Prognosis of Neurological Disorders

Prognosis of Neurological Disorders

Edited by

Randolph W. Evans, M.D.
David S. Baskin, M.D.
Frank M. Yatsu, M.D.

New York   Oxford
OXFORD UNIVERSITY PRESS
1992

Oxford University Press

Oxford New York Toronto
Delhi Bombay Calcutta Madras Karachi
Kuala Lumpur Singapore Hong Kong Tokyo
Nairobi Dar es Salaam Cape Town
Melbourne Auckland

and associated companies in
Berlin Ibadan

Library of Congress Cataloging-in-Publication Data
Prognosis of neurological disorders / edited by
Randolph W. Evans, David S. Baskin, Frank M. Yatsu.
p. cm. Includes bibliographical references and index.
ISBN 0-19-505699-X
1. Nervous system–Diseases–Prognosis.
I. Evans, Randolph W.
II. Baskin, David S.
III. Yatsu, Frank M.
[DNLM: 1. Nervous System Diseases–diagnosis.
2. Prognosis. WL 100 P964]
RC346.P76 1992 616.8–dc20
92-1171

9 8 7 6 5 4 3 2 1

Printed in the United States of America
on acid-free paper

Preface

There are three fundamental questions that we address in clinical neurology and neurosurgery: What is the diagnosis? What is the treatment, medical and/or surgical? What is the prognosis? We deal with prognosis on a daily basis. Our patients and colleagues frequently request prognostic information which is also critical for developing treatment protocols. In years to come, outcome analysis will be of increasing importance as beleaguered health care providers and payors attempt to provide the most cost-effective treatment possible. In medical-legal cases, we are often asked as experts to predict the future with reasonable medical probability.

This volume is designed as a compendium of neurological disorders discussed from the viewpoint of prognosis. The information it provides can be found scattered among various other sources. For example, any of the standard textbooks on neurology and neurosurgery address prognosis to some extent. However, their coverage is not always complete or well-referenced. For many diseases, it is difficult to find any systematic, comprehensive discussions of prognosis. Since there is no single reference that covers neurological disorders from this standpoint, we hope that this book will be useful to neurologists and neurosurgeons.

The book begins with an introductory section discussing the ethical implications of prognosis from the viewpoint of a medical ethicist. A psychologist explores the effect on the patient of giving a prognosis. A critique of natural history studies completes the overview section.

In each chapter, the natural history of a given disease and its subgroups is explored. Clinical prediction rules and prognostic factors, as appropriate, are reviewed. A discussion then follows of how medical and/or surgical therapy alters natural history. Each chapter has full references of the important literature in that area.

This book is a practical guide for the neurologist and neurosurgeon that can be used on a daily basis as a reference source when discussing prognosis with patients and their families, colleagues, and other interested parties. It may also be of interest to psychiatrists, internists, physiatrists, neuropsychologists, and attorneys as well. Looking at the whole of neurological disease, the many inadequacies in our predictive abilities are apparent. We hope that this volume will act as an impetus to acquire increasingly complete prognostic information.

We are grateful to our many outstanding contributors. We appreciate the encouragement and advice of our editor at Oxford University Press, Jeffrey W. House. The support of the Departments of Neurology and Neurosurgery at Baylor College of Medicine and the Department of Neurology at the University of Texas Medical School at Houston has been invaluable. Dr. Yatsu appreciates the ongoing support of the Clayton Foundation for Research. Finally, we wish to give special thanks to our spouses, Marilyn Evans, Juli Baskin, and Mich Yatsu, and our children, Elliott, Rochelle, and Jonathan Evans, and Libby Yatsu.

Houston, Texas   R.W.E.
May, 1992   D.S.B.
                                                                         F.M.Y.

Contents

Contributors

Vincent T. Andriole, M.D.
Professor of Medicine
Yale University School of Medicine
Section of Infectious Diseases
New Haven, Connecticut

Louise V. Appel, R.N.
Instructor
Department of Neurology
Baylor College of Medicine
Houston, Texas

Stanley H. Appel, M.D.
Professor and Chairman
Department of Neurology
Baylor College of Medicine
Houston, Texas

James Ashe, M.D.
Assistant Professor
Department of Neurology
The Johns Hopkins University School of
  Medicine
Baltimore, Maryland

Robert W. Baloh, M.D.
Professor
Department of Neurology and Division of
  Head and Neck Surgery
UCLA School of Medicine
Los Angeles, California

Albert Bandura, Ph.D.
David Starr Jordan Professor of Social
  Science in Psychology
Stanford University
Stanford, California

Michele Barry, M.D.
Associate Professor of Medicine
  Co-Director, Tropical Medicine and
  International Traveler's Clinic
Yale University School of Medicine
New Haven, Connecticut

David S. Baskin, M.D., F.A.C.S.
Chief, Neurosurgery Section
Houston Veterans Affairs Medical Center
Associate Professor
Department of Neurosurgery
Baylor College of Medicine
Houston, Texas

Frank J. Bia, M.D., M.P.H.
Associate Professor of Medicine and
*   Laboratory Medicine*
Co-Director, Tropical Medicine and
*   International Traveler's Clinic*
Yale University School of Medicine
New Haven, Connecticut

John Booss, M.D.
Neurovirologist
VA Medical Center
West Haven, Connecticut; and
Associate Professor
Departments of Neurology and Laboratory
*   Medicine*
Yale University School of Medicine
New Haven, Connecticut

Stephen L. Boswell, M.D.
Instructor
Department of Medicine
Harvard Medical School;
Director, HIV Clinical Services
Infectious Disease Unit and General
*   Medicine Unit*
Massachusetts General Hospital
Boston, Massachusetts

Baruch A. Brody, PhD.
Leon Jaworski Professor of Biomedical
*   Ethics*
Director, Center for Ethics, Medicine, and
*   Public Issues*
Baylor College of Medicine
Houston, Texas

Marc I. Chimowitz, M.D.
Assistant Professor
Department of Neurology
University of Michigan Medical Center
Ann Arbor, Michigan

P.K. Coyle, M.D.
Associate Professor
Department of Neurology
Health Sciences Center
SUNY at Stony Brook
Stony Brook, New York

Thomas J. DeGraba, M.D.
Assistant Professor
Department of Neurology
University of Texas Medical School at
*   Houston*
Houston, Texas

James O. Donaldson, M.D.
Professor of Neurology
University of Connecticut
School of Medicine
Farmington, Connecticut

William H. Donovan, M.D.
Professor of Physical Medicine and
*   Rehabilitation*
Baylor College of Medicine
Medical Director
The Institute for Rehabilitation and
*   Research*
Houston, Texas

Jerrold J. Ellner, M.D.
Professor of Medicine and Pathology
Director, Division of Infectious Diseases
Department of Medicine
Case Western Reserve University
University Hospitals
Cleveland, Ohio

Randolph W. Evans, M.D.
Clinical Assistant Professor
University of Texas Medical School at
*   Houston; and*
Baylor College of Medicine
Houston, Texas

Simon J. Farrow, M.D.
Clinical Assistant Professor
Department of Neurology
Baylor College of Medicine
Houston, Texas

Duncan K. Fischer, M.D., Ph.D.
Resident, Department of Neurosurgery
Baylor College of Medicine
Houston, Texas

Gerhard H. Fromm, M.D.
Professor
Department of Neurology
University of Pittsburgh
School of Medicine
Pittsburgh, Pennsylvania

Anthony J. Furlan, M.D.
Head, Section of Adult Neurology
Director of Cerebrovascular Center
The Cleveland Clinic Foundation
Cleveland, Ohio

Douglas Galasko, M.D.
Assistant Clinical Professor
Department of Neurosciences
School of Medicine
University of California, San Diego
La Jolla, California

Richard E. George, M.D.
Assistant Professor
Section of Pediatric Neurosurgery
Department of Neurosurgery
Baylor College of Medicine
Houston, Texas

Michael J. Glantz, M.D.
Associate in Neurology
Duke University Medical Center
Durham, North Carolina

Christopher G. Goetz, M.D.
Professor
Department of Neurological Sciences
Rush Medical College of Rush University
Rush-Presbyterian-St. Luke's Medical
*   Center*
Chicago, Illinois

James C. Grotta, M.D.
Professor
Department of Neurology
University of Texas Medical School at
*   Houston*
Houston, Texas

Sandra K. Hanson, M.D.
Clinical Fellow in Cerebral Vascular
*   Disease*
Department of Neurology
University of Texas Medical School at
*   Houston*
Houston, Texas

Harold J. Hoffman, M.D.
Professor and Chairman
Department of Neurosurgery
Hospital for Sick Children
Toronto, Ontario

Joseph Jankovic, M.D.
Professor of Neurology
Baylor College of Medicine
Houston, Texas

Bryan Jennett, M.D., F.R.C.S.
Professor
Department of Neurosurgery
University of Glasgow
Institute of Neurological Sciences
Southern General Hospital
Glasgow, Scotland

John L. Johnson, M.D.
Assistant Professor
Department of Medicine
Case Western Reserve University
Cleveland Veterans Administration
*   Hospital*
Cleveland, Ohio

Robert Katzman, M.D.
Professor
Department of Neurosciences
School of Medicine
University of California, San Diego
La Jolla, California

Andrew Kertesz, M.D., F.R.C.P.
Professor of Neurology
University of Western Ontario
Lawson Research Institute
St. Joseph's Hospital
London, Ontario

David G. Kline, M.D.
Professor and Chairman
Department of Neurosurgery
Louisiana State University Medical Center
Charity Hospital and Ochsner Hospital
New Orleans, Louisiana

Thomas D. Koepsell, M.D., M.P.H.
Professor
Departments of Epidemiology and Health
Services
School of Public Health and Community
Medicine
University of Washington
Seattle, Washington

Victor A. Levin, M.D.
Professor and Chairman
Department of Neuro-Oncology
University of Texas M.D. Anderson
Cancer Center
Houston, Texas

David E. Levy, M.D.
Adjunct Clinical Associate Professor
Department of Neurology
Cornell University Medical College
New York, New York

W.T. Longstreth, Jr., M.D., M.P.H.
Associate Professor
Departments of Medicine (Neurology) and
Epidemiology
School of Medicine and School of Public
Health and Community Medicine
University of Washington
Seattle, Washington

Carl E. Lowder, M.D.
Former Senior Resident
Department of Neurosurgery
Louisiana State University Medical Center
Charity Hospital and Ochsner Hospital
New Orleans, Louisiana;
Private Practice
Fort Myers, Florida

Thomas J. Mampalam, M.D.
Former Chief Resident
Department of Neurological Surgery
School of Medicine
University of California, San Francisco
San Francisco, California;
Private Practice
Richmond, California

Elliott L. Mancall, M.D.
Professor and Chairman
Department of Neurology
Hahnemann University School of Medicine
Philadelphia, Pennsylvania

Alan W. Martin, M.D.
Clinical Assistant Professor
University of Texas Southwestern Medical
Center
Dallas, Texas

Richard P. Moser, M.D.
Associate Professor
Department of Neurosurgery
University of Texas M.D. Anderson
Cancer Center
Houston, Texas

Hamilton Moses, III, M.D.
Associate Professor
Department of Neurology
The Johns Hopkins University School of
Medicine
Baltimore, Maryland

Gordon Murray, Ph.D.
Senior Lecturer in Medical Statistics
Department of Neurosurgery
University of Glasgow
Glasgow, Scotland

Lilian Murray, Ph.D.
Statistician
Department of Neurosurgery
University of Glasgow
Glasgow, Scotland

Stephen E. Nadeau, M.D.
Associate Professor of Neurology
University of Florida College of Medicine;
Staff Neurologist
Geriatric Research, Education and Clinical
*   Center*
Veteran's Affairs Medical Center
Gainesville, Florida

Bradford A. Navia, M.D.
Instructor
Department of Neurology
Harvard Medical School and
*   Massachusetts General Hospital*
Boston, Massachusetts

Lorene M. Nelson, Ph.D.
Department of Epidemiology
School of Public Health and Community
*   Medicine*
University of Washington
Seattle, Washington

Eugenie A.M.T. Obbens, M.D.
Clinical Associate Professor
Department of Neurology
University of Arizona College of Medicine
Tucson, Arizona

Gary D. Overturf, M.D.
Professor
Department of Pediatrics
University of New Mexico
Albuquerque, New Mexico

Andrew R. Pachner, M.D.
Associate Professor
Department of Neurology
Georgetown University School of Medicine
Washington, D.C.

Gareth J. Parry, M.D.
Professor
Department of Neurology
Louisiana State University Medical Center
School of Medicine in New Orleans
New Orleans, Louisiana

Thomas F. Patterson, M.D.
Assistant Professor of Medicine
Yale University School of Medicine
Section of Infectious Diseases
New Haven, Connecticut

Isabelle Rapin, M.D.
Professor
Saul R. Korey Department of Neurology
*   the Department of Pediatrics, and*
*   the Rose F. Kennedy Center for*
*   Research in Mental Retardation and*
*   Human Development*
Albert Einstein College of Medicine
Bronx, New York

Vincent M. Riccardi, M.D.
Medical Director
Alfigen, The Genetics Institute
Pasadena, California

Steven P. Ringel, M.D.
Professor of Neurology
University of Colorado School of Medicine
Denver, Colorado

Loren A. Rolak, M.D.
Associate Professor
Department of Neurology
Baylor College of Medicine
Houston, Texas

Mark L. Rosenblum, M.D.
Professor
Department of Neurological Surgery
School of Medicine
University of California, San Francisco
San Francisco, California

Joel R. Saper, M.D.
Director
Michigan Head Pain and Neurological
*   Institute*
Ann Arbor, Michigan;
Clinical Professor
Neurology, Michigan State University
East Lansing, Michigan

S. Clifford Schold, Jr., M.D.
Professor of Neurology
Duke University Medical Center
Durham, North Carolina

Richard K. Simpson, Jr., M.D., Ph.D.
Assistant Professor
Department of Neurosurgery
Baylor College of Medicine
Houston, Texas

Barney J. Stern, M.D.
Director
Division of Neurology
Sinai Hospital of Baltimore
Associate Professor of Neurology
The Johns Hopkins University School of
*   Medicine*
Baltimore, Maryland

David A. Stumpf, M.D.
Professor and Chairman
Department of Neurology
Professor of Pediatrics
Northwestern University Medical School
Chicago, Illinois

Austin J. Sumner, M.D.
Professor and Chairman
Department of Neurology
Louisiana State University Medical Center
School of Medicine in New Orleans
New Orleans, Louisiana

Rosa A. Tang, M.D.
Clinical Associate Professor
Departments of Ophthalmology and
*   Neurology*
University of Texas Medical School at
*   Houston*
Houston, Texas

Graham Teasdale, F.R.C.S.
Professor
Department of Neurosurgery
University of Glasgow
Glasgow, Scotland

Gerald van Belle, Ph.D.
Professor
Departments of Environmental Health and
*   Biostatistics*
Chairman of Department of Environmental
*   Health*
School of Public Health and Community
*   Medicine*
University of Washington
Seattle, Washington

C. Peter N. Watson, M.D., F.R.C.P. (C)
Assistant Professor
Department of Medicine
The Irene Eleanor Smythe Pain Clinic
University of Toronto
Toronto, Ontario

Stuart M. Weil, M.D.
Clinical Assistant Professor
Department of Neurosurgery
Baylor College of Medicine
Houston, Texas

L. James Willmore, M.D.
Professor of Neurology
University of Texas Medical School at
*   Houston*
Houston, Texas

Anthony J. Windebank, B.M., B.Ch.,
*   M.R.C.P. (U.K.)*
Professor of Neurology
Mayo Medical School
Dean Mayo Graduate School
Mayo Clinic and Mayo Foundation
Rochester, Minnesota

Maureen P. Wooten, M.D.
Formerly Fellow, Movement Disorders
Baylor College of Medicine
Houston, Texas;
Neurology Associates of Dallas
Dallas, Texas

Frank M. Yatsu, M.D.
Professor and Chairman
Department of Neurology
University of Texas Medical School at
*   Houston*
Houston, Texas

Prognosis of Neurological Disorders

1

Ethical Issues Raised by the Clinical Use of Prognostic Information

BARUCH BRODY

The development of prognostic indicators and of reliable prognostic information is valuable for many reasons. First, this information can be used in quality assurance activities. If the outcome of the management for a certain type of patients in an institution is significantly worse than the expected outcome based on reliable prognostic information derived from general experience, then that institution has reason to examine carefully its management of those patients to see whether the management is substandard. For example, HCFA recently has funded efforts to develop reliable prognostic indicators for stroke patients for quality assurance purposes (Daley et al. 1988). Second, the information can be used to help assess the impact of new interventions when a randomized test of the impact is impossible or inappropriate. Thus, if the outcome of the management of a certain type of patients is significantly better after a new intervention than the expected outcome based on reliable prognostic information derived from experience prior to the new intervention, then this would be the best form of a historically controlled study, a type of study that the Food and Drug Administration (FDA) has ap-

proved (Kessler 1989). Finally, the information can be used to help make clinical decisions about the management of patients. For example, reliable prognostic information about persistent vegetative patients has been used since the Quinlan decision as the basis for withholding or withdrawing care from such patients (Matter of Quinlan 1976). It is this use (particularly as applied to the decision to withhold or withdraw care) that is examined in this chapter, because it raises the most pressing ethical issues.

What degree of prognostic certainty is required before decisions about withholding or withdrawing care are appropriate? Is this prognostic certainty about mortality or severe morbidity? What decisional processes must be undergone, and how should the information about prognosis be introduced into those decisional processes? These are the ethical questions that must be confronted before reliable information about prognosis can be used as a basis for clinical decisions about withdrawing or withholding care.

It is difficult to assess these issues in the abstract; therefore, a particular example (prognostic information about the neurological outcome of postarrest patients and/or patients suffering from hypoxic–ischemic coma) will be used to make the discussion more concrete. In the first section, two approaches are presented, one used by a Belgian collaborative group (Mullie et al. 1988) and the other used by a group based in New York (Levy et al. 1985). In the next two sections, fundamental features common to both approaches are identified, and some of these features are challenged. In the final section, the ways in which a more appropriately constructed approach could be used clinically are discussed. The goal is not primarily to criticize these two approaches (which have been chosen precisely because of their sophisticated results). Instead, it is to develop an understanding of the use of reliable prognostic information in ethically appropriate clinical decision-making about the level of care provided to patients who have experienced severe neurological damage.

Two Approaches

The Belgian group used the Glasgow coma score as a basis for its prognostic scheme. Its scheme was developed solely for out-of-hospital cardiac arrest patients who were successfully resuscitated (an initial restoration of spontaneous circulation) and fully treated in accordance with accepted standards for advanced cardiac life support and cardiopulmonary resuscitation. A patient's treatment was judged successful if he or she returned to their prearrest status, were only moderately disabled (less than previous status but independent in daily living), or were severely disabled (retained cognition if not independence) within 14 days after their arrest. A patient's treatment was judged a failure if he or she died or remained vegetative without regaining consciousness at the end of 14 days. This group developed its predictive rule based on an analysis of 216 patients seen in 1983 to 1984 and tested the rule in a cohort of 133 patients seen in 1985.

The predictive rule used the patient's best Glasgow Coma Score in the first 2 days as its

basis for initial predictions. Patients with a score of four or less were predicted to be treatment failures while patients with a score of 10 or better were predicted to be treatment successes. Patients whose best score was five to nine were considered not yet predictable, and predictions were deferred for 4 days. At that point, a prediction was made based on their best Glasgow coma score. If it was less than eight, the patient was predicted to be a treatment failure; if it was eight or greater, the patient was predicted to be a treatment success.

The results are quite impressive. Of the 73 patients predicted to be treatment failures (54 on day 2 and 19 on day 6), only two did better by day 14, one died on day 35, and the other survived with "substantial neurological damage." Of the 60 patients predicted to be treatment successes (49 on day 2 and 11 on day 6), 14 actually were failures (12 deaths and two vegetative patients). While the negative predictive value (97%) clearly was better than the positive predictive value (77%), the authors feel that this is appropriate. "Whereas an incorrect positive prediction has no adverse consequences for the patient, an incorrect negative prediction is unacceptable" (Mullie et al. 1988). The authors conclude that their results are encouraging, but that a prospective study in a larger number of patients is needed before their rule can be used safely in clinical practice.

The New York Group used specific neurological findings as a basis for its prognostic scheme. Its scheme was developed for patients suffering from hypoxic–ischemic coma as a result of cardiac arrest, respiratory failure, or profound hypotension. A patient's outcome was judged as best if they recovered some independent function (26 patients of 210 or 13%), poorer if they were severely disabled and dependent on others for daily living activities (20 of 210 or 10%), and poorest if they remained comatose or vegetative (154 of 210 or 77%). This was assessed in terms of best recovery in the first year regardless of length of survival because more than 90% of the patients died in the first year.

The prognostic rules emerging from this study are based on the presence or absence of specific neurological findings at specified periods of time after the coma-inducing event. For example, absence of pupillary reflexes at initial examination is prognostic of the poorest neurological outcome because none of the 52 patients meeting that description ever became independent, and only three (6%) regained consciousness. By contrast, at initial examination presence of preserved pupillary reflexes, motor responses that were extensor or better, and roving conjugate or better spontaneous eye movements are predictive of a much better neurological outcome because 11 of the 27 patients (41%) who met these criteria regained independent function. Similar rules are developed for neurological findings at specified times after the coma-inducing event.

No attempt was made in the New York group's published work to test on a different cohort of patients the rules derived from a retrospective review of one set of patients, so negative or positive predictive values are not available. The authors do suggest, however, that both types of errors may be costly. Just as patients suffer if they are inappropriately predicted to do badly (because they may die as a result of withheld care when they could have survived with a good neurological outcome), they also suffer if they are inappropriately predicted to do well (because they may survive as a vegetative patient for a prolonged period of time, which they may not have wanted).

Common Features of the Two Approaches

Having introduced the two approaches, the following is a discussion of the features they have in common. These common features are usually—but not always—present in other prognostic systems. These features are as follows: They are based on a review of previous cases; they involve a definition of successful and unsuccessful care; and they yield predictions, based on less than 100% certainty, as to whether care in future patients will be successful.

The first feature, that both systems are built on the results of treating patients in the past, is most obvious in the case of the New York group. This group's study base was patients admitted to several hospitals from September, 1973 to June, 1977, although the analysis was published in 1985. It is also true, however, of the Belgian group whose study base was patients admitted to their hospitals from 1983 to 1985. The Belgian group divided their patient base into two groups, developed their indicators by analyzing the results in one group, and tested their indicators on the second group. Moreover, the Belgian group rightly pointed out that such indicators could be tested prospectively as well as retrospectively. Nevertheless, until now all such systems have been developed by retrospective analysis and testing. Moreover, the published results of a prospective test would still describe a test of a prognostic indicator on patients who were ill at an earlier date than the date of publication. The significance of this point will be discussed later.

The second feature of these systems is that the authors of both studies have built in a definition of treatment success and of treatment failure. For the Belgian group, a patient who survived until day 14 and recovered consciousness is considered a treatment success; a patient who failed to survive until day 14 or survived until then as a vegetative patient is considered a treatment failure. The New York group had a more elaborate account of failures and successes because it subdivided the successes into greater or lesser successes depending on best neurological outcome; however, it judged treatment to be a success even if the patient died (from neurological or other problems) shortly after attaining that neurological outcome. Much can be said about the merits of these different approaches, but the crucial point is that the definitions of success and of failure are built into these systems. The significance of this point also will be discussed later.

The third common feature of these systems is that they are designed to yield prognostic indicators with less than 100% certainty. This is true even for these two studies, which initially appear close to certainty. In the Belgian study, even with the deliberately chosen conservative approach, two out of 73 predictions of treatment failure were incorrect. Because they were extremely conservative in predicting treatment failure, they had few errors in predicting failure, but they had a considerable percentage of errors (14 of 60 or 23%) when predicting treatment success. This trade off will be discussed later, but it should be noted that their approach is not 100% certain. It might appear that the New York study is better in this respect because of their 52 patients with no pupillary reflexes at initial examination, none fully recovered and became independent in their activities of daily living. However, three of these 52 patients did regain consciousness (so treatment was successful to some degree). Moreover, the authors of these studies note that their data mean only that there is a 95% confidence that the percentage of patients with this prognostic indicator who will recover independent function is between 0% and 7%. Moreover, like the Belgian study, their indicators of success are much less certain. Only 41% of patients with best neurological findings at initial examination were a treatment success. Therefore, these are not certain indicators.

Three crucial features of these systems of prognostic indicators have been addressed. In the next section, the implications of these features for clinical use as prognostic indicators are discussed.

Clinical Implications of the Common Features

The first of the three features (the use of past data) is the most inevitable, and unfortunately it raises problems of variable magnitude from one prognostic indicator to the next. To understand why this is so, a review of some elementary observations about outcomes is necessary.

Clinical outcomes for a given class of patients are a function of two features: the condition of the patients when they present and the efficaciousness of the care given to

these patients after they present. Over time, changes in medical knowledge (and resulting changes in care) clearly can affect both. As a result, the clinical outcome for patients with the same presenting features can vary over time. Systems of prognostic indicators inevitably use past outcome data to predict future outcomes; therefore, they are vulnerable to the possibility that changes in treatment may cause future outcomes to be different from past outcomes. The degree of vulnerability depends on the importance of treatment changes from the time the system was developed to the time it is being used, which varies from one indicator to another. Therefore, an inevitable uncertainty about the usefulness of prognostic indicators must be recognized when approaching a new patient.

Returning to the Belgian study (similar points also could be made about the New York study), the class of patients they examined were out-of-hospital cardiac arrests between 1983 and 1985 who were successfully resuscitated and admitted to a hospital. The management of these patients has changed since 1985, both in the period immediately following their admission and after. What does this mean for their proposed prognostic indicator, the Glasgow Coma Score, on day 2 or day 6? This is difficult to assess. Does a given Glasgow coma score for a postarrest patient mean the same thing clinically in 1992 that is meant in 1985? Do the changes in follow-up treatment also make a difference? Do improvements in care mean that there are now more survivors who are seriously disabled, thereby producing more incorrect predictions of treatment failure, according to the study's definition of treatment failure?

The authors of the study are not unaware of this problem. At one crucial point, they make the following observation (Mullie et al. 1988):

To exclude the possibility that secular changes during the study could have affected the data, and to show the timeless character of the predictive rule, we also analyzed the patient data in randomly selected subsets; we constructed the predictive rule on a randomly selected subset of two thirds of the whole data set and tested it on the remaining third. The results obtained were identical.

Unfortunately, none of this shows the timeless character of the predictive rule. It merely shows that changes in 1983 to 1985 were not very important. It says nothing about changes since then, or about changes in the period in which the rule might be used to guide patient management.

Conclusions from this first point should be drawn cautiously. The correct conclusion is not that prognostic indicators cannot be used because future outcomes may be different than past outcomes. Rather, it is that there is always a degree of uncertainty in the use of such indicators, which varies from one use to another, and that a correct decisional process must take that uncertainty into account. This will be discussed further in the section, Decisional Processes for the Clinical Use of Prognostic Information.

The second feature (the presence of a definition of success) gives rise to another set of issues but leads to a similar conclusion. The importance of the concept of appropriate decisional processes is highlighted again.

The authors of both systems understand that quality of life, as opposed to mere survival, is crucial in adequately defining success and failure. The Belgians treat survival without regaining consciousness as just as much a treatment failure as the death of the patient, while the New York group defines levels of success totally in terms of regained functioning (and therefore regained quality of life) without reference to length of survival. In this respect, both studies are improvements over other prognostic systems, such as APACHE II, which define success and failure solely in terms of in-hospital survival or death (Knaus et al. 1985). Nevertheless, fundamental difficulties exist with both systems' definitions of success.

One way to approach these issues and difficulties is to raise a series of questions concerning the differences and similarities between the definitions offered by the two systems. For example, a patient who regains consciousness and some independent functioning but dies shortly thereafter (at day 10) from underlying cardiac illness is a relative treatment success (second best) according to the New York group but a failure according to the Belgians. Which is correct? Was treatment a success because the patient recovered some functioning for a period of time, even if it was a short period? Was it a failure because the patient died so soon? Another example is a patient who regains consciousness but lives for many years dependent on others for all activities of daily living. Would this patient be a treatment success as the Belgians would classify it, a minimal success (second lowest) as the New York group would classify it, or a failure as many might think because of the long period of dependence for the patient and suffering for the patient's family? This might even be considered worse than a vegetative state because of the patient's suffering caused by an awareness of dependency. Many such questions could be raised, and it is difficult to see how they can be answered.

This leads to a more in-depth approach to these issues, one which challenges this second feature at an even more fundamental level. Can there be a universal objective resolution of this issue that would enable a study to classify objectively some outcomes as successes and others as failures or that would enable a study to classify objectively some outcomes as greater successes and others as lesser successes? Or would it be better to consider that successes and failures, or greater and lesser successes are subjective, individually determined decisions? Thus, for some people, the possibility of being restored to some normalcy of functioning, even for a short period of time in a hospital, is of considerable value (perhaps because they have agendas to complete and affairs to put in order or because they would enjoy that period of life), while others would find it of less value. For some, survival as a conscious individual for a prolonged period is of great value, even if functionally dependent; for others, prolonged dependence on others is a horrendous prospect (perhaps out of a love for those others or a sense of self-dignity). In the extreme example of a vegetative patient, many would judge this continued existence to be of no value, and some would judge it to be a negative value; however, others would judge it to be of some positive value. For example, a recent important statement about vegetative patients, signed by Catholics and others, argues that continued biological value is a positive value (May et al. 1987).

Again, one needs to be careful about the conclusions that are drawn from these observations. Certainly, quality of life should not be eschewed in an attempt to define successful outcomes and failures; any return to a definition based solely on survival would be a mistake. Rather, the conclusion is that there is a fundamentally subjective character to judgments about quality of life and a correct decisional process must take this into account when determining the success or failure of treatment. This will be examined more closely in the section, Decisional Processes for the Clinical Use of Prognostic Information.

The final feature (prognoses based on less than certainty) gives rise to a final set of issues and leads to similar observations about appropriate decisional processes.

The authors of both systems recognize that there is a fundamental problem with different types of risks that must be confronted when developing a prognostic system. The system mistakenly may predict a bad outcome, which may lead to the death of a patient (if care is withheld on the basis of the prediction) who would have otherwise survived with a good outcome. A high negative predictive value means that there is a low likelihood that this will occur. The system may mistakenly predict a good outcome, and this may lead to a long painful dying process (because aggressive care is provided) or to the survival of a patient in very bad condition (severely disabled or vegetative). A high positive predictive value means that there is a low likelihood that this will occur. All other things being equal, it is desirable to have as high of positive and negative predictive values as possible. In an ideal world, it would be possible to develop a system of prognostic indicators with perfect negative and positive predictive values. In the real world, often one must settle for less, making prognostications with known error rates. The decision must be made as to which rate to improve at the cost of the other rate. A better negative predictive rate is possible by only predicting treatment failures in cases in which the evidence of failure is overwhelming. This will avoid more false predictions of failure but usually results in a lower positive predictive value because many more treatment successes are falsely predicted. On the other hand, a better positive predictive value is possible by predicting failures when the evidence of failure is less overwhelming, but this will result in a lower negative predictive value because many more treatment failures are falsely predicted. How is the choice between these two options made?

The view of the Belgian group that the highest priority must be given to attaining a high negative predictive value because an incorrect negative prediction is unacceptable, while an incorrect positive prediction has no adverse consequences for the patient has been quoted. Such a position cannot serve as the basis for developing a prognostic system in a world of uncertainty. In a world in which a false-negative prediction is unacceptable while a false-positive prediction has no adverse effects, a negative outcome should never be predicted, and one should wait until it occurs before acting on it; until then, only positive predictions should be made. It is precisely because false-positive predictions have adverse effects on patients, families, and the health care system that we try to make predictions, both negative and positive. The following is a quote from the New York group (Levy et al. 1985):

Perhaps the best answer is that if one waits until patients either awaken or die, their medical condition will have stabilized. The result can be prolonged survival for vegetative patients who might otherwise have died. For most of us this is an undesirable state. . . . The ability to predict outcome could also spare families the emotional and financial burden of prolonged care of patients with a hopeless prognosis.

Therefore, it will be necessary to make trade-offs between improvements in negative predictive values and in positive predictive values. The advantages and disadvantages must be determined when lowering one predictive value and raising the other. To do that, however, a value comparison must be made between the two types of mistakes. The more one agrees with the New York group about the harmful effects of false-positive predictions, the more willing one will be to accept some additional false-negative predictions to get an improvement in the positive predictive rate. The more one agrees with the Belgian group about the insignificance of such effects in comparison to the harmful effects of the false-negative prediction, the more one will demand a high negative predictive rate, even if the positive predictive rate is considerably lowered. The trade-offs are value choices, and they are just as much subjective, individually determined choices as are other value choices. There will be no objectively correct trade-offs.

Once again, conclusions should be drawn cautiously from these final observations. The conclusion is not that such trade-offs should be eschewed because they are based on subjective choices; in a world of uncertainty, any prognostic system must involve such trade-offs. Rather, it is that correct decisional processes involving prognostications of outcomes must take into account the subjectivity of the trade-offs. Therefore, as mentioned previously, the conclusion is that the decisional process will be crucial for success when using prognostic information.

Decisional Processes for the Clinical Use of Prognostic Information

The medical ethics and medical jurisprudence literature has been dominated in the last 20 years by a picture of medical decision-making that involves a far more active role of patients or of the surrogates who speak for them. This is central to the legal doctrine of informed consent, which implies that competent patients (or the surrogates of incompetent patients) can refuse care recommended by physicians (Faden and Beauchamp 1986), and to discussions of the ethics of decision making found in such central writings as the reports of the President's Commission (President's Commission for the Study of Ethical Problems in Medicine 1982).

This is based on two fundamental moral themes. The first is the rights of patients. Physicians may carry out procedures on a patient only with the consent of that patient because it is the patient's body and he or she has a right to determine what is done. The second theme is that medical decisions are based on medical facts and on individual values, that the physician has expertise only about the former while the patient has expertise only about the latter, and that decision making must grow out of this joint expertise rather than from unilateral physician recommendations. Both themes are valid, and they often produce the same consequences for medical decision making. Of course there are occasions which the consequences are different, which raises some of the most difficult questions for medical ethics. (For additional information on this topic, please see Brody 1988.) The second theme is used here as the basis for developing a proper process for decision making using prognostic information.

Consider a patient who has suffered an out-of-hospital arrest, has been successfully resuscitated, and is in a coma 12 to 24 hours after the arrest. Obviously, the patient cannot now participate in any decisional process. Any necessary decisions will have to be made by the physicians caring for the patient and the surrogates of that patient (a guardian, someone with durable power of attorney, or a close family member). The crucial decision that has to be made is whether to treat the patient vigorously to avoid further arrests and to produce the best neurological outcome or to keep the patient comfortable and allow nature to take its course. There are many variations on each of the basic approaches, and much can be said about the decisions among these variations, but the focus here is on the crucial fundamental decision.

The physician brings considerable expertise to this decision even before the development of prognostic indicators. He or she is aware of the details of the patient's condition and of what is required to give the patient the best chance of surviving with improved neurological functioning. With prognostic indicators, he or she is better aware of the patient's likelihood of surviving and of recovering varying levels of neurological functioning. The physician brings this expertise to the decisional process. However, none of this expertise answers certain crucial questions: What outcomes would this patient judge as a success, and what outcomes would this patient judge as a failure? Would this patient prefer living for a few months in a severely disabled fashion to dying or would this be something that the patient would want to avoid, even at the cost of dying? Given these preferences, how would this patient want to make the trade-off between the possibility of erring and being alive but disabled versus the possibility of erring and dying when he or she could have recovered? These are value questions, and the answers will vary from individual to individual. Of course, the patient is the best authority on these matters, and any advance directives left by the patient should be consulted for guidance. When directives are not available or when the patient was silent on these issues, the next best source is surrogates who know the patient and the patient's values. They must use their expertise for the decisions that must be made.

Prognostic information is more information that the physician can use in the decisional process. The goal of developing prognostic schemes is to develop information that will be relevant, when combined with the patient's values, to decisions made about levels of care. Prognostic systems should describe outcomes and assign probabilities to them, but they should neither judge some to be successes and others failures nor assign cut-off points at which predictions of success or failures are made.

A physician using the Belgian system needs to convey to the surrogate something like the following:

The extent of a patient's coma is measured by a Glasgow coma score. Your patient has a score of x. A careful study of patients with this score reveals that if that patient is treated vigorously, the likelihood of surviving and recovering to former status is $n_1\%$; of surviving but being moderately disabled is $n_2\%$; of surviving but being seriously disabled is $n_3\%$; of surviving but never regaining consciousness is $n_4\%$; and of dying is $n_5\%$. This is the best information we now have concerning outcome, but it is based on past experience, and things have changed since then. My hunch is that this change means that certain outcomes are now more likely. In particular, there may be fewer deaths but more survivals without regaining consciousness or with severe disabilities. We have a choice between helping the patient to survive and recover some functioning or keeping the patient comfortable and letting nature take its course. I cannot tell you what to decide because what we should do is determined in large measure by the values of the patient, and you know those better than I. What would the patient have judged as an acceptable outcome, and what would he or she have preferred not to undergo, even at the cost of death? For example, how did he or she feel about dependence? With your help, we can jointly try to determine the best choice for this patient in light of his or her values.

Similar remarks, suitably modified, should be made by physicians using the New York system.

Many questions need to be explored when using this approach. Some questions concern the most effective way of conveying this information, asking the questions, and coming to mutually agreed on conclusions. Other questions concern assessing the ability of surrogates to understand and participate and determining what to do if they cannot understand or will not participate. Finally, other questions concern the horrendous difficulties that arise when multiple surrogates are in conflict. The many difficulties involved in adopting this approach should not be underestimated, but it is preferable to the alternative—physicians using prognostic schemes to recommend strongly or to make decisions. The latter approach requires objective assessments of which outcomes are successes and which are failures and of what trade-offs among uncertainties are appropriate. These objective assessments are, as has been discussed earlier in this chapter (see the section, Clinical Implications of the Common Features), nonexistent.

A conclusion can now be drawn about the general clinical use of prognostic information. In the real world, prognostic information does not lead to universal treatment decisions because such decisions are a function of both objective information and subjective values. Systems of prognostic information should therefore be developed in a way that recognizes this fact, eschewing decisions about successful outcomes and trade-offs among predictive values. However, this does not undermine the significance of the development of prognostic information systems. If prognostic information without values leads nowhere, values without prognostic information lead everywhere and yield no grounded clinical decisions. Reliable prognostic information is a crucial component, if not the whole basis, of sound decision making.

Finally, the focus of this discussion of clinical decisions based on prognostic information has been on the implications for the patients, their families, and their care-givers. The implications for other patients (in situations of triaging scarce resources) or for society as a whole (in times of cost-containment) have not been considered. There is a growing recognition that cost-containment pressures may lead to more instances of scarce resources being used, which requires triaging decisions, and that these triaging decisions will need to take into account poor prognoses. An example of this growing recognition is the recent guidelines of the Task Force of the Society of Critical Care Medicine on intensive care unit admissions and discharges (Task Force on Guidelines 1988). This will impact on the decisional processes previously indicated because these processes involve no mechanism that considers these societal considerations. The mechanisms and principles that will be involved in factoring in these social considerations are, however, unclear, so the process by which prognostic indicators will be used clinically will probably be modified in the future in ways that cannot now be predicted.

References

Brody, B. Life and death decision making. New York: Oxford University Press; 1988.

Daley, J.; Jencks, S.; Draper, D.; Lenhart, G.; Thomas, N.; Walker, J. Predicting hospital-associated mortality for medicare patients. JAMA. 260:3617–3624; 1988.

Faden, R.; Beauchamp, T. A history and theory of informed consent. New York: Oxford University Press; 1986.

Kessler, D. The regulation of investigational drugs. N. Engl. J. Med. 320:281–288; 1989.

Knaus, W.; Draper, E.; Wagner, D.; Zimmerman, J. APACHE II: a severity of disease classification system. Crit. Care Med. 13:818–829; 1985.

Levy, D.; Caronna, J.; Singer, B.; Lapinski, R.; Frydman, H.; Plum, F. Predicting outcome from hypoxic–ischemic coma. JAMA. 253:1420–1426; 1985.

Matter of Quinlan 355 A2d. 647 (1976).

May, W.; Barry, R.; Griese, O.; Grisez, G.; Johnstone, B.; Marzen, T.; McHugh, J.; Meilaender, G.; Siegler, M.; Smith, W. Feeding and hydrating the permanently unconscious and other vulnerable persons. Issues in Law and Medicine. 3:203–217; 1987.

Mullie, A.; Verstringe, P.; Buylaert, W.; Houbrechts, H.; Michem, N.; Delooz, H.; Verbruggen, H.; Van Den Broeck, L.; Corne, L.; Lauwaert, D.; De Cock, R.; Weeghmans, M.; Mennes, J.; Bossaert, L.; Quets, A.; Lewi, P. Predictive value of Glasgow coma score for awakening after out-of-hospital cardiac arrest. Lancet. I:137–140; 1988.

President's Commission for the Study of Ethical Problems in Medicine. Making health care decisions. Washington, D.C.: U.S. Government Printing Office; 1982.

Task Force on Guidelines. Recommendations for intensive care unit admission and discharge criteria. Crit. Care Med. 16:807–808; 1988.

2

Psychological Aspects of Prognostic Judgments

ALBERT BANDURA

It is now widely acknowledged that the level of health functioning is governed by biopsychosocial processes rather than solely by biological factors (Engel 1977). Psychological determinants contribute to physical and functional status by their impact on health-related behavior and on the biological systems that mediate health and physical dysfunction. Because psychosocial factors account for some of the variability in the course of health functioning, their inclusion in prognostic schemes can enhance the predictive power. Prognostic judgments activate psychosocial processes that can influence health outcomes, rather than simply serving as nonreactive forecasters. This chapter examines some of the psychological mechanisms through which prognostic judgments and clinical interventions can alter the probabilities of health outcomes.

Psychosocial determinants of health status operate largely through the exercise of personal agency. Among the mechanisms of personal agency, none is more central or pervasive than people's beliefs in their ability to exercise some control over their own health. This self-belief of personal control is called perceived self-efficacy. Evidence from diverse research studies shows that perceived self-efficacy operates as an important psychological mechanism linking psychosocial influences to physical and functional status (Bandura 1991, 1992).

Perceived self-efficacy has diverse effects, each of which can influence health outcomes and the extent to which people use their physical and cognitive capabilities (Bandura 1986). Such self-beliefs affect what people choose to do. They avoid activities they believe exceed their capabilities and as a result, experience declines through disuse. However, they readily undertake tasks they judge themselves capable of handling in ways that enhance their capabilities. Self-efficacy beliefs also play a central role in the self-regulation of motivation. They determine how much effort people will exert in an endeavor and how long they will persevere in the face of difficulties and setbacks. The stronger the belief in their capabilities, the greater and more persistent are their efforts. When faced with obstacles and disabilities, people who are beset by self-doubts about their capabilities slacken their efforts or give up quickly. A resilient sense of personal efficacy thus provides the needed staying power for surmounting difficulties. People's beliefs in their efficacy affect the amount of stress and depression they experience in taxing situations as well as their level of motivation. Stress and depression take their toll on the quality of health functioning.

Impact of Self-Efficacy Beliefs on Health Functioning

There are two levels of research on the psychosocial determinants of health outcomes in which perceived self-efficacy plays an influential role. The more basic level of research examines how psychosocial factors affect the biological systems that mediate health and disease through the self-efficacy mechanism. Stress has been implicated as an important contributing factor to many physical dysfunctions. Controllability appears to be a key organizing principle regarding the nature of these biological stress effects. Exposure to stressors with a concomitant strong sense of coping efficacy has no adverse physiological effects. Exposure to the same stressors with weak coping efficacy, however, heightens autonomic arousal, activates neurotransmitters and stress-related hormones, and impairs cellular components of the immune system (Bandura 1991; Bandura et al. 1985; Maier et al. 1985). Similarly, depression has been shown to produce immunosuppressive effects (Coe and Levine in press; Adler and Cohen 1985).

Physiological sytems are highly interdependent. The types of physiological reactions that accompany weak coping efficacy are involved in the regulation of immune systems. For example, weak efficacy in exercising control over stressors activates endogenous opioid systems (Bandura et al. 1988; Maier 1986). Evidence shows that some of the immunosuppressive effects of weak efficacy in controlling stressors are mediated by the release of endogenous opioids. When opioid mechanisms are blocked by an opiate antagonist, the stress of weak coping efficacy loses its immunosuppressive power (Shavit and Martin 1987).

The second level of research is concerned with modifying habits that enhance or impair health and functional status. Self-efficacy beliefs affect every phase of behavioral change (Bandura 1991). They determine whether people will consider changing their health-related behavior, whether they will enlist the motivation and perseverance needed to succeed if they choose to do so, and how well they will maintain the changes they have achieved. Each of these change processes is discussed briefly in the sections that follow.

People's beliefs that they can improve their physical well-being play a crucial role in determining whether they will consider changing their health habits or pursuing rehabilitative activities. Those who believe they cannot exercise any control over their health outcomes see little point in trying (Beck and Lund 1981). If they make an attempt, they give up easily in the absence of quick results. Effective self-regulation is not achieved through an act of will. It requires development of self-regulatory skills. To build a sense of controlling efficacy, people must develop skills to influence their own motivation and behavior. In such programs, they learn how to monitor the behavior they seek to change; how to set short-range, attainable subgoals to motivate and direct their efforts; and how to inlist incentives and social supports to sustain the effort needed to succeed (Bandura 1986). Once equipped with the skills and self-belief in their capabilities, people are better able to adopt behaviors that promote health and to eliminate those that impair it. They benefit more from treatments for physical disabilities and their psychological well-being is less adversely affected by chronic impairments.

A growing body of evidence reveals that the impact of different therapeutic interventions on health outcomes is partly mediated through their effects on perceived self-efficacy. The stronger the perceived efficacy the interventions instill, the more likely people are to enlist and sustain the effort needed to adopt and maintain health-promoting behavior. This has been shown in studies in such diverse areas of health as level of postcoronary recovery (Ewart et al. 1983; Taylor et al. 1985); enhancement of pulmonary function in patients suffering from chronic pulmonary disease (Kaplan et al. 1984); reduction in pain and dysfunction in rheumatoid arthritis (Lorig et al. 1989; O'Leary et al. 1988); reduction of the pain of childbirth (Manning and Wright 1983); elimination of tension headaches (Holroyd et al. 1984); management of chronic low back, neck, and leg pain and impairment (Council et al. 1988; Dolce 1987); maintenance of diabetic self-care (Crabtree 1986) modification of eating habits and disorders (Glynn and Ruderman 1986; Jeffrey et al. 1984; Love et al. 1985; Maibach et al. 1989); reduction of cholesterol through dietary means (McCann et al. 1988); adherence to prescribed rehabilitative activities (Ewart et al. 1986); adoption and adherence to programs of physical exercise (McAuley 1992; Desharnais et al. 1986; Sallis et al. 1986); control of sexual practices that pose a high risk for transmission of acquired immune deficiency syndrome (AIDS) (McKusick et al. 1990); and control of addictive habits that impair health, such as alcohol abuse, smoking, and use of opiate drugs (Annis and Davis 1989; Condiotte and Lichtenstein 1981; Gossop et al. 1985; Devins and Edwards 1988; DiClemente 1986).

It is one thing to get people to change their health-related behavior; it is another thing to maintain those changes over time. Treatments can be arranged to produce behavioral changes, but if they are to generalize and endure, people must be active agents in their own motivation and behavior. Because they preside constantly over their own behavior, they are in the best position to exercise influence over it. Much of the research on self-directed change has centered on the skills needed to regulate one's own behavior. There is a difference between possessing skills and being able to use them effectively and consistently in difficult circumstances. Self-efficacy beliefs determine whether people abort their efforts prematurely or persevere with whatever means and effort necessary to succeed (Bandura 1988). If they are not fully convinced of their personal efficacy, they rapidly abandon the skills they have been taught when they fail to get quick results or when they suffer setbacks. The strategies for strengthening perceived self-efficacy to enhance maintenance of health-promoting behavior and to reduce vulnerability to relapse will be considered later in this chapter.

As the research amply documents, health outcomes are not governed solely by biologically rooted factors. Psychological determinants also contribute through their impact on both health-related behavior and biological systems that mediate the level of health functioning. Strength of perceived self-efficacy is a psychological prognostic indicator of the course that health outcomes are likely to take. Results of a program of research on enhancement of perceived self-efficacy for postcoronary recovery may illustrate several general issues regarding prognosis of health outcomes and the course they are likely to take.

About half the patients who experience myocardial infarctions have uncomplicated ones (DeBusk et al. 1983). The heart heals rapidly, and they are physically capable of resuming an active life. However, psychological and physical recovery is slow for patients who believe they have an impaired heart. They avoid physical exertion; they fear they cannot handle the strains in their vocational and social life; they give up recreational activities; and they fear that sexual activities will do them in. The recovery problems stem more from patients' beliefs that their cardiac system has been impaired than from physical debility. The rehabilitative task is to convince patients that they have a sufficiently robust cardiovascular system to lead productive lives.

The initial study in this program of research demonstrated that having patients master increasing workloads on the treadmill strengthened patients' beliefs in their physical capabilities (Ewart et al. 1983). The stronger their perceived physical efficacy, the more active they become in their everyday life. Maximal treadmill attainment is a weak predictor of a patient's level and duration of activity. Treadmill experiences, thus, exert their influence indirectly, facilitating recovery by raising patients' beliefs about their physical and cardiac capabilities. Enhanced perceived efficacy, in turn, fosters more active pursuit of everyday activities.

Ewart and his colleagues have further shown that patients' beliefs about their physical efficacy predicts compliance with prescribed exercise programs, whereas actual physical capability does not (Ewart et al. 1986). This corroborates the earlier findings that the effect of treadmill experiences on activity level is largely mediated by changes in perceived self-efficacy. Patients who have a high sense of efficacy tend to overexercise, whereas those who doubt their physical efficacy underexercise at levels that provide little cardiovascular benefits.

Psychological recovery from a heart attack is a social, rather than solely individual, matter. Virtually all patients were males. The wives' judgments of their husbands' physical and cardiac capabilities can aid or retard the recovery process. The direction that social support takes is partly determined by perceptions of efficacy. Spousal support is likely to be expressed in curtailment of activity if the husband's heart function is regarded as impaired, and as encouragement of activity if his heart function is judged to be robust. In the program designed to enhance postcoronary recovery (Taylor et al. 1985), the treadmill was used to raise and strengthen spousal and patients' beliefs in their cardiac capabilities.

Several weeks after patients had a heart attack their beliefs about the amount of strain their heart could withstand were measured. They then performed a symptom-limited treadmill, mastering increasing workloads with three levels of spouse involvement in the treadmill activity. The wife was either uninvolved in the treadmill activity; was present to observe her husband's stamina as he performed the treadmill under increasing workloads; or observed her husband's performance, whereupon she performed the treadmill exercises to gain first-hand experience of the physical stamina required. It was reasoned that having the wives experience the strenuousness of the task and seeing their husbands match or surpass them would convince them that their husband had a robust heart.

After the treadmill activities, couples were fully informed by the cardiologist about the patients' level of cardiac functioning and their capacity to resume activities of daily life. If the treadmill is interpreted as an isolated task, its impact on perceived cardiac and physical capability may be limited. To achieve a generalized impact of enhanced self-efficacy on diverse domains of functioning, the stamina on the treadmill was presented as a generic indicant of cardiovascular capability—that the patients' level of exertion exceeded whatever strain everyday activities might place on their cardiac system. This would encourage them to resume activities in their everyday life that placed weaker demands on their cardiac system than the heavy workloads on the treadmill. The patient's and spouse's beliefs concerning his physical and cardiac capabilities were measured before and after the treadmill activity and again after the medical counseling.

Figure 2–1 shows the patterns of change in perceptions of the patients' physical and cardiac capacities at different phases of the experiment and with varying degrees of spousal involvement in the treadmill activity. Treadmill performances increased patients' beliefs in their physical and cardiac capabilities. Initially, the beliefs of wives and their husbands were highly discrepant—husbands judged themselves moderately hearty, whereas wives judged their husbands' cardiac capability as severely impaired and incapable of withstanding physical and emotional strain. Spouses who were either uninvolved in, or merely observers of, the treadmill activity did not change their considerable doubts about their husbands' physical and cardiac capabilities. Even the detailed medical counseling by the cardiology staff did not alter their preexisting beliefs of their husbands' cardiac debility. However, wives who had personally experienced the strenuousness of the treadmill were persuaded that their husbands had sufficiently robust hearts to withstand the nor-

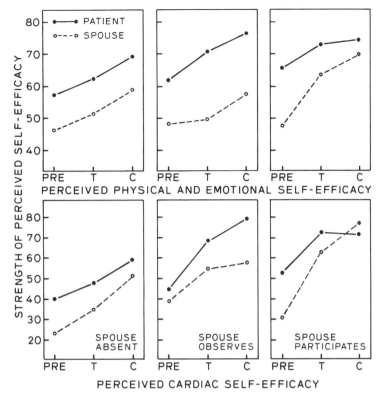

Figure 2-1. Changes in perceived physical and cardiac efficacy as a function of level of spouse involvement in the treadmill activity, patients' treadmill exercises, and the combined influence of treadmill exercises and medical counseling. Perceived efficacy was measured before the treadmill activity (*Pre*), after the treadmill activity (*T*), and after the medical counseling (*C*). One set of efficacy scales measured beliefs about the patient's capability to bear physical stressors (e.g., physical exertion, sexual activity) and strain of emotional stressors (e.g., anger arousal, social discord); cardiac efficacy measured beliefs about how much strain the patient's heart could withstand (Taylor et al. 1985).

mal strains of everyday activities. The participant experience apparently altered spousal cognitive processing of treadmill information, giving greater weight to indicants of cardiac robustness than to symptoms of cardiac debility. The change in perceived efficacy made the wives more accepting of the medical counseling. Following the medical counseling, couples in the participant spouse group had congruently high perceptions of the patients' cardiac capabilities.

The findings further show that beliefs of cardiac capabilities can affect the course of recovery from myocardial infarction. The higher the patients' and the spouses' beliefs in the patients' cardiac capabilities, the greater was the patients' cardiovascular functioning as measured by peak heart rate and maximal workload achieved on the treadmill 6 months later. The joint belief in the patients' cardiac efficacy proved to be the best predictor of cardiac functional level. Initial treadmill performance did not predict level of cardiovascular functioning in the follow-up assessment when perceived efficacy was partialled out. But perceived cardiac efficacy did predict the level of cardiovascular functioning when initial treadmill performance was partialled out.

Wives who believe that their husbands have a robust heart are more likely to encourage them to resume an active life than those who believe their husbands' heart is impaired and vulnerable to further damage. The positive relation between the wife's perceptions of her husband's cardiac capability and his treadmill accomplishments months later is, in all likelihood, partly mediated by spousal encouragement of activities during the interim period. Pursuit of an active life improves the patient's physical ability to engage in activities without overtaxing the cardiovascular system.

Prognostic judgments are not simply nonreactive forecasts of a natural history of a disease. Except in extreme pathologies that may be overwhelmingly determined by biological factors, the nature and course of clinical outcomes is partly dependent on psychological sources of influence. Strong belief in the capability to exercise some control over one's physical condition serves as a psychological prognostic indicator of the level of health functioning. Thus, people with similar levels of physical impairment can achieve different functional outcomes depending on their self-beliefs of efficacy (Holman and Lorig 1992; Kaplan et al. 1984; Lorig et al. 1989; O'Leary et al. 1988). Even in the case of severe permanent impairment in which only partial recovery is possible, psychosocial factors will affect how much of the possible functional capacity is realized. Because prognostic information can affect patients' beliefs of their physical efficacy, diagnosticians not only foretell, but may partly influence the course of recovery from disease. This effect will be examined in greater detail later in this chapter.

Mode of Conveying Prognostic Information

Another important issue in the clinical management of patients concerns the way in which prognostic information is conveyed to them. Usually, possible outcomes and the probabilities associated with them are described. However, verbal prognostications alone may not have the intended impact, especially when they contradict strong preexisting beliefs. This is true even for positive prognostications if patients associate prescribed restorative activities with grave risks. For example, in the study of postcoronary rehabilitation, wives were not at all reassured of their husbands' hardiness by

the positive prognostic judgments of the medical staff unless they had benefit of direct confirmatory experiences.

To increase their persuasive influence, clinicians may have to convey positive prognostic information to their patients by word and by structuring performance tasks for them that provide self-convincing experiences. This issue will be explored later when discussing strategies for instilling a sense of personal efficacy and reducing vulnerability to relapse.

Psychological Impact of Diagnostic Procedures

The manner in which diagnostic tests are conducted also can influence patients' beliefs about their efficacy, as in the cognitive processing of somatic information from the treadmill test. Treadmill activity produces many negative signs, such as fatigue, pain, shortness of breath, and other exercise-induced symptoms that increase as the task continues. Patients who focus on their physical stamina as they master increasing workloads will judge their cardiac system as more robust than patients who selectively attend to and remember the negative somatic signs. Positive indicants of capability can be more salient if patients receive ongoing feedback of their performance attainments as they master heavier workloads. Judgment of cardiac efficacy will vary depending on how this diverse symptom information and the indicants of cardiac robustness are weighted and integrated.

This is illustrated in a study of a group of healthy men and women who completed a symptom-limited treadmill before entering an exercise program (Juneau et al. 1986). Half the participants received concurrent feedback of the workloads they mastered on the treadmill task. The other half received the feedback about their physical attainments just after they had completed the treadmill task. Their perceived cardiac efficacy was measured before and after the treadmill performance. They also recorded the physical signs they recall having experi-

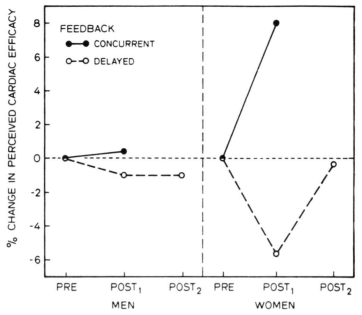

Figure 2-2. Impact of treadmill diagnostic performances on judgment of cardiac efficacy under conditions in which participants received concurrent feedback of the workloads they mastered or the feedback about their attainments was delayed until after the treadmill test was completed (Juneau et al. 1986).

enced during the treadmill activity. Figure 2–2 shows how treadmill performances with and without concurrent feedback affect self-beliefs of cardiac capabilities.

Without feedback of positive indicants of capability, exercise-induced symptoms completely dominate attention and memory representation of the treadmill experience. For healthy men who have a more resilient conception of their cardiac capabilities, a taxing treadmill test without feedback did not alter the beliefs that they had a robust cardiac system. However, positive feedback that makes physical attainments on the treadmill more noticeable raised women's judgments of their cardiac capabilities. In the absence of such feedback, women read the mounting negative physiological sensations accompanying increasing exertion on the treadmill as indicants of cardiac limitations and lowered their judgments of their cardiac capabilities. Women did not experience any more negative physiological sensations than men. The adverse impact of treadmill experiences without positive feedback stemmed from negative cognitive pro-

cessing of symptom information rather than from greater amounts or salience of such symptoms.

Preconceptions tend to bias how information is weighted and integrated (Bandura 1986; Nisbett and Ross 1980). A similar process is indicated in women's reactions to delayed positive feedback regarding their treadmill performances. When told of their notable physical attainments, they raised their perceived cardiac efficacy to the pre-treadmill level, but they achieved no net gain from the treadmill experience. Positive signs of cardiac capability are difficult to assimilate after conceptions of one's efficacy have already been formed under conditions in which negative signs clearly dominate. A coronary can markedly undermine beliefs concerning one's cardiac efficacy. A strong preconception of physical impairment makes negative physiological reactions to performance tests highly salient and recallable. Therefore, concurrent positive feedback of physical stamina would be especially important for countering beliefs of a frail cardiac capability in postcoronary patients who have not suffered clinical complications.

Any diagnostic procedure that gauges impairments and capabilities by testing the upper limits of performance will create a pattern of experiences reflecting both strengths and deficiencies. Patients who selectively notice and recall their performance deficiencies will judge their capabilities as lower than those who notice their strengths as well. As shown in the treadmill example, the adverse impact on perceived self-efficacy of diagnostic procedures that generate negative experiences can be reduced or counteracted by structuring performance tests in ways that give salience to one's remaining strengths. In addition to the type and timing of verbal feedback given to patients, some evidence suggests that diagnostic tasks that create mounting failure by an ascending order of difficulty produce more adverse effects than if tasks of different levels of difficulty are intermixed to maintain a sense of attainment (Zigler and Butterfield 1968). Analysis of how the structure of diagnostic procedures and preconceptions of personal efficacy bias attention to, and cognitive processing of, somatic and behavioral information is clinically important and theoretically interesting (Cioffi 1991). The knowledge gained from these types of microanalytical studies would add greatly to the understanding of the psychological impact of experiences of diagnostic procedures.

Scope of Prognostic Schemes

Another issue regarding prognosis concerns the range of factors included in a prognostic scheme, specifically whether health outcomes are viewed from an individualistic biomedical perspective or from a broader biopsychosocial perspective. As mentioned previously, the level of health functioning is determined not only by the patients' physical status, but also by a system of social influences that can enhance or impede the progress they make. For example, in the study of recovery from uncomplicated myocardial infarction, wives' beliefs in their husbands' cardiac robustness were better predictors of level of recovery of cardiac function than physical indicants of cardiovascular status as measured by the treadmill.

To the extent that interpersonal influences contribute to health outcomes, giving these factors some weight in prognostic schemes will enhance their predictive usefulness. If they are not considered, a puzzling variability remains in the courses that health changes take and differential functional attainments of people who are equally physically impaired are unexplained.

Self-Validating Potential of Prognostic Judgments

Health outcomes are related to predictive factors in complex, multidetermined, and probabilistic ways. Prognostic judgments, therefore, involve some degree of uncertainty. The predictiveness of a given prognostic scheme depends on the number of relevant predictors it encompasses, the relative validities and inter-relations of the predictors, and the adequacy with which they are measured. There is always leeway for expectancy effects to operate because prognostic schemes rarely include all of the relevant biological and psychosocial predictors, and the predictors that are singled out usually have less than perfect validity. Based on selected sources of information, diagnosticians form expectations about the probable course of a disease. The more confident they are in the validity of their prognostic scheme, the stronger are their prognostic expectations.

Prognostic expectations are conveyed to patients by attitude, word, and the type and level of care provided to them. As alluded to previously, prognostic judgments have a self-confirming potential. Expectations can alter patients' self-beliefs and behaviors in ways that confirm the original expectations. Evidence indicates that the self-efficacy mechanism operates as one important mediator of self-confirming effects. This is most clearly revealed in laboratory studies in which arbitrary information of personal capabilities is conveyed to people and its effects on their perceived self-efficacy and behavior are measured. Self-management of pain provides a relevant example.

There are several ways by which perceived coping efficacy can facilitate the personal management of pain. People who believe they can alleviate pain will likely enlist whatever ameliorative skills they have learned and will persevere in their efforts. Those who judge themselves as inefficacious make no effort to reduce their level of experienced pain. A sense of coping efficacy also reduces distressing anticipations that create aversive reactions and bodily tension, which exacerbate pain sensations and discomfort. Consciousness has a very limited capacity. It is hard to keep more than one thing in mind at the same time. If pain sensations are supplanted in consciousness, they are felt less. Dwelling on pain sensations only makes them more noticeable and thus more difficult to bear. Perceived self-efficacy can lessen the extent to which painful stimulation is experienced as conscious pain by diverting attention from pain sensations to competing engrossments or by nondistressing construal of the sensations (McCaul and Malott 1984).

In a study conducted by Litt (1988), people given bogus feedback that they were good pain controllers raised their perceived self-efficacy and tolerance of cold pressor pain; those led to believe that they were poor pain controllers lowered their perceptions of their efficacy and found it hard to bear pain. Instated self-efficacy belief was a better predictor of pain tolerance than past level of actual pain tolerance. A low sense of efficacy constrained efforts to ameliorate pain even when the opportunity to exercise some personal control existed. Arbitrarily altered perceptions of efficacy also affected preference for personal or external control of pain. Those whose efficacy was raised preferred a strong personal role in the management of their pain. Those whose efficacy was lowered wanted external interventions to stop their pain.

Holroyd and his colleagues (1984) demonstrated with sufferers of recurrent tension headaches that the benefits of biofeedback training stem more from boosts in perceived coping efficacy than from the muscular exercises. Perceived self-efficacy, created by bogus feedback that one is a skilled relaxer for controlling pain, predicted reduction in tension headaches, whereas the actual amount of change in muscle activity achieved in treatment was unrelated to the incidence of subsequent headaches.

These experiments should not be interpreted to mean that arbitrary persuasive influence is a good way of enhancing efficacy beliefs to reduce functional impairments associated with clinical conditions. Rather, these studies have special bearing on the self-confirming potential of prognostic judgments because efficacy beliefs are altered independently of actual physical status. In clinical practice, personal efficacy is strengthened by providing patients with the knowledge and means to make optimal use of their capabilities. In the amelioration of clinical pain, the skill development includes behavioral and cognitive pain control techniques. Results of this research indicate that changes in perceived self-efficacy mediate the analgesic potency of different modes of treatment (Dolce 1987; Lorig et al. 1989; Manning and Wright 1983; O'Leary et al. 1988).

The preceding analysis of self-confirming processes focused solely on how people's self-beliefs of efficacy and behavior are affected by what they are told about their capabilities. Other evidence suggests that prognostic judgments may bias the treatment people receive and the information they are given. In these experiments, individuals are arbitrarily led to form high or low expectations for others. The studies reveal that individuals treat others differently when they have high or low expectations in ways that confirm the original expectations (Jones 1977; Jussim 1986). Although there is some variation in the results, the findings generally show that with induced high expectations, individuals pay more attention to those in their charge, provide them with more emotional support, create greater opportunities for them to build their competences, and give them more positive feedback than with induced low expectations.

Differential care that promotes in patients different levels of self-efficaciousness and skill in managing health-related behavior can exert stronger impact on the trajectories of health functioning than just conveying prognostic information. The effects of verbal prognostications alone may be short-lived if they are repeatedly contradicted by personal experiences due to deficient capabilities. However, a sense of personal efficacy rooted in enhanced competencies fosters functional attainments that create their own experiential validation. Clinical transactions operate bidirectionally to shape the course of change. The functional improvements fostered by positive expectancy influences further strengthen clinicians' beneficial expectations and their sense of efficacy to aid progress. In contrast, negative expectations that breed functional declines can set in motion a downward course of mutual discouragement.

Conception of Capability

In recent years, major changes have occurred in the conception of human ability (Bandura 1990; Sternberg and Kolligian 1990). Ability is not a fixed property in one's behavioral repertoire. Rather, it involves a generative capability in which cognitive, emotional, and motivational factors govern the translation of knowledge and skills into performance attainments. Thus, with the same set of capabilities, people may perform poorly, adequately, or extraordinarily depending on their thinking patterns, emotional states, and level of motivation.

The variable use of capabilities is illustrated in research on the impact of self-efficacy beliefs on level of memory functioning with advancing age (Bandura 1989; Berry et al. 1989; Lachman et al. 1987). Human memory is an active constructive process in which information is semantically elaborated, transformed, and reorganized into meaningful cognitive representations that aid recall. People who view memory as a cognitive skill that they can improve are more likely to exert the effort needed to convert the experiences into recallable symbolic forms than those who regard it as an inherent capacity that inevitably declines with aging. Consistent with this expectation, the more that older adults believe in their memory capabilities, the more time they devote to cognitively processing memory tasks (Berry 1987). Higher processing effort, in turn, produces better memory performance. Perceived self-efficacy affects actual memory performance both directly and indirectly through level of cognitive effort. Those who regard memory as simply a biologically shrinking capacity have little reason to try to exercise control over their memory functioning. They are quick to read instances of normal forgetting as indicants of declining cognitive capacity. The less they believe in their memory capabilities, the poorer use they make of their cognitive capabilities. Mnemonic training that improves memory performance does not lead to generalized use of the memory aids unless the training also raises people's beliefs in their memory efficacy (Rebok and Balcerak 1989).

The undermining efforts of disbelief in one's capabilities also may be mediated through depression. A low sense of personal efficacy to fulfill desired goals and to secure things that bring satisfaction to one's life creates depression. Despondent mood further diminishes self-beliefs of capability (Kavanagh and Bower 1985) in ways that can debilitate memory functioning. Indeed, West, Berry, and Powlishta (1989) found that depression is accompanied by a low sense of memory efficacy which in turn is associated with deficient memory performances.

Self-efficacy beliefs similarly contribute to level of physical functioning. This is most strikingly revealed in experiments in which beliefs of physical efficacy are raised in some people and lowered in others by information unrelated to their actual physical capabilities (Weinberg et al. 1979). The higher the induced beliefs in one's physical efficacy, the greater the physical attainments are. Deficient performances spur those with a high sense of efficacy to even greater physical effort, but further impair the performances of those whose efficacy had been undermined. Self-beliefs of physical efficacy arbitrarily heightened in females and arbitrarily weakened in males obliterate large preexisting sex differences in physical strength. The nonability determinants of functional attainments have been amply documented in diverse domains of activity (Bandura 1990, 1992).

Ways of Instilling Resilient Self-Efficacy

People's beliefs about their efficacy can be altered in four principal ways (Bandura 1986): The most effective way of instilling a strong sense of efficacy is through *mastery experiences*. Successes build a robust sense of efficacy. Failures undermine it, especially if failures occur early in the course of developing competencies. Self-efficacy is best developed through a series of subgoals that serve to expand competencies. Subgoal attainments provide indicants of mastery for enhancing a sense of personal efficacy and sustaining motivation along the way. If subgoal challenges are set too high, most performances prove disappointing and reduce motivation to continue the pursuit. People who have a low sense of efficacy are easily discouraged by failure and are quick to attribute it to personal incapacities.

Neurological injuries that produce severe permanent impairments can be devastatingly demoralizing to patients and their families. Patients have to reorganize their perspective to learn alternative ways of regaining as much control as possible over their activities. Goals need to be restructured in ways that capitalize on remaining capacities. Ozer (1988) illustrates effective ways of structuring goals couched in functional terms to minimize disabilities created by chronic neurological impairment. Focus on achievement of functional improvements rather than on degree of organic impairments helps to counteract self-demoralization. Making difficult activities easier by

breaking them down into graduated subtasks of attainable steps helps to prevent self-discouragement of rehabilitative efforts.

Development of resilient self-efficacy requires some experience in mastering difficulties through perseverant effort. If people experience only easy successes, they expect quick results; their sense of efficacy is easily undermined by failure. Some setbacks and difficulties in human pursuits are useful in teaching that success usually requires sustained effort. After people become convinced that they can succeed, they persevere in the face of adversity and quickly rebound from setbacks. By sticking it out through tough times, they emerge from adversity with a stronger sense of efficacy.

The second way of enhancing personal efficacy is through *modeling*. People partly judge their capabilities by comparison with others (Bandura 1990). Seeing people similar to oneself regain, by perseverant effort, some control over their activities despite impairment raises observers' beliefs about their own abilities to regain some control. The failures of others coping with similar problems instill self-doubts about one's ability to manage similar tasks. Having expatients exemplify the active lives they are leading can be especially influential in strengthening beliefs that functional improvements are realizable. Seeing how others manage difficult conditions can alter self-beliefs of efficacy through ways other than social comparison. Efficacious models can teach competencies and effective strategies for dealing with taxing situations. Adoption of serviceable strategies raises perceived self-efficacy.

Social persuasion is the third mode of infuence. People try to talk others into believing they possess the capabilities to achieve what they seek. Realistic boosts in efficacy can lead to a greater exertion of effort, which increases the chances of success. However, raising unrealistic beliefs of personal capability creates the risk of inviting failure. Successful efficacy builders, however, do more than convey positive appraisals. In addition to raising patients' beliefs in their capabilities, they structure tasks for them in ways that are likely to bring improvements and avoid placing them in situations where they are likely to fail. By maintaining an efficacious attitude that functional gains are attainable when patients are beset with self-doubts, clinicians can help patients to sustain their coping efforts in the face of reverses and discouraging obstacles. Through these various means, clinicians can help patients to make the best use of their capacities.

People also rely partly on their physiological state when judging their capabilities. They read their anxiety arousal and tension as signs of vulnerability to dysfunction. In activities involving strength and stamina, people interpret their fatigue, aches, and pains as indicants of physical inefficacy. The fourth way of modifying self-beliefs of efficacy is to equip patients with skills to reduce aversive physiological reactions or alter their interpretation of somatic information. The meanings assigned to bodily sensations and states can have significant health consequences (Bandura 1991; Cioffi 1991).

The health benefits of a sense of personal efficacy do not arise simply from the incantation of capability. Saying something should not be confused with believing it. Simply saying that one is capable is not necessarily self-convincing, especially when it contradicts firm preexisting beliefs. Self-efficacy beliefs are the product of a complex process of self-persuasion that relies on cognitive processing of diverse sources of efficacy information conveyed behaviorally, vicariously, socially, and physiologically. Their strength is affected by the authenticity of the efficacy information on which they are based. Self-efficacy beliefs that are firmly established are resilient to adversity. In contrast, weakly held self-beliefs are highly vulnerable to change and negative experiences readily reinstate disbelief in one's capabilities.

Reduction of Vulnerability to Relapse

Each of the methods for enhancing efficacy can be used to develop the resilient sense of perceived efficacy needed to override difficulties that inevitably arise. With regard to the performance mode, a resilient belief in one's personal efficacy is built through demonstration trials in the exercise of control over progressively more difficult tasks. For example, as part of instruction in cognitive pain control strategies, arthritic patients were given efficacy demonstration trials in which they performed pain-producing activities with and without cognitive control and rated the level of pain they experienced (O'Leary et al. 1988). Explicit evidence that they achieved substantial reduction in pain by cognitive means persuaded the patients that they could exercise some control over pain by enlisting cognitive control strategies. Self-efficacy validating trials not only serve as efficacy builders, but put to trial the value of the techniques being taught.

Modeling influences, in which other patients demonstrate their ability to cope with difficulties and setbacks and show that success usually requires tenacious effort, can further strengthen perceived self-efficacy. Moreover, modeled perseverant success can alter the diagnosticity of failure experiences as partly reflecting difficult situational predicaments rather than solely inherent personal limitations. Difficulties and setbacks prompt redoubling of efforts rather than provoke self-discouraging doubts about one's capabilities. For example, pain threshold and tolerance is affected by modeling influences (Craig, 1983). Thus, people who have seen others persevere despite pain function much more effectively when they are in pain than if they had seen others give up quickly (Turkat and Guise 1983; Turkat et al. 1983).

Persuasive influences that instill self-beliefs conducive to optimal use of skills also can contribute to staying power. As a result, people who are persuaded that they have what it takes to succeed and are told that the gains achieved in treatment verify their capability are more successful in sustaining their altered health habits than those who undergo the same treatment without the efficacy-enhancing component (Blittner et al. 1978).

Self-Management of Chronic Disability

Chronic disease has become the dominant form of illness and the major cause of disability. The treatment of chronic disease must focus on self-management of physical conditions over time rather than on cure. This requires, among other things, pain amelioration, enhancement and maintenance of functioning with growing physical disability, and development of self-regulative compensatory skills. Holman and Lorig (1992) have devised a prototypical model for the self-management of different types of chronic diseases. The self-management skills include (1): cognitive pain control techniques; (2) self-relaxation; (3) proximal goal setting (to increase both the level of activity and the use of self-incentives as motivators); (4) problem solving and self-diagnostic skills (for monitoring and interpreting fluctuation in one's health status and skills in locating community resources); (5) managing medication programs; and (6) effective ways of dealing with other aspects of health care systems to optimize health benefits. Participants are taught how to exercise control over their physical condition through modeling of self-management skills, guided mastery practice, informative feedback, and efficacy demonstration trials.

The effectiveness of this self-regulative approach has been tested extensively for ameliorating the debility and chronic pain of arthritis (Lorig, Seleznick et al. 1989). Patients suffering from rheumatoid arthritis substantially improved their psychophysical functioning following treatment compared to matched controls who received an arthritis helpbook describing self-management techniques and were encouraged to be more active (O'Leary et al. 1988). The self-management program increased patients' perceived self-efficacy to reduce pain and other debilitating aspects of arthritis, and to pursue potentially painful activities. The treated patients reduced their pain and inflammation in their joints, and were less debilitated by their arthritic condition. The higher their perceived coping efficacy, the less pain they experienced, the less they were disabled by their arthritis, and the greater the reduction they achieved in joint impairment. The more efficacious were also less depessed, less stressed, and they slept better.

In follow-up assessments conducted four years later, arthritis patients who had the benefit of self-management training displayed increased self-efficacy (+17%), reduced pain (−19%), experienced slower biological progression of their disease (+9%) over the four-year period and reduced the number of physician visits substantially (−43%). Enhancement of functioning despite some biological progression of the disease provides further testimony that functional limitations may be governed more by self-beliefs of capability than by degree of actual physical impairment (Baron, Dutil, Berkson, Lander, and Becker 1987). Tests of alternative mediating mechanisms reveal that neither increases in knowledge nor degree of change in health behaviors are appreciable predictors of health functioning (Lorig, Chastain et al. 1989; Lorig, Seleznick et al. 1989). However, both baseline-perceived self-efficacy and changes in perceived self-efficacy to exercise some control over one's arthritic condition instilled by treatment explain the variance in pain. When patients are equated for degree of physical debility, those who believe they can exercise some influence over how much their arthritic condition affects them lead more active lives and experience less pain (Shoor and Holman 1984).

Conclusion

This chapter addressed the issue of prognosis from a biopsychosocial perspective on health and human capability. Converging lines of evidence indicate that self-belief of efficacy operates as an influential prognostic indicator of level of functioning. Strength of perceived self-efficacy can influence the course of health outcomes and functional status through its intervening effects on cognitive, motivational, affective, and biological processes. Prognostic schemes that encompass psychosocial determinants will have greater predictive power than those that ignore them. Prognostic evaluations have a self-confirming potential. Whether patients are expected to do well or to do poorly can affect their clinical management and beliefs in their capabilities in ways that confirm the original expectations. Patients are best served by prognosticians that enable them to realize their potential.

References

Ader, R.; Cohen, N. CNS-immune system interactions: Conditioning phenomena. Behav. and Brain Sci. 8:379–394; 1985.

Annis, H. M.; Davis, C. S. Relapse prevention. In: Hester, R. K.; Miller, W. R., eds. Handbook of alcoholism treatment approaches. New York: Pergamon Press; 1989: p. 170–182.

Bandura, A. Social foundations of thought and action: a social cognitive theory. Englewood Cliffs, NJ: Prentice-Hall, Inc.; 1986.

Bandura, A. Self-regulation of motivation and action through goal systems. In: Hamilton, V.; Bower G. H.; Frijda, N. H.; eds. Cognitive perspectives on emotion and motivation. Dordrecht: Kluwer Academic Publishers; 1988: p. 37–61.

Bandura, A. Regulation of cognitive processes through perceived self-efficacy. Dev. Psych. 25:729–735; 1989.

Bandura, A. Reflections on nonability determinants of competence. In: Sternberg, R. J.; Kolligian, J. Jr., eds. Competence considered. New Haven, CT: Yale University Press; 1990: p. 315–362.

Bandura, A. Self-efficacy mechanism in physiological activation and health-promoting behavior. In: Madden, J., IV, ed. Neurobiology of learning, emotion and affect. New York: Raven; 1991: p. 229–270.

Bandura, A. Self-efficacy mechanism in psychobiological functioning. In: Schwarzer, R., ed. Self-efficacy: Thought control of action. Washington, D.C.: Hemisphere; 1992, in press.

Bandura, A.; Cioffi, D.; Taylor, C. B.; Brouillard, M. E. Perceived self-efficacy in coping with cognitive stressors and opioid activation. J. Pers. Soc. Psychol. 55:479–488; 1988.

Bandura, A.; Taylor, C. B.; Williams, S. L.; Mefford, I. N.; Barchas, J. D. Catecholamine secretion as a function of perceived coping self-efficacy. J. Consult. Clin. Psychol. 53:406–414; 1985.

Baron, M.; Dutil, E.; Berkson, L.; Lander, P.; Becker, R. Hand function in the elderly: Relation to osteoarthritis. J. Rheumatology 14:815–819; 1987.

Beck, K. H.; Lund, A. K. The effects of health threat seriousness and personal efficacy upon intentions and behavior. J. Appl. Soc. Psychol. 11:401–415; 1981.

Berry, J. M. A self-efficacy model of memory performance. Paper presented to the American Psychological Association meetings, New York, NY; 1987.

Berry, J. M.; West, R. L.; Dennehey, D. Reliability and validity of the memory self-efficacy questionnaire (MSEQ). Dev. Psychol. 25:701–713; 1989.

Blittner, M.; Goldberg, J.; Merbaum, M. Cognitive self-control factors in the reduction of smoking behavior. Behav. Ther. 9:553–561; 1978.

Cioffi, D. Beyond attentional strategies: A cognitive-perceptual model of somatic interpretation. Psychol. Bull. 109:25–41; 1991.

Coe, C. L.; Levine, S. Psychoimmunology: an old idea whose time has come. In: Barchas P. R., ed. Social physiology of social relations. Oxford: Oxford University Press; in press.

Coletti, G.; Supnick, J. A.; Payne, T. J. The smoking self-efficacy questionnaire (SSEQ): preliminary scale development and validation. Behav. Assess. 7:249–260; 1985.

Condiotte, M. M.; Lichtenstein, E. Self-efficacy and relapse in smoking cessation programs. J. Consult. Clin. Psychol. 49:648–658; 1981.

Council, J. R.; Ahem, D. K.; Follick, M. J.; Kline, C. L. Expectancies and functional impairment in chronic low back pain. Pain. 33:323–331; 1988.

Crabtree, M. K. Self-efficacy beliefs and social support as predictors of diabetic self-care. San Francisco: University of California; 1986.

Craig, K. D. A social learning perspective on pain experience. In: Rosenbaum, M.; Franks, C. M.; Jaffe Y., eds. Perspectives on behavior therapy in the eighties. New York: Springer; 1983: p. 311–327.

DeBusk, R. F.; Kraemer, H. C.; Nash, E. Stepwise risk stratification soon after acute myocardial infarction. Am. J. Cardiol. 12:1161–1166; 1983.

Desharnais, R.; Bouillon, J.; Godin, G. Self-efficacy and outcome expectations as determinants of exercise adherence. Psychol. Rep. 59:1155–1159; 1986.

DiClemente, C. C. Self-efficacy and the addictive behaviors. J. Soc. Clin. Psychol. 4:302–315; 1986.

Dolce, J. J. Self-efficacy and disability beliefs in behavioral treatment of pain. Behav. Res. Ther. 25:289–300; 1987.

Engel, G. L. The need for a new medical model: a challenge for biomedicine. Science. 196:129–136; 1977.

Ewart, C. K.; Stewart, K. J.; Gillilan, R. E.;

Kelemen, M. H. Self-efficacy mediates strength gains during circuit weight training in men with coronary artery disease. Med. Sci. Sports Exerc. 18:531–540; 1986.

Ewart, C. K.; Stewart, K. J.; Gillilan, R. E.; Kelemen, M. H.; Valenti, S. A.; Manley, J. D.; Kalemen, M. D. Usefulness of self-efficacy in predicting overexertion during programmed exercise in coronary artery disease. Am. J. Cardiol. 57:557–561; 1986.

Ewart, C. K.; Taylor, C. B.; Reese, L. B.; DeBusk, R. F. Effects of early post-myocardial infarction exercise testing on self-perception and subsequent physical activity. Am. J. Cardiol. 51:1076–1080; 1983.

Glynn, S. M.; Ruderman, A. J. The development and validation of an eating self-efficacy scale. Cog. Ther. Res. 10:403–420; 1986.

Gossop, M.; Green, L.; Phillips, G.; Bradley, B. Factors predicting outcome among opiate addicts after treatment. Brit. J. Clin. Psych. 29:209–216; 1990.

Holman, H.; Lorig, K. Perceived self-efficacy in self-management of chronic disease. In: Schwarzer, R., ed. Self-efficacy: Thought control of action. Washington, D.C.: Hemisphere; 1992.

Holroyd, K. A.; Penzien, D. B.; Hursey, K. G.; Tobin, D. L.; Rogers, L.; Holm, J. E.; Marcille, P. J.; Hall, J. R.; Chila, A. G. Change mechanisms in EMG biofeedback training: cognitive changes underlying improvements in tension headache. J. Consult. Clin. Psychol. 52:1039–1053; 1984.

Jeffrey, R. W.; Bjornson-Benson, W. M.; Rosenthal, B. S.; Lindquist, R. A.; Kurth, C. L.; Johnson, S. L. Correlates of weight loss and its maintenance over two years of follow-up among middle-aged men, Prev. Med. 13:155–168; 1984.

Jones, R. A. Self-fulfilling prophesies: Social, psychological, and physiological effects of experiences. Hillsdale, NJ: Erlbaum; 1977.

Juneau, M.; Rogers, F.; Bandura, A.; Taylor, C. B.; DeBusk, R. Cognitive processing of treadmill experiences and self-appraisal of cardiac capabilities. Stanford, CA: Stanford University. Unpublished manuscript.

Jussim, L. Self-fulfilling prophecies: a theoretical and integrative review. Psychol. Rev. 93:429–445; 1986.

Kaplan, R. M.; Atkins, C. J.; Reinsch, S. Specific efficacy expectations mediate exercise compliance in patients with COPD. Health Psychol. 3:223–242; 1984.

Kavanagh, D. J.; Bower, G. H. Mood and self-efficacy: impact of joy and sadness on perceived capabilities. Cog. Ther. Res. 9:507–525; 1985.

Kolligian, J., Jr.; Sternberg, R. J., eds. Competence considered: Perceptions of competence and incompetence across the lifespan. New Haven, CT: Yale University Press; 1989.

Lachman, M. E.; Steinberg, E. S.; Trotter, S. D. Effects of control beliefs and attributions on memory self-assessments and performance. Psychology and Aging. 2:266–271; 1987.

Litt, M. D. Self-efficacy and perceived control: cognitive mediators of pain tolerance. J. Pers. Soc. Psychol. 54:149–160; 1988.

Lorig, K.; Chastain, R. L.; Ung, E.; Shoor, S.; Holman, H. Development and evaluation of a scale to measure perceived self-efficacy in people with arthritis. Arthritis Rheum. 32:37–44; 1989.

Lorig, K.; Seleznick, M.; Lubeck, D.; Ung, E.; Chastain, R. L.; Holman, H. R. The beneficial outcomes of the arthritis self-management course are not adequately explained by behavior change. Arthritis and Rheumatism 32:91–95; 1989.

Love, S. Q.; Ollendick, T. H.; Johnson, C.; Schlezinger, S. E. A preliminary report of the prediction of bulimic behavior: A social learning analysis. Bull. Soc. Psychol. Addict. Behav. 4:93–101; 1985.

Maibach, E. W.; Flora, J.; Nass, C. Changes in self-efficacy and health behavior in response to a minimal contact community health campaign. Health Communication 3:1–15; 1991.

Maier, S. F. Stressor controllability and stress-induced analgesia. In: Kelly, D. D., ed. Stress-induced analgesia. Annals of the New York Academy of Sciences. New York: New York Academy of Sciences; 1986: p. 55–72.

Maier, S. F.; Laudenslager, M. L.; Ryan, S. M. Stressor controllability, immune function, and endogenous opiates. In: Brush F. R.; Overmier J. B., eds. Affect, conditioning, and cognition: Essays on the determinants of behavior. Hillsdale, NJ: Erlbaum; 1985: p. 183–201.

Manning, M. M.; Wright, T. L. Self-efficacy expectancies, outcome expectancies, and the persistence of pain control in childbirth. J. Pers. Soc. Psychol. 45:421–431; 1983.

McCann, B. S.; Follette, W. C.; Driver, J. L.; Brief, D. J.; Knopp, R. H. Self-efficacy and adherence in the dietary treatment of hyperlipidemia. Paper presented at the 96th Annual Convention of the American Psychological Association. Atlanta, Georgia; 1988.

McCaul, K. D.; Malott, J. M. Distraction and coping with pain. Psychol. Bull. 95:516–533; 1984.

McAuley, E. Understanding exercise behavior: A self-efficacy perspective. In Roberts, G. C. ed. Understanding motivation in exercise and sport. Champaign, IL: Human Kinetics; 1992.

McKusick, L.; Coates, T. J.; Morin, S. F.; Pollack, L.; Hoff, C. Longitudinal predictors of reductions in unprotected anal intercourse among gay men in San Francisco: the AIDS behavioral research project. Am. J. Public Health. 80:978–983; 1990.

Nisbett, R.; Ross, L. Human inference: strategies and shortcomings of social judgment. Englewood Cliffs, NJ: Prentice-Hall, Inc.; 1980.

O'Leary, A.; Shoor, S.; Lorig, K.; Holman, H. R. A cognitive–behavioral treatment for rheumatoid arthritis. Health Psychol. 7:527–544; 1988.

Ozer, M. N. The management of persons with spinal cord injury. New York: Demos; 1988.

Rebok, G. W.; Balcerak, L. J. Memory self-efficacy and performance differences in young and old adults: effect of mnemonic training. Dev. Psychol. 25:714–721; 1989.

Sallis, J. F.; Haskell, W. L.; Fortmann, S. P.; Vranizan, M. S.; Taylor, C. B.; Solomon, D. S. Predictors of adoption and maintenance of physical activity in a community sample. Prev. Med. 15:331–341; 1986.

Shavit, Y.; Martin, F. C. Opiates, stress, and immunity: Animal studies. Ann. Behav. Med. 9:11–20; 1987.

Shoor, S. M.; Holman, H. R. Development of an instrument to explore psychological mediators of outcome in chronic arthritis. Trans. Assoc. Amer. Physicians 97:325–331; 1984.

Taylor, C. B.; Bandura, A.; Ewart, C. K.; Miller, N. H.; DeBusk, R. F. Exercise testing to enhance wives' confidence in their husbands' cardiac capabilities soon after clinically uncomplicated acute myocardial infarction. Am. J. Cardiol. 55:635–638; 1985.

Turkat, I. D.; Guise, B. J. The effects of vicarious experience and stimulus intensity of pain termination and work avoidance. Behav. Res. Ther. 21:241–245; 1983.

Turkat, I. D.; Guise, B. J.; Carter, K. M. The effects of vicarious experience on pain termination and work avoidance: a replication. Behav. Res. Ther. 21:491–493; 1983.

Weinberg, R. S.; Gould, D.; Jackson, A. Expectations and performance: an empirical test of Bandura's self-efficacy theory. J. Sport Psychol. 1:320–331; 1979.

West, R. L.; Berry, J. M.; Powlishta, K. K. Self-efficacy and prediction of memory task performance. Submitted for publication, 1989.

Zigler, E.; Butterfield, E. C. Motivational aspects of changes in IQ test performance of culturally deprived nursery school children. Child Dev. 39:1–14; 1968.

3

Prognosis: Keystone of Clinical Neurology

W. T. LONGSTRETH, JR., THOMAS D. KOEPSELL, LORENE M. NELSON, AND GERALD VAN BELLE

Classic epidemiology deals with groups of people, risk factors, and causation, while clinical epidemiology concentrates on groups of patients, prognostic factors, and outcomes (Weiss 1986). Central to understanding clinical epidemiology is the study of prognosis. Prognosis is a set of outcomes and the associated probabilities following the occurrence of a condition that can be a symptom (amaurosis fugax), a sign (asymptomatic bruit), a laboratory finding (high-grade stenosis of the internal carotid artery), or a disease (ischemic stroke) (Longstreth et al. 1987b).

Clinical neuroepidemiology addresses important questions in clinical neurology that practitioners and researchers face regularly. The first issue is diagnosis (Longstreth et al. 1987a). A diagnosis that has no prognostic implications does little more than describe a constellation of patient characteristics. Prognosis links diagnoses to outcomes and identifies the diseases that warrant treatment. Treatment becomes an intervention intended to modify prognosis. Thus, in clinical neurology, the concepts of diagnosis, prognosis, and treatment are inseparable, with prognosis as the keystone.

This review describes a set of principles and guidelines that can be useful in designing, analyzing, and evaluating studies of prognosis. It first critiques the different types of study designs used to address questions of prognosis. The cohort design is the strongest and is discussed in greatest detail. The next section reviews the approach to data analysis in studies of prognosis, specifically addressing the time-linked nature of clinical observations and the necessity of examining multiple variables simultaneously. Often factors in the design and analysis that strengthen a study and the conclusions that can be drawn from it about the subjects actually investigated (internal validity) paradoxically limit the ability to generalize the results to other groups of patients (external validity) (Moses 1985). The question of internal and external validity is explored in several sections.

Design of Studies of Prognosis

Except in the restricted domain of randomized trials of therapy, studies of prognosis do not lend themselves to an experimental design in which factors of interest can be randomly assigned. Many patient character-istics of potential prognostic importance are not subject to manipulation or to random as-signment. The typical study of prognosis in-volves patients to whom various diagnostic and therapeutic maneuvers are applied in the course of clinical care. The ideal study of prognosis would seek to describe the nat-ural history of a condition from its biological start to its end without interventions that could influence outcome.

The two broad classes of nonexperimental (or observational) studies used to investi-gate prognosis are descriptive and analyti-cal. Descriptive studies tally the outcomes for a condition. Analytical studies attempt to test hypotheses about the association of certain potential prognostic factors with par-ticular outcomes. The types of clinical stud-ies most commonly encountered in the neu-rological literature are the case report and case series. Unfortunately, they have the weakest design. They are important for de-scribing events that can occur or character-istics that may coexist, but they do not nec-essarily provide information on frequency. They indicate possibilities rather than prob-abilities. Case reports also can provide im-portant exceptions to prognostic rules. An example would be the importance of a single well-documented case of a patient diag-nosed with brain death who subsequently underwent full neurological recovery. Less dramatic examples exist, such as cases of recovery after cardiac arrest when prognos-tic factors indicated otherwise (Rosenberg et al. 1977; Snyder et al. 1983; Ringel et al. 1988).

In the next two sections, designs that are stronger than case reports and case series are addressed, and the cohort and the case-control studies are discussed.

Cohort Design

The most natural design for an investigation of prognosis is the cohort study. It involves a group of patients with a specific condition. Information on potential prognostic factors at baseline and on the occurrence of one or more outcomes over time is collected or is available for review. The essential ingredi-

Table 3-1. Design and Evaluation of Studies of Prognosis

Define condition of study explicitly
Assemble inception cohort
Describe referral pattern
Define prognostic factors explicitly
Assess prognostic factors equally in all groups
Achieve complete follow-up
Define outcomes explicitly
Assess outcomes equally in all groups
Adjust for extraneous prognostic factors

Modified from Sackett (1985).

ents of a cohort study are listed in Table 3-1 and are similar to the components of a good clinical trial. A cohort study enables the in-vestigator to determine the absolute risk of an outcome. This absolute risk is termed *in-cidence* and is the frequency of specific out-come events over time in a defined popula-tion of patients (Rothman 1986). An important prognostic factor will split the co-hort into groups whose incidence for a par-ticular outcome differs. One measure of this difference is the *relative risk*. This is the ra-tio of the incidence in patients with the prognostic factor to the incidence in patients without the factor.

For investigation of prognosis, cohort studies are used for two main purposes. First, they can be used to compare people with the condition of interest to people with-out the condition to determine whether the observed incidence of adverse outcomes ex-ceeds what would be expected in the ab-sence of the condition. For example, a com-parison of stroke incidence among people with and without carotid bruits helped char-acterize the prognostic implications of a bruit (Heyman et al. 1980; Wolf et al. 1981). Second, cohort studies can be used to iden-tify subgroups of people with a certain con-dition who may face different prognoses. For example, the incidence of unprovoked seizures was compared between two groups of patients following recovery from enceph-alitis and meningitis. One group consisted of patients who had early seizures complicat-ing their infections and the other group con-sisted of patients who did not have early seizures. Patients with early seizures were at greater risk for future unprovoked seizures (Annegers et al. 1988).

Cohort studies can be prospective or retrospective. In a prospective study, patients with a certain condition are identified and characterized at the start of the study and are followed for the outcomes of interest as the study proceeds. None of the outcomes to be measured has occurred at the start of a prospective cohort study. The Framingham study is a prime example of an ongoing prospective cohort study (Wolf et al. 1978). A retrospective cohort study is similar in design, except the outcome events have already occurred when the study is initiated. Its success depends on the information on diagnosis, prognostic factors, and outcome having been collected in a complete and comparable fashion and being available for review. The investigator should strive to achieve the same completeness of prognostic information on patients who develop the outcome of interest as on those who do not. Many retrospective cohort studies have been possible with the records linkage system maintained at the Mayo Clinic, serving Olmsted County, Minnesota (Kurland and Brian 1978).

Assembly of Inception Cohort

A cohort study of prognosis entails the investigation of a well-defined group of patients. A specific diagnosis defines who will comprise the cohort. To ensure homogeneity and to allow comparison with other groups of patients, the criteria for diagnosis must be explicitly stated. Often this is not a trivial task, as, for example, when defining primary lateral sclerosis, multiple sclerosis, headache, or carpal tunnel syndrome.

Once the criteria for diagnosis are established, patients meeting these criteria are assembled into an *inception cohort*. This means that the study groups should consist of patients followed from a comparable point in the course of their condition, usually the onset or diagnosis. The beginning of the disease must be defined in the same way for all members of the cohort. Otherwise, the prognosis may differ substantially depending on when in the clinical course the cohort is assembled. In some situations, the prognosis may improve. For example, a cohort of patients assembled 3 months after their subarachnoid hemorrhage or transient ischemic attack has a relatively good outcome compared to a cohort assembled immediately following the onset of first symptoms. In some conditions, the onset of the disease may be difficult to define. For example, the advances in neuroimaging have had an effect on the diagnosis of brain tumors in that the diagnosis may be made earlier; therefore, the amount of time to certain outcomes, such as death, will be longer (Miles et al. 1981). The same may be true for imaging of multiple sclerosis. Because of early imaging, the apparent change in prognosis reflects the change in the definition of when the disease began and is sometimes referred to as *lead time bias* (Weiss 1986). Feinstein and colleagues (1985) have referred to this ''zero-time shift'' and ''stage migration'' produced by new diagnostic techniques as the ''Will Rogers phenomenon.''

Imaging not only leads to an earlier diagnosis, but also to a different diagnosis, which can alter apparent prognosis. Such an effect is demonstrated with primary intracerebral hemorrhage (Drury et al. 1984). Prior to the widespread availability of computed tomographic scanning of the head, the diagnosis of intracerebral hemorrhage rested on a devastating clinical picture and often autopsy confirmation. Not surprisingly, with the introduction of computed tomographic scanning, more benign bleeds have been identified that probably would have been classified as ischemic strokes. Thus, the introduction of computed tomographic scanning has resulted in an apparent decrease in fatality rates for intracerebral hemorrhages. More mild cases are now included in the denominator of these rates but not in the numerator because they survive.

Designs that include patients in a certain population without regard to the beginning of the condition, called prevalent cases, risk yielding misleading results. This is because patients with the condition have different probabilities of being included in the study, depending on the chronicity of their illness.

Figure 3-1 illustrates a hypothetical study of patients with malignant brain tumors. Choosing a prevalent sample involves selecting all such patients whose disease is active at a particular time, as indicated by the vertical line in Figure 3-1. The consideration of prevalent cases, rather than all newly diagnosed or incident cases, results in an over-representation of patients with long survivals. Patients with short survivals are less likely to be included in a sample of prevalent cases because they may have died before or been diagnosed after the sampling is done. If incident cases of the disease had been included at the time of diagnosis, then an inception cohort would have been assembled and individual cases of short survival would not have been missed.

In all cohort studies of prognosis, the selection of patients has a profound effect on prognosis. A bias can result when more severe cases are referred to academic centers where they are more likely to be described in the literature (Motulsky 1978). Examples of this referral bias would include the higher rate of epilepsy following febrile convulsions in studies from referral centers rather than population-based studies, (Fig. 3-2) (Ellenberg and Nelson 1980), the higher rates of mental retardation with tuberous

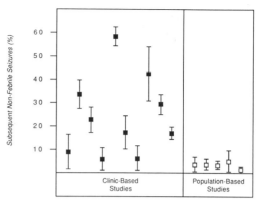

Figure 3-2. Rate of non-febrile seizures following febrile convulsion. Closed squares represent studies from referral centers and open squares from population-based surveys. Rates and their 95% confidence intervals are shown. Modified from Ellenberg (1980), Sackett (1985), and Longstreth (1987b).

sclerosis reported from referral centers (Nagib et al. 1984), the more malignant course of multiple sclerosis reported from referral centers (Nelson et al. 1988), and the shorter survival of Friedreich's disease reported from referral centers (Leone et al. 1988).

Other biases that can affect how the cohort is assembled have been described by Sackett and colleagues (1985). Specific cases being referred to a specific expert leads to *centripetal* bias; interesting cases being more completely evaluated than more mundane cases is *popularity* bias; only certain cases being able to pass through several screening procedures is *referral filter* bias; and the types of diagnostic evaluations varying from institution to institution is *diagnosis access* bias. These biases can make the cohort being studied so unique that even if the study if internally valid, the results cannot be applied to other patients; that is, external validity is lacking. The challenge for a clinician is to find a prognosis study in which the cohort most closely resembles the type of patients that he or she sees or is likely to see. In this respect, a complete ascertainment of cases for a defined population, as in the door-to-door search for patients with strokes in Copiah County, Mississippi (Schoenberg et al. 1986), may be less useful to the clinician than to the public health offi-

Figure 3-1. Incident versus prevalent cases. The start of the horizontal line for each patient with a malignant brain tumor represents the time of diagnosis, and the end of the line, the time of death.

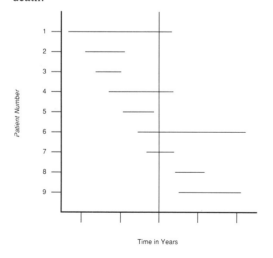

Time in Years

cial because 20% of stroke survivors in that study were never hospitalized. The outcomes in that cohort are unlikely to reflect a neurologist's experience with stroke patients.

Maximizing Internal Validity

Aside from an inception cohort, the internal validity of a cohort study can be strengthened with explicit and standardized definitions of potential prognostic factors and outcomes. Ideally in cohort studies, the outcomes should be determined in such a way that outcome data cannot be biased by knowledge of prognostic factors. Even when blinding or other measures are used to avoid the biased assessment of outcome, the possibility still exists, especially with potentially fatal illnesses, that decisions about medical support will be based on the prognostic factors of interest and that a self-fulfilling prophecy will result. Thus, if physicians consider a large clot and reduced level of consciousness to indicate a poor outcome after intracerebral bleed, they may limit medical support in patients with these factors. Not surprisingly, a retrospective study of prognostic factors in intracerebral hemorrhage may find these two factors important predictors of outcome, namely survival (Tuhrim et al. 1988). Instead, they may be important predictors of a physician's behavior, namely withdrawing medical support, which then determines outcome.

Prospective studies would ideally keep information about prognosis from physicians making medical decisions that could influence outcome. In many clinical settings, important prognostic information cannot deliberately be withheld; physicians cannot be made ignorant of the findings of the physical examination or the imaging studies. An alternative, to provide maximal support to all patients, is neither practical nor ethical in most clinical situations. Sometimes when a new prognostic test is introduced, it can be evaluated without the results influencing researchers' or practitioners' determination of outcome. For example, investigators of cerebrospinal fluid enzymes after cardiac arrest withheld the results of the test from the treating physicians and from themselves until the outcome was determined (Longstreth et al. 1984).

The choice of outcome to assess depends in part on the disease being studied and the information that is available. Life or death is not always an appropriate outcome to measure, but other outcomes dealing with morbidity may be difficult to quantify. For example, Fletcher and coworkers (1988) describe six categories of outcome: disease, death, disability, discomfort, dissatisfaction, and destitution. As discussed by Patrick and Deyo (1989), some outcomes are disease specific, such as the Glasgow Coma Outcome scale (Jennett and Bond 1975), the Kurtzke Disability Status Scale (Kurtzke 1989) or the Mathew Stroke Score (Mathew et al. 1972). Other instruments are designed as broader measures of health status and quality of life, such as the Sickness Impact Profile (Bergner et al. 1981) or the Medical Outcome Study Short Form (Stewart et al. 1988). These generic measures of outcome are only rarely used in studies of neurological illnesses. Their use in clinical neurology is likely to increase because of patients', physicians', and society's growing concerns with medicine's ability to prolong life at the expense of quality. As Fries and colleagues (1989) stated, "add life to your years, not years to your life." One example is the vegetative or severely impaired survivors of a serious head injury, cardiac arrest, or subarachnoid hemorrhage. Instead of these generic measures, outcomes in studies of prognosis are most commonly measured by diagnostic test results, physical signs, or symptoms (Fletcher and Fletcher 1979).

A complete follow-up of the cohort also is required and is one of the most challenging aspects of a cohort study. The more patients lost from the cohort, the greater is the compromise of the study's internal validity and of the conclusions that can be drawn from the study. Depending on which patients were lost to follow-up, the effect on the results can be positive or negative. For example, patients with paralysis in the distribution of the facial nerve might not return for follow-up if their condition spontaneously improves (Katusic et al. 1986). Alternatively, patients with brain tumors who are doing poorly might be lost to follow-up because they seek medical treatment elsewhere or because they die. If the rate at which patients are lost differs substantially among the different comparison groups, then the results of the study may be subject to bias. The magnitude of these effects can sometimes be estimated by alternately assuming the worst and the best outcomes for the patients who are lost to follow-up.

Clinical Trials of Treatment and Prognosis

A unique and increasingly important type of prospective cohort study of prognosis is a randomized, controlled clinical trial of treatment. Particularly when one arm of the trial is a no-treatment or placebo control group, the variation in outcomes within this group may provide insight into the prognostic importance of factors measured at the beginning of the trial. Except for the treatment being randomized, inferences about potential prognostic factors are based on observational comparisons that do not involve random allocation. On the other hand, control and standardization of other treatments in randomized trials help to minimize the mixing of treatment effects with the effects of other potential prognostic factors. Also, information about patient characteristics and outcomes is compiled in a prospective and standardized fashion, which is the exception in most studies dealing solely with issues of prognosis. The reason that information on prognostic factors typically is collected as part of treatment trials is because often these factors have a greater potential effect on outcomes than the treatments being tested (Sather 1986). Knowledge of prognosis is essential for clinical trials to assure that randomization has resulted in treatment groups that are balanced with respect to prognostic factors, to obtain a more precise estimate of treatment effects by controlling for prognostic factors in the analysis, and to examine treatment effects in different prognostic subgroups.

In a clinical trial, the treatment groups can be examined separately to assess potential prognostic factors that may be modified by the treatment given. Information measured at baseline or around the time of randomization is the most useful. For example, important information about prognosis in Guillain-Barré syndrome has emerged from a multicenter clinical trial (McKhann et al. 1988), and data about prognosis after cardiac arrest have come from the Brain Resuscitation Clinical Trials (Edgren 1987). Because of the difficulty in funding studies that deal solely with prognosis, clinical trials will likely remain an important source of prospective information on prognosis.

An important limitation of clinical trials is that only a select group of patients is studied. Typically clinical trials have specific inclusion and exclusion criteria to select patients who are most likely to benefit from the intervention being tested or who are the least likely to suffer adverse effects of the treatment. In one trial of heparin after ischemic stroke (Duke et al. 1986), more than 3000 patients were screened to enroll 225 subjects for the study. In another trial of a calcium channel blocker for ischemic stroke (Gelmers et al. 1988), 20% of the placebo group died in the first month following the stroke, suggesting that the patients included in this trial may have had a more grave prognosis than if all patients with ischemic strokes had been considered. In this trial, more patients with infarctions in the left side of the brain ($n = 116$) than in the right side of the brain ($n = 70$) were included, reflecting the greater ease by which a dysphasic patient could meet entrance criteria for stroke severity.

Case-Control Design

The case-control design is another nonexperimental analytical method often used in epidemiology. In the context of prognosis, the case-control study compares a group of patients known to have had a poor outcome from a condition with another group known to have had a good outcome from the same condition. It relates these differences in outcome to prognostic factors measured earlier in the clinical course of each patient. Advantages of the case-control study inclusive being relatively inexpensive, requiring a relatively brief time to perform, being especially efficient for rare outcomes, and being suitable to study multiple potential predictors of an outcome. More commonly, it is used to address questions of causation in classic epidemiology by identifying risk factors for disease. Its use to address questions of prognosis in clinical epidemiology, is unusual. In one example, investigators identified one group of patients with multiple sclerosis whose disease ran a malignant course and another group whose course was more benign (Clark et al. 1982). They then sought to determine the factors in the patients' past that were different between the groups and that might be important prognostic factors. A more recent study used a similar design to try to identify important prognostic factors in Parkinson's disease (Goetz et al. 1988).

One major disadvantage of the case-control study is that it involves retrospective assessment of the prognostic factors of interest and may be susceptible to bias, which is often difficult to detect or control. Also, it does not yield data on the absolute risk, or incidence, of particular outcomes as the cohort study does. Although the cohort study yields actual relative risks (discussed in the previous section), the case-control study yields only estimates of the relative risk of good versus poor outcomes in people with or without a particular characteristic. In addition, the odds ratio, which is computed from case-control data, is a good estimate of the relative risk only if the outcome of interest is rare (Feinstein 1985). In this context, "rare" usually indicates about 10% or less. For many outcomes of disease, this assumption may be violated, as with a malignant course in patients with multiple sclerosis or Parkinson's disease. Thus, the major advantage of case-control studies, the ability to investigate rare outcomes, makes this design infeasible for common outcomes for certain conditions.

Clinical situations do exist, however, in which the case-control study is a valid and efficient method to investigate prognosis. Its uncommon use in studies of prognosis may indicate that this important study design is under-used. For example, sudden death is a rare outcome that can complicate the clinical course of patients with epilepsy (Jay and Leestma 1981). A case-control design would be appropriate to try to identify the factors associated with sudden death. After establishing specific diagnostic criteria, a group of epileptic patients with sudden death (cases) and a group of epileptic patients without sudden death (controls) would be compared with respect to potential prognostic factors. Information on the factors would be obtained retrospectively by reviewing medical records or by interviewing people who knew the patients. Such a study would not address the question of the absolute risk of sudden death in patients with epilepsy. A cohort study would be needed to answer this question but would be extremely difficult to perform because the outcome of interest, sudden death, is so rare. Other examples of uncommon outcomes suited to the case-control designs include studies of very good outcomes after a diagnosis of glioblastoma multiforme, amyotrophic lateral sclerosis, or severe head injury or studies of very poor outcomes after Guillain-Barré syndrome, unruptured aneurysms, or aseptic meningitis. Finally, depending on how the controls are sampled, the rare disease assumption of a case-control study may not be required for valid estimation of the relative risk (Greenland and Thomas 1982).

Analysis of Studies of Prognosis

The goal of data analysis is usually to describe the incidence of one or more outcomes and to identify associations between potential predictors and outcomes that are not accidental. The potential predictors or prognostic factors are sometimes termed *independent variables,* and the outcomes, *dependent variables.*

The most important prognostic factors to identify are those that may have a causal link to the outcomes. Such a causal relationship can reveal the pathophysiology of a condition and suggest possible avenues of treatment, if that factor can be modified. For instance, if some immunologic change predicts progression in a patient with multiple sclerosis (Trotter et al. 1989), could manipulation of the change prevent the progression? Elevated blood glucose levels may exacerbate brain ischemia and result in worse outcomes for ischemic stroke (Woo et al. 1988). Is this findings a cause or an effect? Could lowering of blood glucose levels result in an improved outcome?

Just because a prognostic factor is associated with an outcome does not mean that a causal relationship exists. Other features of the association are relevant when establishing causality. Some features concern the study data, such as whether a dose-response relationship can be demonstrated and whether the temporal sequence is correct. Other considerations go beyond the study data, such as consistency of findings with other clinical studies and biological plausibility. Even if evidence for a causal link is lacking, a strong association between predictor and outcome identifies an important factor capable of predicting outcomes. Cerebrospinal fluid creatine kinase isoenzymes are strongly related to outcome after cardiac arrest (Vaagenes et al. 1988). No one would argue that the enzymes are causing brain damage; rather, their leakage into the cerebrospinal fluid is a result of brain damage. The lack of causal relation does not obviate their usefulness as predictors of outcome.

The analysis of studies of prognosis aims to describe outcomes and identify predictors. Often this involves striking a balance between deriving an understandable summary measure of a prognostic factor and obtaining as much information about the disease as possible. Clinicians find simple summaries most appealing, but much information can be lost. The simplest approach is to pick some dichotomous outcome (e.g., life or death, ambulatory or not, recurrent seizure or not) and ignore time altogether, assuming that enough time has transpired for all the outcomes of interest to have occurred. Progression to a certain stage, remission, or recurrence can define the endpoint. For example, in 14% of patients with myasthenia gravis, their disease remains clinically localized to the extraocular mus-

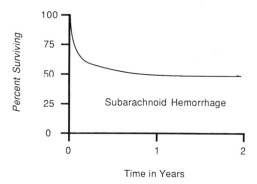

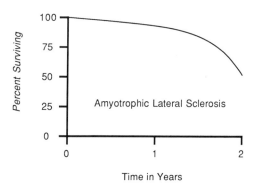

Figure 3-3. Survival curves for a hypothetical cohort of patients with subarachnoid hemorrhage and another with amyotrophic lateral sclerosis.

cles (Grob et al. 1987). Alternatively, a single time may be selected at which to assess outcome. For example, the 5-year survival of patients with an oligodendroglioma is 34% (Mørk et al. 1985). If the outcomes are examined more than once, then more information is captured. For instance, the 2-year survival after the diagnosis of subarachnoid hemorrhage and amyotrophic lateral sclerosis is similar at about 50%, but the shape of the two survival curves might be very different as shown in Figure 3-3. In these hypothetical curves, the time at which 50% of the cohort has died is the median survival. For both illnesses, the median survival is 2 years. Calculation of median survival does not require that all members of the cohort have died, as would be necessary to calculate mean survival.

Survival Analysis

Survival analysis is a powerful technique to study prognosis (Peto et al. 1977). To be applicable, the outcome must be dichotomous, or made to be dichotomous, and it must be an event that a patient can experience only once. Any dichotomous outcome can be used, not just life and death. Recurrent events such as seizures can be studied by considering only the first occurrence. Separate curves can be constructed for patients with particular prognostic factors, and statistical tests can be applied to decide whether the curves are significantly different. The information from the curve can be summarized numerically by using a specific time at which the heights of the curves are compared, such as 5-year survival, or by determining the time required for a given survival curve to reach a certain height, such as median survival time. Alternatively, the curves can be assessed visually or quantified by calculating the area under the curve or the person–years lived by the cohort.

Ideally, the survival curve would include information on the entire cohort followed until all its members had experienced the outcome of interest. Rarely is such an ideal achieved in a clinical study. Instead, members of the cohort are often lost to follow-up or are *censored* for some other reason. One of the strengths of survival analysis is that it uses as much of the information available on each patient as possible. Assumptions need to be made about those censored or lost to follow-up. Consequently, the curves are estimates and become increasingly so as the number of people under observation continues to decrease with movement from left to right. Central to the use of information on all cases, even those subsequently censored or lost to follow-up, is the assumption that those who are lost from the cohort share the same survival experience as the entire cohort with respect to the outcome of interest. In many situations, this assumption is violated. As suggested previously, patients with a brain tumor who are doing poorly may seek help elsewhere and may be lost to follow-up. Their survival experience may be much worse than the remainder of the cohort. A survival curve that assumes that their survival experience is the same will give an overly optimistic picture of the clinical course of this cohort. A simple method

for setting bounds on the extent of the bias is to reconstruct the curves under the pessimistic assumption that all those lost to follow-up had a poor outcome at the time of last contact.

An example is a hypothetical study in which patients with a diagnosis of epilepsy who have been free of any seizures for 2 years are advised to stop their antiseizure medications. The study runs for 4 years, during which time 10 patients who want to stop their medications are enrolled and followed as indicated in Figure 3-4A. Patient number 3 and 5 stop coming to the clinic for follow-up, and neither can be located after the times indicated. Patient number 9 was killed in a motor vehicle accident. He was the driver, but otherwise the details of the accident were unknown. The next step in the analysis is to change the time scale in Figure 3-4A to 3-4B so that all cases line up against the vertical axis at time zero. This is tantamount to establishing an inception cohort, as described previously, in which all patients are followed from an equivalent temporal milestone. Now a survival curve can be constructed. It begins at 100% on the vertical axis, indicating that all the patients are initially seizure free. It continues horizontally until one of the patients experiences his or her first recurrent seizure. Dropouts, such as patients number 3, 5, and 9 with incomplete follow-up information, have no immediate effect on the height of the curve. When a patient suffers his or her first recurrent seizure, the curve drops. The new height of the curve, the new percent remaining seizure-free, is given by the following formula:

(old percent remaining seizure-free)

$$\times \frac{(N-1)}{N}$$

N is the number of subjects under observation just before the time of the seizure, and the old percent refers to the height of the curve just before the time of the seizure. Figure 3-4C shows the resulting survival curve and the appropriate calculations. For this example, based on so few patients, the survival curve is made up of a few large

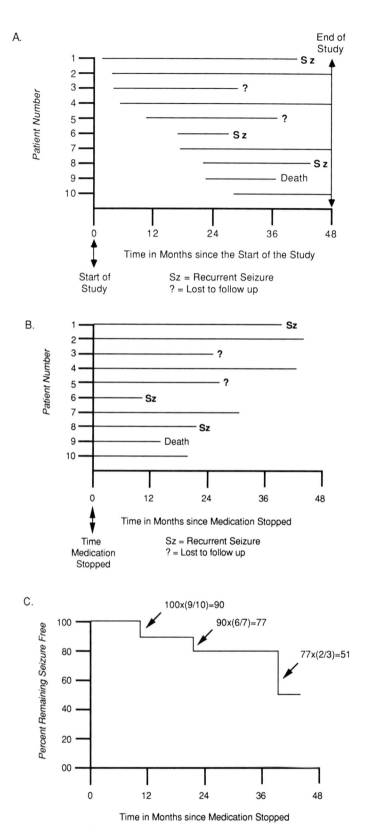

Figure 3-4. Example of survival analysis for recurrence of seizures after stopping anti-seizure medications. See text for complete explanation.

steps; for larger cohorts, it adopts a smoother shape.

The curve in Figure 3-4C assumes that the two patients who were lost to follow-up and the one patient who was censored because of death experienced the same rate of seizure recurrence as the entire cohort. A critic might argue that the two patients lost to follow-up probably had seizures, but because they were disgruntled about the advice they received in the clinic did not return and sought treatment for their epilepsy elsewhere. The patient who died in the motor vehicle accident also could have had a seizure, which could have been the cause of the accident. With these considerations, the curve can be remade under the possibly extreme assumption that all three patients with missing information experienced a recurrent seizure around the time of their last contact for patients number 3 and 5 or at the time of the motor vehicle accident for patient number 9.

Time-linked analysis of outcomes that are not dichotomous are not as well developed and are less commonly used. Often an outcome that is measured as a continuous variable in a study will need to be made a dichotomous variable for survival analysis. For example, a device is used to assess grip strength in a patient with amyotrophic lateral sclerosis. The results are read from the device as a continuous variable. For purposes of a survival analysis, however, a particular level of weakness has to be used to dichotomize the continuous variable, and then a survival curve can be constructed from this dichotomous outcome. Growth curve analysis allows a continuous outcome measure to be examined over time (Zerbe 1979). It is used most commonly to examine change in height of children. It is rarely applied to studies of outcome but seems to be well suited to many situations in clinical medicine.

Effect Modification and Confounding Factors

Other efforts in the analysis of prognosis studies are directed at dealing with confounding factors and effect modifiers (Miet-tinen 1974; Hennekens and Buring 1987). Effect modification occurs when the strength of the association between a prognostic factor and an outcome differs among subgroups formed according to a third factor. For example, the prognosis of patients with cryptococcal meningitis is worsened by concomitant infection with human immunodeficiency virus (HIV) (Rosenblum et al. 1988). To summarize the prognosis of cryptococcal meningitis and not consider HIV status is to ignore an important effect modifier. When effect modification is present, the groups should not be mixed together.

Confounding factors can cause a relationship to be overestimated or underestimated; in an extreme situation, it can cause an association when one is lacking, or vice versa. This occurs because the factor is related to both the predictor and the outcome. An obvious example is one in which gray hair is proposed to be associated with a worse outcome from head trauma. A cohort of patients with head trauma is separated into those with and without gray hair and outcomes in the two groups are compared. If a statistically significant association is found, one should hesitate to conclude that the presence of gray hair is linked to outcome from head trauma without looking at the effects of age. Age is likely related to both gray hair and to outcome from head trauma and therefore must be taken into account in the analysis. Confounding is typically more subtle and can easily invalidate conclusions drawn from a study that is trying to establish a causal link between a prognostic factor and an outcome. If causation is not an issue, then neither is confounding. Thus in the previous example, regardless of any confounding by age, gray hair may still be an important predictor of outcome.

An important and difficult issue in many cohort studies is a phenomenon called confounding by indication. This problem arises when an objective of the study is to determine the effects of a particular form of treatment from observational data or to determine the effect of a patient characteristic that is itself an indication (or contraindication) for a certain form of intervention. In these circumstances, the form of treatment a patient receives may be so closely linked to a particular feature of his or her clinical status that the effect of treatment cannot be separated from the effect of the underlying indication (or contraindication) for it. For example, the case fatality rate for patients in an intensive care unit who have a neurology consultation may be higher than for those who do not have a neurology consultation. The consultation does not kill the patients. Rather, the indication for a neurology consultation is what places a patient at high risk for death. Again, confounding applies only to the attempt to assign causality. Although a causal relationship does not exist, a neurology consultation on a patient in the intensive care unit may be a marker for a poor outcome.

Several techniques exist to search for confounding, but the most generally useful analytical methods are stratification and multivariate modeling. Stratification involves forming subgroups of patients who are similar with respect to the confounding factor. The association between the potential prognostic factor and outcome is then evaluated in each subgroup or stratum. Techniques to combine data across strata include direct and indirect standardization (Fleiss 1986; Hennekens and Buring 1987) and the Mantel-Haenszel method (1959). Such approaches to control confounding factors are most useful when the number of factors is small. Unfortunately, a clinical setting often becomes complicated with several potential predictors and confounders. The question arises as to which predictors are the most important. The importance of a predictor can be established on the basis of statistical significance, clinical usefulness, or both. Patients who regain consciousness and those who do not regain consciousness after out-of-hospital cardiac arrest have admission temperatures that differ by an average of 0.2°C (Longstreth et al. 1983). Although this is a statistically significant difference ($p = .026$), it lacks clinical usefulness. For a prognostic factor to be useful, the relation between predictor and outcome must be clinically convincing, not just statistically significant.

Multivariate Models

Multivariate models allow the investigator to look at the relation of many variables to outcome (Rothman 1986). These models can be used to help identify the most important predictors and create clinical predictive rules (Wasson et al. 1985). Such models have been used to predict outcomes after cardiac arrest (Levy et al. 1985), intracranial hemorrhage (Tuhrim et al. 1988), acute stroke in general (Fullerton et al. 1988), febrile convulsions (Annegers et al. 1987), acute idiopathic neuropathy (Winer et al. 1988), Bell's palsy (Katusic et al. 1986), Alzheimer's disease (Heyman et al. 1987), and many other conditions. Most of the multivariate techniques listed in Table 3-2 are available in statistical packages that are widely distributed. Logistical regression is the multivariate method of choice when prognostic data have been collected using a case-control design. Logistical regression is a favorite of epidemiologists because adjusted odds ratio estimates can be calculated from the coefficients of the predictor vari-

Table 3-2. Multivariate Models

Method	Nature of Predictor Variables	Nature of Outcome Variable
Multiple linear regression	Categorical or continuous	Continuous
Discriminant analysis	Continuous, with normal distribution	Dichotomous
Multiple logistic regression*	Categorical or continuous	Dichotomous
Log–linear modeling*	Categorical	Categorical, with two or more categories
Cox proportional hazards model*	Categorical or continuous	Survival time

* These methods can be used to obtain adjusted relative risk estimates for predictor variables.

ables. These estimates of relative risk are useful to summarize the strength of the relation between a predictor and an outcome, controlling for other potential predictors, and confounders that are included in the model. As with survival analysis, the outcome variable in logistical regression must be dichotomous.

Often little understanding of multivariate modeling is needed to perform the analysis, but because many pitfalls exist, a biostatistician experienced in these techniques should be consulted. The danger exists of finding some idiosyncrasy of the data set rather than true variations in outcomes for the disease in question. This danger increases as the number of predictor variables examined increases. A general rule is to have at least 10 cases for each variable that is included in the model. Often the validity of the model can be tested. One simple approach is to generate prognostic rules on one randomly chosen half of the data set and then test them on the remaining half. As mentioned previously, an internally valid study or model may still have limited external validity as a result of the characteristics of the cohort under study.

One of the more sophisticated models used in studies of outcome is the Cox proportional hazards model (Cox 1972). This model allows multiple variables that can be time-dependent to be considered simultaneously in an analysis of survival. As in logistical regression, the coefficients from the Cox proportional hazards model can be converted to adjusted relative risk estimates. For example, when this type of analysis was applied to seizure recurrence following a febrile convulsion (Annegers et al. 1987), the adjusted relative risk for seizure recurrence was 3.6 if the febrile convulsion had focal features. Thus, patients whose febrile convulsions had focal features would be 3.6 times as likely to have a seizure recurrence per unit of time as those whose febrile convulsions lacked focal features. A relative risk of 1.0 would indicate no relation between potential predictors and outcome. A relative risk of less than 1.0 would indicate a favorable prognosis with the factor present.

From the standpoint of data analysis, clinical trials of treatments can be seen as studies of prognosis in which the treatments are another potential predictor of outcome. Many prognostic factors cannot be modified, such as age, gender, and race. Treatments are simply prognostic factors that can be modified. Examples of this are the cooperative study investigating the treatment of Guillain-Barré syndrome (McKhann et al. 1988) and the one studying treatment of coma after cardiac arrest (Brain Resuscitation Clinical Trial I Study Group 1986). Despite both being randomized, double-blind trials, analyses included the Cox proportional hazards model with several potential predictors of outcome, including treatment. After controlling for the effects of the other predictors of outcome, the analyses yielded a coefficient for the treatment variable that was significant for plasmaparesis in Guillain-Barré syndrome and insignificant for pentobarbital in coma after cardiac arrest.

Another type of model that has been used in studies of prognosis is the Markov process (Beck and Pauker 1983). Various stages of disease progression are defined, and probabilities are assigned for moving from one stage to another. This technique has been used to model the clinical course of patients with headaches (Leviton et al. 1980) and multiple sclerosis (Wolfson and Confavreux 1987). Other models have been used to try to characterize the nonrandom occurrence of seizures in patients with epilepsy (Hopkins et al. 1985).

Conclusions

One of the major tasks of a neurologist is to render an accurate prognosis. Many of the patients that neurologists care for have diseases for which treatments are lacking or are ineffective, so prognosis becomes even more important. With the nervous system at risk, patients and families insist that neurologists "avoid expressing prognosis with vagueness when it is unnecessary, and with certainty when it is misleading" (Fletcher et al. 1988). Armed with an understanding of prognosis, the neurologist can appropriately counsel patients, families, and other physicians. Even with this understanding, the neurologist is challenged by having to translate probabilities that range from 0 *to* 1 in groups of patients to 0 *or* 1 in a specific patient. If the prognosis is known, the clinician can identify situations in which aggressive or dangerous treatments are indicated because of the poor outcomes that follow the usual clinical course. If the prognosis is certain enough, such as with brain death or a persistent vegetative state, decisions to withdraw therapy can be made.

Sometimes studies of prognosis are all clinicians have to judge the external validity of a trial of a treatment (Weiss 1986). If lowering blood pressure in white and African-American men results in reduced cardiovascular disease as outcome events (Hypertension and Detection Follow-up Program Cooperative Group 1979), will the same conclusions apply to women or Asians? If the clinical course of all of these patients with hypertension is similar, then the results of the trial are assumed to apply to women and Asians. If a drug could be shown to prevent post-traumatic epilepsy, it would probably be used in other situations in which a brain injury can lead to epilepsy, such as following meningitis or brain surgery.

For investigators, studies of prognosis may indicate the diseases that have an important impact on function and thus require further investigation. In addition to indicating the diseases that require treatments, the search for prognostic factors may yield clues as to the pathophysiology of the disease and to its treatment.

Many studies of prognosis in neurology fall far short of the ideal study presented here. Nevertheless, each study provides some information and helps to reduce the mystery that surrounds many neurological illnesses. Ignorance about clinical course and prognosis can potentially cause more harm than good in some patients. An example is the discovery of an asymptomatic bruit on physical examination or unidentified bright objects on a magnetic resonance imaging study. Sometimes in the haste to intervene and treat, the clinical course or

prognosis of certain conditions remains ill-defined. Detailed studies of prognosis should precede studies of treatment, but such a sequence often is reversed.

In this chapter, the basic elements of studies of prognosis have been demonstrated. Some skills have been outlined for the researcher and clinician to master to perform and interpret such studies. Finally, the reader has been shown that prognosis is the keystone to clinical neurology. This chapter has dealt in generalities and methodologies; subsequent chapters turn to the prognosis of specific conditions.

References

Annegers, J. F.; Hauser, W. A.; Beghi, E.; Nicolosi, A.; Kurland, L. T. The risk of unprovoked seizures after encephalitis and meningitis. Neurology 38:1407–1410; 1988.

Annegers, J. F.; Hauser, W. A.; Shirts, S. B.; Kurland, L. T. Factors prognostic of unprovoked seizures after febrile convulsions. N. Engl. J. Med. 316:493–498; 1987

Beck, J. R.; Pauker, S. G. The Markov process in medical prognosis. Med. Decis. Making 3:419–458; 1983.

Bergner, M.; Bobbitt, R. A.; Carter, W. B.; Gilson, B. S. The Sickness Impact Profile: development and final revision of a health status measure. Med. Care 19:787–805; 1981.

Brain Resuscitation Clinical Trial I Study Group. Randomized clinical study of thiopental loading in comatose survivors of cardiac arrest. N. Engl. J. Med. 314:397–403; 1986.

Clark, V. A.; Detels, R.; Visscher, B. R.; Valdiviezo, N. L.; Malmgren, R. M.; Dudley, J. P. Factors associated with a malignant or benign course of multiple sclerosis. J.A.M.A. 248:856–860; 1982.

Cox, D. R. Regression models and life tables (with discussion). J. R. Stat. Soc. [B] 34:187–202; 1972.

Drury, I.; Whisnant, J. P.; Garraway, W. M. Primary intracerebral hemorrhage: impact of CT on incidence. Neurology 34:653–657; 1984.

Duke, R. J.; Bloch, R. F.; Turpie, A. G. G.; Trebilcock, R.; Bayer, N. Intravenous heparin for the prevention of stroke progression in acute partial stable stroke: a randomized controlled trial. Ann. Intern. Med. 105:825–828; 1986.

Edgren, E. The prognosis of hypoxic–ischaemic brain damage following cardiac arrest. Acta Univ. Ups. (Comprehensive summaries of Uppsala dissertations from the faculty of medicine) 89:1–44; 1987.

Ellenberg, J. H.; Nelson, K. B. Sample selection

and the natural history of disease: studies of febrile seizures. J.A.M.A. 243:1337–1340; 1980.

Feinstein, A. R. Clinical epidemiology: the architecture of clinical research. Philadelphia: W.B. Saunders Co.; 1985.

Feinstein, A. R.; Sosin, D. M.; Wells, C. K. The Will Rogers phenomenon: stage migration and new diagnostic techniques as a source of misleading statistics for survival in cancer. N. Engl. J. Med. 312:1604–1608; 1985.

Fleiss, J. L. The design and analysis of clinical experiments. New York: John Wiley and Sons; 1986.

Fletcher, R. H.; Fletcher, S. W. Clinical research in general medical journals: a 30-year perspective. N. Engl. J. Med. 301:180–183; 1979.

Fletcher, R. H.; Fletcher, S. W.; Wagner, E. H. Clinical epidemiology: the essentials. 2nd ed. Baltimore: Williams and Wilkins; 1988.

Fries, J. F.; Green, L. W.; Levine, S. Health promotion and the compression of morbidity. Lancet 1:481–483; 1989.

Fullerton, K. J.; Mackenzie, G.; Stout, R. W. Prognostic indices in stroke. Q. J. Med. 66:147–162; 1988.

Gelmers, H. J.; Gorter, K.; de Weerdt, C. J.; Wiezer, H. J. A. A controlled trial of nimodipine in acute ischemic stroke. N. Engl. J. Med. 318:203–207; 1988.

Goetz, C. G.; Tanner, C. M.; Stebbins, G. T.; Buchman, A. S. Risk factors for progression in Parkinson's disease. Neurology 38:1841–1844; 1988.

Greenland, S.; Thomas, D. C. On the need for the rare disease assumption in case-control studies. Am. J. Epidemiol. 116:547–553; 1982.

Grob, D.; Arsura, E. L.; Brunner, N. G.; Namba, T. The course of myasthenia gravis and therapies affecting outcome. Ann. N.Y. Acad. Sci. 505:472–499; 1987.

Hennekens, C. H.; Buring, J. E. Epidemiology in Medicine. Boston: Little, Brown and Company; 1987.

Heyman, A.; Wilkinson, W. E.; Heyden, S.; Helms, M. J.; Bartel, A. G.; Karp, H. R.; Tyroler, H. A.; Hames, C. G. Risk of stroke in asymptomatic persons with cervical arterial bruits: a population study in Evans County, Georgia. N. Engl. J. Med. 302:838–841; 1980.

Heyman, A.; Wilkinson, W. E.; Hurwitz, B. J.; Helms, M. J.; Haynes, C. S.; Utley, C. M.; Gwyther, L. P. Early-onset Alzheimer's disease: clinical predictors of institutionalization and death. Neurology 37:980–984; 1987.

Hopkins, A.; Davies, P.; Dobson, C. Mathematical models of patterns of seizures: Their use in the evaluation of drugs. Arch. Neurol. 42:463–467; 1985.

Hypertension and Detection Follow-up Program Cooperative Group. Five-year findings of the hypertension detection and follow-up program: II. Mortality by race, sex and age. J.A.M.A. 242:2572–2577; 1979.

Jay, G. W.; Leestma, J. E. Sudden death in epilepsy: a comprehensive review of the literature and proposed mechanism. Acta. Neurol. Scand. 63(suppl. 82):1–66; 1981.

Jennett, B.; Bond, M. Assessment of outcome after severe brain damage: a practical scale. Lancet 1:480–484; 1975.

Katusic, S. K.; Beard, C. M.; Wiederholt, W. C.; Bergstralh, E. J.; Kurland, L. T. Incidence, clinical features, and prognosis in Bell's palsy, Rochester, Minnesota, 1968–1982 Ann. Neurol. 20:622–627; 1986.

Kurland, L. T.; Brian, D. D. Contributions to neurology from records linkage in Olmsted County, Minnesota. Adv. Neurol. 19:93–105; 1978.

Kurtzke, J. F. The Disability Status Scale for multiple sclerosis: apologia pro DSS sua. Neurology 39:291–302; 1989.

Leone, M.; Rocca, W. A.; Rosso, M. G.; Mantel, N.; Schoenberg, B. S.; Schiffer, D. Friedreich's disease: survival analysis in an Italian population. Neurology 38:1433–1438; 1988.

Leviton, A.; Schulman, J.; Kammerman, L.; Porter, D.; Slack, W.; Graham, J. R. A probability model of headache recurrence. J. Chron. Dis. 33:407–412; 1980.

Levy, D. E.; Bates, D.; Caronna, J. J.; Singer, B. H.; Lapinski, R. H.; Frydman, H.; Plum, F. Predicting outcome from hypoxic–ischemic coma. J.A.M.A. 253:1420–1426; 1985.

Longstreth, W. T., Jr.; Clayson, K. J.; Chandler, W. L.; Sumi, S. M. Cerebrospinal fluid creatine kinase activity and neurologic recovery after cardiac arrest. Neurology 34:834–837; 1984.

Longstreth, W. T., Jr.; Diehr, P.; Inui, T. S. Prediction of awakening after out-of-hospital cardiac arrest. N. Engl. J. Med. 308:1378–1382; 1983.

Longstreth, W. T., Jr.; Koepsell, T. D.; van Belle, G. Clinical neuroepidemiology: I. diagnosis. Arch. Neurol. 44:1091–1099; 1987a.

Longstreth, W. T., Jr.; Koepsell, T. D.; van Belle, G. Clinical neuroepidemiology: II. outcomes. Arch. Neurol. 44:1196–1202; 1987b.

Mantel, N.; Haenszel, W. Statistical aspects of the analysis of data from retrospective studies of disease. J. Natl. Cancer Inst. 22:719–748; 1959.

Mathew, N. T.; Meyer, J. S.; Rivera, V. H.; Charney, J. Z., Hartmann, A. Double-blind evaluation of glycerol treatment in acute cerebral infarction. Lancet 2:1327–1329; 1972.

McKhann, G. M.; Griffin, J. W.; Cornblath, D. R.; Mellits, E. D.; Fisher, R. S.; Quaskey, S. A.; Guillain-Barré Syndrome study group. Plasmapheresis and Guillain-Barré syndrome:

analysis of prognostic factors and the effect of plasmapheresis. Ann. Neurol. 23:347–353; 1988.

Miettinen, O. S. Confounding and effect modification. Am. J. Epidemiol. 100:350–353; 1974.

Miles, I.; Mitchelson, M.; Morgan, R.; Michaelides, C.; Stuart, G.; Jayasinghe, L.; Baddeley, H. Improved survival of patients with glioma in the CT era. Diagn. Imaging 50:313–320; 1981.

Moses, L. E. Statistical concepts fundamental to investigations. N. Engl. J. Med. 312:890–987; 1985.

Motulsky, A. G. Biased ascertainment and natural history of disease. N. Engl. J. Med. 298:1196–1197; 1978.

Mørk, S. J.; Lindegaard, K-F.; Halvorsen, T. B.; Lehmann, E. H.; Solgaard, I.; Hattevoll, R.; Harvei, S.; Ganz, J. Oligodendroglioma: incidence and biological behavior in a defined population. J. Neurosurg. 63:881–889; 1985.

Nagib, M. G.; Haines, D. J.; Erickson, D. L.; Mastri, A. R. Tuberous sclerosis: a review for the neurosurgeon. Neurosurgery 14:93–98; 1984.

Nelson, L. M.; Franklin, G. M.; Hamman, R. F.; Boteler, D. L.; Baum, H. M.; Burks, J. S. Referral bias in multiple sclerosis research. J. Clin. Epidemiol. 41:187–192; 1988.

Patrick, D. L.; Deyo, R. A. Generic and disease-specific measures in assessing health status and quality of life. Med. Care 27(3 suppl):S217–S232; 1989.

Peto, R.; Pike, M. C.; Armitage, P.; Breslow, N. E.; Cox, D. R.; Howard, S. V.; Mantel, N.; McPherson, K.; Pete, J.; Smith, P. G. Design and analysis of randomized clinical trials requiring prolonged observation of each patient: II. analysis and examples. Br. J. Cancer 35:1–39; 1977.

Reding, M. J.; Ernesto, P. Rehabilitation outcome following initial unilateral hemispheric stroke: life table analysis approach. Stroke 19:1354–1358; 1988.

Ringel, R. A.; Riggs, J. E.; Brick, J. F. Reversible coma with prolonged absence of pupillary and brainstem reflexes: an unusual response to a hypoxic–ischemic event in MS. Neurology 38:1275–1278; 1988.

Rosenberg, G. A.; Johnson, S. F.; Brenner, R. P. Recovery of cognition after prolonged vegetative state. Ann. Neurol. 2:167–168; 1977.

Rosenblum, M. L.; Levy, R. M.; Bredesen, D. E. AIDS and the nervous system. New York: Raven Press; 1988.

Rothman, K. J. Modern Epidemiology. Boston: Little, Brown and Company; 1986.

Sackett, D. L.; Haynes, R. B.; Tugwell, P. Clinical epidemiology: a basic science for clinical medicine. Boston: Little, Brown and Company; 1985.

Sather, H. N. The use of prognostic factors in clinical trials. Cancer 58:461–467; 1986.

Schoenberg, B. S.; Anderson, D. W.; Haerer, A. F. Racial differentials in the prevalence of stroke: Copiah County, Mississippi. Arch. Neurol. 43:565–568; 1986.

Snyder, B. D.; Cranford, R. E.; Rubens, A. B.; Bundlie, S.; Rockswold, G. E. Delayed recovery from post anoxic persistent vegetative state (abstract). Ann. Neurol 14:152; 1983.

Stewart, A. L.; Hays, R. D.; Ware, J. F. The MOS short-form general health survey. Med. Care 26:724–735; 1988.

Trotter, J. L.; Clifford, D. B.; McInnis, J. E.; Griffeth, R. C.; Bruns, K. A.; Perlmutter, M. S.; Anderson, C. A. Collins, K. G.; Banks, G.; Hicks, B. C. Correlation of immunological studies and disease progression in chronic progressive multiple sclerosis. Ann. Neurol. 25:172–178; 1989.

Tuhrim, S.; Dambrosia, J. M.; Price, T. R.; Mohr, J. P.; Wolf, P. A.; Heyman, A.; Kase, C. S. Prediction of intracerebral hemorrhage survival. Ann. Neurol. 24:258–263; 1988.

Vaagenes, P.; Safar, P.; Diven, W.; Moosy, J.; Rao, G.; Cantadore, R.; Kelsey, S. Brain enzyme levels in CSF after cardiac arrest and resuscitation in dogs: markers of damage and predictors of outcome. J. Cereb. Blood Flow Metab. 8:262–275; 1988.

Wasson, J. H.; Sox, H. C.; Neff, R. K.; Goldman, L. Clinical prediction rules: applications and methodological standards. N. Engl. J. Med. 313:793–799; 1985.

Weiss, N. S. Clinical epidemiology: the study of the outcome of illness. New York: Oxford University Press; 1986.

Winer, J. B.; Hughes, R. A. C.; Osmond, C. A prospective study of acute idiopathic neuropathy: I. clinical features and their prognostic value. J. Neurol. Neurosurg. Psychiatry 51:605–612; 1988.

Wolf, P. A.; Kannel, W. B.; Dawber, T. R. Prospective investigations: the Framingham study and the epidemiology of stroke. Adv. Neurol. 19:107–120; 1978.

Wolf, P.A.; Kannel, W. B.; Sorlie, P.; McNamara, P. Asymptomatic carotid bruit and risk of stroke: the Framingham study. J.A.M.A. 245:1442–1445; 1981.

Wolfson, C.; Confavreux, C. Improvements to a simple Markov model of the natural history of multiple sclerosis: I. short-term prognosis. Neuroepidemiology 6:101–115; 1987.

Woo, E.; Ma, J. T. C.; Robinson, J. D.; Yu, Y. L. Hyperglycemia is a stress response in acute stroke. Stroke 19:1359–1364; 1988.

Zerbe, G. O. Randomization analysis of the completely randomized design extended to growth and response curves. J. Am. Statistical Assoc. 74:215–221; 1979.

4

Stroke

THOMAS J. DEGRABA, SANDRA K. HANSON,
FRANK M. YATSU, AND JAMES C. GROTTA

Because of the brain's exquisite vulnerability to ischemic insults, the prognosis of strokes due to vascular occlusion or rupture depends in large measure on the degree of brain damage sustained initially. Depending on the pathological process causing the stroke, eventual prognosis also is influenced by the likelihood of stroke recurrence and how effectively it can be minimized. In this review, the common prognoses of various stroke syndromes are discussed with these considerations in mind.

Effects of Therapy on Prognosis After Transient Ischemic Attack (TIA)

Medical therapy has had a significant impact on prognosis after TIA. It is now accepted, after numerous prospective studies involving thousands of patients, that aspirin reduces the subsequent incidence of stroke after TIA by approximately 20% to 30% and that this effect is most striking in men (Antiplatelet Trialists' Collaboration 1988). No evidence shows that dipyridamole or sulfinpyrazone provides additional benefit; one aspirin (325 mg) is as effective as higher doses and has fewer side effects. The value

of doses lower than 325 mg remains unproven. Recent studies with ticlopidine indicate that it may be 30% more effective than aspirin and equally effective in both sexes (Hass et al. 1989). Although there is theoretical reason to speculate that platelet antiaggregant drugs might affect the severity or pathophysiology of strokes that occur in patients while on treatment, there is no evidence that these agents have any effect other than reducing the number of strokes that occur.

Anticoagulation with Coumadin still has not been studied in a randomized prospective controlled study with enough patients to provide a conclusive answer; however, a recent study demonstrating a 75% reduced incidence of stroke in patients with atrial fibrillation treated with Coumadin (Petersen et al. 1989) suggests that such therapy might affect outcome when TIAs have a cardioembolic basis.

It is difficult to determine the effect of surgical therapy on prognosis after TIA. A large randomized prospective study of extracranial to intracranial bypass proved that this operation had no effect on outcome in patients with occlusion or inaccessible stenosis of their carotid arteries. Only one such study has evaluated carotid endarterectomy and showed that perioperative morbidity canceled the long-term benefits. Two ongoing studies are evaluating carotid endarterectomy after TIA. Hopefully, these studies will identify subgroups of patients that might benefit from this operation, although the results probably will not be known for at least 5 years. The major factor determining the prognosis of patients undergoing endarterectomy is the perioperative morbidity and mortality, which can range from 1% to 20% depending on the series.

Analysis of the overall TIA population shows a 5% yearly incidence of stroke with medical therapy and a 2% yearly incidence of stroke after endarterectomy. Calculations reveal, therefore, that perioperative morbidity and mortality must be kept less than 4% after 5 years to achieve a statistically significant advantage for carotid endarterectomy (Whisnant et al. 1987). The greatest advantage of endarterectomy probably will be for selected patients at very high risk of stroke (i.e., greater than 5% yearly) with medical therapy. This subpopulation is yet to be defined.

Completed Stroke

Short-Term Mortality After Stroke

The 30-day mortality for various types of strokes, based on a review of 1073 consecutive stroke admissions reported by Silver and colleagues (1984), is 15% for supratentorial infarct, 58% for supratentorial hemorrhage, 18% for infratentorial infarct, and 31% for infratentorial hemorrhage.

Much of what is known about recent changes in these mortality rates derives from experience at the Mayo Clinic, where records have been kept for 5-year intervals from 1945 to 1979 (Garraway et al. 1983). These data provide clues about how diagnosis and therapy have affected stroke mortality. For instance, there was marked reduction in mortality in the 1970s for intracerebral hemorrhage, probably due to the

Table 4-1. Mean Survival Times for Patients With Various Causes of Death

Cause	Mean Survival (Days)	Number of Cases
Herniation	3.7	31
Pneumonia	13.7	29
Cardiac	6.9	17
Pulmonary embolus	5.2	13
Other	11.8	10
Total	8.3	100

advent of computed tomography (CT) and consequent identification and inclusion of smaller nonfatal hematomas.

The immediate causes of death after cerebral infarct provide useful clues about how to improve this parameter of stroke outcome. Bounds and associates (1981), also working at the Mayo Clinic, report that 60% of deaths occur in the first week, usually due to herniation, cardiac complications, or pulmonary emboli. Approximately 40% of deaths occur in the second and third weeks, due to pneumonia and sepsis (Table 4-1).

Long-Term Recovery After Stroke

Approximately 80% of stroke patients will achieve independent ambulation and 67% will become fully independent in activities of daily living (ADL) (Feigenson et al. 1977). More specifically, using the Barthel score as a measure of functional recovery, 35% to 40% of patients ultimately will regain a perfect score (100 points), and most will achieve a score greater than 70. The final score depends on the severity of the initial disability (Wade et al. 1983).

Considering outcome in terms of stroke severity and excluding fatalities, conventional practice has been to divide strokes into the following categories: mild, little or no reduction of ADL; moderate, reduction of ADL but ambulatory and independent in some functions; severe, nonambulatory and totally dependent.

Studies often are not comparable because of variability in patient selection; however, representative studies over the past 30 years demonstrate that approximately 60% to 70%

Table 4-2. Stroke Severity in Representative Surveys From 1950 to 1980

Study	Mild	Moderate	Severe
Rankin 1957	64	17	19
Matsumoto 1955–1969 (1973)	74	23	3
Dombovy 1975–1979 (1987)	56	37	7
Dove 1977–1981 (1984)	70	22	8
Kotila 1978–1980 (1984)	81	12	7
Mean	69	24	7

of patients eventually are in the mild category, 20% to 30% are moderate, and 10% are severe (Table 4-2).

Caronna and Levy (1983) report somewhat worse outcomes, after conducting a prospective survey of 107 patients with cerebral infarct. Twenty-one percent of their patients died, and of those surviving at follow-up, 29% had mild, 25% had moderate, and 46% had severe residual deficit. These results might be explained, however, by a high incidence of depressed sensorium and coma, indicating that these researchers were dealing with a particularly devastated population of stroke patients.

Factors in Recovery

Recovery after stroke depends on numerous factors. Data from the University of Mississippi (Haerer and Woosley 1975) and the Mayo Clinic (Turney et al. 1984) demonstrate that survivors of brain stem infarcts walk better and have better functional outcome than those with hemispheric infarcts. This is probably because impaired cognitive function, depression, and personality changes are seen more frequently after hemispheric damage.

Increasing age and presence of comorbidities, especially cardiovascular disease, arthritis, and other neurological conditions, adversely affect outcome (Dombovy et al. 1987). However, the presence of a spouse or family often is more important than many biological factors determining whether a patient will be able to return home (Kotila et al. 1984).

Other markers can be used for prognosticating outcome in a patient. In the first 12 hours after stroke, altered level of consciousness, dysphasia, gaze paresis, and visual field defect (but not paralysis) are predictors of poor outcome (Caronna and Levy 1983).

In cases of hemiplegia, lower extremity strength and function will be greater than in upper extremities. All patients regaining normal arm and hand function will have some voluntary arm movement 1 week and some voluntary hand movement 2 weeks after the stroke. Tendon reflexes usually return within 2 weeks, but if not accompanied by return of some voluntary movement, usually useful function is unlikely.

Different functions show variable recovery. Brain stem dysfunction, especially dysarthria, dysphagia, and vertigo, usually improves, but hearing loss usually is permanent. Hemianopia rarely resolves completely if fixed for more than 1 week, but conjugate gaze abnormalities always disappear. Recovery of language function is extremely difficulty to predict, but receptive and global aphasias generally have a worse prognosis than expressive aphasias (Millikan et al. 1987).

Course of Recovery

The rate of recovery varies among patients and even among functions in a patient. However, most clinically detectable functional improvement occurs within 6 months (Dombovy et al. 1987).

In a recent study of 48 patients with acute hemispheric infarct followed for 3 months at the University of Texas Medical School, gaze palsy was the only function that predictably improved in the first 72 hours. Neglect of the affected half of the body and visual field, visual construction, vision, and sensation showed marked recovery during the first month. All motor functions also showed gradual improvement in the first month, but facial strength improved more than proximal arm and leg strength. Distal arm and leg strength improved the slowest and least. Fluency and comprehension showed gradual partial improvement over 3 months.

In a study of 154 stroke patients, Kotila and coworkers (1984) found that 12 months after the stroke, emotional and neuropsychological disturbances were as prevalent and influenced functional outcome as much as hemiparesis or dysphasia. Also, little functional improvement occurred between 3 months and 1 year.

Right Hemispheric Function

Recovery of right hemispheric function is difficult to quantitate, but Hier and associates (1983) have detailed the return of various functions in 41 patients with unilateral right hemispheric strokes. As in the overall stroke population, motor and visual improvement were rapid but often incomplete, whereas recovery of denial and neglect were both rapid and complete.

On the other hand, extinction and apraxia, while eventually totally resolved, often took up to 1 year to recover. The authors hypothesize that because movement and sensation are mediated by discrete neural pathways, injury usually means incomplete recovery. However, higher cortical functions are more diffusely represented, so recovery of denial, neglect, and apraxia are eventually complete.

Extinction and neglect improved faster after hemorrhage than infarct, perhaps due to the more frequent subcortical location and the dissection rather than destruction of hemorrhage.

Depression After Stroke

Significant depression has been documented in 30% to 50% of stroke survivors. However, while 67% are still depressed 6 months after the onset of depression, most recover spontaneously by 1 year. The period of greatest prevalence and severity of depression is 6 months to 2 years after the stroke (Robinson and Price 1982).

The occurrence of depression correlates with the location and size of the stroke and with "premorbid personality" but does not correlate with demographic factors, functional status, or any component of the neu-

rological examination, including paresis or aphasia (Robinson and Price 1982).

Correlations found between location of the stroke and incidence and type of depression are as follows: Depression has been found to be more frequent and severe with left hemispheric lesions compared to right hemispheric or brainstem strokes, independent of associated aphasia (Robinson and Price 1982). Furthermore, the closer the lesion is to the left frontal pole, the greater the incidence of depression, but this association has been debated (Feibel and Springer 1982; Sinyor et al. 1986).

Sexual Dysfunction After Stroke

Abnormal sexual function after stroke is a problem about which little has been written and which is often ignored or misunderstood by patients and physicians. In one study of 24 male (ages 40–75 years) and 11 female (ages 33–80) stroke survivors, there was no change in their sexual interest or desire compared to their premorbid state; however, decreased erection and ejaculation occurred in men and decreased menstruation in women (Bray et al. 1981).

Cardiogenic Emboli

The prognosis of strokes due to cardiogenic emboli depends on the initial insult, resulting brain damage, and underlying cardiac condition. Although the patient's neurological deficits due to emboli may improve, a major concern is the menacing problem of re-embolization, which may be devastating to the patient, and to a great extent, it may determine the patient's ultimate prognosis if initial survivorship is satisfactory. As indicated in Table 4-3, causes of cardiogenic emboli are multiple, and therefore no single form of therapy will avert all causes of re-embolization. However, the most common causes of cardiogenic emboli are atrial fibrillation, accounting for over 50%, and postmyocardial infarction state; in aggregate, cardiac emboli account for approximately 15% (6% to 23%) of all strokes. Re-embolization may be averted with anticoagulation, which is mentioned because of the predomi-

Table 4-3. Causes of Embolic Strokes of Cardiac Origin

Arrhythmias

Atrial fibrillation
Sick sinus syndrome

Cardiac Muscle Pathology

Myocardial infarction (akinetic segment, ventricular aneurysm)
Alcohol cardiomyopathy
Amyloid cardiomyopathy
Sarcoid cardiomyopathy

Valvular Pathology

Mitral valve prolapse
Mitral annulus calcification
Infective endocarditis
Calcific aortic stenosis
Nonbacterial thrombotic endocarditis
Prosthetic valve replacement

Anatomical or Tumor Pathology

Patent foramen ovale
Atrial myxoma

nance of conditions noted previously. Because a discussion of the pathologies noted in Table 4-3 is beyond the scope of this chapter, only the conditions of atrial fibrillation and postmyocardial states primarily are discussed. (For a comprehensive review of the subject of cardiogenic emboli, the reader is referred to the reports of Dr. Robert G. Hart and his collaborators for the First and Second Embolism Task Force [1986; 1989]).

In a comprehensive study emphasizing the prognostic features of embolic strokes of cardiac origin, Berlit and colleagues (1986) retrospectively analyzed 211 cases. Their well studied cases can serve as a point of departure for further investigations and can provide prognostic data from a modern series. In their series, 21 patients had only TIAs, 39 had Reversible Ischemic Neurological Deficits (RINDs), and 151 had cerebral infarctions. Of the 151 with cerebral infarction, 60 were mild and 91 were severe, 38 of whom died during the hospitalization.

Various risk factors for embolic strokes, such as hypertension, diabetes mellitus, or cigarette smoking, do not influence outcome; however, TIAs and RINDs occur more commonly in individuals who are an average of 5 years younger than the whole

group. This finding suggests that with aging, resilience and response of collateral circulation with vascular occlusion may be impaired.

Prognosis also is influenced by the underlying cardiac disease. For example, patients with myocardial infarction fare better on average than those with atrial fibrillation, which in part may be due to the larger size of the thrombus that can develop in the left atrium in the latter condition. Furthermore, with valvular diseases, mitral causes of emboli have a worse prognosis than aortic pathology, perhaps also due to left atrial thrombus size (Johannessen et al. 1988).

Of the underlying multiple etiologies for emboli, bacterial endocarditis was the worst in the degree of neurological impairment and in causing death.

Greater deficits resulting from the cardiac emboli are associated with a worse prognosis for disability and death. For example, with the most common stroke syndrome of middle cerebral artery occlusion, patients with homonymous hemianopsia had a poor prognosis, as did patients with reduced sensorium, such as stupor or coma. Stupor and coma indicate a poor outcome regardless of stroke etiology and frequently may reflect the poor medical condition of the patient. The common complication of seizures does not adversely affect outcome; seizures reflect cerebral cortical involvement and are not dependent on the size of the embolic infarct.

Findings of greater brain damage, such as CT evidence of infarct or angiographic evidence of an occluded artery with impaired cerebral circulation, were associated with poor prognosis. However, a hemorrhagic infarct on CT scan did not worsen outcome, although bloody cerebrospinal fluid (CSF) worsened the prognosis, perhaps reflecting the extent of brain injury.

Atrial Fibrillation

The occurrence of emboli with atrial fibrillation (AF) is 15% (from 13% to 35%) or approximately 5% per year, but it decreases after first detection of AF (Cerebral Embolism Task force 1989; Harrison et al. 1984; Kempster et al. 1988; Sherman et al. 1986; Wolf et al. 1978). Diagnosis of embolic strokes due to AF has been complicated by the advent of CT scans because clinically silent infarcts presumed to be from AF are now detected. The prediction of stroke occurrence in these patients is unknown, but in those who are symptomtic of strokes, it is 8% (from 10% to 20%) (Cerebral Embolism Task Force 1989). With re-embolization, the second event may be devastating, and efforts to reduce re-embolization with antiplatelet drugs, such as aspirin, and anticoagulation are under active investigation. Re-embolization may be reduced by 50% in both symptomatic and asymptomatic patients with AF (Petersen et al. 1989).

For the prognosis of stroke patients with cardiac emboli due to AF, the results of Berlit and colleagues (1986) reviewed previously, offer reasonably good guidelines to determine probable outcome.

Acute Myocardial Infarction

The occurrence of strokes following myocardial infarction (MI) is 2.5% within the first 2 to 4 weeks, which is the peak period (Cerebral Embolism Task Force 1989). Emboli are more common with anterior MIs than inferior MIs; they occur in 6% (4% to 12%) of the former and 1% of the latter. The occurrence of emboli with anterior MIs is increased if two-dimensional echocardiography shows a left ventricular thrombus and an akinetic segment (Rem et al. 1985). The frequency of left ventricular thrombi is not affected if thrombolytic agents are used for treating the acute MI. Use of anticoagulation may reduce re-embolization up to 50%, but ongoing studies will clarify the true effectiveness and indications for anticoagulation more precisely.

As with cardiac emboli due to AF, the prognostic guidelines from Berlit and co-workers (1986) for post-MI patients were reviewed previously. Collectively, post-MI patients have fewer neurological deficits from emboli than AF patients, which is due to the larger thromboemboli in the former patients.

Lacunar Infarctions

The first large series describing lacunar infarctions was Pierre Marie's (1901) account of 50 cases of capsular infarction. C. Miller Fisher has made important recent contributions to the understanding of this entity. He created a body of literature spanning the early 1960s to today, which includes detailed pathological descriptions of the involved vessels. However, many questions regarding pathophysiology of small-vessel diseases and the etiology of lacunar stroke remain unanswered. For this reason, the discussion of the natural course and prognosis of this disease is incomplete.

Lacunar infarctions occur most frequently in the lenticular nuclei, pons, thalamus, caudate, and internal capsule, that is, in the distribution of the lenticulostriate and thalamoperforant vessels and paramedian branches of the basilar artery (Fisher 1965). Lacunar infarcts range in size from 2 to 15 mm. "Giant lacunas" involving portions of the internal capsule and striatum may be different entities because of their association with large-vessel disease and embolism in some cases (Araki 1978; Fisher 1979; Rascol et al. 1982). However, smaller lacunae are thought to result from small-vessel disease, including the lesions of fibrinoid necrosis, lipohyalinosis, and microatheroma, as described by Fisher. He found a predominance of vessels with lipohyalinosis in a series of patients with smaller infarcts (less than 9 mm), and in a second series, he reported six of 11 larger capsular infarcts occurring in association with microatheroma (Fisher 1969; Fisher 1979). It is unclear whether these different abnormalities represent distinct pathological entities or a continuum of a single pathological process. Likewise, the significance of Charcot-Bouchard aneurysms in the development of lacunar stroke and in the course of small-vessel disease has not been established. Further insights into this pathological process are required to understand the natural history, prognosis, and treatment of lacunar stroke.

Hypertension is a major risk factor for small-vessel disease. It is present in 44% to 75% of patients with lacunar stroke (Gross et al. 1984; Kannel et al. 1970; Lodder et al. 1990; Mohr et al. 1978; Norrving and Cronqvist 1989; Pullicino et al. 1980; Tuszynski et al. 1989). A clear causal relationship, however, has not been demonstrated; in fact, lacunar infarcts occur in the absence of hypertension. Also, the arteriopathy described in autopsy series, although more extensive with hypertension, occurs in its absence as well (Tuszynski et al. 1989). Recently, a large scale community-based study showed no difference between lacunar and nonlacunar strokes with regard to a pre-stroke diagnosis of hypertension (44% and 47%, respectively). Likewise, although diabetes mellitus is commonly associated with lacunar infarct (in 10% to 34%), it is equally common in the setting of cortical stroke, particularly if cardioembolism is excluded (Mohr et al. 1978; Norrving and Cronqvist 1989).

Some authors have suggested that lacunar stroke may result from cardioembolism. Bladin and Berkovic (1984) describe large striatocapsular infarcts associated with an embolic source. These lesions, however, represented occlusion of the proximal middle cerebral artery with involvement of multiple, rather than a single, perforating artery. Because of the angle at which the penetrating vessels arise from their parent arteries, it is unlikely that cardioembolism is a significant cause of lacunar stroke. Likewise, studies comparing lacunar and nonlacunar ischemic strokes suggest that carotid bifurcation and artery-to-artery embolus are much less frequent with lacunar infarction (Kappelle et al. 1988; Norrving and Cronqvist 1989).

Fisher describes more than 20 different presentations of lacunar strokes, the most frequent of which are pure motor hemiparesis, pure sensory stroke, sensorimotor stroke, ataxic hemiparesis, and clumsy-hand dysarthria (Fisher 1982). In clinical se-

ries, lacunar strokes make up between 19% and 27% of first-ever strokes compared with other stroke subtypes (subarachnoid hemorrhage, intracerebral hemorrhage, cardioembolic stroke, and atherothrombotic stroke) (Mohr et al. 1982; Rascol et al. 1982; Sacco et al. 1989a). The annual incidence of lacunar strokes noted in the Oxfordshire Community Stroke Project was 0.33 per thousand. Case fatality rates were 1% at 1 month, which contrasts sharply with that of cerebral hemorrhage (40% to 84%) and cerebral infarction (15% to 33%) (Bamford et al. 1987; Silver et al. 1984). Recovery from lacunar infarction is good and shows the best functional prognosis for all stroke subtypes according to the National Institutes of Neurological and Communicative Disorders and Stroke (NINCDS) Stroke Data Bank population. In patients with motor weakness consisting of incomplete hemiparesis, substantial functional improvement occurred in 85% within 1 month (Mohr et al. 1982). In Rascol's study of 30 patients with pure motor hemiplegia, recovery related to size and location of infarction by CT. Those with large capsulo–putamino–caudate infarcts, or giant lacunas, showed the least improvement, exhibiting in major sequelae three of the four patients with this lesion (Rascol et al. 1982). In Fisher's series of patients with pure sensory stroke, the clinical course in 39 patients is described. In 15 patients, the symptoms has resolved within 2 to 3 days. Although symptoms persisted in 22 patients for weeks to months, there were no serious complaints after 1 year. Only two patients developed a chronic problem with a thalamic pain syndrome (Fisher 1982). Information from the NINCDS Stroke Data Bank reveals an early recurrence risk of lacunar stroke of 2.2% by 30 days. This is significantly lower than other stroke subtypes; the greatest risk for early recurrence occurs in atherothrombotic infarction (Sacco et al. 1989a). On the other hand, Bamford and associates (1987) found the recurrence risk in 1 year to be 11.8%, similar to other stroke subtypes.

In summary, the natural history of lacunar

stroke and small-vessel disease must be studied further. Although the prognosis for recovery is good, the risk for late recurrence approaches that of atherothrombotic stroke, displaying a significant morbidity that could be preventable. For this reason, the pathophysiology of small-vessel disease and the difference between lacunar stroke and other stroke subtypes regarding risk factors, natural history, and response to therapy must be discovered.

Intracerebral Hemorrhage

Intracerebral hemorrhage (ICH) accounts for approximately one fourth of all strokes, and the majority of ICHs are due to rupture of arteriole Charcot-Bouchard aneurysms resulting from longstanding systemic hypertension (Cole and Yates 1967; Fisher 1961; Mohr et al. 1978; Toffol et al. 1987). Other causes of ICH include bleeding diatheses, trauma, arteriovenous malformations, tumor metastases, arteritides, and amyloid angiopathy (Butler et al. 1972; Finelli et al. 1984; Gildersleve et al. 1977; Haerer and Woosley 1975).

The diagnosis of ICH is now made accurately and simply with CT scan or magnetic resonance imaging (MRI) of brain, but the exact underlying cause may not be clarified with neuroimaging techniques (Fieschi et al. 1988; McCormick and Rosenfield 1973). Clinical presentations of ICH syndromes are beyond the scope of this discussion, but briefly, they reflect the tempo, size, and location of ICHs (Fisher 1961). Prognosis for ICHs depends on the initial neurological status reflecting these qualitative and quantitative factors, but also on the underlying cause of ICH and its associated morbidity, potential treatability, and likelihood of recurrence (Douglas and Haerer 1982; Fieschi et al. 1988; Furlan et al. 1979; Garraway et al. 1983).

Primary ICH due to hypertension occurs in six cerebral areas; the most common is the putamen, reflecting the peculiar vulnerability of sharply angulated lenticulostriate vessels to form Charcot-Burchard aneurysms under the pressure of hypertension.

The five other areas are thalamus, cortical white matter ("polar hemorrhages"), pons, cerebellum, and other areas such as caudate nucleus (Fisher 1961; Fisher 1965; Hier et al. 1977; Mizukami et al. 1983; Stein et al. 1984). Amyloid or "congophilic" angiopathy may be multiple and occurs predominantly in the cortical white matter, and particularly in the occipital and temporal lobes (Finelli et al. 1984; Gilbert and Vinters 1983; Giles et al. 1984; Okazaki et al. 1979).

The prognosis of strokes has more changed recently for two reasons, as pointed out by Tuhrim and coworkers (1988) from the Pilot Stroke Data Bank. First, with better surveillance and control of hypertension, the frequency and severity of strokes have decreased. Second, with the advent of CT scans of the brain, small bleeds that were previously believed to be ischemic strokes on clinical examination are now diagnosed. This expands the spectrum of ICHs, not primarily massive hemorrhages with associated major neurological deficits and with death as a frequently inevitable outcome (Fieschi et al. 1988; Longstreth et al. 1989).

In an attempt to identify critical prognostic factors, Portenoy and colleagues (1987) analyzed 112 patients with ICH who were diagnosed with CT scans. By applying multivariate analyses to many factors likely to impact independent and dependent outcome, they found that the factors correlating best with outcome from ICH were level of consciousness, pupillary abnormality, ataxic respiration, and acute hypertension. In a similar study from the Pilot Stroke Data Bank, Tuhrim and associates (1988) analyzed a group of 94 supratentorial ICH patients out of 938 stroke patients. More than 70 variables were examined from the patients' histories, demographics, stroke courses, physical and neurological examinations, and laboratory studies (including blood, electroencephalogram [EEG], electrocardiogram [EKG], CT scan, and angiographic examinations). Eight univariate factors associated with outcome were Glasgow Coma Score, systolic blood pressure, pulse pressure, horizontal and vertical gaze palsies, severity of weakness, brain stem and cerebellar deficits, interval stroke course, and ICH size. To derive a logistical regression model, a step-down variable selection procedure using these factors was used to predict outcome. With this method, 92% of the patients could be predicted as dead or alive after 30 days using the Glasgow Coma Scale, pulse pressure, and hemorrhage size. The authors note that unavailable factors, such as glucose levels and intraventricular hemorrhage, also may predict outcome. Nonetheless, with supratentorial ICHs, 92% accuracy to predict death within 30 days on the basis of sensorium, pulse pressure, and ICH size, reflecting massive impairment and increased intracranial pressure, can be used at the bedside to identify a potentially lethal stroke. Thus, massive hemorrhages of more than 50 cc and the occurrence of coma were lethal insults uniformly resulting in death.

Because of its predominance, only the prognosis of primary ICH is discussed here. As noted previously, the prognosis of ICH depends on the factors surrounding the acute ICH and on the issue of ICH recurrence. The presence of comorbid conditions, such as debilitating cardiac, pulmonary, renal, or other brain diseases, affects activities of daily living (ADL) and mortality.

For acute hypertensive ICH, the most important determinants for prognosis, both functional recovery and survival, are largely dependent on the location, tempo, and quantity of ICH. The six major sites of ICH noted previously are discussed separately for immediate outcome and the value of potential interventional therapies, and collectively for questions of recurrence, risk factor modification, and rehabilitation.

Putaminal ICHs account for up to 60% of primary ICHs and are the most common. The initial symptoms reflect the bleeding tempo and quantity because a massive ICH in the putamen, easily detected by CT scan, was previously misdiagnosed as a TIA due to platelet emboli or hemodynamic factors. Because of the wide spectrum of symptoms, signs, and prognoses, putaminal ICHs cannot, as with cerebral areas, be aggregated together as one syndrome or as displaying uniform prognoses (Giles et al. 1984; Hier et al. 1977; Mizukami et al. 1983).

In all ICHs, if the patient presents in coma, the prognosis for viable outcome is negligible, and because of this, avoidance of therapeutic intervention, such as surgical decompression, should be considered. With putaminal ICHs, in addition to the importance of sensorium, the degree of improvement during the first week is critical in predicting outcome (Fisher 1961; Portenoy et al. 1987). One fourth of patients with ICH will die; another fourth will recover completely to return to their premorbid state; and the remaining half will have moderate to major disabilities for motor, sensory, speech, and higher cortical impairment (Fisher 1961; Mizukami et al. 1983). Neurological improvement can continue over the first year, but often the secondary complications of depression or impaired social interactions and insight are the most disabling (Garraway et al. 1983).

For white matter or "polar" hemorrhages, prognosis for outcome depends on the initial neurological status. With an alert sensorium without increased intracranial pressure or herniation, patients will usually stabilize (Kase et al. 1982). With herniation or reduced sensorium, surgical decompression can be life-saving. Patients may be left only with deficits from the initial ICH, but reasons for postsurgical deterioration remain uncertain (Lipton et al. 1987; Masdeu and Rubino 1984; Ropper and Davis 1980; Tanaka et al. 1986).

Thalamic ICHs may account for up to one third of ICHs (Kwak et al. 1983; Lipton et al. 1987; Mizukami et al. 1983). Prognosis depends on the size of hemorrhage and its location. Hemorrhages measuring greater than 3 cm are usually associated with coma, and the prognosis for recovery is poor (Fisher 1961). Smaller hemorrhages of less than 2 cm are associated with good recovery, but the deficits depend on location of thalamic hemorrhage. For example, lateral hemorrhages are associated with hemiparesis or hemiplegia, while dorsomedial ones occur with memory impairment (Fisher

1961; Mizukami et al. 1983). With dominant thalamic ICHs, a "thalamic aphasia" can occur, which may not recover over time; it may simulate the various cortical aphasias, such as transcortical motor, global, or transcortical sensory aphasias (Fisher 1961).

Pontine ICHs are usually fatal in patients who are comatose. Patients with intracerebellar hemorrhages and in coma also as with other areas of brain injury, do not usually recover, even with surgical decompression. However, patients presenting increasing lethargy and focal brain stem signs, such as gaze palsy, or evidence of fourth ventricular shift and enlarging ventricles, can survive with good recovery, depending on the degree of cerebellar dysfunction at onset (Fisher et al. 1965; Mizukami et al. 1983; Ott et al. 1974).

Subarachnoid Hemorrhage

The management of subarachnoid hemorrhage (SAH) has evolved in the past 2 decades. It is important to appreciate the continuing improvements to understand the natural history and prognosis of SAH, particularly its initial morbidity and mortality and the development of secondary complications of rebleeding, vasospasm, and hydrocephalus.

Saccular aneurysms occur at arterial bifurcations in approximately 5% to 6% of the adult population (McCormick 1984); 26,000 cases of ruptured aneurysms occur annually in the United States. Approximately one fourth of all mortality and morbidity associated with SAH is due to rebleeding. Studies show the risk to be approximately 20% in the first 10 days, and the cumulative risk to be 30% in the first month (Hijda et al. 1987) with a peak rate of 4% in the first 24 hours (Adams 1986; Kassell and Torner 1983). Inigawa and coworkers (1987) report up to a 19% rebleed rate in the first 24 hours. The recurrent rebleed rate drops after 6 months to 2.2% per year for the next 9.5 years. The rate is then reduced to 0.9% per year for the next 10 years (Nishioka et al. 1984). The lethal nature of a recurrent bleed is displayed in a large cooperative aneurysm

study (Nishioka et al. 1984) in which 378 of 568 patients (67%) were followed for 20 years and treated conservatively (no surgery). They died at the end of the period. Forty percent died in the first 6 months after the initial bleed, and fatality was 78% in patients who rebled. This fatality rate is slightly higher than the 66% rebleed mortality reported by Winn and colleagues (1973). In either case, these statistics are formidable. Unlike patients with demonstrable aneurysms, the 477 patients with no aneurysm on angiography had a more favorable prognosis (Nishioka et al. 1984). For patients with a subarachnoid hemorrhage of unknown origin who survived the first 6 months, the recurrent rebleed rate was a maximum of 0.86% per year. Of those who were normotensive, patients surviving the first 6 months had a life expectancy equal to the age- and sex-matched population in the United States. Hypertensive patients with SAH and no demonstrable aneurysm, however, have a 10% to 20% decreased survival rate compared to that of the expected life span of matched U.S. control group (Nishioka et al. 1984).

Patients with multiple aneurysms have an initially high mortality rate of 47% compared with 33% to 38% for single aneurysms. However, after 6 months' survival, their long-term survival parallels patients with single aneurysms (Nishioka et al. 1984).

Studies indicate that level of consciousness and CT scan findings help predict mortality and good outcome. Adams and associates (1985) demonstrate that patients admitted with SAH who were alert on admission ($n = 808$) had a mortality rate of 11%, while comatose patients ($n = 165$) displayed a 71.3% mortality rate. The rate of good outcome was approximately reversed. In addition, impairment of orientation among alert patients had a negative impact on outcome; the alert and oriented patients' mortality rate was 10.5%. Alert but disoriented patients had a mortality of 18.2% with a concomitant drop in good recovery from 77.8% to 52.7%.

From the same study (Adams et al. 1985), evidence indicated that intraventricular hemorrhage (IVH), ICH, and localized, thick, or diffuse collections of blood on CT significantly increased mortality.

The clinical syndrome of delayed cerebral ischemic deficit occurred in 20% to 36% of patients with SAH (Sekhar et al. 1988), with approximately 50% resulting in death or permanent disability.

CT findings have a positive predictive value in determining if significant vasospasm will occur with SAH. Fisher and coworkers (1980) investigated 47 patients with ruptured aneurysms who underwent clinical examinations, CT within 5 days of onset, and cerebral angiographic studies between days 7 and 17. Of the 24 patients with focal clot or greater than 1-mm vertical layering of blood in the cisterns or interhemispheric fissures, 23 had severe and one had moderate vasospasm. Twenty-three of 24 patients had focal neurological deficits, with three deaths, five permanent major deficits, seven permanent minor deficits, and eight recoveries. Of 18 patients with no blood or only thinly diffuse subarachnoid blood on CT scan, only two had severe vasospasm and five had moderate vasospasms, but none developed clinically delayed focal neurological deficits.

Adams and colleagues (1987) also showed a strong correlation between ischemia and focal thick collection of subarachnoid blood on CT.

The role of hydrocephalus secondary to SAH on clinical outcome has been debated, primarily because of the inconsistencies noted in its course. Hasan and colleagues (1989) present data from a study of 473 patients with SAH admitted within 72 hours of onset, and showed that hydrocephalus occurred in 206 (44%). They showed that approximately 33% of these patients were asymptomatic and 50% of those with hydrocephalus and impairment of consciousness improved spontaneously. These results conflict with Vassilouthis and Richardson (1979) who report only one of nine patients with symptomatic hydrocephalus showing spontaneous resolution. Although the accurate prediction of the course of hydrocephalus is elusive, complication due to intracranial

pressure is a significant factor in the outcome of patients with SAH. Other factors that influence prognosis are age of the patient, aneurysm size, and poor medical condition. In a recent international cooperative study (Kassell et al. 1990b) of overall outcome based on aneurysm size, 66% of those with aneurysms less than 12 mm, 52% with aneurysms 12 to 24 mm, and 39% with aneurysms larger than 24 mm showed good recovery. Poor outcome (severely disabled, vegetative state, or death) was seen in 24% of patients with aneurysms less than 12 mm, 37% in those with 12- to 24-mm aneurysms, and 51% in those with larger than 24-mm aneurysms. Outcome based on age revealed a worsening of recovery with each increase in decade. Patients aged 18 to 29 showed an 86% good and a 10% poor outcome; patients aged 70 to 87 who showed a 25% good and a 64% poor outcome.

Timing of neurosurgical intervention has been debated for years. However, the most recent study (Kassell et al. 1990b) has determined that recovery and mortality were not statistically different between the early surgery group (1–3 days post-SAH) and the late surgery group (15–32 days post-SAH), with an overall recovery rate of 63% and an overall mortality of 20% in both groups. Patients receiving surgery between days 4 and 14 did worse than the early and late surgical group. Also, patients who were alert and received early surgery did significantly better than any other group, with a good recovery of 78% and mortality of 11% (see Table 4–1).

It must be noted, however, that this study did not use calcium channel blockers in the medical management. It has been well established (Petruk et al. 1988; Pickard 1989) that the use of calcium channel blockers within 96 hours of the onset of SAH decreases the incidence of delayed neurological deficit due to vasospasm. Because a higher percentage of patients in the early surgery group worsened due to vasospasm, agents preventing this phenomenon could improve further the outcome of the early surgery group. Also the use of vasopressor agents to induce hypertensive therapy with vasospasm is beneficial (Kassell et al.

1990a,b; Muizelaar and Becker 1986). Vigorous use of this therapy generally is undertaken only after clipping of the aneurysm. Unclipped aneurysms require a more cautious approach.

Daily transcranial dopplers may be extremely useful for determining the degree of vasospasm, and they can help regulate the use of antivasospasm and prophylactic hypervolemic hypertensive therapy.

Despite the significant decrease in rebleeding with the use of antifibrolytic agents, an increase in delayed neurological deficits (Adams 1986) and hydrocephalus (Lindsay 1987) make these agents of no benefit in improving overall outcome of SAH. Their usefulness in combination with calcium channel blockers remains to be determined.

Clearly, the treatment and prognosis of SAH depends on multiple variables, and careful attention to each of the factors is crucial for changing outcome.

Arteriovenous Malformation

With cerebral angiomas, the natural history and prognosis vary greatly with the type of vascular malformation. The majority of this discussion focuses on arteriovenous malformations (AVM) in which the size, location, complexity of feeding vessels, and venous drainage pattern of AVM, as well as the patient's age are critical in determining long-term management and prognosis.

Interval angiography performed during the first 2 decades of life has shown that progressive growth of AVMs correlates with brain growth involving enlargement of feeder vessels and increased complexity of venous drainage (Luessenhop and Rosa 1984). These changes correspond with time of onset and relative frequency of first symptoms (Fig. 4–1) (Luessenhop and Rosa 1984). Approximately 80% of the cases that will become symptomatic do so by the end of the fourth decade, with the risk of bleeding decreasing after middle age (Heros and Tu 1987). In their study, Graf and associates (1983) show that approximately 60% of

AVM rupture occurs between the ages of 11 and 35 and that 76% bleed for the first time by the age of 50. Sex of the patient and incidence of seizure are not statistically significant factors in predicting the hemorrhage rate; however, smaller AVMs are believed to have higher incidence of bleeding (Graf et al. 1983; Luessenhop and Rosa 1984). One study (Graf et al. 1983) shows the frequency of hemorrhage in large AVMs (greater than 3 cm) to be 0% at 1 year and 10% at 5 years, while that of small AVMs (less than 3 cm) is 10% at 1 year and 52% at 5 years. They found a cumulative risk of a first hemorrhage to be 2% at 1 year, 14% at 5 years, 31% at 10 years, and 39% at 20 years. An average yearly risk of 2% to 3% was found.

Reported morbidity and mortality due to hemorrhage vary with studies (Fisher 1989; Graf et al. 1983; Luessenhop and Rosa 1984; Svien and McRae 1965). Luessenhop and Rosa (1984) report a morbidity of 30% and mortality of 10%. A review of 11 studies by Fisher (Fisher 1989) places the morbidity and mortality as high as 41.2% and 15.6%, respectively. Perret and Nishioka (1966) also report a 10% mortality with hemorrhage.

After the initial hemorrhage, the risk of rebleed is approximately 6% in the first year and decreases to 2% to 3% for each subsequent year (Fisher 1989; Graf et al. 1983).

The risks and benefits of surgical intervention must be individualized for each case. Studies calculating the morbidity and mortality of surgical obliteration clearly show a relationship to AVM size, location in the brain, pattern of feeding vessels, and complexity of draining veins (Luessenhop and Rosa 1984; Shi and Chen 1986; Spetzler and Martin 1986). Luessenhop and Rosa (1984), using a grading scale based on size, found that AVMs smaller than 4 cm had a surgical morbidity and mortality of 4% and 0%, respectively, whereas AVMs of 4 cm or greater showed a 44% morbidity and a 12.5% mortality.

Spetzler and Martin (1986) incorporated size, eloquence of surrounding brain tissue, and venous drainage into a grading scale.

They found that low-grade AVMs (i.e., small AVMs with superficial draining veins and in the proximity of no vital cerebral structure) had a low morbidity–mortality (less than 3%) and those with high grades (i.e., large AVMs with deep venous drainage and adjacent to eloquent brain) demonstrated approximately 30% neurological deficit.

Heros and Tu (1987), reporting on 49 cases of unruptured AVMs, demonstrated a surgical morbidity of 14.2% with 0% mortality.

Embolization, stereotactic radiosurgery, anticonvulsants, electrothrombosis, and proton beam have all been used with some success in ameliorating symptoms caused by the AVM (Aminoff 1987; Graf et al. 1983). However, methods short of total obliteration do not reduce the incidence of hemorrhage (Aminoff 1987; Luessenhop and Rosa 1984).

Although there are dissenting opinions (Aminoff 1987), a majority of the literature supports early surgery when possible on both ruptured and unruptured AVMs, particularly in patients younger than 50. Beyond the age of 50, the 2% to 3% per year complication rate may not exceed the surgical risks, given the patient's overall medical condition. Serious consideration must be given to conservative therapy if the AVM is positioned in eloquent brain or has a complex vascular supply and drainage.

For the natural course of the other angiomas, a review by Wilkins (1985) reveals that capillary telangiectasias are generally clinically silent and radiographically occult. Cavernous hemangiomas frequently clot and have a low incidence of hemorrhage; however, they can cause seizures. Venous angiomas occurring in 2% to 6% of the population are most often silent, and surgical intervention generally is not recommended if asymptomtic. One exception, however, is based on data that venous angiomas of the cerebellum may cause a high incidence of hemorrhage. Rothfus and coworkers (1984), when reviewing the literature, found that 36% of cerebellar venous angiomas bleed,

and therefore surgery should be considered more seriously in these cases.

Carotid Dissection

Spontaneous cervical carotid artery dissection is well recognized and accounts for 1% to 2.5% of all strokes (Bogousslavsky et al. 1987; Hart and Easton 1986) and as much as 5% of cerebral infarctions in young adults (Hart and Easton 1983). Seventy percent of patients with carotid dissection are between the ages of 35 and 50 (approximately one third to one half developing cerebral ischemia). The most common presenting symptom is headache or neck pain, seen in 76% to 92% of patients (Hart and Easton 1986; Mokri et al. 1986), often preceding cerebral ischemia by several hours to 2 weeks (Hart and Easton 1986; Mokri et al. 1986). Mokri and associates (1986) report that focal ischemia may be found on initial examination in 67% of cases, and Biller and coworkers (1986) report it in 85% of their patients. Oculosympathetic paresis is the third most common presenting sign. Pathological studies reveal that subintimal hemorrhage leads to luminal stenosis, while subadvantitial hemorrhage results in pseudoaneurysms. Thromboembolization from the site of dissection is the most likely cause of ischemia, supported by visualization of distal emboli by angiography (Biller et al. 1986; Bogousslavsky et al. 1987; Hart and Easton 1986; Mokri et al. 1986).

Conflicting reports exist on the natural history of carotid dissections. A review of 140 cases by Hart and Easton (1983) confirms early beliefs that the outcome is generally favorable by reporting an 8% to 10% incidence of major deficits or death. Mokri and coworkers (1986) present similar findings in 36 cases of dissection in which large infarcts were seen in three patients (8%) who displayed only moderate neurological deficits. However, the benign nature of the process is challenged by Bogousslavsky and colleagues (1987) who report 30 cases of carotid occlusion secondary to spontaneous dissection. In their series, seven patients

(23%) died in the first week. Of the survivors, 11 of 23 (48%) had severe deficits, with the remaining 12 (52%) showing good outcomes. Pozzati and associates (1990) report an overall mortality and major morbidity of 32% in patients with occlusive cervical carotid dissection.

Recurrence of dissection is rare. In 36 cases followed for an average of 58.5 months, Mokri and associates (1986) report no recurrence of carotid events, although one patient experienced a renal artery dissection 8 years after bilateral carotid dissection.

Factors that predispose to dissection are believed to be fibromuscular dysplasia (15% to 20% of all spontaneous dissections) (Bogousslavsky et al. 1987; Hart and Easton 1986; Mokri et al. 1986) and hypertension (Mokri et al. 1986; Ojemann et al. 1972). Hypertension was found in one study in 36% of patients, well above the control population. The significance of migraine headaches, atherosclerosis, and oral contraceptives, seen in sporadic cases, remains uncertain.

The evolution of carotid stenosis and occlusion can be followed by serial Doppler examination, which may aid in determining the duration of management required. Bogousslavsky and coworkers (1987) report that 13 of 23 (57%) carotid occlusions returned to normal in the first 6 months, and 11 of 13 (85%) were normal in the first month. No resolution of an occluded vessel was noted after 6 months. Pozzati and colleagues (1990) show a 47% carotid recanalization rate within the first 6 months, although half of those still had vessel wall defects on repeat angiography.

Good outcome from carotid dissection appears to be associated with early recanalization and small or no cerebral infarction on presentation (Bogousslavsky et al. 1987; Mokri et al. 1986). Poor outcome occurs in patients with distal embolization on angiogram, large cerebral infarction, persistent internal carotid artery (ICA) occlusion, and a decreased level of consciousness (Bogousslavsky et al. 1987; Pozzati et al. 1990).

No prospective study has been undertaken to evaluate the benefits of anticoagulation and antiplatelet therapy on outcome. Reviews of the literature show support for the use of anticoagulation in carotid dissection, citing improved neurological outcome and recanalization of the intracranial ICA (Bogousslavsky et al. 1987; Fisher et al. 1965; Hart and Easton 1983; McNeill et al. 1980; Pozzati et al. 1990). Others report spontaneous resolution without anticoagulation therapy (Biller et al. 1986; Ehrenfeld and Wylie 1976; Gee et al. 1980; Mokri et al. 1986.)

Surgical intervention, including emergent removal of the dissected intima, intraluminal thrombus, carotid artery ligation, dilation with a catheter, endarterectomy, and pseudoaneurysm repair (Biller et al. 1986; Ehrenfeld and Wylie 1976; Fisher et al. 1978; Ojemann et al. 1972), have been used in individual cases with variable but generally discouraging results. Good outcome has been reported with superficial temporal artery to middle cerebral artery anastomosis (Pozzati et al. 1990; Welling et al. 1983).

Despite reports that this entity has a benign outcome (Hart and Easton 1983; Mokri et al. 1986), others have demonstrated that the sequela of carotid dissection is catastrophic in a significant minority of the cases (Bogousslavsky et al. 1987; Ehrenfeld and Wylie, 1976; Ojemann et al. 1972; Pozzati et al. 1990). Though each case mut be individualized, the authors support the use of early anticoagulation for at least 6 to 12 weeks. Serial Doppler may assist in continuous monitoring or repeat angiography can document the resolution of dissection. Complete resolution of the intimal defect would prompt the discontinuation of anticoagulation. Residual luminal defects and recurrent symptoms would dictate longer use of effective medical management. If anticoagulation is contraindicated, then antiplatelet therapy should be used. Because the method of infarction is believed to be thromboembolic, initial anticoagulation may preclude surgical intervention, although recurrent embolization despite medical therapy with an accessible lesion may warrant surgical intervention.

Intracranial arterial dissections generally present with neurological deficits simultaneously with other symptoms (Hart and Easton 1986). The resulting neurological deficits in many cases are severe. Anticoagulation generally is not recommended because intracranial dissections have a higher incidence of subarachnoid hemorrhage (Bogousslavsky et al. 1987).

Vertebral Artery Dissection

The natural history and prognosis of vertebral artery dissections are not as well known as carotid dissections. Case studies suggest an association between migraine headaches, past and current use of oral contraceptives, hypertension, fibromuscular dysplasia (FMD), and atherosclerosis and the occurrence of vertebral dissection (Caplan et al. 1985; Chiras et al. 1985; Hart 1988; Mas et al. 1987). However, no case-controlled studies have established these factors as definite variables in predicting outcome or recurrence of dissection.

The most common mechanisms of brain injury resulting from vertebral dissection are due to severe stenosis or occlusion (Hart 1988), intraluminal thrombosis leading to artery emboli (Katirji et al. 1985; Levine et al. 1989; Levine and Welch 1988; Mas and Henin 1989), dissection extending to the basilar artery and obstructing flow, and subadventitial dissection causing SAH.

Neurological symptoms of head and neck pains (the most common presenting complaints) can precede other symptoms and signs by hours to weeks or possibly by as long as 1 month (Farhry et al. 1988; Mas et al. 1987).

Dissections of extracranial vertebral arteries are associated with a favorable prognosis. A review of 46 cases demonstrates good outcome (absent or mild deficits) in 78%; moderate residual deficits in 18%; and death in 4% of the cases (Alexander et al. 1986; Biller et al. 1986; Caplan et al. 1988; Chiras et al. 1985; Farhry et al. 1988; Hart

1988; Katirji et al. 1985; Laterra et al. 1988; Levine et al. 1989; Levine and Welch 1988; Mas et al. 1987; Mas and Henin 1989). One study with well documented initial angiography and follow-up study (average 3 months) showed that 63% returned to normal, 26% displayed marked improvement, and 11% advanced to occlusion. Significant anatomical improvement was seen as early as 7 days on repeat angiography (Mas et al. 1987).

These percentages do not represent the natural history and may lead to false reassurance. In the study by Mas and associates (1987), 11 of 13 patients with vertebral artery dissection had a favorable outcome, but as the author points out, 10 of 11 were on anticoagulation therapy for 2 to 36 months (average 9.6 months), and the 11th on aspirin (ASA) for 1 year. Three of 10 patients had progression or recurrent neurological deficits prior to anticoagulation. Chiras and colleagues (1985) report 10 patients with favorable outcomes. Five of 10 patients who had recurrent or progressive neurological deficits experienced TIAs prior to their infarcts. Six did well after anticoagulation therapy and two after surgical intervention. The remaining two of 10 did well with supportive therapy. Two others in this series with no reported medical or surgical intervention had residual neurological deficits. Although no controlled study is available, residual anatomical abnormalities on follow-up angiography are believed to demonstrate a persistently increased risk for neurological events, as demonstrated by Katirji and coworkers (1985). They report one patient who experienced recurrent vertebral basilar ischemia with cessation of coumadin 16 months after vertebral dissection. Angiography revealed persistent arterial irregularities, and reinstatement of anticoagulation therapy lead to no further symptoms.

Whereas the initial presentation of extracranial vertebral dissection appears to be mild with head and neck pain or mild ischemic syndromes (though some present with infarction), patients with intradural vertebral dissection present with more catastrophic events, predominantly characterized by severe posterior circulation ischemia or SAH (Caplan et al. 1985). Brain stem infarctions are usually due to subintimal dissection extending into the basilar artery. They are more commonly seen in younger patients who often present with a single fatal event (Caplan et al. 1988). Intradural arterial morphology predisposes to the greater likelihood of a subadventitial or transmural dissection (Wilkinson 1972) and subsequent SAH. Caplan and coworkers (1985) report that 10 of 21 patients with SAH after dissection had a normal outcome; five died; one had moderate to severe deficits; and four had mild deficits. Of the ten presenting with acute posterior circulation ischemia, nine died. Thrombosis or flow obstruction due to dissection of the intima or media was seen on autopsy studies. The high propensity for SAH as a result of intradural vertebral artery dissection demands that a CT lumbar puncture (LP) be performed before considering anticoagulation therapy.

Angiographic series have reported the occurrence of multiple arterial involvement with FMD to be as high as 66% (Mokri et al. 1988). The authors review of the recent literature in which four-vessel angiography was reported revealed that 42% of patients (22 of 52) presenting with symptoms of vertebral artery dissection had multiple vessel involvement. The prognostic significance of these findings is not well defined.

Using the existing data, patients with symptomatic angiographically proven extracranial vertebral artery dissection should be treated with anticoagulants or antiplatelet therapy with a repeat angiography at 3 months. If repeat angiography demonstrates persistent significant intimal defects or stenosis, anticoagulation therapy should be continued.

Fibromuscular Dysplasia

FMD is a vascular disease characterized by a nonatherosclerotic, noninflammatory process, involving the smooth muscle, and fibrous and elastic tissues of small and medium arteries. In a review of more than 1400 cases of FMD, 60% to 75% have renovascu-

lar, 25% to 30% cerebrovascular, 9% mesenteric or subclavian artery, and 5% ileal or femoral artery involvement. Twenty-four percent have multiple vessel distribution. The coronary arteries are rarely involved (Corrin et al. 1981; Luscher et al. 1987; Mettinger 1982).

More than 33,000 consecutive angiographic studies reveal the radiological diagnosis of FMD in 0.67% of cases (Corrin et al. 1981; Healton 1986) with a female predominance of 66% to 94% (Corrin et al. 1981; Mettinger 1982).

Because of its relative selectivity with renovascular and cerebrovascular complications, prognosis is influenced by end-organ diseases within these vascular territories. Although renal prognosis is beyond the scope of this chapter, Corrin et al. (1981), and Pohl and Novick (1985), have demonstrated up to 33% progression of renal stenosis with FMD. However, the clinical correlation of elevated blood pressure and impaired renal function in FMD is unclear.

With cerebral vascular FMD, the natural history of the disease appears benign. Corrin and coworkers (1981) report a follow-up of 79 patients with angiographically demonstrated FMD for an average of 60 months (1–231 months) and noted three subsequent cerebral ischemic events. Only one of the three had an FMD lesion by prior angiography corresponding to the distribution of the infarct. Wells and Smith (1982) report 16 women with angiographically demonstrated FMD of the internal carotid artery (ICA) and an average follow-up of 3.8 years. Two patients had subsequent strokes, but both had coincident atherosclerotic disease.

Many series of surgical interventions have been reported in FMD with graduated intraluminal dilation (the procedure of choice) that showed 0% operative mortality with good long-term clinical results. Effeney and associates (1983) report 150 carotid procedures in 101 patients (22% with completed strokes prior to surgery). Three patients had operative strokes (no operative mortality); follow-up to 17 years demonstrated only two subsequent strokes, and repeat angiography up to 5 years later demonstrated only one previously operated vessel to be occluded. Potential complications of FMD are spontaneous ICA dissection, as reported by Grotta and coworkers (1982a) and others; vertebral artery dissections, which are less common; traumtic dissection; thromboembolic carotid cavernous fistulas; and intracranial aneurysms.

Hieshima and colleagues (1986) report four cases of spontaneous arteriovenous fistulas in association with FMD. In addition, Mettinger (1982) reports 284 patients with FMD, 61 (21%) with intracranial aneurysms and 20 of those with multiple aneurysms.

Because of the benign nature of FMD, the recommendation for a conservative approach to therapy generally is accepted. Though the efficacy of medical therapy is unproven, use of antiplatelets and anticoagulation theoretically are appropriate in symptomatic patients (Corrin et al. 1981; Effeney et al. 1983; Stewart et al. 1986; Wells and Smith 1982). However, therapy must be individualized, as demonstrated by reports that anticoagulation therapy is more effective than aspirin in recanalization of spontaneous dissected cerebral artery (Luscher and Lie 1987). Although surgical intervention has been associated with a low incidence of neurological events, (Corrin et al. 1981) it should be reserved for those with progressive cerebral events and not for those with FMD as an incidental finding.

Sickle Cell Anemia

Sickle cell disease, an autosomal recessive disorder seen in the black population, predisposes the patient to an increased risk of cerebrovascular complications. Vascular compromise is seen in large-vessel distributions, with arterial wall ischemia secondary to obstructed flow in the vasovasorum by sickled cells, causing intimal proliferation and obstruction (Grotta et al. 1986) and in small vessels due to increased viscosity and stasis (Osuntokum 1981). The CNS complication rate is 8% to 17% hemoglobin (HbS) disease, 2% to 5% in HbSC, and 1.5% to 2% in patients with HbSA or HbAA (Grotta et al. 1986). The presentation of strokes appears to differ with the age of the patient. Thrombotic infarction is predominant in children, while hemorrhage infarction (SAH and ICH) are more common in adults (Powars et al. 1978). Powars and associates (1978) report the relative risk for stroke in all age groups to be approximately 6% in patients with HbSS and SC disease. In 537 patients in a natural history study, the actuarial risk of stroke for the first 2 decades of life is determined to be 0.761 per 100 person-years, and for those older than 20 years, it is 0.542 per 100 person-years. Of those who had strokes, the recurrence rate was 67% over a 9-year period, with approximately 80% occurring in the first 36 months. The mean age of onset of the first stroke was 7.7 years with a mode of 5 years. As a complication, Powars and colleagues found that approximately 41% of those with strokes developed seizure disorders, half of which were difficult to control. Wood (1978) found in a study of 86 patients with sickle cell disease that 70% of those strokes had generalized or focal seizures at the onset. A majority of these subsequently developed a chronic seizure disorder requiring long-term anticonvulsant therapy.

A decline in cognitive function also is seen with acute strokes. Powars and coworkers (1978) report that in the evaluation of 10 children with five strokes and HbSS, all had some intellectual impairment, and seven of 10 exhibited significant mental retardation (I.Q. score less than 70). Two of these patients demonstrated some improvement with time. Wilimas and colleagues (1980) followed patients with serial neuropsychological testing. They note low I.Q. scores after the initial stroke. Improvement was seen in the majority of these patients, but further decline in cognitive skills was seen with subsequent strokes. These data, and other reports of clinical outcome, show that intellectual capacity is dependent on the number of infarcts sustained.

Although reports show that children with sickle cell disease have poorer musculoskeletal and intellectual development (McCormack et al. 1975; Osuntokum 1981), a study of 200 Jamaican children with HbSA suggests that it does not affect growth or mental development (Ashcroft et al. 1976).

The clinical severity of sickle cell disease (i.e., crisis frequency and so forth) does not predict cerebral infarction. However, patients with strokes have a significantly higher incidence of hospitalization, not only for stroke-related problems, but also for priapism, pneumonia, pyelonephritis, and sickle cell crisis, which occurs twice as often as for nonstroke HbSS patients (Grotta et al. 1986; Powars et al. 1978). EEG and CT Scans also are not useful in predicting cerebral infarctions in these patients (Wilimas et al. 1980). A fetal hemoglobin level of more than 13% is the only laboratory finding that can predict the reduction of risk for cerebral vascular accidents (Powars et al. 1978).

With the known pathogenesis of strokes in sickle cell disease, its natural progression may be altered if HbS concentration could be reduced.

Stockman and coworkers (1972) demonstrate that reduction of HbS to less than 40% prevented strokes during angiography in patients with HbSS disease. Russel and associates (1976) follow five children with strokes and internal carotid artery stenosis with angiography. Three of five received exchange transfusions for 1 year to achieve an HbS of less than 30%. Two of the three had total resolution of the ICA stenosis, and the third showed improvements with no new neurological symptoms. The two untreated patients had angiographically demonstrable progression of their vascular disease.

Wilimas and colleagues (1980) used exchange transfusion of HbS to less than 20% for 2 years in 10 children after their first infarction. No new cerebral events occurred during therapy. However, 7 of the 10 had recurrent strokes in the same vascular territory within the first 5 weeks to 11 months after stopping therapy. In a study of thirty patients with HbSS, Russel and associates (1984) used prolonged transfusion therapy for 1 to 9 years in 19 patients and demonstrated a 75-fold decrease in clinical recurrence of stroke and marked improvement of luminal abnormalities on angiography. The degree of stenosis doubled in four patients

who were not transfused. Of the seven patients with smooth abnormalities of a single vessel, nonocclusive changes, or normal angiography, only one had a recurrent infarct, implying that angiographic change may be useful for determining which patients will require long-term transfusion therapy.

Because prolonged use of transfusion carries risks, medical therapies to alter the percentage of hemoglobin also can be used. The most promising is hydroxyurea, which increases the concentration of fetal Hgb and decreases the rate of hemolysis, intracellular polymerization, and irreversibility of the sickle cell (Rodgers et al. 1990). Further work remains to be done.

The course of sickle cell disease is much more benign. However, ischemic episodes have been precipitated at high altitudes and in patients with concurrent migraine headaches (Osuntokum 1981; Powars et al. 1978).

Patients with HbSC have been reported to be asymptomatic until adulthood, at which time they may present with a stroke. As with sickle cell trait, they are more susceptible to decreased oxygenation tension, as is seen at high altitudes.

Coagulation Disturbance

In recent years, a growing body of literature has recognized coagulation abnormalities as a major factor in the etiology of strokes, particularly in young people. Increased incidence of thrombosis has been noted in patients with antiphospholipid antibodies; deficiencies of antithrombin III (AT III), protein C, and protein S; and increases of fibrinogen, lipoprotein (a) (Lp [a]), and homocysteine.

Antiphospholipid Antibodies

A family of circulating antibodies directed against phospholipids, antiphospholipid antibodies (APLAb), has been associated with a prothrombotic state that can lead to a spectrum of neurological diseases, including cerebral ischemia (Asherton et al. 1989; Levine and Welch 1987). Recent studies have shown an increased prevalence of anti-cardiolipin antibodies (aCL) and lupus anti-coagulant (LA) in patients with cerebral infarctions (Brey et al. 1990; Briley et al. 1989; Kushner 1990; Lechner and Pabiner-Fasching 1985; Levine et al. 1990) and an increased stroke risk with APLAb. Levine and colleagues (1990) studied 48 patients (mean age 43 years) who presented with cerebral or ocular ischemia and had elevated APLAb. During their prospective follow-up, a combined rate for stroke, TIA, systemic thrombosis, and death was 0.54 events per patient year. For patients with prior strokes, the average recurrence rate in the first year is 6.75%. The rate of new thrombotic events in asymptomatic patients with APLAb is unknown with prospective studies. Lechner and Pabinger-Fasching (1985) reviewed 259 cases of patients with LA and found that 33% had thrombotic events, 25% of which were cerebral.

Brey and associates (1990) studied 46 consecutive patients with stroke who were younger than 50 years and found that 21 (46%) were positive for APLA$_b$, while 54% were negative. An age-matched control group with other central nervous system (CNS) disease had only an 8% (two of 26) prevalence of APLAb. In the follow-up period (1.2 years), the recurrence rate was 14% per year (three of 21) in the APLAb-positive group and no recurrence in the APLAb-negative group.

Not only the presence of APLAb, but the elevation of titers appears to correlate with clinical disease. Briley and coworkers (1989) demonstrate that of the patients with aCL and neurological deficit, the highest titers correlated with multiple infarctions and encephalopathy.

Though anecdotal reports show antiplatelet therapy, anticoagulation, steroids, and immunosuppressant therapy to be somewhat effective, prospective clinical trials must be performed.

Protein C

Protein C is a vitamin K-dependent plasma protein that inhibits the ability of factors Va and VIIIa to facilitate clot formation. Along with protein S and AT III, it may prevent intravascular clotting and limit the extension of an existing clot. Protein C deficiency is an autosomal dominant disease with incomplete penetrance. Homozygotes in many cases experience severe thrombotic disease soon after birth and die in infancy if early plasma (i.e., protein C) replacement and long-term anticoagulation are not instituted (D'Angelo et al. 1988). Heterozygotes are at increased risk of thromboembolic disease (Brockman and Conrad 1988; D'Angelo et al. 1988; Kohler et al. 1990) when the protein C concentration falls below 50% of normal (Brockman and Conrad 1988). In a study of French and Dutch patients, Brockman and Conrad (1988) and Comp (1986) report that 60% to 65% of heterozygotes (type I deficiency) had thrombotic episodes at the time of diagnosis and that approximately 50% are symptomatic by age 30. Brockman and Conrad showed that only 3% of those who were symptomatic had cerebral vein thrombosis. These data, however, are not undisputed. Miletich and colleagues (1987) report no thrombotic events in 79 patients found to have protein C level consistent with heterozygote deficiency. Still, arterial thrombosis has been associated with protein C deficiency (Griffin et al. 1981; Kohler et al. 1990). Kohler and associates (1990) report a 32-year-old man who suffered a middle cerebral artery thrombotic infarct and was found to have protein C functional activity of 36%.

The prognostic factor of protein C after strokes was evaluated by D'Angelo and coworkers (1988) who measured fibrinopeptide A (FA) and protein C. In 37 stroke patients, 11 of 37 died in the first 15 days and were found to have a protein C level only 63% that of the survivors. Concomitant elevation in FA suggests that massive thrombosis may activate protein C and decrease its concentration in the blood. This supports trials of protein C supplementation in patients with acute stroke.

A review of acquired protein C deficiency (Comp 1986) reveals that 50% to 96% of patients with disseminated intravascular coagulation (DIC) have a significant decrease in protein C possibly due to consumption. This raises the possibility of plasma protein C supplementation for therapy. Other causes of acquired protein C deficiency are liver disease, acute leukemia, acute respiratory distress syndrome (ARDS), coumadin use, and others.

Treatment for acute thrombosis in protein C deficiency is heparin anticoagulation, which appears to prevent recurrent thrombosis. These patients should be followed with long-term anticoagulation. Other therapies remain to be explored.

Protein S

Protein S, vitamin-K dependent plasma protein, is a cofactor for the anticoagulant activity of protein C. As in protein C deficiency, the onset of thrombotic events occurs at an early age—an average of 27 years in one study (Comp 1986)—and is associated with recurrent thrombosis of the venous system (High 1988). A study by Sacco and colleagues (1989b) shows that 21 of 103 patients with stroke had a decrease in free protein S by quantitative analysis, although in a select group, its prevalence was greater than expected.

The course of therapy from several studies with familial protein S deficiency shows good outcome with long-term anticoagulation (Comp 1986).

Antithrombin III

AT III is a plasma inhibitor that inhibits vitamin K-dependent clotting factor; therefore, it is a natural anticoagulant. Approximately 55% of patients with AT III deficiency have at least one thrombotic event (Comp 1986), and 36% of the first events develop spontaneously. Venous sites are the most predominant although cases of cerebral arterial thrombosis have been reported (Ernerhudh et al. 1990; Vomberg et al. 1987). Prevalence studies of stroke and prospective therapeutic trials have not been done. Case reports reveal that early warfarin therapy in combination with plasma is prudent for severe cases, whereas intravenous heparin is sufficient for good outcome in mild cases (Comp 1986; Filip et al. 1976).

Fibrinogen

Data show that elevated plasma fibrinogen levels are an important risk factor for myocardial infarction and stroke. Wilhelmsen and coworkers (1984) demonstrate in a random sample of 792 men, 54 years of age, that fibrinogen levels were significantly higher in men who had strokes than in those who did not (3.70 g/L vs. 3.30 g/L; $p < 0.01$, respectively). In addition, when combined with hypertension, the risk of stroke was three times greater than the risk with elevated fibrinogen level alone. Fibrinogen increases the risk of stroke by increasing blood viscosity and decreasing cerebral blood flow (Grotta et al. 1982b). In addition, fibrinogen affects atherosclerosis. Grotta and associates (1989) followed 38 patients with asymptomatic carotid stenosis. Eight patients had rapidly progressing atherosclerotic plaques. Multifactorial analysis revealed low-density lipoprotein and fibrinogen levels to be significantly elevated in those eight patients vs. 30 nonprogressing patients. This, combined with coronary artery disease, will predict with 88% accuracy progressive carotid stenosis. Treatment of increased fibrinogen levels remains to be determined.

Homocysteine

Cerebral infarctions in both adults and children have been noted in patients with homocysteine elevation. An international survey by Mudd and coworkers (1985) was conducted to define the natural history of homocystinuria due to cystathionine B-synthase deficiency. From these data, the estimate of thromboembolic events occurring before the age of 20 is 30%, and the probability before the age of 40 is approximately 60% (Boers et al. 1985). Boers and coworkers (1985) studied excess homocysteine accumulation after methionine loading in 75 patients presenting with occlusive arterial disease before age 50. Results showed that 28% of those patients with occlusive peripheral vascular disease (seven of 25) and occlusive cerebrovascular disease (seven of 25) were heterozygous for homocystinuria. The detection of one heterozygote among 40 healthy controls explains the high frequency of symptomatic vascular disease and supports the increased risk of occlusive arterial disease in the carriers of homocystinuria. Coull and associates (1990) compared plasma levels of homocysteine in 99 patients with acute stroke, TIAs, or risk factors with ischemic symptoms to 31 normal controls. Levels were moderately but significantly higher in 30% of the patients vs. the control group, showing that moderate levels of homocysteine are independent risk factors for cerebrovascular disease.

Pyridoxine administration is the most widely used therapy for patients who are B_6-responders (i.e., those who show a decrease in plasma homocysteine with large doses of vitamin B_6). A study of B_6-responders on therapy (3744 months of follow-up) compared to those off therapy (2028 months) displayed thromboembolic event rate of 0.04 and 0.08 events per year, respectively. Of the B_6-nonresponders, methionine restriction vs. no therapy yielded a 40% reduction in events (0.06 and 0.10 events per year) (Mudd et al. 1985).

Screening for homocystinuria in population younger than 50 years appears to be useful for the prevention of future events.

Polycythemia Vera

Polycythemia vera (PCV) is a myeloproliferative disease characterized by a hyperplasia of erythrocyte, leukocyte and megakaryocyte cell lines in the bone marrow. Cerebrovascular symptoms arising from PCV are related to direct arterial or venous occlusion, primarily due to increased whole-blood viscosity. This results from increased red cell mass, which significantly decreases cerebral blood flow (CBF) (Grotta et al. 1982b; Hart and Kanter 1990; Thomas et al. 1977a; Thomas et al. 1977b). Thomas and associates (1977) measured CBF in patients ($n = 15$) with PCV (mean HCT 53.6%) and found it to be significantly less than the normal controls at 37.9 ml per 100 g/min vs. 69.1 ml per 100 g/min, respectively. They also showed that a reduction of the mean hematocrit with venesection from 53.6% to 45.5% resulted in a 73% increase in CBF. In contrast to earlier beliefs, hematocrits in the upper limit of normal are associated with an increase in cerebral occlusive disease and a decrease in CBF (Pearson and Wetherley-Mein 1978; Thomas et al. 1977b). Therapies used to reduce the hematocrit also carry risks. Phlebotomy has been noted to cause early thrombotic events, particularly in the first 3 years (Tartaglia et al. 1986). Attempts to decrease early thrombotic complications with an antiaggregating agent (e.g., ASA, dipyridimole) substantially increase the risk of serious hemorrhage without any benefit (Hart and Kanter 1990; Tartaglia et al. 1986). With myelosuppressive agents, benefits are negated by increased incidences of acute leukemia and other neoplasms (Kaplan et al. 1986). Kaplan and coworkers (1986) show that hydroxyurea added to phlebotomy therapy may be effective in reducing hematocrit without increasing side effects; hydroxyurea reduced the occurrence of thromboembolic events by approximately 25%. Further studies need to be done. Decreasing the hematocrit with isovolemic phlebotomy results in increased CBF and decreased symptoms (Grotta et al. 1982b; Thomas et al. 1977a). Secondary polycythemia has a lower association with ischemic events than PCV, although this may be a function of a lower mean hematocrit in these patients (Hart and Kanter 1990).

Thrombocythemia

Essential thrombocythemia is a myeloproliferative disorder that increases the risk of thrombohemorrhagic infarcts due to an elevated platelet count and qualitative platelet dysfunction. Neurological symptoms, including strokes, are commonly seen with a peak onset age of 50 to 60 years (Jabaily et al. 1983; Lahverta-Palacios et al. 1988). Jabaily (1983) reports that 21 of 33 patients with unequivocal or essential thrombocythemia displayed neurological manifestations at presentation.

Lahverta-Palacios and colleagues (1988) report 35 ischemic events in 12 of 19 essential thrombocythemia patients during a 421 patient-month follow-up. Five of 12 patients had clinical onset of vascular occlusion with platelets less than 1,000,000/ml. The remaining seven had first attacks at lower platelet counts. Of the 35 occlusive events, 22 (68%) occurred with platelet counts between 650,000 to 900,000/ml. No events were recorded with platelets less than 650,000/ml. When comparing nontreated to treated patients, cerebral ischemic events were seen in nine patients (47.3%) in the nontreated, with no events occurring in the treated group. Treatment consisted of melphalon in the early phase of the study but was switched to hydroxyurea for the remainder of the study. The severe natural history of the disease was demonstrated in this study because 63% had ischemic events and 21% had hemorrhagic diathesis. Symptoms can present at platelet counts lower than originally believed, and early management may prevent catastrophic events.

It has been commonly accepted that the course of essential thrombocythemia is relatively benign (Hoagland and Silverstein 1978) in those younger than 30. Millard and coworkers (1990), however, recently reviewed 13 patients' cases (aged 22–35) and found 11 of 13 to be symptomatic at diagnosis, three with life-threatening vaso-occlusive disease and six with cerebrovascular disease. The series clearly demonstrates the potential for poor outcome and should redirect thinking to long-term therapy in this patient population.

No definitive therapy exists; however, hydroxyurea is the treatment of choice for long-term use in minor to moderate vaso-occlusive disease. Platelet plasmaphoresis has been used in emergency settings in combination with hydroxyurea (Jabaily et al. 1983). Further prospective studies must be carried out.

Other factors influence intravascular coagulation that are beyond the scope of this chapter. Lp(a), a lipoprotein that inhibits the thrombolytic action of plasminogen; plasminogen activator inhibitor-1 (PAI-1), a tissue plasminogen activator (tPA) inhibitor; and dysfibrinogenemias all play a role in coagulation. Their natural history must be studied carefully, and a treatment plan to maximize recovery must be tested.

References

Adams, H., Jr. Antifibrinolytic therapy for prevention of recurrent aneurysmal subarachnoid hemorrhage. Semin. Neurol. 6:309–315; 1986.

Adams, H. P. Jr.; Kassel, N. E.; Turner, J. C.; Haley, E. C. J. Predicting cerebral ischemia after aneurysmal subarachnoid hemorrhage. Influence of clinical condition, CT results, and antifibrinolytic therapy. A report of the Cooperative Aneurysm Study. Neurology 37:1586–1591; 1987.

Adams, H. P. Jr.; Kassell, N. F.; Turner, J. C. Usefulness of computed tomography in predicting outcome after aneurysmal subarachnoid hemorrhage. A preliminary report of the Cooperative Aneurysm Study. Neurology 35:1263–1267; 1985.

Alexander, J. J.; Glacou, J.; Zarins, C. K. Repair of a vertebral artery dissection. J. Neurosurg. 64:662–665; 1986.

Aminoff, M. J. Treatment of unruptured cerebral arteriovenous malformations. Neurology 37: 815–819; 1987.

Antiplatelet Trialists' Collaboration. Secondary prevention of vascular disease by prolonged antiplatelet treatment. Br. Med. J. 296:320–331; 1988.

Araki, G. Small infarcts of the basal ganglia with special reference to transient ischemic attacks. Recent Adv. Gerontol. 469:161–162; 1978.

Ashcroft, M. T.; Desai, P.; Richardon, S. A.; Serjeant, G. R. Growth, behavior, and educational achievement of Jamaican children with sickle cell trait. Br. Med. J. 1:1371–1373; 1976.

Asherton, R. A.; Khamashta, M. A.; Gil, A.; Vazquez, J.; Chan, O.; Baguley, E.; Hughes, G. R. Cerebrovascular disease and antiphospholipid antibodies in systemic lupus erythematosis, lupus-like disease, and the primary antiphospholipid syndrome. Am. J. Med. 86:391–399; 1989.

Bamford, J.; Sandercock, P.; Jones, L.; Warlow, C. The natural history of lacunar infarction: the Oxfordshire community Stroke Project. Stroke 18:545–551; 1987.

Berlit, P.; Eckstein, H.; Krause, K. H. Prognosis of cardiogenic cerebral embolism. Fortschr. Neurol. Psychiatr. 54:205–215; 1986.

Biller, J.; Hungtgen, W. L.; Adams, H. P.; Smoker, W. R. K.; Godersky, J. C.; Toffel, G. J. Cervicocephalic arterial dissections. Arch. Neurol. 43:1234–1238; 1986.

Bladin, P. F.; Berkovic, S. F. Striato-capsular infarction: large infarcts in the lenticulo-striate arterial territory. Neurology 34:1423–1430; 1984.

Boers, G. H. J.; Smals, A. G. H.; Trijbels,

F. J. M.; Fowler, B.; Bakkeren, J. A.; Schoonderwaldt, H. C.; Kleijer, W. J.; Kloppenborg, P. W. Heterozygosity for homocystinuria in premature peripheral and cerebrovascular occlusive arterial disease. N. Engl. J. Med. 313:709–715; 1985.

Bogousslavsky, J.; Despland, P. A.; Regli, F. Spontaneous carotid dissection with acute stroke. Arch. Neurol. 44:137–140; 1987.

Bounds, J. V.; Wiebers, D. O.; Whisnant, J. P.; Okazaki, H. Mechanisms and timing of deaths from cerebral infarction. Stroke 12(4):474–477; 1981.

Bray, G. P.; DeFrank, R. S.; Wolfe, T. L. Sexual functioning in stroke survivors. Arch. Phys. Med. Rehabil. 62:286–288; 1981.

Brey, R. L.; Hart, R. G.; Sherman, D. G.; Tegeler, C. H. Antiphospholipid antibodies and cerebral ischemia in young people. Neurology 40:1190–1196; 1990.

Briley, D. P.; Coull, B. M.; Goodnight, S. H. Neurological disease associated with antiphospholipid antibodies. Ann. Neurol. 25:221–227; 1989.

Brockmann, A. W.; Conrad, J. Hereditary protein C deficiency. In: Bertina, R. M., ed. Protein C and related proteins. Edinburgh: Churchill Livingstone; 1988: p. 160–181.

Butler, A. B.; Partian, R. A.; Netsky, M. G. Primary intraventricular hemorrhage. A mild and remediable form. Neurology 22:675–687; 1972.

Caplan, L. R.; Zarins, C. K.; Hemmat, M. Spontaneous dissection of the extracranial vertebral arteries. Stroke 16:1030–1038; 1985.

Caplan, L. R.; Baquis, G. D.; Pessin, M. S.; D'Alton, J.; Achelman, L. S.; DeWitt, C. D.; Ho, K.; Izukawa, D.; Kwan, E. S. Dissection of the intracranial vertebral artery. Neurology 38:868–877; 1988.

Caronna, J.; Levy, D. Clinical predictors of outcome in ischemic stroke. In: Barnett, H. J. M., ed. Neurologic clinics: cerebrovascular disease, Philadelphia: W. B. Saunders Co.; 1983: p. 103–117.

Cerebral Embolism Task Force. Cardiogenic brain embolism. Arch. Neurol. 43:71–84; 1986.

Cerebral Embolism Task Force. Cardiogenic brain embolism. Arch. Neurol. 46:727–743; 1989.

Chiras, J.; Marciano, S.; Molina, J. V.; Touboul, J.; Poirier, B.; Bories, J. Spontaneous dissecting aneurysms of the extracranial vertebral artery (20 cases). Neuroradiology 27:327–333; 1985.

Cole, F. M.; Yates, P. O. The occurrence and significance of intra-cerebral microaneurysms. J. Pathol. Bacteriol. 93:393–411; 1967.

Comp, P. C. Hereditary disorders predisposing to thrombosis. In progress in hemostasis and thrombosis. New York: Grune & Stratton; 1986: p. 71–102.

Corrin, L. S.; Sander, B. A.; Houser, O. W. Cerebral ischemic events in patients with carotid artery fibromuscular dysplasia. Arch. Neurology 38:616–618; 1981.

Coull, B. M.; Malinow, M. R.; Beamer, N.; Sexton, G.; Nordt, F.; deGarmo, P. Elevated plasma homocysteine concentration as a possible independent risk factor for stroke. Stroke 21:572–576; 1990.

D'Angelo, A.; Landi, G.; D'Angelo, S. V.; Orazio, E. N.; Boccardi, E.; Candelise, L.; Mannucci, P. M. Protein C in acute stroke. Stroke 19:579–583; 1988.

Dombovy, M. L.; Basford, J. R.; Whisnant, J. P.; Bergsthralh, E. J. Disability and use of rehabilitation services following stroke in Rochester, MN, 1975–1979. Stroke 18:830–836; 1987.

Douglas, M. A.; Haerer, A. F. Long-term prognosis of hypertensive intracerebral hemorrhage. Stroke 13:488–491; 1982.

Dove, H. G.; Schnieder, K. C.; Wallace, J. D. Evaluating and predicting outcome of acute cerebral vascular accident. Stroke 15:858–864; 1984.

Effeney, D. J.; Krupski, W. C.; Stoney, R. J.; Ehrenfeld, W. K. Fibromuscular dysplasia of the carotid artery. Aust. N.Z. J. Surg. 53:527–531; 1983.

Ehrenfeld, W. K.; Wylie, E. J. Spontaneous dissection of the internal carotid artery. Arch. Surg. 111:1294–1300; 1976.

Ernerudh, J.; Olsson, J. E.; von Schenck, H. Antithrombin III deficiency in ischemic stroke. Stroke 21:967; 1990.

Farhry, J. M.; Jaques, P. E.; Proctor, H. J. Cervical vessel injury after blunt trauma. J. Vasc. Surg. 8:501–508; 1988.

Feibel, J. H.; Springer, C. J. Depression and failure to resume social activities after stroke. Arch. Phys. Med. Rehabil. 63:276–278; 1982.

Feigenson, J. S.; McDowell, F. H.; Meese, P.; McCarthy, M. I.; Greenberg, S. D. Factors influencing outcome and length of stay in a stroke rehab unit. Part I. analysis of 248 unscreened patients—medical and functional prognostic indicators. Stroke 8:651–656; 1977.

Fieschi, C.; Carolei, A.; Fiorelli, M.; Argentino, C.; Bozzao, L.; Fazio, C.; Salvetti, M.; Bastianello, S. Changing prognosis of primary intracerebral hemorrhage: results of a clinical and computed tomographic follow-up study of 104 patients. Stroke 19:192–195; 1988.

Filip, D. J.; Eckstein, J. D.; Veltkemp, J. J. Hereditary antithrombin III deficiency and thromboembolic disease. Am. J. Hematol. 1:343–349; 1976.

Finelli, P. F.; Kessimian, N.; Bernstein, P. W. Cerebral amyloid angiopathy manifesting as recurrent intracerebral hemorrhage. Arch. Neurol. 41:330–333; 1984.

Fisher, C. M. Clinical syndromes in cerebral hemorrhage. In: Fields, W. S., ed. Pathogenesis and treatment of cerebrovascular disease. Charles C Thomas; Springfield, IL: 1961a; p. 318–334.

Fisher, C. M. The pathology and pathogenesis of intracerebral hemorrhage. In: Fields, W. S., ed. Pathogenesis and treatment of cerebrovascular disease. Springfield, IL: Charles C Thomas; 1961b; p. 295–317.

Fisher, C. M.; Kistler, J. P.; Davis, J. M. Relation of cerebral vasospasm to SAH visualized by computerized tomography scanning. Neurosurgery 6:1–9; 1980.

Fisher, C. M.; Picard, E. H.; Polak, A; Dalal, P.; Ojeman, R. G. Acute hypertensive cerebellar hemorrhage: diagnosis and surgical treatment. J. Nerv. Ment. Dis. 140:38–57; 1965.

Fisher, C. M.; Ojemann, R. G.; Roberson, G. H. Spontaneous dissection of cervico-cerebral arteries. Can. J. Neurol. Sci. 5:9–19; 1978.

Fisher, C. M. Pure sensory stroke and allied conditions. Stroke 13:434–447; 1982a.

Fisher, C. M. The arterial lesions underlying lacunes. Acta. Neuropathol. (Berlin) 12:1–15; 1969.

Fisher, C. M. Lacunar strokes and infarcts: a review. Neurology 32:871–876; 1982b.

Fisher, C. M. Capsular infarcts. Arch. Neurol. 36:65–73; 1979.

Fisher, C. M. Lacunes—small, deep cerebral infarcts. Neurology 15:774–784; 1965.

Fisher, W. S. Decision analysis: a tool of the future: an application to unruptured arteriovenous malformations. Neurosurgery 24:129–135; 1989.

Furlan, A. J.; Whisnant, J. P.; Elveback, L. R. The decreasing incidence of primary intracerebral hemorrhage: a population study. Ann. Neurol. 5:367–373; 1979.

Garraway, W. M.; Whisnant, J. P.; Drury, I. The changing pattern of survival following stroke. Stroke 14(5):699–704; 1983.

Gee, W.; Kaupp, H. A.; McDonald, K. M.; Lin, F. Z.; Curry, J. L. Spontaneous dissection of internal carotid arteries. Arch. Surg. 115:944–949; 1980.

Gilbert, J. J.; Vinters, H. V. Cerebral amyloid angiopathy: incidence and complications in the aging brain. I. cerebral hemorrhage. Stroke 14:915–923; 1983.

Gildersleve, N.; Koo, A. H.; McDonald, C. J. Metastatic tumor presenting as intracerebral hemorrhage. Radiol. 124:109–112; 1977.

Giles, C.; Brucher, J. M.; Khoubesserian, P.; Vanderhaeghen, J. J. Cerebral amyloid angiopathy as a cause of multiple intracerebral hemorrhages. Neurology 34:730–735; 1984.

Graf, C. J.; Perret, G. E.; Torner, J. C. Bleeding from cerebral arteriovenous malformations as part of their natural history. J. Neurosurg. 58:331–337; 1983.

Griffin, J. H.; Evatt, B.; Zimmerman, T. S.; Kleiss, A. J. Deficiency of protein C in congenital thrombotic disease. J. Clin. Invest. 68:1370–1373; 1981.

Gross, C. R.; Kase, C. S.; Mohr, J. P.; Cunningham, S. C.; Baker, W. E. Stroke in South Alabama: incidence and diagnostic features—a population based study. Stroke 15:249–255; 1984.

Grotta, J. C.; Yatsu, F. M.; Pettigrew, L. C.; Rhoades, H.; Bratina, P.; Vital, D.; Alam, R.; Earls, R.; Picone, C. Prediction of carotid stenosis progression by lipid and hematologic measurements. Neurology 39:1325–1331; 1989.

Grotta, J. C.; Ward, R. E.; Flynn, T. C.; Cullen, M. L. Spontaneous internal carotid artery dissection associated with fibromuscular dysplasia. J. Cardiovasc. Surg. 23:512–514; 1982a.

Grotta, J. C.; Manner, C.; Pettigrew, L. C.; Yatsu, F. M. Red blood cell disorder and stroke. Stroke 17:5, 811–817; 1986.

Grotta, J.; Ackerman, R.; Correia, J.; Fallick, G.; Chang, J. Whole body viscosity parameters and cerebral blood flow. Stroke 13:296–301; 1982b.

Haerer, A. F.; Woosley, P. C. Prognosis and quality of survival in a hospitalized stroke population from the South. Stroke 6:543–548; 1975.

Hardy, R. C.; Williams, R. G. Moyamoya disease and cerebral hemorrhage. Surg. neurol. 21:507–510; 1984.

Harrison, M. J. G.; Marshall, J. Atrial fibrillation, TIAs, and completed strokes. Stroke 15:441–442; 1984.

Hart, R. G.; Kanter, M. C. Hematologic disorders and ischemic stroke, a selection review. Stroke 21:1111–1121; 1990.

Hart, R. G.; Easton, J. D. Dissection and trauma of cervico-cerebral arteries. In: Barnett, H. J. M.; Mohr, J. P.; Stein, B. M.; Yatsu, F. M. eds. Stroke: pathophysiology, diagnosis, and management. New York: Churchill Livingstone; 1986:p. 775–803.

Hart, R. G.; Easton, T. D. Dissection of cervical and cerebral arteries. Neurol. Clinics 1:155–182; 1983.

Hart, R. G. Vertebral artery dissection. Neurology 38:987–989; 1988.

Hasan, D.; Vermeulen, M; Wisdick, E. F. M; Hijdra, A.; Van Gijn, J. Management problems in acute hydrocophalus after subarachnoid hemorrhage. Stroke 20:747–753; 1989.

Hass, W. K.; Easton, J. D.; Adams, H. P.; Pryse-Phillips, W.; Molony, B. A.; Anderson, S.; Kamm, B. A randomized trial comparing the efficacy of ticlopidine hydrochloride to aspirin for the prevention of stroke in high risk patients. N. Engl. J. Med. 321:501–507; 1989.

Healton, E. B. Fibromuscular dysplasia. In: Barnett, H. J.; Mohr, J. P.; Stein, B. M.; Yatsu, F. M. eds. Stroke, pathophysiology, diagnosis, and management. New York: Churchill Livingstone; 1986:p. 831–843.

Heros, R. C.; Tu, Y. K. Is surgical therapy needed for unruptured arteriovenous malformations? Neurology 37:279–286; 1987.

Hier, D. B.; Mondlock, J.; Caplan, L. R. Behavioral abnormalities after right hemisphere stroke. Neurology 3:377–344; 1983.

Hier, D. B.; Davis, K. R.; Richardson, E. P.; Mohr, J. P. Hypertensive putaminal hemorrhage. Ann. Neurol. 1:152–159; 1977.

Hieshima, G. B.; Cahan, L. D.; Mehringer, C. M.; Bentson, J. R. Spontaneous arteriovenous fistulas of cerebral vessels in association with fibromuscular dysplasia. Neurosurg. 18:454–458; 1986.

High, K. Antithrombin III, protein C and protein S. Arch. Pathol. Lab. Med. 112:28–36; 1988.

Hijda, A.; Vermeulen, M.; Van Gijn, J.; Van Crevel, H. Rerupture of intracranial aneurysms: a clinicoanatomic study. J. Neurosurg. 67:29–33; 1987.

Hoagland, H. C.; Silverstein, M. N. Primary thrombocythemia in the young patient. Mayo. Clin. Proc. 53:578–580; 1978.

Inigawa, T.; Kamiya, K.; Ogasawara, H.; Yano, T. Rebleeding of ruptured intracranial aneurysms in the acute stage. Surg. Neurol. 28:93–99; 1987.

Jabaily, J.; Iland, H. J.; Laszlo, J.; Massey, E. W.; Faquet, G. B.; Briere, J.; Landaw, S. A.; Pisciotta, A. V. Neurologic manifestations of essential thrombocythemia. Ann. Intern. Med. 99:513–518; 1983.

Johannessen, K. A.; Nordrehaug, J. E.; von der Lippe, G.; Vollset, S. E. Risk factors for embolization in patients with left ventricular thrombi and acute myocardial infarction. Br. Heart J. 60:104–110; 1988.

Kannel, W. B.; Wolf. P. A.; Verter, J.; McNamara, P. M. The role of blood pressure in stroke: the Framingham study. J.A.M.A. 214:301–310; 1970.

Kaplan, M. E.; Mack, K.; Goldberg, J. D.; Donovan, P. B.; Berk, P. D.; Wasserman, L. R. Long-term management of polycythemia vera with hydrozyurea: a progress report. Semin. Hematol. 23:167–171; 1986.

Kappelle, L. J.; Koudstaal, P. J.; Van Gijn, J.; Ramos, L. M. P.; Keunen, J. E. E. Carotid angiography in patients with lacunar infarction: a prospective study. Stroke 19:1093–1096; 1988.

Kase, C. S.; Williams, J. P.; Wyatt, D. A.; Mohr,

J. P. Lobar intracerebral hematomas: clinical and CT analysis of 22 cases. Neurology 32:1146–1150; 1982.

Kassell, N. F.; Torner, J. C.; Jane, J. A.; Haley, E. C.; Adams, H. P. The international cooperative study on the timing of aneurysm surgery. Part II, surgical results. J. Neurosurg. 73:37–47; 1990a.

Kassell, N. F.; Peerless, S. J.; Durward, Q. J.; Beck, D. W.; Drake, C. G.; Adams, H. P. Treatment of ischemic deficits from vasospasm with intravascular volume expansion and induced arterial hypertension. Neurosurgery 11:337–343; 1982.

Kassell, N. F.; Torner, J. C.; Haley, E. C.; Jane, J. A.; Adam, H. P.; Kongable, G. L. The international cooperative study on the timing of aneurysm surgery. Part I: J. Neurosurg. 73:18–36; 1990b.

Kassell, N. F.; Torner, J. C. Aneurysmal rebleeding: a preliminary report from the Cooperative Aneurysm Study. Neurosurgery 13:479–481; 1983.

Katirji, M. B.; Reinmuth, O. M.; Latchaw, R. E. Stroke due to vertebral artery injury. Arch. Neurol. 42:242–248; 1985.

Kempster, P. A.; Gerraty, R. P.; Gates, P. C. Asymptomatic cerebral infarction in patients with chronic atrial fibrillation. Stroke 19:955–958; 1988.

Kohler, J.; Kasper, J.; Witt, I.; vanReutern, G. M. Ischemic stroke due to protein C deficiency. Stroke 21:1077–1080; 1990.

Kotila, M.; Waltimo, O.; Niemi, M. L.; Laaksonen, R.; Lempinen, M. The profile of recovery from stroke and factors influencing outcome. Stroke 15:1039–1044; 1984.

Kushner, M. J. Prospective study of anticardiolipin antibodies in stroke. Stroke 21:295–298; 1990.

Kwak, R.; Kadoya, S.; Susuki, T. Factors affecting the prognosis in thalamic hemorrhage. Stroke 14:493–500; 1983.

Lahverta-Palacios, J. J.; Bornstein, R.; Fernandez-Debora, F. J.; Gutierrez-Rivas, E.; Ortiz, M. C.; Larrelgla, S.; Calandre, L.; Montero-Castillo, J. Controlled and uncontrolled thrombocytosis. Its clinical role in essential thrombocythemia. Cancer 61:1207–1212; 1988.

Laterra, J.; Gebarski, S.; Sackellares, J. C. Transient amnesia resulting from vertebral artery dissection. Stroke 19:98–101; 1988.

Lechner, K.; Pabinger-Fasching, I. Lupus anticoagulants and thrombosis: a study of 25 cases and review of the literature. Haemostasis 15:254–262; 1985.

Levine, S. R.; Welch, K. M. A. The spectrum of neurologic disease associated with antiphospholipid antibodies. Lupus anticoagulants and anticardiolipin antibodies. Arch. Neurol. 44:876–883; 1987.

Levine, S. R.; Deegan, M. J.; Futrell, N; Welch, K. M. A. Cerebrovascular and neurologic disease associated with antiphospholipid antibodies; 48 cases. Neurology 40:1181–1189; 1990.

Levine, S. R.; Quint, D. J.; Pessin, M. S.; Boulos, R. S.; Welch, K. M. A. Intraluminal clot in the vertebrobasilar circulation. Neurology 39:515–522; 1989.

Levine, S. R.; Welch, K. M. A. Superior cerebellar artery infarction and vertebral artery dissection. Stroke 19:1431–1434; 1988.

Lindsay, K. Antifibrinolytic agents in subarachnoid hemorrhage. J. Neurol. 234(1):1–8; 1987.

Lipton, R. B.; Berger, A. R.; Lesser, M. L.; Portenoy, R. K. Lobar vs thalamic and basal ganglion hemorrhage: clinical and radiographic features. J. Neurol. 234:86–90; 1987.

Lodder, J.; Bamford, J. M.; Sandercock, D. M. Are hypertension or cardiac embolism likely causes of lacunar infarction. Stroke 21:375–381; 1990.

Longstreth, W. T. Jr.; Koepsell, T. D.; van Belle, G. Predictors of outcome after intracerebral hemorrhage. Ann. Neurol. 26:105–106; 1989.

Luessenhop, A. J.; Rosa, L. Cerebral arteriovenous malformations. J. Neurosurg. 60:14–22; 1984.

Luscher, T. F.; Lie, J. T.; Stanton, A. W.; Houser, O. W.; Hollier, L. H.; Sheps, S. G. Arterial fibromuscular dysplasia (subject review). Mayo Clin. Proc. 62:931–952; 1987.

Marie, P. Des foyers lacunaire de disintegration et de differents autres etats cavitaires du cerveau. Rev. Med. 21:181–198; 1901.

Mas, J. L.; Bousser, M. G.; Hasbourn, D; Laplane, D. Extracranial vertebral artery dissections: a review of 13 cases. Stroke 18:1037–1047; 1987.

Mas, J. L.; Henin, D. Dissecting aneurysm of the vertebral artery and cervical manipulation. A case report with autopsy. Neurology 39:512–515; 1989.

Masdeu, J. C.; Rubino, F. A. Management of lobar intracerebral hemorrhage, medical or surgical. Neurology 34:381–383; 1984.

Matsumoto, N,; Whisnant, J. P.; Kurland, H. Natural history of stroke in Rochester, MN, 1955 through 1969; an extension of a previous study 1945 to 1954. Stroke 4:20–29; 1973.

McCormack, M. K.; Scarr-Salapatek, S.; Polesky, H.; Thompson, W.; Katz, S. H.; Barker, W. B. A comparison of the physical and intellectual development of black children with and without sickle cell trait. Pediatrics 56:1021–1025; 1975.

McCormick, W. F. Pathology and pathogenesis of intracranial saccular aneurysms. Semin. Neurol. 4:291–303; 1984.

McCormick, W. F.; Rosenfield, D. B. Massive brain hemorrhage: a review of 144 cases and an examination of their causes. Stroke 4:946–954; 1973.

McNeill, D. H.; Dreisbach, J.; Marsden, R. J. Spontaneous dissection of the internal carotid artery. Arch. Neurology 37:54–55; 1980.

Mettinger, K. L. Fibromuscular dysplasia and the brain. Current concept of the disease. Stroke 13:53–58; 1982.

Miletich, J.; Sherman, L; Broze, G. Absence of thrombosis in subjects with heterozygous protein C deficiency. N. Engl. J. Med. 317:991–996; 1987.

Millard, F. E.; Hunter, C. S.; Anderson, M.; Edelman, M. J.: Kosty, M. P.; Luiken, G. A.; Marino, G. G. Clinical manifestations of essential thrombocythemia in young adults. Am. J. Hematol. 33:27–31; 1990.

Millikan, C. H.; McDowell, F.; Easton, J. D.; eds. Stroke. Philadelphia: Lea and Febiger;1987:p. 201–222.

Mizukami, M.; Kogure, K.; Kanaya, H.; Yamori, Y. Hypertensive intracerebral hemorrhage. New York: Raven Press; 1983.

Mizukami, M.; Nishijima, M.; Kin, H. Computed tomographic findings of good prognosis for hemiplegia in hypertensive putaminal hemorrhage. Stroke 12:648–652; 1981.

Mohr, J. P.; Kase, C. S.; Wolf, P. A.; Price, T. A.; Heyman, A.; DAmbrosia, J. H.; Kunitz, S. Lacunes in the NINCDS Pilot Stroke Data Bank. Ann. Neurol. 12:84; 1982.

Mohr, J. P.; Caplan, L. R.; Melski, J. W.; Goldstein, R. J.; Duncan, G. W.; Kistler, J. P.; Pessin, M. S.; Bleich, H. L. The Harvard Cooperative Stroke Registry: a prospective registry. Neurology 28:754–762; 1978.

Mokri, B.; Sundt, T. F.; Houser, D. W.; Piepgras, D. G. Spontaneous dissection of the cervical internal carotid artery. Ann. Neurol. 19:126–138; 1986.

Mokri, B.; Houser, O. W.; Sandok, B. A.; Piepgras, D. G. Spontaneous dissection of the vertebral artery. Neurology 38:880–885; 1988.

Mudd, S. H.; Skovby, F.; Levy, H. L.; et al. The natural history of homocystinura due to cystathionine β-synthase deficiency. Am. J. Hum. Genet. 37:1–31; 1985.

Mudd, S. H. Vascular disease and homocysteine metabolism. N. Engl. J. Med 313:751–753, 1985.

Muizelaar, J. P.; Becker, D. P. Induced hypertension for the treatment of cerebral ischemia after subarachnoid hemorrhage. Surg. Neurol. 25:317–325; 1986.

Nishioka, H.; Torner, J. C.; Graf, C. J.; Kassell, N. F.; Sahs, A. L.; Goettle, L. C. Cooperative study of intracranial aneurysms and subarachnoid hemorrhage: a long term prognostic study. Arch. Neurol. 41:1142–1151; 1984.

Norrving, B.; Cronqvist, S. Clinical and radiologic features of lacunar versus nonlacunar minor stroke. Stroke 20:59–64; 1989.

Ojemann, R. G.; Fisher, C. M.; Rich, J. C. Spontaneous dissecting aneurysm of the internal carotid artery. Stroke 3:434–440; 1972.

Okazaki, H.; Reagan, T. J.; Campbell, R. J. Clinicopathologic studies of primary cerebral amyloid angiopathy. Mayo Clin. Proc. 54:22–31; 1979.

Osuntokum, B. O. Neurological syndromes, management and prognosis in sickle cell anemia. Tropical Doctor 11:2–7; 1981.

Ott, K. H.; Kase, C. S.; Ojemann, R. G.; Mohr, J. P. Cerebellar hemorrhage: diagnosis and treatment. Arch. Neurol. 31:160–167; 1974.

Pearson, T. C.; Wetherley-Mein, G. Vascular occlusive episodes and venous hematocrit in primary proliferative polycythemia. Lancet 2:1219–1222; 1978.

Perret, G.; Nishioka, H. Report on the Cooperative Study of Intracranial Aneurysms and SAH. Sec. VI. Arteriovenous malformations. J. Neurosurg. 25:467–490; 1966.

Petersen, P.; Godtfredsen, J.; Boysen, G. Placebo-controlled, randomized trial of warfarin and aspirin for prevention of thromboembolic complications of chronic atrial fibrillation. Lancet 1:175–178; 1989.

Petruk, N. K.; West, M.; Mohr, G.; Weir, B.; Benoit, B.; Gentici, F.; Disney, L.; Khan, M.; Grace, M.; Holness, R.; Karwon, M.; Ford, R.; Cameron, G. S.; Tucker, W. S.; Purves, G. B.; Miller, J. D. R.; Hunter, K. M.; Richard, M. T.; Durity, F. A.; Chan, R.; Clein, L. J.; Maroun, F. B.; Godon, A. Nimodipine treatment in poor-grade aneurysm patients— results of a multicenter double-blind placebo controlled trial. J. Neurosurg. 68:505–517; 1988.

Pickard, J. D. Effects of oral nimodipine on cerebral infarction and outcome after subarachnoid hemorrhage. British aneurysm nimodipine trial. Br. Med. J. 298:636–642; 1989.

Pohl, M. A.; Novick, A. C. Natural history of atherosclerotic and fibrous renal artery disease: clinical implications. Am. J. Kidney Dis. 5:A120–A130; 1985.

Portenoy, R. K.; Lipton, R. B.; Berger, A. R.; Lesser, M. L.; Lantos, G. Intracerebral hemorrhage: a model for the prediction of outcome. J. Neurol. Neurosurg. Psych. 50:976–979; 1987.

Powars, D. R.; Schroeder, W. A.; Weiss, J. N.; Chan, C. S.; Azen, S. P. Lack of influence of fetal hemoglobin levels or erythrocyte indices on the severity of sickle cell anemia. J. Clin. Invest. 65:732–740; 1980.

Powars, D.; Wilson, B.; Imbus, C.; Peglow, C.; Allen, J. The natural history of stroke in sickle cell disease. Am. J. Med. 65:461–471; 1978.

Pozzati, E.; Givliani, G.; Acciarri, N.; Giacomo, N. Long-term follow-up of occlusive cervical carotid dissection. Stroke 21:528–531; 1990.

Pullicino, P.; Nelson, R. F.; Kendall, B. E.; Marshall, J. Small deep infarcts diagnosed on computed tomography. Neurology 30:1090–1096; 1980.

Rankin, J. Cerebral vascular accidents in patients over age of 60. II. Prognosis. Scot. Med. J. 2:200–215; 1957.

Rascol, A.; Clanet, M.; Manelfe, C.; Guiraud, B.; Bonafe, A. Pure motor hemiplegia: CT Study of 30 cases. Stroke 13:11–17; 1982.

Rem, J. A.; Hachinski, V. C.; Boughner, D.R.; Barnett, H. J. M. Value of cardiac monitoring and echocardiography in TIA and stroke patients. Stroke 16:950–956; 1985.

Robinson, R. G.; Price, T. R. Post-stroke depressive disorders: a follow-up study of 103 patients. Stroke 13(5):635–641; 1982.

Robinson, R. G.; Szetela, B. Mood change following left hemispheric brain injury. Ann. Neurol. 9:447–453; 1981.

Rodgers, G. P.; Dorer, G. J.; Noguchi, C. T.; Schechter, A. N.; Nienhuis, A. W. Hematologic response of patients with sickle cell disease to treatment with hydroxyurea. N. Engl. J. Med. 322:1037–1044; 1990.

Ropper, A. H.; Davis, K. R. Lobar cerebral hemorrhages: acute clinical syndromes in 26 cases. Ann. Neurol. 8:141–147; 1980.

Rothfus, W. E.; Albright, A. L.; Casey, K. F.; Latchaw, R. E.; Roppolo, H. M. Cerebellar venous angiomas "benign" entity? A.J.N.R. 5:61–66; 1984.

Russell, M. O.; Goldberg, H. I.; Hodson, A.; Kim, H. C.; Halus, J.; Reivich, M.; Schwartz, E. Effect of transfusion therapy on arteriographic abnormalities and on recurrence of stroke in sickle cell disease. Blood 63:162–169; 1984.

Russell, M. O.; Goldberg, H. I.; Reis, L; Friedman, S.; Slater, R.; Reivich, M.; Schwartz, E. Transfusion therapy for cerebrovascular abnormalities in sickle cell disease. J. Pediatr. 88:382–387; 1976.

Sacco, R. L.; Foulkes, M. A.; Mohr, J. P.; Wolf, P. A.; Hier, D. B.; Price, T. R. Determinants of early recurrence of cerebral infarction. Stroke 20:983–989; 1989a.

Sacco, R. L.; Owen, J.; Mohr, J. P.; Tatemichi, T. K.; Grossman, B. A. Free protein S deficiency: a possible association with cerebrovascular occlusion. Stroke 20:1657–1661; 1989b.

Sekhar, L. N.; Weschler, L. R.; Yonas, H.; Luyckx, K.; Obrist, W. Value of transcranial doppler examination in the diagnosis of cerebral vasospasm after SAH. Neurosurg. 22:813–821; 1988.

Sherman, D. G.; Hart, R. G.; Easton, J. D. The secondary prevention of stroke in patients with atrial fibrillation. Arch. Neurol. 43:68–70; 1986.

Shi, Y.; Chen, X. A proposed scheme for grading intracranial arteriovenous malformations. J. Neurosurg. 65:484–489; 1986.

Silver, F. L.; Norris, J. W.; Lewis, A. J.; Hachinski, V. C. Early mortality following stroke: a prospective review. Stroke 15:492–496; 1984.

Sinyor, O.; Jacques, P.; Kaloupek, D.; Becker, R.: Goldberg, M.; Oppersmith, H. Post-stroke depression and lesion location. Brain 109:537–546; 1986.

Spetzler, R. F.; Martin, N. A. A proposed grading system for arteriovenous malformations. J. Neurosurg. 65:476–483; 1986.

Stein, R. W.; Kase, C. S.; Hier, D. B.; et al. Caudate hemorrhage. Neurology 34:1549–1554; 1984.

Stewart, M. T.; Moritz, M. W.; Smith, R. B.; et al. The natural history of carotid fibromuscular dysplasia. J. Vasc. Surg. 3:305–310; 1986.

Stockman, J. A.; Nigro, M. A.; Mishkin, M. M.; Oski, F. A. Occlusion of large cerebral vessels in sickle cell anemia. N. Engl. J. Med. 287:846–849; 1972.

Svien, H. J.; McRae, J. A. Arteriovenous anomalies of the brain. J. Neurosurg. 23:23–28; 1965.

Tanaka, Y.; Furuse, M.; Isawa, H.; Masuzawa, T.; Saito, K.; Sato, F.; Mizuno, Y. Lobar intracerebral hemorrhage: etiology and a long-term follow-up study of 32 patients. Stroke 17:51–57; 1986.

Tartaglia, A. P.; Goldberg, J. D.; Berk, P. D.; Wasserman, L. R. Adverse effects of antiaggregating platelet therapy in the treatment of polycythemia vera. Semin. Hematol. 23:172–176; 1986.

Thomas, D. J.; Marshal, J.; Russell, R. W. R.; Wetherley-Mein, G.; duBoulay, G. H.; Pearson, T. C.; Symon, L.; Zilka, E. Cerebral blood flow in polycythemia. Lancet 2:161–163; 1977a.

Thomas, D. J.; Marshal, J.; Russell, R. W. R.; Wetherley-Mein, G.; duBoulay, G. H.; Pearson, T. C.; Symon, L.; Zilka, E. Effects of hematocrit on cerebral blood flow in man. Lancet 2:941–943; 1977.

Toffol, G. J.; Biller, J.; Adams, H. P. Jr. Nontraumatic intracerebral hemorrhage in young adults. Arch. Neurol. 44:483–485; 1987.

Tuhrim, S.; Dambrosia, J. M.; Price, T. R.; Mohr, J. P.; Wolf, P. A.; Heyman, A.; Kase, C. S. Prediction of intracerebral hemorrhage survival. Ann. Neurol. 24:258–263; 1988.

Turney, T. M.; Garraway, W. M.; Whisnant, J. P. The natural history of hemispheric and brainstem infarction in Rochester, MN. Stroke 15(5):790–794; 1984.

Tuszynski, M. H.; Petito, C. K.; Levy, D. E. Risk factors and clinical manifestations of pathologically verified lacunar infarctions. Stroke 20:990–999; 1989.

Vassilouthis, J.; Richardson, A. E. Ventricular dilation and communicating hydrocephalus following spontaneous SAH. J. Neurosurg. 51:341–351; 1979.

Vomberg, P. P.; Brecerveld, C.; Pleury, P.; Arts, W. F. M. Cerebral thromboembolism due to antithrombin III deficiency in two children. Neuropediatrics 18:42–44; 1987.

Wade, D. T.; Stilbeck, C. E.; Hewer, R. L. Predicting Barthel ADL score at 6 months after an acute stroke. Arch. Phys. Med. Rehabil. 64:24–28; 1983.

Welling, R. E.; Taha, A.; Goel, T.; Cranley, J.; Krause, R.; Hafner, C.; Tew, J. Extracranial carotid artery aneurysms. Surgery 93:319; 1983.

Wells, R. P.; Smith, R. R. Fibromuscular dysplasia of the internal carotid artery: a long term follow up. Neurosurgery 10:39–43; 1982.

Whisnant, J. P.; Fisher, L.; Robertson, J. T.; Scheinbert, P. Does carotid endarterectomy decrease stroke and death in patients with transient ischemic attacks? Ann. Neurol. 22(1):72–76; 1987.

Wilhelmsen, L.; Svärdsudd, K.; Korsan-Bengtsen, K.; Larsson, Bo; Welin, L.; Tibblin, G. Fibrinogen as a risk factor for stroke and myocardial infarction. N. Engl. J. Med. 311:501–505; 1984.

Wilimas, J.; Goff, J. R.; Anderson, H. R.; Langston, J. W.; Thompson, E. Efficacy of transfusions therapy for one to two years in patients with sickle cell disease and cerebrovascular accidents. J. Pediatr. 96:205–208, 1980.

Wilkins, R. H. Natural history of intracranial vascular malformations: a review. Neurosurgery 16:421–430; 1985.

Wilkinson, I. M. The vertebral artery, extracranial and intracranial structure. Arch. Neurol. 27:392–396; 1972.

Winn, H. R.; Richardson, A. E.; Jane, J. A. Late morbidity and mortality in cerebral aneurysms. Trans. Am. Neurol. Assoc. 98:23–24; 1973.

Wolf, P. A.; Dawber, T. R.; Thomas, H. E.; Kannel, W. B. Epidemiologic assessment of chronic atrial fibrillation and risk of stroke: the Framingham Study. Neurology 28:973–977; 1978.

Wood, D. H. Cerebrovascular complications of sickle cell anemia. Stroke 9:73–75; 1978.

5

Neurological Complications Related to Open Heart Surgery

MARC I. CHIMOWITZ AND ANTHONY J. FURLAN

Central and peripheral nervous system complications are major causes of morbidity following open heart surgery (OHS). This chapter reviews the neurological complications of coronary artery bypass graft (CABG) surgery and focuses on identifying the high-risk patient. Related topics include the neurological complications of cardiac catheterization, percutaneous transluminal coronary angioplasty (PTCA), and cardiac transplantation.

Cardiac Catheterization

Cardiac catheterization has been performed since the 1960s. In an early study of 46,904 patients undergoing this procedure at 173 hospitals, stroke occurred in 106 patients (0.23%) (Adams et al. 1973). Since then, systemic heparinization, aspirin pretreatment, avoidance of entry into cerebral vessels, modifications in catheter design and original techniques, and acquisition of skill and experience have contributed to the decrease in the incidence of stroke associated with this procedure. In a recent study of 30,000 cardiac catheterizations performed at the Cleveland Clinic Foundation during a 5-year

period, 35 patients (.11%) sustained a focal deficit (15 had carotid territory, 20 had vertebrobasilar) (Lockwood et al. 1983). The deficit resolved within 48 hours in 19 patients but persisted in the other 16, two of whom died. No clear etiology was identified in the majority of these patients, although many had a history of stroke or potential cardioembolic sources, such as ventricular hypokinesia, mural thrombus, or valvular disease.

The preponderance of vertebrobasilar strokes in most series (Dawson and Fischer 1977; Kosmorsky et al. 1988) strongly suggests the stroke mechanism is technical factors and local trauma to the aorta, or origins of the vertebral, innominate, or subclavian arteries with release of embolic material.

Diffuse encephalopathy also occurs following cardiac catheterization, although the frequency has never been documented. Risk factors for encephalopathy include advanced age, dehydration, and injection of large volumes of contrast material.

Peripheral nerve injury also is relatively common during cardiac catheterization (Dawson and Fischer 1977). The causes of nerve trauma include cut-downs, tight arm-

Table 5-1. Central Nervous System Complications in Patients Undergoing Various Types of Open Heart Surgery

Percent with	CABG ($n = 421$)	Left Heart Valves ($n = 80$)	Ventricular Aneurysm ($n = 19$)	Valves and Aneurysms ($n = 99$)
Stroke	5	8	0	6
Encephalopathy	12	10	5	9
Death	2	9	0	7

Furlan, A. J.; Breuer, A. C. Central nervous system complications of open heart surgery. Stroke 15:912–915; 1984.

boards, positioning of the hand, and inadvertent catheterization of the nerve sheath.

Percutaneous Transluminal Coronary Angioplasty

The use of PTCA to dilate proximal stenoses of coronary arteries has increased markedly during the past decade. The National Heart, Lung, and Blood Institute (Dorros et al. 1983) reported the complications in 1500 patients undergoing PTCA. Five patients (0.3%) suffered a central nervous system (CNS) complication: one cerebrovascular accident, three transient neurological deficits, and one anoxic encephalopathy. At the Cleveland Clinic Foundation, 1829 patients underwent a total of 1968 PTCA procedures between 1981 and 1985 (Galbreath et al. 1986). Four patients (0.2%) suffered stroke-related problems: two cerebral hemisphere infarcts, one hemisphere transient ischemic attack, and one brain stem infarct. Two patients experienced strokes while the angiographer was searching the ascending aorta for aortosaphenous vein graft anastomatic sites. One patient suffered a stroke during a hypotensive episode after successful PTCA, and in the patient with the TIA, a small quantity of air was injected through the guiding catheter.

CABG Surgery

Most of the data pertaining to the neurological complications of OHS are derived from studies of this operation. In our institution, the risk of stroke or encephalopathy during CABG surgery did not differ from the risk in other forms of cardiac surgery not involving transplantation (Table 5-1).

Stroke

In the 1960s, stroke[*] was a common complication of OHS (Gilman 1965). Since then, the stroke risk has been substantially lowered because of improved monitoring and surgical techniques and the introduction of membrane oxygenators and in-line filtration, which decreases the release of microaggregates into the circulation. Nonetheless, several studies since the 1970s have shown little change in the stroke rate associated with OHS, which ranges between 2% and 5% (Table 5-2). These studies indicate that of the 150,000 patients who undergo CABG surgery annually in the United States, at least 5000 will suffer a stroke.

Various studies have attempted to identify specific risk factors for the development of stroke during CABG surgery. In a prospective study of 421 patients undergoing CABG surgery at the Cleveland Clinic, 5% developed focal brain or ocular infarction, and 2% suffered major neurological disability (Breuer et al. 1983). The authors were unable to correlate a large number of preoperative, intraoperative, and postoperative variables with the risk of stroke (Table 5-3). However, these results should be interpreted cautiously because major stroke occurs in less than 2% of all patients undergoing this procedure, making it difficult to

* In the setting of open heart surgery, stroke generally refers to focal brain infarction.

Table 5-2. Summary of Recent Data on CNS Complications of Open Heart Surgery*

Authors	Year†	Study Design	Type of OHS	Number of Patients	Percent with CNS Deficit		Method of Ascertainment
Hill et al.	1968	Pathological analysis of autopsy cases	Mixture	133	** (fat emboli in 62% of brains, nonfat emboli in 31% of brains)	**	Pathological analysis of brains
Heller et al.	1969	Prospective	Mixture	100	** (postoperative delirium in 24%)	9%	Psychiatric interview, psychological testing
Javid et al.	1969	Prospective	Mixture	100	13+%	35%	Clinical examination, psychological testing
Tufo et al.	1969	Prospective	Mixture	100	13+%	43%	Clinical examination, psychological testing
Branthwaite	1970	Retrospective, % prospective	Mixture	417	9.4%	10.1%	Chart review, clinical examination
Hansotia et al.	1972	Prospective	Mixture	177	** (51% with persistently abnormal EEG at discharge; includes 11 patients who died)	**	Serial EEG
Cannon et al.	1972	Retrospective	Mixture	400	1%††	**	Chart review
Hutchinson et al.	1972	Retrospective	CABG alone	376	0.3%††	**	Chart review
Branthwaite	1973	Prospective	Mixture	140	** (7.1% had clinical neurologic damage)	**	Intraoperative use of cerebral function monitor
Branthwaite	1973	Retrospective	Mixture	538	3%	4.8%	Chart review
Hodgman et al.	1974	Retrospective	Mixture	100	** (20% had minor psychiatric problems)††	**	Chart review
Kolkka et al.	1977	Prospective	Mixture	204	2.9%	17.2%	Clinical examination
Lee et al.	1978	Retrospective	Mixture	943	0.7%	**	Chart review
Gonzalez-Scarano et al.	1978	Retrospective case control	CABG alone	1427	1%	0.4%	Chart review
Loop et al.	1978	Retrospective	CABG alone	8741	†† 1.3% to 2%	**	Chart review, computerized cardiovascular information registry
Muraoka et al.	1979	Prospective	Congenital heart	57	0% (10.5% had persistent CT scan changes)	0%	Chart review
Turnipseed et al.	1979	Prospective	CABG alone	170	4.7%	**	Clinical examination
Breuer et al.	1980	Prospective	CABG alone	421	5.2% (total) 2% (severe)	11.6%	Clinical examination
Coffey et al.	1981	Retrospective	CABG alone	1669	0.8%	3.4%	Chart review

* Modified from Breuer, A. C.; Furlan, A. J.; Hanson, M. R.; Lederman, R. J.; Loop, F. D.; Cosgrove, D. M.; Greenstreet, R. L.; Estafanous, F. G. Central nervous system complications of coronary artery bypass graft surgery: prospective analysis of 421 patients. Stroke 14:682–687; 1983.

† Latest year patients in study underwent surgery reflecting technology of that time (i.e., not year study published)

** No clinical examination data available

†† These retrospective studies are highly devoted to analyses of non-neurological issues and complications and mention CNS dysfunction

demonstrate that one variable significantly alters that risk. Stroke etiology, however, could be determined in many cases (Table 5-4).

Other studies have suggested various factors for increased stroke risk during CABG surgery, such as carotid occlusive disease, prior stroke, perioperative arrythmias, and pump time.

Table 5-3. Potential Risk Factors Not Significant for Stroke or Prolonged Encephalopathy in CABG Surgery

Preoperative Variables

age, sex, race, weight, height, history (hx) of prior stroke, TIA, migraine, smoking hx, hx of noncoronary arteriography, calf claudication, other vascular syndrome, admission BP, cholesterol level, triglycerides, abnormal lipoproteins, glucose intolerance, carotid bruit, use of medication (ASA, Persantine, Coumadin, anti-CHF, anti-BP, antidysrhythmia, antiangina), hx of AMI, cardiac arrhythmia, NYHA functional class, CHF, old MI, LVH by EKG, degree of coronary artery disease by angiography, prior cardiac or vascular surgery, preoperative spirometry (predicted or found), abnormal chest x-ray, abnormal arterial blood gases, abnormal neurological preoperative examination

Operative Variables

surgical team, day of week, time of day, first or second time for myocardial revascularization, anesthetic agent used, type of oxygenator (bubble vs. membrane), duration of total bypass, aortic clamp time, various extracorporeal circulation pump parameters, use of hypothermia, lowest hematocrit during bypass, use of Haemonetics autologous blood recovery, intraoperative blood gases, hypertension, intraoperative use of drugs before, during, or after bypass

Postoperative Variables

hypertension, ventilatory parameters, plasma-free hemoglobin level, total units of blood transfused, use of Sorensen autotransfusion device, reoperation within 48 hours, duration of intubation, postoperative use of morphine, Pantopon, Valium, Pavulon, postoperative cardiac enzymes, postoperative BB-CPK

Key: ASA, acetylsalicylic acid; CHF, congestive heart failure; BP, blood pressure; AMI, acute myocardial infarction; MI, myocardial infarction; LVH, left ventricular hypertrophy.

Breuer, A. C.; Furlan, A. J.; Hanson, M. R.; Lederman, R. J.; Loop, F. D.; Cosgrove, D. M.; Greenstreet, R. L.; Estafanous, F. G. Central nervous system complications of coronary artery bypass graft surgery: prospective analysis of 421 patients. Stroke 14:682–687, 1983.

Table 5-4. Myocardial Revascularization: Possible Etiology of Focal CNS Deficits*

Cause	Number of Patients
Cardiac arrhythmia intraoperatively and perioperatively	6
Internal carotid artery atherosclerosis	5
Air embolism from left ventricle	2
Carotid artery trauma during internal jugular vein cannulation	1
Aortic atherosclerosis at site of clamping	1
Prolonged intraoperative blood pressure decrease	1

* Cause identified in 16 of 22 patients (73%)

Furlan, A. J.; Breuer, A. C. Central nervous system complications of open heart surgery. Stroke 15:912–915, 1984.

Carotid Artery Disease and Stroke During CABG

Extracranial carotid artery disease often is implicated as an important cause of stroke during CABG surgery; therefore, it is common practice in some hospitals to perform a prophylactic staged or combined carotid endarterectomy in asymptomatic patients discovered to have carotid stenosis (usually through a bruit) prior to OHS. Most studies have failed to demonstrate a correlation between stroke during CABG and asymptomatic carotid bruits (Breuer et al. 1983; Hart and Easton 1983; Ropper et al. 1982), although a recent retrospective study (Reed et al. 1988) demonstrates a small but significant increased risk. The results of bruit studies should be interpreted cautiously. They are of limited use because bruits are neither specific nor sensitive for the presence or degree of carotid occlusive disease (Ivey et al. 1984; Barnes et al. 1982), and often they do not correlate with the location of the stroke.

Studies using carotid noninvasive tests to screen asymptomatic patients before CABG surgery have provided conflicting data regarding stroke risk in patients with carotid stenosis. Most studies have not shown an increased risk of stroke (Barnes and Marszalek 1981; Breslau et al. 1981; Turnipseed et al. 1980), but others have (Brener et al. 1984; Kartchner and McRae 1981).

In a retrospective angiographic study, Furlan and Craciun (1985) identified 155 stenotic (≥50%) or occluded carotid arteries in 144 patients undergoing CABG surgery. Ipsilateral stroke occurred in one of (1.1%) with 50% to 90% stenosis, one of 16 (6.2%) with greater than 90% stenosis, and one of 49 (2.06%) with occlusion. In a prospective study, Brener and colleagues (1987) identified 193 stenotic or occluded carotid arteries in 153 out of 4047 patients undergoing CABG surgery using noninvasive screening. Angiography confirmed the diagnosis in 95 of the 153 patients. Definite TIA or stroke ipsilateral to these arteries occurred in eight of the 47 (17%) that were occluded, two of 89 (2.2%) that were stenotic (≥50%) and not operated on during CABG surgery, and two of 57 (3.5%) that were stenotic (≥50%) and operated on during CABG surgery.

In a recent study at the Cleveland Clinic, Hertzer and associates (1989) randomized 129 patients with angiographically proven asymptomatic carotid stenosis (>70%) into two treatment subsets: one group of 71 patients underwent combined carotid endarterectomy and CABG and the other group of 58 patients underwent CABG followed by delayed carotid endarterectomy. There were two perioperative strokes (2.8%) in the group undergoing the combined procedure and four (6.9%) in the group undergoing CABG alone. The difference in stroke rate between these two groups was not statistically significant.

There are virtually no data on the risk of stroke during CABG in patients with symptomatic high-grade unilateral carotid stenosis or multiple-vessel occlusive disease because most of these patients undergo staged or combined carotid endarterectomy. Hertzer and colleagues (1989) report a group of 23 patients with symptomatic or bilateral carotid stenosis (≥70%) who underwent "unprotected" CABG because their unstable coronary disease required urgent revascularization. Two (8.7%) of these 23 patients developed a stroke perioperatively.

The results of these studies should be interpreted cautiously because each has at least one of the following serious design flaws: insufficient numbers of patients, retrospective analyses, lack of randomization, lack of information regarding degree of carotid disease and relation of stroke to side of stenosis, different definitions of what constitutes a stroke, and incomplete information regarding investigation of other potential stroke mechanisms. However, it is possible to make some conclusions on the basis of these and other studies.

There is no evidence that asymptomatic patients with less than 90% stenosis of one carotid artery are at increased risk of stroke during CABG surgery.

The majority of patients with asymptomatic unilateral stenosis of 90% or more or carotid occlusion do not sustain a stroke during CABG surgery. However, a small subset of these patients are probably at increased risk of stroke, but there are no means to identify them preoperatively or intraoperatively. Recently, intraoperative transcranial doppler studies have failed to demonstrate ipsilateral blood flow velocity changes during CBAG in patients with high-grade internal carotid artery stenosis (von Reutern et al. 1988).

No evidence shows that either staged or combined endarterectomy lowers stroke risk in these asymptomatic patients.

There are insufficient data on stroke risk in symptomatic patients with high-grade stenosis or patients with multiple-vessel occlusive disease. The authors know of patients (e.g., bilateral carotid occlusion) who have successfully undergone CABG, but patients are rarely not cleared for nonemergent CABG in the face of multiple, symptomatic, surgically inaccessible vascular lesions.

The current approach to patients with carotid disease undergoing CABG surgery is to view the carotid disease independently. For symptomatic carotid disease, the authors prefer endarterectomy before CABG surgery if the coronary status is stable, or combining the procedures if both diseases are unstable. In most asymptomatic patients, prophylactic endarterectomy is not preferred. On rare occasions in patients with bilateral severe carotid disease (≥90% stenosis or occlusion), especially without collateral supply from the posterior circulation, unilateral endarterectomy on the side of the dominant hemisphere is recommended. If the results of the ongoing asymptomatic carotid artery stenosis (ACAS) study favor the surgically treated group, combined carotid endarterectomy and CABG may prove a safe, cost-effective alternative for some patients.

Prior Stroke and Risk of Subsequent OHS

The data on this subject are limited. Studies have provided discrepant estimates of stroke risk in patients with prior stroke undergoing OHS, varying from no increased risk (Breuer et al. 1983; Gonzalez-Scarano and Hurtig 1981) to six times the risk of control patients (Reed et al. 1988). However, these studies lack sufficient statistics to permit definite conclusions.

In a recently completed retrospective study of 127 patients with a history of stroke who underwent OHS, 17 (13.4%) had a new stroke or worsening of prior deficits (Rorick and Furlan 1989). This rate is significantly higher ($p < .005$) than the expected rate of 5.2% predicted by a previous prospective study from the same institution. Almost all patients had no or minimal neurological deficit at the time of OHS. The stroke rate was higher (16.7%) in those patients whose previous strokes were recent (<3 months before OHS) compared to those with remote strokes (12.3%), although this was not statistically significant. Sixty percent of the complications in the recent stroke group involved worsening of deficits, whereas 75% of the complications in the remote stroke group involved new deficits. This suggests different mechanisms of strokes in the two groups. Extracranial carotid disease was not a factor. In the remote group, perioperative cardiogenic embolus was considered the most likely mechanism because new areas of brain were affected, and 50% had atrial fibrillation or flutter prior to their strokes. Patients with recent stroke exhibited periinfarction hemodynamic vulnerability, which led to infarct extension when mean arterial pressure fell during OHS. The authors suggest delaying OHS for 3 months in patients with recent stroke, unless the patient has preinfarction angina, bacterial endocarditis, myxoma, or another urgent cardiac condition. In these situations, the benefits of removing the embolic source or improving cardiac hemodynamics outweigh the risk of stroke extension during surgery (Zisbrod et al. 1987). Intraoperative blood pressure should be maintained above average in these cases.

Postoperative Arrhythmia and Stroke

Transient cardiac arrhythmias occur in 5% to 40% of patients after OHS (Michelson et al. 1979; Ghosh and Pakrashi 1972). Supraventricular arrhythmias are particularly common, prompting some authors to suggest prophylactic treatment with Inderal or digoxin after OHS (Roffman and Feldman 1981). Data on the incidence of stroke associated with atrial fibrillation after OHS are scanty. Taylor and associates (1987) and Reed and colleagues (1988) report a threefold to fivefold increase in stroke risk after CABG surgery in these patients. However, in a prospective study of 200 patients undergoing CABG surgery at the Cleveland Clinic Foundation, the stroke rate among patients with postoperative atrial fibrillation was 3.6%, compared to 3.5% in patients without cardiac dysrhythmias (O'Neill et al. 1983). The authors' experience suggests that prophylactic antiarrhythmic agents would not lower the stroke rate following CABG surgery. Similarly, the authors do not routinely use anticoagulation for post-OHS atrial fibrillation unless cardioversion is attempted or the atrial fibrillation is persistent. Ventricular arrhythmias have not been associated with an increased stroke risk after CABG surgery (Taylor et al. 1987; O'Neill et al. 1983).

Pump Time and Stroke Risk

Pump time more than 2 hours has been associated with an increased frequency of stroke during CABG (Branthwaite 1975; Reed et al.

1988; Tufo et al. 1970). However, Breuer and coworkers (1983) report that the stroke rate in 36 patients undergoing prolonged bypass (>2 hours) was not increased. Patients undergoing prolonged bypass tend to have compromised preoperative cardiorespiratory function and unstable hemodynamics intraoperatively, and it is likely that the reported increased stroke risk in patients undergoing prolonged bypass relates instead to these factors.

Encephalopathy and OHS

The incidence of encephalopathy following OHS varies from 3% to 12% (Breuer et al. 1983; Coffey et al. 1983; Shaw et al. 1985). This variability is due partly to the timing of the neurological examination and whether neuropsychological testing is performed. In a study of 59 patients undergoing CABG at the Cleveland Clinic, four of 59 patients showed transient evidence of delirium in the intensive care unit, while none had evidence of delirium on postoperative day 6. However, subtle but significant cognitive deficits were noted in these patients after neuropsychological testing on postoperative day 6 compared to preoperative testing (Calabrese et al. 1987).

In the study by Breuer and coworkers (1983) no preoperative or intraoperative risk factors emerged for the development of prolonged encephalopathy, but two postoperative variables correlated with confusion— use of an intra-aortic balloon pump and use of pressor drugs, both markers for patients with severe hypotension. In most patients, multiple potential causes for encephalopathy were recognized, including medications, hypoxia, fever or sepsis, metabolic derangements, hemodynamic instability, and intensive care psychosis. These findings suggest that meticulous medical management of these patients should decrease the incidence of encephalopathy following OHS.

Although 80% of patients with encephalopathy after OHS in the study by Breuer and associates (1983) had normal bedside mental status examinations by the time of discharge, subtle long-term changes in mental functioning detected by neuropsychological testing may occur in up to 30% of patients undergoing OHS (Sotaniemi et al. 1986).

Rare CNS Complications of OHS

Several other rare CNS complications of OHS have been reported, including coma (Furlan and Breuer, 1984), intracranial hemorrhage (Humphreys et al. 1975; Hungxi et al. 1982), pituitary apoplexy (Cooper et al. 1986), and hearing loss (Plasse et al. 1981). The hemorrhages may be due to anticoagulation associated with OHS. Nonmetabolic coma occurs in less than 1% of patients undergoing OHS. Of 34 patients seen by the stroke service at the Cleveland Clinic Foundation over a 4-year period for failure to awaken after OHS, the causes were global ischemia or hypoxia (seven patients), hemisphere infarct with herniation (five), multifocal infarcts (five), and unknown (19). Prognosis in this group of patients is very poor—29 died, four evolved into a persistent vegetative state, and only one had useful neurological recovery (Furlan and Breuer 1984).

Peripheral Nervous System Complications of OHS

The incidence of peripheral nervous system (PNS) complications of OHS varies between 2.6% and 13% (Keates et al. 1975; Lederman et al. 1982). In a study by Lederman and coworkers (1982), 55 of 421 patients (13%) undergoing CABG surgery at the Cleveland Clinic Foundation developed 63 new PNS complications. These included brachial plexopathy (23 patients), saphenous neuropathy (13), common peroneal palsy (eight), ulnar mononeuropathy (five), phrenic nerve lesion (six), recurrent laryngeal nerve paralysis (five), radial sensory neuropathy (one), facial neuropathy (one), and Horner's syndrome (one). Most PNS deficits were transient, but two patients, one with brachial plexopathy and another with peroneal palsy, had persistent significant weakness. The plexopathy can be severe and associated with causalgia, which may respond to carbamazepine or sympathetic block.

Risk factor analysis indicates that male sex and the use of hypothermia during surgery were significantly correlated with PNS injury. The predominance of men is difficult to explain, but hypothermia is a recognized cause of neuropathy (Stephens and Appleby 1955). Diabetes was not significantly associated with PNS complications despite the well-recognized risk of stretch and compressive nerve injuries in this subgroup (Aguayo 1975).

The mechanisms of brachial plexus injury were attributed to trauma from jugular vein cannulation and plexus stretching, especially during internal mammary artery dissection, which requires more chest wall retraction. Ulnar and peroneal palsies were attributed to compression at susceptible sites, and the saphenous neuropathies were related to the injury to the nerve during harvesting of the donor vessels for bypass. Special attention to positioning the patient, improving techniques of percutaneous cannulation of the jugular vein, harvesting of the saphenous vein, and using guards to protect against compression neuropathies at the elbows and knees will lower the incidence of PNS injury associated with OHS (Britt and Gordon 1964).

Cardiac Transplantation

Cardiac transplantation has become an accepted treatment for patients with terminal heart disease. Neurological complications frequently are associated with this procedure (Hotson and Pedley 1976; Schober and Herman 1973). In a study of 83 patients undergoing cardiac transplantation at Stanford University Medical Center, 50 patients (60%) developed neurological complications, and eight others sustained vertebral compression fractures without neurological impairment. Ten patients (12%) showed impaired neurological function in the immediate postoperative period. This was attributed to focal cerebral infarction (four patients), diffuse encephalopathy (three),

severe anoxic damage following postoperative cardiac arrest (one), worsening of a pre-existing hemiparesis (one), and multiple cerebral infarcts from fat emboli (one). The deficits of the patients with focal infarcts and diffuse encephalopathy resolved within 2 weeks of the transplantation. A comparison of these results to other forms of OHS suggests that the incidence of immediate postoperative neurological complications is higher in the transplant group, but the mechanisms of brain injury are similar. Unfortunately, an even higher incidence of neurological complications occurs after the first 10 days of transplantation—forty out of 83 patients (48%) developed delayed neurological dysfunction attributed to CNS infection (20 patients), stroke (five), metabolic encephalopathy (seven), acute psychosis (seven), and cerebellar lymphoma (one). The infections and lymphoma were related to immunosuppressive therapy, and the strokes were attributed to cardioembolism in two, hypoperfusion in one, and intracerebral hemorrhage in two patients.

An unusual complication, delayed intracerebral hemorrhage (ICH), occurred in three of 58 patients on days 7, 16, and 38 after cardiac transplantation at the Cleveland Clinic Foundation (Sila et al. 1989). All three patients were young, had cardiomyopathy, had a small body surface area, and developed a vascular type headache several days before the ICH. Vascular headache without ICH and often responsive to beta blockers occurred in an additional 30.6% of the patients. The improvement in cardiac index after transplantation was higher for patients with ICH (median 331%) or vascular headache (median 96%) than for those without these complications (median 58%), although this was not statistically significant ($p = 0.09$). ICH and a vascular headache complicating cardiac transplantation probably occur because of cerebral hyperperfusion and breakthrough of cerebral autoregulation due to an abrupt improvement in cardiac index. The authors advise strict control of blood pressure not only in the immediate postoperative period, but also for a few months after transplantation to prevent delayed ICH and to treat vascular headaches with beta blockers.

Conclusion

The risk of stroke associated with cardiac catheterization or PTCA is extremely low (0.11% to .3%) is the procedures are performed by experienced angiographers. Stroke associated with CABG surgery occurs in 2% to 5% of patients, whereas encephalopathy is more frequent (3% to 12%). The potential mechanisms of these two complications are numerous and incompletely understood. In studies, no consistent risk factors for the development of these two complications have emerged, which makes it difficult to institute prophylactic measures. Brachial plexopathy and mononeuropathy occur in 2% to 13% of patients undergoing CABG surgery and are related to stretching of, or direct trauma to, these nerves during the procedure. Cardiac transplantation carries the greatest risk of neurological complications of all OHS procedures (60% of patients). Most of these occur after the immediate postoperative period and frequently are due to CNS infection related to immunosuppressive therapy.

References

Adams, D. R.; Fraser, D. B.; Abrams, H. L. The complications of coronary arteriography. Circulation 48:609–617; 1973.

Aquayo, A. J. Neuropathy due to compression and entrapment. In: Dyck, P. J.; Thomas, P. K.; Lambert, E. H. eds. Peripheral neuropathy. Philadelphia: W.B. Saunders; 1975; p. 688–713.

Barnes, R. W.; Rittgers, S. E.; Putney, W. W. Real-time Doppler spectrum analysis. Arch. Surg. 117:52–57; 1982.

Barnes, R. W.; Marszalek, P. B. Asymptomatic carotid disease in the cardiovascular surgical patient: is prophylactic endarterectomy necessary? Stroke 12:497–500; 1981.

Branthwaite, M. A. Neurological damage related to open-heart surgery: a clinical study. Thorax 27:748–753; 1972.

Branthwaite, M. A. Cerebral blood flow and metabolism during open-heart surgery. Thorax 29:633–638; 1974.

Branthwaite, M. A. Prevention of neurological damage during open-heart surgery. Thorax 30:258–261; 1975.

Brener, B. J.; Brief, D. K.; Alpert, J.; Goldenkranz, R. J.; Parsonnet, V.; Feldman, S.; Gielchinsky, I.; Abel, R. M.; Hochberg, M.; Hussain, M. A four-year experience with preoperative noninvasive carotid evaluation of two thousand twenty-six patients undergoing cardiac surgery. J. Vasc. Surg. 1:326–338; 1984.

Brener, B. J.; Brief, D. K.; Alpert, J.; Goldenkranz, R. J.; Parsonnet, V. The risk of stroke in patients with asymptomatic carotid stenosis undergoing cardiac surgery: a follow-up study. J. Vasc. Surg. 5:269–279; 1987.

Breslau, P. J.; Fell, G.; Ivey, T. D.; Bailey, W. W.; Miller, D. W.; Strandness, D. E. Jr. Carotid arterial disease in patients undergoing coronary artery bypass operations. J. Thorac. Cardiovasc. Surg. 82:765–767; 1981.

Breuer, A. C.; Furlan, A. J.; Hanson, M. R.; Lederman, R. J.; Loop, F. D.; Cosgrove, D. M.; Greenstreet, R. L.; Estafanous, F. G. Central nervous system complications of coronary artery bypass graft surgery: prospective analysis of 421 patients. Stroke 14:682–687; 1983.

Britt, B. A.; Gordon, R. A. Peripheral nerve injuries associated with anaesthesia. Can. Anaesth. Soc. J. 11:514–536; 1964.

Calabrese, J. R.; Skwerer, R. G.; Gulledge, A. D.; Gill, C. G.; Mullen, J. D.; Rodgers, D. A.; Taylor, P. C.; Golding, L. A.; Lytle, B. W.; Cosgrove, D. M.; Bazarel, M. G.; Loop, F. D. Incidence of post-operative delirium following myocardial revascularization: a prospective study. Cleve. Clin. J. Med. 54:29–32; 1987.

Cannon, D. S.; Miller, D. C.; Shumway, N. E.; Fogarty, T. J.; Daily, P. O.; Hu, M.; Brown, B. Jr.; Harrison, D. C. The long-term follow-up of patients undergoing saphenous vein bypass surgery. Circulation 49:77–85; 1974.

Coffey, C. E.; Massey, E. W.; Roberts, K. B.; Curtis, S.; Jones, R. H.; Pryor, D. B. Natural history of cerebral complications of coronary artery bypass graft surgery. Neurology 33:1416–1421; 1983.

Cooper, D. M.; Bazaral, M. G.; Furlan, A. J.; Seuilla, E.; Ghatias, M. A.; Sheeler, L. R.; Little, J. R.; Hahn, J. F.; Sheldon, W. C.; Loop, F. D. Pituitary apoplexy: a complication of cardiac surgery. Ann. Thorac. Surg. 41:547–550; 1986.

Dawson, D. M.; Fischer, E. G. Neurologic complications of cardiac catheterization. Neurology 27:496–497; 1977.

Dorros G.; Cawley, M. J.; Simpson, J.; Bentivaglio, L. G.; Block, P. C.; Bourassa, M.; Detre, K.; Gosselin, A. J.; Grunteig, A. R.;

Kelsey, S. E.; Kent, K. M.; Mock, M. B.; Mullin, S. M.; Meyer, R. K.; Passamani, E. R.; Stertzer, S. H.; Williams, D. O. Percutaneous transluminal coronary angioplasty: report of complications from the National Heart, Lung and Blood Institute PTCA registry. Circulation 67:723–730; 1983.

Furlan, A. J.; Breuer, A. C. Central nervous system complications of open heart surgery. Stroke 15:912–915; 1984.

Furlan, A. J.; Craciun, A. R. Risk of stroke during coronary artery bypass graft surgery in patients with internal carotid artery disease documented by angiography. Stroke 16:797–799; 1985.

Galbreath, C.; Salgado, E. D.; Furlan, A. J.; Hollman, J. Central nervous system complications of percutaneous transluminal coronary angioplasty. Stroke 17:616–619; 1986.

Ghosh, P.; Pakrashi, B. C. Cardiac dysrhythmias after thoracotomy. Br. Heart. J. 374–376; 1972.

Gilman, S. Cerebral disorders after open heart operations. N. Engl. J. Med. 272:489–498; 1965.

Gonzalez-Scarano, F.; Hurtig, H. I. Neurologic complications of coronary artery bypass grafting: Case control study. Neurology 31:1032–1035; 1981.

Hansotia, P. L.; Myers, W. O.; Ray, J. F. III; Greehling, C.; Sautter, R. D. Prognostic value of electroencephalography in cardiac surgery. Ann. Thoracic Surg. 19:127–134; 1975.

Hart, R. G.; Easton, J. D. Management of cervical bruits and carotid stenosis in preoperative patients. Stroke 14:290–297; 1983.

Heller, S. S.; Frank, K. A.; Malm, J. R.; Bowman, F. O.; Harris, P. D.; Charlton, M. H.; Kornfeld, D. S. Psychiatric complications of open-heart surgery. N. Engl. J. Med. 283:1015–1020; 1970.

Hertzer, N. R.; Loop, F. D.; Beven, E. G.; O'Hara, P. J.; Krajewski, L. P. Surgical staging for simultaneous coronary and carotid disease: a study including prospective randomization. J. Vasc. Surg. 9:455–463; 1989.

Hill, J. D.; Aguilor, M. J.; Baranco, A.; DeLanerolle, P.; Gerbode, F. Neuropathological manifestations of cardiac surgery. Ann. Thorac. Surg. 7:409–419; 1969.

Hodgman, J. R.; Cosgrove, D. M. Post-hospital course and complications following coronary bypass surgery. Cleve. Clin. Q. 43:125–129; 1976.

Hotson, J. R.; Pedley, T. A. The neurological complications of cardiac transplantation. Brain 99:673–694; 1976.

Humphreys, R. P.; Hoffman, H. J.; Mustard, W. T.; Trusler, G. A. Cerebral hemorrhage following heart surgery. J. Neurosurg. 43:671–675; 1975.

Hungxi, S.; Xiaogin, H.; Kongsoon, L.; Jiaquiang, G.; Letien, X. Intracranial hemorrhage and hematoma following open heart surgery. In: Becker, R.; Katz, J.; Polonius, M. J.; Speidel, H. eds. Psychopathological and neurological dysfunctions following open heart surgery. New York: Springer-Verlag; 1982:p. 293–299.

Hutchinson, J. E. III; Green, G. E.; Mekhjian, H. A.; Kemp, H. G. Coronary bypass grafting in 376 consecutive patients, with three operative deaths. J. Thorac. Cardiovasc. Surg. 67:7–16; 1974.

Ivey, T. D.: Standness, D. E. Jr.; Williams, D. B.; Langlois, Y.; Misbach, G. A.; Delegans, A. D. Management of patients with carotid bruit undergoing cardiopulmonary bypass. J. Thorac. Cardiovasc. Surg. 87:183–189; 1984.

Javid, H.; Tufo, H. M.; Nafaji, H.; Dye, W. S.; Hunter, J. A.; Julian, O. C. Neurological abnormalities following open-heart surgery. J. Thorac. Cardiovasc. Surg. 58:502–509; 1969.

Kartchner, M. M.; McRae, L. P. Guidelines for noninvasive evaluation of asymptomatic carotid bruits. Clin. Neurosurg. 28:418–428; 1981.

Keates, J. R. W.; Innocenti, D. M.; Ross, D. N. Mononeuritis multiplex: a complication of open heart surgery. J. Thorac. Cardiovasc. Surg. 69:816–819; 1975.

Kolkka, R.; Hilberman, M. Neurologic dysfunction following cardiac operation with low-flow, low-pressure cardiopulmonary bypass. J. Thorac. Cardiovasc. Surg. 79:432–437; 1980.

Kosmorsky, G.; Hanson, M. R.; Tomsak, R. L. Neuroophthalmologic complications of cardiac catheterization. Neurology 38:483–485; 1988.

Lederman, R. J.; Breuer, A. C.; Hanson, M. R.; Furlan, A. J.; Loop, F. D.; Cosgrove, D. M.; Estafanous, F. G.; Greenstreet, R. L. Peripheral nervous system complications of coronary artery bypass graft surgery. Ann. Neurol. 12:297–301; 1982.

Lee, M. C.; Geiger, J.; Nicoloff, D.; Klassen, A. C.; Resch, J. A. Cerebrovascular complications associated with coronary artery bypass (CAB) procedure. Stroke 10:13, 1979.

Lockwood, K. I.; Capraro, J.; Hanson, M.; Conomy, J. Neurologic complications of cardiac cathetherization. Neurology 33(suppl 2):143; 1983.

Loop, F. D.; Cosgrove, D. M.; Lytle, B. W.; Thurer, R. L.; Simpendorfer, C.; Taylor, P. C.: Proudfit, W. L. An 11-year evolution of coronary arterial surgery (1967–1978). Ann. Surg. 190:444–445; 1979.

Michelson, E.; Morganroth, J.; MacVaugh, H. Post-operative arrythmias after coronary artery and cardiac valvular surgery detected by long term electrocardiographic monitoring. Am. Heart J. 97:442–448; 1979.

Muraoka, R.; Yokota, M.; Aoshima, M.; Kyoku, I.; Momoto, S.; Kobayashi, A.; Nakano, H.; Ueda, K.; Saito, A.; Hojo, H. Subclinical changes changes in brain morphology following cardiac operations as reflected by computed tomographic scans of the brain. J. Thorac. Cardiovasc. Surg. 81:364–369; 1981.

O'Neill, B. J. III.; Furlan, A. J.; Hobbs, R. D. Risk of stroke in patients with transient postoperative atrial fibrillation/flutter. Stroke 14:133; 1983.

Plasse, H. M.; Mittleman, M.; Frost, J. O. Unilateral sudden hearing loss after open heart surgery: a detailed study of seven cases. Laryngoscope 91:101–109; 1981.

Reed, G. L. III: Singer, D. E.; Picard, E. H.; DeSanctis, R. W. Stroke following coronary artery bypass surgery: a case control estimate of the risk from carotid bruits. N. Engl. J. Med. 319:1246–1250; 1988.

Roffman, J.; Fieldman, A. Digoxin and Propanolol in the prophylaxis of supraventricular tachydysrythmias after coronary artery bypass surgery. Ann. Thorac. Surg. 31:496–501; 1981.

Ropper, A. H.; Wechsler, L. R.; Wilson, L. S. Carotid bruit and the risk of stroke in elective surgery. N. Engl. J. Med. 307:1388–1390; 1982.

Rorick, M.; Furlan, A. J. Risk of heart surgery in patients with prior stroke. Neurology 40:835–837; 1990

Schober, R.; Herman, M. M. Neuropathology of cardiac transplantation: survey of 31 cases. Lancet 1:962–967; 1973.

Shaw, P. J.; Bates, D.; Cartlidge, N. E. F.; Heaviside, D.; Julian, D. G.; Shaw, D. A. Early neurological complications of coronary artery bypass surgery. Br. Med. J. 291:1384–1387; 1985.

Sila, C. A.: Furlan, A. J.; Stewart, R. W. Intracerebral hemorrhage complicating cardiac transplantation: hyperperfusion breakthrough. Neurology 39(suppl 1):161; 1989.

Sotaniemi, K. A.; Mononen, H.; Hokkanen, T. E. Long-term cerebral outcome after open heart surgery. A five-year neuropsychological follow-up study. Stroke 17:410–416; 1986.

Stephens, J.; Appleby, S. Polyneuropathy following induced hypothermia. Trans. Am. Neurol. Assoc. 80:102–104; 1955.

Taylor, G. J.; Malik, S. A.; Colliver, J. A.; Dove, J. T.; Moses, H. W.; Mikell, F. L.; Batchelder, J. E.; Schneider, J. A.; Wellons, H. A. Usefulness of atrial fibrillation as a predictor of stroke after isolated coronary artery bypass grafting. Am. J. Cardiol. 60:905–907; 1987.

Turnipseed, W. D.; Berkoff, H. A.; Belzer, F. O. Post-operative stroke in cardiac and peripheral vascular disease. Ann. Surg. 192:365–368; 1980.

Tufo, H. M.; Ostfeld, A. M.; Shekelle, R. Central nervous system dysfunction following open heart surgery. J.A.M.A. 212:1333–1340; 1970.

von Reutern, G-M.; Hetzel, A.; Birnbaum, D.; Schlosser, V. Transcranial doppler ultrasonography during cardio-pulmonary bypass in patients with severe carotid stenosis or occlusion. Stroke 19:674–680; 1988.

Zisbrod, Z.; Rose, D. M.; Jacobwitz, I. J.; Kramer, M.; Acinapura, A. J.; Cunningham, J. N. Jr. Results of open heart surgery in patients with recent cardiogenic embolic stroke and central nervous system dysfunction. Circulation 76(suppl V):V109–V112; 1987.

6

Head Injury

BRYAN JENNETT, GRAHAM TEASDALE, GORDON MURRAY, AND
LILIAN MURRAY

Prognosis After Head Injury

The uncertainty of what will happen after a
head injury, whether mild or severe, causes
great concern not only to the patient and
family, but also to doctors in several special-
ties who must make decisions about management. The capriciousness of head injury
was summarized by Hippocrates: "No head
injury is too serious to despair of, nor too
trivial to ignore." This reflects the fact that
some apparently severely injured patients
make a good recovery, while others with
mild injury can develop life-threatening
complications.

Several questions arise soon after a head
injury. If it was mild, what is the risk of
complications? If it was serious, what are
the chances of survival. For survivors, what
is the likelihood of recovery without permanent disability?

Uses of Prognosis

When doctors could do little to influence the
course or outcome of disease or injury, their
reputations rested largely on their ability to
give a correct prognosis. Many diagnostic,
monitoring, and therapeutic technologies

can now be used; however, some are hazardous, others are in short supply, and all
use health-care resources. Selecting patients
for these technologies should depend on the
probability of benefit vs. the possibility of
harm—an exercise in prediction.

Prognosis also is essential for assessing
the effectiveness of methods of management, as judged by significantly better outcomes than had been predicted without their
use. Prospective randomization is the recommended way to select comparable groups
of patients who will be treated differently,
but for sudden life-threatening conditions,
this can be difficult. An alternative is to ex-
ploit the wide variations in practice between
clinicians (or hospitals), and use predictive
models to identify patients whose outcomes
were expected to be similar, but who had
been managed in different ways.

Difficulty of Prognosis After Head Injury

Traditionally, clinicians base prognoses on
comparing current patients with supposedly similar previously encountered patients. What is remembered about the early
features and late outcomes of previous patients, however, is always incomplete and is

often unduly influenced by more recent cases and by those whose outcomes were unexpected. After head injury, the limited number of comparable cases that one doctor encounters also makes a reliable prognosis difficult. Mild injuries are common, but complications seldom occur, and severe injuries are infrequent. Even after many years in practice, a clinician is likely to have dealt with only a few truly similar patients with complications or severe injuries. Only since computers have made it possible to store and analyze large amounts of data has it been possible to express predictions as estimates of the probability that certain events will occur. Prognostic data may emerge as by-products of randomized trials, but some systematic collections of data on head injuries have been specifically designed to discover predictive criteria.

Establishing a prognostic system for head injuries has four stages. It begins with defining the outcome event(s) to be predicted and the time after injury when outcome is to be assessed. Next, the factors considered to be of predictive value are identified, usually based on a consensus among experienced clinicians. The long and tedious process of collecting data prospectively from large number of patients follows. The data consist of the pattern of occurrence of putative predictive factors and of outcomes. Statistical techniques are then applied to determine the relative predictive power of different factors alone and in various combinations.

Three examples are given of prognosis after head injury: the risk of an early complication, acute intracranial haematoma; the risk of a late complication, traumatic epilepsy; and the likelihood of different outcomes after severe head injury.

The Risk of Acute Intracranial Hematoma

A hematoma requiring surgery within 14 days of injury is a major cause of potentially avoidable mortality and morbidity after head trauma. Early detection and evacuation, before irreversible secondary brain damage has occurred, improves the outcome (Teasdale et al. 1982a; Bullock and Teasdale 1990). Large numbers of mildly in-

jured patients are admitted to hospitals for observation in the hope of achieving this early detection, but an acute hematoma occurs in less than one in 5000 of those who attend hospitals after head injury. Some controversy exists about the relative predictive importance of impaired consciousness and skull fracture and therefore of their relevance as indications for admission. Radiologists have repeatedly expressed concern about the number of negative skull x-rays carried out in emergency rooms, and some assert that the presence of a fracture is of little significance in identifying patients at risk (Royal College of Radiologists 1983; Masters et al. 1987; Cooper 1983).

The authors' studies on the risk factors of patients admitted to the hospital (i.e., in the emergency room) were initially confined to adults, in whom hematoma is more common (Mendelow et al. 1983). However, about one half of all recent head injury patients are children (younger than 15 years), and risk factors have now been calculated for them as well (Teasdale et al. 1990). Because a hematoma is so uncommon, it was considered impractical to collect sufficient data on patients in emergency departments with this complication to allow valid statistical analysis. Instead, the occurrence of suspected risk factors was analyzed separately in a sample of 4767 adults and 3599 children. These data were collected in a systematic survey of head injuries in Scottish hospitals (Strang et al. 1978). The frequency of risk factors was also recorded for 861 adults and 99 children who had been operated on for acute hematoma in the Glasgow neurosurgical unit.

The risk of a hematoma in adults proved to be six times greater than in children, but in both age groups, a lowered conscious state and a skull fracture increased the risk that a hematoma would develop (Table 6–1). In patients who were fully conscious when first seen at the hospital, a fracture increased the risk of hematoma 176 times in adults and 80 times in children. Impaired consciousness (short of coma) with no fracture increased the risk only 44 times in adults and 22 times in children; and the risk

Table 6-1. Absolute Risk of Hematomas in A/E Patients

	Children*			Adults†		
	Fully Conscious	Impaired Consciousness	Coma	Fully Conscious	Impaired Consciousness	Coma
No fracture	13,000 (95%)	580	65	7900 (92%)	180	27
Skull fracture	160	25	12	45	5	4
Increased risk if fracture	×81	×23	×5	×176	×36	×7
Increased risk if impaired (not coma)						
No fracture		×22			×44	
With fracture		×6			×9	

* Figures indicate risk of one in 2100.

† Figures indicate risk of one in 348.

  Based on data from Teasdale, G. M.; Murray, G.; Anderson, E.; Mendelow, A. D.; Brookes, M. Risks of a traumatic intracranial haematoma in children and adults: implications for the management of head injuries. Br. Med. J. 300:363–367; 1990.

was increased only 9 and 6 times, respectively, in those with a fracture. Skull fracture was therefore a more powerful predictor of this complication than impairment of consciousness in both adults and children.

As well as comparing the relative predictive power of different factors, this analysis indicated that the absolute risk for patients with different combinations of risk factors ranged from about one in 6000 to one in four. Guidelines based on these data were published in Britain in 1984, which recommended the types of adult patients who should be admitted and those who should have a computed tomography (CT) scan or neurosurgical consultation (Group of Neurosurgeons 1984). Their suggestions have affected clinical practice and have improved patient outcome (Teasdale and Bullock 1990; Brocklehurst et al. 1987).

The Risk of Late Epilepsy After Head Injury or Intracranial Surgery

Epilepsy occurring more than 1 week after injury is another uncommon complication. It can spoil an otherwise good recovery, particularly because it limits a person's eligibility to hold a driver's licence. Therefore, there is a premium on the ability to estimate this risk.

The overall risk for nonmissile head injuries who have been admitted to the hospital is about 5% but it ranges from 1% to more than 60%, according to the type of injury and early complications (Jennett 1975). The

risk is increased by an early fit (within 1 week of injury), an acute intracranial hematoma that has been evacuated surgically, and certain types of depressed fractures (Table 6-2, Fig. 6-1). Patients with none of these risk factors can be reassured that epilepsy is unlikely. These patients include many with a depressed fracture and some with severe injuries—both types of injuries that were previously believed to create a risk of late epilepsy. Modern imaging techniques can now show sizable intradural clots in patients whose clinical condition does not justify surgery, and 23% of these patients develop late epilepsy, compared with 45% for operated intradural hematomas (Crandon et al. 1990). The higher rate after operated hematoma may be because the clots in themselves have done more damage, as evidenced by their need for surgery, but the surgery also may impose epileptogenic brain damage.

Table 6-2. Residual Risk of Late Epilepsy With Increasing Fit-Free Interval Since Injury

Initial Risk	Time Since Injury Without a Fit Since First Week			
	1 Year	2 Years	3 Years	4 Years
10%	5	3	2.5	2
20%	10	7	5	4.5
30%	16	12	9	7.5
40%	23	17	13	11
50%	31	24	19	16

If early epilepsy, the decay rate is more rapid; therefore residual risk would be approximately $\frac{2}{3}$ of figure shown.

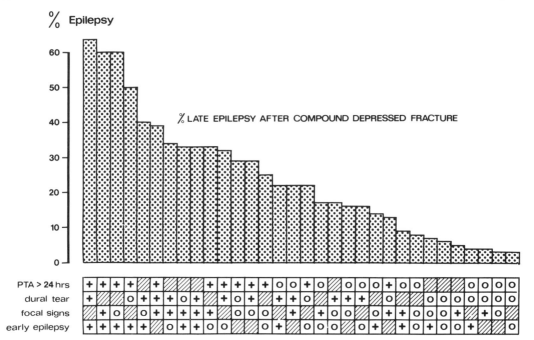

Figure 6-1. Proportion of patients developing late epilepsy after compound depressed fracture (at least 1 year follow-up). Different groups of patients are shown, according to four features, with the calculations based on knowledge of three of these four features (as information is often missing about one of them). (From Jennett, with permission.)

Predicting Outcome After Severe Injury

Predicting outcome is complicated after severe injury by several possible outcomes and many determining factors. It has long been recognized that the older the patient and the more severe the brain damage, the worse the outcome. But how old is old? How and when should severity be measured? How do these two predictors combine to influence outcome? Which can radiology and other investigations contribute to prediction?

Systematic attempts to discover the influence of age and depth or duration of coma on outcome were first reported between 1968 and 1973 from Scandinavia (Carlsson et al. 1968; Heiskanan and Sipponen 1970; Vapalahti and Troupp 1971; Overgaard et al. 1973). These studies used two or three age groups, complex definitions of coma, and three outcomes (death, severe deficits, and recovery). Jennett (1972) stressed the importance of clear definitions of coma and outcome, presaging the Glasgow Coma Scale (Teasdale and Jennett 1974) and the

Glasgow Outcome Scale (Jennett and Bond 1975). Both the Scandinavians and Jennett noted that reliable predictions soon after injury should make it possible to minimize efforts to rescue patients who eventually either die or survive with severe deficits.

Outcomes

Most predictions now focus on the outcome on the Glasgow Scale 6 months after injury (Jennett and Bond 1975). For most purposes, three outcomes are sufficient: dead or vegetative, severe disability, and moderate or good. These categories reflect the objective of management—to increase the proportion of patients who become independent. When only two outcomes are predicted, severe disability should be counted with dead and vegetative or with moderate and good.

Selection of Predictive Criteria

Prospective data collection to identify the features that have predictive power for severely injured patients was first reported in

1976 (Jennett et al. 1976). This experience with 600 cases from Glasgow–Netherlands data bank was expanded to include a center from California (Jennett et al. 1979). Until the minimum data required for prediction has been identified, it is necessary to analyze an excess of data. One advantage of the efficient manipulation of data by computers is that eventually the amount of required data can be reduced. The Glasgow system began with almost 300 separate indicants, which eventually was reduced to 17, with most data provided by only eight. The most powerful predictors are clinical criteria that can be reliably observed and readily recorded by practicing clinicians and nurses. A predictive system based on these criteria should be applicable to the wide range of hospitals where head injuries may be treated, many of which have neither specialist staff nor the special investigations that are available in regional neurosurgical units.

There are two kinds of data. Those that do not change include the age of the patient, cause and type of injury, skull fracture, and major extracranial injuries. Those that change with time reflect the state of brain dysfunction—depth and duration of coma, pupil reactions, and eye movements. The changing state of a patient can be captured by recording the best and worst state during different periods after injury.

Selection of Patients for Prediction

There can be no absolute definition of a severe head injury. Entry to the Glasgow-based data bank required at least 6 hours of coma, an interval during which initial resuscitation was completed, the effects of alcohol or other drugs diminished, and moribund patients either died or became brain dead.

Other studies have defined patients who are in coma on admission, usually after initial resuscitation, as severely injured. The interval between injury and assessment may vary considerably, however, and many accounts fail to state whether "admission" applies to arrival at the first hospital or at a neurosurgical or intensive care unit after transfer. Many patients emerge from coma in the first few hours, and this proportion is

Table 6-3. Comparison of Features of Head Injury in Three Countries

	Glasgow $n = 593$	Netherlands $n = 239$	Los Angeles $n = 168$
Mean age	35 yr	32 yr	35 yr
Lucid interval	32%	25%	23%
Intracranial hematoma	54%	28%	56%
Extracranial injury	32%	51%	51%
Responsiveness (24 hours, best)	21%	21%	21%
Coma sum 3/4	65%	58%	66%
Coma sum 5/6/7/			
Pupils not reacting	19%	29%	31%
Eye movements Absent/impaired	45%	37%	40%

From Jennet, B.; Teasdale, G.; Braakman, R.; Minderhoud, J.; Heiden, J.; Kurze, T. Prognosis of patients with severe head injury. Neurosurgery 4:283–288; 1979.

greater in children. When patients are sedated, paralyzed, and mechanically ventilated soon after injury, there is little option but to classify them as in a coma, although some of these patients may not be injured severely and, without this intervention, would have come out of coma within a few hours. Within the category of severe injury, there can be considerable variation between series; for example, the proportion of patients whose coma scores are 3 or 4 as distinct from those whose scores are 7 or 8.

The Glasgow Data Bank has now collected more than 3000 cases of severe injury, about two thirds from Glasgow and the rest from two centers in the Netherlands and two in California. The clinical features of patients from different centers (Table 6-3) are similar, as are the outcomes (Table 6-4).

A multicenter U.S. study classified more than 1100 severely injured patients accord-

Table 6-4. Outcome at 6 Months in Three Countries

	Glasgow $n = 593$ (%)	Netherlands $n = 239$ (%)	Los Angeles $n = 168$ (%)
Dead	48	50	50
Vegetative	2	2	5
Severe disability	10	7	14
Moderate disability	18	15	19
Good recovery	23	26	12

From Jennett, B.; Teasdale, G.; Braakman, R.; Minderhoud, J.; Heiden, J.; Kurze, T. Prognosis of patients with severe head injury. Neurosurgery 4:283–288; 1979.

ing to various anatomical and pathological lesions. This produced a range of mortality for various subgroups from 9% to 74% (Gennarelli et al. 1982). Whatever the lesion, however, the coma scale still had a dominant influence on prognosis. It can be useful to consider predictions after specific types of injury, for example, in patients who have had an operated intracranial hematoma, according to their state prior to operation, and when assessed again 24 hours after operation. At the other extreme, it is possible to predict mortality in subgroups of large series of injuries admitted to hospital, of which approximately 80% are mild. A study of 41 hospitals in three separate metropolitan areas in the United States identified a low-risk group with an expected mortality of less than 10%; these accounted for 68% of head injury admissions (Klauber et al. 1989). The mortality of these low-risk patients was much lower than predicted in the best hospitals compared to the worst hospitals. There was little variation in the outcome of severe injuries between hospitals, however.

Factors Strongly Related to Outcome

Age has long been correlated closely with mortality rate, and with survival with disability. Most publications show poorer results above or below certain arbitrary ages, such as 20 or 40 or 60 years of age. In the data bank series, a continuous relationship was found between increasing age and bad outcome (death or vegetative survival), and an inverse relationship was related to independent recovery (Teasdale et al. 1979; Teasdale et al. 1982). However, mortality was higher in those younger than age 5 than in those age 5 to 10. A similar continuous relationship has been found not only for severe, but also for moderate injuries in the study of 41 U.S. hospitals (Luerssen et al. 1988). In the authors' experience, the effect of age on outcome is independent of the state of coma, eye movements, or pupil reactions. However, the slope relating age to outcome is similar for severe injuries of differing degrees of severity, but set at a different point on the graph (Fig. 6-2). A similar

slope was shown for all severities in the study by Luerssen and colleagues (1988), but age had less influence on outcome after milder injuries.

Aspects of coma are important for determining prognosis, including depth and duration of coma and spontaneous and reflex pupil reactions and eye movements (Table 6-5). For patients in a coma, the motor response alone can be as good a predictor as the coma score, which frequently cannot be calculated because intubation prevents a verbal response or by drugs or local injury affect eye opening.

Autonomic abnormalities are less strongly related to outcome. These include periodic respiration, tachypnea, tachycardia, systemic hypotension, and temperature higher than 39°C. Although these signs are usually associated with a worse recovery than in patients without them, they occur relatively seldom even among severe injuries. Moreover, other factors, such as extracranial in-

Figure 6-2. Log odds on death after severe head injury. Lines are parallel but set at different levels according to degree of severity of brain damage. (From Teasdale et al., 1982.)

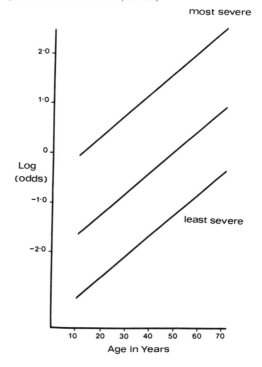

Table 6-5. Outcomes Associated With Best Level of Responsiveness in First 24 Hours After Coma

	n	Dead or Vegetative (Percent)	Moderate Disability, Good Recovery (Percent)
Coma response sum			
>11	57	12	87
8/9/10	190	27	68
5/6/7	525	53	34
3/4	176	87	7
Pupils			
Reacting	748	39	50
Nonreacting	226	91	4
Eye movements			
Intact	463	33	56
Impaired	143	62	25
Absent/bad	186	90	5
Motor response pattern (any limb)			
Normal or weak	568	36	54
Abnormal	393	74	16
Motor response pattern (best limb)			
Obey/localizes	395	31	58
Withdraws/flexes	402	54	35
Extensor/nil	191	85	8

From Jennett, B.; Teasdale, G.; Braakman, R.; Minderhoud, J.; Heiden, J.; Kurze, T. Prognosis of patients with severe head injury. Neurosurgery 4:283–288; 1979.

juries and complications, may affect these functions, and this reduces their value as indicators of the severity of brain damage.

Intracranial hematoma was excluded from the seminal study on prognosis by Carlsson and associates (1968) because it was considered a confusing factor. However, a complication that occurs in one third to one half of the patients with severe injuries can hardly be ignored in a predictive system. Most series show that patients with intracranial hematoma do worse than other comparably severe patients. Once the patient with a hematoma has entered a coma, which is not immediately improved by operation, the outcome is poor. A hematoma is more common with increasing age, and this interaction between age and hematoma affects outcome adversely.

Special Investigations

With the evolution of CT scanning, intracranial pressure (ICP) monitoring, multimodality evoked potentials (MEP), and various biochemical markers of brain damage, it was natural that their predictive value should be sought (Narayan et al. 1981). However,

these tests are normal in about one half of patients with a severe injury. When they are abnormal, the severity of injury is always obvious from clinical features. Other problems include the difficulty of standardizing the equipment and skills from place to place and the impossibility of repeating the tests as frequently as clinical examinations. As a result, the time after injury for which test results are available varies from place to place, making comparisons difficult. Moreover, these investigations are not available in all hospitals managing head injuries. Even the center that pioneered the predictive value of these features omitted them from their most recent predictive system "in order to make the model more universally useful" (Choi et al. 1988).

Predictions Based on Combinations of Factors

Since the early publications of predictive data from Glasgow, several other centers have explored alternative statistical means of selecting predictive variables and of predicting outcome (Narayan et al. 1981; Choi et al. 1988; Braakman et al. 1980; Young et al. 1981; Habbema et al. 1979). The same factors have been selected by a succession of studies. A comparison of 19 statistical methods applied to the same data base showed little difference between several of the most effective methods (Titterington et al. 1981). The main determinant of the strength of a predictive system is the validity of the data entered. The two main methods used are logistical regression and an independence model based on Bayes' theorem.

The reliability of a predictive system using different variables can be compared in different ways. Predictions should be considered correct only if made with some degree of confidence. In the Virginia study, this was 90% (Narayan et al. 1981). Taking into account optimistic and pessimistic predictions, this study calculated a modified prediction rate based on the correct figure minus the optimistic mistakes, which was used to compare the power of single indicants and of various combinations (Table 6-

6). There was little difference between correct predictions and the modified prediction rate for different combinations, but some features predicted more cases at more than 90% confidence. The best three factors were age, motor response, and pupillary response (Choi et al. 1988). The addition of other factors improved the correct prediction rate by less than 0.5%. Mortality was the outcome most accurately predicted, while moderate or severe disability were the most difficult. This was also true in the Glasgow studies.

The number of confident and correct predictions increases as time passes during the first week after injury (Teasdale et al. 1979). The level of probability at which a prediction is regarded as "confident" also affects the number that can be made. A predictive system should seek to minimize pessimistic errors because they could lead to decisions to limit treatment on the grounds of a poor outcome. Some optimistic predictions are inevitable because some patients whose brain damage is recoverable develop com-

Table 6-6. Modified Prediction Rate Associated with Severe Head Injury Patients

Single Indicants	Correct Prediction		Wrong Prediction		Modified Prediction Rate
	>90%	Total Confidence	Optimistic	Correct— Pessimistic	Optimistic
Clinical*					
GCS only	80	25	9	9	93
Best three	84	34	7	9	91
Best five	79	43	11	9	90
All six	82	43	9	9	91
MEP	91	25	9	0	100
CT	64	0	7	29	71
ICP	75	0	11	14	86
Combined indicants					
Clinical+					
MEP	89	64	7	4	96
Best five	86	64	7	7	93
All nine	86	61	7	7	93
Clinical +					
ICP	80	55	11	9	91
Clinical +					
CT	77	52	14	9	91

Clinical indicants: age, GCS, pupils = best three; + eye movements, surgery = best five; + motor posturing = best six.

Best five indicants: MEP, age, ICP, GCS, pupils.

From Narayan, R. K.; Greenberg, R. P.; Miller, J. D.; et al. Improved confidence of outcome prediction in severe head injury. A comparative analysis of clinical examination, multimodality evoked potentials, CT scanning and intracranial pressure. J. Neurosurg. 54:751–762; 1981.

plications, either intracranial or extracranial, that are inherently unexpected.

Uses Made of Predictions

Reliable predictions make it possible to advise families of head-injured victims more confidently about the outcome. Rational decisions about the use of different methods of management in the acute stage should be based on predictions of the likely course and outcome, regardless of technological packages, such as aggressive intensive care, or individual therapies, such as steroids or controlled ventilation. Methods that are potentially hazardous and those that use substantial scarce resources may not be justified if the patient is not sufficiently seriously brain damaged to require the method or is too severely affected to benefit (Jennett 1988).

Not only may certain measures be initiated based on predictions, but treatment may be limited if the outcome is expected to be poor. Doctors, patients, and society are becoming more aware of the importance of quality of life, particularly when severe brain damage is involved. Therefore, there is increasing pressure to consider carefully the use or maintenance of life-sustaining technologies, once it is clear that a reasonable recovery is extremely unlikely (Wanzer et al. 1989; Stanley et al. 1989). More than 50 neurosurgeons from different countries were asked to indicate the level of probability of a poor outcome at which they would withhold or withdraw ventilation from a patient or decide not to evacuate an acute traumatic intracranial hematoma. As expected, they required a high degree of probability of a poor outcome before making such a decision (Barlow and Teasdale 1986). However, when the same surgeons were asked to consider themselves as the victims of brain damage, most of them considerably reduced the level of probability of a poor outcome at which they would want their own treatment to be limited.

Research with the Glasgow system is now focused on its clinical impact. A 3-year prospective multicenter study will soon be completed. The key design features have been described by Murray and coworkers (1986). The main thrust of the study is to examine the relationship between predicted prognosis and resource use. The authors of this chapter hypothesize that the routine use of the predictive system should lead to resources being concentrated on patients whose outcomes are most uncertain. These individuals have the greatest potential to benefit from intensive neurosurgical care.

Predictions also can be valuable for assessing the efficacy of alternative regimes of management for acute life-threatening conditions, for which randomized controlled trials are difficult. The wide variation in the use of methods of management makes it feasible to compare outcomes in groups of patients who are well-matched for age and severity but whose management was different. The question to be asked with a combination of treatment or an individual component is how much better the outcome with this method is than in patients not treated in this way. For example, the Glasgow-based data bank showed no difference in the actual and predicted deaths whether or not steroids were used (Table 6-7). Subsequent controlled trials in other centers confirmed this finding (Todd and Teasdale 1989). This method of technology assessment for severe injuries has been described by Murray (1986) and by Jennett (1987; 1980). A similar study of the outcome of head injuries in a wide range of severities conducted in 41 U.S. hospitals showed a striking difference between the best and worst hospitals in the

Table 6-7. Three Countries' Data Banks from 1970 to 1977*

Deaths	
No Steroids	Steroids
Expected: 232.4	Expected: 193.6
Observed: 232	Observed: 194
49% of 558	51% of 447

* Stratified for age, GCS, pupils, haematoma, country.

From Jennett, B.; Teasdale, G.; Fry, J.; Braakman, R.; Minderhoud, J.; Heiden, J.; Kurze, T. Treatment for severe head injury. J. Neurol. Neurosurg. Psychiat. 43:289–295; 1980.

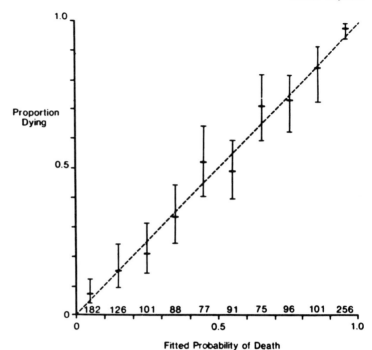

Figure 6-3. Fitted probability of death compared with observed proportion of deaths with 95% confidence limits. Based on 763 cases from Glasgow and 430 from San Francisco — n for each point on abscissa. (From Murray, 1986.)

outcome of patients at low risk, defined as having a predicted mortality of less than 10% (Klauber et al. 1989).

For such an approach to be credible, it is imperative that the predictive model be well calibrated (i.e., the model leads to reliable statements relating to prognosis). For example, of patients predicted to survive with a probability of 0.9, 90% should survive, as with any other predicted probability. Good calibration relies on careful statistical modelling, a feature that is noticeably lacking in some approaches to prognosis (e.g., those based on "expert systems" methodology). The standard of calibration that can be achieved in the context of head injury prognosis is illustrated by Murray (1986). With a suitable logistic regression model, he demonstrated that excellent calibration can be obtained over the entire probability range (Fig. 6-3).

Another application of the predictive method is to define patients whose outcome is uncertain, for whom formal prospective trials of new methods of treatment would be most appropriate. If prerandomized stratification by a predictive model is carried out, a desired effect is likely to be demonstrated with a much smaller number of cases than when randomization is applied to all cases. This is because the predictive system would exclude patients whose outcome was already highly probable and therefore was unlikely to be affected by the new method.

Conclusion

The rational management of head injuries, both mild cases in emergency rooms and severe cases in intensive care units, should be based on estimates of the likelihood of complications or recovery. Computers as tools and statistics as a method have transformed prognosis from physicians' experience and intuition to a technology—in many ways, the key technology of medicine.

References

Barlow, P.; Teasdale, G. Prediction of outcome in the management of severe head injuries: the attitudes of neurosurgeons. Neurosurgery 19:989–991; 1986.

Braakman, R.; Gelpke, G. J.; Habbema, J. D. F.; et al. Systematic selection of prognostic features in patients with severe head injury. Neurosurgery 6:362–370; 1980.

Brocklehurst, G.; Gooding, M.; James, G. Comprehensive care of patients with head injuries. Br. Med. J. 294:345–347; 1987.

Bullock, R.; Teasdale, G. Management of intracranial haematomas. In: Vinken, P. J.; Bruyn, G. W.; Klowans, H. L. eds. Handbook of clinical neurology. Amsterdam; Elsevier; 1990.

Carlsson, C. A.; Von Essen, C.; Lofgren, J. Factors affecting the clinical course of patients with severe head injuries. J. Neurosurg. 29:242–251; 1968.

Choi, S. C.; Narayan, R. K.; Anderson, R. L.; et al. Enhanced specificity of prognosis in severe head injury. J. Neurosurg. 69:381–385; 1988.

Cooper, P. R. Role of emergency skull X-ray films in the evaluation of the head injured patient: a retrospective study. Neurosurgery 13:136–140; 1983.

Crandon, I.; Jennett, B.; Kay, M. Traumatic epilepsy after unoperated acute haematomas. Unpublished manuscript.

Gennarelli, T. A.; Spielman, G. M.; Langfitt, T. W.; et al. Influence of the type of intracranial lesion on outcome from severe head injury. J. Neurosurg 56:26–32; 1982.

Habbema, D. F.; Braakman, R.; Avezaat, C. J. J. Prognosis of the individual patient with severe head injury. Acta. Neurosurg. 28(suppl):158–160; 1979.

Heiskanan, O.; Sipponen, P. Prognosis of severe brain injury. Acta. Neurol. Scan. 46:343–348; 1970.

Jennett, B. Prognosis after severe head injury. Clin. Neurosurg. 19:200–207; 1972.

Jennett, B. Epilepsy after non-missile head injury. 2nd ed. London: Heinemann; 1975.

Jennett, B. Assessment of a technological package using a predictive tool. Int. J. Tech. Assess. Health Care 3:335–338; 1987.

Jennett, B. Assessment of clinical technologies; importance for provision and use. Int. J. Tech. Assess. Health Care 4:435–445; 1988.

Jennett, B.; Bond, M. R. Assessment of outcome after severe brain damage. Lancet 1:480–484; 1975.

Jennett, B.; Teasdale, G.; Braakman, R.; Minderhoud, J.; Knill-Jones, R. Predicting outcome in individual patients after severe head injury. Lancet 1:1031–1034; 1976.

Jennett, B.; Teasdale, G.; Braakman, R.; Minderhoud, J.; Heiden, J.; Kurze, T. Prognosis of patients with severe head injury. Neurosurgery 4:283–288; 1979.

Jennett, B.; Teasdale, G.; Fry, J.; Braakman, R.; Minderhoud, J.; Heiden, J.; Kurze, T. Treatment for severe head injury. J. Neurol. Neurosurg. Psychiat. 43:289–295; 1980.

Klauber, M. R.; Marshall, L. F.; Leurssen, T. G.; et al. Determinants of head injury mortality: importance of the low risk patient. Neurosurgery 24:31–36; 1989.

Luerssen, T. G.; Klauber, M. R.; Marshall, L. F. Outcome from head injury related to patient's age: a longitudinal prospective study of adult and paediatric head injury. J. Neurosurg. 68:409–416; 1988.

Masters, S. J.; McClean, P. M.; Arcarese, J. S.; et al. Skull X-ray examinations after head trauma. N. Engl. J. Med. 316:85–91; 1987.

Mendelow, A. D.; Teasdale, G.; Murray, G. Risks of intracranial haematoma in head injured adults. Br. Med. J. 287:1629–1631; 1983.

Murray, G. D. Use of an international data bank to compare outcome following severe head injury in different centres. Stats. in Med. 5:103–112; 1986.

Murray, G. D.; Murray, L. S.; Barlow, P.; Teasdale, G. M.; Jennett, B. Assessing the performance and clinical impact of a computerised prognostic system in severe head injury. Stats. in Med. 5:403–410; 1986.

Narayan, R. K.; Greenberg, R. P.; Miller, J. D.; et al. Improved confidence of outcome prediction in severe head injury. A comparative analysis of clinical examination, multimodality evoked potentials, CT scanning and intracranial pressure. J. Neurosurg. 54:751–762; 1981.

Overgaard, J.; Christensen, S.; Hvid-Jansen, O. Prognosis after head injury based on early clinical examination. Lancet 2:631–635; 1973.

Royal College of Radiologists. Patient selection for skull radiography in uncomplicated head injury. Lancet 1:115–118; 1983.

Stablein, D. M.; Miller, J. D.; Choi, S. C.; et al. Statistical methods for determining prognosis in severe head injury. Neurosurgery 6:243–248; 1980.

Stanley, J. M. The Appleton Consensus—suggested international guidelines for decisions to forego medical treatment. J. Med. Ethics 15:129–136; 1989.

Strang, I.; MacMillan, R.; Jennett, B. Head injuries in accident/emergency departments in Scottish hospitals. Injury 10:154–159; 1978.

Teasdale, G.; Galbraith, S.; Murray, L.; et al. Management of traumatic intracranial haematoma. Br. Med. J. 285:1695–1697; 1982a.

Teasdale, G.; Jennett, B. Assessment of coma and impaired consciousness: a practical scale. Lancet 2:81–84; 1974.

Teasdale, G. M.; Murray, G.; Anderson, E.; Mendelow, A. D.; Brookes, M. Risks of a traumatic intracranial haematoma in children and adults: implications for the management of head injuries. Br. Med. J. 300:363–367; 1990.

Teasdale, G.; Parker, L.; Murray, G.; et al. Predicting the outcome of individual patients in the first week after severe head injury. Acta. Neurochir. 28(suppl):161–164; 1979.

Teasdale, G.; Skene, A.; Parker, L.; Jennett, B. Age, severity and outcome of head injury. Acta. Neurochir. 28(suppl):140–143; 1979.

Teasdale, G.; Skene, A.; Spiegelhalter, D.; et al. Age, severity and outcome of head injury. In: Grossman, R. G.; Gildenberg, P. L. eds. Head injury: basic and clinical aspects. New York: Raven Press; 1982b.

Titterington, D. M.; Murray, G. D.; Murray, L. S.; et al. Comparison of discrimination techniques applied to a complex data set of head injured patients. J. R. Stat. Soc. (A) 144:145–175; 1981.

Todd, N. V.; Teasdale, G. M. Steroids in human head injury: clinical studies. In: Capildeo, R. ed. Steroids and disease of the central nervous system. London: John Wiley; 1989.

Vapalahti, M.; Troupp, H. Prognosis for patients with severe brain injuries. Br. Med. J. 3:404–407; 1971.

Wanzer, S. H.; Federman, D. D.; Adelstein, S. J.; Cassel, C. K.; Cassem, E. H.: Cranford, R. E.; Hook, E. W.; Lo, B.; Moertel, C. G.; Safar, P.; Stone, A.; van Eys, J. The physician's responsibility toward hopelessly ill patients. A second look. N. Engl. J. Med. 320:844–849; 1989.

Young, B.; Rapp, R. P.; Norton, J. A.; et al. Early prediction of outcome in head-injured patients. J. Neurosurg. 54:300–303; 1981.

7

The Post-Concussion Syndrome

RANDOLPH W. EVANS

The post-concussion syndrome refers to a large number of symptoms and signs that may occur alone or in combination usually following mild head trauma (Evans 1987; 1992). Loss of consciousness does not have to occur for the post-concussion syndrome to develop. The following symptoms and signs are associated with the syndrome: headaches, dizziness, vertigo, tinnitus, hearing loss, blurred vision, diplopia, convergence insufficiency, light and noise sensitivity, diminished taste and smell, irritability, anxiety, depression, personality change, fatigue, sleep disturbance, decreased libido, decreased appetite, memory dysfunction, impaired concentration and attention, slowing of reaction time, and slowing of information processing speed. Rare sequelae of mild head injury include subdural and epidural hematomas, seizures, transient global amnesia, tremor, and dystonia (Table 7-1). Headaches, dizziness, fatigue, irritability, anxiety, insomnia, loss of concentration and memory, and noise sensitivity are the most common complaints (Rutherford et al. 1977; Minderhoud 1980; Dikmen et al. 1986; Edna 1987).

Epidemiology

Head trauma is a cause of significant morbidity and mortality in all societies. The an-nual incidence of mild head injury per 100,000 population has been estimated to be 149 for Olmsted County, Minnesota (Annergers et al. 1980), 131 for San Diego County, California (Kraus et al. 1984), and 511 for Auckland, New Zealand (Wrightson 1989). Mild head injury accounts for 75% or more of all brain injuries (Krause and Nourjah 1988). In an industrialized country such as the United States, the relative causes of head trauma are approximately as follows: motor vehicle accidents, 45%; falls, 30%; occupational accidents, 10%; recreational accidents, 10%; and assaults, 5% (Jennett and Frankowski 1990). By comparison, in the People's Republic of China, traffic-related accidents accounted for 31.7% of head trauma cases, but most were bicycle accidents (Wang et al. 1986). About one-half of all patients with mild head injury are between the ages of 15 and 34. Motor vehicle accidents are more common in the young and falls more common in the elderly (Roy 1986).

Historical Aspects

The post-concussion syndrome has been controversial for over a century (Strauss and Savitsky 1934; Trimble 1981). One interesting historical case involved a 26-year-old-maid servant who had been hit over the head

Table 7-1. Sequelae Of Mild Head Injury

Headaches
Muscle contraction type
Migraine
Cluster
Occipital neuralgia
Supraorbital and infraorbital neuralgia
Secondary to neck injury
Secondary to temporomandibular joint syndrome
Due to scalp laceration or local trauma
Mixed

Cranial Nerve Symptoms and Signs
Dizziness
Vertigo
Tinnitus
Hearing loss
Blurred vision
Diplopia
Convergence insufficiency
Light and noise sensitivity
Diminished taste and smell

Psychological and Somatic Complaints
Irritability
Anxiety
Depression
Personality change
Fatigue
Sleep disturbance
Decreased libido
Decreased appetite

Cognitive Impairment
Memory dysfunction
Impaired concentration and attention
Slowing of reaction time
Slowing of information processing speed

Rare Sequelae
Subdural and epidural hematomas
Seizures
Transient global amnesia
Tremor
Dystonia

with a stick and reported symptoms of retrograde amnesia. Six months later she still complained of headaches, dizziness, tinnitus, and tiredness. In 1694, a judge requested a specialists' report of the Swiss physician, J. J. Wepfer and two other surgeons who stated, "We can't say anything definite, but it is certain that there has been a grave contusion of the head and that this will leave its mark in the form of an impediment." (DeMorsier 1943).

Rigler raised the issue of "post-compensation neurosis" in 1879. He described the increased incidence of post-traumatic invalidism after a system for financial compensation was established for accidental injuries on the Prussian railways in 1871. The London surgeon John Erichsen (1882) took an opposing viewpoint. He argued that minor injuries to the head and spine could result in severe disability due to "molecular disarrangement" and/or anemia of the spinal cord.

In more recent times, Henry Miller (1961) summarized the viewpoint of those who believe that the post-concussion syndrome is really a compensation neurosis: "The most consistent clinical feature is the subject's unshakable conviction of unfitness for work. . . ." Sir Charles Symonds (1962) took an equally strong opposing viewpoint when he wrote, "It is, I think, questionable whether the effects of concussion, however slight, are ever completely reversible."

Evidence for Organicity

Many physicians, lay persons, defense attorneys, and agents of insurance companies have serious doubts about the existence of the post-concussion syndrome (McMordie 1988). A group of Canadian college students did not find the symptoms of post-concussion syndrome credible (Aubrey 1989). Among their doubts were that the patients had sustained only minor head injuries, they appeared normal, and they had multiple vague complaints. The group suspected that they wanted to get out of school or work responsibilities or enhance a compensation claim. Other than not being familiar with the literature, another explanation for these doubts is the Hollywood head-injury myth, which Robertson (1988) has called the "Three Stooges model."

The Hollywood Head Injury Myth

Most people's knowledge of the sequelae of mild head injuries is largely the product of movie magic. Some of the funniest scenes in slapstick comedies and cartoons depict the character sustaining single or multiple head injuries, looking dazed, and then recovering immediately. In cowboy movies, detective and action stories, and boxing and Kung Fu films, seemingly serious head trauma is often inflicted by blows from guns and heavy objects, falls, motor vehicle injuries, fists, and kicks, all without lasting sequelae. Our experience is miniscule compared to the thousands of simulated head injuries witnessed in the movies and on television. Because of this compelling mythology, the physician has a difficult job educating patients, their families, and others in the realities of mild head injuries.

In addition to summarizing the literature on mild head injury, the physician can counter the myths by the use of vivid examples from boxing. The public is quite familiar with the punch-drunk syndrome of cumulative head trauma in boxers (Martland 1928). The examples of two successful boxers, Joe Louis and Muhammad Ali, are well-known. Many have witnessed powerful punches resulting in dazed, disoriented boxers or knockouts. Considering the common incidence of head injuries, the example of family members, friends, or acquaintances with sequelae from mild head injuries sometimes can be used as well.

In 1835 Gama wrote, "Fibers as delicate as those of which the organ of mind is composed are liable to break as a result of violence to the head." Over the last 30 years, a growing body of evidence strongly supports the organic theory. Diffuse axonal injury has been documented with mild head injury in neuropathological studies of humans and animals (Oppenheimer 1968; Blumbergs et al. 1989; Povlishock and Coburn 1989). Neuroimaging studies, particularly magnetic resonance imaging (MRI) scans of the brain, have demonstrated clinically occult contusions and white matter changes (Hesselink et al. 1988; Yokota et al. 1991). Although not prognostically helpful, in some cases auditory brain stem responses have been reported as abnormal in mild head injury

(Schoenhuber et al, 1988). Neuropsychological studies have revealed deficits in information processing, memory, reaction time, and attention (Levin 1989; 1992). Prognostic studies have shown the persistence of symptoms over time as compared to controls.

Variables in Prognostic Studies

Although many prognostic studies have been reported during the last 60 years, comparison among the studies is difficult. There are significant differences including the definition of mild head injury, use of testing, study design, and subject variables (Table 7-2) (Levin et al. 1987a).

Table 7-2. Variables in Prognostic Studies

Definition of Mild Head Injury
Loss of consciousness and if so, duration
Duration, if present, of post-traumatic amnesia
Glasgow Coma Scale score
Inclusion of skull fractures/or cerebral contusions?

Use of Testing
Radiological
Neurophysiological
Neuropsychological

Study Design
Prospective vs. retrospective
Length of follow-up
Spontaneous volunteering of symptoms vs. responding to a checklist
Face to face interview vs. a mailed questionnaire
Use of matched controls
Number of subjects
Symptoms assessed

Subject Variables
Cause of head injury
Hospital vs. outpatient presentation
Geographic and cultural differences
Age and sex
Socioeconomic and educational level
Pre-morbid personality and psychopathology
Prior head trauma
Use of alcohol and drugs
Multiple trauma
Pending or completed litigation
Attrition rate of subjects

Definition of Mild Head Injury

Although the terms mild and minor head injury are often used interchangeably, "mild head injury" is preferred in delineating the continuum of mild, moderate, and severe. Mild head injury is nonuniformly defined in different studies by different criteria including: loss of consciousness, and if present, duration of the loss; duration of post-traumatic amnesia, if present; and a Glasgow Coma Scale of 13 to 15. In some studies, patients with skull fractures and/or cerebral contusions are included. Strict criteria used in recent studies (Dikmen et al. 1986; Levin et al. 1987b) are crucial to ensure studying similar types of injuries without confounding variables. For future studies, I would propose the following criteria for mild head injury: loss of consciousness does not have to occur; if it does occur, a duration of 30 minutes or less; an initial Glasgow Coma Scale rating of 13 to 15 without subsequent deterioration; and absence of focal neurological deficits without evidence of depressed skull fracture, intracranial hematomas, or other neurosurgical disease.

The persistence and severity of symptoms and neuropsychological deficits are not predicted by a loss of consciousness of less than one hour as compared to a patient being just dazed (Denker 1944; Leininger et al. 1990). The consideration of post-traumatic amnesia is problematic as a criteria for mild head injury since the duration is frequently not reliably reported by the patient or observers. In addition, the duration of post-traumatic amnesia has been variably reported as being predictive (Minderhoud 1980) and not predictive (Edna 1987) of post-concussion sequelae. Wrightson (1981) found no correlation between duration of post-traumatic amnesia and time off work following the injury.

Testing Variables

Imaging studies, as well as neurophysiological and neuropsychological testing have been used to variable degrees in prognostic studies. EEG studies have frequently been used to assess patients with mild head injury. Since there is a significant incidence of premorbidly abnormal electroencephalograms (EEGs) (Lorenzoni 1970), an abnormal EEG cannot be stated unequivocally to be caused by mild head injury without a pre-injury study to compare. For the specific patient, EEG results do not have predictive value. Dow, Ulett, and Raaf (1944) found that an EEG taken immediately following head injury was borderline or abnormal in 43% of patients compared to 38% of controls. An abnormal EEG study was less predictive than clinical judgment in predicting time lost from work. However, for certain population groups, there may be some predictive information. For example, an increased incidence of abnormal EEGs has been found in former soccer players with chronic post-concussion symptoms (Tysvaer et al. 1989). A significant correlation has also been reported between EEG abnormalities and the number of bouts fought by ex-boxers (Ross et al. 1983). Although auditory brain stem responses have been reported as being abnormal in mild head injury, there is no correlation between abnormal results and postconcussion symptoms (Schoenhuber et al. 1988).

MRI of the brain is more sensitive than CAT scan in evaluating mild head injury. Levin, Amparo et al. (1987), in a comparison study between MRI and computerized axial tomography (CAT) scan in evaluating mild to moderate head injury, found that MRI detected lesions in 85% of patients which were not detected by CAT scan. Most of the lesions detected were in the frontal and temporal regions. Lesions present on the MRI scan had prognostic value for deficits of frontal lobe functioning and memory. Follow-up scans at 1 and 3 months showed marked reduction of lesion size with improvement in cognition and memory.

Study Design Variables

Study designs also vary. Some studies have prospective designs (Rutherford et al. 1977, 1978; Cartlidge 1977; Rimel et al. 1981; Dikmen et al. 1986; Edna 1987; Levin et al.

1987b), while others are retrospective (Denker 1944; Denny-Brown 1945). Retrospective studies may have incomplete baseline data and have varying lengths of follow-up for individual subjects. In studies of both types the length of follow-up varies from weeks to months to years. Techniques for assessing the presence of symptoms and signs differ. Some studies have used spontaneous volunteering of symptoms, where others have had subjects respond to a checklist. Directed interviews and physical exams have been used, although other studies have obtained follow-up only by a questionnaire sent through the mail. More recent studies have used controls matched by sex, education, geographic location, and socioeconomic level. This is particularly important since so many of the symptoms of post-concussion syndrome are frequent in the general population. The symptoms assessed vary as well. Finally, the number of subjects used is a notable variable.

Subject Variables

Multiple subject variables influence the usefulness of the prognostic studies. The cause is important because of different mechanisms of head injury and the circumstances which may influence the outcome. For example, compare a motor vehicle accident involving an inertial-loading injury where another person is at fault with a contact phenomena sports injury. Subjects who are hospitalized might have different outcomes than patients seen in outpatient clinics and sent home with milder injuries. The cultural background and geographical location of subjects may be significant.

Sir Charles Symonds observed, "The symptom picture depends, not only upon the kind of injury, but upon the kind of brain." (Symonds 1937). Preexisting psychopathology and premorbid personality are certainly important (Keshaven et al. 1981). The socioeconomic status, educational level, age, and sex of subjects are critical variables. Rimel et al. (1981) reported that significant predictors for return to work by

three months included: older age; higher level of education, employment, and socioeconomic status; and greater income. By three months, 100% of executives and business managers had returned to work compared to 68% of skilled laborers and 57% of unskilled laborers.

Although the degree of initial impairment of information processing speed is not related to intelligence, high-IQ patients recover faster than low-IQ patients (Gronwall 1976). This may illustrate the greater motivation of high achievers. Older age (over 40 years) has been reported as a risk factor for increased duration and number of post-concussion symptoms (Denker 1944; Hernesniemi 1979; Edna 1987), and slower recovery of cognitive deficits (Barth et al. 1983; Gronwall 1989). The gender of the patient is an important variable since late symptoms occur more often in women (Rutherford 1977, 1978; Edna 1987).

A history of prior head injuries or prior use of alcohol and illicit drugs compounds the effects of even mild head injury. Prior head injury is a risk factor for persistence and number of post-concussion symptoms (Gromwall and Wrightson 1975; Carlsson 1987) and is consistent with the neuropathological concept of cumulative diffuse axonal injuries and contusions. A history of alcohol abuse may increase the number of post-traumatic sequelae and may also be related to additional slowing of reaction time (Carlsson 1987). Alcohol intoxication makes the initial assessment of patients with head injury more difficult (Brismar et al. 1983) and is also a risk factor for neurosurgical sequelae.

Multiple trauma can cause additional functional impairment, depression, anxiety, and stress (Carlsson, 1987; Berrol, 1989). Multiple trauma with associated orthopedic/soft tissue injuries also contributes to the persistence and frequency of post-concussion symptoms (Dikmen, et al, 1986).

Litigation that is pending or completed is a significant variable and is discussed in a subsequent section. The attribution of subjects can skew results. Patients without persisting symptoms may not wish to participate in serial evaluations.

Prognosis of Post-Concussion Symptoms

Because of the many variables in prognostic studies, the percentage of patients with symptoms following mild head injury varies often (Table 7-3). The percentage of patients with headaches at 1 month varies from 31.3% (Minderhoud et al. 1980) to 90% (Denker 1944) and at 3 months from 47% (Levin et al. 1987b) to 78% (Rimel et al. 1981). Edna (1987) found that 24% of the patients had persisting headaches at 4 years. The frequency of dizziness also varies considerably, ranging from 19% (Cartlidge 1978) to 53% (Levin et al. 1987b) at 1 week. At 2 years, 18% of patients report continuing dizziness (Denker 1944; Cartlidge 1978) and 18% at 4 years (Edna 1987). Memory problems are reported in 18.8% at 1 month (Minderhoud et al. 1980), 59% at 3 months (Rimel et al. 1981), 15.3% at 6 months (Minderhoud 1980), and 19% at 4 years (Edna 1987). Irritability is reported in 24.7% of patients at 1 month (Minderhoud 1980) and 5.3% at 1 year (Rutherford 1978).

Since the symptoms of post-concussion syndrome are so common in the general population, comparison to controls is very important. Dikmen, McLean, and Tempkin (1986) studied 20 consecutive patients hospitalized with mild head injury and compared them to carefully matched controls. At 1 month postinjury, the following symptoms were endorsed by subjects and controls, respectively: headaches 51%, 38%; memory difficulties 52%; 6%; dizziness 41%; 11%; fatigue 68%, 41%; noise sensitivity 52%, 10%; light sensitivity 32%, 28%; difficulty concentrating 42%, 21%; and irritability 68%, 42%. These results demonstrate that although these nonspecific symptoms are common in controls, they are still more frequent in patients with mild head injury.

Neuropsychological Deficits

Following mild head injury, deficits in cognitive functioning have been described including a reduction in information processing speed, attention, reaction time, and memory for new information. For information processing speed, recovery is seen within 3 months in most patients (Gronwall and Wrightson 1974; Levin et al. 1987b; Hugenholtz et al. 1988). Memory for new information also recovers within 1 to 3 months (Dikmen et al. 1986; Levin et al. 1987b), although persisting impairment in visual memory and performance of digit span has been

Table 7-3. Percentage of Patients With Persistence Of Symptoms After Mild Head Injury

	One Week	One Month	Six Weeks	Two Months	Three Months	Six Months	One Year	Two Years	Three Years	Four Years	Five Years
Headache	71[i] 36[e]	90[a] 31.3[f] 56[i]	24.8[d]	31.5[b]	78[g] 47[i]	21.6[f] 27[e]	35[a] 8.4[d] 18[e]	22[a] 24[e]	20[a]	24[h]	12[c] (Headaches and/or Dizziness)
Dizziness	53[i] 19[e]	12[a] 21.9[f] 35[i]	14.5[d]	23[b]	22[i]	13.1[f] 22[e] 14[e]	26[a] 4.6[d]	18[a] 18[e]	16[a]	18[h]	
Memory Problems		18.8[f]	8.3[d]		59[g]	15.3[f]	3.8[d]			19[h]	
Irritability		24.7[f]	9[d]			19.6[f]	5.3[d]				

[a] Denker (1944).
[b] Dennÿ-Brown (1945).
[c] Steadman and Graham (1969).
[d] Rutherford et al. (1978).
[e] Cartlidge (1977).
[f] Minderhoud et al. (1980).
[g] Rimel et al. (1981).
[h] Edna (1987).
[i] Levin et al. (1987).

noted (Levin et al. 1987b). Reaction time has been reported to be abnormal between 6 weeks (MacFlynn 1984) and 3 months (Hugenholtz et al. 1988) with recovery occurring within about 6 months (MacFlynn 1984). Attention deficits show persisting impairment at 3 months (Levin et al. 1987b; Gentilini et al. 1989). Dikmen et al. (1986) found recovery of cognitive impairment by 1 year compared to controls, although a minority had ongoing impairment preventing return to normal activities.

Resolution of cognitive impairment on testing does not necessarily imply resolution of subjective symptoms. In the Levin et al. study (1987b), although testing showed nearly complete neuropsychological recovery at 3 months postinjury, headaches were still present in 47% of subjects, dizziness in 22%, and decreased energy in 22%.

Even in patients with complete resolution of cognitive deficits and subjective symptoms, there may be residual brain impairment. Ewing et al. (1980) compared university students who made a full recovery from mild head injury 1 to 3 years previously to matched controls. The two groups were tested at ground level and at a simulated altitude of 12 500 feet. Although both groups performed similarly at ground level, the mild head injury group showed significant impairment in tests of memory and vigilance performed with the mild hypoxia. This study suggests that patients considered completely recovered may demonstrate impairment with environmental stress. Other physical and psychosocial stressors may cause similar impairment. As noted previously, the effects of mild head injury even in persons fully recovered are cumulative, which also suggests permanent brain damage.

Although most patients with mild head injury do not have neuropsychological sequelae after 3 months, a significant subgroup exists with persistent post-concussion difficulties and neuropsychological deficits. Leininger et al. (1990) compared 53 patients with mild head injury who had persisting symptoms after 1 or more months to matched controls. Exclusion criteria included prior history of significant head trauma, substance abuse, or low academic achievement. Thirty-two percent of the patients were dazed without loss of consciousness and 58% were unconscious for 20 minutes or less. Testing was performed from 1 to 22 months following injury. Tests of reasoning, information processing, and verbal learning revealed the most deficits. Controls performed better than the patients on reproduction of a complex geometric design. The loss of consciousness, per se, compared to being dazed, was not predictive for neuropsychological sequelae. Test results were similar in those patients assessed within 3 months of the injury as compared to the others tested after 3 months.

The Effect of Litigation

Litigation as a cause of the post-concussion syndrome has been a topic of debate during the last century. As a matter of routine, defense attorneys still cite Miller's work (1961) and invariably raise questions of secondary gain and malingering. However, studies of the last 20 years demonstrate that secondary gain and malingering certainly exist, but are usually minor elements of the overall picture.

Compensation Neurosis and Malingering

Miller (1961) reported on 200 consecutive cases seen for medico-legal examination in Newcastle-upon-Tyne, England with an average interval between the head injury and the first examination of 14 months. Forty-seven of the 200 patients were perceived to have gross and unequivocally psychoneurotic complaints. The patients exhibited characteristic behaviors during the consultation. They frequently arrived late and were accompanied by a family member who took an active part in the interview process. The patients displayed an attitude of "martyred gloom" and were very defensive. An obvious dramatization of symptoms was reported to be present in more than one half of the patients such as avoiding the opthalmoscope, performing grip testing inconsistently, or slumping forward while holding their face in their hands. Miller (1961) reports: "The most consistent feature is the subject's unshakable conviction of unfitness for work, a conviction quite unrelated to overt disability, even if his symptomatology is accepted at its face value. At a later stage, the patient will declare his fitness for light work, which is often not available. . . . Another cardinal feature is an absolute refusal to admit any degree of symptomatic improvement."

Miller's study can be criticized for several reasons. He describes a biased sample; the patients were referred specifically for medico-legal consultation, Although his behavioral observations may have some validity, they are by necessity quite subjective and judgmental. Miller dismisses symptoms as being minor but does not give information on the percentage of patients with various complaints such as headaches, dizziness, and memory problems at the time of the examination. In the same paper Miller states, "The consistency of the post-concussional syndrome of headache, postural dizziness, irritability, failure of concentration, and intolerance of noise, argues a structural or at least a pathophysiological basis." Miller's study has stimulated many other investigators to further explore issues of compensation.

Guthkelch (1980) in Pittsburgh, Pennsylvania, reported on 398 consecutive head injury patients he examined in connection with a claim for compensation. All were employed at the time of the injury. Accident neurosis was often defined by bizarre and inconsistent complaints, exaggeration of length of initial unconsciousness, and attention-seeking behaviors. About one half of the patients returned to work but left work within a few days complaining of headaches and noise intolerance. About one half of the patients did not return to work until their compensation claim had been settled or they were turned down for disability. Accident neurosis was more common in manual workers sustaining accidents at work than in nonmanual workers. Psychiatric treatment was not found to be helpful. Guthkelch concluded, "Accident neurosis is not particularly common; even in this series, which

was exclusively composed of patients with a compensation problem, it was identified in only 6.8% of patients."

Are Litigants Different from Nonlitigants?

Patients with litigation are quite similar to those without litigation. Both groups have similar symptoms improving with time (Merskey and Woodford 1972; McKinlay et al. 1983; Leininger et al. 1990) and similar cognitive test results (McKinlay et al. 1983; Leininger et al. 1990). Both groups also respond similarly to appropriate treatment as suggested by a study on post-traumatic migraine (Weiss et al. 1991).

The end of litigation does not mean the end of symptoms or return to work for many claimants. Fee and Rutherford (1988) reported that 34% of patients were still symptomatic 1 year after settlement of claims. Many litigating patients, particularly those who are older or employed in more dangerous occupations, do not return to work after settlement (Kelly and Smith 1981).

Pending litigation may increase the level of stress for some claimants and may result in an increased frequency of symptoms after settlement (Mendelson 1982; Fee and Rutherford 1988). Skepticism of many physicians about persisting symptoms may accentuate this level of stress.

Miller (1961) and Guthkelch (1980) make several excellent observations about patients who exaggerate or malinger with which clinicians can readily agree. Certainly physicians should consider a patient's motivation when litigation is involved. Patients with premorbid neuroticism or psychosocial problems may exaggerate symptoms. In a medico-legal setting, subjective criteria for making a diagnosis of post-concussion syndrome should be clearly delineated from objective findings. However, since some patients with seemingly hysterical signs and symptoms may actually have underlying organic disease (Gould et al. 1986), the diagnosis of accident neurosis, malingering, or conversion neurosis should be made with a great deal of caution.

In 1934, Strauss and Savitsky came to the following conclusion which is still warranted: "The harshness, injustice, and brutal disregard of complaints shown by the physicians and representatives of insurance companies and their ready (assumptions) of intent to swindle do not foster wholesome patterns of reaction in injured persons. The frequent expression of unjustifiable skepticism on the part of examiners engenders resentment, discouragement, and hopelessness and too often forces these people to resort to more primitive modes of response (hysterical)."

Summary

The post-concussion syndrome refers to a large number of symptoms and signs that may occur alone or in combination following usually mild head trauma. Headaches, dizziness, fatigue, irritability, anxiety, insomnia, loss of concentration and memory, and noise sensitivity are the most common complaints. Mild head injury and the post-concussion syndrome are major public health concerns since mild head injury has an annual incidence of about 150 per 100 000 population and accounts for 75% or more of all head injuries.

Many physicians, lay persons, and interested third parties have serious doubts about the existence of the post-concussion syndrome. An important contributor to these doubts has been the frequent portrayal of head injury in motion pictures and on television as innocuous without sequelae or as humorous; as with the Hollywood head injury myth. Educational efforts can include: examples of such media misinformation; familiar counterexamples from boxing, such as technical knockouts, knockouts, and punch-drunk syndrome; and summarizing the evidence demonstrating organicity.

A growing body of evidence documents the organicity of the post-concussion syndrome. Abnormalities have been reported in neuropathological, neuroimaging, neuropsychological, and neurophysiological studies. Despite differences in definition of mild head injury, use of testing, study design, selection of subjects, and use of controls, prognostic studies clearly substantiate the existence of a post-concussion syndrome and the persistence of symptoms over time as compared to controls. For future studies, the use of a standardized definition of mild head injury such as the one proposed earlier in the chapter is critical to ensure studying similar types of injuries without confounding variables.

The symptoms and signs of the post-concussion syndrome are common and resolve in most patients by 3 to 6 months after the injury. However, a distinct minority of patients may have persisting symptoms and cognitive deficits for additional months or years.

Risk factors for persisting sequelae include: age over 40 years; lower educational, intellectual, and socioeconomic level; female gender; alcohol abuse; prior head injury; and multiple trauma. Duration of post-traumatic amnesia, EEG, and auditory brain stem response studies are not predictive of sequelae. Lesions present on MRI scans of the brain do have predictive value for cognitive impairment.

Compensation neurosis or malingering in patients with compensation claims is quite uncommon. Patients with persisting symptoms frequently seek redress because they do have persisting symptoms. Physicians should educate patients, their families, and other interested parties at all stages following injury. As our knowledge increases, hopefully treatments will be discovered which favorably improve outcome.

Acknowledgments

The suggestions of Richard I. Evans, PhD and Harvey S. Levin, PhD for manuscript revision are appreciated.

References

Annegers, J. F.; Grabow, J. D.; Kurland, L. T.; Laws, E.R. The incidence, causes, and secular trends of head trauma in Olmsted County, Minnesota, 1935–1974. Neurology 30:912–919; 1980.

Aubrey, J.; Dobbs, A. R.; Rule, B. G. Laypersons' knowledge about the sequelae of minor head injury and whiplash. J. Neurol. Neurosurg. Psychiatry 52:842–846; 1989.

Barth, J. T.; Macciocchi, S. N.; Giordani, B.; Rimel, R.; Jane, J.A.; Boll, T.J. Neuropsychological sequelae of minor head injury. Neurosurgery 13:529–533; 1983.

Berrol, S. Other factors: age, alcohol, and multiple injuries. In: Hoff, J. T.; Anderson, T. E.; Cole, T. M., eds. Mild to moderate head injury. Boston: Blackwell; 1989: p. 135–142.

Blumbergs, P. C.; Jones, N. R.; North, J. B. Diffuse axonal injury in head trauma. J. Neurol. Neurosurg. Psychiatry 52:838–841; 1989.

Brismar, B.; Engstrom, A.; Rydberg, U. Head injury and intoxication: a diagnostic and therapeutic dilemma. Acta. Chir. Scand. 149:11–14; 1983.

Carlsson, G. S.; Svardsudd, K.; Welin, L. Long-term effects of head injuries sustained during life in three male populations. J. Neurosurg. 67:197–205; 1987.

Cartlidge, N. E. F. Postconcussional syndrome. Scottish Med. J. 23:103; 1977.

de Morsier, G. Les encephalopathies traumatiques. Etude neurologique. Schweiz. Arch. Neurol. Neurochir. Psychiat. 50:161; 1943.

Denker, P. G. The postconcussion syndrome: prognosis and evaluation of the organic factors. N.Y. State J. Med. 44:379–384; 1944.

Denny-Brown, D. Disability arising from closed head injury. J.A.M.A. 127:429–436; 1945.

Dikmen, S.; McLean, A.; Temkin, N. Neuropsychological and psychosocial consequences of minor head injury. J. Neurol. Neurosurg. Psychiatry 49:1227–1232, 1986.

Dow, R. S.; Ulett, G.; Raak, J. Electroencephalographic studies immediately following head injury. Am. J. Psychiatry 101:174–183; 1944.

Edna, T.-H.; Cappelen, J. Late postconcussional symptoms in traumatic head injury. An analysis of frequency and risk factors. Acta. Neurochir. (Wien) 86:12–17; 1987.

Edna, T.-H. Disability 3-5 years after minor head injury. J. Oslo City Hosp 37:41–48; 1987.

Erichsen, J. E. On concussion of the spine: nervous shock and other obscure injuries of the nervous system in their clinical and medicolegal aspects. London: Longmans Green; 1882.

Evans, R. W. Postconcussive syndrome: an overview. Texas Med. 83:49–53; 1987.

Evans, R. W. The postconcussion syndrome and the sequelae of mild head trauma. Neurologic Clinics 10:4, 1992 (in press).

Ewing, R.; McCarthy, D.; Gronwall, D.; Wrightson, P. Persisting effects of minor head injury observable during hypoxic stress. J. Clin. Neuropsychol. 2:147–155; 1980.

Fee, C. R. A.; Rutherford, W. H. A study of the effect of legal settlement on postconcussion symptoms. Arch. Emergency Med. 5:12–17; 1988.

Gama, J. H. P. Traite des plaies de tete et de l'encephalite. Paris, 1835.

Gentilni, M.; Nichelli, P.; Schoenhuber, R. Assessment of attention in mild head injury. In: Levin, H. S.; Eisenberg, H. M.; Benton, A. L. eds. Mild head injury. New York: Oxford University Press; 1989: p. 163–175.

Gould, R.; Miller, B. L.; Goldberg, M. A.; Benson, D.F. The validity of hysterical signs and symptoms. J. Nervous Mental Disease 174:593–597; 1986.

Gronwall, D. Concussion: does intelligence help? NZ Psychologist 5:72–78; 1976.

Gronwall, D. Cumulative and persisting effects of concussion on attention and cognition. In: Levin, H. S.; Eisenberg, H. M.; Benton, A. L. eds. Mild head injury. New York: Oxford University Press; 1989: p. 153–162.

Gronwall, D., Wrightson, P. Delayed recovery of intellectual function after minor head injury. Lancet. 2:605–609; 1974.

Gronwall, D.; Wrightson, P. Cumulative effects of concussion. Lancet 2:995–997; 1975.

Guthkelch, A. N. Postraumatic amnesia, postconcussional symptoms, and accident neurosis. Eur. Nerol. 19:91–102; 1980.

Hernesniemi, J. Outcome following head injuries in the aged. Acta Neurochirurgica 49:67–79; 1979.

Hesselink, J. R.: Dowd, C. F.: Healy, M. E.; et al. MR imaging of brain contusions: a comparative study with CT. A.J.R. 150:1133–1142; 1988.

Hugenholtz, H.; Stuss, D. T.; Stethem, L. L.; et al. How long does it take to recover from a mild concussion? Neurosurgery 22:853–858; 1988.

Jennett, B.; Frankowski, R. F. The epidemiology of head injury. In: Brinkman, R. ed. Handbook of Clinical Neurology. Vol. 13. New York: Elsevier, 1990: p. 1–16.

Kelly, R.; Smith, B. N. Posttraumatic syndrome: another myth discredited. J. R. Soc. Med. 74:275–277; 1981.

Kraus, J. F.; Black, M. A.; Hessol, N. The incidence of acute brain injury and serious impairment in a defined population. Am. J. Epidemiol. 119:186–201; 1984.

Kraus, J. F.; Nourjah, M. S. The epidemiology of mild, uncomplicated brain injury. J. Trauma. 28:1637–1643; 1988.

Leininger, B.E.; Gramling, S.E.; Farrell, A. D.; Kreutzer, J. S.; Peck, E. A. Neuropsychological deficits in symptomatic minor head injury patients after concussion and mild concussion. J. Neurol. Neurosurg. Psychiatry. 53:293–296; 1990.

Levin, H. S.; Amparo, E. G.; Eisenberg, H. M.; Williams, D. H.; High, W. M.; McArdle, C. B.; Weiner, R. L. Magnetic resonance imaging and computerized tomography in relation to the neurobehavioral sequelae of mild and moderate head injuries. J. Neurosurg. 66:706–713; 1987.

Levin, H. S.; Gary, H. E.; High, W. M.; et al. Minor head injury and the postconcussional syndrome: methodological issues in outcome studies. In: Levin, H. S.; Grafman, J.; Eisenberg, H. M. eds. Neurobehavioral recovery from head injury. New York: Oxford University Press; 1987a: p. 262–275.

Levin, H. S.; Mattis, S.; Ruff, R. M.; et al. Neurobehavioral outcome following minor head injury: a three-center study. J. Neurosurg. 66:234–243; 1987b.

Levin, H. S. Neurobehavioral outcome of mild to moderate head injury. In: Hoff, J. T.; Anderson, T. E.; Cole, T. M. eds. Mild to moderate head injury. Boston: Blackwell; 1989: p. 153–185.

Levin, H. S. Neurobehavioral sequelae of mild head injury. Neurologic Clinics 10:4;1992 (in press).

Lorenzoni, E. Electroencephalographic studies before and after head injuries. Electroencephalogr. Clin. Neurophysiol. 28:216, 1970.

MacFlynn, G.; Montgomery, E. A.; Fenton, G. W.; et al. Measurement of reaction time following minor head injury. J. Neurol. Neurosurg. Psychiatry 47:1326–1331; 1984.

Martland, H. S. Punch-drunk. J.A.M.A. 19:1103–1107; 1928.

McKinlay, W. W.; Brooks, D. N.; Bond, M. R. Postconcussional symptoms, financial compensation, and outcome of severe blunt head injury. J. Neurol. Neurosurg. Psychiatry 46:1084–1091; 1983.

McMordie, W. R. Twenty-year follow-up of the prevailing opinion on the posttraumatic of postconcussional syndrome. Clin. Neuropsych. 2:198–212; 1988.

Mendelson, G. Not "cured by a verdict:" effect of legal settlement on compensation claimants. Med. J. Aust. 2:132–134; 1982.

Merskey, H.; Woodforde, J. M. Psychiatric sequelae of minor head injury. Brain. 95:521–528; 1972.

Miller, H. Accident neurosis. B.R. Med. J. 1:919–925; 992–998; 1961.

Minderhoud, J. M.; Boelens, M. E. M.; Huizenga, J.; Saan, R. J. Treatment of minor head injuries. Clin. Neurol. Neurosurg. 82:127–140; 1980.

Oppenheimer, D. R. Microscopic lesions in the brain following head injury. J. Neurol. Neurosurg. Psychiatry 31:299–306; 1968.

Povlischock, J. T.; Coburn, T. H. Morphopathological change associated with mild head injury. In: Levin, H. S.; Eisenberg, H. M.; Benton, A. L. eds. Mild head injury. New York: Oxford University Press; 1989: p. 37–53.

Rigler, J. Veber die Verletzungen auf Eisen-

bahnen Insbesondere der Verletzungen des Rueckenmarks. Berlin: Reimer, 1879.

Rimel, R. W.; Giordani, B.; Barth, J. T., et al. Disability caused by minor head injury. Neurosurgery 9:221–228; 1981.

Robertson, A. The postconcussional syndrome then and now. Australian and N.Z. J. Psychiatry 22:396–402; 1988.

Roy, C. W.; Pentland, B.; Miller, J. D. The causes and consequences of minor head injury in the elderly. Injury 17:220–223; 1986.

Rutherford, W. H.; Merrett, J. D.; McDonald, J. R. Sequelae of concussion caused by minor head injuries. Lancet 1:1–4; 1977.

Rutherford, W. H.; Merrett, J. D.; McDonald, J. R. Symptoms at 1 year following concussion from minor head injuries. Injury 10:225–230; 1978.

Schoenhuber, R.; Gentilini, M.; Orlando, A. Prognostic value of auditory brain stem responses for late postconcussion symptoms following minor head injury. J. Neurosurg. 68:742–744; 1988.

Steadman, J. H.; Graham, J. G. Rehabilitation of the brain-injured. Prog. Royal Soc. Med. 63:23–28; 1969.

Strauss, I.; Savitsky, N. Head injury: neurologic and psychiatric aspects. Arch. Neurol. Psychiatry 31:893–955; 1934.

Symonds, C. The assessment of symptoms following head injury. Guys Hospital Gazette 51:464; 1937.

Symonds, C. Concussion and its sequelae. Lancet 1:1–5; 1962.

Trimble, M. Post-traumatic Neurosis: From Railway Spine to the Whiplash. Chichester: Wiley, 1981.

Tysvaer, A. T.; Storli, O. V.; Bachen, N. I. Soccer injuries to the brain. A neurologic and electroencephalographic study of former players. Acta. Neurol. Scand. 80:151–156; 1989.

Wang, C., Schoenberg, B. S.; Li Sc, Yang, Y.; Cheng, X.; Bolis, C.L. Brain injury due to head trauma. Epidemiology in urban areas of the People's Republic of China. Arch. Neurol. 43:570–572; 1986.

Weiss, H. D.; Stern, B. J. Goldberg, J. Posttraumatic migraine: chronic migraine precipitated by minor head or neck trauma. Headache 31:451–456; 1991.

Wrightson, P. Management of disability and rehabilitation services after mild head injury. In: Levin, H. S.; Eisenberg, H. M.; Benton, A. L. eds. Mild head injury. New York: Oxford University Press; 1989; p. 245–256.

Wrightson, P.; Gronwall, D. Time off work and symptoms after minor head injury. Injury 12:445–454; 1981.

Yokota, H.; Kurokawa, A.; Otsuka, T.; et al. Significance of magnetic resonance imaging in acute head injury. J. Trauma 31:351–357; 1991.

8

Spinal Cord Injury

WILLIAM H. DONOVAN

Etiology

Spinal cord injury is a malady that affects approximately 7000–10 000 individuals per year in the United States. Its incidence also is quoted as between 20 and 40 per million people. It affects men four times more than women and although it occurs in all age groups, the mean age is 29, the median age 25 and the mode age 19. Etiologies of this condition vary from country to country but in the developed countries, the distribution is as follows: motor vehicle (48%), falls (21%), sports (14%), acts of violence (15%), and other (3%) (Stover and Fine 1986).

Pathophysiology

The presenting findings are anesthesia or hypesthesia and paralysis or paresis below the level of the spinal cord injury. The neurological level of injury can be determined by physical examination; the skeletal level of injury is determined by radiographic examination. Magnetic resonance imaging (MRI) also has proven useful in recent years by providing a visual image of the nature of the injury to the spinal cord as well as the surrounding soft tissues.

Whether an individual will recover any or all neurological function after a spinal cord injury basically depends on three factors: 1) the magnitude of force impacted on the spinal cord and the resultant destruction of its cells and axons: (2) the extent to which secondary injury is prevented, including the controlling of any movement of the unstable spine during the rescue: and (3) the degree to which the complications known to affect the function of neural tissue, such as hypotension and hypoxia, are prevented. Physicians are able to influence the second and third factors by careful handling and by supporting the circulation and respiration during the rescue and acute treatment. Nothing, however, can completely reverse the disastrous effects of the violence imparted directly to the spinal cord. Recent investigations have indicated that methylprednisolone in some cases can improve outcomes if given within eight hours after an injury (Bracken et al. 1990). In addition, gangliosides such as GM-1 may also enhance neurologic recovery if administered within 72 hours of injury and over a four-week period (Geisler 1991). Investigations of other drugs have been conducted on animals and humans to determine whether these agents can influence the extent of necrosis of neural tissue. Many investigators believe that the reactive edema and release of cellular degradation products in the region of the spinal cord injury can cause further neural compromise. Results from clinical trials using high doses of steroids, while initially disappointing (Bracken et al. 1984) now appear more promising (Bracken et al. 1990). Other substances, such as opiate antagonists and thyrotropic-releasing hormone, have been shown to be of no benefit (Bracken et al. 1990; Faden et al. 1984). Surgical interventions have also been tried such as the use of pedicled omentum, spinal cord cooling, myelotomy, laminectomy, and anterior decompression (Donovan 1986). None of these procedures, however, has yet been proven effective in reversing the extent of spinal cord injury.

Similarly, efforts to restore neurologic function by interposing an autograft from a peripheral nerve after resecting the damaged cord have been tried but have not been successful. Recently, however, Aguayo (1989) (1991) has achieved promising success in one location of the Central Nervous System (CNS) by interposing a nerve graft in the optic nerve through which CNS axons have grown and reached their ultimate destination. Previously considered impossible, Dr. Aguayo has accomplished this in rats when action potentials from retinal cells reached the superior colliculis through the graft (Aguayo 1989, Aguayo 1991). Also, Schwab has demonstrated that at least two inhibiting factors released by mature oligodendrocytes are responsible for the failure of mammalian CNS tissue to grow through a nerve graft (Schwab 1991).

Prognosis for further recovery is usually better in those with incomplete lesions. It is generally believed that in recent years the percent of incomplete spinal injuries has increased, owing to improved rescue and retrieval, emergency care, expert delivery of definitive care at the trauma center and early rehabilitation (Stover and Fine 1986). Further, it is the experience of most centers throughout the United States that the over-all incidence of spinal cord injury has dropped because of active preventive measures, including a heightened awareness of the dangers of driving while intoxicated, the use of passive restraints in automobiles, and the reduction of and enforcement of speed limits.

It would be useful if predictive factors could be identified that would allow an absolute prognostic statement after the initial examination of the spinal cord-injured patient. Attempts to do so appear in the literature (Holdsworth 1970), but too many exceptions have been noted to make them universally applicable. While the failure of motor or sensory function to reappear following a complete injury to the spinal cord within 24 to 48 hours must be considered a grave sign, this condition does not always mean that no neurological function will ever return (Bracken et al. 1990; Donovan et al. 1991). Some complete cervical and lumbar lesions can show some recovery and are more likely to do so than the thoracic injuries (Donovan et al. 1987; Bedbrook and Clark 1981). Incomplete lesions are even far more difficult, if not impossible, to prognosticate in the early stages. Further neurological recovery often occurs although it is unusual to regain full motor and sensory function unless the injury was quite incomplete to begin with.

Whether the spinal cord is subjected to a minor, though sufficient force, or a major force, its function becomes impaired and normal conduction through the area of injury does not occur. Therefore, in the early hours following spinal injury, it is impossible to predict how serious the tissue damage is. However, after minor insults, function begins to return soon. After major insults leading to cell and axon death, it will never return. Between these extremes, gradations of the seriousness of the direct injury, the edema, and the ischemia following the trauma will affect the extent of remaining action potential transmission through the injured area. Current technology cannot assess how many nerve fibers have undergone necrosis compared with those that still have membranes, myelin, and axons that are only partly damaged and can recover. Only the passage of time will reveal this and ulti-mately how severe the injury was. Hence, the patient and family are forced to live with uncertainty regarding the ultimate neurolog-ical outcome following most incomplete le-sions and even some that initially are com-plete. A rule of thumb that the author has found useful is: when the lesion is complete and no function returns after 3 months, it is likely that none will ever return. Whereas, if some function does occur during that time, it is possible that more will return, so one must wait about 18 months just as for incom-plete lesions, before it can be determined that maximum recovery has been achieved.

Functional Expectations

The spinal cord readily lends itself to predic-tions of functional outcome or functional ex-pectation because each neurological level (or segment) controls specific muscles and transmits sensory information from its der-matome. Distribution of myotomies and dermatomes are quite similar among in-dividuals. When considering functional ex-pectations, it is useful to group several lev-els together. It is understood that the lower level within each group will perform better than the higher levels, but the expectations for each group are nearly the same. The fol-lowing expectations are for musculoskele-tal, bowel, bladder, and sexual function.

Levels Below S2

Below S2, only bowel, bladder, and sexual function are affected. Patients in this group have acontractile bladders, lax sphincters and impotence. Intermittent catheterization often is used, unless patients are easily able to overcome sphincter resistance by strain and Crede's maneuver. Some urinary and fecal incontinence may occur with a Valsal-va's maneuver. For this reason, it is pre-ferred to avoid stool softeners, and let the stool consistency be firm. Men and women may need to wear incontinence pads if drib-bling is a problem. Artificial sphincters may be appropriate for some patients in this group. Usually, no motor deficits exist in the extremities, except perhaps to the foot in-

trinsics, which could lead to hammer-toe de-formities.

Levels L4, L5, S1, S2

In the L4 to S2 group, indwelling catheters rarely are needed, while intermittent cathe-terization often is used. These patients also often have compromised sphincters and are-flexic, acontractile detrussors. Therefore, elimination also may be accomplished by in-creasing abdominal pressure (strain and crede). When the sphincteric mechanism is weakened, residual urine volumes close to zero generally are achievable. However, as with the first group, some dribbling may oc-cur, particularly with a Valsalva's maneu-ver. If dribbling is significant, men may use an external collecting system consisting of a condom, a tube, and a collecting bag, and women may wear absorbent pads. In some instances, an artificial sphincter can also be surgically implanted. This requires the pa-tient to inflate and deflate a periurethral cuff by squeezing hydraulic fluid through tubes and valves implanted under the skin. When the patient's sphincter is not lax, intermit-tent catheterization provides for satisfactory bladder emptying while the sphincter tone assures continence. Bowel elimination usu-ally is accomplished by strain, Crede's ma-neuver, and manual removal of feces from the rectal vault. Suppositories may be used but they often are ineffective because of the inability to induce reflex emptying. Because the anal sphincters also may be lax, some fecal soiling may occur when straining; therefore, patients generally prefer a firm stool.

As in the first group, to circumvent the male patient's inability or difficulty in ob-taining an erection, a penile prosthesis, con-sisting of semirigid silicone or an inflatable mechanism (Scott et al. 1973) can be im-planted surgically. The erectile mechanism also can be influenced by the injection of papaverine or phentolamine into one of the corpora cavernosa (Brindley 1983). More re-cently, prostaglandin E has been used suc-cessfully in this regard (Lee et al. 1989). Thirdly, an erection may be achieved by an external condom device which exerts a negative pressure or vacuum on the penis. Competent sexual counseling should be offered before any procedures or devices are recommended so that the patients and their partners can assess their expectations and feelings toward each other to be sure they are realistic.

Transfers, eating, dressing, and personal hygiene pose no problem. The patient's ambulation often is quite good even without orthoses, although the L4 and L5 lesion patients will benefit from the ankle stability provided by two ankle–foot orthoses (AFO) that essentially lock the ankle joint, (i.e., prevent both dorsiflexion and plantarflexion). Two canes or crutches are required by some when the hip abductors and extensors are weak, although at the lower levels, one cane may be sufficient, even for long distances. Wheelchairs are not needed in most cases. At the lower levels, driving without special hand controls can be achieved, although hand controls may be preferred by some patients. No attendant care is necessary.

Levels L1, L2, L3

At these levels, urethral and anal sphincters are inclined to be hyperreflexic, and involuntary tone will keep them in the closed position. Bladder elimination still may be accomplished by abdominal pressure if the sphincteric pressures are not too great. If they are, it may be necessary to use a procedure called anal stretch (Kiviat et al. 1975) which consists of manually dilating the anal sphincter, which relaxes the urethral sphincter, permitting bladder emptying when the individual bears down. Alternately, intermittent catheterization may be used. External collecting systems for the man and absorbent pads for the women are often not necessary because the sphincter tone prevents leaking and the bladder can usually store enough urine for the patient to remain dry between catheterizations or voidings. However, if the bladder is too hyperreflexic to store a sufficient volume of urine to allow intermittent catheterizations every 4 hours, anticholinergics usually will improve blad-

der capacity. Bowel leakage is less of an issue, although if diarrhea occurs, incontinence would be a problem. Reflex stimulation of peristalsis to enhance bowel emptying is more successful. Timing (i.e., after a meal taking advantage of the gastrocolic reflex), use of a suppository, and digital stimulation are more effective than in the lower lesions. Transfers, eating, dressing, and personal hygiene are independent after training.

Bipedal ambulation usually will require knee–ankle–foot orthoses (KAFO) and two crutches or even a walker. The gait pattern used may be either the four-point, swing-through, or swing-to depending on the ability of hip flexors (L2, L3) to advance the limb. Wheelchairs at these levels usually are not discarded, particularly at the higher levels because they are needed for ambulating long distances. Automobile driving requires the use of hand controls, but no attendant is necessary.

Levels T7-T12

Bladder and bowel function is similar to the more rostral lumbar levels. However, the bladder may be more hyperreflexic. Men and women may be managed with intermittent catheterization, but they often will require anticholinergics to suppress the detrusor. If a man wants to stop catheterizing and allow spontaneous voiding to empty his bladder, a sphincterotomy may be necessary to allow him to void with acceptable pressures and with sufficiently low residual volumes because the sphincter may be too hypertonic and the dyssynergia between the detrusor and sphincter, which often occurs at these and higher levels, may prevent effective voiding. If successful voiding is achieved, external collecting devices will be needed and are now practical to use because newer materials cause less skin irritation.

The higher the level between T7 and T12, the greater the loss of control over the abdominal and back musculature. Because these muscles are used for coughing, the higher the lesion, the less effective the cough. Therefore, these patients and those at higher levels are at greater risk for developing respiratory infections, particularly smokers and those with intrinsic lung disease.

Transfers, eating, dressing, and personal hygiene are easily achieved through training. Bipedal ambulation with KAFOs and crutches at these levels is less functional particularly at the higher levels. While walking in orthoses is good for exercise, the wheelchair is the main mode of ambulation. Therefore, use of public buildings, public toilets, and so forth is restricted to those that were wheelchair accessible. Hand controls are required for automobile driving.

Levels T1 to T6

Bladder elimination is essentially the same as for T7 to T12. Control of leakage between intermittent catheterization usually requires anticholinergics and maintenance of sterile urine. Some men may avoid catheterizations and wear external collecting systems at all times, providing that whether they have had a sphincterotomy or not, they have acceptable voiding pressures and volumes. Some women may find that leakage is not controlled sufficiently, regardless of anticholinergics and sterile urine so they may choose to use an indwelling catheter with drainage to a leg bag. Bowel function control is as described for the previous group. Transfers, eating, dressing, and personal hygiene, although achievable, require much training.

Bipedal ambulation is no longer practical, even for exercise, because of poor trunk control; however, some patients might choose to attempt it. The wheelchair is the main mode of ambulation. Hand controls and sometimes external trunk support are required for automobile driving. No attendant is necessary and living alone is possible. However, a spouse or roommate can be very helpful.

Levels C7, C8

At these levels, the hands are involved. Training takes longer because special adaptive equipment is needed, and manual manipulative techniques must be learned.

An external collector is preferred by many male patients since impairment of finger dexterity makes intermittent catheterization difficult, particularly the C7 patients. If voiding pressures and residual volumes are acceptable, bladder elimination can be facilitated by stimulating the micturition reflex. A sphincterotomy may be required to lower the voiding pressure to an acceptable range (less than 80 cm of H_2O). But this will be helpful only if the detrusor has sufficient contractility to expel the urine when the resistance is reduced. Many women will use a continuous indwelling Foley catheter. Autonomic dysreflexia may be a significant problem, particularly if the catheter becomes obstructed. Bowel function, particularly the insertion of suppositories and the use of digital stimulation of the anus, is more difficult for the patient to perform but can be achieved with a finger splint or performed by an attendant.

Due to grip weakness the patient may require wheelchair modifications to make propulsion easier, such as applying friction tape, tubing, or projections on the hand wheel rims. Transfers usually require a sliding board, and electric hospital beds usually are necessary. Clothing modifications, such as the use of snaps or Velcro, will circumvent the difficulty these patients have with buttons. Dressing and personal hygiene may require an attendant for donning pants and for total body bathing. Special adaptations of bathroom systems are required. Hand controls and sometimes steering wheel attachments are necessary for automobile driving.

The C7 level is the most rostral level for which living without an attendant is possible, but not all achieve this goal particularly if they have any coexisting medical problems.

Level C6

Bladder and bowel function is the same as for the C7-C8 levels but intermittent catheterization is not practical for most patients because it must be performed by an attendant. Most women and many men will require an indwelling urethral or suprapubic catheter. The only remaining hand and wrist function at this level is extension or dorsiflexion of the wrist. Triceps function is lost; therefore, the patient cannot forcibly extend his or her elbow. Elevating the trunk from the sitting position is more difficult because only the shoulder depressors can be used and even then, only if the elbows are free of flexion contractures and can be placed in full extension. Therefore, preservation of full range of elbow motion is essential.

Eating skills are achieved using adaptive equipment applied to the hands, such as a wrist-driven flexor hinge splint that creates opposition of the thumb to the second and third fingers by dorsiflexing the wrist. A cuff worn over the hand with a slot for a spoon or fork (universal cuff) also may be used.

Ambulation in a manually operated wheelchair with modifications of the wheel rims as previously described, is still possible on level, smooth surfaces. However electric wheelchairs also are needed for more difficult terrain. Total dressing and personal hygiene usually require an attendant. Some patients can achieve this independently with the aid of loops sewn onto the clothes into which they can hook a finger, but this is often too time consuming. Pull-over garments for the upper torso usually are preferable because of the difficulty with manipulation of buttons, snaps, and so forth. If the arms are long enough, a sliding board is used for transfers. Transfers in and out of bed and on and off of a toilet can be achieved; car transfers also can be achieved by certain patients without assistance. A van with a lift makes driving easier and more practical for most people in this group. An attendant is required to assist with dressing, personal hygiene, and in some cases transfers. Some of the bladder care (e.g., application of the external collector and bowel care, such as suppository insertion and clean up) also require assistance. These patients are rarely able to live alone.

Level C5

At this level, the patient has shoulder and elbow flexion function but no wrist and hand function.

Bladder management is essentially the same as for C6. Male patients cannot apply an external collector independently. For bowel function, an attendant must provide a suppository, and digital stimulation and clean up functions. Eating, except perhaps for cutting, can be accomplished with set up (i.e., food already cut, cartons opened, and straw inserted), but transfers, dressing, and personal hygiene are impossible without an attendant.

Ambulation is possible using an electric wheelchair, which the patient controls by placing his or her hand on a stick mounted on the control box of the wheelchair on the side of his or her stronger arm. Automobile driving capability depends on residual shoulder motion, but a van is absolutely necessary. A full-time attendant is necessary to assist with most functions but does not need to be at the patient's side at all times. With appropriate seat cushions, on a power controlled reclining wheelchair, these patients can sit for 8 to 10 hours a day. Some pressure relief, provided by the attendant, may be needed.

Other adaptive equipment used by some patients with lesions at C5 include battery-powered and cable-operated wrist-driven flexor hinge splints and ball bearing feeders (mobile arm supports) anchored to the wheelchair. These allow the patient to feed himself or herself and do facial grooming. To operate the mobile arm supports, the patient places the arms in special troughs attached to movable hinged bars. By using the shoulder muscles, the patient can make the troughs swivel to bring the arms toward the body and move them out again. The troughs also can be tipped downward at the elbow to raise the hands.

Levels C1, C2, C3, C4

Acutely, the C4 level is the highest that is still compatible with life, unless at the time of the accident, assisted respiration was provided. All C1, C2, and many C3 patients invariably require permanent assisted ventilation, as do some of the C4 patients. A battery-powered portable respirator can, for some, be fitted to the wheelchair. The C1 to

C3 patients usually require a permanent tracheostomy and special techniques to control lung and throat secretions (Vaughn and Syers 1989).

C1 to C4 patients can use a chin, mouth, or breath control device and perhaps a voice-controlled device to operate electric wheelchairs. They may have greater control of their environment through "environmental control systems." In these systems, the patient can control a telephone, radio, television, lights, buzzer, and so forth from one place. Selection of one of these options is made by using breath, tongue, sound, or residual muscle power to activate the system. A full-time trained attendant is required for this patient, and that attendant probably should always be available.

Sexual Function

Sexual functioning is another area affected by spinal cord injury that deserves separate discussion regarding prognosis. Although many aspects of sexuality, including production of sex hormones, feelings of sexual attraction to potential partners, and the ability to be sexually attractive to others, are essentially undisturbed by spinal cord injury, certain functions will be altered by a spinal cord injury. Modifications will be needed in the spinal cord-injured person's techniques of behaving sexually. While in some cases the libido decreases, the psychic, hormonal, and many physical aspects of a loving relationship are still present and available to the spinal cord-injured person (Trieschmann 1989).

Physical aspects of the love relationship such as the involvement of the senses of vision, smell, and touch by way of oral or manual contact, still remain. Areas in which sensation is preserved that were previously erogenous to the person, e.g. the lips will remain so. Such areas might, in fact, have heightened erotic significance after the injury. Often the transition zone between normal sensation and no sensation acquires a heightened erotic feeling, and if the lesion is incomplete, some sensory areas below the main level of paralysis also may have heightened erotic feeling. Sensation in the genital

area requires an intactness of the second, third, and fourth sacral nerves and conduction past the area of injury. Thus, unless incomplete, the spinal cord-injured person will not have direct appreciation of touch or contact in these areas. Some of the higher lumbar nerves around the upper thigh approach the genital area and might convey genital-like sensations. Spinal cord injury does alter physical mobility and therefore will alter the physical aspects of sexual interactions. This will necessitate modification of techniques.

Genital sexual functions of the man may be broken down into erection, seminal emission into the urethra, and forceful ejaculation of the semen. Erection is controlled by the parasympathetic nerves, emission by the sympathetic nerves, and ejaculation by the somatic motor nerves. The achievement of a full orgasmic experience requires normal sensation and a coordination and integration of these three nervous system functions. This coordination occurs in the spinal cord and depends on integration from higher centers. In general, erections are more likely to be sustained in the higher lesions, while ejaculation, or at least emission, is more likely to occur in the lower lesions. The fertility rate for men with lumbar lesions is higher than for men with cervical lesions, because ejaculation may occur, but in both cases, it is less than 10% (Bors and Comarr 1960). The reason for this remains unclear, but maturation arrest has been noted on testicular biopsy (Stemmermann 1950).

Reflex erection is usually possible for all spinal cord injured men with lesions at about T12 and above. Mechanical stimulation can produce and often sustain an erection. In some men with lower thoracic and upper lumbar level lesions, weak erections can occur. Lumbar lesions, particularly if they involve the cauda equina, will interfere with the development of a full reflex erection but may allow for a less intense psychic erection, which may or may not be strong enough for vaginal intromission. Since seminal emission depends on sympathetic nerve function, it may appear without intense ejaculation in men with lumbar lesions. Ejaculation rarely appears in men with complete lesions at any level. Recently, however, ejaculation has been induced by electrical or vibratory means and the quantity and quality of sperm has been found to improve upon repeated ejaculations (VerVoort 1987). Despite the absence of ejaculation, some sexually active spinal cord-injured men report an orgasm-like experience associated with a certain amount of release of sexual tension.

As with spinal cord-injured men, women with spinal cord injuries may develop sensory areas that convey heightened erotic meaning. Lubrication of the vaginal area will be inconsistent and may necessitate additional lubrication. While women will exhibit reflex engorgement of the clitoris (equivalent to erection in men), they will not exhibit the spasmodic contractions of the perineal and vaginal musculature (equivalent to ejaculation). The absence of the latter orgasmic experience is sometimes replaced, as in the man, by an orgasm-like experience associated with a release of sexual tension. Female fertility is generally undisturbed by spinal cord injury. Immediately after an injury, menstruation may cease temporarily but then return in a regular pattern. Thus, spinal cord-injured women can conceive and carry a fetus to term. A spinal cord-injured woman who chooses to do this needs careful medical attention during pregnancy, because of an increased risk of autonomic dysreflexia during pregnancy and particularly during delivery for those with lesions above T6. Uterine contractions and lactation are unaffected by the spinal cord injury. For women who do not wish to conceive, birth control methods must be considered carefully because the problems associated with some of them may be increased in spinal cord-injured women (e.g., estrogen and the risk of deep venous thrombosis and pulmonary and cerebral thromboembolism) (Donovan 1985).

Incomplete Lesions

The previous considerations related to functional expectations apply to complete lesions. When incomplete lesions are encountered, the expectations may be greater

depending on how useful the motor function and sensory function are that return. Alternately, faced with significant medical problems, additional injuries, or age, the previous expectations would have to be lessened.

Frankel originally classified spinal cord lesions as follows:

A—Complete
B—Sensory sparing only
C—Motor function preserved but not useful
D—Motor function preserved useful
E—Fully recovered

The American Spinal Injury Association (ASIA) has modified this (American Spinal Injury Association 1989) and is currently updating it.

A. Complete

All motor and sensory function is absent below the zone of partial preservation (ZPP). The ZPP is defined as the region below the neurological level that retains some sensory and/or motor function. The neurologic level is the lowest or most caudal level where both motor and sensory innervation is normal.

B. Incomplete, Preservation Sensation Only

Any demonstrable, reproducible sensation, extending into the sacral segments, excluding phantom sensations are preserved. Voluntary motor functions are absent.

C. Incomplete, Preserved Motor Nonfunctional

Voluntary motor function is preserved, but it is minimal and performs no useful purpose. Minimal is defined as preserved voluntary motor ability below the neurological level in which the majority of the key muscles tests less than a grade of 3.

D. Incomplete, Preserved Motor Function

Voluntary motor function is preserved, and it is useful functionally. This is defined as preserved voluntary motor ability below the level of injury in which the majority of the key muscles tests at least a grade of 3.

E. Fully Recovered

All motor and sensory function is returned completely, but one may still have abnormal reflexes (American Spinal Injury Association 1989).

Although this is a crude classification, it has enjoyed clinical acceptance because of its simplicity and ease of application. In making prognostications, regardless of the neurological level of injury, one needs to note and follow the degree of incompleteness because visceral and somatic functional expectations will improve accordingly.

Summary

The level of function following spinal cord injury is predictable, particularly in complete lesions. Incomplete lesions may show further improvement. There are no absolute prognostic features that one can rely on in the early stages following spinal injuries, except that incompleteness at the outset is encouraging for further recovery. In general, the sooner recovery is observed, the more complete it is likely to be. Optimal treatment depends on careful handling and stabilization of the spine as well as prevention of complications known to impair neurological function. Early rehabilitation is absolutely necessary in order to reach the patient's functional expectations as early as possible and to avoid the serious complications that are known to occur in the spinal injury population, such as pressure ulcers, infection of the urinary tract, severe spasticity, and others. Finally, this malady requires long-term follow up for the purposes of health maintenance, equipment replacement, preservation of skills and psychosocial adjustment.

References

Aguayo, A. J. Regrowth and connectivity of regenerating axons from the adult mammalian CNS. Presented at the Seventh Annual Neurotrauma Symposium. Phoenix, AZ Oct. 28, 1989.

Aguayo, A. J.; Raminsky, M.; Bray, G. M.; Carbonetto, S.; McKerracher, L.; Villegas-Perez, M. P.; Vidal-Sanz, M.; Carter, D. A. Degenerative and regenerative responses of injured neurons in the central nervous system of adult mammals. Philos. Trans. R. Soc. Lond. Ser. B: Biol. Sci; 1991; 331:337–343.

American Spinal Injury Association. Standards for Neurological Classification of Spinal Injury Patients. 1989.

Bedbrook, G.; Clark W. Thoracic spine injuries with spinal cord damage. J. Royal College Surg. Edinburgh 26:264–271; 1981.

Bors, E.; Comarr, A. E. Neurologic disturbances of sexual function with special reference to 529 patients with spinal cord injury. Urological Survey 10:191–222; 1960.

Bracken, M. B.; Collins, W. F.; Freeman, D. F.; Shepard, M. J.; Wagner, F. W.; Silten, R. M.; Hellenbrand, K. G.; Ranshoff, J.; Hunt, W. E.; Perot, P. L.; Grossman, R. G.; Green, B. A.; Eisenberg, H. M.; Rifkinson, N.; Goodman, J. H.; Meagher, J. N.; Fischer, B.; Clifton, G. L.; Flamm, E. S.; Rawe, S. E. Efficacy of methylprednisolone in acute spinal cord injury. J.A.M.A. 251:45–52; 1984.

Bracken, M. B.; Shepard, M. J.; Collins, W. F.; Holford, T. R.; Young, W.; Baskin, D. S.; Eisenberg, H. M.; Flamm, E.; Leo-Summers, L.; Maroon, J.; Marshall, L. F.; Perot, Jr., P. L.; Piepmeier, J.; Sonntag, V. K. H.; Wagner, F. C.; Wilberger, J. E.; Winn, H. R. A randomized, controlled trial of methylprednisolone or naloxone in the treatment of acute spinal-cord injury. N. Engl. J. Med. 322:20:1405–1411; 1990.

Brindley, G. S. Cavernosal alpha-blockade: a new technique for investigating and treating erectile impotence. Br. J. Psychiat. 143:332–337; 1983.

Donovan W. H. Sexuality and sexual function, in lifetime care of the paraplegic patient. In: Bedbrook, G. M. ed. Edinburgh: Churchill Livingstone; 1985:149–161.

Donovan, W. H. New developments in spinal cord injury. Rehabilitation Report 2:1–2; 1986.

Donovan, W. H.; Kopaniky, D.; Stolzmann, E.; Carter, R. E. The neurological and skeletal outcome in patients with closed cervical spinal cord injury. J. Neurosurg. 66:690–694; 1987.

Donovan, W. H.; Cifu, D. X.; Schotte, D. E. Neurological and skeletal outcomes in 113 patients. Paraplegia. (In press.)

Faden, A. L.; Jacobs, T. P.; Smith, M. T.; Zivin, J. A. Naloxone in experimental spinal cord ischemia: dose-response studies. Eur. J. Pharmacol. 103:115–120, 1984.

Geisler, F. H.; Dorsey, F. C.; Coleman, W. P. Recovery of motor function after spinal-cord injury–a randomized, placebo-controlled trial with GM-1 ganglioside. N. Engl. J. Med. 324:26:1829–1838; 1991.

Hallstead, L. S.; VerVoort, S.; Seager, S. W. Rectal probe electrostimulation in treatment of anejaculatory spinal-cord injured men. Paraplegia. 25:120–129; 1987.

Holdsworth, F. Review article: Fractures, dislocations, and fracture dislocations of the spine. J. Bone Joint Surg. (Am) 52:1534–1551; 1970.

Kiviat, M. D.; Zimmerman, T. A.; Donovan, W. H. Sphincter stretch: a new technique resulting in continence and complete voiding in paraplegics. J. Urol. 114:895–897; 1975.

Lee, L. M.; Stevenson, R. W. D.; Szaz, G. Prostaglandin E versus phentolamine/Papavarine for the treatment of erectile impotence: a double blind comparison. J. Urol. 141:549–550, 1989.

Schwab, M. E. Regeneration of lesioned central nervous system axons by neutralisation of neurite growth inhibitors: a short review. Paraplegia. 29:294–298; 1991.

Scott, F. B.; Bradley, W. E.; Timm, G. W. Management of erectile impotence. Urology. 2:80–82; 1973.

Stemmermann, G. N., Weiss, L., Overbach, O., Friedmann, M. A study of germinal epithelium in male paraplegics. Amer. J. of Clin. Path. 20:24–34; 1950.

Stover, S. L.; Fine, P. R. Spinal cord injury: the facts and figures. In: Stover, S. L.; Fine, P. R., eds. University of Alabama at Birmingham; Birmingham: 1986.

Trieschmann, R. B. Loving, sexuality and relationships. In: Spinal cord injuries: psychological, social, and vocational rehabilitation. Trieschmann, R. B., ed. Demos Publications; New York: 1989, 158–185.

9

Viral Diseases

JOHN BOOSS

Aseptic Meningitis

For the majority of cases, aseptic meningitis is a benign disease. In fact, one of the criteria established by Wallgren (1925) for its diagnosis was a short benign course. Beghi and coworkers (1984) found that of 283 patients with aseptic meningitis, 95% recovered completely, and only mild residua were found in the remainder. Ambiguity may arise in cases of meningoencephalitis, that is, when the brain parenchyma is involved in addition to signs of meningeal irritation. In these circumstances, the prognosis is adversely affected by the encephalitic component. As pointed out by Oldfeldt (1949) in her review of the sequelae of mumps meningoencephalitis, the risks of permanent ill effects appeared to be greater in cases demonstrating symptoms of acute encephalitis.

In the United States, the most common viral causes of aseptic meningitis are the nonpolio enteroviruses. Following the introduction of the vaccine, there has been a marked decline in aseptic meningitis due to mumps. Numerous other viruses, such as lymphocytic-choriomeningitis, Epstein-Barr, and Herpes simplex type 2, also cause

this syndrome. In addition, viruses more commonly associated with encephalitis, such as the arboviruses, occasionally can cause aseptic meningitis. Although clues from season, geographic locality, regional virus activity, and associated systemic findings can help, few clinical observations can distinguish one type of viral meningitis from another (Adair et al. 1953). Furthermore, usually no specific therapy is indicated. Exceptions occur when the disease extends beyond the meninges to involve the brain, roots, or spinal cord as in various types of meningoencephalitis, herpes simplex type 2 meningitis, and enteroviral parenchymal involvement in patients with agammaglobulinemia. Recurrence of aseptic meningitis is rare, found in only three of 283 cases reviewed by Beghi and associates (1984). Recurrence should raise a concern over other diagnostic possibilities.

While the long-term outlook is benign, the convalescence period may be prolonged following the acute phase of aseptic meningitis. Lepow and colleagues (1962) found that two thirds of patients at a 3-month follow-up had subjective symptoms or minor objective findings. Complaints included fatigue, irritability, reduced concentration, and muscle pain. Other less common symptoms included headache, parasthesias, dizziness, and clumsiness. The most common findings were tightness and muscle weakness. However, 95% of patients were fully recovered at 1 year. As a consequence of the prolonged recovery phase, Lepow and co-workers (1962) recommend a careful program of graded activity and follow-up, postponing full activity until an evaluation 1 to 2 months following discharge. Evaluation also should include testing for potential hearing loss. In contrast to viral encephalitis, the long-term risk for seizures is no greater in patients recovering from aseptic meningitis than in the general population (Annegers et al. 1988).

Infection of the central nervous system (CNS) by enteroviruses in infants younger than 1 year is an apparent exception to the generally benign prognosis of aseptic meningitis. Sells and colleagues (1975) found a smaller mean head circumference, lower mean I.Q., and impaired language and speech skills in children whose enterovirus infection occurred at less than 1 year. Wilfert and coworkers (1981) compared infants who experienced enteroviral meningitis in the first 3 months of life with matched controls. They found no difference in head circumference, I.Q., or sensorineural hearing. However, receptive language function was impaired in the aseptic meningitis group. There are at least two potential explanations for the worsened prognosis in this group. First, there may be parenchymal involvement during the acute phase. Ten of the 19 infants reviewed by Sells and associates (1975) had clinical evidence of parenchymal involvement during the acute phase. Brain parenchymal lesions were found at autopsy in many infants dying of fatal coxsackie B infections reported and reviewed by Kaplan and colleagues (1983). Kitamoto and colleagues (1989) used repeat EEGs and evoked potential studies to assess subclinical parenchymal involvement in neonates and infants with aseptic meningitis. Second, maturation of the brain may be impeded. In any case, children who

have had aseptic meningitis in the first year of life deserve serial developmental and sensorineural hearing evaluations to facilitate adjustment.

Encephalitis

There is a wide range in the reported incidence of morbidity and mortality following viral encephalitis (Booss and Esiri 1986; Whitley 1990). In small prospective studies, the combined percentage of significant morbidity and mortality is as high as 80% in brain biopsy-confirmed herpes simplex encephalitis (HSE) (Whitley et al. 1977) and as low as 7% in encephalitis associated with systemic but not direct infection of the brain (Kennedy et al. 1988). In retrospective studies, there is also a wide variation. Mortality in children younger than 1 year with encephalitis studied by Wong and Yeung (1987) was 42%; in Olmsted County, Minnesota (Nicolosi et al. 1984) it was 3.8% among all age groups. Within certain limits, these differences can be understood. In the examples cited, the nature of the agent and the age of the patient were important factors. Other factors that relate to the interaction of virus and host (infected patient) can be analyzed to help predict the outcome of acute viral encephalitis (Table 9-1).

Host

Age and immunocompetence are the host factors most important to outcome. In addition to the study by Wong and Yeung (1987), Rautonen and colleagues (1991) also found that an age of less than 1 year predicted a poor outcome. In certain infections, age appears to be a particularly important variable. In Eastern equine encephalitis, both those younger than 10 years and those older than 50 years suffer the highest mortality, while those of an intermediate age fare much better (Przelomski et al. 1988). In Herpes simplex encephalitis, Miller and Ross (1968) found that none of five patients younger than 15 years died from HSE, whereas six of seven patients older than 17 years succumbed. The studies of antiviral therapy for HSE confirmed age to be an important fac-

Table 9-1. Factors Influencing Prognosis in Viral Encephalitis

Host
  Age
  Immunocompetence
Virus
  Identity
  Virulence
Pathogenesis
  Disease mechanism
    Direct infection
    Immune-mediated attack
  Distribution of lesions
  Reversibility of CNS injury
  Secondary complications
    Cerebral edema—raised intracranial pressure
      Brain stem compression
      Impaired cerebral perfusion
    Vascular occlusion or hemorrhage
Intervention
  Specific
    Antivirals, corticosteroids
  Monitoring and support
    Neuroimaging studies, intensive care unit management

tor, with those younger than 30 years faring better than those older than 30 (Whitley 1988).

Herpes simplex infection under one month (herpes neonatorum), is an exception. Infection may be classified as disseminated, encephalitis, or restricted to skin, eyes, or mouth (Whitley et al. 1991a). The collaborative anti-viral study group found that morbidity was related to the extent of involvement at the start of antiviral therapy (Whitley et al. 1991b). The mortality rate was 57% with disseminated infection, 15% with encephalitis, and no deaths occurred in restricted infection.

With regard to immunocompromise, the most common effect on viral encephalitis is an increased susceptibility to usually well handled agents. In AIDS, for example, cytomegalovirus (CMV) is the most common viral superinfection of the brain (Morgello et al. 1987), whereas CMV is virtually unknown as a pathogen for the CNS in immunocompetent adults. Progressive multifocal leukoencephalopathy (see HIV-1 Infection, chapter 17) and varicella zoster virus (VZV) infections are other opportunistic infections found in AIDS patients (Petito et al. 1986). Bone marrow transplant recipi-

ents are another immunosuppressed population in which relatively benign viral agents can produce fatal meningoencephalitis.

Agent and Mechanism
It has long been recognized that the identity of the infecting agent is a major determinant of outcome (Meyer et al. 1960). Certain strains of virus also are more likely to involve the CNS, such as ECHO-9 of the ECHO viruses. The mechanism of injury to the CNS is important, and it is useful to classify acute viral encephalitis by type: sporadic, epidemic, or parainfectious (Booss and Esiri 1986). Direct infection of the brain can occur in sporadic or epidemic infections, such as HSE or arboviral encephalitis. In general, the worst outcome is associated with certain direct infections of the brain (Table 9-2). Among the sporadic encephalitides, rabies, with only very rare survival (Hattwick et al. 1972), and untreated HSE, with a combined morbidity and mortality rate of 80% (Whitley et al. 1977), are exceptionally lethal. In contrast, viral encephalitis due to other agents that mimic HSE, has a significantly better outcome than HSE itself (Whitley et al. 1989). However, with the wider application of the polymerase chain reaction (PCR) assay to CSF to diagnose HSE, atypical, mild, and recurrent forms may be defined. The course and prognosis of such syndromes await the accumulation of PCR data. Of the arboviruses found in the United States, Eastern equine

Table 9-2. Types of Acute Encephalitis Associated With Significant Mortality and Sequelae

Direct Infection	Parainfectious
Sporadic	Acute hemorrhagic
Rabies virus	leukoencephalopathy of
Herpes simplex virus	Hurst
Epidemic	
Eastern equine encephalitis virus	
Japanese B encephalitis virus	
Far eastern subtype of tick-borne encephalitis virus	

encephalitis is the most deadly, with 68% mortality found in earlier studies and 31% found in a more recent study (Przelomski et al. 1988). Worldwide, other arboviruses with significant morbidity and mortality are Japanese B encephalitis and the Far Eastern subtype of tick-borne encephalitis.

While immune-mediated parainfectious encephalitidies associated with systemic infections usually have a better prognosis than direct viral infection of the brain, acute hemorrhagic leukoencephalopathy of Hurst is an exception (see Table 9-2). A very rare condition, this illness frequently follows an upper respiratory infection, is associated with massive cerebral swelling and hemorrhage, and has a mortality of about 70% (Behan and Currie 1978). While early aggressive measures to manage increased intracranial pressure may be facilitated by serial computed tomography (CT) studies and intracranial pressure monitoring, there is insufficient experience to determine if these will significantly improve the outcome. In a biopsy-confirmed case, recovery followed management of raised intercranial pressure and prolonged immunosuppression (Seales and Greer 1991).

The more common manifestation of parainfectious encephalitis is characterized clinically by multifocal areas of dysfunction and is termed acute disseminated encephalomyelitis (ADEM). Previously, ADEM had been most commonly identified following clinically recognizable childhood illnesses such as measles. In countries in which widespread immunization against measles, mumps, and rubella is practiced, encephalomyelitis associated with these agents has virtually dissappeared (Koskiniemi and Vaheri 1989; Beghi et al. 1984). Local outbreaks of measles in the United States, including CNS complications (Pearl et al. 1990), however, raise concern about the duration of protection. Of the recognizable childhood illnesses, chicken pox remains a cause of ADEM. Whether treatment of chickenpox with acyclovir (Dunkle et al. 1991) will alter the incidence or manifestations of ADEM has not been determined. Many cases of encephalitis occur following relatively nonspecific respiratory or other viral illness symptoms in a biphasic disease pattern compatible with ADEM. Wider application of magnetic resonance imaging (MRI), which has great sensitivity to white matter lesions, may clarify the diagnosis in these cases. Kennedy and colleagues (1987) found that encephalitis in response to non-CNS systemic infection had a lower incidence of long-term sequelae than that with evidence for direct infection of the brain.

Acute Phase

Of the several factors in the clinical evaluation of the patient during the acute phase of encephalitis, level of consciousness is the most critical. Many studies have reported that coma augurs poorly for full recovery. In prospective studies of antiviral therapy in HSE, both the Swedish (Skoldenberg et al. 1984) and the U.S. (Whitley, 1988) collaborative study groups found that level of consciousness correlated with outcome. In the U.S. study, level of consciousness was assessed by the Glasgow coma scale (Jennett et al. 1977). Scores in the range of normal were associated with excellent recovery, whereas scores indicative of coma had a poor outcome. Other significant variables in the outcome of HSE are age and duration of symptoms before therapy; those treated before 4 days have the best outcome (Whitley 1988). In treated herpes neonatorum, level of consciousness at entry was a significant prognostic factor, as were the development of disseminated intravascular coagulopathy or pneumonitis, age of gestation and type 1 virus (Whitley et al. 1991b). In a study of encephalitis in children, Wong and Yeung (1987) found that both the development of deep coma within the first 24 hours after admission and the persistence of coma for more than 1 week were associated with death or major complications. These investigators report that death usually is preceded by a deterioration of the electroencephalogram (EEG). Depressed level of consciousness or coma has been identified by Kennedy and associates (1987) and by Kennard and Swash (1981) as associated with a poor outcome. Status epilepticus was associated with a poor prognosis in the report by Wong and Yeung (1987).

There is little concensus on the value of any single laboratory study as predictive of outcome. However, if various tests demonstrate a significant amount of tissue destruction, the prognosis is less favorable. An example is the finding of a swollen hemorrhagic region on CT or MRI as in HSE or acute hemorrhagic leukoencephalopathy of Hurst. Buttner and Dorndorf (1988) found normal CT studies in 19 of 26 patients who experienced complete recovery, whereas pathological CTs were found in six of eight patients who died. Elevated CSF protein, the presence of significant numbers of RBCs, and polymorphonuclear cells in the CSF may suggest significant tissue damage. The EEG almost always is abnormal in encephalitis (Kennedy et al. 1987) and may deteriorate significantly before death (Wong and Yeung 1987).

Effect of Specific Therapy
The outcome of HSE is markedly improved by the prompt use of the antiviral compound acyclovir. In untreated patients, mortality is approximately 70%. Mortality was reduced to 19% in the Swedish (Skoldenberg et al. 1984) and to 28% in the U.S. (Whitley et al. 1986) collaborative studies. The quality of survival also was significantly enhanced; 56% and 38% of the survivors in the Swedish and U.S. studies respectively returned to normal life. Some of the differences between the two studies may be accounted for by the strict entry criteria of the U.S. study in which all patients underwent brain biopsy (Whitley et al. 1986). Of 13 patients younger than 30 years and not in coma, none died, and eight had no or only minor residua (Whitley et al. 1986). However, the outcome is still daunting: 46% of patients were left with severe impairment or died (Whitley 1988). No difference in outcome of neonatal HSV infection between acyclovir or vidarabine was found on controlled trial (Whitley et al. 1991a).

Acyclovir has been used in disseminated VZV. However its effectiveness in CNS VZV is unclear, resulting in part from a mixture of types of pathogenesis: angiitis, viral replication, and demyelination. Other confounding factors are the immunocompromised setting in which VZV infection of the CNS often occurs (Jensek et al. 1983) and the capacity for spontaneous recovery of VZV-associated infections of the CNS. In reviewing their own and others' experience with acyclovir for Zoster-associated encephalitis, Peterslund (1988) found that treatment with acyclovir decreased the length of neurological symptoms and that a reduction of mortality that was not statistically significant occurred.

Ganciclovir and foscarnet have been used in serious CMV infections in acquired immune deficiency syndrome (AIDS) patients, particularly for CMV retinitis. Treatment of CMV in the central nervous system in AIDS patients is complicated by the difficulty in establishing the diagnosis of a CMV-related disorder (See HIV-1 Infection, chapter 17).

Following the work of Pasteur, there has been general acceptance of postexposure immunization as a means of preventing rabies. When the rabies vaccine is prepared from brain material, semple vaccine, a variable percentage will experience neuroparalytic accidents. Hemachudha and colleagues (1987) report that in Bangkok, the incidence of this in 1983 was one per 220 vaccinees. They cited data from Vejjajiva indicating a mortality of 17% and a morbidity of 23%. Hemachudha and associates (1987) were unable to identify any indicators of prognostic value. In regions where the vaccine is prepared in human diploid cells, the immunization is safe and effective when used as part of the recommended protocol (Baer and Fishbein 1987).

The use of corticosteroids or ACTH in the therapy of parainfectious ADEM has been based on their capacity to interfere with immunological mechanisms (Miller 1953). Except for clinical reports on progressive ADEM (Pasternak et al. 1980) and relapsing ADEM (Yahr and Lobo-Antunes 1972), the effect on the clinical course has been difficult to document. With MRI's sensitivity in detecting white matter lesions, case selection may be improved, facilitating controlled clinical observations on the effectiveness of corticosteroid therapy, particularly high doses IV.

Relapse and Chronicity

Relapse or recurrence of encephaltis is distinctly unusual, but there are recognized cases. Following antiviral therapy, HSE has been noted to recur within a few days to several weeks in less than 5% of cases (Whitley et al. 1981; Skoldenberg et al. 1984). The mechanism underlying recurrence is important for therapy. Skoldenberg and colleagues (1984) treated such patients with antiviral drug therapy and corticosteroids. This combined therapy covered the possibilities of the reemergence of a suppressed viral infection and a secondary immune-mediated ADEM. A feared potential complication of antiviral therapy of HSE is the development of drug resistance, requiring an alternative drug. In the Collaborative study of neonatal herpes, three or more skin recurrences with type 2 virus within six months correlated with long-term neurological sequelae (Whitley et al. 1991b). This finding raised the consideration of treatment of the skin lesions by oral acyclovir to protect the nervous system (Whitley et al. 1991b). Relapse also has been recorded in two of 61 cases of VZV-associated encephalitis reviewed by Peterslund (1988). When relapses are suspected in ADEM, the distinction from multiple sclerosis (MS) must be made. Separately, certain acute demyelinating syndromes may be the harbinger of multiple sclerosis. These include optic neuritis, brain stem and spinal cord syndromes. Miller and colleagues (1989) found that the risk of progression to MS in brain stem or spinal cord syndromes was increased by the presence of oligoclonal bands of immunoglobulin in the cerebrospinal fluid, and by disseminated brain lesions on MRI in the spinal cord syndrome. Immunosuppression creates the conditions for many attacks on the host by opportunistic infections. In agammaglobulinemia, chronic enteroviral infection of the brain can occur, most commonly by ECHO viruses (McKinney et al. 1987). Systemic treatment with immuno-

globulins is used, and intraventricular delivery of immunoglobulins through a reservoir has produced improvement in some patients. Recurrence of symptoms in three children who had recovered from Japanese encephalitis eight to nine months earlier was associated with virus isolation from blood mononuclear cells. Virus could also be isolated from some children not experiencing recurrence of symptoms (Sharma et al. 1991).

Residua

Following the acute phase of encephalitis, the rate of recovery and the degree of resolution of symptoms vary greatly among survivors. Some patients dramatically improve during the acute course, others recover over weeks to months, and some retain a fixed deficit without significant improvement. Among children, impaired development, mental retardation, seizures, personality changes, behavioral disorders, motor or coordination defects, and language and hearing impairment are observed (Kennedy et al. 1987; Rantakallio et al. 1986; Wong and Yeung 1987). Behavioral disorders may be transient, appearing and resolving in the recovery period, or they may remain as permanent sequelae. Because the plasticity and capacity for recovery in the nervous system of children is greater than that of adults, repeated assessments over a 2-year period following encephalitis are needed to determine the degree of disability and the appropriate education level. Among 41 children surviving acute viral encephalitis, evaluated 1 to 10 years later, Wong and Yeung (1987) report that 76% appeared to be normal.

Following HSE with damage to the temporal lobes, dysphasia, or aphasia, cognitive defects, motor disability (notably hemiparesis), and seizures are reported among survivors. For example, of 15 survivors judged to have moderate disability and followed for 1 year, 13 had dysphasia, eight had seizures and hemiparesis, and six had minimal cognitive impairment (Whitley et al. 1981). Kennedy et al. (1988) found some type of residual disability, including epilepsy, weakness, emotional lability, reduced intelligence, personality change, and memory difficulties in patients characterized as having a useful survival. Long-term neuropsychological testing has confirmed that even mild cases of HSE treated rapidly with acyclovir experience long-lasting cognitive defects (Gordon et al. 1990). Behavioral disturbances following HSE have included several patients with symptoms compatible with a human Klüver-Bucy syndrome (Greewood et al. 1983) and a Tourette's-like disorder, which responded to haloperidol therapy (Turley 1988). Particularly striking in some HSE survivors is an amnesic syndrome (Kapur 1988).

Hearing loss and vestibular disturbances, motor weakness, intellectual impairment, and visual acuity decrease have been found in survivors of CNS mumps virus infection, (Julkunen et al. 1985; Oldfelt 1949). Headache, irritability, and dizziness also are reported. In a follow-up of 47 patients with a history of mumps encephalitis 1 to 15 years earlier, Julkunen and colleagues (1985) found 12 to be experiencing moderate clinical sequelae, while 35 had mild sequelae or were normal.

Encephalitis following measles, rubella, and varicella is parainfectious, involving immune-mediated mechanisms. Miller and associates (1956) characterized the neurological complications following infections by whether the involvement was primarily encephalitic, myelitic, or polyradicular. Following measles, 95% of cases were classified as primarily encephalitic, with a mortality of 27%. However, mortality was found to be 15% after the introduction of antibiotics. On discharge, 42% had sequelae, including motor defects, intellectual impairment, psychiatric disability, and seizures. More recently, Johnson and coworkers (1984) studied 19 patients with ADEM after measles, all of whom survived and were able to return home. However, of the 10 patients examined from 6 weeks to 2 years later, only one patient was without sequelae. Severity of the acute phase did not have prognostic value for the type or degree of sequelae. Following rubella, Miller and associates (1956) found that 74 of 80 cases were primarily encephalitic. Mortality was 20%, but re-

covery was rapid, and sequelae were found in only 8.5% of survivors. ADEM following chicken pox remains a significant clinical problem because vaccination against VZV is not in general use. Miller and coworkers (1956) report that 90% of the parainfectious neurological problems following varicella were encephalitic, with a mortality of 11%. Sequelae, including seizures, retardation, ataxia, and psychiatric symptoms, were found in 16% of survivors. Johnson and Milbourne (1970) analyzed 57 cases and divided them into cerebral and cerebellar types. The cerebral group, which included patients in coma and semi-coma, experienced a mortality of 35%, and 13% had residua. The cerebellar group contained no patients in coma or semi-coma, and most had ataxia or incoordination in the acute phase. There were no deaths, and all patients recovered.

Mortality and the incidence of sequelae vary greatly among the arboviruses (Booss and Esiri 1986). The most frequent arbovirus in the United States, St. Louis encephalitis, has a mortality of less than 10%. In epidemic form, the disease is most severe in the elderly. Sequelae include a convalescent fatigue syndrome in up to half of the survivors. Tremor, but not parkinsonism, also has been described. Intellectual impairment is apparently uncommon. Mortality in Western equine encephalitis is less than 3%. The disease is most disabling in infants and children in whom sequelae include cognitive and behavioral abnormalities. Objective sequelae in adults are rare. La Crosse virus, a member of the California serogroup, is the second most common arbovirus infection in the United States. It infects children predominantly and has a mortality of less than 1%. Rapid reversal of neurological dysfunction occurs in the majority of cases with convalesence lasting up to 3 months. Seizures, hemiparesis, and intellectual impairment have been found in a few patients with minimal motor and coordination difficulties, which were reported in other patients. The greatest mortality among arbovirus infections in the United States is found in Eastern equine encephalitis, with children and the elderly most severely affected. Severe residual impairment is found in many children; return to normal function is more common in adults. Worldwide, Japanese B encephalitis and the Far Eastern subtype of tick–borne encephalitis have been associated with significant mortality and sequelae in survivors. About 37% of Indian children with laboratory confirmed Japanese encephalitis succumbed while in hospital (Kumar et al. 1990). An extrapyramidal syndrome was seen in about 22% of children who had a prolonged type of illness. Other sequelae previously noted have been motor, coordination, and speech defects as well as mental retardation (or dementia) and psychological impairment (Shoji et al. 1990). The Central European subtype of tick-borne encephalitis is widely distributed in Europe but is less destructive than the Far Eastern subtype. Mortality has varied in different studies but is less than 1%. Because of occasional cervical spinal cord involvement, sequelae include atrophy and weakness of the arms, shoulders, or neck. Other sequelae include facial weakness and Kozhevnikov's focal continuous epilepsy.

Seizures commonly are noted among the sequelae of many types of encephalitis. Annegers and colleagues (1988) found the 20-7 year risk of unprovoked seizures was 22% in patients who experienced seizures during acute encephalitis and 10% for patients who were without seizures during the acute phase. Rasmussen's encephalitis is a rare syndrome which includes intractable partial epilepsy. Recent data suggest that CMV may be a potential etiology (Power et al. 1990). Finally, two reports of rare sequelae of encephalitis are of interest. Recurrent bouts of morbid hunger and periodic somnolence (Kleine-Levin syndrome) were observed in a patient surviving encephalitis of undetermined type (Merriam 1986). In the second report, recurrent failure of automatic respiration (Ondine's curse) was described in a patient who died following encephalitis of undetermined etiology (Giangaspero et al. 1988). At autopsy, findings compatible with encephalitis were found in the hypothalamus, midbrain, pons, and medulla.

Myelitis

Various syndromes of spinal cord dysfunction with differing prognoses can result from virus infection. Polio, for example, produces a syndrome of asymmetric lower motor neuron disease. In contrast, both acute transverse myelitis (ATM) and ascending myelitis reflect involvement of long tract motor, sensory, and sphincter dysfunction as well as focal involvement. ATM and ascending myelitis can each be caused by a parainfectious immune-mediated mechanism or direct infection. These acute syndromes are a contrast to the slowly evolving myelopathies associated with human T-lymphotropic virus type I (HTLV-I) infection.

Poliomyelitis

Polio remains a significant cause of paralysis in countries without effective or thorough vaccine distribution. Because of the massive epidemics of polio, a great deal is known of its clinical expression, including factors that influence prognosis (Bodian and Horstmann 1965; Modlin 1985). Ninety percent of infected patients have inapparent infection, and less than 8% have a minor systemic illness. Of the remainder, some develop aseptic meningitis only (nonparalytic), and others progress to paralysis. Overall, one in 1000 cases of infection develop into paralysis. Virulence of the virus strain, genetic susceptibility of the host, and certain host factors influence the severity of the disease. Polio is more severe, and death is more common in adults. Paralysis is more common in boys up to 15 years old. Other factors that increase the risk of paralytic disease include pregnancy, tonsillectomy, recent inoculations, trauma, and exercise. Tonsillectomy, particularly within the preceding 2 weeks, was found to increase the likelihood of bulbar involvement. During the early phase of infection, vigorous exercise and trauma are associated with more severe paralysis. In oral vaccine-associated cases, the presence of immunodeficiency should be determined (Nkowane et al. 1987).

Different clinical forms of the illness have

been associated with widely differing rates of mortality, depending on the incidence of respiratory failure (Bodian and Horstmann 1965; Modlin 1985). Thus, while the overall rate of mortality varies from 5% to 10%, the rate increases markedly in the spinobulbar and pure bulbar forms. Mortality from respiratory failure also can occur in the spinal form following neural compromise of the diaphragm and intercostal muscles.

The acute evolution of paralysis usually is terminated 3 to 7 days after onset, along with the fever (Bodian and Horstmann 1965; Modlin 1985). Death is most likely during the first week of paralysis. Strength returns within the first month following reversible lesions, and most recovery is accomplished by 6 to 9 months. However, some recovery can continue up to 2 years. It has been estimated that one third of patients will have residua. Those with the greatest degree of paralysis acutely are most likely to suffer permanent residua. Other viruses, such as mumps virus and other enteroviruses, can cause a polio-like syndrome. In general, resolution in these cases is more benign than in polio.

Two to 4 decades following acute poliomyelitis, some patients develop a slowly progressive syndrome termed progressive post-poliomyelitis muscular atrophy (PPMA). Weakness may occur in muscles recovered from or uninvolved in the original illness. Dalakas and coworkers (1986), found the process to be focal, slow, and not life-threatening, with an average decline in strength of 1% per year. PPMA may result from the death of individual nerve terminals rather than the motor neuron itself. Klingman and associates (1988) found that the risk of developing PPMA was associated with relatively greater functional recovery following paralysis in the acute phase. It is possible that the increased metabolic demands placed on surviving neurons that had extended their fields of innervation ultimately are too great (Dalakas et al. 1986; Klingman et al. 1988). Intrathecal oligoclonal IgM bands specific for polio virus in certain patients suggest possible viral persistence as an alternative explanation (Sharief et al. 1991).

Parainfectious Myelitis

A myelitic component may occur in a clinical constellation of lesions as part of ADEM following a recognized viral illness. Myelitis, ATM, or ascending myelitis also can occur in isolation following viral infections. Miller and coworkers (1956) found myelitic forms of ADEM in 3% or less of cases following measles, rubella, and chicken pox. Mortality ranged from 22% following measles to 0% following chicken pox. Death was primarily associated with respiratory and bulbar muscle compromise. Full functional recovery occurred in one half of the cases following measles and in all of the cases following chicken pox.

Acute transverse myelitis is a clinical syndrome with diverse causes and variable outcomes (Lipton and Teasdall 1973). About one third of patients with this syndrome have a recent history of nonspecific viral illness, most commonly an upper respiratory infection. A history of preceding infection is more commonly obtained in younger patients (Paine and Byers 1953). Three clinical courses were observed by Ropper and Poskanzer (1978). The most rapid form was established within less than 1 to 12 hours. The most common course stabilized within 2 weeks. A few patients had a more prolonged course with stuttering progression up to 4 weeks.

The majority of patients with an explosive onset had a poor outcome in Ropper and Poskanzer's (1978) study. Overall, 33% had a good outcome (no or little disfunction), 42% had a fair outcome (functional and ambulatory but major symptoms of spinal cord disfunction), and 25% a poor outcome (chair- or bed-bound), including 8% who died. Sudden onset and back pain were predictive of a poor outcome. Similar outcomes were observed by Berman and colleagues (1981). These authors found that recovery usually took place between 4 weeks and 3 months of onset. Recovery continued to 1 year and up to 2 years in some. Except for rapid onset and back pain, Ropper and Poskanzer (1978) found no factors that predict outcome. Thus, age, sex, history of preceding infectious illness, CSF pleocytosis and protein elevation, and type of therapy were not found to influence the outcome. Lipton and Teasdall (1973) found that patients with spinal shock had a less favorable prognosis than those who retained posterior column function and tendon reflexes. The late development of multiple sclerosis is unusual, found in only 7% in a review of the literature by Ropper and Poskanzer (1978).

Myelitis Following Direct Infection

In some cases of transverse or ascending myelitis, the clinical circumstances suggest direct infection of the spinal cord. Zoster myelitis, for example, may occur in the context of a characteristic rash. Although vasculitis and other demyelinating mechanisms may exist, direct infection appears to play a prominent role (Cho et al. 1989). The rationale for treating such cases with an antiviral compound, such as acyclovir or adenine arabinoside, appears sound; however, there is insufficient reported experience. Similarly, myelitis, radiculomyelitis, or sacral myeloradiculitis associated with genital or sacral herpes simplex infection merit antiviral therapy, but there are scant data on which to base a prognosis. In AIDS, an acute clinical picture of ascending myeloradiculitis should suggest CMV-related progressive radiculopathy and treatment with ganciclovir instituted (Miller et al. 1990). This entity and the vacuolar myelopathy of AIDS are considered in Chapter 17 in this book.

HTLV-I

Two clinical syndromes recognized in association with infection by the retrovirus HTLV-I have clinical courses very similar to the spinal form of chronic multiple sclerosis. Long recognized in the tropics, endemic tropical spastic paraparesis (TSP) has been associated with HTLV-I (Gessain et al. 1985). This sparked the recognition and description of a similar syndrome in Japan, HTLV-I–associated myelopathy (HAM) (Osame et al. 1986). Predominently affecting

the spinal cord, TSP is of insidious onset and gradual progression. Sensory and sphincter dysfunction occur, but the predominent disability is weakness and spasticity of gait. Use of ambulatory aids often are required, and some patients are bed-ridden after 1 decade. One third of patients become incapacitated in 2 years or less (Roman et al. 1987). Immunosuppression, particularly with steroids, has been tried. Recent preliminary results with danazol merit further study (Harrington et al. 1991). Anti-viral therapy with zidovudine, ineffective in the short term, merits further study (Gout et al. 1991). Despite the association of HTLV-I and adult T-cell leukemia–lymphoma (ATL-L). The association between ATL-L and TSP or HAM is, with rare exception, not found.

Reye-Johnson Syndrome

Reye-Johnson syndrome, or Reye's syndrome as it is commonly known, deserves separate attention in the context of cerebral disorders following acute viral infection. It appears to be a metabolic encephalopathy in children triggered by viral infection, with an epidemiological link to aspirin ingestion. Two recent developments influence the understanding of its incidence. First, there has been a marked decline in the incidence of Reye's syndrome in the United States during a time when aspirin use by children also has declined (Hurwitz 1988). Second, there has been an increasing recognition that certain inborn errors of metabolism can mimic the clinical appearance of Reye's syndrome, which may have been misdiagnosed in the past (Rowe et al. 1988).

Some factors are clearly linked to outcome in Reye's syndrome. Clinical staging, or stages of coma, use four or five levels, depending on the classification, to describe clinical progression and response to therapy. One of the clearest prognostic factors is the speed of disease recognition, hospitalization, and institution of therapy. With institution of intravenous (IV) glucose and electrolyte therapy at an early stage, outcome is almost uniformly benign (Lovejoy et al. 1974; Dezateux et al. 1986; De Vivo

1985). In contrast, rapid progression into deeper stages of coma and high blood ammonia levels (Lovejoy et al. 1974; Fitzgerald et al. 1982) indicate severity and potential mortality. Once a patient demonstrates progression into deeper stages and has raised intracranial pressure, the goal in therapy becomes maintenance of cerebral arterial perfusion pressure. Many centers use invasive monitoring to intervene accurately (Shaywitz et al. 1986; Jenkins et al. 1987). Using such a protocol, Shaywitz and associates (1986) achieved an overall survival rate of 93% in 55 cases, with 80% recovering completely. Three children had severe residua, whereas another three had mild defects. Jenkins and coworkers (1987) found that patients who failed to maintain a perfusion pressure above 40 mm Hg died or had severe impairment.

During the acute and postacute phases, Goff and coworkers (1983) found that the return of somatic evoked potentials correlates with return of neurological function. During the recovery phase, recuperation may last up to 2 years. Shaywitz and colleagues (1982) observed irritability, agitation, and impulsiveness during this period.

Slow Virus Encephalopathies

The concept of "slow virus" disease as articulated by Sigurdsson (1954) includes a long incubation period and a protracted course, resulting in serious disease or death. The human slow virus encephalopathies, a heterogeneous set of conditions caused by diverse agents (Table 9-3), share these features. Two of the illnesses, progressive multifocal leukoencephalopathy and the AIDS dementia complex, are discussed in Chapter 17 in this book. This section focuses on subacute sclerosing panencephalitis (SSPE) and Creutzfeldt-Jakob disease (CJD).

SSPE

This uncommon disease, which usually strikes children between the ages of 5 and 15, is associated with prior measles virus infection. Although SSPE can follow vacci-

Table 9-3. Slow Virus Encephalopathies

DNA Virus
Progressive multifocal leukoencephalopathy
    Papovavirus
RNA Viruses
Subacute sclerosing panencephalitis
    Measles virus
Progressive rubella panencephalitis
    Rubella virus
Retrovirus
AIDS encephalopathy (AIDS dementia complex)
    Human immunodeficiency virus
Spongiform Agents
    Kuru
    Creutzfeldt—Jakob disease
    Gerstmann-Straussler syndrome (certain cases)

nation, the widespread application of measles vaccination has been associated with a fall in the average number of recorded SSPE cases in the United States from 48.6 per year for the years 1967 to 1971 to 4.2 per year for the years 1982 to 1986 (Dyken et al. 1989). Although the clinical course may vary considerably, the majority succumb between 9 months and 3 years. Unusually rapid courses leading to death or a vegetative state within weeks (Gilden et al. 1975) and prolonged remissions of up to several years (Risk et al. 1978) have been reported. Progression of the disease can be plotted through four stages (Jabbour et al. 1969). The onset is associated with altered mental function, regression, and behavioral changes and may persist for 1 to 2 months. The second stage is characterized by motor abnormalities of myoclonus, choreoathetosis, seizures, and tremors and also may last 1 to 2 months. Stage three is associated with coma and opisthotonos and may last 1 to 3 months. In stage four, hypertonia and myoclonus diminish, and roving eye movements and pathological laughter and crying occur. Since interruption in the progress of the disease can occur spontaneously, including a few cases with prolonged improvement (Risk et al. 1978), it is difficult to assess the effect of therapy. However, two large nonrandomized studies, using historical controls, reached the conclusion that inosiplex slows the progression of the disease or prolongs life in some cases (Jones et al. 1982;

Fukuyama et al. 1987). Continued attention also has been directed to interferon therapy, but the results have been variable (Crols and Lowenthal 1987).

Progressive Rubella Panenencephalitis

This exceedingly rare illness, resulting from congenital or childhood infection with rubella virus, is a slowly progressive disorder of childhood (Waxham and Wolinsky 1984). Learning disability, ataxia, seizures, and spasticity evolve to a fatal conclusion in 10-years. No therapy has been demonstrated to retard the progress of the illness.

Creutzfeldt-Jakob Disease

One of a group of illnesses of animals and humans transmitted by atypical agents (spongiform agents, prions), CJD is a progressive, dementing disorder. The course ends fatally for 90% of patients with 1 year and 95% within 2 years (Brown et al. 1986). Long-term survival up to 13 years has been reported in a small percentage of cases (Brown et al. 1984). Mental deterioration has been found as the first symptom in two thirds of patients (Brown et al. 1986). Ultimately, three fourths of patients evolve to a clinical state of dementia and myoclonus with a periodic EEG. The mean duration of illness was 7.6 months, and the youngest patients had the longest courses (Brown et al. 1986). The two other transmissible encephalopathies of humans are kuru, which is disappearing from the highlands of New Guinea with the cessation of cannabalism, and certain cases classified as the Gerstmann-Straussler syndrome. The latter syndrome, characterized clinically by ataxia and dementia, is associated with widespread amyloid plaques in the brain (Masters et al. 1981). These cases tend to have an earlier onset in life than CJD and an average survival time of almost 5 years. Bovine spongiform encephalopathy has received considerable recent attention in the United Kingdom because of human consumption of beef (Collee 1990). No therapy has been demonstrated to alter the course of any of the transmissible spongiform encephalopathies.

References

Adair, C. V.; Gauld, R. L.; Smadel, J. E. Aseptic meningitis, a disease of diverse etiology: clinical and etiologic studies on 854 cases. Ann. Intern. Med. 39:675–704; 1953.

Annegers, J. F.; Hauser, W. A.; Beghi, E.; Nocolosi, A.; Kurland, L. T. The risk of unprovoked seizures after encephalitis and meningitis. Neurology 38:1407–1410; 1988.

Appelbaum, E.; Rachelson, M. H.; Dolgolpol, V. B. Varicella encephalitis. Am. J. Med. 15:223–30; 1953.

Baer, G. M.; Fishbein, D. B. Rabies post-exposure prophylaxis. N. Engl. J. Med. 316:1270–1272; 1987.

Beghi, E.; Nicolosi, A.; Kurland, L. T.; Mulder, D. W.; Hauser, W. A.; Shuster, L. Encephalitis and aseptic meningitis, Olmsted County, Minnesota, 1950–1981: I. Epidemiology. Ann. Neurol. 16:283–294; 1984.

Behan, P. O.; Currie, S. Acute haemorrhagic leucoencephalitis. Clinical neuroimmunology. Philadephia: W. B. Saunders Co.; 1978: p. 34–48.

Berlit, P. The prognosis and long term course of viral encephalitis. J. Neuroimmunol. 20:117–125; 1988.

Berman, M.; Feldman, S.; Alter, M.; Zilber, N.; Kahana, E. Acute transverse myelitis: incidence and etiologic considerations. Neurology 31:966–971; 1981.

Bodian, D.; Horstmann, D. M. Polioviruses. In: Horsfall, F. L. Jr.; Tamm, I. eds. Viral and Rickettsial Infections of Man. 4th ed. Philadelphia: J. B. Lippincott; 1965: p. 430–473.

Booss, J.; Esiri, M. M. Viral encephalitis. Pathology, diagnosis and management. Oxford: Blackwell Scientific Publications; 1986.

Brown, P.; Cathala, F.; Castaigne, P.; Gajdusek, D. C. Creutzfeldt-Jakob disease: clinical analysis of a consecutive series of 230 neuropathologically verified cases. Ann. Neurol. 20:597–602; 1986.

Brown, P.; Rodgers-Johnson, P.; Cathala, F.; Gibbs, C.J. Jr.; Gajdusek, D. C. Creutzfeldt-Jakob disease of long duration: clinicopathological characteristics, transmissibility and differential diagnosis. Ann. Neurol. 16:295–304; 1984.

Buttner, T.; Dorndorf, W. Prognostic value of computed tomography and cerebrospinal fluid analysis in viral encephalitis. J. Neuroimmunol. 20:163–164; 1988.

Cho, E. S.; Petito, C. K.; Devinsky, O.; Price, R. W. Neuropathology of varicella zoster virus (VZV) myelitis (Abstract). J. Neuropathol. Exp. Neurol. 48:314; 1989.

Collee, J. G. Foodborne Illness. Bovine spongiform encephalopathy. Lancet 336:1300–1303; 1990.

Crols, R.; Lowenthal, A. Long-term intramuscular recombinant DNA interferon alpha-2 therapy in subacute sclerosing panencephalitis: reduction of serum measles antibodies without clinical improvement. Eur. Neurol. 27:72–77; 1987.

Dalakas, M. C.; Elder, G.; Hallett, M.; Ravits, J.; Baker, M.; Papadopoulos, N.; Albrecht, P.; Sever, J. A long-term follow-up study of patients with post-poliomyelitis neuromuscular symptoms. N. Engl. J. Med. 314:959–963; 1986.

De Vivo, D. C. Reye syndrome. Neurol. Clin. 3:95–115; 1985.

Dezateux, C. A.; Dinwiddie, R.; Helms, P.; Matthew, D. J. Recognition and early management of Reye's Syndrome. Arch. Dis. Child. 61:647–651; 1986.

Dunkle, L. M.; Arvin, A. M.; Whitley, R. H.; Rotbart, H. A.; Feder, H. M. Jr.; Feldman, F.; Gershon, A. A.; Levy, M. L.; Hayden, G. F.; McGuirt, P. V.; Harris, J.; Balfour, H. H. Jr. A controlled trial of acyclovir for chicken pox in normal children. N. Engl. J. Med. 325:1539–1544; 1991.

Dyken, P. R.; Cunningham, S. C.; Ward, L. C. Changing character of subacute sclerosing panencephalitis in the United States. Pediatr. Neurol. 5:339–341; 1989.

Fitzgerald, J. F.; Clark, J. H.; Angelides, A. G.; Wyllie, R. The prognostic significance of peak ammonia levels in Reye syndrome. Pediatrics 70:997–1000; 1982.

Fukuyama, Y. (Chairman); the Inosiplex–SSPE Research Committee. Clinical effects of MND-19 (Inosiplex) on subacute sclerosing panencephalitis: a multi-institutional collaborative study. Brain Dev. 9:270–282; 1987.

Gessain, A.; Barin, F.; Vernant, J. C.; Gout, O.; Maurs, L.; Calendar, A.; De The, G. Antibodies to human T-lymphotropic virus type-1 in patients with tropic spastic paraparesis. Lancet 2:407–410; 1985.

Giangaspero, F.; Schiavina, M.; Sturani, C.; Mondini, S.; Cirignotta, F. Failure of automatic control of ventilation (Ondine's Curse) associated with viral encephalitis of the brainstem: a clinicopathologic study of one case. Clin. Neuropathol. 7:234–237; 1988.

Gilden, D. H.; Rorke, L. B.; Tanaka, R. Acute SSPE. Arch. Neurol. 32:644–646; 1975.

Goff, W. R.; Shaywitz, B. A.; Goff, G. D.; Reisenhauer, M. A.; Jasiorkowski, J. G.; Venes, J. L.; Rothstein, P. T. Somatic evoked potential evaluation of cerebral status in Reye Syndrome. Electroenceph. Clin. Neurophysiol. 55:388–398; 1983.

Gordon, B.; Selnes, O. A.; Hart, J., Jr.; Hanley, D. F.; Whitley, R. J. Long-term cognitive sequelae of acyclovir-treated herpes simplex encephalitis. Arch. Neurol. 47:646–647; 1990.

Gout, O.; Gessain, A.; Iba-Zizan, M. T.; Kouzan, S.; Bolgert, F.; deThé G.; Lyon-Caen, O. The effect of zidovudine on chronic myelopathy associated with HTLV-1. J. Neurol. 238:108–110, 1991.

Greenwood, R.; Bhalla, A.; Gordon, A.; Roberts, J. Behavior disturbances during recovery from herpes simplex encephalitis. J. Neurol. Neurosurg. Psychiat. 46:809–817; 1983.

Harrington, W. J. Jr.; Sheremata, W. A.; Snodgrass, S. R.; Emerson, S.; Phillips, S.; Berger, J. R. Tropical spastic parapanosesis/HTLV-1 associated myelopathy (TSP/HAM): treatment with the anabolic steroid danazol. AIDS research and human retroviruses 7:1031–1034; 1991.

Hattwick, M. A. W.; Weis, T. T.; Stechschulte, C. J.; Baer, G. M.; Gregg, M. B. Recovery from rabies. A case report. Ann. Intern. Med. 76:931–942; 1972.

Hemachudha, T.; Phanuphak, P.; Johnson, R. T.; Griffin, D. E.; Ratanavongsiri, J.; Siriprasomsup, W. Neurologic complications of semple-type rabies vaccine: clinical and immunologic studies. Neurology 37:550–556; 1987.

Heubi, J. E.; Daugherty, C. C.; Partin, J. S.; Partin, J. C.; Schubert, W. K. Grade I Reye's syndrome—outcome and predictors of progression to deeper coma grades. N. Engl. J. Med. 311:1539–1542; 1984.

Hurwitz, E. S. The changing epidemiology of Reye's syndrome in the United States: further evidence for a public health success. J.A.M.A. 260:3178–3180; 1988.

Jabbour, J. T.; Garcia, J. H.; Lemmi, H.; Ragland, J.; Duenas, D. A.; Sever, J. L. Subacute sclerosing panencephalitis. A multidisciplinary study of eight cases. J.A.M.A. 207:2248–2254; 1969.

Jemsek, J.; Greenberg, S. B.; Taber, L.; Harvey, D.; Gershon, A.; Couch, R. B. Herpes zoster-associated encephalitis: clinicopathologic report of 12 cases and review of the literature. Medicine 62:81–97; 1983.

Jenkins, J. G.; Glasgow, J. F. T.; Black, G. W.; Fannin, T. F.; Hicks, E. M.; Keilty, S. R.; Crean, P. M. Reye's Syndrome: assessment of intracranial monitoring. Br. Med. J. 294:337–338; 1987.

Jennett, B.; Teasdale, G.; Galbraith, S.; Pickard, J.; Grant, H.; Braakman, R.; Avezaat, C.; Maas, A.; Minderhoud, J.; Vecht, C. J.; Heiden, J.; Small, R.; Caton, W.; Kurze, T. Severe head injuries in three countries. J. Neurol. Neurosurg. Psychiat. 40:291–298; 1977.

Johnson, R.; Milbourn, P. E. Central nervous system manifestations of chickenpox. Can. Med. Assoc. J. 102:831–834; 1970.

Johnson, R. T.; Griffin, D. E.; Hirsch, R. L.; Wolinsky, J. S.; Roedenbeck, S.; de Soriano, I. L.; Vaisberg, A. Measles encephalomyelitis—clinical and immunologic studies. N. Engl. J. Med. 310:137–141; 1984.

Jones, C. E.; Dyken, P. R.; Huttenlocher, P. R.; Jabbour, J. T.; Maxwell, K. W. Inosiplex therapy in subacute sclerosing panencephalitis. A multicentre, non-randomized study in 98 patients. Lancet 1:1034–1037; 1982.

Julkunen, I.; Koskiniemi, M.; Lehtokoski-Lehtiniemi, E.; Sainio, K.; Veheri, A. Chronic mumps virus encephalitis. Mumps antibody levels in cerebrospinal fluid. J. Neuroimmunol. 8:167–175; 1985.

Kaplan, M. H.; Klein, S. W.; McPhee, J.; Harper, R. G. Group B coxsackievirus infections in infants younger than three months of age: a serious childhood illness. Rev. Inf. Dis. 5:1019–1032; 1983.

Kapur, N. Selective sparing of memory functioning in a patient with amnesia following herpes encephalitis. Brain Cogn. 8:77–90; 1988.

Kennard, C.; Swash, M. Acute viral encephalitis. Its diagnosis and outcome. Brain 104:129–148; 1981.

Kennedy, C. R.; Duffy, S. W.; Smith, R.; Robinson, R. O. Clinical predictors of outcome in encephalitis. Arch. Dis. Child. 62:1156–1162; 1987.

Kennedy, P. G. E.; Adams, J. H.; Graham, D. I.; Clements, G. B. A clinico-pathological study of herpes simplex encephalitis. Neuropath. Appl. Neurobiol. 14:395–415; 1988.

Kitamoto, I.; Nakayama, M.; Miyazaki, C.; Minami, T.; Kurokawa, T.; Ueda, K. Evoked potentials in neonates and infants with aseptic meningitis. Pediatric Neurology 5:342–346; 1989.

Klingman, J.; Chui, H.; Corgiat, M.; Perry, J. Functional recovery. A major risk factor for the development of postpoliomyelitis muscular atrophy. Arch. Neurol. 45:645–647; 1988.

Koskiniemi, M.; Vaheri, A. Effect of measles, mumps, rubella vaccination on pattern of encephalitis in children. Lancet 1:31–34; 1989.

Kumar, R.; Mathur, A.; Kumar, A.; Sharma, S.; Chakraborty, S.; Chaturvedi, U. C. Clinical features and prognostic indicators of Japanese encephalitis in children in Lucknow (India). Indian J. Med. Res. (A)91:321–327; 1990.

Lepow, M. L.; Coyne, N.; Thompson, L. B.; Carver, D. H.; Robbins, F. C. A clinical, epidemiologic and laboratory investigation of aseptic meningitis during the four year period, 1955–1958. II. The clinical disease and its sequelae. N. Engl. J. Med. 266:1188–1193; 1962.

Lipton, H. L.; Teasdall, R. D. Acute transverse myelopathy in adults. Arch. Neurol. 28:252–257; 1973.

Lovejoy, F. H. Jr.; Smith, A. L.; Bresnan, M. J.; Wood, J. N.; Victor, D. I.; Adams, P. C. Clinical staging in Reye syndrome. Am. J. Dis. Child. 128:36–41; 1974.

Masters, C. L.; Gajdusek, D. C.; Gibbs, C. J. Jr.; Creutzfeldt-Jakob disease virus isolations from the Gerstmann-Straussler syndrome. With an analysis of the various forms of amyloid plaque deposition in the virus-induced spongiform encephalopathies. Brain 104:559–588; 1981.

McKinney, R. E.; Katz, S. L.; Wilfert, C. M. Chronic enteroviral meningoencephalitis in agammaglobulinemic patients. Rev. Inf. Dis. 9:334–356; 1987.

Merriam, A. E. Kleine-Levin syndrome following acute viral encephalitis. Biol. Psychiat. 21:1301–1304; 1986.

Meyer, H. M.; Johnson, R. T.; Crawford, I. P.; Dascomb, H. E.; Rogers, N. G. Central nervous system syndromes of "viral" etiology. A Study of 713 Cases. Am. J. Med. 129:334–347; 1960.

Miller, D. H.; Ormerod, I. E. C.; Rudge, P.; Kendall, B. E.; Moseley, I. F.; McDonald, W. I. The early risk of multiple sclerosis following isolated syndromes of the brainstem and spinal cord. Ann. Neurol. 26:635–639; 1989.

Miller, H. G. Acute disseminated encephalomyelitis treated with A.C.T.H. Br. Med. J. 1:177–183; 1953.

Miller, H. G.; Stanton, J. B.; Gibbons, J. L. Para-infectious encephalomyelitis and related syndromes. A critical review of the neurological complications of certain specific fevers. Q. J. Med. 25:427–505; 1956.

Miller, J. D.; Ross, C. A. C. Encephalitis. A four year study. Lancet 1:1121–1126; 1968.

Miller, R. G.; Storey, J. R.; Greco, C. M. Ganciclovir in the treatment of progressive AIDS-related polyradiculopathy. Neurology 40:569–574; 1990.

Modlin, J. F. Poliovirus. In: Mandell, G. L.; Douglas, R. G. Jr.; Bennett, J. E. eds. Principles and practice of infectious diseases. 2nd ed. New York: John Wiley and Sons; 1985: p. 806–814.

Morgello, S.; Cho E-S.; Nielsen, S.; Devinsky, O.; Petito, C. K. Cytomegalovirus encephalitis in patients with acquired immunodeficiency syndrome: an autopsy study of 30 cases and a review of the literature. Hum. Pathol. 18:289–297; 1987.

Nicolosi, A.; Hauser, W. A.; Beghi, E.; Kurland, L. T. Epidemiology of central nervous system infections in Olmsted County, Minnesota, 1950–1981. J. Inf. Dis. 154:399–408; 1986.

Nkowane, B. M.; Wassilak, S. G. F.; Orenstein, W. A.; Bart, K. J.; Schonberger, L. B.; Hinman, A. R.; Kew, O. M. Vaccine-associated paralytic poliomyelitis. United States: 1973 through 1984. J.A.M.A. 257:1335–1340; 1987.

Oldfelt, V. Sequelae of mumps-meningoencephalitis. Acta. Medica. Scand. 84:405–414; 1949.

Osame, M.; Usuku, K.; Izumo, S.; Ijich, N.; Amitani, H.; Igata, A.; Matsumoto, M.; Tara, M. HTLV-I associated myelopathy, a new clinical entity. Lancet 1:1031–1032; 1986.

Paine, R. S.; Byers, R. K. Transverse myelopathy in childhood. Am. J. Dis. Child 85:151–163; 1953.

Pasternak, J. F.; DeVivo, D. C.; Prensky, A. L. Steroid-responsive encephalomyelitis in childhood. Neurology 30:481–486; 1980.

Pearl, P. L.; Abu-Farsakh, H.; Starke, J. R.; Dreyer, Z.; Louis, P. T.; Kirkpatrick, J. B. Neuropathology of two fatal cases of measles in the 1988–1989 Houston epidemic. Pediatr. Neurol. 6:126–130; 1990.

Peterslund, N. A. Herpes zoster associated encephalitis: clinical findings and acyclovir treatment. Scand. J. Infect. Dis. 20:583–592; 1988.

Petito, C. K.; Cho, E.-S.; Lemann, W.; Navia, B. A.; Price, R. W. Neuropathology of acquired immunodeficiency syndrome (AIDS); an Autopsy Review. J. Neuropathol. Exp. Neurol. 45:635–646; 1986.

Power, C.; Poland, S. D.; Blume W. T.; Girvin, J. P.; Rice, G. P. A. Cytomegalovirus and Rasmussen's encephalitis. Lancet 336:1282–1284; 1990.

Przelomski, M. M.; O'Rourke, E.; Grady, G. F.; Berardi, V. P.; Markley, H. G. Eastern equine encephalitis in Massachusetts: a report of 16 cases, 1970–1984. Neurology 38:736–739; 1988.

Rantakallio, P.; Leskinen, M.; Von Wendt, L. Incidence and prognosis of central nervous system infectious in a birth cohort of 12,000 children. Scand. J. Infect. Dis. 18:287–294; 1986.

Rautonen, J.; Koskiniemi, M.; Vaheri, A. Prognostic factors in childhood acute encephalitis. Pediatr. Inf. Dis. 10:441–446; 1991.

Risk, W. S.; Haddad, F. S.; Chemali, R. Substantial spontaneous long-term improvement in subacute sclerosing panencephalitis. Six cases from the middle east and a review of the literature. Arch. Neurol. 35:494–502; 1978.

Roman, G. C.; Spencer, P. S.; Schoenberg, B. S.; Hugon, J.; Ludolph, A.; Rodgers-Johnson, P.; Osuntokun, B. O.; Shamlaye, C. F. Tropical spastic paraparesis in the Seychelles Islands: a clinical and case-control neuroepidemiologic study. Neurology 37:1323–1328; 1987.

Ropper, A. H.; Poskanzer, D. C. The prognosis of acute and subacute transverse myelopathy based on early signs and symptoms. Ann. Neurol. 4:51–59; 1978.

Rowe, P. C.; Valle, D.; Brusilow, S. W. Inborn errors of metabolism in children referred with Reye's Syndrome. A changing pattern. J.A.M.A. 260:3167–3170; 1988.

Seales, D. and Greer, M. Acute hemorrhagic leukoencephalitis: A successful recovery. Arch. Neurol. 48:1086–1088; 1991.

Sells, C. J.; Carpenter, R. L.; Ray, C. G. Sequelae of central nervous system enterovirus infections. N. Engl. J. Med. 293:1–4; 1975.

Sharief, M. K.; Hentges, R.; Chiardi, M. Intrathecal immune response in patients with the post-polio syndrome. N. Eng. J. Med. 325:749–755; 1991.

Sharma, S.; Mathur, A.; Prakash, V.; Kulshreshtha, R.; Kumar, R.; Chaturvedi, U. C. Japanese encephalitis virus latency in peripheral blood lymphocytes and recurrence of infection in children. Clin. Exp. Immunol. 85:85–89; 1991.

Shaywitz, B. A.; Lister, G.; Duncan, C. C. What is the best treatment for Reye's syndrome. Arch. Neurol. 43:730–731; 1986.

Shaywitz, S. E.; Cohen, P. M.; Cohen, D. J.; Mikkelson, E.; Morowitz, G.; Shaywitz, B. A. Long-term consequences of Reye Syndrome: a sibling-matched, controlled study of neurologic, cognitive, academic and psychiatric function. J. Pediat. 100:41–46; 1982.

Shoji, H.; Murakami, T.; Murai, I.; Kida, H.; Sato, Y.; Kojima, K.; Abe, T.; Okudera, T. A follow-up study by CT and MRI in 3 cases of Japanese encephalitis. Neuroradiology 32:215–219; 1990.

Sigurdsson, B. Rida, a chronic encephalitis of sheep. With general remarks on infections which develop slowly and some of their special characteristics. Br. Vet. J. 110:341–354; 1954.

Skoldenberg, B.; Forsgren, M.; Alestig, K.; Bergstrom, T.; Burman, L.; Dahlgqvist, E.; Forkman, A.; Fryden, A.; Lovgren, K.; Norlin, K.; Norrby, R.; Olding-Stenkvist, E.; Stiernstedt, G.; Uhnoo, I.; deVahl, K. Acyclovir versus vidarabine in herpes simplex encephalitis. Randomized multicenter study in consecutive Swedish patients. Lancet 2:707–71; 1984.

Turley, J. M. Tourette-like disorder after herpes encephalitis. Am. J. Psychiat. 145:1604–1605; 1988.

Wallgren, A. Une novelle maladie infectieuse du systeme nerveux central? Acta. Paediatrica. 4:158–182; 1925.

Waxham, M. N.; Wolinsky, J. S. Rubella virus and its effects on the nervous system. Neurol. Clin. 2:367–385; 1984.

Whitley, R. J. Viral encephalitis. N. Engl. J. Med. 323:242–250; 1990.

Whitley, R. J. Herpes simplex virus infections of the central nervous system. A Review. Am. J. Med. 85(suppl 2A):61–67; 1988.

Whitley, R.; Arvin, A.; Prober, C.; Burchett, S.; Corey, L.; Powell, D.; Plotkin, A.; Starr, S.; Alford, C.; Connor, J.; Jacobs, R.; Nahmias, A.; Soong, S.-J., and The National Institute of Allergy and Infections Diseases Collaborative Antiviral Study Group. A controlled trial comparing vidarabine with acyclovir in neonatal herpes simplex virus infection. N. Engl. J. Med. 324:444–449; 1991a.

Whitley, R.; Arvin, A.; Prober, C.; Corey, L.; Burchett, S.; Plotkin, S.; Starr, S.; Jacobs, R.; Powell, D.; Nahmias, A.; Sumaya, C.; Edwards, K.; Alford, C.; Caddell, G.; Soong, S.-J. and The National Institute of Allergy and Infectious Diseases Collaborative Antiviral Study Group. Predictors of morbidity and mortality in neonates with herpes simplex virus infections. N. Engl. J. Med. 324:450–454; 1991b.

Whitley, R. J.; Alford, C. A.; Hirsch, M. S.; Schooley, R. T.; Luby, J. P.; Aoki, F. Y.; Hanley, D.; Nahmias, A. J.; Soong, S.-J.; the NIAID Collaborative Antiviral Study Group. Vidarabine versus acyclovir therapy in herpes simplex encephalitis. N. Engl. J. Med. 314:144–149; 1986.

Whitley, R. J.; Cobbs, G.; Alford, C. A. Jr.; Soong, S.-J.; Hirsch, M. S.; Connor, J. D.; Corey, L.; Hanley, D. F.; Levin, M.; Powell, D. A.; The NIAID Collaborative Antiviral Study Group. Diseases that mimic herpes simplex encephalitis. Diagnosis presentation, and outcome. J.A.M.A. 262:234–239; 1989.

Whitley, R. J.; Soong, S.-J.; Hirsch, M. S.; Karchmer, A. W.; Dolin, R.; Galasso, G.; Dunwick, J. K.; Alford, C. A.; NIAID Antiviral Study Group. Herpes simplex encephalitis. Vidarabine therapy and diagnostic problems. N. Engl. J. Med. 304:313–318; 1981.

Whitley, R. J.; Soong, S.-J.; Dolin, R.; Galasso, G. J.; Ch'ien, L. T.; Alford, C. A.; the Collaborative Study Group. Adenine arabinoside therapy of biopsy proved herpes simplex encephalitis. National Institutes of Allergy and Infectious Diseases Collaborative Antiviral study. N. Engl. J. Med. 297:289–294; 1977.

Wilfert, C. M.; Thompson, R. J. Jr.; Sunder, T. R.; O'Quinn, A.; Zeller, J.; Blacharsh, J. Longitudinal assessment of children with enteroviral meningitis during the first three months of life. Pediatrics 67:811–815; 1981.

Wong, V.; Yeung, C. Y. Acute viral encephalitis in children. Aust. Paeditr. J. 32:339–342; 1987.

Yahr, M. D.; Lobo-Antunes, J. Relapsing encephalomyelitis following the use of influenza vaccine. Arch. Neurol. 27:182–183; 1972.

10

Congenital Viral Infections

P. K. COYLE

Rubella

Natural History

Rubella is an enveloped single-stranded ribonucleic acid (RNA) togavirus (Waxham and Wolinsky 1984). Humans are the only known host for this ubiquitous agent, which invades the upper respiratory tract by way of infected aerosol droplets. Typical rubella is a very mild exanthematous illness lasting a few days. The maculopapular rash is centrifugal, spreading from face and neck to trunk and extremities. Postauricular lymphadenopathy and low-grade fever are frequent accompaniments of symptomatic infection, and arthralgias are noted in about 20% of adults. Diagnosis requires laboratory confirmation because clinical suspicion is incorrect in up to 35% of cases. At least two thirds of infections are subclinical, manifested solely by seroconversion.

Unlike rubella in children and adults, fetal infection is not benign (Cooper 1985). Intrauterine infection may be lethal. Pregnant women with confirmed rubella have a spontaneous abortion twice that of normal (4% to 5%) and a stillborn rate slightly above normal (1% to 2%). Congenital rubella is a chronic infection. Virus seeds the placenta, infects multiple fetal organs, and then persists throughout gestation and postnatally. Virus may be cultured for weeks to months and even years in the pharynx, urine, conjunctiva, feces, cerebrospinal fluid (CSF), bone marrow, and circulating leukocytes of infected infants. Rubella interferes with normal organ development. Infection may result in a normal infant or in an infant with congenital malformations, transient neonatal problems, or who appears healthy but develops problems months or years later (Table 10-1). The overall mortality rate for congenitally infected infants with rubella defects is 10% to 20% in the first year of life.

Prognosis

Fetal risk for infection and damage is predicted by two factors: the immune status of the mother and the timing of maternal rubella infection (Dudgeon 1975). Maternal infection requires a lack of specific viral immunity. Despite a vaccination program instituted in 1969, 10% to 20% of women of childbearing age remain susceptible to rubella. Among vaccinees, 1% to 3% have antibody titers that decrease over time,

135

Table 10-1. Abnormalities Associated With
Congenital Rubella

Cardiac
Patient ductus arteriosus, pulmonary stenosis,
septal defects
Ocular
Cataracts, glaucoma, microphthalmia, retinopathy
Hearing loss
Sensorineural
Neurological
Behavioral disturbances, cerebral palsy, meningo-
encephalitis, mental retardation, microcephaly
Other
Growth retardation, hepatosplenomegaly, osteitis,
purpura
Delayed
"Late-onset disease," endocrinopathies, hearing
loss, neurological, ocular, vascular

making reinfection possible. Primary or symptomatic maternal infection has significant fetal risk; secondary (recurrent) or asymptomatic maternal infection has a much lower risk. In a series that examined 269 infants, mothers with symptomatic infections transmitted the virus to 45% of infants, compared to 0% transmission from mothers with asymptomatic infections (Miller et al. 1982). In another series of 60 pregnant women with documented rubella, mothers with symptomatic infections transmitted the virus to 53% of fetuses and produced defects in 27% (Peckham 1974). Subclinical infections resulted in the virus being transmitted to 19% of fetuses and caused defects in only 4%. In a third study of 118 infants, mothers with symptomatic infections had a 31% transmission rate and 26% of these had defects, compared to a 13% transmission and no defects with asymptomatic infections (Grillner et al. 1983). Although maternal reinfection may result in a congenitally infected infant, it is unusual because most reinfections are subclinical.

The timing of maternal rubella infection is crucial (Table 10-2). Rubella shortly before conception (before the last menstrual period) carries almost no fetal risk (Enders et al. 1988). Rubella during the first 12 weeks of pregnancy carries a high risk of fetal infection and the greatest risk of fetal damage. Fetal defects are unusual when maternal rubella occurs after 17 weeks, despite an in-

creasing transmission rate during the last trimester. It is not clear why maternal rubella late in pregnancy is likely to produce an infected but healthy infant, except that organogenesis virtually is complete.

Permanent sequelae of congenital rubella include senorineural hearing loss (58%), cardiac disease (13%), ocular damage (13%), and neurological damage (8%). The type of rubella defect depends in large part on the timing of fetal infection (Ueda et al. 1979). Cataracts and cardiac disease are seen with infection in the first 8 weeks of pregnancy when these organs are being formed, but they are unusual after that time. Deafness, retinopathy (generally asymptomatic), and neurological damage occur with infection in the first 17 weeks. Multiple defects are common in the first 8 weeks and unusual after that.

Certain problems, such as mild hearing loss or brain damage, may be difficult to detect at the time of birth and may only become apparent later. In addition, because congenital rubella is a chronic infection, new defects may develop months or years after birth. An uncommon syndrome called "late-onset disease" may occur in congenitally infected infants who were either normal at birth or showed typical rubella defects (Tardieu et al. 1980). By 6 months of age, these infants develop interstitial pneu-

Table 10-2. Risk of Fetal Infection and Damage With Maternal Rubella Infection

Pregnancy (Weeks)	Fetal Infection (%)*	Rubella Defects (%)*
0–11	90%	87%
11–16	55%	60%
17–30	34%	<6%
≥31	70%	0%

* Based on 258 infants born to women with symptomatic confirmed rubella during pregnancy. (Miller E.; Cradock-Watson, J. E.; Pollock, T. M. Consequences of confirmed maternal rubella at successive stages of pregnancy. Lancet 1:781–784; 1982.)
† Based on 208 infected infants born to women with symptomatic confirmed rubella during pregnancy. (Miller E.; Cradock-Watson, J. E.; Pollock, T. M. Consequences of confirmed maternal rubella at successive stages of pregnancy. Lancet 1:781–784; 1982. Munro, N. D.; Sheppard, S.; Smithells, R. W.; Holzel, H.; Jones, G. Temporal relations between maternal rubella and congenital defects. Lancet 1:201–204; 1987.)

monia, rash, chronic diarrhea, and meningo-encephalitis. Laboratory accompaniments include low immunoglobulin (Ig) G, high Ig M, and circulating immune complexes. Mortality rate is high, and survivors may show severe neurological damage. A variety of other delayed manifestations have been reported in more than 20% of infants with congenital rubella who show defects at birth (Sever et al. 1985). The most frequent of these is type 1 diabetes mellitus, which has developed as late as 33 years of age. Late-onset hearing loss has been reported up to 10 years of age. A very rare slow virus infection, progressive rubella panencephalitis, may develop years after birth and cause fatal neurological disease.

Therapy

There is no definitive treatment for established congenital rubella. Therapeutic abortion may be performed based on the calculated risk to the fetus and informed parental choice. In the late-onset disease syndrome, there are anecdotal reports that steroids and plasmapheresis may be beneficial (Tardieu et al. 1980; Verder et al. 1986). Congenital infection must be confirmed by viral isolation in the first weeks of life and by documented rubella IgM antibodies in cord blood. Infected infants receive careful ophthalmological, medical, and neurological assessments that continue through early childhood. The definitive treatment for congenital rubella rests with preventing maternal infection through vigorous support of the immunization program. Women who do not have confirmed vaccination or rubella antibody titer are vaccinated at a time when they are unlikely to become pregnant for the next 3 months. Even if there is accidental exposure to a fetus, risk of damage from the attenuated vaccine virus is negligible.

Cytomegalovirus

Natural History

Cytomegalovirus (CMV) is a double-stranded deoxyribonucleic acid (DNA) herpesvirus. After initial infection, it persists in the host in latent form. CMV is ubiquitous, particularly in lower socioeconomic groups, with antibody prevalence rates of 60% to 100% (Zaia and Lang 1984). It is transmitted through body fluids and intimate contact. Primary infection generally occurs early in childhood or during the reproductive years. In normal hosts, infection is subclinical, although a few individuals develop a heterophile-negative mononucleosis syndrome. In immunocompromised hosts, infection is much more likely to be symptomatic. CMV infection cannot be diagnosed clinically and requires laboratory confirmation. Unlike rubella, primary infection does not protect a person from secondary infection. Most secondary infections are reactivations of latent virus, but reinfections with different strains also occur. Infection results in virus excretion from multiple sites, even in the presence of antibody. Excretion may continue for years, especially in neonatal infections.

CMV is the most common congenital viral infection and the leading infectious cause of mental retardation and sensorineural hearing loss (Demmler 1991). Infection occurs in 0.2% to 2.2% of all live births (Stagno and Whitley 1985). Like rubella, congenital infection with CMV may be asymptomatic or symptomatic at birth or result in delayed problems. Congenital infection is symptomatic at birth in 5% to 10% of infants; approximately one half show multisystem abnormalities characteristic of CMV inclusion body disease. Infants develop lethargy, respiratory distress, and seizures at birth or shortly thereafter. Characteristic features include thrombocytopenia with petechiae, jaundice, hepatosplenomegaly, and intrauterine growth retardation. Neurological abnormalities include psychomotor retardation, microcephaly, spasticity, hearing loss, seizures, chorioretinitis or optic atrophy, and hydrocephalus. CMV has a particular affinity for the rapidly growing subependymal cells and produces calcifications localized to this region and diffusely in brain in approximately 26% of infected infants. Calcifications are present at birth or develop during the first year of life.

Earlier series reported that more than 90% of symptomatic infants who survive developed late sequelae that were either not apparent or not present at birth (Stagno and Whitley 1985). However, two European studies suggest that the true rate may be much lower, in the range of 25% (Ahlfors et al. 1984). Because congenital CMV is a chronic infection, ongoing damage is not surprising, particularly within the central nervous system (CNS), which continues to develop after birth. Among infants with asymptomatic infection, 10% to 17% also develop late sequelae (Demmler 1991). The major late sequela is sensorineural hearing loss (bilateral in almost half), which develops in 10% to 24% of symptomatic and asymptomatic infections (Kumar et al. 1984; Preece et al. 1984; Saigal et al. 1982). Serious neurological sequelae (cerebral palsy, mental retardation, brain damage, behavioral problems, and lowered IQ) develop in 6% to 12% (Ahlfors et al. 1984; Griffiths and Baboonian 1984; Peckham et al. 1983; Preece et al. 1984). Late chorioretinitis and dental problems are rare but have been reported.

Intrapartum or postpartum CMV infection is probably more common than intrauterine infection (Alford et al. 1990). CMV can be transmitted to neonates by passage through an infected cervix, by virus excreted in breast milk, or by transfusion of infected blood. CMV is present in the cervical secretions of 0% to 2% of women in the first trimester, 6% to 10% in the second trimester, and 11% to 28% in the third trimester (Ho 1985). Transmission occurs in 12.5% of neonates born to women who excreted virus during the first two trimesters, 37% who excreted CMV during the third trimester, and 57% who excreted virus around the time of delivery (Ho 1985). When CMV is shed in breast milk, 58% to 69% of breast-fed babies become infected (Zaia and Lang 1984). When seropositive blood is used to transfuse infants, 14% to 30% become infected with CMV.

Prognosis

Prognosis of congenital CMV infection relates to the type of maternal infection and the presence or absence of symptoms at birth (Demmler 1991). Intrauterine infection occurs with primary and secondary maternal infection. Unlike congenital infections with rubella or toxoplasma, many studies have found that maternal antibody is an important disease marker. A high prevalence of maternal antibody implies a high rate of congenital infection with CMV (Demmler 1991). This is assumed to be on the basis of reactivation of endogenous maternal virus, although reinfection with a new strain is also possible. However, maternal immunity does provide a degree of fetal protection, because almost all symptomatic congenital infections occur with primary maternal infection (Zaia and Lang 1986). In the rare sequential pregnancies involving CMV, the first infant is more severely affected. Among susceptible women, 0.7% to 4.1% seroconvert during pregnancy. Seroconversion is most likely in younger, primiparous women from lower socioeconomic groups. In upper socioeconomic groups, the at-risk seronegative woman tends to be white, younger than age 30, college educated, bottle-fed as an infant, with no children at home (Walmus et al. 1988). The average role for fetal transmission primary maternal CMU infections is 40%, with reported rates in the range of 25% to 75%. Studies suggest that defective cell-mediated immunity to CMV in the mother may correlate with transmission (Stern et al. 1986). Transmission also may be more likely in mothers who develop CMV mononucleosis and mothers who have particularly intense prolonged antibody production (Alford et al. 1988). Unlike rubella, the timing of maternal infection does not affect the transmission rate. Congenital CMV occurs with infections in all trimesters, but fetal damage may be more likely when infection occurs early in pregnancy (Britt and Vugler 1990).

A recent animal study found that maternal and fetal genetic factors may modulate congenital CMV infection and its sequelae (Fitzgerald and Shellam 1991).

Sequelae are common when a congenitally infected infant is symptomatic at birth: 90% to 95% will suffer sequelae, including death in 10% to 20%. In contrast, only 10%

to 17% of asymptomatic infected infants will show later sequelae (Demmler 1991).

Inclusion body disease is the most severe form of congenital CMV, and early studies suggested a poor prognosis with a mortality rate of 20% to 30%. Severe hematologic or liver abnormalities, as well as ocular involvement, have correlated roughly with severity of infection. Hearing loss does not correlate. Laboratory accompaniments of more severe symptomatic infection include low platelets, elevated SGOT, hyperbilirubinemia, elevated cord IgM, and increased CSF protein (Pass et al. 1980).

Neurological sequelae are three times more likely to occur with symptomatic infections (Ahlfors et al. 1984), but they may develop in infants who appear normal at birth without microcephaly or CSF abnormalities (Pass et al. 1980). A recent series examined 22 congenital cases for laboratory and clinical correlates of poor outcome (Bale et al. 1988). Only abnormal neuroimaging studies (skull x-rays, sonograms, computed tomography [CT]), predicted adverse outcome; microcephaly, jaundice, organomegaly, thrombocytopenia, retinitis, and hearing loss did not predict adverse outcome.

With regard to intrapartum or postpartum infection, risk factors include women with a history of venereal disease and with a young age. Cervical viral excretion has been found to be unusual after age 31 (Knox et al. 1979). Transfusion-associated CMV in the neonate correlates with lack of maternal antibody, multiple transfusions, transfusions from seropositive donors, and low-birth weight infants (Zaia and Lang 1984). Unlike fetal infection, intrapartum or postpartum infection carries no immediate or long-term sequelae, except in premature infants. The neonate of low birth weight (less than 1200 g) is an immunocompromised host. By age 3 months, infants infected postnatally who survive the initial complications of prematurity may develop a syndrome characterized by fever, pallor, respiratory distress with pneumonia, hepatitis, organomegaly, hypoalbuminemia, and atypical lymphocytosis. The syndrome is self-limited over several weeks but may be fatal (Ballard et al. 1979).

Therapy

Treatment of asymptomatic congenital infection is not currently justified due to the risks associated with available antiviral drugs. Gancyclovir is being evaluated for treatment of symptomatic congenital infection. Gancyclovir is an acyclic nucleoside analog of acyclovir. It is virostatic at relatively low concentrations against all human herpesviruses, including CMV. Preliminary data available on the first 25 babies treated indicates that the drug markedly decreases but does not clear viral excretion. Neutropenia developed in several cases, but platelet counts improved and no significant toxicity was noted (Demmler 1991). Newer antiviral drugs, such as foscarnet, are potential therapies for severely affected infants. Other possible treatment modalities could combine antiviral drugs with steroids, intravenous or hyperimmune immunoglobulin, or specific monoclonal antibodies directed against major CMV antigens. Hyperimmune globulin can also be used to treat the mother when there is documented maternal infection, but its efficacy is unproven.

Current therapy is based on prevention of fetal infection. The best treatment is acquired immunity, and there is ongoing research to develop an effective live attenuated or subunit vaccine. No vaccine is yet available, however. At present, it is important for young women at risk to be aware of CMV, to know their serologic status, and to use hygienic measures when they are in at-risk surroundings. Infected newborns need to be identified as early as possible.

Congenital fetal infection is confirmed by virus isolation (generally from urine) in the first few days of life and by documenting CMV-specific IgM or IgE in cord blood (Nielsen et al. 1987). Infants are closely monitored at birth for CMV sequelae, and they receive aggressive symptomatic treatment as needed. Asymptomatic infected infants should receive careful longitudinal follow-up for the delayed problems that may require specific therapy. In the premature infant at risk for postnatal infection, transmission can be avoided by using only seronegative blood for transfusions and by screening the cervical and breast secretions of high-risk mothers.

Herpes Simplex Virus

Natural History

Herpes simplex virus (HSV) is another ubiguitous DNA herpesvirus. There are two distinct antigenic types, HSV-1 and HSV-2, which have different clinical and epidemiological patterns (Whitley and Hutto 1985). HSV-1 produces oropharyngeal lesions, is spread mainly by oral secretions, and has an antibody prevalence rate of 30% to 90%, depending on the group surveyed. HSV-2 produces genital lesions, is spread mainly by sexual contact, and has an antibody prevalence rate of 10% to 60%. Both HSV types are more common in lower socioeconomic groups, and HSV-2 is more common in sexually active groups. Following primary infection, HSV becomes latent, particularly in sensory nerve ganglia. Secondary infections mainly are due to reactivation of latent virus rather than reinfection.

Neonatal HSV infection now occurs in 0.02% to 0.05% of all newborns (Whitley and Hutto 1985). The numbers are increasing as genital HSV becomes more common. HSV-2 accounts for 80% of neonatal infections, and the remainder is due to HSV-1 (Whitley et al. 1980). The virus is acquired *in utero*, intrapartum, or after birth. Intrauterine infection is rare but has been reported most often with primary maternal infections involving HSV-2 (Hutto et al. 1987). Fetal infection may result in abortion. Surviving infants have cutaneous lesions and CNS and ocular damage similar to severe congenital CMV or rubella syndrome. The majority of neonatal infections are intrapartum (approximately 80%) or postpartum. Exposure occurs as HSV-2 ascends from an infected cervix after rupture of the membranes or during delivery through an infected genital tract. Infants also may be exposed to HSV-1 shortly after birth in infected secretions or, rarely, through breast-feeding. Neonatal HSV almost always is symptomatic and takes one of three forms (Whitley and Hutto 1985). Approximately 50% of infants present around day 11 with mucocutaneous vesicular lesions localized to the eye, mouth, or skin. Skin and eye lesions periodically recur over several months, and 30% eventually develop neurological impairment. Approximately 30% of infants have infection restricted to the CNS. These infants present around day 16 with irritability, poor feeding, seizures, bulging fontanelle, and corticospinal tract signs. Two thirds also have vesicular lesions. About 20% of infants have a disseminated infection. They present around day 10 with irritability, poor feeding, seizures, bleeding diathesis, and shock. The characteristic vesicles are present in approximately 80%, and infection also involves the CNS in 60% to 75%. Regardless of the form, infection is present approximately 4 to 5 days before diagnosis (Whitley et al. 1988). Neonatal HSV infection may produce late sequelae, particularly with regard to ocular defects and neurological problems.

Prognosis

Risk factors for intrauterine infection are unknown. Both primary and recurrent maternal infections can involve the fetus, as well as ascending or transplacental infections. Risk factors for intrapartum infection include primary (as opposed to recurrent) maternal infection, maternal antibody status, duration of rupture of the membranes (more than 6 hours), prematurity, vaginal delivery, and placement of a fetal scalp monitor in a woman who is excreting virus (Whitley 1991).

Mothers who are young, unmarried, and from lower socioeconomic groups are at greater risk. A maternal history of genital HSV or exposure from sexual contacts also increases fetal risk because 1% of women with a history of genital HSV shed the virus at the time of delivery (Yeager and Arvin 1984). Unfortunately, up to 81% of pregnant women exposed to HSV-2 will not give a suggestive history (Whitley et al. 1988). Virus shedding tends to be asymptomatic. Recent series have documented 0.2% (0.01% to 0.4%) of pregnant women excrete genital

HSV at birth (Prober et al. 1988). Preterm infants comprise more than one fourth of those infected and are most likely to develop mucocutaneous or disseminated disease (Sullender et al. 1987). Neonatal risk is significantly greater with primary maternal infection. Approximately 35% of pregnant women with primary genital HSV have a preterm delivery. Up to 50% of infants will be infected by HSV (Nahmias et al. 1971; Arvin 1991), compared to less than 8% of infants born to mothers with secondary infection (Prober et al. 1987). Several factors contribute to this difference. Primary infections, unlike recurrent infections, commonly involve the cervix and are associated with excretion of large amounts of virus for extended periods. Transplacental maternal antibodies also appear to offer a degree of protection. In a recent study, 86% of exposed infants who remained well had specific antibodies to HSV-2, compared to only 12% of infants who became infected (Sullender et al. 1988). Lack of neonatal neutralizing antibody has been associated with symptomatic infection in the first week and disseminated infection (Sullender et al. 1987). The specificity of antiviral antibody may be prognostically useful. Neonates with high titers of antibody to a specific HSV polypeptide were reported to have more severe infections and poorer long-term outcomes (Kahlon and Whitley 1988).

The risk of asymptomatic vs. symptomatic primary maternal infection is poorly defined. A recent study found that symptomatic primary genital HSV has a 40% incidence of perinatal complications, including abortion, prematurity, and intrauterine growth retardation in addition to neonatal HSV (Brown et al. 1987). Primary infection in the first two trimesters has a 20% complication rate. In the third trimester the rate is 80%, including a 40% neonatal infection rate. Symptomatic recurrent HSV, in marked contrast has no perinatal morbidity.

Untreated neonatal HSV has an overall mortality of 50%, and more than half of the survivors sustain permanent damage. Disseminated infection carries the worst prognosis, with a mortality rate of 85% and an even higher morbidity rate. Death typically occurs in the first week due to coagulopathy and respiratory dysfunction. HSV pneumonitis is a particularly ominous sign (Barnes and Whitley 1986). Untreated CNS disease also has a poor prognosis. The mortality rate is 50% and the morbidity rate is 83%. Death occurs in the third to fourth week and is related to brain stem involvement. Localized infection, although it carries morbidity, is not fatal. However, more than 70% of untreated neonates with localized disease will progress within the first week to more severe involvement.

In a recent study of treated neonatal HSV infection, predictors of mortality included disseminated infection, coma or near-coma status at initiation of therapy, disseminated intravascular coagulation, prematurity, and pneumonitis complicating a disseminated infection. (Whitley et al. 1991b). In this same study, predictors of morbidity in treated survivors included encephalitis, disseminated infection, seizures, and infection with type 2 HSV. For neonates with treated local HSV infection, there was an increased risk for neurologic involvement when vesicles recurred more than three times.

Early diagnosis of neonatal HSV is crucial to improve the prognosis through prompt treatment. Diagnosis can be delayed because the clinical features are nonspecific, but the presence of cutaneous vesicles is an extremely helpful clue. Diagnosis is confirmed by isolating HSV from skin lesions, CSF, stool, urine, throat, nose, conjunctiva, or duodenal aspirates. Cytology is positive in 50% to 60% of specimens. CNS involvement is confirmed by CSF, EEG, and neuroimaging abnormalities. When pregnancy has been complicated by genital HSV, absent or low titers of HSV-2–neutralizing antibodies in the newborn is a definite risk factor for neonatal infection. Neuroimaging or serial studies late in the disease course are prognostically useful (Noorbehesht et al. 1987). Severe CT abnormalities (three or more lobes involved) correlate with marked neurological deficts. Serial CT findings reflect a stable or worsening clinical condition.

It was believed that HSV type made no difference because mortality rates were similar. Recently, 24 cases of treated neonatal HSV encephalitis were studied. Nine were due to HSV-1 and 15 due to HSV-2 (Corey et al. 1988). There was one mortality among treated patients, a neonate with HSV-2 who presented late. However, HSV-2 encephalitis was a more severe disease; patients had more lethargy and seizures, higher CSF cell and protein counts, and greater evidence of structural CT brain abnormalities. The long-term prognosis was excellent for HSV-1 patients who were treated early; all nine children were normal when followed up at 6 months to 3 years of age. In marked contrast, only 23% of the HSV-2 group were normal at follow-up. Sequelae included cerebral palsy (64%), ocular defects (64%), mental retardation (57%), seizure disorder (57%), and microcephaly (50%). Increased morbidity associated with HSV-2 compared to HSV-1 was confirmed in 2 subsequent analyses (Malm et al. 1991; Whitley et al. 1991b).

Therapy

Newer antiviral agents have had a dramatic impact on neonatal HSV infection. Both acyclovir and adenine arabinoside (AraA; vidarabine) decrease overall mortality from 50% to 19% (Whitley et al. 1991a; Whitley and Hutto 1985; Whitley and the NIAID Collaborative Antiviral Study Group 1983). With these drugs, the mortality rate of disseminated infection is reduced from 80% to 50% to 60% (although morbidity is 40%), and the mortality of CNS disease is reduced from 50% to 14% (with a 60% to 70% morbidity rate). Only 5% (vs. 70%) of localized infections progress if treated with 30 mg/kg of acyclovir. Although acyclovir and vidarabine appear equally effective, acyclovir is the preferred drug because it can be given in smaller fluid volumes over short periods 3 times daily (Arvin et al. 1987; Whitley 1991). Approximately 2% of treated infants will develop recurrent CNS infection that requires retreatment, but no predictive factors for relapse have emerged (Whitley and Hutto 1985).

Preventive therapy involves screening all women who present in labor for genital HSV by history and examination. If genital lesions are noted, a cesarean is performed. If the history is positive, genital cultures are obtained at the time of delivery to identify neonates at higher risk. Postnatal infection by HSV-1 can also be reduced by awareness and proper hygienic measures. Ultimately, subunit vaccines for HSV may provide definitive preventive therapy.

Human Immunodeficiency Virus Type 1

Natural History

Human immunodeficiency virus type 1 (HIV-1) is the RNA retrovirus responsible for acquired immune deficiency syndrome (AIDS). Infants and children comprise approximately 2% of reported AIDS cases, and AIDS is now the ninth leading cause of death for those younger than 5 years. It has been estimated that by 1992, nearly a million children will be infected worldwide (Chin 1990). Most cases are believed to be due to transplacental passage of maternal virus. A small number have resulted from postpartum exposure to virus in breast milk. Although no cases have been unequivocally documented, intrapartum exposure to virus in maternal blood or genital secretions also may occur. The natural history of neonatal infection with HIV-1 is not fully delineated, but preliminary studies indicate significant morbidity and mortality. In one prospective African study, children born to infected mothers had a mortality rate in the first year almost 6 times higher than control infants (Ryder et al. 1988). Neonates infected with HIV-1 range from the asymptomatic seropositive child with normal or abnormal laboratory immune parameters, to the critically ill child with AIDS. In large prospective series, at least 67% to 80% of infected neonates develop symptoms. Symptomatic infection includes nonspecific findings (fever, failure to thrive, thrush, organomegaly, lymphadenopathy, parotitis, diarrhea), progressive neurological disease, lymphoid interstitial pneumonitis, and secondary infections

Table 10-3. Characteristics of HIV-1 Encephalopathy

Clinical features
  Acquired microcephaly
  Cognitive impairment
  Developmental delay
  Loss of milestones
  Progressive corticospinal tract deficits
  Cerebellar deficits (uncommon)
  Movement disorders (uncommon)
  Seizures (uncommon)
Pathological features
  Cerebral atrophy, ventricular dilation
  Subacute encephalitis with perivascular inflammation, microglial nodules, multinucleated giant cells (deep white and gray matter)
  Progressive calcific vasculopathy (basal ganglia)
  White matter gliosis, pallor
  Corticospinal tract degeneration (spinal cord)

or malignancies (Belman 1990). Fifty percent to 90% of children develop a devastating encephalopathy, which is due to brain invasion by the virus (Table 3). The incubation period to clinical disease appears shorter in children with perinatal infection. The median age at onset of symptoms is 9 months, and over 80% will be ill by age 3 years (Pizzo 1990; Scott et al. 1989). In the recent report of the European Collaborative Study, most children showed clinical or laboratory evidence of infection by age 6 months. By 1 year of age, 26% of infected children develop AIDS and 17% die from HIV-related diseases (European Collaborative Study 1991). However, after age one many of the infected children seemed to stabilize or even improve. At the same time, there are some children who have remained healthy for 7 to 12 or more years (Burger et al. 1990; Pizzo 1990). Such variable incubation periods indicate that virus strain, immunological and genetic factors, and other cofactors are likely to affect prognosis for the individual infected child.

Prognosis

In the United States, 79% of women with AIDS are in their child bearing years. HIV-1 prevalence in pregnant women has ranged from 0.09% in rural areas (Hoff et al. 1988) to more than 3% in inner cities (Falloon et al. 1989). The geographic and racial distribution of neonatal infection parallels that of adult cases. Although most infected neonates have been born in New York, New Jersey, Florida, California, and Texas, other geographic areas are now seeing cases. In particular, urban cities with large numbers of intravenous drug users, such as New York City, Newark, and Miami, are particular foci. The majority of children are Afro-Americans (58%) and less commonly hispanic (21%) or white (24%) (Gayle et al. 1990). Mothers with a history of intravenous drug use or with a sexual partner with a history of drug use, bisexuality, or hemophilia are at particular risk. Other maternal risk factors include a history of prostitution, multiple sex partners, residence in an area where heterosexual HIV-1 transmission is high (Haiti or central Africa), and prior infection by contaminated blood. No study has noted any effect of sex on pediatric AIDS, but the recent Italian multicenter study reports that both infection rate and mortality were higher in female infants.

Studies have suggested maternal transmission rates of 25% to 30% (Pizzo and Butler 1990) with a low rate of 13% (European Collaborative Study 1988) and a high rate of 45% from Kenya. One problem with vertical transmission studies is the difficulty in determining an infected neonate (Pizzo 1990). HIV-1 antibody cannot be used because maternal antibody may persist to 15 months of age. Specific IgM is very difficult to measure. Antibody negative infections also are well documented. Neonatal diagnosis requires documentation of virus in blood or tissues, HIV-1 antibody in the setting of immune abnormalities and symptomatic infection, or symptoms meeting Centers for Disease Control (CDC) criteria for AIDS (CDC 1987). Besides virus culture, other tests that may be helpful include polymerase chain reaction detection of nucleic acids, antigen detection, and in vitro antibody production (De Rossi et al. 1991).

Transmission may be more likely when the mother shows symptomatic infection or has laboratory evidence of advanced immunosuppression (Curran et al. 1988; Jonckheer et al. 1986). In an early study, children of such mothers were nine times more likely to develop AIDS or symptomatic HIV infection (Mok et al. 1987). However, the larger follow-up study was unable to confirm a link between symptomatic maternal infection and neonatal infection (European Collaborative Study 1988). Certain maternal antibodies with neutralizing activity may be protective (Devash et al. 1990). The precise timing of infection is unclear, but can probably occur early in gestation (Pizzo, 1990). The precise risk of infection intrapartum during vaginal delivery and postpartum from breast milk is unknown. The Italian multicenter study reported significantly increased transmission to infants delivered vaginally and then breast-fed. Using several different modalities to measure virus components, a Swedish study reported intrapartum rather than congenital infection was likely in the cases they studied (Ehrnst et al. 1991). A recent prospective study from Africa suggests that colostrum and breast milk may indeed transmit virus and account for a proportion of infected children (Van de Perre et al. 1991).

Early onset of clinical disease in the infected neonate correlates with a rapidly progressive course and a poor outcome (Jonckheer et al. 1986). Clinical predictors of poor outcome include *Pneumocystis carinii* pneumonia, opportunistic infections in the first year of life, hepatitis, and progressive neurological disease, but not lymphoid interstitial pneumonitis (Italian Multicenter Study 1988; Pizzo et al. 1988a). In particular, progressive encephalopathy has been associated with a mortality rate of more than 75% (Belman et al. 1988; Epstein et al. 1988b). Death generally occurs within 1 year. Although the specificity for HIV-1 is somewhat controversial, craniofacial abnormalities have been described, including hypertelorism, box-like forehead, and a flat nasal bridge. More severe dysmorphism has been correlated with earlier onset of symptomatic infection (Marion et al. 1987). Laboratory findings that have suggested poor prognosis in small series have included hypogammaglobulinemia (Pahwa et al. 1987), absent serum neutralizing or cytotoxic antibodies

to HIV-1 (Robert-Guroff et al. 1987; Ljunggren et al. 1990), HIV-1 antigen in CSF or blood (Epstein et al. 1987; Borkowski et al. 1989), elevated serum Beta 2(B2)-microglobulin and neoopterin (Chan et al. 1990) and basal ganglia calcification on neuroimaging studies (Belman et al. 1986). These findings need to be confirmed in larger studies. A recent study tested for HIV-1 antigen in the serum of 43 infected children followed for a median of 17 months (Epstein et al. 1988a). Children who had persistent antigen were three times more likely to deteriorate and almost five times more likely to die.

Therapy

Zidovudine (AZT) is an antiretroviral agent that has been available since 1987. Continuous intravenous treatment (via a Hickman-Broviac catheter) in infected children resulted in significant increases in IQ, improved language and social skills, improved motor and coordination skills, and in specific cases improved brain CT and PET scans (Pizzo et al. 1988b). Improvement occurred within 3 to 4 weeks and continued over several months of therapy. IQ rose even in children without apparent encephalopathy, suggesting that HIV-1 may be associated with subtle cognitive deficits. Intermittent AZT therapy resulted in significantly less improvement than with constant infusion therapy (Pizzo 1990).

Oral sustained-release and subcutaneous AZT therapies are currently under investigation, and AZT is now approved for use in children. However, prolonged use of AZT in children is likely to be limited by bone marrow toxicity, just as it is in adults. New antiretroviral agents are currently being evaluated. These include 2',3'-dideoxycytidine (ddC), 2',3'-dideoxyinosine (ddI), and recombinant CD4 (Pizzo 1990). Dideoxyinosine given orally to a group of 43 infected children was able to lower their p24 antigen levels, raise their CD4 cell counts, and improve their IQ scores (Butler et al. 1991). These trials are being expanded, along with a continued search for new as well as multimodality therapy. Prophylactic intravenous immunoglobulin has been reported to decrease the frequency of bacterial infections, with fewer episodes of fever and sepsis in treated children. A multicenter study is now in progress, in which children on AZT are being randomized to receive or not receive intravenous immunoglobulin. Research on a potential AIDS vaccine, the definitive preventive therapy for at-risk populations, also continues.

References

Alford, C. A.; Stagno, S.; Pass, R. F.; Britt, W. J. Congenital and perinatal cytomegalovirus infections. Rev Infect Dis 12:S745–S753; 1990.

Alford, C. A.; Hayes, K.; Britt, W. Primary cytomegalovirus infection in pregnancy: comparison of antibody responses to virus-encoded proteins between women with and without intrauterine infection. J. Infect. Dis. 158:917–924; 1988.

Ahlfors, K.; Ivarsson, S-A.; Harris, S.; Svanberg, L.; Holmqvist, R.; Lernmark, B.; Theander, G. Congenital cytomegalovirus infection and disease in Sweden and the relative importance of primary and secondary maternal infections. *Scand. J. Infect. Dis.* 16:129–137; 1984.

Arvin, A. M. Relationships between maternal immunity to herpes simplex virus and the risk of neonatal herpesvirus infection. Rev. Infect. Dis. 13(S11):S953–S956; 1991.

Arvin, A. M.; Johnson, R. T.; Whitley, R. T.; Nelson, J. D.; McCracken, G. H. Consensus: management of the patient with herpes simplex encephalitis. Pediatr. Infect. Dis. J. 6:2–5; 1987.

Bale, J. F.; Blackman, J. A.; Sato, Y. Predicting outcome in infants with symptomatic congenital cytomegalovirus infection (Abstr). Ann. Neurol. 24:341; 1988.

Ballard, R. A.; Drew, W. L.; Hufnagle, K. G.; Riedel, P. A. Acquired cytomegalovirus infection in preterm infants. Am. J. Dis. Child. 133:484–485; 1979.

Barnes, D. W.; Whitley, R. J. Varicella zoster virus and HSV infection. Neurol. Clin. 4:265–283; 1986.

Belman, A. L. AIDS and pediatric neurology. Neurologic Clinics 8:571–603; 1990.

Belman, A. L.; Diamond, G.; Dickson, D.; Horoupian, D.; Llena, J.; Lantos, G.; Rubinstein, A. Pediatric acquired immunodeficiency syndrome. Am. J. Dis. Child. 142:29–35; 1988.

Belman, A. L.; Lantos, G.; Horoupian D,; Novick, B. E.; Ultmann, M. H.; Dickson, D. W.; Rubinstein, A. AIDS: calcification of the basal ganglia in infants and children. Neurology 36:1192–1199; 1986.

Borkowsky, W.; Krasinski, K.; Paul, D.; Holzman, R.; Moore, T.; Bebenroth, D.; Lawrence, R. ; Chandwani, S. Human immunodeficiency virus type 1 antigenemia in children. J. Ped. 114:940–945; 1989.

Britt, W. J.; Vugler, L. G. Antiviral antibody responses in mothers and their newborn infants with clinical and subclinical congenital cytomegalovirus infection. J. Infect. Dis. 161:214–219; 1991.

Brown, Z. A.; Vontver, L. A.; Benedetti, J.; Critchlow, C. W.; Sells, C. J.; Berry, S.; Corey, L. Effects on infants of a first episode of genital herpes during pregnancy. N. Engl. J. Med. 317:1246–1251; 1987.

Burger, H.; Belman, A. L.; Grimson, R.; Kaell, A., Flaherty, K.; Gulla, J.; Gibbs, R. A.; Nguyun, P.-N.; Weiser, B. Long HIV-1 incubation periods and dynamics of transmission within a family. Lancet 336:134–136; 1990.

Butler, K. M.; Husson, R. N.; Balis, F. M.; Brouwers, P.; Eddy, J.; El-Amin, D.; Gress, J.; Hawkins, M.; Jarosinski, P.; Moss, H.; Poplack, D.; Santocroce, S.; Venzon, D.; Wiener, L.; Wolters, P.; Pizzo, P.A. Dideoxyinosine in children with symptomatic human immunodeficiency virus infection. N. Engl. J. Med. 324:137–144; 1991.

CDC. Classification system for human immunodeficiency virus (HIV) infection in children under 13 years of age. M.M.W.R. 36:225–236; 1987.

Chan, M. M.; Campos, J. M.; Josephs, S.; Rifai, N. β_2-microglobulin and neopterin: predictive markers for human immunodeficiency virus type 1 infection in children? J. Clin. Microbiol. 28:2215–2219; 1990.

Chin, J. Current and future dimensions of the HIV/AIDS pandemic in women and children. Lancet 336:221–224; 1990.

Cooper, L. Z. The history and medical consequence of rubella. *Rev. Inf. Dis.* 7:S2–S10; 1985.

Corey, L.; Whitley, R. J.; Stone, E. F.; Mohan, K. Difference between herpes simplex virus type 1 and type 2 neonatal encephalitis in neurological outcome. Lancet 1:1–4; 1988.

Curran, J. W.; Jaffe, H. W.; Hardy, A. M.; Morgan, W. M.; Selik, R. M.; Dondero, T. J. Epidemiology of HIV infection and AIDS in the United States. Science 239:610–616; 1988.

Demmler, G. J. Summary of a workshop on surveillance for congenital cytomegalovirus disease. Rev. Infect. Dis. 13:315–329; 1991.

DeRossi, A.; Ades, A. E.; Mammano, F.; DelMistro, A.; Amadori, A.; Giaquinto, C.; Chieco-Bianchi, L. Antigen detection, virus culture, polymerase chain reaction, and in vitro antibody production in the diagnosis of vertically transmitted HIV-1 infection. AIDS 5:15–20; 1991.

DeVash, Y.; Calvelli, T. A.; Wood, D. G.; Reagan, K. J.; Rubinstein, A. Vertical transmission of human immunodeficiency virus is correlated with the absence of high-affinity/avidity maternal antibodies to the gp120 principal neutralizing domain. Proc. Natl. Acad. Sci. 87:3445–3449; 1990.

Dudgeon, J. A. Congenital rubella. J. Pediatr. 87:1078–1086; 1975.

Editorial. Vertical transmission of HIV. Lancet 2:1057–1058; 1988.

Ehrnst, A.; Lindgren, S.; Dictor, M.; Johansson, B.; Sonnerborg, A.; Czajkowski, J.; Sundin, G.; Bohlin, A.-B. HIV in pregnant women and their offspring: evidence for late transmission. Lancet 338:203–207; 1991.

Enders, G.; Nickerl-Pacher, U.; Miller, E.; Cradock-Watson, J. E. Outcome of confirmed periconceptional maternal rubella. Lancet 2:1445–1447; 1988.

Epstein, L. G.; Boucher, C. A. B.; Morrison, S. H.; Connor, E. M.; Oleske, J. M.; Lange, J. M. A.; Van der Noordaa, J.; Bakker, M.; Dekker, J.; Scherpier, H.; Van Den Berg, H.; Boer, K.; Goudsmit, J. Persistent human immunodeficiency virus type 1 antigenemia in children correlates with disease progression. Pediatrics 82:919–924; 1988a.

Epstein, L. G.; Goudsmit, J.; Paul, D. A.; Morrison, S. H.; Connor, E. M.; Oleske, J. M.; Holland, B. Expression of human immunodeficiency virus in cerebrospinal fluid of children with progressive encephalopathy. Ann. Neurol. 21:397–401; 1987.

Epstein, L. G.; Sharer, L. R.; Goudsmit, J. Neurological and neuropathological features of human immunodeficiency virus infection in children. Ann. Neurol. 23S:S19–S23; 1988b.

European Collaborative Study. Children born to women with HIV-1 infections: natural history and risk of transmission. Lancet 337:253–260; 1991.

European Collaborative Study. Mother-to-child transmission of HIV infection. Lancet 2:1039–1043; 1988.

Fitzgerald, N. A., Shellam, G. R. Host genetic influences on fetal susceptibility to murine cytomegalovirus after maternal or fetal infection. J. Infect. Dis. 163:276–281; 1991.

Gayle, J. A.; Selik, R. M.; Chu, S. Y. Surveillance for AIDS and HIV infection among black and hispanic children and women of childbearing age, 1981–1989. MMWR 39:23–30; 1990.

Griffiths, P. D.; Baboonian, C. A prospective study of primary cytomegalovirus infection during pregnancy: final report. 91:307–315; 1984.

Grillner, L.; Forsgren, M.; Barr, B.; Bottiger, M.; Danielsson, L.; De Verdier, C. Outcome of rubella during pregnancy with special reference to the 17th-24th weeks of gestation. Scand. J. Infect. Dis. 15:321–325; 1983.

Ho, M. Cytomegalovirus. Principles and Practice of Infectious Disease. 2nd ed. New York; John Wiley & Sons: 1985.

Hoff, R.; Berardi, V. P.; Weiblen, B. J.; Mahoney-Trout, L.; Mitchell, M. L.; Grady, G. F. Seroprevalence of human immunodeficiency virus among childbearing women. N. Engl. J. Med. 318:525–530; 1988.

Hutto, C.; Arvin, A.; Jacobs, R.; Steele, R.; Stagno, S.; Lyrene, R.; Willett, L.; Powell, D.; Andersen, R.; Werthammer, J.; Ratcliff, G.; Nahmias, A.; Christy, C.; Whitley, R. Intrauterine herpes simplex virus infections. J. Pediatr. 110:97–101; 1987.

Italian Multicenter Study. Epidemiology, clinical features, and prognostic factors of pediatric HIV infection. Lancet 2:1043–1046; 1988.

Jonckheer, T.; Levy, J.; Van De Perre, P.; Thiry, L.; Henrivaux, P.; Sacre, J. P.; Schepens, G.; Mees. N.; Dab, I.; Taelman, H.; Mascart-Lemone, F.; Zissis, G.; Clumeck, N.; Butzler, J. P.; Sprecher-Goldberger, S. LAV/HTLV-III infection in children of African origin: experience in Belgium. Eur. J. Pediatr. 145:511–516; 1986.

Kahlon, J.; Whitley, R. J. Antibody response of the newborn after herpes simplex virus infection. J. Infect. Dis. 158:925–933; 1988.

Knox, G. E.; Pass, R. F.; Reynolds, D. W.; Stagno, S.; Alford, C. A. Comparative prevalence of subclinical cytomegalovirus and herpes simplex virus infections in the genital and urinary tracts of low-income, urban women. J. Infect. Dis. 140:419–422; 1979.

Kumar, M. L.; Nankervis, G. A.; Jacobs, I. B.; Ernhart, C. B.; Glasson, C. E.; McMillan, P. M.; Gold, E. Congenital and postnatally acquired cytomegalovirus infections: long-term follow-up. J. Pediatr. 104:674–679; 1984.

Libman, M. D.; Dascal, A.; Kramer, M. S.; Mendelson, J. Strategies for the prevention of neonatal infection with herpes simplex virus: a decision analysis. Rev. Infect. Dis. 13:1093–1104; 1991.

Ljunggren, K.; Moschese, V.; Brolinden, P.-A.; Giaquinto, C.; Quinti, I.; Fenyo, E.-M.; Wahren, B.; Rossi, P.; Jondal, M. Antibodies mediating cellular cytotoxicity and neutralization correlate with a better clinical stage in children born to human immunodeficiency virus-infected mothers. J. Infect. Dis. 161:198–202; 1990.

Malm, G.; Forsgren, M.; Elazazi, M.; Persson, A. A follow-up study of children with neonatal herpes simplex virus infections with particular regard to late nervous disturbances. Acta. Paediatr. Scand. 80:226–234; 1991.

Marion, R. W.; Wiznia, A. A.; Hutcheon, R. G.;

Rubinstein, A. Fetal AIDS syndrome score, Am. J. Dis. Child. 141:429–431; 1987.

Miller, E.; Cradock-Watson, J. E.; Pollock, T. M. Consequences of confirmed maternal rubella at successive stages of pregnancy. Lancet 1:781–784; 1982.

Mok, J. Q.; Giaquinto, C.; De Rossi, A.; Grosch-Worner, I.; Ades, A. E.; Peckham, C. S. Infants born to mothers seropositive for human immunodeficiency virus. Preliminary findings from a multicenter European Study. Lancet 1:1164–1168; 1987.

Munro, N. D.; Sheppard, S.; Smithells, R. W.; Holzel, H.; Jones, G. Temporal relations between maternal rubella and congenital defects. Lancet 1:201–204; 1987.

Nahmias, A. J.; Josey, W. E.; Naib, Z. M.; Freeman, M. G.; Fernandez, R. J.; Wheeler, J. H. Perinatal risks associated with maternal genital herpes simplex virus infection. Am. J. Obstet. Gynecol. 110:825–833; 1971.

Nielsen, S. L.; Ronholm, E.; Sorensen, I.; Jaeger, P.; Andersen, H. K. Improvement of serological diagnosis of neonatal cytomegalovirus infection by simultaneously testing for specific immunoglobulins E and M by antibody-capture enzyme-linked immunosorbent assay. J. Clin. Microbiol. 25:1406–1410; 1987.

Noorbehesht, B.; Enzmann, D. R.; Sullender, W.; Bradley, J. S.; Arvin, A. M. Neonatal herpes simplex encephalitis: correlation of clinical and CT findings. Radiology 162:813–819; 1987.

Pahwa, R.; Good, R. A.; Pahwa, S. Prematurity, hypogammaglobulinemia, and neuropathology with HIV infection. Proc. Natl. Acad. Sci. 84:3826–3830; 1987.

Pass, R. F.; Stagno, S.; Myer, G. J.; Alford, C. A. Outcome of symptomatic congenital cytomegalovirus infection: Results of long term longitudinal follow-up. Pediatrics 66:758–762; 1980.

Peckham, C. S. Clinical and serological assessment of children exposed in utero to confirmed maternal rubella. Br. Med. J. 1:259–261; 1974.

Peckham, C. S.; Chin, K. S.; Coleman, J. C.; Henderson, K.; Hurley, R.; Preece, P. M. Cytomegalovirus infection in pregnancy: preliminary findings from a prospective study. Lancet 1:1352–1355; 1983.

Pizzo, P. A. Pediatric AIDS: problems within problems. J. Infect. Dis. 161:316–325; 1990.

Pizzo, P. A.; Butler, K. M. In the vertical transmission of HIV, timing may be everything. N. Engl. J. Med. 324:652–654; 1991.

Pizzo, P. A.; Eddy, J.; Falloon, J. Acquired immune deficiency syndrome in children. Am. J. Med. 85(S2A):195–202; 1988a.

Pizzo, P. A.; Eddy, J.; Falloon, J.; Balis, F. M.; Murphy, R. F.; Moss, H.; Wolters, P.; Brouwers, P.; Jarosinski, P.; Rubin, M.; Broder, S.; Yarchoan, R.; Brunetti, A.; Maha, M.; Nusinoff-Lehrman, S.; Poplack, D. G. Effect of continuous intravenous infusion of zidovudine (AZT) in children with symptomatic HIV infection. N. Engl. J. Med. 319:889–896; 1988b.

Preblud, S. R.; Alford, C. A. Rubella. In: Remington JS, Klein JO, eds. Infectious Diseases of the Fetus and Newborn Infant. Philadelphia: WB Saunders; 1990: p. 196–240.

Preece, P. M.; Pearl, K. N.; Peckham, C. S. Congenital cytomegalovirus infection. Arch. Dis. Child. 59:1120–1126; 1984.

Prober, C. G.; Hensleigh, P. A.; Boucher, F. D.; Yasukawa, L. L.; Au, D. S.; Arvin, A. M. Use of routine viral cultures at delivery to identify neonates exposed to herpes simplex virus. N. Engl. J. Med. 318:887–891; 1988.

Prober, C. G.; Sullender, W. M.; Yasukawa, L. L.; Au, D. S.; Yeager, A. S.; Arvin, A. M. Low risk of herpes simplex virus infections in neonates exposed to the virus at the time of vaginal delivery to mothers with recurrent genital herpes simplex virus infections. N. Engl. J. Med. 316:240–244; 1987.

Ramsay, M. E. B.; Miller, E.; Peckham, C. S. Outcome of confirmed symptomatic congenital cytomegalovirus infection. Arch. Dis. Child. 66:1068–1069; 1991.

Robert-Guroff, M.; Oleske, J. M.; Connor, E. M.; Epstein, L. G.; Minnefor, A. B.; Gallo, R. C. Relationship between HTLV-III neutralizing antibody and clinical status of pediatric AIDS and AIDS-related complex cases. Pediatr. Res. 21:547–550; 1987.

Ryder, R. W.; Nsa, W.; Behets, F.; Vercauteren, G.; Baende, E.; Lubaki, M.; Baudox, M.; Quinn, T.; Piot, P. Perinatal HIV transmission in two African hospitals: one year follow-up. Fourth International Conference on AIDS, Stockholm. Swedish Ministry of Health and Social Affairs, abstract 4128; 1988.

Saigal, S.; Lunyk, O.; Bryce Larke, R. P.; Chernesky, M. A. The outcome in children with congenital cytomegalovirus infection. Am. J. Dis. Child. 136:896–901; 1982.

Sever, J. L.; South, M. A.; Shaver, K. A. Delayed manifestations of congenital rubella. Rev. Inf. Dis. 7:S164–S169; 1985.

Scott, G. B.; Hutto, C.; Makuch, R. W.; Mastrucci, M. T.; O'Connor, T.; Mitchell, C. D.; Trapido, E. J.; Parks, W. P. Survival in children with perinatally acquired human immunodeficiency virus type 1 infection. N. Engl. J. Med. 321:1791–1796; 1989.

Stagno, S.; Whitley, R. J. Herpesvirus infections of pregnancy part I: cytomegalovirus and Epstein-Barr virus infections. N. Engl. J. Med. 313:1270–1274; 1985.

Stern, H.; Hannington, G.; Booth, J.; Moncrieff, D. An early marker of fetal infection after primary cytomegalovirus infection in pregnancy. Br. Med. J. 292:718–720; 1986.

Sullender, W. M.; Miller, J. L.; Yasukawa, L. L.; Bradley, J. S.; Black, S. B.; Yeager, A. S.; Arvin, A. M. Humoral and cell-mediated immunity in neonates with herpes simplex virus infection. J. Infect. Dis. 155:28–37; 1987.

Sullender, W. M.; Yasukawa, L. L.; Schwartz, M.; Pereira, L.; Hensleigh, P. A.; Prober, C. G.; Arvin, A. M. Type-specific antibodies to herpes simplex virus type 2 (HSV-2) glycoprotein G in pregnant women, infants exposed to maternal HSV-2 infection at delivery, and infants with neonatal herpes. J. Infect. Dis. 157:164–171; 1988.

Tardieu, M.; Grospierre, B.; Durandy, A.; Griscelli, C. Circulating immune complexes containing rubella antigens in late-onset rubella syndrome. J. Pediatr. 97:370–373; 1980.

Ueda, K.; Nishida, Y.; Oshima, K.; Shepard, T. H. Congenital rubella syndrome: correlation of gestational age at time of maternal rubella with type of defect. J. Pediatr. 94:763–765; 1979.

Van De Perre, P.; Simonon, A.; Msellati, P.; Hitimana, D.-G.; Vaira, D.; Bazubagira, A.; VanGoethem, C.; Stevens, A.-M.; Karita, E.; Sondag-Thull, D.; Dabis, F.; LePage, P. Postnatal transmission of human immunodeficiency virus type 1 from mother to infant. N. Engl. J. Med. 325:593–598; 1991.

Verder, H.; Dickmeiss, E.; Haahr, S.; Kappelgaard, E.; Leerboy, J.; Moller-Larsen, A.; Nielsen, H.; Platz, P.; Koch, C. Late-onset rubella syndrome: coexistence of immune complex disease and defective cytotoxic effects all functions. Clin. Exp. Immunol. 63:367–375; 1986.

Walmus, B. F.; Yow, M. D.; Lester, J. W.; Leeds, L.; Thompson, P. K.; Woodward, R. M. Factors predictive of cytomegalovirus immune status in pregnant women. J. Infect. Dis. 157:172–177; 1988.

Waxham, M. N.; Wolinsky, J. S. Rubella virus and its effects in the central nervous system. Neurol. Clin. 2:367–385; 1984.

Whitley, R. J. Herpes simplex virus infections of the central nervous system. Drugs 42:406–427; 1991.

Whitley, R.; Arvin, A.; Prober, C.; Burchett, S.; Corey, L.; Powell, D.; Plotkin, S.; Starr, S.; Alford, C.; Connor, J.; Jacobs, R.; Nahmias, A.; Soong, S.-J.; the NIAID Collaborative Antiviral Study Group. A controlled trial comparing vidarabine with acyclovir in neonatal herpes simplex virus infection. N. Engl. J. Med. 324:444–449; 1991a.

Whitley, R.; Arvin, A.; Prober, C.; Corey, L.; Burchett, S.; Plotkin, S.; Starr, S.; Jacobs, R.; Powell, D.; Nahmias, A.; Sumaya, C.; Edwards, K.; Alford, C.; Caddell, G.; Soong, S.-J., the NIAID Collaborative Antiviral Study Group. Predictors of morbidity and mortality in neonates with herpes simplex virus infections. N. Engl. J. Med. 324:450–454; 1991b.

Whitley, R. J.; Corey, L.; Arvin, A.; Lakeman, F. D.; Sumaya, C. V.; Wright, P. F.; Dunkle, L. M.; Steele, R. W.; Soong, S.-J.; Nahmias, A. J.; Alford, C. A.; Powell, D. A.; San Joaquin, V.; the NIAID Collaborative Antiviral Study Group. Changing presentation of herpes simplex virus infection in neonates. J. Infect. Dis. 158:109–116; 1988.

Whitley, R. J.; Hutto, C. Neonatal herpes simplex virus infections. Pediatr. Rev. 7:119–126; 1985.

Whitley, R. J.; Nahmias, A. J.; Visintine, A. M.; Fleming, C. L.; Alford, C. A. The natural history of herpes simplex virus infection of mother and newborn. Pediatrics 66:489–501; 1980.

Whitley, R. J.; the NIAID Collaborative Antiviral Study Group. Interim summary of mortality in herpes simplex encephalitis and neonatal herpes simplex virus infections: vidarabine versus acyclovir. J. Antimicrob. Chemother. 125B:105–112; 1983.

Yeager, A. S.; Arvin, A. M. Reasons for the absence of a history of recurrent genital infections in mothers of neonates infected with herpes simplex virus. Pediatrics 73:188–193; 1984.

Zaia, J. A.; Lang, D. J. Cytomegalovirus infection of the fetus and neonate. Neurol. Clin. 2:387–410; 1984.

11

Spirochetal Infections, Lyme Disease, and Neurosyphilis

ANDREW R. PACHNER

History

The story of Lyme disease begins with the persistence of concerned mothers in Lyme, Connecticut. After observing that many children in Lyme, including their own, had been diagnosed with juvenile rheumatoid arthritis, which is relatively uncommon, and suspicious that the diagnoses were incorrect, Polly Murray and Judith Mensch contacted state health authorities. The evaluation and characterization of Lyme arthritis by Dr. Allen Steere (Steere et al. 1977) established this arthritis as "epidemic" and set the stage for clinical and basic research in Lyme borreliosis. The subsequent identification by Willy Burgdorfer (Burgdorfer et al. 1982) of a spirochete as the causative organism was somewhat serendipitous because the *Ixodes dammini* ticks he was studying were actually being examined for *Ricketssiae ricketsii*. When he saw the spirochetes, he surmised that they might be the organisms associated with the tick-borne arthritis described by Steere. The characteristic skin rash, erythema chronicum migrans (ECM), however, was reported in Sweden by Afzelius (Afzelius 1910) in 1910. The clinical manifestations of Lyme borreliosis have been reviewed extensively (Benach and Bosler 1988) and include neurological, cardiac, and skin involvement. The causative organism has been named *Borrelia burgdorferi* and its fate *in vivo,* its relationship to the host immune system, and mechanisms of latency are subjects of intensive study.

Epidemiology and Microbiology

B. burgdorferi is transmitted to humans primarily by the bite of ixodid ticks, such as *I. dammini* in the Northestern United States, *Ixodes pacificus* in the Western United States, and *Ixodes ricinus* in Europe. It is generally accepted that the tick requires 24 to 72 hours on a human host prior to transmission of the spirochete. It is probable that a number of host factors are involved when a tick bites, such as host chemicals and reactions to tick salivary proteins. Most tick bites occur in the warm months of late spring, summer, and early fall, although in warm climates such as California, the tick bite season can last throughout the year.

There are many strains of *B. burgdorferi,* which have been isolated from ticks, animals, human skin lesions, or cerebrospinal fluid (CSF). Protein content, as determined by sodium dodecyl sulfate-polyacrylamide gel electrophoresis (SDS-PAGE) of bacterial sonicates, varies from strain to strain, although most strains have prominent 60-, 41-, 34-, 31-, and 22-kd proteins. The antigenicity of some of these proteins is highly variable (Barbour and Schrumpf 1986), especially in the 31- and 34-kd proteins which are outer surface proteins. Some of these variant proteins are encoded by DNA in extrachromosomal plasmids (Barbour 1988), and these plasmids vary in size and number from strain to strain. Some of the most dramatic differences between strains have been noted when American organisms have been compared to those from Europe; some of the differences in clinical manifestations between the two continents may be related to characteristics determined by these plasmids.

Fate of *B. burgdorferi in vivo* (Animal Models)

Much of what is known about *B. burgdorferi* infection has been learned from animals models. Experimental models of Lyme borreliosis have been induced (Johnson et al. 1984; Johnson et al. 1986; Burgess 1986; Burgdorfer 1984; Kornblatt et al. 1984; Benach et al. 1984; Barthold et al. 1988; Barthhold et al. 1988; Burgdorfer and Gage 1987; Schmitz et al. 1988; Hejka et al. 1989; Donahue et al. 1987; Levine et al. 1985; Stanek et al. 1986; Burgess et al. 1986; Schwan et al. 1988) in hamsters, dogs, rabbits, and rats, generally by injection of cultured spirochetes. Hamsters developed generalized infection but developed arthritis only when irradiated with local injection into the joint. Rat pups inoculated with the strain N40 developed arthritis without irradiation; the brain was culture-positive at 14 days, but not 90 days, after inoculation. *B. burgdorferi* could be cultured from the kidney, spleen, and blood of experimentally infected *Peromyscus leucopus* most frequently from the urinary bladder (94%). The bladder from a mouse with a 47-week-old infection appeared necrotic and was culture positive. It has been demonstrated that

chronic infection occurs despite high anti-*B. burgdorferi* antibody titers in inbred mice injected intraperitoneally with the N40 strain; organisms cultured from the brain may have different plasmid profiles than the original infecting strain; and the organism can be passed *in vivo* by injection of brain isolates intraperitoneally and subsequent culture of brain (Pachner and Itano 1990).

Clinical Manifestations

After inoculation of *B. burgdorferi* into human skin by the tick, ECM occurs frequently; in some patients, there is early hematogenous spread of organism leading to multiple skin lesions. Seeding into the central nervous system (CNS) can occur early because meningitis may occur within a few weeks of the tick bite; the nervous system is the most common site of involvement, and arthritis is very unusual in Europe (Stanek et al. 1988) where the infection is more common than in the United States. Infection may remain latent for years with subsequent reactivation within the joints, skin, or CNS. Late manifestations of Lyme borreliosis are due to direct infection on the basis of successful culture from the late skin manifestation, acrodermatitis chronica atrophicans (ACA) (Asbrink and Hovmark 1985). However CSF cultures are positive in less than 10% of patients with Lyme meningitis, and culture positivity generally is present early rather than late in the course of the infection (Karlsson 1990). Neither the CSF nor joint have been culture-positive in late neurological manifestations or Lyme arthritis, although spirochetes have been identified in affected tissue (Pachner et al. 1989; DeKoning and Hoogkamp-Korstanje 1986). Thus, the possible mechanism for arthritis or late neurological sequelae must include postinfective immune reaction, such as in Reiter's syndrome because there is no proof the organisms are still alive in the brain or joint.

Because of the wide range of clinical manifestations of the infection, it has been hypothesized that there is a genetic component in the susceptibility to the infection and the development of disease (Steere 1988). In four studies, three in Europe and one in the United States, there has been a lack of consensus on histocompatibility locus (HLA) association with Lyme borreliosis, although the studies may not actually be contradictory since the American study dealt with arthritis and the European studies primarily with neurological disease (Steere et al. 1979; Kristoferitsch and Mayr 1984; Van Doorn et al. Pfluger et al. 1988). These early studies, though limited in number of patients, indicate the potential importance of the major histocompatibility complex (MHC) in human Lyme borreliosis.

Fate of the Organism After Infection

B. burgdorferi is hematogenously spread and establishes infections in many organs; its fate within target organs is poorly understood. Its ability to bind to and penetrate endothelial cells (Comstock and Thomas 1989; Thomas and Comstock 1989) may mean that it can take up residence within the endothelium or other macrophage-like cells. This perivascular localization is supported by the frequent occurrence of vasculitis or perivascular inflammation in human disease, although when spirochetes are seen, they are extracellular (Duray and Steere 1988). In most manifestations of Lyme borreliosis in humans that occur after ECM, spirochetes are visualized in affected organs or are extremely rare. This is true whether they are studied by the Warthin-Starry or Dieterle techniques (Steere et al. 1988) or monoclonal antibodies to major proteins of *B. burgdorferi*. In experimentally infected hamsters, spirochetes were in very low numbers in culture-positive organs (Duray and Johnson 1986). The Bosma modification of the Steiner technique, in which the protective nonantigenic "slime layer" of the borrelia is removed prior to staining, has demonstrated larger numbers of organisms in affected skin joints, and hearts in humans; therefore, organisms may be much more prevalent in infected tissue than previously thought. There also is controversy regarding inflammatory responses to the organism in experimental disease, with some investiga-

tors finding little pathology (Duray and Johnson 1986), while Schaible and coworkers have found that infection in scid mice resulted in giant cells and perivascular infiltrates in brains, hearts, spleens, and other organs (Schaible et al. 1989a; Schaible et al. 1989b).

Most studies of experimental animals have been done within a few weeks after infection when organs are culture positive. Yet the neurological manifestations of Lyme borreliosis frequently become manifest many months to years after initial infection; *B. burgdorferi* was cultured from the heart in a patient with cardiomyopathy 3 years, and from acrodermatitis lesions more than 10 years, and after initial infection (Stanek et al. 1989; Asbrink and Hovmark 1985). Prior to the age of freezers and *in vitro* culture media, relapsing fever borelia were kept in the brains of rodents and passed once or twice a year. The demonstration of latency of other pathogenic spirochetes within the CNS in these situations contrasts to a dearth of information in Lyme borreliosis. Results from the laboratory and from the rat and hamster models have documented culture positivity in brain tissue within the first few weeks of infection, but late culture shows negativity. Although these data imply that early infection of brain by *B. burgdorferi* in the mouse may be cleared, culture negativity does not prove absence of the organism, given the insensitivity of the culture technique. In *Treponema pallidum* infections in mice, histopathological identification of the spirochete and culture are known to be erratic or useless, and investigators have documented infection by transfer of infected tissue into susceptible animal hosts (i.e., rabbits [Klein and Monjan 1983]). Optimal means of documenting late infection in Lyme borreliosis have not been carefully studied in an animal model.

The means by which *B. burgdorferi* evades host defenses also is unknown. There are many possibilities, all of which are testable in the mouse model. The possibility of a "slime" layer around *B. burgdorferi in vivo* already has been discussed; histopathological techniques that remove this layer may increase the ability to visualize *B. burgdorferi in vivo.* Intracellular residence of the organism is another possibility, which is more likely because the spirochete can penetrate endothelial cells (Comstock and Thomas 1989). The organism may rearrange DNA *in vivo* in response to the host's immune attack. This strategy is used by *B. hermsii,* in which both expression-linked and silent copies of the genome for outer surface proteins exist; change in serotype is associated with transposition of a silent copy of the gene to an expression site on a plasmid (Barstad et al. 1985; Meier et al. 1985; Plasterk et al. 1985).

In the case of *B. burgdorferi,* outer surface proteins change during *in vitro* cultivation (Schwan and Burgdorfer 1987), and a new IgM antibody against outer surface protein B late Lyme disease in humans has been documented (Craft et al. 1986).

Prognosis

Because of the lack of knowlege about the organism and the short history of awareness of the disease, it is not surprising that little information is available about prognosis. It is likely that the most common event after initial infection is spontaneous, probably immune-mediated, resolution and clearance of organism. In some patients, ECM occurs as a manifestation of local infection and spread. There is little doubt that oral antibiotic treatment ECM is highly effective in curing the infection (Steere et al. 1986). In patients in whom ECM does not occur, is not noticed, or is misdiagnosed, later manifestations such as Lyme meningitis, Lyme arthritis, or late neurological manifestations can occur after infection; the frequency with which this happens is unknown, but is probably low. In the early stages of infection within the brain, the disease is highly curable with 2 weeks of intravenous penicillin (Steere et al. 1983). After prolonged periods of infection, the organism can become localized to the brain parenchymia or cause a chronic subarachnoid inflammation (Pach-

ner et al. 1989). These late manifestations within the CNS are unusual, but they can be diagnostic and therapeutic conundra. Active inflammation, even in these late stages of infection, can be eradicated with intravenous antibiotics, but fixed lesions will not be cured.

Neurosyphilis

Most neurological manifestations of neurosyphilis occur years after the initial infection. Classically, neurosyphilis can present in a meningovascular form or as tabes dorsalis or general paresis. More focal manifestations are optic neuritis or eighth nerve involvement.

The diagnosis, as in Lyme disease, is made by a combination of clinical evaluation and serology because *T. pallidum* cannot be cultured from infected tissues. Increasingly, neurosyphilis is presenting in less classical ways, so that the differential diagnosis of almost any inflammation within the CNS must include syphilis. Thus, serology has become as crucial in the diagnosis of neurosyphilis.

Only a small percentage (5% to 10%) of those infected by *T. pallidum,* as manifested by initial lesions, develops neurological disease. A smaller percentage develops subclinical involvement with an abnormal CSF. It is generally accepted that symptoms and signs in neurosyphilis are due to direct infection with the organism and that adequate antibiotic treatment will, in most cases, cure the disease and, in the others, prevent progression. Treatment for neurosyphilis requires the maintenance of treponemicidal levels of antibiotics for an extended time within the CNS, but the ideal way of achieving this has not been determined. Twenty million units of penicillin intravenously (IV) per day for 2 to 3 weeks are most commonly used. Treatment failure is uncommon and is defined as a lack of improvement in the CSF pleocytosis, lack of diminution of serum VDRL, and persistence of CSF VDRL. Successful treatment can sometimes be effected by retreatment. The frequency with which intravenous antibiotics are unsuccessful in neurosyphilis has not been carefully studied, but it is quite low.

Prognosis of Neurosyphilis

This topic has been reviewed by Pachner (1986). It is generally accepted that the clinical manifestations of neurosyphilis are different in the postantibiotic vs. preantibiotic era, with more dementia and personality change and less tabes (Hooshmand et al. 1972). Of patients with neurosyphilis in a more recent study, 17% had received recommended doses of antibiotics for treatment of their primary lesions (Nordenbo and Sorensen 1981). Unlike Lyme disease in which there is no serological test that is useful to indicate activity of the disease, the VDRL in serum or CSF serves this purpose and can be used to follow the response to treatment. The generally accepted treatment for established neurosyphilis is 24 million units of IV penicillin G for 2 to 3 weeks. Although there are no data about frequency of failure of this regimen, retreatment is accepted in instances in which clinical findings progress without another cause, especially in the presence of persistent CSF pleocytosis or CSF VDRL. The course of neurosyphilis is sometimes accelerated or more malignant in patients with immunodeficiency, such as in human immunodeficiency virus infection (Johns et al 1987). As in Lyme disease, the vast majority of untreated patients never develop CNS infection. However, the factors, other than immunodeficiency, that determine the likelihood of CNS infection are not understood.

The Role of the Immune System to Protect Against Spirochetal Infection

It is clear that CNS infection with *B. burgdorferi* or *T. pallidum* results in a balance between the needs of the host to destroy the offending bacterium and the needs of the spirochete to grow. There is little doubt that the immune system maintains an attack against the spirochete that is not com-

pletely effective. The strategies that the spirochete uses to escape this attack are not known. The mechanisms by which previously latent infection changes into disease also are unknown. The recent research in the basic and clinical aspects of Lyme borreliosis should provide insight into the strategies used by host and pathogenic spirochete.

References

Afzelius, A. Verhandlungen der dermatologischen Gesellschaft zu Stockholm. Arch. Dermatol. Syph. 101:404; 1910.

Asbrink, E.; Hovmark, A. Successful cultivation of spirochetes from skin lesions of patients with erythema chronicum migrans Afzelius and acrodermatitis chronica atrophicans. Acta Pathol. Microbiol. Immunol. Sect. B 93:161–163; 1985.

Asbrink, E.; Hovmark, A. Successful cultivation of spirochetes from skin lesions of patients with erythema chronicum migrans Afzelius and acrodermatitis chronica atrophicans, Acta Pathol. Microbiol. Immunol. Sect. B. 93:161–163; 1985.

Barbour, A. G.; Ramona, A. H.; Howe, T. R. Heterogeneity of major proteins in Lyme disease Borreliae: a molecular analysis of North American and European isolates. J. Infect. Dis. 152:478–484; 1985.

Barbour, A. G.; Schrumpf, M. E. Polymorphisms of major surface proteins of Borrelia burgdorferi. Zbl. Bakt. Hyg. A. 263:83–91; 1986.

Barbour, A. G. Plasmid analysis of the Borrelia burgdorferi, the Lyme disease agent, J. Clin. Microbiol. 26:475–478; 1988.

Barstad, P. A.; Coligan, J. E.; Raum, M. G.; Barbour, A. G. Variable major proteins of *Borrelia hermsii:* epitope mapping and partial sequence analysis. J. Exp. Med. 161:1302–1314; 1985.

Barthold, S. W.; Moody, K. D.; Terwilliger, G. A.; Steere, A. C.; Duray, P. H.; Jacoby, R. O. Experimental Lyme arthritis in rats infected with Borrelia burgdorferi. J. Infect. Dis. 157:842–846; 1988a.

Barthold, S. W.; Moody, K. D.; Terwilliger, G. A.; Jacoby, R. O.; Steere, A. C. An animal model for Lyme arthritis. Ann. N.Y. Acad. Sci. 539:264–273; 1988b.

Benach, J. L.; Bosler, E. M.; Coleman, J. L.; Habicht, J. Experimental transmission of the Lyme disease spirochete to rabbits (letter). J. Infect. Dis. 150:786–787; 1984.

Benach, J.; Bosler, E. M. Lyme disease and related disorders, Ann. N.Y. Acad. Sci. 539:1–513; 1988.

Burgdorfer, W. A.; Barbour, A. G.; Hayes, S. F.; Benach, J.; Grunwaldt, E.; Davis, J. P. Lyme disease—a tick-borne spirochetosis? Science 216:1317–1319, 1982.

Burgdorfer, W. The New Zealand rabbit: an experimental host for infecting ricks with Lyme disease spirochetes. Yale J. Biol. Med. 57:609–612; 1984.

Burgdorfer, W.; Gage, K. L. Susceptibility of the hispid cotton rat to the Lyme disease spirochete. Am. J. Trop. Med. Hyg. 37:624–628; 1987.

Burgess, E. C. Experimental inoculation of dogs with Borrelia burgdorferi. Zb. Bakt. Hyg. A 263:49–54; 1986.

Burgess, E. C.; Amundson, T. E.; Davis, J. P.; Kaslow, R. A.; Edelman, R. Experimental inoculation of Peromyscus spp. with Borrelia burgdorferi: evidence for contact transmission. Am. J. Trop. Med. Hyg. 35:335–339; 1986.

Comstock, L. E.; Thomas, D. D. Penetration of endothelial cell monolayers by Borrelia burgdorferi. Infect. Immun. 57:1626–1628; 1989.

Craft, J. E.; Fischer, D. K.; Shimamoto, G. T.; Steere, A. C. Antigens of Borrelia burgdorferi recognized during Lyme disease: appearance of a new immunoglobulin M response and expansion of the immunoglobulin G response late in the illness. J. Clin. Invest. 78:934–939; 1986.

DeKoning, J.; Hoogkamp-Korstanje, J. A. A. Diagnosis of Lyme disease by demonstration of spirochetes in tissue biopsies. ZBl. Bakt. Hyg. A 263:179–188; 1986.

Donahue, J. G.; Piesman, J.; Spielman, A. Reservoir competence of white-footed mice for Lyme disease spirochetes. Am. J. Trop. Med. Hyg. 36:92–96; 1987.

Duray, P. H.; Johnson, R. C. The histopathology of experimentally infected hamsters with the Lyme disease spirochete, Borrelia burgdorferi. Proc. Soc. Exp. Biol. Med. 181:263–269; 1986.

Duray, P. H.; Steere, A. C. Clinical pathologic correlations of Lyme disease by stage. Ann. N.Y. Acad. Sci. 539:65–79; 1988.

Hejka, A.; Schmitz, J. L.; England, D. M.; et al. Histopathology of Lyme arthritis in LSH hamsters. Am. J. Pathol. 134:1113–1123; 1989.

Hooshmand, H.; Escobar, M. R.; Kopf, S. W. Neurosyphilis: a study of 241 patients. J.A.M.A. 219:726–729; 1972.

Johns, D. R.; Tierney, M.; Felsenstein, D. Alteration in the natural history of neurosyphilis by concurrent infection with HIV. N. Engl. J. Med. 316:1569–1572; 1987.

Johnson, R. C.; Marek, N.; Kodner, C. Infection of Syrian hamsters with Lyme disease spirochetes. J. Clin. Microbiol. 20:1099–1101; 1984.

Johnson, R. C.; Kodner, C.; Russell, M. Active immunization of hamsters against experimental infection with Borrelia burgdorferi. Infect. Immun. 54:897–898; 1986.

Johnston, Y. E.; Duray, P. H.; Steere, A. C.; Kashgarian, M.; Buza, J.; Malawista, S. E.; Askenase, P. W. Lyme arthritis: spirochetes found in synovial microangiopathic lesions. Am. J. Pathol. 118:26–34; 1985.

Karlsson, M.; Hovind-Hougen, K.; Svenungsson, B.; Stiernstedt, G. Cultivation and characterization of spirochetes from the cerebrospinal fluid of patients with Lyme borreliosis. J. Clin. Microbiol. 28:473–479; 1990.

Klein, J. R.; Monjan, A. A. Delayed-type hypersensitivity response in mice to Treponema pallidum. Immunological Communications 12:25–30; 1983.

Kornblatt, A. N.; Steere, A. C.; Brownstein, D. G. Experimental Lyme disease in rabbits: spirochetes found in erythema migrans and blood. Infect. Immun. 46:220–223; 1984.

Kristoferitsch, W.; Mayr, W. R. HLA-DR in meningopolyneuritis of Garin-Bujadoux-Bannwarth: contrast to Lyme disease? J. Neurol. 231:271–272; 1984.

Levine, J. F.; Wilson, M. L.; Spielman, A. Mice as reservoirs of the Lyme disease spirochete. Am. J. Trop. Med. Hyg. 34:355–360; 1985.

Meier, J. T.; Simon, M. I.; Barbour, A. G. Antigenic variation is associated with DNA rearrangements in a relapsing fever borrelia. Cell 41:403–409; 1985.

Nordenbo, A. M.; Sorensen, P. S. The incidence and clinical presentation of neurosyphilis in greater Copenhagen, 1974–8. Acta. Neurolog. Scand. 63:237–246; 1981.

Pachner, A. R. Spirochetal diseases of the CNS. Neurologic Clinics, 4:207–222; 1986.

Pachner, A. R.; Duray, P. H.; Steere, A. C. Central nervous system manifestations of Lyme disease. Arch. Neurol. 46:790–795; 1989.

Pachner, A. R.; Duray, P. H.; Steere, A. C. CNS manifestations of Lyme disease, Arch. Neurol. 46:790–795; 1989.

Pachner, A. R.; Itano, A. Borrelia burgdorferi infection of the brain—characterization of the organism and response to antibiotics and immune sera in the mouse model. Neurology 40:1535–1540; 1990.

Pfluger, K. H.; Reimers, C. D.; Neubert, U.; et al. Erythema migrans borreliosis: an HLA-associated disease? Ann. N.Y. Acad. Sci. 539:414–417; 1988.

Plasterk, R. H.; Simon, M. I.; Barbour, A. G. Transposition of structural genes to an expression sequence on a linear plasmid causes antigen variation in the bacterium Borrelia hermsii. Nature (London) 318:257–263; 1985.

Schaible, U. E.; Kramer, M. D.; Justus, C. W. E.; et al. Demonstration of antigen-specific T cells and histopathological alterations in mice experimentally inoculated with Borrelia burgdorferi. Infect. Immun. 57:41–47; 1989a.

Schaible, U. E.; Kramer, M. D.; Museteanu, C., et al. The severe combined immunodeficiency mouse: a laboratory model for the analysis of Lyme arthritis and carditis. J. Exp. Med. 170:1427–1432; 1989b.

Schmitz, J. L.; Schell, R. F.; Hejka, A.; et al., Induction of Lyme arthritis in LSH hamsters. Infect. Immun. 56:2336–2342; 1988.

Schwan, T. G.; Burgdorfer, W. Antigenic changes of Borrelia burgdorferi as a result of *in vitro* cultivation. J. Infect. Dis. 156:852–853; 1987.

Schwan, T. G.; Burgdorfer, W.; Schrumpf, D. A.; Karstens, R. H. The urinary bladder, a consistent source of *Borrelia burgdorferi* in experimentally infected white-footed mice (*Peromyscus leucopus*). J. Clin. Microbiol. 26:893–895; 1988.

Stanek, G.; Burger, I.; Hirschl, A.; et al. Borrelia transfer by ticks during their life cycle. Zbl. Bakt., Hyg. A. 263:29–33, 1986.

Stanek, G.; Pletschette, M.; Flamm, H.; et al. European Lyme borreliosis, Ann. N.Y. Acad. Sci. 539:274–282; 1988.

Stanek, G.; Klein, J.; Bittner, R.; et al. Isolation of Borrelia burgdorferi from the myocardium of a patient with longstanding cardiomyopathy. N. Engl. J. Med. 322:249–252; 1989.

Steere, A. C.; Malawista, W. E.; Snydman, D. R.; Shope, R. E.; Andiman, W. A.; Ross, M. R.; Steele, F. M. Lyme arthritis: an epidemic of oligoarticular arthritis in children and adults in three Connecticut communities. Arthritis Rheum. 20:7–17; 1977.

Steere, A. C.; Gibofsky, A.; Patarroya, M. E.; et al. Chronic Lyme arthritis. Ann. Intern. Med. 90:896–901; 1979.

Steere, A. C.; Pachner, A. R.; Malawista, S. E. Neurologic abnormalities of Lyme disease: successful treatment with high-dose intravenous Penicillin. Ann. Intern. Med. 99:767–772; 1983.

Steere, A. C.; Green, J.; Pachner, A. R.; et al. Treatment of Lyme disease. Zbl. Bakt. Hyg. A 263:352–356; 1986.

Steere, A. C. Pathogenesis of Lyme arthritis. Ann. N.Y. Acad. Sci. 539:87–92; 1988.

Steere, A. C.; Duray, P. H.; Butcher, E. C. Spirochetal antigens and lymphoid cell surface markers in Lyme synovitis: comparison with rheumatoid synovium and tonsillar lymphoid tissue. Arthritis Rheum. 31:487–495; 1988.

Thomas, D. D.; Comstock, L. E. Interaction of Lyme disease spirochetes with cultured eucaryotic cells. Infect. Immun. 57:1324–1326; 1989.

VanDoorn, P. A.; Brand, A.; Vermeulen, M.; et al. Antibodies against neural tissue and DR2 antigen in patients with Bannwarth's syndrome (Abstr). Presented at the Lyme Borreliosis Update Conference, Baden, June 2–4; 1989.

12

Parasites

FRANK J. BIA AND MICHELE BARRY

Understanding the prognosis for parasitic infections occurring within the central nervous system (CNS) is critical to several medical and surgical disciplines. Those who care for travelling students, tourists, missionaries, and career relief workers must be familiar with disease entities often considered obscure by others. In many instances, the prognoses for parasitic infections can be markedly altered by the administration of immunosuppressive agents, as might be used for prevention of transplant rejection. Immunosuppressive diseases, such as acquired immunodeficiency syndrome (AIDS), may dramatically alter the natural history of parasitic diseases within the CNS. Prognosis, therefore, often depends on several important factors, including prior exposure, natural immunity, and the underlying immune status of the human host (Bia and Barry 1986). An understanding of the parasite's life cycle and the pathogenesis of human infection often will aid in predicting neurological outcome.

Toxoplasmosis

Prior to the AIDS epidemic, which is associated with human retrovirus infections, the CNS manifestations of human toxoplasmosis were largely confined to congenital CNS infections and retinitis occurring in the fetus or adult (McCabe and Remington 1984). Today, toxoplasmosis is included in the differential diagnosis of most mass lesions discovered in the CNS. Diagnosis is critical because the prognosis of such infections can be markedly altered by antimicrobial therapy.

The obligate intracellular parasite, *Toxoplasma gondii*, is consumed by humans as tissue cysts in undercooked meat or as oocysts passed in infected cat feces, both forms ultimately releasing infectious organisms into the gastrointestinal tract. Although usually asymptomatic, acute infection also may present as a self-resolving mononucleosis-like syndrome requiring no specific antibiotic therapy. Following acute infections, tissue cysts containing living bradyzoites survive in many tissues, including eye and brain. Later in life, immunosuppression may cause a decrease in cell-mediated immunity (e.g., lymphomas, chemotherapy, graft rejection, or AIDS). This results in cyst rupture, release of invasive tachyzoites, and tissue necrosis.

Prior to the AIDS era, a review of toxo-

157

plasmosis at Memorial Sloan-Kettering Cancer Center from 1972 to 1978 revealed only eight cases of CNS toxoplasmosis in patients with underlying malignancies, mostly lymphomas (Hakes and Armstrong 1983). All were fatal despite therapy. In contrast, at the same institution and at New York Hospital, AIDS was highly associated with CNS toxoplasmosis, which was found in 25% of all AIDS patients with CNS infections (Snider et al. 1983).

When the neurological complications in more than 300 cases of AIDS or generalized lymphadenopathy at University of California at San Francisco were reviewed, results were similar (Levy et al. 1985). Nearly one third of all CNS complications of any type were associated with toxoplasmosis (103 cases in 315 patients).

A diagnosis of toxoplasmosis in AIDS patients often is presumptive (i.e., based on the presence of ring-enhancing lesions on computed tomography [CT] scan, often in the basal ganglia, occasionally associated with chorioretinitis, and usually is responsive to specific antibiotics within weeks). Regardless of the serological method used to document infection, several principles apply. CNS toxoplasmosis represents a reactivation of latent infection under conditions of immunosuppression. A rise in immunoglobulin M (IgM) antibody should not be expected, but a test for *T. gondii* IgG antibodies in serum should be positive despite an immunosuppressed state (Luft et al. 1984). Demonstration of IgG antibodies against *T. gondii* in the CNS was useful in a study showing local production of antibodies in 11 of 16 patients with AIDS and toxoplasma-associated encephalitis (Potasman et al. 1988). Methods for diagnosis of toxoplasmic encephalitis are shown in Table 12-1.

Primary therapy for toxoplasmic encephalitis relies on a combination of folic acid antagonists, usually pyrimethamine and sulfa drugs. Within 2 weeks of initiating therapy, clinical improvement generally is apparent (Israelski and Remington 1988). The level of consciousness, the number of lesions on computed tomography (CT) scan, cerebrospinal fluid (CSF) abnormalities, and serum

Table 12-1. Usual Methods for Diagnosis of Encephalitis Associated With *T. gondii* Infections in AIDS Patients

---

Presence of IgG, and occasionally IgM, antibodies in serum
Demonstration of intrathecal antibody production against *T. gondii*
Demonstration of ring-enhancing lesions on CT scan or MRI
Brain biopsy for histopathology, antigen detection (immunoperoxidase staining), and isolation of *T. gondii* in tissue culture

---

toxoplasma antibody titers do not correlate with outcome (Navia et al. 1986).

For patients allergic to sulfa drugs, clindamycin when used with pyrimethamine, has been shown to be a useful alternative (Gellin and Soave 1992). Within 1 month, CT scans should demonstrate a decrease in size and number of lesions, or additional diagnoses must be considered. Folinic acid (leucovorin) is administered concurrently to prevent marrow toxicity. Folic acid should not be used as a substitute or it will counter the effects of these agents. An 80% to 90% rate of clinical response is expected, despite the severity of this illness. However, if patients who respond are not maintained on a regimen of lifelong toxoplasmosis suppression, most will have relapses of necrotizing encephalitis, often beginning at the sites of original infection. Whether the newer macrolides, such as azithromycin and clarithromycin, will be useful in combination with pyrimethamine is currently under evaluation. Azithromycin will likely be approved for opportunistic infections soon. Until then it can be obtained on a compassionate use basis from Pfizer Incorporated (Groton, CT. 800 742-3029).

Malaria

With the bite of a plasmodium-infected female anopheline mosquito, one of four species of human malaria can be transmitted. Following human infection and asexual parasite multiplication in the liver, erythrocytes are invaded by merozites, initiating an attack of clinical malaria. However, cerebral malaria occurs only as a consequence of infection caused by *Plasmodium falciparum*

(Maegraith 1980). Parasitized erythrocytes develop surface projections or knobs that adhere to the endothelial surface of blood vessels. When more than 5% to 10% of erythrocytes are infected, the resulting capillary blockage becomes severe enough to threaten cerebral function, however, peripheral blood parasitemia does not always reflect the degree of CNS infection. High fevers, chills, headaches, myalgias, icterus, hepatosplenomegaly, and anemia are common during acute falciparum malaria.

In terms of prognosis, renal failure, mesenteric ischemia, and acute respiratory distress syndrome result from capillary blockage, and they may portend cerebral malaria. However, when cerebral malaria does occur, it may be missed because clinical findings are not specific. The range of disturbances in consciousness varies from lethargy and stupor to coma (Daroff et al. 1967).

Organic findings include confusion, disorientation, and intellectual impairment, not necessarily associated with depressed consciousness. Other neurological findings include movement disorders, such as chorea, myoclonus, and tremors; focal signs, such as seizures, hemiplegias, and sensory disturbances; and acute personality changes (Anderson 1927). Retinal hemorrhages occur in 15% of cases.

The diagnosis of cerebral malaria is still made using a peripheral blood smear showing *P. falciparum*-infected erythrocytes in association with the clinical syndromes described previously, usually during a period of intense parasitemia and fever. In fatal cases, cerebral vasculature show capillaries and venules packed with parasitized erythrocytes. Hyperparasitemia is defined as a density of asexual forms of *P. falciparum* exceeding 5% of erythrocytes (i.e., more than 250,000 parasites per μl). *Prognosis directly relates to the density of parasitemia.* Mortality is approximately 1% at counts below 100,000/μl. It exceeds 50% with counts above 500,000/μl. In a recent study of pediatric patients in Ghana, the initial Glasgow coma score was a better prediction of length of stay than the *initial* parasitemia level.

High parasitemia levels were associated with deep levels of coma, but only when both parameters were assessed *throughout* hospital stay (Wolf–Gould et al.). CSF analysis is indicated to rule out another infection. In most cases, opening pressures and CSF analysis are normal (Daroff et al. 1967). Once malaria is confirmed and treatment is rapidly initiated, full neurological recovery is the expected outcome. For children presenting with cerebral malaria, poor prognostic indicators include age of younger than 3 years, hypoglycemia, unresponsiveness to painful stimuli, and convulsions (Taylor et al. 1988).

The mortality of cerebral malaria in children is approximately 10% to 40%. The incidence of sequelae in African children who survive has been approximately 10%, and they are associated particularly with hypoglycemia (Warrell et al. 1990). In Gambia, hemiplegias were the most common events, but cortical blindness, ataxia, and behavior disturbances have also been seen. Cranial nerve palsies, neuropathies and extrapyramidal tremors are rarely observed. Cerebellar signs have been observed in Sri Lankan patients weeks after an attack of cerebral malaria. Among adults in Thailand, greater than 95% of survivors are said to show no neurological sequelae (Warrell et al. 1990).

Several issues must be considered carefully to achieve a desirable outcome. Because *P. falciparum* often is chloroquine-resistant, alternative antimalarials may be required (Medical Letter 1990). These include quinine, pyrimethamine, sulfadoxine, tetracycline, and newer agents, such as the quinolinemethanol, mefloquine (CDC 1990). Parenteral therapy generally is required for initial therapy of cerebral malaria (Phillips et al. 1985). Quinine therapy may cause hypoglycemia and mimic neurological symptoms. This must be considered in patients apparently responding poorly to therapy.

Quinidine, the well-known antiarrythmic agent and isomer of quinine, may be even more effective than quinine against some strains of resistant malaria. More importantly, its wide availability hastens therapy, which markedly improves prognosis. Corticosteroids do not improve prognosis and are *not* indicated during the therapy of cerebral malaria (Warrell et al. 1982). Rarely, exchange transfusions can be performed to rapidly decrease parasitemia. Hence, full neurological recovery from cerebral malaria generally can be expected if the diagnosis is considered early, blood smears are obtained, and therapy is initiated quickly with the possibility of chloroquine-resistance always kept in mind.

Primary Amebic Meningoencephalitis

When warm-water temperatures and organic food sources in freshwater swimming areas reach optimum levels for growth of the ameboflagellate, *Naegleria fowleri*, the stage is set for primary amebic meningoencephalitis (PAM) (Fowler and Carter 1965).

This infection is found throughout the world and is associated with several species of free-living amebae, with *N. fowleri* most commonly reported. A large proportion of cases are reported from the United States (John 1982).

Access of *N. fowleri* to the CNS is direct (i.e., passage through the nasal mucosa, cribiform plate, and along the olfactory nerves). When they reach the CNS, fulminant meningoencephalitis occurs; death ensues within a few days. Fever, headache, olfactory disturbances, nausea, and vomiting initially are associated with purulent meningitis (John 1982).

Progression to meningoencephalitis is marked by invasion of gray matter, subarachnoid, and perivascular spaces. CSF becomes grossly purulent (Carter 1972). The poor prognosis and rapid progression of illness caused by *N. fowleri* infections must be distinguished from the chronic progressive illness caused by *Acanthamoeba* spp, often found in debilitated individuals. The latter is associated with a chronic, granulomatous, necrotizing encephalitis with minimal involvement of the leptomeninges. Death follows after weeks to months of CNS infection. Eyes, skin, and respiratory tract are

the likely entrance sites (Barry and Bia 1989).

Diagnosis of PAM is based on visualization of the living or strained organism in CSF (Wright's or Giemsa stains). Routine Gram stains may not demonstrate organisms because of the effects of heat fixation. Culturing techniques are available. *Acanthamoeba* spp has not been observed during CSF examination but may be found in tissue sections or corneal scrapings. Serology for *Acanthamoeba* may be more helpful than for rapidly progressive *Naegleria* infections.

There is no satisfactory therapy for primary amebic meningoencephalitis associated with *N. fowleri*. One patient was successfully treated with the antifungal agents amphotericin B and miconazole in combination with rifampin (Seidel et al. 1982). The fourth documented survivor was treated with high-dose amphotericin B after motile amebae were visualized in CSF four hours after hospital admission (Brown 1991). For infections caused by *Acanthamoeba* sp, sulfadiazine has been used successfully only for experimental murine infections. The usual course of human infection is rapid progression to coma and death.

Entamoeba Histolytica and Amebic Brain Abscesses

Entamoeba histolytica is the parasitic organism associated with classic amebic dysentery. However, amebic infections range from the asymptomatic carrier state to one of dissemination throughout many organ systems, including brain. Organisms vary in their pathogenicity, which is a major factor in prognosis. Amebiasis is acquired from human carriers by ingestion of cysts in fecally contaminated food or drink. Active intestinal infection results in colonic ulcerations harboring amebic trophozoites. When trophozoites penetrate mucosa, they may reach the portal system, ultimately causing abscess formation in liver, lungs, pleural, or pericardial spaces. Brain abscesses are not common. They carry a poor prognosis because they are associated with severe, widespread infection. In one autopsy series, all 17 proven cases of cerebral amebiasis were associated with hepatic abscess formation, indicating advanced disease (Lombardo et al. 1964). The reported incidence of brain abscess has ranged between 0.6% to approximately 5%. In the autopsy series noted previously, the rate of brain involvement was approximately 8%. Hepatic abscesses usually are singular, but brain lesions can be multiple. In 53 such lesions, basal ganglia and frontal lobes were frequently involved with occasional meningeal infections.

In the AIDS era, suspected brain lesions must be distinguished from lymphomas or other infections, such as toxoplasmosis, fungal, and mycobacterial disease, which are far more common. Stool examinations for ova and parasites generally are not useful in documenting extraintestinal amebiasis, because they are usually negative.

Serological examination for antibodies to *E. histolytica* is useful in nonendemic regions where the prevalence of such antibodies is low. In extraintestinal disease, which has already disseminated to brain, serology would be positive in most, if not all, cases. Unlike the dismal prognosis for primary amebic meningoencephalitis, soft-tissue infections caused by *E. histolytica*, including those in brain, can now be successfully treated with metronidazole and, when necessary, surgery (Hughes et al. 1975). Hopefully, this will lead to a reduction in the mortality rate of greater than 90%. Prognosis depends on the degree of dissemination, the pathogenicity of the infecting organism, and local complicating factors for each lesion in the CNS. Immunocompromised patients predispose to dissemination of pathogenic organisms.

Neurocysticercosis

Human ingestion of pork tapeworm (*Taenia solium*) ova predisposes to neurocysticercosis, the most common parasitic disease of the CNS (Pawlowski 1984). Human exposure occurs with fecal contamination of water and food by these ova (Nash and Neva 1984). Humans actually require the adult *T. solium* by eating poorly cooked pork.

In this form of infection, the adult tapeworm's head, or scolex, is attached to the small bowel along with several meters of additional segments, which are the ova-producing proglottids. Ova are passed in the stool of the infected host. When these ova are ingested by humans, encysted larvae spread from the gut to the brain (neurocysticercosis), eyes, or skeletal muscle (Del Brutto and Sotelo 1988; Keane 1984).

The several patterns of neurocysticercosis are based on resulting pathology. The major patterns of active disease are: (1) parenchymal cysts, which cause seizures, focal neurological deficits (McCormick et al. 1982), or slow mental deterioration; (2) meningeal inflammation, which causes arachnoiditis and obstructive hydrocephalus; (3) intraventricular cysts, which intermittently obstruct the flow of CSF (Shanley and Jordon 1980); and (4) brain infarcts secondary to vasculitis. Occasionally, spinal neurocysticercosis will present with progressive motor deficits, root pain, sensory level sphincter disturbances, or hydrocephalus. Most patients have mixed forms of the disease, but approximately 25% of all patients with neurocysticercosis have apparently normal neurological function (Sotelo et al. 1985)

Untreated larva-containing parenchymal brain cysts expand slowly. Ultimately, the larvae will die. Inflammation occurs and seizures result; neurocysticercosis is the most common cause of seizures in endemic areas. To assess prognosis, *untreated* parenchymal disease was followed with CT scan in 17 patients for more than 8 months before initiating treatment. Lesions showed a 12% increase in the number of cysts and a 33% increment in the mean diameter of these cysts (Sotelo et al. 1984).

Usually *T. solium* ova are not present in stools of patients with neurocysticercosis; that is, concomitant cysticercosis and taeniasis (adult intestinal tapeworm infection) are uncommon (Richards and Schantz 1985). Brain biopsy, which demonstrates the three-layered wall of a cysticercus and its invaginated scolex, is definitive. Brain biopsy may be avoided because approximately half of patients with neurocysticercosis will have subcutaneous and soft-tissue cysts available for biopsy.

In the absence of soft-tissue cysts or calcifications serological tests are useful in demonstrating an antibody response to a variety of antigens. Such tests are more helpful in areas of low endemicity, where positive results have more significance, and cross-reactions from other parasitic infections are unlikely. CSF findings are nonspecific in leptomeningeal infections; they include elevated pressure, lymphocytic pleocytosis, increased protein concentration, and decreased glucose concentration (Nash and Neva 1984). Diagnosis is more dependent on imaging techniques such, as CT and MRI scans (Byrd et al. 1982), to demonstrate the characteristic calcifications, cysts, hydroencephaly, and postcontrast lesion enhancement (Minguetti and Ferreira 1983). Occasionally the characteristic invaginated scolex of neurocysticercosis is visualized.

The antiparasitic agent, praziquantel, is an isoquinoline used since 1980 to treat human cysticercosis, and it has had a major effect on prognosis (Sotelo et al. 1984). Sotelo and associates demonstrated satisfactory results in 26 patients with active parenchymal neurocysticercosis within 3 months of praziquantel therapy. There was a disappearance of 101 out of 152 original cysts and a reduction in the mean diameter of all cysts by 72%.

Repeat CT or MRI evaluation before 3 months following therapy may be misleading when continued inflammation is present (Sotelo et al. 1985). Chronic arachnoiditis and intraventricular cysts (Apuzzo et al. 1984) are less likely than parenchymal cysticercosis to show resolution. Large intraventricular cysts may require shunting or surgical removal; however, one patient with both subarachnoid and ventricular cysticercosis has responded to therapy with albendazole (Del Brutto and Sotello 1990).

During praziquantel therapy, most patients experience clinical exacerbations of their disease, including nausea, vomiting, and seizures. Unrelated to drux toxicity, these findings represent an inflammatory response to parasite degeneration. Use of corticosteroids and antiseizure medications to prevent these responses during praziquantel therapy is the accepted practice (DeGhetaldi et al. 1983). The new imidazole, albendazole, is less expensive than praziquantel and may be slightly more efficacious (DelBrutto and Sotelo 1988). A 30-day course of albendazole (15 mg/kg/d) is now an accepted alternative to praziquantel (15 mg/kg/d for 14 d in 3 divided doses each day) (Medical Letter 1990); Albendazole may be administered with dexamethasone to prevent inflammatory reactions (Santoyo et al. 1991).

Echinococcal (Hydatid) CNS Disease

Echinococcal disease of the CNS can cause large cystic parenchymal brain lesions (Tiberin et al. 1984) or spinal cord compression due to vertebral infection (Pamir et al. 1984). In addition, medical therapy does not alter the prognosis of CNS disease to the same extent that it does for neurocysticercosis.

Patient contact with dogs and sheep are important historical points because *Enchinococcus granulosis* is a tapeworm of dogs that they acquire by eating the infected viscera of sheep. Ova shed by canines are infective for humans, in whom the parasite's cycle is analogous to that of cysticercosis. After ova are ingested, larval development in the human host may produce large hydatid cysts.

These cysts cause symptoms by their expansion and compression of adjacent viscera, including brain. Cyst rupture spreads infection further and is associated with anaphylactoid reactions. When echinococcal cysts invade bone, vertebral fractures may cause epidural spinal cord compression (Schantz 1984).

A more aggressive form of echinococcal disease is seen in the Northern hemisphere. This "alveolar" infection is caused by *Echinococcus multilocularis* and is transmitted by dogs and foxes. Whereas *E. granulosis* produces single large cystic structures, *E. multilocularis* maintains an invasive proliferative growth stage. The parasite often invades liver relentlessly and can metastasize to other tissues, including vertebral bodies and brain (Honma et al. 1982).

Serology is limited in making a diagnosis of echinococcal disease. Infected people often are seronegative, and cross-reactivity due to other disease, such as cysticercosis, can occur. The CT scan has become an important tool in diagnosis. Unlike neurocysticercosis, in which parenchymal brain lesions tend to be small and multiple, echinococcal cysts tend to be single and grow larger without much surrounding edema or inflammatory reaction. In geographic areas where the disease is not endemic, a positive serology and CT appearance of a typical large cyst indicate the presumptive diagnosis (Schantz 1984).

The prognosis for cystic hydatid disease is one of steady progression in the absence of surgical, and possibly medical, therapy. Surgical extirpation of cysts has been the treatment of choice with provisions made to avoid spillage and spread of viable protoscolices from within the cysts. Cardiac cysts are particularly dangerous in this regard because rupture of a cyst may cause dissemination of protoscolices. For alveolar hydatid infection, surgery to remove all larval tissue is essential and must be accompanied by long-term therapy.

Antiparasitic chemotherapy is evolving. If cyst rupture occurs during surgery, praziquantel and albendazole can be used to kill the Protoscolices that spread infection. A trial of chemotherapy prior to surgery can be attempted. Chemotherapy without surgery has also been undertaken. The recommended course of adult antimicrobial therapy is albendazole 800 mg/d for 28 days, which can be repeated as deemed necessary (Medical Letter 1990). Mebendazole and flubendazole therapy are under evaluation. Even with long-term therapy, the prognosis for many treated patients is one of slow progression and recurrences with occasional spontaneous resolution (Schantz 1984).

Trichinosis

Infections caused by the nematode *Trichinella spiralis* are found wherever trichina-infected, undercooked pork or horse meat is consumed. Cysts ingested by humans release larvae that develop into adults in the small bowel. Adults are excreted leaving behind large numbers of new larvae that migrate to striated human muscle. Cooking or freezing pork for human consumption is the major means of preventing human infection (Kazura 1984).

When infection first occurs, a minority of patients experience abdominal symptoms. With larval migration, fever, myalgias, headache, subconjunctival hemorrhages, and occasionally myocarditis occur. Larvae eventually may reach the CNS where their migration causes various syndromes, including meningoencephalitis, thrombosis of vessels, focal deficits, and seizure disorders. Hemorrhagic infarction of brain appears to underlie much of the CNS pathology. Linear necrosis of cerebral tissue likely is secondary to larval migration (Most and Abeles 1937).

Diagnosis is dependent on a history of raw pork ingestion, eosinophilia (present in approximately 90%), and elevated muscle enzymes (50%). Serological conversion lags behind clinical presentation. Muscle biopsy demonstrates larvae in more than 90% of cases and is preferred over brain biopsy. In CNS infections, larvae may be observed in centrifuged CSF in 28% of cases (VanCott and Lintz 1914; Kramer and Aita 1972). CT scan may show nonspecific findings that suggest emboli or hemorrhagic infarction (Gay et al. 1982). MRI scanning has confirmed findings of multifocal areas of hypoxia and infarction (Ellrodt et al. 1987).

The prognosis for CNS trichinosis that results in severe neurological deficits in poor, and it is unlikely to be influenced solely by the available antiparasitic agents. Thiabendazole has little effect on migrating larvae. Corticosteroids are the suggested mainstay of therapy, but their benefits in CNS infection are not uniform. A report from France suggests efficacy of corticosteroids used in combination with the me-bendazole derivative, flubendazole, in three patients with CNS trichinosis (Ellrodt et al. 1987). Albendazole may also prove helpful (Medical Letter 1990). Analgesics, such as aspirin, which could contribute to further hemorrhage, should be avoided.

Paragonimiasis

Although not often encountered by most clinicians outside of Asia, the diagnosis of cerebral paragonimiasis is suggested readily by a history of raw crustacean ingestion, the presence of chronic hemoptysis, and impairment of vision. Chest radiographs simulate chronic cavitary tuberculosis, and skull films often are abnormal, showing unusual parenchymal calcifications (Bunnag and Harinasuta 1984). The recent influx of immigrants from Southeast Asia to North America has underscored the need to be aware of this disease (Coleman and Barry 1982).

Cerebral paragonimiasis evolves from lung fluke infections caused by *Paragonimus westermani* or related species (Higashi et al. 1971). Infective metacercaria are ingested by eating undercooked freshwater crabs or crayfish. Young flukes migrate from intestine to lung where they reside as adults in abscess cavities and mimic tuberculosis and cause hemoptysis. Early during infection, flukes often will migrate to brain, particularly in children, and the majority of patients with cerebral disease present prior to age 20 (Oh 1967).

Major symptoms are variable and include headache, nausea, vomiting, mental deterioration, and occasionally acute meningitis. Epilepsy is common, and a majority of patients have visual disturbances, including impairment of acuity, homonymous hemianopsia, optic atrophy, papilledema, nystagmus, and pupillary abnormalities.

To confirm a diagnosis of paragonimiasis, serological evaluations of blood and CSF for antibodies are useful. Ova of *P. westermani* often are found in stool and sputum. Eosinophilia may be present. CSF findings are nonspecific. Mononuclear pleocytosis and elevated protein concentration may suggest tuberculosis. CSF eosinophilia is uncommon. Skull films should be performed because they may demonstrate a variety of cerebral calcifications, including those with a characteristic "soap-bubble" appearance (Oh 1968). CT findings can provide additional information on ventricular size and possible ocular disease (Yoshida et al. 1982).

Praziquantel has replaced the dichlorophenol, bithionol, for treatment because the latter has been associated with treatment failures (Coleman and Barry, 1982). Praziquantel has the potential to treat pulmonary paragonimiasis as its earliest stages and to halt CNS spread. Once established in the CNS, surgical therapy for removal of established lesions can be combined with praziquantel to control residual disease. Clearly, ultimate control of cerebral paragonimiasis depends on interruption of the organism's life cycle and alterations in methods of food preparation. In the absence of preventive measures, early therapy of pulmonary disease in children should alter prognosis favorably by at least preventing CNS spread.

Strongyloidiasis

Two major CNS complications result from disseminated infection by the intestinal nematode *Strongyloides stercoralis* (i.e., direct CNS invasion by larval forms and meningitis secondary to larva-induced bacteremia). How often does this occur as a consequence of a relatively common nematode infection with world-wide distribution? Fortunately, the need for specific clinical circumstances makes such complications relatively uncommon (Scowden et al. 1978).

Free-living larvae of *S. stercoralis* that are present in soil can penetrate the skin of humans, ultimately migrating to and maturing within, the upper gastrointestinal tract where the adults lay eggs. The ova hatch into larvae before their passage in stool. If full maturation into infectious filariform larvae occurs in the gastrointestinal tract before elimination, larvae may autoinfect their human host by penetration of the gut wall (Faust and DeGroat 1940). Hematogenous spread carries the larvae to the CNS during disseminated infection (Owor and Wamukota 1976). These events, however, do not constitute the usual course and prognosis for *S. stercoralis* infections except under conditions of severe immunosuppression (Neefe et al. 1973).

Because of autoinfection, strongyloidiosis may persist in the human host for many years. It has been documented that prisoners in southeast Asia during World War II maintained persistent infection for more than 40 years. Such infections could go unnoticed except for occasional epigastric discomfort, unexplained skin eruptions, and eosinophilia (Berenson et al. 1987).

However, should such individuals undergo immunosuppression as a result of lymphoma, chemotherapy, or transplantation and its required course of immunosuppression, the risk of dissemination and neurological disease increases.

Under these circumstances, larvae migrating within brain may cause capillary obstruction and microinfarcts with severe mental status changes, including coma. Disseminated infection also is associated with gram-negative bacillary meningitis. This may be caused by hematogenous spread of bacteria from the injured gut wall or direct invasion by larvae carrying such bacterial organisms.

Patients with CNS infection often are immunosuppressed individuals from endemic areas who were not properly screened for *S. stercoralis*. By the time the diagnosis has been considered, these patients may be moribund from gram-negative bacteremia, pneumonitis, and meningitis. Eosinophilia may have been masked by corticosteroids, and nonspecific CSF findings may only reflect bacillary meningitis.

All potentially immunosuppressed patients who have travelled to endemic areas should be evaluated with stool examinations and duodenal aspirates for characteristic larvae. Intestinal carriage must be eliminated prior to initiating immunosuppression. Serological examination for specific antibodies may not be as useful as for other parasitic infections reviewed in this chapter. In patients from endemic areas, such antibodies may only reflect previous infections. However, even this is important to alert the clinician who manages a potentially immunosuppressed patient.

To improve the poor prognosis for disseminated infection, strongyloidiasis must be considered seriously in any immunocompromised patient who develops gram-negative meningitis, bacteremia, or pneumonitis, particularly if eosinophilia is present. However, eosinophilia may be suppressed by concomitant administration of corticosteroids.

Thiabendazole therapy is effective against *S. stercoralis* infections (Grove 1982). However, the standard regimen of 2 days is not adequate for disseminated infections. Therapy for disseminated infection should be prolonged until stools are free of larvae, which may take weeks. The prognosis for such infections, even with therapy, is poor unless the immunocompromised state can be limited. Prognosis is best improved by prevention of dissemination (i.e., appropriate screening and elimination of asymptomatic carriage). Ivermectin, a semisynthetic macrocyclic lactone which causes nematode paralysis, is effective against a variety of nematodes, including *S. stercoralis* (Naquira et al. 1989). There may be a future role for this agent in the treatment of disseminated infections.

Schistosomiasis

Several species of schistosomes can cause severe neurological disease in humans. The major organisms are *Schistosoma mansoni*, *Schistosoma haematobium*, and *Schistosoma japonicum*. Adult worms (approximately 2 cm) usually live within human vesical and mesenteric vessels. Ova are deposited by adult worms within blood vessels and appear in stool and urine. Human fecal or urine contamination of fresh water with ova permits them to hatch and infect snails. Infections are acquired from exposure to infective cercariae released by snails into fresh water. Acute syndromes relating to human infection by cercariae include cercarical dermatitis and a febrile illness (Katayama fever) characterized by lymphadenopathy, diarrhea, urticaria, and splenomegaly (Mahmoud 1984). The pathophysiology of neurological and other organ manifestations of schistosomiasis are related to the human immune response to the ova shed by these parasites.

These responses range from negligible immunological reactions to intense granulomatous inflammatory responses (Bambirra et al. 1984) and myelonecrosis. Hence, prognosis depends on intensity of infection, degree of host immunity, and allergic inflammatory response to infection (Querioz et al. 1979). In addition, each major species of schistosome causes an infection with a different neurological syndrome as discussed below.

S. mansoni

A recent outbreak of acute schistosomiasis and transverse myelitis among 15 American students in Kenya reminded clinicians that *S. mansoni* can cause CNS disease. Autopsy studies in Puerto Rico (Marcial-Rojas and Fiol 1963) had demonstrated a few cases associated with reactions to ova in the spinal cord. More than 20 additional cases were noted in the literature, largely involving the spinal cord. The American students had bathed in a dammed stream and acquired acute schistosomiasis (Centers for Disease Control 1984). Two students developed acute transverse myelitis, both associated with eosinophilia, CSF pleocytosis, and evidence of swelling on CT scan. One student improved following specific chemotherapy and dexamethasone. The constellation of a lower cord syndrome, a known exposure to schistosomes, and no other obvious cause for transverse myelitis raised the diagnosis of schistosomiasis.

S. hematobium

Adult worms of this species reside in the vesical plexus around the bladder. Hematuria and obstructive uropathy characterize the non-neurological presentation. Spinal cord syndromes are similar to those seen with *S. mansoni* infections, including occlusion of the anterior spinal artery, granulomatous involvement of multiple nerve roots, intrathecal granuloma formation, and transverse myelitis. *S. hematobium* ova are more likely than *S. mansoni* to be deposited in brain (Scrimgeour and Gajdusek 1985).

S. japonicum

This schistosome is endemic in China, the Philippines, Indonesia, and Southeast Asia, and cerebral complications are more common than with other schistosome infections (Warren 1978). This has been more closely evaluated since the end of World War II, and several features characteristic of *S. japonicum* have been described. These include monoplegias, hemiplegia and quadriplegia, diffuse encephalitis or meningitis, and syndromes that simulate brain tumor (Kane and Most 1948). Seizure disorders and abnormal EEG findings are more common during *S. japonicum* infections (Marcial-Rojas and Fiol 1963).

Diagnosis of schistosomal infection rests on history and the demonstration of characteristic ova in tissues such as rectum, bladder, or CNS. Eosinophilia suggests acute infection as does CSF pleocytosis and elevated CSF protein concentration. Serology may not be reliable for acute infection, but it is useful during chronic syndromes for documenting prior exposure.

Therapy with praziquantel has altered the prognosis considerably for all forms of schistosomiasis to the point at which an 80% cure rate could be expected after a single dose. However, acute neurological syndromes involve an immunological response, and corticosteroids are used in conjunction with antiparasitic agents to suppress this response. In addition, praziquantel has been associated with acute exudative effusive polyserositis and respiratory failure in one patient following treatment of schistosomiasis (Azher et al. 1990).

The importance of corticosteroids is best illustrated in a patient with Katayama fever and CNS involvement shown by CT scan. This patient showed complete neurological resolution after a 10-week course of dexamethasone alone, prior to any antiparasitic therapy (Kirchoff and Nash 1984). Mass lesions may require surgical removal to assure return of neurological function. Antiparasitic therapy should be undertaken since mating worms can live for many years, releasing large numbers of ova. For the glomerulopathy associated with chronic schistomiasis, antiparasitic therapy does not appear to influence the clinical course of the disease once the lesion is established (Martinelli et al. 1987).

Acknowledgment The authors wish to sincerely thank Ms. Lori Hudnall for her many contributions to manuscript preparation.

References

Alarcon, F.; Escalante, L.; Duenas, G.; Montalvo, M.; Roman, M. Neurocysticercosis. Short course of treatment with albendazole. Arch. Neurol. 46:1231–1236; 1989.

Anderson, W. K. Malarial Psychosis and Neurosis. Edinburgh, Oxford University Press; 1927: pp 1–395.

Apuzzo, M. L. J.; Dobkin, W. R.; Chi-Shing Zee; Chan, J. C.; Giannotta, S. L.; Weiss, M. H. Surgical considerations in treatment of intraventricular cysticercosis. J. Neurosurg. 60:400–407; 1984.

Azher, M.; El-Kassimi, F. A.; Wright, S. G.; Mofti, A. Exudative polyserositis and acute respiratory failure following praziquantel therapy. Chest. 98:241–243; 1990.

Bambirra, A. E.; DeSouza-Andrade, J.; Cesarini, I.; Rodriques, P. A.; Drummond, C. A. S. A. The tumoral form of schistosomiasis: report of a case with cerebellar involvement. Am. J. Trop. Med. Hyg. 33:76–79; 1984.

Barry, M.; Bia, F. J. Acanthamoeba meningoencephalitis. In: Goldsmith, R. S.; Heyneman, D., eds. Tropical medicine and medical parasitology. Norwalk, CT: Appleton and Lange; 1989; p. 253–255.

Brown, R. L. Successful treatment of primary amebic meningoencephalitis. Arch. Intern. Med. 151:1201–1201; 1991.

Berenson, C. S.; Dobuler, K.; Bia, F. J. Fever, petechiae and pulmonary infiltrates in an immunocompromised Peruvian man. Yale J. Biol. Med. 60:437–445; 1987.

Bia, F. J.; Barry, M. Parasitic infections of the central nervous system. Neurologic Clinics 4:171–206; 1986.

Bunnag, D.; Harinasuta, T. Opisthorchiasis, clonorchiasis, paragonimiasis. In: Warren, K. S.; Mahmoud, A. A. F. eds. Tropical and geographical medicine. New York: McGraw-Hill Book Co.; 1984: p. 461–469.

Byrd, S. E.; Locke, G. E.; Biggers, S.; Percy, A. K. The computed tomographic appearance of cerebral cysticercosis in adults and children. Radiology 144:819–823; 1982.

Carter, R. Primary amoebic meningo-encephalitis: an appraisal of present knowledge. Trans. R. Soc. Trop. Med. Hyg. 66:193–213; 1972.

Centers for Disease Control. Acute schistosomiasis with transverse myelitis in American students returning from Kenya. M.M.W.R. 33:445–447; 1984.

Centers for Disease Control. Recommendations for prevention of malaria in travelers. M.M.W.R. 39: Recommendations and reports 3:1–10; 1990.

Coleman, D.; Barry, M. Relapse of *Paragonimus westermani* lung infection after biothional therapy. Am. J. Trop. Med. Hyg. 3:71–74; 1982.

Daroff, R. B.; Deller, J. J. Jr.; Kastl, A. J. Jr.; Blocker, W. W. Jr. Cerebral malaria. J.A.M.A. 202:679–682; 1967.

DeGhetaldi, L. D.; Norman, R. M.; Douville, A. W. Jr. Cerebral cysticercosis treated biphasically with dexamethasone and praziquantel. Ann. Intern. A. Med 99:179–181; 1983.

DelBrutto, O. H.; Sotelo, J. Albendazole therapy for subarachnoid and ventricular cysticercosis. J. Neurosurg. 72:816–817; 1990.

DelBrutto, O. H.; Sotelo, J. Neurocysticercosis: an updated. Rev. Infect. Dis. 10:1075–1087; 1988.

Ellrodt, A.; Halfon, P.; LeBras, P.; Halimi, D.; Bouree, P.; Desi, M.; Caquet, R. Multifocal central nervous system lesions in three patients with trichinosis. Arch. Neurol. 44:432–434; 1987.

Faust, E. C.; DeGroat, A. Internal autoinfection in human strongyloidiasis. Am. J. Trop. Med. 20:359–375; 1940.

Fowler, M.; Carter, R. F. Acute pyogenic meningitis probably due to Acanthamoeba sp. A preliminary report. Br. Med. J. 2:740–742; 1965.

Gay, T.; Pankey, G. A.; Beckman, E. N.; Washinton, P.; Bell, K. A. Fatal CNS trichinosis. J.A.M.A. 247:1024–1025; 1982.

Gellin, B. G.; Soave, R.; Coccidian infections in AIDS. Medical Clinics 76:205–234; 1992.

Grove, D. I. Treatment of strongyloidiasis with thiabendazole: an analysis of toxicity and effectiveness. Trans. R. Soc. Trop. Med. Hyg. 76:114–118; 1982.

Hakes, T. B.; Armstrong, D. Toxoplasmosis: problems in diagnosis and treatment. Cancer 52:1535–1540; 1983.

Higashi, K.; Aoki, H.; Tatebayashi, K.; Morioka, H.; Sakata, Y. Cerebral paragonimiasis. J. Neurosurg. 34:515–527; 1971.

Honma, K.; Sasano, N.; Andoh, N.; Iwai, S.

Hepatic alveolar echinococcus invading pancreas vertebrae and spinal cord. Hum. Pathol. 13:944–946; 1982.

Hughes, F. B.; Faehnle, S. T.; Simon, J. L. Multiple cerebral abscesses complicating hepatopulmonary amebiasis. J. Pediatr. 86:95–96; 1975.

Israelski, D. M.; Remington, J. S. Toxoplasmic encephalitis in patients with AIDS. Infect. Dis. Clinics. of North Am. 2:429–445; 1988.

John, D. T. Primary amebic meningoencephalitis and the biology of *Naegleria fowleri*. Ann Rev. Microbiol. 36:101–123; 1982.

Kane, C. A.; Most, H. Schistosomiasis of the central nervous system: experiences in World War II and a review of the literature. Arch. Neurol. Psychiatry 59:141–183; 1948.

Kazura, J. W. Trichinosis. In: Warren, K. S.; Mahmoud, A. A. E., eds. Tropical and geographical medicine. New York: McGraw-Hill Book Co.; 1984: p. 427–430.

Keane, J. R. Death from cysticercosis: seven patients with unrecognized obstructive hydrocephalus. West J. Med. 140:787–789; 1984.

Kirchhoff, L. V.; Nash, T. E. A case of Schistosomiasis japonica: resolution of CAT-scan detected cerebral abnormalities without specific therapy. Am. J. Trop. Med. Hyg. 33:1155–1158; 1984.

Kramer, M. D.; Aita, J. F. Trichinosis with central nervous system involvement. Neurology 22:485–491; 1972.

Levy, R. M.; Bredesen, D. E.; Rosenblum, M. L. Neurological manifestations of the acquired immunodeficiency syndrome (AIDS): experience at UCSF and review of the literature. J. Neurosurg. 62:475–495; 1985.

Lombardo, L.; Alonso, P.; Arroyo, L. S.; Brandt, H.; Mateos, J. H. Cerebral amebiasis: report of 17 cases. J. Neurosurg. 21:704–708; 1964.

Luft, B. J.; Brooks, R. G.; Conley, F. K.; McCabe, R. E.; Remington, J. S. Toxoplasmic encephalitis in patients with acquired immune deficiency syndrome. J.A.M.A. 252:913–917; 1984.

Maegraith, B.; Malaria. In: Adams, A. R. D.; Maegraith, B., eds. Clinical Tropical Diseases. Oxford: Blackwell Scientific Publications, Alden Press; 1980: p. 240–291.

Mahmoud, A. A. F. Schistosomiasis. In: Warren, K. S.; Mahmoud, A. A. F. Tropical and geographical medicine. New York: McGraw-Hill Book Co., 1984: p. 443–457.

Marcial-Rojas, R. A.; Fiol, R. E. Neurological complications of schistosomiasis. Ann. Intern. Med. 59:215–230; 1963.

Martinelli, R.; Pereira, L. J.; Rocha, H. The influence of anti-parasitic therapy on the course of schistosomiasis associated with *Schistoso-miasis mansoni*. Clin. Nephrol. 27:229–232, 1987.

McCabe, R. E.; Remington, J. S. Toxoplasmosis. In: Warren, K. S.; Mahmoud, A. A. F., eds. Tropical and geographical medicine. New York: McGraw-Hill Book Co.; 1984: p. 281–292.

McCormick, G. F.; Chi-Shing Zee; Heiden, J. Cysticercosis cerebri. Review of 127 cases. Arch. Neurol. 39:534–539; 1982.

Medical Letter. Drugs for parasitic infections. 32:23–32; 1990.

Minguetti, G.; Ferreira, M. V. C. Computed tomography in neurocysticercosis. J. Neurol. Neurosurg. Psychiatry. 46:936–942; 1983.

Most, H.; Abeles, M. M. Trichiniasis involving the central nervous system. Arch. Neurol. Psychiatry 37:589–616; 1937.

Naquira, C.; Jimenez, G.; Guerra, J. G.; Bernal, R.; Nalin, D. R.; Neu, D.; Aziz, M. Ivermectin for human strongyloidiasis and other intestinal helminths. Am. J. Trop. Med. Hyg. 40:304–309; 1989.

Nash, T. E.; Neva, F. A. Recent advances in the diagnosis and treatment of cerebral cysticercosis. N. Engl. J. Med. 311:1492–1496; 1984.

Navia, B. A.; Petito, C. K.; Gold, J. W. M.; Cho, E. S.; Jordan, B. D.; Price, R. W. Cerebral toxoplasmosis complicating the acquired immunodeficiency syndrome: clinical and neuropathological findings in 27 patients. Ann. Neurol. 19:224–238; 1986.

Neefe, L. I.; Pinilla, O.; Garagusi, V. F.; Bauer, H. Disseminated strongyloidiasis with cerebral involvement: A complication of corticosteroid therapy. Am. J. Med. 55:832–838; 1973.

Oh, S. J. Cerebral paragonimiasis. Trans. Am. Neurol. Assoc. 92:275–277; 1967.

Oh, S. J. Ophthalmological signs in cerebral paragonimiasis. Trop. geogr. Med. 20:13–20; 1968.

Owor, R.; Wamukota, W. M. A fatal case of strongyloidiasis with Strongyloides larvae in the meninges. Trans. R. Soc. Trop. Med. Hyg. 70:497–499; 1976.

Pamir, M. N.; Akalan, N.; Ozgen, T.; Erbengi, A. Spinal hydatid cysts. Surg. Neurol. 21:53–57; 1984.

Patterson, T. P.; Patterson, J. E.; Barry, M.; Bia, F. J. Parasitic infections of the central nervous system. In: Schlossberg, D. ed. Infections of the nervous system. New York: Springer-Verlag, 1990: p 234–261.

Pawlowski, Z. S. Cestodiases: Taeniasis, Diphyllobothriasis, Hymenolepiasis and others. In: Warren, K. S.; Mahmoud, A. A. F. eds. Tropical and geographical medicine. New York: McGraw-Hill Book Co.; 1984: p. 471–479.

Phillips, R. E.; Warrell, D. A.; White, N. J.; Looaresuwan, S.; Karbwang, J. Intravenous quinidine for the treatment of severe falciparum malaria. N. Engl. J. Med. 312:1273–1278; 1985.

Potasman, I.; Resnick, L.; Luft, B. J.; Remington, J. S. Intrathecal production of antibodies against T. gondii in patients with toxoplasmic encephalitis and AIDS. Ann. Intern. Med. 108:49–51; 1988.

Querioz, L.; de S.; Nucci, A.; Facure, N. O.; Facure, J. J. Massive spinal cord necrosis in schistosomiasis. Arch. Neurol. 36:517–519; 1979.

Richards, F. Jr.; Schantz P. M. Cysticercosis and taeniasis. N. Engl. J. Med. 312:787–788; 1985.

Santoyo, H.; Corona, R.; Sotelo, J. Total recovery of visual function after treatment of cerebral cysticercosis. N. Engl. J. Med. 324:1137–1139; 1991.

Schantz, P. M. Echinococcosis (Hydatidosis). In: Warren, K. S.; Mahmoud, A. A. F., eds. Tropical and geographical medicine. New York: McGraw-Hill Book Co.; 1984:p. 487–497.

Scowden, E. B.; Schaffner, W.; Stone, W. J. Overwhelming strongyloidiasis: An unappreciated opportunistic infection. Medicine 57:527–544; 1978.

Scrimgeour, E. M.; Gajdusek, D. C. Involvement of the central nervous system in *Schistosoma mansoni* and *S. haematobium* infection. A review. Brain 108:1023–1038; 1985.

Seidel, J. S.; Harmatz, P.; Visvesvara, G. S.; Cohen, A.; Edwards, J. Successful treatment of primary amebic meningoencephalitis. N. Engl. J. Med. 306:346–348; 1982.

Shanley, J. D.; Jordan, M. C. Clinical aspects of CNS cysticercosis. Arch. Intern. Med. 140:1309–1313; 1980.

Snider, W. D.; Simpson, D. M.; Nielson, S.; Gold, J. W. M.; Metroka, C. E.; Posner, J. B. Neurological complications of acquired immune deficiency syndrome: analysis of 50 patients. Ann. Neurol. 14:403–418; 1983.

Sotelo, J.; Escobedo, F.; Rodriquez-Carbajal, J.; Torres, B.; Rubio-Donnadieu, F. Therapy of parenchymal brain cysticercosis with praziquantel. N. Engl. J. Med. 310:1001–1007; 1984.

Sotelo, J.; Torres, B.; Rubio-Donnadieu, F.; Escobedo, F.; Rodriquez-Carbajal, J. Praziquantel in the treatment of neurocysticercosis: long-term follow-up. Neurology 35:752–755; 1985.

Sotelo, J.; Guerreror, V.; Rubio, F. Neurocysticercosis: a new classification based upon active and inactive forms. A study of 753 cases. Arch. Intern. Med. 145:442–445; 1985.

Taylor, T. E.; Molyneux, M. E.; Wirima, J. J.; Fletcher, K. A. & Morris, K. Blood glucose levels in Malawian children before and during the administration of intravenous quinine for severe falciparum malaria. N. Engl. J. Med., 319:1040–1047; 1988.

Tiberin, P.; Heilbronn, Y. D.; Hirsch, M.; Barmeir, E. Giant cerebral *Echinococcus* cysts with galactorrhea and amenorrhea. Surg. Neurol. 21:505–506; 1984.

Van Cott, J. M.; Lintz, W. Trichinosis. J.A.M.A. 62:680–684; 1914.

Warrell, D. A.: Molyneux, M. E.; Beales, P. F. Severe and complicated malaria. Trans. R. Soc. Trop. Med. Hyg. 84 (suppl. 2): p 1–65; 1990.

Warrell, D. A.; Looareesuwan, S.; Warrell, M. J.; Kasemsarn, P.; Intraprasert, R.; Bunnag, D.; Harinasuta, T. Dexamethasone proves deleterious in cerebral malaria. N. Engl. J. Med. 306:313–319; 1982.

Warren, K. S. Schistosomiasis japonica. Clin. Gastroenterol. 7:77–85; 1978.

Wolf-Gould, C.: Osei, L.; Commey, J. O. O.; Bia, F. J. Pediatric cerebral malaria in Accra, Ghana (In Press). J. Trop. Pediat.

Yoshida, M.; Moritaka, K.; Kuga, S.; Anegawa, S. CT findings of cerebral paragonimiasis in the chronic state. J. Comput. Assist Tomogr. 6:195–196; 1982.

13

Fungal Disease

THOMAS F. PATTERSON AND VINCENT T. ANDRIOLE

Central nervous system (CNS) fungal infection has been relatively uncommon (Lyons and Andriole 1986), confined largely to immunosuppressed patients in specific epidemiological settings (Hooper et al. 1982; Salaki et al. 1984). However, with acquired immune deficiency syndrome (AIDS) and with an expanding population of immunocompromised cancer and transplant patients, the incidence of CNS fungal disease has increased dramatically (Conti and Rubin 1988, Macher et al. 1988). The prognosis of CNS fungal disease depends on the underlying immune status of the patient because treatment in persistently immunosuppressed hosts often is unsuccessful (Patterson et al. 1988). The prognosis of most fungal infections is affected favorably by early diagnosis and treatment (Aisner et al. 1977). Amphotericin B has been the cornerstone of antifungal therapy for CNS fungal infections for more than 3 decades; however, its effectiveness is limited by poor CNS penetration, resistance of some fungi, and drug toxicity (Graybill 1988). New antifungals, such as fluconazole and itraconazole, appear to be a

major advance in improving the prognosis of CNS fungal infection (Graybill 1988).

Cryptococcosis

Cryptococcus neoformans causes meningitis, meningoencephalitis, and rarely, a localized granulomatous CNS lesion (Patterson and Andriole 1989a). The respiratory route is the usual means of primary infection, but it can be followed by disseminated disease, usually in patients with compromised cell-mediated immunity (Rippon 1989). Patients with dissemination of disease commonly have CNS involvement because the organism has a predilection for neural tissues (Diamond 1985).

In recent years, the incidence of cryptococcosis has risen due to the increased cases of cancer, transplant, and AIDS (Kovacs et al. 1985). The prognosis of cryptococcal infection depends on whether the underlying predisposing condition can be reversed (Perfect 1989). Kaplan and coworkers (1977) report a 100% fatality rate within 2 years of the diagnosis of cryptococcosis in cancer patients in whom host defense defects were persistent. Similarly, steroid therapy or lymphoreticular malignancy predicts a poor outcome (Diamond and Bennett 1974).

Severity of disease at time of presentation is an important determinant of outcome. An abnormal mental status at the time of diagnosis correlates with a poor outcome (Dismukes et al. 1987). Headache is a favorable symptom presumably because it leads to an earlier diagnosis. Higher mortality rates are seen in patients who present with an initial cerebrospinal fluid (CSF) leukocyte count of less than 20 cells/mm³, low CSF glucose, elevated CSF opening pressure, positive India ink stain for the organism, and CSF or serum cryptococcal antigen titers of 1 : 32 or more (Diamond and Bennett 1974). Isolation of cryptococci at extraneural sites and absence of serum cryptococcal antibody also are predictive of treatment failure or relapse (Diamond and Bennett 1974).

In non-AIDS patients, higher rates of relapse are seen with persistently low CSF glucose concentrations, lack of decrease in CSF antigen titers during therapy, and a post-treatment CSF or serum antigen of 1 : 8 or more (Diamond and Bennett 1974). Elevated pretreatment antigen titers (≥1 : 32) also were associated with treatment failures (Dismukes et al. 1987). However, changes in CSF antigen titers during therapy do not correlate with outcome. Cryptococci can be detected by india ink with negative cultures for years after therapy in 15% to 20% of the patients but do not correlate with relapse (Sarosi et al. 1969).

Cryptococcemia has an extremely poor prognosis, with no survivors in 40 patients in one study (Diamond and Bennett 1974). Perfect et al. (1983) found a survival rate of four out of 15 (29%) in patients with cryptococcemia, but no patients died from uncontrolled cryptococcemia. They suggest that the prognosis of cryptococcemia, while poor, was better than previous predictions.

In patients with AIDS, the outcome of acute infection is poor but may not differ significantly from that of other persistently immunosuppressed hosts (Kovacs et al. 1985; Zuger et al. 1986). AIDS patients at highest risk for mortality include those with an abnormal mental status at presentation, CSF antigen titers more than 1 : 1024, and an absent CSF inflammatory response (less than 20 white blood cells/mm³). Mortality rates in those high risk patients range as high as 40% even with antifungal therapy (Saag et al. 1992). Other risk factors have been less predictive of outcome but Chuck and Sande (1989) found the presence of extraneural cryptococci and a low serum sodium to be the only two dependent factors predictive of survival.

Although cryptococcal meningitis is almost universally fatal without therapy (Lyons and Andriole 1986), it is notable for its chronic nature. Beeson (1952) reports one patient with cryptococcosis who was untreated for 16 years before the disease was fatal. Campbell and colleagues (1982) report on 13 patients who survived more than 2 years without therapy, but these patients are exceptional. In a review of 178 patients in 1951, 86% died within 1 year without treatment (Carton and Mount 1951).

Amphotericin B has been the standard treatment for cryptococcosis for more than 30 years. In early reports, amphotericin B was given intravenously in high doses (1.0–1.5 mg/kg daily or every other day) (Butler et al. 1964; Sarosi et al. 1969; Spickard et al. 1963). A single course of amphotericin B achieved a 52% cure rate, which increased to 64% when patients that relapsed were successfully treated (Sabetta and Andriole 1985a). Death from cryptococcal meningitis occurred in 25% of the patients.

Flucytosine is ineffective when used alone; relapses, development of drug resistance, and significant bone marrow toxicity are common (Kauffman and Frame 1970). However, a cooperative trial (Bennett et al. 1979) compared amphotericin B alone (at a dose of 0.4 mg/daily) to a combination regimen of low-dose amphotericin B (0.3 mg/kg daily) and flucytosine. Combination therapy resulted in cure or improvement in 16 of 24 patients (67%) as compared to 11 of 27 (41%) with amphotericin B alone. In addition, less nephrotoxicity, fewer relapses, and more rapid sterilization of cultures were noted in the group receiving combination therapy. The authors conclude that combination therapy is the regimen of choice for treating cryptococcal meningitis. However, the cure rates seen with low doses of amphotericin B alone are lower than previous studies using higher doses of amphotericin B (Sabetta and Andriole 1985a). Further, toxicity, mainly due to flucytosine, occurred in almost one third of patients receiving combination therapy.

A recent study (Dismukes et al. 1987) showed that a 4-week combination regimen was as effective as a 6-week combination course in patients with no underlying disease who were not receiving immunosuppressive therapy or who had an uncomplicated course of disease. Unfortunately, these criteria for a short course of therapy exclude most patients with cryptococcal meningitis. Although combination therapy offers some advantages over amphotericin B alone, particularly in more rapid sterilization of cultures, the toxicity of combination therapy is significant (Stamm et al. 1987), and amphotericin B can be successfully used alone.

In AIDS patients with cryptococcal meningitis initial responses to therapy have ranged from 50% to 70% using combination or single-drug therapy (Grant and Armstrong 1988; Larson et al. 1990; Saag et al. 1992; Sugar et al. 1990). Virtually all of these patients will suffer relapse, which may be recalcitrant to further antifungal therapy without maintenance therapy (Bozzette et al. 1991a; Zuger et al. 1986). A single weekly dose of 50 to 100 mg of amphotericin B has been effective in suppressing the disease and is surprisingly well tolerated (Zuger et al. 1986). However, recent studies have shown maintenance therapy with fluconazole improves outcome as compared to amphotericin and is less toxic (Powderly et al. 1990). Persistent prostatic infection often precedes CNS release and is a predictor of CNS infection (Bozzette et al. 1991b; Larsen et al. 1989.

Ketoconazole is not useful for CNS disease (Perfect et al. 1982). Fluconazole, which has excellent CSF penetration, and itraconazole have been successfully used to treat cryptococcal meningitis in AIDS patients who failed combination therapy (Byrne and Wajszczuk 1988; DeGans et al. 1988; Denning et al. 1989). These drugs appear to offer a tremendous improvement in the therapy of cryptococcal meningitis.

Intraventricular amphotericin therapy generally has been reserved for patients with progressive disease despite systemic amphotericin B therapy (Bennett 1981). However, a recent study shows increased survival in seriously ill patients treated with both intraventricular and systemic amphotericin B (Polsky et al. 1986).

Surgical therapy in cryptococcal disease is used for complications such as hydrocephalus (Young et al. 1985) or cryptococcomas (Salaki et al. 1984). The development of hydrocephalus may cause visual loss and cause early deaths due to cryptococcal meningitis (Denning et al. 1991a). Optimal treatment for this complication is not established but includes ventricular shunting, high volume lumbar punctures and perhaps steroids (Denning et al. 1991a). Cryptococcomas are rare and can be managed medically in many cases (Perfect 1989). In cases of large lesions (greater than 3 cm), surgical therapy may be required, which has a survival rate of 33% in reported cases of cerebral cryptoccoma (Salaki et al. 1984).

Coccidioidomycosis

Coccidioidal meningitis is the most serious complication of disseminated disease due to *Coccidioides immitis,* which is endemic to the Southwestern United States, Mexico, and areas of Central and South America (Bouza et al. 1981). Disseminated disease occurs in less than 1% of patients with primary pulmonary infections, and one third to one half of those with disseminated disease develop meningeal involvement (Einstein 1974). Risk of dissemination and the prognosis of infection depends on several factors, including race, gender, and immune status. African-American people, Filipinos, and Native Americans have an increased risk of dissemination and higher mortality rates as compared to white people (Drutz and Catanzaro 1978). Meningitis in these patients tends to present acutely with multiple organ systems involved and often is rapidly fatal (Rippon 1989). Meningeal disease in white people presents more commonly as chronic meningitis, which may become symptomatic long after the primary infection (Rippon 1989). In patients presenting with chronic symptoms, the disease course may be protracted, with long periods of remission.

Dissemination is less common in women except during pregnancy, which is a poor prognostic factor (Bouza et al. 1981). In one study, dissemination occurred in 37 of 65 pregnant women and was fatal in 78% (Smale and Waechter, 1970). The risk of dissemination and of fatality is greater later in pregnancy (Rippon 1989). Disseminated disease is associated with greater than 50% fetal loss primarily due to premature birth (Van Bergen et al. 1976). Untreated, the infection is fatal, but successful treatment with amphotericin B has been reported (Rippon 1989).

Opportunistic meningeal infection occurs in immunosuppressed patients, including those with malignancies (Deresinski and Stevens 1974), steroid treatment (Bayer et al. 1976a), organ transplantation, and AIDS (Roberts 1984). In these patients, disseminated infection appears to result from reactivation of disease, and it carries a poor prognosis if the underlying condition continues (Deresinki and Stevens 1974). Coccidioidomycosis in AIDS patients appears to occur as an acute infection and has been associated with a high fatality rate (Fish et al. 1991). Life-long suppressive treatment has been required in these patients (Roberts 1984).

Meningeal involvement usually occurs within the first 3 months of the illness, although it has been reported to remain latent for intervals up to 12 years (Bouza et al. 1981). The diagnosis of CNS infection often is not made for months after CNS involvement occurs, which adversely affects response to therapy (Bronnimann and Galgiani 1989). Stevens (1983) reports responses in five of nine patients with disease of less than 1 month duration, but only seven responses in 30 patients with the disease for more than 6 months. In nine patients presenting with symptoms of advanced neurological disease, including coma, none responded to treatment.

Spinal fluid findings are nonspecific and generally do not predict prognosis. However, in one study, only one of twelve patients with a CSF protein greater than 200 responded to therapy (Stevens 1983). Even in cases of fulminant disease with positive CSF cultures, the CSF formula has been unremarkable (Bouza 1981). The organism is cultured in less than one third of patients but is not predictive of outcome (Salaki et al. 1984). Coccidioidal fungemia reflects severe disseminated disease and carries a poor prognosis. Death occurred in 11 of 16 patients within 1 month of a positive *C. immitis* blood culture (Ampel et al. 1986).

Serological testing is useful in establishing an early diagnosis of disseminated coccidioidomycosis and correlates with outcome. Serum complement fixing (CF) antibodies in titers of more than 1 : 16 are predictive of disseminated disease and are associated with a poorer prognosis (Smith et al. 1950). CSF CF antibodies are found in 75% to 95% of patients with meningeal disease (Drutz and Catanzaro 1978). Patients who lack CSF antibodies early in the course of disease may have a favorable prognosis if therapy is initiated rapidly. However, patients with immunodeficiencies also may fail to develop CF antibodies in serum or CSF, and their prognosis is poor. Serial CF titers can be used to assess response to therapy and to evaluate for relapse after therapy is discontinued (Rippon 1989; Smith et al. 1950).

Skin testing is useful in assessing prognosis (Cox and Vivas 1977). In widely disseminated disease, the skin test frequently is negative and is associated with a poor prognosis. The skin test was negative in 19 of 27 patients with coccidioidal meningitis but did not correlate with outcome (Bouza et al. 1981). Conversion of the skin test to positive during therapy may be associated with an improved prognosis (Smith et al. 1950).

Computed tomography (CT) offers a significant improvement in the diagnosis of hydrocephalus, a complication of the disease that adversely affects the prognosis (McGahan et al. 1982). Prior to CT, hydrocephalus was reported to occur in 0% to 19% of cases, but in recent series 29% to 45% required a shunting procedure (Labadie and Hamilton 1986).

The mortality rate of coccidioidal meningitis prior to the use of amphotericin B approached 100% (Einstein 1974). Although exceptional cases surviving untreated for up to 25 years are reported (Drutz and Catanzaro 1978), the fatality rate at 1 year for untreated disease is 86% (Stevens 1983). Sobel and colleagues (1984) found a mean survival of 3.6 months in untreated patients, compared to 21.1 months in patients treated with amphotericin B. However, they concluded that while amphotericin B prolonged survival, it did not alter the pathological findings in CNS coccidioidal infection.

Prolonged therapy with intrathecal (IT) and intravenous (IV) amphotericin B is required, but relapses are common, which requires extensive periods of maintenance therapy (Bennett 1981). Survival rates with combined IT (at doses of 0.5 mg three times weekly) and IV amphotericin B range from 51% to 83% (Labadie and Hamilton 1986). Bouza and colleagues (1981) show that in patients treated with minimal total amounts of IT amphotericin B (less than 10 mg), the mortality rates were 67%, compared to mortality rates of only 12% in patients receiving more than 20 mg of IT amphotericin B. Labadie and Hamilton (1986) used high doses of IT amphotericin B (1.0–1.5 mg per dose; mean total dose 82 mg) and achieved overall survival rate of 91%.

Ketoconazole at doses up to 400 mg/day fails to achieve detectable CSF concentrations (Galgiani 1983) and has not been successful in treating meningeal disease. Higher doses (up to 1200 mg/day) suppress the infection, but toxicity is common at those doses (Craven et al. 1983; Goodpasture et al. 1985). Miconazole has been used for coccidioidal meningitis, but its use is limited because response rates are only 31% with relapse rates as high as 78% (Stevens 1983). A combination regimen of oral ketoconazole with IT miconazole was successful in nine children with coccidiodial meningitis (Shehab et al. 1988). Fluconazole has been used successfully to treat coccidioidal meningitis, and it may improve the prognosis of this disease (Classen et al. 1988; Tucker et al. 1988; 1990).

Surgical therapy in coccidioidal disease is reserved for placement of reservoirs for IT amphotericin and for shunting of hydrocephalus (Young et al. 1985). Aggressive shunting procedures combined with antifungal treatment can improve the prognosis of patients with advanced neurological symptoms related to uncontrolled hydrocephalus (Winston et al. 1983).

Histoplasmosis

Histoplasmosis is a granulomatous disease caused by the pathogenic fungus *Histoplasma capsulatum,* which has a worldwide distribution and is common in the Central United States (Goodwin and DesPrez 1978; Wheat 1989a). CNS disease generally occurs in the setting of disseminated infection from a pulmonary focus (Salaki et al. 1984). Disseminated histoplasmosis is uncommon, occurring in patients with some defect of cellular immunity including patients with AIDS (Goodwin and DesPrez 1978; Johnson et al. 1988; Wheat et al. 1990a). CNS histoplasmosis occurs in 10% to 24% of patients with disseminated disease (Cooper and Goldstein 1963; Salaki et al. 1984).

Dissemination can occur shortly after initial infection or can develop years later, depending on the immune status of the patient, which is important in determining the prognosis of infection. An acute presentation of overwhelming disseminated disease with granulomatous CNS involvement occurs in patients with serious underlying immune defects, such as underlying malignancies and AIDS (Anaissie et al. 1988; Goodwin and DesPrez 1978). In patients with less severe immune defects, chronic or subacute presentations of meningitis or focal cerebral histoplasmomas are seen (Shapiro et al. 1955). In these patients, the course may be characterized by long asymptomatic periods with relapses of clinical illness (Gelfand and Bennett 1975). In patients with focal histoplasmomas and occasionally chronic meningitis, the CNS process may present as an isolated lesion, without evidence of active disseminated infection (Couch et al. 1978; Gilden et al. 1974; LeBourgeois 1979; White and Fritzlen 1962). Strong clinical suspicion is required for appropriate diagnosis and treatment.

CNS histoplasmosis is not common in cancer and transplantation patients, but both dissemination of primary infection and reactivation of latent infection occurs. In a review of immunosuppressed patients with disseminated histoplasmosis (Kaufmann et al. 1978), all patients who had CNS involvement developed extensive disease in the brain and meninges. In patients with a delay in diagnosis, the mortality rate was 100%.

In patients with AIDS, histoplasmosis usually results from reactivation of infection (Minamoto and Armstrong 1988). The presentation can be one of persistent fever and weight loss, but an overwhelming disseminated infection resembling gram-negative sepsis has been described. CNS involvement occurs in at least 18% of disseminated histoplasmosis in AIDS (Wheat et al. 1990b).

Clinical symptoms of CNS histoplasmosis generally occur late in the course of infection and indicate extensive neurological damage (Karalakulasingam et al. 1976). Cooper and Goldstein (1963) report neurological manifestations, including mental status changes, chronic meningeal signs, and focal defects in seven of 26 (26.9%) patients, all of whom had extensive CNS infection. However, in the remaining 19 patients, five had extensive CNS involvement but did not exhibit any neurological finding.

Spinal fluid findings in CNS histoplasmosis generally are indistinguishable from other chronic meningititis and are not predictive of the prognosis of disease. Normal CSF can be found in some patients with proven CNS disease and are positive in less than 50% of the cases (Salaki et al. 1984).

Serological testing for serum antibodies by CF or immunodiffusion are not specific or sensitive for either disseminated or CNS disease (Wheat 1989a). Higher titers of CF antibody are present in symptomatic patients, but only 11 of 26 (42%) immunosuppressed patients with disseminated disease developed CF antibody (Kauffman et al. 1978). Wheat and colleagues (1986) detected CF antibody in the CSF in eight of nine patients with CNS histoplasmosis, but CSF antibodies were positive in 23% of patients with chronic meningitis of other etiologies. A decrease in CSF antibody titer may be useful in predicting a favorable response to therapy (Wheat 1989a).

H. capsulatum polysaccharide antigen detected in serum and urine has been successfully used to diagnose active disseminated infection (Wheat 1989a). Recently, histoplasma antigen has been detected in the CSF and appears specific for the disease (Wheat 1989b). Serial antigen levels have decreased following successful therapy and risen with relapse of infection (Wheat et al. 1991).

In patients with focal signs, CT and EEG are useful in localizing lesions and in directing neurosurgical biopsy to allow prompt diagnosis and treatment (Walpole and Gregory 1987).

The mortality rates of untreated disseminated histoplasmosis ranges from 83% to 100%, compared to 5% to 23% when treated with amphotericin B (Wheat 1989a). Relapses occur in up to 23% of patients and are associated with underlying immunosuppression, treatment with less than 2 g of amphotericin B, and meningitis (Bradsher et al. 1982). Mortality rates in untreated CNS disease have been estimated at more than 70% (Karalakulasingam et al. 1976), but the course of histoplasma meningitis can vary from an acute overwhelming infection to an indolent course with survival for years, even without therapy (Tynes et al. 1963; Gelfand and Bennett 1975). Goodwin and colleagues (1980) reviewed seven cases of severe meningitis that were treated with amphotericin B. Five patients were cured, and two improved with therapy. Wheat et al. (1990a) reported overall cure rates of 42%. Intrathecal amphotericin B does not appear to improve survival, although experience is limited (Salaki et al. 1984; Wheat et al. 1990a). Ketoconazole is not useful for CNS histoplasmosis (Craven et al. 1983).

In patients with AIDS, therapy has been less successful, with mortality rates of more than 50% (Minamoto and Armstrong 1988). Suppressive therapy with antifungals in patients with AIDS is required because relapse occurs in more than 90% after therapy is discontinued (McKinsey et al. 1989).

Optimal therapy for CNS histoplasmoma is anecdotal, but biopsy of suspected lesions should be used to establish a diagnosis if necessary because several patients have been treated for brain tumors rather than histoplasma infection (Venger et al. 1987; Walpole and Gregory 1987). Successful therapy of cerebral histoplasmoma usually includes surgical resection combined with amphotericin B (Allo et al. 1979; Greer et al. 1964; Schochet et al. 1980).

Aspergillosis

The CNS is involved in 50% to 70% of patients with disseminated invasive aspergillosis, primarily a result of hematogenous spread from a pulmonary focus (Bodey and Vartivarian 1989; Denning and Stevens 1990), but it can occur from direct extension of a sinus infection (Jinkins et al. 1987), iatrogenic contamination, or intravenous injection in drug addicts (Feely and Steinberg 1977). Occasionally, an isolated CNS lesion occurs with no known portal of entry (DeWyngaert et al. 1986).

Most CNS aspergillosis occurs in immunocompromised cancer and transplant patients in whom the disease usually is associated with fatal disseminated infection (Conti and Rubin 1988; Greene et al. 1991). The mortality rates in patients with persistent immune defects is extremely high (Boon et al. 1990). CNS aspergillosis was fatal in all 36 patients with cardiac and renal transplantation and with other chronically immunosuppressed conditions (Britt et al. 1981; Hooper et al. 1982; Weiland et al. 1983). CNS aspergillosis occurs in patients with AIDS, usually during advanced stages of HIV-infection (Minamoto et al. 1992; Pursell et al. 1992). CNS aspergillosis in these patients is usually fatal even with antifungal therapy (Woods and Goldsmith 1990; Denning et al 1991b).

In patients with no predisposing conditions and in patients in whom immunosuppression is temporary, CNS aspergillosis is more likely to present as a result of local extension or as an isolated CNS lesion (Mukoyama et al. 1969). In patients without persistent underlying immune defects, CNS aspergillosis has an improved prognosis. In a review of 25 patients with CNS aspergillomas, 12 had no identifiable underlying condition (Yanai et al. 1985). Although the infection was fatal in 24 of those 25 patients, symptoms were chronic in some patients, lasting up to 14 years. Several cases of long-term survival in patients without underlying diseases have been described, including isolated pituitary abscess (Ramos-Gabatin and Jordan 1981), arachnoiditis (DeWyngaert et al. 1986), and erosive sinusitis (Sekhar et al. 1980).

The duration of symptoms in immunosuppressed patients with CNS aspergillosis from one series was 5.3 weeks but has ranged from days to months (Young et al. 1970). The most common clinical presentation of CNS aspergillosis is an acute neurological deterioration, which usually is associated with rapidly fatal disease (Hooper et al. 1982). Focal neurological deficits, most commonly hemiparesis caused by hemorrhagic infarction, were seen in 10 of 17 patients reported by Walsh and colleagues (1985). In these patients, death occurred 1 to 17 days after the neurological event.

Prompt diagnosis is critical for the successful outcome of CNS aspergillosis. Hematogenous dissemination in the CNS occurs in more than 50% of patients by the time the pulmonary lesion is identified (Conti and Rubin 1988). The CSF formula is nonspecific, and CSF cultures rarely are positive for the organism (Salaki et al. 1984). CT may be useful for isolating the CNS lesion, but early lesions may be missed even when EEG and neurological examinations show localizing signs (Weiland et al. 1983).

Serological techniques have been used to facilitate the early diagnosis of invasive aspergillosis (Patterson et al. 1988; Sabetta et al. 1985; Weiner 1980). Serum antibodies are not useful in predicting invasive disease because immunosuppressed patients at highest risk for infection often do not have a positive antibody response. However, in one patient, CSF antibodies to *Aspergillus* were detected at the time of diagnosis and became undetectable with successful therapy (De-Wyngaert et al. 1986). Serum antigen detection has been useful in the early diagnosis of invasive disease (Sabetta et al. 1985b). Serial measurement of antigen is useful for allowing response to therapy (Patterson et al. 1989c).

Treatment of invasive aspergillosis prior to CNS seeding offers the best possibility for cure (Aisner et al. 1977) because the prognosis is extremely poor in patients with dissemination to the CNS, particularly if underlying immune deficits remain (Patterson and Andriole 1989d). Amphotericin B remains the treatment of choice for CNS disease (Bodey and Vartivarian 1989). Anecdotal cases of cure are reported with the addition of either flucytosine or rifampin or both to amphotericin B (DeWyngaert et al. 1986; Sekhar et al. 1980; Weinstein et al. 1982). One report describes cure with flucytosine alone, although flucytosine is expected to have little activity as a single agent (Atkinson and Israel 1973). Liposomal amphotericin B and the new triazoles appear to offer improvements in treatment of invasive aspergillosis but remain experimental (Graybill 1988; Patterson et al. 1989b). Empiric amphotericin B therapy at the time of reinduction chemotherapy has been shown to improve the prognosis in patients with invasive aspergillosis who require continued immunosuppressive therapy (Karp et al. 1988).

Surgical therapy is used in cases of solitary brain abscess or granuloma but usually is not successful (Salaki et al. 1984). However, cases of cure with surgery alone (Venugopal et al. 1977) or more commonly followed by antifungal therapy (Ramos-Gabatin and Jordan 1981) have been reported.

Candidiasis

The incidence of CNS candidal disease has risen in recent years due to increased numbers of patients receiving immunosuppressive therapies, steroids, broad-spectrum antibiotics, and parenteral hyperalimentation (Lipton et al. 1984). *Candida* has become the predominant mycosis detected at autopsy, even in areas endemic for other types of mycoses (Parker et al. 1978). In some series, most CNS disease has resulted from disseminated candidiasis, which usually produces multiple cerebral microabscesses (Edwards 1978). Parenchymal CNS involvement occurs in 30% to 90% of cases of disseminated candidiasis (Crislip and Edwards 1989; Meunier 1989) and is usually not diagnosed or suspected until autopsy (Louria et al. 1962). In 12 patients with cerebral candidiasis reported by Lipton and colleagues (1984), survival after detection of *Candida* ranged from 2 days to 2 months. Only one patient in that series had a preterminal diagnosis of CNS infection.

In the past, meningitis was reported as the most common manifestation of CNS candidal infection (DeVita et al. 1966). However, in more recent series, less than 15% of patients with CNS candidiasis have meningitis (Lipton et al. 1984). Candidal meningitis is reported to present as a complication of disseminated candidiasis in half the patients but can occur with endocarditis (Andriole et al. 1962), direct inoculation (Chmel 1973), and local extension (Crislip and Edwards 1989). CNS meningitis also can occur as an isolated event, usually from hematogenous spread that may have occurred days or months previously from an indwelling catheter (Salaki et al. 1984). In patients with CNS meningitis, the presentation is usually one of subacute or chronic disease with symptoms lasting for a mean duration of 2 months prior to diagnosis (Crislip and Edwards 1989). In addition, candida meningitis has occurred following therapy for bacterial meningitis (Gelfand et al. 1990).

Candidiasis in the newborn is associated with a high incidence of CNS involvement, usually meningitis (Faix 1984). Risk factors for development of CNS disease include prematurity, immunocompromised condition, antibiotics, and parenteral nutrition (Lilien et al. 1978). Mortality rates in newborns are higher than in older individuals, perhaps due to the subacute course of disease, lack of clinical signs, and variable CSF culture results (Chesney et al. 1978). Long-term neurological sequelae of the disease are seen most commonly in premature infants younger than 4 months (Smego et al. 1984).

Bayer and colleagues (1976b) report that a delay of more than 2 weeks from onset of symptoms to diagnosis, a CSF glucose of less than 35 mg/dl, and the development of intracranial hypertension and focal neurological deficits are associated with increased mortality. In contrast, age, sex, underlying diseases, prior use of antibiotics, and presence of disseminated candidiasis does not correlate with outcome.

Candida meningitis is usually fatal, but a surprising number of spontaneous cures occur (Smego et al. 1984). In the series reported by Buchs and Pfister (1983), 39% of patients survived without specific antifungal therapy. In the preamphotericin B era, 69% of patients reported by DeVita and colleagues (1966) survived with therapy that would be expected to have little antifungal activity (such as iodides and sulfonamides). Similar results of 63% and 67% survival without antifungal therapy were reported (Bayer et al. 1976b; Buchs and Pfister 1983).

Therapy with IV amphotericin B improves cure rates of candidal meningitis in adult patients to 83% to 89% (Smego et al. 1984). Cure rates reported in neonatal disease, using amphotericin B alone, ranged from 71% to 100% (Smego et al. 1984). However, 56% of surviving neonates had psychomotor retardation, and hydrocephalus occurred in 50%.

Combination therapy with amphotericin B and flucytosine produces cure rates for candidal meningitis of 88%, similar to that achieved by amphotericin B alone (Smego et al. 1984). Combination therapy has been recommended by some investigators as the treatment of choice for candidal meningitis (Bennett 1981). More rapid sterilization of the CSF does seem to occur, and combination therapy has been suggested in one study to reduce the long-term neurological sequelae in newborns (Smego et al. 1984). However, not all strains are sensitive to flucytosine, and the drug is associated with significant toxicity (Bennett 1981).

Miconazole and flucytosine used as single agents have been successful, but experience with these regimens is anecdotal (Nordstrom et al. 1977; Morison et al. 1988). Fluconazole with its excellent CNS penetration may prove useful in treating the disease, but it remains experimental (Crislip and Edwards 1989).

Mucormycosis

Mucormycosis is a relatively uncommon opportunistic fungal infection caused by broad, rarely septate hyphal organisms of the order Mucoraceae, usually *Rhizopus arrhizus* and less commonly *Mucor* and *Absidia* spp (Rinaldi 1989a). The most common CNS presentation of mucormycosis (also referred to as zygomycosis) is a rhinocerebral form, in which the disease may initially involve only the paranasal area and sinuses but progress through the cribiform plate to involve the frontal lobes (Abedi et al. 1984). Rhinocerebral mucormycosis occurs most commonly in patients with diabetes mellitus, lymphoma, and leukemia (Parfrey 1986), organ transplantation (Morduchowicz et al. 1986) and has been reported in AIDS (Blatt et al. 1991). In addition, deferoxamine therapy in hemodialysis patients has been associated with risk for infection (Daly et al. 1989). Survival rates of diabetic patients in a series of 179 patients was 60% as compared to only 20% in persistently immunosuppressed patients with other diseases, such as hematological malignancies (Blitzer et al. 1980).

CNS mucormycosis also occurs as a focal cerebral lesion, which is most commonly seen from hematogenous seeding in patients with IV drug use (Stave et al. 1989). Disseminated mucormycosis is seen in patients with underlying host defects and is commonly associated with CNS involvement (Abedi et al. 1984). The development of cerebral involvement from disseminated disease carries an extremely poor prognosis (Ochi et al. 1988).

Clinical findings of hemiplegia are present in up to one third of patients and reflect vascular invasion of the large arteries. Hemiplegia and facial necrosis are poor prognostic signs (Blitzer et al. 1980) that reflect a delay in diagnosis. Coma occurs in about two thirds of the patients with cerebral involvement and is present with far-advanced intracerebral disease (Rinaldi 1989a).

CT is useful for defining the extent of lesions prior to surgical resection and for documenting cerebral extension (Centeno et al. 1981; Maniglia et al. 1982). However, some reports have shown that CT scans underestimate the clinical extent of disease (Van Johnson et al. 1988). CT scans also can be used to localize focal cerebral mucormycosis lesions, which is important in determining prognosis in patients with isolated cerebral disease. Magnetic resonance imaging may be superior to CT in localizing the pathological process, which may improve survival in this disease (Galetta et al. 1990). Focal cerebral lesions in patients with IV drug use are most commonly located in the basal ganglia (Stave et al. 1989); only two patients with basal ganglia lesions have survived.

In the past, survival for patients with rhinocerebral mucormycosis treated without surgery or amphotericin B was 6% to 12% (Parfrey 1986), and most cases were discovered at autopsy (Esakowitz et al. 1987). Prompt recognition of the disease, aggressive surgical resection, and high doses of amphotericin B have improved the survival rates to 50% to 80% (Eisenberg et al. 1977). However, few patients with intracranial extension have survived. High dose amphotericin delivered via liposomes may improve survival (Fisher et al. 1991). Reversal of the underlying condition, such as treatment of ketoacidosis in diabetics or limitation of the immunosuppressive regimen in transplant patients, is critical to successful therapy (Parfrey 1986). Extensive surgical intervention is necessary for successful therapy, but early surgery appears more important in minimizing the ultimate extent of débridement than in improving rates of survival (Ochi et al. 1988).

Recent studies have evaluated the use of hyperbaric oxygen in improving survival, particularly in cases of cerebral extension of disease (Couch et al. 1988; Ferguson et al. 1988). The mortality rate for patients treated with adjunctive hyperbaric oxygen (along with amphotericin B and surgical debridement) was 33%, compared to 57% in patients who did not receive hyperbaric therapy. More encouraging was the fact that three patients with intracerebral extension, a condition that is usually fatal, were successfully treated. While hyperbaric oxygen therapy offers theoretical advantages of inhibiting fungal growth and relieving tissue ischemia, its effect on the prognosis of patients with rhinocerebral mucormycosis remains to be determined.

Blastomycosis

Blastomycosis of the CNS is a rare complication of infection due to the dimorphic fungus *Blastomyces dermatitis*. The disease is endemic to the Ohio, Mississippi, and St. Lawrence River valleys, but has been reported in a world-wide distribution (Rippon 1989). The CNS is involved in 3% to 10% of cases but is present in up to 33% of cases of systemic disease at autopsy (Witorsch and Utz 1968). CNS lesions usually result from hematogenous seeding from a pulmonary source, but direct extension occurs from osseous lesions in the skull or spine (Tenenbaum et al. 1982). CNS blastomycosis presents as intracranial or spinal abscesses or as meningitis (Gonyea 1978).

Disseminated blastomycosis uncommonly is seen in immunocompromised patients, such as patients with lymphoreticular malignancies and pregnancy (Salaki et al. 1984). More rapid dissemination of disease is reported in these patients, but response to therapy is good if promptly instituted (Tenenbaum et al. 1982). Disseminated blastomycosis is an uncommon infection in patients with AIDS, but it occurs as a late complication of HIV-infection and is associated with a mortality rate of over 50% (Pottage et al. 1990).

The presentation of blastomycotic meningitis includes an acute, chronic, self-limited form of disease (Gonyea 1978). In one series (Kravitz et al. 1981), all four patients with mild symptoms of chronic meningitis were cured. In contrast, three of five patients with severe obtundation or focal neurological findings died despite therapy. Two of three patients with hydrocephalus died, perhaps reflecting a delay in diagnosis of the disease.

The diagnosis of CNS blastomycosis is difficult, which often delays initiation of appropriate therapy (Benzel et al. 1986). The presence of organisms on smear or CSF culture is not predictive of outcome because less than half the smears and one fourth of the lumbar CSF cultures yield the organism. Culturing ventricular fluid may improve markedly the CSF yield (Bradsher 1988). Radiographic techniques are useful in delineating intracranial and spinal lesions but are nonspecific in their appearance. Antibodies to blastomycosis detected by immunodiffusion have been detected in patients with blastomycosis (Tang et al. 1984) but have limited prognostic use in CNS disease. In one series (Gonyea 1978), seven of 10 patients with blastomycosis had detectable antibody, but all three patients negative for antibodies had disseminated disease.

Before amphotericin B became available, the mortality rate of CNS blastomycosis was as high as 77% (Buechner and Clawson 1967), although cures with 2-hydroxystilbamidine were reported. The use of amphotericin B has reduced overall mortality for blastomycosis to approximately 10% (Tenenbaum et al. 1982). However, in a review of blastomycotic meningitis, Kravitz and colleagues (1981) report a mortality rate of 40% in patients treated with amphotericin B. The use of intrathecal amphotericin B is not established, and its use is controversial. Relapse after amphotericin B treatment has occurred (Loudon and Lawson 1961) and appears more common when total doses less than 1.5 g are used (Tenenbaum et al. 1982).

Other antifungal therapies, such as 2-hydroxystilbamidine or ketoconazole, should not be used for CNS disease (Bradsher 1988). CNS lesions have developed in patients with ketoconazole for cutaneous and pulmonary infection (Pitrak and Anderson 1989; Yancey et al. 1991).

Other Mycoses

In recent years, a wide variety of fungal organisms has emerged as opportunistic pathogens that involve the CNS in immunosuppressed patients (Rinaldi 1989b). CNS infections with these emerging fungi range from mass legions in the brain to chronic meningitis (Garcia et al. 1990). Some of these organisms, such as *Sporothrix*, are potential pathogens in normal hosts but are less virulent than the pathogenic organisms discussed previously (Winn 1988). Other fungi, including *Fusarium, Trichosporon, Xylohypha, Drechslera, Curvularia, Geotrichum*, and *Pseudallescheria*, are largely nonpathogenic and cause systemic disease, including CNS infection in compromised hosts (Anaissie et al. 1989; Richardson et al. 1988). The prognosis of infection with these pathogens primarily depends on the immunocompromised state of the host, but prompt diagnosis and therapy with amphotericin B improve the outcome of most.

The mycological identification of these organisms is important because specific antifungal therapy other than amphotericin B improves the prognosis in some of these infections. Therapy with potassium iodide improves the outcome of disseminated sporotrichosis, although its efficacy in CNS disease is limited (Winn 1988). The specific identification of *Pseudallescheria* in a brain abscess is important because it is usually resistant to amphotericin B and sensitive to miconazole (Gari et al. 1985; Patterson et al. 1990). The prognosis of CNS pseudallescheriasis is poor, but cures with miconazole have been reported (Anderson et al. 1984; Berenguer et al. 1989; Kershaw et al. 1990).

References

Abedi, E.; Sismanis, A.; Choi, K.; Pastore, P. Twenty-five years' experience treating cerebro-rhino-orbital mucormycosis. Laryngoscope 94:1060–1062; 1984.

Aisner, J.; Schimpff, S. C.; Wiernik, P. H. Treatment of invasive aspergillosis: relation of early diagnosis and treatment to response. Ann. Intern. Med. 86:539–543; 1977.

Allo, M. D.; Silva, J.; Kauffman, C. A.; Dicks, R. E. Enlarging histoplasmomas following treatment of meningitis due to *Histoplasma capsulatum*. J. Neurosurg. 51:242–244; 1979.

Ampel, N. M.; Ryan, K. J.; Carry, P. J.; Wieden, M. A.; Schifman, R. B. Fungemia due to *Coccidioides immitis*. An analysis of 16 episodes in 15 patients and a review of the literature. Medicine 65:312–321; 1986.

Anaissie, E.; Fainstein, V.; Samo, T.; Bodey, G. P.; Sarosi, G. A. Central nervous system histoplasmosis. An unappreciated complication of the acquired immunodeficiency syndrome. Am. J. Med. 84:215–217; 1988.

Anaissie, E.; Bodey, G. P.; Kantarjian, H.; Ro, J.; Vartivarian, S. E.; Hopfer, R.; Hoy, J.; Rolston, K. New spectrum of fungal infections in patients with cancer. Rev. Infect. Dis. 3: 369–378; 1989.

Anderson, R. L.; Carroll, T. F.; Harvey, R. T.; Myers, M. G. *Petriellidium* (*Allescheria*) *boydii* orbital and brain abscess treated with miconazole. Am. J. Ophthalmol. 97:771–775; 1984.

Andriole, V. T.; Kravetz, H. M.; Roberts, W. C.; Utz, J. P. Candida endocarditis. Clinical and pathological studies. Am. J. Med. 32:251–285; 1962.

Arndt, C. A.; Walsh, T. J.; McCully, C. L.; Balis, F. M.; Pizzo, P. A.; Poplack, D. G. Fluconazole penetration into cerebrospinal fluid: implications for treating fungal infections of the central nervous system. J. Infect. Dis. 157: 178–179; 1988.

Atkinson, G. W.; Israel, H. L. 5-fluorocytosine treatment of meningeal and pulmonary aspergillosis. Am. J. Med. 55:496–504; 1973.

Bayer, A. S.; Yoshikawa, T. T.; Galpin, J. E.; Guze L. B. Unusual syndromes of coccidioidomycosis: diagnostic and therapeutic considerations. Medicine 55:131–152; 1976a.

Bayer, A. S.; Edwards, J. E.; Seidel, J. S.; Guze, L. B. Candida meningitis. Report of seven cases and review of the English literature. Medicine 55:447–486; 1976b.

Beeson, P. B. Cryptococcal meningitis of nearly sixteen years' duration. Arch. Intern. Med. 89: 797–801; 1952.

Bennett, J. E. Treatment of cryptococcal, candidal, and coccidioidal meningitis. In: Remington, J. S.; Swartz, M. N., eds. Current clinical topics in infectious diseases. 2nd ed. New York: McGraw-Hill; 1981:p. 54–88.

Bennett, J. E.; Dismukes, W. E.; Duma, R. J.; Medoff, G.; Sande, M. A.; Gallis, H.; Leonard, J.; Fields, B. T.; Bradshaw, M.; Haywood, H.; McGee; Z. A.; Cate, T. R.; Cobbs, C. G.; Warner, J. F.; Alling, D. W. A comparison of amphotericin B alone and combined with flucytosine in the treatment of cryptococcal meningitis. N. Engl. J. Med. 301:126–131; 1979.

Benzel, E. C.; King, J. W.; Mirfakhraee, M.; West, B. C.; Misra, R. P.; Hadden, T. A. Blastomycotic meningitis. Surg. Neurol. 26:192–196; 1986.

Berenguer, J.; Diaz-Mediavilla, J.; Urra, D.; Muñoz, P. Central nervous system infection caused by *Pseudallescheria boydii:* Case report and review. 11:890–896; 1989.

Blatt, S. P.; Lucey, D. R.; DeHoff, D.; Zellmer, R. B. Rhinocerebral zygomycosis in a patient with AIDS. J. Infect. Dis. 164:215–216; 1991.

Blitzer, A.; Lawson, W.; Meyers, B. R.; Biller, H. F. Patient survival factors in paranasal sinus mucormycosis. Laryngoscope 90:635–648; 1980.

Bodey, G. P.; Vartivarian, S. Aspergillosis. Eur. J. Clin. Microbiol. Infect. Dis. 8:413–437; 1989.

Boon, A. P.; Adams, D. H.; Buckels, J.; McMaster, P. Cerebral aspergillosis in liver transplantation. J. Clin. Pathol. 43:114–118; 1990.

Bouza, E.; Dreyer, J. S.; Hewitt, W. L.; Meyer, R. D. Coccidiodal meningitis. An analysis of thirty-one cases and review of the literature. Medicine 60:139–172; 1981.

Bozzette, S. A.; Larsen, R. A.; Chiu, J.; Leal, M. A. E.; Jacobsen, J.; Rothman, P.; Robinson, P.; Gilbert, G.; McCutchan, J. A.; Tilles, J.; Leedom, J. M.; Richman, D. D.; the California collaborative treatment group. A placebo-controlled trial of maintenance therapy with fluconazole after treatment of cryptococcal meningitis in the acquired immunodeficiency syndrome. N. Eng. J. Med. 324:580–584; 1991a.

Bozzette, S. A.; Larsen, R. A.; Chiu, J.; Leal, M. A. E.; Tilles, J. G.; Richman, D. D.; Leedom, J. M., McCutchan, J. A. Fluconazole treatment of persistent *Cryptococcus neoformans* prostatic infection in AIDS. Ann. Intern. Med. 115:285–286; 1991b.

Bradsher, R. W.; Alford, R. H.; Hawkins, S. S.; Spickard, W. A. Conditions associated with relapse of amphotericin B-treated disseminated histoplasmosis. Johns Hopkins Med. J. 150:127–131; 1982.

Bradsher, R. W. Blastomycosis. Infect. Dis. Clin. North Am. 2:877–898; 1988.

Britt, R. H.; Enzmann, D. R.; Remington, J. S.

Intracranial infection in cardiac transplant recipients. Ann. Neurol. 9:107–119; 1981.

Bronnimann, D. A.; Galgiani, J. N. Coccidioidomycosis. Eur. J. Clin. Microbiol. Infect. Dis. 8:466–473; 1989.

Buchs, S.; Pfister, D. Candida meningitis: course, prognosis, and mortality before and after introduction of the new antimycotics. Mycoses 26:73–81; 1983.

Buechner, H. A.; Clawson, C. M. Blastomycosis of the central nervous system. II. A report of nine cases from the Veterans Administration cooperative study. Am. Rev. Respir. Dis. 95:820–826; 1967.

Butler, W. T.; Alling, D. W.; Spickard, A.; Utz, J. P. Diagnostic and prognostic value of clinical and laboratory findings in cryptococcal meningitis. A follow-up study of forty patients. N. Engl. J. Med. 270:59–67; 1964.

Byrne, W. R.; Wajszczuk, C. P. Cryptococcal meningitis in the acquired immunodeficiency syndrome (AIDS). Successful treatment with fluconazole after failure of amphotericin B. Ann. Intern. Med. 108:384–385; 1988.

Campbell, G. D.; Currier, R. D.; Busey, J. F. Survival in untreated cryptococcal meningitis. Neurology 31:1154–1157; 1982.

Carton, C. A.; Mount, C. A. Neurosurgical aspects of cryptococcosis. J. Neurosurg. 8:143–155; 1951.

Centeno, R. S.; Bentson, J. R.; Mancuso, A. A. CT scanning in rhinocerebral mucormycosis and aspergillosis. Radiology 140:383–389; 1981.

Chesney, P. J.; Justman, R. A.; Bodganowicz, W. M. Candida meningitis in newborn infants: a review and report of combined amphotericin B-flucytosine therapy. Johns Hopkins Med. J. 142:155–160; 1978.

Chmel, H. Candida albicans meningitis following lumbar puncture. Am. J. Med. Sci. 266:465–467; 1973.

Chuck, S. L.; Sande, M. A. Infections with Cryptococcus neoformans in the acquired immunodeficiency syndrome. N. Engl. J. Med. 321:794–799; 1989.

Classen, D. C.; Burke, J. P.; Smith, C. B. Treatment of coccidioidal meningitis with fluconazole. J. Infect. Dis. 158:903–904; 1988.

Conen, P. E.; Walker, G. R.; Turner, J. A.; Field, P. Invasive primary aspergillosis of the lung with cerebral metastasis and complete recovery. Dis. Chest 42:88–94; 1962.

Conti, D. J.; Rubin, R. H. Infection of the central nervous system in organ transplant recipients. Neurologic Clinics 6:241–260; 1988.

Cooper, R. A.; Goldstein, E. Histoplasmosis of the central nervous system. Report of two cases and review of the literature. Am. J. Med. 35:45–57; 1963.

Couch, J. R.; Abdou, N. I.; Sagawa, A. Histoplasma meningitis with hyperactive suppressor T cells in cerebrospinal fluid. Neurology 28:119–123; 1978.

Couch, L.; Theilen, F.; Mader, J. T. Rhinocerebral mucormycosis with cerebral extension successfully treated with adjunctive hyperbaric oxygen therapy. Arch Otolaryngol. Head Neck Surg. 114:791–794; 1988.

Cox, R. A.; Vivas, J. R. Spectrum of in vivo and in vitro cell-mediated immune responses in coccidioidomycosis. Cell. Immunol. 31:130–141; 1977.

Craven, P. C.; Graybill, J. R.; Jorgensen, J. H.; Dismukes, W. E.; Levine, B. E. High-dose ketoconazole for treatment of fungal infections of the central nervous system. Ann. Intern. Med. 98:160–167; 1983.

Crislip, M. A.; Edwards, J. E. Candidiasis. Infect. Dis. Clin. North Am. 3:103–133; 1989.

Daly, A. L.; Velazquez, L. A.; Bradley, S. F.; Kauffman, C. A. Mucormycosis: Association with deferoxamine therapy. Am. J. Med. 87:468–471; 1989.

Denning, D. W.; Tucker, R. M.; Hanson, L. H.; Hamilton, J. R.; Stevens, D. A. Itraconazole therapy for cryptococcal meningitis and cryptococcosis. Arch. Intern. Med. 149:2301–2308; 1989.

Denning, D. W.; Stevens, D. A. Antifungal and surgical treatment of invasive aspergillosis: Review of 2,121 published cases. Rev. Infect. Dis. 12:1147–1201; 1990

Denning, D. W.; Armstrong, R. W.; Lewis, B. H.; Stevens, D. A. Elevated cerebrospinal fluid pressures in patients with cryptococcal meningitis and acquired immunodeficiency syndrome. Am. J. Med. 1:267–272; 1991a.

Denning, D. W.; Follansbee, S. E.; Scolaro, M.; Norris, S.; Edelstein, H.; Stevens, D. A. Pulmonary aspergillosis in the acquired immunodeficiency syndrome. N. Engl. J. Med. 324:654–662; 1991b.

DeGans, J.; Schattenkerk, J. K. M. E.; VanKetal, R. F. Itraconazole as maintenance treatment for cryptococcal meningitis in the acquired immune deficiency syndrome. Br. Med. J. 297:339; 1988.

Deresinski, S. C.; Stevens, D. A. Coccidioidomycosis in compromised hosts. Experience at Stanford University hospital. Medicine 54:377–395; 1974.

DeVita, V. T.; Utz, J. P.; Williams, T.; Carbone, P. P. Candida meningitis. Arch. Intern. Med. 117:527–535; 1966.

DeWyngaert, F. A. V.; Sindic, C. J. M.; Rousseau, J. J.; Xavier, F. G. F.; Brucher, J. M.; Laterre, E. C. Spinal arachnoiditis due to aspergillus meningitis in a previously healthy patient. J. Neurol. 233:41–43; 1986.

Diamond, R. D. Cryptococcus neoformans. In: Mandell, G. L.; Douglas, R. C.; Bennett, J. E., eds. Principles and practice of infectious diseases. 2nd ed. New York: John Wiley; 1985:p. 1460–1468.

Diamond, R. D.; Bennett, J. E. Prognostic factors in cryptococcal meningitis. A study in 111 cases. Ann. Intern. Med. 80:176–181; 1974.

Dismukes, W. E.; Cloud, G.; Gallis, H. A.; Kerkering, T. M.; Medoff, G.; Craven, P. C.; Kaplowitz, L. G.; Fisher, J. F.; Gregg, C. R.; Bowles, C. A.; Shadomy, S.; Stamm, A. M.; Biasio, R. G.; Kaufman, L.; Soong, S.; Blackwelder, W. C.; the National Institute of Allergy and Infectious Diseases Mycoses Study Group. Treatment of cryptococcal meningitis with combination amphotericin B and flucytosine for four as compared with six weeks. N. Engl. J. Med. 317:334–341; 1987.

Drutz, D. J.; Catanzaro, A. Coccidioidomycosis. Part II. Am. Rev. Respir. Dis. 117:727–771; 1978.

Edwards, J. E. Severe candidal infections. Clinical perspective, immune defense mechanisms, and current concepts of therapy. Ann. Intern. Med. 89:91–106; 1978.

Einstein, H. E. Coccidioidomycosis of the central nervous system. Adv. Neurol. 6:101–105; 1974.

Eisenberg, L.; Wood, T.; Boles, R. Mucormycosis. Laryngoscope 87:347–356; 1977.

Eng, R. H. K.; Bishburg, E.; Smith, E. M.; Kapila, R. Cryptococcal infections in patients with acquired immune deficiency syndrome. Am. J. Med. 81:19–23; 1986.

Esakowitz, L.; Cook, S. D.; Adams, J.; Doyle, D.; Grossart, K. W. M.; Macpherson, P.; McFadzean, R. M. Rhino-orbital-cerebral mucormycosis—a clinico-pathological report of two cases. Scott. Med. J. 32:180–182; 1987.

Faix, R. G. Systemic Candida infections in infants in intensive care nurseries: high incidence of central nervous system involvement. J. Pediatr. 105:616–622; 1984.

Feely, M.; Steinberg, M. Aspergillus infection complicating transphenoidal yttrium-90 pituitary implant. Report of two cases. J. Neurosurg. 46:530–532; 1977.

Ferguson, B. J.; Mitchell, T. G.; Moon, R.; Camporesi, E. M.; Farmer, J. Adjunctive hyperbaric oxygen for treatment of rhinocerebral mucormycosis. Rev. Infect. Dis. 10:551–559; 1988.

Fish, D. G.; Ampel, N. M.; Galgiani, J. N.; Dols, C. L.; Kelly, P. C.; Johnson, C. H.; Pappagianis, D.; Edwards, J. E.; Wasserman, R. B.; Clark, R. J.; Antoniskis, D.; Larsen, R. A.; Englender, S. J.; Petersen, E. A. Coccidioidomycosis during human immunodeficiency virus infection. A review of 77 patients. Medicine 69:384–391; 1990.

Fisher, E. W.; Toma, A.; Fisher, P. H.; Chees-

man, A. D. Rhinocerebral mucormycosis: use of liposomal amphotericin B. J. Laryngol. Otol. 105:575–577; 1991.

Galetta, S. L.; Wule, A. E.; Goldberg, H. I.; Nichols, C. W.; Glaser, J. S. Rhinocerebral mucormycosis: Management and survival after carotid occlusion. Ann. Neurol. 28:103–107; 1990.

Galgiani, J. N. Ketoconazole in the treatment of coccidioidomycosis. Drugs 26:355–363; 1983.

Garcia, J. A.; Ingram, C. W.; Granger, D. Persistent neutrophilic meningitis due to Pseudallescheria boydii. Rev. Infect. Dis. 12:959–960; 1990.

Gari, M.; Fruit, J.; Rousseaux, P.; Garnier, J. M.; Trichet, C.; Baudrillart, J. C.; Comte, P.; Feucheres, P.; Pinon, J. M. Scedosporium (Monosporium) apiospermum: multiple brain abscesses. Sabouradia. 23:371–376; 1985.

Gelfand, J. A.; Bennett, J. E. Active Histoplasma meningitis of 22 years' duration. J.A.M.A. 233:1294–1295; 1975.

Gelfand, M. S.; McGee, Z. A.; Kaiser, A. B.; Tally, F. P.; Moses, J. Candidal meningitis following bacterial meningitis. South. Med. J. 83:567–570; 1990.

Gilden, D. H.; Miller, E. M.; Johnson, W. G. Central nervous system histoplasmosis after rhinoplasty. Neurology 24:874–877; 1974.

Gonyea, E. F. The spectrum of primary blastomycotic meningitis: a review of central nervous system blastomycosis. Ann. Neurol. 3:26–39; 1978.

Goodpasture, H. C.; Hershberger, R. E.; Barnett, A. M.; Peterie, J. D. Treatment of central nervous system fungal infection with ketoconazole. Arch. Intern. Med. 145:879–880; 1985.

Goodwin, R. A.; DesPrez, R. M. Histoplasmosis. Am. Rev. Respir. Dis. 117:929–956; 1978.

Goodwin, R. A.; Shapiro, J. L.; Thurman, G. H.; Thurman, S. S.; DesPrez, R. M. Disseminated histoplasmosis: clinical and pathologic correlations. Medicine 59:1–33; 1980.

Grant, I. H.; Armstrong, D. Fungal infections in AIDS. Cryptococcosis. Infect. Dis. Clin. North Am. 2:457–464; 1988.

Graybill, J. R. Therapeutic agents. Infect. Dis. Clin. North Am. 2:805–825; 1988.

Green, M.; Wald, E. R.; Tzakis, A.; Todo, S.; Starzel, T. E. Aspergillosis of the CNS in a pediatric liver transplant recipient: case report and review. Rev. Infect. Dis. 13:653–657; 1991.

Greer, H. D.; Geraci, J. E.; Corbin, K. B.; Miller, R. H.; Weed, L. A. Disseminated histoplasmosis presenting as a brain tumor and treated with amphotericin B: report of a case. Mayo Clin. Proceed. 39:490–494; 1964.

Hooper, D. C.; Pruitt, A. A.; Rubin, R. H. Central nervous system infection in the chronically immunosuppressed. Medicine 61:166–188; 1982.

Jinkins, J. R.; Sisqueira, E.; Al-Kwai, M. Z. Cranial manifestations of aspergillosis. Neuroradiology 29:181–185; 1987.

Johnson, P. C.; Khardori, N.; Najjar, A. F.; Butt, F.; Mansell, G. A.; Sarosi, G. A. Progressive disseminated histoplasmosis in patients with acquired immunodeficiency syndrome. Am. J. Med. 85:152–156; 1988.

Kaplan, M. H.; Rosen, P. P.; Armstrong, D. Cryptococcus in a cancer hospital. Cancer 39: 2265–2274; 1977.

Karalakulasingam, R.; Arora, K. K.; Adams, G.; Serratoni, F.; Martin, D. G. Meningoencephalitis caused by *Histoplasma capsulatum*. Arch. Intern. Med. 136:217–220; 1976.

Karp, J. E.; Burch, P. A.; Merz, W. G. An approach to intensive antileukemia therapy in patients with previous invasive aspergillosis. Am. J. Med. 85:203–206; 1988.

Kauffman, C. A.; Frame, P. T. Bone marrow toxicity associated with 5-fluorocytosine therapy. Antimicrob. Agents Chemother. 11:244–247; 1977.

Kauffman, C. A.; Israel, K. S.; Smith, J. W.; White, A. C.; Schwarz, J.; Brooks, G. F. Histoplasmosis in immunosuppressed patients. Am. J. Med. 64:923–932; 1978.

Kershaw, P.; Freeman, R.; Templeton, D.; DeGirolami, P. C.; DeGirolami, U.; Tarsy, D.; Hoffmann, S.; Eliopoulos, G.; Karchmer, A. W. *Pseudallescheria boydii* infection of the central nervous system. Arch. Neurol. 47:468–472; 1990.

Kovacs, J. A.; Kovacs, A. A.; Wright, W. C.; Gill, V. J.; Tuazon, C. V.; Gelman, E. P.; Lanc, H. C.; Longfield, R.; Parrillo, J. E.; Bennett, J. E.; Magur, H.; Polis M. Cryptococcosis in the acquired immunodeficiency syndrome. Ann. Intern. Med. 103:533–538; 1985.

Kravitz, G. R.; Davies, S. F.; Eckman, M. R.; Sarosi, G. A. Chronic blastomycotic meningitis. 71:501–505; 1981.

Labadie, E. L.; Hamilton, R. H. Survival improvement in coccidioidal meningitis by high-dose intrathecal amphotericin B. Arch. Intern. Med. 146:2013–2018; 1986.

Larsen, R. A.; Bozzette, S.; McCutchan, J. A.; Chiu, J.; Leal, M. A.; Richman, D. D.; the California collaborative treatment group. Persistent *Cryptococcus neoformans* infection of the prostate after successful treatment of meningitis. Ann. Intern. Med. 111:125–128; 1989.

Larsen, R. A.; Leal, M. A. E.; Chan, L. S. Fluconazole compared with amphotericin B plus flucytosine for cryptococcal meningitis in

AIDS. A randomized trial. Ann. Intern. Med. 113:183–187; 1990.

LeBourgeois, P. A. Isolated central nervous system histoplasmosis. South. Med. J. 72:1624–1625; 1979.

Lilien, L. D.; Ramamurthy, R. S.; Pildes, R. S. *Candida albicans* meningitis treated with 5-fluorocytosine and amphotericin B: a case report and review of the literature. Pediatrics 61: 57–61; 1978.

Lipton, S. A.; Hickey, W. F.; Morris, J. H.; Loscalzo, J. Candidal infection in the central nervous system. Am. J. Med. 76:101–108; 1984.

Loudon, R. G.; Lawson, R. A. Systemic blastomycosis. Recurrent neurological relapse in a case treated with amphotericin B. Ann. Intern. Med. 55:139–147; 1961.

Louria, D. B.; Stiff, D. P.; Bennett, B. Disseminated moniliasis in the adult. 21:307–333; 1962.

Lyons, R. W.; Andriole, V. T. Fungal Infections of the CNS. Neurologic Clinics 4:159–170; 1986.

Macher, A. M.; DeVinatea, M. L.; Tuur, S. M.; Angritt, P. AIDS and the mycoses. Infect. Dis. Clin. North Am. 2:827–839; 1988.

Maniglia, A. J.; Mintz, D. H.; Novak, S. Cephalic phycomycosis: a report of eight cases. Laryngoscope 92:755–760; 1982.

McGahan, J. P.; Graves, D. S.; Palmer, P. E. S.; Stadalnik, R. S.; Dublin, A. B. Classic and contemporary imaging of coccidioidomycosis. A. J. R. 136:393–404; 1982.

McKinsey, D. S.; Gupta, M. R.; Riddler, S. A.; Driks, M. R.; Smith, D. L.; Kurtin, P. J. Long-term amphotericin B therapy for disseminated histoplasmosis in patients with the acquired immunodeficiency syndrome (AIDS). Ann. Intern. Med. 111:655–659; 1989.

Meunier, F. Candidiasis. Eur. J. Clin. Microbiol. Infect. Dis. 8:438–447; 1989.

Minamoto, G.; Armstrong, D. Fungal infections in AIDS. Histoplasmosis and coccidioidomycosis. Infect. Dis. Clin. North Am. 2:447–456; 1988.

Minamoto, G. Y.; Barlam, T. F.; Vander Els, N. J. Invasive aspergillosis in patients with AIDS. Clin. Infect. Dis. 14:66–74; 1992.

Morduchowicz, G.; Shmueli, D.; Shapira, Z.; Cohen, S. L.; Yussim, A.; Block, C. S.; Rosenfeld, J. B.; Pitlik, S. D. Rhinocerebral mucormycosis in renal transplant recipients: report of three cases and review of the literature. Rev. Infect. Dis. 8:441–446; 1986.

Morison, A.; Erasmus, D. S.; Bowie, M. D. Treatment of *Candida albicans* meningitis with intravenous and itrathecal miconazole. South Afr. Med. J. 74:235–236; 1988.

Mukoyama, M.; Gimple, K.; Posner, C. M. Aspergillosis of the central nervous system. Report of a brain abscess due to *A. fumigatus* and review of the literature. Neurology 19:967–974; 1969.

Nordstrom, L.; Oistamo, S.; Olmebring, F. Candida meningoencephalitis treated with 5-fluorocytosine. Scand. J. Infect. Dis. 9:63–64; 1977.

Ochi, J. W.; Harris, J. P.; Feldman, J. I.; Press, G. A. Rhinocerebral mucormycosis: results of aggressive surgical debridement and amphotericin B. Laryngoscope. 98:1339–1342; 1988.

Parfrey, N. A. Improved diagnosis and prognosis of mucormycosis. A clinicopathologic study of 33 cases. Medicine 65:113–123; 1986.

Parker, J. C.; McCloskey, J. J.; Lee, R. S. The emergence of candidosis. The dominant postmortem cerebral mycosis. Am. J. Clin. Pathol. 70:31–36; 1978.

Patterson, T. F.; Miniter, P.; Ryan, J. L.; Andriole, V. T. Effect of immunosuppression and amphotericin B on aspergillus antigenemia in an experimental model. J. Infect. Dis. 158:415–422; 1988.

Patterson, T. F.; Andriole, V. T. Current concepts in cryptococcosis. Eur. J. Clin. Micro. Infect. Dis. 8:457–465; 1989a.

Patterson, T. F.; Miniter, P.; Dijkstra, J.; Szoka, F. C.; Ryan, J. L.; Andriole, V. T. Treatment of experimental invasive aspergillosis with novel amphotericin B/cholesterol-sulfate complexes. J. Infect. Dis. 159:717–724; 1989b.

Patterson, T. F.; Miniter, P.; Andriole, V. T. Predictive value of serum aspergillus antigen detection in the diagnosis of invasive aspergillosis (F-2). In: Abstracts of the 89th Annual Meeting of the American Society for Microbiology, New Orleans, LA, 1989c.

Patterson, T. F.; Andriole, V. T. The role of liposomal amphotericin B in the treatment of systemic fungal infections. European J Clin Oncology 25:S63–S68; 1989d.

Patterson, T. F.; Andriole, V. T.; Zervos, M. J.; Therasse, D.; Kauffman, C. A. The epidemiology of pseudallescheriasis complicating transplantation: Nosocomial and community-acquired infection. Mycoses 33:297–302; 1990.

Perfect, J. R. Cryptococcosis. Infect. Dis. Clin. North Am. 3:77–102; 1989.

Perfect, J. R.; Durack, D. T.; Hamilton, J. D.; Gallis, H. A. Failure of ketoconazole in cryptococcal meningitis. J.A.M.A. 247:3349–3351; 1982.

Perfect, J. R.; Durack, D. T.; Gallis, H. A. Cryptococcemia. Medicine 62:98–109; 1983.

Perfect, J. R.; Savani, D. V.; Durack, D. T. Comparison of itraconazole and fluconazole in treatment of cryptococcal meningitis and candida pyelonephritis in rabbits. Antimicrob. Agents Chemother. 29:579–583; 1986.

Pierce, N. F.; Millan, J. C.; Bender, B. S.; Curtis, J. L. Disseminated *Curvularia* infection.

Additional therapeutic and clinical considerations with evidence of medical cure. Arch. Pathol. Lab. Med. 110:959–961; 1986.

Pitrak, D. L.; Anderson, B. R. Cerebral blastomycoma after ketoconazole therapy for respiratory tract blastomycosis. Am. J. Med. 86:713–714; 1989.

Plouffe, J. F.; Fass, R. J. Histoplasma meningitis: diagnostic value of cerebrospinal fluid serology. Ann. Intern. Med. 92:189–191; 1980.

Polsky, B.; Depman, M. R.; Gold, J. W. M.; Galicich, J. H.; Armstrong, D. Intraventricular therapy of cryptococcal meningitis via a subcutaneous reservoir. Am. J. Med. 81:24–28; 1986.

Pottage, J. C.; Tapper, M. L.; Shafer, R. W.; Blum, R. N.; Bonebrake, F. C.; Klein, B.; Chmel, H.; Stratton, C. W.; Powderly, W. G.; Dismukes, W. E. Blastomycosis in AIDS patients (abstract 1166). In: Abstracts of the 30th Interscience Conference on Antimicrobial Agents and Chemotherapy, Atlanta, GA, 1990.

Powderly, W.; Saag, M.; Cloud, G.; Dismukes, W.; Meyer, R.; Robinson, P.; NIAID AIDS Clinical Trials Group, NIAID Mycosis Study Group, and Pfizer Central Research. Fluconazole versus amphotericin B as maintenance therapy for prevention of relapse of AIDS-associated cryptococcal meningitis (abstract 1162). In: Abstracts of the 30th Interscience Conference on Antimicrobial Agents and Chemotherapy, Atlanta, GA, 1990.

Pursell, K. J.; Telzak, E. E.; Armstrong, D. *Aspergillus* species colonization and invasive disease in patients with AIDS. Clin. Infect. Dis. 14:141–148; 1992.

Ramos-Gabatin, A.; Jordan, R. M. Primary pituitary aspergillosis responding to transphenoidal surgery and combined therapy with amphotericin-B and 5-fluorocytosine. J. Neurosurg. 54:839–841; 1981.

Richardson, S. E.; Bannatyne, R. M.; Summerbell, R. D.; Milliken, J.; Gold, R.; Weitzman, S. S. Disseminated fusarial infection in the immunocompromised host. Rev. Infect. Dis. 10:1171–1181; 1988.

Rinaldi, M. G. Invasive aspergillosis. Rev. Infect. Dis. 5:1061–1077; 1983.

Rinaldi, M. G. Zygomycosis. Infect. Dis. Clinics North Am. 3:19–41; 1989a.

Rinaldi, M. G. Emerging opportunists. Infect. Dis. Clin. North Am. 3:65–76; 1989b.

Rippon, J. W. Medical Mycology. The pathogenic fungi and the pathogenic actinomycetes. 3rd ed. Philadelphia: W. B. Saunders, 1989.

Roberts, C. J. Coccidioidomycosis in acquired immune deficiency syndrome. Depressed humoral as well as cellular immunity. Am. J. Med. 76:734–736; 1984.

Rogers, T. R.; Haynes, K. A.; Barnes, R. A. Values of antigen detection in predicting invasive aspergillosis. Lancet 336:1210–1213; 1990.

Roos, K. L.; Bryan, J. P.; Maggio, W. W.; Jane, J. A.; Scheld, W. M. Intracranial blastomycoma. Medicine 66:224–235; 1987.

Saag, M. S.; Powderly, W. G.; Cloud, G. A.; Robinson, P.; Grieco, M. H.; Sharkey, P. K.; Thompson, S. E.; Sugar, A. M.; Tuazon, C. U.; Fisher, J. F.; Hyslop N.; Jacobsen, J. M.; Hafner, R.; Dismukes, W. E.; the NIAID Mycoses Study Group and the AIDS Clinical Trial Group. Comparison of amphotericin B with fluconazole in the treatment of acute AIDS-associated cryptococcal meningitis. N. Engl. J. Med. 326:83–89; 1982.

Sabetta, J. R.; Andriole, V. T. Cryptococcal infection of the central nervous system. Med. Clin. North Am. 69:333–344; 1985a.

Sabetta, J. R.; Miniter, P.; Andriole, V. T. The diagnosis of invasive aspergillosis by an enzyme-linked immunosorbent assay for circulating antigen. J. Infect. Dis. 152:946–952; 1985b.

Salaki, J. S.; Louria, D. B.: Chmel, H. Fungal and yeast infections of the central nervous system. A clinical review. Medicine 63:108–132; 1984.

Sarosi, G. A.; Parker, J. D.; Doto, I. L.; Tosh, F. E. Amphotericin B in cryptococcal meningitis. Long-term results of treatment. Ann. Intern. Med. 71:1079–1087; 1969.

Schochet, S. S.; Mohammad, S.; Kelly, P. J.; Masel, B. E. Symptomatic cerebral histoplasmoma. Case report. J. Neurosurg. 52:273–275; 1980.

Sekhar, L. N.; Dujovny, M.; Rao, G. R. Carotid-cavernous sinus thrombosis caused by Aspergillus fumigatus. J. Neurosurg. 52:120–125; 1980.

Shapiro, J. L.; Lux, J. J.; Sprofkin, B. E. Histoplasmosis of the central nervous system. Am. J. Path. 31:319–330; 1955.

Shehab, Z. M.; Britton, H.; Dunn, J. H. Imidazole therapy of coccidioidal meningitis in children. Pediatr. Infect. Dis. J. 7:40–44; 1988.

Smale, L. E.; Waechter, K. G. Dissemination of coccidioidomycosis in pregnancy. Am. J. Obstet Gynecol. 107:356–359; 1970.

Smego, R. A.; Perfect, J. R.; Durack, D. T. Combined therapy with amphotericin B and 5-fluorocytosine for candida meningitis. Rev. Infect Dis. 6:791–801; 1984.

Smith, C. E.; Saito, M. T.; Beard, R. R.; Kepp, R. M.; Clark, R. W.; Eddie, B. U. Serological tests in the diagnosis and prognosis of coccidioidomycosis. Am. J. Hyg. 52:1–21; 1950.

Sobel, R. A.; Ellis, W. G.; Nielsen, S. L.; Davis, R. L. Central nervous system coccidioidomycosis: a clinical pathologic study of treatment with and without amphotericin B. Hum. Pathol. 15:980–995; 1984.

Spickard, A.; Butler, W. T.; Andriole, V.; Utz, J. P. The improved prognosis of cryptococcal meningitis with amphotericin B therapy. Ann. Intern. Med. 58:66–83; 1963.

Stamm, A. M.; Diasio, R. B.; Dismukes, W. E. Toxicity of amphotericin B plus flucytosine in 194 patients with cryptococcal meningitis. Am. J. Med. 83:236–242; 1987.

Stave, G. M.; Heimberger, T.; Kerkering, T. M. Zygomycosis of the basal ganglia in intravenous drug users. Am. J. Med. 86:115–117; 1989.

Stevens, D. A. Miconazole in the treatment of coccidioidomycosis. Drugs 26:347–354; 1983.

Sugar, A. M.; Stern, J. J.; Dupont, B. Overview: treatment of cryptococcal meningitis. Rev. Infect. Dis. 12:S338–S348; 1990.

Tang, T. T.; Marsik, F. J.; Harb, J. M.; Williams, J. E.; Frommell, G. T.; Dunn, D. K. Cerebral blastomycosis: an immunodiagnostic study. Am. J. Clin. Pathol. 82:243–246; 1984.

Tenenbaum, M. J.; Greenspan, J.; Kerkering, T. M. Blastomycosis. CRC Crit. Rev. Microbiol. 9:139–163; 1982.

Tucker, R. M.; Williams, P. L.; Arathoon, E. G.; Levine, B. E.; Harstein, A. I.; Hanson, L. H.; Stevens, D. A. Pharmacokinetics of fluconazole in cerebrospinal fluid and serum in human coccidioidal meningitis. Antimicrob. Agents Chemother. 32:369–373; 1988.

Tucker, R. M.; Galgiani, J. N.; Denning, D. W.; Hanson, L. H.; Graybill, J. R.; Sharkey, K.; Eckman, M. R.; Salemi, C.; Libke, R.; Klein, R. A.; Stevens, D. A. Treatment of coccidioidal meningitis with fluconazole. Rev. Infect. Dis. 12:S380–S389; 1990.

Tynes, B. S.; Crutcher, J. C.; Utz, J. P. Histoplasma meningitis. Ann. Intern. Med. 59:615–621; 1963.

Utz, J. P.; Shadomy, S.; McGeenhee, R. F. Flucytosine. N. Engl. J. Med. 286:777–778; 1972.

Van Bergen, W. S.; Fleury, F. J.; Cheatle, E. L. Fatal maternal disseminated coccidioidomycosis in a nonendemic area. Am. J. Obstet. Gynecol. 124:661–663; 1976.

Van Johnson, E.; Kline, L. B.; Julian, B. A.; Garcia, J. H. Bilateral cavernous sinus thrombosis due to mucormycosis. Arch. Ophthalmol. 106:1089–1092; 1988.

Venger, B. H.; Landon, G.; Rose, J. E. Solitary histoplasmoma of the thalamus: case report and literature review. Neurosurgery 20:784–786; 1987.

Venugopal, P.; Venugopal, T. V.; Thiruneelakantan, K.; Subramanian, S.; Shetty, B. M. V. Cerebral aspergillosis: report of two cases. J. Med. Vet. Mycol. 15:225–230; 1977.

Viviani, M. A.; Tortorano, A. M.; Giani, P. C.; Ariei, C.; Goglio, S. A.; Crocchiolo, P.; Almavira, M. Itraconazole for cryptococcal infection in the acquired immunodeficiency syndrome. Ann. Intern. Med. 106:166; 1987.

Walpole, H. T.; Gregory, D. W. Cerebral histoplasmosis. South. Med. J. 80:1575–1577; 1987.

Walsh, T. J.; Hier, D. B.; Caplan, L. R. Aspergillosis of the central nervous system: clinicopathological analysis of 17 patients. Ann. Neurol. 18:574–582; 1985.

Weiland, D.; Ferguson, R. M.; Peterson, P. K.; Snover, D. C.; Simmons, R. L.; and Najarian, J. S. Aspergillosis in 25 renal transplant patients. Epidemiology, clinical presentation, diagnosis, and management. Ann. Surg. 198:622–629; 1983.

Weiner, M. H. Antigenemia detected by radioimmunoassay in systemic aspergillosis. Ann. Intern. Med. 92:793–796; 1980.

Weinstein, J. M.; Sattler, F. A.; Towfighi, J.; Sassani, J.; Page, R. B. Optic neuropathy and paratrigeminal syndrome due to *Aspergillus fumigatus*. Arch. Neurol. 39:582–585; 1982.

Wheat, L. J.; French, M. L. V.; Kamel, S.; Tewari, R. P. Evaluation of cross-reactions in *Histoplasma capsulatum* serologic tests. J. Clin. Microbiol. 23:493–499; 1986.

Wheat, L. J. Diagnosis and management of histoplasmosis. Eur. J. Clin. Microbiol. Infect. Dis. 8:480–490; 1989a.

Wheat, L. J.; Kohler, R. B.; Tewari, R. P.; Garten, M.; French, M. L. V. Significance of *Histoplasma* antigen in the cerebrospinal fluid of patients with meningitis. Arch. Intern. Med. 149:302–304; 1989b.

Wheat, L. J.; Batteiger, B. E.; Sathapatayavongs, B. *Histoplasma capsulatum* infections of the central nervous system. Medicine. 69:244–260; 1990a.

Wheat, L. J.; Connolly-Stringfield, P. A.; Baker, R. L.; Curfman, M. F.; Eads, M. E.; Israel, K. S.; Norris, S. A.; Webb, D. H.; Zeckel, M. L. Disseminated histoplasmosis in the acquired immune deficiency syndrome: Clinical findings, diagnosis and treatment, and review of the literature. Medicine. 69:361–374; 1990b.

Wheat, L. J.; Connolly-Stringfield, B. S.; Blair, R.; Connolly, K.; Garringer, T.; Katz, B. P. Histoplasmosis relapse in patients with AIDS: Detection using *Histoplasma capsulatum* variety *capsulatum* antigen levels. Ann. Intern. Med. 115:936–941; 1991.

White, H. H.; Fritzlen, T. J. Cerebral granuloma caused by *Histoplasma capsulatum*. J. Neurosurg. 19:260–263; 1962.

Winn, R. E. Sporotrichosis. Infect. Dis. Clin. North Am. 2:899–911; 1988.

Winston, D. J.; Kurtz, T. O.; Fleischmann, J.; Morgan, D.; Batzdorf, U.; Stern, W. E. Successful treatment of spinal arachnoiditis due to coccidioidomycosis. Case report. J. Neurosurg. 59:328–331; 1983.

Witorsch, P.; and Utz, J. P. North American blastomycosis: a study of 40 patients. Medicine 47:169–200; 1968.

Woods, G. L.; Goldsmith, J. C. *Aspergillus* infection of the central nervous system in patients with acquired immunodeficiency syndrome. Arch. Neurol. 47:181–184; 1990.

Yanai, Y.; Wakao, T.; Fukamachi, A.; Kunimine, H. Intracranial granuloma caused by *Aspergillus fumigatus*. Surg. Neurol. 23:597–604; 1985.

Yancey, R. W., Jr.; Perlino, C. A.; Kaufman, L. Asymptomatic blastomycosis of the central nervous system with progression in patients given ketoconazole therapy: a report of two cases. J. Infect. Dis. 164:807–810; 1991.

Young, R. C.; Bennett, J. E.; Vogel, C. L.; Carbone, P. P.; DeVita, V. T. Aspergillosis. The spectrum of the disease in 98 patients. Medicine 49:147–173; 1970.

Young, R. F.; Gade, G.; and Grinnell, V. Surgical treatment for fungal infections in the central nervous system. J. Neurosurg. 63:371–381; 1985.

Zuger, A.; Louie, E.; Holzman, R. S.; Simberkoff, M. S.; Rahal, J. J. Cryptococcal disease in patients with the acquired immunodeficiency syndrome. Ann. Intern. Med. 104:234–240; 1986.

14

Bacterial Meningitis

GARY D. OVERTURF

Bacterial meningitis is caused by a variety of bacterial pathogens and attacks newborn infants to the elderly. Not surprisingly, the prognosis greatly depends on two major determinants: the age of the patient and the microbiological pathogen. In general, the very young and the very old fare the worst, regardless of the cause. However, other factors that are determined during the acute course, such as the intensity and duration of infection or the associated intensity of the inflammatory reaction, also can predict the quality of survival. Because meningitis is predominantly a disease of children and therefore often occurs in a neurologically developing individual, most of the data on prognosis are applicable to the three major community-acquired pathogens, *Haemophilus (H.) influenzae*, *Neisseria (N.) meningitidis*, and *Streptococcus (S.) pneumoniae*. Similarly, meningitis is nearly epidemic in the neonatal period, and relatively large numbers of infants have been observed following meningitis caused by common neonatal pathogens, such as the enteric rods and group B streptococci. In contrast, adult cases are relatively rare; community-acquired cases frequently are due to *S. pneumoniae* and often occur in a patient with some immune compromise (e.g., alcoholism). In general, adult patients fare worse after meningitis, but specific risk data by etiology are less available. For rarely encountered pathogens, only minimal information is available. Thus, this chapter focuses on the prognosis of bacterial meningitis after infection with the primary causes of the disease. It refers to the rarer forms of the disease, such as listeriosis, brucellosis, and rickettsial infection when appropriate.

Neonatal Meningitis

The neonatal period extends from birth through the second month. Bacterial pathogens at this age predominantly include the gram-negative enteric rods (30% to 60%), the group B streptococci (12% to 48%), and occasionally other gram-positive cocci (5% to 10%) and *Listeria monocytogenes* (1%). Mortality is high, and gram-negative rods account for the highest death rate. Combined rates of mortality and significant se-

189

quelae often exceed 50% to 75% for infants with neonatal meningitis (Bush 1971; Dyggve 1962).

Group B Streptococci

Mortality rates of group B streptococcal infections vary from 18% to 60%, and sequelae have varied from 36% to 44% (Barton et al. 1973). Age and prematurity are significant factors in increasing mortality for infants with birth weights less than 1500 g and in disease severity as indicated by positive blood cultures, pulmonary involvement, and early postnatal onset. In studies of long-term survivors, 29% had severe sequelae, 21% had moderate deficits, and 50% were functionally normal (Edwards and Baker 1982). Predictive features for death or long-term sequelae at admission to the hospital include a comatose or semicomatose state, decreased peripheral perfusion, total peripheral white blood count (WBC) less than 5000/mm^3, and cerebrospinal fluid (CSF) protein less than 300 mg/dl (Chin and Fitzhardinge 1985). Major sequelae have included global mental retardation, relapse, uncontrolled seizures, cortical blindness, microcephaly or hydrocephaly, quadriparesis, central diabetes insipidus, and mild mental retardation. Mild or moderate sequelae have included unilateral sensioneural deafness, mild mental retardation, monoparesis, arrested hydrocephalis, controlled seizure disorders, language delay, porencephalic cyst, frontal atrophy, and deficits in visual or auditory memory (Edwards et al. 1985). In cohort-controlled studies, rates of major sequelae have been as low as 12% (Wald et al. 1986). Thus, for the majority of children who survive their disease, outcome may be relatively good for this pathogen as opposed to other neonatal pathogens.

Gram-Negative Enteric Rods

In contrast to disease with group B streptococci, infants with infections caused by gram-negative pathogens have much worse outcomes. These infections are largely confined to the first 14 days of life and are associated with other known risk factors for severe infection and neurological injury, such as prematurity, male sex, and obstetrical or perinatal complications (premature or prolonged rupture of the membranes, unsterile or precipitous delivery, maternal bleeding, twin birth, cesarian section, and maternal sepsis) (Groover et al. 1961; Ziai and Haggerty 1985). These complications may be present in 56% to 94% of births. Mortality rates in the early antibiotic or preantibiotic era ranged from 64% to 83%, and sequelae rates were 73% to 83% among survivors. The highest mortality occurred in premature infants. Some investigators have suggested that fewer infants in shock or with a poor neurological response survive, whereas the presence of fever (as opposed to no fever or hypothermia) is a good prognostic sign (Groover et al. 1961; Yu and Grauaug 1963). Survivors also tend to have significantly lower initial CSF proteins. Both subdural effusions and hydrocephaly occur more often in infants with CSF protein greater than 300 mg/dl (29% vs. 83%, respectively) (Yu and Grauaug 1963). The use of corticosteroids has been suggested to improve the outcome in neonatal meningitis; however, this has not been tested in a controlled trial or shown clear statistical significance in treated versus untreated groups. Hydrocephalus is a frequent, severe, and persistent sequelae among survivors of gram-negative enteric neonatal meningitis. Hydrocephalus is characteristically observed in 25% to 75% of survivors.

The Neonatal Meningitis Cooperative Study Groups have provided more insight into the outcome of neonatal meningitis caused by gram-negative enteric bacteria in today's antimicrobial era. In particular, these investigators have related the variables of treatment modality to prognosis (McCracken and Mize 1976; McCracken et al. 1980). Overall, the case fatality rate in these studies has been about half of those reported from the early treatment era, varying from 12% to 45%. Also, these studies have stressed the striking differences between term (18%) and low–birth-weight (48%) mortality. In the first report, 64% of survivors were assessed as normal 4 years after illness (McCracken and Mize 1976).

However, there were no statistically significant differences in the incidence of mortality or morbidity or in the number of days that CSF fluid cultures remained positive among infants receiving no intrathecal or three sequential daily intrathecal injections of gentamicin. The routine use of intraventricular therapy also did not improve outcome and actually adversely affected survival (42.9% died with intraventricular therapy vs. 12.5% without) (McCracken et al. 1980). In these latter studies, the CSF WBC was significantly greater for infants who died (2370 ± 900 vs. 990 ± 240) as was CSF protein (1282 ± 626 vs. 517 ± 120); therefore, these two CSF findings used as prognostic indicators were consistent with earlier studies.

Certain etiologic agents may be specifically associated with a more adverse prognosis. *Escherichia coli* with K1 capsular antigens are more invasive (Mulder et al. 1984; McCracken et al. 1974). Approximately 75% to 80% of the strains of *E. coli* that cause neonatal meningitis carry the K1 capsular antigen, which is related to the capsular polysaccharide of group B *N. meningitidis*. The high incidence of K1 *E. coli* serotypes in neonatal meningitis suggests that it is a significant virulence factor. Certain other infections due to enteric organisms other than *E. coli* appear to be associated with an increased mortality in neonatal meningitis, particularly *Proteus spp* (Cussen and Ryan 1967), *Citrobacter spp* and *Salmonella spp* (Choen et al. 1987). In part, this may be related to more frequent resistance to effective antimicrobial therapy, or it may be related to specific undefined virulence factors. A higher incidence of cerebral hemorrhagic necrosis has been associated with infections due to *Proteus spp* (Cusson and Ryan 1967) but also has been noted less frequently in cases caused by *Aerobacter, Pseudomonas, E. coli,* and *Serratia. Citrobacter diversus* may cause sporadic or epidemic neonatal meningitis (Kline 1988). One third of infected infants die, and 75% of survivors have severe neurological impairment. Brain abscess is found in 75% of infants, and 82% of meningeal *Citrobacter* strains possess a 32,000 molecular weight (MW) outer membrane protein (compared to 21% of strains isolated from other body sites). This protein may constitute a specific virulence protein. Epidemics of meningitis among newborn infants due to *Flavobacterium meningosepticum* have had a particularly poor outcome. In one series of 14 infants, infection resulted in nine deaths (George et al. 1961). This high mortality was caused by early toxemia and necrotizing meningitis. Only one of the 14 infants survived without apparent sequelae, while four other survivors developed hydrocephalus.

Recovery from salmonella meningitis appears to be comparable to other enteric meningitides, but relapse is more common. More than half of all cases of salmonella meningitis occur within the first year of life (Choen et al. 1987). In early studies, untreated salmonella meningitis had a mortality of 94% compared to 52% in those treated with sulfonamides (Henderson 1948). The prognosis remained poor even when treated with conventional antibiotics (i.e., chloramphenicol and tetracycline), with a mortality of 83% (Beene et al. 1951). Most of those who survive were left with significant neurological residue. However, more recent series have noted a better outcome. In three cases of salmonella meningitis in infants treated by Rabinowitz and McLeod (1972) all three survived, two without sequelae. The third infant developed massive hydrocephalus.

Community-Acquired Meningitis

Three major pathogens account for more than 95% of all cases of community-acquired bacterial meningitis: *H. influenzae, N. meningitidis,* and *S. pneumoniae* (Underman et al. 1978). *Haemophilus* infections account for approximately 60% of the cases, and thus more long-term data on outcome are available for this pathogen than for any other. In the preantibiotic era, meningeal disease due to these pathogens was nearly universally fatal, with rates of mortality for haemophilus, pneumococcal, and meningococcal infection of 95% to 100%, 98% to 100%, and 72% to 99%, respectively. With the advent of effective chemotherapy, mortality rapidly fell to rates near those seen today.

Age and clinical severity at admission to the hospital have been significant predictors of survival or sequelae from the time of the earliest studies in the chemotherapeutic era. Thus, in large studies by Wilson and Lerner (1964), Mathies (1972), and Geisler and coworkers (1980), older patients consistently had a poorer prognosis for survival and neurological recovery regardless of etiology, and pneumococcal-diseased patients had poorer prognosis regardless of age. Thus, mortality rates for children (younger than 14 years) averaged 5.0% (0% to 25%) and for adults 38% (33% to 68%). In the series by Wilson and Lerner (1964), all adult patients with pneumococcal meningitis who were in a coma or were older than sixty years died. The presence of pneumonia in adults also was nearly as predictive of a poor prognosis. In the series by Mathies (1972), 40.2% of all adults and children with pneumococcal disease who entered with severe disease died, as opposed to rates of 18.2% and 22% for haemophilan and meningococcal disease, respectively. However, even in the mildest disease category, rates of meningococcal and haemophilan mortality were 1.0%, whereas pneumococcal disease was nearly four times as high at 3.8%. In the most recent large series (1316 patients) (Giesler et al. 1980) covering the years from 1954 to 1976, the mean annual fatality rate was 7.1% excluding neonates who had the highest rate (38.4%). Beyond the neonatal period, people older than 40 years had the highest fatality rate (21.5%), while children 1 to 4 years old had the lowest rate (5.6%). Pathogen-specific rates of fatality were 3.4%, 6.0%, and 17.9% for Haemophilus, meningococcal, and pneumococcal infections, respectively.

Haemophilus influenzae

H. influenzae type B meningitis was, until recently, the single most common cause of bacterial meningitis with rates approximating 3.0 in 100,000 or approximately 8000 per year in the United States. Ninety-four percent of all cases occur before 6 years of age. Thus, the incidence within the first 5 years of life is one in 400 to 2000 children. A few cases occur due to nontypable *Haemophilus spp,* but more than 98% are caused by the type B encapsulated organism.

Early studies revealed that only 35% of children escaped neurological sequelae (Sproles et al. 1969), and 10% of survivors were incapable of independent function. The first comprehensive review of sequelae and prognosis in haemophilus meningitis was completed in 1972 by Sell and coworkers (1972). Eighty-six patients were followed after acute episodes occurring from 1950 to 1964. Eleven of 86 (12.7%) died; eight died within the first 24 hours of illness. Follow-up appraisals (including history, physical examination, neurological examination, audiometry, and Wechsler Intelligence Scale for Children) were completed for 56 survivors when they were school aged. Of the survivors, 29% had severe handicaps, 14% had possible residuals, and only 43% were free of detectable deficits. Significant or "severe" handicaps were defined as an IQ of 50 to 69, seizures requiring medication, hearing loss, failure in school, paralysis or paresis (i.e., motor deficits), or sight loss. "Possible" residuals were defined as failure in school without other problems, IQ of 70 to 89, mild hearing or speech difficulty, or behavioral problems with an IQ more than 90.

Taylor and coworkers (1984) completed a sibling-controlled study of neuropsychological achievement in 24 children 6 to 8 years after recovery from haemophilus meningitis. After meningitis, children consistently scored worse on performance and full scale IQ tests, and on a panel of neuropsychological tasks. However, the groups did not differ in verbal IQ, and they performed comparably on all academic measures used in the study. Unlike the study by Sell and coworkers (1987), these investigators did not find notable differences in behavioral problems or temperament.

A significantly increased risk of morbidity and mortality in Haemophilus meningitis has been correlated with seizures, coma, hypothermia shock, age younger than 12 months, hemoglobin less than 11 g/100 ml, pretreatment symptoms for longer than 3 days, a CSF WBC less than 1000/mm³, or CSF glucose less than 20 mg/dl (Herson and Todd 1977). Feldman and coworkers (1982) have shown that certain microbiological features of Haemophilus meningitis correlate with sequelae and are the probable determinants of the severity of presenting symptoms and prognosis. Patients with CSF colony counts of *H. influenzae* 10^7/ml or more had significantly greater frequencies of speech impairment, hearing loss, and moderate or severe neurological sequelae; similarly, CSF glucose less than 10 mg/dl correlated with an increased incidence of hearing loss, speech impairment, and high colony counts. However, partially effective antibiotic therapy, CSF protein concentrations, and the number of CSF polymorphonuclear leukocytes did not predict severe outcome. High CSF colony counts also were correlated with CSF *Haemophilus* antigen concentrations of more than 1.0 µg/ml. Pretreatment concentrations of bacteria varied from 400 to 1.7×10^9 CFU/ml with a mean of 4×10^7 CFU/ml.

Data regarding invasive disease due to *H. influenzae* in adults are limited. In one review, six of 31 adult patients (older than 16 years) had meningitis due to type B strains, and only one patient died (a patient with multiple cranial fractures) (Trollfors et al. 1984). In the same series, four of 15 had disease due to nontypable strains, and one died. Thus, the overall mortality was approximately 20% or about fourfold that observed in children. In a series from the Netherlands, 35 patients older than 6 years with *H. influenzae* meningitis were reviewed; this group represented about 4% of all Haemophilus meningitis cases. Fifty-four percent had a predisposing cause, particularly CSF leaks or skull fractures, and 54% of the strains were nontypable. No patients died, and only 14% had sequelae. However, among the 16 patients without predisposing causes, five of 16 (31%) had sequelae. In another review of 15 cases of

Haemophilus meningitis in adults, sequelae were infrequent among survivors, although mortality was 27% (Spagnuolo et al. 1982). Prognosis was not affected by age or low CSF WBC; however, the occurrence of coma on admission was predictive of a poor outcome. Similar to other studies, patients with CSF rhinorrhea, head trauma, and infection with nontypable *Haemophilus* strains seemed to have a better prognosis.

Pneumococcal and Meningococcal Meningitis

Both pneumococcal and meningococcal disease occur with sufficient frequency to make some conclusions regarding their severity. In a review of meningococcal infection in children from 1947 to 1962 (Stiehm and Damrosch 1966), a 42.8% mortality rate was noted in meningococcemia without meningitis, whereas meningococcal meningitis had a 5.9% mortality. Features associated with a poor prognosis were onset of petechiae within 12 hours before admission, the absence of meningitis, shock, normal or low WBC (10,000 or less), and a normal or low ESR (10 or less). The presence of three or more of these factors was associated with a mortality of greater than 85%. A similar correlation with the poor prognosis with meningococcal septicemia without meningitis has been reported by other investigators (Halstensen et al. 1987). These latter investigations also noted significant differences in survival between patients who reported between 7 and 11 AM (mortality was 33%) and those who reported during the rest of the day (20.7%). They felt that this might be due to overnight delays before seeking medical attention.

Recently, it has been recognized that as many as 50% of children with meningococcemia have clinically evident myocarditis (i.e., depressed left ventricular ejection fraction on ECHO cardiography). In children without myocarditis, no deaths occurred, while three of seven with myocarditis died (Boucek et al. 1984). Tumor necrosis factor (TNF) also has been associated with a poor outcome (Waage et al. 1987). This factor was detected in the serum in 10 of 11 patients who died but in only eight of 68 survivors. All patients with levels greater than 440 U/ml died.

In direct comparisons of meningococcal and haemophilus infections, meningococcal disease generally is associated with fatality rates of about twofold higher. Thus, in Finnish epidemiological studies, overall mortality rates were 3.1% and 7.9% for haemophilus versus meningococcal infections, respectively; again, meningococcemia without meningitis had a 33% mortality (Valmari et al. 1987). However, when comparing meningitis only, mortality was 4.1% and 5.3%, respectively. A long duration of prediagnostic symptoms (3 days or more) was associated with development of neurological sequelae among survivors (14 of 34 or 41% vs. 13 of 74 or 17%).

Pneumococcal meningitis is the most severe disease in terms of mortality and neurological sequelae. Higher rates of mortality apply even if age is taken into account; thus, at all ages, rates of mortality for pneumococcal infection are at least two to three times higher than meningococcal and three to four times higher than haemophilus meningitis. Dodge and Swartz (1965) performed detailed study of sequelae following pneumococcal infections. In this series, 16 pediatric and 17 adult survivors of pneumococcal meningitis were reviewed; six of 27 (22%) patients had major neurological sequelae, whereas only three of 30 (10%) of haemophilus and one of 25 (4%) meningococcal patients had sequelae.

In a series of 42 patients with pneumococcal meningitis, the mortality was 48%, despite treatment with penicillins; 32 of these patients were older than 10 years (Tugwell et al. 1976). Higher mortality was positively correlated with impaired consciousness, a CSF WBC of less than 1000/mm³, and a CSF antigen titer of more than 1 : 100 (CIE). No significant correlation was found with age, sex, duration of symptoms, presence of a focal infection, or focal neurological signs. In a follow-up study (Baird et al. 1976), these investigators continued to find significant associations and poor outcome with

high CSF protein (3.0 g/L or more) and CSF lactate. Feldman (1977) showed that both meningococcal and pneumococcal infections were associated with lower CSF bacterial counts than *H. influenzae* (1.27×10^5, 3.87×10^5, and 2.03×10^7, pneumococcal, meningococcal, and *Haemophilus* respectively). Thus, in contrast to haemophilus infections, higher CSF colony counts did not appear to be the reason for increased morbidity due to these pathogens.

In the large series by Geisler and coworkers (1980), 17.5% of all patients were alert at admission, 51.1% were irritable or lethargic, 19.9% were stuporous or semicomatose, and 9.5% were comatose or convulsing. No deaths occurred among patients with haemophilus or meningococcal disease who were alert on admission, whereas patients who were admitted comatose or convulsing had the highest mortality. In those younger than 10 years, incidence of significant morbidity was 9.8%, 47.3%, and 31.6% for meningococcal, haemophilus, and pneumococcal infection, respectively. The mortality in those older than 10 years was 7.2%, 0%, and 47.4%, for each etiology, respectively.

In an attempt to assess the risk of sequelae with abbreviated (7 days) antibiotic therapy, Jadavji and coworkers (1986) evaluated the outcome of 235 children (mean age 26.4 months) treated with ampicillin or chloramphenicol. The overall mortality was 6.4%. Of 220 survivors, 171 were examined at one year. No neurological, developmental, or audiological abnormalities were detected in 80% of the children. The frequency of identifying at least one handicap during the first year after disease was 57% in pneumococcal meningitis, compared to 14.5% in *H. influenzae* and 0% in *N. meningitides*. Sensorineural hearing deficit was observed in 12.9% (bilateral, 8.2%; unilateral 4.7%). Incidence of hearing loss was 43.3% with pneumococcal meningitis, compared to 7.3% with *H. influenzae*. Incidence of other abnormalities for pneumococcal versus haemophilus disease included speech delay (16.6% vs. 2.4%), developmental delay (13.3% vs. 4%), motor deficits

(10% vs. 1.6%), optic atrophy (10% vs. 0%), and seizure disorder (6.6% vs. 0%). Movement disorders characterized by ataxia, athetosis, choreoathetosis, and hemiballismus rarely are observed in acute episodes of childhood meningitis and may persist for months (Burstein and Breiningstall 1986). Lastly, a Scandinavian group has attempted to correlate prognosis in bacterial meningitis with a multivariate analysis. Risk factors taken into account included sex, CSF glucose, CSF WBC, otitis media, hemoglobin, serum potassium, CSF gram-stain, focal neurological sign, peripheral WBC, CSF percentage of granulocytes, platelet count, petechiae, and duration of symptoms. The predictive value of a calculated good prognosis was 94%; sensitivity in predicting death or mild to severe sequelae was 83% (Valmari et al. 1987).

The findings and prognostic significance of computerized tomography and electrophysiologic studies has been examined (Pike, 1190; Cabral 1987). In forty-one patients older than two months of age at the time of bacterial meningitis (*H. influenzae:* 29 patients, *N. meningitides:* 6 patients, *S. pneumoniae:* 6 patients) CT abnormalities were frequent including subdural effusion, focal infarction, purulence in the basal cisterna, and transient mild dilatation of the subarachnoid space (Cabral 1990). Early increased ventricular size was observed in 29 of 36 patients, but decreased to normal on the third scan in 30 of 33 patients. Clinical management was not influenced by the CT. A second study was specifically designed to assess the prognostic significance of CT, EEG, brain-stem evoked auditory response, and visual evoked potential mapping during acute bacterial meningitis in 32 young children (Pike 1987). Focal or generalized EEG suppression was associated with a poor outcome as well as CT evidence for cerebral infarction and edema, whereas enlarged ventricles or subarachnoid spaced were not. However, neither CT or electrophysiologic studies were better than physical exam in making this prognosis. Subdural effusions may develop in up to 39% of young children with bacterial meningitis (Snedeker 1990).

Young age, rapidity of onset of illness, low peripheral white blood count, and high CSF protein and bacterial antigen were associated with the occurrence of subdural effusion. Although patients with effusions were more likely to have neurologic abnormalities and seizure both at admission and at the completion of therapy, there did not appear to be a greater incidence of seizures, hearing loss, neurologic deficits, or developmental delay on long-term follow-up.

Hearing Impairment

Among the best studied sequelae of acute bacterial meningitis is hearing impairment. Frank deafness or some degree of hearing impairment is among the most commonly recognized serious complications of bacterial meningitis in children and infants. Retrospective and prospective surveys suggest an incidence of 5% to 30% (Berlow et al. 1980; Edwards and Baker 1982; Feigin and Dodge 1976; Gamstorp and Klockhoff 1974; Jones and Hansen 1977; Munoz et al. 1983; Nylen and Rosenhall 1979; Ozdumar et al. 1983; Richner et al. 1979; Teng Y-C et al. 1962). Kaplan and coworkers (1984) have pointed out that the onset of neurological injury leading to hearing loss occurs very early in the disease. All four children in their report had hearing loss at admission, and in two children, hearing loss was profound and persistent. In other studies, as many as 20% of children have transient hearing abnormalities, and only 10% have persistent abnormalities (Vienny et al. 1984). Thus, early testing identifies patients who may have persistent hearing loss, although about one third to two thirds of patients will have an improvement in their hearing deficit. A crucial period of 2 weeks after diagnosis may be necessary to define which patients will have persistent loss and which will experience improvement. In a prospective study of (185 children) by Dodge and coworkers (1984), 10.3% of children with acute bacterial meningitis suffered sensorineural bilateral or unilateral hearing loss. Hearing loss was much more common with pneumococcal (31%) than with meningococcal (10.5%) or haemophilus (6%) infection. In contrast to Vienny's study, these workers did not find improvement in sensorineural hearing loss at any time. Further, in contrast to most other neurological sequelae, the number of days of illness (symptoms) before hospitalization or institution of antibacterial treatment did not correlate with the development of sensorineural deafness.

Earlier studies suggesting a rate of neurologic sequelae in excess of 20% to 25% following bacterial meningitis may have been flawed by too early an assessment after recovery. Two recent studies (Taylor 1990; Pomeroy 1990) have found that the high incidence of neurologic complications during the acute course (37%) or neurologic abnormalities one month following meningitis, did not persist; in both studies only 14% of children had persistent neurologic abnormalities after 1 year, which contrasts with higher sequelae rates in previous series. Sensorineural hearing loss (4–11%) were prominent sequelae in both series, whereas seizures occurred in 2% to 7%. Pomeroy (1990) noted that only the presence of neurologic deficits indicative of cerebral injury was predictive of late afebrile seizures. In the combined total of these 2 series of 282 children, hemiparesis or quadraparesis was observed in 9, with concurrent mental retardation in 8 of these 9. Taylor and colleagues (1990) found only minuscule differences in IQ scores (102 for cases vs. 109 for siblings) among 41 children who experienced neurologic complications compared to unaffected siblings. Both of these series were based on children treated from 1972–1984; therefore, "improvements" in antibiotic therapy and supportive care occurring during the last decade are unlikely explanations for the seemingly better outcome observed in these latter studies.

Effect of Steroids on Neurological Sequelae

Early trials and reviews of steroid use in bacterial meningitis and their effects on the incidence of sequelae and mortality were flawed or inconclusive (Harbin and Hodes 1979). However, recent experience in animal models of steroid-treated bacterial meningitis has shown a significant reduction of either total brain water content or intracranial pressure (Sande et al. 1987). In two prospective studies conducted by Lebel and coworkers (1988), dexamethasone therapy (0.15 mg/kg every 5 hours for 4 days) was associated with a more rapid increase in 24-hour CSF glucose and a decrease in CSF lactate and protein levels. Patients who received dexamethasone became afebrile earlier and were less likely to acquire moderate or severe bilateral sensorineural hearing loss (15.5% vs. 3.3%, treated vs. untreated). Twelve of 98 (13%) patients in the placebo group had severe or profound hearing loss requiring a hearing aid, whereas only one of 102 (1%) dexamethasone-treated patients required a hearing aid. There were no statistically significant differences in the incidence of other neurological abnormalities at 6 weeks or 1 year in steroid-treated or -untreated children, although the outcome was slightly better for steroid-treated children. Other investigators have noted a decline in mortality (but not sequelae) with the routine use of other new modalities, like artificial ventilation (Rasmussen et al. 1988).

Gram-Negative Bacillary Meningitis in Adults

Only a few reviews have been completed regarding meningitis in the elderly. However, in a review by Behrman and colleagues (1989), of 57 episodes of central nervous system infection in patients 65 years and older, 50 cases of meningitis were described. S. pneumonae accounted for 43% of all isolates, but 25% were gram-negative organisms. Sixty-five percent of the patients survived. Increased mortality was associated with altered mental status at diagnosis, inappropriate initial antibiotic therapy (73% vs 27% survival) and low-CSF glucose. Also, patients with associated urinary tract infection or pneumonia also had survival rates of 1/2, compared to 2/3 in patients without such associated infections. Fifteen of twenty-one (66%) patients with gram-positive organisms survived compared to only two of seven patients with gram-negative infections. Surprisingly, delay in diagnosis, the presence of underlying disease, or bacteremia did not significantly alter outcome.

Similar to neonates, gram-negative bacillary meningitis occurring among adult patients carries mortality rates from 30% to 80% (Rajal 1972). Most cases have been associated with severe underlying chronic disease, central nervous system (CNS) trauma, surgery, or spinal anesthesia. Thus, in part, the poor prognosis for survival may be due to the severe life-threatening associated conditions. Gram-negative bacillary infections may constitute up to 69% of the etiologic agents in postoperative neurosurgical infections (Mangi et al. 1975). In the series by Mangi and coworkers (1975), *E. coli* was the major pathogen in children younger than 1 year and had a mortality of 62%. *Klebsiella spp* was the major pathogen in those older than 1 year (usually older than 13 years) and had a mortality of 59%. Associated conditions such as alcoholism increased mortality, whereas burns, cervical infection, and mastoiditis did not. The most significant variables predicting an increased mortality (61.8%) were the occurrence of neurological complications (seizure, coma, focal neurological signs, cranial nerve palsies, brain abscess, or infected foreign body) compared to those patients without such complications (22.7%). In this series, etiologic agent was not associated with a poor outcome. However, other investigators have noted higher mortality with certain pathogens, such as *Serratia spp* (75%) or *Pseudomonas spp* (84%); both agents are typically opportunistic pathogens (Cherubin et al. 1981). Similarly, other investigators have noted a much higher mortality after infection with hospital-acquired infections compared with community-acquired infection (70% vs. 33%, respectively) (Crane and Lerner 1978). Cherubin and coworkers (1981) recorded an overall mortality of 71.3% in 158 patients and noted the more frequent involvement of *Klebsiella* among adult cases. These investigators also noted two major groups: septic elderly adults older than 60 years and preceding cranial trauma

in patients 20 to 60 years of age. Sepsis was a significant factor in death rates (76% with sepsis vs. 54% without sepsis) as was an initial glucose of less than 50 mg/dl. In addition, variations in antibiotic therapy appeared to be a significant factor in mortality; high mortality rates (83%) were seen in patients treated with bacteriostatic antibiotics (chloramphenicol) vs. those treated with bactericidal combinations (51% with ampicillin or ampicillin plus aminoglycosides). Thus, the variables affecting survival with enteric meningitis in adult patients have been age, the pathogen (nosocomial), the occurrence of concomitant sepsis, underlying disease, and the use of bacteriostatic antibiotics.

Rajal and coworkers (1974) studied the effect of intrathecal gentamicin on the outcome of gram-negative meningitis. Although meningitis was clinically or bacteriologically eradicated from 16 of 19 patients, 12 died of other causes during hospitalization. The use of newer cephalosporins in gram-negative bacillary infections in adults is associated with a much better prognosis for survival. Bacteriologic cure in infections due to *E. coli* and *Klebsiella* (in 37 patients) has been as high as 96% in early studies (Cherubin et al. 1982); most of these cases were treated with cefotaxime. In studies by Rajal and coworkers (1974), 16 adult patients were treated with moxalactam. Twelve patients had neurosurgical procedures for head trauma or neoplasm, while other possible predisposing conditions included intravenous (IV) catheter sepsis, hepatic cirrhosis, alcoholism and diabetes, gram-negative pneumonia, and mastoiditis. Infections were due to *Proteus mirabilis* (one), *K. spp* (nine), *E. coli* (three), *Serratia marcescens* (one), and *Acinetobacter* (one). One *Klebsiella* infection was mixed with *Pseudomonas*. Nine of 16 (56%) patients were cured. Six additional patients died of meningitis or complications of antibiotic therapy (bleeding), but achieved a microbiological cure; only one patient with *Serratia* infection did not achieve a microbiological cure and died of hemorrhage and complications of the disease. Other investigators have noted the more frequent failures of treatment associated with *Serratia* and *Enterobacter* meningitis (Eng et al. 1987).

Few adult studies have focused on the long-term prognosis of their patients. This may be because survival has been so hard to achieve. Further, many of the adult patients have neurological morbidity related to their primary disease (e.g., head trauma, neoplasms). However, the intrinsic pathology of adult gram-negative bacillary meningitis does not apper to be as severe as in neonates because diffuse necrotization is infrequent and hydrocephalus appears to be far less frequent.

The understanding of the importance of inflammatory mediators in bacterial meningitis has become greater in recent years (Sande 1987). Interleukin-1β[a] (IL-1β) and cachectin (tumor-necrosis factor i.e., TNF), are thought to be primary mediators in the immune response to microbial invasion. Mustafa and colleagues (1989) have shown correlations between poor neurological outcome in patients with IL-1β concentrations $\geq$500 pg/ml, which also correlated with higher CSF white count, lactate protein and TNF concentrations, and lower glucose concentrations. Arditi and colleagues (1990) have shown similar associations between high CSF TNF and platelet-activating factor, and greater risks of dying, seizures, the number of febrile days and severity scores, as well as higher CSF lipopolysaccharide levels. Although these factors are useful research tools, they are not generally available to the clinician, but serve to extend our knowledge of the pathogenesis of bacterial meningitis. However, they add little to the predictive features of prognosis provided by the clinical symptoms and usual CSF factors associated with adverse prognosis.

Brucellosis

Brucellosis is a common infection in underdeveloped communities. Although reliable incidence figures are difficult to obtain, true neurobrucellosis may be an uncommon disease, occurring in 5% or less of patients with brucellosis (Bouza et al. 1987). However,

neurological complaints are frequent, particularly in chronic infections (i.e., 3 months or more). Neurobrucellosis may evolve at anytime during the course of the disease and may manifest as encephalitis, meningoencephalitis, radiculitis, myelitis, and neuritis (Fincham et al. 1963). Subclinical involvement of the meninges may be more common than clinically appreciated. In general, the prognosis for most neurobrucellosis is good, and the disease is relatively benign if recognized and treated promptly. Increased morbidity is perhaps related most to the delays in treatment and to prolonged symptoms. Less than 30% of patients have positive CSF or blood cultures. Finally, it has been long recognized that confusion regarding toxic neurological and psychiatric symptoms occurs in 75% of patients, and true infection must be differentiated from actual invasion of the nervous system (De Jong 1936).

Seven patients with brucellar meningitis were reported by Bouza and coworkers (1987), and an additional 17 cases were reviewed. Two of the initial seven patients experienced sequelae (flacid paraparesis and gait difficulties), although both of these patients had only minimal CSF changes. A third patient (untreated) presented with a 1-year history of transient ischemic attacks and was admitted for evaluation of dementia and low-pressure hydrocephalus; this patient succumbed to disease after clear evidence of meningitis was obtained (75 cells/mm^3, CSF glucose 5 mg/dl, protein 2.5 g/L, and isolation of *Brucella melitensis* from CSF). In the subsequent review of all 24 confirmed cases of brucellar meningitis, only three were younger than 3 years and only two had *Brucella* sp isolated from CSF. Only one of the 24 confirmed cases of brucellar meningitis died (4.2%). Data on morbidity were included only for the first seven cases, and a detailed analysis was not available for the remaining 17 cases. However, there was one case of epidural abscess complicating the disease, which required surgery, and three patients (12.5%) had detectable nonspecific neurological sequelae. Two of twenty-four cases (8.3%) relapsed.

In another review of 48 cases of neurobrucellosis ranging in age from 2 to 55 years, 10 (21%) cases died (Nichols 1951). The associated pathology included vascular and perivascular inflammation and diffuse meningeal inflammation, including granulomatous cellular infiltration, perineural (nerve root) inflammation, and cerebral cortical edema in some cases. At onset, complications included paraplegia (six), facial paralysis (one), seizures (14), and variable cranial nerve deficits (24) of cranial nerves I, III, VI, VIII. From this report, the prognosis for each of these acute phenomena was not clear, but it seems likely that many were persistent and most occurred in patients who had prolonged symptoms (i.e., more than 3 months) before diagnosis and treatment.

Similar to other central nervous infections, neurobrucellosis in children appears to have a better prognosis than in adults (Lubani et al. 1989). Of nine children with neurobrucellosis, eight had a meningeal component, six of whom had meningitis alone. No relapses, mortality, or sequelae occurred among any of these children, despite a duration of symptoms of 1 to 5 weeks before administration of effective therapy. Coma, convulsions, and cranial nerve involvement occurred in a 1-year-old girl who recovered without sequelae.

Listeriosis

Listeriosis is primarily a disease of the neonate or the geriatric population older than 70 years (Ciesielski et al. 1988). Overall, reported mortality rates have been approximately 19.1% with increasing mortality associated with advanced age (e.g., mortality is 30% in those aged 60 years or older). The presence or absence of meningitis is not a determining factor of mortality; approximately the same death rates occur in both groups of patients, except in infants (six deaths of 35 without meningitis, 0 of 107 with meningitis). However, the incidence of meningitis is far more common in infants and in those older than 40 years (75.3%, less

than 1 month; 52%, 1 month to 13 years; 14.4%, 16 to 39 years; and 49.6%, older than 40 years). Fatality rates for listeriosis among adults and children with underlying malignancy are high (33%), although only one half develop meningitis (Louria et al. 1967).

In an extensive review of 63 neonatal cases by Ray and Wedgewood (1964), 54% were fatal. Fatality was much higher in "early-onset" (i.e., younger than 4 days of age) than "late-onset" disease. Of those who recovered, two developed hydrocephalus (6%). These fatality rates are higher than those observed in more recent series and may be related to the use of less effective or timely antimicrobial therapy. However, some recent outbreaks have continued to be associated with perinatal death rates as high as 46.7% (Evans et al. 1983).

Data on morbidity prognosis are not as readily available as mortality data. Lavetter and coworkers (1971) report the outcome of 25 cases, including nine neonates and 11 patients older than 55 years. In this series, 8 of 25 (32%) died; a 3-week-old infant was left with a residual hydrocephalus and convulsive disorder, whereas all 16 of the remaining survivors had no detectable residua at the time of discharge. Other data regarding long-term prognosis following listerial meningitis are not available.

Miscellaneous Bacterial Pathogens of Low Incidence

Meningitis due to *S. aureus* is infrequent, although it constitutes about 9% to 10% of postneurosurgical infections (Mangi et al. 1975). Gordon and coworkers (1985) describe the course and outcome of 10 adult patients with community-acquired *S. aureus* meningitis. A poor prognosis was associated with severe underlying disease, hyponatremia at presentation, development of seizures, failure of nuchal rigidity to develop, or persistent or recurrent bacteremia. The degree of mental status changes or CSF glucose, protein, or pleocytosis did not influence outcome. Treatment with semisynthetic penicillins combined with rifampin was associated with survival in six; whereas three of four patients treated with other regimens died. Death occurred 4, 20, and 23 days after admission. The quality of survival of the six survivors was not detailed.

In a more recent series of 8 children and 20 adults with meningitis due to *S. aureus,* central nervous system trauma was the precedent cause in 46% (75% of children vs 35% of adults), (Kim 1989). Overall mortality was 37%, but one half of the adults died, whereas no children died. No patient with simple shunt infection died. Nine of eleven patients with foci outside the CNS succumbed. Also, treatment with penicillinase-resistant penicillin resulted in death in only one patient, whereas 6 of 12 patients not so treated died.

Bacteroides fragilis is a normal bowel anaerobic bacterial species and generally is associated with intracerebral abscess when it involves the CNS. Meningitis due to this organism is rare. Only nine patients have been reported; seven of these nine occurred in premature infants or neonates (Feder 1987). Two patients died, and four survived with severe neurological sequelae (hydrocephalus in all four). Two neonates with gastrointestinal problems and one 83-year-old woman with chronic otitis media survived without sequelae.

As mentioned previously, *Pseudomonas spp* is a rare cause of meningitis, but may occasionally cause meningitis in immunocompromised patients (Cherubin et al. 1981). Previously, mortality exceeded 85% in children and adults, whereas the newer cephalosporins (i.e., ceftazidime) may reduce this to less than 20% (Norrby 1984).

Meningitis is a relatively rare complication of typhoid fever (*Salmonella typhi*), occurring in 0% to 5% of patients. In the review by Rowland (1961), only four cases occurred among 530 patients (0.75%); three patients survived, two with neurological residua (gait disturbance and mental retardation). Most neurological manifestations observed during typhoid are not a direct result of bacterial CNS invasion, although delirium, coma, and seizures may be observed in 5% to 12% of patients (Hoffman et al. 1975).

Similarly, in a large series from India (500 cases), meningism occurred in 9.8%, delirium in 6.8%, and inappropriate behavior in 2.4%, but frank meningitis occurred in only one patient (0.2%) (Samantray et al. 1977). Early neurological symptoms may mislead the clinician to a diagnosis of meningitis in up to 8% of children (Johnson and Aderle 1981). Scragg and coworkers (1969) list multiple neurological complications occurring in 316 cases of typhoid in children in South Africa, including seizures (16), aphasia or mutism (nine), ataxia (eight), hemiplegia (two), facial nerve palsy (two), meningitis (one), cerebellar hemorrhage (one), and peripheral neuritis (one). Seizures and ataxia uniformly resolved. Aphasia or mutism also resolved, although one child who had aphasia in conjunction with a left-sided hemiplegia had persistent speech difficulties. In two other children with hemiplegia, one child recovered, and the other was left with residual weakness. Facial palsy occurred once in isolation and once in association with meningitis. Encephalitis (or acute encephalopathy) also occurred in 20 children (1%), all of whom made an uneventful recovery. Transient sensorineural hearing loss also has been described. This complication almost always resolves without residua (David and Tolaynat 1978).

Rickettsia

Central nervous symptoms (particularly headache) are frequent in many rickettsial infections, but invasion of the CNS appears to be relatively unusual for most syndromes and changes in the CSF are also unusual, except in Rocky Mountain Spotted Fever (RMSF), suggesting that the predominant rickettsial-associated syndrome is an encephalitis. Severe headache is a cardinal symptom of RMSF (*Rickettsia rickettsii*), which is present in more than 60% of patients within the first few days (Helmick et al. 1985). Similarly, headache, described as retro-orbital or "constant pressure," is common (87%) in murine (endemic) typhus (Miller et al. 1974). Photophobia and meningismus may be experienced in both diseases

Table 14-1. Range of Morbidity and Mortality Observed in Common Syndromes of Bacterial Meningitis and CNS Rickettsial Infections

Age	Etiology	Morbidity*	Mortality (Mean)
Neonate	Group B streptococci	14% to 50%	18% to 60% (22%)
	Gram-negative rods	23% to 75% (term/neonatal)	18% to 50% (25%)
		50% to 80% (premature)	48% to 83% (56%)
	Listeria monocytogenes	0% to 10%	0% to 63% (37.1%)
Children	*Salmonella pneumoniae*	20% to 57%	10.5% to 19.8% (12.5%)
(10 yrs old)	*Neisseria meningitidis*	0% to 10%	1.0% to 18% (7.9%)
	Haemophilus influenzae	0% to 43%	1.0% to 18.2% (3.1%)
Adults	Gram-negative rods	50%	22.7% to 84% (59%)
	Listeria monocytogenes	0% to 5%	28% to 37% (31%)
All ages	*Brucella spp*	0% to 28%	0% to 4.2%
All ages	*Rickettsia rickettsii* (RMSF)	5.5% to 27%	7% to 46%
All ages	*Rickettsia prowazecki*	Rare	1.3% to 59.4%

* Morbidity includes any severe or moderate sequelae; severe = bilateral sensorineural hearing loss, uncontrolled or controlled seizures, cortical blindness, microcephaly or hydrocephaly, central diabetes insipidus, paralysis or paresis; moderate–mild developmental retardation (IQ > 70), unilateral hearing loss, speech delay, behavioral, school, or work problems.

(Sah and Hornick 1979). However, evidence of distinct neurological involvement is frequent only in RMSF. Approximately 10% experience coma, and 8% have seizures; stupor also is a frequent sign (26%), as is ataxia (18%) (Helmick et al. 1984). Miller and Price (1972) note prominent neurological involvement in 54 (72%) of their 75 cases, which included lethargy or confusion in 22 cases (29%), coma in 8 cases (11%), delirium or hallucinations in 12 cases (16%), convulsions in 2 cases (2.6%), and focal deficits in 8 cases (11%). Decreased hearing was noted as a sequelae of RMSF in 7%, and the mortality with meningeal involvement (occurring in 18%) was approximately 4% to 5%. In review of 260 cases of RMSF by Vianna and Hinman (1971), initial misdiagnosis included viral encephalitis (11), meningococcal meningitis (11), and aseptic meningitis (10); therefore, 32 of 260 (12%) patients had signs suggesting primary CNS infections. Early treatment with an effective regimen of tetracycline or chloramphenicol was an important prognostic variable; only six of 108 (5.5%) patients treated within the first 4 days of illness suffered major sequelae (neurological, cardiac, or pulmonary), whereas 38 of 140 (27%) treated after the fifth day experienced serious sequelae. The overall mortality was 7%. Fifteen of the 18 patients who died had nuchal rigidity and frequent episodes of generalized convulsions, followed by coma. In four patients,

postmortem examinations demonstrated a meningoencephalitis, and one patient who had an ascending flaccid paralysis died of respiratory failure.

The characteristic pathological lesion of patients dying with RMSF is focal vasculitis that involves multiple organs but particularly the heart, kidney, brain, liver, and testes (Bradford and Hawkins 1977). This necrotizing vasculitis involves capillaries, arterioles, and venules with perivascular infiltrates of both polymorphonuclear leukocytes and lymphocytes. Intimal necrosis leads to thrombosis and disseminated intravascular coagulation. Abnormal CSF may be obtained in as many as 30% to 50% of patients submitted to lumbar puncture, with 11 to 300 CSF leukocytes per ml (usually lymphocytic), elevated protein to 41 to 200 mg/dl, but normal CSF glucose (Haynes et al. 1970).

Thus, involvement of the CNS appears to decrease the prognosis for survival in RMSF. Advanced age appears to be a less important prognostic variable, whereas a delay in treatment leads to a higher mortality and higher rates of apparent sequelae. A single prospective study has specifically evaluated the neurological prognosis in children with RMSF (Gorman 1981). The presence of sequelae was more common in children with initially severely impaired states of consciousness. Behavioral disturbances and learning disabilities were the most common sequelae. Seizures, although a common presenting symptom did not occur as a sequelae in any of the 42 children examined. However, other authors have noted seizures as sequelae in a small number of children (i.e., two of 78 or 2.6%) (Haynes et al. 1970). Seven of 42 children had coma for longer than 3 hours, and seven experienced stupor and hallucinations. Scores for mean IQ (Weschler), reading, math, spelling, and standardized test of behavior were significantly less for children presenting in coma. Other deficits included perceptual motor deficits (85%), behavioral adjustment problems (57%), and deviant finger tapping performance (42%).

Other Rickettsial Infections

In murine typhus, headache also is a prominent symptom (Woodward 1988). Mild stupor, prostration, and lethargy may be apparent during the second week. In louse-borne (i.e., epidemic) typhus, delirium, coma, and combative episodes are common. Children experience far fewer neurological symptoms than adults, and coma in elderly patients almost always is followed by death. Although nuchal rigidity may prompt lumbar puncture, the CSF almost always is normal in most infections of the typhus fever group of rickettsia. Despite the clinical neutropism of typhus (pathologically) and related rickettsia, grossly abnormal localizing physical or central neurological signs are rare. Transient partial deafness or weakness may occur in murine and louse-borne typhus, but precise incidence data from the modern literature are not available. The basic typhus

lesion is the result of invasion of the small vessel endothelia with subsequent necrosis, as well as lymphocytic and plasma cell infiltration (Manson-Bahr and Apted 1982), forming the "typhus nodule." These are found in brain, particularly the basal ganglia, medulla, and cerebral cortex (Wolbach 1948). In patients who progress to coma and meningitis, 80% may suffer deafness. Hemiplegia and persistent paresis also has been reported, but specific incidence figures in the modern literature are difficult to find.

Linnemann (1989) and coworkers describe a case of probable murine typhus caused by either R. typhi or R. canada which they link to the previously described syndrome of "acute febrile cerebrovasculitis"; this syndrome occurred in five patients with fever, headache, altered mentation, multifocal neurological signs, and CSF pleocytosis. In the case described, the patient had similar symptoms and CSF with greater than 3000 cells—97% lymphocytes. The patient experienced hallucinations, decreased right corneal reflex, a positive jaw and snout reflex, and impaired two-point discrimination, associated with electroencephalographic abnormalities. Over two weeks of hospitalization, myoclonus and tremors decreased and her sensorium cleared. Recently Sanra (1989) and coworkers (1989) described two similar patients with probable murine typhus. Both resolved their acute symptoms of encephalopathy, but in both, recovery was prolonged from several weeks up to 18 months. These cases again emphasize the vasculitic encephalopathy of rickettsial agents and the variability of involvement and subsequent recovery.

Table 14-2. Factors Associated With Poor Prognosis in Bacterial Meningitis in Neonatal Infants*

Clinical Signs and Symptoms	Etiology	Laboratory Values
Prematurity	*Escherichia coli;* K1 antigen	CSF protein > 300 mg/dl
Birth weight ≤ 1500 g	*Proteus spp*	CSF WBC < 500/ml
Early onset (≤7 days of life)	*Citrobacter diversus*	Positive blood cultures
Shock, decreased peripheral perfusion	*Salmonella spp*	Total WBC < 5000/mm^3
Hypothermia or afebrile	*Flavobacterium spp*	
Associated pulmonary infection†		
Seizures		

* Infants are younger than 2 months.

† Group B streptococcal infection; Listeriosis.

Table 14-3. Factors Associated With a Poor Prognosis in Bacterial Meningitis in Patients Older Than 2 Months

Clinical Signs and Symptoms	Laboratory Values
Shock or hypotension	Initial CSF CFU $> 10^7$ bacteria/ml*
Coma	Initial CSF glucose < 20 mg/dl
Purpura	Initial CSF protein > 1.0 g/dl
Convulsions at presentation	Initial CSF WBC < 1000/ml
Etiology†	Antigen (polysaccharide) > 1.0 μg/ml**
Bacteremia without meningitis††	ESR ≤ 10 mm/hr
Clinical symptoms greater than 3 days before treatment	Elevated tumor necrosis factor††
"Delayed" therapy	CIE antigen titer $> 1 : 100$§
Age younger than 12 months	Elevated CSF lactate
Age older than 60 years	Relative or absolute leuckopenia (<5000 or $<10,000$/ml)
	Anemia (Hgb < 11.0 g/dl)**

* haemophilus only

† Etiology: Prognosis worst for pnemococcal > meningococcal > haemophilus

†† meningococcal only

§ pneumococcal only

Mortality data are widely variable, but epidemic louse-borne typhus (*Rickettsia prowazeckii*) has the worst outcome; deaths from Brill-Zinsser and murine typhus variants are rare. In epidemics reported from Naples during World War II, mortality varied from 1.3% to 2.6% in children younger than 10 years, to 50% to 59.4% in those older than 55 years (Snyder 1965). The higher death rates in older people have been consistent in most epidemics. Deaths have been rare in the antibiotic era.

Mycoplasma pneumoniae

A large variety of neurological complications of infection with *Mycoplasma pneumoniae* have been reported. Despite the diversity of neurological syndromes, neurological involvement with this infection is unusual. In a prospective survey of 560 hospitalized patients with serologically confirmed mycoplasmal infection, 27 (4.8%) had CNS manifestations (Ponka 1979). Half of these 27 patients were younger than 10 years old. Because hospitalization for this infection is rare, the absolute incidence of CNS involvement also must be rare. Of the 27 patients, 18 had encephalitis or meningoencephalitis, eight had aseptic meningitis, and one had polyradiculitis. Of the patients with meningoencephalitis, four died, and three

had permanent sequelae. All cases of aseptic meningitis were benign. The patient with polyradiculitis survived and recovered without sequelae. In this review, nine deaths were reported out of a total of 87 neurological cases reported before 1980 (33 meningoencephalitis, 24 polyradiculitis, 10 "aseptic" meningitis, eight cerebellar ataxas, five cranial neuropathies, three psychosis, three transverse myelitis, and one phrenic nerve paralysis). Most deaths occurred in the meningoencephalitis group (six out of nine).

Meningitis, meningoencephalitis, peripheral and cranial nerve neuropathy, and cerebellar ataxia have been described following mycoplasmal CNS infection as noted previously (Murray et al. 1973). Among those prospectively described, five of 11 patients with meningoencephalitis suffered sequelae (expressive aphasia, sensory deficits, nystagmus, gait disturbances, and focal EEG-abnormalities). Peripheral neuropathy and severe Guillain-Barré syndrome also have been described in five patients, often with prolonged involvement after the initial episode. Cerebellar ataxia has been described in three children, all of whom fully recovered within 6 weeks after acute infection (Steele et al. 1972). With all these syndromes, onset usually is within the first 14 days of the presenting respiratory symptoms. Recovery may take up to 5 months but usually is complete. Through 1973, no deaths had been confirmed due to neurological involvement in mycoplasmal disease in this series. Antimicrobial therapy appeared to have no influence on the severity or duration of the morbidity due to CNS complications.

Finally, sudden transient sensorineural deafness has been described in a few patients with acute mycoplasmal infection (Shanon et al. 1982). The pathogenesis and outcome of this complication have been poorly defined. Transverse myelitis and ascending polyneuritis also have been described in association with bilateral optic neuritis (Rothstein and Kenny 1979) in a 28-year-old woman; she suffered severe persistent visual loss, but motor function returned. Conversely, isolated transverse myelitis usually has had a good outcome (Westenfelder et al. 1981) with return of relatively normal function.

References

Arditi, M.; Manogue, K. R.; Caplan, M.; Yogev, R. Cerebrospinal fluid cachectin/tumor necrosis factor-α and platelet-activating factor concentrations and severity of bacterial meningitis in children. J. Infect. Dis. 162:139–147; 1990.

Baird, D. R.; Whittle, H. C.; Greenwood, B. M. Mortality from pneumococcal meningitis. Lancet 2:1344–1346; 1976.

Barton, L. L.; Feigin, R. D.; Lins, R. Group B beta hemolytic streptococcal meningitis in infants. J. Pediatr. 82:719–723; 1973.

Beene, M. L.; Hansen, A. E.; Fulton, N. Salmonella meningitis: recovery from meningitis due to Salmonella spp with consideration of the problem of Salmonella meningitis. Am. J. Dis. Child. 82:567–573; 1951.

Behrman, R. E.; Meyers, B. R.; Mendelson, M. H.; Sachs, H. S.; Hirschman, S. Z. Central nervous system infections in the elderly. Arch. Intern. Med. 149:1596–1599; 1989.

Berlow, S. J.; Caldarelli, D. D.; Matz, G. I.; Meyer, D. H.; Harsch, G. G. Bacterial meningitis and sensorineural hearing loss: a prospective investigation. Laryngoscope 90:1445–1452; 1980.

Bol, P.; Spanjaard, L.; Van Alphen, L.; Baner, H. C. Epidemiology of Haemophilus influenzae meningitis in patients more than 6 years of age. J. Infect. 14:81–84; 1987.

Boucek, M. M.; Bowerth, R. C.; Artman, M.;

Graham, I. P.; Boucek, R. J. Myocardial dysfunction in children with acute meningococcemia. J. Pediatr. 105:5388–542; 1984.

Bouza, E.; Garcia de la Torre, M.; Parras, F.; Guerrero, A.; Rodriguez-Creixems, M.; Gobernado, J. Brucellar meningitis. Rev. Infect. Dis. 9:810–822; 1987.

Bradford, W. D.; Hawkins, H. K. Rocky mountain spotted fever in children. Am. J. Dis. Child. 131:1228–1232; 1977.

Burstein, L.; Breiningstall, G. N. Movement disorders in bacterial meningitis. J. Pediatr. 109:260–264; 1986.

Bush, R. T. Purulent meningitis of the newborn: a survey of 28 cases. N. Z. Med. J. 73:278–282; 1971.

Cabral, D. A.; Flodmark, O.; Farrell, K.; Speert, D. P. Prospective study of computed tomography in acute bacterial meningitis. J. Pediatrics 111:201–205; 1987.

Chin, K. C.; Fitzhardinge, P. N. Sequelae of early-onset group B hemolytic streptococcal meningitis. J. Pediatr. 106:819–822; 1985.

Cherubin, C. E.; Marr, J. S.; Sierra, M. F.; Becker, S. Listeria and gram-negative bacillary meningitis in New York City, 1972–1979. Am. J. Med. 71:199–209; 1981.

Cherubin, C. E.; Corrado, M. L.; Nair, S. R.; Gombert, M. E.; Landesman, S.; Humbert, G. Treatment of gram-negative bacillary meningitis: role of the new cephalosporin antibiotics. Rev. Infect. Dis. 4(suppl):5453–5464; 1982.

Choen, J. I.; Bartlett, J. A.; Corey, G. R. Extraintestinal manifestations of Salmonella infections. Medicine 66:349–388; 1987.

Ciesielski, C. A,; Hightower, A. W.; Parsons, S. K.; Bloome, C. V. Listeriosis in the United States 1980–82. Arch. Intern. Med. 148:1416–1419; 1988.

Crane, L. R.; Lerner, A. M. Non-traumatic gram-negative bacillary meningitis in the Detroit Medical Center 1964–1974. Medicine 57:197–209; 1978.

Cussen, L. J.; Ryan, G. B. Hemorrhagic cerebral necrosis in neonatal infants with enterobacterial meningitis. J. Pediatr. 71:771–776; 1967.

David, C. B.; Tolaymat, A. Typhoid fever: an unusual presentation. J. Pediatr. 93:533; 1978.

De Jong, R. N. Central nervous system involvement in undulant fever, with a report of a case and a survey of the literature. J. Nerv. Ment. Dis. 83:430–442; 1936.

Dodge, P. R.; Davis, H.; Feigin, R. D.; Holmes, S. J.; Kaplan, S. L.; Jubelirer, D. P.; Stechenberg, B. W.; Hirsh, S. K. Prospective evaluation of hearing impairment as a sequela of acute bacterial meningitis. N. Engl. J. Med. 311:869–874; 1984.

Dodge, R.; Swartz, M. H. Bacterial meningitis—a review of selected aspects. II Special neurologic problems, postmeningitic complications and clinicopathological correlations. N. Engl. J. Med. 272:954–1010; 1965.

Dyggve, H. Prognosis in meningitis neonatorum. Acta. Pediatr. 51:303–312; 1962.

Edwards, M. S.; Baker, C. J. Complications and sequelae of meningococcal infections in children. J. Pediatr. 99:540–545; 1982.

Edwards, M. S.; Rench, M. A.; Haffar, A. A. M.; Murphy, M. A.; Desmond, M. M.; Baker, C. J. Long-term sequelae of group B streptococcal meningitis in infants. J. Pediatr. 106:717–722, 1985.

Eng. R. H. K.; Cherubin, C. E.; Pechere, J-C; Bean, T. Treatment failures of cefotaxime and latamoxef in meningitis caused by Enterobacter and Serratia spp J. Antimicrob. Chemother. 20:903–911; 1987.

Evans, J. R.; Allen, A. C.; Stinson, D. A.; Bortolussi, R.; Peddle, L. J. Perinatal listeriosis: report of an outbreak. Pediatr. Infect. Dis. J. 4:237–241; 1983.

Feder, H. M. Bacteroides fragilis meningitis. Rev. Inf. Dis. 9:783–786; 1987.

Feigin, R. D.; Dodge, P. R. Bacterial meningitis: newer concepts of pathophysiology and neurologic sequelae. Pediatr. Clin. North Am. 23:541–556; 1976.

Feldman, W. E.; Ginsberg, C. M.; McCracken, G. H.; Allen, D.; Ahmann, P.; Graham, J.; Graham, L. Relation of concentrations of Haemophilus influenzae type b in cerebrospinal fluid to late sequelae of patients with meningitis. J. Pediatr. 100:209–212; 1982.

Feldman, W. E. Relation of concentrations of bacteria and bacterial antigen in cerebrospinal fluid to prognosis in patients with bacterial meningitis. N. Engl. J. Med. 296:433–435; 1977.

Fincham, R. W.; Saks, A. L.; Joynt, R. J. Protean manifestations of nervous system brucellosis. J.A.M.A. 184:269–275; 1963.

Gamstorp, I., Klockhoff, I. Bilateral severe sensorineural hearing loss after Haemophilus influenzae meningitis. Neuropaediatric 5:121–124; 1974.

George, R. M.; Cochran, C. P; Wheeler, W. E. Epidemic meningitis of the newborn caused by Flavobacteria. Am. J. Dis. Child. 123:259–262; 1961.

Geisler, P. J.; Nelson, K. E.; Levin, S.; Reddi, K. T.; Moses, V. K. Community-acquired purulent meningitis: a review of 1,316 cases during the antibiotic era, 1954–1976. Rev. Infect. Dis. 2:725–745; 1980.

Gordon, J. J.; Harter, D. H.; Phair, J. P. Meningitis due to Staphylococcus aureus. Am. J. Med. 78:965–970; 1985.

Gorman, R. J.; Saxon, S.; Snead, O. C. Neurotic sequelae of Rocky Mountain spotted fever. Pediatrics 67:354–357; 1981.

Groover, R. V.; Sutherland, J. M.; Landing, B.

Purulent meningitis of newborn infants. Eleven-year experience in the antibiotic era. N. Engl. J. Med. 264:115–121; 1961.

Halstensen, A.; Pedersen, S. H. J.; Haneberg, B.; et al. Case fatality of meningococcal disease in western Norway. Second. J. Infect. Dis. 19:35–42; 1987.

Harbin, L.; Hodes, G. R. Corticosteroids as adjunctive therapy for acute bacterial meningitis. So. Med. J. 72:977–980; 1979.

Haynes, R. E.; Sanders, D. Y.; Cramblett, H. G. Rocky mountain spotted fever in children. J. Pediatr. 76:683–693; 1970.

Helmick C. G.; Bernard, K. W.; D'Angelo, L. J. Rocky mountain spotted fever: clinical, laboratory and epidemiological features of 262 cases. J. Infect. Dis. 150:480–488; 1985.

Henderson, L. L. Salmonella meningitis. Am. J. Dis. Child. 75:351–358; 1948.

Herson, V. C.; Todd, J. K. Prediction of morbidity in Haemophilus influenzae meningitis. Pediatr. 59:35–39; 1977.

Hoffman, T. A.; Ruiz, C. J.; Counts, G. W.; Sachs, J. M.; Nitzkin, J. L. Waterborne typhoid fever in Dade County Florida. Am. J. Med. 59:481–508; 1975.

Jadavji, T.; Biggar, W. B.; Gold, W. R.; Probes, C. G. Sequela of acute bacterial meningitis in children treated for seven days. Pediatrics 78:21–25; 1986.

Johnson, A. O. R.; Aderle, W. I. Enteric fever in childhood. J. Trop. Med. Hyg. 84:29–35; 1981.

Jones, F. E., Hansen, D. R. H. influenzae meningitis treated with ampicillin or chloramphenicol, and subsequent hearing loss. Dev Med Child Neurol 19:593–597.

Kaplan, S. L.; Catlin, F. I.; Weaver, T.; Feigin, R. Onset of hearing loss in children with bacterial meningitis. Pediatrics 73:573–578; 1984.

Kim, J. H.; vanderHorst C.; Mulrow, C. D.; Corey, G. R. Staphylococcus aureus meningitis: A review of 28 cases. Rev. of Infect. Dis. 11:698–706; 1989.

Klein, J. O.; Feigin, R. D.; McCracken, G. H. Report of the task force on diagnosis and management of meningitis. Pediatrics 78(suppl): 5959–5982; 1986.

Kline, M. W. Citrobacter meningitis and brain abscess in infancy: Epidemiology, pathogenesis and treatment. J. Pediatr. 113:430–434; 1988.

Kline, M. W.; Mason, E. O.; Kaplan, S. L. Characterization of Citrobacter diversus strains causing neonatal meningitis. J. Infect. Dis. 137:101–105; 1988.

Lavetter, A.; Leedom, J. M.; Mathies, A. W.; Ivler, D.; Wehrle, P. F. Meningitis due to listeria monocytogenes. A review of 25 cases. New. Eng. J. Med. 285:598–603; 1971.

Lebel, M. H.; Freej, B. J.; Syroginannopoulos,

G. A.; Chrane, D. E.; Hoyt, M. J.; Stewart, S. M.; Kennard, B. D.; Olsen, K. D.; McCracken, G. H. Dexamethasone therapy for bacterial meningitis. Results of two double blind, placebo-controlled trials. N. Engl. J. Med. 319:964–971; 1988.

Linnemann, C. C.; Pretzman, C. I.; Peterson, E. D. Acute febrile cerebrovasculitis. A non-spotted fever group of rickettsial disease. Arch Intern Med 149:1682–1684; 1989.

Louria, D. B.; Hensle, T.; Armstrong, D.; Collins, H. S.; Blevins, A.; Krugman, D.; Buse, M. Listeriosis complicating malignant disease. A new association. Ann. Intern. Med. 67:261–281; 1967.

Lubani, M. M.; Dudin, K. I.; Araj, G. F.; et al. Neurobrucellosis in children. Pediatr. Infect. Dis. J. 8:79–82; 1989.

Mangi, R. J.; Quintiliani, R.; Andriole, V. T. Gram-negative bacillary meningitis. Am. J. Med. 59:829–836; 1975.

Mathies, A. W. Penicillins in the treatment of bacterial meningitis. J. Roy. Coll. Phys. (London) 5:139–146; 1972.

Manson-Bahr, P. E. C.; Apted, F. I. C. Manson's tropical diseases. London: Bailliere and Tindall; 1982.

McCracken, G. H.; Mize, S. G. A controlled study of intrathecal antibiotic therapy in gram-negative enteric meningitis of infancy. J. Pediatr. 89:66–72; 1976.

McCracken, G. H.; Mize, S. G.; Threlkeld, N. Intraventricular gentamicin therapy in gram-negative bacillary meningitis of infancy. Report of the second neonatal meningitis cooperative study group. Lancet 1:787–791; 1980.

McCracken, G. H.: Jr.; Sarff, L. D.; Glode, M. P.; Mize, S. G.; Schifter, M. S.; Robbins, J. B.; Gotschlich, E. C.; Orskov, J.; Orskov, F. Relation between E. coli K1 capsular polysaccharide antigen and clinical outcome in neonatal meningitis. Lancet 2:246–250, 1974.

Miller, J. Q.; Price, T. R. Involvement of the brain in rocky mountain spotted fever. South Med. J. 65:437–439; 1972.

Miller, M. B.; Bratton, J. L.; Hunt, J. Murine typhus in Vietnam. Military Medicine 139:184–186; 1974.

Mulder, C. J. J.; Van Alphen, L.; Zanen, H. C. Neonatal meningitis caused by Escherichia coli in the Netherlands. J. Infect. Dis. 130:935–940; 1984.

Munoz, O.; Benitez, L.; Martinez, M. C.; Guiscafre, H. Hearing loss after Haemophilus meningitis: followup study with auditory brainstem potentials. Ann. Otol. Rhinol. Laryngol. 92:272–275; 1983.

Murray, H. W.: Masur, H.; Senterfit, L. B.; Roberts, R. B. The protean manifestations of Mycoplasma pneumoniae infection in adults. Am. J. Med. 58:229–242; 1973.

Mustafa, M. M.; Lebel, M. H.; Ranilo, O.; Olsen, K. D.; Reisch, J. S.; Beutler, B.; McCracken, G. H. Correlation of interleukin-1β and cachectin concentrations in cerebrospinal fluid and outcome from bacterial meningitis. J. Pediatr. 115:208–213; 1989.

Nichols, E. Meningo-encephalitis due to brucellosis with the report of a case in which B abortus was recovered from the cerebrospinal fluid, and a review of the literature. Ann. Intern. Med. 35:673–693; 1951.

Norrby, S. R. The place of cephalosporins in the treatment of bacterial meningitis in adults. Proceeding of Symposium: Advances in Cephalosporin Therapy, Glaxco, Inc.; 1984.

Nylen, O.; Rosenhall, U. Haemophilus influenzae meningitis and hearing. Int. J. Pediatr. Otorhino/Laryngol. 1:997–1001; 1979.

Ozdamar, O.; Kraus, N.; Stein, L. Auditory brainstem responses in infants recovering from bacterial meningitis: audiologic evaluation. Arch. Otolaryngol. 109:13–18; 1983.

Pike, M. G.; Wong, P. K. H.; Bencivenga, R.; Flodmark, O.; Cabral, D. A.: Speert, D. P.; Farrell, K. Electrophysiologic studies, computed tomography, and neurological outcome in acute bacterial meningitis. J. Pediatrics 116:702–706; 1990.

Pomeroy, S. L.; Holmes, S. J.; Dodge, P. R.; Fergin, R. D. Seizures and other neurologic sequelae of bacterial meningitis in children. New Engl. J. Med. 32:1651–1656; 1990.

Ponka, A. The occurrence and clinical picture of serologically verified Mycoplasma pneumoniae infections with emphasis on central nervous system, cardiac and joint manifestations. Ann. Clin. Res. 11(suppl 24):1–60; 1979.

Rabinowitz, S. G.; McLeod, N. B. Salmonella meningitis. A report of three cases and review of the literature. Am. J. Dis. Child. 123:259–262; 1972.

Rajal, J. J. Treatment of Gram-negative bacillary meningitis in adults. Ann. Int. Med. 77:295–302; 1972.

Rajal, J. J.; Hyams, P. J.: Simberkoff, M. S.; Rubinstein, E. Combined intrathecal and intramuscular gentamicin for gram-negative meningitis. N. Engl. J. Med. 290:1394–1398; 1974.

Rajal, J. J.; Simberkoff, M. S.; Landesman, S. H.; the Metropolitan Meningitis Study Group. Prospective evaluation of moxalactam therapy for gram-negative bacillary meningitis. J. Infect. Dis. 149:562–567; 1984.

Rasmussen, N.; Hansen, B.; Bohr, H.; Kristensen, H. S. Artificial ventilation and prognostic factors in bacterial meningitis. Infection 16:158–162; 1988.

Ray, C. G.; Wedgwood, R. J. Neonatal Listeriosis. Six case reports and a review of the literature. Pediatrics 34:378–390; 1964.

Richner, B.; Hof, E.; Prader, A. Hearing impairment following therapy of *Haemophilus influenzae* meningitis. Helv. Pediatr. Acta. 34:443–447; 1979.

Rothstein, T. L.; Kenny, G. E. Cranial neuropathy, myeloradiculopathy and myositis. Complications of *Mycoplasma pneumoniae* infection. Arch. Neurol. 36:476–477; 1979.

Rowland, H. A. K. The complications of typhoid fever. Trop. Med. Hyg. 64:143–152; 1961.

Sah, A. J.; Hornick, R. B. Rickettsiosis, in *Principles and practices of infectious diseases*. In: Mandell, G. L.; Douglas, R. G.; Bennett, J. E., eds. New York: John Wiley and Sons; 1979:1081–1082.

Samantray, S. K.; Johnson, S. C.; Chakraborti, A. K. Enteric fever: an analysis of 500 cases. Practitioner 218:400–408; 1977.

Sande, M. A.; Scheld, W. M.; McCracken, G. H. Jr. Summary of a workshop: pathophysiology of bacterial meningitis—implications for new management strategies. Pediatr. Infect. Dis. J. 6:1167–1171; 1987.

Sanra, Y.; Shaked, Y.; Maier, M. K. Delayed neurologic display in murine typhus. Arch Intern Med 149:949–951; 1989.

Scragg, J.; Rubidge, C.; Wallace, H. L. Typhoid fever in African and Indian children in Durban. Arch. Dis. Child. 44:18–20; 1969.

Sell, S. H. *Haemophilus influenzae* type b meningitis: manifestations and long term sequelae. Pediatr. Infect. Dis. 6:775–778; 1987.

Sell, S. H.; Merrill, R. E.; Doyne, E. O.; Zimsky, E. P. Long-term sequelae of *Haemophilus influenzae* meningitis. Pediatrics 49:206–217; 1972.

Shanon, E.; Redianoui, C.; Zikk, D.; Eylom, E. Sudden deafness due to infection by *Mycoplasma pneumoniae* Am. Otol. Rhinol. Laryngol. 91:163–165; 1982.

Snedeker, J. D.; Kaplan, S. L.; Dodge, P. R.; Holmes, S. J.; Fergin, R. D. Subdural effusions and its relationship with neurologic sequelae of bacterial meningitis in infancy: A prospective study. Pediatrics 86:163–270; 1990.

Snyder, J. C. Typhus fever rickettsia. In Horsfall, F. L.; Tamm, I., eds. Viral and rickettsial infections of man. Philadelphia: J.B. Lipincott Co.; 1965:p. 1059–1094.

Spagnuolo, P. J.; Ellner, J. J.; Lerner, P. I.; McHenry, M. C.; Flatauer, F.; Rosenberg, P.; Rosenthal, M. S. *Haemophilus influenzae* meningitis: the spectrum of disease in adults. Medicine 61:74–85; 1982.

Sproles, E. T.; Azarrad, J.; Williamson, C.; Merril, R. E. Meningitis due to *Haemophilus influenzae*. Long term sequelae. J. Pediatr. 75:782–787; 1969.

Steel, J. C.; Gladstone, R. M.; Thanasophon, S.; Flemming, P. C. Severe cerebellar ataxia and

concomitant infection with *Mycoplasma pneumoniae* J. Pediatr. 80:467–469; 1972.

Stiehm, E. R.; Damrosch, D. S. Factors in the prognosis of meningococcal infection. A review of 63 cases with emphasis on recognition and management of the severely ill patient. J. Pediatr. 68:457–467; 1966.

Taylor, H. G.; Mills, E.; Ciampi, A.; deBerger, R.; Watters, G. V.; Gold, R.; MacDonald, N.; Michaels, R. H. The sequelae of *Haemophilus influenzae* meningitis in school-age children. New Engl. J. Med. 323:1657–1663; 1990.

Taylor, G. H.; Michaels, R. H.; Mazur, P. M.; Bauer, R. E.; Linden, C. B. Intellectual, neuropsychological, and achievement outcomes in children six to eight years after recovery from *Haemophilus influenzae* meningitis. Pediatrics 74:198–205; 1984.

Teng, Yy-C.; Lice, J. H.; Hsu, Y. H. Meningitis and deafness: report of 337 cases of deafness due to cerebrospinal meningitis. Chin. Med. J. 81:127–130; 1962.

Trollfors, B.; Claesson, B.; Lagergard, T.; Sandberg, T. Incidence, predisposing factors and manifestations of invasive *Haemophilus influenzae* infections in adults. Eur. J. Clin. Microbiol. 3:180–184; 1984.

Tugwell, P.; Greenwood, B. M.; Warrell, D. A. Pneumococcal meningitis: a clinical and laboratory study. Quart. J. Med. 180:583–601; 1976.

Underman, A. E.; Overturf, G. D.; Leedom, J. M. Bacterial meningitis. Disease A Month 24:7–63; 1978.

Valmari, P.; Kataja, M.; Peltola, H. Multivariate prognostication in bacterial meningitis of childhood. Scand. J. Infect. Dis. 19:29–34; 1987

Viannnna, J. J.; Hinman, A. R. Rocky mountain spotted fever on Long Island. Epidemiologic and clinical aspects. Am. J. Med. 51:725–730; 1971.

Vienny, H.; Despland, P. A.; Lutschg, J.; Deonna, T.; Dutoit–Marco, M. L.; Gander, C. Early diagnosis and evolution of deafness in childhood bacterial meningitis: a study using brainstem auditory evoked potentials. Pediatrics 73:579–586; 1984.

Waage, A.; Halstenesen, A.; Esperik, T. Association between tumor necrosis factor in serum and fatal outcome in patients with meningococcal disease. Lancet 1:355–357; 1987.

Wald, E. R.; Bergman, I.; Taylor, G.; Chiponis, D.; Porter, C.; Kubek, K. Long-term outcome of group B streptococcal meningitis. Pediatrics 77:217–221; 1986.

Westenfelder, G. O.; Akey, T.; Corwin, S. J.; Vick, N. A. Acute transverse myelitis due to *Mycoplasma pneumoniae* infection. Arch. Neurol. 38:317–318; 1981.

Wilson, F. M.; Lerner, A.M. Etiology and mortality of purulent meningitis at the Detroit Receiving Hospital. N. Engl. J. Med. 271:1245–1248; 1964.

Wolbach, S. B. The pathology of the rickettsial diseases of man. In: Soule, M. H., ed. Rickettsial diseases of man. Washington, DC: American Association of Advancement of Science; 1948: p. 118–123.

Woodward, T. E. Murine typhus: its clinical and biologic similarity to epidemic typhus. In:

Walker, D. H., ed. Biology of rickettsial disease, Vol. I. Boca Raton, FL: CRC Press, Inc.; 1988: p. 79–92.

Yu, J. S.; Grauaug, A. Purulent meningitis in the neonatal period. Arch. Dis. Child. 38:391–396; 1963.

Ziai, M.; Haggerty, R. J. Neonatal meningitis. N. Engl. J. Med. 259:314–320; 1958.

15

Tuberculous Meningitis

JOHN L. JOHNSON AND JERROLD J. ELLNER

Tuberculous meningitis is the most life-threatening manifestation of extrapulmonary tuberculosis. The mortality of tuberculous meningitis was essentially 100% in the prechemotherapy era; the fact that meningeal spread was most common in infants and children younger than 5 years compounded the tragedy. Tuberculous meningitis remains the leading cause of death and chronic disability in infants and children with tuberculosis (Ramachandran et al. 1986).

The impact of effective antituberculous chemotherapy on survival in tuberculous meningitis has been enormous. In a group of 365 cases of meningitis without miliary disease, the U.S. Veterans Administration—Armed Forces Cooperative trial demonstrated that survival increased from 31% to 81% during the period from 1946–1957. This was largely due to the beneficial effects of treatment with isoniazid (INH)-containing drug regimens (Falk 1965).

In developing countries, the prevalence of tuberculous meningitis remains high; the main problems are limited access to medical care, clinical diagnosis without adequate laboratory support, and limited access to effective drugs. In nations in which the overall prevalence of tuberculosis is low, the major problem lies in recognition of this uncommon medical problem. Tuberculous meningitis accounted for only 3.3% of all pediatric tuberculosis cases in the United States in 1985 (Snider et al. 1988).

In the developing world, there is a high prevalence of tuberculosis in which the infant or child develops tuberculous meningitis in the setting of progressive primary infection or miliary spread; however, in the more developed nations, tuberculous meningitis is now more frequently seen in adults and the elderly in association with reactivation or late generalized (miliary) tuberculosis. The pandemic of human immunodeficiency virus (HIV) infection has been associated with high rates of progression to active tuberculosis in HIV infected individuals with prior tuberculous infection. Approximately one half of these patients have extrapulmonary tuberculosis, and it is likely that an increased incidence of CNS tuberculosis will be seen.

Tuberculous meningitis resembles many other forms of chronic meningitis in its clinical manifestations. Tuberculous meningitis occurs without other evidence of extracranial tuberculosis in a significant minority of cases, further increasing the difficulty of establishing the diagnosis. Nevertheless, early diagnosis and institution of appropriate antituberculous chemotherapy are vitally important; they are probably the main factors under the control of the physician that impact on good clinical outcome, including neurological sequelae (Kennedy and Fallon 1979).

Pathogenesis and Pathology

Tuberculous meningitis sometimes is a manifestation of miliary disease. Early authors believed that it arose from direct hematogenous seeding of the meninges; the fact that many cases with widely disseminated disease had no central nervous system (CNS) involvement at autopsy conflicted with this hypothesis. In fact, experimental tuberculous meningitis was produced readily by intrathecal injection of mycobacteria, but it developed only in a minority of animals given intravenous infusions of tubercle bacili. The classic work by Rich and McCordock (1933) clarified the pathogenesis of the disease. The presence of 1 mm to 1 cm subependymal tubercles in the brain parenchyma, meninges, or both was demonstrable in 93% of 82 autopsy-confirmed cases of tuberculous meningitis. Rich and McCordock postulated that the development of tuberculous meningitis depended on the unfortunate but infrequent "chance" event whereby an old subependymal tuberculous focus (the "Rich focus") ruptured into the subarachnoid space. The rupture of the subependymal tubercle resulted in viable tubercle bacilli and mycobacterial antigens spilling into the cerebrospinal fluid (CSF), provoking an inflammatory and immunological response that limited the growth of the microorganisms (Rich and McCordock 1933; Auerbach 1951). Unfortunately, the severe inflammatory responses elicited in the subarachnoid space produce much of the symptomatology and ultimate morbidity of tuberculous meningitis.

The pathology of tuberculous meningitis is characterized by a thick gelatinous basilar exudate capable of entrapping cranial nerves and blood vessels and blocking normal CSF resorptive pathways; a severe arteriolitis and perivasculitis with subsequent endarteritis and thrombosis of nutrient vessels to the brain parenchyma, resulting in ischemia and infarction; and severe comorbidity due to debility and secondary infection resulting from a decreased level of consciousness and loss of other neurological functions. At autopsy,

"tuberculous meningitis is characterized especially by its exudative inflammatory nature, and by the tendency to widespread necrosis of the inflammatory exudate and of contiguous meningeal tissues . . . there is a very definite tendency for the infection to be particularly severe in and about the walls of the meningeal arterioles; the media and adventitia are frequently infected and may become completely caseous; or a perfectly normal vessel may be surrounded by a sheath of inflammatory cells with necrosis which encroaches upon the adventitia only" (Rich and McCordock 1933).

Clinical Aspects

In children, tuberculous meningitis frequently is associated with evidence of pulmonary tuberculosis, manifested as primary tuberculosis with hilar adenopathy or as progressive miliary disease (Auerbach 1951; Idriss et al. 1976). In 136 cases of pediatric tuberculous meningitis in which the time between the first diagnosis of tuberculosis and the onset of meningitis was known, the onset of meningitis in 75% of the cases was within 1 year of their initial diagnosis (Lincoln et al. 1960).

This is not always the case in adults. Adults may develop tuberculous meningitis in the setting of progressive primary disease or miliary tuberculosis similar to children. However, they also may develop tuberculous meningitis as a result of the late meningeal seeding and subependymal tubercle formation arising from bacilli spread from small smoldering or previously dormant extracranial tuberculous foci. Fifty-four percent of 63 fully autopsied cases of late generalized tuberculosis in adults had evidence of tuberculous meningitis (Slavin et al. 1980).

Approximately three fourths (48% to 79%) of adults with tuberculous meningitis have radiographic evidence of pulmonary tuberculosis. Most of the remaining cases have other clinical or radiographic stigmata of tuberculosis in other extracranial sites. A few have no other visible evidence of tuberculosis outside the CNS. The source of late meningeal involvement in these cases is cryptic. Many probably arise either from a longstanding parameningeal focus (Rich focus), which spontaneously discharges infectious material into the subarachnoid space, or by hematogenous spread from small renal or osseous foci (Auerbach 1951). Therefore, the absence of stigmata of extracranial tuberculosis cannot be used reliably to exlude the diagnosis of tuberculous meningitis.

There are usually three stages in the natural history of the illness. In most patients there is a history of a variable period (1 week to several months) of fever, malaise, and vague mental status changes, including apathy or intense irritability, difficulty with memory, and subtle personality changes. At the end of this period, definite signs and symptoms of neurological illness appear. Meningism usually is present at some point, often late, during the course of the illness in adults but is less common in infants (Smith and Daniel 1947; Illingworth 1956). Focal neurological findings are present, including signs of elevated intracranial pressure (ICP) or hydrocephalus resulting from arachnoiditis which blocks the basilar cisterns, hemiplegia, and unilateral or bilateral cranial nerve palsies (most commonly cranial nerve VI, but also III, IV, and the optic nerve in descending order of frequency). The onset of stupor and coma mark the final stage of the disease. Death often occurs within 6 to 8 weeks after the onset of the second stage of tuberculous meningitis.

Many neurological syndromes have been described in tuberculous meningitis. The syndromes can be considered in five groups: meningeal irritation, increased ICP, mental status changes, epilepsy, and focal neurological signs (Smith and Daniel 1947). Atypical forms include isolated cranial neuropathies, hemiballismus, tremors, myoclonus, a midline cerebellar syndrome, and a spinal form of the disease in adults that include radicular pain, lower extremity weakness, and absent deep-tendon reflexes, resembling cord compression or the Guillian-Barré syndrome (Udani et al. 1971; Cybulska and Rucinski 1988).

Clinical manifestations commonly seen at presentation include headache, fever, irritability, and in children, episodes of protracted vomiting. Fever is present in 47% to 100% of the cases, meningismus in 57% to 90% of cases, and decreased level of consciousness, papilledema, and cranial nerve palsies are seen less frequently (Tables 15-1 and 15-2).

In adults, the clinical syndrome of tuberculous meningitis resembles that of many other forms of chronic meningitis with fever, meningismus, headache, and abnormalities in the level of consciousness (Ellner and Bennett 1976). These clinical features suggest tuberculous meningitis, and a high clinical suspicion for the disease must be maintained in the appropriate setting. The combination of neuromeningeal signs, especially new or evolving cranial nerve palsies, a chest x-ray consistent with tuberculosis, and a CSF profile of hypoglycorrhachia with lymphocytic pleocytosis should suggest this diagnosis.

The duration of symptoms before hospitalization varies from a few days to 6 months. Today, more cases present with an acute or subacute onset, possibly reflecting better access to health care facilities or the availability of therapy that prevents tuberculous spread to the meninges. CNS symptoms were present for less than 2 weeks in 50% of the cases in two recent series (Haas et al. 1977; Hinman 1967).

Rarely, tuberculous meningitis may present as an acute transient aseptic meningitis followed by complete clinical recovery over 2 to 4 weeks without specific therapy (Emond and McKendrick 1973). Lincoln also reported an earlier series of 12 similar pediatric cases (Lincoln 1947). All of these children had concomitant active primary pulmonary tuberculosis. The CSF in

Table 15-1. Clinical Symptoms of Tuberculous Meningitis

	Percent of Cases*— Pediatric Series ($n = 697$)	Percent of Cases— Adult Series ($n = 142$)	Percent of Cases— Combined Series ($n = 1147$)
Headache	41 (20–73)	61 (28–100)	55 (20–100)
Vomiting	60 (30–80)	28	55 (28–80)
Anorexia	52 (27–67)	35 (25–44)	46 (25–67)
Cough or respiratory symptoms	35 (22–48)	33	34 (28–48)
Irritability	42 (24–76)	—	36 (17–76)
Weight loss	24 (18–30)	29 (25–35)	25 (18–35)
Seizures	12 (7–20)	11 (8–13)	11 (7–20)
History of fever	67 (38–98)	82 (63–100)	65 (19–69)
Photophobia	8 (7–9)	—	10 (7–13)
Abdominal pain	10	—	10
Constipation	22 (10–34)	—	27 (10–37)
Apathy	51 (30–72)	52	50 (30–72)
Night sweats	—	32	32

* Mean (range).

Adult series = Barrett-Connor 1967; Haas et al. 1977; Crocco et al. 1980; Stockstill and Kauffman 1983; Traub et al. 1984; Klein et al. 1985; Phuapradit and Vejjajva 1987

Pediatric series = Illingworth 1956; Lincoln 1960; Smith 1975; Idriss et al. 1976; Delage and Dusseault 1979; Escobar et al. 1975

Combined series = adult and pediatric plus. Lepper and Spies 1963; Hinman 1967; Kennedy and Fallon 1979; Swart et al. 1981; Ogawa et al. 1987

these children showed pleocytosis but normal chemistries.

Laboratory Tests

Skin Testing

The initial skin test with old tuberculin or purified protein derivative (PPD) was positive in 31% to 61% of adults and 50% to 96% of children (Table 15-3). The delayed type hypersensitivity skin test response to intermediate strength PPD (5 TU/0.1 ml.) usually was positive. Severely debilitated patients may be anergic. Anergy also was common in patients who presented at later clinical stages of the disease (Illingworth 1956).

Chest X-Ray

The chest x-ray often is helpful diagnostically—44% to 88% of adult and pediatric patients have radiographic abnormalities compatible with pulmonary tuberculosis. Approximately 20% of these patients will have a miliary pattern (see Table 15-3). The

Table 15-2. Clinical Signs of Tuberculous Meningitis

	Percent of Cases*— Pediatric Series ($n = 697$)	Percent of Cases— Adult Series ($n = 142$)	Percent of Cases— Combined Series ($n = 1147$)
Fever	67 (47–81)	83 (69–100)	77 (47–100)
Meningismus	75 (74–77)	67 (57–75)	73 (57–90)
Decreased level of consciousness, stupor, coma	36 (30–40)	64 (22–86)	52 (22–86)
Cranial nerve palsy	19 (12–33)	27 (13–42)	27 (12–73)
Papilledema	16 (4–41)	9 (4–13)	14 (4–41)
Choroidal tubercles	11 (9–12)	—	9 (2–12)

* Mean (range).

Adult series = Barrett-Connor 1967; Haas et al. 1977; Crocco et al. 1980; Stockstill and Kauffman 1983; Traub et al. 1984; Klein et al. 1985; Phuapradit and Vejjajva 1987

Pediatric series = Illingworth 1956; Lincoln 1960; Smith 1975; Idriss et al. 1976; Delage and Dusseault 1979; Escobar et al. 1975

Combined series = adult and pediatric plus. Lepper and Spies 1963; Hinman 1967; Kennedy and Fallon 1979; Swart et al. 1981; Ogawa et al. 1987

Table 15-3. Skin Testing and Chest X-Ray Data in Tuberculous Meningitis

	Percent of Cases*— Pediatric Series ($n = 970$)	Percent of Cases— Adult Series ($n = 141$)	Percent of Cases— Combined Series ($n = 1438$)
Initial positive tuberculin or PPD skin test	76 (50–96)	43 (31–61)	66 (31–96)
Abnormal admission chest x-ray (includes patterns consistent with TB including miliary disease)	72 (55–88)	60 (48–79)	64 (44–88)
Miliary chest x-ray pattern	21 (16–24)	22 (10–29)	22 (10–29)

* Mean (range).

Adult series = Barrett-Connor 1967; Haas et al. 1977; Crocco et al. 1980; Stockstill and Kauffman 1983; Traub et al. 1984; Klein et al. 1985; Phuapradit and Vejjajva 1987

Pediatric series = Lincoln 1960; Lee and Brown 1968; Zarabi et al. 1971; Steiner and Portugaleza 1973; Smith 1975; Sumaya et al. 1975; Idriss et al. 1976; Delage and Dusseault 1979, Rahajoe et al. 1979

Combined series = adult and pediatric plus. Lepper and Spies 1963; Hinman 1967; Kennedy and Fallon 1979; Ogawa et al. 1987

chest x-ray should be examined carefully. Early miliary disease may be very subtle and cases will be missed unless the radiologist is aware of the clinical suspicion of tuberculosis.

Adjunctive Tests

Sputum smears and cultures or gastric aspirate cultures often are positive (14% to 51% of cases) (Klein et al. 1985; Hinman 1967) and provide important supportive evidence for the diagnosis of tuberculous meningitis. Bone marrow and liver biopsies also potentially yield the diagnosis in miliary disease. The peripheral white blood cell (WBC) count usually is less than 10,000 WBC/mm³. Elevated erythrocyte sedimenation rates greater than 20 mm per hour were present in 42% to 83% of cases (Kennedy and Fallon 1979; Clark et al. 1986). Hyponatremia occurs in tuberculous meningitis in 37% to 73% of patients (Haas et al. 1977; Karandanis and Shulman 1976) due to the syndrome of inappropriate antidiuretic hormone secretion (Smith and Godwin-Austen 1980). Hyponatremia is more common in tuberculous meningitis than in pyogenic or aseptic meningitis (Karandanis and Shulman 1976). The combination of CSF pleocytosis, a negative CSF gram-strain, and hyponatremia should suggest the diagnosis. Metabolic alkalosis may be present in patients with protracted vomiting.

CSF Findings

CSF opening pressures frequently are elevated early in the course of the disease (Cairns 1951); however, this does not correlate with the presence or severity of hydrocephalus (Leiguarda et al. 1988). Early ICP elevations may be due to cerebral edema alone (Udani et al. 1971). The opening pressure in cases of spinal block may be low or normal with absence of augmentation of the CSF pressure when the neck veins are compressed.

The usual CSF findings (Table 15-4) consist of hypoglycorrhachia (CSF glucose less than 40 mg/dl) in 32% to 88% (Haas et al. 1977; Lincoln et al. 1960), elevated protein levels, and a mononuclear pleocytosis with 100 to 500 cells/ml (Fishman 1980). Eighty-five percent of patients in an older series of 84 cases had a CSF WBC between 50 to 500 cells/ml (Merritt and Fremont-Smith 1938). This pattern of abnormalities is generally the same across all age groups (Merritt and Fremont-Smith 1938). The majority of the cells are lymphocytes. The CSF usually is clear but may have a "ground glass" or translucent appearance. A delicate web-like clot or pellicle will form in roughly one-half of the cases (Idriss et al. 1976) if the fluid is allowed to stand. Pellicle formation reflects the presence of increased fibrinogen in the CSF due to generalized meningeal inflammation with increased permeability to serum

Table 15-4. Spinal Fluid Findings in Tuberculous Meningitis

Review	Number of Cases	CSF WBC/ml Range (mean)	Percent of Patients With PMN Predominance on Initial LP	Percent of Patients With CSF Glucose <40 mg/dl	Percent of Patients With Positive CSF AFB Smears	Percent of Patients With Positive CSF Mycobacterial Culture
Foord and Forsyth 1933	87	0–1120 (209)	19	—	67	—
Merritt and Fremont-Smith 1935	84	50–2021	11	—	—	—
Lincoln et al. 1960	241	—	—	88	—	53
Hinman 1967	35	—	14	66	22	75
Barrett-Connor 1967	18	29–1030	45	83	11	39
Mackay 1967	130	24–1200	7	65	50	—
Smith 1975	43	10–950	28	77	—	—
Karandanis and Shulman 1976	11	3–3900	36	64	0	55
Haas et al. 1977	19	8–900 (244)	—	32	37	47
Kennedy and Fallon 1979	52	—	27	—	87	83
Klein et al. 1985	21	0–960	6	50	19	81
Ogawa et al. 1987	45	0–8600	32	54	10	40

proteins (Fremont-Smith 1932; Fishman 1980). These findings are nonspecific and may be seen in other chronic meningitides. Markedly elevated protein levels (greater than 500 mg/dl) and xanthochromic fluid occasionally may be seen in severe cases that are complicated by hydrocephalus with spinal subarachnoid block.

The fluid rarely is hemorrhagic except in rare cases of advanced disease with degeneration of cortical or meningeal vessels (Merritt and Fremont-Smith 1938). A predominance of polymorphonuclear (PMN) leucocytes (rather than the usual lymphocytic pleocytosis) may be seen early in the course of tuberculous meningitis. This phenomenon occurred in 11% to 45% of reported cases (see Table 15-4). The presence of a PMN predominance was not related to the total CSF WBC count (Merritt and Fremont-Smith 1938). Serial lumbar punctures often demonstrate a subsequent shift to a lymphocytic pleocytosis over time.

The CSF glucose characteristically decreases as the disease progresses. A rising CSF protein may signal the development of spinal block due to worsening basilar or spinal arachnoiditis. With appropriate treatment, the CSF findings normalize in the following order: glucose, CSF protein, and

CSF cell counts (Hinman 1967). CSF abnormalities take 3 to 6 months to resolve completely in most patients (Lepper and Spies 1963).

Direct CSF smears for acid fast bacilli (AFB) are positive in a highly variable number of cases. In 16 studies, the yield from direct CSF smears ranged from 0% to 87% with mean of 30%. CSF mycobacterial cultures are positive in 40% to 83% (see Table 15-4). Examination of fluid from up to four taps increases the diagnostic yield (Kennedy and Fallon 1979). Examination of fluid from the last tube collected at lumbar puncture or staining of the clot (pellicle), if one forms, may increase the yield from direct AFB smears (Merritt and Fremont-Smith 1938). Examination of larger volumes of CSF (5–20 ml) also may be useful. Stewart (1953) describes a method that resulted in positive smears in 91 of 100 consecutive proven cases of tuberculous meningitis. Ten to 20 ml of CSF were centrifuged at 2500 rpm for 30 minutes. The supernatant was decanted and the pellet agitated. A drop of fluid from the resuspended pellet was placed on a slide. The edge of the drop was heated gently to decrease spreading and to keep the size of the preparation to 5 to 10 mm. in diameter. Several drops of fluid were placed over the first drop and allowed to dry serially. The slides were stained and examined for 30 to 90 minutes by light microscopy.

A normal or minimally abnormal CSF profile rarely may be seen in tuberculous involvement of the CNS. Patients may have multiple tuberculous lesions of the brain or spinal cord before a subependymal tubercle ruptures into the CSF producing the typical CSF picture. These lesions (abscesses or small tuberculomas) may cause seizures, hemiplegia, or paraplegia before any CSF abnormalities are present. Thus, normal CSF findings do not exclude the possibility of CNS tuberculosis in cases in which such involvement is suspected clinically.

Serial lumbar punctures performed over a few days may help clarify the clinical picture in difficult cases. Anergic patients, debilitated alcoholics, and patients receiving treatment with corticosteroids for other systemic disorders also may not show an early CSF pleocytosis. This is because the CSF pleocytosis may be likened to a tuberculin reaction in the subarachnoid space, and agents such as corticosteroids that interfere with cell-mediated immune responses may blunt such responses (Kocen and Parsons 1970; Wilkinson et al. 1971).

Newer techniques have been applied to the diagnosis of tuberculous meningitis (Daniel 1987). Assessment of the lymphocyte-derived enzyme, adenosine deaminase, and measurement of the partition of bromide ions in CSF compared to serum are sensitive diagnostic tests; however, a significant number of false-positive tests have been described, especially in CSF from patients with pyogenic meningitis (Coovadia et al. 1986).

Tests for CSF antibodies to mycobacterial antigens have had poor specificity in geographic areas with a high prevalence of tuberculosis. In one series, low levels of antibodies to mycobacterial antigens were present in 18% of controls and in 37% of proven cases of pyogenic meningitis in people from such an area (Chandramuki et al. 1985).

Tests that detect growing mycobacteria in the CSF directly are more likely to offer the high specificity required for routine clinical use in areas with a low prevalence of tuberculous meningitis (Daniel 1987). These techniques focus on the detection of mycobacterial constituents in the CSF. The use of electron capture gas chromatography–mass spectral identification of 3-(2'-ketohexyl) indoline (Brooks et al. 1977) and tuberculostearic acid (TBSA) (Mardh et al. 1983; Brooks et al. 1987) in the CSF of patients with tuberculous meningitis has been described. TBSA is a component of mycobacteria and other members of the *Actinomycetales*, including *Nocardia*. Detection of TBSA in the CSF appears to be both sensitive and specific for the diagnosis of tuberculous meningitis and requires only 0.5 to 1 ml of CSF to perform. A recent study detected TBSA in the CSF of all of 13 proven and eight of nine suspected cases of tuberculous meningitis (French et al. 1987). TBSA was present in the CSF of only one of 87 patients with nontuberculous meningitis or other diseases. The single false-positive result in this study was thought to be due to intrathecal therapy with amikacin in a case of gram-negative meningitis. Amikacin is derived from an *Actinomyces* spp and produces peaks similar to TBSA in the assay. TBSA was identified in the CSF of proven cases of tuberculous meningitis for up to 8 months after initiation of treatment, making the assay potentially valuable for confirming the diagnosis in patients previously started on empiric therapy. Frequency-pulsed electron capture gas–liquid chromatographic detection of TBSA and other related carboxylic acids in CSF for the rapid diagnosis of tuberculous meningitis is available on special request from the laboratories at the Centers for Disease Control (CDC) in Atlanta. This test has had a specificity of 91% and a sensitivity of 95% in studies of 41 coded and 75 clinical cases at the CDC (Brooks et al. 1990).

Direct detection of mycobacterial antigens in the CSF by latex agglutination (Krambovitis et al. 1984), radioimmunoassay (RIA) (Kadival et al. 1987), or enzyme-linked immunosorbent assay (ELISA) (Watt et al. 1988) techniques has been associated with high sensitivity and a low false-positive rate. The few false-positive results are probably due to the fact that mycobacteria share common antigens with other bacteria. Further research to develop more specific tests for CSF mycobacterial antigens is ongoing. Use of the polymerase chain reaction to amplify and detect mycobacterial DNA in the CSF is attractive but untested.

Cerebral Imaging Techniques in Tuberculous Meningitis

Computed tomography (CT) scanning is the major neurodiagnostic imaging technique used in the study and management of patients with chronic meningitides. It has largely replaced the older and less sensitive techniques of radionucleotide brain scanning and ventriculography. The availability of high-resolution studies that can be performed serially with no appreciable risk to the patient has increased our understanding of the role of basal arachnoiditis and hydrocephalus in the pathophysiology of the disease.

Hydrocephalus is common in tuberculous meningitis and usually is of the communicating type due to blockage of CSF flow at the level of the basilar cisterns (especially the cisterna ambiens) by the arachnoiditis (Schoemen et al. 1988). Rarely, noncommunicating hydrocephalus may occur due to obstruction by proteinaceous debris or mass effect at the level of the aqueduct of Sylvius or the foramina of Magendie and Luschka. Eighty-three percent of adult and pediatric cases of tuberculous meningitis in one large series had hydrocephalus (Bhargava et al. 1982). Hydrocephalus was more common in children younger than 10 years and more likely to be associated with severe degrees of obstruction in the pediatric age group. Advanced grades of hydrocephalus were present in cases in which the duration of illness was more than 1 month prior to diagnosis. Contrast enhancement of the basilar exudate at the time of presentation was associated with a poorer prognosis for neurological recovery and represented more extensive basilar arachnoiditis (Bhargava et al. 1982).

CT scanning also has confirmed that ischemic cerebral infarction is common in tuberculous meningitis. An earlier autopsy series found evidence of ischemic stroke in 41% of cases (Dastur et al. 1970). More recent CT studies have revealed evidence of ischemic infarction in 28% to 38% of cases (Bhargava et al. 1982; Leiguarda et al. 1988). Ischemic stroke occurs more commonly in pediatric patients, in the distribution of the anterior cerebral circulation (especially middle cerebral artery), and in patients with more severe grades of hydrocephalus. It is associated with a higher mortality and more extensive neurological residuals (Leiguarda et al. 1988).

Nuclear magnetic resonance imaging (MRI) may add additional information regarding brain stem lesions not visible on CT scans. In a study of 27 children with advanced tuberculous meningitis, focal brain stem lesions were present in 37% of cases (Schoeman et al. 1988). Most of the children with these lesions died or survived in a vegetative state. Ischemic strokes in the basal ganglion were often associated with brain stem lesions. Although the presence of these lesions on MRI correlated well with clinical findings, the prognostic value of the MRI findings was limited. Many patients who did not have brain stem lesions also died. MRI studies have confirmed the earlier clinicopathological concepts (Tandon 1978) that neurological deterioration in advanced stages of tuberculous meningitis is due to the effects of a combination of parenchymal brain inflammation, basilar meningitis with vasculitis and ischemia, and elevated ICP.

Vasospasm and irregularity involving the large cerebral arteries at the base of the brain have been demonstrated by angiography (Greitz 1964; Lehrer 1966). Tuberculous arteritis narrows vessels due to symmetrical or asymmetrical endothelial proliferation, leading to ischemia and infarction. The inflammatory exudate often entraps the smaller thalamostriate and nutrient-perforating branches at the base of the brain, which may account for the large number of lacunar strokes of the basal ganglia (Dastur et al. 1970; Kingsley et al. 1987). During treatment, a dense, woody, fibrous exudate may form at the base of the brain in the regions of the optic chiasm and interpeduncular cisterns. This can lead to visual loss and opthalmoplegia due to optochiasmatic arachnoiditis (Scott et al. 1977; Navarro et al. 1981; Kingsley et al. 1987).

Diagnosis and Differential Diagnosis

In patients with a history of recent tuberculous infection or active pulmonary or extrapulmonary tuberculosis, the diagnosis of tuberculous meningitis is easy. The presence of new neurological signs or symptoms should provoke a prompt search for evidence of meningeal involvement.

However, a high level of clinical suspicion is necessary to diagnose many cases of tuberculous meningitis. Only one fourth to one third of the patients will have a history of exposure to a known case of tuberculosis, although this information should be sought. The clinical signs and symptoms described early in the course of tuberculous meningitis are nonspecific, although a longer history of symptoms and the evolution of cranial neuropathy are suggestive. The physical examination should be meticulous to define the clinical stage of neurological involvement, localize other sites of extracranial disease, and look for potential sites of sampling or biopsy, such as enlarged lymph nodes or an enlarged liver. The chest x-ray (if consistent with tuberculosis) provides an important diagnostic clue.

Acid-fast smears and cultures of secretions and tissues from other organs may provide important supportive evidence. The yield of sputum and gastric aspirate smears and cultures in cases of tuberculous meningitis varies widely (14% to 51%). Examination of three early morning deep sputum specimens is adequate for the diagnosis of pulmonary tuberculosis in most patients with pulmonary involvement. Cultures of gastric aspirates are useful alternatives in children, who often have difficulty producing acceptable sputum specimens for study.

Smears of gastric aspirate material are less helpful because up to one third of the positive results may be false-positive due to contamination by commensal mycobacteria. However, one study of gastric aspirate specimens demonstrated that when more than six AFB per high-power field were present, all of the patients had disease due to pathogenic mycobacteria (Strumpf et al. 1976). A rapid diagnosis also may be facilitated by histological examination and culture of tissue from enlarged lymph nodes or from liver, lung, or bone marrow in suspected cases of miliary disease with meningeal spread.

Lumbar puncture and examination of the CSF form the cornerstones of diagnosis in tuberculous meningitis. Caution is required when interpreting CSF findings in suspected cases of tuberculous meningitis. The differential diagnosis of subacute or chronic lymphocytic meningitis, with or without hypoglycorrhachia, is broad, and it includes other treatable infectious and noninfectious causes of meningitis. Fungal meningitis (especially cryptococcal meningitis), "aseptic" viral meningitis, parasitic infestations of the CNS, acute syphilitic meningitis, brain abscesses, subdural hematomata, carcinomatous meningitis, and other rare diseases, such as CNS sarcoidosis, Behçet's disease, and CNS vasculitis, may have a comparable CSF picture (Ellner and Bennett 1976). These findings are consistent with partially treated bacterial meningitis as well.

Much of the immediate diagnostic value of the lumbar puncture lies in the rapid exclusion of the entities of bacterial meningitis and subarachnoid hemorrhage and in finding cell counts and chemistries consistent with tuberculous meningitis. Multiple lumbar punctures (up to three to four taps of 10 to 20 ml of CSF) are indicated if the diagnosis is in doubt because the yields from cultures and smears increases with serial taps, and the evolution of CSF chemistries and cell counts (progressive decrease in glucose and lymphocytic pleocytosis) can be followed. CSF gram-stains are negative in tuberculous meningitis and positive, if properly performed, in more than 80% of cases of pyogenic meningitis (Karandanis and Shulman 1976). Occasional difficulty may arise in cases of partially treated pyogenic meningitis in which the CSF findings may be atypical. CSF latex agglutination tests for the detection of microbial polysaccharide antigens of pneumococcus, meningococcus, *Haemophilus influenzae*, and group B streptococcus are rapid, sensitive, and available in most larger hospital laboratories to assist in such cases. The CSF india ink preparation is positive in about 50% of cases of cryptococcal meningitis, the most common cause of chronic meningitis in the immunocompromised host. It can be performed rapidly and reliably within minutes. Latex agglutination tests for cryptococcal polysaccharide antigen in the serum and CSF are readily available, and each are positive in greater than 85% of cases of cryptococcal meningitis (Bennett and Bailey 1971). In 94% of patients, serum or CSF will contain cryptococcal antigen. Serologic tests for syphilis performed on CSF should allow rapid diagnosis of CNS involvement in that disease.

CNS involvement by tuberculosis has been reported in a few patients with HIV infection. Tuberculomas, multiple intracerebral abscesses, and tuberculous meningitis have been reported. All patients have been anergic. CSF findings have been comparable to those in other immunocompetent patients. *Mycobacterium tuberculosis* has been isolated from CSF and brain biopsy tissue; however, direct smears of the CSF have been negative (Bishburg et al. 1986).

In summary, the clinical, laboratory, and radiographic findings are variable in tuberculous meningitis. Many syndromes have been described. Compatible chest x-rays, positive skin tests, and detection of AFB in other tissues or secretions are strongly supportive, though not directly diagnostic of, tuberculous meningitis. Negative direct AFB smears of the CSF, negative PPD skin tests, or negative chest x-rays cannot be used reliably to rule out the diagnosis of tuberculous meningitis. The long clinical course and the evolution of changes in the CSF (lymphocyte predominance and falling CSF glucose) may be helpful in suggesting the diagnosis in some cases, but a high clinical suspicion must be maintained. Because direct CSF smears for AFB often are negative and mycobacterial cultures may take 4 to 6 weeks for growth, patients with the presumptive diagnosis of tuberculous meningitis should be started on antituberculous chemotherapy without delay to avoid unnecessary morbidity.

Treatment

The initial United States Veterans Administration–Armed Forces studies (Falk 1965) clearly demonstrated the positive effects of isoniazid (INH) on survival in tuberculous meningitis. Initial improvement followed by relapse and death was commmon during treatment with streptomycin (STM) alone before 1952 (preisoniazid era). The mortality in tuberculous meningitis fell from 85% to 33% after the addition of INH to treatment regimens (Lepper and Spies 1963). In general, INH-containing regimens adequate for extrapulmonary tuberculosis are effective for the treatment of tuberculous meningitis. The number of microorganisms present in the CNS in tuberculous meningitis is likely to be smaller than in cavitary pulmonary tuberculosis, and the potential for the development of resistant organisms is less (Sheller and DesPrez 1986). In the presence of meningeal inflammation, INH achieves levels in the CSF comparable to those in the blood (DesPrez and Boone 1961). Rifampicin (RMP) levels in CSF are 20% to 50% of those in the peripheral blood and exceed the reported minimal inhibitory concentrations for *M. tuberculosis* (D'Oliveira 1972; Sippel et al. 1974; Tuchman et al. 1984). Ethambutol (EMB), pyrazinamide (PZA), and ethionamide (ETH) penetrate into the CNS relatively well. The notable exceptions are the aminoglycosides, which penetrate even inflamed meninges poorly.

The following are recommended and are highly effective for treating tuberculous meningitis: combinations of INH (5 mg/kg per day with a maximum dose of 300 mg/day) and RMP (10–20 mg/kg per day with a maximum dose of 600 mg/day) administered for 9 months; the newer 6-month regimens of INH/RMP/EMB (15–25 mg/kg per day with a maximum dose of 2.5 g/day)/PZA (25 mg/kg per day with a maximum of 2 g/day); or INH/RMP/PZA for 2 months followed by INH/RMP for 4 months (American Thoracic Society 1986; Phuapradit and Vejjajiva 1987; Snider et al, 1988). The use of the four-drug regimen (INH/RMP/EMB/PZA) should be strongly considered in cases of patients in areas with high rates of primary isoniazid resistance or who have received earlier treatment for tuberculosis. PZA is mycobactericidal, achieves excellent and sustained levels in the CSF (Ellard et al. 1987), and is unique in its efficacy in killing intracellular bacilli at acidic pH (Girling 1984). PZA clearly improves the efficacy of regimens for the treatment of pulmonary tuberculosis of less than 9 months' duration. The use of 2 months of PZA in short-course regimens for pulmonary tuberculosis resulted in greater than 80% cure rates even when only 3 months of treatment were received (Fox 1983). This makes the drug particularly valuable in cases in which compliance problems are anticipated. However, one clinical trial documented no improved outcome when 2 initial months of PZA were added to 12-month regimens containing INH and RMP (Ramachandran et al. 1986).

Longer regimens using INH/STM/RMP for 2 months followed by INH/EMB for 10 months are highly efficacious in children (Ramachandran et al. 1986). Continuing STM beyond the first 2 months of treatment does not improve efficacy and adds increased risk of ototoxicity (American Thoracic Society 1986). EMB is less expensive than RMP but is difficult to use in infants and young children because ocular toxicity cannot be monitored. Compliance with therapy remains a major determinant of clinical success. Essentially no mycobacterial relapses were noted in compliant patients during clinical trials with the previous regimens.

The duration of antituberculous chemotherapy for the treatment of patients with acquired immunodeficiency syndrome (AIDS) and tuberculous meningitis is not known. Many of these patients respond to therapy for tuberculosis but die of other less treatable opportunistic infections during their course of antituberculous chemotherapy. CNS relapse of tuberculosis has been described in a patient with HIV infection who had been treated 1 year earlier for disseminated tuberculosis (Bishburg et al. 1986). Ultimately, some patients with AIDS may require lifelong maintenance therapy.

The usefulness and indications for adjunctive corticosteroid treatment in tuberculous meningitis are unclear. While many series include significant numbers of patients who were treated with corticosteroids to prevent the severe basilar inflammatory processes and the "vasculitis" characterized earlier, few comparative controlled trials are available. The best data come from 3 series (Lepper and Spies 1963; O'Toole et al. 1969; Escobar et al. 1975). Two studies were prospective randomized blind studies.

O'Toole and coworkers (1969) studied 23 cases of tuberculous meningitis from Calcutta, India. Only patients with moderately severe and advanced disease were chosen for the study because of "study bed limitations." Seventy percent of these cases had eventual cultural confirmation of *M. tuberculosis*, and all patients were started on therapy within 72 hours of presentation. The treatment protocol was INH and STM *plus* placebo *or* dexamethasone (2.25 mg every 6 hours for 1 week, 1.5 mg every 6 hours for 1 week, 0.75 mg every 6 hours for 1 week, and 0.375 mg every 6 hours for 1 week). Two control patients and one patient who received dexamethasone developed spinal block. Dexamethasone-treated patients showed decreased CSF pressures at 4 days; more rapid normalization of CSF glucose, protein, and leucocyte counts; and decreased evidence of "herniation syndrome" during the first 72 hours of treatment. Mortality was lower in the corticosteroid-treated group (six of 11 versus nine of 12 in the control group), but the difference was not statistically significant.

Another series studied the usefulness of adjunctive corticosteroids in the treatment of 99 Columbian children (ages 1–14 years) with tuberculous meningitis (Escobar et al. 1975). The patients had moderate to severe disease (25% comatose on admission), and the overall mortality rate was 51%. The investigators used a randomized, sequential, matched-pairs analysis (to control for age and severity of disease) and studied the efficacy of antituberculous therapy (INH/STM/para-amino salicylic acid [PAS]) with or without corticosteroids in two phases. Phase 1 examined the use of antituberculous drug therapy plus high- (10 mg/kg per day) or low- (1 mg/kg per day) dose prednisone therapy. Phase 2 examined the use of antituberculous drug therapy plus high- or low-dose corticosteroids versus no corticosteroids.

The use of adjunctive corticosteroids along with conventional antituberculous chemotherapy decreased mortality compared to antituberculous drug therapy alone. Prednisone given 1 mg/kg per day for 30 days was as effective as higher-dosage regimens. No differences in neurological sequela were seen between the corticosteroid- and noncorticosteroid-treated groups; however, only a few patients had greater than 2-year follow-up.

A review of the available evidence suggests that the administration of corticosteroids (given as prednisone 1 mg/kg per day with a maximum of 60 mg/day or its equivalent for the first 4–6 weeks of treatment) may be beneficial in cases of severe tuberculous meningitis with depressed levels of consciousness, hydrocephalus, stroke, cranial nerve palsies, or other focal neurological abnormalities. The corticosteroid dosage is tapered after 4 to 6 weeks depending on the clinical response of the patient. Antituberculous chemotherapy alone probably is adequate in cases of lesser clinical severity.

As described previously, elevated ICP and hydrocephalus are common in tuberculous meningitis, especially in children. In acute tuberculous meningitis, cerebral edema may cause increased ICP without hydrocephalus (Udani et al. 1971). In later stages, hydrocephalus due to basilar meningitis with CSF block at the cisterns around the rostral brain stem is the most common cause of elevated ICP and may cause death due to herniation (Singhal et al. 1975). Corticosteroids usually provide only temporary relief of symptoms related to elevated ICP due to hydrocephalus. In some cases, the communicating hydrocephalus may resolve spontaneously with chemotherapy alone. The role of neurosurgical management of hydrocephalus in tuberculous meningitis with ventriculoatrial or ventriculoperitoneal shunting procedures is controversial and decisions are best made on a case-by-case basis. Not all patients benefit from the procedure but in some instances, it may be life-saving (Bullock and Van Dellen 1982). There is a high risk of shunt dysfunction in the setting of chronic meningitis. Recommended indications for shunting include neurological deterioration despite appropriate antituberculous chemotherapy and ICP-reducing measures, progressively increasing ICP and the presence of severe hydrocephalus on CT scan (Bullock and Welchman 1982; Murray et al. 1981; Clark et al. 1986; Wilkinson et al. 1971; Newman et al. 1980; Schoeman et al. 1985).

Neurosurgical intervention also may be useful for restoring vision in patients with blindness due to optochiasmatic arachnoiditis. The formation of scar tissue at the level of the optic chiasm may cause chronic compression and ischemia of the optic nerves. This syndrome is characterized by progressively worsening visual acuity with optic nerve pallor and poor or absent pupillary light reflexes during therapy for tuberculous meningitis. Microsurgical dissection and lysis of adhesions in these areas may partially or fully restore vision (Navarro et al. 1981; Scott et al. 1977). Late paraplegia during treatment may be due to spinal epidural granuloma formation, which also is treatable surgically (Kocen and Parsons 1970).

Prognosis

Today, with antituberculous chemotherapy, the overall mortality from tuberculous meningitis is approximately 10% to 33%

(Lorber 1960; Mackay 1967; Phuaprandit and Vejjajva 1987; Rahajoe et al. 1979; Ramachandran et al. 1986; Sumaya et al. 1975). The prognosis for patients with tuberculous meningitis is determined by the severity of neurological involvement at the time of presentation and the duration of illness (Idriss et al. 1976; Kennedy and Fallon 1979; Kilpatrick et al. 1986; Lincoln et al. 1960; Ramachandran 1986; Smith 1975). Comatose patients have the highest mortality and the highest risk of permanent neurological sequelae. These sequelae are the result of the severe inflammatory process mentioned previously and include hydrocephalus, cerebral infarction due to arteriolitis, hemiparesis or monoparesis, permanent learning disabilities, and chronic cranial nerve damage leading to blindness, deafness, and opthalmoplegia. Rarely, Cushing's syndrome, obesity, diabetes insipidus, and excessive lethargy may occur due to permanent hypothalamic or pituitary damage (Udani et al. 1971). In one series with an 85% 5-year follow-up rate, 25% of patients had severe neurological residuals (Falk 1965). Severe neurological sequelae are more common in children. In more recent series, significant neurological residua have been present in 31% of adult cases (Haas et al. 1977) and in 51% of pediatric cases (Idriss et al. 1976).

The British Medical Research Council (1948) clinical staging system and modifications of this system have been used extensively in studies of tuberculous meningitis (Gordon and Parsons 1972). This system reiterates the three stages seen in the natural history of the disease. Stage 1 includes patients who are mildly ill with no signs of altered consciousness, focal neurological findings, or signs of hydrocephalus. Stage 1 patients may have meningeal signs. Stage 2 patients are moderately ill, are often confused but not comatose, or have focal neurological signs, including single cranial nerve palsies, paraparesis, or hemiparesis. Stage 3 patients are severely ill, stuporous or comatose, or have multiple cranial nerve palsies, complete hemiplegia, or paraplegia. This system and its modifications have been useful because the prognosis for recovery correlates with the clinical stage at the time specific antituberculous chemotherapy is initiated (Crocco et al. 1980; Kennedy and Fallon 1979; Kilpatrick et al. 1986; Klein et al. 1985; Swart et al. 1981).

Other prognostic factors (Table 15-5) that have been identified include the severity and presence of contrast enhancement of basilar exudates on head CT scans, the presence of ischemic stroke on head CT scans (threefold increase in mortality), and the presence and severity of hydrocephalus (Bhargava et al. 1982; Leiguarda et al. 1988). Higher mortality rates also have been reported in children younger than 5 years and in adults older than 50 years (Hinman 1967; Kennedy and Fallon 1979; Weiss and Flippin 1965). The existence of other chronic medical illnesses is a strong contributing factor to the high mortality of the disease in the elderly (Weiss and Flippin 1965). The presence of miliary disease, especially in infants, may worsen the prognosis (Delage and Dusseault 1969; Lepper and Spies 1963). Extremely high CSF protein levels (greater than 300 mg/dl) also carry a worse prognosis (Haas et al. 1977; Weiss and Flippin 1965).

Table 15-5. Prognostic Factors in Tuberculous Meningitis

Well established
Clinical stage at time of admission to hospital
Clinical stage at time antituberculous chemotherapy begun
Delay in initiation of antituberculous chemotherapy
Age (poorer prognosis in infants and children <5 years old, adults >50 years old)
Presence of associated chronic medical disease, alcoholism
Length of illness before hospitalization
Presence of cranial nerve palsy
CSF protein level >300 mg/dl at initial lumbar puncture

Possible
Miliary disease on chest x-ray
Active pulmonary tuberculosis
Active extracranial tuberculosis
Acute neurological complications during treatment
Ischemic stroke

Conclusion

Tuberculosis meningitis remains an uncommon but highly lethal disease with mortality rates of 20% to 33%, even in more recent series. Despite modern antituberculous chemotherapy and supportive care, a significant number of patients treated die or suffer permanent neurological disability. The prognosis is very good, however, for most patients who are diagnosed in the early stages of the illness. An awareness of the signs and symptoms of the disease, rapid diagnosis, and initiation of early therapy (often on empirical grounds) remain the principal means of optimizing clinical outcome.

Note Added in Proof

Forty-five culture proven cases of tuberculous meningitis in HIV-infected individuals from Madrid have recently been reported (Berenguer et al. 1992). Two percent of non-HIV-infected patients with tuberculosis had meningeal involvement, as compared to 10 percent of HIV-infected patients. Fifty-four percent of the HIV-infected patients with tuberculous meningitis had chest x-ray findings consistent with tuberculosis; 29% of patients tested had positive tuberculin skin tests (greater than 10 mm induration). CSF AFB smears were positive in 22% of HIV-infected patients who were tested. Presenting signs and symptoms, CSF findings, and inpatient mortality were comparable between HIV-infected and non-HIV-infected individuals. Peripheral blood CD4 lymphocyte counts less than 200/mm^3 and the presence of symptoms longer than 2 weeks before admission were adverse prognostic signs in HIV-infected patients with tuberculous meningitis.

References

American Thoracic Society. Treatment of tuberculosis and tuberculosis infection in adults and children. Am. Rev. Resp. Dis. 134:355–363; 1986.

Auerbach, O. Tuberculous meningitis: correlation of therapeutic results with the pathogenesis and pathologic changes. I. General considerations and pathogenesis. Am. Rev. Tuberc. 64:408–418; 1951.

Barrett-Connor, E. Tuberculous meningitis in adults. South Med. J. 60:1061–1067; 1967.

Bennett, J. E.; Bailey, J. W. Control for Rheumatoid Factor in the Latex Test for Cryptococcosis. Am. J. Clin. Pathol. 56:360–365; 1971.

Berenguer, J.; Moreno, S.; Laguna, F.; Vincente, T.; Adrados, M.; Ortega, A.; Gonzalez-LaHoz, J.; Bouza, E. Tuberculous meningitis in patients infected with the human immunodeficiency virus. N. Engl. J. Med. 326:668–672, 1992.

Bhargava, S.; Gupta, A. K.; Tandon, P. N. Tuberculous meningitis—a CT study. Br. J. Radiol. 55:189–196; 1982.

Bishburg, E.; Sunderam, G.; Reichman, L. B.; Kapila, R. Central nervous system tuberculosis with the acquired immunodeficiency syndrome and its related complex. Ann. Int. Med. 105:210–213; 1986.

Brooks, J. B.; Choudhary, G.; Craven, R. B.; Alley, C. C.; Liddle, J. A.; Edman, D. C.; Converse, J. D. Electron capture gas chromatography detection and mass spectrum identification of 3-(2'-Ketohexyl) indoline in spinal fluids of patients with tuberculous meningitis. J. Clin. Microbiol. 5:625–628; 1977.

Brooks, J. B.; Daneshvar, M. I.; Fast, D. M.; Good, R. C. Selective procedures for detecting femtomole quantities of tuberculostearic acid in serum and cerebrospinal fluid by frequency-pulsed electron capture gas-liquid chromatography. J. Clin. Microbiol. 25:1201–1206; 1987.

Brooks, J.; Daneshuvar, M. I.; Haberberger, R. L.; Mikhail, I. A. Rapid diagnosis of tuberculous meningitis by frequency pulsed electron capture gas-liquid chromatography detection of carboxylic acids in cerebrospinal fluid. J. Clin. Microbiol. 28:989–997; 1990.

Bullock, M. R. R.; VanDellen, J. R. The role of cerebrospinal fluid shunting in tuberculous meningitis. Surg. Neurol. 18:274–277; 1982.

Bullock, M. R.; Welchman, J. M. Diagnostic and prognostic features of tuberculous meningitis on CT scanning. J. Neurol. Neurosurg. Psych. 45:1098–1101; 1982.

Cairns, H. Neurosurgical methods in the treatment of tuberculous meningitis. Arch. Dis. Child. 26:373–386; 1951.

Chandramuki, A.; Allen, P. R. J.; Keen, M.; Ivanyi, J. Detection of mycobacterial antigen and antibodies in the cerebrospinal fluid of patients with tuberculous meningitis. J. Med. Microbiol. 20:239–247; 1985.

Clark, W. C.; Metcalf, J. C.; Muhlbauer, M. S.; Dohan, F. C.; Robertson, J. H. Mycobacterium tuberculosis meningitis: a report of twelve cases and a literature review. Neurosurg. 18:604–610; 1986.

Coovadia, Y. M.; Dawood, A.; Ellis, M. E.;

Coovadia, H. M.; Daniel, T. M. Evaluation of adenosine deaminase activity and antibody to Mycobacterium tuberculosis antigen 5 in cerebrospinal fluid and the radioactive bromide partition test for the early diagnosis of tuberculosis meningitis. Arch. Dis. Child. 61:428–435; 1986.

Crocco, J. A.; Blecker, M. J.; Rooney, J. J.; Kramer, S.; Lyons, H. A. Tuberculous meningitis in adults. N.Y. State J. Med. 80:1231–1234; 1980.

Cybulska, E.; Rucinski, J. Tuberculous meningitis. Br. J. Hosp. Med. 39:63–66; 1988.

Daniel, T. M. New approaches to the rapid diagnosis of tuberculous meningitis. J. Infect. Dis. 155:599–602; 1987.

Dastur, D. K.; Lalitha, V. S.; Udani, P. M.; Parekh, U. The brain and meninges in tuberculous meningitis—gross pathology in 100 cases and pathogenesis. Neurol. India 18:86–100; 1970.

Delage, G.; Dusseault, M. Tuberculous meningitis in children: a retrospective study of 79 patients with an analysis of prognostic factors. Can. Med. Assoc. J. 120:305–309; 1979.

DesPrez, R. M.; Boone, I. U. Metabolism of C[14]-isoniazid in humans. Am. Rev. Resp. Dis. 84:42–51; 1961.

DesPrez, R. M.; Heim, C. R. Tuberculosis. In: Mandell, G. L.; Douglas, R. G., Jr.; Bennett, J. E., eds. Principles and practice of infectious disease. 3rd ed. New York: J. Wiley and Sons; 1989:p. 1897–1898.

D'Oliveira, J. J. G. Cerebrospinal fluid concentrations of rifampin in meningeal tuberculosis. Am. Rev. Resp. Dis. 106:432–437; 1972.

Ellard, G. A.; Humphries, M. J.; Gabriel, M.; Teoh, R. Penetration of pyrazinamide into the cerebrospinal fluid in tuberculous meningitis. Br. Med. J. 294:284–285; 1987.

Ellner, J. J.; Bennett, J. E. Chronic meningitis. Medicine 55:341–369; 1976.

Emond, R. T. D.; McKendrick, G. D. W. Tuberculosis as a cause of transient aseptic meningitis. Lancet 2:234–236; 1973.

Escobar, J. A.; Belsey, M. A.; Dueñas, A.; Medina, P. Mortality from tuberculous meningitis reduced by steroid therapy. Pediatrics 56:1050–1055; 1975.

Fallon, R. J. The treatment of tuberculous meningitis. J. Antimicrob. Chemother. 4:1–2; 1978.

Falk, A. U.S. Veterans Administration–Armed Forces Cooperative study on the chemotherapy of tuberculosis. XIII. tuberculous meningitis in adults, with special reference to survival, neurologic residuals, and work status. Am. Rev. Resp. Dis. 91:823–831; 1965.

Fishman, R. A. Cerebrospinal fluid in diseases of the nervous system. Philadelphia: W.B. Saunders; 1980.

Foord, A. G.; Forsyth, A. The laboratory diagnosis of tuberculous meningitis. Am. J. Clin. Path. 3:45–54; 1933.

Fox, W. Compliance of patients and physicians: experience and lessons from tuberculosis—II. Br. Med. J. 287:101–105; 1983.

Fremont-Smith, F. Pathogenesis of the changes in cerebrospinal fluid in meningitis. Arch. Neurol. Psych. 28:778–788; 1932.

French, G. L.; Chan, C. Y.; Cheung, S. W.; Teoh, R.; Humphries, M. J.; O'Mahony, G. Diagnosis of tuberculous meningitis by detection of tuberculostearic acid in cerebrospinal fluid. Lancet 2:117–119; 1987.

Girgis, N. I.; Yassin, M. W.; Sippel, J. E.; Sorenson, K.; Hassan, A.; Miner, W. F.; Farid, Z.; Abu El Ella, A. The value of ethambutol in the treatment of tuberculous meningitis. J. Trop. Med. Hyg. 79:14–17; 1976.

Girling, D. J. The role of pyrazinamide in primary chemotherapy for pulmonary tuberculosis. Tubercle 65:1–4; 1984.

Gordon, A.; Parsons, M. The place of corticosteroids in the management of tuberculous meningitis. Br. J. Hosp. Med. 7:651–655; 1972.

Greitz, T. Angiography in tuberculous meningitis. Acta. Radiol. [Diagn] (Stockholm) 2:369–378; 1964.

Haas, E. J.; Madhavan, T.; Quinn, E. L.; Cox, F.; Fisher, E.; Burch, K. Tuberculous meningitis in an urban general hospital. Arch. Intern. Med. 137:1518–1521; 1977.

Hinman, A. R. Tuberculous meningitis at Cleveland Metropolitan General Hospital 1959 to 1963. Am. Rev. Resp. Dis. 95:670–673; 1967.

Idriss, Z. H.; Sinno, A. A.; Kronfol, N. M. Tuberculous meningitis in childhood. Forty-three cases. Am. J. Dis. Child. 130:364–367; 1976.

Illingworth, R. S. Miliary and meningeal tuberculosis: difficulties in diagnosis. Lancet 2:646–649; 1956.

Kadival, G. V.; Samuel, A. M.; Mazarelo, T. B.; Chaparas, S. D. Radioimmunoassay for detecting Mycobacterium tuberculosis antigen in cerebrospinal fluids of patients with tuberculous meningitis. J. Infect. Dis. 155:608–611; 1987.

Karandanis, D.; Shulman, J. A. Recent survey of infectious meningitis in adults: review of laboratory findings in bacterial, tuberculous and aseptic meningitis. South Med. J. 69:449–457; 1976.

Kennedy, D. H.; Fallon, R. J. Tuberculous meningitis. J.A.M.A. 241:264–268; 1979.

Kilpatrick, M. E.; Girgis, N. I.; Yassin, M. W.; Abu El Ella, A. A. Tuberculous meningitis—clinical and laboratory review of 100 patients. J. Hyg. (London) 96:231–238; 1986.

Kingsley, D. P. E.; Hendrickse, W. A.; Kendall, B. E.; Swash, M.; Singh, V. Tuberculous meningitis: role of CT in management and prognosis. J. Neurol. Neurosurg. Psych. 50:30–36; 1987.

Klein, N. C.; Damsker, B.; Hirschman, S. Z. Mycobacterial meningitis: retrospective analysis from 1970–1983. Am. J. Med. 79:29–34; 1985.

Kocen, R. S.; Parsons, M. Neurological complications of tuberculosis: some unusual manifestations. Q. J. Med. 39:17–30; 1970.

Krambovitis, E.; McIllmurray, M. B.; Lock, P. E.; Hendrickse, W.; Holzel, H. Rapid diagnosis of tuberculous meningitis by latex particle agglutination. Lancet 2:1229–1231, 1984.

Lee, T. Y.; Brown, H. W. Tuberculous meningitis patients as index cases in the epidemiology of tuberculosis. Am. J. Pub. Health 58:1901–1909; 1968.

Lehrer, H. The angiographic triad in tuberculous meningitis. A radiographic and clinicopathologic correlation. Radiology 87:829–835; 1966.

Leiguarda, R.; Berthier, M.; Starkstein, S.; Nogues, M.; Lylyk, P. Ischemic infarction in 25 children with tuberculous meningitis. Stroke 19:200–204; 1988.

Lepper, M. H.; Spies, H. W. The present status of the treatment of tuberculosis of the central nervous system. Ann. N. Y. Acad. Sci. 106:106–123; 1963.

Lincoln, E. M. Tuberculous meningitis in children with special reference to serous meningitis. Part II. serous tuberculous meningitis. Am. Rev. Tuberc. 56:95–109; 1947.

Lincoln, E. M.; Sordillo, S. V. R.; Davies, P. A. Tuberculous meningitis in cihldren. A review of 167 untreated and 74 treated patients with special reference to early diagnosis. J. Pediatr. 57:807–823; 1960.

Lorber, J. Treatment of tuberculous meningitis. Br. Med. J. 1:1309–1312; 1960.

Mackay, J. B. Tuberculous meningitis: a 25 year survey in the Wellington area. N. Z. Med. J. 66:82–89; 1967.

Mardh, P.; Larsson, L.; Hoiby, N.; Engbaek, H. C.; Odham, G. Tuberculostearic acid as a diagnostic marker in tuberculous meningitis. Lancet 1:367; 1983.

Medical Research Council. Streptomycin treatment of tuberculous meningitis: report of the committee of streptomycin in tuberculosis trial. Lancet 1:582–596; 1948.

Merritt, H. H.; Fremont-Smith, F. Cerebrospinal fluid in tuberculous meningitis. Arch. Neurol. Psych. 33:516–536; 1935.

Merritt, H. H.; Fremont-Smith, F. The cerebrospinal fluid. Philadelphia: W.B. Saunders; 1938.

Murray, H. W.; Brandstetter, R. D.; Lavyne, M. H. Ventriculoatrial shunting for hydrocephalus complicating tuberculous meningitis. Am. J. Med. 70:895–898; 1981.

Navarro, I. M.; Peralta, V. H.; Leon, J. A.;

Varela, E. A.; Cabrera, J. M. Tuberculous optochiasmatic arachnoiditis. Neurosurg. 9:654–660; 1981.

Newman, P. K.; Cumming, W. J. K.; Foster, J. B. Hydrocephalus and tuberculous meningitis in adults. J. Neurol. Neurosurg. Psychiat. 43:188–190; 1980.

Ogawa, S. K.; Smith, M. A.; Brennessel, D. J.; Lowy, F. J. Tuberculous meningitis in an urban medical center. *Medicine* 66:317–326; 1987.

O'Toole, R. D.; Thornton, G. F.; Mukherjee, M. K.; Nath, R. L. Dexamethasone in tuberculous meningitis. Relationship of cerebrospinal fluid effects to therapeutic efficacy. Ann. Intern. Med. 70:39–48; 1969.

Phuapradit, P.; Vejjajiva, A. Treatment of tuberculous meningitis: role of short course chemotherapy. Q. J. Med. 62:249–258; 1987.

Rahajoe, N. N.; Rahajoe, N.; Boediman, I.; Said, M.; Lazuardi, S. The treatment of tuberculous meningitis in children with a combination of isoniazid, rifampin, and streptomycin—preliminary report. Tubercle 60:245–250; 1979.

Ramachandran, P.; Duraipandian, M.; Nagarajan, M.; Prabhakar, R.; Ramakrishnan, C. V.; Tripathy, S. P. Three chemotherapy studies of tuberculous meningitis in children. Tubercle 67:17–29; 1986.

Rich, A. R.; McCordock, H. A. The pathogenesis of tuberculous meningitis. Bull. J. Hopkins Hosp. 52:5–38; 1933.

Schoeman, J. F.; Roux, D. L.; Bezuidenhout, P. B.; Donald, P. R. Intracranial pressure monitoring in tuberculous meningitis: clinical and computerized tomographic correlation. Dev. Med. Child Neurol. 27:644–654; 1985.

Schoeman, J.; Hewlett, R.; Donald, P. MR of childhood tuberculous meningitis. Neuroradiol. 30:473–477; 1988.

Scott, R. M.; Sonntag, V. K.; Wilcox, L. M.; Adelman, L. S.; Rockel, T. H. Visual loss from optochiasmatic arachnoiditis after tuberculous meningitis. J. Neurosurg. 46:524–526; 1977.

Sheller, J. R.; DesPrez, R. M. CNS tuberculosis. Neurol. Clin. 4:143–158; 1986.

Singhal, B. S.; Bhagwati, S. N.; Syed, A. H.; Lud, G. W. Raised intracranial pressure in tuberculous meningitis. Neurol. India 23:32–39; 1975.

Sippel, J. E.; Mikhail, I. A.; Girgis, N. I. Rifampin concentrations in cerebrospinal fluid of patients with tuberculous meningitis. Am. Rev. Resp. Dis. 109:579–580; 1974.

Slavin, R. E.; Walsh, T. J.; Pollack, A. D. Late generalized tuberculosis: a clinical pathologic analysis and comparison of 100 cases in the preantibiotic and antibiotic eras. Medicine 59:352–366; 1980.

Smith, A. L. Tuberculous meningitis in childhood. Med. J. Austral. 1:57–60; 1975.

Smith, H. V.; Daniel, P. Some clinical and pathological aspects of tuberculosis of the central nervous system. Tubercle 28:64–80; 1947.

Smith, J.; Godwin-Austen, R. Hypersecretion of antidiuretic hormone due to tuberculous meningitis. Postgrad. Med. J. 56:41–44; 1980.

Snider, D. E.; Rieder, H. L.; Combs, D.; Bloch, A. B.; Hayden, C. H.; Smith, M. H. D. Tuberculosis in children. Pediatr. Inf. Dis. J. 7:271–278; 1988.

Steiner, P.; Portugaleza, C. Tuberculous meningitis in children. A review of 25 cases observed between the years 1965 and 1970 at the Kings County Medical Center of Brooklyn with special reference to the problem of infection with primary drug-resistant strains of M. tuberculosis. Am. Rev. Resp. Dis. 107:22–29; 1973.

Stewart, S. M. The bacteriological diagnosis of tuberculous meningitis. J. Clin. Path. 6:241–242; 1953.

Stockstill, M. T.; Kauffman, C. A. Comparison of cryptococcal and tuberculous meningitis. Arch. Neurol. 40:81–85; 1983.

Strumpf, I. J.; Tsang, A. Y.; Schork, M. A.; Weg, J. A. The reliability of gastric smears by auramine-rhodamine staining technique for the diagnosis of tuberculosis. Am. Rev. Resp. Dis. 114:971–976; 1976.

Sumaya, C. V.; Simek, M.; Smith, M. H. D.; Seidemann, M. F.; Ferriss, G. S.; Rubin, W. Tuberculous meningitis in children during the isoniazid era. J. Pediatr. 87:43–49; 1975.

Swart, S.; Briggs, R. S.; and Millac, P. A. Tuberculous meningitis in Asian patients. Lancet 2:15–16; 1981.

Tandon, P. N. Tuberculous meningitis. In: Vinken, P. J.; Bruyn, G. W., eds. Handbook of clinical neurology. Amsterdam: Elsevier/North Holland; 1978: vol. 33 p. 195–262.

Traub, M.; Colchester, A. C. F.; Kingsley, D. P. E.; Swash, M. Tuberculosis of the central nervous system. Q. J. Med. 53:81–100; 1984.

Tuchman, A. J.; Burger, S.; Daras, M. Cerebrospinal fluid penetration of anti-infective agents. South Med. J. 77:1443–1445; 1984.

Udani, P. M.: Parekh, U. C.: Dastur, D. K. Neurological and related syndromes in CNS tuberculosis. Clinical features and pathogenesis. J. Neurol. Sci. 14:341–357; 1971.

Watt, G.; Zaraspe, G.; Bautista, S.; Laughlin, L. W. Rapid diagnosis of tuberculous meningitis by using an enzyme-linked immunosorbent assay to detect mycobacterial antigen and antibody in cerebrospinal fluid. J. Inf. Dis. 158:681–686; 1988.

Weiss, W.; Flippin, H. F. The changing incidence and prognosis of tuberculous meningitis. Am. J. Med. Sci. 250:80–93; 1965.

Wilkinson, H. A.; Ferris, E. J.; Muggid, A. L.; Cantu, R. C. Central nervous system tuberculosis: a persistent disease. J. Neurosurg. 34:15–22; 1971.

Zarabi, M.; Sane, S.; Girdany, B. R. The chest roentgenogram in the early diagnosis of tuberculous meningitis in children. Am. J. Dis. Child. 121:389–392; 1971.

16

Brain Abscesses

THOMAS J. MAMPALAM AND MARK L. ROSENBLUM

The prognosis of patients with brain abscesses may be considered in terms of mortality rate statistics and neurological sequelae. The most important determinant of the morbidity and mortality rates is the condition of the patient at the start of treatment. Early diagnosis and prompt initiation of appropriate therapy are necessary to ensure a satisfactory therapeutic outcome.

Mortality

There is no evidence that standard demographic characteristics, such as sex or ethnic origin, influence the mortality rate among patients with brain abscess. Although age at diagnosis is an important determinant of long-term neurological status, it has no consistent relationship with the mortality rate (Bhatia et al. 1973; Britt et al. 1981; Carey et al. 1971; Carey et al. 1972; Garfield 1969; Jooma et al. 1951; Joubert and Stephanov 1977; Loeser and Scheinberg 1957; Mampalam and Rosenblum 1988; Morgan et al. 1973), except among neonates with gram-negative abscesses (Renier et al. 1988). The high mortality rate in such patients may be related to immaturity, the virulence of gram-negative organisms, and coexistent systemic disease. In the University of California, San Francisco (UCSF), series, which included patients aged 6 weeks to 78 years (mean, 34.9 years), the mortality rate did not correlate with age (Mampalam and Rosenblum 1988).

The risk of death from brain abscess depends on the neurological condition at diagnosis (Table 16-1) (Alderson et al. 1981; Bhatia et al. 1973; Britt et al. 1981; Carey et al. 1971; Carey et al. 1972; Garfield 1969; Jooma et al. 1951; Joubert and Stephanov, 1977; Karandanis and Shulman 1975; Loeser and Scheinberg 1957; Mampalam and Rosenblum 1988; Morgan et al. 1973). The mortality rate exceeds 60% if signs of brain herniation are present (Yang 1981) and is approximately 90% among comatose patients (Alderson et al. 1981; Bhatia et al. 1973; Garfield 1969; Karandanis and Shulman 1975). Among alert patients, however, the mortality rate is less than 20%, even in older studies (Alderson et al. 1981; Karandanis and Shulman 1975; Mampalam and Rosenblum 1988; Svanteson et al. 1988). Misdiagnosis and delays in diagnosis were the main reasons for the high mortality rate

Table 16-1. Factors Affecting the Prognosis of Patients With Brain Abscesses

Factor	Mortality Rate (%)
Before advent of CT scanning	30–50
After advent of CT scanning	0–15
Neurological condition at diagnosis	
Alert, no deficit	0–20
Comatose	25–90
Etiology	
Local infection	6–50
Metastatic	12–100
Number of abscesses	
Solitary	0–45
Multiple	8–100
Type of treatment	
Aspiration	0–45
Excision	0–30
Nonoperative	0–13

from brain abscesses in previous decades (Bhatia et al. 1973; Garfield 1969; Loeser and Scheinberg 1957; Rosenblum et al. 1986; Samson and Clark, 1973).

The advent of computerized tomography (CT) has been the most important factor in reducing the mortality rate among patients with brain abscesses (Joubert and Stephanov 1977; Renier et al. 1988; Rosenblum et al. 1978; Svanteson et al. 1988). The prompt and accurate localization of the lesions by CT has reduced the mortality rate from 30% to 50% to less than 15% in most recent series (Alderson et al. 1981; Bhatia et al. 1973; Britt et al. 1981; Carey et al. 1971; Carey et al. 1972; Garfield 1969; Jooma et al. 1951; Joubert and Stephanov 1977; Karandanis and Shulman 1975; Mampalam and Rosenblum 1988; Rosenblum et al. 1978; Rosenblum et al. 1986; Samson and Clark 1973; Stephanov 1988; Svanteson et al. 1988; Van Alphen and Dreissen 1976; Yang 1981). Since CT scanning became a routine diagnostic test at UCSF, the diagnosis of brain abscess has not been missed in any patient, and the mortality rate has decreased from 40.9% to 4.3% (Mampalam and Rosenblum 1988). During this period, there were no significant changes in culture identification of organisms or in antibiotic regimens that could account for the remarkable drop in mortality rate.

Differences in perioperative mortality rates between aspiration and excision procedures seem to be related to patient selection criteria rather than to the safety of the respective treatments. Stephanov (1988) recently reviewed the literature on surgical treatment of brain abscesses and found mortality rates of 38% (356 of 941) after aspiration and 19% (283 of 1505) after excision. Most of these patients, however, were treated before CT scanning became available, and in most series, the initial neurological or general medical condition was worse among patients whose lesions were aspirated than among those who underwent excision. Given similar initial conditions, the results of aspiration and excision appear to be comparable. In the UCSF series, the perioperative mortality rate was 6.1% after aspiration and 8.7% after excision; the difference was not statistically significant (Mampalam and Rosenblum 1988).

Certain characteristics of brain abscesses have been associated with higher mortality rates. Abscesses arising from adjacent otogenous infections or previous craniotomy seem to carry lower mortality rates than those caused by metastatic spread (Bhatia et al. 1973; Carey et al. 1972; Karandanis and Shulman 1975; Loeser and Scheinberg 1957; Van Alphen and Dreissen 1976). Metastatic abscesses tend to be deep and multiple and occur more often in patients with serious systemic disease, such as endocarditis or pulmonary infections. In addition, the diagnosis of metastatic abscess may be delayed because the index of suspicion for abscess is higher when a patient has an adjacent infection than when the source is remote from the brain.

Multiple abscesses also have been associated with high mortality rates, up to 100% in older studies (Garfield, 1969; Karandanis and Shulman 1975). Liske and Weikers (1964) found a 70% mortality rate in patients with multiple abscesses, compared with only 45% in those with solitary abscesses. CT scanning, more effective antibiotic regimens and stereotactic surgical procedures, have markedly diminished the difficulty of managing multiple lesions; in several recent series, all patients with multiple abscesses were treated successfully (Mampalam and Rosenblum 1988; Rosenblum et al. 1978; Rosenblum et al. 1986).

In the UCSF series, the size of the abscess was the only factor other than the initial neurological grade that influenced the mortality rate; patients who died tended to have larger abscesses than the survivors (Mampalam and Rosenblum 1988). Rupture of an abscess into the ventricle can cause ventriculitis, acute hydrocephalus, and death if a large volume of purulent material enters the ventricle and if there is a significant delay in diagnosis. If only a small amount of pus reaches the ventricle and if a ventriculostomy is performed without delay, the prognosis is better. Immunocompromised patients and those with coexisting septicemia or cardiac disease have higher mortality rates than other patients (Britt et al. 1981; Kagawa et al. 1983; Rosenblum et al. 1986).

Neurological Sequelae

Despite successful eradication of the infection, many patients with brain abscess in early series suffered persistent neurological sequelae, probably because of delayed diagnosis, poor initial neurological condition, and larger abscesses. The surgical technique also might have contributed to the higher morbidity rate because before CT scanning was introduced, there was no way to localize the lesions precisely before and during the operation. We believe that earlier diagnosis will be reflected in lower long-term morbidity. The morbidity from brain abscess has already decreased markedly; at UCSF, for example, none of 43 patients treated successfully between 1981 and 1986 had significant and neurological deficits when discharged from the hospital (Mampalam and Rosenblum 1988).

The most common neurological sequelae of brain abscesses have been cognitive deficits, hemiparesis, and seizures. Earlier studies reported permanent neurological deficits, such as hemiparesis and dysphasia, in 27% to 50% of patients with brain abscesses (Carey et al. 1971; Carey et al. 1972; Nielsen et al. 1983). Infants and children have a worse outcome than adults. Most neonates with brain abscess subsequently have IQs less than 80 (Renier et al. 1988). In one study, 70% of the children had difficulties with school performance, and 38% had severe hemiparesis, whereas 87% of the adults were able to continue their normal work, and none had severe hemiparesis (Carey et al. 1971). Abscess location also influences neurological sequelae. Patients with lesions near the sensorimotor strip had much worse therapeutic outcomes than patients with abscesses in other locations (Carey et al. 1971).

Seizure disorders occur in about 50% of patients with brain abscess (range, 15% to 72%) (Jooma et al. 1951; Legg et al. 1973; Nielsen et al. 1983; Northcraft and Wyke 1957), usually within 3 to 4 years after treatment (Legg et al. 1973). Early studies suggested that seizures might be more common after aspiration than excision, but later reports did not confirm this (Jooma et al. 1951; Legg et al. 1973; Nielsen et al. 1983). Children are more likely than adults to develop epilepsy after brain abscess (Carey et al. 1971). Seizures are most common in patients with lesions of the frontal and temporal lobes; seizures after occipital lobe abscesses are unusual, and cerebellar abscesses do not predispose to seizures (Carey et al. 1971; Northcraft and Wyke 1957). Prolonged administration of anticonvulsant drugs is recommended for all patients with brain abscess.

In conclusion, the outcome after brain abscess depends largely on the clinical presentation at the time of diagnosis. The development of techniques that permit more rapid and accurate diagnosis, especially CT scanning, has resulted in earlier diagnosis, when the abscess is still small and few neurological deficits have occurred. As a result, the prognosis has improved markedly. The long-term follow-up of patients treated since CT became available will undoubtedly reveal both a lower mortality rate and fewer serious neurological sequelae than in the past.

Acknowledgment

The authors thank Stephen Ordway for editorial assistance.

References

Alderson, D.; Strong, A. J.; Ingham, H. R.; Selkon, J. B. Fifteen-year review of the mortality of brain abscess. Neurosurgery 8:1–5; 1981.

Bhatia, R.; Tandon, P. N.; Banerji, A. K. Brain abscess—an analysis of 55 cases. Int. J. Surg. 58:565–568; 1973.

Britt, R. H.; Enzmann, D. R.: Remington, J. S. Intracranial infection in cardiac transplant recipients. Ann. Neurol. 9:107–119; 1981.

Carey, M. E.; Chou, S. N.; French, L. A. Long-term neurological residua in patients surviving brain abscess with surgery. J. Neurosurg. 34:652–656; 1971.

Carey, M. E.; Chou, S. N.; French, L. A. Experience with brain abscesses. J. Neurosurg. 36:1–9; 1972.

Garfield, J. Management of supratentorial intracranial abscess: a review of 200 cases. Br. Med. J. 2:7–11; 1969.

Jooma, O. V.; Pennybacker, J. B.; Tutton, G. K. Brain abscess: aspiration, drainage, or excision? J. Neurol. Neurosurg. Psychiat. 14:308–313; 1951.

Joubert, M. J.; Stephanov, S. Computerized tomography and surgical treatment in intracranial suppuration. Report of 30 consecutive unselected cases of brain abscess and subdural empyema. J. Neurosurg. 47:73–78; 1977.

Kagawa, M.; Takeshita, M.; Yato, S.; Kitamura, K. Brain abscess in congenital cyanotic heart disease. J. Neurosurg. 58:913–917; 1983.

Karandanis, D.; Shulman, J. A. Factors associated with mortality in brain abscess. Arch. Intern. Med. 135:1145–1150; 1975.

Legg, N. J.; Gupta, P. C.; Scott, D. F. Epilepsy following cerebral abscess: a clinical and EEG study of 70 patients. Brain 96:259–268; 1973.

Liske, E.; Weikers, N. J. Changing aspects of brain abscesses. Review of cases in Wisconsin 1940 through 1962. Neurology 14:294–300; 1964.

Loeser, E., Jr.; Scheinberg, L. Brain abscess: a review of ninety-nine cases. Neurology 7:601–609; 1957.

Mampalam, T. J.; Rosenblum, M. L. Trends in the management of bacterial brain abscesses: a review of 102 cases over 17 years. Neurosurgery 23:451–458; 1988.

Morgan, H.; Wood, M. W.; Murphey, F. Experience with 88 consecutive cases of brain abscess. J. Neurosurg. 38:698–704; 1973.

Nielsen, H.; Harmsen, A.; Gyldensted, C. Cerebral abscess: a long term follow-up. Acta Neurol. Scand. 67:330–337; 1983.

Northcraft, G. B.; Wyke, B. D. Seizures following surgical treatment of intracranial abscesses: a clinical and electroencephalographic study. J. Neurosurg. 14:249–263; 1957.

Renier, D.; Flandin, C.; Hirsch, E.; Hirsch, J-F. Brain abscesses in neonates. A study of 30 cases. J. Neurosurg. 69:877–882; 1988.

Rosenblum, M. L.; Hoff, J. T.; Norman, D.; Weinstein, P. R.; Pitts, L. Decreased mortality from brain abscesses since advent of computerized tomography. J. Neurosurg. 49:658–668; 1978.

Rosenblum, M. L.; Mampalam, T. J.; Pons, V. Controversies in the management of brain abscesses. Clin. Neurosurg. 33:603–632; 1986.

Samson, D. S.; Clark, K. A current review of brain abscess. Am. J. Med. 54:201–210; 1973.

Stephanov, S. Surgical treatment of brain abscess. Neurosurgery 22:724–730; 1988.

Svanteson, B.; Nordstrom, C. H.; Rausing, A. Non-traumatic brain abscess. Acta Neurochir. (Wien) 94:57–65; 1988.

Van Alphen, H. A. M.; Dreissen, J. J. R. Brain abscess and subdural empyema: factors influencing mortality and results of various surgical techniques. J. Neurol. Neurosurg. Psychiat. 39:481–490; 1976.

Yang, S-Y. Brain abscess: a review of 400 cases. J. Neurosurg. 55:794–799; 1981.

17

HIV-1 Infection

STEPHEN L. BOSWELL AND BRADFORD A. NAVIA

The peripheral and central nervous systems (CNS) can be profoundly affected by human immunodeficiency virus type-1 (HIV-1). The effects of this virus can result from direct infection of or injury to nervous system tissues or from opportunistic infections and malignancies that frequently accompany immunological suppression. These complications can occur early in HIV infection, as with Bell's palsy, or late, as with the AIDS dementia complex (ADC). Several of the diseases associated with HIV infection, such as ADC and vacuolar myelopathy, have never been seen before. The spectrum of HIV-induced neurological disorders is as perplexing to researchers as it is devastating to many persons living with HIV infection (Table 17-1).

It has been estimated that CNS symptoms are the initial manifestation of AIDS in about 10% of cases, and that more than 60% of patients with AIDS have neurological symptoms (Levy, R. M. et al. 1985; Navia et al. 1986; Petito et al. 1985; Snider et al. 1983). Postmortem evidence of CNS disease can be found in 80% of patients, while radiographic evidence of abnormalities occur in 60% of individuals diagnosed with AIDS

(Levy et al. 1986; Navia et al. 1986). Several studies have revealed that between 73% and 87% of individuals diagnosed with AIDS will experience nervous system abnormalities (Jordon et al. 1985; Levy et al. 1986; Navia et al. 1986a; Moskowitz et al. 1984; Urmacher and Nielsen, 1985). The neurological manifestations of AIDS are being observed with a frequency that parallels the improvements in antiretroviral therapy and *Pneumocystis carinii* pneumonia prophylaxis. A growing collection of evidence indicates that as these advances continue to prolong life, the risk of neurological complications will likely increase (Harris 1990).

The prevalence of certain neurological complications associated with HIV infection appears to vary widely. For example, estimates of the prevalence of *Toxoplasma gondii* encephalitis varies from 2.6% to 30.8%, with most estimates between 5% and 15% of patients (Haverkos 1987; Moskowitz et al. 1984; Navia et al. 1986c; Wong et al. 1984). Geographic location, risk group, stage of disease, and other factors determine the prevalence of these neurological complications. In the United States, cryptococcal meningitis is reported approximately twice

Table 17-1. Neurological Disorders Associated With HIV-1 Infection

Central Nervous System

Focal Brain
  Toxoplasmosis
  Primary CNS lymphoma
  Progressive multifocal leukoencephalopathy
  Vascular disorders
  VZV encephalitis
  Tuberculoma
  Cryptococoma
Diffuse Brain
  AIDS dementia complex
  Cytomegalovirus encephalitis
  Herpes encephalitis
  Acute HIV-associated encephalitis
  Metabolic encephalopathies
Meningitides
  Aseptic meningitis
  Cryptococcal meningitis
  Tuberculous meningitis
  Lymphomatous meningitis
  Syphilitic meningitis
Myelopathies
  Vacuolar myelopathy
  HSV-2 myelopathy
  VZV myelitis
  CMV myelopathy
  Epidural or intradural lymphoma

Peripheral Nervous System

Myopathies
  Polymyositis
  Zidovudine-associated myopathy
Peripheral Neuropathies
  Chronic sensory-motor polyneuropathy
  Mononeuritis multiplex
  Mononeuropathy associated with aseptic meningitis
  Acute demyelinating polyneuropathy (Guillain-Barré)
  Chronic idiopathic demyelinating polyneuropathy
  CMV polyradiculopathy
  Autonomic neuropathy
  Toxic neuropathies

as frequently in New Jersey (7.8%) as in the rest of the country (4.1%) (Levy et al. 1988). Florida has reported *Toxoplasmosis* approximately three times more frequently (4.9%) than the remainder of the country. CNS lymphoma has been reported uniformly around the country (approximately 0.4%), with the exception of Texas where it has been reported in 0.8% of cases. A more uniform distribution of the prevalence of progressive multifocal leukoencephalopathy has been reported.

A similar variation in disease prevalence can be seen for different risk groups. Individuals born in Haiti are approximately 3.7 times more likely to be reported as having neurological complications as other risk groups (21.4% vs. 6.9%). Intravenous drug users also seem to be at increased risk of the neurological complications of HIV infection, with a 12.5% prevalence reported.

Focal Brain Disorders

Toxoplasmosis

The incidence of CNS toxoplasmosis increases with age and exhibits marked geographic variation. Immunoglobulin G (IgG) antibodies to *T. gondii* are detected in 40% to 50% of healthy young adults in the United States. This compares with approximately 20% of British adults and 90% of French adults (World Health Organization 1984). Within the United States, there is great geographical variation in the prevalence of toxoplasmosis; approximately three times more cases are reported in Florida than in the rest of the country (Levy et al. 1988). People with AIDS who were born in Haiti are at increased risk of developing cerebral toxoplasmosis compared with other risk groups, 12.4% vs. 1.7% ($p < 0.001$) in one study (Levy et al. 1988). Hispanics also appear to be at increased risk of developing toxoplasmosis when compared to non-Haitian, non-Hispanic patients (Levy et al. 1988). It has been suggested that the increased incidence of toxoplasmosis among these populations is in large part due to the higher prevalence of this disease in tropical areas (Feldman 1974). Several studies have estimated the risk of active toxoplasmosis in those who are HIV seropositive. The risk of developing toxoplasmosis has been estimated to be as high as 30% among those who are seropositive for *T. gondii* (Grant et al. 1986).

Serological techniques to diagnose toxoplasmosis are of great value when applied to immunocompetent individuals. However, when these techniques are applied to those who are HIV seropositive, their value diminishes markedly. With available techniques, the diagnosis of toxoplasmosis rarely can be confirmed or excluded by the results of a single examination. IgM antibody is found in less than 3% of AIDS patients with documented acute infection, while IgG antibody titers infrequently show the fourfold rise that is diagnostic in immunocompetent patients (Luft et al. 1984; Navia et al. 1986c; Wong et al. 1984). Several researchers have argued that a negative IgG titer excludes the diagnosis (Elder and Sever 1988). While this may be true in the majority of cases, physicians should remember that several exceptions to this rule have been reported (Leoung et al. 1985; Wanke et al. 1984).

Antifolates, such as pyrimethamine in combination with sulfadiazine, are the only therapeutic regimen of proven value in the treatment of toxoplasmosis. Although sulfadiazine is the most commonly used sulfonamide, the trisulfapyrimidine can be substituted. Pyrimethamine is given orally at a loading dose of 3–4 mg/kg (usually 200 mg in an averaged size adult) followed by 1 to 1.5 mg/kg/d (usually 75 to 100 mg) in a single daily dose; 50 to 100 mg/kg/day of sulfadiazine (usually 4 to 6 gm) are given in divided doses. Folinic acid (usually 10–20 mg/d) is provided to avoid pyrimethamine-induced bone marrow toxicity. The majority of patients will begin to show improvement in neurological symptoms within 7 to 10 days of initiating therapy. However, some patients may improve much more gradually, requiring 4 to 6 weeks. Seventy-five percent of patients with toxoplasmosis respond to this regimen (Leoung et al. 1985; Navia et al. 1986c; Pitchenik et al. 1983; Wanke et al. 1984). Computed tomography (CT) scan and magnetic resonance imaging (MRI) abnormalities are reversible in the majority of cases following prompt institution of treatment (Navia et al. 1986c). The likelihood of response appears not to correlate with the level of consciousness, number of lesions on CT scan, abnormalities in CSF, or serum titer, but it is significantly reduced with treatment delay (Navia et al. 1986). Response to treatment correlates histologically with abscess organization and eradication of free tachyzoites. Therapy needs to be continued indefinitely, however, because relapse occurs in almost all cases following discontinuation of treatment (Luft and Remington 1988; Cohn et al. 1989). The presence of encysted organisms in brains of patients who had been on chronic treatment underscores the importance of continued therapy (Navia et al. 1986). In a prospective study of 35 people with AIDS and toxoplasmic encephalitis, 31 individuals (89%) had a complete or partial response to 4 to 6 weeks of therapy (Leport et al. 1988). Twenty-four of these patients were followed for a mean of 8 months (range 2.5–28 months) while on maintenance pyrimethamine and sulfadiazine. Six (25%) of these individuals had 10 relapses. The majority of these relapses occurred within a short period (5–7 weeks) of discontinuation of primary therapy. Increasing the doses resulted in complete resolution of signs and symptoms in eight (80%) of the episodes of relapse and a partial response in one patient. Another patient died of the relapse before therapy could be reinitiated. The high relapse rate for this infection has lead to the development of chronic maintenance therapy. Pyrimethamine, 0.5–0.75 mg/kg/d (usually 25–50 mg/d); folinic acid, 5–20 mg/d; and sulfadiazine 25–50 mg/kg/d (usually 2 gm/d) is the most common regimen.

Cryptococcosis

The most common deep-seated fungal infection in patients with AIDS is that caused by *Cryptococcus neoformans,* a common soil fungus found throughout the world. It is believed to enter the body through the respiratory tract. Although meningitis and meningoencephalitis are the most frequent manifestations of this infection among persons with AIDS, it can affect other tissues as well, including skin, lungs, pericardium, retina, joints, and the peritoneum (Borton and Wintroub 1984; Gal et al. 1986; Kovacs et al. 1985; Ricciardi et al. 1986; Rico and

Penneys 1984; Schuman and Friedman 1983; Zuger et al. 1986).

Cryptococcosis occurs with increased frequency among individuals who are immunosuppressed, especially those with defects in cell-mediated immunity. This makes individuals with advanced HIV infection particularly susceptible to this organism. Data from several sources indicate that between 5% and 10% of people with AIDS will experience cryptococcosis (Centers for Disease Control 1986; Zuger et al. 1986). Most estimates of the prevalence of cryptococcal meningitis among patients with AIDS in the United States have placed the figure between 5% and 6% (Koppel et al. 1985; Levy, R. M. et al. 1985). Among those who are black or who have a history of intravenous drug use, the prevalence of cryptococcal meningitis may be as high as 10% (Selik et al. 1984).

The manifestations of cryptococcal disease are similar in both HIV-seropositive and -seronegative individuals. Typically, the early signs and symptoms of disease are deceptively mild, and their rate of progression is slow. Often individuals are ill for several weeks or months before the diagnosis is made. In one study of initial cryptococcal infection in people with AIDS, documentation of meningeal involvement could be made in 84% of cases (Chuck and Sande 1989). In 13% of these patients, documentation of meningitis by positive cerebrospinal fluid (CSF) culture could not be made. In the remainder of patients, cryptococcal meningitis could not be excluded.

Amphotericin B therapy dramatically reduced the mortality rate for those who suffer from cryptococcal infection. Prior to the introduction of amphotericin B for treatment of cryptococcal meningitis in 1957, this disease was almost uniformly fatal. Three out of four patients with cryptococcal meningitis died in the first year. With the use of amphotericin B, the mortality rate has decreased to between 25% and 30%. The mortality rate, however, varies greatly based on the state of the immune system. For those who are immunocompetent, the rate has been estimated to be 15% or less compared to 85% among those who are immunosup-pressed for reasons other than HIV infection (Kaplan et al. 1977; Sarosi et al. 1969; Spickard et al. 1963). What is surprising, given their immunocompromised state, is that people with AIDS have a relatively low rate of mortality during initial therapy. In one study, only three of 24 patients died of cryptococcal disease during initial therapy (Zuger et al. 1986).

There is no standard treatment of cryptococcal disease among persons with AIDS. Amphotericin B alone and in combination with flucytosine have been used to treat AIDS-associated cryptococcal disease (Zuger et al. 1986; Karsen et al. 1990). It remains controversial whether the addition of flucytosine improves outcome among individuals who experience AIDS-associated cryptococcal disease (Chuck and Sande 1989; Sugar et al. 1990). In approximately one third of patients, flucytoxine must be stopped because of leukopenia, thrombocytopenia, or liver function abnormalities (Kovacs et al. 1985). Fluconazole may be an effective alternative to amphotericin B for the treatment of cryptococcal meningitis among those patients who have no alteration of mental status (lethargy, somnolence or obtundation) prior to initiation of treatment. (see Table 17-2) (Saag et al. 1992).

The principle problem associated with cryptococcal disease in people with AIDS is the high rate of recurrence. Zuger and co-workers (1986) found that 50% of patients with meningitis who received no maintenance therapy after their initial course had a relapse within 6 months compared with no patients out of seven who received weekly amphotericin B therapy. This has led researchers to investigate other methods of maintenance therapy for cryptococcal meningitis. The drug that has received the most attention in this role has been fluconazole, a triazole antifungal agent. A study of 205 patients with AIDS-associated cryptococcal meningitis compared fluconazole, 200 mg taken orally as a single daily dose, with amphotericin B, 1 mg/kg, intravenously (IV) administered once weekly. It concluded that 200 mg of oral fluconazole is at least as effective at suppressing recurrences and may be superior to maintenance with weekly IV amphotericin B at a dose of 1 mg/kg (NIAID 1990). As a result of this study, oral fluconazole has become standard maintenance therapy for AIDS-associated cryptococcal meningitis.

Research has focused on assessing the likelihood of responding to acute therapy and of relapse. These prognostic factors do not seem to be the same for patients with AIDS as for non-AIDS patients. Zuger and associates (1986) found that high CSF cryptococcal antigen level (more than 1:10,000) was associated with a 100% mortality rate compared to a 22% mortality rate for those with CSF cryptococcal antigen levels below 1:10,000. Other studies have been unable to demonstrate this association (Chuck and Sande 1989; Gal et al. 1987). Zuger and colleagues also found that a positive india ink smear at diagnosis was associated with a mortality rate of 47%, but there were no recurrences among those with a negative smear ($0.05 < p < 0.10$). Relapse also appears to be a significant predictor of mortality, while CSF opening pressure, low CSF glucose, low CSF leukocyte count, and isolation of cryptococci outside the CNS do not. In a recent retrospective review of cryptococcal infections among people with AIDS, Chuck and Sande found that positive cryptococcal cultures of extrameningeal specimens (median survival of 147 days vs. 265 days) and hyponatremia (median survival of 113 days vs. 214 days) were statistically significant predictors of earlier mortality (Chuck and Sande 1989). Age, symptoms, titers of cryptococcal antigen, and CSF characteristics were not associated

with shorter survival. The factor that seems to be one of the best predictors of relapse is a CSF cryptococcal antigen titer of 1:8 or more at the end of acute therapy (Chuck and Sande 1989).

In a recently completed clinical trial comprising fluconazole and amphotericin B for treatment of acute cryptococcal meningitis, several markers correlate with early deterioration and death (Table 17-2). It has been suggested that these mrkers can be used to identify patients in whom acute treatment with fluconazole (200 mg/day) may be preferred over amphotericin B (with or without 5-flucytosine) therapy. The study was unable to demonstrate a statistically significant difference in outcome between low-risk patients treated with fluconazole and those treated with amphotericin B (0.5 mg/kg per day). However, because the number of patients involved in this study was small, it is difficult to determine clinically significant differences in outcome between these two groups.

CNS Lymphoma

The incidence of primary CNS lymphoma has increased significantly among people with AIDS. In a recent review of 1286 adults with AIDS, 1.9% developed primary CNS lymphomas (Levy et al. 1988). Primary CNS lymphoma is the second most frequent CNS mass lesion in adults with AIDS and the most frequent in pediatric AIDS (3%) (Levy et al. 1988). It has been predicted that during 1991, there will be 1800 cases reported in the United States

Table 17-2. Risk of Death from AIDS-associated Cryptococcal Meningitis in Adults*

	Risk Factor		Estimated Probability of Survival During Therapy
Mental Status	CSF Cryptococcal Antigen Titer > 1:1024	CSF White Blood Cell Count > 0.02×10^{-9} per liter	
Normal	No	No	0.94 ± 0.02
Normal	No	Yes	1.0 ± 0.00
Normal	Yes	Yes	1.0 ± 0.00
Abnormal	Yes	Yes	0.56 ± 0.08

* Adapted from Saag et al. (1992).

(Baumgartner et al. 1990). According to CDC surveillance data, there does not seem to be a regional or risk group variation in the prevalence of primary CNS lymphomas, except for a slight increase in reported cases from Texas (Levy et al. 1988). Pluda and coworkers (1988) have suggested that lymphomas among HIV-seropositive individuals may be increased as a consequence of antiretroviral therapy. Such a link, if confirmed, could imply a much larger increase in the incidence of non-Hodgkin's lymphomas among HIV-seropositive individuals in the future. Further, an increase in the incidence in non-Hodgkin's lymphoma may be a consequence of prolonged life expectancy among HIV-seropositive individuals.

The natural history of CNS lymphomas among people with AIDS differs markedly from the natural history among immunocompetent patients. In general, HIV-seropositive individuals present with more advanced disease (Ziegler et al. 1984). High-grade lymphomas are disproportionately prevalent among AIDS patients with non-Hodgkin's lymphomas (Levy et al. 1985). In general, primary CNS lymphoma has an extremely poor prognosis. Untreated, it is rapidly fatal, frequently within 1 month. When radiation therapy is used, the mean duration of survival is significantly increased (median 27 days vs. 119 days, $p < 0.5$) (Baumgartner et al. 1990). Necropsy results suggest that patients who do not receive radiation therapy die from tumor progression, whereas those given radiation therapy tend to die of opportunistic infections (Baumgartner et al. 1990). Adjuvant chemotherapy has been used in some patients with primary CNS lymphoma. However, one study using procarbazine, lomustine, and vincristine in patients with AIDS was unable to demonstrate a difference between those who received chemotherapy and radiation therapy and those who received radiation therapy alone (Levy et al. 1988).

Progressive Multifocal Leukoencephalopathy

Progressive multifocal leukoencephalopathy (PML) is a subacute, demyelinating disease of the CNS that is caused by JC virus. It has long been associated with defects in cell-mediated immunity, such as those caused by myeloproliferative disorders, chronic inflammatory diseases, and immunosuppressive therapy. The first report of PML in association with HIV infection was in 1982 (Miller et al. 1982). It remains a relatively rare complication among AIDS patients, with estimates ranging from 0.5% to 4% (Berger et al. 1987; Levy et al. 1988; Snider et al. 1983). However, in 25% of afflicted patients, it can be the presenting manifestation of AIDS. Prevalence does not seem to vary geographically or by risk group. Prognosis generally is poor with a mean survival of nearly 4 months in afflicted patients, although rare instances of spontaneous sustained remissions have been reported (Berger et al. 1987; Berger and Muck 1988). No single therapy has proven effective for this disorder. Although several cases have now been treated with cytosine arabinoside (ARA-C), results have varied (Krupp et al. 1985; Snider et al. 1983). Patients with mild or slowly progressive impairment associated with a single lesion may be more likely to respond (Carolyn Britton, personal communication).

Cytomegalovirus

Cytomegalovirus (CMV) is a common pathogen among people with AIDS. It infects a wide variety of organs, including eyes, lungs, esophagus, colon, adrenal glands, liver, and brain. Data suggest that 80% to 90% of patients develop active CMV infection during their illness. Viral inclusions consistent with CMV infection have been reported in approximately 25% of brains (Navia et al. 1986b). It has been estimated that clinical manifestations of CMV chorioretinitis occur in 5% to 10% of HIV-infected individuals. At necropsy, up to 30% of those who are HIV infected have evidence of CMV chorioretinitis.

Less is known about CMV infection of the brain and spinal cord. Neuropathological studies have found great variability in the pathological appearance of CMV in the nervous system. The most frequent finding is microglial nodule encephalitis but leptomeningitis, ependymitis, focal necrotizing encephalitis, demyelination, and vaculitis also occur (Vinters et al. 1989). Among HIV-1-seropositive patients, HIV-1 and CMV are simultaneously found in brain tissue with varying frequency (Navia et al. 1986b; Wiley et al. 1988). This finding has led to the hypothesis that dual infection may contribute to progressive ADC (Wiley et al. 1988; Vinters et al. 1989). However, the presence of both viruses is not a necessary condition for the development of severe AIDS dementia complex (ADC) (Navia et al. 1986). In general, little correlation exists between CNS neuropathology and neurological impairment (Navia et al. 1986b; Vinters et al. 1989). Although the pathogenetic role of CMV in the CNS remains unclear, it is increasingly recognized as an important agent in the peripheral nervous system (see below).

CMV chorioretinitis is the most common ocular infection in patients with AIDS. Evidence supports the notion that this disorder is occurring with increasing frequency among HIV-seropositive individuals. According to the CDC surveillance data, the incidence with which CMV chorioretinitis has been reported as an initial AIDS-defining illness has increased from 0.2% in 1984 to 2.4% in 1989. The cause of this apparent increase is not well understood but several investigators have postulated that increased longevity among persons with AIDS may play a significant role.

In spite of the large increase in the incidence of CMV chorioretinitis in the past decade, many questions remain regarding natural history of this disease. If left untreated, clinically evident progression occurs in more than 90% of patients (Drew et al. 1990). Retinitis usually begins at the periphery of the fundus and progresses centrally to involve the macula and the optic disc with bilateral involvement noted in 50% of cases (Pepose et al. 1985). Visual field defects associated with CMV chorioretinitis do not reverse significantly with therapy; however, the loss of visual acuity that is believed to result from edema of the macula may improve with treatment.

The only drug approved by the Food and Drug Administration for the treatment of CMV chorioretinitis is ganciclovir (9-(1,3-dihydroxy-2-propoxymethyl guanine). Treatment usually lasts for 14 days and results in an initial response rate of approximately 80% (Pepose et al. 1985). Maintenance therapy however, needs to be continued indefinitely because the virus reactivates in a significant percentage of patients. Even with the use of high-dose (25–35 mg/kg per week) maintenance therapy, approximately 40% of patients will relapse within 120 days (Pepose et al. 1985). Dose escalation is thus frequently required to prevent reactivation but frequently is limited by drug toxicity, especially neutropenia. Research is ongoing to assess the value of intravitreal ganciclovir injection to circumvent this particular complication. In general, however, maintenance therapy using ganciclovir alone eventually fails, resulting in further progression of disease. Foscarnet, a selective inhibitor of CMV DNA polymerase, is being evaluated for treatment and maintenance therapy. Although preliminary data have shown efficacy, use of this drug, as with ganciclovir, may be limited by toxicity.

AIDS Dementia Complex

Background

Since its initial description as a distinct syndrome related to HIV-1 infection (Navia et al. 1986a), the AIDS dementia complex (ADC) remains an important and common problem in the HIV-1 infected population (Navia et al. 1986a; Price et al. 1988a). Although the epidemiology of this disorder has yet to be fully defined, it has been estimated to occur in 40% to 60% of patients with AIDS (Navia et al. 1986; Price et al. 1988b) and in 10% to 30% of patients with the AIDS-related complex (ARC) (Janssen et al. 1989; Miller et al. 1990; Tross et al. 1988). Its occurrence among asymptomatic seropositive individuals has remained a controversial issue, but it appears that a significant smaller percentage of these patients may also be affected (Goethe et al. 1989; Tross et al. 1988; Janssen et al. 1989; Miller et al. 1990; Selnes et al. 1990). In the majority of patients, neurological impairment consistent with ADC occurs months after the diagnosis of AIDS, but in nearly 5% to 10%, it can be the presenting or the predominant clinical manifestation of their HIV-1 infection (Navia and Price 1987). The clinical presentation is characterized by a stereotypical constellation of symptoms and signs that reflect cognitive, motor, or behavioral impairment (Navia et al. 1986a; Price et al. 1988a,b) (Table 17-3). Neuropsychological abnormalities most often involve areas of sustained attention, verbal fluency, mental flexibility, and psychomotor speed (Miller et al. 1990; Tross et al. 1988). The clinical course usually is insidious and slowly progressive over months, but it can be static or less often fulminant, leading to severe dementia within weeks. Patients with severe ADC show diffuse cognitive and motor impairment with little capacity for social or intellectual function. Progression of disease often is accompanied by a global reduction in cerebral glucose metabolism (Rottenberg et al. 1987). Mean survival, previously reported to be 4 months, can now exceed 1 year in patients receiving AZT (Navia et al. 1986a). A similar encephalopathy occurs in 30% to 50% of children with HIV-1 infection and is characterized by loss of developmental milestones, intellectual impairment, and motor deficits. Encephalopathy can be static or progressive. As with the adult form of this disease, onset of progression is a poor prognostic

Table 17-3. AIDS Dementia Complex: Clinical Features

Symptoms	
Cognitive	Slowness, impaired attention and concentration, forgetfulness, confusion
Motor	Clumsiness, deterioration in fine motor tasks, tremor, loss of balance
Behavioral	Reduced spontaneity, apathy, social withdrawal
Signs	
Mental status	Psychomotor slowing, impaired word reversal, and serial subtraction, blunted affect, organic psychosis
Motor	Impaired rapid movements, sustention tremor, hyperreflexia, limb paresis, impaired tandem gait

sign, and if left untreated, it is almost invariably fatal (Belman et al. 1986; Epstein et al. 1985b; Epstein et al. 1987).

Brain pathology is confined to the deep white and subcortical grey matter regions (basal ganglia and thalamus) with relative sparing of the cortex (Navia et al. 1986b). Pathological abnormalities most commonly include pallor and vacuolation of the white matter, gliosis, and collections of perivascular or parenchymal inflammatory cells composed of macrophages, lymphocytes, and multinucleated cells (Navia et al. 1986b; de la Monte et al. 1987; Budka et al. 1987; Budka 1991). Early brain changes noted in either asymptomatic or mildly demented patients consist of mild myelin pallor and gliosis, associated with perivascular inflammation (Navia et al. 1986b; McArthur et al. 1988). Multinucleated cells represent the histological hallmark of HIV-1 infection of the brain and probably results from viral-induced fusion of brain macrophages or microglia (Navia et al. 1986; Sharer et al. 1986; Watkins et al. 1990). Further, it usually is associated with severe brain pathology, progressive disease, and a poor outcome (Navia et al. 1986b).

Early studies of this disorder noted that cortical abnormalities were relatively sparse, consisting primarily of gliosis and mild neuronal loss (Navia et al 1986b, de la Monte et al. 1987). Notable exceptions were found in brains of patients with end-stage dementia. More recent studies, however, using quantitative morphometry have demonstrated significant neuronal loss associated with HIV-1 infection of the brain when compared to control brains (Kezler et al. 1990; Everall et al. 1991; Wiley et al. 1991). The importance of this finding although intriguing remains unclear. It nonetheless has lent further support to the hypothesis that neurotoxic mechanisms may be involved in this disorder (Lipton, 1991).

Accumulating evidence indicates that HIV-1 is the most likely cause of this disorder. Since its initial detection in brain by Southern analysis (Shaw et al. 1985), HIV-1 has been identified by a variety of other methods, including electron microscopy

Table 17-4. Clinical, Pathological, and Virological Correlations of the AIDS Dementia Complex

Approximate Percentage of AIDS Patients	Clinical Severity of AIDS Dementia Complex	Neuro-pathological Severity	Multi-nucleated Cells	Neuroimaging (CT and MRI)	HIV Detection (Antigen and Southern Blot)
25	Severe	Severe	+	Atrophy/increased white matter signal	+
50	Mild to moderate	Mild to moderate	–	Atrophy	–
25	None/subclinical	Mild	–	Normal/atrophy	–

(Epstein et al. 1985), viral culture (Ho et al. 1985; Levy et al. 1986), immunohisto-chemistry (Gabuzda et al. 1986; Pumorola-Sune et al. 1987; Wiley et al. 1986), and *in situ* hybridization (Koenig et al. 1986). These studies also established that the most commonly infected cell in the brain is the macrophage. Although virus has been localized to other cells, including endothelial cells, glia, and rarely neurons, it is now widely believed that the macrophage or cells of similar lineage play a key role in the pathogenesis of ADC (Giulian et al. 1990; Price et al. 1988a,b; Puliam et al. 1991). Further, the capacity of this virus to infect macrophages, which varies among different strains, may correlate with its capacity to cause brain injury.

Clinical–pathological studies revealed that nearly one third of AIDS patients developed severe ADC, often associated with white-matter abnormalities on MRI and a multinucleated cell encephalitis with evidence of HIV-1 in brain (Table 17-4). Thus, in the majority of demented patients, brain pathology appeared disproportionately mild to the degree of neurological impairment, and virus was rarely identified. In the remaining patients, despite the absence of overt neurological impairment, brain abnormalities consistent with a possibly mild form of HIV-1 infection were nonetheless found, suggesting that HIV-1 enters the CNS of almost all AIDS patients but in a few does not cause significant brain injury. Additional molecular and virological studies also showed that in a few patients with severe ADC, the brain appeared to be the major target of infection when compared to peripheral tissues (Shaw et al. 1985). In addition,

genetically distinct but related HIV-1 strains were identified in the brains of these patients, suggesting that differences in clinical and pathological expression of ADC (i.e., differences in natural history and outcome) may be partly related to the existence of such variants (Shaw et al. 1985). The observed discrepancies between clinical disease, pathology, and infection also have led to the hypothesis, now receiving wide support, that neurological dysfunction associated with HIV-1 infection may result from physiological alterations of brain cells rather than from direct infection (Navia et al. 1986b; Price et al. 1988b).

Despite recent clinical and biological advances, the factors that affect the natural history and prognosis of this disorder remain to be fully defined. Nonetheless, emerging evidence suggests that several host- and viral-related factors may be important in shaping the clinical course and outcome (Table 17-5). Each of these are discussed separately.

Table 17-5. Factors Associated With Progressive AIDS Dementia Complex

Clinical
  AIDS (systemic CMV, MAI)
  Lack of zidovudine therapy, immunosuppression
Biological
  Host immune responses (cytotoxic T lymphocytes, IgG antibodies)
  Neurovirulent strains (macrophage-tropic)
Laboratory Markers
  CSF
    HIV-1 antigen
    Beta-2 microglobulin
    Quinolinate
    Tumor necrosis factor
    Cytotoxic T lymphocytes
  Neuroimaging
    White matter abnormalities

Clinical Studies

Several studies have shown that the risk of developing ADC increases with progression of systemic disease toward AIDS. Neuropsychological impairment was found in nearly 70% of patients with established AIDS compared to 30% of patients at an earlier stage in their disease (early AIDS) (Tross et al. 1988). Patients with systemic CMV or mycobacterium avium intracellulare (MAI) appear to be at greater risk for developing progressive ADC compared to patients without these complications (Navia et al. 1986a). Among asymptomatic seropositive subjects, rates of 12% to 44% were initially reported (Grant et al. 1987; Tross et al. 1988). More recent studies, however, including the multicenter AIDS cohort study (MACS), detected significant neuropsychological impairment only in patients with AIDS and ARC and not in asymptomatic patients when compared to seronegative controls (Janssen et al. 1989; Miller et al. 1990; Selnes et al. 1990). In addition, neuropsychological impairment did not correlate with immunological suppression or with the duration of HIV-1 infection.

Although ADC is a relatively late complication of HIV-1 infection, it appears that HIV-1 enters the CNS during the early stages of infection. This is primarily based on studies of asymptomatic seropositive individuals, 30% to 40% of whom have been reported to show CSF abnormalities, including pleocytosis and intrathecal HIV-1 antibodies (Goudsmit et al. 1987; McArthur et al. 1988), cerebral perfusion defects detected by single photon emission tomography (SPECT) (Schielke et al. 1990), and various electrophysiological abnormalities (Koralnik et al. 1990; Ollo et al. 1991). Further, it appears that several otherwise asymptomatic patients have developed mild to moderate ADC, associated with SPECT cerebral perfusion defects (Navia and Worth, unpublished).

In summary, the risk for cognitive impairment increases with progression of disease; outcome is generally worse in patients with AIDS, but abnormalities may occur in 10% to 30% of patients with ARC and in a smaller percentage of asymptomatic patients. Further, in some, either mild or progressive ADC may occur as the primary or predominant clinical manifestation of HIV-1 infection. The fact that the incidence of ADC appears to be highest in AIDS patients when compared to other HIV-infected groups suggests that additional factors, such as immunosuppression, coinfection with other viruses, and advancing sytemic disease, may contribute further to risk and poor outcome. However, there are patients who develop progressively severe dementia with no systemic signs of HIV-1 infection other than immunosuppression. Further, nearly 25% of AIDS patients will develop minimal or mild neurological impairment (Table 17-6). These differences in clinical course and outcome may be attributed to several factors, including infection by neurovirulent strains of HIV-1 with biological properties distinct from isolates that do not cause CNS infection, differences in host immunological responses to infection, or the production of toxins believed to contribute to brain injury caused by HIV-1.

Table 17-6. Peripheral Neuropathies

Type	Course	Etiology	Treatment
Distal sensory	Mildly progressive	HIV-1	Zidovudine
Mononeuritis Multiplex	Progressive	HIV-1? Vasculitis	None
ADIP CDIP	Progressive	HIV-1? Autoimmune	Plasmapheresis Steroids
Multifocal Motor/sensory	Progressive	CMV	Ganciclovir
Polyradiculopathy	Progressive	CMV	Ganciclovir

Laboratory Studies

Various nonspecific CSF abnormalities have been reported in patients with ADC, including pleocytosis, elevated protein, and less often oligoclonal banding (Navia et al. 1986a; Elovaara et al. 1988). A progressive rise in protein is seen in some patients with advancing disease, but the significance of this finding remains unclear (Navia et al. 1986a).

Several CSF markers that may predict risk or progression recently have been described. Two studies have shown that the detection of intrathecal HIV antigen and not antibodies correlates with the development of progressive neurological impairment. Intrathecal synthesis of HIV-1 specific antibodies and antigen were assayed in 27 children with HIV-1 infection (11 with progressive encephalopathy and nine with static encephalopathy) (Epstein et al. 1987). CSF antigen was detected in eight patients with PE but in none of the children with SE. In another study, antigen was found in six of seven patients with PE but in only one of 18 without neurological impairment (Goudsmit et al. 1987). On the other hand, CSF antibodies were found in the majority of infected patients, including asymptomatic subjects (Goudsmit et al. 1987). The presence of HIV antigen in the CSF, reflecting viral expression in the CNS, may therefore serve as a sensitive and specific marker for ADC. However, because it's rarely detectable in patients receiving AZT, its clinical usefulness may be limited (Portegies et al. 1989). It should also be realized that in some patients HIV antigen may be bound to immune complexes and would not be detected by conventional assays (Epstein et al. 1987).

Beta-2 microglobulin is a low molecular-weight molecule present on activated lymphocytes. Increased serum levels have been associated with an increased risk for AIDS in seropositive patients (Bhalla et al. 1985) and recently have been correlated with the severity of ADC (Brew et al. 1989). Tumor necrosis factor (TNF), a monokine produced by activated macrophage and lymphocytes, recently has been shown to act as a myelin toxin in culture (Selmaj and Raine 1988). Significantly increased CSF levels of TNF have been reported in HIV-infected patients with active CNS infection, including HIV-1 (Grimaldi et al. 1991). In addition, increased serum, but not CSF, levels have been correlated with the development of progressive encephalopathy in children (Mintz et al. 1989). Quinolinate, a tryptophan metabolite and a potent neurotoxin in the mammalian brain, also is increased with AIDS and particularly in patients with severe ADC (Heyes et al. 1991). Additional longitudinal and quantitative studies correlating levels of these different molecules with severity of disease and response to antiviral therapy are needed to confirm their usefulness as markers of CNS infection or progression.

Neuroimaging

Cerebral atrophy is the most common neuroradiological finding associated with ADC (Navia et al. 1986a; Post et al. 1988); its relationship to dementia, however, remains unclear except that it is seen in most demented patients and may antedate the onset of cognitive impairment (Navia et al. 1986a). Various white-matter abnormalities also have been described in patients with ADC (Grant et al. 1987; Navia et al. 1986a; Post et al. 1988). The presence of diffuse white-matter attenuation on CT scan or a patchy increased signal on MRI may correlate with risk for progression and the presence of HIV-1 in brain. Before AZT, this finding was almost invariably seen in patients with severe dementia and a multinucleated cell or HIV-1 encephalitis (Navia and Price unpublished). High-signal lesions in the splenium of the corpus callosum and the crura of the fornices have also been reported in HIV-1 infected patients with early cognitive impairment (Kieburtz et al. 1990). The prognostic significance of this finding remains to be determined.

SPECT provides a sensitive method of detecting brain abnormalities associated with HIV-1 infection, particularly during its early stages. Perfusion defects are seen in almost all demented patients (Johnson et al. 1991; Pohl et al. 1988) and in many asymptomatic seropositive patients (Schielke et al. 1990). Defects show a frontal and temporal lobe predominance, but the pathological patterns are not specific for this disease (Johnson et al. 1991; Schielke et al. 1990). Similar defects have been noted in patients with chronic cocaine abuse (Holman et al. 1992). It remains to be determined whether asymptomatic patients with SPECT abnormalities on SPECT are at greater risk for developing ADC.

Positron emission tomography (PET) using fluorodeoxyglucose has identified altered metabolic patterns associated with different stages of ADC. Early impairment is characterized by relative subcortical (thalamus and basal ganglia) hypermetabolism, while disease progression is associated with diffuse cortical and subcortical hypometabolism (Rottenberg et al. 1987). Using a novel mathematical approach, Rottenberg and coworkers (1987) further determined that disease progression in the subcortical regions was the best predictor of motor performance (grooved pegboard), whereas cortical disease correlated best with tests of frontal lobe function (verbal fluency and trail making). HIV-1 has been identified in the brains of patients with primary severe ADC and a global reduction in cerebral glucose metabolism, suggesting that HIV-1 is the likely cause of these changes (Navia and Price unpublished observations).

Electrophysiology

Various nonspecific electrophysiological abnormalities have been described in AIDS patients with and without ADC and in asymptomatic seropositive individuals. Recent findings indicate subclinical brain involvement during the early stages of infection. Both EEG (theta slowing, abnormal reactivity) and otoneurological abnormalities (e.g., prolonged brain stem auditory evoked responses) have been reported in as many as 40% of asymptomatic seropositive patients (Koralnik et al. 1990). Abnormal event-related potentials (ERP) also have

been reported in different HIV-1 infected groups. ERP provides a sensitive measurement of various aspects of information processing in response to auditory or visual stimuli. Specifically, the P300 ERP component was significantly altered in both AIDS–ARC and asymptomatic patients, suggesting that subclinical cognitive impairment may occur early in the course of HIV-1 infection (Ollo et al. 1991).

Biological Studies

Patients with mild ADC or static neurological impairment frequently showed a vigorous humoral response in the CSF characterized by an increased total intrathecal IgG synthesis and the presence of oligoclonal bands in contrast to patients with severe ADC who tend to show a declining B cell response (Elovaara et al. 1988). These results suggest that differences in clinical course may be partly related to host humoral defenses operable within the CNS. Host Cytotoxic T lymphocytes (CTL) represent an important host defense against viral infection, presumably by responding to cell surface antigens in an HLA-restricted manner (Walker et al. 1987). CTL targeted against specific regions of the HIV-1 gene have been identified in the serum of most seropositive patients but infrequently in patients with AIDS, suggesting that they may serve a protective role during the early stages of infection. It also has been postulated that CTL may contribute to the pathogenesis of certain neurological disorders. CTL recently has been identified in the serum of HTLV-I infected patients with, but not in those without, tropical spastic paraparesis (Jacobsen et al. 1990) and in the CSF of patients with HIV-1 related encephalopathy (Sethi et al. 1988). More recently, a vigorous and broadly directed HIV-1 specific CTL response has been found in the CSF of most patients with ADC (Jassoy et al. 1991). In addition, CTL have been isolated with significantly greater frequency in the CSF than in the peripheral blood, excluding the possibility that the CSF-derived clones resulted from simple trafficking across the blood-brain barrier. CTL may therefore contribute to the pathogenesis of ADC either through direct immunological mechanisms or the production of soluble factors such as cytokines.

Virus

HIV-1 is remarkably polymorphic, particularly in the envelope region. Consequently, distinct but related strains of HIV-1 are frequently isolated from infected patients as well as from different tissues of the same patient. Such variants differ in their capacity to replicate in culture and cause cytopathic change (Cheng-Mayer et al. 1988). Clinical progression to AIDS has been correlated with the emergence of variants that exhibit greater in vitro cytopathogenicity compared to the original isolate (Cheng-Mayer et al. 1988).

It has been postulated that neurotropic or neurovirulent variants may be implicated in ADC (Price et al. 1988b). Noncytocidal variants of HIV-1 have been isolated in some patients with ADC and may partly explain differences in their clinical course (Anand et al. 1987). In addition genetically distinct variants have been identified in brains of demented patients (Shaw et al. 1985) and have been isolated from CSF, brain, and peripheral blood of the same patient (Koyanagi et al. 1987). Such isolates appear to infect macrophages more efficiently than peripheral blood strains. Recent observations have shown that neurovirulence may be related to infection of these cells: Macrophage-, but not lymphocyte-, derived isolates productively infect glial cell cultures (Watkins et al. 1990); produce a toxin that reduces neuronal survival in culture (Giulian et al. 1990) and cause histological abnormalities in human fetal brain cells (Puliam et al. 1991). The genetic basis for this tropism has been recently identified and mapped to a domain within gp120, outside the CD4-binding region (O'Brien et al. 1990; Shioda et al. 1991). It is therefore possible that ADC, particularly its progressive form, may be causally linked to infection by macrophage tropic variants of HIV-1. Additional isolates from patients with and without ADC, however, need to be characterized to show that risk for ADC is greatest in patients infected with such strains.

Treatment

It is now recognized that AZT decreases the morbidity and mortality associated with HIV-1 infection (Fischl et al. 1987; Volberding et al. 1990). Further, it has been shown that AZT can ameliorate or stabilize neuropsychological impairment in adults and children with HIV-1 encephalopathy (Pizzo et al. 1988; Schmitt et al. 1988; Yarchoan et al. 1987). In a randomized, double-blind, placebo-controlled study of 281 patients, AZT significantly improved neuropsychological performance in AZT-treated patients compared to placebo (Schmitt et al. 1988). Further, it appears that the incidence of progressively severe ADC has declined significantly following the use of AZT (Portegies et al. 1989). The majority of severe ADC cases are now seen in patients not receiving treatment, most of whom appear to be AZT intolerant or resistant (Navia and Boswell unpublished observations). Abrupt neurological deterioration also can occur in some patients with rapidly advancing ADC following discontinuation of AZT. Hence, it appears that AZT may not only improve neurological function in patients with ADC, but may also be neuroprotective. The effective treatment dose has not been established. Although AZT at 500 mg/day is effective for AIDS (Volberding et al. 1990), treatment of ADC may require higher doses (more than 1000 mg/day). AZT, by intravenous (IV) or oral administration, has been shown to penetrate the blood–brain barrier to provide an adequate inhibitory concentration of more than 1 μmol/L (Klecker et al. 1987). However, a high degree of intersubject variability has been noted, and peak concentrations are lower than those achieved in plasma. Increasing the dose in patients with progressive neurological impairment may result in clinical improvement (Navia, Worth, and Boswell unpublished observations). Additional studies correlating plasma and CSF pharmacokinetics with treatment response in patients with ADC are needed. Other possible therapies include dideoxyinosine (ddI), an inhibitor of HIV-1 replication, and nimodipine, a calcium channel antagonist that has been shown to prevent the neurotoxic effects of gp120 in culture (Dreyer et al. 1990). The clinical efficacy of either treatment in patients with ADC remains to be shown.

Spinal Cord

The most common spinal cord disorder associated with HIV-1 infection is a vacuolar myelopathy, with a reported autopsy incidence of 25% (Petito et al. 1985). It pathologically resembles subacute combined degeneration and is characterized by vacuolation of the posterior and anterolateral columns, presumably related to intramyelin swelling associated with lipid-laden macrophages. Clinical presentation consists of spastic paresis and ataxia, initially involving the lower extremities; in severe cases, quadriparesis can develop. Moderate to severe disease occurred in nearly 60% of patients in the original series and generally correlated with histological severity (Petito et al. 1985). Progressive weakness is a poor prognostic sign. Etiology remains uncertain, but it is likely related to HIV-1 infection (Eilbott et al. 1989).

Myelopathies associated with herpes simplex type 2, herpes zoster, and CMV infection have been reported but are rare (Britton et al. 1985; Vinters et al. 1989; Wiley et al. 1987). The clinical course may be fulminant or slowly progressive, but progression generally is inexorable and prognosis poor. Antiviral therapy does not improve outcome.

Peripheral Neuropathies

Peripheral nerve manifestations are fairly common in HIV-1 infected patients, accounting for 5% to 20% of all neurological complications (Dalakas and Pezeshkpour 1988; Lange et al. 1988; Parry 1988; Snider et al. 1983). Several forms have been reported to occur during different stages of HIV-1 infection; outcome depends on the type of neuropathy, its underlying etiology, and treatment (Table 17-6). HIV-1 is believed to be the cause for most of these disorders through direct or autoimmune mechanisms (de la Monte et al. 1988). Several recent studies, however, indicate that CMV may be the pathogenetic agent in some cases.

Based on clinical and electrophysiological criteria, the following peripheral nerve disorders have been described: Distal sensory neuropathy is the most common, seen typically in patients with AIDS, the majority of whom show concomitant signs of ADC (Bailey et al. 1988). Clinical findings include painful paresthesias associated with sensory loss and depressed ankle reflexes; motor deficits occur infrequently. Electrophysiological findings are consistent with a predominant axonal neuropathy and include slowed conduction velocities and reduced amplitudes of distal and proximal evoked responses. The clinical course is either static or mildly progressive. Symptoms often are debilitating but may respond to amitriptyline or AZT.

A rapidly progressive, multifocal, necrotizing neuropathy recently has been described with evidence that CMV is the likely cause (Said et al. 1991). It usually is seen in patients with advanced AIDS, but in some, it can antecede systemic complications including CMV. It is predominantly a motor or mixed neuropathy distinguishable from the more common distal sensory form by its fulminant course and association with systemic CMV or CMV retinitis. Prognosis generally is poor, but based on recent reports, it may be improved with ganciclovir (Said et al. 1991).

Acute or chronic demyelinating inflammatory neuropathy is clinically and histologically indistinguishable from Guillain-Barré syndrome (Cornblath et al. 1987). It has been described in patients who have recently seroconverted or have been otherwise symptomatic. The clinical picture consists of diffuse progressive weakness with mild sensory symptoms of acute or subacute onset, progressing in some cases to flaccid paralysis and areflexia. CSF shows a mononuclear pleocytosis and an elevated protein. Tempo and degree of clinical improvement appear to be slower in patients with acute onset and rapid progression. The majority of patients improve spontaneously over several months or in response to steroids, plasmapheresis, or both. It appears, however, that HIV-1 infected patients do not recover as well as uninfected patients with this disorder (Cornblath 1988).

Mononeuritis multiplex is characterized by multiple sensory or motor deficits that follow the distribution of multiple nerves (spinal, cranial, or peripheral), and it pathologically resembles a necrotizing vasculitis (Dalakas and Pezeshkpour 1988). It has been seen in otherwise asymptomtic patients, ARC, and AIDS. Clinical symptomology may remit spontaneously or progress to a diffuse sensorimotor polyneuropathy. Prognosis is considered generally poor.

Polyradiculoneuropathy represents a progressive inflammatory disorder of ventral and dorsal roots caused by direct CMV infection (Behar et al. 1987; Eidelberg et al. 1986; Miller et al. 1990). It is most often seen in patients with advanced AIDS and frequently in association with other manifestations of CMV infection, particularly CMV retinitis. Typical clinical findings include a progressive flaccid paraparesis, urinary retention, and an ascending sensory level. Progression generally is relentless, and prognosis poor. However, early diagnosis with prompt institution of ganciclovir may result in clinical improvement (Miller, R. G. et al. 1990).

Myopathy

Myopathy is comparatively an infrequent and poorly understood complication that can occur in any of the HIV-1 infected groups, including otherwise asymptomatic seropositive patients. It can antecede the development of ARC or AIDS by several months (Dalakas and Pezeshkpour 1988; Lange et al. 1988; Simpson and Bender 1988; Snider et al. 1983). Several forms have been described, including a disorder identical to polymyositis, noninflammatory myopathy, and nemaline rod myopathy. Clinical findings consist of progressive proximal weakness, myalgias, and an elevated creatine phosphokinase. Muscle biopsy typically shows fiber necrosis and regenerating fibers associated in some, but not all, cases with inflammatory infiltrates. Nemaline rods, indicative of Z-band involvement, also have been described, and suggests that HIV-1 can directly or through immunopathogenetic mechanisms alter the integrity of specific muscle proteins. Most patients improve spontaneously or respond to immunosuppressive therapy (e.g., prednisone, azathioprine).

A myopathy associated with AZT has been recently described. Although clinically indistinguishable from the other myopathies, it is characterized by a distinct histological appearance, recently reported by Dalakas and coworkers. Twenty patients with myopathy were studied, 15 of whom had been treated with AZT. Muscle biopsy showed "ragged-red" fibers indicative of mitochondrial dysfunction only in the AZT-treated group (Dalakas et al. 1990). In addition, a T-cell mediated inflammatory myopathy was noted in all patients. Most patients improved in several days following discontinuation of treatment, while the remaining patients responded either to prednisone or nonsteroidal anti-inflammatory drugs. These findings indicate that AZT can cause a reversible toxic mitochondrial myopathy. It is conceivable that patients with an underlying HIV-1 associated myopathy are at greater risk for developing this complication. Because it appears that immune-mediated and toxic mechanisms may be involved, treatment response may vary depending on which mechanism is responsible for the clinical state (Dalakas et al. 1990). Given the potential adverse effects of immunosuppressive therapy, it has been suggested that initial treatment consist of use of nonsteroidal anti-inflammatory drugs and, if appropriate, discontinuation of AZT.

References

Anand, R.; Siegal, F.; Reed, C.; et al. Noncytocidal natural variants of human immunodeficiency virus isolated from AIDS patients with neurological disorders. Lancet 2:234–238; 1987.

Bailey, R. O.; Baltch, A. L.; Venkatesh, R.; et

al. Sensory motor neuropathy associated with AIDS. Neurology 38:886–891; 1988.

Baumgartner, J. E.; Rachlin, J. R.; Beckstead, J. H.; Meeker, T. C.; Levy, R. M.; Wara, W. M.; Rosenblum, M. L. Primary central nervous system lymphomas: natural history and response to radiation therapy in 55 patients with acquired immunodeficiency syndrome. J. Neurosurg. 73:206–211; 1990.

Behar, R.; Wiley, C.; McCutchan, J. A. Cytomegalovirus polyradiculoneuropathy in acquired immune deficiency syndrome. Neurology 37:557–561; 1987.

Belman, A. L.; Ultman, M. H.; Horoupian, D.; et al. Neurological complications in infants and children with acquired immunodeficiency syndrome. Ann. Neurol. 18:560–566; 1985.

Berger, J. R.; Kaszovitz, B.; Post, M. J. D.; Dickinson, G. Progressive multifocal leukoencephalopathy associated with human immunodeficiency virus infection. Ann. Intern. Med. 107:78–87; 1987.

Berger, J. R.; Muck, E. L. Prolonged survival and partial recovery in AIDS-associated PML. Neurology 38:1060–1065; 1988.

Bhalla, R. B.; Safai, B.; Pahwa, S.; Schwartz, M. K. Beta 2 microglobulin as a prognostic marker for development of AIDS. Clin. Chem. 31:1411–1412; 1985.

Borton, L. K.; Wintroub, B. U. Disseminated cryptococcosis presenting as herpetiform lesions in a homosexual man with acquired immunodeficiency syndrome. J. Am. Acad. Dermatol. 110:387–390; 1984.

Brenneman, D. E.; Westbrook, G. L.; Fitzgerald, S. P.; et al. Neuronal cell killing by the envelope protein of HT and its prevention by vasoactive intestinal peptide. Nature 335:639–642; 1988.

Brew, B. J.; Bhalla, R. B.; Fleisher, M.; et al. Cerebrospinal fluid B2 microglobulin in patients infected with human immunodeficiency virus. Neurology 39:830–834; 1989.

Britton, C. B.; Mesa-Tejada, R.; Fengolio, C.; et al. A new complication of AIDS: thoracic myelitis caused by herpes simplex virus. Neurology 35:1071–1074; 1985.

Brosnan, C. F.; Selmaj, K.; Raine, C. S. Hypothesis: a role for tumor necrosis factor in immune-mediated demyelination and its relevance to multiple sclerosis. J. Neuroimmun. 18:87–94; 1988.

Budka, H.; Costanzi, G.; Cristina, S.; et al. Brain pathology induced by infection with the human immunodeficiency virus (HIV). A histological, immunocytochemical, and electron microscopical study of 100 autopsy cases. Acta. Neuropathol. (Berl.) 75:185–198; 1987.

Budka, H. Neuropathology of human immunodeficiency virus infection. Brain Pathology 1:163–175; 1991.

Centers for Disease Control. Update: acquired immunodeficiency syndrome—United States. Morbid. Mortal. Week. Rep. 35:542, 757–760, 765–6; 1986.

Cheng-Mayer, C.; Seto, D.; Tateno, M.; et al. Biological features of HIV-1 that correlate with virulence in the host. Science 240:80–82; 1988.

Chuck, S. L.; Sande, M. A. Infections with Cryptococcus neoformans in the acquired immunodeficiency syndrome. N. Engl. J. Med. 321:794–799; 1989.

Cohn, J. A.; McMeeking, A.; Cohen, W.; et al. Evaluation of the policy of empiric treatment of suspected toxoplasma encephalitis in patients with the acquired immunodeficiency syndrome. Am. J. Med. 186:521–527; 1989.

Cornblath, D. R.; McArthur, J. C.; Kennedy, P. G. E.; et al. Inflammatory demyelinating peripheral neuropathie associated with human T-cell lymphotropic virus type III infection. Ann. Neurol. 21:32–40; 1987.

Cornblath, D. R. Treatment of the neuromuscular complications of human immunodeficiency virus infection. Ann. Neurol. 23(suppl):S88–S91; 1988.

Dalakas, M. C.; Pezeshkpour, G. H. Neuromuscular diseases associated with human immunodeficiency virus infection. Ann. Neurol. 23(suppl):S38–S46; 1988.

Dalakas, M. C.; Illa, I; Pezeshkpour, G. H.; et al. Mitochondrial myopathy caused by long term zidovudine therapy. N. Engl. J. Med. 322:1098–1105; 1990.

de la Monte, S. M.; Gabuzda, D. H.; Ho, D. D.; et al. Peripheral neuropathy in the acquired immunodeficiency syndrome. Ann. Neurol. 23:485–492; 1988.

Drew, W. L.; Buhles, W.; Dworkin, R. J.; Erlich, K. S. Management of herpes virus infections (CMV, HSV, VZV). In: Sande, M. A.; Volberding, P. A., eds. The medical management of AIDS. Philadelphia: W. B. Saunders Company; 1990: p. 318.

Dreyer, E. B.; Kaiser, P. K.; Offermann, I. T.; et al. HIV-1 coat protein neurotoxicity prevented by calcium channel antagonists. Science 248:364–367; 1990.

Eidelberg, D.; Sotrel, A.; Vogel, H.; et al. Progressive polyradiculopathy in acquired immune deficiency syndrome. Neurology 36:912–916; 1986.

Eilbott, D. J.; Peress, N.; Burger, H.; et al. Human immunodeficiency virus type 1 in spinal cords of acquired immunodeficiency syndrome patients with myelopathy: expression and replication in macrophages. PNAS 86:3337–3341, 1989.

Elder, G. A.; Sever, J. L. Neurologic disorders associated with AIDS retroviral infection. Rev. Infec. Dis. 10:286–302; 1988.

Elovaara, I.; Seppala, I.; Poutiainen, M. A.; Suni, J.; Valle, S-L. Intrathecal humoral immunologic response in neurologically symptomatic and asymptomatic patients with human immunodeficiency virus infection. Neurology 38:1451–1456; 1988.

Epstein, L. G.; Goudsmit, J.; Paul, D. A.; et al. Expression of human immunodeficiency virus in cerebrospinal fluid of children with progressive encephalopathy. Ann. Neurol. 21:397–401; 1987.

Epstein, L. G.; Sharer, L. R.; Cho, E.-S.; et al. HTLV-III/LAV-like retrovirus particles in brains of patients with AIDS encephalopathy. AIDS Res. 1:447–454; 1985.

Epstein, L. G.; Sharer, L. R.; Joshi, V. V.; et al. Progressive encephalopathy in children with acquired immune deficiency syndrome. Ann. Neurol. 17:488–496; 1985b.

Everall, I. P.; Luthert, P. J.; Lantos, P. L. Neuronal loss in the frontal cortex in HIV infection. Lancet 337:1119–1121; 1991.

Feldman, H. A. Toxoplasmosis: an overview. Bull. N.Y. Acad. Med. 50:110–127; 1974.

Fischl, M. A.; Richman, D. D.; Grieco, M. H.; et al. The efficiency of aziodothymidine (AZT) in the treatment of patients with AIDS and AIDS-related complex: a double-blind, placebo-controlled trial. N. Engl. J. Med. 317:185–191; 1987.

Gabuzda, D. H.; Ho, D. D.; de la Monte, S. M.; et al. Immunohistochemical identification of HTLV-III antigen in brains of patients with AIDS. Ann. Neurol. 20:289–295; 1986.

Gal, A. A.; Evans, S.; Meyer, P. R. The clinical and laboratory evaluation of cryptococcal infections in the acquired immunodeficiency syndrome. Diagn. Microbiol Infect. Dis. 7:249–254; 1987.

Gal, A. A.; Koss, M. N.; Hawkins, J.; et al. The pathology of pulmonary cryptococcal infections in the acquired immunodeficiency syndrome. Arch. Pathol. Lab. Med. 110:502–507; 1986.

Giulian, D.; Vaca, K.; Noonan, C. A. Secretion of neurotoxins of mononuclear phagocytes infected with HIV-1. Science 250:1593–1596; 1990.

Goethe, K. E.; Mitchell, J. E.; Marshall, D. W.; et al. Neuropsychological and Neurological function of human immunodeficiency virus seropositive asymptomatic individuals. Arch. Neurol. 46:129–133; 1989.

Goudsmit, J.; Epstein, L. G.; Paul, D. A.; et al. Intra-blood–brain barrier synthesis of human immunodeficiency virus antigen and antibody in humans and chimpanzees. Proc. Natl. Acad. Sci. USA 84:3876–3880; 1987.

Grant, I.; Atkinson, J. H.; Hesselink, J. R.; et al. Evidence for early central nervous system involvement in the acquired immunodeficiency syndrome (AIDS) and other human immunodeficiency virus (HIV) infection. Ann. Intern. Med. 107:828–836; 1987.

Grant, I. H.; Gold, J. M. W.; Armstrong, D. Risk of CNS toxoplasmosis in patients with AIDS. Presented at the 26th Interscience Conference on Antimicrobial Agents and Chemotherapy. New Orleans, September–October, 1986.

Grimaldi, L. M. E.; Martino, G. V.; Franciotta, D. M.; et al. Elevated alpha-tumor necrosis factor levels in spinal fluid from HIV-1-infected patients with central nervous system involvement. Ann. Neurol. 29:21; 1991.

Harris, J. E. Improved short-term survival of AIDS patients initially diagnosed with Pneumocystis carinii pneumonia, 1984 through 1987. J.A.M.A. 263:397–401; 1990.

Haverkos, H. Assessment of therapy for toxoplasma encephalitis. Am. J. Med. 82:907–914; 1987.

Heyes, M. P.; Brew, B. J.; Martin, A.; et al. Quinolinic acid in cerebrospinal fluid and serum in HIV-1 infection: relationship to clinical and neurological status. Ann. neurol. 29:202–209; 1991.

Ho, D. D.; Rota, T. R.; Schooley, R. T.; et al. Isolation of HTLV-III from cerebrospinal fluid and neural tissues patients with neurological syndromes related to the acquired immunodeficiency syndrome. N. Engl. J. Med. 1:493–497; 1985.

Ho, D. D.; Bredesen, D. E.; Vinters, H. V.; Daar, E. S. The acquired immunodeficiency syndrome (AIDS) dementia complex. Ann. Intern. Med. 111:400–409; 1989.

Holman, B. J.; Gerada, B.; Johnson, K. A.; et al. A comparison of brain perfusion SPECT in cocaine abuse and AIDS dementia complex. J. Nucl. Med. in press, 1992.

Important information on results of a control clinical trial of fluconazole vs. amphotericin B for suppression of cryptococcal meningitis. A note to physicians from the Division of AIDS, National Institute of Allergy and Infectious Diseases, National Institutes of Health. Bethesda, Maryland; April 30, 1990.

Jacobson, S.; Shida, H.; McFarlin, D. E.; et al. Circulating CD8 cytotoxic T lymphocytes specific for HTLV-1 pX in patients with HTLV-1 associated neurological disease. Nature 348:245–248; 1990.

Janssen, R. S.; Saykin, A. J.; Cannon, L.; et al. Neurological and neuropsychological manifestations of HIV-1 infection: association with AIDS-related complex but not asymptomatic HIV-1 Infection. Ann. Neurol. 26:592–600; 1989.

Jassoy, C.; Navia, B. A.; Johnsen, R. P.; Worth, J.; Walker, B. D. Frequent detection of HIV-1-specific cytotoxic lymphocytes in cerebrospinal fluid of persons with AIDS dementia complex. Neurology (abstract), in press, 1992.

Johnson, K.; Worth, J.; Holman, L.; Navia, B. SPECT in AIDS dementia complex. Neurology (abstract), 1991.

Jordan, B. D.; Navia, B. A.; Petito, C.; Cho, E. S.; Price, R. W. Neurological syndromes complicating AIDS. Front. Radiat. Ther. Oncol. 19:82–87; 1985.

Kaplan, L. D. The malignancies associated with AIDS. In: Sande, M. A.; Volberding, P. A. The medical management of AIDS. Philadelphia: W. B. Saunders Company; 1990.

Kezler, S.; Weis, S.; Budka, H. Loss of neurons in the frontal cortex in AIDS brains. Acta. Neuropathol. 80:92–94; 1990.

Kaplan, M. H.; Rosen, P. P.; Armstrong, D. Cryptococcosis in a cancer hospital. Clinical and pathological correlates in forty-six patients. Cancer 39:2265–2274; 1977.

Kieburtz, K. D.; Ketonen, L.; Zettelmaier, A. E.; et al. Magnetic resonance imaging findings in HIV cognitive impairment. Arch. Neurol. 47:643–645; 1990.

Klecker, R. W.; Collins, J. M.; Yarchoan, R.; et al. Plasma and cerebrospinal fluid pharmacokinetics of 3'-azid 3'deoxythymidine: a novel pyrimidine analog with potential application for the treatment of patients with AIDS and related diseases. Clin. Pharmacol. Ther. 41:407–412; 1987.

Koenig, S.; Gendelman, H. E.; Orenstein, J. M.; et al. Detection of AIDS virus in macrophages in brain tissue from AIDS patients with encephalopathy. Science 233:1089–1093, 1986.

Koppel, B. S.; Wromser, G. P.; Tuchman, A. J.; et al. Central nervous system involvement in patients with acquired immune deficiency syndrome (AIDS). Acta. Neurol. Scand. 71:337–353; 1985.

Koralnik, I. J.; Beaumanoir, A.; Hausler, R.; et al. A controlled study of early neurologic abnormalities in men with asymptomatic human immunodeficiency virus infection. N. Engl. J. Med. 323:864–870; 1990.

Kovacs, J. A.; Kovacs, A. A.; Polis, M.; et al. Cryptococcosis in the acquired immunodeficiency syndrome. Ann. Intern. Med. 103:533–538; 1985.

Koyanagi, Y.; Miles, S.; Mitsuyasu, R. T.; et al. Dual infection of the central nervous system by AIDS viruses with distinct cellular tropisms. Science 236:819–822; 1987.

Krupp, L. B.; Lipton, R. B.; Swerdlow, M. L.; Leeds, N. E.; Llena, J. Progressive multifocal leukoencephalopathy: clinical and radiographic features. Ann. Neurol. 17:344–349; 1985.

Lange, D. J.; Britton, C. B.; Younger, D. S.; et al. The neuromuscular manifestations of hu-

man immunodeficiency virus infections. Arch. Neurol. 45:1084–1088; 1988.

Larsen, R. A.; Leal, M. A. E.; Chan, L. S. Fluconazole compared with amphotericin B plus flucytosine for cryptococcal meningitis in AIDS. A randomized trial. Ann. Intern. Med. 113:183–187; 1990.

Leoung, G. S.; Mills, J.; Hadley, W. K.; Remington, J. S.; Wofsy, C. Cerebral toxoplasmosis in AIDS patients: clinical presentation with laboratory, radiographic and histologic correlations. Proceedings of the International Conference on AIDS, Atlanta, GA, 1985 (abstr).

Leport, C.; Raffi, K.; Katlama, C.; et al. Treatment of central nervous system toxoplasmosis with pyrimethamine/sulfadiazine combination in 35 patients with acquired immunodeficiency syndrome. Am. J. Med. 84:94–100; 1988.

Levy, J. A.; Shimabukuro, J.; Hollander, H.; Mills, J.; Kaminsky, L. Isolation of AIDS-associated retroviruses from cerebrospinal fluid and brains of acquired immune deficiency syndrome patients. Lancet 2:586–588; 1985.

Levy, R. M.; Bredesen, D. E.; Rosenblum, M. L. Neurological manifestations of the acquired immunodeficiency syndrome (AIDS): experience at UCSF and review of the literature. J. Neurosurg. 162:475–495; 1985.

Levy, R. M.; Rosenbloom, S.; Perrett, L. V. Neurological findings in AIDS: a review of 200 cases. A. J. R. 7:833–839; 1986.

Levy, R. M.; Janssen, R. S.; Bush, T. J.; et al. Neuroepidemiology of acquired immunodeficiency syndrome. In: Rosenblum, M. L.; Levy, R. M.; Bredesen, D. E., eds. AIDS and the nervous system. New York: Raven Press; 1988: p. 13–27.

Lipton, S. A. HIV-related neurotoxicity. Brain Pathology 1:193–199; 1991.

Luft, B. J.; Brooks, R. B.; Conley, F. K.; McCabe, R. E.; Remington, J. W. Toxoplasmic encephalitis in patients with acquired immune deficiency syndrome. J.A.M.A. 252:913–917; 1984.

Luft, B. J.; Remington, J. S. Toxoplasmic encephalitis. J. Infect. Dis. 157:1–6; 1988.

McArthur, J. C.; Cohen, B. A.; Farzedegan, H.; et al. Cerebrospinal fluid abnormalities in homosexual men with and without neuropsychiatric findings. Ann. Neurol. 23(suppl):S34–S37; 1988.

McArthur, J. C.; Cohen, B. A.; Selnes, O. A.; et al. Low prevalence of neurological and neurospsychological abnormalities in otherwise healthy HIV-1-infected individuals: results from the Multicenter AIDS Cohort Study. Ann. Neurol, 26:601–611; 1989.

McArthur, J. C.; Becker, P. S.; Pansi, J. E.; et al. Neuropathological changes in early HIV-1 dementia. Ann. Neurol. 26:681–684; 1989.

Miller, E. N.; Selnes, O. A.; McArthur, J. C.; et al. Neuropsychological performance in HIV-1-infected homosexual men: the Multicenter AIDS Cohort Study (MACS). Neurology 40:197–203; 1990.

Miller, J. R.; Barrett, R. E.; Britton, C. B.; et al. Progressive multifocal leukoencephalopathy in a male homosexual with T-cell immune deficiency. N. Engl. J. Med. 307:1436–1438; 1982.

Miller, R. G.; Storey, J. R.; Greco, C. M. Ganciclovir in the treatment of progressive AIDS-related polyradiculopathy. Neurology 40:569–574; 1990.

Mintz, M. M.; Rapaport, R.; Oleske, J. M.; et al. Elevated serum levels of tumor necrosis factor are associated with progressive encephalopathy in children with acquired immunodeficiency syndrome. A.J.D.C. 143:771–774; 1989.

Moskowitz, L. B.; Hensley, G. T.; Chan, J. C.; Gregorios, J.; Conley, F. K. The neuropathology of the acquired immune deficiency syndrome. Arch. Pathol. Lab. Med. 108:867–872; 1984.

Navia, B. A.; Jordan, B. D.; Price, R. W. The AIDS dementia complex: I. clinical features. Ann. Neurol. 19:517–524, 1986a.

Navia, B. A.; Cho, E.-S.; Petito, C. K.; Price, R. W. The AIDS dementia complex: II. neuropathology. Ann. Neurol. 19:525–535; 1986b.

Navia, B. A.; Petito, C. K.; Gold, J. W. M.; et al. Cerebral toxoplasmosis complicating the acquired immune deficiency syndrome: clinical and neuropathological findings in 27 patients. Ann. Neurol. 19:224–238; 1986c.

Navia, B. A.; Price, R. W. The acquired immunodeficiency syndrome dementia complex as the presenting or sole manifestation of human immunodeficiency virus infection. Arch. Neurol. 44:65–69; 1987.

NIAID communication, April 30, 1990.

O'Brien, W. A.; Koyanagi, Y.; Namazie, A.; et al. HIV-1 tropism for mononuclear phagocytes can be determined by regions of gp120 outside the CD4-binding domain. Nature 348:69–72; 1990.

Ollo, C.; Johnson, R.; Grafman, J. Signs of cognitive change in HIV disease: an event-related brain potential study. Neurology 41:209–215; 1991.

Parry, G. J. Peripheral neuropathies associated with human immunodeficiency virus infection. Ann. Neurol. 23(suppl):S49–S53; 1988.

Pepose, J. S.; Holland, G.N.; Nestor, M.S.; et al. Acquired immune deficiency syndrome: pathogenic mechanisms of ocular disease. Ophthalmology 92:472–484; 1985.

Petito, C. K.; Navia, B. A.; Cho, E.-S.; et al. Vacuolar myelopathy pathologically resembling subacute combine degeneration in pa-tients with the acquired immunodeficiency syndrome. N. Engl. J. Med. 312:874–879; 1985.

Pitchenik, A. E.; Fischl, M. A.; Dickinson, G. M.; et al. Opportunistic infections and Kaposi's sarcoma among Haitians: evidence of a new acquired immunodeficiency state. Ann. Intern. Med. 98:277–284; 1983.

Pizzo, P. A.; Eddy, J.; Falloon, J.; et al. Effect of continuous intravenous infusion of zidovudine (AZT) in children with symptomatic HIV infection. N. Engl. J. Med. 319:879–896; 1988.

Pluda, J. M.; Yarchoan, R.; Jaffe, E. S.; et al. Development of non-Hodgkin lymphoma in a cohort of patients with severe human immunodeficiency virus (HIV) infection on long-term antiretroviral therapy. Ann. Intern. Med. 113:276–282; 1990.

Pohl, P.; Vogl, G.; Fill, H.; et al. Single photon emission computed tomography in AIDS dementia complex. J. Nucl. Med. 29:1382–1386; 1988.

Portegies, P.; deGans, J.; Lange, J. M. A.; et al. Declining incidence of AIDS dementia complex after introduction of zidovudine treatment. Br. Med. J. 299:819–821; 1989.

Post, M. L. D.; Tate, L. G.; Quencer, R. M.; et al. CT, MR and pathology in HIV encephalitis and meningitis. A. J. R. 151:373–380; 1988.

Price, R. W.; Brew, B.; Sidtis, J. J.; et al. The brain in AIDS: central nervous system HIV-1 infection and AIDS dementia complex. Science 239:586–592, 1988a.

Price, R. W.; Sidtis, J. J.; Navia, B. A.; et al. The AIDS dementia complex. In: Rosenblum, M. L.; Levy, R. M.; Bredesen, D. E., eds. AIDS and the nervous system. New York: Raven Press; 1988b:203–219.

Puliam, L.; Herndier, B. G.; Tang, N. M.; Mcgrath, M. S. Human immunodeficiency virus-infected macrophages produce soluble factors that cause histological and neurochemical alterations in cultured human brains. J. Clin. Inv. 87:503–512; 1991.

Pumarola-Sune, T.; Navia, B. A.; Condon-Cardi, C.; et al. HIV antigen in the brains of patients with the AIDS dementia complex. Ann. Neurol. 490–496; 1987.

Resnick, L.; DiMarzo-Veronese, F.; Schupbach, J.; et al. Intra-blood–brain barrier synthesis of HTLV-III-specific IgG in patients with neurologic symptoms associated with AIDS or AIDS-related complex. N. Engl. J. Med. 313:1498–1504; 1985.

Ricciardi, D. D.; Sepkowitz, D. V.; Berkowitz, L. B.; et al. Cryptococcal arthritis in a patient with acquired immunodeficiency syndrome. Case report and review of the literature. J. Rheumatol. 13:455–458; 1986.

Rico, M. J.; Penneys, N. S. Cutaneous cryptococcosis resembling molluscum contagiosum in a patient with AIDS. Arch. Dermatol. 121(7):901–2; 1985.

Rottenberg, D. A.; Moeller, J. R.; Strother, S. C.; et al. The metabolic pathology of the AIDS dementia complex. Ann. Neurol. 22:700–706; 1987.

Saag, M. S.; et al. Comparison of amphotericin B with flucanozole in the treatment of acute AIDS associated cryptococcal meningitis. The NIAID Mycoses Study Group and the AIDS Clinical Trials Group. N. Engl. J. Med. 326(2):83–89; 1992.

Said, G.; Lacroix, C.; Chemoulli, P.; et al. Cytomegalovirus neuropathy in acquired immunodeficiency syndrome: a clinical and pathological study. Ann. Neurol. 29:139–146; 1991.

Sarosi, G. A.; Parker, J. D.; Doto, II.; Tosh, F. E. Amphotericin B in cryptococcal meningitis. Long-term results of treatment. Ann. Intern. Med. 71:1079–1087; 1969.

Schielke, E.; Tatsch, K.; Pfister, H. W.; et al. Reduced cerebral blood flow in early stages of human immunodeficiency virus infection. Arch. Neurol. 47:1342–1345; 1990.

Schmitt, F. A.; Bigley, J. W.; McKinnis, R.; et al. Neuropsychological outcome of zidovudine (AZT) treatment of patients with AIDS and AIDS-related complex. N. Engl. J. Med. 319:1573–1578; 1988.

Schuman, J. S.; Friedman, A. H. Retinal manifestations of the acquired immune deficiency syndrome: cytomegalovirus, Candida albicans, cryptococcus, toxoplasmosis, and Pneumocystis carinii. Trans. Ophthal. Soc. U.K. 103:177–190; 1983.

Selik, R. M.; Haverkos, H. W.; Curran, J. W. Acquired immune deficiency syndrome (AIDS). Trends in the United States, 1078–1982. Am. J. Med. 76:493–500; 1984.

Selmaj, K. W.; Raine, C. S. Tumor necrosis factor mediates myelin and oligodendrocyte damage in vitro. Ann. Neurol. 23:339–346, 1988.

Selnes, O. A.; Miller, E.; McArthur, J: et al. HIV-1 infection: no evidence of cognitive decline during the asymptomatic stages. Neurology 40:204–208; 1990.

Sethi, K. K.; Naher, H.; Stroehmann, J. Phenotypic heterogeneity of cerebrospinal fluid-derived HIV-specific and HLA-restricted cytotoxic T-cell clones. Nature 335:178–181; 1988.

Sharer, L. R.; Cho, E. S.; Epstein, L. G. Multinucleated giant cells and HTLV-III in AIDS encephalopathy. Hum. Pathol. 17:271–284; 1986.

Shaw, G. M.; Harper, M. E.; Hahn, B. H.; et al. HTLV-III in brains of children and adults with AIDS encephalopathy. Science 227:177–182; 1985.

Shioda, T.; Levy, J. A.; Cheng-Mayer, C. Macrophage and T cell-line tropisms of HIV-1 are determined in specific regions of the envelope gp120 gene. Nature 349:167–169; 1991.

Simpson, D. A.; Bender, A. N. Human immunodeficiency virus-associated myopathy: analysis of 11 patients. Ann. Neurol. 24:79–84; 1988.

Snider, W. D.; Simpson, D. M.; Nielsen, S.; et al. Neurological complications of acquired immune deficiency syndrome: analysis of 50 patients. Ann. Neurol. 14:403–418; 1983.

Spickard, A.; Butler, W. T.; Andrcole, V.; Utz, J. P. The improved prognosis of cryptococcal meningitis with amphotericin B therapy. Ann. Intern. Med. 58:66–83; 1963.

Sugar, A. M.; Stern, J. J.; Dupont, B. Overview: treatment of cryptococcal meningitis. Rev. Infec. Dis. 12:S338–S348; 1990.

Tross, S.; Price, R. W.; Navia, B.; et al. Neuropsychological characterization of the AIDS dementia complex: preliminary report. AIDS 2:81–88; 1988.

Urmacher, C.; Nielsen, S. The histopathology of the acquired immune deficiency syndrome. Pathol. Ann. 20:197–220; 1985.

Vinters, H. V.; Kwok, M. K.; Ho, H. W.; et al. Cytomegalovirus in the nervous system of patients with acquired immune deficiency syndrome. Brain 112:245–268; 1989.

Volberding, P. A.; Lagakos, S. W.; Koch, M. A.; et al. Zidovudine in asymptomatic human immunodeficiency virus infection. N. Engl. J. Med. 322:941–949; 1990.

Walker, B. D.; Chakrabarti, S.; Moss, B.; et al. HIV-specific cytotoxic T lymphocytes in seropositive individuals. Nature 328:345–348; 1987.

Wanke, C. A.; Tuazon, C. U.; Levy, C.; et al. Clinical course of tissue documented cerebral toxoplasmosis in patients with acquired immune deficiency syndrome (AIDS). Proceedings of the Interscience Conference on Antimicrobial Agents and Chemotherapy, p. 230 (abstr), 1984.

Watkins, B. A.; Dorn, H. H.; Kelly, W. B.; et al. Specific tropism of HIV-1 for microglial cells in primary human brain cultures. Science 249:549–553; 1990.

Wiley, C. A.; VanPatten, P. D.; Carpenter, P. M.; et al. Acute ascending necrotizing myelopathy caused by herpes simplex virus type 2. Neurology 37:1791–1794; 1987.

Wiley, C. A.; Nelson, J. A. Role of human immunodeficiency virus and cytomegalovirus in AIDS encephalitis. Am. J. Pathol. 133:73–81; 1988.

Wiley, C. A.; Masliah, E.; Morey, M.; et al. Neocortical Damage during HIV infection. Ann. Neurol. 29:651–657; 1991.

Wong, B.; Gold, J. W. M.; Brown, A. E.; et al. Central nervous system toxoplasmosis in homosexual men and parenteral drug abusers. Ann. Intern. Med. 100:36–42; 1984.

World Health Organization: Toxoplasmosis surveillance. Weekly Epidemiol. Rec. WHO 59:162–164; 1984.

Yarchoan, R.; Berg, G.; Brouwers, P.; et al. Response of human immunodeficiency-virus-associated neurological disease to 3-azido-deoxythymidine. Lancet 1:132–135; 1987.

Ziegler, J.; Beckstead, J.; Volberding, P.; et al. Non-Hodgkin's lymphoma in 90 homosexual men. Relation to generalized lymphadenopathy and the acquired immunodeficiency syndrome. N. Engl. J. Med. 311:565–570; 1984.

Zuger, A.; Louie, E.; Holzman, R. S.; Simberkoff, M. S.; Rahal, J. J. Cryptococcal disease in patients with the acquired immunodeficiency syndrome. Diagnostic features and outcome of treatment. Ann. Intern. Med. 104:234–240; 1986.

18

Movement Disorders

MAUREEN P. WOOTEN AND JOSEPH JANKOVIC

Parkinson's Disease

Parkinson's disease, a common neurological disease, was first described by James Parkinson in 1817. He noted that the disease was characterized by "involuntary tremulous motion, with lessened muscular power, in parts not in action and even when supported; with a propensity to bend the trunk forward, and to pass from a walking to a running pace, the senses and intellect being uninjured" (Parkinson 1817). However, he failed to describe rigidity, bradykinesia, sensory complaints, dementia, and other motor and nonmotor findings, which are now recognized as important clinical features of Parkinson's disease. He did recognize the disease as a distinct clinical entity and suggested that it was a result of a dysfunction in the brain.

The etiology of Parkinson's disease is unknown. All parkinsonian cases were once thought to be the sequela of the encephalitis lethargica epidemics of 1917 to 1926. It was proposed that once the cohort of these patients died, Parkinson's disease would disappear (PosKanzer et al. 1969). However, the incidence of Parkinson's disease has remained the same in the past several decades (Rajput et la. 1984). The prevalence of Parkinson's disease from population studies in North America and Europe has been estimated to be one in 1000, with a rise in incidence in the fifth and sixth decades. In the population older than 60 years, the estimated prevalence is 1%, with peak age at onset in the sixth decade (Rajput et al. 1984).

Most studies have shown approximately 1.5:1 male preponderance (Kessler 1978). Population studies have found a 4:1 ratio of whites to blacks in Parkinson's disease (Kessler 1978), but more recent data suggest only slightly increased frequency in whites (Schoenberg et al. 1985). The prevalence of Parkinson's disease varies from country to country; the disease is much less common in Japan and China than in Europe and North America (Kondon 1984; Li et al. 1985).

Trauma, emotional upset, overwork, exposure to cold, and other stresses have been suggested as precipitating factors for Parkinson's disease, but there is no convincing evidence in humans to support this. Snyder and colleagues (1985) tested this hypothesis in an animal model of subclinical parkinsonism and found that glucose deprivation, cold exposure, and electrical shock precipitated the emergence of parkinsonian symptoms. They concluded that the 6-hydroxydopamine-induced damage to the dopaminergic nigrostriatal pathway had been concealed in the preclinical stage due to compensatory changes of the intact dopaminergic neurons.

Because of a 15% to 20% occurrence of tremor in first-degree relatives, an autosomal dominant pattern of inheritance in some families, and a possible relationship between familial essential tremor and Parkinson's disease, a genetic predisposition has been suggested at least for a subtype of parkinsonian patients (Barbeau 1984). However, while occasional concordant twin pairs for Parkinson's disease have been reported, (Jankovic and Reches 1986) larger studies of twins have found only rare instances of concordance (Ward et al. 1983; Marttila et al. 1986). Although the methods used in these studies have been challenged, it is difficult to prove that Parkinson's disease is a genetic disease.

A more recent theory involves environmental neurotoxins in the pathogenesis of Parkinson's disease. This hypothesis was conceived after a report of a severe parkinsonian syndrome in heroin addicts exposed to 1-methyl-4-phenyl-1,2,3,6-tetrahydropyridine (MPTP) (Langston et al. 1983). Subsequent experiments found that MPP^+, the oxidation product of MPTP, can induce a syndrome resembling Parkinson's disease in animals that is associated with destruction of substantia nigra neurons (Langston et al. 1983; Snyder and D'Amato 1986). The discovery of MPTP-induced parkinsonism lead to a search for possible environmental toxins. One case control study in China revealed an increased risk for Parkinson's disease in those expected to industrial chemicals (Tanner et al. 1989). A study in Canada found a correlation between agricultural pesticides and an increased risk for Parkinson's disease (Barbeau et al. 1986). These early findings have been supported by later epidemiological studies (Ho et al. 1989). Parkinson's disease also is one of the few diseases that has been found to have an inverse relationship to smoking and cancer (Jansson and Jankovic 1985; Rajput et al. 1987).

Parkinson's disease is characterized pathologically by degeneration of pigmented brain stem nuclei, primarily the pars compacta of the substantia nigra, with an associated formation of eosinophilic cytoplasmic inclusions called Lewy bodies. The cardinal signs of Parkinson's disease result from loss of dopaminergic neurons in the nigrostriatal pathway. These clinical signs (i.e., resting tremor, cogwheel, rigidity, akinesia, and postural instability) usually do not appear until about 80% of these neurons are lost (Bernheimer et al. 1973; Iversen 1984).

Resting tremor is the most common presenting symptom in Parkinson's disease, occurring in 70% of patients at onset. It is frequently accompanied by an action postural tremor. Other motor symptoms include cogwheel rigidity and bradykinesia. A relatively late sign in Parkinson's disease is loss of postural reflexes, which probably accounts for the frequent falls in these patients (Jankovic and Calne 1987; Jankovic 1988).

Autonomic dysfunction, with orthostatic hypotension, constipation, bladder and sexual dysfunction, and thermal dysregulation also can occur in Parkinson's disease (Martignoni et al. 1986; Mathers et al. 1989). Various sensory symptoms, which include limb paresthesias, aches, and pains may be experienced by approximately half of all patients (Snider et al. 1976; Koller 1984; Sandyk and Bamford 1988). Olfactory and vestibular dysfunction, seborrhea, pedal edema, fatigue, and weight loss are some other common nonmotor abnormalities associated with Parkinson's disease.

Recent studies have drawn attention to a variety of neurobehavioral abnormalities in Parkinson's disease. In the early stage, personality changes may occur, which include withdrawal, apathy, or more dependency. This may lead to significant depression in approximately one half of patients (Gotham et al. 1986) and dementia in one third of patients (Brown and Marsden 1984). Depression is more common in the young-onset

Parkinson's disease group (Starkstein et al. 1989).

The marked clinical heterogeneity in Parkinson's disease suggests the possibility of Parkinson's disease subgroups with distinct clinical patterns and possibly different pathogenetic mechanisms. Many studies have indicated that older patients deteriorate more rapidly than those with young-onset Parkinson's disease (Scott and Brody 1971; Birkmayer et al. 1979; Goetz et al. 1988; Hoehn and Yahr 1967), while others found similar rates of progression (Hoehn and Yahr 1967; Diamond et al. 1989; Gibband and Lees 1988). Young-onset Parkinson's disease patients and usually improve initially with levodopa but seem to develop "on–off" fluctuations and levodopa-induced dyskinesias sooner than the older onset patients. (Pederzoli et al. 1983; Quinn et al. 1987; Gershanik 1988).

Aside from the subgroups that can be differentiated by their age at onset, several studies show that parkinsonian patients may be categorized according to distinct clinical patterns (Hoehn and Yahr 1967; Zetusky et al. 1985). One study identified at least two major subgroups—one group with tremor as the predominant symptom and the other, postural instability and gait difficulty (Zetusky et al. 1985). Most patients fit in one group or the other. The tremor group patients tended to have an earlier age of onset with a slower progression, and some had a family history of the disease. They seemed to have little bradykinesia or postural instability, and cognitive impairment was uncommon. Overall, the tremor group patients had a better prognosis than the postural instability and gait difficulty group patients, who tended to progress more rapidly and had more bradykinesia and dementia.

Another study of 800 patients with early untreated Parkinson's disease enrolled in the DATATOP study supports these clinical subtypes (Jankovic et al.) A number of different subsets of patients were compared. First, a group of young-onset patients (onset at age 40 or younger) was compared to old-onset (onset at age 70 or older). While both groups had similar degrees of disability at the time of entry into the study, the young-onset group had longer duration of symptoms than the old-onset, suggesting slower progression of the disease in the young-onset group. The young-onset group also had more tremor and better cognitive functioning than the old-onset group, which had more difficulties with postural instability and cognitive impairment. Overall, the young-onset group of Parkinson's disease patients had a more favorable prognosis.

Another comparison involved patients with a benign course (symptoms for more than 4 years) compared to those with a malignant course (symptoms for less than 1 year—Hoehn/Yahr stage of 2.5). The benign group had a younger age at onset, while the malignant group had symptoms of bradykinesia, postural instability, and gait difficulty more frequently at onset. Finally, patients with the tremor-predominant form of Parkinson's disease were compared to those with predominant postural instability and gait difficulty (PIGD). In general, the PIGD group had more severe symptoms than the tremor group. The study concluded that an old age at onset, presenting symptoms of bradykinesia, and the PIGD form of parkinsonism are indicative of a more malignant course than onset at a young age with tremor.

Treatment of Parkinson's disease involves correction of the dopamine depletion with levodopa, activation of the intact post-synaptic receptors with dopamine agonists, or blocking of the relatively overactive cholinergic system with anticholinergics. The most effective of these various antiparkinsonian drugs is levodopa, which usually is combined with carbidopa (Sinemet), a peripheral dopa decarboxylase inhibitor. The addition of carbidopa blocks the conversion of levodopa to dopamine in the periphery.

With the advent of levodopa therapy, the mortality of Parkinson's disease patients has changed dramatically. Hoehn and Yahr (1967) in a classic study of patients with parkinsonism prior to the introduction of levodopa therapy, provided data on mortality before levodopa. They noted an observed-to-expected death ratio of 2.9 in this group of Parkinson's disease patients. As a result of levodopa therapy, this ratio dropped remarkably. A collaborative multi-center study of levodopa in the treatment of parkinsonism in 1978 reported an observed-to-expected death ratio of 1.03 (Joseph et al. 1978). Other studies showed similar improvements in life expectancy with levodopa therapy (Rajput et al. 1984; Hoehn 1986; Sweet and McDowell 1975; Cedarbaum and McDowell 1986). Cause of death in Parkinson's patients is not from the disease itself, but rather from complications of such a debilitating disease, such as aspiration pneumonia, other infections, and sacral decubiti. Levodopa helps prevent these complications, but it does not alter the progression of the underlying pathological process.

For patients treated with levodopa, four general responses have been noted. About 15% of patients fail to respond, even at high doses. These are usually patients with a postsynaptic form of parkinsonism (parkinsonism plus syndromes). About one third of patients respond to levodopa for 7 to 8 years, and another one third revert to pre-levodopa disability after 4 to 5 years of treatment. However, in up to one third of all parkinsonian patients, levodopa provides a lifelong improvement (Jankovic and Marsden 1988).

The usual starting dose of Sinemet is the 25/100 tablet, one-half tablet twice a day and increasing by one-half tablet every third day until improvement is seen or side effects develop. Early side effects, which include nausea, vomiting, and orthostatic hypotension, usually can be managed with dosage adjustment and addition of carbidopa (Jankovic and Marsden 1988). Major side effects, which tend to occur later in treatment, can be more difficult to manage. These include levodopa-induced dyskinesias, clinical fluctuations, and psychiatric disturbances. Although levodopa dosage can be reduced to alleviate these complications, often this causes a worsening of parkinsonian symptoms. A new controlled-release formulation of carbidopa and levodopa (Sinemet CR4) can decrease the daily number of doses but

usually at the expense of increased "on" time dyskinesia (Jankovic et al. 1989). Antipsychotic drugs should be avoided in these patients because they block the striatal dopamine receptors and may exacerbate parkinsonian symptoms. When all other measures fail, the patient may benefit from a closely supervised levodopa withdrawal in the hospital. The drug is restarted after a few days off levodopa. This form of therapy, called a "drug holiday," is somewhat controversial, and its benefits have been shown to be sustained in only 50% to 60% of patients 6 months later (Kaye and Feldman 1986).

In addition to levodopa, other dopaminergic drugs include those that improve dopaminergic transmission by enhancing the release of dopamine presynaptically, by blocking the action of its metabolizing enzymes, or by direct interaction with the postsynaptic receptors (dopamine agonists) (Jankovic 1985). At this time, only bromocriptine and pergolide are commercially available dopamine agonists. Both have a long duration of action, which makes them useful in treating clinical fluctuations. Pergolide is thought to have a longer duration and less side effects than bromocriptine, but comparative studies have shown no clinical difference in efficacy (Goetz et al. 1985; Goetz et al. 1989). Side effects include psychiatric disturbances, erythromelalgia, and pulmonary and retroperitoneal fibrosis (Goetz et al. 1985). Dopamine agonists are useful in the treatment of patients with clinical fluctuations and in early stages of antiparkinsonian therapy. When used early, they may delay the onset of levodopa-induced complications, including dyskinesias and fluctuations.

Anticholinergic drugs are useful primarily in relieving the tremor or Parkinson's disease. The most commonly used drugs of this class include trihexyphenidyl (Artane), benzotropine mesylate (Cogentin), and ethopropazine (Parsidol). All are considered equally effective. Side effects include dry mouth, blurred vision, constipation, urinary retention, glaucoma, memory disturbance, and confusional states.

Another medication used in the treatment of Parkinson's disease is amantadine. Its exact mechanism of action is unknown but may include enhancement of dopamine release and synthesis, inhibition of dopamine reuptake, and an anticholinergic effect (Gerlak et al. 1970; Stone 1977). Amantadine has improved tremor and bradykinesia. Side effects include dry mouth, urinary retention, constipation (from its anticholinergic effects), livedo reticularis, ankle edema, and exacerbation of congestive heart failure.

Some of the most promising therapeutic strategies in Parkinson's disease include drugs that may halt the progression of the disease. As noted previously, a leading hypothesis to explain the pathogenesis of Parkinson's disease implicates an environmental substance (like MPTP) that becomes neurotoxic when oxidized in the brain (Langston 1988). Selegiline, a monoamine oxidase-B (MAO-B) inhibitor, prevents MPTP-induced parkinsonism and has been shown in a retrospective survey to increase life span in a group of patients with Parkinson's disease (Birkmayer et al. 1985). In a double-blind, placebo-controlled study of patients with early Parkinson's disease, Tetrud and Langston (1989) showed that selegiline delayed the need for levodopa therapy by at least 8 months when compared to placebo. They also noted that disease progression, as monitored by five different assessment scales, was slowed 40% to 83% per year. They theorized that selegiline achieves its effects by blocking the oxidation of exogenous or endogenous MPTP-like substances or by preventing oxidation of dopamine by MAO-B. Preliminary results from a much larger, multicenter, DATATOP study (Parkinson Study Group 1989a; Parkinson Study Group 1989b), with 800 subjects enrolled, provide further evidence that selegiline, alone or in combination with vitamin E, can slow the progression of Parkinson's disease.

Finally, there has been a resurgence of interest on neurosurgical treatment of Parkinson's disease. While initial reports of adrenal medulla and fetal substantia nigra grafting in patients with Parkinson's disease were encouraging, recent studies show that only a modest improvement in function can be achieved with this procedure (Jankovic 1988; Madrazo et al. 1987; Goetz et al. 1989; Jankovic et al. 1989).

Parkinsonism Plus Syndromes

When patients with parkinsonian symptoms present without tremor and with a poor response to levodopa, one of the parkinsonism plus syndromes should be considered in the differential diagnosis (Jankovic 1989; Quinn 1989). The term "parkinsonism plus syndrome" is used to describe a heterogenous group of disorders that includes olivopontocerebellar atrophy (OPCA), striatonigral degeneration (SND), Shy-Drager syndrome (SDS), and progressive supranuclear palsy (PSP). Pathologically, this group of disorders is categorized as multisystem degenerations. These disorders share several pathological features, including neuronal degeneration in the striatum, cerebellar cortex, pigmented brainstem nuclei, pontine nuclei, and inferior olives.

The additional ("plus") symptoms that should alert the examiner of a diagnosis of parkinsonism plus syndrome include eye findings (i.e., vertical gaze paralysis in PSP and nystagmus in OPCA); autonomic dysfunction in SDS; ataxic, broad-based gait in OPCA; pyramidal tract findings in SND; respiratory abnormalities in SND and SDS.

Progressive Supranuclear Palsy

PSP, also known as the Steele-Richardson-Olszewski syndrome, is frequently under-recognized, and its etiology is unknown (Jankovic 1989; Quinn 1989; Steele et al. 1963; Richardson et al. 1963). Recent studies of PSP patients have delineated its natural history (Friedman and Jankovic 1989; Golbe et al. 1988; Maher and Lees 1986). The mean age at onset averaged 63 years; the mean interval from onset of symptoms to PSP diagnosis was 3.6 to 3.9 years, and the mean survival from time of onset was 6 years (Golbe et al. 1988; Maher and Lees, 1986). PSP often was misdiagnosed as Parkinson's disease, and it accounted for approximately 6% of patients referred to a Parkinson's disease clinic (Friedman et al. 1989). In a survey of neurologists serving a population of approximately 800,000, the sex-adjusted prevalence for PSP was estimated at 1.53 per 100,000 men and 1.23 per 100,000 women for a combined ratio of 1.39 out of 100,000 (Golbe et al. 1988).

The predominant features of PSP include parkinsonian signs of bradykinesia, postural and gait disorders, axial rigidity, and minimal or no resting tremor. Other characteristics are supranuclear gaze palsies, with vertical gaze palsy most common. Pseudobulbar signs of dysarthria and dysphagia, hyperactive gag reflex, and emotional incontinence also are seen (Friedman and Jankovic 1989; Jankovic et al. 1990). The most common presenting symptoms include unsteadiness of gait, frequent falls, monotonous speech, loss of eye contact, slowness of movement and mentation, sloppy eating habits, blurred vision, and difficulty focusing (Golbe et al. 1988; Golbe and Davis 1988).

The diagnosis of PSP is made from the signs and symptoms that characterize this syndrome. Atrophy is the most common abnormality seen on computed tomography (CT) or magnetic resonance imaging (MRI) of the brain. Other less common abnormalities include single and multiple infarcts, midbrain atrophy, or cerebellar atrophy. The EEG frequently is abnormal, showing diffuse bitemporal, bifrontal, or focal slowing (Jankovic et al. 1990). Neuropathological findings include neurofibrillary tangles, granulovacular degeneration, gliosis, and cell loss in multiple subcortical structures (Jankovic et al. 1990).

No clear etiology or risk factors have been identified for PSP (Friedman and Jankovic 1989; Golbe et al. 1988), but some cases are due to multiple infarcts or other vascular lesions (Dubinsky and Jankovic 1987). Because of the loss of postsynaptic dopamine receptors, the PSP patients responded poorly to dopaminergic drugs, even at high doses (Golbe and Davis 1988). Other medications, including amitriptyline, methysergide, and anticholinergics, have been tried without success (Friedman and Jankovic 1989; Golbe et al. 1988; Jankovic et al. 1990). Death usually results from the complications of immobility and aspiration.

Striatonigral Degeneration

The average age at onset of SND is 54, and there is no male or female predominance. The syndrome resembles an akinetic–rigid parkinsonism syndrome and may be unilateral or bilateral. Tremor usually is less prominent than rigidity. Distinguishing features include autonomic dysfunction with orthostatic hypotension, incontinence, and impotence; pyramidal tract findings with hyperreflexia and upgoing toes; arrhythmias, central respiratory abnormalities, aspiration, and stridor; sleep abnormalities; dysarthria, cerebellar speech, or aphonia; and cerebellar findings (Jankovic 1989; Pollinsky 1984; Adams et al. 1964).

The pathological features include neuronal loss and depigmentation, seen primarily in the putamen and substantia nigra, but also in the caudate, globus pallidus, locus coeruleus, and dentate nuclei (Jankovic 1989; Pollinsky 1984; Adams et al. 1964). CT or MRI of the head may show atrophy in the brain stem or cerebellum (Jankovic 1989). Treatment primarily is supportive, and response to levodopa therapy is poor (Jankovic 1989).

Olivopontocerebellar Atrophy

The term olivopontocerebellar atrophy designates a heterogeneous group of ataxic disorders that are either familial or sporadic. The male to female ratio of familial OPCA patients is 1.8 : 1 and that of sporadic is 1 : 1 (Friedman and Jankovic 1989; Berciano 1988). The age at onset ranges from birth to 66 years with average age at onset for familial OPCA approximately 28 and 50 for sporadic OPCA. The duration of the disease is 15 years for familial OPCA and 6 years for sporadic OPCA (Jankovic 1989; Berciano 1988; Harding 1987).

OPCA usually presents with signs of cerebellar dysfunction, such as gait or limb ataxia, and intention tremor. Parkinsonian findings also are present in many patients. Other clinical findings include pyramidal findings with spastic paraparesis and pseudobulbar abnormalities; sphincter disturbances; dementia; involuntary movements to include dystonia, chorea, athetosis, and myoclonus; ophthalmoplegic and visual (optic nerve) disturbances, particularly in familial OPCA; and spinal cord findings reflecting corticospinal and anterior horn cell damage, including fasciculations, amyotrophy, hyperreflexia or areflexia, and pes cavus (Berciano 1988; Harding 1987).

The major pathological findings are cerebellar atrophy with loss of Purkinje's cells, pontine atrophy, depigmentation of the substantia nigra, and neuronal loss in the inferior olivary nuclei and spinal cord. CT and MRI scanning can show atrophy of cerebellar hemispheres, brain stem, and vermis (Harding 1987). The parkinsonian symptoms of OPCA respond variably to dopaminergic therapy.

Shy-Drager Syndrome

SDS was first described in 1960 as a neurological syndrome associated with orthostatic hypotension (Shy and Drager 1960). SDS usually begins after age 50 and affects men twice as frequently as women. The most common presenting symptoms include orthostatic light-headedness, impotence, and difficulty with micturition. Autonomic findings predominate in this disorder and are manifested chiefly by postural hypotension, bowel and bladder dysfunction, impotence, defective sweating, pupillary abnormalities, and sleep disturbances (Polinsky 1988; Cohen et al. 1987). SDS can overlap with symptoms of OPCA, SND, and Parkinson's disease.

Two subgroups of pathological findings are noted in SDS. The first subgroup is characterized by the presence of Lewy bodies in the substantia nigra, locus coeruleus, and sympathetic ganglion cells. Cell loss in the dorsal nuclei of the vagus and intermediolateral columns also is seen. The second subgroup has neuropathological findings similar to those seen in OPCA and SND (Polinsky 1988; Cohen et al. 1987).

Therapy for SDS consists of symptomatic treatment of the orthostatic symptoms. This includes fluorocorticsone, increased salt consumption, indomethacin and other nonsteroidal antiinflammatory drugs, midodrine, compressive garments, and sleeping with the head up. Dopaminergic agents may exacerbate the orthostatic symptoms and must be used cautiously (Polinsky 1988; Cohen et al. 1987).

Other Syndromes

Lower body parkinsonism denotes a subgroup of parkinsonian patients with predominant lower body involvement characterized by shuffling gait and freezing (Fitzgerald and Jankovic 1989). It differs from Parkinson's disease by its presenting symptoms of gait disturbance, lack of tremor, and predominant leg involvement. It has a relatively rapid course and responds poorly to levodopa. The etiology is uncertain, but a vascular etiology has been proposed partly because of the increased incidence of hypertension in these patients (Fitzgerald and Jankovic 1989).

Corticobasal degeneration is a syndrome with clinical features similar to PSP and pathological features similar to Pick's disease (Gibb et al. 1989). Patients may exhibit parkinsonian features, mild cerebellar signs, supranuclear gaze palsies, chorea, or focal dystonias. The disease usually progresses to death in 4 to 6 years and responds poorly to levodopa. Pathological features include frontoparietal atrophy with cortical cell loss, gliosis, and Pick cells and cell loss and gliosis in the thalamus, lentiform nucleus, subthalamic nucleus, red nucleus, substantia nigra, and locus coeruleus. Lewy bodies and neurofibrillary tangles generally were absent (Gibb et al. 1989).

Tremors

A tremor is an involuntary oscillation of a part of the body produced by alternating or synchronous contractions of reciprocally innervated antagonistic muscles. It can result from normal or pathological processes and is characterized in terms of amplitude and frequency.

Most tremors can be categorized as occurring at rest or with action. Action tremor can be further subdivided as a postural tremor or intention (kinetic) tremor. A postural tremor occurs with a sustained posture against gravity, while an intention tremor is produced during goal-directed movement. Common examples of rest tremor are those seen in Parkinson's disease. Physiological tremor, essential tremor (ET), and the postural tremors of Parkinson's disease are classified as postural tremors (Findley 1988; Capildeo and Findley 1984). Another method to classify tremor is by its distribution. Physiological tremor primarily involves the distal part of the limb, as does essential tremor, but essential tremor also involves the neck, head, and voice. Parkinsonian rest tremor can affect the face, tongue, and lips as well as distal limbs but does not involve the head or voice (Findley 1988; Capildeo and Findley 1984; Elbe 1986; Sabra and Hallett 1984). Frequency also can be used to classify tremor. The tremor of cerebellar disease has a frequency of 2.5 to 4 Hz, parkinsonian rest tremor is 4 to 5 Hz, essential tremor is 4 to 12 Hz, and physiological tremor is 6.5 to 12 Hz (Findley 1988; Jankovic and Fahn 1980).

Physiological tremor is the most common form of postural tremor and consists of small-amplitude, high-frequency movements that are primarily distal and do not usually require medical attention. The frequency of physiological tremor in the hands usually is very rapid (8–12 Hz), but it can be as low as 6.5 Hz in other body parts. The tremor can be accentuated by various factors, including anxiety, stress, fatigue, exercise, thyrotoxicosis, hypoglycemia, hypothermia, pheochromocytoma, alcohol withdrawal, beta-adrenergic and dopaminergic drugs, lithium, valproate, neuroleptics, tricyclics, caffeine, and other drugs and toxins (Findley 1988; Capildeo and Findley 1984; Elbe 1986). Any one or a combination of the previous factors increases the amplitude of physiological tremor by synchronizing motor unit discharges and therefore increasing muscle spindle afferent activity from the contracting muscles (Logigian et al. 1988). The amplitude can be improved

with use of beta-adrenergic blocking drugs (Kruse et al. 1986), and the frequency can be decreased by loading the extended limb with weights (Homberg et al. 1987).

ET is present during activity or a sustained posture but is absent with rest. It typically involves the distal portion of limbs with flexion–extension oscillation of the hands at the wrist or adduction–abduction movements of the fingers when the arms are outstretched. ET also may involve the neck, head, and voice. The tremor is inherited as an autosomal dominant disorder with high penetrance. It can be associated with dystonia, myoclonus, and Charcot-Marie-Tooth disease (Jankovic and Fahn 1980; Salisachs 1976). Prevalence rates for individuals older than 40 range from 0.4% to 5.6% (Rautakorpi et al. 1984; Haerer et al. 1982), with a positive family history noted in about 30% of cases (Findley 1988).

ET differs from the rest tremor of Parkinson's disease because it occurs with goal-directed movements or sustained postures, and it frequently involves the head and voice (Lou and Jankovic 1991). However, ET patients also may have a tremor at rest and parkinsonian patients may exhibit a postural tremor. The overlapping features and the higher-than-expected coexistence of ET and PD has suggested to some that the two disorders may be related. Some studies have shown an increased association between the two diseases (Geraghty et al. 1985), while others have not (Cleeves et al. 1988). This controversy may be solved by defining the disorders more clearly and obtaining a more accurate family history (Jankovic 1989).

The frequency of ET is 4 to 12 Hz and is produced by alternating or synchronous contractions of antagonistic muscles. The more rapid type of ET (frequency 8–12 Hz) can be similar in appearance to physiological tremor, but there are a few important differences. Physiological tremor improves with intravenous propranolol and worsens with alpha-adrenergic blockade, whereas ET improves with both (Abila et al. 1985). Also, ET is not affected by mechanical loading as is physiological tremor (Elbe 1986). Although ET is frequently referred to as benign, it can be disabling, interfering with writing, eating, and drinking. The slower frequency ET (4–6 Hz) usually has a higher amplitude and seems to be the most disabling.

More recently, a marked clinical variability in clinical expression of ET has been noted (Findley and Koller 1987). Examples of different types of ET include the task-specific tremor that occurs only during specific activities, such as writing, holding objects, and playing instruments. Similar to other types of ET, task-specific tremors often are associated with focal dystonia (Rosenbaum and Jankovic 1988). Another form of task-specific ET is orthostatic tremor, which affects the trunk and legs and occurs primarily in the standing position (Fitzgerald and Jankovic 1989).

The most effective treatment for ET is beta-adrenergic blocking drugs, especially propranolol. Long-acting propranolol has been found to be as effective as standard propranolol (Cleeves and Findley 1988), but because it readily crosses the blood–brain barrier, it can have prolonged undesirable side effects, such as sedation. Arotinolol, a new peripherally acting beta blocker, has been effective in the treatment of ET and has less side effects (Kuroda et al. 1988). Other drugs effective in the treatment of ET include primidone (Sasso et al. 1988), (Huber and Paulson 1988), and possibly glutethimide (McDowell 1989). An investigation into the acute and chronic effects of propranolol and primidone in essential tremor revealed acute adverse reactions in 8% of patients treated with propranolol and 32% treated with primidone. The drug was ineffective in 30% of the propranolol-treated patients and in 32% of primidone treated patients. Significant chronic side effects occurred in 17% of patients taking propranolol but were minimal in the primidone-treated patients (Koller and Vetere-Overfield 1989).

The tremor of Parkinson's disease primarily is distal, occurs at rest, and usually is asymmetrical at onset. It can diminish or disappear with voluntary movement. Characteristically, the tremor can be an alternating supination–pronation or "pill rolling" type movement of the forearm, adduction–abduction of the thumb, flexion–extension of the foot, or opening–closing of the jaw. Treatment of this tremor is discussed in the Parkinson's disease section. As noted in the section entitled Parkinson's disease, anticholinergic and dopaminergic drugs usually are successful in the treatment of tremor. Additionally, long-acting propranolol (but not primidone or clonazepam) often are effective in the treatment of this tremor (Koller and Herbster 1987). In cases with medically intractable parkinsonian tremor, CT-guided stereotactic ventralis lateralis thalamotomy may be helpful (Andrew 1984).

Tremor associated with lesions in the cerebellar outflow pathways often is slow (2.5–4 Hz) and proximal in its distribution. It can be ipsilateral or contralateral to the lesion and often is associated with other cerebellar symptoms. Cerebellar tremor occurs primarily during goal-directed movement but may have a coarse, postural ("wing-beating," "rubral") component (Findley 1988; Cole et al. 1988). There is no satisfactory drug treatment for cerebellar tremor; however, physostigmine, isoniazid, and carbamazepine have been used with variable results (Findley 1988). Thalamotomy has been shown to decrease the amplitude of kinetic cerebellar tremors, but patients may continue to be disabled because of persistent dysmetria (Speelman and Van Manen 1984). Application of limb weights may be of limited help (Sanes et al. 1988).

Tremors also can occur due to hysteria or other psychogenic causes (Koller et al. 1989). The clinical features of psychogenic tremor include an abrupt onset, static course, and spontaneous remissions. These tremors often are complex or unclassifiable with associated inconsistent symptomatology, selective disabilities, and absence of other neurological signs. The tremor often increases with attention and decreases with distraction and typically does not respond to antitremor drugs. These patients also may form of adult-onset focal dystonia (Jankovic et al. 1991). Onset is usually insidious in the fourth to sixth decade with initial symptoms of neck pain or muscle stiffness. They may exhibit symptoms of psychiatric illness and may recover with psychotherapy (Koller et al. 1989).

Dystonia

In 1984, an ad hoc committee of the Dystonia Medical Research Foundation proposed the following definition of dystonia: "Dystonia is a syndrome dominated by sustained muscle contractions frequently causing twisting and repetitive movements or abnormal postures." Dystonia can be classified by its severity, clinical characteristics, distribution, age at onset, or etiology. Because of its variable clinical expression and because it is often exacerbated by stress, dystonia is frequently misdiagnosed (Fahn 1984).

A review of medical records at the Mayo Clinic in Rochester, Minnesota, estimated the incidence of generalized dystonia at two per 1 million people per year, while that of focal dystonias was 24 per 1 million per year (Nutt et al. 1988). An estimated prevalence of all forms of dystonia in the same population was 329 per 1 million (Nutt et al. 1988).

Generalized dystonia often is inherited in an autosomal dominant pattern with variable clinical severity. The Ashkenazi Jewish population seems to have higher prevalence of dystonia than the non-Jewish population, but inheritance patterns are similar (Elridge 1970; Zilber et al. 1984; Bressman et al. 1988; Fletcher et al. 1988). The gene for torsion dystonia has been located on chromosome 9q32-q34 (Ozelius et al. 1989). There are no genetic studies on focal dystonia, but a family history of dystonia has been noted in 10% of patients with blepharospasm (Grandas et al. 1988), and about one third of blepharospasm patients have a family history of a movement disorder (Jankovic and Nutt 1988).

One method of classifying dystonia is by its severity (i.e., whether it occurs with rest or with activity or "action"). Action dystonia may be task-specific, that is, present only during a certain activity. The best example of this is writer's cramp or graphospasm (Rosenbaum and Jankovic 1988). As symptoms progress, simple but not necessarily specific tasks, such as cutting food, using utensils, or buttoning clothes can elicit the dystonia. Later, actions in one part of the body can induce involuntary movements in another part of the body (the overflow phenomena) (Fahn 1984). Eventually, the dystonic movements occur at rest and produce sustained dystonic postures that could progress to actual contractures.

Most dystonic movements are present as long as the patient is awake, but the timing and intensity of movements can be influenced by fatigue, relaxation, motor activity, and emotion. In some, dystonic movements fluctuate in a diurnal pattern; they are more severe in the evening and mild to nonexistent in the morning (Segawa et al. 1976; Segawa et al. 1986). This type of dystonia may be associated with parkinsonian features and may improve with levodopa or anticholinergic therapy (Nygaard and Duvoisin 1986; Nygaard et al. 1988).

Paroxysmal dystonia refers to a sudden onset of dystonic movements lasting for seconds to hours. This dystonia can be subdivided as to whether it is induced by movement. Kinesigenic paroxysmal dystonia can be induced by sudden movements and often is associated with choreoathetosis. It is best treated with anticonvulsants, such as phenytoin, carbamazepine, or barbituates (Lance 1977). Nonkinesigenic paroxysmal dystonia usually is precipitated by alcohol, caffeine, fatigue, stress, or exercise and often is associated with choreoathetosis and ataxia. It is best treated with clonazepam, oxazepam, acetazolamide, valproate, or carbamazepine (Bressman et al. 1988).

A useful method of classifying dystonia is by its distribution (Jankovic and Fahn 1988). Focal dystonia involves a single body part. Examples of focal dystonia include blepharospasm, oromandibular dystonia, laryngeal dystonia, and pharyngeal dystonia, the combination of which can be classified as segmental, cranial–cervical dystonia. Other examples of focal dystonia are writer's cramp, torticollis, and foot dystonia (Jankovic and Fahn 1988; Marsden 1986; Fahn 1986).

In blepharospasm, the eyelids close involuntarily due to contraction of the orbicularis oculi muscles. If it occurs alone, it is called essential blepharospasm. More commonly, blepharospasm is associated with dystonic movements of other facial, pharyngeal, laryngeal, or cervical muscles. In this case the term cranial–cervical dystonia (sometimes referred to as Meige's syndrome) is more appropriate. Blepharospasm initially can affect only one eye, but the asymmetrical contractions of the eyelids usually become bilateral and symmetrical within a few weeks after onset. Patients frequently complain of eye irritability or a feeling of sand in the eyes with increased blinking. This can later develop into more sustained eyelid closure, at times so severe that patients become legally blind (Marsden 1986). Blepharospasm can be triggered by light or stress and is frequently the worst while the patient is trying to read, watch television, or drive (Jankovic and Fahn 1988; Marsden 1986). It commonly begins in the fourth to sixth decade of life (Marsden 1986; Tolosa and Marti 1988; Jankovic and Orman 1984).

The pathogenesis of blepharospasm is unclear. Postmortem studies on three brains of patients with idiopathic dystonia showed elevated levels of norepinephrine in the red nucleus (Hornykiewicz et al. 1988; Jankovic et al. 1987). This finding coupled with abnormal blink reflexes recorded in patients with blepharospasm and other focal dystonias suggests an abnormality in the rostral brain stem (Tolosa et al. 1988). This also is suggested by reports of secondary blepharospasm occurring after an acute lesion in the brain stem (Jankovic and Patel 1983). In addition to focal brain stem lesions, such as stroke and multiple sclerosis, blepharospasm has been associated with Parkinson's disease, PSP, multiple system atrophy, and other neurodegenerative disorders (Jankovic 1988).

Many patients use dark glasses or sensory "tricks," such as pulling on an upper eyelid, pinching the neck, talking, humming, or singing to help relieve the symptoms of blepharospasm, at least transiently (Grandas et al. 1988; Jankovic and Fahn 1988). Long-term management of blepharospasm

with drug therapy generally is not successful (Grandas et al. 1988; Jankovic and Fahn 1988; Marsden 1986). Anticholinergics occasionally are useful, but baclofen, benzodiazepines, and tetrabenazine also are sometimes used (Grandas et al. 1988; Marsden 1986; Tolosa and Marti 1988). Surgical therapy to destroy peripheral branches of the facial nerve or to resect the orbicularis oculi muscle may be successful, but recurrence secondary to reinnervation occurs frequently; therefore surgery is rarely needed and is used only when all other treatments have failed and the patient continues to be disabled (Anderson and Patrinely 1988).

Treatment of blepharospasm with botulinum toxin injections has largely replaced pharmacological therapy and surgery. Botulinum toxin blocks the release of acetylcholine at the neuromuscular junction and when injected into the oribicularis oculi muscle, leads to paralysis and relief of blepharospasm. The benefit of injections lasts about 3 to 4 months. Most studies have found botulinum toxin injections to be effective in more than 90% of patients (Jankovic and Orman 1987; Jankovic et al. 1991).

Oromandibular dystonia is a focal dystonia manifested by spasmodic contractions of the muscles of the mouth, jaw, and tongue. This may cause involuntary jaw opening with associated protrusion of the tongue or jaw closure with trismus, bruxism, and pursing of the lips. Laryngeal dystonia can cause dystonic adduction of the vocal cords that results in a strained voice (spasmodic dysphonia); dystonic abduction of the vocal cords produces a soft voice (whispering dysphonia). Pharyngeal dystonia, an uncommon form of focal dystonia, can cause dysphagia. Patients may present with one focal dystonia, and later adjacent body segments may be involved (cranial–cervical dystonias). Botulinum toxin has been used successfully in the treatment of oromandibular dystonia and spasmodic dysphonia (Jankovic and Orman 1987; Jankovic et al. 1990; Brin et al. 1987; Brin et al. 1989).

Cervical dystonia, in the form of torticollis, retrocollis, anterocollis, laterocollis, or a combination of these is the most common be a history of trauma prior to the advent of symptoms, but in a majority of cases, no specific etiology for the dystonia can be found. The neck spasms may be intermittent, causing jerking movements of the head, or continuous, producing a sustained deviation of neck posture. The muscles involved in the abnormal movements usually can be palpated easily and often are hypertrophied. Like blepharospasm, torticollis in some cases can be relieved temporarily by a variety of sensory tricks including touching the chin, face, or back of the head (Weiner and Nora 1984).

The natural history of spasmodic torticollis has not been well studied, but the neck spasms and abnormal posture usually increase during the first few years with minimal or no progression 5 years after onset. A study of 24 patients with idiopathic spasmodic torticollis followed for more than 1 year found three outcome groups (Lowenstein and Aminoff 1988). Three patients (13%) underwent complete to almost complete remission at a median of 3 years into the illness. Eight patients (33%) had partial remissions up to 9 years into the illness, and the remaining 13 patients (54%) had no improvement. For the combined patient population, the interval from the onset of symptoms to maximum disability ranged from 1 month to 18 years and did not correlate with the final outcome. A history of trauma was present in 13% of patients, and tremor was found in 33%. Pain occurred initially in 21% of patients and increased to 75% of patients when the torticollis was the most severe; the presence of pain at onset did not predict outcome. The progression of symptoms from onset to maximum severity ranged from 1 month to 18 years, with more than 75% without remission and reaching their maximum severity of symptoms within 5 years of onset. The investigators found that in patients with a complete or partial recovery, onset of symptoms occurred at a younger age, and they were less likely to have constant deviation of the head.

Treatment of torticollis can be difficult. As with the other cranial–cervical dysto-nias, response to drugs is poor. Anticholinergics, benzodiazepines, levodopa, tetrabenazine, and haloperidol, among others, have been used with minimal success (Marsden 1986). Surgical treatment such as posterior rhizotomy may provide partial improvement (Bertrand). Botulinum toxin injections of the affected muscles provide the most effective, albeit transient, relief and is now considered the treatment of choice for most patients with cervical dystonia (Stell et al. 1988; Jankovic and Schwartz, 1990 & 1991; Jankovic and Brin 1991).

Limb dystonia is manifested frequently as writer's cramp or other occupational cramps, such as typist's, musician's and sports-related cramps (Marsden 1986; Sheehy and Marsden 1982; Newmark and Hochberg 1987). Essential tremor can coexist with the dystonia. At onset, the involuntary movement is task-specific, but with time, the dystonia may occur with other actions or may become generalized. Some forms of peripheral focal dystonia follow an injury to the affected body part, and this post-traumatic focal dystonia may later become segmental and even generalized in distribution (Jankovic and Van Der Linden 1988). Treatment with medications usually is unsuccessful, but local injections of botulinum toxin usually provides a satisfactory relief.

Other categories of dystonia include segmental, multifocal and generalized dystonia, and hemidystonia (Jankovic and Fahn 1988). Segmental dystonia involves contiguous body parts; examples include cranial–cervical dystonia, axial dystonia (neck and trunk musculature), brachial dystonia (one or both arms involved with or without axial muscles), and crural dystonia (one leg and trunk or both legs). Multifocal dystonia involves two or more noncontiguous body parts. Generalized dystonia is a combination of segmental cervical dystonia, cervical dystonia, and dystonia of another body part. Hemidystonia affects one half of the body and is usually associated with a lesion in the contralateral basal ganglia (Pettigrew and Jankovic 1985).

The causes of dystonia can be divided into two major classes: idiopathic (primary) and symptomatic (secondary). The idiopathic dystonias are either familial or sporadic. Symptomatic dystonias are associated with neurodegenerative, metabolic, or due to some other known specific cause (Jankovic and Fahn 1988). Idiopathic dystonia has no associated symptoms of weakness, spasticity, ataxia, abnormal eye movements, dementia, reflex changes, or seizures (Jankovic and Fahn 1988). The onset of idiopathic dystonia usually is focal and task-specific and later spreads to other body parts (Marsden 1986; Marsden et al. 1976). A study of torsion dystonia among Jews in Israel revealed that approximately 35% of patients reached maximum disability within 10 years of onset (Inzelberg et al. 1988). Of patients that reached maximum disability within 10 years, the greatest proportion were those patients with young-onset (younger than age 16) dystonia. A rapid deterioration also was more common in this group of young-onset patients. Stabilization (lack of deterioration for more than 5 years) usually occurred 10 years after onset. The initial site of involvement was most commonly the upper limb (47%), followed by the lower limb (28%) and axial musculature (21%). Initial lower limb involvement occurred most often in the young-onset patients. Therefore, poor prognostic factors for idiopathic torsion dystonia include juvenile onset and lower limb involvement, both of which predict progression to a generalized dystonia.

Symptomatic dystonia often begins with dystonia at rest and may have a sudden onset. It can progress quickly to fixed dystonic postures (Jankovic and Fahn 1988). Usually it is associated with other nondystonic neurological signs, such as dementia, seizures, spasticity, and ataxia (Calne and Lang 1988). There are many causes of symptomatic dystonia. Neurodegenerative disorders causing dystonia include Wilson's, Huntington's, and Parkinson's disease (Poewe et al. 1988); PSP; progressive pallidal degeneration; Hallervorden-Spatz disease; Joseph disease; ataxia–telangiectasia;

multiple sclerosis; neuroacanthocytosis; Rett syndrome; intraneuronal inclusion disease; and familial basal ganglia calcifications (Nygaard et al. 1988). Metabolic disorders, such as amino acid disorders, lipid storage diseases, and Lesch-Nyhan syndrome, also should be considered, particularly when dystonia is associated with some atypical features. Other cases of symptomatic dystonia include head trauma and peripheral trauma with nerve injury, infection, perinatal injury, drugs, and toxins (Calne and Lang 1988). It is important to screen for Wilson's disease in patients with symptomatic dystonia because early treatment can improve prognosis. Wilson's disease is discussed more thoroughly in the section entitled Heredodegenerative Disorders.

The pathogenetic mechanisms underlying idiopathic torsion dystonia are unknown. However, abnormalities of the basal ganglia, particularly putamen, are seen frequently in symptomatic cases (Rothwell and Obeso 1987). Also, brain stem pathology has been found in some patients with cranial dystonia (Jankovic and Patel 1983; Gibb et al. 1988; Zweig et al. 1988). Few biochemical changes have been found in brains of dystonia patients at autopsy (Hornykiewicz et al. 1988; Jankovic et al. 1987). These include reduction in norepinephrine levels in areas of the hypothalamus, mammillary bodies, subthalamic nucleus, and locus coeruleus, with elevated levels in the red nucleus, system thalamus, colliculi dorsal raphe nuclei, and substantia nigra.

The treatment of dystonia with medication is usually difficult. Anticholinergic drugs when used in high doses produce at least a modest improvement in most patients (Burke et al. 1986). However, sedation, forgetfulness, blurred vision, and other anticholinergic side effects can be intolerable. Drug response may be better if started within 5 years of onset of symptoms (Greene et al. 1988). Baclofen and tetrabenazine also may provide some improvement in selected patients (Greene et al. 1988; Jankovic and Orman 1988). Botulinum A toxin injections provide the most effective treatment, particularly in focal dystonias (Jankovic and Orman 1987; Jankovic et al. 1990; Brin et al. 1987; Brin et al. 1989; Stell et al. 1988; Jankovic and Schwartz 1990; Jankovic and Brin 1991).

Tics and Tourette's Syndrome

Tics are rapid, abrupt, involuntary, brief, coordinated jerks that involve discrete muscle groups (usually facial, neck, shoulder, abdominal, and respiratory musculature) and often are associated with involuntary noises or words. The patient can suppress them for a few minutes, but this can result in emotional distress, which is relieved by the occurrence of the tic. Tics are classified as motor, phonic, and sensory (Jankovic and Fahn 1986; Jankovic 1986; Lees and Tolosa 1988). The motor and phonic tics may be simple or complex (Jankovic and Fahn 1986; Jankovic 1986; Lees and Tolosa 1988).

Gilles de la Tourette's syndrome (TS), first described more than 100 years ago, is probably the most common cause of tics. The spectrum of TS symptoms ranges from mild, almost unnoticeable motor tics to disabling, complex motor and phonic tics (Kurlan 1988; Kurlan 1989). The following criteria are required for the diagnosis of definite Tourette's syndrome: multiple motor and one or more phonic tics present at some time during the illness; the tics can occur from many times a day to intermittently for more than 1 year; the location, number, frequency, complexity, type, or severity of tics change over time; onset is before age 21; involuntary movements and noises cannot be explained by other medical conditions; and tics must be witnessed by a reliable examiner or recorded on film (Jankovic 1989).

Because of the variable expression of TS, the disorder is frequently missed or misdiagnosed. Therefore, the true prevalence of TS is unknown but is estimated to be from 0.29 to 0.5 per 1000 (Shapiro and Shapiro 1982; Caine et al. 1988). The disorder occurs in one per 2000 men and in one per 10,000 women (Pauls et al. 1988). It is estimated that one in 83 people carry the gene for TS (Comings 1987).

Motor tics can be characterized as simple or complex. Simple tics are brief muscle jerks (clonic tics) or more sustained contractions (dystonic tics) that occur in an isolated body segment, usually the eyelids, neck, or shoulders. Complex motor tics are well-coordinated acts that can appear as purposeful or nonpurposeful movements. Examples of complex motor tics include hair twiddling, jumping, kicking, squatting, and obscene gesturing (Jankovic and Fahn 1986). Both simple and complex motor tics can be exacerbated by stress.

Phonic tics involve involuntary vocalization and also can be classified as simple or complex. Simple phonic tics include throat clearing, sniffling, and grunting. Examples of complex phonic tics include echolalia, palilalia, and coprolalia. Coprolalia, the shouting of obscenities, is sometimes thought to be synonymous with the diagnosis of TS but is actually present in less than 50% of cases (Jankovic and Rohaidy 1987).

Sensory tics are described as abnormal sensations of pressure, pain, cold, warmth, or tickle in the skin, bones, muscles, and joints. These findings can be localized to a specific area, and the patient can temporarily relieve them by certain movements, usually involving muscle tightening or stretching (Shapiro et al. 1988).

The behavioral abnormalities associated with TS involve a spectrum of disorders, including obsessive–compulsive, impulsive, and self-destructive behavior; attention deficits; and sleep disorders (Pauls et al. 1986; Glaze et al. 1983). Approximately 50% of TS patients show signs of obsessive–compulsive disorder (Frankel et al. 1986). A similar percentage of patients show evidence of attention deficit and hyperactivity.

A variety of medications are available for treatment of TS, and their choice must be tailored to each patient's needs. Mild cases of TS usually require no treatment except education for the patient, family, and school personnel and perhaps restructuring of the school environment. If the symptoms of TS are socially or occupationally disabling, medical therapy should be considered.

Haloperidol is probably the most common

medication used for TS; however, some other neuroleptics may be more effective and have fewer side effects. Patients and families should be warned about the risk of developing tardive dyskinesia and tardive dystonia with any and all of the neuroleptics (Singh and Jankovic 1988). Other side effects include sedation, depression, and weight gain (Jankovic 1986). Fluphenazine has an effectiveness similar to that of haloperidol, but seems to have less side effects. The daily dosage is between 4 and 8 mg/day (Shapiro et al. 1987). Pimozidine is the newest neuroleptic approved for treatment of TS (Shapiro et al. 1987). The initial dosage is 1 mg/day, gradually increased to an average dose of 6 to 16 mg/day. Pimozide also appears to be less sedating than haloperidol but can cause a prolonged Q–T interval. Clonidine also has been used in the treatment of TS, but its effects are variable, and it seems more efficacious in treating behavioral abnormalities of TS than motor tics (Leckman et al. 1988). Other medications that may have tic-suppressing effects include clonazepam, reserpine, and tetrabenazine (Jankovic and Orman 1988).

For treatment of the obsessive–compulsive disorders associated with TS, neuroleptics and clonidine have limited success (Leckman et al. 1988). However, the antidepressant fluoxetine (Prozac), which blocks serotonin reuptake, appears promising (Burrows et al. 1988). In addition to pharmacotherapy, psychotherapy and behavioral therapy play an important role in treating TS.

TS is a lifelong condition, but tic severity has been known to wax and wane throughout its course. In addition, approximately 20% of affected patients achieve permanent remission after age 20. In a study of 43 members of a family with TS or chronic tics, 30% were unaware of tics and only 18.5% sought medical treatment for their tics (Kurlan et al. 1987). In another study of 58 patients with TS, the majority indicated they had fewer tics as they reached adolescence or young adulthood. The majority also reported that they were coping well with the associated behavior or learning problems (Erenberg et al. 1987). Therefore, it is important to emphasize that although TS may be lifelong, the symptoms may not be severe, and pharmacological therapy provides adequate control of symptoms in most cases.

Tardive Dyskinesia

In 1957, the first description of orofacial dyskinetic movements after chlorpromazine treatment was reported in the German literature (Schonecker 1957). However, the existence of chronic movement disorders secondary to neuroleptic treatment remained controversial for many years (Waddington 1984). Some of the difficulties in characterizing these chronic movement disorders, or tardive dyskinesias (TD), are due to its variable presentation and clinical course. TD may be present only a short time after discontinuation of neuroleptics, or it may be suppressed by higher doses (Sigwald et al. 1959; Klawans 1973). TD also is similar to spontaneous stereotypic movements that occur in some psychotic patients (Marsden et al. 1975). The preponderance of clearly documented cases of involuntary movement disorders after neuroleptic use has established TD as an important iatrogenic disorder. In 1980, the American Psychiatric Association task force proposed the following definition of TD: an abnormal involuntary movement, which does not include tremor, resulting from treatment with a neuroleptic drug for 3 months. The patient also must have no other identifiable cause for the movement disorder (Baldessarini et al. 1980). Drugs other than the major tranquilizers that cause TD include the antiemetics (e.g., perphenazine, metoclopramide) (Miller and Jankovic 1989) and some calcium channel blockers, such as cinnarizine and fluphenazine.

A variety of movement disorders resulting from chronic neuroleptic use has been described. The majority of movements are stereotypical or choreic, but dystonia, myoclonus, tics, and akathisia also have been reported (Miller and Jankovic 1990). A common way to classify TD is by these predominant movement disorders (Gardos et al. 1987; Tanner 1986).

Some investigators have suggested that TD is a combination of disorders with different etiologies and pathophysiological mechanisms (Kidger et al. 1980). In a study of 228 patients with TD, two subgroups were identified: orofacial and nonorofacial dyskinetic movements, each with a different set of prognostic and possibly etiologic determinants (Glazer et al. 1988). Severe orofacial movements were associated with increasing age, a diagnosis of schizoaffective or affective disorder, and living alone. Poor prognostic factors for nonorofacial movements included current neuroleptic medication dose, living alone, and no psychiatric medications used other than neuroleptics.

The stereotypic movements found in TD most often affect the oral region, with characteristic tongue protrusion, lip smacking and puckering, grimacing, and chewing (Druckman et al. 1962; Schmidt and Jarcho 1966). These may be accompanied by chorea of the hands, arms, and toes, head nodding and pelvic rocking movements (Tanner 1986; Druckman et al. 1962; Schmidt and Jarcho 1966). Tardive dystonia is now recognized as an important and relatively common sequela of neuroleptic drug use. In one study, 64% of patients with tardive dystonia had segmental dystonia (64%), 21% had focal dystonias (primarily of the head and neck), and 14% had generalized dystonia (Burke et al. 1982). Other movement disorders, including chorea and myoclonus, also were seen with tardive dystonia. Tardive dystonia also has been reported in two TS patients after treatment with neuroleptics (Singh and Jankovic 1988).

Other less common sequelae of chronic neuroleptic treatment include multifocal motor and vocal tics, which resemble those of TS (Klawans et al. 1978; Stahl 1980) and tardive myoclonus (Little and Jankovic 1987). Tardive akathisia also may result from antipsychotic drug use (Gardos et al. 1987; Weiner and Luby 1983; Burke et al. 1989). These patients feel an uncomfortable internal need to move, usually the lower extremities. Movements include stereotypical truncal rocking, complex hand movements, hair and face rubbing, scratching, and respiratory grunting, shouting, and moaning (Burke et al. 1989). While sitting, the legs often show repetitive crossing and uncrossing, abduction and adduction, and pumping up and down. While standing, the patient may march in place, shift weight, or pace (Burke et al. 1989). Tardive akathisia has been reported to result in increased psychosis and violent behavior in schizophrenics (Herrara et al. 1988). Tardive tremor has been recently reported (Stacy and Jankovic 1992).

The prevalence of TD, estimated in various studies to be 0.5% to 56%, has been difficult to determine because TD is diagnosed only by clinical criteria, and its signs and symptoms are variable (Jus et al. 1976). Prevalence rate for TD of 10–20% is probably a reasonable estimate of the true prevalence (Baldessarini et al. 1980; Crane and Smith, 1980). Females and those patients older than 40 years seem to be at increased risk for developing TD (Balderssarini et al. 1980; Jus et al. 1976). Duration of neuroleptic therapy also has been reported to be an important factor in developing TD (Jeste et al. 1979), as has the use of depot neuroleptics or frequent interruption of neuroleptic treatment (Jus et al. 1986; Jeste et al. 1979; Csernansky et al. 1981).

Recently, Waddington and colleagues (1988) have found a relationship between cognitive state and risk of developing TD. In a study of 42 patients with bipolar affective disorder, they found that patients with involuntary movements had lower cognitive scores, fewer depressive episodes, and less prolonged exposure to lithium than those with dyskinesias (Waddington and Youssef 1988). Another study found a greater degree of cognitive impairment in mentally handicapped and epileptic patients with involuntary movements complicating neuroleptic therapy (Youseff and Waddington 1988). An impairment of memory in schizophrenic patients may predispose them to the develop TD to a greater degree than schizophrenics with normal memory (Sorokin et al. 1988).

The pathophysiology of TD is unknown.

The most plausible explanation for this syndrome is the phenomenon of denervation hypersensitivity of dopamine receptors, resulting from the use of dopamine-receptor blockers, such as neuroleptics. Although this is an oversimplification, the hypothesis is supported by the observation that withdrawal of the offending agents or addition of a dopamine agonist will worsen the dyskinesias (Iverson 1975). However, the denervation supersensitivity theory remains speculative (Jenner and Marsden 1986; Fibiger and Lloyd 1986; Waddington 1986). One argument against the theory is that dopamine-depleting agents, such as reserpine and tetrabenazine, do not cause TD in animals, despite their ability to increase sensitivity of dopamine receptors (Hong et al. 1987). A more recent study showed little change in dopamine function in rats with chronic treatment of reserpine, but chronic tetrabenazine treatment did show evidence of dopamine-receptor hypersensitivity (Hong et al. 1987). Dopamine receptors were investigated *in vivo* in eight neuroleptic-free patients with persistent tardive dyskinesia using a positron emission tomography (PET) scanner (Blin et al. 1989). In these patients, the density of dopamine receptors in the striatum was not elevated. However, the striatal dopamine-receptor density was positively correlated with the severity of orofacial dyskinesia.

It is not necessary to emphasize that the best treatment of TD is prevention and early recognition. Whenever possible, withdrawal of the causative agent should be attempted. It has been estimated that up to 33% of patients with TD achieve spontaneous remission within 3 months of cessation of neuroleptics (Jeste et al. 1983). Other prognostic factors for remission of TD after discontinuation of neuroleptics include the speed at which neuroleptics are stopped after the first signs of dyskinesias and the age of the patient (Klawans and Tanner 1983; Quitkin et al. 1977). The sooner the cessation of neuroleptics is after onset of dyskinesias, the more likely the remission, and the older the patient, the less likely remission.

If the symptoms of TD persist and remain disabling, or if neuroleptic withdrawal is not feasible, pharmacologic therapy may be required. The dopamine-depleting drugs, such as reserpine and tetrabenazine, are the most efficacious, and, in contrast to dopamine-receptor blocking drugs, they have not been reported to cause TD (Fahn 1983). Complete remission of TD have been noted with reserpine (Fahn 1983). In a review of 217 patients with various movement disorders, Jankovic and Orman (1988) report an excellent or very good response in 71% of patients with TD and an overall moderate response in patients with tardive dystonia. The main side effects of these drugs include orthostatic hypotension, parkinsonism, and depression. Doses are introduced gradually over weeks to months to help diminish these side effects.

Gamma-aminobutyric acid (GABA) is thought to have an inhibitory effect on the nigrostriatal dopaminergic activity (Penney and Young 1983). Therefore, a number of agents that enhance GABA activity has been used to treat TD with only minimal effects. These agents include sodium valproate (Fisk and York 1987), baclofen (Stewart et al. 1982), and benzodiazepines, such as clonazepam, alprazolam, and diazepam (Bobruff et al. 1981; Csernansky et al. 1988). Results with other agents, such as diltiazem (Bobruff et al. 1981), pindolol (Greendyke et al. 1988) and naloxone (Blum et al. 1987) also have been disappointing. Several well-documented cases of TD improved after electroconvulsive therapy (Gosek and Weller 1988; Chacko and Root 1983).

Heredodegenerative Disorders

Huntington's Disease

HD, initially described in 1872 by George Huntington in residents of Huntington County, New York, has an estimated prevalence of five to 10 per 100,000. It is inherited as an autosomal dominant disorder with complete penetrance. Therefore, all people who carry the gene for HD eventually develop symptoms of the disease, and each offspring has a 50% chance to inherit it.

The average age at onset is from the mid-30s to mid-40s (Martin 1984; Hayden 1981), but it can range as early as infancy to as late as 80 years (Penney and Young 1988). Therefore, most patients have had children before the development of symptoms, unfortunately ensuring continued passage of the HD gene.

Although patients with HD may present with abnormal movements (i.e., chorea, athetosis), behavioral changes frequently precede these motor manifestations. Emotional disturbances can occur up to 1 decade or more prior to the onset of movement disorders (Brackenridge 1971). Symptoms of emotional disturbance can include depression, impulsiveness, fits of rage, and erratic behavior. Impaired memory also may be an early sign of HD (Martin 1984). As the disease progresses, the classis choreiform movements associated with HD become more evident, and signs of dementia become more obvious. Later manifestations include eye movement abnormalities, dysarthria, rigidity, ataxia, myoclonus, dystonia, and weight loss. In the juvenile (symptom onset younger than age 20) form of HD, a parkinsonian type tremor with rigidity, cerebellar signs, and seizures can be seen (Penney and Young 1988; Martin 1984; Hayden 1981).

Diagnosis is based on recognition of the signs and symptoms along with a history of HD in the family. MRI and CT scans of the head may show generalized atrophy, particularly involving the head of the caudate. More recent data demonstrate that cognitive impairment is a clear characteristic of early HD and that it is closely linked to the extent of caudate atrophy as measured by CT scan (Bramford et al. 1989). Using PET, Kuhl and coworkers (1982) showed glucose hypometabolism in the basal ganglia in HD patients. Other investigators used the PET technique to correlate the movement and cognitive dysfunction of HD with cerebral metabolic abnormalities (Young et al. 1986). They found that all HD patients studied had decreased metabolic activity in the caudate, while patients in the early stages of HD had normal putamen metabolism. The level of caudate hypometabolism correlated highly with patients' overall functional capacity and bradykinesia and rigidity. Putamen metabolism correlated with motor functions (i.e., chorea, oculomotor abnormalities, and fine motor coordination). Other studies used PET scans to identify individuals at risk for HD but found the technique relatively unreliable for presymptomatic diagnosis (Young et al. 1987; Young et al. 1988). While some investigators have found abnormalities in the at-risk relatives (Hayden et al. 1987), PET scans remain controversial for presymptomatic HD testing.

A more accurate method of testing at-risk patients was made possible by the discovery of a genetically linked DNA marker near the HD gene on the p terminal segment of chromosome 4 (Gusella et al. 1983). To improve the 5% error rate due to recombination with this marker, there is an ongoing search for a marker closer to the HD gene (Smith et al. 1988; Wasmoth et al. 1988).

The characteristic pathological abnormalities in HD include degeneration and astrogial proliferation of the caudate nucleus, putamen, and to a lesser extent the nucleus accumbens and their projection zones, the globus pallidus and substantia nigra pars reticulata (Von Sattel et al. 1985). Most of the degeneration occurs in the medium-sized spiny neurons that are primarily located in the striatum (Von Sattel et al. 1985). These neurons contain GABA, enkephalins, substance P, and dynorphin and account for a large portion of striatal output projections. Postmortem biochemical analysis of HD-affected brains show decreased GABA neurons and spared somatostatin–neuropeptide gamma-containing neurons (Beal et al. 1988). Some investigators have found the N-methyl-D-aspartate subtype of glutamate receptors to be reduced by 93% in the putamen and cerebral cortex in HD patients, consistent with the hypothesis that neuronal degeneration in HD is due to some excitotoxicity by these receptors (Young et al. 1988).

Maternal inheritance of the HD gene usually is associated with the later-onset HD compared to the juvenile-onset form, which has a paternal inheritance in about 75% of cases (Breckenridge, 1971). This raises the possibility of a maternally transmitted protective factor that relates to the expression of the HD gene. The duration of the disease ranges from 10 to 30 years, depending on the quality of care the patient receives. Death usually occurs from secondary complications such as aspiration pneumonia. The course of juvenile-onset HD is usually more rapid, which may represent an inability of the developing brain to adjust to the associated cell death and atrophy. No difference in disease onset, severity, or progression has been found between homozygous and heterozygous patients (Young et al. 1986).

Dopaminergic drugs that may be used to treat some of the rigid forms of the disease can exacerbate the chorea. Phenothiazines, butyrophenones, and tetrabenazine can help suppress the chorea and alleviate some of the behavioral disorders. Although these antidopaminergic agents are effective in suppressing chorea, they often precipitate or exacerbate underlying parkinsonism.

Wilson's Disease

Wilson's disease (WD), first described by Kinnier Wilson in 1912, is an uncommon inborn error of copper metabolism transmitted as an autosomal recessive trait and localized to chromosome 13 (Wilson 1912; Bowcock et al. 1987). The neurological and systemic manifestations result from an accumulation of copper in various organs but mainly in the brain and liver. Its prevalence has been estimated at 20 to 30 per million, but the estimated prevalence of heterozygous gene carriers is one in 100 to one in 500 (Patten 1988; Scheinberg and Sternlieb 1984). Heterozygotes do not develop the disease but may exhibit mild abnormalities of copper metabolism that can cause confusion in the diagnosis. Siblings of patients with WD have a 25% chance of developing the disease and thus must be screened carefully for WD because early diagnosis and treatment is essential for preventing permanent neurological, hepatic, and renal complications.

Patients with WD present with neurological, psychiatric, or hepatic disease. About one third to one half of WD patients present with neurological abnormalities, usually in the second or third decade. The three neurological syndromes most commonly seen are (1) an akinetic–rigid syndrome resembling parkinsonism; (2) a dystonic syndrome with abnormal involuntary movements and sustained dystonic postures; and (3) postural and intention tremors often with ataxia of limbs and gait, dysarthria, and head titubation (Scheinberg and Sternlieb 1984; Marsden 1987; Walshe 1988). Seizures occur in WD in a minority of cases but with a prevalence that is higher than in the normal population (Dening et al. 1988). Cognitive changes and dementia also occur.

Ophthalmological abnormalities include disturbances of eye movements (Marsden 1987) and the well-characterized Kayser–Fleischer (K–F) rings. K–F rings are found in 95% of WD patients and virtually all WD patients with neurological or psychiatric abnormalities, but they are not specific for WD. They also may be seen in primary biliary cirrhosis, silver intoxication, chronic active hepatitis, and cryptogenic cirrhosis.

Four psychiatric symptom clusters have been described in WD. They include (1) affective disorders; (2) behavioral or personality disorders; (3) schizophrenia-like symptoms; and (4) cognitive disorders, but affective and behavioral or personality are the most common (Dening 1985).

The hepatic manifestations of WD are the most common presenting symptoms in childhood WD. These manifestations include signs and symptoms of acute hepatitis (i.e., nausea, vomiting, malaise, jaundice, anorexia, weakness). These may progress to fulminant liver failure or cirrhosis.

The two simplest methods for diagnosing WD are examination for K–F rings and measurement of serum ceruloplasmin, which will be decreased (less than 200 mg/L) in 95% of WD patients. Also, 24-hour urinary copper excretion will be increased (100–1000 μg/24 hours) in WD. If there is doubt, a liver biopsy can confirm the diagnosis. In WD, liver copper concentrations will be markedly increased (more than 250 μg/g dry weight) (Bowcock et al. 1987; Patten 1988; Scheinberg and Sternlieb 1984).

CT or MRI scans often show ventricular dilation, cortical atrophy, brain stem atrophy, or hypodense areas in the basal ganglia (Walshe 1988). EEG findings in WD include generalized or focal abnormalities that are usually mild and nonspecific (Dening et al. 1988).

Treatment of WD consists of increasing copper excretion but decreasing intake of foods high in copper also is advisable. Penicillamine has been the gold standard for treatment of WD since the late 1950s (Walshe 1988). The recommended dose for older children and adults is 1 g in four divided doses. Because penicillamine has antipyridoxine effects, it also is recommended that patients take 25 mg pyridoxine daily. With treatment, the patient may improve dramatically, have little to no improvement, or even worsen (Brewer et al. 1987). Usually, the shorter the duration of symptoms are, the better the response to penicillamine will be. Prophylactic therapy with penicillamine in presymptomatic individuals is similar to that of symptomatic patients. The side effects of penicillamine are many, including hypersensitivity reactions, bone marrow suppression, and renal problems. This may limit its usefulness in treatment.

Triethylenetetramine dihydrochloride (also called trien, TETA, and Cuprid) provides a useful alternative to penicillamine. The recommended adult dose is 400 to 800 mg three times a day. Trien has been shown to be an effective decoppering agent due to chelation, but experience with long-term usage is not available. Zinc has been shown to block copper absorption in the small intestine, and also has been recommended in the treatment of WD. The usual dose is 150 mg in three divided doses. Side effects are minimal, but again, experience with long-term usage is limited (Brewer et al. 1987).

The prognosis of WD depends on the amount of damage prior to treatment. A normal life span is possible for patients treated prior to onset of neurological defects, psychiatric abnormalities, or liver disease. This emphasizes the need for screening and early diagnosis in the siblings of patients with WD. Even symptomatic patients have a good chance for good recovery and for a normal life, particularly if treatment is instituted early.

Myoclonus

Myoclonus has been defined as brief, sudden, shock-like movement caused by involuntary muscle contraction (positive myoclonus) or a brief inhibition of muscle contraction (negative myoclonus) (Fahn et al. 1986; Hallet et al. 1986; Patel and Jankovic 1988). Myoclonus can be confused with other jerk-like movements. Simple tics may be indistinguishable from myoclonus, especially if they are not accompanied by vocalizations or other features of TS (Jankovic and Fahn 1986). Tics, however, consist of more coordinated movements, can be voluntarily suppressed for at least a brief period, and often are associated with a feeling of tension released by the occurrence of the tic (Jankovic and Fahn 1986; Lees and Tolosa 1988). Chorea also is characterized by quick, random, involuntary muscle contractions, but in contrast to myoclonus, choreic movements are continuous and tend to flow from one muscle to another (Shoulson 1986). The sustained contractions of agonist and antagonist muscles producing twisting movements, typical of dystonia, are easy to differentiate from myoclonus. However, rapid dystonic movements may prevent a diagnostic problem. When these rapid movements predominate, the term "myoclonic dystonia" has been used (Obeso et al. 1983). Several families have been reported with the combination of myoclonus and dystonia (Kurlan et al. 1987; Kurlan et al. 1988). Fasciculations, which is the spontaneous firing of a single motor unit, may be confused with myoclonus, especially when motor units are enlarged, as in motor neuron disease (Hallet et al. 1986). Minipolymyoclonus, which can be seen in benign forms of spinal muscular atrophy, appears as irregular tremulousness in the outstretched hands (Spiro 1970). Asterixis is a form of a negative myoclonus that appears as an arrhythmic, involuntary brief lapse of posture during tonic contraction (Patel and Jankovic 1988). It can be elicited in awake patients with outstretched arms, hands dorsiflexed, and fingers extended and abducted. In comatose patients, flapping movements of the hips can be elicited by holding down the patient's feet and flexing and abducting the hips to 60 to 90 degrees (Nodo et al. 1985).

Myoclonus can be classified in many ways: by its distribution (focal, segmental, or generalized); by its etiology; by its periodicity (rhythmic, arrhythmic, or oscillatory); by its synchronization (synchronous vs. asynchronous); by its relationship to motor activity (at rest or with action, with intention of external stimuli, such as sudden noise, visual threat, pain, muscle stretch, or certain types of movement); and by whether it is produced by muscle contraction (positive myoclonus) or by muscle relaxation (negative myoclonus) (Fahn et al. 1986; Hallet et al. 1986; Patel and Jankovic 1988).

Focal myoclonus is a myoclonus that involves a localized group of muscles and can be rhythmic or arrhythmic. It differs from tremor because it is produced by repetitive contractions of agonists only and may persist during sleep (Hallet et al. 1986; Patel and Jankovic 1988). Focal cortical myoclonus also involves a discrete group of muscles, typically more distal than proximal and more flexor than extensor. It can be elicited with volitional movement or by cutaneous or stretch receptor stimulation, such as flicking a fingernail (Obeso et al. 1985). Focal cortical myoclonus also may occur as a manifestation of epilepsia partialis continua or simple motor seizures (Obeso et al. 1985; Thomas et al. 1977; Hallett et al. 1979).

Segmental myoclonus involves muscles or groups of muscles supplied by contiguous segments of the spinal cord (spinal myoclonus) or brain stem (branchial myoclonus), without associated EEG changes (Jankovic and Pardo 1986). The frequency is usually between 1 and 3 Hz but can vary in rate from 1 to 600 Hz/min (Jankovic and Pardo 1986).

Segmental myoclonus has been seen after root, plexus, and peripheral nerve lesions (Marsden et al. 1984; Banks et al. 1985; So-

taniemi 1985). If branchial structures are involved, the segmental myoclonus may appear as palatal movements and rhythmic contractions of the lower facial and jaw muscles (oculomasticatory myorrhythmia) (Schwartz et al. 1986). This is typically seen in cerebral Whipple's disease (Jankovic and Pardo 1986; Schwartz et al. 1986).

Multifocal myoclonus involves focal myoclonic jerks occurring asynchronously in noncontiguous parts of the body, while generalized myoclonus consists of synchronous, widespread involvement. Generalized myoclonus is due to involvement of reticular mechanisms and hyperexcitability of the caudal brain stem reticular formation or other neuronal motor control circuits (Patel and Jankovic 1988).

One of the best methods used to study myoclonus involves video-monitored EEG combined with time-locked EEG–electromyography measurements and back averaging (Kelly et al. 1981; Shibasaki et al. 1978). High-voltage somatosensory-evoked potentials (SEP) have been noted in patients with stimulus-sensitive cortical myoclonus (Hallett et al. 1979; Kelly et al. 1981; Shibasaki et al. 1978).

There are four major etiologic categories of myoclonus: physiological, essential, epileptic, and symptomatic. Physiological myoclonus occurs in healthy individuals and often is manifested as a startle response or while falling asleep. These sleep or hypnic jerks are quick (10–100 msec), myoclonic movements that tend to occur in stage I and REM sleep (Lugarsi et al. 1986). Sleep starts (also known as massive myoclonic jerks or sudden body jerks), another type of physiological myoclonus, are sporadic, involuntary muscle contractions of primarily axial and proximal limb musculature. They usually last longer than 1 second and occur while a patient is falling asleep or in stage I or II of sleep. They may be precipitated by external stimuli, such as sudden noise (Lugarsi et al. 1986; Walters and Hening 1987). Patients frequently report a feeling of falling into a void or a feeling of shock (Lugarsi et al. 1986). Other forms of physiological myoclonus include benign neonatal sleep myoclonus (Resnick et al. 1986), anxiety-induced myoclonus, exercise-induced myoclonus, hiccups (Lewis 1985), and the normal startle reflex (Wilkens et al. 1986).

Essential myoclonus is a term used for patients with persistent myoclonus and, except for occasional coexistence with essential tremor and dystonia, no other associated neurological deficit or epilepsy (Fahn et al. 1986). Hereditary essential myoclonus is a subset of this group. It is inherited in an autosomal dominant pattern, with onset younger than age 20 and variable severity (Bressman and Fahn 1986; Korten et al. 1974). It generally has a benign course and no associated EEG abnormalities (Korten et al. 1974).

The other disorders grouped with essential myoclonus are heterogenous with no definite age of onset and with a sporadic occurrence. These movements often are increased by stimuli such as bright light or movement (Bressman and Fahn 1986). Examples of essential myoclonus include ballistic movement overflow myoclonus (Hallett et al. 1977), oscillatory myoclonus (Fahn and Singh 1981; Obeso et al. 1983), rhythmic essential segmental myoclonus (Jankovic and Pardo 1986), nonrhythmic essential segmental myoclonus, and nonrhythmic essential multifocal myoclonus (Bressman and Fahn 1986). These disorders do not show progressive disability and can improve with medications (Bressman and Fahn 1986). Ballistic movement overflow myoclonus may improve with alcohol or propranolol (Hallett et al. 1977), while oscillatory myoclonus responds to alcohol and occasionally clonazepam (Fahn and Singh 1981; Obeso et al. 1983). Rhythmic essential segmental myoclonus can improve with clonazepam or valproate, and nonrhythmic essential segmental and multifocal myoclonus responds to clonazepam (Bressman and Fahn 1986).

Nocturnal and symptomatic nocturnal myoclonus are other forms of essential myoclonus. Nocturnal myoclonus is a periodic movement, occurring once every 20 to 40 seconds and involving dorsiflexion of the toes or feet. It is associated with flexion of the legs and occasionally the arms (Lugarsi et al 1986). It is usually associated with EEG signs of arousal and occurs most often in stage I and II of sleep (Lugarsi et al. 1986). Symptomatic nocturnal myoclonus often occurs while the patient is awake and may be associated with restless leg syndrome (Lugarsi et al. 1986; Walters and Hening 1987). The patient often complains of poorly defined paresthesias in the legs that improve with motor activity of the legs. Many medications have been successful in the treatment of symptomatic nocturnal myoclonus with restless leg syndrome. These include levodopa, bromocriptine, baclofen, diazepam, 5-hydroxytryptophan, phenoxybenzamine, clonidine, carbamazepine, and various opioids (Lugarsi et al. 1986; Walters and Hening 1987; Martinelli et al. 1987; Montplasir et al. 1986; Akpinar 1987).

Epileptic myoclonus describes myoclonic movements associated with seizures but without significant encephalopathy. This disorder is discussed in more detail in the section entitled ''Seizure Disorders.''

Symptomatic myoclonus refes to a group of disorders with myoclonus associated with progressive or static encephalopathies. A major category in this group is progressive myoclonus epilepsy (PME). Unverricht-Lundborg disease (Baltic myoclonus) was one of the first types of PME to be described (Unverricht 1891; Lundborg 1903). It is usually inherited in an autosomal recessive pattern, with an incidence of one per 20,000 in Finland, but it is less common outside the Baltic region (Koskiniemi 1986; Norio and Koskiniemi 1979). The criteria for diagnosis of Unverricht-Lundborg disease include stimulus-sensitive myoclonus, age at onset between 6 and 15 years, generalized tonic–clonic seizures, EEG abnormalities, and a progressive clinical course (Koskiniemi 1986; Berkovic et al. 1986). Myoclonus is usually severe and precipitated by movement, stress, and sensory stimuli. Neurological signs are minimal to absent at onset, but ataxia, dysarthria, and intention tremor usually develop and are accompanied by gradual intellectual decline. Progression of the disease varies, but most patients survive into adulthood. Death usually is due to complications of the seizures, anticonvulsant toxicity, or infections (Eldridge et al. 1983; Iivanainen and Himberg 1982).

Lafora-body disease was described approximately 20 years after Unverricht's initial clinical description of PME. Interneuronal inclusion bodies were found in the brain of a patient with myoclonus, seizures, and dementia (Lafora and Glueck 1911). Age of onset is 11 to 18 years, with patients showing progressive neurological deterioration, dementia, apraxia, and cortical blindness (Berkovic et al. 1986; Rapin 1986). Myoclonic seizures are not as prominent, and approximately half of the patients have focal occipital seizures. Most patients become severely disabled or die before age 25 or about 5 to 8 years after onset (Patel and Jankovic 1988; Berkovic et al. 1986; Rapin 1986). Diagnosis is based on the clinical findings and on the detection of the characteristic periodic acid–Schiff-positive inclusions that are present in brain, liver, skeletal muscle, and cardiac muscle (Patel and Jankovic 1988; Berkovic et al. 1986; Rapin 1986). The condition is inherited as an autosomal recessive trait.

Ramsay Hunt syndrome was first described in 1921 as PME with associated cerebellar ataxia (Hunt 1921). The syndrome described by Hunt has no specific pathology and a variety of degenerative disorders can produce it, including spinocerebellar degenerations, PSP, cerebellar telangiectasias, mitochondrial and metabolic encephalopathies (Berkovic et al. 1986), and celiac disease (Lu et al. 1986). A more recent report of 13 patients with Ramsay Hunt syndrome found no evidence of mitochondrial abnormalities on muscle biopsies (Tassinari et al. 1989). The designation of Ramsay Hunt as a true syndrome is controversial (Marsden and Obeso 1989). PME also can be seen in various storage diseases, including neuronal ceroid lipofuscinosis, sialidoses, and mitochondrial encephalomyelopathies (Berkovic et al. 1986; Rapin 1986), most with autosomal recessive inheritance.

One of the most frequent causes of symptomatic myoclonus is acute and prolonged hypoxia and ischemia (Lance–Adams syndrome). In a recent study of 114 comatose adult survivors of cardiopulmonary resuscitation, 50 (44%) had seizures or myoclonus, and 40 (35%) had myoclonus alone (Krumholz et al. 1988). Previous studies have suggested impaired serotonin transmission as the cause of posthypoxic myoclonus (Fahn 1986; Von Woert et al. 1986), but evidence of serotonin hyperactivity has been noted in one patient with this disorder (Gimenez-Roldan et al. 1988).

Myoclonus can be seen as a characteristic and early feature of Creutzfeldt-Jakob disease (Davanipour et al. 1986) and Alzheimer's disease (Hauser 1986). It also can occur in viral encephalopathies (most commonly herpes simplex encephalitis); various metabolic encephalopathies, including hepatic failure and renal failure (Hallett et al. 1986; Patel and Jankovic 1988), toxin exposure, drug-use, and trauma (Hallett et al. 1986; Patel and Jankovic 1988). Neuroleptic drugs have recently been reported to cause tardive myoclonus (Little and Jankovic 1987).

The pathology of generalized myoclonus is unknown, but clinical studies indicate an abnormality in the brain stem reticular formation. Serotonergic (Gascon et al. 1988), cholinergic (Chokroverty et al. 1987), gamma-aminobutyric acidergic (Crossman et al. 1988), glycinergic (Truong et al. 1988), and peptidergic (Walters et al. 1986) systems have been implicated in the pathogenesis of myoclonus.

Improvement of myoclonic symptoms is possible in many patients, but the underlying cause will determine the progression of symptoms. Thus, the myoclonus caused by progressive or static encephalopathies has a much poorer prognosis than myoclonus with other causes. Because various biochemical systems are implicated in the pathogenesis of myoclonus, many drugs have variable efficacies. The CSF concentration of 5-hydroxyindoleacetic acid (5-HIAA), the main metabolite of serotonin, has been low in some patients with postanoxic and post-traumatic myoclonus (Van Woert et al. 1986). Thus, these patients may improve with 5-hydroxytryptophan (5-HT), a serotonin precursor (Van Woert et al. 1986). Other drugs that are effective in the treatment of myoclonus include clonazepam, sodium valproate, primidone, anticholinergics, and opioids (Walters et al. 1986; Obeso et al. 1989). Piracetam, a nootropic drug in which the mechanism of action is unknown, has been shown to improve cortical myoclonus (Obeso et al. 1989; Obeso et al. 1988), but more controlled trials are needed.

Chorea, Athetosis, and Ballism

The term chorea is derived from the Latin choreus (dancing). It refers to brief, involuntary, irregular, purposeless, arrhythmic, abrupt, random, rapid, unsustained, and continuous movements that flow from one body part to another and vary in amplitude (Padberg and Bruyn 1986; Shoulson 1986; Fahn 1989). Mild forms of chorea can appear as semipurposeful movements. Large amplitude movements, especially if involving one limb or one side of the body, are termed ballismus. Slow choreatic movements or chorea superimposed on dystonic postures is termed choreoathetosis. Many consider ballism as a severe form of chorea and athetosis as a slow form of chorea. The movements may be generalized or may involve only the extremities, trunk, face, or oral–buccal–lingual region. Orofacial chorea is more stereotypical, particularly if it is due to tardive dyskinesia. It may be symmetrical, asymmetrical, or unilateral (Padberg and Bruyn 1986; Shoulson 1986; Fahn 1989).

Patients with chorea frequently have difficulty maintaining sustained postures and tend to drop things. This motor impersistence can be manifested, for example, by an inability to keep the tongue protruded or to maintain a tight grip ("mild-maid" grip). The movements of Huntington's disease are classic examples of chorea (Tan et al. 1976).

Marsden and colleagues (1983) found no characteristic pattern of EMG activity in chorea. They did find continuous changes in activation order of muscles and in length of the EMG bursts. They also noted considerable overlap in EMG patterns of chorea with those of myoclonus and dystonia.

There are many causes of chorea, but the brunt of pathology usually is subcortical, most notably the caudate, putamen, globus pallidus, and subthalamic nucleus (Marsden 1984; Marsden 1984; Penney and Young 1983). The mechanism for the chorea in Huntington's disease is thought to be inadequate striatal output of GABA, resulting in tonic inhibition of the thalamus and loss of ability to maintain normal motor behaviors and release of choreic movements (Shoulson 1986). In patients with subthalmic nucleus infarcts, chorea is thought to arise from defective disinhibition of thalamic cells by the subthalamic nucleus (Johnson and Fahn 1977; Dewey and Jankovic 1989). Heightened dopaminergic activity may induce chorea from inadequate collateral inhibition of output neurons from the striatum and globus pallidus to the thalamus. This can be seen in levodopa-treated parkinsonian patients and in patients with tardive dyskinesia (Shoulson 1986).

Numerous disorders have been described in which chorea is one chief neurological abnormality. Shoulson (1986) listed more than 150 diseases, drugs, and toxins. The causes of chorea can be divided into two major groups—primary and secondary. The primary group includes idiopathic, hereditary, and paroxysmal causes; the secondary group includes infectious, immunological, vascular, drugs and toxins, and metabolic and endocrine disorders.

The best known cause of hereditary chorea, Huntington's disease, was discussed previously in this chapter. Most of the hereditary choreas that occur in infancy, childhood, or early adolescence are either X-linked or recessive (Gordon 1980; Adams and Lyon 1982). Examples include inborn errors of metabolism affecting the CNS, phakomatoses, lipofuscinoses, gangliosidoses, Hallervorden-Spatz, myoclonic epilepsy, Lesch–Nyhan syndrome (Jankovic et al. 1988), and TS (Shoulson 1986).

The hereditary choreas that are late adolescent or adult onset usually are inherited in an autosomal dominant pattern. Examples of hereditary chorea include Huntington's disease, benign familial chorea, and neuroacanthocytosis. Benign familial chorea (also called hereditary nonprogressive chorea and hereditary chorea without dementia) is not associated with dementia or any other neurological deficits beyond chorea, and it is nonprogressive (Haerer et al. 1967; Harper 1978; Schady and Meara 1988). PET scans have shown striatal glucose hypometabolism in some patients (Suchowersky et al. 1986). The gene for benign familial chorea is not in the same region as the gene for Huntington's disease (Quarrel et al. 1988).

Neuroacanthocytosis, a slowly progressive disease, commonly presents in young adulthood and is manifested by tics, seizures, amyotrophy, absent deep tendon reflexes, high serum creatine phosphokinase, feeding dystonia, and self-mutilation, particularly lip and tongue biting (Brin et al. 1988; Sakai et al. 1981). PET scans show glucose hypometabolism in the caudate nucleus, along with a striatal degeneration that is similar to Huntington's disease (Philips et al. 1987). Spitz and coworkers (1985) found that the hyperkinetic movement disorder in the early stages of the disease may be replaced gradually by parkinsonism in more advanced stages. Decreased dopamine levels have been found in many areas of the brain; additionally low substance P in the substantia nigra and striatum and increased norepinephrine in the putamen and pallidum have been noted in a postmortem examination (De Yebenes et al. 1988). The paroxysmal choreas are similar to the paroxysmal dystonias and can be classified as kinesigenic or nonkinesigenic. A discussion of this subject can be fund in the dystonia section.

Wilson's disease, as previously mentioned, also can present as many types of movement disorders, including chorea. Thus, any patient presenting with chorea, especially in childhood or as a young adult, should be evaluated for Wilson's disease. Wilson's disease is treatable and early diagnosis and treatment can prevent and possibly reverse neurological deficits (Walshe 1988).

Sydenham's chorea, a form of secondary chorea, has been pathogenetically linked to beta-hemolytic streptococcus infection and can occur in acute, remitting, or persistent patterns (Nausieda et al. 1980; Stollerman 1975). The incidence of Sydenham's chorea, has decreased with the introduction of antibiotics (Nausieda et al. 1980). In a study of 1000 patients with rheumatic fever between 1920 to 1950, 52% of patients had associated Sydenham's chorea (Bland and Jones 1951). Later studies showed that after 1968, this number dropped to 8% of patients (Stollerman 1975). Other neurological findings include speech impairment, gait disturbance, reflex changes, weakness, encephalopathy, and seizures. The age of onset primarily is between ages 5 and 15, with duration ranging from 1 to 22 weeks and a recurrence rate of 20% (Nausieda et al. 1980; Stollerman 1975). There have been rare cases of adult onset (Gordon 1988) and persistence throughout life (Gibb et al. 1985). Residual persistent behavioral and EEG changes are common (Bird et al. 1976). Women with a history of Sydenham's chorea may be more susceptible to chorea gravidarum and are at higher risk to develop chorea when using oral contraceptives (Nausieda et al. 1983). Antineuronal IgG antibodies to human caudate and subthalamic nucleus have been found in the sera of children with Sydenham's chorea (Husby et al. 1976). Other infections have been shown to cause chorea, including mononucleosis (Friedland and Yahr 1977), HIV infection (Nath et al. 1987), ECHO virus type 25 (Peters et al. 1979), tuberculosis (Riela and Roach 1982), bacterial endocarditis (Medley 1963), Lyme disease (Reik et al. 1979), and legionnaires' disease (Bamford and Hakin 1982).

Systemic lupus erythematosus (SLE) can cause an intermittent chorea that is thought to be immunological (Lusins 1975; Bruyn and Padberg 1984). Cardiolipin antibodies and antiphospholipid antibodies have been associated with SLE chorea (Asherson et al. 1988). The male to female ratio is 1:9, with chorea occurring in 1% to 4% of SLE cases (Grigor et al. 1978). Chorea also has been reported as a remote effect of carcinoma (Albin et al. 1988).

Chorea can be caused by a variety of drugs and toxins, including neuroleptics (Goetz and Klawans 1982); dopaminergic drugs; anticholinergics (Warne and Grubbay 1979); anticonvulsants, such as phenytoin and carbamazepine (Chadwick et al. 1976; Weaver et al. 1988); noradrenergic stimulants, such as amphetamines (Lundh and Iunving 1981); methylphenidate (Weiner et al. 1978); oral contraceptives (Dove); and many other drugs (Miller and Jankovic 1990). Alcohol use (Mullin et al. 1970) and carbon monoxide poisoning (Davous et al. 1986) also can cause chorea. There are many other causes of chorea, including hyperthyroidism (Fischbeck and Layzer 1979; Drake 1987), various metabolic disorders (Padberg and Bruyn 1986; Shoulson 1986), multiple sclerosis (Mao et al. 1988), tuberous sclerosis (Evans and Jankovic 1983), and CNS lymphoma (Poewe et al. 1988).

Generalized chorea is a rare consequence of cerebrovascular disease, usually due to bilateral basal ganglia lacunes (Tabaton et al. 1985; Sethi et al. 1987). Earlier studies found that the most common cause of hemiballism–hemichorea was vascular, usually infarction of hemorrhage, in the contralateral subthalamic nucleus or caudate nucleus (Johnson and Fahn 1977). A more recent study of 21 patients with hemiballism–hemichorea found a cause other than stroke for almost half the patients (Dewey and Jankovic 1989). Hemiballism–hemichorea is usually self-limited.

Treatment of chorea typically rests on correcting the underlying disorder. Thus, patients with Syndenham's chorea, SLE-induced chorea, metabolic or endocrine disorder, or drug intoxications have a good chance of improvement with proper treatment. Unfortunately, many of the underlying causes are not amenable to therapy, such as most of the inherited choreas. In these cases, antidopaminergic agents, such as phenothiazines, butyrophenones, reserpine, and tetrabenazine, may provide symp-

tomatic relief (Shoulson 1982; Jankovic and Orman 1983). While these antidopaminergic agents may improve the chorea, multiple side effects can occur, including dystonic reactions, parkinsonism, and tardive dyskinesias. The prognosis is poor in patients with heredodegenerative diseases such as Huntington's disease.

Peripherally Induced Movement Disorders

CNS damage or dysfunction is the usual cause for involuntary motor movements. However, lesions in spinal roots, cervical or lumbar plexus, or peripheral nerves also can cause involuntary movements. These peripherally induced movement disorders are becoming more recognized (Marsden et al. 1984; Schott 1985; Schott 1986; Scherokman et al. 1986; Brin et al. 1986).

Segmental myoclonus, while often caused by spinal cord or brain stem lesions, also may be produced by compression or other injury to the cranial nerves (e.g., hemifacial spasm), roots, plexi, or peripheral nerves (Luttrell and Bang 1958; Luttrell et al. 1959; Kao and Crill 1972). Palatal myoclonus, a segmental myoclonus, has been reported following head and neck trauma (Jacobson and Garmen 1949; Matuso and Ajaz 1979). Hemifacial spasm usually is attributed to compression of the facial nerve by aberrant, anomalous, or arteriosclerotic arteries in the posterior fossa or by a mass in the cerebellopontine angle (Digre and Corbett 1988). It usually begins in the fourth to fifth decade with clonic contractions around the eye. These can spread to muscles of one side of the face, including the platysmal muscle. The movements usually occur in bursts and can occur in sleep. Botulinum toxin injections into the contracting muscles or neurosurgical decompression of the aberrant arteries (or lesion) compressing the facial nerve provide satisfactory relief in most cases (Jannetta et al. 1977).

Many types of spinal cord lesions can produce spinal myoclonus, including trauma, tumors, ischemia, infection, demyelinating disease, cervical spondylosis, and arteriovenous malformations (Luttrell et al. 1959;

Kao and Crill 1972; Jankovic and Pardo 1986; Garcin et al. 1968; Davis et al. 1981; Dhaliwal and McGreal 1974; Hoehn and Cherington 1977; Levy et al. 1983). A focal myoclonus of the legs has been reported with lumbar radiculopathy and after lumbar laminectomy for lumbar stenosis and root lesions (Jankovic and Pardo 1986). A more in depth discussion on spinal myoclonus can be found in the section on myoclonus. Focal myoclonus of the arm following radiotherapy and trauma has been attributed to brachial plexus involvement (Banks et al. 1985).

Injury to peripheral nerves or lumbar roots may cause another peripherally induced movement disorder called painful legs and moving toes (Spillane et al. 1971; Nathan 1978; Montagna et al. 1983; Schott 1981). This syndrome consists of a combination of pain in the feet and writhing or dystonic movements of the toes and sometimes the feet. In addition to lumbar spondylosis with radiculopathy, similar symptoms have been described with L5 herpes zoster infection, S1 root compression and cauda equina lesions (Nathan 1978), peripheral neuropathy (Montagna et al. 1983), and minor leg trauma (Schott 1981). "Painful arm and moving fingers" have been described with a brachial plexus lesion (Verhagen et al. 1985).

Marsden and colleagues described 5 patients who developed reflex sympathetic dystrophy, Sudeck's atrophy and abnormal movements after limb injury (Marsden et al. 1985). The abnormal movements included myoclonus, muscle spasms, and dystonia of the injured limb. Similar sequelae have been associated with amputations (Steiner et al. 1974; Marion et al. 1989) and phantom tardive dyskinesia of the amputated arm (Jankovic and Glass 1985). These and other reports indicate a relationship between peripheral nerve injury, with or without reflex sympathetic dystrophy, and various movement disorders, including myoclonus, tremor, and dystonia.

In one study of 23 patients with onset of abnormal movements within 1 year of peripheral nerve injury, focal dystonia of the involved body part was found in 18, nine of whom had associated reflex sympathetic dystrophy (Jankovic and Van Der Linden 1988). One of five patients with peripherally induced tremor had reflex sympathetic dystrophy. In 65% of the patients studied, possible predisposing factors for these peripherally induced movement disorders, such as perinatal problems, use of neuroleptics, family history or patient history of essential tremor, and AIDS-related complex, were found. Even torticollis and other forms of cervical dystonia have been attributed to head, neck, and VIII or IX cranial nerve injury (Sheehy and Marsden 1980; Drake 1987; Bronstein et al. 1987).

The mechanisms underlying these peripherally induced movement disorders are not completely understood. The frequent association of pain and peripherally induced movement disorders suggests that the motor symptoms are analogous to, and share common pathogenetic mechanisms with, peripherally induced sensory disturbances, such as phantom pain (Jankovic and Glass 1985) and causalgia (Schott 1986; Schwarttman and McLellan 1987). It has been suggested that an altered afferent input into the spinal cord or brain stem may lead to reorganization of the local neuronal circuitry and enhancement of evoked and spontaneous motor output, perhaps mediated by hyperexcitable gamma neurons (Eccles et al. 1962; Loeser and Ward 1967; Loeser et al. 1968).

Treatment of peripherally induced movement disorders is difficult. Effective, although temporary, relief of dystonic movements may be achieved with botulinum A toxin injections (Jankovic and Brin, In press). Various analgesic treatments, including sympathectomy, may improve the causalgic pain but usually do not improve the movement disorder.

References

Parkinson's Disease

Barbeau, A. Etiology of Parkinson's disease: A research strategy. Can. J. Neurol. Sci. 11:24–28; 1984.

Barbeau, M.; Cloutier, T.; Plasse, L.; et al. Environmental and genetic factors in the etiology of Parkinson's disease. In: Yahr, M. D.; Bergmann, K. J., eds. Advances in neurology: parkinson's disease, Vol. 45. New York: Raven Press; 1986: p. 299–306.

Bernheimer, H.; Birkmayer, W.; Hornykiewicz; et al. Brain dopamine and the syndromes of Parkinson and Huntington. J. Neurol. Sci. 20:415–455; 1973.

Birkmayer, W.; Reiderer, P.; Youdim, J. B. H. Distinction between benign and malignant type of Parkinson's disease. Clin. Neurol. Neurosurg. 81:158–164; 1979.

Birkmayer, W.; Knoll, J.; Riederer, P.; et al. Increased life expectancy resulting from addition of L-deprenyl to madopar treatment in Parkinson's disease. A long-term study. J. Neurol. Transm. 64;113–127; 1985.

Brown, R. G.; Marsden, C. D. How common is dementia in Parkinson's disease? Lancet 2:1262–1265; 1984.

Cedarbaum, J. M.; McDowell, F. H. Sixteen year follow-up of 100 patients begun on levodopa in 1968: emerging problems. In: Yahr, M. D.; Bergmann, K. J., eds. Advances in neurology: Parkinson's disease, Vol. 45. New York: Raven Press; 1986: p. 469–472.

Diamond, S. G.; Markham, C. H.; Hoehn, M. M.; et al. Effect of age at onset of progression and mortality in Parkinson's disease. Neurology 39:1187–1190; 1989.

Gerlak, R. P.; Clark, R.; Stump, J. M.; et al. Amantadine-dopamine interaction. Science 169;203–204; 1970.

Gershanik, O. S. Parkinsonism of early onset. In: Jankovic, J.; Tolosa, E., eds. Parkinson's disease and movement disorders. Baltimore-Munich: Urban and Schwarzenberg, Inc.; 1988: p. 191–204.

Gibb, W. R. G.; Lees, A. J. A comparison of clinical and pathological features of young- and old-onset Parkinson's disease. Neurology 38:1402–1406; 1988.

Goetz, C. G.; Tanner, C. M.; Glantz, R. H.; et al. Chronic agonist therapy for Parkinson's disease: a 5 year study of bromocriptine and pergolide. Neurology 35:749–751; 1985.

Goetz, C. G.; Tanner, C. M.; Stebbins, G. T.; et al. Risk factors for progression in Parkinson's disease. Neurology 38:1841–1844; 1988.

Goetz, C. G.; Olanow, C. W.; Koller, W. C.; et al. Multi-center study of autologous adrenal medullary transplantation to the corpus striatum in patients with advanced Parkinson's disease. N. Engl. J. Med. 320:337–341; 1989.

Goetz, C. G.; Shannon, K. M.; Tanner, C. M.; et al. Agonist substitution in advanced Parkinson's disease. Neurology 39:1121–1122; 1989.

Gotham, A. M.; Brown, R. G.; Marsden, C. D. Depression in Parkinson's disease: a quantitative and qualitative analysis. J. Neurol. Neurosurg. Psychiatry 49:381–389; 1986.

Ho, S. H.; Woo, J.; Lee, C. M. Epidemiologic study of Parkinson's disease in Hong Kong. Neurology 38:1314–1318; 1989.

Hoehn, M. M.; Yahr, M. D. Parkinsonism: onset, progression and mortality. Neurology 17:427–442; 1967.

Hoehn, M. M. M. Parkinson's disease: Progression and mortality. In: Yahr, M. D.; Bergmann, K. J., eds. Advances in neurology: Parkinson's disease, Vol. 45. New York: Raven Press; 1986: p. 457–461.

Iversen, S. Dopamine depletion and replacement. In: Evered, D.; O'Connor, M., eds. Functions of the basal ganglia: CIBA foundation symposium 107. London: Pitman; 1984: p. 216–221.

Jankovic, J. Long-term use of dopamine agonists in Parkinson's disease. Clin. Neuropharmacol. 8:131–140; 1985.

Jankovic, J.; Reches, A. Parkinson's disease in monozygotic twins. Ann. Neurol. 19:403–408; 1986.

Jankovic, J.; Calne, D. B. Parkinson's disease: etiology and treatment. In: Appel, S. H., ed. Current neurology. Chicago: Year Book Medical Publishers, Inc.; 1987: p. 193–234.

Jankovic, J. Parkinson's disease: recent Advances in Therapy. South Med. J. 81(8):1021–1027; 1988.

Jankovic, J. Neural transplants in the treatment of Parkinson's disease and other neurodegenerative disorders. In: Jankovic, J.; Tolosa, E., eds. Parkinson's disease and movement disorders. Baltimore-Munich: Urban and Schwarzenbarg, Inc.; 1988: p. 471–480.

Jankovic, J.; Grossman, R.; Goodman, C.; et al. Clinical, biochemical and neuropathologic findings following transplantation of adrenal medulla to the caudate nucleus for treatment of Parkinson's disease. Neurology 39:1227–1234; 1989.

Jankovic, J.; Marsden, C. D. Therapeutic strategies in Parkinson's disease. In: Jankovic, J.; Tolosa, E., eds. Parkinson's disease and movement disorders. Baltimore-Munich: Urban and Schwarzenberg, Inc.; 1988: p. 95–119.

Jankovic, J.; Schwartz, K.; Van der Linden, C. Comparison of Sinemet CR4 and standard Sinemet: double-blind and long-term open trial in parkinsonian patients with fluctuations. Movement Disorders 4:303–309; 1989.

Jankovic, J.; McDermott, M.; Carter, J.; et al. Variable expression of Parkinson's disease. A baseline analysis of the DATATOP cohort. Neurology 40:1529–1534; 1990.

Jansson, B.; Jankovic, J. Low cancer rates among patients with Parkinson's disease. Ann. Neurol. 17:505–509; 1985.

Joseph, C.; Chassan, J. B.; Koch, M. L. Levodopa in Parkinson's disease: a long-term appraisal of mortality. Ann. Neurol. 3:116–118; 1978.

Kaye, J. A.; Feldman, R. G. The role of L-dopa holiday in the long-term management of Parkinson's disease. Clin. Neuropharmacol. 9:1–13; 1986.

Kessler, I. Parkinson's disease in epidemiologic perspective. Adv. Neurol. 19:355–384; 1978.

Koller, W. C. Sensory symptoms in Parkinson's disease. Neurology 34:957–959; 1984.

Kondo, K. Epidemiological clues for the etiology of Parkinson's disease. In: Hassler, R. G.; Christ, J. F., eds. Advances in neurology, Vol. 40. New York: Raven Press; 1984: p. 345–361.

Langston, J. W.; Ballard, P. A.; Tetrud, J. W.; et al. Chronic parkinsonism in humans due to a product of meperidine—analog synthesis. Science 219:979–980; 1983.

Langston, J. W. The etiology of Parkinson's disease: new directions for research. In: Jankovic, J.; Tolosa, E., eds. Parkinson's disease and movement disorders. Baltimore-Munich: Urban and Schwarzenberg, Inc.; 1988: p. 85–86.

Li, S. C.; Schoenberg, B. S.; Wang, C.; et al. A prevalence survey of Parkinson's disease and other movement disorders in the Peoples Republic of China. Arch. Neurol. 42:655–657; 1985.

Madrazo, I.; Drucker-Colin, R.; Diaz, V.; et al. Open microsurgical antograft of adrenal medulla to the right caudate nucleus in two patients with intractable Parkinson's disease. N. Engl. J. Med. 316:831–834; 1987.

Martignoni, E.; Micieli, G.; Cavallini, A.; et al. Autonomic disorders in idiopathic parkinsonism. Neurol. Transm. 22 (suppl.):149–161; 1986.

Marttila, R. J.; Kaprio, J.; Koskenvuo, M.; Rinne, U. K. Finnish Parkinson's disease in a nationwide twin cohort. Neurology 38:1217–1219; 1988.

Mathers, S. E.; Kempster, P. A.; Law, P. J.; et al. Anal sphincter dysfunction in Parkinson's disease. Arch. Neurol. 46:1061–1064; 1989.

Parkinson, J. An essay on the shaking palsy. London: Sherwood, Neely and Jones; 1817.

Parkinson Study Group. Effect of deprenyl on the progression of disability in early Parkinson's disease. N. Engl. J. Med. 321:1364–1371; 1989a.

Parkinson Study Group. DATATOP: a multicenter controlled clinical trial in early Parkinson's disease. Arch. Neurol. 46:1052–1060; 1989b.

Pederzoli, M.; Girotti, F.; Scigliano, G.; et al. L-dopa long term treatment in Parkinson's disease: age-related side effects. Neurology 33:1518–1522; 1983.

Poskanzer, D. C.; Schwab, R. S.; Fraser, D. W. The cohort phenomenon in Parkinson's syndrome. In: Gillingham, F. J.; Donaldson, I. M., eds. Third symposium on parkinson's disease. Edinburgh: E. & S. Livingstone; 1969: p. 8–12.

Quinn, N.; Critchley, P.; Marsden, C. D. Young-onset Parkinson's disease. Movement Disorders 2:73–91; 1987.

Rajput, A. H.; Offord, K. P.; Beard, C. M.; et al. Epidemiology of parkinsonism: Incidence, classification, and mortality. Ann. Neurol. 16:278–282; 1984.

Rajput, A. H.; Offurd, K. P.; Beard, C. M.; et al. A case-control study of smoking habits, dementia, and other illnesses in idiopathic Parkinson's disease. Neurology 37:226–228; 1987.

Sandyk, R.; Bamford, C. R. Pain and sensory symptoms in Parkinson's disease. Intern. J. Neurosci. 39:15–25; 1988.

Schoenberg, B. S.; Anderson, D. W.; Haerer, A. F. Prevalence of Parkinson's disease in the birural population of Copiah County, Mississippi. Neurology 35:841–845; 1985.

Scott, R. M.; Brody, J. A. Benign early onset of Parkinson's disease: a syndrome distinct from classical postencephalitic parkinsonism. Neurology 21:366–368; 1971.

Snider, S. R.; Fahn, S.; Isgreem, W. P.; et al. Primary sensory symptoms in parkinsonism. Neurology 26:423–429; 1976.

Snyder, A. M.; Stricker, E. M.; Zigmond, M. J. Stress-induced neurologic impairments in an animal model of parkinsonism. Ann. Neurol. 18:544–551; 1985.

Snyder, S. H.; D'Amato, R. J. MPTP: a neurotoxin relevant to the pathophysiology of Parkinson's disease. Neurology 36:250–258; 1986.

Starkstein, S. E.; Berthier, M. L.; Boldoc, P. L.; et al. Depression in patients with early versus late onset of Parkinson's disease. Neurology 38:1441–1445; 1989.

Stone, T. W. Evidence for a nondopaminergic action of amantadine. Neurosci. Lett. 4:343–346; 1979.

Sweet, R. D.; McDowell, F. H. Five years treatment of Parkinson's disease with levodopa: therapeutic results and survival of 100 patients. Ann. Intern. Med. 83:456–463; 1975.

Tanner, C. M.; Chen, B.; Wang, W.; et al. Environmental factors and Parkinson's disease: A case-control study in China. Neurology 39:660–663; 1989.

Tetrud, J. W.; Langston, J. W. The effect of deprenyl (selegiline) on the natural history of Parkinson's disease. Science 245:519–522; 1989.

Ward, C. D.; Duvoisin, R. C.; Ince, S. E.; et al. Parkinson's disease in 65 pairs of twins and in a set of quadruplets. Neurology 33:815–824; 1983.

Zetusky, W. J.; Jankovic, J.; Pirozzolo, F. J. The heterogeneity of Parkinson's disease: clinical and prognostic implications. Neurology 35:522–526; 1985.

Parkinson's Plus Syndromes

Adams, R. D.; Van Bogaert, L.; Van der Eecken, H. Striatonigral degeneration. J. Neuropath. Exp. Neurol. 23:548–608; 1964.

Berciano, J. Olivopontocerebellar atrophy. In: Jankovic, J.; Tolosa, E., eds. Parkinson's disease and movement disorders. Baltimore-Munich: Urban and Schwarzenberg, Inc.; 1988: p. 131–151.

Cohen, J.; Low, P.; Fealey, R.; et al. Somatic and autonomic function in progressive autonomic failure and multiple system atrophy. Ann. Neurol. 22:692–699; 1987.

Dubinsky, R. M.; Jankovic, J. Progressive supranuclear palsy and a multi-infarct state. Neurology 37:570–576; 1987.

FitzGerald, P. M.; Jankovic, J. Lower body parkinsonism: evidence for vascular etiology. Movement Disorders. 4:249–260; 1989.

Friedman, D. I.; Jankovic, J. Progressive supranuclear palsy. A quarter century of progress. In: Appel, S. H., ed. Current neurology. Chicago: Year Book Medical Publishers, Inc.; 1989: p. 191–217.

Gibb, W. R. G.; Luthert, P. J.; Marsden, C. D. Corticobasal degeneration. Brain 112:1171–1192, 1989.

Golbe, L. I.; Davis, P. H. Progressive supranuclear palsy: recent advances In: Jankovic, J.; Tolosa, E., eds. Parkinson's disease and movement disorders. Baltimore-Munich: Urban and Schwarzenberg, Inc.; 1988: p. 121–130.

Golbe, L. I.; Davis, P. H.; Schoenberg, B. S.; et al. Prevalence and natural history of progressive supranuclear palsy. Neurology 38:1031–1034; 1988.

Harding, A. E. Commentary: olivopontocerebellar atrophy is not a useful concept. In: Marsden, C. D.; Fahn, S., eds. Movement disorders II. London: Butterworths; 1987: p. 269–271.

Jankovic, J. Parkinsonism-plus syndromes. Movement Disorders 4(suppl. 1):S95–S119; 1989.

Jankovic, J.; Friedman, D.; Pirozzolo, F. J.; McCrary. J. A. Progressive supranuclear palsy: Clinical, neurobehavioral, and neuro-ophthalmic findings. In: Streifler, M.; Korczyn, A. D.; Melamed, E.; Youdim, M. B. H., eds. Advances in neurology, Vol. 59. New York: Raven Press; 1990: p. 293–304.

Maher, E. R.; Lees, A. J. The clinical features and natural history of the Steele-Richardson-Olszewski syndrome (progressive supranuclear palsy). Neurology 36:1005–1008; 1986.

Polinsky, R. J. Multisystem atrophy: clinical aspects, pathophysiology and treatment. Neurol. Clin. 2:473–486; 1984.

Polinsky, R. J. Shy-Drager syndrome. In: Jankovic, J.; Tolosa, E., eds. Parkinson's disease and movement disorders. Baltimore-Munich: Urban and Schwarzenberg, Inc.; 1988: p. 153–166.

Quinn, N. Multiple system atrophy—the nature of the beast. J. Neurol. Neurosurg. Psychiatry S78–S89; 1989.

Richardson, J. C.; Steele, J.; Olszewski, J. Supranuclear ophthalmoplegia, pseudobulbar palsy, nuchal dystonia, and dementia. Trans. Am. Neurol. Assoc. 88:25–27; 1963.

Shy, G. M.; Drager, G. A. A neurological syndrome associated with orthostatic hypotension: a clinical-pathological study. Arch. Neurol. 2:511–527; 1960.

Steele, J. C.; Richardson, J. C.; Olszewski, J. Progressive supranuclear palsy: a heterogenous degeneration involving the brainstem, basal ganglia and cerebellum with vertical gaze and pseudobulbar palsy, nuchal dystonia and dementia. Arch. Neurol. 2:473–486; 1963.

Tremors

Abila, B.; Wilson, J. F.; Marshall, R. W.; et al. Differential effects of alpha-adrenoceptor blockade on essential physiological and isoprenaline-induced tremor: evidence for a central origin of essential tremor. J. Neurol. Neurosurg. Psychiat. 48:1031–1036; 1985.

Andrew, J. Surgical treatment of tremor. In: Findley, L. J.; Capildea, R., eds. Movement disorders: tremor. London: MacMillan; 1984: p. 339–351.

Capildeo, R.; Findley, L. J. Classification of tremor. In: Findley, L. J.; Capildeo, R., eds. Movement disorders: tremor. London: MacMillan Press; 1984: p. 3–13.

Cleeves, L.; Findley, L. J. Propranolol and propranolol-LA in essential tremor: a double-blind comparative study. J. Neurol. Neurosur. Psychiat. 51:379–384; 1988.

Cleeves, L.; Findley, L. J.; Koller, W. Lack of association between essential tremor and Parkinson's disease. Ann. Neurol. 24:23–26; 1988.

Cole, J. D.; Philip, H. I.; Sedgewick, E. M. Stability and tremor in the fingers associated with cerebellar hemisphere and hemisphere tract lesions in man. J. Neurol. Neurosurg. Psychiat. 51:1558–1568; 1988.

Elbe, R. Physiologic and essential tremor. Neurology 36:225–231; 1986.

Findley, L. J.; Koller, W. C. Essential tremor: a review. Neurology 37:1194–1197; 1987.

Findley, L. J. Tremors: differential diagnosis and pharmacology. In: Jankovic, J.; Tolosa, E., eds. Parkinson's disease and movement disorders. Baltimore-Munich: Urban and Schwarzenberg, Inc.; 1988: p. 243-261.

FitzGerald, P.; Jankovic, J. Orthostatic tremor: an association with essential tremor. Movement Disorders. 6:60-64; 1991.

Geraghty, J. J.; Jankovic, J.; Zetusky, W. J. Association between essential tremor and Parkinson's disease. Ann. Neurol. 17:329-333; 1985.

Haerer, A. F.; Anderson, D. W.; Schoenberg, B. S. Prevalence of essential tremor: Results from the Copian county study. Arch. Neurol. 39:750-755; 1982.

Homberg, V.; Heffner, H.; Reiners, K.; et al. Differential effects of changes in mechanical limb properties on physiological and pathological tremor. J. Neurol. Neurosurg. Psychiat. 50:568-579; 1987.

Huber, S. J.; Paulson, G. W. Efficacy of alprazolam for essential tremor. Neurology 38:241-243; 1988.

Jankovic, J.; Fahn, S. Physiologic and pathologic tremors. Diagnosis, mechanism and management. Ann. Intern. Med. 93:460-465; 1980.

Jankovic, J. Essential tremor and Parkinson's disease. Ann. Neurol. 25:211-212; 1989.

Koller, W. C.; Herbster, G. Adjuvant therapy of parkinsonian tremor. Arch. Neurol. 44:921-923; 1987.

Koller, W.; Lang, A.; Vetere-Overfield, B.; et al. Psychogenic tremors. Neurology 39:1094-1099; 1989.

Koller, W. C.; Vetere-Overfield, B. Acute and chronic effects of propranolol and primidone in essential tremor. Neurology 39:1587-1588; 1989.

Kruse, P.; Ladefoged, J.; Nielsen, U.; et al. β blockade used in precision sports: effect on pistol shooting performance. J. Appl. Physiol. 61:417-420; 1986.

Kuroda, Y.; Ryusuke, K.; Shibasaki, H. Treatment of essential tremor with arotinolol. Neurology 38:650-652; 1988.

Logigian, E. L.; Wierzbicka, M. M.; Bruyninckx, F.; et al. Motor unit synchronization in physiologic, enhanced physiologic and voluntary tremor in man. Ann. Neurol. 23:242-250; 1988.

Lou, J. S.; Jankovic, J. Essential tremor: Clinical correlates in 350 patients. Neurology 41:234-238, 1991.

McDowell, F. H. The use of glutethimide for treatment of essential tremor. Movement Disorders 4(1):75-80; 1989.

Rautakorpi, I.; Marttila, R. J.; Rinne, U. K. Epidemiology of essential tremor. In: Findley, L. J.; Capildeo, R., eds. Movement disorders: tremor. London: MacMillan Press; 1984: p. 211-218.

Rosenbaum, F.; Jankovic, J. Focal task-specific tremor and dystonia: categorization of occupa-

tional movement disorders. Neurology 38:522-527; 1988.

Sabra, A. F.; Hallett, M. Action tremor with alternating activity in antagonist muscles. Neurology 34:151-156; 1984.

Salisachs, P. Charcot-Marie-Tooth disease associated with "essential tremor." J. Neurol. Sci. 28:17-40; 1976.

Sanes, J. N.; LeWitt, P. A.; Mauritz, K. H. Visual and mechanical control of postural and kinetic tremor in cerebellar system disorders. J. Neurol. Neurosurg. Psychiat. 51:934-943; 1988.

Sasso, E.; Perucca, E.; Calzetti, S. Double-blind comparison of primidone and phenobarbital in essential tremor. Neurology 38:808-810; 1988.

Speelman, J. D.; Van Manen, J. Stereotactic thalamotomy for the relief of intention tremor of multiple sclerosis. J. Neurol. Neurosurg. Psychiat. 47:596-599; 1984.

Dystonia

Anderson, R. L.; Patrinely, J. R. Surgical management of blepharospasm. In: Jankovic, J.; Tolosa, E., eds. Advances in neurology: facial dyskinesias, Vol. 49. New York: Raven Press; 1988: p. 501-520.

Bertrand, C. M. Peripheral versus central surgical approach for the treatment of spasmodic torticollis. In: Marsden, C. D.; Fahn, S., eds. Movement disorders. London: Butterworths; 1982: p. 315-318.

Bressman, S. B.; DeLeon, D.; Brin, M. F.; et al. Inheritance of the idiopathic torsion dystonia in Ashkenazi Jews. In: Fahn, S.; Marsden, C. D.; Calne, D. B., eds. Advances in neurology: dystonia, Vol. 50. New York: Raven Press; 1988: p. 45-56.

Bressman, S. B.; Fahn, S.; Burke, R. E. Paroxysmal non-kinesigenic dystonia. In: Fahn, S.; Marsden, C. D.; Calne, D. B., eds. Advances in neurology: dystonia 2, Vol. 50. New York: Raven Press; 403-413; 1988.

Brin, M. F.; Fahn, S.; Moskowitz, C. B.; et al. Localized injections of botulinum toxin for the treatment of focal dystonia and hemifacial spasm. Movement Disorders 2:237-254; 1987.

Brin, M. F.; Butzer, A.; Fahn, S.; et al. Adductor laryngeal dystonia (spastic dysphonia): treatment with local injections of botulinum toxin (botox). Movement Disorders 4:287-296; 1989.

Burke, R. E.; Fahn, S.; Marsden, C. D. Torsion dystonia: a double-blind, prospective trial of high-dosage trihexyphenidyl. Neurology 36:160-164; 1986.

Calne, D. B.; Lang, A. E. Secondary dystonia. In: Fahn, S.; Marsden, C. D.; Calne, D. B., eds. Advances in neurology: dystonia 2, Vol. 50. New York: Raven Press; 1988: p. 9-33.

Elridge, R. The torsion dystonias: literature review and genetic and clinical studies. Neurology 20(pt. 2):1–78; 1970.

Fahn, S. The varied clinical expressions of dystonia. In: Jankovic, J., ed. Neurologic clinics: movement disorders, Vol. 2. Philadelphia: W. B. Saunders; 1984: p. 541–554.

Fahn, S. Generalized dystonia: concept and treatment. Clinical Neuropharmacology 9(suppl. 2):S37–S48; 1986.

Fletcher, N. A.; Harding, A. E.; Marsden, C. D. A genetic study of primary torsion dystonia in the UK (abstract). J. Neurol. 235(suppl.):S6; 1988.

Gibb, W. R. G.; Lees, A. J.; Marsden, C. D. Pathological report of four patients presenting with cranial dystonias. Movement Disorders 3:211–221; 1988.

Grandas, F.; Elston, J.; Quinn, N.; et al. Blepharospasm—a review of 264 patients. J. Neurol. Neurosurg. Psychiat. 51:767–772; 1988.

Greene, P.; Sahle, H.; Fahn, S. Analysis of open-label trials in torsion dystonia using high dosages of anti-cholinergics and other drugs. Movement Disorders 3:46–60; 1988.

Hornykiewicz, O.; Kish, S.; Becker, L. E.; et al. Biochemical evidence for brain neurotransmitter changes in idiopathic torsion dystonia (Dystonia Musculorum Deformans). In: Fahn, S.; Marsden, C. D.; Colne D. B., eds. Advances in neurology: Dystonia 2, Vol. 50. New York: Raven Press; 1988: p. 157–165.

Inzelberg, R.; Kahana, E.; Korczyn, A. D. Clinical course of idiopathic torsion dystonia among Jews in Israel. In: Fahn, S.; Marsden, C. D.; Calne, D. B., eds. Advances in neurology: dystonia 2, Vol. 50. New York: Raven Press; 1988: p. 93–100.

Jankovic, J.; Patel, S. Blepharospasm associated with brainstem lesions. Neurology 33:1237–1240; 1983.

Jankovic, J.; Orman, J. Blepharospasm: demographic and clinical survey of 250 patients. Ann. Opthal. 16:371–376; 1984.

Jankovic, J.; Orman, J. Botulinum A toxin for cranial–cervical dystonia: a double-blind, placebo-controlled study. Neurology 37:616–623; 1987.

Jankovic, J.; Svendsen, C. N.; Bird, E. P. Brain neurotransmitters in dystonia. N. Engl. J. Med. 316:278–279; 1987.

Jankovic, J. Etiology and differential diagnosis of blepharospasm and oromandibular dystonia. In: Jankovic, J.; Tolosa, E., eds. Advances in neurology: facial dyskinesias, Vol. 49. New York: Raven Press; 1988: p. 103–116.

Jankovic, J.; Fahn, S. Dystonic syndromes. In: Jankovic, J.; Tolosa, E., eds. Parkinson's disease and movement disorders. Baltimore–Munich: Urban and Schwarzenberg, Inc.; 1988: p. 283–314.

Jankovic, J.; Nutt, J. G. Blepharospasm and cranial–cervical dystonia (Meige's syndrome): Familial occurrence. In: Jankovic, J.; Tolosa, E., eds. Advances in neurology: facial dyskinesias, Vol. 49. New York: Raven Press; 1988: p. 117–124.

Jankovic, J.; Orman, J. Tetrabenazine therapy of dystonia, chorea, tics, and other dyskinesias. Neurology 38:391–394; 1988.

Jankovic, J.; Van Der Linden C. Dystonia and tremor induced by peripheral trauma: predisposing factors. J. Neurol. Neurosurg. Psychiat. 51:1512–1519; 1988.

Jankovic, J.; Schwartz, K. Botulinum toxin injection for cervical dystonia. Neurology. 40:277–280; 1990.

Jankovic, J.; Schwartz, K.; Donovan, D. T. Botulinum toxin treatment of cranial-cervical dystonia, spasmodic dysphonia, other focal dystonias, and hemifacial spasm. J. Neurol. Neurosurg. Psychiatry. 53:633–639; 1990.

Jankovic, J.; Leder, S.; Warner, D.; Schwartz, K. Cervical dystonia: Clinical findings and associated movement disorders. Neurology 41:1088–1091; 1991

Jankovic, J.; Schwartz, K. S. Clinical correlates of response to botulinum toxin injections. Arch. Neurol. 48:1253–1256; 1991.

Jankovic, J.; Brin, M. Therapeutic uses of botulinum toxin. N. Engl. J. Med. 324:1186–1194, 1991.

Lance, J. W. Familial paroxysmal dystonic choreoathetosis and its differentiation from related syndromes. Ann. Neurol. 2:285–293; 1977.

Lowenstein, B. H.; Aminoff, M. J. The clinical course of spasmodic torticollis. Neurology 38:530–532; 1988.

Marsden, C. D.; Harrison, M. J. G.; Bundley, S. Natural history of idiopathic torsion dystonia. Adv. Neurol. 14:177–187; 1976.

Marsden, C. D. The focal dystonias. Clinical Neuropharmacology 9(suppl. 2):S49–S60; 1986.

Newmark, J.; Hochberg, F. H. Occupational focal dystonias: painless incoordination in 59 musicians. J. Neurol. Neurosurg. Psychiat. 50:291–295; 1987.

Nutt, J. G.; Muenter, M. D.; Aronson, A.; et al. Epidemiology of focal and generalized dystonia in Rochester, Minnesota. Movement Disorders 3:188–194; 1988.

Nutt, J. G.; Muenter, M. D.; Melton, J.; et al. Epidemiology of dystonia in Rochester, Minnesota. In: Fahn, S.; Marsden, C. D.; Calne, D. B., eds. Advances in neurology: dystonia, Vol. 50. New York: Raven Press; 1988: p. 361–365.

Nygaard, T. G.; Duvoisin, R. C. Hereditary dystonia–parkinsonism syndrome of juvenile onset. Neurology 36:1424–1428; 1986.

Nygaard, T. G.; Marsden, C. D.; Duvoisin, R. C.

Dopa-responsive dystonia. In: Fahn, S.; Marsden, C. D.; Calne, D. B., eds. Advances in neurology: dystonia 2, Vol. 50. New York: Raven Press; 1988: p. 377–384.

Ozeliuys, L.; Kramer, P. L.; Moskowitz, C. B.; et al. Human gene for torsion dystonia located on chromosome 9q32–q34. Neuron 2:1427–1434; 1989.

Pettigrew, L. C.; Jankovic, J. Hemidystonia: a report of 22 patients and a review of the literature. J. Neurol. Neurosurg. Psychiat. 48:650–657; 1985.

Poewe, W. H.; Lees, A. J.; Stern, G. M. Dystonia in Parkinson's disease: clinical and pharmacological features. Ann. Neurol. 23:73–78; 1988.

Rosenbaum, F.; Jankovic, J. Focal task-specific tremor and dystonia: categorization of occupational movement disorders. Neurology 38:522–527; 1988.

Rothwell, J. C.; Obeso, J. A. The anatomical and physiologic basis of torsion dystonia. In: Marsden, C. D.; Fahn, S., eds. Movement disorders 2. London: Butterworth; 1987: p. 313–314.

Segawa, M.; Hosaka, A.; Miyagawa, F.; et al. Hereditary progressive dystonia with marked diurnal fluctuation. In: Eldridge, R.; Fahn, S., eds. Advances in neurology: dystonia, Vol. 14. New York: Raven Press; 1976: p. 215–233.

Segawa, M.; Nomura, Y.; Kase, M. Diurnally fluctuating hereditary progressive dystonia. In: Vinken, P. J.; Bruyn, G. W.; Klawans, H. L., eds. Extrapyramidal disorders: handbook of clinical neurology, Vol. 5(49). Amsterdam: Elsevier; 1986:529–540.

Sheehy, M. P.; Marsden, C. D. Writers' cramp: a focal dystonia. Brain 105:461–480; 1982.

Stell, R.; Thompson, P. D.; Marsden, C. D. Botulinum toxin in spasmodic torticollis. J. Neurol. Neurosurg. Psychiat. 51:920–923; 1988.

Tolosa, E.; Marti, M. J. Blepharospasm–oromandibular dystonia syndrome (Meige syndrome): clinical aspects. In: Jankovic, J.; Tolosa, E., eds. Advances in neurology: facial dyskinesias, Vol. 49. New York: Raven Press; 1988: p. 73–84.

Tolosa, E.; Montserrat, L.; Bayes, A. Blink reflex studies in focal dystonias: enhanced excitability of brainstem interneurons in cranial dystonia and spasmodic torticollis. Movement Disorders 3:61–69; 1988.

Weiner, W. J.; Nora, L. M. "Trick" movements in facial dystonia. J. Clin. Psychiatry 45:519–521; 1984.

Zilber, N.; Korczy, A. D.; Kahane, E.; et al. Inheritance of idiopathic torsion dystonia among Jews. J. Med. Genet. 21:13–20; 1984.

Zweig, R. M.; Hedreen, J. C.; Jankel, W. R.; et al. Pathology in brainstem regions of individuals with primary dystonia. Neurology 38:702–706; 1988.

Tics and Tourette's Syndrome

Burrows, G. D.; McIntyre, I. M.; Judd, F. K.; Norman, T. R. Clinical effects of serotonin reuptake in the treatment of depressive illness. J. Clin. Psychiatry 49(suppl. 8):18–22; 1988.

Caine, E. D.; McBride, M. C.; Chiverton, P.; et al. Tourette's syndrome in Monroe county school children. Neurology 38:472–475; 1988.

Comings, D. E. A controlled study of Tourette syndrome. VII. Summary: a common genetic disorder causing disinhibition of the limbic system. Am. J. Hum. Genet. 41:839–866; 1987.

Erenberg, G.; Cruse, R. P.; Rothner, A. D. The natural history of Tourette syndrome: A follow-up study. Ann. Neurol. 22:383–385; 1987.

Frankel, M.; Cummings, J. L.; Robertson, M. M.; et al. Obsessions and compulsions in Gilles de la Tourette syndrome. Neurology 36:378–382; 1986.

Glaze, D. G.; Frost, J. D.; Jankovic, J. Sleep in Gilles de la Tourette's syndrome: disorder of arousal. Neurology 33:586–592; 1983.

Jankovic, J. Recent advances in the management of tics. Movement Disorders 1(suppl. 2):S100–S110; 1986.

Jankovic, J.; Fahn, S. The phenomenology of tics. Movement Disorders 1:17–26; 1986.

Jankovic, J.; Rohaidy, H. Motor, behavioral and pharmacologic findings in Tourette's syndrome. Canad. J. Neurol. Sci. 14:541–546; 1987.

Jankovic, J.; Orman, J. Tetrabenazine therapy of dystonia, chorea, tics, and other dyskinesias. Neurology 38:391–394; 1988.

Jankovic, J. Tremor, tic and myoclonic disorders. Current Opinion in Neurology and Neurosurgery 2:324–329; 1989.

Kurlan, R.; Behr, J.; Medved, L.; et al. Severity of Tourette's syndrome in one large kindred: implication for determination of disease prevalence rate. Arch. Neurol. 44:268–269; 1987.

Kurlan, R. The spectrum of Tourette's syndrome. Current Opinion in Neurology and Neurosurgery 1:294–298; 1988,

Kurlan, R. Tourette's syndrome: current concepts. Neurology 39:1625–1630, 1989.

Leckman, H.; Cohen, D. J.; Bruun, R. D. Tourette's syndrome and tic disorders. New York: John Wiley and Sons; 1988: p. 291–302.

Lees, A. J.; Tolosa, E. Tics. In: Jankovic, J.; Tolosa, E., eds. Parkinson's disease and movement disorders. Baltimore-Munich: Urban and Schwarzenberg. Inc.; 1988: p. 275–281.

Pauls, D. L.; Hurst, C. R.; Kruger, S. D.; et al. Gilles de la Tourette's syndrome and attention deficit disorder with hyperactivity: evidence against a genetic relationship. Arch. Gen. Psychiatry 43:1177–1179; 1986.

Pauls, D. L.; Cohen, D. J.; Kidd, K. K.; et al. (Letter to the editor) Tourette syndrome and neuropsychiatric disorders: Is there a genetic relationship? Am. J. Hum. Genet. 43:206–209; 1988.

Shapiro, A. K.; Shapiro, E. An update on Tourette's syndrome. Am. J. Psychother. 36:379–89; 1982.

Shapiro, A, K.; Shapiro, E.; Fulop, G. Pimozide treatment of tic and Tourette disorders. Pediatrics 79:1032–1039; 1987.

Shapiro, A. K.; Shapiro, E. S.; Young, J. G.; et al. Sensory tics. In: Shapiro, E. S.; Young J. G.; Feinberg, T. E., eds. Gilles de la Tourette syndrome. New York: Raven Press; 1988: p. 356–360.

Singh, S. K.; Jankovic, J. Tardive dystonia in patients with Tourette's syndrome. Movement Disorders 3:274–280; 1988.

Tardive Dyskinesias

Baldessarini, R. J.; Cole, J. O.; Davis, J. M.; et al. Tardive dyskinesia: summary of a task force report of the American Psychiatric Association. Am. J. Psychiat. 137:1163–1172; 1980.

Blin, J.; Baron, J. C.; Cambon, H.; et al. Striatal dopamine D2 receptors in tardive dyskinesia: PET study. J. Neurol. Neurosurg. Psychiat. 52:1248–1252; 1989.

Blum, I.; Nisipeanu, P. F.; Roberts, E. Naloxone in tardive dyskinesia. Psychopharmacol. 93:538; 1987.

Bobruff, A.; Gardos, G.; Tarsy, D.; et al. Clonazepam and phenobarbital in tardive dyskinesia. Am. J. Psychiat. 138:189–193; 1981.

Burke, R. E.; Fahn, S.; Jankovic, J.; et al. Tardive dystonia: late-onset and persistent dystonia caused by antipsychotic drugs. Neurology 32:1335–1341; 1982.

Burke, R. E.; Kang, U. J.; Jankovic, J.; et al. Tardive akathisia: an analysis of clinical features and response to open therapeutic trials. Movement Disorders 4(2):157–175; 1989.

Chacko, R. C.; Root, L. ECT and tardive dyskinesia: two cases and a review. Jnl. Clin. Psychiatry 44:254–266; 1983.

Crane, G.; Smith, R. C. The prevalence of tardive dyskinesia. In: Fahn, W. E.; Smith, R. C.; Davis, J. M.; Domino, E. F., eds. Tardive dyskinesia: research and treatment. New York: SP Medical and Scientific Books; 1980: p. 269–280.

Csernansky, J. G.; Grabowski, K.; Cervantes, J.; et al. Fluphenazine decanoate and tardive dyskinesia: a possible association. Am. J. Psychiatry 138:1362–1365; 1981.

Csernansky, J. G.; Tacke, U.; Rusen, D.; et al. The effect of benzodiazepines on tardive dys-

kinesia. J. Clin. Psychopharmacol. 8:154–155; 1988.

Druckman, R.; Seelinger, D.; Thulin, B. Chronic involuntary movements induced by phenothiazines. J. Nerv. Ment. Dis. 135:69–75; 1962.

Fahn, S. Treatment of tardive dyskinesia with dopamine depleting agents. Clin. Neuropharmacol. 6:151–158; 1983.

Fibiger, H. C.; Lloyd, K. G. Reply to letter to the editor. Trends in Neurosciences 9:259–260; 1986.

Fisk, G. G.; York, S. M. The effect of sodium valproate on tardive dyskinesia—revisited. Br. J. Psychiat. 150:542–549; 1987.

Gardos, G.; Cole, J. O.; Salomon, M.; et al. Clinical forms of severe tardive dyskinesia. Am. J. Psychiat. 144:895–902; 1987.

Glazer, W. M.; Morgenstein, H.; Niedzwiecki, D.; et al. Heterogeneity of tardive dyskinesia: a multivariate analysis. Br. J. Psychiat. 152:253–259; 1988.

Gosek, E.; Weller, R. A. Improvement of tardive dyskinesia associated with electroconvulsive therapy. J. Nerv. Ment. Dis. 176:120–122; 1988.

Greendyke, R. M.; Webster, J. C.; Kim, J.; et al. Lack of efficacy of pindolol in tardive dyskinesia. Am. J. Psychiat. 145:1318–1319; 1988.

Herrara, J. N.; Sramek, J. J.; Costa, J. F.; et al. High potency neuroleptics and violence in schizophrenics. J. Nerv. Ment. Dis. 176:558–561; 1988.

Hong, M.; Jenner, P.; Marsden, C. D. Comparison of the acute actions of amine-depleting drugs and dopamine receptor antagonists on dopamine function in the brain in rats. Neuropharmacology 26:237–245; 1987.

Hong, M.; Kilpatrick, G. J.; Jenner, P.; et al. Effects of continuous administration for 12 months of amine-depleting drugs and chlorpromazine on striatal dopamine function in the rate. Neuropharmacology 26:1061–1069; 1987.

Iverson, I. L. Dopamine receptors in the brain. Science 188:1084–1089; 1975.

Jankovic, J.; Orman, J. Tetrabenazine therapy of dystonia, chorea, tics, and other dyskinesias. Neurology 38:391–394; 1988.

Jenner, P.; Marsden, C. D. Is the dopamine hypothesis of tardive dyskinesia completely wrong? (Letter to the editor) Trends in Neurosciences 9:259; 1986.

Jeste, D. V.; Potkin, S. G.' Sinha, S.; et al. Tardive dyskinesia: reversible and persistent. Arch. Gen. Psychiatry 36:585–590; 1979.

Jeste, D. V.; Jeste, S. D.; Wyatt, R. J. Reversible tardive dyskinesia: implications for therapeutic strategy and prevention of tardive dyskinesia. Mod. Probl. Pharmacopsychiatry 21:34–48; 1983.

Jus, A,; Pineau, R.; Lachance, R.; et al. Epidemiology of tardive dyskinesia. Part 1. Dis. Nerv. Syst. 37:210–213; 1976.

Kidger, T.; Barnes, T. R. E.; Trauer, T.; et al. Subsyndromes of tardive dyskinesia. Psychological Medicine 10:513–520; 1980.

Klawans, H. L. The pharmacology of extrapyramidal movement disorders. Basel: Kager; 1973.

Klawans, H. L.; Falk, D. K.; Nausieda, P. E.; et al. Gilles de la Tourette syndrome after long term chlorpromazine therapy. Neurology 28:1058–1064; 1978.

Klawans, H. L.; Tanner, C. M. The reversibility of permanent tardive dyskinesia. Neurology 33(suppl. 2):163; 1983.

Leys. D.; Vemersch, P.; Danel, T.; et al. Diltiazem for tardive dyskinesia (letter to the editor) Lancet 1:250–251; 1988.

Little, J. T.; Jankovic. J. Tardive myoclonus. Movement Disorders 2(4):307–311; 1987.

Marsden, C. D.; Tarsy, D.; Baldessarini, R. J. Spontaneous and drug-induced movement disorders in psychotic patients. In: Benson, D. F.; Blumer, D., eds. Psychiatric aspects of neurologic disease. New York: Grune & Stratton; 1975: p. 219–265.

Miller, L. G.; Jankovic, J. Metoclopramide-induced movement disorders. Arch. Intern. Med. 149:2486–2492; 1989.

Miller, L.; Jankovic, J. Drug-induced dyskinesias. In: Appel, S. H., ed. Current Neurology, Vol. 10. Chicago: Yearbook Medical Publisher; 1990: p. 321–355.

Penney, J. B. Jr.; Young, A. B. Speculations on the functional anatomy of basal ganglia disorders. Ann. Rev. Neurosci. 6:73–94; 1983.

Quitkin, F.; Rifkin, A.; Gochfeld, L.; et al. Tardive dyskinesia: are first signs reversible? Am. J. Psychiat. 134:84–87; 1977.

Schmidt, W. R.; Jarcho, L. W. Persistent dyskinesias following phenothiazine therapy. Arch Neurol. 14:369–377; 1966.

Schonecker, M. Ein eigentumliches syndrom in oralen Bereich bei Megphen applikation. Nervenarzt 28:35–36; 1957.

Sigwald, J.; Bouttier, D.; Raymondeaud, C.; et al. Quatre cas de dyskinesie facie-buccal-lingo-masticatrice a l'evolution pronlongee secondaire a un traitement par les neuroleptiques. Revue Neurol. 100:750–755; 1959.

Singh, S. K.; Jankovic, J. Tardive dystonia in patients with Tourette's syndrome. Movement Disorders 3(3):274–280; 1988.

Sorokin, J. E.; Giordani, B.; Mohs, R. C.; et al. Memory impairment in schizophrenic patients with tardive dyskinesia. Biol. Psychiat. 23:129–135; 1988.

Stacy, M.; Jankovic, J.; Tardive tremor. Mov. Disor. 7:53–57; 1992.

Stahl, S. M. Tardive Tourette syndrome in an autistic patient after long-term neuroleptic administration. Am. J. Psychiat. 137:1267–1269; 1980.

Stewart, R. M.; Rollins, J.; Beckham, B. Baclofen in tardive dyskinesia patients maintained on neuroleptics. Clin. Neuropharmacol. 5:365–373; 1982.

Tanner, C. M. Drug-induced movement disorders (tardive dyskinesia and dopa-induced dyskinesia). In: Vinken, P. J.; Bruyn, G. W.; Klawans, H. L., eds. Handbook of clinical neurology: extrapyramidal disorders, Vol. 5(49). Amsterdam: Elsevier Science Publishers BV; 1986: p. 185–204.

Waddington, J. L. Tardive dyskinesia: a critical reevaluation of the causal role of neuroleptics and of the dopamine receptor supersensitivity hypothesis. In: Callaghan, N.; Calvin, R., eds. Recent researches in neurology. London: Pitman Publishing; 1984: p. 34–48.

Waddington, J. L. Reply to letter to the editor. Trends in Neurosci. 9:261; 1986.

Waddington, J. L.; Youssef, H. A. Tardive dyskinesia in bipolar affective disorder—aging, cognitive dysfunction, course of illness, and exposure to neuroleptics and lithium. Am. J. Psychiat. 145:613–616; 1988.

Weiner, W. J.; Luby, E. D. Persistent akathisia following neuroleptic withdrawal. Ann. Neurol. 13:466–467; 1983.

Youssef, H. A.; Waddington, J. L. Involuntary orofacial movements in hospitalized patients with mental handicap or epilepsy—relationship to developmental/intellectual deficit and presence or absence of long-term exposure to neuroleptics. J. Neurol. Neurosurg. Psychiat. 51:863–865; 1988.

Heredodegenerative Disorders

Bamford, K. A.; Caine, E. D.; Kido, D. K.; et al. Clinical–pathologic correlation in Huntington's disease: a neuropsychological and computed tomography study. Neurology 38:796–801; 1989.

Beal, M. F.; Mazurek, M. F.; Ellison, D. W.; et al. Somatostatin and neuropeptide γ concentrations in pathologically graded cases of Huntington's disease. Ann. Neurol. 23:562–569; 1988.

Bowcock, A. M.; Farrer, L. A.; Cavalli-Sforza, L. L.; et al. Mapping the Wilson's disease locus to a cluster of linked polymorphic markers on chromosome 13. Am. J. Hum. Genet. 41:27–35; 1987.

Brackenridge, C. J. The relation of type of initial symptoms and line of transmission to ages at onset and death in Huntington's disease. Clin. Genet. 2:287–297; 1971.

Brewer, D. J.; Yuzbasiyan-Gurkan, V.; Young, A. B. Treatment of Wilson's disease. Seminars in Neurology 7(2):209–220; 1987.

Dening, T. R. Psychiatric aspects of Wilson's disease. Br. J. Psychiat. 147:677–682; 1985.

Dening, T. R.; Berrios, G. E.; Walshe, J. M. Wilson's disease and epilepsy. Brain 111:1139–1155; 1988.

Gusella, J. F.; Wexler, N. S.; Conneally, P. M.; et al. A polymorphic DNA marker genetically linked to Huntington's disease. Nature 306:234–238; 1983.

Hayden, M. R. Huntington's chorea. New York: Springer-Verlag; 1981.

Hayden, M.; Hewitt, J.; Stoessl, A.; et al. The combination of positron emission tomography and DNA polymorphisms for preclinical detection of Huntington's disease. Neurology 37:1441–1447; 1987.

Huntington, G. On chorea. Med. Surg. Reporter 26:317–321; 1972. Reprinted In: Barbeau, A.; Chase, T. N.; Paulson, G. W., eds. Advances in neurology: huntington's chorea, Vol. 1. New York: Raven Press; 1972: p. 33–35.

Kuhl, D. E.; Phelps, M. E.; Markham, C. H.; et al. Cerebral metabolism and atrophy in Huntington's disease determined by [18]FDG and computed tomographic scan. Ann. Neurol. 12:425–434; 1982.

Marsden, C. D. Wilson's disease. Q. J. Med. 248:959–966; 1987.

Martin, J. B. Huntington's disease: new approaches to an old problem. Neurology 34:1059–1072; 1984.

Patten, B. M. Wilson's disease. In: Jankovic, J.; Tolosa, E., eds. Parkinson's disease and movement disorders. Baltimore-Munich: Urban and Schwarzenberg, Inc.; 1988: p. 179–190.

Penney, J. B. Jr.; Young, A. B. Huntington's disease. In: Jankovic, J.; Tolosa, E., eds. Parkinson's disease and movement disorders. Baltimore-Munich: Urban and Schwarzenberg, Inc.; 1988: p. 167–178.

Scheinberg, I. H.; Sternlieb, I. Wilson's disease. Philadelphia: W. B. Saunders; 1984.

Smith, B.; Skarecky, D.; Bengtsson, U.; et al. Isolation of DNA markers in the direction of the Huntington disease gene from the G8 locus. Am. J. Human Genet. 42:335–344; 1988.

Von Sattel, J. P.; Myers, R. H.; Stevens, T. J.; et al. Neuropathological classification of Huntington's disease. J. Neuropath. Exp. Neurol. 44:559–577; 1985.

Walshe, J. M. Wilson's disease: yesterday, today and tomorrow. Movement Disorders 3(1):10–29; 1988.

Wasmoth, J. J.; Hewitt, J.; Smith, B.; et al. A highly polymorphic locus very tightly linked to the Huntington's disease gene. Nature 332:734–736; 1988.

Wilson, S. A. K. Progressive lenticular degeneration: a familial nervous disease associated with cirrhosis of the liver. Brain 34:295–507; 1912.

Young, A. B.; Penney, J. B.; Starosta-Rubinstein, S.; et al. PET scan investigations of Huntington's disease: cerebral metabolic correlates of neurologic features and functional decline. Ann. Neurol. 20:296–303; 1986.

Young, A. B.; Shoulson, I.; Penney, J. B.; et al. Huntington's disease in Venezuela: neurologic features and functional decline. Neurology 36:244–249; 1986.

Young, A. B.; Penney, J. B.; Starosta-Rubinstein, S.; et al. Normal caudate glucose metabolism in persons at-risk for Huntington's disease. Arch. Neurol. 44:254–257; 1987.

Young, A. B.; Greenamyre, J. T.; Hollingsworth, Z.; et al. NMDA receptor losses in putamen from patients with Huntington's disease. Science 241:981–983; 1988.

Young, A. B.; Penney, J. B.; Markel, D. S.; et al. Genetic linkage analysis, glucose metabolism, and neurologic examination: comparison in persons at risk for Huntington's disease (HD). Neurology 38(suppl. 1):359; 1988.

Myoclonus

Akpinar, S. Restless legs syndrome treatment with dopaminergic drugs. Clin. Neuropharmacol. 10:69–71; 1987.

Andermann, F.; Berkovic, S.; Carpenter, S.; et al. The Ramsay Hunt syndrome is no longer a useful clinical entity. Movement Disorders 4(1):13–17; 1989.

Banks, G.; Nielsen, V. K.; Short, M. P.; et al. Brachial plexus myoclonus. J. Neurol. Neurosurg. Psychiatry 48:582–584; 1985.

Berkovic, S. F.; Andermann, F.; Carpenter, S.; et al. Progressive myoclonic seizures: specific causes and diagnoses. N. Engl. J. Med. 315:296–305; 1986.

Bressman, S.; Fahn, S. Essential myoclonus. In: Fahn, S.; Marsden, C. D.; and Van Woert, M., eds. Advances in neurology: myoclonus, Vol. 43. New York: Raven Press; 1986: p. 287–294.

Chokroverty, S.; Manocha, M. K.; Duvoisin, R. C. A physiologic and pharmacologic study in anticholinergic-responsive essential myoclonus. Neurology 37:608–615; 1987.

Crossman, A. R.; Mitchell, I. J.; Sambrook, M. A.; et al. Chorea and myoclonus in the monkey induced by gamma-aminobutyric acid antagonism in the lentiform complex. Brain 111:1211–1234; 1988.

Davanipour, Z.; Alter, M.; Sobel, E. Creutzfeldt-Jakob Disease. Neurologic Clinics 4(2):415–426; 1986.

Eldridge, R.; Iivanainen, M.; Stern, R.; et al. "Baltic" myoclonus epilepsy: hereditary disorder of childhood made worse by phenytoin. Lancet 2:838–842; 1983.

Fahn, S.; Singh, N. An oscillating form of essential myoclonus. Neurology 31:80; 1981.

Fahn, S. Posthypoxic action myoclonus: review of the literature and report of two new cases with response to valproate and estrogen. In: Fahn, S.; Marsden, C. D.; Van Woert, M., eds. Advances in neurology: myoclonus, Vol. 43. New York: Raven Press; 1986: p. 49–84.

Fahn, S.; Marsden, C. D.; Van Woert, M. H. Definition and classification of myoclonus. In: Fahn, S.; Marsden, C. D.; Van Woert, M. H., eds. Advances in neurology: myoclonus, Vol. 43. New York: Raven Press; 1986: p. 1–5.

Gascon, G.; Wallenberg, B.; Daif, A. K.; et al. Successful treatment of cherry red spot-myoclonus syndrome with 5-hydroxytryptophan. Ann. Neurol. 24:453–455; 1988.

Gimenez-Roldan, S.; Mateo, D.; Muradas, V.; et al. Clinical, biochemical, and pharmacological observation in a patient with postasphyxic myoclonus: association to serotonin hyperactivity. Clin. Neuropharmacol. 11:151–160; 1988.

Hallett, M.; Chadwick, D.; Marsden, C. D. Ballistic movement overflow myoclonus. A form of essential myoclonus. Brain 100:299–312; 1977.

Hallett, M.; Chadwick, D.; Marsden, C. D. Cortical reflex myoclonus. Neurology 29:1107–1125; 1979.

Hallet, M.; Marsden, C. D.; Fahn, S. Myoclonus. In: Vinken, P. J.; Bruyn, G. W.; Klawans, H. L., eds. Handbook of clinical neurology: extrapyramidal disorders, Vol. 5(49). Amsterdam: Elsevier Science Publishers BV; 1986: p. 609–625.

Hauser, W. A.; Morris, M. L.; Heston, L. L.; et al. Seizures and myoclonus in patients with Alzheimer's disease. Neurology 36:1226–1230; 1986.

Hunt, J. R. Dyssynergia cerebellaris myoclonica-primary atrophy of the dentate system: a contribution to the pathology and symptomatology of the cerebellum. Brain 44:490–538; 1921.

Iivanainen, M.; Himberg, J. J. Valproate and clonazepam in the treatment of severe progressive myoclonus epilepsy. Arch. Neurol. 39:236–238; 1982.

Jankovic, J.; Fahn, S. The phenomenology of tics. Movement Disorders 1:17–26; 1986.

Jankovic, J.; Pardo R. Segmental myoclonus: clinical and pharmacologic study. Arch. Neurol. 43:1025–1031; 1986.

Kelly, J. J.; Sharborough, F. W.; Daube, J. R. A clinical and electrophysiological evaluation of myoclonus. Neurology 31:581–589; 1981.

Korten, J. J.; Notermans, S. L. H.; Frenken, C. W. G. M.; et al. Familial essential myoclonus. Brain 97:131–138; 1974.

Koskiniemi, M. L. Baltic myoclonus. In: Fahn, S.; Marsden, C. D.; Van Woert, M., eds. Advances in neurology: myoclonus, Vol. 43. New York: Raven Press; 1986: p. 57–64.

Krumholz, A.; Stern, B. J.; Weiss, H. D. Outcome from coma after cardio-pulmonary resuscitation: Relation to seizures and myoclonus. Neurology 38:401–405; 1988.

Kurlan, R.; Behr, J.; Shoulson, I. Hereditary myoclonus and chorea: the spectrum of hereditary nonprogressive hyperkinetic movement disorders. Movement Disorders 2:301–306; 1987.

Kurlan, R.; Behr, J.; Medved, L.; et al. Myoclonus and dystonia: a family study. In: Fahn, S.; Marsden, C. D.; Calne, D. B., eds. Advances in neurology: dystonia 2, Vol. 50. New York: Raven Press; 1988: p. 391–401.

Lafora, G. R.; Glueck, B. Beitrag zur Histopathologie der myoklonischen Epilepsie. Z. Gesamte. Neurol. Psychiatr. 6:1–14; 1911.

Lees, A. J.; Tolosa, E. Tics. In: Jankovic, J.; Tolosa, E., eds. Parkinson's disease and movement disorders. Baltimore-Munich: Urban and Schwarzenberg, Inc.; 1988: p. 275–281.

Lewis, J. H. Hiccups: causes and cures. J. Clin. Gastroenterol. 7:538–552; 1985.

Little, J. T.; Jankovic, J. Tardive myoclonus. Movement Disorders 2(4):307–311; 1987.

Lu, C. S.; Thompson, P. D.; Quinn, N. P.; et al. Ramsay Hunt syndrome and celiac disease: a new association? Movement Disorders 1:209–219; 1986.

Lugarsi, E.; Cirignotta, F.; Coccagna, G.; et al. Nocturnal myoclonus and restless legs syndrome. In: Fahn, S.; Marsden, C. D.; Van Woert, M., eds. Advances in neurology: myoclonus, Vol. 43. New York: Raven Press; 1986: p. 295–307.

Lundborg, H. Die progressive Myoklonus-Epilepsie (Unverricht's Myoklonie). Uppsala: Almqvist and Wiksell; 1903.

Marsden, C. D.; Obeso, J. A.; Traub, M. M.; et al. Muscle spasms associated with Sudeck's atrophy after injury. Br. Med. J. 288:173–176; 1984.

Marsden, C. D.; Obeso, J. A. Viewpoints on the Ramsay Hunt syndrome. 1. The Ramsay Hunt syndrome is a useful clinical entity. Movement Disorders 4(1):6–12; 1989.

Martinelli, P.; Coccagna, G.; Lugaresi, E. Nocturnal myoclonus, restless legs syndrome, and abnormal electrophysiological findings. Ann. Neurol. 21:515; 1987.

Montplasir, J.; Godboat, R.; Poirer, G.; et al. Restless legs syndrome and periodic movements in sleep: physiopathology and treatment with L-dopa. Clin. Neuropharmacol. 9:456–463; 1986.

Nodo, S.; Ito, H.; Umezaki, H.; et al. Hip flexion–abduction to elicit asterixis in unresponsive patients. Ann. Neurol. 18:96–97; 1985.

Norio, R.; Koskiniemi, M. Progressive myo-clonic epilepsy: genetic and nosological as-pects with special reference to 107 Finnish pa-tients. Clin. Genet. 15:382–398; 1979.

Obeso, J. A.; Lang, A. E.; Rothwell, J. C.; et al. Postanoxic symptomatic oscillatory myo-clonus. Neurology 33:240–243; 1983.

Obeso, J. A.; Rothwell, J. C.; Lang, A. E.; et al. Myoclonic dystonia. Neurology 33:825–830; 1983.

Obeso, J. A.; Rothwell, J. C.; Marsden, C. D. The spectrum of cortical myoclonus. Brain 108:193–224; 1985.

Obeso, J. A.; Artieda, J.; Quinn, N.; et al. Piracetam in the treatment of different types of myoclonus. Clin. Neuropharmacol. 11:529–536; 1988.

Obeso, J. A.; Artieda, J.; Rothwell, J. C.; et al. The treatment of severe action myoclonus. Brain 112:765–777; 1989.

Patel, V. M.; Jankovic, J. Myoclonus. In: Appel, S. H., ed. Current neurology. Chicago: Year Book Medical Publishers; 1988: p. 109–156.

Rapin, I. Myoclonus in neuronal storage and La-fora diseases. In: Fahn, S.; Marsden, C. D.; Van Woert, M., eds. Advances in neurology: myoclonus, Vol. 43. New York: Raven Press; 1986: p. 65–85.

Resnick, T. J.; Moshe, S. L.; Perotta, L.; et al. Benign neonatal sleep myoclonus. Relation-ship to sleep states. Arch. Neurol. 43:266–268; 1986.

Schwartz, M. A.; Selhorst, J. B.; Ochs, A. L.; et al. Oculomasticatory myorrhythmia: a unique movement disorder occurring in Whipple's dis-ease. Ann. Neurol. 20:677–683; 1986.

Shibasaki, H.; Yamashita, Y.; Kuroiwa, Y. Elec-troencephalographic studies of myoclonus. Brain 191:447–460; 1978.

Shoulson, I. On chorea. Clin. Neuropharmacol. 9(suppl. 2):S85–S99; 1986.

Sotaniemi, K. A. Paraspinal myoclonus due to spinal root lesions. J. Neurol. Neurosurg. Psy-chiat. 48:722–723; 1985.

Spiro, A. J. Minipolymyoclonus: a neglected sign in childhood spinal muscular atrophy. Neurol-ogy 20:1124–1126; 1970.

Tassinari, C. A.; Michelucci, R.; Genton, P.; et al. Dyssynergia cerebellaris myoclonica (Ramsay Hunt syndrome): a condition unre-lated to mitochondrial encephalomyopathies. J. Neurol. Neurosurg. Psychiat. 52:262–265; 1989.

Thomas, J.; Reggan, J.; Klass, D. Epilepsia par-tialis continua. A review of 32 cases. Arch. Neurol. 34:266–275; 1977.

Truong, D. D.; de Yebenes, J. C.; Pezzoli, G.; et al. Glycine involvement in DDT-induced myo-clonus. Movement Disorders 3(1):77–87; 1988.

Unverricht, H. Die myoclonie. Leipzig: Franz Deuticke; 1891.

Van Woert, M. H.; Rosenbaum, D.; Chung, E. Biochemistry and therapeutics of posthypoxic myoclonus. In: Fahn, S.; Marsden, C. D.; Van Woert, M. H., eds. Advances in neurology: myclonus, Vol. 43. New York: Raven Press; 1986: p. 171–182.

Walters, A. S.; Hening, W. Clinical presentation and neuropharmacology of restless legs syn-drome. Clin. Neuropharmacol. 10:235–237; 1987.

Walters, A.; Hening, W.; Cote, L.; et al. Domi-nantly inherited restless legs with myoclonus and periodic movements of sleep: a syndrome related to endogenous opiates? In: Fahn, S.; Marsden, C. D.; Van Woert, M. H., eds. Ad-vances in neurology: myoclonus, Vol. 43. New York: Raven Press; 1986: p. 309–319.

Wilkens, D. E.; Hallett, M.; Wess, M. M. Audio-genic startle reflex of man and its relationship to startle syndromes: a review. Brain 109:561–573; 1986.

Chorea, Athetosis and Ballism

Adams, R. D.; Lyon, G. Neurology of hereditary metabolic diseases of children. New York: McGraw–Hill; 1982: p. 226–235.

Albin, R. L.; Bromberg, M. B.; Penney, J. B.; et al. Chorea and dystonia: a remote effect of car-cinoma. Movement Disorders 3:162–169; 1988.

Asherson, R. A.; Hughes, G. R. V.; Gledhill, R.; et al. Absence of antibodies to cardiolipin in patients with Huntington's chorea, Syden-ham's chorea and acute rhematic fever. J. Neurol. Neurosurg. Psychiatry 51:1458–1488; 1988.

Bamford, J. M.; Hakin, R. N. Chorea after le-gionnaire's disease. Br. Med. J. 284:1232–1233; 1982.

Bird, M. T.; Palkes, H.; Prensky, A. L. A follow-up study of Sydenham's chorea. Neurology 26:601–606; 1976.

Bland, E. F.; Jones, T. D. Rheumatic fever and rheumatic heart disease. A twenty year report on 1000 patients followed since childhood. Cir-culation 4:836–843; 1951.

Brin, M. F.; de Yebenes, J.; Fahn, S. Neuro-acanthocytosis (Levine syndrome). Curr. Opin. Neurol. Neurosurg. 1:332–334; 1988.

Bruyn, G. W.; Padberg, G. Chorea and systemic lupus erythematosis: a critical review. Eur. Neurol. 23:278–290; 1984.

Chadwick, D.; Reynolds, E. H.; Marsden, C. D. Anti-convulsant-induced dyskinesias: a com-parison with dyskinesias induced by neurolep-tics. J. Neurol. Neurosurg. Psychiat. 39:1210–1218; 1976.

Davous, P.; Rondot, P.; Marion, M. H.; et al. Severe chorea after acute carbon monoxide poisoning. J. Neurol. Neurosurg. Psychiat. 49:206–208; 1986.

De Yebenes, J. G.; Brin, M. F.; Mena, M. A.; et al. Neurochemical findings in neuroacanthocytosis. Movement Disorders 3:300–312; 1988.

Dewey, R. B. Jr.; Jankovic, J. Hemiballism–hemichorea. Clinical and pharmacologic findings in 21 patients. Arch. Neurol. 46:862–867; 1989.

Dove, D. J. Chorea associated with oral contraceptive therapy. Am. J. Obstet. Gynecol. 137:740–742; 1980.

Drake, M. E. Jr. Paroxysmal kinesigenic choreoathetosis in hyperthyroidism. Postgrad. Med. J. 63:1089–1090; 1987.

Evans, B. K.; Jankovic, J. Tuberous sclerosis and chorea. Ann. Neurol. 13:106–107; 1983.

Fahn, S. Choreic disorders. Curr. Opin. Neurol. Neurosurg. 2:319–323; 1989.

Fischbeck, K. I. T.; Layzer, R. B. Paroxysmal choreoathetosis associated with thyrotoxicosis. Ann. Neurol. 6:453–454; 1979.

Friedland, R.; Yahr, M. D. Meningoencephalopathy secondary to infectious mononucleosis. Unusual presentation with stupor and coma. Arch. Neurol. 34:186–188; 1977.

Gibb, W. R. G.; Lees, A. J.; Scadding; J. W. Persistent rheumatic chorea. Neurology 35:101–102; 1985.

Goetz, C. G.; Klawans, H. L. Tardive dyskinesia. In: Benson, D. F.; Blumer, D., eds. Psychiatric aspects of neurologic disease II. New York: Grune and Stratton; 1982: p. 195–218.

Gordon, N. Choreoathetosis of genetic origin. Dev. Med. Child Neurol. 22:521–524; 1980.

Gordon, M. F. New onset choreiform disorder in an adult with recent Group A beta hemolytic streptococcal pharyngitis. J. Neurol. Neurosurg. Psychiat. 51:448–419; 1988.

Grigor, R.; Edmonds, J.; Lewkonia, R.; et al. Systemic lupus erythematosus. Ann. Rheum. Dis. 37:121–128; 1978.

Haerer, A. F.; Currier, R. D.; Jackson, J. F. Hereditary nonprogressive chorea of early onset. N. Engl. J. Med. 27:1220–1224; 1967.

Harper, P. S. Benign hereditary chorea: clinical and genetic aspects. Clin. Genet. 13:85–95; 1978.

Husby, G.; Van der Rijn, I.; Zabriskie, J. B.; et al. Antibodies reacting with cytoplasm of subthalamic and caudate nuclei neurons in chorea and acute rhematic fever. J. Exp. Med. 144:1094–1110; 1976.

Jankovic, J.; Orman, J. Tetrabenazine therapy of dystonia, chorea, tics, and other dyskinesias. Neurology 38:391–394; 1983.

Jankovic, J.; Caskey, T. C.; Stout, J. T.; et al. Lesch-Nyhan syndrome: a study of motor behavior and cerebrospinal fluid neurotransmitters. Ann. Neurol. 23:466–469; 1988.

Johnson, W. G.; Fahn, S. Treatment of vascular hemiballism and hemichorea. Neurology 27:634–636; 1977.

Johnson, W. G.; Fahn, S. Treatment of vascular hemiballism and hemichorea. Neurology 27:634–636; 1977.

Lundh, H.; Iunving, K. An extrapyramidal choreiform syndrome caused by amphetamine addiction. J. Neurol. Neurosurg. Psychiat. 44:728–730; 1981.

Lusins, J. O. Clinical features of chorea associated with systemic lupus erythematosus. Am. J. Med. 58:857–861; 1975.

Mao, C-C.; Gancher, S. T.; Herndon, R. M. Movement disorders in multiple sclerosis. Movement Disorders 3:109–116; 1988.

Marsden, C. D.; Obeso, J. A.; Rothwell, J. C. Clinical neurophysiology of muscle jerks. In: Desmedt, J. E., ed. Motor control mechanisms in health and disease. New York: Raven Press; 1983: p. 865–881.

Marsden, C. D. The pathophysiology of movement disorders. Neurol. Clin. 2:435–459; 1984a.

Marsden, C. D. Motor disorders in basal ganglia disease. Hum. Neurobiol. 2:245–250; 1984b.

Medley, D. R. K. Chorea and bacterial endocarditis. Br. Med. J. 1:861–862; 1963.

Mullin, P. J.; Kershaw, P. W.; Bolt, J. M. W. Choreoathetotic movement disorder in alcoholism. Br. Med. J. 4:278–281; 1970.

Nath, A.; Jankovic, J.; Pettigrew, L. C. Movement disorders and AIDS. Neurology 37:37–41; 1987.

Nausieda, P. A.; Grossman, B. J.; Koller, W. C.; et al. Sydenham's chorea: an update. Neurology 30:331–334; 1980.

Nausieda, P. A.; Bieliauskas, L. A.; Bacon, L. D.; et al. Chronic dopaminergic sensitivity after Sydenham's chorea. Neurology 33:750–754; 1983.

Padberg, G. W.; Bruyn, G. W. Chorea—differential diagnosis. In: Vinken, P. J.; Bruyn, G. W.; Klawans. H. L., eds. Handbook of clinical neurology: extrapyramidal disorders, Vol. 5 (49). Amsterdam: Elsevier Science Publishers BV; 1986: p. 549–564.

Penney, J. B.; Young, A. B. Speculations on the functional anatomy of basal ganglia disorders. Ann. Rev. Neurosci. 6:73–94; 1983.

Peters, A. C. B.; Vielvoye, G. J.; Versteeg, J.; et al. ECHO 25 focal encephalitis and subacute hemichorea. Neurology 29:676–681; 1979.

Philips, P. C.; Brin, M. F.; Fahn, S.; et al. Abnormal regional cerebral glucose metabolism in choreoacanthocytosis: an 18F-fluorodeoxyglucose positron emission tomographic study. Neurology 37(suppl. 1):211; 1987.

Poewe, W. H.; Kleedorfer, B.; Willeit, J.; et al. Primary CNS lymphoma presenting as a choreic movement disorder followed by segmental dystonia. Movement Disorders 3:320–325; 1988.

Quarrell, O. W. J.; Youngman, S.; Sarfarazi, M.; et al. Absence of close linkage between benign hereditary chorea and the locus D4S10 (probe G8). J. Med. Genet. 25:191–194; 1988.

Reik, L.; Steere, A. C.; Bartenhagen, W. H.; et al. Neurologic abnormalities of Lyme disease. Medicine 58:281–294; 1979.

Riela, A.; Roach, E. S. Choreoathetosis in an infant with tuberculous meningitis. Arch. Neurol. 39:526; 1982.

Sakai, I.; Mawatari, S.; Iwashita, H.; Goto, I.; et al. Choreoacanthocytosis: clues to clinical diagnosis. Arch. Neurol. 38:335–338; 1981.

Sethi, K. D.; Nichols, F. T.; Yaghmai, F. Generalized chorea due to basal ganglia lacunar infarcts. Movement Disorders 2:61–66; 1987.

Schady, W.; Meara, R. J. Hereditary progressive chorea without dementia. J. Neurol. Neurosurg. Psychiat. 51:295–297; 1988.

Shoulson, I. (1982) Pharmacotherapy of chorea. Neurol. Neurosurg. Update Series 39:1–7; 1982.

Shoulson, I. On chorea. Clin. Neuropharmacol. 9(suppl. 2):S85–S99; 1986.

Shoulson, I. Huntington's disease. In: Asbury, A. K.; McKhann, G. M.; McDonald, W. I., eds. Diseases of the nervous system. Philadelphia: W. B. Saunders; 1986: p. 1258–1267.

Spitz, M. C.; Jankovic, J.; Killian, J. M. Familial tic disorder, parkinsonism, motor neuron disease and acanthocytosis. Neurology 35:366–370; 1985.

Stollerman, G. H. Rheumatic fever and streptococcal infection. New York: Grune and Stratton, Inc.; 1975.

Suchowersky, O.; Hayden, M. R.; Martin, W. R. W.; et al. Cerebral metabolism of glucose in benign hereditary chorea. Movement Disorders 1:33–44; 1986.

Tabaton, M.; Mancardi, G.; Loeb, C. Generalized chorea due to bilateral small, deep cerebral infarcts. Neurology 35:588–589; 1985.

Tan, B. K.; Mastebroek, H. A. K.; Zaagman, W. H. Huntington's chorea: a random process. Clin. Neurol. Neurosurg. 79:215–221; 1976.

Walshe, J. M. Wilson's disease: yesterday, today and tomorrow. Movement Disorders 3(1):10–29; 1988.

Warne, R. W.; Grubbay, S. S. Choreiform movements induced by anticholinergic therapy. Med. J. Aust. 1:465; 1979.

Weaver, D. F.; Camfield, P.; Fraser, A. Massive carbamazepine overdose: clinical and pharmacologic observations in five episodes. Neurology 38:755–759; 1988.

Weiner, W. J.; Nausieda, P. A.; Klawans, H. L. Methyl-phenidate-induced chorea: case report

and pharmacologic implications. Neurology 29:1041–1044; 1978.

Peripherally Induced Movement Disorders

Banks, G.; Nielsen, V. K.; Short, M. P.; et al. Brachial plexus myoclonus. J. Neurol. Neurosurg. Psychiat. 48:582–584; 1985.

Brin, M. F.; Fahn, S.; Bressman, S. B.; et al. Dystonia precipitated by trauma. Neurology 36(suppl. 1):119; 1986.

Bronstein, A. M.; Rudge, P.; Beechey, A. H. Spasmodic torticollis following unilateral VIII nerve lesions: neck EMG modulation in response to vestibular stimuli. J. Neurol. Neurosurg. Psychiat. 50:580–586; 1987.

Davis, S. M.; Murray, N. M. F.; Diengdoh, J. V.; et al. Stimulus-sensitive spinal myoclonus. J. Neurol. Neurosurg. Psychiat. 44:884–888; 1981.

Dhaliwal, G. S.; McGreal, D. A. Spinal myoclonus in association with herpes zoster infection: two case reports. Can. J. Neurol. Sci. 1:239–241; 1974.

Digre, K.; Corbett, J. J. Hemifacial spasm: differential diagnosis, mechanism and treatment. In: Jankovic, J.; Tolosa, E., eds. Advances in neurology: facial dyskinesias, Vol. 49. New York: Raven Press; 1988: p. 151–176.

Drake, M. E. Spasmodic torticollis after closed-head injury. J. Natl. Med. Assoc. 79:561–563; 1987.

Eccles, R. M.; Kozak, W.; Westerman, R. A. Enhancement of spinal monosynaptic reflex responses after denervation of synergic hindlimb muscles. Exp. Neurol. 6:451–464; 1962.

Garcin, R.; Rondot, P.; Guiot, G. Rhythmic myoclonus of the right arm as the presenting symptom of a cervical cord tumor. Brain 91:75–84; 1968.

Hoehn, M. M.; Cherington M. Spinal myoclonus. Neurology 27:942–946; 1977.

Jacobson, M. P.; Garmen, V. I. Palatal myoclonus and primary nystagmus following trauma. Arch. Neurol. Psychiatry 62:798–801; 1949.

Jankovic, J.; Glass, J. P. Metoclopramide-induced phantom dyskinesia. Neurology 35:432–435; 1985.

Jankovic, J.; Pardo, R. Segmental myoclonus: clinical and pharmacological study. Arch. Neurol. 43:1025–1031; 1986.

Jankovic, J.; Van Der Linden, C. Dystonia and tremor induced by peripheral trauma: predisposing factors. J. Neurol. Neurosurg. Psychiat. 51:1512–1519; 1988.

Jankovic, J.; Schwartz, K.; Donovan, D. T. Botulinum toxin treatment of cranial-cervical dystonia, spasmodic dysphonia, other focal dystonias, and hemifacial spasm. J. Neurol. Neurosurg. Psychiat. 53:633–639; 1990.

Janetta, P. J.; Abbasy, M.; Maroon, J. C.; et al. Aetiology and definitive microsurgical treatment of hemifacial spasms. J. Neurosurg. 47:321–328; 1977.

Kao, L. I.; Crill, W. E. Penicillin-induced segmental myoclonus. I. Motor responses and intracellular recording from motoneurons. Arch. Neurol. 26:156–161; 1972.

Koller, W. C.; Wong, G. F.; Lang, A. Posttraumatic movement disorders: a review. Movement Disorders 4(1):20–36; 1989.

Levy, R.; Plassche, W.; Riggs, J.; et al. Spinal myoclonus related to an arteriovenous malformation: response to clonazepam therapy. Arch. Neurol. 40:254–255; 1983.

Loeser, J. D.; Ward, A. A. Some effects of deafferentiation on neurons of the cat spinal cord. Arch. Neurol. 17:629–636; 1967.

Loeser, J. D.; Ward, A. A.; White, L. E. Chronic deafferentation of human spinal cord neurons. J. Neurosurg. 29:48–50; 1968.

Luttrell, C. N.; Bang, C. B. Newcastle disease encephalomyelitis in cats. I. Clinical and pathological features. Arch. Neurol. 79:647–657; 1958.

Luttrell, C. N.; Bang, C. B.; Luxenberg, K. Newcastle disease encephalomyelitis in cats. II. Psychologic studies on rhythmic myoclonus. Arch. Neurol. 81:285–291; 1959.

Marion, M. H.; Gladhill, R. F.; Thompson, P. D. Spasms of amputation stumps: a report of 2 cases. Movement Disorders 4:354–358; 1989.

Marsden, C. D.; Obeso, J. A.; Traub, M. M.; et al. Muscle spasms associated with Sudeck's atrophy after injury. Br. Med. J. 288:173–176; 1984.

Matuso, F.; Ajaz, E. T. Palatal myoclonus and denervation supersensitivity in the central nervous system. Ann. Neurol. 5:72–78; 1979.

Montagna, P.; Cirignotta, F.; Sacquegna, T.; et al. "Painful legs and moving toes" associated with polyneuropathy. J. Neurol. Neurosurg. Psychiat. 46:399–403; 1983.

Nathan, P. W. Painful legs and moving toes: evidence on the site of the lesion. J. Neurol. Neurosurg. Psychiat. 41:934–939; 1978.

Scherokman, B.; Husain, F.; Cuetter, A.; et al. Peripheral dystonia. Arch. Neurol. 43:830–832; 1986.

Schott, G. D. "Painful legs and moving toes:" the role of trauma. J. Neurol. Neurosurg. Psychiat. 44:344–346; 1981.

Schott, G. D. The relationship of peripheral trauma and pain to dystonia. J. Neurol. Neurosurg. Psychiat. 48:698–701; 1985.

Schott, G. D. Induction of involuntary movements by peripheral trauma: an analogy with causalgia. Lancet 2:712–716; 1986.

Schott, G. D. Mechanisms of causalgia and related clinical conditions. The role of the central nervous and sympathetic nervous systems. Brain 109:717–738; 1986.

Schwarttman, R. J.; McLellan, T. L. Reflex sympathetic dystrophy. Arch. Neurol. 44:555–561; 1987.

Sheehy, M. P.; Marsden, C. D. Trauma and pain in spasmodic torticollis. Lancet 1:777–778; 1980.

Spillane, J. D.; Nathan, P. W.; Kelly, R. E.; et al. Painful legs and moving toes. Brain 94:541–556; 1971.

Steiner, J. L.; DeJesus, P. V.; Mancall, E. L. Painful jumping amputation stumps: pathophysiology of a "sore circuit." Trans. Am. Neurol. Assoc. 99:253–255; 1974.

Verhagen, W. I. M.; Horstink, M. W. I. M.; Notermans, S. L. H. Painful arm and moving fingers. J. Neurol. Neurosurg. Psychiat. 48:384–389; 1985.

19

Multiple Sclerosis

LOREN A. ROLAK

Multiple sclerosis (MS), the most common nontraumatic disabling neurological disease of adults younger than 40 years, affects 250,000 Americans and costs the American economy 3 billion dollars per year (Johnson et al. 1979). MS is an inflammatory disease of the central nervous system (CNS) myelin, the lipoprotein sheath that surrounds axons and enhances nerve conduction. The precise etiology remains unknown, but considerable evidence suggests an autoimmune cause, possibly triggered by a viral infection, which leads to intermittent episodes of inflammation and destruction of myelin in the optic nerves, brain, and spinal cord. The average age of onset is 30 years, and attacks occur at variable intervals but approximately once per year. It affects more women than men by a 3 : 2 ratio. Because it is confined to the white matter, the most common symptoms involve optic nerve blindness, pyramidal tract weakness, spinothalamic numbness, cerebellar ataxia, and myelopathic bowel and bladder dysfunction. Although no effective therapy exists to alter the natural history of MS, its presumed autoimmune etiology has encouraged the use of anti-inflammatory and immune suppressant drugs; therefore, steroids have become the standard treatment for attacks of the disease. Cyclophosphamide (Cytoxan) and other immune suppressive treatments also are used occasionally.

When people learn that they have MS, their immediate concern turns to the prognosis of future disability. Patients anxiously desire a complete revelation of all aspects of the disease as soon as the diagnosis is considered probable (Elian and Dean 1985). Aside from the patient's concern about prognosis, the outcome has become more important because proliferating experimental therapies allow treatment of an increasing number of patients with drugs that carry significant risks. If it were possible to identify at an early stage patients destined to have a benign course, they could be spared the toxic effects of experimental and conventional therapies. Similarly, if it were clear which patients would do poorly, more aggressive measures would be justified. Determining the prognosis of MS has been the subject of an increasing number of research studies, with disparate and conflicting results. In part, this is because different MS populations have been analyzed by different

295

methodologies. Despite these discrepancies, some general principles have emerged that allow reasonable predictions about the prognosis of MS.

Clinical Course

The hallmark of MS is its variability. Figure 19-1 indicates the many different patterns the illness may take, ranging from transient isolated neurological symptoms to chronic progressive multisymptomatic disease (McAlpine 1954). For the sake of simplicity, MS usually is divided into two broad clinical types: exacerbating–remitting and chronic progressive.

The majority of patients present with exacerbating–remitting disease, characterized by "attacks" of MS in which symptoms ap-

pear abruptly. The onset of an MS symptom may be instantaneous in up to 10% of patients, take several minutes in 20%, several hours in 40%, and several days in 30%. The symptom will then persist for an average of 4 to 8 weeks and then remit, often recovering completely. An attack of MS, from the onset to the return to baseline, may last 2 or 3 months. Attacks may occur at any time throughout the course of the disease (Goodkin et al. 1989) with variable frequency. On average, most patients have one attack per year. The majority of recurring attacks mimic previous ones and reproduce old symptoms, but approximately one third strike a new area of the nervous system and produce additional clinical deficits. Most attacks cause pyramidal, cerebellar, or brain stem deficits (Kurtzke 1989).

Figure 19-1. The course of multiple sclerosis.

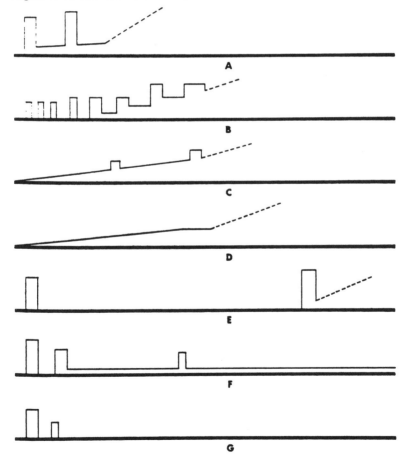

After several years, the exacerbating–remitting pattern gradually gives way to the chronic progressive form, in which symptoms appear more insidiously and persist more indefinitely without the sudden onset and eventual resolution. Although patients may remain in the exacerbating–remitting phase, most will ultimately enter the chronic progressive form. The mean time from first symptoms to onset of the chronic progressive phase is 6 or 7 years, and one half of all patients become chronic progressive within 10 years (Compston 1987; Bernardi et al. 1987; Weinshenker et al. 1989). In one study, 90% of patients had chronic progressive MS after 25 years (Weinshenker et al. 1989). Approximately 15% to 25% of MS patients begin with chronic progressive symptoms, never experiencing a preceding exacerbating–remitting phase, and these usually are patients whose symptoms appear after age 40 (Compston 1987). Their symptoms are more likely to involve the pyramidal tracts, especially the spinal cord.

Recent large studies, using modern computer databases, have revealed the oversimplification of this traditional categorization of MS patients. Patients with chronic progressive disease may suddenly suffer attacks, while those with exacerbating–remitting disease may show chronic progression of some symptoms. Patients who are stable for years may suddenly worsen, while rapidly deteriorating patients may stabilize.

During any period of time, there is a great deal of activity within an MS population; patients "shift" from exacerbating to progressive forms and vice versa, which highlights the extreme variability of this disease. Magnetic resonance imaging (MRI) has shown that even during clinically quiet periods between attacks when patients may be asymptomatic and have normal neurological examinations, lesions within the brain appear and disappear. This suggests that MS is always active but only occasionally does the damage, similar to the tip of an iceberg, appear "above the surface" to cause symptoms (Paty 1987).

The variability of MS is an essential issue in treatment and counseling. Patients find it extremely difficult to come to grips with a disease that is so unpredictable.

Prognosis

The most important factor for predicting disability is the onset of the progressive phase of the disease (Minderhoud et al. 1988). Most studies show that patients have little permanent disability until the progressive phase begins (Compston 1987; Bernardi et al. 1987; Poser et al. 1986; Lauer and Firnhaber 1987; Thompson et al. 1986; Minderhoud et al. 1988). However, while progressive MS portends a disappointing prognosis, caution is necessary because even the chronic progressive form of the disease is

Table 19-1. Prognostic Factors in Multiple Sclerosis

Good Prognosis	Bad Prognosis	No Apparent Influence
Exacerbating–remitting course	Progressive course	Gender
Early onset (younger than age 40)	Late onset (older than age 40)	Rate or number of attacks
Optic neuritis or sensory symptoms	Motor, cerebellar, or sphincter symptoms	Monosymptomatic or polysymptomatic onset
Long first remission (delayed second attack)		
Benign course first 5 years		Magnetic resonance imaging changes
		Visual evoked responses
		CSF immunoglobulins
		Oligoclonal bands
		HLA type
		Lymphocyte subpopulations

Reprinted from Rolak, L.A. Multiple sclerosis. Curr. Neurol. 9:109–148, 1989, by permission of Year Book Medical Publishers, Inc.

highly variable. Although progressive disease often leads to disability, it is too variable to forecast the outcome accurately in any patient (Goodkin et al. 1989). A prospective 1-year study of chronic progressive MS patients showed that 21% stabilized with no neurological deterioration, and many showed spontaneous improvement (Miller et al. 1988). This study could not identify any prognostic features in patients with progressive MS and concluded that the course of the illness cannot be predicted in an individual case (Table 19-1).

Onset of MS late in life carries a discouraging prognosis (Bernardi et al. 1987; Confavreux et al. 1980; Detels et al. 1982; Lauer and Firnhaber 1987; Phadke 1987; Poser et al. 1986; Poser et al. 1989; Riise et al. 1988; Thompson et al. 1986; Visscher et al. 1984). This is only partly because patients older than 40 years often have the chronic progressive form of the disease; independent of other variables, the prognosis worsens with increasing age of onset.

Surprisingly, the number and frequency of attacks has little value in determining the future course of MS (Bernardi et al. 1987; Compston 1987; Liebowitz and Alter 1970; Weinshenker and Ebers 1987). Throughout the course of the disease, the rate of attacks may vary widely (Goodkin et al. 1989). Although the frequency of attacks does not seem predictive, evidence suggests that a prolonged interval between the first and second attack indicates a more benign disease (Bernardi et al. 1987; Confavreux et al. 1980; Phadke 1987; Thompson et al. 1986; Wolfson and Confavreux 1987). Although not all investigators agree (Confavreux et al. 1980; Lauer and Firnhaber 1987; Wolfson and Confavreux 1987) most analyses have shown that patients with sensory symptoms at onset, including optic neuritis, fare significantly better than patients with cerebellar, motor, or sphincter changes (Bernardi et al. 1987; Compston 1987; Phadke 1987; Poser et al. 1986; Thompson et al. 1986; Visscher et al. 1984). This may be because sensory loss is less disabling than motor loss. It has long been assumed that women do better than

Table 19-2. Average Disability 15 Years After the Onset of Multiple Sclerosis

60% ambulatory without assistance
20% require cane, crutches, or wheelchair
20% bedridden or dead

Modified from Weinshenker, B.G.; Ebers, G.C. The natural history of multiple sclerosis. Can. J. Neurol. Sci. 14:255–261; 1987.

men (Detels et al. 1982), but the most recent studies do not confirm any role for gender in the prognosis of MS (Bernardi et al. 1987; Poser et al. 1986; Weinshenker and Ebers 1987).

Unfortunately, few data exist about the rate of accumulation of disability. Patients with benign disease 5 to 10 years after their first symptoms (at which time, the progressive phase of the disease has usually begun if it is going to occur) tend to have a better outcome than the average MS patient (Compston 1987; Phadke 1987; Weinshenker and Ebers 1987). The same does not necessarily apply to patients with moderate or severe disability after 5 years because in this group, past performance does not predict future performance (Table 19-2).

MS victims should understand that their disease is seldom fatal and that many patients do well. In one study of 400 patients followed for 10 years, 42% had slight or no disability (Thompson et al. 1986). Certainly, MS rarely is a fatal disease, and the number of deaths in excess of the natural baseline mortality is only 19 per 1000 MS patients per year (Poser et al. 1986). Only 1% of patients die of causes directly related to their MS (Weinshenker et al. 1989).

Laboratory Studies

Some studies have suggested that various laboratory parameters can help predict the clinical course of MS, such as an increased number of oligoclonal bands (Livrea et al. 1987) or the appearance of confluent lesions around the ventricles on magnetic resonance imaging (MRI) (Paty 1987). However, other studies have disagreed, and there is presently no predictive value to MRI, evoked potentials, spinal fluid, immunoglobulins or oligoclonal bands, CNS immunoglobulin synthesis, HLA type, or lymphocyte subpopulations (Compston 1987; Confavreux et al. 1980; Thompson et al. 1986; Phadke 1987). Although these laboratory tests may provide support for the diagnosis and give researchers important clues about the nature of the disease, they offer no help in predicting the outcome of MS at onset or in subsequent evaluations (Rolak 1989).

Treatment

No therapy can alter the prognosis of MS. Steroids probably shorten the symptoms of attacks and possibly minimize the disability from an episode. However, no data show that patients treated with repeated courses of steroids ultimately acquire less disability than those not treated. Thus, steroids are useful to manage flares of MS but probably do not prevent final disability. For chronic progressive patients, some steroid regimens may temporarily slow or even improve progressive symptoms but without clear evidence that ultimate disability can be avoided. Although many forms of steroids have been used, such as prednisone, ACTH, and cortisone, increasing emphasis is being placed on extremely high doses of intravenous steroids, such as methylprednisolone.

Immune-altering drugs also are widely accepted therapies for MS, but there is no convincing evidence that they alter the natural history of the disease. The most common is cyclophosphamide, which some unblinded studies have suggested stabilizes symptoms and improves disability, sometimes for 1 year or longer. However, blinded controlled trials have not shown much success, so it remains unclear whether this therapy alters the course of MS. Similarly, other immune suppressive treatments, such as azathioprine, cyclosporine, plasmaphoresis, and lymphoid irradiation, have shown no or minimal impact on the disease. Because of the immune abnormalities that accompany MS, there is a strong belief that immune-altering

therapies will prove successful, but given the rudimentary understanding of the immune functions critical to MS, an appropriate therapy has not yet been devised.

Conclusion

The course of MS is unpredictable, possibly because it is determined by an unpredictable event—the chance location of autoimmune attacks on the nervous system (Compston 1987). The purely random location of lesions, rather than any fundamental differences in the disease process, may determine ultimate disability.

References

Bernardi, S.; Buttinelli, C.; Grasso, M. G.; Millefiorini, E.; Pace, A.; Prencipe, M.; Fieschi, C. Evolution and severity markers in 233 multiple sclerosis patients. Riv. Neurol. 57:197–200; 1987.

Compston, A. Can the course of multiple sclerosis be predicted? In: Warlow, C.; Garfield, J., eds. More dilemmas in the management of the neurological patient. Edinburgh: Churchill Livingstone Inc.; 1987:p. 45–50.

Confavreux, C.; Aimaro, G.; Devic, M. Course and prognosis of multiple sclerosis assessed by the computerized data processing of 349 patients. Brain 103:281–300; 1980.

Detels, R.; Clark, V.A.; Valdiviezo, N. L.; Visscher, B. R.; Malmgren, R. M.; Dudley, J. P. Factors associated with a rapid course of multiple sclerosis. Arch. Neurol. 39:337–341; 1982.

Elian, M.; Dean, G. To tell or not to tell the diagnosis of multiple sclerosis. Lancet II:27–28; 1985.

Goodkin, D. E.; Hertsgaard, D.; Rudick, R. A. Exacerbation rates and adherence to disease type in a prospectively followed-up population with multiple sclerosis. Arch. Neurol. 46:1107–1112; 1989.

Johnson, R. T.; Katzmann, R.; McGeer, E.; Price, D.; Shooter, E. M.; Silberberg, D. Report of the panel on inflammatory, demyelinating and degenerative disease. NIH Publication 79-1916. Washington, DC: U.S. Department of Health, Education, and Welfare. 1979.

Kurtzke, J. F. Patterns of neurologic involvement in multiple sclerosis. Neurology 39:1235–1238; 1989.

Lauer, K.; Firnhaber, W. Epidemiological investigations into multiple sclerosis in Southern

Hesse: V. Course and prognosis. Acta. Neurol. Scand. 76:12–17; 1987.

Liebowitz, U.; Alter, M. Clinical factors associated with increased disability in multiple sclerosis. Acta. Neurol. Scand. 46:53–70; 1970.

Livrea, P.; Simone, I. L.; Trojano, M.; Pissichio, L.; Logroscino, G.; Rosato, A. Cerebral spinal fluid parameters and clinical course of multiple sclerosis. Riv. Neurol. 57:189–195; 1987.

McAlpine, D. Multiple sclerosis: a reappraisal. London: Churchill Livingstone; 1954.

Miller, A.; Drexler, E.; Keilson, M.; Slagle, S.; Bornstein, M.; Rolak, L.; Appel, S. Spontaneous stabilization in patients with chronic progressive multiple sclerosis. Neurology 38:194; 1988.

Minderhoud, J. M.; Van der Hoeven, J. H.; Prange, A. J. A. Course and prognosis of chronic progressive multiple sclerosis. Acta. Neurol. Scand. 78:10–15; 1988.

Paty, D. W. Multiple sclerosis: assessment of disease progression and effects of treatment. Can. J. Neurol. Sci. 14:518–520; 1987.

Phadke, J. G. Survival pattern and cause of death in patients with multiple sclerosis: results from an epidemiological survey in Northeast Scotland. J. Neurol. Neurosurg. Psychiat. 50:523–531; 1987.

Poser, S.; Poser, W.; Schlaf, G.; Firnhaber, W.; Lauer, K.; Wolter, M.; Evers, P. Prognostic indicators in multiple sclerosis. Acta. Neurol. Scand. 74:387–392; 1986.

Poser, S.; Kurtzke, J. F.; Poser, W.; Schlaf, G. Survival in multiple sclerosis. J. Clin. Epidemiol. 42:159–168; 1989.

Riise, T.; Gronning, M.; Aarli, J. A.; Nyland, H.; Larslen, J. P.; Edland, A. Prognostic factors for life expectancy in multiple sclerosis analyzed by Cox-models. J. Clin. Epidemiol. 41:1031–1036; 1988.

Rolak, L. A. Multiple sclerosis. Curr. Neurol. 9:109–148; 1989.

Thompson, H. A.; Hutchinson, M.; Brazil, J.; McAlpine, D. A clinical and laboratory study of benign multiple sclerosis. Quart. J. Med. 225:69–80; 1986.

Visscher, B. R.; Liu, K. S.; Clark, V. A.; Detels, R.; Malmgren, R. M.; Dudley, J. P. Onset symptoms as predictors of mortality and disability in multiple sclerosis. Acta. Neurol. Scand. 70:321–328; 1984.

Weinshenker, B. G.; Ebers, G. C. The natural history of multiple sclerosis. Can. J. Neurol. Sci. 14:255–261; 1987.

Weinshenker, B. G.; Bass, B.; Rice, G. P. A.; Noseworthy, J.; Carriere, W.; Baskerville, J.; Ebers, G. C. The natural history of multiple sclerosis: a geographically based study. I. Clinical course and disability. Brain 112:133–146; 1989.

Wolfson, C.; Confavreux, C. Improvements to a simple Markov model of the natural history of multiple sclerosis. Neuroepidemiology 6:101–115; 1987.

20

Neuro-Ophthalmology

ROSA A. TANG

Optic Neuritis

Optic neuritis is the broad term for an inflammatory condition affecting the function of the optic nerve. Visual loss is almost always present, usually accompanied with pain (Rizzo and Lessell 1991).

The presentation and the etiology seems to be different in children than in adults. In children, a postviral or parainfectious process is the most common cause; examination reveals a neuroretinitis that usually affects both optic nerves.

This type of optic neuritis is thought to be due to an immunological process that produces demyelination of the optic nerve and is occasionally accompanied by an immunologically mediated meningoencephalitis. Visual recovery usually is excellent, within 4 to 6 weeks of onset, regardless of the initial degree of visual impairment (Maitland and Miller 1984). Kriss and coworkers (1988) followed 39 children for a mean number of 8.8 years and found that 8% had recurrences, while 15% developed demyelinating disease.

A few cases of optic neuritis have been associated with Guillain-Barré syndrome following a flu-like illness. In the reported cases, the patients progressed to blindness, and the visual recovery was minimal. Steroid therapy did not appear to change the disease progression or recovery (Pal and Williams 1987). Optic neuritis can be associated with human immunodeficiency virus (HIV) infection and opportunistic organisms, such as cytomegalovirus (CMV) infection of the retina and optic nerve. In these cases, the optic neuritis usually causes blindness with poor response to treatment, because of irreversible optic nerve necrosis including intravitreal or intravenous ganciclovir (Grossniklaus et al. 1987).

Syphilitic perineuritis, when associated with HIV infection, is usually bilateral and presents with profound loss of vision, leading to blindness. High-dose penicillin therapy helps visual recovery when the treatment is instituted early (Sombrono and Smith 1987).

Autoimmune optic neuritis can be seen in patients that present with severe visual loss (20/200 or less) and progressive disease. Laboratory evidence of autoimmune disease is present, such as antinuclear antibody, high sedimentation rate, and circulating lupus anticoagulant levels. Occasionally these patients have manifestations of SLE. Treatment with systemic corticosteroids is highly successful (Openheimer and Hoffbrand 1986). Kuppersmith and Burde (1988) described that skin biopsy may be a valuable tool in the diagnosis of autoimmune optic neuritis.

Radiation optic neuritis can result from conventional courses of proton beam radiation when the treated field includes, or is adjacent to, the optic nerve. Optic neuritis presenting as painless loss of vision, with or without disc edema is seen 6 months to 3 years after radiation therapy with a peak incidence between 8 to 13 months after radiation therapy. Visual loss may be progressive, resulting in blindness (Fitzgerald et al. 1981). This condition is unresponsive to steroid treatment.

Hyperbaric oxygen therapy has been used with questionable success in isolated situations in which the patient's clinical picture is compatible with the diagnosis of radiation optic neuritis (Guy and Schatz 1986). Histological studies have shown ischemic demyelination and an obliterative endarteritis with necrosis, which usually progress to irreversible visual loss.

Optic neuritis may be associated with many systemic diseases with infectious and parainfectious processes such as sinusitis, some of which are amenable to specific treatment. The evaluation should thus seek evidence of connective tissue disorder (i.e., systemic lupus erythematosus), infections (i.e., syphillis, HIV, Lyme disease), or inflammatory diseases (i.e., sinusitis and sarcoidosis). Certain forms of optic neuritis appear to be unrelated to significant systemic diseases, such as Leber's idiopathic stellate neuroretinitis, in which the visual prognosis is good and recurrence is unusual. It therefore requires no treatment.

Optic neuritis occurs frequently in patients with diagnosed demyelinating disease, but more frequently, it occurs as an idiopathic disorder in patients who develop other neurological manifestations that suggest multiple sclerosis. What is the risk of developing multiple sclerosis after one isolated episode of optic neuritis with no other neurological manifestations? In a review of 20 studies that have addressed this topic, the conversion has ranged from 15% to 85% (Kurtzke 1985).

Rizzo and Lessell (1988) followed 60 patients for 15 years, and 58% with optic neuritis developed multiple sclerosis. Using life-table analysis to examine this series with projected features for 20-year follow-up, the study suggests that 74% of the women and 34% of the men will develop multiple sclerosis within 15 years of the initial attack of isolated optic neuritis. After 20 years of follow-up, it was predicted that 91% of women and 45% of men will develop multiple sclerosis. The risk of developing multiple sclerosis does not appear to decline with time. However, the proportion of patients at risk for developing multiple sclerosis seems to increase with time, and women are at higher risk than men.

No study has shown that the recurrence of optic neuritis increases the risk of multiple sclerosis if the effects of age and sex are taken into account. Francis and colleagues (1987), studied a population with optic neuritis, and found that 15% converted to demyelinating disease in 11.6 years of follow-up. In these patients, the risk of demyelination was higher in patients with oligoclonal bands in the cerebrospinal fluid (CSF) and in those who were human leukocyte antigen (HLA)-DR3 positive. This study based the diagnosis of optic neuritis on clinical data alone, so it may underestimate the true frequency of multiple sclerosis in patients with optic neuritis. A prospective study by Amarkrud and Slettnes (1989) showed that in 30 patients with retrobulbar neuritis, 50% developed demyelinating disease within 2 to 11 years. In patients with oligoclonal bands in the CSF, the risk of developing multiple sclerosis increased to 79%. When the CSF was normal, the risk decreased to 10%. In most patients, the diagnosis of multiple sclerosis was made within 3 years of the initial episode of retrobulbar optic neuritis.

Jacobs and associates (1986) used nuclear magnetic resonance imaging (MRI) to diagnose and assess the likelihood of developing multiple sclerosis. Of the 16 patients with isolated optic neuritis and no other neurological complaints, 50% had abnormal MRI findings with periventricular lesions, similar to patients with demyelinating disease. The MRI scan can detect silent white-matter lesions in patients with multiple sclerosis, but it cannot detect lesions in the optic nerves of patients with optic neuritis. Miller and coworkers (1988) found high-signal regions in 8% of asymptomatic patients. Johns and associates (1986) found 70% of positive MRI brain lesions in patients with optic neuritis.

The number of patients with optic neuritis and MRI findings who develop demyelinating disease is not known. It is of interest that the optic neuritis study group (1991) found abnormal MRI findings suggesting demyelination in 48.7% of patients studied. The existence of silent brain lesions in many patients with optic neuritis may eventually show relationship between optic neuritis and multiple sclerosis. Recovery of visual acuity (Snellen optotype), in most cases, is good. However, Fleishman and coworkers (1987) demonstrated that 85% of patients may be left with deficits in other parameters of visual function, such as color sensation, contrast sensitivity, light brightness sense, and stereopsis. A recent report on the clinical profile of optic neuritis may be helpful in making a more firm clinical diagnosis of this entity (Optic neuritis study group 1991).

Therapy for optic neuritis remains controversial. The effects of high-dose intravenous steroid therapy administered in a pulse fashion compared to oral corticosteroids have not been demonstrated to be valuable. Therapeutic intervention does not influence the course of idiopathic optic neuritis, but The Optic Neuritis Treatment Trial, a multicenter, randomized study sponsored by The National Eye Institute should provide better guidelines. The role of high-dose corticosteroid therapy in optic neuritis remains to be elucidated.

Ischemic Optic Neuropathy

Ischemic optic neuropathy (ION) is a condition characterized by optic nerve dysfunction secondary to hypoperfusion of the nutrient vessels in the optic nerve head, or less commonly, thrombosis or embolic occlusion of the posterior ciliary artery (Hayreh 1981). It can be divided into two types: anterior and posterior.

Anterior ischemic optic neuropathy (AION) is the most common type of ION. There is sudden visual loss in one eye with an inferior nasal sectorial or attitudinal field deficit. Pain may or may not be present. The ophthalmoscopic examination reveals hyperemic or pale disc edema with peripapillary hemorrhages.

The less common form, posterior ischemic optic neuropathy (PION), is characterized by hypoxic swelling of the orbital portion of the optic nerve and compression of the centripetal pial vessels without evidence of disc edema (Rizzo and Lessell 1987).

ION, whether anterior or posterior, also can be divided into arteritic and nonarteritic types. The nonarteritic type usually occurs in patients 45 to 65 years old. The association with disorders that cause vaso-occlusive disease, such as diabetes and arterial hypertension is debatable. The incidence of bilateral nonarteritic AION is between 10% and 73%. In a prospective study by Beri and colleagues (1987), 388 patients with nonarteritic AION were followed for 28.6 months. The incidence of bilateral AION continued a linear rise during the 28.6 months of the study. This study also stated that the time to develop bilateral AION in nonarteritic ION was 32.4 months, using the estimated 25 percentile time statistical analysis. Also, in nonarteritic AION, the risk of bilaterality was significantly greater in men, patients under 45 years old, and patients with diabetes.

Although most patients with nonarteritic AION present with moderate acute monocular loss of vision (i.e., 20/50 to 20/200), there are cases of disc swelling without visual loss, the "asymptomatic AION" (Levine 1987). Kline (1988) has reported cases of nonarteritic AION with progression of vision loss over a period of up to 7 months. The cause of nonarteritic AION is not well defined, but it is most likely multifactorial, encompassing unusually small cups, crowded discs, arteriosclerotic vascular changes, and dysfunction in the optic nerve circulation. Several studies have emphasized that cup to disc ratios are significantly smaller in both eyes of patients with nonarteritic ION (Beck et al. 1987). These findings suggest that patients with nonarteritic AION may have anatomical factors at the disc that predispose them to the disease. Edematous axons in a smaller space within the scleral canal cause compression of the fine vessels on the nerve fiber bundles of the optic nerve, causing ischemia, axoplasmic flow stasis, disc edema, and infarction.

There is little histopathological evidence to support the concept of irreversible infarction. Most patients experienced sudden deficits of vision, but approximately 25% developed progressive deterioration of visual function 1 to 4 weeks after the onset of symptoms. The progressive nature of visual loss suggest that in some cases, AION may be an evolving process amenable to surgical intervention. Optic nerve sheath decompression (fenestration) reverses some of these progressive cases (Sergot et al. 1989). When considering a patient for optic nerve sheath fenestration, it is important to evaluate the patient's visual acuity, visual fields, and color vision every 10 days for 1 month after making the diagnosis. If there is progressive loss of vision, surgery may be indicated. There is a tremendous controversy as to the use of this treatment modality. Pharmacological intervention with corticosteroids, anticoagulants, and other medications has not been beneficial (Sergot et al. 1989).

Nonarteritic ION also has been described in association with posthemorrhage cases accompanied by arterial hypotension (Hayreh 1987). Coronary artery bypass, chronic hypotension, platelet transfusion, general anesthesia, and coagulopathies also have been implicated as etiologies of ION.

In cases of nonarteritic ION, there is usually secondary optic atrophy after the resolution of the disc edema. Recurrences of AION of the same eye are extremely rare. However, AION may develop in the fellow eye in 25% to 40% of cases, but this happens months to years after the initial involvement (Beri et al. 1987). When this occurs, pseudo Foster Kennedy syndrome occurs. The examination reveals nonarteritic AION with disc edema in one eye and optic atrophy in the other eye where a previous episode of AION had occurred. Consecutive AION in both eyes occurs in approximately 25% of cases (Repka et al. 1983; Boghen and Glaser 1975).

Arteritic AION presents with devastating loss of vision, that is, hand motion, light perception, or no light perception. The prevalence of arteritic AION in the general population is not known. However, in a selected population of 438 patients with AION, 50 patients had arteritic AION (Beri et al. 1987). This study found that the risk of bilaterally in patients with arteritic AION was 1.9 times higher than in those with nonarteritic AION. This study also found that the development of bilateral AION occurred much sooner in patients with arteritic AION (0.4 months) compared with patients with nonarteritic AION (32.4 months). The cup to disc ratios were studied in patients with arteritic AION; larger cups were present in the fellow eye of patients with this type of AION. Also, the optic nerve atrophy resulting from arteritic AION was different than the small cup-to-disc ratios in nonarteritic AION (Beck et al. 1987).

The diagnosis of temporal arteritis (TA) or giant cell arteritis (GCA) usually is made clinically when the patient presents with an early massive acute loss of vision due to AION. Amaurosis tugax is an important symptom preceeding visual loss in arteritic cases. AION associated with a chalky white optic disc, cilioretinal artery occlusion or massive infarction of the choroidal layer is extremely suggestive of arteritic involvement (Hayreh 1974). The abrupt loss of vision often is but not always (occult form) accompanied by systemic symptomatology, such as headaches, jaw claudication, malaise, weight loss, and fever. Serological indicators of GCA are an elevated erythrocyte sedimentation rate (ESR) calculated by the Westergren method (Love et al. 1988) high

C-reactive protein, and an elevated serum fibrinogen level (Keltner 1985). It must be stressed that a normal ESR value is seen in 2% to 10% of patients with GCA. This is more common if the patient is taking a nonsteroidal anti-inflammatory agent such as aspirin (Tang 1988). The C-reactive protein, or the fibrinogen level may be more helpful than the ESR in monitoring the progress of the disease, the success of therapy, and any possible recurrences (Bates et al. 1989).

Histological confirmation of the clinical diagnosis of GCA must be obtained when possible. This is because of the long-term treatment that these patients will require and the complications inherent in long-term steroid therapy. In many cases, the vessels may be affected in a discontinuous fashion; that is, one section of the artery may show the typical inflammatory invasion of the vessel wall, while another section of the same specimen may look normal. For this reason, the biopsy specimen should be at least 2 cm in length, and serial cross sections of the whole specimen are needed (McDonnell et al. 1986).

In a review of 250 biopsies by McDonnell and associates (1986), the "skipped" areas were infrequently found. Active arteritis was seen up to 7 weeks after initiation of treatment, suggesting that the biopsy should not be performed more than 2 months after treatment is instituted. Cases have been reported in which the biopsy has been positive many months, and even years, after treatment (Smith 1988). Chambers and Bernardino (1988) have suggested that with proper sampling techniques and examined correctly, the likelihood of a negative biopsy is less than 1%.

If tissue confirmation is vitally important and the first biopsy is negative, a second biopsy may be required for patients with relative contraindications to long-term steroid therapy. McDonnell and coworkers (1986) found no definite advantage to bilateral biopsy in 13 cases in which this was done.

The pathological features of GCA include a granulomatous reaction involving the media of the artery. If GCA is suspected, immediate treatment should be instituted because of the threat of permanent visual damage. Daily high-dose corticosteroid therapy appears to be the only effective treatment (Hunder et al. 1975). Modern immunosuppressive therapy has not been as beneficial as corticosteroids in the treatment of GCA (Keltner 1982). Patients who present within 36 hours of the visual loss or who have severe systemic symptoms should be considered for intravenous corticosteroids (500 mg of methylprednisolone every 6 hours is recommended). Once the severe symptoms are controlled, oral prednisone may be given, beginning with 100 mg daily and gradually decreasing the dosage as the patient's symptoms and ESR abate. In most patients with GCA treated with adequate doses of steroids, the laboratory abnormalities return to the baseline levels within 1 week of treatment (Gordy 1983; Rosenfeld et al. 1986).

When the symptoms and the ESR C-reactive protein and/or fibrinogen are under control, monthly ESRs may be obtained. Giant cell arteritis is not a self-limiting disease, and the symptomatology may reappear when the steroids are tapered off. The patient should be warned of this and should be taught how to monitor their steroid intake based on any recurrence of the clinical symptoms.

Fauchald (1972) found a relapse rate of 26% in patients with biopsy-proven GCA. These patients had been treated with steroids for less than 2 years.

Patients with polymyalgia rheumatica have a higher risk of developing GCA. High-dose steroid treatment may be indicated in these patients when symptoms of temporal artery involvement, such as temple headaches, jaw claudication, transient blindness, diplopia, and scalp tenderness, appear.

This disease leads to an ocular disaster when left untreated. High-dose corticosteroid treatment must be instituted immediately when the clinical suspicion is high to maintain vision, relieve the serious systemic manifestations, and improve the overall health of elderly patients with this condition. Ancillary tests, such as an elevated sedimentation rate, elevated fibrinogen or C-reactive protein, and a positive biopsy, are of secondary importance to the clinical presentation of this disease.

Idiopathic Intracranial Hypertension (or Pseudotumor Cerebri)

The incidence of idiopathic intracranial hypertension (IIH) in the general population has been reported to be 0.9 per 100,000 people (Durcan et al. 1988). Up to 10% of patients with IIH have no symptoms and are discovered during a routine examination, so the incidence may be skewed. The incidence is higher (14.8 per 100,000) in obese women, with a female to male ratio of 8:1.

To meet the diagnosis, the following four criteria were used:

1. Documented elevation of intracranial pressure (ICP) more than 250 ml of water
2. Normal CSF composition
3. Normal results of neuroimaging studies, with normal or small ventricles
4. Normal neurological examination results except for papilledema and nerve VI paresis

The symptoms of IIH are related to the elevated ICP and include headaches, transient visual obscurations, and diplopia.

Numerous medications have been implicated in the pathogenesis of IIH. These include vitamin A, retinoic acid, indomethacin, lithium, tetracycline, nalidixic acid, and corticosteroids.

Thrombosis or obstruction of the dural venous sinuses, primary and secondary hypothyroidism, and other endocrinopathies have been implicated in the pathophysiology of IIH (Fishman 1984).

This disease seems to occur in obese women of childbearing age, and severe visual impairment occurs in up to 25% of patients (Wall et al. 1983). In children, the incidence of IIH is equal among boys and girls and may be related to nutritional deprivation (Couch et al. 1985). Children may present with irritable behavior or ataxia rather than with headache, which is more commonly

found in adults (Couch et al. 1985). In the study by Couch and associates (1985), 12 of the 38 children had no clinically significant papilledema. Papilledema is almost always found in adults with IIH. The prevalence of visual loss in children with IIH is not known (Lessell and Rosman 1986). In children, the majority of cases may require nonspecific means of lowering ICP. Dexamethasone may be started at the time of diagnosis if the child has a mild to moderate abnormality of the vision or severe papilledema. Resolution of the visual abnormalities and decreased severity of the papilledema should appear within 2 weeks of therapy; otherwise, an alternative form of therapy should be instituted. If medical therapy fails to prevent visual loss, surgical therapy should not be delayed (Baker et al. 1985).

Although several series have reported visual loss in 10% to 25% of patients, the incidence may be as high as 49% (Orcutt et al. 1984). Patients with high-grade papilledema, glaucoma, high myopia, and peripapillary subretinal neovascularization were more prone to lose visual function. Wall and George (1987) studied 300 patients with IIH and found visual field deficits in 75%. This suggests that visual fields should be carefully monitored because early recognition of visual field loss may be the key to successful management of these patients. In the largest follow-up series regarding visual loss in IIH, 57 patients were followed from 5 to 41 years (Corbett and Thompson 1989). Severe, permanent visual loss was found in approximately 25% of these patients. The authors reported that visual loss may occur early or late in the disease and the patient may be asymptomatic for long periods of time. For this reason, the patient should be monitored with serial visual acuity and visual fields.

Arterial hypertension was considered a statistically significant risk factor for visual loss, although not all authors agree. Of patients with systemic hypertension, 61% had severe unilateral or bilateral visual loss when reexamined. Corticosteroid administration has the potential to raise intraocular pressure and may be an additional risk factor for associated visual loss in these patients. A 10% recurrence rate is the average in most series and may occur even years after the first episode (Rush 1980).

Men with IIH tend to be less obese than women. Black men with hypertension seem more likely to lose vision than other groups of male patients. Surgical intervention seemed to be required more commonly in men with IIH than in women (Digre and Corbett 1988).

Management should be directed toward lowering the ICP, controlling the headache, weight loss, and prevention of visual loss. Careful monitoring of visual fields by perimetry (Wall 1991) is essential, and visual fields as well as visual acuity must be performed periodically. Medical treatment should include weight reduction for obese patients and carbonic anhydrase inhibitors or diuretics to depress CSF production. Corticosteroids are not recommended as long-term therapy because the associated weight gain is an undesirable side effect in these patients. Other drugs, such as glycerol and digoxin, have been used, but no prospective randomized study has demonstrated that one medication is better than others for reducing elevated ICP. In asymptomatic patients, no treatment other than weight loss is necessary. Repeated lumbar punctures may be tried in cases in which pharmaceutical contraindications exist.

When the patient does not respond to medical treatment, the decision to proceed with surgery depends on the quantitative change in the visual field examination and visual acuity or on the presence of intractable headaches (Wall 1991). Indications for surgical intervention are recent visual loss, progressive visual field loss in a preexisting visual field defect, reduction in visual acuity not due to macular edema (Corbett and Thompson 1989), and severe headaches. The surgical modalities of treatment are lumboperitoneal shunt and optic nerve sheath fenestration (Hupp et al. 1987; Corbett and Thompson 1989). These modalities may be usd safely in children and adults.

IIH is not self-limiting as some early studies suggest. The ICP may be elevated for months to years.

Patients with loss of vision at the time of diagnosis have a worse prognosis, but otherwise symptoms are not useful in predicting potential loss of vision (Lessell and Rosman 1986). CSF pressure may remain clinically elevated in most patients regardless of whether the clinical syndrome recurs. Lessell and Rosman (1986) recommend that visual fields and visual acuity be monitored, probably for life. Recurrences are not uncommon, and visual loss may occur early or late, depending on the evolution in each patient.

A prospective study in Denmark followed 24 patients for 49 months with regular neurological and ophthalmological examinations. Medical treatment with diuretics induced a rapid relief of symptoms in 75% of cases. Twenty-five percent had a more protracted course with persistent headache, and five patients required surgical intervention. Clinical changes in the optic discs were seen in 2% of the patients, but only one patient developed optic atrophy with blindness. Serial measurements of the ICP were taken, and they showed that elevated CSF pressures persisted for long periods in many cases, even though the patients were asymptomatic (Renson et al. 1988).

IIH is a diagnosis of exclusion in which diagnostic criteria includes elevation of CSR pressure and disturbed cerebrovenous outflow, which may result in significant ancillary clinical findings (Wessell et al. 1987). Approximately 45% of cases have enlarged optic nerve sheaths and empty sellas, which represents a consequence of long-term increased pressure within the CSF spaces. A disturbance of CSF pressure regulation may be the pathophysiological mechanism in IIH (Corbett and Thompson 1989).

Cranial Neuropathies

Sixth Cranial Nerve Palsy

After an abduction deficit is determined to be caused by a lesion of the nerve and not the muscle or neuromuscular junction, the most important step is to determine the level of involvement. If it is at the brain stem level, usually there are other neurological complaints and vascular or neoplastic disease should be considered.

As it enters the subarachnoid space, the sixth nerve can be stretched as a result of increased ICP. This may be caused by a number of events, including lumbar puncture, postmyelography, and treatment for cervical fractures (McCrary 1987).

Lesions involving the sixth nerve in the cavernous sinus usually affect other cranial nerves (III, IV, V), with variable degrees of proptosis and pupillary involvement.

Cluster headache can produce a sixth cranial nerve palsy, which may be recurrent and responsive to steroids, simulating the Tolosa-Hunt syndrome (Peatfield 1985).

Cranial nerve VI paresis is the most frequent cranial nerve to involve the eye. Trauma, postviral illness, and brain stem neoplasias account for most of the cranial nerve palsies in children (Rush 1981). The lack of improvement or development of additional signs within 6 weeks to 3 months of onset is reason to proceed with neuroimaging and CSF studies.

In patients 20 to 40 years of age, a neoplastic etiology must be considered and appropriate diagnostic studies used. In patients older than 40 years, a sixth nerve paresis usually represents an ischemic infarction of the nerve due to diabetes, vasculitis, or arterial hypertension with a good prognosis for recovery within 3 months of onset (Rush and Young 1981).

Treatment with botulinum A toxin for chemodenervation of the antagonistic extraocular muscle has been found to be valuable in the management of patients with unresolved sixth nerve paresis. The etiology must be sought prior to treatment because of the obvious changes in the natural history that prevent evaluation after the use of botulinum A toxin (Biglan et al. 1989). Surgery usually is considered for stable deviations that have been present for 6 months or more.

Bilateral sixth nerve paresis is usually of neoplastic origin in any age group and thus demands a thorough evaluation.

Most sixth cranial nerve pareses present with an incomitant esotropia at distance,

and need to be differentiated from the acute comitant esotropia reported in children with brain tumors (Williams and Hoyt 1989).

Fourth Cranial Neve Palsy

The trochlear nerve is most commonly affected by closed head injury. Approximately 50% of patients recover, but 50% require surgical intervention (Sydnor et al. 1982).

In children, a postviral etiology is the most common cause; CNS tumors rarely are a cause. In the 20- to 40-year age group, a decompensated congenital fourth nerve palsy is common and is diagnosed by the finding of large vertical fusional amplitudes. Rarely, a superior cerebellar artery aneurysm, posterior fossa tumor, or cavernous sinus meningioma may produce a fourth nerve paresis.

In patients older than forty years, most fourth nerve palsies are due to vascular ischemia from diabetic or hypertensive disease (Young and Sutula 1977). When no cause is found, the fourth nerve palsy improves with time, so temporizing treatment with prisms or patching may be necessary.

Third Cranial Nerve Palsy

In the majority of patients presenting with a third nerve paresis, the most important question is whether the patient has a life-threatening compressive lesion (aneurysm) or an ischemic process. A rule of thumb is to determine if the pupils are dilated, which suggests a compressive lesion. This is true, but there have been reports of 5% to 14% of patients with no pupillary involvement and with underlying aneurysms of the posterior communicating artery (Nadeau and Trobe 1983).

Guy and Day (1989) recently reviewed pupillary sparing in aneurysms and recommend cerebral angiography to be considered in the initial evaluation of patients with third nerve palsies. Recently the use of noninvasive MRI angiography has been described as an additional tool (Tomsak 1991).

Brain stem lesions producing oculomotor palsies are diagnosed more frequently because MRI is being used. Gliomas, abscess, infarction, and demyelination are common etiologies in this group (Kashiwagi et al. 1987).

In children, oculomotor palsies usually are congenital, traumatic, and migraine-related. Occasionally they can present with cyclic spasm. These have a good prognosis for life, but not for improvement in the ophthalmoplegia (Miller 1977).

Aberrant regeneration of the third cranial nerve is characterized by lid retraction or pupillary miosis in attempted adduction or downgaze and is seen after palsies caused by compressive, traumatic, and not ischemic lesions (Lepore and Glaser 1980). It has been described as a primary phenomenon, without the preceding acute oculomotor palsy, in cavernous sinus lesions (Schatz 1977).

Studies using horseradish peroxidase in laboratory animals have shown that the underlying mechanism in these cases is an aberrant reinnervation and not ephaptic transmission (Sibony et al. 1986).

Internuclear Ophthalmoplegia

Internuclear ophthalmoplegia (INO) is characterized by slow adduction of the affected eye and associated horizontal jerk nystagmus of the abducting eye (Cogan 1970). The adduction weakness is due to a dysfunction of the medial longitudinal fasciculus (Baloh et al. 1978).

Most cases of unilateral and bilateral INO are due to demyelinating disease, and the prognosis depends on the degree of neurological involvement (Burde 1987).

Trauma, arteriovenous malformations, brain stem tumors or ischemia, SLE, and migraines are rare etiologies.

Unilateral INO has a guarded prognosis if it does not clear completely within 2 to 3 months of onset, as stated in the study by Smith and Cogan (1957). Bilateral INO almost always indicates multiple sclerosis as the etiology. The prognosis in these cases is as unpredictable as MS in general. The syndrome may clear and not reappear, although Smith and Cogan (1957) reported six patients who died within 83 months after the INO had been diagnosed.

Bilateral INO also is a frequent finding in toxic coma (7 of 70 cases). Such an association is a strong argument for the toxic etiology of coma; however, it has no value in determining the source of intoxication and is not a good prognostic indicator for recovery or outcome (Barrett et al. 1983).

References

Aziz, S.; Abdul-Rahim, A. S.; Savino, P. J.; Zimmerman, R. A.; Sergott, R. C.; Bosley, T. M. Cryptogenic oculomotor nerve palsy. The need for repeated neuro-imaging studies. Arch. Ophthalmol. 107:387–390; 1989.

Armarkrud, N.; Slettnes, O. N. Uncomplicated retrobulbar optic neuritis and the development of multiple sclerosis. Acta. Ophthalmol. 67(3):306–309; 1989.

Baker, R. S.; Carter, D.; Hendrick, E. D.; Bonsick, J. R. Visual loss in pseudotumor cerebri in childhood: a follow-up study. Arch. Ophthalmol. 103:1681–1686; 1985.

Baloh, R. W.; Yee, R. D.; Honrubia, V. Internuclear ophthalmoplegia I. Saccades and dissociated nystagmus. Arch. Neurol. 35:484–489; 1978.

Barrett, L. G. Internuclear ophthalmoplegia in patients with toxic coma: frequency, prognostic value and diagnostic significance. J. Toxicol. Clin. Toxicol. 20:373–379; 1983.

Bates, J. H.; Tomsac, R. L. Serum fibrinogen concentration in the diagnosis and management of biopsy proven giant cell arteritis. Program Abstracts, American Academy of Ophthalmology, New Orleans; 1989.

Beck, R. W.; Servais, G. E.; Hayreh, S. S. Anterior ischemic optic neuropathy IX. Cup to disc ratios and its role in pathogenesis. Opthalmology 94:1503–1508; 1987.

Beri, M.; Klugman, M. R.; Kohler, J. A. Anterior ischemic optic neuropathy: incidence of bilaterality and various influencing factors. Opthalmology 94:1020–1028; 1987.

Biglan, A. W.; Burnstine, R. A.; Rogers, G. L.; Saunders, R. A. Management of strabismus with botulinum A toxin. Ophthalmology 96:935–943; 1989.

Boghen, D. R.; Glaser, J. S. Ischemic optic neuropathy: the clinical profile and natural history. Brain 98:689–692; 1975.

Burde, R. M. Internuclear ophthalmoplegia. (letter) Ann. Neurol. 22:668–669; 1987.

Chambers, W. A.; Bernardino, V. B. Specimen length in temporal artery biopsies. J. Clin. Neurol. Ophthalmol. 8:121–125; 1988.

Cogan, D. Internuclear ophthalmoplegia: typical and atypical. Arch. Ophthalmol. 84:583–589; 1970.

Corbett, J. J.; Thompson, H. S. The rational management of idiopathic intracranial hypertension. Arch. Neurol. 46:1049–1051; 1989.

Couch, R.; Camfield, P. R.; Tibbles, J. A. R. The changing picture of pseudotumor cerebri in children. Can. J. Neurol. Sci. 12:48–50; 1985.

Digre, K. B.; Corbett, J. J. Pseudotumor cerebri in men. Arch. Neurol. 45:866–872; 1988.

Durcan, F. J.; Corbett, J. J.; Wall, M. The incidence of Pseudotumor cerebri. Population studies in Iowa and Louisiana. Arch. Neurol. 45:875–877; 1988.

Fauchald, P.; Rygvolg, O.; Oysdes, D. B. Temporal arteritis and polymyalgia rheumatica clinical and biopsy findings. Ann. Intern. Med. 77:845–852; 1972.

Fishman, R. A. The pathophysiology of pseudotumor cerebri: an unsolved puzzle. Arch. Neurol. 41:257–258; 1984.

Fitzgerald, C. R.; Enoch, J. M.; Tamma, L. A. Radiation therapy in and above the retinal, optic nerve, and anterior visual pathway: psychophysical assessment. Arch. Ophthalmol. 99:611–623; 1981.

Fleishman, J. A.; Beck, R. W.; Linares, O. A.; Klein, A. W. Deficits in visual function after resolution of optic neuritis. Ophthalmol. 94:1029–1035; 1987.

Francis, B. A.; Compstond, A. S.; Bachelor, J. R.; McDonald, W. I. Reassessment of the risk of multiple sclerosis developing in patients with optic neuritis after extended follow-up. J. Neurol. Neurosurg. Psychiat. 57:758–765; 1987.

Gordy, R. M. Treatment of giant cell arteritis. Proceedings of the American Association of Ophthalmology, North Central Region Update Course, Section 3, Course Outline 29. A.A.O. 135–138; 1983.

Grossniklaus, H. E.; Frank, K. E.; Tomsac, R. L. Cytomegalovirus retinitis, and optic neuritis in acquired immunodeficiency syndrome: report of a case. Ophthalmol. 94:1601–1604; 1987.

Guy, J.; Day, A. Intracranial aneurysms with superior division paresis of the oculomotor nerve. Ophthalmology 96:1075–1076; 1989.

Guy, K.; Schatz, N. J. Hyperbaric oxygen in the treatment of radiation induced optic neuropathy. Ophthalmol. 93:1083–1088; 1986.

Hayreh, S. S.; Anterior ischemic optic neuropathy II. Fundus on ophthalmology and fluorescein angiography. Br. J. Ophthalmol. 8:964–980; 1974.

Hayreh, S. S.; Anterior ischemic optic neuropathy. Arch. Neurol. 38:675–678, 1981

Hayreh, S. S. Anterior ischemic optic neuropathy VIII. Clinical features and pathogenesis of posthemorrhagic amaurosis. Ophthalmology 94:1488–1502; 1987.

Hunder, G. G.; Sheps, S. G.; Allen, G. L. Daily and alternate day corticosteroid regimens in treatment of giant cell arteritis: comparison in a prospective study. Ann. Intern. Med. 82:613–618; 1975.

Hupp, S. L.; Glaser, J. S.; Byrne, S. F. Optic nerve sheath decompression. Review of 17 cases. Arch. Opthalmol. 105:386–389; 1987.

Jacobs, L.; Kinkle, B. R.; Kinkle, W. R. Silent brain lesions in patients with isolated idiopathic optic neuritis. Arch. Neurol. 43:452–455; 1986.

Johns, K.; Levine, P.; Elliott, J. H.; Partain, G. L. Magnetic resonance imaging of the brain in isolated optic neuritis. Arch. Ophthalmol. 104:1486–1488; 1986.

Kashiwagi, S.; Abiko, S.; Aoki, H. Brainstem abscess. Surg. Neurol. 28:63–66; 1987.

Keltner, J. L. Giant cell arteritis: signs and symptoms. Ophthalmology 88:1101–1110; 1982.

Kline, N. L. Progression of visual defects in ischemic optic neuropathy. Am. J. Ophthalmol. 106:199–203; 1988.

Kriss, A.; Francis, D. A.; Cuendet, F.; Halliday, A. M.; Taylor, D. S.; Wilson, J.; Keast-Butler, J.; Batchelor, J. R.: McDonald, W. T. Main recovery after optic neuritis in childhood. J. Neurol. Neurosurg. Psychiatr. 51(10):1253–1258; 1988.

Krogsaa, B.; Soelberg-Srensen, P.; Seedorff, H. H. Ophthalmologic prognosis in benign intracranial hypertension. Acta. Ophthalmol. Suppl. 173:62–64; 1985.

Kupersmith, M. J.: Burde, R. M.; Warren, F. A. Autoimmune optic neuropathy: Evaluation and treatment. J. Neurol. Neurosurg. Psychiatry 51:1381–1386; 1988.

Kurtzke, J. S. Optic neuritis or multiple sclerosis. Arch. Neurol. 42:704–710; 1985.

Lepore, F. E.; Glaser, J. S. Misdirection revisited: a critical appraisal of acquired oculomotor nerve synkinesis. Arch. Ophthalmol. 98:2206–2209; 1980.

Lessell, S.; Rosman, M. P. Permanent visual impairment in childhood pseudotumor cerebri. Arch. Neurol. 43:801–804; 1986.

Levine, M. L. Chronic asymptomatic ischemic optic neuropathy: a report of two cases in adults with diabetes mellitus. J. Clin. Neuroophthalmol. 7:198–201; 1987.

Love, D. C.: Berler, D. K.; O'Dowd, D. J.; Brooks, J. Z.; Love, R. A. Erythrocyte sedimentation rate and its relationship to giant cell arteritis. Arch. Ophthalmol. 106:309–310; 1988.

Maitland, C. G.; Miller, N. R. Neuroretinitis. Arch. Opthalmol. 102:1146–1150; 1984.

McCrary, M. E.; Nabors, M. W.; Fischer, B. A.; Kobrine, A. I. Delayed abducens nerve palsies associated with cervical spine fractures. Neurol 37:1565–1567; 1987.

McDonnell, P. J.; Moore, G. W.; Miller, N. R.; Hutchins, G. M.; Green, W. R. Temporal arteritis: clinical pathology. Ophthalmology 33:518–530; 1986.

Miller, D. H.; Newton, M. R.; Van der Poel, J. C.; duBouky, E. P.; Halliday, A. M.; Kendall, B. E.; Johnson, G.; MacManus, D. G.; Mosely, I. F.; McDonald, W. I. Magnetic resonance imaging of the optic nerve and optic neuritis. Neurology 38:175–179; 1988.

Miller, N. R. Solitary oculomotor nerve palsy in childhood. Am. J. Ophthalmol. 83:106–111; 1977.

Nadeau, S. E.; Trobe, J. D. Pupil sparing in oculomotor palsy: a brief review. Ann. Neurol. 13:143–148; 1983.

Openheimer, S.; Hoffbrand, B. I. Optic neuritis and myelopathy in systemic lupus erythematosus. Can. J. Neurol. Sci. 13:129–132; 1986.

Optic Neuritis Study Group. The clinical profile of optic neuritis: experience of the optic neuritis treatment trial. Arch. Ophthalmol. 109:1673–1678; 1991.

Orcutt, J. C.; Page, N. G. R.; Sanders, M. D. Factors affecting visual loss in benign intracranial hypertension. Ophthalmology 91(11):1302–1312; 1984.

Pal, H. D.; Williams, A. C. Subacute polyradiculopathy with optic and auditory nerve involvement. Arch. Neurol. 44:885–887; 1987.

Peatfield, R. C. Recurrent sixth nerve palsy in cluster headache. Headache 25:325–327; 1985.

Renson, P. S.; Krogsaa, B.; Gjerris, F. Clinical course and prognosis of pseudomotor cerebri: a prospective study of 24 patients. Acta. Neurol. Scan. 77:164–172; 1988.

Repka, M. X.; Savino, P. J.; Schatz, N. J.; Sergott, R. C. Clinical profile and long term complications of anterior ischemic optic neuropathy. Am. J. Ophthalmol. 96:478–483; 1983.

Rizzo, J. F. III; Lessell, S. Posterior ischemic optic neuropathy during general surgery. Am. J. Ophthalmol 103:808–811; 1987.

Rizzo, J. F. III; Lessell, S. Optic neuritis and ischemic optic neuropathy: overlapping clinical profiles. Arch. Ophthalmol. 109:1668–1672; 1991.

Rizzo, J. S.; Lessell, S. Risk of developing multiple sclerosis after uncomplicated optic neuritis: a long term prospective study. Neurology 38:185–190; 1988.

Rosenfeld, S. I.; Kosmorsky, G. S.; Klingele, T. G. Treatment of temporal arteritis with ocular involvement. Am. J. Med. 80:143–145; 1986.

Rush, J. H. Pseudotumor cerebri: a clinical profile and visual outcome in 63 patients. Mayo Clin. Cross. 55:541–546; 1980.

Schatz, N. J.; Savino, P. J.; Corbett, J. J. Primary aberrant oculomotor regeneration. A sign of intracavernous meningioma. Arch. Neurol. 34:29–32; 1977.

Sergott, R. C.; Cohen, M. S.; Bosley, T. M.; Savino, P. J. Optic nerve decompression may improve the progressive form of non-arteritic ischemic optic neuropathy. Arch. Ophthalmol. 107:1743–1754; 1989.

Sibony, P. A.; Evinger, C.; Lessell, S. Retrograde horseradish peroxidase transport after oculomotor nerve injury. Invest. Ophthalmol. Vis. Sci. 27:975–980; 1986.

Smith, J. L.; Cogan, D. Internuclear Ophthalmoplegia: a review of fifty-eight cases. Arch. Ophthalmol. 39:687–694; 1957.

Sombrono, W. G. M.; Smith, J. L. Acute syphilitic blindness in AIDS. J. Clinical Ophthalmol. 7:1–5; 1987.

Sydnor, C. F.; Seaber, J. H.; Buckley, E. G. Traumatic superior oblique palsies. Ophthalmology 89:134–138; 1982.

Tang, R. A. Giant cell arteritis diagnosis and management. Semin. Ophthalmol. 3(4):244–248; 1988.

Tomsak, R. L.; Masaryk, T. J.; Bates, J. H. Magnetic resonance angiography (MRA) of isolated aneurysmal third nerve palsy. J. Clin. Neuro. Ophthalmol. 11:16–18; 1991.

Wall, M. Idiopathic intracranial hypertension. Neurol. Clinics Vol 9: No. 1:73–95; 1991.

Wall, M.; George, C. Visual loss in pseudotumor cerebri: the incidence and defects related to visual field discrepancy. Arch. Neurol. 44:170–175; 1987.

Wall, M.; Hart, W. M.; Burde, R. M. Visual field defects in idiopathic intracranial hypertension. Am. J. Ophthalmol. 96:664–669; 1983.

Wessel, K.; Thron, A.; Linden, D.; Petersen, D.; Dichgan, J. J. Pseudotumor cerebri: clinical and neuroradiological findings. Eur. Arch. Psychiatr. Neurol. Sci. 237:54–60; 1987.

Williams, A. S.; Hoyt, C. S. Acute comitant esotropia in children with brain tumors. Arch. Ophthalmol. 107:376–378; 1989.

Young, B. R.: Sutula, F. Analysis of trochlear nerve palsies. Mayo Clin. Proc. 52:11—18; 1977.

21

Neuro-Otology

ROBERT W. BALOH

Vestibular Neuritis

Vestibular neuritis (vestibular neuronitis) presents with a gradual onset of vertigo, nausea, and vomiting for several hours (Dix and Hallpike 1952; Coats 1969; Schuknecht and Kitamura 1981; Silvoniemi 1988). The symptoms usually reach a peak within 24 hours and then gradually resolve after several days. During the first day, the patient is off-balance and has difficulty focusing because of spontaneous nystagmus. These acute symptoms typically resolve within 2 to 3 days, but often it takes several weeks before the patient feels well enough to return to normal activities. About two thirds of patients have a benign course with complete recovery within 1 to 3 months. There are important exceptions to this rule, however. Occasionally, patients, particularly the elderly, will have intractable dizziness that persists for years. At least one recurrent bout of vertigo (usually less severe than the initial bout) occurs in 20% to 30% of patients. This may represent reactivation of a latent virus because it is often associated with a systemic viral illness. A small percentage of patients will have multiple recurrent episodes of vertigo, leading to a profound bilateral vestibulopathy (called bilateral sequential vestibular neuritis) (Schuknecht and Witt 1985). The episodic vertigo eventually is replaced by a persistent dysequilibrium and oscillopsia (Baloh et al. 1989).

Management

The optimum management of patients with vestibular neuritis is controversial because the pathophysiology is uncertain. Unless there is convincing evidence to suspect a vascular or nonviral infectious cause, the patient should be managed as a presumed viral neuritis (i.e., symptomatic treatment of the acute vertigo with antivertiginous medication). Although steroids have been recommended for their anti-inflammatory effect (Adour et al. 1981), there have been no controlled studies to assess the risk to benefit ratio for these drugs. Although approximately one third of patients with vestibular neuritis have a permanent loss of vestibular function (as documented by serial caloric examinations), the central nervous system (CNS) is able to adapt to the vestibular loss, and residual symptoms usually are minimal once the compensation has occurred. Vestibular exercises should be started immediately after the acute vomiting and nausea subside, and they should be continued until the dizziness and imbalance are minimal (Baloh 1983).

Bacterial Labyrinthitis

Two types of labyrinthitis are associated with bacterial infections of the temporal bone: (1) serous or toxic labyrinthitis in which bacterial toxins or chemical products invade the inner ear and (2) suppurative labyrinthitis in which bacteria invade the inner ear. The former often leads to only subtle symptoms, such as an insidious high-frequency sensory neural hearing loss, whereas the latter typically leads to a profound combined auditory and vestibular loss with little or no recovery.

Serous labyrinthitis is probably the most common complication of acute or chronic middle ear infection. The toxins or inflammatory cells penetrate the basilar membrane and invade the endolymph at the basal turn of the cochlea. Damage to this region of the cochlea explains the high incidence of high-frequency sensorineural hearing loss in patients with chronic otitis media (Paparella et al. 1980). Acute suppurative labyrinthitis is manifested by the sudden onset of severe vertigo, nausea, vomiting, and unilateral deafness. The infection originates in the middle ear or the cerebrospinal fluid (CSF). When the labyrinthitis is a direct complication of middle ear disease, it is more likely to occur from chronic otitis media and mastoiditis than from an acute middle ear infection. The most common port of entry of bacteria into the inner ear, however, is from the spinal fluid in patients with meningitis (Neely 1986). Meningogenic bacterial labyrinthitis is usually bilateral, whereas direct invasion from a chronic otitic infection is almost always unilateral. Endolymphatic hydrops can be a sequela of both serous and suppurative labyrinthitis (Schuknecht 1974).

Management

Management of bacterial labyrinthitis is directed at the associated infection of the middle ear, mastoid, and, if present, meninges. A patient with acute or chronic bacterial ear disease associated with sudden or rapidly progressive inner ear symptoms should be hospitalized and treated with local cleansing and topical antibiotic solutions to the affected ear and parenteral antibiotics capable of penetrating the blood–brain barrier (Neely 1986). Surgical intervention to eradicate the middle ear and mastoid infection usually is required after a few days of antibiotic treatment. If the labyrinthitis is secondary to a primary meningitis, it is best to treat the underlying meningitis. A resistant or recurrent meningitis may result from an unrecognized posterior fossa epidural abscess with dural perforation or from a congenital direct communication with the CSF.

Viral Labyrinthitis

Of the thousands of infants born deaf every year, approximately 20% are thought to be the result of congenital viral infections of the inner ear (Pappas 1983). More than 4000 people are stricken each year with "sudden deafness," a unilateral, infrequently bilateral, sensorineural hearing loss of acute onset, presumed to be of viral origin in most cases (Wilson et al. 1983). An acute onset of intense vertigo (some due to viral vestibular neuritis, others to labyrinthitis) strikes a similar number of people. The most convincing evidence for a viral cause of these isolated auditory and vestibular syndromes comes from the temporal bone studies of Schuknecht (1985) and colleagues (1986) in Boston. These pathological studies are supported by experimental studies in animals in which it has been shown that several viruses will selectively infect the labyrinth or the eighth nerve (Davis and Johnson 1983; Nomura et al. 1985).

Viral labyrinthitis can present with sudden deafness (usually unilateral and rarely bilateral), acute vertigo (with associated autonomic symptoms), or a combination of auditory and vestibular symptoms. Although the term sudden deafness is commonly used, the hearing loss due to viral infection usually takes several hours and may evolve over several days (Wilson et al. 1983). The hearing loss reverses, at least partially, in most cases. It returns to normal in more than 50% of patients (with or without treatment). Tinnitus and fullness in the involved ear also are common.

Management

As in vestibular neuritis, management of viral labyrinthitis is directed at the symptoms. Vestibular exercises are begun as soon as possible to accelerate the compensation process. Antiviral agents, such as cytosine arabinase and acyclovir, have been used for treating systemic viral illnesses in children, however, it is unclear whether the hearing loss often associated with disorders such as cytomegalovirus and rubella infections is altered by this treatment. There have been no reports of the efficacy of antiviral agents in adults with presumed viral labyrinthitis.

Syphilitic Labyrinthitis

Syphilitic infections remain an important cause of vertigo and hearing loss despite the general availability of penicillin. Syphilitic infections produce auditory and vestibular symptoms by two mechanisms: meningitis with involvement of the eight nerve and osteitis of the temporal bone with associated labyrinthitis. The former typically occurs as an early manifestation of acquired syphilis, whereas the latter occurs as a late manifestation of both congenital and acquired syphilis. With early acquired syphilis, the predominant pathological finding is basilar meningitis affecting the eighth nerve, particularly the auditory branch. The hearing loss typically occurs with the rash and lymphadenopathy of secondary syphilis (Saltiel et al. 1983). Usually it is abrupt in onset, tends to be bilateral, and is rapidly progressive. Vestibular symptoms often are absent. Patients may demonstrate symptoms and signs of meningitis, including headache, stiff neck, cranial nerve palsies, and optic neuritis.

Both congenital and acquired syphilitic infections produce temporal bone osteitis and labyrinthitis as a late manifestation. The congenital variety is approximately three time as common as the acquired variety (Morrison 1975). The time of onset of congenital syphilitic labyrinthitis is anywhere from the first to seventh decades, with the peak incidence in the fourth and fifth decades. Acquired syphilitic labyrinthitis rarely occurs before the fourth decade and has a peak incidence in the fifth and sixth decades. The natural history of syphilitic labyrinthitis is a slow, relentless progression to profound or total bilateral loss of vestibular and auditory function (Morrison 1975; Steckelberg and McDonald 1984). This progression is marked by episodes of sudden deafness, vertigo, and fluctuation in the magnitude of hearing loss and tinnitus.

Management

Penicillin is the treatment of choice for the otological manifestations of syphilis. Because CSF infection accompanies the early manifestations of acquired syphilis, high-dose intravenous penicillin is recommended. If the penicillin is begun early, the prognosis is excellent; complete recovery of hearing and vestibular function usually occurs (Saltiel et al. 1983). For the late manifestations of congenital and acquired syphilitic labyrinthitis, the combination of steroids and penicillin is superior to penicillin alone (Morrison 1975; Steckelberg and McDonald 1984). Numerous penicillin regimens have been used, but the most popular is benzathine penicillin (2.4 million U) given weekly for 6 weeks to 3 months. Along with the penicillin, prednisone, beginning at a dose of 60 mg/pday on an alternate day regimen, is given for 3 months followed by slow tapering. If symptoms recur during the tapering, a maintenance dose of prednisone may be required. Most patients can be expected to stabilize or improve on this therapeutic regimen.

Benign Positional Vertigo

Patients with benign positional vertigo develop brief episodes of vertigo with position change, typically turning over in bed, getting in and out of bed, bending over and straightening up, and extending the neck to look up. So called "topshelf" vertigo, in which a patient experiences an episode of vertigo while reaching for something on a high shelf, is nearly always due to benign positional vertigo. The syndrome is important to recognize because in the majority of patients, the symptoms spontaneously remit and extensive diagnostic procedures are not warranted. The diagnosis is based on the finding of the characteristic torsional upbeat paroxysmal positional nystagmus with rapid positional testing (the Hallpike maneuver) (Baloh et al. 1979).

In a review of 240 cases of benign positional vertigo (Baloh et al. 1987) no episodes of positional vertigo lasted longer than 1 minute. Often after a flurry of episodes, however, patients complained of prolonged nonspecific dizziness (light-headed, swimming sensation) and nausea that lasted for hours to days. Typically, bouts of positional vertigo were intermixed with variable periods of remission. The mean age of onset was 54 years with a range of 11 to 84 years. At the time of examination, one third of the patients reported benign positional vertigo (including remissions) lasting longer than 1 year, and seven reported them lasting longer than 10 years.

In slightly more than one half of the cases, a likely diagnosis was determined. The two largest diagnostic categories were post-traumatic and postviral labyrinthitis or viral neuritis. In patients with the former, onset of positional vertigo occurred immediately after major head trauma. Patients with viral neuritis or labyrinthitis reported a prior episode of acute vertigo gradually resolving over 1 to 2 weeks. Episodes of benign positional vertigo began as soon as 1 week and as long as 8 years after the acute attack. Women outnumbered men by a ratio of 1.6 : 1, combining all diagnostic categories.

This ratio was approximately 2 : 1 if only the idiopathic and miscellaneous groups were considered. Others have reported an even higher female to male preponderance with idiopathic benign positional vertigo (Katsarkas and Kirkham 1978). The age of onset peaked in the sixth decade in the idiopathic group and in the fourth and fifth decades in the postviral group, and it was evenly distributed during the second to sixth decade in the post-traumatic group.

Management

It is important to recognize that the course of benign positional vertigo can be protracted. In the study mentioned previously, nearly every patient had at least one exacerbation after the initial remission. The likelihood of a recurrence should be explained to patients so that they are not unduly frightened if it occurs. Positional exercises can accelerate remissions in most cases of benign positional vertigo (Brandt and Daroff 1980). The patient is instructed to sit on the edge of the bed and then rapidly assume a lateral position to induce positional vertigo. After the vertigo subsides, he or she returns to the upright position, usually experiencing less vertigo. These positional exercises are repeated in each session until the vertigo fatigues, and the sessions are repeated three times a day until the vertigo no longer occurs. Patients who show the most prominent fatigue on the standard diagnostic positional test will derive the most benefit from positional exercise. Antivertiginous medications, such as meclizine or phenergan (25 mg) can be used for symptomatic treatment of vertigo during an acute exacerbation while the patient is performing positional exercises.

For rare patients with prolonged intractible benign positional vertigo unresponsive to conventional therapy, a surgical procedure in which the ampullary nerve is sectioned from the posterior semicircular canal crista, may be considered. This operation, known as a singular neurectomy, has been effective in relieving benign positional vertigo in more than 90% of patients (Gacek

1978). The main complication is a sensori-neural hearing loss, which occurs in approximately 8% of patients.

Meniere's Disease

Meniere's disease (endolymphatic hydrops) is characterized by fluctuating hearing loss and tinnitus, episodic vertigo, and a sensation of fullness or pressure in the ear. The underlying problem is progressive endolymphatic hydrops caused by impaired resorption of the endolymph; all symptoms and signs result from this increased endolymph pressure. Typically, the patient develops a sensation of fullness, pressure, and decreased hearing and tinnitus in one ear. Vertigo rapidly follows, reaching a maximum intensity within minutes and then slowly subsiding during the next several hours. The patient usually is left with a sense of unsteadiness and dizziness for days after the acute vertiginous episode. In the early stages, the hearing loss is completely reversable, but in later stages, a residual hearing loss remains. Tinnitus may persist between episodes but usually increases in intensity immediately before or during the acute episode. It is typically described as a roaring sound (the sound of the ocean or a hollow sea shell). The patient prefers to lie in bed without eating until the acute symptoms pass. The episodes occur at irregular intervals for years, with periods of remission unpredictably intermixed (Eggermont and Schmidt 1985). Approximately 50% will have at least one major remission within the first 2 years of onset (Silverstein et al. 1989). Eventually, severe permanent hearing loss develops and the episodic nature spontaneously disappears ("burned-out phase"). In about one third of patients, bilateral involvement eventually will occur (Wladislavosky-Waserman et al. 1984).

Delayed endolymphatic hydrops develops in an ear that has been damaged years before usually by viral or bacterial infection (Nadol et al. 1975; Schuknecht, 1978). With this disorder, the patient reports a history of hearing loss since early childhood, followed many years later by typical symptoms and signs of endolymphatic hydrops. If the hearing loss is profound, as it often is, the episodic vertigo will not be accompanied by fluctuating hearing levels and tinnitus. Delayed endolymphatic hydrops can be unilateral or bilateral, depending on the extent of damage of the original insult.

Management

Medical management of Meniere's disease consists of symptomatic treatment of the acute spells with antivertiginous medications, such as meclizine and phenergan (25–50 mg), and long-term prophylaxis with salt restriction and diuretics. The mechanism by which a low-salt diet decreases the frequency and severity of attacks with Meniere's disease is unclear, but there is extensive empiric evidence for its efficacy (Boles et al. 1975; Jackson et al. 1981). The author recommends salt restriction in the range of 1 g of sodium per day with a minimal therapeutic trial of 2 to 3 months. Diuretics (acetazolamide, 250 mg or hydrochlorthiazide, 50 mg) provide additional benefit in some patients, but generally they cannot replace a salt-restriction diet.

Two types of surgery have been used for treating Meniere's disease: endolymphatic shunts and ablative procedures. Although shunts are logical because of the presumed pathophysiology of Meniere's syndrome, it is difficult to achieve a long-standing functional shunt (Schuknecht 1986). Furthermore, recent clinical trials have questioned seriously the efficacy of these surgical shunt procedures (Silverstein et al. 1989; Thomsen et al. 1981). The rationale for ablative surgery in the treatment of Meniere's disease is that the nervous system is better able to compensate for complete loss of vestibular function than for loss that fluctuates in degree. Ablative procedures are most effective in patients with unilateral involvement with no functional hearing on the damaged side. Severe vertigo is expected during the immediate postoperative period, but most patients who follow a structured vestibular exercise program can return to normal activity within 1 to 3 months. Ablative surgical procedures generally should be avoided in elderly patients because they have great difficulty adjusting to a vestibular imbalance.

Perilymph Fistula

The classic presentation of an acute perilymph fistula is a sudden audible pop in the ear immediately followed by hearing loss, vertigo, and tinnitus. The key to diagnosis is to identify the characteristic precipitating factors: head trauma, barotrauma, cough, sneeze, straining, vigorous exercise, chronic otitis with cholesteatoma, post-stapedectomy surgery, and congenital malformations. Nonspecific imbalance and disequilibrium aggravated by quick head movements or sudden turning may result from a chronic perilymph fistula. Patients may prefer to sleep on one side rather than the other to avoid an ill-defined uncomfortable dizzy sensation. The latter feature may suggest benign positional vertigo, although the positional dizziness with perilymph fistula is not as intense and is more persistent than that associated with benign positional vertigo.

Management

The majority of perilymph fistulae spontaneously heal without innervation. For this reason, most authors advocate conservative management with an initial period of bedrest, sedation, head elevation, and measures to decrease straining (Mattox 1986). One exception to this conservative approach is acute barotrauma, in which immediate exploration has been advocated (Pullen et al. 1979). Persistent fluctuating auditory and vestibular symptoms indicate exploration of the middle ear after an initial trial of conservative management. Even in these cases, however, only about one half to two thirds of ears are found to have fistulae. The goal of surgery for chronic fistulae is to stabilize hearing loss and relieve vestibular symptoms. The middle ear typically is explored through a posterior tympanotomy. Often the fistula is in the area of the oval window. Recurrence of symptoms after repair occurs

in at least 10% of cases; rarely, intractible symptoms will necessitate ablative surgery with labyrinthectomy or nerve section.

Acoustic Neuromas

The most common symptom associated with an acoustic neuroma is slowly progressive unilateral hearing loss (Erickson et al. 1965; Mattox 1987). Often patients will complain of an inability to understand speech when using the telephone before they are aware of a loss in hearing. Unilateral tinnitus is the next most common symptom. Vertigo occurs in less than 20% of patients, although approximately one half will complain of some mild impairment of balance. Next to the auditory nerve, the most commonly involved cranial nerves (by compression) are the seventh and fifth, producing facial weakness and numbness, respectively. Involvement of the sixth, ninth, tenth, 11th, and 12th nerves occurs only in the late stages of disease with massive tumors. Large acoustic neuromas also may produce increased intracranial pressure from obstruction of CSF outflow, resulting in severe headaches and vomiting.

Management

With few exceptions, management of an acoustic neuroma is surgical. Occasionally, a patient with a small acoustic neuroma may be followed, particularly if the patient is elderly or has underlying medical problems (Wazen et al. 1985). These tumors may remain confined to the internal auditory canal for years, and symptoms may be restricted to those of the eighth nerve.

There are three general surgical approaches to the cerebellopontine angle: translabyrinthine, suboccipital, and middle fossa (Mattox 1987). The translabyrinthine approach destroys the labyrinth, but often it allows complete removal of the tumor within endangering nearby neural structures, particularly the facial nerve. This is the procedure of choice for a patient with severe hearing loss and a tumor of less than 3 cm. With the suboccipital and middle fossa approaches, residual hearing can be saved because the labyrinth is not destroyed (Tator and Nedzelski 1985). Traditionally, the suboccipital approach is performed by neurosurgeons, while the middle fossa approach is performed by otologic surgeons.

References

Adour, K. K.; Sprague, M. A.; Hilsinger, R. L. Vestibular vertigo. A form of polyneuritis. J.A.M.A. 246:1564; 1981.

Baloh, R. W.; Honrubia, V.; Jacobson, K. Benign positional vertigo. Clinical and oculographic features in 240 cases. Neurology 37:371; 1987.

Baloh, R. W.; Jacobson, K.; Honrubia, V. Idiopathic bilateral vestibulopathy. Neurology 39:272; 1989.

Baloh, R. W.; Sakala, S. M.; Honrubia, V. Benign paroxysmal position nystagmus. Am. J. Otolaryngol. 1:1; 1979.

Baloh, R. W. The dizzy patient. Symptomatic treatment of vertigo. Postgrad. Med. 73:317; 1983.

Boles, R.; Rice, D. H.; Hybels, R.; Work, W. P. Conservative management of Meniere's disease: Furstenberg regimen revisited. Ann. Otol. Rhinol. Laryngol. 84:513; 1975.

Brandt, T.; Daroff, R. B. Physical therapy for benign paroxysmal positional vertigo. Arch. Otolaryngol. 106:484; 1980.

Coats, A. C. Vestibular Neuronitis. Acta. Otolaryngol. 251(suppl.):1; 1969.

Davis, L. E.; Johnsson, L. G. Viral infections of the inner ear: Clinical, virologic and pathologic studies in humans and animals. Am. J. Otolaryngol. 4:347; 1983.

Dix, M.; Hallpike, C. The pathology, symptomatology and diagnosis of certain common disorders of the vestibular systems. Ann. Otol. Rhinol. Laryngol. 61:987; 1952.

Eggermont, J. J.; Schmidt, P. H. Meniere's disease: a long-term follow-up study of hearing loss. Ann. Otol. Rhinol. Laryngol. 94:1; 1985.

Erickson, L.; Sorenson, G.; McGavran, M. A review of 140 acoustic neurinomas (neurilemmomas). Laryngoscope 75:601; 1965.

Gacek, R. R. Further observations on posterior ampullary nerve transection for positional vertigo. Ann. Otol. Rhinol. Laryngol. 87:300; 1978.

Jackson, C. G.; Glasscock, M. E.; Davis, W. E.; Hughes, G. B.; Sismanis, A. Medical management of Meniere's disease. Ann. Otol. 90:142; 1981.

Katsarkas, A.; Kirkham, T. H. Paroxysmal positional vertigo: a study of 255 cases. J. Otolaryngol. 7:320; 1978.

Mattox, D. E. Perilymph fistulas. In: Cummings,

C. W.; Fredrickson, J. M.; Harker, L. A.; Krause, C. J.; Schuller, D. E., eds. Otolaryngology—head and neck surgery. St. Louis: C. V. Mosby; 1986: p. 3113.

Mattox, D. E. Vestibular schwannomas. Otolaryngol. Clin. North Amer. 20:149; 1987.

Morrison, A. W. Late syphilis. In: Morrison, A. ed. Management of sensorineural deafness. Boston: Butterworths; 1975: p. 109–144.

Nadol, J. B.; Weiss, A. D.; Parker, S. W. Vertigo of delayed onset after sudden deafness. Ann. Otol. 84:841; 1975.

Neely, J. G. Complications of temporal bone infection. In: Cummings, C. W.; Fredrickson, J. M.; Harker, L. A.; Krause, C. J.; Schuller, D. E., (eds.) Otolaryngology—head and neck surgery. St. Louis: C. V. Mosby Co.; 1986:2988.

Nomura, Y.; Kurata, T.; Saito, K. Sudden deafness: human temporal bone studies and an animal model. In: Nomura, Y. ed. Hearing loss and dizziness. Tokyo: Sgaku-Shoin; 1985: p. 58.

Paparella, M. M.; Goycoolea, M. V.; Meyerhoff, W. L. Inner ear pathology and otitis media: a review. Ann. Otol. Rhinol. Laryngol. 89:249; 1980.

Pappas, D. G. Hearing impairments and vestibular abnormalities among children with subclinical cytomegalovirus. Ann. Otol. Rhinol. Laryngol. 92:552; 1983.

Pullen, F. W.; Rosenberg, G. H.; Cabeza, C. H. Sudden hearing loss in divers and fliers. Laryngoscope 84:1373; 1979.

Saltiel, P.; Melmed, C. A.; Portnoy, D. Sensorineural deafness in early acquired syphilis. Can. J. Neurol. Sci. 10:114; 1983.

Schuknecht, H. F.; Kitamura, K. Vestibular neuritis. Ann. Otol. Rhinol. Laryngol. 90(suppl.):1; 1981.

Schuknecht, H. F.; Witt, R. L. Acute bilateral sequential vestibular neuritis. Am. J. Otolaryngol. 6:255; 1985.

Schuknecht, H. F.; Kimura, R. R.; Nanfal, P. M. The pathology of idiopathic sensorineural hearing loss. Arch. Otorhinolaryngol. 243:1; 1986.

Schuknecht, H. F. Delayed endolymphatic hydrops. Ann. Otol. 87:743; 1978.

Schuknecht, H. F. Endolymphatic hydrops: can it be controlled? Ann. Otol. Rhinol. Laryngol. 95:36; 1986.

Schuknecht, H. F. Neurolabyrinthitis. Viral infections of the peripheral auditory and vestibular systems. In: Nomura, Y., ed. Hearing loss and dizziness. Tokyo: Igaku Shoin; 1985: p. 1.

Schuknecht, H. F. Pathology of the ear. Cambridge: Harvard University Press; 1974: p. 241.

Silverstein, H.; Smouha, E.; Jones, R. Natural history vs surgery for Meniere's disease. Otolaryngol. Head Neck Surg. 100:6; 1989.

Silvoniemi, P. Vestibular neuronitis. Acta. Otolaryngol. 453(suppl.):9; 1988.

Steckelberg, J. M.; McDonald, T. J. Otologic involvement in late syphilis. Laryngoscope
94:753; 1984.

Tator, C. H.; Nedzelski, J. M. Preservation of
hearing in patients undergoing excision of
acoustic neuromas and other cerebellopontine
angle tumors. J. Neurosurg. 63:168; 1985.

Thomsen, J.; Brettan, P.; Tos, M.; Johnsen,
N. J. Placebo effect of surgery for Meniere's
disease. Arch. Otolaryngol. 107:271; 1981.

Wazen, J.; Silverstein; Norrell, H.; Besse, B.

Preoperative and postoperative growth rates in
acoustic neuromas documented with CT scanning. Otolaryngol. Head Neck Surg. 93:151;
1985.

Wilson, W. R.; Veltri, R. W.; Larid, N. et al.
Viral and epidemiologic studies of idiopathic
sudden hearing loss. Otolaryngol. Head Neck
Surg. 91:653; 1983.

Wladislavosky-Waserman, P.; Facer, G. W.;
Mokri, B.; Kurland, L. T. Meniere's disease: a
30-year epidemiologic and clinical study in
Rochester, MN, 1951–1980. Laryngoscope
94:1098; 1984.

22

Dementia

DOUGLAS GALASKO AND ROBERT KATZMAN

The average life expectancy has increased dramatically in developed countries, and therefore diseases associated with aging, such as dementia, have become major public health issues. Physicians caring for elderly individuals frequently must advise patients and their families about the anticipated course and prognosis of dementia.

Dementia is not a unitary syndrome; it represents the clinical expression of many disease processes. Alzheimer's disease (AD), vascular dementia, and a combination of these ("mixed dementia") are the most common, accounting for more than 80% of the dementia cases in several recent large studies of patients (Wade et al. 1987; Thal et al. 1988). The remaining 10% to 20% comprises a large number of conditions, some amenable to treatment (Table 22-1). Case reports or small series often are the only source of information on which to base prognoses in patients with less common dementing illnesses. Therefore, an important subset of patients with these conditions are considered under the rubric of reversible dementia, and the focus of this chapter is mainly on prognosis in AD, vascular dementia, and mixed dementia.

The term prognosis used in dementing illnesses has several dimensions. When the diagnosis is initially made, a major issue is the likelihood of the dementia being treatable or reversible. In progressive dementias, such as AD, the patient and family may be concerned about how long the patient is likely to remain functional and able to live in the community, and they may want to know the likely life expectancy of dementia. The interaction between biological and social factors frequently influences the functional ability of an individual with dementia and complicates the question of prognosis. Furthermore, the prognosis in a single dementing disorder may show marked variability; for instance, life expectancy following the onset of AD may vary from 1 to 20 years (Merritt 1975). Consequently, the general prognosis for dementia that can be gleaned from the literature may need to be modified in clinical practice.

Reversibility of Dementia in General

The search for treatable causes of dementia stems from reports in which 10% to 40% of patients with dementia had potentially reversible underlying illnesses (Marsden and

Table 22-1. Causes of Dementia and Reversible Dementia

Etiology of Dementia*	Percent	Reversible Dementia†	Percent
Alzheimer's disease	57	Depression	26.2
Vascular dementia	14.1	Normal pressure hydrocephalus	10.7
Depression	4.5	Thyroid	6.8
Alcohol	4.2	Subdural hematoma	5.8
Drugs	1.5	Neoplasm	4.0
Normal pressure hydrocephalus	1.6	Calcium	1.9
Metabolic	1.5	Liver failure	1.9
CNS neoplasm	1.5	B_{12} deficiency	1.0
Parkinson's disease	1.2	Other	3.9
Huntington's disease	0.9		
Infections	0.6		
Subdural hematoma	0.4		
Other	7.1		
Nondemented	3.7		

* Twenty-five studies.

† Eleven out of the 25 studies, $n = 103$ of 1051.

From Clarfield, A. M. The reversible dementias: do they reverse? Ann. Intern. Med. 109:476–486; 1988.

Harrison 1972; Smith and Kiloh 1981). In an extensive review of this issue, Clarfield (1988) considers whether treatable dementias actually improved, based on studies that supplied follow-up data on patients after treatment was instigated. Of 25 studies concerning treatable dementias, only 11 provided adequate follow-up data to allow the effectiveness of treatment to be determined. These were pooled to yield a total of 1051 patients with dementia, of whom 11% were thought to have treatable underlying illnesses. With treatment, 8% recovered partially and 3% completely. The most frequent reversible causes of dementia were medication, depression, thyroid disease, normal pressure hydrocephalus, brain tumor, and subdural hematoma. A portion of treatable dementias, however, represented the co-occurrence of subclinical AD and metabolic or intracranial disease; for example, some patients developed deterioration consistent with AD after initial improvement following treatment of an apparent cause of dementia (Larson et al. 1984). A thorough history and examination and a judicious laboratory workup of a patient with dementia, including a computed tomography (CT) scan or magnetic resonance imaging (MRI) are important despite the relatively low yield (Larson et al. 1986). This is because the discovery of a potentially reversible etiology may change the outcome completely.

Prognosis in Alzheimer's Disease

Numerous studies have reported that AD shortens the life span (Katzman 1976). Several factors need to be considered in interpreting these findings. In general, AD patients in studies from the 1970s survived longer than patients in series from the 1950s, which reflects more active treatment of complications of AD, such as infections and malnutrition. In addition, some studies have included actuarial life expectancy data for elderly individuals as historical controls, whereas others have used cognitively normal controls matched for age. Actuarial data should be based on recent surveys of people who have similar demographic characteristics to the patients with AD to allow a fair comparison. Because of the large sample size, such population-based life expectancy data avoid many potential biases that may occur in the recruitment of normal elderly controls. Controls usually are volunteers who lead healthy lifestyles, are highly motivated, and may have above-average life expectancy. Another difficulty arises in estimating accurately the interval from onset of dementia to death because the date of onset of initial symptoms of dementia is difficult to determine.

A group of studies from 1950 to 1969 reviewed by Wang (1978) showed that patients with presenile dementia had an average life expectancy of 6.9 years after the onset of dementia, compared to the anticipated life expectancy of 22.3 years based on actuarial tables. For senile dementia, the average life expectancy was 6.0 years, compared to anticipated survival for elderly individuals of 11.9 years. During the past 10 years, the issue of prognosis in AD has been reexamined because diagnostic criteria for AD have been refined. Table 22-2 includes five recent prospective studies of prognosis in AD, all of which documented a reduction of life expectancy for AD compared to actuarial data- or age-matched controls. Within 5 years of presentation, 20% to 30% of AD patients in one of these studies had died, and the median survival from the estimated date of onset of AD until death was 8.1 years (Barclay et al. 1985).

The shortened survival of AD patients results from complications due to severe mental decline. Pneumonia and other infections, dehydration, and malnutrition occur frequently in advanced stages, when patients are bed-bound, incontinent, and unable to communicate or feed themselves.

Compared to other elderly individuals, AD patients are not predisposed to cancer or cerebrovascular or cardiovascular disease (Mölsa et al. 1986). It is therefore unlikely that survival in AD can be increased unless a specific form of treatment is found that can slow or reverse the degeneration of neurons.

Many patients with AD reside in nursing homes or institutions late in the course of their illness. The time between the diagnosis of AD and the need for institutional placement is not easy to predict for several reasons. First, AD patients are clinically heterogeneous and show marked variability in their rate of cognitive and functional decline. A recent study evaluated the annual rate of decline of scores on the Blessed Information–Memory–Concentration test for patients with AD in a private referral practice, in a community study of 80-year-olds in New York, in a research panel in San Diego, and in a nursing home (Katzman et al. 1988). Regardless of initial cognitive status, patients lost an average of 4.4 points per year on the test, until a ceiling effect was reached in advanced stages of dementia (Table 22-3). The rate of decline was not affected by residence, geography, age, or educational status (Table 22-4). However, there was a wide range of cognitive deterioration; one patient

Table 22-2. Studies of Survival in Dementia

Study	Methods	Outcome
Wang 1978	Review of six studies: 1950–1969	Survival from onset: AD at age 65 or older: 6.0 years vs. 11.1 expected AD younger than 65: 6.9 years vs. 22.3 expected Vascular: 3.8 years vs. 13.4 expected
Barclay et al. 1985	Prospective: 199 AD, 69 MID, 43 MIX	Median survival from onset: AD: 8.1 years MID: 6.7 years MIX: 6.2 years
Mölsa et al. 1986	Prospective: 218 AD, 115 MID	Survival at 6 years: AD: 21% vs. 49% expected MID: 12% vs. 45% expected
Heyman et al. 1987	Prospective: 92 AD, onset before age 70	5-year outcome: 24% died, 63% institutionalized
Martin et al. 1987	Prospective: 134 AD, 41 MID	Survival at 3 years: 74% for dementia vs. 80% for controls
Berg et al. 1988	Prospective: 48 AD, 58 controls	5-year outcome: AD: 30% died, 73% institutionalized

AD = Alzheimer's disease; MID = multi-infarct dementia; MIX = mixed dementia.

Table 22-3. Rate of Cognitive Change in AD as a Function of Initial Mental Status Score*

IMC Score	Number of Subjects	Decline Mean ± S.E.M.
0–7	51	4.29 ± 0.45
8–15	46	4.46 ± 0.53
16–23	45	4.16 ± 0.60
24–33	19	0.40 ± 0.29
Overall	161	3.84

AD = Alzheimer's disease; IMC = information–memory–concentration test.

* The number of errors, ranging from 0 to 33, reflects the severity of dementia. The mean decline on the IMC test was calculated over 1 year.

From Katzman, R.; Brown, T.; Thal, L. J.; Fuld, P.; et al. Comparison of rate of annual change of mental status score in four independent studies of patients with Alzheimer's disease. Ann. Neurol. 24:384–389; 1988.

declined by 15 points in 1 year, and several patients' scores did not change.

A second variable that may greatly influence the need for institutional placement of a demented patient independent of cognition is social support. Many elderly individuals live alone in society and may rapidly reach a point at which cognitive impairment interferes with essential activities, such as shopping for food. In contrast, patients with care-givers—spouses, relatives, or friends—often are able to remain in the community until the late stages of dementia. A high degree of devotion by the care-giver makes home care feasible under such adverse circumstances. The care-giver's physical health and ability to cope with the demands and stresses involved in round-the-clock care of a demented patient may weigh heavily in the decision to place the patient in a nursing home.

Also apart from cognitive impairment, behavioral symptoms of demented patients, in particular agitation or aggression, may precipitate nursing home placement. A study of 14 married men with AD found that the determinants of institutionalization were urinary or fecal incontinence, inability to speak coherently, and loss of skills needed for bathing and grooming (Hutton et al. 1985). In a longitudinal study of 92 patients with the onset of AD before age 70, 54 were admitted to nursing homes during 5 years of follow-up (Heyman et al. 1987). Although lower scores on cognitive screening tests or aphasia on entry increased the risk of institutionalization, nursing home placement occurred most frequently because the patients had become almost completely helpless and required 24-hour supervision; for 10 patients, the decision was reached because of death or serious illness of their care-givers.

As discussed previously, the rate of progression of AD is highly variable. Several studies have attempted to identify subgroups of patients with AD who are predisposed to a more rapid course of deterioration. In a longitudinal study of 43 community-dwelling AD patients, the presence of aphasia and relatively poor performance on a psychometric test distinguished patients

Table 22-4. Annual Rate of Change of IMC Test Over All Years—Initial Score Less than 24

			Mean			
Location	Number of Patients	Age (Year)	Education (Year)	Score	Standard Deviation	Range
NH	28	85.1	9.7	4.0	±3.8	−0.5 to −14
PP	38	67.8	12.4	4.7	±3.0	+0.5 to −12
BAS	32	81.0	9.2	4.8	±3.5	+1 to −15
ADRC	44	71.3	13.4	3.8	±4.0	−5 to −15
All locations	142	74.9	11.5	4.4*	±3.6	−5 to −15

IMC = information–memory–concentration test; NH = nursing home; PP = private practice; BAS = Bronx Aging Study; ADRC = Alzheimer's Disease Research Center.

* Differences between sites not significant, $p = 0.43$

Modified from Katzman, R.; Brown, T.; Thal, L. J.; Fuld, P.; et al. Comparison of rate of annual change of mental status score in four independent studies of patients with Alzheimer's disease. Ann. Neurol. 24:384–389; 1988.

whose cognition deteriorated 1 year later (Berg et al. 1984). Laboratory measures, such as CT, electroencephalogram (EEG), and visual evoked potentials, were not predictive. The same group of patients were studied after a longer period of follow-up to identify factors influencing survival in AD (Berg et al. 1988). Age at onset of dementia did not affect survival, which contradicts the clinical belief that patients with early-onset AD (formerly referred to as presenile dementia) progress more rapidly (Seltzer and Sherwin 1984). This study and the study on rate of cognitive change in AD cited previously (Katzman et al. 1988) provide compelling evidence against the division of AD into presenile and senile forms based on the arbitrary age of 65.

Men with AD had a higher mortality rate than women in the study by Berg and co-workers, which was attributed to inherent gender differences in life span. When mortality rather than cognitive change was taken as the endpoint, none of the initial clinical or laboratory findings correlated with the length of survival. A high level of education did not slow the rate of progression of dementia, which agrees with an earlier study (Filley et al. 1985).

There is evidence that certain clinical features may mark a more rapid course of AD. Patients with extrapyramidal signs or psychosis reached a predetermined level of cognitive impairment faster than patients without these features (Mayeux et al. 1985; Stern et al. 1987). In a cross-sectional study, AD patients with extrapyramidal features, psychosis, or myoclonus had lower scores on psychometric tests than when these findings were absent (Chui et al. 1985). A longitudinal study found that patients with AD who developed extrapyramidal signs had significantly shorter survival (Morris et al. 1987). However, myoclonus is uncommon in AD, and extrapyramidal findings are infrequent early in the course of AD, so these features rarely are useful as prognostic indicators early in the course of AD.

Although there are many observations about the natural history of AD, the variability of the clinical course of AD has not been explained adequately, nor has the validity of subgroups been established clinically or verified in neuropathological terms. No form of treatment influences the progression of AD or improves cognition, although the clinical course may be affected by treating intercurrent illness or symptoms, such as depression.

Prognosis in Vascular Dementias

Multiple bilateral cerebral infarcts may produce a clinical picture of dementia with cognitive deficits resembling those found in AD. Diagnostic accuracy for vascular dementia has improved with the introduction of clinical scales, such as the Hachinski ischemic index (Hachinski et al. 1975) or its modification (Rosen et al. 1980), and imaging techniques, such as CT and MRI, approaching 80% to 85% accuracy in recent clinicopathological series (Erkinjuntti et al. 1988). When differentiating vascular dementia from AD, helpful features are the abrupt onset of deficits, stepwise deterioration, focal neurological symptoms or signs, and infarcts or areas of white-matter attenuation on cranial CT. Autopsy series of patients with vascular dementia have found coexisting Alzheimer pathology in many cases, referred to as mixed dementia. Mixed dementia is difficult to clinically diagnose correctly, and the relative contribution of AD and vascular pathology to dementia in these cases is not always clear. Consequently, studies of vascular dementia include patients with the mixed form, which may have a different clinical course and may progress in spite of control for vascular risk factors.

There are several major patterns of vascular dementia, in addition to numerous less common diseases that may produce multiple strokes. In a large study of 175 patients (Meyer et al. 1988), the most common finding was bilateral lacunar infarcts located at subcortical sites, including the basal ganglia, thalamus, upper brain stem, or white matter. This group of patients had a high prevalence of hypertension. The next largest group of patients had severe stenosis of the carotid or vertebrobasilar arteries and reduced cerebral blood flow ("misery perfusion" syndrome). Lacunar or cortical infarcts, frequently in a watershed distribution, often were present. The third most common pattern associated with vascular dementia was multiple embolic strokes, thought to be cardiogenic in many instances. Many less common conditions may cause multiple strokes, which lead to dementia. Several of these conditions are treatable—in particular vasculitis due to inflammatory diseases, such as giant cell and granulomatous arteritis.

Studies of the natural history of vascular dementia have not divided patients into subgroups based on presumed etiology, so survival data apply to the syndrome as a whole. Based on studies from the 1960s, Wang (1978) concluded that vascular dementia markedly reduced life span because mean life expectancy after onset was 3.8 years compared to the anticipated life expectancy of 13.4 years. In a 5-year prospective study of dementia, Barclay and colleagues (1985) found that patients with vascular dementia and AD had similar rate of cognitive decline; however, vascular dementia carried a higher mortality rate. Four years after entry to this study, survival for vascular dementia was less than 50% of that for a control group. After estimating the date of onset of dementia, the authors calculated that 50% survival was 6.7 years for vascular dementia and 6.2 years for mixed dementia. Another longitudinal study confirmed the shortened life expectancy due to vascular dementia and noted that the most common causes of death were dementia-related illnesses, such as pneumonia in 38% and acute stroke in 33% (Mölsa et al. 1986). It has not been established whether subtypes of vascular dementia carry different prognoses.

Because the increased morbidity and mortality in vascular dementia mainly are due to repeated strokes, attempts have been made to improve the prognosis by treating the underlying risk factors. Treatment of hypertension or cessation of smoking can lead to improved cerebral blood flow and autoregulation, as measured by positron emission tomography (PET) (Meyer et al. 1985), and may reduce the risk of developing vascular dementia. In a recent prospective study, 52 patients with multiple lacunes or misery perfusion (Meyer et al. 1986) were treated for risk factors for atherosclerosis in vascular dementia. For hypertensive patients in this series, systolic blood pressure lowered to between 135 and 150 mm Hg correlated with improved or stabilized cognition. In normotensive patients, improved cognition was associated with abstinence from smoking. Several patients with impaired perfusion due to bilateral internal carotid stenosis improved on cognitive testing and cerebral blood flow following anastomosis of the superficial temporal to the middle cerebral artery or carotid endarterectomy. These encouraging results need to be confirmed by larger studies.

The inception of CT and MRI has resulted in the discovery of abnormalities in the deep white matter of many elderly patients, which appear as lucencies on CT and bright signals on MRI. The radiological disease called leukoaraiosis has been proposed (Hachinski et al. 1987) and is supported by studies that correlate mild cognitive deficits with white-matter lucencies in elderly subjects. Further, it was suggested that these changes represent Binswanger's disease, a dementia formerly considered rare, in which demyelination of periventricular white matter is accompanied by severe sclerosis and hyalinization of small arteries and arterioles (Caplan and Schoene 1978). Recent studies have noted that up to 5% of nondemented elderly subjects have some degree of white-matter change on CT (Goto et al. 1981; Kinkel et al. 1985), and up to 20% have abnormalities on MRI. Although numerous theories have been proposed to account for these radiological findings, their neuropathological identity as infarcts has not been demonstrated convincingly. The relationship between the extent of deep white-matter lesions and cognitive deterioration in elderly individuals and whether these radiological findings have a vascular basis remains to be determined.

In summary, vascular dementia appears to reduce life expectancy more than AD. Within the spectrum of vascular dementias, subgroups of patients with markedly different prognoses have not been identified. Medical treatment of vascular risk factors, in particular hypertension, and surgical treatment of severe carotid stenosis may improve the outcome of patients with vascular dementia.

References

Barclay, L. L.; Zemcov, A.; Blass, J. P.; Sansone, J. Survival in Alzheimer's disease and vascular dementias. Neurology 35:834–840; 1985.

Berg, L.; Danziger, W. L.; Storandt, M.; Coben, L. A.; Gado, M.; Hughes, C. P.; Knesevich, J. W.; Botwinick, J. Predictive features in mild senile dementias of the Alzheimer type. Neurology 34:563–569; 1984.

Berg, L.; Miller, J. P.; Storandt, M.; Duchek, J.; Morris, J. C.; Rubin, E. H.; Burke, W. J.; Cohen, L. A. Mild senile dementia of the Alzheimer type: 2. Longitudinal assessment. Ann. Neurol. 23:477–484; 1988.

Caplan, L. R.; Schoene, W. C. Clinical features of subcortical arteriosclerotic encephalopathy (Binswanger's disease). Neurology 28:1206–1215; 1978.

Chui, H. C.; Teng, E. T.; Henderson, V. W.; Moy, A. C. Clinical subtypes of dementia of the Alzheimer type. Neurology 35:1544–1550; 1985.

Clarfield, A. M. The reversible dementias: do they reverse? Ann. Intern. Med. 109:476–486; 1988.

Erkinjuntti, T.; Haltia, M.; Palo, J.; Sulkava, R.; Paetau, A. Acuracy of the clinical diagnosis of vascular dementia: a prospective clinical and post-mortem pathological study. J. Neurol. Neurosurg. Psychiat. 51:1037–1044; 1988.

Filley, C. M.; Brownell, H. H.; Albert, M. L. Education provides no protection against Alzheimer's disease. Neurology 35:1781–1784; 1985.

Goto, K.; Ishii, N.; Fukasawa, H. Diffuse white matter disease in the geriatric population. Radiology 141:687–695; 1981.

Hachinski, V. C.; Lassen, N. A.; Marshall, J. Multi-infarct dementia: a cause of mental deterioration in the elderly. Lancet 2:207–209; 1974.

Hachinski, V. C.; Potter, P.; Merskey, H. Leuko-araiosis. Arch. Neurol. 44:21–23; 1987.

Heyman, A.; Wilkinson, W. E.; Hurwitz, B. J.; Helms, M. J.; et al. Early-onset Alzheimer's disease; clinical predictors of institutionalization and death. Neurology 37:980–984; 1987.

Hutton, J. T.; Dippel, R. L.; Loewenson, R. B.; Mortimer, J. A.; Christians B. L. Predictors of nursing home placement of patients with Alzheimer's disease. Texas Med. 81:40–43; 1985.

Katzman, R. The prevalence and malignancy of Alzheimer disease: a major killer. Arch. Neurol. 33:217–218; 1976.

Katzman, R.; Brown, T.; Thal, L. J.; Fuld, P.; et al. Comparison of rate of annual change of mental status score in four independent studies of patients with Alzheimer's disease. Ann. Neurol. 24:384–389; 1988.

Kinkel, W. R.; Jacobs, L.; Poklachini, I.; Bates, V.; et al. Subcortical arteriosclerotic encephalopathy (Binswanger's disease). Arch. Neurol. 42:951–959; 1985.

Larson, E. B.; Burton, M. P. H.; Reifler, F.; Featherstone, H. J.; English, D. R. Dementia in elderly outpatients: a prospective study. Ann. Intern. Med. 100:417–423; 1984.

Larson, E. B.; Reifler, B. V.; Sumi, S. M.; Canfield, C. G.; et al. Diagnostic tests in the evaluation of dementia. Arch. Intern. Med. 146:1917–1922; 1986.

Marsden, C.; Harrison, M. Outcome of investigation of patients with presenile dementia. Br. Med. J. 2:249–252; 1972.

Martin, D. C.; Miller, J. K.; Kapoor, W.; Arena, V. C.; Boller, F. A controlled study of survival with dementia. Arch. Neurol. 44:1122–1126; 1987.

Mayeux, R. T.; Stern, Y.; Spanton, S. Heterogeneity in dementia of the Alzheimer type: evidence of subgroups. Neurology 35:453–461; 1985.

Merritt, H. H. A textbook of neurology, 5th ed. Philadelphia: Lea & Febiger; 1975.

Meyer, J. S.; Judd, B. W.; Tawakina, T.; Rogers, R. L.; Mortel, K. F. Improved cognition after control of risk factors for multi-infarct dementia. J.A.M.A. 256:2203–2209; 1986.

Meyer, J. S.; Mc Clintic, K. L.; Rogers, R. L.; Sims, P.; Mortel, K. F. Aetiological considerations and risk factors for multi-infarct dementia. J. Neurol. Neurosurg. Psychiatr. 51:1489–1497; 1988.

Meyer, J. S.; Rogers, R. L.; Mortel, K. F. Prospective analysis of long-term control of mild hypertension on cerebral blood flow. Stroke 16:985–989; 1985.

Mölsa, P. K.; Marttila, R. J.; Rinne, U. K. Survival and cause of death in Alzheimer's disease and multi-infarct dementia. Acta. Neurol. Scand. 74:103–107; 1986.

Morris, J. C.; Drazner, M.; Fulling, K.; Berg, L. Parkinsonism in senile dementia of the Alzheimer type. In: Wurtman R. J.; Corkin,

S. H.; Growdon, J. H.; eds. Alzheimer's disease: advances in basic research and therapies. Cambridge: Cambridge Press; 1987: p. 499–505.

Rosen, W. G.; Terry, R. D.; Fuld, P. A.; Katzman, R.; Peck, A. Pathological verification of ischemic score in differentiation of dementias. Ann. Neurol. 7:486–488; 1980.

Seltzer, B.; Sherwin, I. A comparison of clinical features in early- and late-onset primary degenerative dementia. Arch. Neurol. 40:143–146; 1984.

Smith, J. S.; Kiloh, L. G. The investigation of dementia: results in 200 consecutive admissions. Lancet 1:824–827; 1981.

Stern, Y.; Mayeux, R.; Sano, M.; Hauser, W. A.; Bush, T. Predictors of disease course in patients with probable Alzheimer's disease. Neurology 37:1649–1653; 1987.

Thal, L. J.; Grundman, M.; Klauber, M. R. Dementia: characteristics of a referral population and factors associated with progression. Neurology 38:1083–1090; 1988.

Wade, J.; Mirsen, T.; Hachinski, V.; Fisman, M.; et al. The clinical diagnosis of Alzheimer disease. Arch. Neurol. 44:24–29; 1987.

Wang, H. S. Prognosis in dementia and related disorders in the aged. In: Bick, K. L.; eds. Aging, Vol. 7: Alzheimer's disease: senile dementia and related disorders. New York: Raven Press; 1978: p. 309–313.

23

Normal Pressure Hydrocephalus

JAMES ASHE AND HAMILTON MOSES, III

A quarter of a century ago the idea of a surgically treatable dementia generated great excitement when it was first described by Hakim and Adams (1965). The dementia assiciated with normal pressure hydrocephalus (NPH) improved following ventricular shunting procedures. The initial enthusiasm gave way to confusion as the number of diagnostic and predictive tests increased and it became clear that a significant number of patients failed to improve after surgery.

Hakim and Adams (1965) described three patients with ventricular enlargement but normal lumbar cerebrospinal fluid (CSF) pressures who exhibited a progressive syndrome of dementia, gait disturbance, and incontinence. The symptoms were insidious and progressive. Psychomotor retardation was a major feature of the dementia. The gait disturbance was difficult to characterize, but postural instability was a prominent component. The incontinence, which tended to be a late feature of the disease was frontal lobe in character, in that the patients were generally unconcerned about the symptom.

The basic abnormality in NPH is thought to be a disturbance of normal CSF flow through a patent ventricular system. This may be due to obstruction around the basal cisterns, in the subarachnoid space over the convexities, or to malfunction of the arachnoid granulations that resorb CSF. NPH may be ideopathic or secondary to such conditions as subarachnoid hemorrhage, intracranial surgery, meningitis, and head trauma (Adams et al 1965; Foltz and Ward 1956; Greitz et al. 1969; Hakim and Adams 1965). Although pressure measurements from the lumbar subarachnoid space are usually normal, continuous pressure monitoring may show elevated mean pressures, plateau waves (A waves), or B waves (Symon et al 1975). The transmantle pressure (i.e. the pressure differential between the subarachnoid space and the ventricles) may also contribute to the pathogenesis of NPH; it is elevated in an experimental model of the condition, and abnormal in some patients (Hoff et al 1974). At least three mechanisms contribute to the signs and symptoms of NPH: (1) a stretching of the long periventricular fibres due to ventriculat dilatation (Greitz et al 1969; Yakovlev 1947); (2) and overall decrease in the cortical blood flow; (3) an increase in the water content and gliosis of the periventricular tissues (Mori et al 1977).

The clinical syndrome (hydrocephalus, normal lumbar CSF pressure, dementia, gait disturbance, incontinence) is well defined; however many patients who fit the clinical description of the syndrome do not respond to surgery. Improvement after surgery can be expected in 35% to 45% of patients with NPH of unknown cause, and in 65% of those with a known cause (Udvarhelyi et al 1975). When one considers the significant surgical mortality rate of 9% (Udvarhelyi et al 1975) and a complication rate between 10% and 50% (Haan and Thomeer 1988, Stein and Langfitt 1974) shunting procedures should not be undertaken lightly in this group of patients. For these reasons great emphasis has been placed on reliable methods for the selection of patients who are most likely to improve following ventriculoatrial of ventriculoperitoneal shunt operations.

The factors on which the selection of patients for surgery is based can be grouped into three categories; clinical indicators, CSF tests, and radiological studies.

Clinical Indicators

In patients with clinical diagnosis of NPH, known cause and a disease history of less than 12 months duration are associated with a good surgical outcome (Thomsen et al 1986). When gait disturbance is the first or the major symptom then the prognosis is also good (Fisher 1982; Graff-Redford and Godensky 1986). If dementia is the most prominent feature the outcome is likely to be less favourable. The age of patient is not related to surgical outcome.

Obviously, every effort should be made to exclude patients with conditions that may resemble NPH clinically, especially Alzheimer's disease and multi-infarct dementia.

CSF Studies

CSF Flow

CSF flow is thought to be abnormal in NPH. In normal circumstances, the CSF formed in the lateral ventricles flows through the foramina into the subarachnoid space and over the convexities to the apex of the brain where it is absorbed by the arachnoid granulations. Radionuclide injected into the cisterna magna has long been used to assess CSF flow (Bannister et al. 1967; Ross et al. 1974). In the typical NPH pattern, the radioisotope refluxes into ventricles where it clears slowly and does not flow over the convexities. This typical pattern is associated with a good response to surgery (Borgesen et al. 1981). However, equivocal or normal patterns are not helpful, and the test seems to be less useful than initially predicted (Laws and Mokri 1977).

CSF Pressure

The lumbar CSF pressure is normal in NPH, but many studies have shown abnormal pressure elevations and patterns of pressure change when intracranial pressure is monitored for 24 to 48 hours (Symon and Dorsch 1975; Chawla et al. 1974). It is clear that patients with elevated mean intracranial pressure and plateau waves (Lundberg A-waves) will improve with surgery (Lamas and Lobato 1979; Lundberg 1960; Symon et al. 1975). Patients with significant B-wave activity (pressure fluctuations with a rate of 1-2 per minute for periods of greater than 10 minutes) also have a good prognosis. Borgesen and Gjerris (1982) found that the extent of B-wave activity was inversely related to CSF absorption. In their study, 13 out of 14 patients with B-wave activity for more than 50% of the recording period improved following surgery. No patient with less than 5% B-wave activity improved, and those who had between 5% and 50% had a mixed outcome.

CSF Drainage

Withdrawal of 40 to 50 cc of CSF through lumbar puncture has led to transient improvement of symptoms in some patients (Adams et al. 1965). When positive, this is a reliable predictive test (Wikkelso et al. 1982); however, it is of little help when equivocal or negative. A modification of this method, called external lumbar drainage, has been described in which 300 ml of CSF are withdrawn daily for 5 days from an external lumbar catheter (Di Lauro et al. 1986; Haan and Thomeer 1988). External drainage correctly predicted the outcome of surgery in the 17 of 22 patients who completed the test (Haan and Thomeer 1988).

CSF Absorption

Impaired CSF absorption may contribute to the pathogenesis of NPH (Borgesen and Gjerris 1982). Two indirect tests of absorption have been developed based on the belief that patients with very poor CSF absorption are most likely to respond to surgery. The constant infusion test (Katzman and Hussey 1970) monitors the rise in CSF pressure in response to a constant infusion of saline through a lumbar catheter. CSF absorption can then be calculated. There have been encouraging results with this method (Lamas and Lobato 1979), but many studies have found that it does not reliably predict the outcome of surgery (Belloni et al. 1976; Stein and Langfitt 1974; Symon and Hinzpeter 1976; Wolinsky et al. 1973). In the lumboventricular perfusion test (Borgesen et al. 1978), the CSF pressure is maintained at a steady level during constant lumbar infusion by varying the outflow from a ventricular catheter. The outflow measurements are then used to calculate the absorption. In a prospective study of 80 patients, this test successfully predicted the outcome in 96% of patients (Borgesen and Gjerris 1982). The results were somewhat less impressive in a study of 40 patients with idiopathic NPH in which strict neuropsychological criteria for improvement were used (Thomsen et al. 1986). Although it seems to be a reliable test, the requirement for ventricular catheterization may limit its usefulness.

Radiological Studies

CT Scan

The CT scan is the most common method used to document hydrocephalus, which is part of the syndrome. To confirm the diagnosis of hydrocephalus radiologically, one can rely on the opinion of an experienced radiologist or calculate the Evans ratio (greater than 3.0). The absolute size of the ventricles has no relation to outcome (Borgesen and Gjerris 1982). Cortical atrophy does not preclude a good response to surgery (Laws and Mokri, 1977; Salmon and Timperman 1972; Vassilouthis, 1984), but small sulci increase the chance of a favorable outcome (Borgesen and Gjerris 1982). The presence of periventricular hypodensities, which probably indicates edema of the white matter, is highly correlated with a good outcome (Mori et al 1977) and usually resolves postoperatively. Evidence of hypertensive vascular disease on CT is not a contraindication to surgery (Graff-Redford and Godensky 1987).

MRI

No definitive studies have been performed to determine the specificity of MRI findings in NPH or the patterns associated with a good response to surgery. There is a suggestion that increased CSF flow void is more likely to be found over the aqueduct in NPH than in other conditions, but this needs to be confirmed (Bradley et al. 1986; Jack et al. 1987).

Blood Flow

Cerebral blood flow often is reduced in patients with NPH and may increase postoperatively (Greitz et al. 1969; Salmon and Timperman 1971). Some studies have documented a regional, but not necessarily characteristic, decrease in blood flow (Graff-Radford and Godensky 1987; Mathew et al. 1975), while others have found a more global disturbance (Jagust et al. 1985). Because of conflicting evidence in various studies, cerebral blood flow has little predictive value at present.

Complications

The complication rate of CSF shunting procedures in patients with NPH is significant. The mortality rate in one study was 9% (Udverhelyi 1975). Overall complication rate varies from 10% to 50% and includes, in order of frequency, shunt malfunctions, shunt infections, subdural effusions, subdural hematomas, and seizures.

Conclusions

Not all patients with NPH will respond to surgery. Because of the morbidity and mortality of the surgical procedure, patient selection is crucial. There is no completely accurate method of predicting which patients will improve following surgery. Those with the full clinical triad of dementia, gait disturbance, and incontinence and patients with gait disturbance as the predominant symptom are likely to do well. A known cause or a short history also are favorable indicators. In one study, 13 of 14 patients (with a combination of known cause, short history, dementia, incontinence, and gait disturbance) had a good result from surgery (Borgesen 1982). Thomsen and associates (1896) found that in patients with three or more indicators (short history, known cause, low CSF output, small sulci, and periventricular hypodensity on CT), 80% had improvement in cognitive function postoperatively.

CSF tap tests and radionuclide cisternography are relatively reliable if predictive tests are unequivocally positive. If the situation is still unclear after the simpler predictive tests, one may need to resort to long-term pressure monitoring or CSF absorption studies.

References

Adams, R. D.; Fisher, C. M.; Hakim, S.; Ojeman, R. G.; Sweet, W. H. Symptomatic occult hydrocephalus with "normal" cerebrospinal fluid pressure. N. Engl. J. Med. 273:117; 1965.

Bannister, R.; Gilford, E. E.; Kocen R. Isotope encephalography in the diagnosis of dementia due to communicating hydrocephalus. Lancet 2:1014; 1967.

Belloni, G.; DiRocco, C.; Focacci, C.; Galli, G.; Maira, G.; Rossi, G. F. Surgical indications in normotensive hydrocephalus. A retrospective analysis of the relations of some diagnostic findings to the results of surgical treatment. Acta. Neurochir. 33:1; 1976.

Borgesen, S. E.; Westergard, L.; Gherris, F. Isotope cisternography and conductance to outflow of CSF in normal pressure hydrocephalus. Acta. Neurochir. 57:67; 1981.

Borgesen, S. E.; Gjerris, F.; Sorensen, S. C. The resistance to cerebrospinal fluid absorption in humans. A method of evaluation by lumboventricular perfusion, with particular reference to normal pressure hydrocephalus. Acta. Neurologica. Scandinavica. 57:88; 1978.

Borgesen, S. E.; Gjerris, F. The predictive value of conductance to outflow of cerebrospinal fluid in normal pressure hydrocephalus. Brain 105:65; 1982.

Bradley, W. G., Jr.; Kortman, K. E.; Borgoyne, B. Flowing cerebrospinal fluid in normal and hydrocephalic states: Appearance on MR images. Radiol. 159(3):611; 1986.

Chawla, J. C.; Hulme, A.; Cooper, R. Intracranial pressure in patients with dementia and communicating hydrocephalus. J. Neurosurg. 40:376; 1974.

Connor, E. S.; Black, P. M.; Foley, L. Experimental normal pressure hydrocephalus is accompanied by increased transmantle pressure. J. Neurosurg. 61:322; 1984.

Crockard, H. A.; Hanlon, K.; Duda, E. E.; Mullan, J. F. Hydrocephalus as a cause of dementia: evaluation by computerized tomography and intracranial pressure monitoring. J. Neurol. Neurosurg. Psychiat. 40:736; 1977.

di Lauro, L.; Mearini, M.; Bollati, A. The predictive value of 5 days diversion for shunting of normal pressure hydrocephalus. J. Neurol. Neurosurg. Psychiat. 49:842; 1986.

Fisher, C. M. Hydrocephalus as a cause of disturbances of gait in the elderly. Neurology 32:1358; 1982.

Fisher, C. M. The clinical picture in occult hydrocephalus. Clin. Neurosurg. 24:270; 1977.

Foltz, E. L.; Ward, A. A. Communicating hydrocephalus from subarachnoid bleeding. J. Neurosurg. 13:546; 1956.

Graff-Redford, N. R.; Godensky, J. C. Normal-pressure hydrocephalus: onset of gait abnormality before dementia predicts good surgical outcome. Arch. Neurol. 43:940; 1986.

Graff-Redford, N. R.; Godensky, J. C. Idiopathic normal pressure and systemic hypertension. Neurology 37:868; 1987.

Greitz, T. V. B.; Grepe, A. O. L.; Kalmar, M. S. F. Pre- and post-operative evaluation of cerebral blood flow in low-pressure hydrocephalus. J. Neurosurg. 31:644; 1969.

Haan, J.; Thomeer, R. T. W. M. Predictive value of temporary external lumbar drainage in normal pressure hydrocephalus. Neurosurgery 22:388; 1988.

Hakim, S.; Venegas, J. G.; Burton, J. D. The physics of the cranial cavity, hydrocephalus and normal pressure hydrocephalus: mechanical interpretation and mathematical model. Surg. Neurol. 5:187; 1976.

Hakim, S.; Adams, R. D. The special clinical problem of symptomatic hydrocephalus with normal cerebrospinal fluid pressure. Observations on cerebrospinal fluid hydrodynamics. J. Neurol. Sci. 2:307; 1965.

Jack, C. R.; Mokri, B.; Laws, E. R., Jr.; Houser, O. W.; Baker, H. L., Jr.; Petersen, R. C. MR findings in normal-pressure hydrocephalus: significance and comparison with other forms of dementia. J. Comput. Assist. Tomogr. 11(6):923; 1987.

Jagust, W. J.; Friedland, R. R. P.; Budinger, T. F. Positron emission tomography with [18F] Fluorodeoxy glucose differentiates normal pressure hydrocephalus from Alzheimer's type dementia. J. Neurol. Neurosurg. Psychiat. 48:1091; 1985.

Katzman, R.; Hussey, F. A simple constant-infusion manometric test for measurement of CSF absorption. Rationale and method. Neurology 20:534; 1970.

Lamas, E.; Lobato, R. D. Intraventricular pressure and CSF dynamics in chronic adult hydrocephalus. Surg. Neurol. 12:287; 1979.

Laws, E. R.; Mokri. B. Occult hydrocephalus: Results of shunting correlated with diagnostic tests. Clin. Neurosurg. 24:316; 1977.

Lundberg, N. G. Continuous recording and control of ventricular fluid pressure in neurosurgical patients. ACTA Psychiatrica Scand. 36:(suppl. 149):1; 1960.

Mathew, N. T.; Meyer, J. S.; Hartmann, A.; Ott, E. O. Abnormal cerebrospinal fluid—blood flow dynamics. Implications in diagnosis, treatment and prognosis in normal pressure hydrocephalus. Arch. Neurol. 32:657; 1975.

Mori, K.; Murata, T.; Makano, Y.; Handa, H. Periventricular lucency in hydrocephalus on computerized tomography. Surg. Neurol. 8:337; 1977.

Ross, G. F.; Gull, G.; diRocco, C.; Maria, G.; Meglio, M.; Troneonne, D. Normotensive hy-drocephalus: The relations of pneumoencephalography and isotope cisternography and isotope to the results of surgical treatment. Acta. Neurochir. 30:69; 1974.

Salmon, J. H.; Timperman, A. L. Cerebral blood flow in post-traumatic encephalopathy. The effect of ventriculoatrial shunt. Neurology 21:33; 1971.

Stein, S. C.; Langfitt, T. W. Normal pressure hydrocephalus. Predicting the results of cerebrospinal fluid shunting. J. Neurosurg. 41:463; 1974.

Symon, L.; Dorsch, N. W. C. Use of long-term intracranial pressure measurement to assess hydrocephalic patients prior to shunt surgery. J. Neurosurg. 42:258; 1975.

Symon, L.; Hinzpeter, T. The enigma of normal pressure hydrocephalus: tests to select patients for surgery and to predict shunt function. Clin. Neurosurg. 24:285; 1976.

Thomsen, A. M.; Borgesen, S. E.; Bruhn, P.; Gjerris, F. Prognosis of dementia in normal pressure hydrocephalus after a shunt operation. Ann. Neurol. 20:304; 1986.

Udvarhelyi, G. B.; Wood, J. H.; James, A. E.; Bartlet, D. Results and complications in 55 shunted patients with normal pressure hydrocephalus. Surg. Neurol. 3:271; 1975.

Vassilouthis, J. The syndrome of normal pressure hydrocephalus. Surg. Neurol. 61:501; 1984.

Wikkelso, C.; Andersson, H.; Blomstrand, C. The clinical effect of lumbar puncture in normal pressure hydrocephalus. J. Neurol. Neurosurg. Psychiat. 45:64; 1982.

Wolinsky, J. S.; Barnes, B. D.; Margolis, M. T. Diagnostic tests in normal pressure hydrocephalus. Neurology 23:706; 1973.

Yakovlev, P. I. Paraplegias of hydrocephalus. Am. J. Ment. Defic. 51:561; 1947.

24

Spinal Spondylosis and Disc Disease

DUNCAN K. FISCHER, RICHARD K. SIMPSON, JR., AND
DAVID S. BASKIN

The prognosis underlying cervical, thoracic, and lumbar spondylosis and disc disease has consistently improved over the last 20 years. Conservative therapy, as well as modern anterior, lateral, and posterior approaches to the spine, often using microsurgical techniques, have been refined and improved. The conservative measures and each surgical approach have unique advantages, and decisions about management often are related to specific patients and particular circumstances. Nevertheless, certain guidelines still apply, based on an understanding of the natural history and prognosis of spinal spondylosis and disc disease, that can aid in determining the optimal therapeutic course for a patient. The significance of refined microsurgical techniques that avoid manipulation of delicate structures is of paramount importance, as is the surgical approach used. Moreover, conservative therapy is important in the management of this disease because only certain patients will benefit from surgical intervention. Some patients have a poor prognosis regardless of intervention, and they are unquestionably important to identify so that ineffective surgery is avoided. An understanding of the prognostic factors most closely related to outcome is increasingly important today when justification and efficiency of medical services are required, and medicolegal issues often are raised.

Historical Overview

Although the ancient Egyptians observed the relationship between spine injury and paralysis, it was not until much later that spondylotic and disc compression of the spinal cord and nerve roots were appreciated. In 1838, ventral spondylotic ridges compromising the spinal canal were described (Key 1838; Wilkinson 1967). Later in that century, von Bechterew (1893) suggested on clinical grounds that spondylosis was an entity separate from ankylosing spondylitis; however, he was severely criticized for his conclusions. The potential role of decompressive surgical intervention was shown by Sir Victor Horsely who performed a cervical laminectomy on a patient with progressive myelopathy following a falling injury. The patient made an excellent recovery (Taylor and Collier 1901), and an anterior transverse ridge of bony material compressing the cord was found at surgery. The reports in the late 1920s of chondromatous, cartilagenous material encroaching on the spinal canal (Dandy 1929; Stookey 1928) were a major step forward in understanding upper and lower extremity radicular symptoms unrelated to tumors.

The landmark paper by Mixter and Barr in 1934 established disc rupture as a major etiology of radiculopathy and myelopathy in lumbar and cervical cases. It included four patients with myelopathy from cervical disease, three of whom benefited from surgical intervention. Others soon appreciated that osteoarthritic changes and foraminal encroachment compressing nerve roots also could lead to radicular complaints (Oppenheimer and Turner 1937). With the advent of Lipiodol myelography (Sicard and Forestier 1926) and its refinement using Pantopaque (Steinhausen et al. 1944) to provide preoperative localization in support of the neurological findings, surgery to decompress disc herniation and foraminal encroachment gradually was accepted (Love 1936; Spurling and Bradford 1939). Although some confusion remained concerning cervical radiculopathy and the scalenus anticus syndrome, most agreed that the latter required a positive true Adson's test and the existence of a supernumerary cervical rib (Adson and Coffey 1927).

The earliest clear distinction between disc rupture and the calcified ridges and osteophytic spurring of cervical spondylosis was provided by Brain and colleagues (1952), who analyzed 45 patients, of whom 84% had myelopathic signs. Surgical decompression yielded improvement in eight of 21 severely affected patients, with two deaths. Their conclusions about cervical spondylosis causing cord compression and mimicking motor neuron disease were soon confirmed by others (Liversedge et al. 1953; Symonds 1953).

Nevertheless, additional time was required for the influence of the size of the cervical and lumbar spinal canal to be appreciated (Ehni 1984). Early efforts to remove spondylotic "hard discs" instead of decompressive laminectomy yielded unacceptable neurological deficits. Verbiest (1949) suggested that the size of the lumbar canal and its foramina had an influence. Subsequently, cervical posture and extension were noted to influence Queckenstedt's manometric test for spinal block in 12 of 294 neurological patients, and six of eight of these patients showed complete resolution of the myelographic block after laminectomy (Kaplan and Kennedy 1950). Verbiest's concepts of the role of spinal canal and intervertebral foraminal size in the development of compression syndromes were soon extended to the cervical spine (Arnold 1955; Payne and Spillane 1957). Research since the 1960s has established the view that multiple factors are involved in the pathogenesis of spondylotic myeloradiculopathies, including motion segment arthropathy, developmental canal shallowness, ligamentum flavum hypertrophy, and vascular ischemia or congestion (Ehni 1984).

Surgical Decision Making

Pain radiating into an extremity that is not resolved with conservative measures (including bedrest, traction, analgesics, antiinflammatories, and muscle relaxants) and progressive myelopathy are clear indications for myelographic evaluation of the cervical, thoracic, or lumbar region. This should be followed immediately by computed tomographic (CT) scanning to assess thecal sac, cord, or nerve root compromise. Surgery for lumbar spondylosis and disc disease usually involves hemilaminotomy, medial facetectomy, and discectomy when appropriate. Clearly defined disc lesions can be removed efficaciously by microsurgical discectomy, but controversy remains concerning the balance between reduced postoperative pain and a higher incidence of recurrent symptoms due to missed disc fragments (Frymoyer 1988). For cervical disease amenable to surgery, the choice of a laminectomy or an anterior approach (Cloward 1958; Dereymaeker and Mulier 1958; Robinson and Smith 1955; Smith and Robinson 1958) depends on the patient's clinical presentation, myelographic defects, judgement, surgeon's experience, and long-term prognostic factors. Sufficient data do not exist to compare directly modifications and refinements of the original approaches, such as the ventrolateral exposure (Verbiest 1968; 1970; 1978) and dorsolateral foraminotomy (Frykholm 1951). The latter is thought to be particularly efficacious for monoradicular disease, although controversy persists as to whether to open the cervical nerve root sheath as Frykholm recommends (Knight 1964).

Sensory levels, long tract signs, hyperreflexia, and sphincter dysfunction, which alone or in combination may herald an acute cervical or thoracic disc, must be immediately and completely evaluated; conservative therapy may cause dangerous neurological deterioration and has no place in the management of patients with these abnormalities. For thoracic disc excision, the choices include costotransversectomy (Hulme 1960), transthoracic (Perot and Munro, 1969; Ransohoff et al. 1969), far lateral laminectomy (Carson et al. 1971), and lateral pedicle (Patterson and Arbit 1978) approaches. A clear understanding of the prognostic factors in spinal spondylosis and disc disease aids in management decisions, particularly for complex cases. Delicate microsurgical technique is difficult to quantitate but is as significant as the operative approach.

Cervical Spine Disease

Natural History and Pathophysiology

Cervical spondylosis and disc disease can present as radiculopathy, myelopathy, or myeloradiculopathy (Brain et al. 1952; Hoff and Wilson 1977). Some of the confusion in the literature probably stems from unilateral cervical disc herniation being grouped with unilateral foraminal spur, or "hard disc," because of the similar clinical presentations (Odem et al. 1958).

An acute presentation in a young person more likely indicates disc extrusion. In a large review of 2035 surgical cases expanded from the initial series by Scoville and cowor-

kers (1976) of 383 patients at Hartford Hospital, the mean patient age was 37 years; 85% were lateral soft-disc ruptures, 11% were hard-disc protrusions, and 4% were soft central discs (Collias and Roberts 1988). The male-to-female ratio was nearly 1:1 (49% and 51%, respectively). The most commonly affected roots were C6 and C7, with 54% of disc ruptures occurring at C6 to C7 and 35% at C5 to C6. Only 6% of disc ruptures were localized to C7 to T1 and 5% to C4 to C5. Herniated cervical discs at C3 to C4 were decidedly rare, with only two of 2035 cases (0.1%) documented. Cervical disc herniations were one-tenth as common as lumbar disc protrusions, but 19% of patients with lateral cervical discs developed disc herniations requiring surgical intervention at other spinal levels. Routine cervical spine roentgenograms were helpful in only 20% to 30% of cases because they showed abnormalities at the interspace adjacent to the myelographically or operatively confirmed ruptured disc (Odom et al. 1958; Scoville et al. 1976). A classic sign included neck pain, often progressing in stepwise fashion to the medial scapula, shoulder, and down the arm; more than 80% of the patients denied a history of antecedent trauma. The involved roots often could be identified by myotomal and dermatomal findings if the patient was cooperative and not in severe pain; the Scoville–Spurling test was a valuable confirmation in the majority of cases (Spurling and Scoville 1944). Independent of whether the disc was hard or soft, 78% had radicular pain, 53% had numbness, and 37% had weakness of one arm (Lunsford et al. 1980a). Single dermatomal and myotomal deficits were noted in 48% and 61% of soft discs, and 39% and 43% of hard discs, respectively; hyporeflexia appropriate for the cervical root was specific in 50% to 64% of cases (Lunsford et al. 1980a). The findings of the extensive Hartford Hospital series on cervical disc herniation (Collias and Roberts 1988) are supported by a similar series of 246 patients at Duke University Hospital (Odom et al. 1958) and 648 cases from Semmes–Murphey Clinic (Murphey et al. 1973a; 1973b).

Spondylotic disease prevalence based on radiographic findings dramatically increases with age, it being present in 50% of non-neurological patients older than 50 years and in approximately 80% younger than 65 years (Pallis et al. 1954; McRae 1960). The most common and usually earliest syndrome includes neck pain and upper extremity radiculopathy following acute or chronic involvement of one or more roots, often C6 and C7, and hemicranial occipital headache in 30% of cases (Mayfield 1965; Wilkinson 1971). In contrast to cervical disc herniation, in which 54% of disc ruptures occur at C6 to C7 (Collias and Roberts 1988), cervical spondylosis most commonly involves two adjacent intervertebral levels, and 51% to 64% of cases include the C5 to C6 interspace (Crandall and Batzdorf 1966; Haft and Shenkin 1963). The mean age of 59 years also is older than the average age of 37 for cervical disc rupture (Collias and Roberts 1988). In a long-term study by Crandall and colleagues at UCLA, greater than 80% of patients admitted with spondylotic radiculopathy had symptoms of local cervical and radicular pain, 60% had extremity numbness, and almost 70% had upper extremity weakness (Gregorius et al. 1976). On physical examination, nearly 90% had restricted cervical range of motion, more than 60% had a dermatomal sensory deficit, and approximately 40% and 55% had distal and proximal upper extremity weakness, respectively (Gregorius et al. 1976). The hallmark of spondylotic cervical radiculopathy amenable to surgical correction is arm pain, often worsened by cervical motion, and root localization by dermatomal sensory loss, hyporeflexia, and muscle weakness. Importantly, the severity of most acute attacks of cervical radiculopathy resolves over time with conservative measures; however, these episodes can progress to a chronic syndrome with an intermittent, dull, aching discomfort and more clear-cut signs of dermatomal sensory dysesthesia, hyporeflexia, and muscle wasting.

The syndrome of cervical spondylotic myelopathy develops gradually, often over years, or can be precipitated by injury or trauma to an underlying stenotic cervical canal. Although bowel and bladder sphincter function usually are not affected early, long sensory and motor tracts characteristically are involved (Brain et al. 1952). This leads to altered proprioception, hyperreflexia, extensor plantar reflexes, and spastic paraparesis. Complaints of gait disturbance, lower extremity weakness, diffuse pains, and sensations of discomfort are present in 68% to 77% of cases (Kugelgen et al. 1980). The long-term study of Crandall and associates at UCLA revealed that 90% of patients with spondylotic myelopathy had gait disturbance, and approximately 70% and 75% had sphincter disturbance and lower extremity weakness, respectively (Gregorius et al. 1976). Physical examination on admission documented that more than 80% of patients had lower extremity spasticity, 70% had lower extremity weakness, and 40% to 50% had spinothalamic tract and posterior column deficits (Gregorius et al. 1976). Lower percentages were reported in an earlier series of 199 patients with cervical spondylotic myelopathy, which was noted to be the most common single cause of spinal cord disease (Crandall and Batzdorf 1966). In this initial UCLA series, five classic syndromes of neurological findings in spondylotic cervical myelopathy were described; these included a transverse lesion syndrome (47%), a motor system syndrome (19%), a central cord syndrome (13%), the Brown–Séquard syndrome (13%), and a brachialgia and cord syndrome (8%) (Crandall and Batzdorf 1966).

The diagnosis of cervical spondylosis often has been confused with amyotrophic lateral sclerosis, but the latter usually can be excluded electromyographically (including paraspinal muscle and tongue recordings) and by the lack of preferential involvement of distal extensors. Lees and Alden-Turner (1963) followed the natural history of cervical spondylosis in 44 myelopathic and 51 nonmyelopathic patients at St. Bartholomew's Hospital for more than 10 years. In most cases, the clinical picture of chronic myelopathy often was dominated by prolonged periods of nonprogressive disability, interrupted by occasional exacerbations. Moreover, they could attribute very few deaths to cervical spondylosis and found that symptoms of myelopathy did not develop in any patient who did not have them at first presentation. They argued that the data supported a very conservative approach in the management of cervical myelopathy. The often benign nature of cervical spondylotic myelopathy subsequently was confirmed in a study of 91 patients at Queen Square, in which only older age at presentation could be associated with progressive deterioration and disability (Nurick 1972b).

Myeloradiculopathy combines cord and root lesions, typically from lower cervical segment spondylosis caudal to C4. The syndrome is one of intermittent progressive deterioration, but acute episodes often remit clinically with conservative measures. Acute presentation can follow a hyperextension injury in a spondylotic canal, compromised by ventral ridging and dorsal infolding ligamentum flavum, resulting in a central cord syndrome (Schneider et al. 1954). It involves upper extremity but spares lower extremity motor function based solely on the lamination pattern of the corticospinal tracts. The deficit can mimic intrinsic cord lesions, such as intramedullary tumors, hematomyelia, and syringomyelia, but often without the dissociated sensory loss seen in the latter.

The pathophysiological mechanisms (Hoff and Wilson 1977) leading to cervical myeloradiculopathy serve as a well-researched model to understand spondylotic disease presentation, progression, and curative approaches at all levels of the spinal canal. For this reason, this subject is discussed in some depth. Although the width of the spinal canal usually has little significance except with major lateral mass lesions, a canal height on lateral films of 13 mm or less at C5 is considered abnormal (Arnold 1955; Epstein et al. 1978). Canal heights or sagittal diameters less than 10 or 11 mm are especially critical (Epstein et al. 1978). The constitutional and acquired sagittal dimensions of the spinal canal are narrower in individuals afflicted with cervical spondylotic myelopathy, and the degree of narrowing is associated with the severity of the paraparesis (Crandall and Batzdorf 1966; Nurick 1972a; Payne and Spillane 1957).

The height of cervical discs is about 45% of the adjacent vertebral body height (Payne 1959). During the normal aging process, the nucleus pulposus progressively shrinks, losing water from 88% content in infancy and childhod to roughly 65% in elderly patients (Coventry et al. 1945). Trauma and degenerative changes can more readily deform and fragment the aging, inelastic intervertebral disc onto adjacent neural structures, which leads to clinical symptoms.

In the normal aging process, as the cervical nuclei pulposi dehydrate and fragment, both intervertebral disc height and elasticity are diminished, placing greater stresses on the articular surfaces of the vertebral endplates. Greater pressures consequently are exerted on a degenerating disc, which can lead to herniation of the nucleus pulposus, often causing foraminal compromise and nerve root impingement. Osteophytic bone formation occurs naturally to stabilize and increase the weight-bearing surface area of adjacent vertebral endplates (Hoff and Wilson 1977). However, osteophytic bone growth is poorly regulated, and excessive formation can lead to ventral spondylotic ridging and foraminal encroachment. The combination of the infolding of ligamentum flavum dorsally and the presence of osteophytic bony ridges ventrally maximizes canal compromise with neck hyperextension (Schneider et al. 1954). It also can precipitate a central cord syndrome acutely or a myelopathy more insidiously. Cord flattening and distortion, readily visualized on modern CT scans following myelogram studies, can lead to demyelination in the lateral columns of the cord (Wilkinson 1960), which explains how undiagnosed cervical spondylosis can mimic amyotrophic lateral sclerosis.

Vascular changes also may contribute to the development of myeloradiculopathy. The radicular arteries accompany the pairs of dorsal and ventral cervical roots through the neural foramina. They come from the vertebral, deep cervical, and thyrocervical vessels, which together lead into the anterior spinal artery that supplies the central gray matter and anterolateral regions of the cord (Turnbull et al. 1966). The smaller, paired posterior spinal arteries primarily supply the dorsal column tracts within the cord. Venous drainage is accomplished through radicular veins that exit through the neural foramina emptying into extravertebral venous plexuses. The epidural veins, which often are a hemostatic annoyance to the neurosurgeon during laminectomy, have little or no role in the vascular drainage of the cord (Turnbull et al. 1966). The radicular arteries and veins may be adversely affected by bone spur growth and foraminal encroachment, possibly leading to arterial spasm, thrombosis (Gooding 1974), edema, and decreased spinal cord blood flow from compromised venous drainage (Brain et al. 1952).

Research has suggested that both compression and ischemia appear to be independent etiologic factors contributing to spondylotic myeloradiculopathy, but an underlying additive synergism can exacerbate progression. Each factor has been hypothesized to play a role, because ischemic pathological changes similar to anterior spinal artery occlusion (Hughes and Brownell 1966) have been noted in severe cases; however, many patients with osteophytic compression improve after adequate laminectomy (Fager 1973). Experimental studies in laboratory dogs have produced an animal model of a cord lesion resembling spondylotic myeloradiculopathy in humans. In these studies, although ischemia and compression can independently produce the deficit, microangiographic and autoradiographic analysis have shown an additive, synergistic effect. Anterior cord blood flow is more severely reduced by both compression and ischemia than by either influence alone (Gooding et al. 1975; 1976; Hukuda and Wilson 1972). The possibility of impaired autoregulation in the ischemic spinal cord has been suggested by studies in dogs (Griffiths 1973) and rhesus monkeys (Kobrine et

al. 1974). This is supported by reports stressing the importance of local vascular abnormalities, including the compression of important feeding radicular arteries, in the predisposition to cervical myelopathy (Taylor 1964).

Prognosis and Treatment

In most cases, the initial therapy for cervical radiculopathy should be nonsurgical and should consist of rest, analgesics, muscle relaxants, and nonsteroidal anti-inflammatory agents. These measures, a rigid cervical collar, and cervical traction several times daily with 10 to 15 lb of weight often are helpful; up to 88% of patients will recover within a few months without surgical intervention (Collias and Roberts 1988; Mayfield 1965). Nonsurgical treatment should be continued as long as the patient is improving. Although time limits are arbitrary, the failure of conservative measures for 3 to 4 weeks is an indication to proceed with water-soluble myelography, followed by spinal CT scanning to evaluate compromise of subarachnoid space or impingement of cervical nerve roots.

Although in some nations, such as Great Britain, conservative therapy is recommended for protracted periods of time, most physicians in the United States proceed with surgery for a myelographic defect seen on the neurological examination after failure of conservative measures for 4 weeks. Most agree, however, with the recommendation that conservative treatment should be continued if the patient is improving (Collias and Roberts 1988). Clearly, conservative therapeutic management is bypassed in acutely worsening myeloradiculopathy, which requires prompt radiological evauation and often surgical decompression.

The surgical approaches to cervical disc excision, traditionally involving laminotomy and foraminotomy, have yielded results superior to those of any other neurosurgical operation, with the possible exception of that for trigeminal neuralgia. In the study by Murphey and colleagues (1973a; 1973b) 84% of 380 surgical patients followed for 1 to 28 years estimated their degree of relief to be greater than 90%. Equally encouraging results were described by others in which 84% of 175 patients with ruptured cervical discs had good to excellent results (Odom et al. 1958), and 90% of cases had similarly graded good to excellent postoperative courses (Haft and Shenkin 1963). The results were only slightly worse for the posterior decompression of foraminal spurs causing radicular symptoms, with 67% (Odom et al. 1958) and 83% of patients (Haft and Shenkin 1963) reporting good to excellent results. Careful case selection, including myelography demonstrating a defect properly accounting for the patient's neurological examination, is crucial. The Hartford Hospital series studied 2035 surgical cases of cervical disc herniation approached posteriorly by foraminotomy and discectomy; 96% achieved good to excellent results (Roberts and Collias 1988). These statistics mirror the earlier Hartford series of 171 cases of lateral cervical disc surgery, in which 95% had good to excellent results; 96% of men and 100% of women returned to the same job within an average of 4.2 weeks (Scoville et al. 1976).

Although initial claims of high success rates by Cloward were considered overly optimistic (Cloward 1963; Connolly et al. 1965), similar postoperative outcomes have been reported when anterior cervical approaches were used for lateral disc herniation in 253 patients at Pittsburgh followed for 1 to 7 years after surgery (Lunsford et al. 1980a). Of these cases, 67% had good to excellent results, and 77% at least initially noted complete relief of symptoms (Lunsford et al. 1980a). As in other studies (Robertson 1973; Martins 1976), surgical results were not affected by whether fusion accompanied anterior discectomy, but the former was associated with a more lengthy hospital stay and a higher incidence of postoperative complications (Lunsford et al. 1980a). Moreover, no significant differences were observed between hard or soft disc cases, the number of operative levels, or Cloward interbody dowel vs. Smith–Robinson wedge-type graft fusions. No statistically significant differences in surgical out-

come were noted with the presence or absence of occipital headache, whiplash, work-related injury, prior surgery, radicular arm pain, myotomal upper extremity weakness, dermatomal sensory deficits, or hyporeflexia (Lunsford et al. 1980a).

In contrast to the uniformly superb results of posterior and anterior surgical approaches for cervical disc disease, the postoperative prognosis for spondylotic myelopathy is considerably less gratifying. A successful outcome appears to depend on underlying etiologic factors rather than surgical approach; in a series of 85 anterior fusions, acute syndromes usually secondary to hernited discs improved significantly more often than chronic syndromes from osteophytic spurring (Hamel et al. 1980). Initial reports suggested that decompressive laminectomy improved 70% to 74% of myelopathic patients (Stoops and King 1962; Symon and Lavender 1967). Other analyses, however, indicate that the 56% improvement rate for laminectomy in 474 patients and the 55% improvement rate for anterior interbody fusion in 123 patients do not differ significantly from to 40% to 53% of improved cases using conservative measures alone (Lees and Aldren-Turner 1963; Nurick 1972b). Of 59 myelopathic patients, including those of Northfield, followed 10 years after decompressive laminectomy, the percentage of improved cases dropped from 61% to 51% (Northfield 1955; Bishara 1971). Further pessimism was provided by a Mayo Clinic series in which only 38% of 47 patients undergoing cervical laminectomy with section of intradural dentate ligaments showed some improvement (Peserico et al. 1962). Nevertheless, some support for the efficacy of anterior interbody fusion was noted by a report of 51% good and very good results in 45 cases with long-term follow-up, which is superior to the group's results with laminectomy (Guidetti and Fortuna 1969). A subsequent study from the same group confirmed their conclusions in 270 cases, in which good to very good results of 64.5% with anterior interbody fusion dropped to 51.6% with extended laminectomy (Guidetti et al. 1980). In contrast, late operative results in 57 myelopathic patients followed an average of 3.5 years revealed no significant difference between posterior and anterior approaches in the percentage of patients improved (24% to 35%) versus their disease progression arrested (54% to 59%) (Arienta et al. 1980).

The Lahey Clinic experience, favoring the posterior approach, revealed that extensive cervical laminectomy, often with dentate ligament sectioning, yielded 66% to 69% of myelopathic cases improved (Fager 1973; 1976; 1989). A subsequent analysis, although retrospective, uncovered no significant differences in outcome when comparing laminectomy alone with laminectomy, dentate sectioning, and dural graft; 43% to 44% of each treatment group improved at follow-up (Piepgras 1977). The Hartford Hospital experience included 84 myelopathic patients followed for an average of more than 2 years after decompressive laminectomy; 69% of cases remained improved and 60% returned to work (Collias and Roberts 1988). Recent additional support for posterior decompression has come from a Japanese study of 191 surgical cases of cervical spondylotic myelopathy followed for an average of 2.6 years (Hukuda et al. 1985). Posterior approaches yielded statistically significant results with superior to anterior operations, particularly for the more severe transverse lesion, Brown–Séquard, and motor syndrome myelopathies as defined by Crandall and Batzdorf's 1966 classification (Hukuda et al. 1985). A long preoperative duration of symptoms, not the age of the patient, was associated with a poor response to surgery. No clear superiority was identified for any specific anterior or posterior approach, including the Cloward, Smith–Robinson, conventional laminectomy, French window laminectomy, or French window laminoplasty procedures. This excellent study in terms of scope and follow-up, concluded that one or two levels without canal stenosis could be addressed anteriorly, but posterior decompression was preferable for three or more levels, particularly when spinal stenosis exists (Hukuda et al. 1985). Other groups have supported the con-

clusions that laminectomy is advisable in the presence of a narrow spinal canal (Fager 1989; Roosen and Grote 1980).

In the initial UCLA series of 62 surgical cases of spondylotic cervical myelopathy, anterior cervical discectomy with interbody fusion and laminectomy with dural patch graft insertion were the best procedures, yielding 71% and 82% excellent and improved results, respectively (Crandall and Batzdorf 1966). In contrast, 69% of laminectomies with the dura closed (i.e., unopened or reapproximated) yielded unchanged or worse postoperative results. Overall, the prognosis for pain and spasticity was best with 84% of these manifestations resolved or improved; approximately 41% of sphincter disturbances and fasciculations were unchanged or worse. Among the five classic spinal cord syndromes described, most improvement was observed for the Brown–Séquard syndrome and least improvement noted for the motor system syndrome (Crandall and Batzdorf 1966). The UCLA data subsequently were supported by an excellent comparison of anterior and posterior approaches, including historical data, demonstrating that anterior interbody fusion yielded a 73% cured or improved rate for myelopathic cases, whereas extensive laminectomy yielded a comparable 70% rate (Gorter 1976). Although it is less commonly performed, many European groups favor decompressive laminectomy with duroplasty, reporting 48% to 60% of myelopathic cases substantially improved (Demirel et al. 1980; Hamer and Kahl 1980). Comparing the number of laminae removed with outcome, better postoperative results have been achieved by a more extensive laminectomy (Thal et al. 1980).

In the long-term follow-up study at UCLA of 96 surgical cases of severe cervical spondylosis (55 patients characterized as myelopathic and 41 primarily with radicular features), improvement in disability was noted following anterior interbody fusion for myelopathy, whereas laminectomy worsened disability (Gregorius et al. 1976). Only 21 of 55 myelopathic cases (38%) had excellent or improved postoperative results, whereas 18% were unchanged, and 44% were worse. In contrast, consistently good operative outcomes were achieved in the radiculopathy group, independent of whether anterior interbody fusion of laminotomy were used. Thirty-one of 41 patients (76%) were excellent or improved. Among the numerous complaints, symptoms, and physical findings in the myelopathic group, only preoperative sphincter disturbance and lower extremity weakness portended a worse outcome (Gregorius et al. 1976). Having symptoms less than 12 months before operative intervention for myelopathy yielded a better prognosis for postoperative walking. Neither race, sex, age at admission or symptom onset, level of CSF protein, length of follow-up, levels involved on spine films, nor presence of myelographic block had any prognostic relationship to postoperative outcome (Gregorius et al. 1976). Other groups have disagreed about the role of age, stating that older myelopathic patients have a worse prognosis than those younger than 50 years (Fischer et al. 1980). The somewhat pessimistic long-term follow-up UCLA study is supported by a more recent Pittsburgh series of 32 myelopathic cases assessed 1 to 7 years after anterior cervical surgery, 84% of which were fused by Cloward interbody dowel or Smith–Robinson wedge grafts in equal proportions; only 50% of patients improved and 50% of cases were classified as poor. The overall results were not superior to conservative therapy alone (Lunsford et al. 1980b). This pessimism is mirrored in an earlier report in which only 36% of 28 surgical cases with long-tract spinal cord signs improved, and this percentage decreased as the follow-up lengthened (Galera and Tovi 1968). No firm statistics or long-term follow-up exist to clearly estabish one surgical procedure as superior to others; therefore, the often adamant opinion concerning anterior vs. posterior approaches remains unwarranted (Lunsford et al. 1980b).

In a large clinical series of 270 cases with good follow-up, data show that chances for an improved surgical prognosis are inversely related to the preoperative duration of spondylotic myelopathy (Guidetti et al. 1980). Postoperative results of good to very good dropped from 64% for a history of less than 6 months, to 30% for histories of more than 1 year. This suggests that it is not prudent to continue conservative treatment of cervical spondylosis after myelopathic features develop. Other studies also have confirmed that the length of history of myelopathy adversely affects the postoperative prognosis (Guidetti and Fortuna 1969; Peserico et al. 1962).

Many of the surgical series summarized previously have attempted to address the efficacy, relative advantages and disadvantages, and prognostic factors associated with traditional laminectomy, often foraminotomy, versus the more recently described anterior approach (Cloward 1958; Robinson and Smith 1955). The data suggest that both approaches are equally effective and achieve excellent results in most cases of cervical radiculopathy, whether due to lateral cervical disc extrusion or osteophytic foraminal spurring. The anterior approach is probably superior, particularly in early cases of myelopathy (Crandall and Batzdorf 1966; Gregorius et al. 1976), but it has a potentially higher complication rate. It can be less effective than posterior decompression in the more advanced myelopathies associated with a multilevel stenotic canal (Hukuda et al. 1985). In complex cases of myeloradiculopathy, each approach yields roughly the same results. Analysis of most reported surgical series reveals that the best results were achieved in patients with pure radiculopathy, and myelopathic features carried a worse prognosis (Ehni 1984). The conclusions that can be drawn from the published analyses of surgical results remain limited because randomized prospective comparisons of surgical approach do not exist; the operative approaches were not separated according to whether the myelographic defect was ventral or dorsal; many authors failed to segregate cases of cervical root and cord compression; and the surgical results of many individual surgeons often were weighted equally, yet their methods of patient selection and extent of operative experience were not equivalent.

There are other advantages and disadvantages for each cervical approach. Transient dysphagia, vocal cord paresis, Horner's syndrome, tracheitis, graft extrusion, donor-site problems, spinal cord malfunction, and vertebral artery perforation have been reported with the anterior approach (Baumann et al. 1980; Cloward 1958; Robinson et al. 1962; Smith and Robinson 1958; Sugar 1981); however, the procedure does not dissect away the paraspinal muscles and thus involves less postoperative pain, more rapid mobilization, a shorter hospital stay, and no fusion if only one level is surgically addressed (Franco and Matricali 1980; Hankinson and Wilson 1975). Minimal amounts of longus coli musculature are dissected free anteriorly, under which the retractors are secured to avoid esophageal or carotid injury. This small extent of muscle dissection has not been associated with postoperative cervical discomfort. The stability of anterior cervical discectomy without fusion, in spite of slight angular kyphosis, has been supported by other studies (Gilsbach and Eggert, 1980; Giombini and Solero, 1980; Martins 1976; Murphy and Gado 1972). Others, however, have reported considerable restriction of sagittal movement of the cervical spine, particularly retroflexion, with fusion of two or more vertebrae and have proposed that the efficacy of the Cloward operation may lie in part on its restriction of cervical movement (Gruninger and Gruss 1980). Additional potential problems with the anterior Cloward procedure include kyphosis within the operated segment, slippage of the fashioned bone plug, and increased stressing leading to premature degeneration of cervical segments above and below the operative site (Gruss et al. 1980). Operative failures and deaths following 100 anterior cervical discectomies without fusion have been associated with a stenotic canal, particularly less than 13 mm, and posterior decompression may be preferable in this subset of cases (Giombini and Solero 1980). Another group, in a series of 132 cervical cases, have noted postoperative deaths only in dorsally operated patients (Thomalske et al. 1980).

Thoracic Disc Disease

Thoracic disc herniations represent only 0.25% to 0.75% of symptomatic disc lesions, and primarily occur betwen T8 and T11 (Simeone and Rashbaum 1988). These disc lesions can acutely or chronically cause radicular pain if they are lateral, but they can lead to long tract signs and paraparesis if the herniation is more central. In approximately 75% of patients, back pain with uncomfortable paresthesias below the level of the lesion is the initial complaint. Often these symptoms have been precipitated by a fall or other trauma. It is difficult to differentiate clinically chronic thoracic disc herniations and other etiologies of thoracic cord compromise, so definitive diagnosis must await the early and timely use of myelography, CT after water-soluble agent myelography, and magnetic resonance imaging (MRI). Fifty-five percent of thoracic disc protrusions are calcified and readily apparent on plain spine films (McAllister and Sage 1976).

Conservative therapy is not appropriate for thoracic disc lesions after long tract signs, sensory levels, proprioceptive changes, hyperreflexia, or sphincter dysfunction develop. Immediate and full radiological evaluation is needed because delay can cause progressive neurological deterioration. After radiological studies have ruled out tumor, metastasis, spinal epidural hematoma, abscess, and other possible etiologies and have established a diagnosis of herniated thoracic disc as the cause of the symptoms, timely operative excision and often immediate initiation of intravenous dexamethasone are indicated to optimize chances for a full recovery. High-quality radiographic studies are imperative because the suspected pathological lesion largely determines the operative approach, and the precise intraoperative localization of thoracic discs remains a challenge.

Initially, the approach to thoracic disc excision was based on laminectomy, but serious postoperative sequelae, including paraplegia, were noted. Three out of four cases reported were paraplegic after surgery (Mueller 1951). In a series of 11 cases of central thoracic disc herniations, all had an increase in neurological deficit after laminectomy, and five suffered a functional cord transection (Logue 1952). Of 91 thoracic intervertebral disc protrusions treated by laminectomy reviewed in the literature, 18 did not improve, 16 (17.6%) were paraplegic, and six died (Perot and Munro 1969). Further analysis of these cases revealed that good results could be obtained with laminectomy if the thoracic disc was lateral, but 12 (35.3%) out of 34 cases with central herniations above T10 to T11 became paraplegic.

Analysis of these cases revealed that factors predisposing to a poor surgical outcome included an anterior, central location of the disc protrusion and a hard, calcified consistency. These characteristics may lead to considerable cord manipulation at the time of laminectomy in an attempt to remove an anterior lesion by a posterior approach. Accordingly, new surgical exposures were developed to gain access to the thoracic intervertebral disc space without compromising or manipulating the thoracic cord. One of the first developed was the lateral extrapleural approach, or costotransversectomy (Hulme 1960). This was an extension of earlier procedures to decompress Pott's disease (Dott 1947) and biopsy the vertebral body (Michele and Krueger 1949).

Other surgical approaches followed. A transthoracic, transpleural exposure requiring a lateral thoracotomy was reported by two groups simultaneously (Perot and Munro 1969; Ransohoff et al. 1969). This was particularly efficacious for the removal of firm, calcified discs because it required the least manipulation of the already compromised thoracic cord. Whether the thoracotomy should be right- or left-sided remains controversial and depends on whether the surgeon's concern for mobilization of the great vessels or avoiding the artery of Adamkiewicz is considered paramount.

Far lateral laminectomy exposures also have been described (Carson et al. 1971) to gain access to the thoracic disc space without risking thoracic cord compromise. A final refinement has been the lateral pedicle approach (Patterson and Arbit 1978), in which the facet joint and pedicle of the vertebral body caudal to the herniated disc are carefully drilled to expose completely the offending intervertebral disc space.

As discussed previously, existing series suggest distinctly improved surgical outcomes using the costotransversectomy and transthoracic approaches, particularly with firm, calcified, centrally located thoracic discs (Patterson and Arbit 1978; Perot and Munro 1969). Although large series with adequate follow-up that detail prognostic factors and outcome, have not been conducted, the transthoracic and lateral extrapleural approaches are reported to achieve cure rates of 75% and 61%, respectively (Patterson and Arbit 1978). In contrast, laminectomy yielded a cure rate of only 32%, and 44% of patients did not improve, were paraplegic, or died (Perot and Munro 1969). The success of the far lateral laminectomy and lateral pedicle exposures suggests that manipulation prior to lateral decompression and lack of attention to careful microsurgical technique may have played a role as significant as the approach in the postoperative deficits initially seen with thoracic laminectomy.

Lumbar Spine Disease

Natural History

Low back pain with or without sciatica is the most common reason that patients present to physicians in the United States and Canada (Long 1987). In spite of conservative therapy and approximately 200,000 lumbar laminectomies performed yearly, 9 million workers remain impaired and 2.4 million remain disabled from the pain, costing an average of 30 to 50 billion U.S. dollars (Long 1987). The costs of conservative measures often are greater than those of surgical intervention (Kelsey and White 1980).

The same pathophysiological mechanisms of disc degeneration, ligamentum hypertrophy, osteophyte formation, foraminal encroachment, and canal stenosis as discussed in the section on cervical spine disease are involved in the lumbar region. The nucleus pulposus, which dehydrates with age and often is exposed to considerable shearing stresses and vertical torsion, can herniate through the annulus fibrosus to compromise an adjacent nerve root. If the disc tissue also ruptures through the posterior longitudinal ligament and into the spinal canal, it is considered an extruded disc. Lumbar spondylosis refers to the entire range of osteophyte formation and facet hypertrophy that occur in association with disc degeneration. Historically, the two most significant breakthroughs in the understanding of lumbar spine disease were the description of disc herniation as a cause of lower extremity radiculopathy (Mixter and Barr 1934), and the appreciation of lumbar stenosis as an etiology of intermittent neurogenic claudication (Ehni, 1969; Verbiest 1949; Wilson 1969). The concentric constriction of lumbar stenosis may diminish the blood supply to the cauda equina, resulting in intermittent ischemic or postural claudication, which can be relieved by rest or increased spinal canal size achieved through flexion (Wilson 1969).

Scandinavian studies, using data from national health insurance programs, have documented the natural history and incidence of lumbar spine disease. Almost 53% of individuals engaged in light labor and 64% of heavy laborers experience back pain (Hult 1954). Nearly 80% of the population will suffer from significant lower back pain at some point, and of these, 35% will develop sciatica (Horal 1969; Nachemson 1976). Sciatica is most common in the 30- to 50-year age group, with a mean age of 42 for patients undergoing a lumbar laminectomy for disc herniation (Spangfort 1972). Almost 98% of lumbar herniations involve the L4 to L5 or L5 to S1 interspaces (Spangfort 1972). In another series primarily representing spondylosis, 64% of patients were 40 to 60 years of age with a male-to-female ratio of 5:3. Twenty-six percent reported prior trauma or exertion (Weinstein et al. 1977). Back pain was the major complaint in 66% of cases, followed by unilateral lower extremity radiculopathy in 36%, leg paresthesias in 26%, leg weakness in 23%, and urinary incontinence in 10%. On examination, muscle weakness, found in 64% of cases, was more typical than the sensory loss noted in only 50% of patients. Deep-tendon reflexes were hypoactive or absent in 70% of cases, most commonly the Achilles reflex.

Part of the difficulty in evaluating the outcome of conservative measures vs. surgical intervention is due to the high rate of spontaneous recovery in the natural history of sciatica. Following the first episode of low back pain with sciatica, 60% to 74% of patients recover within 1 month, and 85% are asymptomatic within 3 months (Long 1987).

Treatment and Prognosis

The best studies evaluating nonsurgical conservative therapy have dealt with spondylosis of the lumbar spine. Typical measures have included bedrest, muscle relaxants, nonsteroidal anti-inflammatory agents, analgesics, traction or brace immobilization, and short courses of dexamethasone. Steroids reduce the nerve root inflammation, which is needed for the patient's perception of pain with root compression (Macnab 1990). Multidisciplinary, integrated rehabilitative programs are considered the most successful of the nonsurgical regimens (Rosomoff and Rosomoff 1987; Seres and Newman 1989). A thorough evaluation of psychosocial factors, secondary gain, and pending litigation or compensation is paramount because the efficacy of any treatment regimen is compromised dramatically by these constraints. In a series of 50 patients with lumbar stenosis treated in such a program for 3 to 4 weeks and followed up to 8 years after therapy, 64% reported pain relief, and an additional 30% noted some decrease in pain (Rosomoff and Rosomoff 1987). Patients with unequivocal lumbar disc herniation were assigned to surgical and conservative therapy randomly in a prospective study. One year later, 92% of surgical cases and 60% of conservative cases were substantially improved; however, there were no statistical differences between the two groups at 4 years and 10 years of follow-up (Weber 1983). This is supported by a retrospective study of 583 patients with unilateral L5 or S1 sciatica, which documented that after 6 months there was no difference between the surgical and nonsurgical groups (Hakelius 1970). The validity of these comparisons is being challenged as more neurosurgeons adopt microsurgical approaches in addressing lumbar disc disease. There is broad consensus among the neurosurgical community that "bulging" lumbar discs, readily visualized on outpatient spine MRI studies, should not be addressed surgically.

The generally accepted surgical standard is open laminotomy and discectomy for herniated nucleus pulposus; further laminectomy, medial facetectomy, and foraminotomy for spondylotic encroachment; and decompressive laminectomy for lumbar stenosis. In general, a thorough discectomy is advantageous in limiting the removal primarily to the offending disc fragment. In a recent study of 257 patients undergoing lumbar microdiscectomy, 84% considered the operation a success and 95% returned to work (Macnab and McCulloch 1990). These results are supported by a newly published German series of 299 patients, of whom 89% returned to the same occupation and 74% had good or excellent results following lumbar microdiscectomy (Caspar et al. 1991). A return to full employment in 78% and a relief of disability in 75% of spondylotic cases has been reported (Weinstein et al. 1977). Not only does chemonucleolysis have a significant incidence of anaphylaxis, but its efficacy is inferior to that of open discectomy. In 151 patients with disc herniation at L4 to L5 or L5 to S1 evenly divided between open discectomy and chemonucleolysis, 22% of the chemonucleolysis group (and 0% in the discectomy group) had increased radicular pain, and 25% required open discectomy after failed chemonucleolysis (Van Alphen et al. 1989). For some of the same reasons, percutaneous discectomy has a high rate of recurrent symptoms due to missed disc fragments and to collapse of the disc space and exacerbations of spondylitic nerve root compression.

Acknowledgments We appreciate the cheerful secretarial assistance provided by Ms. Anne Fischer. This work was supported in part by a scholarship from the American College of Surgeons (D.K.F), and by Grants from the Department of Veterans' Affairs Merit Review Board and the Henry J. N. Taub Fund for Neurosurgical Research (D.S.B.).

References

Adson, A. W.; Coffey, J. R. Cervical rib, method of anterior approach for relief of symptoms by division of scalenus anticus. Ann. Surg. 85:839–857; 1927.

Arienta, C.; De Benedittis, G.; Infuso, L.; Villani, R. Operative results in the treatment of cervical spondylotic myelopathy. In: Grote, W.; Brock, M.; Clar, H. E.; Klinger, M.; Nau, H. E., eds. Advances in neurosurgery 8: Surgery of cervical myelopathy. Berlin: Springer–Verlag; 1980: p. 64–68.

Arnold, J. G., Jr. The clinical manifestations of spondylochondrosis (spondylosis) of the cervical spine. Ann. Surg. 141:872–889; 1955.

Baumann, H.; Samii, M.; Von Wild, K. Contribution to ventral microsurgical foraminotomy in case of cervical nerve root compression. In: Grote, W.; Brock, M.; Clar, H. E.; Klinger, M.; Nau, H. E., eds. Advances in neurosurgery 8: surgery of cervical myelopathy. Berlin: Springer–Verlag; 1980: p. 287–290.

Bishara, S. N. The posterior operation in treatment of cervical spondylosis with myelopathy: a long-term follow-up study. J. Neurol. Neurosurg. Psychiat. 34:393–398; 1971.

Brain, W. R.; Northfield, D.; Wilkinson, M. The neurological manifestations of cervical spondylosis. Brain 75:187–225; 1952.

Carson, J.; Gumpert, J.; Jefferson, A. Diagnosis and treatment of thoracic intervertebral disc protrusions. J. Neurol. Neurosurg. Psychiat. 34:68–77; 1971.

Caspar, W.; Campbell, B.; Barbier, D. D.; Kretschmmer, R.; Gotfried, Y. The Caspar microsurgical discectomy and comparison with a conventional standard lumbar disc procedure. Neurosurgery 28:78–87; 1991.

Cloward, R. B. The anterior approach for removal of ruptured cervical discs. J. Neurosurg. 15:602–617; 1958.

Cloward, R. B. Lesions of the intervertebral disks and their treatment by interbody fusion methods. The painful disk. Clin. Orthoped. 27:51–77; 1963.

Collias, J. C.; Roberts, M. P. Posterior operations for cervical disc herniation and spondylotic myelopathy. In: Operative neurosurgical techniques, 2nd ed. Schmidek, H.; Sweet, W., eds. Orlando: Grune & Stratton: 1988: p. 1347–1358.

Connolly, E. S.; Seymour, R. J.; Adams, J. E. Clinical evaluation of anterior cervical fusion for degenerative cervical disc disease. J. Neurosurg. 23:431–437; 1965.

Coventry, M. B.; Ghormley, R. K.; Kernohan, J. W. The intervertebral disc: its microscopic anatomy and pathology. Part I. Anatomy, development, and physiology. J. Bone Joint Surg. 27A:105–112; 1945.

Crandall, P. H.; Batzdorf, U. Cervical spondylotic myelopathy. J. Neurosurg. 25:57–66; 1966.

Dandy, W. E. Loose cartilage from intervertebral disk simulating tumor of the spinal cord. Arch. Surg. 19:660–672; 1929.

Demirel, T.; Maksoud, M.; Braun, W. Therapeutic results following laminectomy and dural enlargement in cases of cervical myelopathy. In: Grote, W.; Brock, M.; Clar, H. E.; Klinger, M.; Nau, H. E., eds. Advances in neurosurgery 8: surgery of cervical myelopathy. Berlin: Springer–Verlag; 1980: p. 120–121.

Dereymaeker, A.; Mulier, J. La fusion vertebrale par voie ventrale dans la discopathie cervicale. Rev. Neurol. 99:597–616; 1958.

Dott, N. M. Skeletal traction and anterior decompression in the management of Pott's paraplegia. Edinb. Med. J. 54:620–627; 1947.

Ehni, G. Significance of the small lumbar spinal canal: cauda equina compression syndromes due to spondylosis. Part 1: Introduction. J. Neurosurg. 31:490–494; 1969.

Ehni, G. Cervical arthrosis: diseases of cervical motion segments. Chicago: Year Book Medical Publishers, Inc.; 1984: p. 11–13, 178–258.

Epstein, B. S.; Epstein, J. A.; Jones, M. D. Anatomicoradiological correlations in cervical spine discal disease and stenosis. Clin. Neurosurg. 25:148–173; 1978.

Fager, C. A. Results of adequate posterior decompression in the relief of spondylotic cervical myelopathy. J. Neurosurg. 38:684–692; 1973.

Fager, C. A. Rationale and techniques of posterior approaches to cervical disk lesions and spondylosis. Surg. Clin. North Am. 56(3):581–592; 1976.

Fager, C. A. Atlas of spinal surgery. Philadelphia: Lea and Febiger; 1989: p. 129–195.

Fischer, D.; Herrmann, H. D.; Loew, F. Chronic spondylogenic myelopathy: analysis of data of 62 patients operated on by the anterior approach. In: Grote, W.; Brock, M.; Clar, H. E.; Klinger, M.; Nau, H. E., eds. Advances in neurosurgery 8: surgery of cervical myelopathy. Berlin: Springer–Verlag; 1980: p. 112–114.

Franco, A.; Matricali, B. Anterior disc surgery in cases of cervical myelopathy. In: Grote, W.; Brock, M.; Clar, H. E.; Klinger, M.; Nau, H. E., eds. Advances in neurosurgery 8: surgery of cervical myelopathy. Berlin: Springer–Verlag; 1980: p. 50–52.

Frykholm, R. Cervical nerve root compression resulting from disc degeneration and root-sleeve fibrosis. A clinical investigation. Acta Chir. Scand. 160:(suppl.)1–149; 1951.

Frymoyer, J. W. Back pain and sciatica. N. Engl. J. Med. 318:291–300; 1988.

Galera, G. R.; Tovi, D. Anterior disc excision with interbody fusion in cervical spondylotic myelopathy and rhizopathy. J. Neurosurg. 28:305–310; 1968.

Gilsbach, J.; Eggert, H. R. Technique and results of the cervical discectomy. In: Grote, W.; Brock, M.; Clar, H. E.; Klinger, M.; Nau, H. E., eds. Advances in neurosurgery 8: surgery of cervical myelopathy. Berlin: Springer–Verlag; 1980: p. 284–286.

Giombini, S.; Solero, C. L. Considerations on 100 anterior cervical discectomies without fusion. In: Grote, W.; Brock, M.; Clar, H. E.; Klinger, M.; Nau, H. E., eds. Advances in neurosurgery 8: surgery of cervical myelopathy. Berlin: Springer–Verlag; 1980: p. 302–307.

Gooding, M. R. Pathogenesis of myelopathy in cervical spondylosis. Lancet 2:1180–1181; 1974.

Gooding, M. R.; Wilson, C. B.; Hoff, J. T. Experimental cervical myelopathy. Effects of ischemia and compression of the canine cervical spinal cord. J. Neurosurg. 43:9–17; 1975.

Gooding, M. R.; Wilson, C. B.; Hoff, J. T. Experimental cervical myelopathy: autoradiographic studies of spinal cord blood flow patterns. Surg. Neurol. 5:233–239; 1976.

Gorter, K. Influence of laminectomy on the course of cervical myelopathy. Acta. Neurochir. 33:265–281; 1976.

Gregorius, F. K.; Estrin, T.; Crandall, P. H. Cervical spondylotic radiculopathy and myelopathy: a long-term follow-up study. Arch. Neurol. 33:618–625; 1976.

Griffiths, I. R. Spinal cord blood flow in dogs: the effect of blood pressure. J. Neurol. Neurosurg. Psychiat. 36:914–920; 1973.

Gruninger, W.; Gruss, P. The influence of the Cloward fusion operation on the motility of the cervical spine. In: Grote, W.; Brock, M.; Clar, H. E.; Klinger, M.; Nau, H. E., eds. Advances in neurosurgery 8: surgery of cervical myelopathy. Berlin: Springer–Verlag; 1980: p. 291–296.

Gruss, P.; Gruninger, W.; Engelhardt. F. R. Differential therapy of cervical radiculopathy and myelopathy in degenerative changes of the cervical vertebral column. In: Grote, W.; Brock, M.; Clar, H. E.; Klinger, M.; Nau, H. E., eds. Advances in neurosurgery 8: surgery of cervical myelopathy. Berlin: Springer–Verlag; 1980: p. 130–137.

Guidetti, B.; Fortuna, A. Long-term results of surgical treatment of myelopathy due to cervical spondylosis. J. Neurosurg. 30:714–721; 1969.

Guidetti, B.; Fortuna, A.; Zamponi, C.; Lunardi, P. P. Cervical spondylosis myelopthy. In: Grote, W.; Brock, M.; Clar, H. E.; Klinger, M.; Nau, H. E., eds. Advances in neurosurgery 8: surgery of cervical myelopathy. Berlin: Springer–Verlag; 1980: p. 104–111.

Haft, H.; Shenkin, H. A. Surgical end results of cervical ridge and disk problems. J.A.M.A. 186:312–315; 1963.

Hakelius, A. Prognosis in sciatica: a clinical follow-up of surgical and non-surgical treatment. Acta Orthop. Scand. 129:(suppl.)1–76; 1970.

Hamel, E.; Frowein, R. A.; Karimi-Nejad, A. Classification and prognosis of cervical myelopathy. In: Grote, W.; Brock, M.; Clar, H. E.; Klinger, M.; Nau, H. E., eds. Advances in neurosurgery 8: surgery of cervical myelopathy. Berlin: Springer–Verlag; 1980: p. 115–119.

Hamer, J.; Kahl, M. Long-term results after decompressive laminectomy in cases of multisegmental cervical spinal stenosis. In: Grote, W.; Brock, M.; Clar, H. E.; Klinger, M.; Nau, H. E., eds. Advances in neurosurgery 8: surgery of cervical myelopathy. Berlin: Springer–Verlag; 1980: p. 95–99.

Hankinson, H. L.; Wilson, C. B. Use of the operating microscope in anterior cervical discectomy without fusion. J. Neurosurg. 43:452–456; 1975.

Hoff, J. T.; Wilson, C. B. The pathophysiology of cervical spondylotic radiculopathy and myelopathy. Clin. Neurosurg. 24:474–487; 1977.

Horal, J. The clinical appearance of low back disorders in the city of Gothenburg, Sweden. Acta. Orthop. Scand. 118:(suppl.)7–109; 1969.

Hughes, J. T.; Brownell, B. Spinal cord ischemia due to arteriosclerosis. Arch. Neurol. 15:189–202; 1966.

Hukuda, S.; Mochizuki, T.; Ogata, M.; Shichikawa, K.; Shimomura, Y. Operations for cervical spondylotic myelopathy: a comparison of the results of anterior and posterior procedures. J. Bone Joint Surg. 67B:609–615; 1985.

Hukuda, S.; Wilson, C. B. Experimental cervical myelopathy: effects of compression and ischemia on the canine cervical cord. J. Neurosurg. 37:631–652; 1972.

Hulme, A. The surgical approach to thoracic intervertebral disc protrusions. J. Neurol. Neurosurg. Psychiat. 23:133–137; 1960.

Hult, L. The Munkfors investigation. Acta Orthop. Scand. 16(suppl.):5–76; 1954.

Kaplan, L.; Kennedy, F. The effect of head posture on the manometrics of the cerebrospinal fluid in cervical lesions: a new diagnostic test. Brain 73:337–345; 1950.

Kelsey, J. L.; White, A. A., III. Epidemiology and impact of low-back pain. Spine 5:133–142; 1980.

Key, C. A. Guy's hospital report, vol. 3, p. 17. (Referenced in Wilkinson, M. Cervical spondylosis. Philadelphia: W. B. Saunders; 1967: p. 1–9.)

Knight, G. Neurosurgical treatment of cervical spondylosis. Proc. Royal Soc. Med. 57:165–168; 1964.

Kobrine, A. I.; Doyle, T. F.; Martins, A. N. Spinal cord blood flow in the rhesus monkey by the hydrogen clearance method. Surg. Neurol. 2:197–200; 1974.

Kugelgen, B.; Liebig, K.; Huk, W. Neurological approach to differential diagnosis and indication for surgery in chronic cervical myelopathy. In: Grote, W.; Brock, M.; Clar, H. E.; Klinger, M.; Nau, H. E., eds. Advances in neurosurgery 8: surgery of cervical myelopathy. Berlin: Springer–Verlag; 1980: p. 41–46.

Lees, F.; Aldren-Turner, J. W. Natural history and prognosis of cervical spondylosis. Br. Med. J. 2:1607–1610; 1963.

Liversedge, L. A.; Hutchinson, E. C.; Lyons, J. B. Cervical spondylosis simulating motor neuron disease. Lancet 2:652–659; 1953.

Long, D. M. Nonsurgical therapy for low back pain and sciatica. Clin. Neurosurg. 35:351–359; 1987.

Love, J. G. Protrusion of the intervertebral disk (fibrocartilage) into the spinal canal. Proc. Staff Meet. Mayo Clin. 11:529–535; 1936.

Lunsford, L. D.; Bissonette, D. J.; Jannetta, P. J.; Sheptak, P. E.; Zorub, D. S. Anterior surgery for cervical disc disease. Part 1: Treatment of lateral cervical disc herniation in 253 cases. J. Neurosurg. 53:1–11; 1980a.

Lunsford, L. D.; Bissonette, D. J.; Zorub, D. S. Anterior surgery for cervical disc disease. Part 2: Treatment of cervical spondylotic myelopathy in 32 cases. J. Neurosurg. 53:12–19; 1980b.

Macnab, I.; McCulloch, J. Backache, 2nd ed. Baltimore: Williams and Wilkins; 1990.

Martins, A. N. Anterior cervical discectomy with and without interbody bone graft. J. Neurosurg. 44:290–295; 1976.

Mayfield, F. H. Cervical spondylosis: observations based on surgical treatment of 400 patients. Postgrad. Med. 38:345–357; 1965.

McAllister, V. L.; Sage, M. R. The radiology of thoracic disc protrusion. Clin. Radiol. 27:291–299; 1976.

McRae, D. L. The significance of abnormalities of the cervical spine. Am. J. Roentgenol. 84:3–25; 1960.

Michele, A. A.; Krueger, F. J. Surgical approach to the vertebral body. J. Bone Joint Surg. 31A:873–878; 1949.

Mixter, W. J.; Barr, J. S. Rupture of the intervertebral disk with involvement of the spinal canal. N. Engl. J. Med. 211:210–215; 1934.

Murphey, F.; Simmons, J. C. H.; Brunson, B. Ruptured cervical discs, 1939 to 1972. Clin. Neurosurg. 20:9–17; 1973a.

Murphey, F.; Simmons, J. C. H.; Brunson, B. Surgical treatment of laterally ruptured cervical disc: review of 648 cases, 1939 to 1972. J. Neurosurg. 38:679–683; 1973b.

Murphy, M. G.; Gado, M. Anterior cervical discectomy without interbody bone graft. J. Neurosurg. 37:71–74; 1972.

Nachemson, A. L. The lumbar spine: an orthopedic challenge. Spine 1:59–71; 1976.

Northfield, D. W. C. Diagnosis and treatment of myelopathy due to cervical spondylosis. Br. Med. J. 2:1474–1477; 1955.

Nurick, S. The pathogenesis of the spinal cord disorder associated with cervical spondylosis. Brain 95:87–100; 1972a.

Nurick, S. The natural history and the results of surgical treatment of the spinal cord disorder associated with cervical spondylosis. Brain 95:101–108; 1972b.

Odom, G. L.; Finney, W.; Woodhall, B. Cervical disk lesions. J. A. M. A. 166:23–28; 1958.

Oppenheimer, A.; Turner, E. L. Discogenic disease of the cervical spine with segmental neuritis. Am. J. Roentgenol. 37:484–493; 1937.

Pallis, C.; Jones, A. M.; Spillane, J. D. Cervical spondylosis, incidence, and implications. Brain 77:274–289; 1954.

Patterson, R. H., Jr.; Arbit, E. A surgical approach through the pedicle to protruded thoracic discs. J. Neurosurg. 48:768–772; 1978.

Payne, E. E. The cervical spine and spondylosis. Neurochirurgia 1:178–196; 1959.

Payne, E. E.; Spillane, J. D. The cervical spine: an anatomico-pathological study of 70 specimens (using a special technique) with particular reference to the problems of cervical spondylosis. Brain 80:571–596; 1957.

Perot, P. L., Jr.; Munro, D. D. Transthoracic removal of midline thoracic disc protrusions causing spinal cord compression. J. Neurosurg. 31:452–458; 1969.

Peserico, L.; Uihlein, A.; Baker, G. S. Surgical treatment of cervical myelopathy associated with cervical spondylosis. Acta. Neurochir. 10:365–375; 1962.

Piepgras, D. G. Posterior decompression for myelopathy due to cervical spondylosis: laminectomy alone versus laminectomy with dentate ligament section. Clin. Neurosurg. 24:508–515; 1977.

Ransohoff, J.; Spencer, F.; Siew, F.; Gage, L., Jr. Transthoracic removal of thoracic disc: report of three cases. J. Neurosurg. 31:459–461; 1969.

Robertson, J. T. Anterior removal of cervical disc without fusion. Clin. Neurosurg. 20:259–261; 1973.

Robinson, R. A.; Smith, G. W. Anterolateral cervical disc removal and interbody fusion for cervical disc syndrome (Abst.). John Hopkins Hosp. Bull. 96:223–224; 1955.

Robinson, R. A.; Walker, A. E.; Ferlic, D. C.; Wiecking, D. K. The results of anterior interbody fusion of the cervical spine. J. Bone Joint Surg. 44A:1569–1587; 1962.

Roosen, K.; Grote, W. Late results of operative treatment of cervical myelopathy. In: Grote, W.; Brock, M.; Clar, H. E.; Klinger, M.; Nau, H. E., eds. Advances in neurosurgery 8: surgery of cervical myelopathy. Berlin: Springer–Verlag; 1980: p. 69–77.

Rosomoff, H. L.; Rosomoff, R. S. Nonsurgical aggressive treatment of lumbar spinal stenosis. Spine: State of the Art Reviews 1:383–400; 1987.

Schneider, R. C.; Cherry, G.; Pantek, H. The syndrome of acute central cervical spinal cord injury with special reference to the mechanisms involved in hyperextension injuries of cervical spine. J. Neurosurg. 11:546–577; 1954.

Scoville, W. B.; Dohrmann, G. J.; Corkill, G. Late results of cervical disc surgery. J. Neurosurg. 45:203–210; 1976.

Seres, J. L.; Newman, R. I. Failed back syndrome: nonsurgical treatment. Curr. Ther. Neurol. Surg. 2:291–293; 1989.

Sicard, J. A.; Forestier, J. Roentgenologic exploration of the central nervous system with iodized oil (lipiodol). Arch. Neurol. Psychiat. 16:420–434; 1926.

Simeone, F. A.; Rashbaum, R. Transthoracic disc excision. In: Operative neurosurgical techniques, 2nd ed. Schmidek, H.; Sweet, W.; eds. Orlando: Grune & Stratton, Inc.; 1988: p. 1367–1374.

Smith, G. W.; Robinson, R. A. The treatment of certain cervical-spine disorders by anterior removal of the intervertebral disc and interbody fusion. J. Bone Joint Surg. 40A:607–624; 1958.

Spangfort, E. V. The lumbar disk herniation: a computer-aided analysis of 2,504 operations. Acta. Orthop. Scand. 142(suppl.):1–95; 1972.

Spurling, R. G.; Bradford, F. K. Neurologic aspects of herniated nucleus pulposus. J.A.M.A. 113:2019–2022; 1939.

Spurling, R. G.; Scoville, W. B. Lateral rupture of the cervical intervertebral discs. A common cause of shoulder and arm pain. Surg. Gynecol. Obstet. 78:350–358; 1944.

Steinhausen, T. B.; Dungan, C. E.; Furst, J. B.; Plati, J. T.; Smith, S. W.; Darling, A. P.; Wolcott, E. C.; Warren, S. L.; Strain, W. H. Iodinated organic compounds as contrast media for radiographic diagnosis: III. Experimental and clinical myelography and iodophenylundecylate (Pantopaque). Radiology 43:230–235; 1944.

Stookey, B. Compression of the spinal cord due to ventral extradural cervical chondromas: diagnosis and surgical treatment. Arch. Neurol. Psychiat. 20:275–291; 1928.

Stoops, W. L.; King, R. B. Neural complications of cervical spondylosis: their response to laminectomy and foramenotomy. J. Neurosurg. 19:986–999; 1962.

Sugar, O. Spinal cord malfunction after anterior cervical discectomy. Surg. Neurol. 15:4–8; 1981.

Symon, L.; Lavender, P. The surgical treatment of cervical spondylotic myelopathy. Neurology 17:117–127; 1967.

Symonds, C. P. The interrelation of trauma and cervical spondylosis in compression of the cervical cord. Lancet 1:451–454; 1953.

Taylor, A. R. Vascular factors in the myelopathy associated with cervical spondylosis. Neurology 14:62–68; 1964.

Taylor, J.; Collier, J. The occurrence of optic neuritis in lesions of the spinal cord: Injury, tumor, myelitis. Brain 24:533–552; 1901.

Teng, P. Spondylosis of the cervical spine with compression of the spinal cord and nerve roots. J. Bone Joint Surg. 42A:392–407; 1960.

Thal, H. U.; Miltz, H.; Bock, W. J.; Kuhlendahl, H. Long-term results after operative treatment of cervical myelopathy by laminectomy. In: Grote, W.; Brock, M.; Clar, H. E.; Klinger, M.; Nau, H. E., eds. Advances in neurosurgery 8: surgery of cervical myelopathy. Berlin: Springer–Verlag; 1980: p. 90–94.

Thomalske, G.; Braunsdorf, W.; Galow, W. Ventral or dorsal surgical approach in chronic cervical myelopathies? In: Grote, W.; Brock, M.; Clar, H. E.; Klinger, M.; Nau, H. E., eds. Advances in neurosurgery 8: surgery of cervical myelopathy. Berlin: Springer–Verlag; 1980: p. 59–63.

Turnbull, I. M.; Breig, A.; Hassler, O. Blood supply of cervical spinal cord in man. J. Neurosurg. 24:951–965; 1966.

Van Alphen, H. A. M.; Braakman, R.; Bezemer, P. D.; Broere, G.; Berfelo, M. W. Chemonucleolysis versus discectomy: a randomized multicenter trial. J. Neurosurg. 70:869–875; 1989.

Verbiest, H. Sur certaines formes rares de compression de la queue de cheval. In: Maloine, ed. Hommage a clovis vincent. Paris; Maloine: 1949: p. 163–174.

Verbiest, H. A lateral approach to the cervical spine: technique and indications. J. Neurosurg. 28:191–203; 1968.

Verbiest, H. La chirurgie anterieure et laterale du radis cervical. Neuro-chirurgie 16:154–158; 1970.

Verbiest, H. From anterior to lateral operations on the cervical spine. Neurosurg. Rev. 1:47–67; 1978.

Von Bechterew, V. Steifigkeit der Wirbelsaule und ihre Verkrummung als besondere Erkrankungsform. Neurologisches Zentralblatt 12:426–434; 1893.

Weber, H. Lumbar disc herniation: a controlled prospective study with 10 years of observation. Spine 8:131–140; 1983.

Weinstein, P. R.; Ehni, G.; Wilson, C. B. Clinical features of lumbar spondylosis and stenosis. In: Lumbar spondylosis: diagnosis, management, and surgical treatment. Chicago: Year Book Medical Publishers, Inc.; 1977: p. 115–133.

Wilkinson, M. Historical introduction. In: Brain, W. R.; Wilkinson, M., eds. Cervical spondylosis and other disorders of the cervical spine. London: Heinemann; 1967: p. 1–9.

Wilkinson, M. The morbid anatomy of cervical spondylosis and myelopathy. Brain 83:589–616; 1960.

Wilkinson, M. Cervical spondylosis: its early diagnosis and treatment. Philadelphia: W. B. Saunders Co.; 1971: p. 1–182.

Wilson, C. B. Significance of the small lumbar spinal canal: Cauda equina compression syndromes due to spondylosis. Part 3: Intermittent claudication. J. Neurosurg. 31:499–506; 1969.

Yuhl, E. T.; Hanna, D.; Rasmussen, T.; Richter, R. B. Diagnosis and surgical therapy of chronic midline cervical disc protrusions. Neurology 5:494–509; 1955.

25

Disorders of Consciousness

DAVID E. LEVY

Like dementia, disorders of consciousness should be considered a syndrome with many causes rather than a specific disease entity. For this reason, the chance of recovery from the condition is governed by etiological considerations and by demographic, clinical, and laboratory factors; the organization of this chapter thus reflects the contributing influence of underlying causes. Three general disorders of consciousness are discussed in this chapter: (1) nontraumatic coma, (2) the chronic vegetative state, and (3) syncope. Also considered is the related topic of transient global amnesia.

Nontraumatic Coma

For centuries, coma has been viewed as a near-death-state with a correspondingly poor prognosis. Because of the perceived hopelessness of the condition, little effort was made to distinguish its causes or to differentiate patients likely to recover from those with a poor prognosis. In recent decades, however, advances have been made in the diagnosis of different causes of coma (Fisher 1969; Plum and Posner 1966), and several groups (Levy et al. 1981; Snyder and

Tabbaa 1988; Snyder et al. 1981) have carefully analyzed factors governing its prognosis.

Overall accuracy may not be the best criterion by which to judge papers about prognosis of coma. The reason is that not all errors of prediction have equivalent consequences. Inappropriately relying on a predicted bleak prognosis could lead to premature withdrawal of care from a patient who might recover; a known low rate for that particular error is the best protection. Because the consequence of acting on an incorrect prediction of poor prognosis is much more serious than the converse error from an incorrect prediction of good prognosis, the number and kind of errors made by the prediction rule are more important than overall accuracy.

One truism that is validated by studies is the belief that the longer the coma lasts, the poorer are the chances of recovery. Bell and Hodgson (1974) initially reported that recovery was exceptional if coma after cardiac arrest lasted at least 3 days. In the study by Levy and coworkers (1981), only one of 38 patients in nontraumatic coma for 1 week ever achieved a good recovery.

The cause is another important consideration governing prognosis. Nearly all patients comatose from exogenous poisoning (e.g., sedative drugs) recover fully if they receive prompt supportive care. Patients in traumatic coma generally do better than those in nontraumatic coma. Jennett and colleagues (1979) reported that 39% of 1000 patients in traumatic coma for at least 6 hours recovered independent function at 6 months, whereas Levy and associates (1981) found comparable recovery at any time within the first year in only 16% of 500 patients in nontraumatic coma. Among patients in nontraumatic coma not caused by drugs (Levy et al. 1981), cerebrovascular disease (cerebral infarct, parenchymal hemorrhage, or subarachnoid hemorrhage) is associated with the poorest prognosis (9% achieving independent function), cardiopulmonary arrest has a better prognosis (12%), and endogenous metabolic factors, such as hepatic encephalopathy, have the best (33%).

Demographic factors also play a role, although their influence is greatest in traumatic coma, in which young age is associated with improved chances of recovery. In the study by Jennett and coworkers cited previously (1979), 78% of patients older than 60 years in traumatic coma either died or were vegetative, as opposed to only 29% of those younger than 20 years. For patients with nontraumatic coma, age generally is not a powerful prognostic indicator, but this may be because so many of these patients with underlying systemic conditions are old, and thus effects of age are not apparent. Clinicians should exercise particular caution discussing the prognosis of young patients.

Clinical signs are powerful predictors of outcome from nontraumatic coma. Abnormalities of multiple brain stem signs identify groups with particularly poor outcomes. In the study by Levy and coworkers (1981), at the end of the first day, the absence of two or more of the following signs identified one fourth of the survivors, only 2% of whom ever regained consciousness: pupillary reaction to light, eye blinking in response to corneal stimulation, eye movements in response to caloric stimulation, and motor responses to pain identified. In contrast, a small group displayed recognizable speech (even isolated words) at the same time; 67% regained independent function, which is nearly a 30-fold better chance of recovery.

Reliance on a single clinical sign can mislead the physician. Willoughby and Leach (1974) reported that the absence of any motor response to pain within the first few hours of coma caused by cardiac arrest reliably identified patients with no chance of regaining independence, but subsequent investigators have presented evidence that contradicts this simplistic rule. Similarly, although absent pupillary reaction to light is an ominous clinical sign, recovery in such patients has been reported (Price and Knill-Jones 1979).

One of the problems of relying on single signs is the possibility of extraneous factors confounding the examination. For example, anticholinergic drugs given during a cardiac arrest can interfere with pupillary reactivity. Paralyzing agents (e.g., pancuronium) that are sometimes used after a cardiac arrest in patients combating mechanical ventilation can interfere with motor responses. Therefore, it is wise to use the full clinical picture in estimating prognosis and to be aware of the possible effects of prior illnesses or interfering medications.

Despite a long search for an ideal laboratory technique to distinguish patients with different prognoses, none has yet been found. Most commonly recommended is electroencephalography (EEG). Scolio-Lavizzari and Bassetti (1987), for example, report that comatose patients with the best of five EEG grades (alpha and theta–delta activity) have a very good prognosis, usually with full recovery, that the intermediate grades of dominant theta–delta with or without detectable alpha are associated with an intermediate prognosis, and that the poorest grades (low-voltage delta sometimes with brief isoelectric periods, nonreactive alpha, periodic phenomena, or isoelectricity) have the worst prognosis.

Considerable interest has focused on the significance of clinical or EEG evidence of paroxysmal activity. Levy and associates (1981) were unable to document an association of clinical seizures with poor outcome; by contrast, Snyder and colleagues (1980) report a survival rate of 32% in comatose patients with seizures after cardiac arrest, as opposed to 43% in patients without seizures. Examining the problem in greater detail, Krumholz and coworkers (1988) report that neither seizures nor myoclonus alone are associated with poor outcome, but status epilepticus and status myoclonus did predict poor outcome in coma after cardiac arrest.

Thought initially to be better prognostically than EEG, evoked potentials have not lived up to their initial promise. Nonetheless, in some circumstances they may add prognostic information to clinical and demographic data. Ganes and Lundar (1988) concluded that in comatose patients, the absence of spontaneous EEG, somatosensory evoked potentials (SSEP), brain stem auditory evoked potentials (BAEP), and visual evoked potentials (VEP) was invariably associated with brain death. DeLecluse and coworkers (1987) partially contradict this, however, because they found that although absent SSEPs in patients with nontraumatic coma predicted no recovery, this was not true in traumatic coma, in which five of 10 patients with absent SSEPs achieved a moderate recovery. Similar discrepancies between SSEP and outcome were noted in brain dead patients by Grigg and associates (1987), who reported some preserved EEG activity in nearly 20% of patients otherwise fulfilling clinical criteria for brain death, and by Belsh and Chokroverty (1987), who noted persistent SSEPs in 10 patients otherwise meeting clinical criteria for brain death.

Longstreth and coworkers (1983) have reported that comatose patients after cardiac arrest who had high initial blood glucose levels did worse than those with lower values. The mean glucose in patients who never awakened was 332 mg/dl, whereas it was only 249 mg/dl in those who awakened. Using a discriminant analysis model, these authors found that a rule incorporating motor response to pain, pupillary light reflex, spontaneous eye movements, and initial blood glucose correctly predicted whether patients would awaken in 81% of patients who were not missing any of the information. Sixteen percent of patients predicted not to awaken, however, did awaken, a potentially serious error rate.

Although there is no single comprehensive prognostic guide that uses all available clinical and laboratory data, the physician can incorporate results of these and other studies to guide families in decision-making for patients with nontraumatic coma. From these studies, it is possible within the first few days of coma to identify subgroups of patients with virtually no chance of meaningful recovery or with more than 50% chance of an independent life.

Chronic Vegetative State

Unlike acute coma, large databases do not exist for the chronic vegetative state. The most extensive published study was based on a survey of hospitals within an entire Japanese prefecture of approximately 1.5 million (Higashi et al. 1977; 1981); 37 vegetative patients were identified for a prevalence of 2.5 per 100,000. An additional 73 patients from other sources brought the total to 110 patients. All were vegetative for at least 3 months, one half had been vegetative for less than 1 year, and four patients had been vegetative more than 10 years. After 5 years of follow-up, only two had regained the ability to communicate and move in a wheelchair, and three responded at times to simple commands. One of the two patients who made the best recovery had remained vegetative for 3 years after a subarachnoid hemorrhage and then gradually recovered; the other had been vegetative for 8 months after cerebral anoxia. Neither lived independently, and both remained physically and mentally disabled. This study suggests partial recovery in 2% to 3% of patients vegetative for at least 3 months, but because nearly one third of the patients showed some "emotional expression" when first examined, it is possible that some might not have been vegetative.

Although the Japanese combined patients vegetative from traumatic and nontraumatic causes, there are two published experiences that permit differentiation of prognosis in accordance with the cause. As with coma, it appears that the vegetative state has a better prognosis when caused by trauma than by cardiac arrest or other medical conditions. Among 23 vegetative patients after nontraumatic coma who had survived 1 month (Levy et al. 1981), none regained independent function, and only three regained consciousness. Two of these three died before the end of the first year, and the third was left severely disabled, capable only of uttering individual, isolated words. Although this study shows a potential for partial recovery after 1 month in a vegetative state, independent function did not occur.

By contrast, independent function was reported in a higher proportion of 140 patients vegetative for 1 month after trauma (Braakman et al. 1988). Fifty-nine (42%) regained consciousness within the first year, most (49) within the first 3 months. However, only 14 patients regained independent function, and none recovered completely. Thus, of 140 patients vegetative at 1 month, none achieved a complete recovery, 10% regained independent function, and 42% regained at least a fragment of conscious behavior. No patient vegetative for 3 months regained independent function.

Older age reduces the chances of recovery from the vegetative state when caused by trauma, but not when caused by other sources. In the experience of Braakman and colleagues (1988), none of 41 patients older than 40 years, and vegetative for 1 month ever regained independence. By contrast, such recovery was achieved by 9% of those between ages 20 and 40 and by 21% of those younger than 20 years. No effect of age emerged when the vegetative state was not caused by trauma, but again, patient numbers and age range are more limited.

Clinical discriminators have not been thoroughly studied in the vegetative state. Nonetheless, trauma patients do better if their pupils react or if they have spontaneous conjugate eye movements than if either or both of these are missing (Braakman et al. 1988). The Japanese study (Higashi et al. 1977) distinguished three groups of patients on the basis of the presence of "emotional reactivity" and sleep–wake cycles (best), sleep–wake cycles alone (intermediate), and neither (worst). Mortality was greatest in the group lacking both sleep–wake cycles and emotional responses.

The press contains occasional reports of people miraculously recovering after prolonged unconsciousness. Published reports of such recoveries are few, however, and must be placed in the context of the number of patients from which they are drawn. The authors know of fewer than 10 published case reports of late "recovery" during a 15-year period. In all but two cases (Arts et al. 1985; Shuttleworth 1983), recovery was only to a level of severe disability, and even the two exceptions were not living at their previous level. No good epidemiological studies establish the reservoir from which these patients might be drawn. The Japanese experience extrapolates to 5000 to 10,000 vegetative patients in American hospitals, but we know that most vegetative patients in the United States are cared for outside of acute hospitals (e.g., nursing homes, head injury centers, and private homes). The author conducted an informal survey of 1989 attendees at a conference dealing with prolonged unconsciousness and found that 50 attendees had an average of four new vegetative patients during the previous year, a figure that extrapolates to 40,000 among the more than 10,000 neurologists in the United States. If the median survival is 1 year, then in 15 years, there would be 300,000 vegetative patients in the United States (not counting those seen by neurosurgeons, internists, and others), from which these 10 published "recoveries" have been derived, which is a very low recovery rate of only 3.3 per 100,000 vegetative patients.

Clearly, more work is needed, especially

in defining the reservoir of vegetative patients, but the likelihood of recovery once the condition has lasted more than 4 to 6 months appears to be minimal, especially for those not related to trauma, regardless of age or clinical signs.

Syncope

For syncope and transient global amnesia, the prediction that is needed does not pertain to recovery from the current episode, which is assured. Rather, the problem is in gauging the likelihood of subsequent serious medical problems, including heart disease and death.

In one study (Houdent et al. 1988), the 1-year mortality of patients older than age 65 with syncope was nearly 20%. Several studies have shown that syncope patients with clinical evidence of prior cardiovascular or cerebrovascular disease have a significantly increased risk of sudden death within 1 to 2 years of the syncopal episode. Electrophysiological study of one group of 70 patients (Bass et al. 1988) identified more than half (37) with abnormal monitoring—ventricular tachycardia (31), supraventricular tachycardia (3), and abnormal conduction (3). Sudden death within 3 years occurred much more frequently in this group (48%) than in the group with normal electrophysiology (9%).

Transient Global Amnesia

Unlike syncope, the major concern in transient global amnesia is that of future cerebrovascular rather than cardiovascular disease. Happily, this is not common. In three recent studies of more than 400 patients (Colombo and Scarpa 1988; Hinge et al. 1988; Miller et al. 1987), stroke risk was not increased in patients with transient global amnesia, although approximately one fourth of patients had recurrent episodes. Another study (Palmer 1986) divided patients into three groups based on (1) history of migraine, (2) associated neurological deficits during the episode risk factors for cerebro-

vascular disease, and (3) no underlying cause. The first group had an increased risk of recurrence, and the second group had a higher incidence of stroke or dementia, but this group also was older.

Summary

All of these conditions are serious, especially coma and the vegetative state. They require understanding and decisions of great importance. The increasing availability of information regarding the prognosis of these conditions is essential in guiding individual physicians and families in some decisions. Moreover, understanding the factors that govern prognosis will facilitate a more efficient design of future treatment trials intended to improve outcome.

References

Arts, W. F. M.; Van Dongen, H. R.; Van Hof-Van Duin, J.; Lammens, E. Unexpected improvement after prolonged posttraumatic vegetative state. J. Neurol. Neurosurg. Psychiatry 48:1300–1303; 1985.

Bass, E. B.; Eison, J. J.; Fogoros, R. N.; Peterson, J.; Arena, V. C.; Kapoor, W. N. Long-term prognosis of patients undergoing electrophysiologic studies for syncope of unknown origin. Am. J. Cardiol. 62:1186–1191; 1988.

Bell, J. A.; Hodgson, H. J. F. Coma after cardiac arrest. Brain 97:361–372; 1974.

Belsh, J. M.; Chokroverty, S. Short-latency somatosensory evoked potentials in brain-dead patients. Electroencephalogr. Clin. Neurophysiol. 68:75–78; 1987.

Braakman, R.; Jennett, W. B.; Minderhoud, J. M. Prognosis of the posttraumatic vegetative state. Acta Neurochir. 95:49–52; 1988.

Colombo, A.; Scarpa, M. Transient global amnesia: pathogenesis and prognosis. Eur. Neurol. 28:111–114; 1988.

DeLecluse, F.; Zegers de Beyl, D.; Brunko, E. The bilateral absence of cortical somatosensory evoked potentials has a different importance in traumatic versus anoxic coma (abst). Ann. Neurol. 22:142; 1987.

Fisher, C. M. The neurological examination of the comatose patient. Acta Neurol. Scand. 45(suppl. 36):4–56; 1969.

Ganes, T.; Lundar, T. EEG and evoked potentials in comatose patients with severe brain damage. Electroenceph. Clin. Neurophysiol. 69:6–13; 1988.

Grigg, M. M.; Kelly, M. A.; Celesia, G. G.; Ghobrial, M. W.; Ross, E. R. Electroencephalographic activity after brain death. Arch. Neurol. 44:948–954; 1987.

Higashi, K.; Hatano, M.; Abiko, S.; Ihara, K.; Katayama, S.; Wakuta, Y.; Okamura, T.; Yamashita, T. Five-year follow-up study of patients with persistent vegetative state. J. Neurol. Neurosurg. Psychiatr. 44:552–554; 1981.

Higashi, K.; Sakata, Y.; Hatano, M.; Abiko, S.; Ihara, K.; Katayama, S.; Wakuta, Y.; Okamura, T.; Ueda, H.; Zenke, M.; Aoki, H. Epidemiological studies on patients with a persistent vegetative state. J. Neurol. Neurosurg. Psychiatr. 40:876–885; 1977.

Hinge, H. H.; Jensen, T. S.; Kjaer, M.; Marquardsen, J.; de Fine Olivarius, B. The prognosis of transient global amnesia: results of a multicenter study. Arch. Neurol. 43:673–676; 1988.

Houdent, C.; Morcamp, D.; Séréni, D.; Conri, C.; Colvez, A.; Delegove, H.; Vinceneux, P.; Pibarot, M. L.; 1-year prognosis of syncope and brief loss of consciousness in patients over 65: a multicenter study of 188 cases. Presse Med. 17:626–629; 1988.

Jennett, B.; Teasdale, G.; Braakman, R.; Minderhoud, J.; Heiden, J.; Kurze, T. Prognosis of patients with severe head injury. Neurosurgery 4:283–289; 1979.

Krumholz, A.; Stern, B. J.; Weiss, H. D. Outcome from coma after cardiopulmonary resuscitation: relation to seizures and myoclonus. Neurology 38:401–405; 1988.

Levy, D. E.; Bates, D.; Caronna, J. J.; Cartlidge, N. E. F.; Knill-Jones, R. P.; Lapinski, R. H.; Singer, B. H.; Shaw, D. A.; Plum, F. Prognosis in nontraumatic coma. Ann. Intern. Med. 94:293–301; 1981.

Longstreth, W. T., Jr.; Diehr, P.; Inui, T. S. Prediction of awakening after out-of-hospital cardiac arrest. N. Engl. J. Med. 308:1378–1382; 1983.

Miller, J. W.; Petersen, R. C.; Metter, E. J.; Millikan, C. H.; Yanagihara, T. Transient global amnesia: clinical characteristics and prognosis. Neurology 37:733–737; 1987.

Palmer, E. P. Transient global amnesia and the amnestic syndrome. Med. Clin. North Am. 70:1361–1374; 1986.

Plum, F.; Posner, J. B. Diagnosis of stupor and coma. Philadelphia: F. A. Davis Company; 1966.

Price, D. J.; Knill-Jones, R. The prediction of outcome of patients admitted following head injury in coma with bilateral fixed pupils. Acta Neurochir. 28(suppl.):179–182; 1979.

Scolio-Lavizzari, G.; Bassetti, C. Prognostic value of EEG in post-anoxic coma after cardiac arrest. Eur. Neurol. 26:161–170; 1987.

Shuttleworth, E. Recovery to social and economic independence from prolonged post-anoxic vegetative state. Neurology 33:372–374; 1983.

Snyder, B. D.; Gumnit, R. J.; Leppik, I. E.; Hauser, W. A.; Loewenson, R. B.; Ramirez-Lassepas, M. Neurologic prognosis after cardiopulmonary arrest: IV. brainstem reflexes. Neurology 31:1092–1097; 1981.

Snyder, B. D.; Hauser, W. A.; Loewenson, R. B.; Leppik, I. E.; Ramirez-Lassepas, M.; Gumnit, R. J. Neurologic prognosis after cardiopulmonary arrest: III. seizure activity. Neurology 30:1292–1297; 1980.

Snyder, B. D.; Tabbaa, M. A. Assessment and treatment of neurological dysfunction after cardiac arrest. Stroke 19:269–273; 1988.

Willoughby, J. O.; Leach, B. G. Relation of neurological findings after cardiac arrest to outcome. Br. Med. J. 3:437–439; 1974.

26

Neuromuscular Junction and Muscle Disease

ALAN W. MARTIN AND STEVEN P. RINGEL

The prognosis of selected neuromuscular disorders has received increasing attention in the last decade by investigators designing therapeutic trials. The CIDD group, a multiuniversity trial group studying Duchenne dystrophy, has emphasized that the natural history of a disease must be clearly established using a standard protocol before any positive therapeutic effect can be demonstrated (Brooke et al. 1983). In Duchenne dystrophy, a relatively homogenous disorder with a linear progression of weakness, this group has demonstrated significant variation in the rate at which weakness develops. The situation is much more complicated for myotonic dystrophy and facioscapulohumeral dystrophy. Within the same family, one person may be severely weakened while another may have only minimal evidence of disease throughout his or her life. This extensive variability is highlighted in this chapter, so that the reader will develop an understanding of the spectrum of each neuromuscular illness.

The emphasis of this chapter is on the more common, chronic disorders that effect the neuromuscular junction and voluntary skeletal muscles, including myasthenia gravis, inflammatory myopathies, muscular dystrophies, congenital myopathies and representative metabolic disorders of muscle. Although the structural and enzymatic defects of many rare metabolic diseases of muscle are being characterized increasingly, their infrequent occurrence precludes an accurate description of natural history in this chapter.

Myasthenia Gravis

Myasthenia gravis involves fluctuating weakness in cranial and skeletal musculature. Numerous immune abnormalities may account for the disorder; in more than 50% of cases, immunoglobulin G (IgG) autoantibodies are directed against the acetylcholine receptor (Engel 1986A) and there is thymic hyperplasia (Jaretzki et al. 1988; Olanow 1987). Diagnosis is confirmed by the presence of serum acetylcholine receptor antibodies, a decrement in response to the compound muscle action potential to repetitive stimulation, increased jitter with single-fiber electromyography, or clinical improvement with intravenous endrophonium (Tensilon). The disease is associated with thymoma in

10% of cases, thyroid disease in 13%, and other autoimmune diseases in at least 3% (Genkins and Kornfeld 1987). Myasthenia is associated with HLA types A1, B7, and DRw3 in patients younger than 40 years and types A3, B7, and DRw3 in those older than 40 years (Compston et al. 1980).

Myasthenia may occur at any age and is more prevalent in women (57%) particularly if the onset is younger than 40 years of age (Grob et al. 1987). After 40 years of age, the disease characteristics shift to a male predominance, increased severity, and different HLA correlates (Compston et al. 1980).

The presenting symptoms of myasthenia gravis are ocular (ptosis or ophthalmoplegia) in approximately 50% of patients. Almost all patients develop ocular symptoms at some point in the course of their disease (Grob et al. 1987). Few patients (14%) have symptoms that remain localized to the eyes; most develop generalized symptoms in the first year after diagnosis (Donaldson et al. 1989; Grob et al. 1987). If weakness remains localized to the eyes for more than 2 years, few of the patients (10%) progress to generalized myasthenia (Grob et al. 1987). Patients with ocular myasthenia have a positive Tensilon test in 95% of cases, a decremental response of limb muscles to repetitive stimulation in 50%, and a detectable acetylcholine receptor antibody in 50% (Evoli et al. 1988; Tindall 1981). Uncontrolled series suggest that these patients will respond to acetylcholinesterase drugs, immunosuppressive agents, and thymectomy (Evoli et al. 1988; Schumm et al. 1985). Most patients with ocular myasthenia improve with treatment, but complete remission rates are only 8% to 14% (Evoli et al. 1988; Grob et al. 1987).

Generalized myasthenia gravis reaches maximum severity during the first 3 to 5 years, with improvement or remission thereafter in 95% of patients (Grob et al. 1987). No good predictors of outcome have been identified (Donaldson et al. 1989). Grob and coworkers (1987) found that 84% of patients reached their first episode of maximum severity within 3 years of diagnosis. This pattern was evident even in an early series in which a small minority of patients had undergone thymectomy or received immunosuppressive medications.

The reported disease remission rate varies from 12% to 81% (Grob et al. 1987; Jaretzki et al. 1988). This may reflect differences in patient selection, treatment regimens, and definitions of remission. Grob and colleagues (1987) found a 12% rate of remission in 220 patients before 1953. They found that this low percentage remained relatively unchanged with the advent of thymectomy and more aggressive immunosuppression. Jaretzki and associates (1988) followed 72 myasthenic patients after maximal thymectomy and found a 45% drug-free remission rate. They estimated an 81% remission rate at 89 months by life-table analysis. Both studies suggest that patients treated with thymectomy or immunosuppression or both have a better outcome than patients treated before 1953 who received neither therapy (Grob et al. 1987). Between the extremes of these two studies, most data suggest that 65% to 85% of treated patients will remit or improve (Rowland 1980).

Assessing the effects of individual therapies is difficult because of the paucity of prospectively controlled trials. Most retrospective series are poorly controlled and individual patients receive multiple treatments. Furthermore, improved management of patients with respiratory failure has played a central role in reducing the high mortality rates of the early 1900s (Grob et al. 1987).

Anticholinesterase medicines improve the strength of some myasthenics, but they are unable to control the symptoms of many other patients with more severe disease. Thymectomy has become standard therapy in many centers. Although many poorly controlled series and anecdotal reports attest to its effectiveness, the benefit from thymectomy has not been demonstrated in a well-controlled study (Rowland 1980). Olanow and coworkers (1987) avoided the use of corticosteroids, performing thymectomy following symptom reduction with plasmapheresis. After 3 years, 80% of patients had an excellent response with 55% off all medicines and 38% never receiving any medical therapy. Others have argued that thymectomy offers no additional benefit to other therapies (McQuillen and Leone 1977).

Corticosteroids reduce the severity of symptoms or induce remission in most patients with generalized myasthenia gravis. Johns (1987) found 80% of patients treated with prednisone were in remission or had minimal symptoms. There was no difference in clinical response or medication requirements among patients who did or did not receive thymectomy. Up to 50% of myasthenic patients on long-term steroids, however, incur significant side effects (Donaldson et al. 1989). Clinical improvement has been observed in patients receiving multiple immunosuppressive medications, including azathioprine, cyclophosphamide, and cyclosporin A (Niaken et al. 1986; Tindall et al. 1987). Plasmapheresis results in short-term improvement lasting 3 to 6 weeks in most patients (Seybold 1987).

Congenital myasthenia results from genetic defects causing structural or biochemical abnormalities at the neuromuscular junction. Patients often develop symptoms at a very young age. Electrophysiological tests confirm a neuromuscular junction defect, but there are no antibodies to acetylcholine receptor. Patients have little disease progression or remission, but their symptoms may fluctuate in severity. Congenital myasthenia does not respond to corticosteroids, but some patients improve with anticholinesterase drugs (Engel 1986).

Transient neonatal myasthenia occurs in approximately 12% of infants born to mothers who have myasthenia gravis, regardless of the mother's disease severity. These infants typically develop poor feeding, weakness, and respiratory difficulty during the first day of life. Passively transferred maternal acetylcholine receptor antibodies and possibly independently produced antibodies are detectable in the infant's sera. Symptoms last an average of 18 days but may persist up to 47 days (Lefvert and Osterman 1983).

Dermatomyositis and Polymyositis

Dermatomyositis and polymyositis are inflammatory diseases of muscle of unknown etiology. Immune responses involving cell-mediated and humoral mechanisms play a central role in muscle injury.

Although the disease may begin at any age, it most frequently occurs between ages 5 and 14 or 45 and 64 years (Bohan et al. 1977; Mastaglia and Ojeda 1985). Most estimates of disease incidence vary from two to five per 1 million population, and polymyositis is more common than dermatomyositis (Medsger 1970; Rose and Walton 1966). Women outnumber men in most series (Bohan et al. 1977; DeVere and Bradley 1975). The clinical symptoms typically are symmetrical proximal muscle weakness with occasional myalgias and dysphagia. Facial, extraocular, and distal skeletal muscles rarely are involved. Greater than 90% of patients with dermatomyositis have a rash at the time of presentation (Bohan et al. 1977). The symptoms of myositis progress insidiously in most patients for weeks to months, although spontaneous remissions have been described (Rosenberg et al. 1988). Rarely, the disease may present acutely over the course of 1 to 3 weeks with profound weakness, severe myalgias, dysphagia, and occasionally myoglobinuria or respiratory insufficiency.

Serum creatine kinase (CPK) is elevated in 66% to 95% of active or untreated cases of myositis (Bohan et al. 1977; DeVere and Bradley 1975). Electromyography demonstrates short duration, small amplitude polyphasic motor unit potentials in 90% of cases, fibrillation potentials and positive sharp waves in 74%, and complex repetitive discharges in 38% (Bohan et al. 1977). Abnormalities occasionally are isolated to paraspinous muscles. The spontaneous activity often decreases after effective movement (Kimura 1983).

During the course of polymyositis and

dermatomyositis, organs other than skeletal muscle may be involved. A review of 10 series demonstrated an average of 21% of patients with another definable connective tissue disorder (Rosenberg et al. 1988). The most common overlapping diseases were systemic lupus erythematosus, progressive systematic sclerosis, Sjögren's syndrome, rheumatoid arthritis, and mixed connective tissue disease (Isenberg 1984).

A characteristic rash with lilac discoloration of the eyelids, facial and neck erythema, and erythematosus, atrophic, scaly changes over extensor surfaces of joints occurs in dermatomyositis. The rash precedes muscle weakness in at least half of patients. Changes in the rash do not correlate with disease activity (Barwick and Walton 1963; Bohan et al. 1977; Bohan 1988; Eaton 1954; Logan et al. 1966).

Many types of malignancy have been associated with the inflammatory myopathies. The data from several series of patients suggest that the frequency of malignancy is 9%, but recent studies have shown lower percentages (Henricksson and Sandstedt 1982; Rosenberg and Ringel 1988a; Rowland et al. 1977). Two thirds of patients with malignancies have dermatomyositis, and almost all are older than 40 years (Bohan 1988; Talbott 1977). Malignancy may occur before, following, or at the time of diagnosis of myositis (Cailen 1982).

Symptomatic cardiac involvement is uncommon (Bohan et al. 1977); however, sophisticated cardiac assessment reveals abnormalities in more than 70% of patients (Denbow et al. 1979; Gottdiener et al. 1978). Creatine kinase (CPK)-MB fractions should be interpreted with caution because they may be elevated in the absence of cardiac pathology (Adornato and Engel 1977; Larca et al. 1981).

Pulmonary function testing can determine interstitial lung disease with restrictive characteristics in more than 30% of patients, but symptomatic disease with radiographic changes occurs in less than 10% (Dickey and Myers 1984). The anti-Jo-1 antibody is a useful marker for interstitial lung disease in inflammatory myopathies (Yoshida et al. 1983). Interstitial lung disease is thought to have a poor progress, but it may respond to immunosuppression (Bohan 1988). Diagnostic confusion of interstitial lung disease may occur with aspiration pneumonia, opportunistic infections, and methotrexate pneumonitis (Rosenberg et al. 1988). Dysphagia from pharyngeal or esophageal dysfunction increases the risk of aspiration and pneumonia (Bohan 1988; Kagen et al. 1985). Respiratory insufficiency secondary to fulminant muscle weakness occurs occasionally and improves if the underlying myositis responds to immunosuppression (Dickey and Myers 1984).

Patients with childhood dermatomyositis develop unique complications, including subcutaneous calcifications and gastrointestinal hemorrhages. Calcinosis occurs in 20% to 50% of children, is refractory to treatment, and may lead to skin ulcerations and secondary infections (Bohan 1988; Sarnat 1988). The necrotizing vasculitis results in symptoms of systematic illness, bowel hemorrhage and infarction, and vasculitic skin and nail bed changes (Banker and Victor 1966). Early contractures of extremities also are common in childhood cases.

The prognosis of untreated dermatomyositis or polymyositis in most cases is poor and has a progressive downhill course. A mortality rate of 50% was recorded in a group of patients before the availability of prednisone or other immunosuppressive medications (O'Leary and Waisman 1940). Most series demonstrate significant clinical improvement with the chronic administration of corticosteroids, but some patients are refractory to treatment or have only a partial return of function (Henriksson and Sandstedt 1982; Rose and Walton 1966; Vignos et al. 1964). This incomplete response, which may be attributed to steroid resistance, insufficient treatment, or permanent muscle injury, is not predictable prospectively (Rosenberg and Ringel 1988). Other effective treatments include azathioprine (Bunch 1981), methotrexate (Fischer et al. 1979), cyclophosphamide (Currie and

Walton 1971), cyclosporin A (Bendtzen et al. 1984), chlorambucil (Ansell 1984), total body radiation (Kelly et al. 1984) and total lymphoid radiation (Rosenberg and Ringel 1988).

The prognostic factor most related to poor outcome is the presence of an associated malignancy. The increased mortality is related to tumor spread and not a response of the myositis to corticosteroids (Bohan et al. 1977; Engel and Askansas 1976). Other factors suggesting a worse prognosis include an acute fulminant course, dysphagia, cardiac involvement, acute pulmonary infiltration, and black race (Rosenberg and Ringel 1988).

Inclusion Body Myositis

Inclusion body myositis is an idiopathic, slowly progressive inflammatory myopathy with characteristic basophilic rimmed vacuoles in muscle fibers and eosinophilic cytoplasmic and nuclear filamentous inclusions. Since its initial description in 1967 (Chou 1967), inclusion body myositis has been increasingly identified, so it may represent a large number of chronic inflammatory myopathies refractive to treatment (Ringel et al. 1987). It is more common in elderly men and typically progresses so insidiously that an average of 6.5 years transpires from first weakness until diagnosis (Chou 1988; Ringel et al. 1987; Sawchek and Kula 1988). Weakness often is distal as well as proximal and may be asymmetric (Ringel et al. 1987). Mixed neurogenic and myopathic features are seen with electromyography. The creatine kinase (CPK) is normal or minimally elevated (Carpenter et al. 1978). In almost all cases, the progressive weakness does not improve with corticosteroids or other immunosuppressive drugs (Carpenter et al. 1978; Ringel et al. 1987). Occasional steroid-responsive cases have been reported, but several of these have had atypical features, such as very high CPK, extensive inflammation on biopsy, and association with other collagen vascular diseases (Levin 1986; Somer et al. 1986).

X-Linked Muscular Dystrophy

Duchenne type muscular dystrophy is a disease causing progressive muscular weakness, pseudohypertrophy of muscles, joint contractures, and respiratory insufficiency by the end of the second decade. It is inherited in an X-linked recessive manner, with the gene localized to the Xp-21 locus on the short arm of the X chromosome (Rowland 1988).

Clinical experience with this prevalent disease and two studies of natural history have provided reliable information concerning the course and prognosis (Scott et al. 1982). The best data are from a comprehensive study by the Clinical Investigations in Duchenne Dystrophy (CIDD) group (Brooke et al. 1983).

The diagnosis of Duchenne type muscular dystrophy is evident in the first days of life by the 20- to 50-fold elevation in CPK. The disease often is not apparent clinically until age 4 or 5 and is characterized by falling, wobbling gait, persistent toe walking, and the need for support to rise from the floor (Brooke 1986). Individual muscles continue to decline in strength at a uniform rate of 0.40 strength units per year on a 0 to 10 strength scale (Brooke et al. 1983) (Fig. 26-1). Functional abilities change less uniformly (Allsop and Ziter 1981). Some affected children appear to improve their ambulation between ages 3 and 6, even while losing objective muscle strength. After age 8, ambulation typically deteriorates rapidly. In one assessment of natural history, all children walked until age 8, and 60% of children at age 12 were using a wheelchair (Brooke et al. 1983) (Fig. 26-2).

Becker's muscular dystrophy has been defined arbitrarily as a patient who has the phenotypic appearance of Duchenne type muscular dystrophy and is able to walk independently after the age of 15. Molecular genetics have determined that Becker's dystrophy is the result of an allelic mutation at the Xp-21 locus and therefore is variant of Duchenne type muscular dystrophy (Rowland 1988). There is great variability in the

AGE PERCENTILES OF MUSCLE STRENGTH SCORE

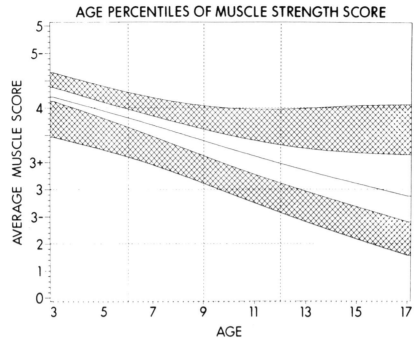

Figure 26-1. Graphic representation of the change in muscle strength with the age of a child. The data comes from a multicenter study of 150 patients with Duchenne muscular dystrophy. Source: Brooke (1983). The "average muscle score" is the mean Medical Research Council (MRC) grade of 34 separate muscles. The center line represents the 50% percentile, the upper hatched area the 75–95 percentiles, and the lower area the 5–25 percentiles.

Figure 26-2. Graphic representation of the ages at which functional milestones are lost by patients with Duchenne muscular dystrophy. Source: Brooke (1989).

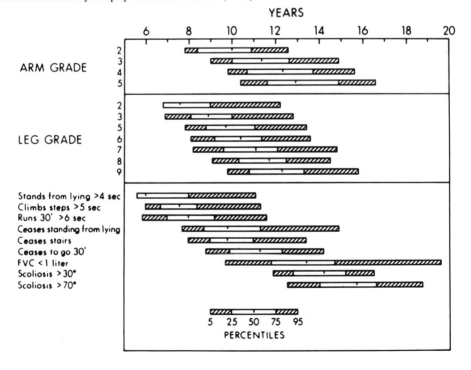

clinical course of patients, with some ambulating until middle-age or beyond (Ringel et al. 1977).

Once patients are wheelchair bound, problems other than skeletal muscle weakness become most apparent in boys with Duchenne type muscular dystrophy (Brooke 1986). Joint contractures, which are detectable by age 6, rapidly increase in severity around the hip, knee, ankle, and elbow joints. Progressive kyphoscoliosis develops in up to 80% of children who are no longer ambulatory and may compromise a weakening respiratory system (Brooke 1986; Smith et al. 1987). Pulmonary function is always below normal, but further deterioration can be documented when the child requires assistance with ambulation. The respiratory decline continues, and death typically occurs in the late second or early third decade from pulmonary infection or respiratory insufficiency (Inkley et al. 1976). A refractory cardiomyopathy occasionally may lead to death.

Although there is no proven therapy to alter the underlying disease process in Duchenne type muscular dystrophy, prednisone may delay weakness (Mendell 1989; Fenichel 1991). Several treatments may improve the quality and length of the patient's life (Smith et al. 1987). Many clinics perform early surgical release of hip, knee, and ankle contractures with early postoperative ambulation in long leg braces to prolong walking for 2 to 4 years (Siegal 1986). Orthotic devices and specialized wheelchair supports will not prevent but may delay the progression of scoliosis. Spinal bracing with polypropylene body jackets decreases vital capacity in many patients and should therefore not be used during respiratory infections or in patients with a very low vital capacity (Noble-Jamieson et al. 1986). Surgical correction of scoliosis using Luque's technique of L-rod segmental stabilization has become popular (Luque 1982). Data conflict as to whether this procedure slows the decline of vital capacity (Kurz et al. 1983; Smith et al. 1987). Inspiratory resistive training has been used to improve temporarily vital capacity and inspiratory pressure in patients with

Duchenne type muscular dystrophy. This technique provides invaluable assistance during weaning from the mechanical ventilation and during preoperative respiratory training (Aldrich and Uhrlass 1987; Ringel and Martin 1985). The effects of all types of respiratory therapy on long-term prognosis are controversial (Smith et al. 1987).

Proper nutrition to maintain a moderate weight is important because obesity and malnutrition impair respiratory function (Smith et al. 1987). Prompt treatment of acute respiratory infections with antibiotics, physiotherapy, postural drainage, and assistive ventilation is imperative. An annual influenza vaccination also is recommended (Smith et al. 1987). Long-term assisted ventilation, often only nocturnally, prolongs life in selected cases late into the third decade (Rideau et al. 1983; Rideau 1986). Although it seems most beneficial in patients who retain significant functional abilities, much controversy remains as to which patients should receive ventilation and how to resolve the ethical and financial questions raised by long-term assistive ventilation of patients with advanced Duchenne's dystrophy (Smith et al. 1987).

Myotonic Dystrophy

Myotonic dystrophy is an autosomal dominant disease that is linked to a locus on the long arm of chromosome 19 (Pericak-Vance et al. 1986). The disease results in progressive muscle weakness and myotonia as well as abnormalities of many other organ systems. A prevalence rate of five per 100,000 makes myotonic dystrophy the most common dystrophy in the general population (Harper 1986).

The age of clinical onset and the severity of symptoms vary greatly. Occasionally, adult carriers of the gene are unaware of their disease, although careful clinical evaluation usually demonstrates signs of myotonic dystrophy by late adolescence (Bundey 1974).

The first signs of myotonic dystrophy often are weakness and atrophy of facial, temporalis, masseter, and sternocleidomastoid muscles, giving the patient a characteristic expressionless facial appearance with ptosis, hollowing (hatchet-face), and a thin, swan-like neck. Early in the course, myotonia and weakness are confined to distal muscles. A recent therapeutic trial that quantified strength during a 12-month period indicated that muscle weakness is very slowly progressive (Griggs et al. 1989). In the later stages of the disease, muscles may be affected more diffusely, and some patients eventually require wheelchairs for mobility (Brooke 1986). Anticipation, or progressively increased severity and earlier onset of symptoms in successive generations, has been suggested, but this may be the consequence of inaccuracy of historical information or bias of observation (Brooke 1986; Harper 1986).

A peculiar form of the disease has a clinical onset during infancy and is almost exclusively a result of maternal inheritance (Harper 1975). These infants are hypotonic with weak facial muscles and feeding difficulties. Respiratory insufficiency may occur and often results in pulmonary infection or death (Harper 1975b). As these children develop, they are more likely to have severe physical problems and mental retardation (Harper 1975b). Myotonia typically is absent in the first few years of life, despite profound weakness.

Involvement of organ systems other than skeletal muscle eventually occurs in most patients with myotonic dystrophy and can cause the most disabling aspects of the disorder. Electrocardiographic (ECG) changes occur in up to 90% of patients (Church 1967). Progressive conduction defects and cardiac arrhythmias are noted in one half of patients tested and may be responsible for occasional sudden deaths (Florek et al. 1989; Griggs 1975; Holt et al. 1964). Pulmonary hypoventilation resulting from respiratory muscle weakness and impaired central ventilatory response frequently is present (Carroll et al. 1977; Caughey and Pachomov 1959; Martin 1984). Although often asymptomatic, these respiratory abnormalities combined with cardiac defects lead to an increased risk with general anesthetics and hypersensitivity to other respiratory depressants (Ravin et al. 1975). Cataracts eventually develop in almost all patients (Brooke 1986). Although mental retardation occurs in a minority of patients (Harper 1975b; Woodward et al. 1982), cerebral evoked potential abnormalities (Thompson et al. 1983) and changes on psychometric and behavioral testing are much more frequent (Bird et al. 1983). These findings may account for the high incidence of unemployment and low motivation described in many patients (Woodward et al. 1982). Other abnormalities noted during the course of the disease include dysphagia, cholelithiasis, hyperinsulinemia, sleep disturbance, testicular atrophy, polyhydramnios, and spontaneous abortions in women (Brooke 1986; Marshall 1959; Stuart et al. 1977; Swick et al. 1981).

Pharmacological therapy may diminish the myotonia, but it does not improve muscle strength (Griggs et al. 1989). Premature death may be prevented by periodic screening for cardiac conduction defects and by taking special precautions when general anesthetics or respiratory depressant medications are used. Proper treatment of swallowing difficulties, sleep disturbance (Martin 1984), and cataracts may improve the patient's quality of life.

Facioscapulohumeral Dystrophy

Facioscapulohumeral dystrophy is a genetically determined disease that is transmitted in an autosomal dominant manner with full penetrance but variable degrees of expression (Padberg 1982). Prevalence is estimated to be one per 100,000 (Morton and Chung 1959).

Weakness is manifested initially in muscles of facial expression with relative sparing of masseter, temporalis, and extraocular muscles. Progressive muscle weakness and laxity lead to the characteristic appearance of an expressionless, smooth, drooping face and protruding lips or "bouche de tapir." Attempts to smile show poor horizontal movement of the angles of the mouth with dimpling (Brooke 1986). Patients often admit to a lifelong history of difficulty blowing up balloons or drinking from straws. Weakness and wasting of scapular muscles also is observed early in the course of the disease, resulting in prominent scapular winging and loss of shoulder abduction. Most patients initially notice shoulder girdle weakness, even though facial weakness is readily demonstrable (Padberg 1982). Later in the course, selective involvement of biceps, triceps, and peronial and anterior tibialis muscles may develop, the latter producing a slapping, footdrop gait. A minority of patients develop more extensive weakness, involving muscles of the pelvic girdle and distal upper extremities. Asymmetric weakness has been well documented (Walton and Nattress 1954).

The onset of symptoms varies greatly in facioscapulohumeral dystrophy. Symptoms are first noted in late adolescence or early adulthood in most patients (Padberg 1982; Tyler and Stephens 1950; Walton and Nattress 1954). Several studies, however, document that at least 30% of affected family members may be unaware of their disease, even during adulthood (Padberg 1982; Tyler and Stephens 1950; Zundel and Tyler 1965). Uncommonly, symptoms may begin during infancy, and these children typically have more severe weakness and occasionally other associated abnormalities, such as nerve deafness and severe lumbar lordosis (Brooke 1986).

The disease is very slowly progressive and may remain confined to the upper extremities in up to 50% of affected people (Munsat 1986). Lifespan is not reduced (Padberg 1982). A few cases that begin with isolated facial weakness and rapidly progress to profound generalized weakness during a period of months have been noted, but they are poorly documented (Brooke 1986).

Facioscapulohumeral dystrophy rarely is associated with Coat's syndrome, an exudative telangiectasia of the retina. Inflammatory muscle biopsies occasionally are seen, particularly in more severe juvenile cases, but these patients with familial disease have not improved with corticosteroid therapy (Munsat et al. 1972; Munsat and Bradley 1977; Wulff, 1982). No pharmacologic therapy alters the course of muscle weakness. Surgical fixation of the scapula to the thoracic wall may allow greater functional use of the arms in selected cases (Ketenjian 1978).

Scapuloperoneal Dystrophy

Scapuloperoneal dystrophy is a poorly defined disease that phenotypically resembles facioscapulohumeral dystrophy, except that facial weakness is less prominent. Many other diseases may mimic this pattern of muscle weakness, including spinal muscular atrophies, polyneuropathies, and other myopathies (Harding and Thomas 1980; Kaeser 1965). The disease may have its onset from infancy to middle-age, and the course is extremely variable (Feigenbaum and Munsat 1970; Kaeser 1965). It may be transmitted in an autosomal dominant, recessive, X-linked, or sporadic manner (Munsat 1986; Thomas et al. 1975). Nerve deafness also may be associated with this syndrome (Ansher et al. 1989).

Emery-Dreifuss disease has a similar distribution of weakness, early joint contractures, X-linked inheritance, and unique prognosis (Emery and Dreifuss 1966; Rowland et al. 1979). Cardiac conduction blocks and arrhythmias requiring electronic pacing are common. Sudden death may occur as a consequence if the cardiac disturbance is not recognized and treated promptly (Brooke 1986).

Limb Girdle Muscular Dystrophy

Limb girdle dystrophy is a syndromic grouping of inherited diseases with progressive weakness of the shoulder and pelvic muscles and relative preservation of facial, pharyngeal, and distal extremity musculature. These grouped diseases show dystrophic changes on muscle biopsy and evade classification into a better defined disorder (Shields 1986). The prevalence of this syndrome has been estimated at 12 per 1 million (Morton and Chung 1959). Great variability exists with inheritance, onset, and progression of disease, but a few patterns may help identify the prognosis. Onset of disease younger than age 15 usually is associated with autosomal recessive inheritance, or it is sporadic. These patients often require wheelchairs for mobility by the fifth decade (Ben Hamida and Fardeau 1980; Jackson and Strehler 1968). Individuals with onset of symptoms later in adulthood typically have less severe and more slowly progressive disease. Their inheritance may be autosomal dominant or recessive (DeCoster et al. 1974; Schneiderman et al. 1969). Weakness limited to the pelvic, femoral, or scapuloperoneal regions also suggests a more benign course (Jackson and Strehler 1968; Moser et al. 1966).

Cardiac involvement with arrhythmias and congestive failure may be seen, but they are typically much milder than involvement in Duchenne type muscular dystrophy (Hoshio et al. 1987; Hunter 1980; Welsh et al. 1963). Diaphragmatic and intercostal muscle weakness may lead to alveolar hypoventilation in a few severely affected patients (Kilburn et al. 1959; Newsom-Davis 1980). Severe low back pain may develop as a consequence of lumbar muscle weakness (Brooke 1986).

Congenital Myopathies

Patients with congenital myopathies present with neonatal weakness and hypotonia and follow a nonprogressive or very slowly progressive course (Bodensteiner 1988). All forms of inheritance have been described. These diseases, grouped according to their morphological appearances on muscle biopsy, have names such as central core disease, nemaline (rod body) myopathy, centronuclear (myotubular) myopathy, congenital fiber type disproportion, reducing body myopathy, and trilaminar myopathy. The severity may vary from rapidly fatal to clinical asymptomatic, even within a single morphological classification, such as nemaline myopathy (Bender and Willner 1978; Tsujihata et al. 1983).

A few generalizations may be useful for prognostic purposes. Most cases present at birth, attain motor milestones, slowly, and have little progression. Central core disease often follows an autosomal dominant pattern of transmission. Although delayed, most of these patients attain motor milestones and become ambulatory (Byrne et al. 1982; Shy and Magee 1956). An association with malignant hyperthermia has been described (Frank et al. 1980). The X-linked form of centronuclear myopathy and one form of nemaline myopathy may result in profound neonatal weakness, with feeding difficulties, skeletal abnormalities, and respiratory insufficiency, often leading to death (Barth et al. 1975; Tsujihata et al. 1983). Adult onset of weakness in a limb girdle distribution has been described with nemaline, central core, and centronuclear myopathies. These cases typically have mild weakness and very slow progression (Meier et al. 1984; Patterson et al. 1979).

Primary Muscle Carnitine Deficiency

Although several myopathies have been described with lipid excess in muscle, these disorders occur so infrequently that a description of the pathogenesis of each is not practical. In one example, primary muscle carnitine deficiency is a clinical syndrome defined by progressive muscle weakness, lipid storage in muscle, and decreased free carnitine levels restricted to muscle (Carroll 1988; Engel and Angelini 1973). The underlying biochemical defect or defects have not been fully defined. The same syndrome may occur as a secondary consequence of other defined metabolic diseases (Willner et al. 1979). An autosomal recessive inheritance has been suggested (DiMarco et al. 1985).

The onset of symptoms typically occurs during childhood but may range from 18 months to 38 years (Engel 1986b). Exercise intolerance is an early symptom in some patients (Willner et al. 1979). A slowly progressive proximal muscle weakness develops in most patients (DiMarco et al. 1985). The weakness may fluctuate and sometimes will progress rapidly (Bradley et al. 1978; Carroll 1988). Several patients have been described with cardiomyopathy, which may lead to early death (Hart et al. 1978). Some patients have been reported to improve with oral carnitine, corticosteroids, riboflavin, or propranolol, while others are unresponsive to treatment (Angelini et al. 1976; Carroll et al. 1980; Carroll et al. 1981; Engel and Siekert 1972; Isaacs et al. 1986).

Myophosphorylase Deficiency

The most common glycogen storage disease of muscle is myophosphorylase deficiency (McArdle's disease), an autosomal recessively inherited disorder with male predominance (DiMauro and Breselin 1986). Rarely, automosomal dominant transmission also has been documented (Chu and Munsat 1976). Because of a deficiency in muscle phosphorylase activity, patients develop muscle contractures (electrically silent cramps) and myoglobinuria after strenuous exercise. Onset of symptoms has been reported during infancy and as late as age 60 (DiMauro and Breselin 1978; Kost and Verity 1980). In one large series, 85% of patients developed symptoms younger than 15 years (DiMauro and Breselin 1986).

Initial symptoms include exercise intolerance, myalgias, and fatigue, which typically progress to painful contractures following strenuous exercise. Only 50% of patients ever develop myoglobinuria, and 60% of these have less than three episodes. Renal failure occurs in approximately 25% of patients with myoglobinuria (DiMauro and Breselin 1986). Most patients learn to prevent severe contractures and myoglobinuria by limiting exercise when they first notice muscle fatigue. More frequent exacerbations have been documented secondary to vascular insufficiency and other unknown factors (Kost and Verity 1980; Wheeler and Brooke 1983).

A minority of patients develop fixed proximal muscle weakness, typically after many years of symptoms (Schmid and Hammer 1961). Cardiac muscle involvement is rare due to preserved activity of the cardiac isoenzyme of phosphorylase (Miranda et al.

1979). Convulsions may precipitate myoglobinuria or contractures but are not clearly etiologically related to McArdle's disease (Salmon and Turner 1965).

Several therapies have been used to improve exercise tolerance, including oral fructose, intravenous glucose, isoproterenol, fasting, and exercise training (DiMauro and Breselin 1986). The effects generally have been mild and not beneficial in all patients.

References

Adornato, B. T.; Engel, W. K. MB-creatine phosphokinase isoenzyme elevation not diagnostic of myocardial infarction. Arch. Intern. Med. 137:1089–1090; 1977.

Aldrich, T. K.; Uhrlass, R. M. Weaning from mechanical ventilation: Successful use of modified inspiratory resistive training in muscular dystrophy. Crit. Care Med. 15(3):247–249; 1987.

Allsop, K. G.; Ziter, F. A. Loss of strength and functional decline in Duchenne's dystrophy. Arch. Neurol. 38:406–411; 1981.

Angelini, C.; Lucke, S.; Cantarutti, F. Carnitine deficiency of skeletal muscle: Report of a treated case. Neurology 26:633–637; 1976.

Ansell, B. M. Management of polymyositis and dermatomyositis. Clin. Rheum. Dis. 10:205–213; 1984.

Ansher, M.; Smith, L. D.; Ringel, S. P. Autosomal dominant scapuloperoneal syndrome with sensorineural hearing loss. Otolaryngol. Head Neck Surg. 100:242–244; 1989.

Banker, B. Q.; Victor, M. Dermatomyositis (systemic angiopathy) of childhood. Medicine. 45:261–288; 1966.

Barth, P. G.; Van Wijngaarden, G. K.; Bethlem, J. X-linked myotubular myopathy with fatal neonatal asphyxia. Neurology 25:531–536; 1975.

Barwick, D. D.; Walton, J. N. Polymyositis. Am. J. Med. 35:646–660; 1963.

Bender, A. N.; Willner, J. P. Nemaline (rod) myopathy: need for histochemical evaluation of affected families. Ann. Neurol. 4:37–42; 1978.

Bendtzen, K.; Tuede, N.; Andersen, V.; Bendixen, G. Cyclosporin for polymyositis. Lancet 1:792–793; 1984.

Ben Hamida, M.; Fardeau, M. Severe autosomal recessive limb-girdle muscular dystrophies frequent in Tunisia. In: Angelini, C.; Danieli, G. A.; Fontanari, D., eds. Muscular dystrophy research: advances and new trends. Amsterdam: Excerpta Medica; 1980: p. 143–146.

Bird, T. D.; Follett, C.; Griep, E. Cognitive and personality function in myotonic muscular dystrophy. 1. Cognitive function 2. Personality profiles. J. Neurol. Neurosurg. Psych. 46:971–980; 1983.

Bodensteiner, J. B. Congenital myopathies. Neurol. Clin. 3:499–518; 1988.

Bohan, A. Clinical presentation and diagnosis of polymyositis and dermatomyositis. In: Dalakas, M., ed. Polymyositis and dermatomyositis. Boston: Butterworth Publishers; 1988: p. 19–36.

Bohan, A.; Peter, J. B.; Bowman, R. L.; Pearson, C. M. A computer-assisted analysis of 153 patients with polymyositis and dermatomyositis. Medicine 56:255–286; 1977.

Bradley, W. G.; Tomlinson, B. E.; Hardy, M. Further studies of mitochondrial and lipid storage myopathies. J. Neurol. Sci. 35:201–210; 1978.

Brooke, M. R.; Fenichel, G. M.; Griggs, R.; Mendell, J. R.; Moxley, R.; Miller, J.; Philip, A. B.; Province, M. A. Clinical investigation in Duchenne Dystrophy: 2. Determination of the "Power" of therapeutic trials based on natural history. Muscle Nerve 6:91–103; 1983.

Brooke, M. H. A clinical view of neuromuscular disease, 2nd ed. Baltimore: Williams & Wilkins; 1986.

Brooke, M. H.; Fenichel, G. M.; Griggs, R. C.; Mendell, J. R.; Moxley, R.; Florence, J.; King, W. M.; Pandya, S.; Robinson, J.; Schierbecker, P. T.; Signore, R. N.; Miller, J. P.; Gilder, B. F.; Kaiser, K. K.; Mandel, S.; Arfken, C. Duchenne muscular dystrophy: Patterns of clinical progression and effects of supportive therapy. Neurology 39:475–481; 1989.

Bunch, T. W. Prednisone and azathioprine for polymyositis: long-term follow-up. Arth. Rheum. 24:45–48; 1981.

Bundey, S. Detection of heterozygotes for myotonic dystrophy. Clin. Genet. 5:107–109; 1974.

Byrne, E.; Blumbery, P. C.; Hallpike, J. F. Central core disease. Study of a family with five affected generations. J. Neurol. Sci. 53:77–83; 1982.

Cailen, J. P. The value of malignancy evaluation in patients with dermatomyositis. J. Am. Acad. Dermatol. 6:253–259; 1982.

Carpenter, S.; Karpati, G.; Heller, J. Inclusion body myositis: a distinct variety of idiopathic inflammatory myopathy. Neurology 28:8–17; 1978.

Carroll, J. E.; Zwillich, C. W.; Weil, J. V. Ventilatory response in myotonic dystrophy. Neurology 27:1125–1128; 1977.

Carroll, J. E.: Brooke, M. H.; DeVivo, D. C.; Shumata, J. B.; Kratz, R. Ringel, S. P.; Hagberg, J. M. Carnitine "deficiency": lack of response to carnitine therapy. Neurology 30:618–626; 1980.

Carroll, J. E.; Shumate, J. B.; Brooke, M. H.; Hagberg, J. M. Riboflavin-responsive lipid myopathy and carnitine deficiency. Neurology 31:1557–1559; 1981.

Carroll, J. E. Myopathies caused by disorders of lipid metabolism. Neurol. Clin. 6(3):563–574; 1988.

Caughey, J. E.; Pachomov, N. The diaphragm in dystrophia myotonic. J. Neurol. Neurosurg. Psychiatry 22:311–313; 1959.

Chou, S. M. Myositis-like structures in a case of human chronic polymyositis. Science 158:1453–1455; 1967.

Chou, S. M. Viral myositis. In: Mastaglia, F. ed. Inflammatory diseases of muscle. Oxford. Blackwell Scientific Publishing Company 1988: p. 125–153.

Chu, L. A.; Munsat, T. L. Dominant inheritance of McArdle syndrome. Arch. Neurol. 33:636–641; 1976.

Church, S. C. The heart in myotonia atrophica. Arch. Intern. Med. 119:176–181; 1967.

Compston, D. A. S.; Vincent, A.; Newsom-Davis, J.; Batchelor, J. R. Clinical, pathological, HLA antigen and immunological evidence for disease heterogeneity in myasthenia gravis. Brain 103:579–601; 1980.

Currie, S.; Walton, J. N. Immunosuppressive therapy in polymyositis. J. Neurol. Neurosurg. Psychiatry 34:447–452; 1971.

DeCoster, W.; DeReuck, J.; Thiery, E. A late autosomal dominant form of limb girdle muscular dystrophy. Eur. Neurol. 12:159–172; 1974.

Denbow, C. E.; Lie, J. T.; Tancredi, R. G.; Bunch, T. W. Cardiac involvement in polymyositis: a clinicopathologic study of 20 autopsied patients. Arth. Rheum. 22:1088–1092; 1979.

DeVere, R.; Bradley, W. G. Polymyositis: its presentation, morbidity and mortality. Brain 98:637–666; 1975.

Dickey, B. F.; Myers, A. R. Pulmonary disease in polymyositis/dermatomyositis. Semin. Arthritis Rheum. 14:60–76; 1984.

DiMarco, A. P.; DiMarco, M. S.; Jacobs, I.; Shields, R.; Altase, M. D. The effects of inspiratory resistive maining of respiratory muscle for patients with muscular dystrophy. Muscle Nerve 8:284–290; 1985.

DiMauro, S.; Hartlage, P. L. Fatal infantile from of muscle phosphorylase deficiency. Neurology 28:1124–1129; 1978.

DiMauro, S.; Breselin, N. Phosphorylase deficiency. In: Engel, A. G.; Banker, B. Q., eds. Myology. New York: McGraw-Hill Book Company; 1986: p. 1585–1601.

Donaldson, D. H.; Ansher, M.; Horan, S.; Rutherford, R. B.; Ringel, S. P. The relationship of age to outcome in myasthenia gravis. Neurology 40:786–790; 1990

Eaton, L. M. The perspective of neurology in regard to polymyositis: a study of 41 cases. Neurology 4:245–263; 1954.

Emery, A. E. H.; Dreifuss, F. E. Unusual type of benign x-linked muscular dystrophy. J. Neurol. Neurosurg. Psychiatry 29:338–342; 1966.

Engel, A. G. Myasthenic syndromes. In: Engel, A. G.; Banker, B. Q., eds. Myology. New York: McGraw-Hill Book Company; 1986a: p. 1955–1992.

Engel, A. G.; Siekert, R. G. Lipid storage myopathy responsive to prednisone. Arch. Neurol. 27:174–181; 1972.

Engel, A. G.; Angelini, C. Carnitine deficiency of human skeletal muscle with associated lipid storage myopathy. Science 179:899–902; 1973.

Engel, W. K.; Askansas V. Remote effects of focal cancer on the neuromuscular system. In: Thompson, R. A.; Green, J. R., eds. Advances in neurology. New York: Raven Press; 1976: p. 119–148.

Engel, A. G. Carnitine deficiency syndromes and lipid storage and myopathies. In: Engel, A. G.; Banker, B. Q., eds. Myology. New York: McGraw-Hill Book Company; 1986b: p. 1663–1696.

Evoli, A.; Tonali, P.; Bestorcian, E.; Manoco, M. L. Ocular myasthenia: diagnosis and therapeutic problems. Acta. Neurol. 77:31–35; 1988.

Feigenbaum, J. A.; Munsat, T. L. A neuromuscular syndrome of scapuloperoneal distribution. Bull. Los Angeles Neurol. Soc. 35:47–57; 1970.

Fenichel, G. M.; Mendell, J. R.; Moxley, R. T.; et al. A comparison of daily and alternate day prednisone therapy in the treatment of Duchenne muscular dystrophy. Arch. Neurol. 48:575–579; 1991.

Fischer, T. J.; Rachelefsky, G. S.; Klein, R. B.; Paulus, H. E.; Steihm, E. R. Childhood dermatomyositis and polymyositis: treatment with methotrexate and prednisone. Am. J. Dis. Child. 133:386–389; 1979.

Florek, R. C.; Triffon, D. W.; Mann, D. E.; Ringel, S. P.; Reiter, M. J. Prevalence and progression of electrocardiographic abnormalities in patients with myotonic dystrophy. West. J. Med. 153:24–27; 1990.

Frank, J. P.; Harat, Y.; Butler, I. J.; Nelson, T. E.; Scott, C. I. Central core disease and malignant hyperthermia syndrome. Ann. Neurol. 7:11–17; 1980.

Genkins, G.; Kornfeld,; Papatestas, A. I.; Bender, A. N.; Mattas, R. J. Clinical experience in more than 2000 patients with myasthenia gravis. Ann. N.Y. Acad. Sci. 505:500–513; 1987.

Gottdiener, J. S.; Sherberg, H. S.; Hawley, R. J.; Engel, W. K. Cardiac manifestations in polymyositis. Am. J. Cardiol. 41:1141–1149; 1978.

Griggs, R. C.; Pandya, S.; Florence, J. M.; Brooke, M. H.; Kingston, W.; Miller, J. P.; Chutkow, J.; Herr, B. E.; Moxley, R. T., III. Randomized controlled trial of testosterone in myotonic dystrophy. Neurology 39:219–222; 1989.

Griggs, R. C. Cardiac conduction in myotonic dystrophy. Am. J. Med. 59:37–42; 1975.

Grob, D.; Arsuie, E. L.; Brunner, N. G.; Namber, T. The course of myasthenia and therapies affecting outcome. Ann. N.Y. Acad. Sci. 505:472–499; 1987.

Harding, A. E.; Thomas, P. K. Distal and scapular peroneal distribution of muscle involvement occurring within a family with type I hereditary motor and sensory neuropathy. J. Neurol. 224(1):17–23; 1980.

Harper, P. S. Congenital myotonic dystrophy in Britain II. Genetic Basis. Arch. Dis. Child. 50:514–521; 1975a.

Harper, P. S. Congenital myotonic dystrophy in Britain I. Clinical aspect. Arch. Dis. Child. 50:505–513; 1975b.

Harper, P. S. Myotonic disorders. In: Engel, A. G.; Banker, B. Q., eds. Myology. New York: McGraw-Hill Book Company; 1986: p. 1267–1295.

Hart, Z. H.; Chang, C. H.; DiMauro, S.; Farooki, Q.; Ayyar, R. Muscle carnitine deficiency and fatal cardiomyopathy. Neurology 28:147–151; 1978.

Henriksson, K. G.; Sandstedt, P. Polymyositis—treatment and prognosis: a study of 107 patients. Acta. Neurol. Scandinav. 65:280–300; 1982.

Holt, J. M.; Lambert, E. N. Heart disease as the presenting feature in myotonia atrophica. Br. Heart J. 26:433–436; 1964.

Hoshio, A.; Kotake, H.; Saitoh, M.; Ogino, K.; Fujimoto, Y.; Cardiac involvement in a patient with limb girdle muscular dystrophy. Heart Lung 16(4):439–441; 1987.

Hunter, S. The heart in muscular dystrophy. Br. Med. Bull. 36:133–134; 1980.

Inkley, S.; Oldenberg, F.; Vignos, P. J. Jr., Pulmonary function in Duchenne muscular dystrophy related to stage of disease. Am. J. Med. 6:297–306; 1976.

Isaacs, H.; Heffron, J. J. A.; Badenhorst, M.; Pickering, A. Weakness associated with the pathological presence of lipid in skeletal muscle: a detailed study of a patient with carnitine deficiency. J. Neurol. Neurosurg. Psychiatry 39:1114–1123; 1976.

Isenberg, D. Myositis in other connective tissue disorders. Clin. Rheum. Dis. 10:151–175; 1984.

Jackson, C. E.; Strehler, D. A. Limb-girdle muscular dystrophy: Clinical manifestations and detection of pre-clinical disease. Pediatrics 41(2):495–502; 1968.

Jaretzki III, A.; Penn, A.; Younger, D.; Wolff, M.; Olarte, M. R.; Lovelace, R. E.; Rowland, L. P. Maximal thymectomy for myasthenia gravis. J. Thorac. Cardiovasc. Surg. 95:747–757; 1988.

Johns, T. R. Long-term corticosteroid treatment of myasthenia gravis. Ann. N.Y. Acad. Sci. 505:568–583; 1987.

Kaeser, H. E. Scapuloperoneal muscular atrophy. Brain 88:407–418; 1965.

Kagen, L. J.; Hochman, R. B.; Strong, E. W. Cricopharyngeal obstruction in inflammatory myopathy (polymyositis/dermatomyositis): report of three cases and review of the literature. Arthritis Rheum. 28:630–636; 1985.

Kelly, J. J.; Madoc-Jones, H.; Adelman, L. R.; Munsat, T. L. Treatment of refractory polymyositis with total body irradiation. Neurology 34(suppl. 1):80; 1984.

Ketenjian, A. Y. Scapulocostal stabilization for scapular winging in facioscapulohumeral muscular dystrophy. J. Bone Joint Surg. 60:476–480; 1978.

Kilburn, K. H.; Eagan, J. T.; Sieker, H. O.; Heyman, A. Cardiopulmonary insufficiency in myotonic and progressive muscular dystrophy. N. Engl. J. Med. 261:1089–1096; 1959.

Kimura, J. Electrodiagnosis of nerve and muscle. Philadelphia: FA Davis; 1983.

Kost, G. J.; Verity, M. A. A new variant of late-onset myophosphorylase deficiency. Muscle Nerve 3:195–201; 1980.

Kurz, L. T.; Mubeersade, S. J.; Schultz, P.; Park, S. M.; Lead, J. Correlation of scoliosis and pulmonary function in DMD. J. Ped. Orthop. 3:347–153; 1983.

Larca, L. J.; Coppola, J. T.; Honig, S. Creatine kinase MS isoenzyme in dermatomyositis: a noncardiac source. Ann. Intern. Med. 94:341–343; 1981.

Lefvert, A. K.; Osterman, P. O. Newborn infants to myasthenic mothers: a clinical study and an investigation of acetylcholine receptor antibodies in 17 children. Neurology 33:133–138; 1983.

Levin, K.; Misumoto, H.; Agamanolis, D. Steroid-responsiveness and clinical variability in inclusion body myositis. Muscle Nerve 9(suppl.): 217; 1986.

Logan, R. G.; Bandera, J. M.; Mikkelsen, W. M. Polymyositis: a clinical study. Ann. Intern. Med. 65:996–1007; 1966.

Luque, E. R. Segmental spinal instrumentation for correction of scoliosis. Clin. Orthop. 163:192–198; 1982.

Marshall, J. Observations on endocrine function in dystrophia myotonica. Brain 82:221–231; 1959.

Martin, R. J. Cardiorespiratory disorders during sleep. N.Y. Futura. 1984.

Mastaglia, F. L.; Ojeda, V. J. Inflammatory myopathies. Ann. Neurol. 17:215–217, 317–323; 1985.

McQuillen, M. P.; Leone, M. G. A treatment carol: thymectomy revisited. Neurology 27:1103–1106; 1977.

Medsger, T. A.; Dawson, W. N.; Masi, A. T. The epidemiology of polymyositis. Am. J. Med. 48:715–723; 1970.

Meier, C.; Voellmy, W.; Gertsch, M.; Zimmerman, A.; Geissbuhler, J. Nemaline myopathy appearing in adults in cardiomyopathy. Arch. Neurol. 41:443–445; 1984.

Mendell, J. R.; Moxley, R. T.; Griggs, R. C.; et al. Randomized double-blind controlled trial of prednisone in Duchenne muscular dystrophy. N. Engl. J. Med. 320:1592–1597; 1989.

Miranda, A. F.; Nette, E. G.; Hartlage, P. L.; DiMauro, S. Phosphorylase isoenzymes in normal and myophosphorylase-deficient human heart. Neurology 29:1538–1540; 1979.

Morton, N. E.; Chung, C. S. Formal genetics of muscular dystrophy. Am. J. Hum. Genet. 11:360–379; 1959.

Moser, Von H.; Wiesmann, U.; Richterich, R.; Rossi, E. Progressive muskeldystrophie, VIII Haufigkeit, Klinik und Genetik der Typen I and II. Schweiz Med Wochenschr. 96:169–174, 205–211; 1966.

Munsat, T. L.; Piper, D.; Cancilla, P.; Mednick, J. Inflammatory myopathy with facioscapulohumeral dystrophy. Neurology 22:335–347; 1972.

Munsat, T. L.; Bradley, W. G. Serum creatine phosphokinase levels and prednisone treated muscle weakness. Neurology 27:96–97; 1977.

Munsat, T. L. Facioscapulohumeral dystrophy and the scapuloperoneal syndrome. In: Engel, A.; Banker, B. Q., eds. Myology. New York: McGraw-Hill Book Co.; 1986: p. 1251–1265.

Newsom-Davis, J. The respiratory system in muscular dystrophy. Br. Med. Bull. 36:135–137; 1980.

Niaken, E.; Hasati, Y.; Rolak, L. Immunosuppressive drug therapy in myasthenia gravis. Arch. Neurol. 43:155–156; 1986.

Noble-Jamieson, C. M.; Heckmett, J. Z.; Dubowitz, V.; Silverman, M. Effects of posture and spinal bracing on respiratory function in neuromuscular disease. Arch. Dis. Child 61:178–181; 1986.

O'Leary, P. A.; Waisman, M. Dermatomyositis: a study of 40 cases. Arch. Dermatol. Syph. 41:1001–1019; 1940.

Olanow, C. W.; Wechsler, A. S.; Sirotkin-Roses,

M.; Stajich, J.; Roses, A. D. Thymectomy as primary therapy in myasthenia gravis. Ann. N.Y. Acad. Sci. 505:595–605; 1987.

Padberg, G. Facioscapulohumeral disease. Doctoral thesis, University of Leiden, Intercontinental Graphics, 1982.

Patterson, V. H.; Hill, T. R.; Fletcher, P. J. Heron, J. R. Central core disease: clinical and pathological evidence of progression within a family. Brain 102:581–594; 1979.

Pericak-Vance, M. A.; Yamaoka, L.; Assinder, R.; Hung, W. Y.; Bartlett, R. J.; Stajich, J. M.; Gaskell, P. C.; Ross, D. A. Sherman, S.; Fey, G. H.; Humphries, S.; Williamson, R.; Roses, A. D. Tight linkage of apolipoprotein C_2 to myotonic dystrophy on chromosome 19. Neurology 36:1418–1423; 1986.

Ravin, M.; Newmark, Z.; Saviello, G. Myotonia dystrophica: an anesthetic hazard: two case reports. Anesth. Analg. 54:216–218; 1975.

Rideau, Y.; Gatin, G.; Bach, J.; Gines, G. Prolongation of life in Duchenne's muscular dystrophy. Acta. Neurol. (Napol.) 5:118–124; 1983.

Rideau, Y. The Duchenne muscular dystrophy child: case of wheelchair dependent patient: death prevention. Muscle Nerve 9(suppl.):86; 1986. (Abstr.)

Ringel, S. P.; Carroll, J.; Schold, C. The spectrum of mild x-linked recessive muscular dystrophy. Arch. Neurol. 34:408–416; 1977.

Ringel, S. P.; Martin, R. J. Respiratory complications and their management in neuromuscular disorders: In: Maloney, F. P.; Burks, J. S.; Ringel, S. P., eds. Interdisciplinary rehabilitation of multiple sclerosis and neuromuscular disorders. Philadelphia: J. B. Lippincott Company; 1985: p. 211–227.

Ringel, S. P.; Kenny, C. E.; Neville, H. E.; Giorno, R.; Carly, M. R. Spectrum of inclusion body myositis. Arch. Neurol. 44:1154–1157; 1987.

Rose, A. L.; Walton, J. N. Polymyositis: a survey of 89 cases with particular reference to treatment and prognosis. Brain 89:747–768; 1966.

Rosenberg, N. L.; Carry, M. R.; Ringel, S. P. Association of inflammatory myopathies with other connective tissue disorders and malignancy. In: Dalakas, M., ed. Polymyositis and Dermatomyositis. Boston: Butterworth Publications; 1988: p. 37–70.

Rosenberg, N. L.; Ringel, S. P. Adult polymyositis and dermatomyositis. In: Mastaglia, F., ed. Inflammatory diseases of muscle. Oxford: Blackwell Scientific Publications: 1988: p. 87–106.

Rowland, L. P.; Clark, C.; Olarte, M. Therapy for dermatomyositis and polymyositis. In: Griggs, R. C.; Moxely, III, R. T., eds. Advances in neurology. New York: Raven Press; 1977: p. 63–95.

Rowland, L. P.; Fetell, M.; Olarte, M.; Hayes, A.; Singh, N.; Wanat, F. E. Emergy–Dreifuss muscular dystrophy. Ann. Neurol. 5:111–117; 1979.

Rowland, L. P. Controversies about the treatment of myasthenia gravis. J. Neurol. Neurosurg. Psychiatry 43(7):644–659; 1980.

Rowland, L. P. Clinical concepts of Duchenne muscular dystrophy. Brain 111:479–495; 1988.

Salmon, S. E.; Turner, C. E. McArdle's disease presenting as convulsions and rhabdomyolysis. Am. J. Med. 39:142–146; 1965.

Sarnat, H. B. Juvenile dematomyositis. In: Mastaglia, F., ed. Inflammatory disease of muscle. Oxford: Blackwell Scientific Publications; 1988: p. 71–86.

Sawcheck, J. A.; Kula, R. W. Inclusion body myositis. In: Dalakas, M., ed. Polymyositis and dermatomyositis. Boston: Butterworth Publications; 1988: p. 121–132.

Schmid, R.; Hammer, L. Hereditary absence of muscle phosphorylase (McArdle's syndrome). N. Engl. J. Med. 264:223–225; 1961.

Schneiderman, L. J.; Sampson, W. I.; Schoene, W. C.; Haydon, G. B. Genetic studies of a family with two unusual autosomal dominant conditions: Muscular dystrophy and Pelger-Huet anomaly. Am. J. Med. 46:380–393; 1969.

Schumm, F.; Wietholter, H.; Fatch-Moghadam, A.; Dichgans, J. Thymectomy in myasthenia with pure ocular symptom. J. Neurol. Neurosurg. Psychiatry 48:332–337; 1985.

Scott, W.; Hyde, S. A.; Goddard, C. M.; Dubowitz, V. Quantitation of muscle function in children: a prospective study in Duchenne muscular dystrophy. Muscle Nerve 5:291–301; 1982.

Seybold, M. Plasmapheresis in myasthenia gravis. Ann. N.Y. Acad. Sci. 505:584–587; 1987.

Shields, R. W., Jr. Limb girdle syndromes. In: Engel, A. G.; Banker, B. Q., eds. Myology. New York: McGraw-Hill Book Company; 1986: p. 1349–1365.

Shy, G. M.; Magee, K. R. A new congenital nonprogressive myopathy. Brain 79:610–621; 1956.

Siegal, I. M. Orthopedic management of muscle disease. In: Maloney, F. P.; Burks, J. S.; Ringel, S. P., eds. Interdisciplinary rehabilitation of multiple sclerosis and neuromuscular disorders. Philadelphia: J.B. Lippincott Company; 1986: p. 277–295.

Smith, E. M.; Calvesley, P. M.; Edwards, R. H.; Evans, G. A.; Campbell, E. J. Practical problems in the respiratory case of patients with muscular dystrophy. N. Engl. J. Med. 316:1197–1204; 1987.

Somer, H.; Haltia, M.; Somer, T. "Inclusion body myositis" in scleroderma: improvement with steroids. Muscle Nerve 9(suppl.):220; 1986.

Stuart, C. A.; Armstrong, R.; Provow, S.; Plishker, G.A. Insulin resistance in patients with myotonic dystrophy. Neurology 33:679–685; 1977.

Swick, H. M.; Werlin, S. L.; Dodds, W. J.; Hogan, W. J. Pharyngo-esophageal motor function in patients with myotonic dystrophy. Ann. Neurol. 10:454–457; 1981.

Talbott, J. H. Acute dermatomyositis-polymyositis and malignancy. Semin. Arthritis. Rheum. 6:305–360; 1977.

Thomas, P. K.; Calne, D. B.; Elliot, C.F. X-linked scapuloperoneal syndrome. J. Neurol. Neurosurg. Psychiatry 35:208–215; 1975.

Thompson, D. S.; Woodward, J. B.; Ringel, S. P.; Nelson, L. M. Evoked potential abnormalities in myotonic dystrophy. EEG and Clin. Neurophysiol. 56:453–456; 1983.

Tindall, R. S. A. Humoral immunity in myasthenia gravis: Biochemical characterization of acquired antireceptor antibodies and clinical correlation. Ann. Neurol. 10:437–447; 1981.

Tindall, R. S. A.; Rollins, J.; Phillips, T.; Greenlee, R. G.; Wells, L.; Belendiuk, G. Preliminary results of a double-blind, randomized, placebo-controlled trial of cyclosporine in myasthenia gravis. N. Engl. J. Med. 361:719–724; 1987.

Tsujihata, M.; Shimomura, C.; Yoshimura, T.; Sato, A.; Ogawa, T.; Tsuji, Y.; Nagataki, S.; Matsuo, T. Fatal neonatal nemaline myopathy: a case report. J. Neurol. Neurosurg. Psychiatry 46:856–858; 1983.

Tyler, F. H.; Stephens, F. E. Studies in disorders of muscles. II. Clinical manifestations and inheritance of facioscapulohumeral dystrophy in a large family. Ann. Intern. Med. 32:640; 1950.

Vignos, P. J.; Bowling, G. F.; Watkins, M. P. Polymyositis: effect of corticosteroids on final results. Arch. Intern. Med. 114:263–2677; 1964.

Walton, J. N.; Nattrass, F. J. On the classification, natural history and treatment of the myopathies. Brain 77:169–231; 1954.

Welsh, J. D.; Lynn, T. N.; Haase, G. R. Cardiac findings in 73 patients with muscular dystrophy. Arch. Intern. Med. 112:199; 1963.

Wheeler, S. D.; Brooke, M. H. Vascular insufficiency in McArdle's disease. Neurology 33(2):249–250; 1983.

Willner, J. H.; DiMauro, S.; Hays, A.; Roohi, F.; Lovelace, R. Muscle carnitine deficiency: genetic heterogeneity. J. Neurol. Sci. 41:235–246; 1979.

Woodward, W. B.; Keaton, R. K.; Simon, D. B.; Ringel, S. P. Neuropsychological defects in myotonic dystrophy. J. Clin. Neuropsych. 4:335–342; 1982.

Wulff, J. D.; Lin, J. T.; Kepes, J. J. Inflammatory facioscapulohumeral muscular dystrophy and Coats syndrome. Ann. Neurol. 12:398–401; 1982.

Yoshida, S.; Akizuki, M.; Mimori, T.; Yamagata, H.; Inada, S.; Homma, M. The precipitating antibody to an acidic nuclear protein antibody, the Jo-1, in connective tissue disease: a marker for a subset of polymyositis with interstitial pulmonary fibrosis. Arthritis Rheum. 26:604–611; 1983.

Zundel, W. S.; Tyler, F. H. The muscular dystrophies. Parts I and II. N. Engl. J. Med. 273:537, 596; 1965.

27

Diseases of the Peripheral Nervous System

GARETH J. PARRY AND AUSTIN J. SUMNER

Several factors influence prognosis in diseases of peripheral nerves. The most important is whether the neuropathy is characterized primarily by segmental demyelination or axonal degeneration. Also important are the etiology, severity, and location of the nerve damage and the acuteness of its development.

Axonal Degeneration Vs. Demyelination

At least in the acquired demyelinating neuropathies, a potential exists for rapid and complete recovery of function by remyelination, an efficient and rapid process. These neuropathies are mainly characterized by a macrophage-mediated, multifocal attack on myelin. The Schwann cell itself is not a target and retains its capacity to proliferate. Following an episode of demyelination, Schwann cells from the demyelinated segment of nerve rapidly proliferate and lay down new myelin in the demyelinated segment. In experimental animals with acute monophasic demyelination, this process begins within a few days and is complete within weeks (Saida et al. 1980). In humans

with Guillain-Barré syndrome (GBS), the quintessential acute demyelinating neuropathy, the time course is not much more protracted: in mild cases with pure demyelination, complete recovery occurs within weeks (Berger et al. 1988). In GBS, prognosis is almost entirely related to the degree of associated axonal degeneration. In the chronic inflammatory demyelinating polyradiculoneuropathies (CIDP), this process of remyelination may be frustrated somehow and recovery is less rapid and less assured. Even so, the potential for recovery with treatment in CIDP is still largely related to the degree of associated axonal degeneration. In the inherited demyelinating neuropathies, demyelination is a much more indolent process and may arise as the result of a pathological process directed at the Schwann cell, rather than at myelin, or may be secondary to axonal disease. Remyelination in this instance is ineffective, although Schwann cell proliferation is still active, and very thinly myelinated axons are surrounded by redundant Schwannn cell processes (onion bulbs). Nonetheless, even in this situation, the functional prognosis may

375

depend primarily on the degree of associated axonal degeneration. Even in early childhood, severe demyelination with onion bulb formation is seen without a clinically detectable abnormality of function. Progressive loss of function later in life is paralleled closely by progressive axonal degeneration, rather than by any change in the state of myelination (Dyck et al. 1989).

In contrast, recovery in neuropathies, in which the primary pathological process is axonal degeneration, is less assured. If the cause of the neuropathy can be removed, axons have the ability to sprout and elongate, if the perikaryon is intact. Therefore, in distal axonopathies, in which axons die back from their receptor or effector organ, some recovery of function by way of axonal regeneration can be expected. However, wallerian degeneration from proximal sites invariably carries a poor prognosis.

Severity

It may appear obvious that more severe neuropathies have a worse prognosis, but this is not inevitably the case. In GBS, patients with complete flaccid quadriplegia with respiratory and bulbar paralysis may recover quickly and completely if there is little associated axonal degeneration. However, in general, prognosis is directly related to severity. In the aforementioned example of GBS, significant associated axonal degeneration usually would occur and recovery likely would be slow and incomplete. With axonal degeneration, prognosis depends on severity. Recovery of function may take place by two mechanisms, namely axonal regeneration and collateral sprouting. If there is only partial loss of axons, recovery may take place by collateral sprouting from the preterminal nodes of Ranvier of surviving axons. In this situation, elongation of these preterminal sprouts occurs over very short distances, before synaptic contact is made. The process of recovery is therefore quick and effective, taking place over a period of weeks to months, and the chance for aberrant innervation is negligible. In muscle, complete restoration of normal function

is possible, even when as much as 70% of the motor axons are lost (Daube 1985). When severe axonal degeneration exceeds the capacity of the motor neuron to reinnervate by collateral sprouting, recovery by way of axonal regeneration is possible, but functional recovery usually is limited. Not only do axons regenerate poorly but the more axons that are damaged, the more likely that misdirected regeneration will occur. Therefore, the primary determinant of recovery in axonal degeneration is the severity of the damage.

Location

When a pathological attack is directed primarily at the proximal segments of peripheral nerves and axonal degeneration is present, potential for recovery is limited for several reasons. First, axonal injury in close proximity to the perikaryon may result in cell death. Second, axons regenerate slowly, elongating at less than 1 mm/day, and they regenerate poorly over long distances. Third, there is an increased probability of misdirected regeneration, which may impair the quality of functional recovery even when significant numbers of axons successfully regenerate. In addition, with an attack on the dorsal root, there will be degeneration of the central processes of the sensory neurons and regeneration into the central nervous system (CNS) will not occur.

Acute Vs. Chronic Neuropathies

With acutely evolving neuropathies, the patient is likely to seek medical advice before pathological changes have become advanced or irreversible. Therefore, if the cause can be treated or removed, potential for recovery is good. In contrast, with chronic neuropathies, the insidious onset and progression enables maximal compensation to occur during the evolution, and the patient may not present clinically until the severity of the neuropathy has exceeded his or her ability to compensate. In this situation, the best that can be expected is prevention of further progression.

Etiology

Recovery from peripheral neuropathies depends on removal of the cause, which often is impossible because 25% to 40% of neuropathies are idiopathic. Even when the etiology is apparent, there may be no effective treatment. In general, idiopathic and untreatable neuropathies usually progress inexorably but at a slow pace. If the cause can be removed or treated, progression usually will halt, and some recovery then occurs, the rapidity and efficacy of which is determined by the factors discussed previously.

The Role of Electrodiagnosis in Determining Prognosis

As stated previously, the primary determinant of prognosis in neuropathies is the relative proportion of demyelination and axonal degeneration. The only reliable means by which to determine this is electrodiagnostic evaluation. In addition, the electrophysiological studies provide invaluable information regarding localization, severity, and acuteness, which may not be available by other means. The presence of conduction block, differential dispersion of responses, and severe conduction slowing point toward a primary demyelinating process and indicate a potential for excellent recovery. Low-amplitude responses without dispersion or conduction block and relatively preserved conduction velocity are characteristic nerve conduction changes seen with axonal degeneration. Furthermore, with needle electromyography (EMG), the presence of fibrillation potentials indicates recent or ongoing denervation. Large amplitude, long duration, polyphasic motor unit potentials indicate the presence of partial denervation with reinnervation by collateral sprouting or regeneration, which indicates the chronicity of the neuropathy. Highly polyphasic, extremely long duration motor unit potentials, usually of low amplitude (nascent units), may be seen in muscles in

which the degree of denervation has exceeded the ability of the nerve to reinnervate by collateral sprouting. These nascent units reflect reinnervation by regeneration and are usually seen following severe wallerian degeneration lesions (trauma or infarction). The electrodiagnostic changes also can be evaluated in a semiquantitative manner to provide objectivity in the assessment of severity and are invaluable in the longitudinal assessment of improvement or deterioration in patients with neuropathy. Finally, the presence of covert changes in both nerve conduction studies and EMG allows for refinement of the clinical localization. In most neuropathies, the clinical manifestations are distally accentuated, particularly in the legs. However, the clinical picture may be similar in distal axonopathy, sciatic neuropathy, or multiple lumbosacral radiculopathies, all of which may cause distally accentuated weakness and sensory loss. In such cases, there may be denervation in muscles innervated by the posterior primary ramus of the ventral root (erector spinae) or of proximal muscles of a myotome or peripheral nerve distribution, even when strength is normal.

Prognosis is different for the wide range of disorders that involve the peripheral nerves. The clinical presentation of neuropathy ranges from acute fulminant paralysis with respiratory impairment to slow onset with evolution over a period of half a century. The manifestations may be predominantly motor, predominantly sensory, or combined motor and sensory or autonomic. Generally, symptoms are most prominent distally, but there are exceptions. Neuropathy may be asymptomatic, demonstrated only by neurophysiological testing. Depression of ankle jerks might be the only finding to alert the clinician of a neuropathy. This chapter separates acutely evolving neuropathies from those that are more likely to present with a subacute or chronic evolution and discusses prognosis in selected examples of each.

The terms polyneuropathy, polyneuritis, peripheral neuropathy, and neuropathy are synonomous. Multiple mononeuropathy or mononeuritis multiplex occurs when multiple focal lesions randomly affect different nerves or roots. These lesions, whether ischemic or demyelinating, generally are asymmetric.

Acutely Evolving Polyneuropathies

Guillain-Barré Syndrome

GBS is an acute inflammatory demyelinating polyradiculoneuropathy that is characterized clinically by acutely evolving paralysis, usually heralded by parasthesias and often accompanied by autonomic instability. Weakness involves proximal and distal muscles but usually is distally accentuated. Severity may range from minimal limitation of function to complete flaccid paralysis, including respiratory and bulbar muscles. Although sensory symptoms often are prominent, objective sensory loss is minimal, but areflexia or hyporeflexia is almost invariable. Approximately 70% of patients report an antecedent viral illness, usually gastroenteritis or an upper respiratory infection. The most commonly identified viruses are cytomegalovirus and Epstein-Barr virus, although the offending organism usually is not identified. Other antecedent events include other infections, surgery, vaccination, and a variety of systemic diseases. Neurological symptoms typically appear within 7 to 10 days of the preceding event and evolve rapidly, reaching a nadir within 2 to 3 weeks. The cardinal feature used to define the syndrome is albuminocytologic dissociation in the cerebrospinal fluid (CSF); however, this is variable. CSF pleocytosis is not uncommon and is the rule with human immunodeficiency virus (HIV)-associated disease. Normal protein may be seen, particularly early. The diagnosis almost always can be confirmed with nerve conduction studies that show multifocal conduction block and conduction slowing. The pronounced pathological changes are lymphoid cellular infiltration associated with segmental foci of demyelination in ventral roots, limb girdle plexuses, and proximal nerve trunks; changes also occur in dorsal roots, autonomic ganglia, and distal peripheral nerves.

GBS is usually monophasic with a low mortality, and complete or partial recovery almost always occurs. In a retrospective study of 40 patients collected for 42 years, Kennedy and coworkers (1978) found complete recovery (judged by return to all former activities), in 31 and partial recovery in seven of the 38 patients who survived the acute illness. Andersson and Siden (1982) also found a high rate or recovery, with 54 of 56 survivors making a complete functional recovery, 30% within 3 months, 73% within 6 months, and the remainder within 18 months. Several other studies (Loffel et al. 1977; Marshall 1963; Ravn 1967; Wiederholt et al. 1964) have reported complete functional recovery in more than 75% of cases. In contrast, McLeod and colleagues (1976) found that less than 50% of 18 patients recovered completely, but only four had significant disability. The difference probably relates to differences in data collection. The larger epidemiological studies rely more on patient reports of recovery or retrospective evaluation of medical records. McLeod and associates (1976) examined their patients prospectively and documented mild abnormalities of neurological examination, such as absent ankle reflexes and impaired two-point discrimination, abnormalities that are unlikely to have functional significance.

Although most patients make an excellent recovery, approximately 15% are left with significant disability. The most important determinant of rate and efficacy of recovery is the degree of axonal degeneration that accompanies the demyelination. Remyelination is an efficient and rapid process that is usually completed within 4 to 6 months of the nadir of the illness. Thus, severe neurological disability resulting solely from demyelination may recover completely within approximately 6 months of the onset of clinical disease. Conversely, axonal regeneration is a slow and inefficient process, with elongation of regenerating sprouts occurring at a rate of less than 1 mm/day. Therefore, recovery occurs over a period of many months and may not be maximal for up to 2 years. Furthermore, even if regeneration occurs, the axon may fail to make synaptic contact with an appropriate receptor or effector organ, so functional recovery may fail. Additionally, synaptic contact may be made with inappropriate end organs (aberrant regeneration), resulting in abnormal function.

Several studies have sought to identify factors that might enable a prediction of outcome at the time of the acute illness. It has been suggested that CSF pleocytosis (Kaeser 1964), severe sensory loss, or a prolonged progressive phase (Loffel et al. 1977; Osler and Sidell, 1960) and papilledema (Morely and Reynolds 1966) carry an ominous prognosis. However, others (Duvoison 1960; Forster et al. 1941; Miller et al. 1988; Pleasure et al. 1968) have disputed these claims. Winer and associates (1985) and Miller and coworkers found that the most powerful and reliable clinical predictor of poor outcome was the need for ventilator assistance during the acute illness. Ravn (1967) and Loffel and colleagues (1977) noted that severe maximum motor impairment carried a poorer prognosis. Children also have a significantly higher incidence of permanent and significant neurological sequelae (Peterman et al. 1959; Ravn 1967) but paradoxically, the mortality in children is lower (Berglund 1954; Peterman et al. 1959).

The most reliable predictor of outcome in GBS is the result of nerve conduction studies and EMG. Miller and coworkers (1988), Cornblath and colleagues (1988), McKhann and associates (1988), and Ropper and colleagues (1990) found that a low-amplitude compound muscle action potential (CMAP), even when recorded during the early stages of the disease, carried an ominous prognosis. In the former study, 60 patients who were bed-ridden, 22 of whom were ventilator dependent, were analyzed retrospectively to see which factors correlated significantly with inadequate recovery. The only electrophysiological parameter that correlated significantly was the low CMAP amplitude. They found no significant correlation between poor outcome and density of fibrillation potentials found with needle EMG. The studies by Cornblath and coworkers and McKhann and colleagues report the results of the North American multicenter

trial of plasmapheresis for GBS, in which no needle EMG data were gathered. The only significant correlation with poor outcome was low-amplitude CMAP. However, even in patients in whom the CMAP amplitude was 0% to 20% of normal, outcome was improved with plasmapheresis.

Pleasure and associates (1968) were the first to suggest that the presence of fibrillation potentials in weak muscles at the nadir of weakness predicted a poor outcome. Their observations were confirmed in more extensive studies by Eisen and Humphreys (1974) and McLeod (1976). However, Miller and coworkers (1988) found no such correlation. Ropper and associates (1990) found widespread fibrillation in 10 of 113 patients, most of whom were studied in the first 3 weeks of illness. Only the two patients in whom the CMAP amplitude was severely reduced did not have complete recovery. Because recovery primarily depends on the degree of axonal degeneration, it is not surprising that patients with electrodiagnostic evidence of severe axonal degeneration have a worse outcome. Fibrillation of muscle indicates loss of the trophic influence of axons on muscle fibers and indicates axonal degeneration. A low-amplitude CMAP may be due to acute axonal degeneration. Conduction block in distal intramuscular nerve twigs can give the same result, but the CMAP usually is dispersed and the distal motor latency prolonged. Nonetheless, if the amplitude is reduced to less than 10% of normal, regardless of whether it is initially due to severe conduction block or axonal degeneration, the result is severe axonal degeneration and carries a poor prognosis.

Although, as strictly defined, GBS is monophasic, a few patients do relapse. Some develop a relapsing chronic demyelinating neuropathy. Overall, approximately 5% of patients have a single relapse and never develop a chronic disorder. Osterman and coworkers (1988) and Ropper and colleagues (1990) suggest that relapses are more likely in patients treated with plasmapheresis. However, neither the North American nor the French multicenter studies of several hundred cases found a significant difference in the relapse rates between treated and untreated patients.

Acute Sensory Neuronopathy

An acutely evolving syndrome of patchy sensory loss, predominantly affecting functions subserved by large myelinated axons, also may occur as an infectious or postinfectious syndrome (Sterman et al. 1980, Windebank et al. 1990). Although often there are complaints of weakness, objective strength testing is normal. Symptoms typically appear within days of the viral illness and are of acute onset and progression, reaching a nadir within 1 week. A subacute disorder, identical except for its rate of progression, may occur as a paraneoplastic syndrome (Horwich et al. 1977) and with Sjögren's syndrome (Malinow et al. 1986). A chronic sensory neuronopathy occurs with pyridoxine intoxication (Parry 1985).

In acute sensory neuronopathy, prognosis is poor, but some patients make a complete recovery. In one series (Windebank et al. 1990) eight of 42 patients made a complete recovery, and 22 recovered sufficiently to return to work. The remainder were severely disabled. The subacute form has an even worse outlook with few, if any, patients improving; this is partly because the cause is difficult or impossible to treat. In the chronic form, associated with pyridoxine intoxication, steady improvement occurs after stopping the pyridoxine. The degree of recovery depends primarily on the severity of the neuropathy. Patients with severe sensory ataxia, particularly if it involves the arms, seldom return completely to normal. The persistent abnormalities may be due to degeneration of the central projections of the sensory neurons in the rostral portion of the posterior columns, where regeneration cannot take place.

Acute Inflammatory Autonomic Neuropathy (Pandysautonomia)

Acute inflammatory autonomic neuropathy is a rare condition with an etiology that is not understood, but it may be the autonomic counterpart of GBS. Postural hypotension,

cramping, abdominal pain, and varying amounts of diarrhea are the principle symptoms. It may run a relapsing course.

Prognosis for full recovery in acute dysautonomia is guarded. Young and associates (1975) showed a good prognosis in their cases who all had excellent recovery. However, Hart and Kanter (1990), in their review, noted that only 40% of reported cases recovered completely, and another 48% had partial recovery. Time course for recovery is several months to 2 years.

Brachial Neuritis

Brachial neuritis is an idiopathic, acute, painful paralysis of the arm. It most commonly involves the muscles of the shoulder girdle, but involvement of more distal (e.g., anterior interosseus) and more proximal (e.g., spinal accessory) nerves may be seen. Occasionally, it is bilateral but it is always markedly asymmetrical. Recurrent and familial forms are seen. An analogous neuropathy involving the lumbosacral plexus occasionally may occur.

The prognosis for recovery is generally good. The pain, which may be severe, almost invariably resolves over a period of weeks. However, treatment with narcotic analgesics usually is needed. Although corticosteroids do not influence the rate or efficacy of recovery from the paralysis, it has been suggested that they may shorten the painful phase (Sumner 1990). Recovery from the paralysis is less predictable but is usually good. In one study (Tsiaris et al. 1972), 80% of patients fully recovered, and 18% were still improving at the time of follow-up. Similarly, Devathasan and Tong (1980) found complete recovery of motor function in 18 of 19 patients followed for 2 years; 12 within 6 months and five by 1 year. In general, the most reliable predictor of outcome is the severity of the paralysis. With mild to moderate weakness and survival of 10% to 30% of motor units on EMG, full functional recovery can occur in 3 to 6 months. With complete or near-complete denervation, prognosis is much more guarded. However, some cases with severe weakness and atrophy have a remarkable recovery, perhaps reflecting the site of pathology. The lesion in brachial neuritis may be located distally (England and Sumner 1987), peerhaps in the intramuscular nerve terminals (Kraft 1969), so that recovery by regeneration over these short distances is possible.

Porphyria

The dominantly inherited hepatic porphyrias (acute intermittent porphyria, variegate porphyria, and coproporphyria) also may present with an acute neuropathy mimicking GBS (Ridley 1984). Manifestations usually begin at puberty. Abdominal, loin, or back pain followed by restlessness, crying, and hysterical or neurotic behavior occur. Barbiturates worsen the situation. In variegate porphyria and hereditary coproperphyria, cutaneous photosensitivity (unrelated to the attacks) may occur. Motor involvement can be acute and symmetrical and may involve respiratory muscles and cranial nerves. Sensory symptoms can be distal or proximal with a "bathing trunk" or "long johns" distribution. Muscle pain and aching may be prominent. Tendon reflexes are unpredictable but are usually diminished or absent. There may be prominent autonomic instability and signs of CNS involvement with encephalopathy and seizures. The CSF is acellular with a modest elevation of protein. The predominant pathology is distal axonopathy with preferential involvement of short motor axons, but there may be some secondary paranodal demyelination (Albers 1991).

Although porphyric neuropathy is primarily an axonal degeneration neuropathy, prognosis for recovery of function is good. There is significant mortality during the acute phase, probably exacerbated by delayed diagnosis in many cases. This early mortality mainly is related to autonomic failure. Patients who survive the acute attack steadily improve, and even those with severe paralysis usually recover functionally after many months, although some weakness may be permanent (Sorensen and With 1971).

Polyarteritis Nodosa

The neuropathy associated with polyarteritis nodosa (PAN) is the quintessential acute, multifocal, axonal neuropathy and results from nerve infarction secondary to necrotizing vasculitis. Similar ischemic neuropathies occur with the other vasculitides (e.g., rheumatoid arthritis, systemic lupus erythematosis, Sjögren's syndrome) but are usually less severe and tend to be more chronic. Although PAN is rare, neuropathy occurs in approximately two thirds of cases and may be the presenting feature (Lovelace 1964). Additional clinical features include abdominal pain, purpura, ulcers, asthma, arthralgia, hypertension, renal sediment abnormalities, and anemia. Extensive arterial involvement precedes neuropathy. The mononeuropathies are abrupt in onset, and pain and numbness occur in the distribution of the affected nerve, quickly followed by weakness. Confluence of multiple nerve lesions can result in a symmetrical sensorimotor polyneuropathy (Parry 1985). Hepatitis B surface antigen has been found in 10% to 50% of patients. Complement levels usually are normal but may be depressed (mixed cryoglobulinemia). Proteinuria or red cell casts suggest renal involvement. Muscle and whole sural nerve biopsies can be useful, and visceral angiography should be considered if biopsy does not show necrotizing vasculitis. An acutely evolving axonal mononeuropathy multiplex is a clear indication for high-dose corticosteroids, which probably should be combined with cyclophosphamide.

Prognosis for recovery from systemic necrotizing vasculitis is guarded. Early recognition and aggressive treatment of the underlying disease limits the extent of neurological damage. Recovery from the associated mononeuropathies is determined primarily by the severity of the axonal degeneration. With partial nerve infarcts, some recovery of function occurs by two mechanisms. First, with mild nerve infarcts, there may be ischemic conduction block, which resolves rapidly over days to weeks

(Parry and Linn 1988). This usually accounts for only a small proportion of the clinical deficit, but some rapid resolution may occur. The second mechanism involves reinnervation by terminal collateral sprouting of surviving motor axons. Up to one third of motor axons may be lost with full strength restored by this mechanism (Daube 1985). Therefore, full recovery of function may occur with mild lesions. Because these distal branches must only elongate a short distance before establishing synaptic contact, recovery of some function may occur within weeks to months. Reinnervation by regeneration is unlikely in ischemic nerve lesions, perhaps because of ischemic damage to Schwann cells and the perineurium. As a result, recovery is poor in more severe lesions. Although recovery of neurological function is poor, the pain associated with these acute nerve infarcts usually resolves spontaneously; however, it may take several months and sometimes as long as 2 years.

Toxic Neuropathies

Toxic neuropathies may be an acute, but delayed, result of massive exposure to a neurotoxin. More commonly, they are subacute or chronic, secondary to more protracted exposure. Causes of acute delayed neuropathy include organophosphates, arsenic, and thallium. Subacute and chronic toxic neuropathies may occur with many toxins, including many medications. An important phenomenon that occurs with some toxic neuropathies is "coasting" (Schaumburg and Spencer 1984). After exposure to the neurotoxin is stopped, there is continuing worsening of the neurological signs and symptoms for days to weeks before recovery begins. This has important implications for prognosis. Most toxic neuropathies are distal axonopathies (Cavanaugh 1985; Schaumburg and Spencer 1984), usually with both central and peripheral axonal degeneration. With some acute, massive exposures, there may be multifocal demyelination as well, presumably from a direct toxic effect on Schwann cells, but severe axonal degeneration always predominates (Dono-

frio et al. 1987). Clinical features of toxic neuropathy seldom are sufficiently specific or characteristic to identify the toxic etiology or the specific toxin. Sensory, motor, and autonomic features occur in varying proportions, but sensory usually predominates, often with severe pain. There may be evidence of involvement of other organs, including the CNS, that may provide a clue to the etiology.

Prognosis following identification of intoxication is related to many factors. Involvement of other organs, particularly liver and kidney, may be fatal or severely disabling. CNS involvement also may limit the extent and rapidity of recovery. With regard to the neuropathic features, because these are mainly distal axonopathies, the principle determinant of recovery is the extent of the axonal degeneration. When minimal "dying-back" has occurred, regeneration with re-establishment of appropriate contact with receptor or effector organs occurs within weeks. As mentioned previously, some continued deterioration of neurological function may occur even after the toxin is removed, so it is important to recognize a toxic neuropathy as soon as possible so that irreversible changes are minimized. With more severe neuropathies, some improvement will occur, but sensory abnormalities may persist indefinitely because of the central degeneration in the rostral portion of the posterior columns.

Bell's Palsy

Bell's palsy is an acute idiopathic facial paralysis that evolves over hours to days. Commonly, it is preceded by pain in or behind the ear, with accompanying subjective numbness of the face but no objective sensory loss. Occasionally, with very proximal facial nerve involvement, there is hyperacusis, xerostomia, and xerophthalmia. It is important to distinguish Bell's palsy from facial paralysis associated with geniculate herpes zoster because the latter has a much worse prognosis.

Prognosis for recovery from Bell's palsy is excellent (Katusic et al. 1986). Without treatment, 75% recover completely within 2 to 4 weeks or occasionally longer. Fifteen percent make an excellent functional recovery with only subtle facial asymmetry apparent on examination. The remainder are left with varying degrees of facial weakness. Even in these patients, slow improvement may continue over many months. However, there is a significant risk of developing aberrant regeneration, resulting in facial synkinesis and occasionally lacrimation induced by eating (crocodile tears). Controlled trials indicate that treatment with high-dose prednisone accelerates recovery and reduces the incidence of aberrant regeneration (Wolf et al. 1978). Prognosis is closely related to the severity of the axonal degeneration. Electrodiagnostic testing 3 to 5 days after the onset of the paralysis can provide useful prognostic information. If there is partial or complete preservation of the evoked motor response, voluntary motor unit potentials and a paucity of fibrillations with needle EMG, prognosis is better. Furthermore, the rate of recovery is proportional to the amplitude of the evoked motor response (Albers and Bromberg 1990).

Subacutely Evolving Neuropathies

Alcoholic Neuropathy

There is a close association between alcoholism and peripheral nerve disease. Symptoms may be subacute or slowly progressive (Tabaraud et al. 1990; Walsh and McLeod 1970). Alcohol neuropathy may be caused by lack of thiamine and other vitamins, but alcohol might be neurotoxic. The neuropathy can be asymptomatic with mild leg thinness, muscle tenderness, ankle reflex depression, and subtle loss of distal sensation. Mild calf ache, discomfort of the soles, distal paresthesia, pain, and weakness are common. The legs invariably are affected before the arms. Motor and sensory symptoms usually occur together, although the sensory features usually predominate. One fourth of patients have distressing pain, paresthesia, and burning feet. The entire leg may become sensitive to touch. The condition usually is accompanied by excessive perspiration.

Distal weakness and atrophy usually are present, but proximal weakness and even bulbar involvement also may occur and may be ascribed erroneously to alcoholic myopathy. Postural hypotension and bladder symptoms are common. Muscle tenderness is characteristic. Lower limb ataxia from alcohol-induced anterior cerebellar vermis atrophy also may be present. CSF usually is normal, but the protein may be elevated. Other signs of alcoholism and nutritional deprivation are common, particularly Korsakoff's psychosis.

Alcoholic neuropathy is primarily a distal axonopathy, although the commonly associated malnutrition may produce significant associated demyelination. Two factors are important in determining prognosis, most importantly, the ability of the patient to stop drinking. If alcohol intake can be stopped, the neuropathy will improve and the amount of improvement that occurs is related to the severity of the neuropathy. Some numbness of the feet usually is permanent with the more severe cases, but the pain usually subsides over several months.

Chronically Evolving Neuropathies

Chronic Inflammatory Demyelinating Polyneuropathy

CIDP may run a relapsing or inexorably progressive course. It shares many clinical and pathological features with GBS. Onset can be at any age. The neuropathy usually evolves over months or years, and relapses or steps with progression can occur. Sensory complaints are as common as weakness. Generally, weakness is out of proportion to atrophy, providing a clinical clue to the underlying pathology. Cranial nerves may be involved. Signs of CNS involvement, such as extensor plantar responses and papilledema, occasionally may be seen. CSF protein usually is elevated and there may be an associated paraproteinemia.

Electrodiagnostic studies show the characteristic features of demyelination, and the amount of associated axonal degeneration is extremely variable. Particularly in the relapsing form, axonal degeneration may be minimal, while in some cases of indolently evolving CIDP, there is so much axonal degeneration that it may be difficult to recognize the primary underlying pathology. In such cases, nerve biopsy can be diagnostically helpful.

CIDP is a potentially treatable neuropathy. Most cases will respond to some form of immune manipulation. High-dose corticosteroids and plasmapheresis have been shown in controlled studies to benefit CIDP more than no treatment (Dyck 1982; Dyck et al. 1986). A host of anecdotal reports also attest to the efficacy of other immunosuppressives, particularly high-dose intravenous hyperimmune globulin (Faed et al. 1989).

Prognosis in CIDP primarily depends on the ability to control the immune-mediated demyelination. If the immune attack on myelin can be halted, dramatic recovery by way of remyelination can occur. Improvement is sometimes seen within hours and continues for days to weeks. If there is significant associated axonal degeneration, recovery is more protracted and often is incomplete. The frequent involvement of nerve roots makes full recovery impossible if there is significant axonal degeneration at this level. There is low mortality from complications of the neuropathy. In the series by Dyck and associates (1975) 11% of patients died from complications, but a lower mortality is usually found (Austin 1958). A few cases eventually go into complete remission, as a result of treatment or spontaneously. Only 4% of the patients in the series by Dyck and coworkers (1975) recovered. However, treatment was not given or was less than optimal. Approximately 60% of the patients will be able to return or continue to work. The remainder will be disabled, some confined to bed or wheelchair. There is no similar long-term follow-up for patients with CIDP who have received optimal treatment, but experience suggests that this gloomy prognosis can be improved. In the authors' experience of more than 50 patients in the last 15 years, there have been no deaths, and

most patients continue in their former activities. The chronic demyelinating neuropathy associated with monoclonal gammopathies may have a worse prognosis. However, chronic demyelinating neuropathy associated with a solitary plasmacytoma can be cured with complete removal of the tumor.

Diabetic Neuropathies

The diabetic neuropathies have considerable clinical heterogeneity and the prognosis differs for different syndromes. Broadly, these neuropathies can be divided into two categories (Dyck et al. 1987). Focal or multifocal neuropathies have an acute or subacute onset, are overtly asymmetrical, have a predilection for proximal nerves, are most likely the result of nerve ischemia, and have a reasonably good prognosis for complete or partial recovery. These include cranial, truncal, and proximal asymmetrical motor neuropathies and mononeuropathies. Distal neuropathies have a slow onset and insidious progression, are largely symmetrical, are distally accentuated, and have an obscure pathogenesis, and are inexorably progressive. It is this more common syndrome which is generally called diabetic neuropathy.

Focal and Multifocal Neuropathies

These are characterized clinically by the acute or subacute onset of pain followed by loss of function in the distribution of the affected nerve(s). The maximal clinical deficit usually is reached within hours, or at the most, days, although a more subacute course may occur. Cranial neuropathy most commonly affects the third nerve but classically spares pupillary function. Truncal neuropathy appears to have a predilection for the thoracic and upper lumbar spinal nerves although cervical and lumbar involvement, if it occurs, is more likely to be attributed to spondylosis or disc disease. Proximal motor neuropathy mainly affects the femoral and obturator nerves. The main symptom is pain, and it may be severe. Motor abnormalities are equally common but may be missed with truncal neuropathies because focal abdominal paralysis produces negligible clinical symptoms and signs (Parry and Floberg 1989).

Prognosis in each of these focal or multifocal neuropathies is similar. After the nadir is reached, the condition persists for several weeks and then slowly improves. With oculomotor neuropathy, complete recovery is the rule in 3 to 5 months, and aberrant regeneration is rare. The other focal neuropathies recover more slowly and incompletely. Pain, which may be severe and require narcotic analgesics, always improves but may take several months and sometimes as long as 2 years. Recovery from the motor deficit also is quite good and occurs in approximately the same time period. More than half the patients regain normal or near-normal function, but the remainder have significant residual disabilities (Casey and Harrison 1972). Experience and common sense indicate that the degree of motor recovery is predicated primarily by the severity of the denervation. Little information is available on the effect of glycemic control on prognosis for recovery but most recommend attempts to lower blood glucose.

Distal Neuropathies

This more common type of neuropathy is almost an inevitable consequence of long-standing diabetes. After 25 years of diabetes, 50% of patients develop symptomatic neuropathy, and the rest have asymptomatic neuropathy of varying severity (Pirart 1978). Surveys of prevalence indicate that one third to one half of patients with diabetes have symptomtic neuropathy and approximately 15% have significant disability attributable to the neuropathy (Dyck et al. 1987). On rare occasions, death may be attributable to neuropathy, presumably due to autonomic instability.

Prognosis for this type of neuropathy is less optimistic. Inexorable progression is the rule with the development of sensory, motor, and autonomic disorders in varying combinations. Apart from duration of diabetes, the only factor related to the development and progression of neuropathy is glycemic control (Dahl-Jorgensen 1987). Age, sex, family history, and obesity have no effect (Pirart 1978). Severe hyperglycemia in recently diagnosed diabetics (usually type I) may be associated with symptoms of neuropathy and abnormal nerve conduction. When this short-term hyperglycemia is controlled, symptoms resolve, and nerve conduction returns to normal. These neuropathic symptoms are probably related to functional metabolic abnormalities of the nerve and are completely reversible. Occasionally, irreversible neuropathy is present at the time of diagnosis of diabetes in older patients with mild type II diabetes, and symptoms of neuropathy may lead to the diagnosis of diabetes. In such cases, mild diabetes usually has been unrecognized for many years. Once established in chronic diabetics, objective evidence of neuropathy cannot be reversed, but symptoms can be improved. Rigid glycemic control, with diet, oral hypoglycemic drugs, insulin, or pancreatic transplantation (Kennedy et al. 1990), can only halt or reduce the rate of progression. None of the other treatments suggested for diabetic neuropathy have been shown to influence progression (Harati 1987).

Genetically Determined Neuropathies

A diverse group of inherited neuropathies affect myelin, axons, or both. Peripheral nerves also are often involved in multisystem degenerations. All inherited neuropathies are inexorably progressive, and no treatment is available. However, the rate of progression is enormous, not only between genetically distinct types of neuropathy, but also within a particular neuropathy type and even within individual kindreds. For example, completely asymptomatic people with hereditary motor sensory neuropathy type I may be discovered only when their children or grandchildren present with severe disease in childhood. Furthermore, siblings with similar age of onset of symptoms may progress at different rates. Thus, it is difficult to anticipate the prognosis of neuropathy at the time of diagnosis.

References

Albers, J. W.; Bromberg, M. B. Bell's Palsy. In: Johnson, R. T., ed. Current therapy in neurologic disease. 3rd ed. Philadelphia: B. C. Decker Inc.; 1990: p 376–380.

Albers, J. W. Porphyric neuropathy. In press.

Andersson, T.; Siden, A. A clinical study of the Guillain-Barré syndrome. Acta. Neurol. Scand. 66:316–327; 1982.

Asbury, A. K. Focal and multifocal neuropathies of diabetes. In: Dyck, P. I.; Thomas, P. K.; Asbury, A. K.; Winegrad, A. I.; Porte, D., eds. Philadelphia: W. B. Saunders; 1987: p 45–55.

Austin, J. H. Recurrent polyneuropathies and their corticosteroid treatment; with five-year observations of a placebo-controlled case treated with corticotrophin, cortisone and prednisone. Brain 81:157–194; 1958.

Berger, A. R.; Logigian, E. L.; Shahani, B. T. Reversible proximal conduction block underlies rapid recovery in Guillain-Barré syndrome. Muscle Nerve 11:1039–1042; 1988.

Berglund, A. Polyradikulonevriter. Nord. Med. 52:1091–1095; 1954.

Casey, E. B.; Harrison, M. J. G. Diabetic amyotrophy: a follow-up study. Br. Med. J. 1:656–659; 1972.

Cavanagh, J. B. Mechanisms of damage by chemical agents. In: Swash, M.; Kennard, C., eds. Scientific basis of neurology. Edinburgh: Churchill Livingstone; 1985: p 631–645.

Cornblath, D. R.; Mellits, E. D.; Griffin, J. W.; McKhann, G. M.; Albers, J. W.; Miller, R. G.; Feasby, T. E.; Quaskey, S. A.; and the Guillain-Barré study group. Motor conduction studies in Guillain-Barré syndrome: description and prognostic value. Ann. Neurol. 23:354–359; 1988.

Dahl-Jorgernsen, K. Near-normoglycemia and late diabetic complications. The Oslo study. Acta. Endocrinol. 115(suppl. 284):1–38; 1987.

Daube, J. R. Electrophysiologic studies in the diagnosis and prognosis of motor neuron diseases. Neurol. Clin. 3:473–481; 1985.

Devathasan, G.; Tong, H. I. Neuralgic amyotrophy: criteria for diagnosis and a clinical with electromyographic study of 21 cases. Aust. N. Z. J. Med. 10:188–191; 1980.

Donofrio, P. D.; Wilbourn, A. J.; Albers, J. W.; Rogers, L.; Salanga, V.; Greenberg, H. S. Acute arsenic intoxication presenting as Guillain-Barré-like syndrome. Muscle Nerve 10:114–120; 1987.

Duvoisin, R. C. Polyneuritis: clinical review of 23 cases of Landry-Guillain-Barré syndrome. U. S. Armed Forces Med. J. 11:1294–1306; 1960.

Dyck, P. J.; Lais, A. C.; Ohta, M.; Bastron, J. A.; Okazaki, H.; Groover, R. V. Chronic inflammatory polyradiculoneuropathy. Mayo Clin. Proc. 50:621–637; 1975.

Dyck, P. J.; O'Brien, P. C.; Oviatt, K. F.; Dinapoli, R. P.; Daube, J. R.; Bartleson, J. D.; Mokri, B.; Swift, T.; Low, P. A.; Windebank, A. J. Prednisone improves chronic inflammatory demyelinating polyradiculoneuropathy more than no treatment. Ann. Neurol. 11:136–141; 1982.

Dyck, P. J.; Daube, J. R.; O'Brien, P.; Pineda, A.; Low, P. A.; Windebank, A. J.; Swanson, C. Plasma exchange in chronic inflammatory demyelinating polyradiculoneuropathy. N. Engl. J. Med. 314:416–465; 1986.

Dyck, P. J.; Karnes, J.; O'Brien, P. C. Diagnosis, staging, and classification of diabetic neuropathy and associations with other complications. In: Diabetic neuropathy. Dyck, P. J.; Thomas, P. K.; Asbury, A. K.; Winegrad, A. I.; Porte, D., eds. Philadelphia: W. B. Saunders; 1987:p 36–44.

Dyck, P. J.; Karnes, J. L.; Lambert, E. H. Longitudinal study of neuropathic defects and nerve conduction abnormalities in hereditary motor and sensory neuropathy Type I. Neurology 39:1302–1308; 1989.

Eisen, A.; Humphreys, P. The Guillain-Barré syndrome. A clinical and electrodiagnostic study of 25 cases. Arch. Neurol. 30:438–443; 1974.

England, J. D.; Sumner, A. J. Neuralgic amyotrophy: an increasingly diverse entity. Muscle Nerve 10:60–68; 1987.

Faed, J. M.; Day, B.; Pollock, M.; Taylor, P. K.; Nukada, H.; Hammond-Tooke, G. D. High-dose intravenous human immunoglobulin in chronic inflammatory demyelinating polyneuropathy. Neurology 39:422–425; 1989.

Forster, F. M.; Brown, M.; Merritt, H. H. Polyneuritis with facial diplegia: a clinical study. N. Engl. J. Med. 225:51–56; 1941.

French Cooperative Group on Plasma Exchange in Guillain-Barré Syndrome. Efficiency of plasma exchange in Guillain-Barré syndrome: role of replacement fluids. Ann. Neurol. 22:753–761; 1987.

The Guillain-Barré Study Group. Plasmapheresis and acute Guillain-Barré syndrome. Neurology 35:1096–1104; 1985.

Harati, Y. Diabetic peripheral neuropathies. Ann. Intern. Med. 107:546–559; 1987.

Hart, R. G.; Kanter, M. C. Acute autonomic neuropathy. Two cases and a clinical review. Arch. Intern. Med. 150:2373–2376; 1990.

Horwich, M. S.; Cho, L.; Porro, R. S.; Posner,

J. B. Subacute sensory neuropathy: a remote effect of carcinoma. Ann. Neurol. 2:7–19; 1977.

Kaeser, H. E. Klinische und elektromyographische verlaufsuntersuchungen beim Guillain-Barré-syndrome. Schweiz. Arch. Neurol. Neurochir. Psychiat. 94:278–286; 1964.

Katusic, S. K.; Beard, C. M.; Wiederholt, W. C.; Bergstralh, E. J.; Kurland, L. T. Incidence, clinical features, and prognosis in Bell's palsy, Rochester, Minnesota, 1968–1982. Ann. Neurol. 20:622–627; 1986.

Kennedy, R. H.; Danielson, M. A.; Mulder, D. W.; Kurland, L. T. Guillain-Barré syndrome: a 42 year epidemiologic and clinical study. Mayo Clin. Proc. 53:93–99; 1978.

Kennedy, W. R.; Navarro, X.; Goetz, F. C.; Sutherland, D. E. R.; Najarian, J. S. Effects of pancreatic transplantation on diabetic neuropathy. N. Engl. J. Med. 322:1031–1037; 1990.

Kraft, G. H. Multiple distal neuritis of the shoulder girdle: an electromyographic clarification of "paralytic brachial neuritis." Electroencephalogr. Clin. Neurophysiol. 27:722; 1969.

Loffel, N. B.; Rossi, L. N.; Mumenthaler, M.; Lutschg, J.; Ludin, H-P. The Landry-Guillain-Barré syndrome. Complications, prognosis and natural history in 123 cases. J. Neurol. Sci. 33:71–79; 1977.

Lovelace, R. E. Mononeuritis multiplex in polyarteritis nodosa. Neurology 14:434–442; 1964.

Malinow, K.; Yannakakis, G. D.; Glusman, S. M.; Edlow, D. W.; Griffin, J.; Pestronk, A.; Powell, D. L.; Ramsey-Goldman, R.; Eidelman, B. H.; Medsger, T. A.; Alexander, E. L. Subacute sensory neuronopathy secondary to dorsal root ganglionitis in primary Sjogren's syndrome. Ann. Neurol. 20:535–537; 1986.

Marshall, J. The Landry-Guillain-Barré syndrome. Brain 86:55–66; 1963.

McKhann, G. M.; Griffin, J. W.; Cornblath, D. R.; Mellits, E. D.; Fisher, R. S.; Quasket, S. A.; and the Guillain-Barré study group. Plasmapheresis and Guillain-Barré syndrome: analysis of prognostic factors and the effects of plasmapheresis. Ann. Neurol. 23:347–353; 1988.

McLeod, J. G.; Walsh, J. C.; Prineas, J. W.; Pollard, J. D. Acute idiopathic polyneuritis. A clinical and electrophysiological follow-up study. J. Neurol. Sci. 27:145–162; 1976.

Miller, R. G.; Peterson, G. W.; Daube, J. R.; Albers, J. W. Prognostic value of electrodiagnosis in Guillain-Barré syndrome. Muscle Nerve 11:769–774; 1988.

Morley, J. B.; Reynolds, E. H. Papilloedema and the Landry-Guillain-Barré syndrome. Brain 89:205–222; 1966.

Osler, L. D.; Sidell, A. D. The Guillain-Barré syndrome. The need for exact diagnostic criteria. N. Engl. J. Med. 262:964–969; 1960.

Osterman, P. O.; Fagius, J.; Safwenberg, J.; Wikstrom, B. Early relapse of acute inflammatory polyradiculoneuropathy after successful treatment with plasma exchange. Acta. Neurol. Scand. 77:273–277; 1988.

Parry, G. J. G. Mononeuropathy multiplex. (AAEE case report #11). Muscle Nerve 8:493–498; 1985.

Parry, G. J.; Bredesen, D. E. Sensory neuropathy with low-dose pyridoxine. Neurology 35:1466–1468; 1985.

Parry, G. J.; Linn, D. J. Conduction block without demyelination following acute nerve infarction. J. Neurol. Sci. 84:265–273; 1988.

Parry, G. J.; Floberg, J. Diabetic truncal neuropathy presenting as abdominal hernia. Neurology 39:1488–1490; 1989.

Peterman, A. F.; Daly, D. D.; Dion, R. F.; Keith, H. M. Infectious neuronitis (Guillain-Barré syndrome) in children. Neurology 9:533–539; 1959.

Pirart, J. Diabetes mellitus and its degenerative complications: a prospective study of 4400 patients observed between 1947 and 1973. Diab. Care. 1:168–188; 252–263; 1978.

Pleasure, D. E.; Lovelace, R. E.; Duvoisin, R. C. The prognosis of acute polyradiculoneuritis. Neurology 18:1143–1148; 1968.

Ravn, H. The Landry-Guillain-Barré syndrome. A survey and a clinical report of 127 cases. Acta Neurol Scand 43(suppl. 30):1–64; 1967.

Ridley, A. Porphyric neuropathy. In: Dyck, P. J.; Thomas, P. K.; Lambert, E. H.; Bunge, R., eds. Philadelphia: W. B. Saunders; 1984:p 1704–1716.

Ropper, A. H.; Albers, J. W.; Addison, R. Limited relapse in Guillain-Barré syndrome after plasma exchange. Arch. Neurol. 45:314–315; 1988.

Ropper, A. H.; Wijdicks, E. F. M.; Shahani, B. T. Electrodiagnostic abnormalities in 113 consecutive patients with Guillain-Barré syndrome. Arch. Neurol. 47:881–887; 1990.

Saida, K.; Sumner, A. J.; Saida, T.; Brown, M. J.; Silberberg, D. H. Antiserum-mediated demyelination: relationship between remyelination and functional recovery. Ann. Neurol. 8:12–24; 1980.

Schaumburg, H. H.; Spencer, P. S. Human toxic neuropathy due to industrial agents. In: Dyck, P. J.; Thomas, P. K.; Lambert, E. H.; Bunge, R., eds. Peripheral neuropathy. Philadelphia: W. B. Saunders; 1984:p 2115–2132.

Sorensen, A. W. S.; With, T. K. Persistent paresis after porphyric attacks. S. Afr. Med. J. 45(special issue):101–103; 1971.

Sterman, A. B.; Schaumburg, H. H.; Asbury,

A. K. The acute sensory neuronopathy syndrome: a distinct clinical entity. Ann. Neurol. 7:354–358; 1980.

Sumner, A. J. Brachial neuritis. In: Johnson, R. T., ed. Current therapy in neurologic disease. 3rd ed. Philadelphia: B. C. Decker Inc.; 1990:p 374–375.

Tabaraud, F.; Vallat, J. M.; Hugon, J.; Ramiandrisoa, H.; Dumas, M.; Signoret, J. L. Acute or subacute alcoholic neuropathy mimicking Guillain-Barré syndrome. J. Neurol. Sci. 97:195–205; 1990.

Tsiaris, P.; Dyck, P. J.; Mulder, D. W. Natural history of brachial plexus neuropathy. Arch. Neurol. 27:109–117; 1972.

Walsh, J. C.; McLeod, J. G. Alcoholic neuropathy: an electrophysiological and histological study. J. Neurol. Sci. 10:457–469; 1970.

Wiederholt, W. C.; Mulder, D. W.; Lambert, E. H. The Landry-Guillain-Barré-Strohl syndrome or polyradiculoneuropathy: historical review, report on 97 patients, and present concepts. Mayo Clin. Proc. 39:427–451; 1964.

Windebank, A. J.; Blexrud, M. D.; Dyck, P. J.; Daube, J. R.; Karnes, J. L. The syndrome of acute sensory neuropathy: clinical features and electrophysiological and pathological changes. Neurology 40:584–591; 1990.

Winer, J. B.; Greenwood, R. J.; Hughes, R. A. C.; Perkin, G. D.; Healy, M. J. R. Prognosis in Guillain-Barré syndrome. Lancet 1:1202–1203; 1985.

Wolf, S. M.; Wagner, J. H.; Davidson, S.; Forsythe, A. Treatment of Bell's palsy with prednisone: a prospective, randomized study. Neurology 28:158–161; 1978.

Young, R. R.; Asbury, A. K.; Corbett, J. L.; Adams, R. D. Pure pandysautonomia with recovery. Brain 98:613–636; 1975.

28

Peripheral Nerve Injury

CARL E. LOWDER AND DAVID G. KLINE

Peripheral nerve injury during war or peace rarely results in significant mortality, but morbidity due to pain and serious loss of function is common (Horowitz and Rizzoli 1967; Woodhall and Beebe 1956). Prognosis for improvement depends on many factors, all of which may aid or hinder the nerve's natural response to injury—regeneration. Regeneration always involves axon regrowth except in the mildest classification of nerve injury, Seddon's neurapraxia (Seddon 1943) or Sunderland's first-degree injury in which axonal continuity is maintained (Sunderland 1968). Thus, the neuron's capacity to restore significant axonal continuity is maximal after most nerve injuries. The pathophysiology of peripheral nerve injury and regeneration has been studied extensively. An initial period of degeneration is followed by a complex regenerative process (Ducker 1985; Sunderland 1981). Thus, when significantly injured, the neuron and its axon are involved in four phases of regrowth: (1) initial delay—the time required for the neuron soma to recover, sprout new axons, and grow to the level of injury; (2) scar delay—the time required for axonal growth through the lesion; (3) intermediate delay—the time required for axons to transverse old or new distal endoneurial tubes to the end-organ level; and (4) terminal delay—the time involved in restoration of function of the end-organ (McQuarrie 1985; Sunderland 1981).

Because regeneration can be delayed or arrested at any of these four stages, there is wide variation in the course of recovery of injuries and the rate of regeneration. The predictive and prognostic factors vary and decisions regarding conservative management approaches vs. surgical intervention can be difficult.

Classification of Injury Related to Prognosis

Axon number and diameter, degree of myelination, fascicular arrangement, and relative amount of connective tissue vary from nerve to nerve and within a nerve, level to level. Thus, the effect of an injury can vary greatly depending on the nerve involved and the level injured. In addition, due to the multiplicity of forces involved in injury, it is not surprising that there is a full range of pathological grades of damage. This extensive variability makes formulating a prognosis challenging, but Seddon's (Seddon 1943) and Sunderland's (Sunderland 1968) classifications of injury can be of prognostic value.

Seddon's neurapraxic injury is characterized by a temporary local conduction block with complete reversibility within hours to 8 weeks. Corresponding to Sunderland's first-degree injury, this lesion has little or no wallerian degeneration and normal distal conduction but may involve a biochemical disturbance and segmental demyelination (Gentili and Hudson 1985; Spinner 1978). Usually this grade lesion causes the greatest loss in motor and proprioceptive functions, a slight loss of touch and pain sensation, and no loss of autonomic function (Sunderland 1968). Neurapraxia is present in "Saturday night palsy," tourniquet palsy, and some entrapment–compression neuropathies; it most often involves nerves at anatomical points where they are relatively superficial and in juxtaposition to bony surfaces (Gentili and Hudson 1985). Because the axon and endoneurial tubes are intact, this is not a surgical lesion, and the prognosis is excellent for full recovery with conservative treatment.

Axonotmesis involves complete loss of axon and myelin continuity with relative preservation of the connective tissue framework. Corresponding to Sunderland's second-degree injury and mild third-degree injury, this lesion results in wallerian degeneration distal to the injury site and a small amount of retrograde reduction of axon and myelin sheath diameter (Cragg and Thomas 1961; Gentili and Hudson 1985). Unlike neurapraxia, this lesion causes acute loss of motor, sensory, and autonomic functions. However, with preservation of the endoneurial tube, spontaneous recovery of most sensory and motor functions occurs within 18 to 36 months (Seddon 1943; Sunderland 1968). This incontinuity lesion is due to moderate stretch and contusion, such as those associated with skeletal fractures. Because the regrowing axons do not stray from intact endoneurial tubes, prognosis is good for eventual recovery. However, a problem arises in clinical assessment with a long delay in recovery. This lesion may be difficult to distinguish from neurotmesis. The physician must use the patient's history and physical examination along with electrophysiological data to discern the most likely classification; with axonotmesis, spontaneous recovery is quantitatively and qualitatively superior to that of neurotmesis and to that seen after resection and suture repair (Kline and Hudson 1989).

Seddon's most serious category, neurotmesis, corresponds roughly to Sunderland's higher grade of injury. Neurotmesis involves loss of connective tissue and axonal integrity. This may be an in-continuity lesion (Sunderland's fourth degree) or an anatomically sectioned nerve (Sunderland's fifth degree). With this lesion, the connective tissue framework of the endoneurial tubes and their axons and myelin coverings are disrupted. Clinical and electrophysiological findings are identical to those seen with axonotmetic injuries. Also, loss of motor, sensory, and autonomic functions occurs. The difference lies in the inability of a neurotmetic lesion to recover spontaneously secondary to the loss of a guiding endoneurial framework across the lesion site. Thus, multiple proximal sprouts are deflected and divide many times as they attempt to re-enter the distal endoneurial tubular system and reestablish the proper end-organ contact (Lundborg 1982; Lundborg 1987). Recent evidence supports a concomitant formation of totally new endoneurial tubes in the distal stump after injury and repair. This process may not always aid direction of axons to proper end-organ sites (Cabaud et al. 1982). Neurotmesis usually is associated with severe stretch, such as that due to high-velocity missile injury and laceration. If identified, these lesions require surgical resection and repair. Even with excellent technical repair, prognosis is guarded. Recovery may require 3 to 5 years, and it may never be complete (Kline 1989; Sunderland 1968).

Injury classifications assist in forming a prognosis, but many lesions are mixed, and detailed distinctions can only be made in retrospect. Sunderland believes the severity of the lesion is the most important prognostic factor shaping both the extent of regeneration and the time it takes.

Etiology and Extent of the Lesion

Closely related to the severity and degree of injury is the force involved because the greater the violence to the nerve, the greater the damage. Neat wounds from glass, knives, and other sharp objects have a better prognosis than those involving diffuse contusion or stretch, such as from a missile wound or a vehicular accident. Explanations for this include nerve involved by stretch or contusion suffers a greater severity and length of damage; the lesions often are more proximal than lacerating injuries, and more extensive retrograde neural damage occurs; the proximal level preponderance requires a greater distance and time of regeneration; and associated extensive soft-tissue and bony damage may delay repair and leave the lesion and repair site in a bed of scar (Frykman and Colly 1988; McQuarrie 1985). Soft-tissue laceration is a low-velocity injury and usually causes complete paralysis secondary to severance of the nerve. Nonetheless, 12% to 15% of these injuries leave the nerve in continuity (Gentili and Hudson 1985; Kline and Judice 1983; Kline and Hackett 1975). Many lacerations are clean and neatly transect the nerve. These benefit from acute or primary repair. Ragged lacerations, such as from a chain saw, fan blade, or jagged auto metal, are best repaired secondarily after several weeks of delay (Kline and Judice 1983; Kline and Hackett 1975). Severity of gunshot wounds varies from high-velocity injuries seen in combat to lower velocity injuries seen in civilian life. Immediate loss of neural function with missile injury, in contrast to other penetrating injuries to nerve, does not necessarily signify a transected nerve. Thus, most nerves injured by gunshot are in continuity. Some recover and some do not. Initial conservative care is best (Gentili and Hudson 1985; Kline 1989; Omer 1980). Approximately 40% of such injuries recover significant function, and the rest require operation (Omer 1982).

Stretch–traction injuries are most commonly associated with extreme movements of the neck and shoulder, bony dislocations, and fractures. Although mild, focal neural damage can occur; these lesions typically are more serious, nonfocal, and lengthy. Surgical repair frequently is required, and prognosis usually is poor. The majority of brachial plexus lesions are due to traction injury, and repair is difficult. Even when cases are carefully selected for surgery, significant recovery occurs in only 50% of patients (Kline et al. 1986; Narakas 1981). Some distal traction injuries to nerves associated with skeletal fracture have a better prognosis for spontaneous recovery. For example, substantial spontaneous recovery occurs in 80% of patients with humeral fracture involving the radial nerve (Bateman 1962; Pollack and Davis 1933). By comparison, many axillary nerve stretch lesions do not recover spontaneously and require operation.

Compression by entrapment begins with paranodal demyelination but can progress to more extensive demyelination and axonal dropout. Unless chronic or severe, these lesions have a good prognosis if decompressed in a timely and proper fashion. Most Saturday night palsies and anesthetic palsies also have a good prognosis; most recover spontaneously, but there are exceptions (Gentili and Hudson 1985). Surgical decompression of the median and ulnar nerves for carpal tunnel syndrome and tardy ulnar palsy usually prevents progression of loss and sometimes reverses the loss, as does release of compression in properly diagnosed thoracic outlet syndrome (Hardy and Wilbourn 1985; Hudson et al. 1987; Kline and Hudson 1984; Rengachary 1985). Conversely, compression secondary to closed compartment syndrome can result in an extremely poor prognosis unless fasciotomies are done acutely. Some traumatic aneurysms, fistulas, or hematomas produce acute neural ischemia as well as progressive compression of already injured elements.

When this occurs, prognosis for recovery is poor.

Although relatively uncommon, thermal, chemical, and electrical injury often involves extensive lengths of nerve. For severe loss, resection and lengthy graft repair are necessary, and prognosis generally is poor. Protection from compression from secondary scar by thorough debridement of associated soft-tissue injuries offers the best chance of recovery when neural loss is partial or incomplete to begin with (Bateman 1962; Rosen 1981; Salisbury and Dingeldein 1982). Drug injection injuries most often involve the radical and sciatic nerves and cover a large range of severity of injury. Conservative therapy is the initial course though intraoperative electrical evaluation, and repair may be necessary when severe loss with or without pain does not reverse. Depending on the extent and level of injury, prognosis is good, but one study revealed significant residual motor deficit in most patients, and another series showed surgical intervention required in 30% to 40% of cases (Clark 1972; Kline and Hudson 1984; Kline and Hudson 1987). With a lengthy damaged segment of nerve, the delay in spontaneous recovery or growth through lengthy grafts needed for repair is protracted.

The cross-sectional extent of a complete or partial lesion can affect the prognosis. A dissociated nerve lesion results when loss of the heavily myelinated large fibers of motor, proprioceptive, vibratory, and touch function are lost out of proportion to the thinly myelinated small fibers for pain, temperature, and autonomic function. The prognosis is slightly better for spontaneous regeneration and recovery in these cases (Spinner 1978; Sunderland 1981; Williams and Jabaley 1986).

In summary, lesion etiology and severity of injury are major factors affecting prognosis for functional recovery.

Type of Nerve Injured

Nerves are composed of sensory, motor, or mixed (sensory and motor) fascicles. Pure sensory nerves, such as the antebrachial cutaneous, superficial sensory radial, and sural nerves, have the best prognosis for spontaneous regeneration when injury is nontransecting. Predominant motor nerves are less likely to recover function spontaneously because of additional difficulty in reinnervation of myoneural junctions, especially if the muscle is small and concerned with fine movements, such as those of the fingers. Prognosis for recovery is the worst with serious injury to mixed motor–sensory nerves. Because regeneration involves random axonal regrowth into distal endoneurial tubes, proper functional alignment may be precarious. A motor axon entering a sensory endoneurial tube most likely fails to connect and make a myoneural junction. Variation of a nerve's fascicular makeup at different levels may affect the prognosis because although the median nerve is 66% sensory above the elbow, it is 94% sensory at the wrist level (Doyle 1980; Kline and Hurst 1984; Millesi 1979). Thus, with regeneration after injury or repair, there is a greater chance for an adverse mix of sensory and motor fibers at a proximal rather than a distal level. Additionally, Sunderland and others have shown that the fascicular pattern of nerves can change over small linear segments. Fascicles can be difficult to realign properly, particularly if injury or resection before repair involves a length of nerve.

Specific Nerve Injured

For many reasons, some nerves recover function better than others, regardless of whether spontaneous regeneration or surgical repair is involved. With the brachial plexus, prognosis is best for the upper trunk outflow and its origins, the C5 and C6 roots. Middle trunk and C7 root offer a variable prognosis, while the lower trunk and its C8 and T1 root origins offer the worst prognosis. As might be expected, lateral cord injuries have a better prognosis than posterior cord lesions, which in turn fare better than medial cord injuries. Kline and Judice (1983; 1989) also report encouraging results with medial and lateral cord to median nerve repairs. Bateman (1962) states that although motor return is poor, sensory return could be good in lower root and trunk repair of lacerating injuries. Overall, the infraclavicular plexus has a better prognosis than the more frequently injured supraclavicular plexus. In either case, the return of intrinsic hand muscle function after complete injury is rare (Alnot 1984; Frykman et al. 1981; Kline 1989).

In the upper extremity, the musculocutaneous and radial nerves have an excellent prognosis for functional recovery, the median nerve has a good prognosis, and the ulnar nerve has the worst prognosis. This probably relates to the regenerative distance required and the nature of the muscle to be innervated. Proper reinnervation and function are more difficult in small, fine movement muscles, such as those in the hand (Miller and Kline 1982; Miller 1987; Millesi 1981).

In the lower extremity, repair or spontaneous regeneration of the tibial nerve has a good prognosis, while that for the peroneal nerve has a poor prognosis (Kline and Hudson 1989; Sunderland 1968). The peroneal's anterior compartment is composed of relatively fine, linear muscles, while the tibial's gastrocnemius and soleus are large, bulky muscles. With sciatic injury or repair, prognosis for return of plantar flexion and sole sensation is good, but dorsiflexion is difficult to recover. As might be expected, return of the foot's intrinsic fine muscle function is uncommon with tibial nerve regeneration. Finally, if femoral nerve repair is possible proximal to its branches, the prognosis for functional quadriceps return is good (Kline 1985; Seddon 1972). In summary, prognosis is in part dependent on the specific nerve injured.

Anatomic Level of the Lesion

Studies of nerve injuries during war and peace, regardless of technique used (neurolysis or epineurial, fascicular, or graft repair), have concluded that the more distal the neural repair, the better the prognosis for functional sensory and motor return (Bateman 1962; Braun 1982; Dellon 1981; Frykman and Colly 1988; Williams and Jabaley 1986). There are several reasons for this. Retrograde changes may irreversibly damage the neuron in proximal lesions, whereas only regenerative chromatolysis of the spinal cord's anterior motoneuron or dorsal root ganglion occurs in distal lesions; in proximal injuries, neurons innervating distal limb regions are more susceptible to retrograde damage than in distal lesions. Also, at a distal level, fascicles are well localized and fixed in arrangement; proximally, the sensory and motor fibers are scattered and change position over short nerve segments. Third, the more proximal the lesion, the longer the denervation interval and the more significant the possible muscle atrophy. Finally, in proximal lesions, axons require a longer period of regenerative growth to reach end-organs (McQuarrie 1985; Sunderland 1968). However, results do not totally depend on the distance between lesion and muscle endpoint. In C5 to C6 and upper trunk lesions, the biceps–brachialis function returns more often and usually earlier than in the deltoid muscle, even though the latter is closer to the lesion. An explanation might be that the deltoid requires more point-to-point reinnervation for synchronous fiber contraction (Doyle 1980; Kline et al. 1986).

Age of the Patient

Studies of recovery patterns in infants and children indicate a significantly improved prognosis compared to adults. Although a few early investigators doubted this, the consensus now shows a marked advantage with youth (Dellon 1981; Kutz et al. 1981; Lindsay et al. 1962; Lundborg 1982; Muller and Grubel 1983; Stellini 1982). Tupper and coworkers (1988) suggest an improved prognosis up to the early 20s in age, and Spinner (1978) notes that distal ulnar return may be fairly complete in children. Seddon's extensive series (1972) revealed differences of

71% to 31% in good functional recovery in patients up to 10 years old vs. 41 to 50 years old, respectively.

This prognostic advantage involves sensory, motor, and mixed nerves and may be due to many factors: regeneration rates in children may be faster, although this is debatable (Sunderland 1968); sprouting of axons in children may be more extensive, although again debatable; regenerating axons have shorter distances to travel in small limbs; children's central nervous systems (CNS) have greater plasticity; and children have more substitutive abilities and less fear in using an impaired limb. Although youth is considered an advantage, whether a patient beyond a certain age has a worse prognosis is unknown (Doyle 1980; Kline and Hurst 1984; Lundborg 1982; Sunderland 1968).

Associated Injuries and Medical Problems

Prognosis for recovery might be expected to be hindered by any complicating factor. Seddon (1972) noted that in World War II, extensive soft-tissue damage precluded repair in 20% of cases, but statistical analysis by Woodhall and Beebe (1956) showed no prognostic difference between isolated nerve injuries and neural lesions in association with bone and joint fractures. Nevertheless, 12% to 20% of humeral shaft fractures have an associated radial nerve injury, of which approximately 80% will recover spontaneously if not caught in callus or between bone edges. Approximately 85% of all neural injuries associated with fracture occur in the upper limb. Radial nerve comprises 60%, ulnar 18%, median 6%, and peroneal 15% of these types of complications. Joint dislocations can cause neural damage 18% of the time at the knee and 13% at the hip. These injuries have a worse prognosis than those associated with fracture. Stretch and neurotmesis over a length is more likely with dislocations than in fractures associated with injury (Gentili and Hudson 1985; Kline and Hudson 1987).

Vascular injuries may involve nerve in many cases, but the prognosis for spontaneous regeneration usually is good (Visser et al. 1980). Nevertheless, acute neural compression by hematomas, traumatic aneurysms and fistulas, and compartment syndromes have a poor prognosis unless emergent decompression is performed. An exception is the usually excellent prognosis for spontaneous regeneration when pelvic plexus is compressed by a retroperitoneal hematoma (Hudson et al. 1987; Kline and Hudson 1987).

Intact nerves are relatively resistant to local soft-tissue infection. Damaged or repaired nerve, however, is not as resistant to infection (Hudson et al. 1987; Kline and Hudson 1987). Longitudinal fibrosis or suture line dehiscence can occur. Injury to a nerve already involved by diabetic or other metabolic neuropathies carries a poor prognosis. Associated injuries or medical problems also may alter prognosis because of a substantial delay in neural repair due to care of the associated condition (Woodhall and Beebe 1956).

Factors Involved in Determining Medical or Surgical Management

History and Physical Examination

The importance of a thorough history and physical examination cannot be overemphasized. The history will show the physician the prognostic values of the many variables involved. Determining the acuteness or chronicity of the nerve injury is extremely important in assessing prognosis because prolonged damage may preclude surgical correction. Sunderland (1968) felt that the capacity for proper muscle reinnervation was not lost within the first year of injury, while Seddon (1972) felt that late and repetitive attempts to gain regeneration might eventually cause neuronal exhaustion.

The regenerating axon does not fully mature in diameter and conduction without end-organ continuity. Experimental studies reveal neuromuscular function breakdown before distal wallerian degeneration (Miller 1987). Distal target atrophy begins early and a delay of 24 months without restitution of axonal continuity to muscle may cause irreversible atrophy and fibrosis. Sensory organs are more resistant, but skin receptor atrophy also occurs (Lundborg 1982; Williams and Jabaley 1986). Attempts to regain regeneration in muscle should begin relatively early. If delay is necessary, denervated muscle should be exercised regularly.

Sequential physical examination is of paramount importance. Prognosis varies according to whether the lesion is partial or complete because a dissociative or partial injury usually has a better outlook. Tinel's sign progressing distal to the lesion is encouraging because it represents regenerating young sensory axons. On the other hand, presence of a Tinel's sign recorded well distal to the injury site does not guarantee recovery of significant function. Months after injury, the absence of a Tinel's sign distal to the initial lesion is a poor prognostic sign. The first sign of sensory regeneration is pain on compression of a denervated muscle belly. Next in order of return is a high threshold hyperpathic response to pinprick, characterized by unpleasant and diffuse discomfort. Temperature perception and focal pain signify regeneration of small-diameter axons. The large-fiber functions of proprioception, vibration, and two-point discrimination appear later. Return of sensation in the autonomous zone and recovery of sweating indicate an increased prognosis for recovery but do not guarantee it (Kline and Hudson 1987; Sunderland 1968).

Except in certain distal nerves, the extent of motor reinnervation is the most important prognostic factor. A thorough motor examination and grading of muscle strength are important. Significant improvement supports conservative management (Groff and Houtz 1945). In brachial plexus injuries, preservation of rhomboids, serratus anterior, supraspinatus, and infraspinatus muscle function is a good sign, and their loss is a poor prognostic sign. Motor recovery seldom is complete, even when prognosis is good due to the nature of the injury, lesion level, age of patient, and nerves involved

(Hudson and Trammer 1985; Kline and Judice 1983; Millesi 1984; Sedel 1984).

Laboratory Studies

Routine chest and neck x-rays may uncover poor prognostic signs, such as concomitant fractures of the spine or paralysis of the diaphragm, denoting proximal brachial plexus pathology. Additionally, myelography may reveal meningoceles, cord edema, or atrophy, which signify severe proximal brachial plexus root damage. Nonetheless, in the author's and other's experience with stretch injuries of the plexus, injury to roots can be proximal without meningoceles or absent at a level where a meningocele is present (Kline and Hudson 1987; Nagano et al. 1984; Sunderland 1968). Angiography is useful in the evaluation of a few thoracic outlet cases, some traumatic aneurysms, and all arteriovenous fistulas.

Electromyography (EMG) is helpful in determining the degree and prognosis of neural injury. There are rare fibrillations with neurapraxic injury but frequent fibrillations and denervation potentials 2 to 4 weeks after the injury with axonotmesis or neurotmesis. Similarly, electroneurography reveals preserved conduction distal to the lesion in neurapraxia but absent conduction distally in axonotmesis unless adequate regeneration is occurring. Successful regeneration in serious injuries will reveal on EMG decreased denervation potentials, increased nascent potentials, and return of insertional activity. Likewise, with successful regeneration, electroneurography or direct stimulation and recording from nerve will reveal nerve action potentials (NAPs) from points progressively distal to the original lesion. Absent NAPs across the lesion 8 to 10 weeks after the injury is a poor prognostic sign that implies a neurotmetic lesion (Kline 1981; Kline and Hackett 1985). Even EMG signs of regeneration can precede voluntary contraction by 2 to 6 weeks; if progression to voluntary movement does not occur in 2 months, this signifies a largely neurotmetic lesion. Presence of rapidly conducting high-amplitude NAPs from a brachial plexus element with absent clinical function suggests a preganglionic lesion and thus sensory fiber sparing. This is a poor prognostic sign for recovery of function in that element's distribution. Thus, both the EMG and electroneurography (NAP recording) are useful prognostic indicators (Gilbert et al. 1988; McGillicuddy 1985; McQuarrie 1985; Spinner 1978; Sunderland 1968).

Rate(s) of Regeneration

Velocity of regeneration is fast when the regenerative axon tips are close to the neuron and slow when they are away from it. Thus, the regenerative rate over the length of the nerve is not constant. Long delays in reinnervation are more common with a proximal lesion, but this is probably due as much to severity of the lesion as level (Lundborg 1982; Sunderland 1968). Several texts have attempted to calculate the regenerative rates for specific nerves and levels for various lesions (Bateman 1962; Groff and Houtz 1945; Sunderland 1968). The tables by Sunderland (1968) and Pollock and Davis (1933) offer some of the best prognostic information for gauging the expected proximal to distal return of voluntary contraction. Generally, the rule of thumb is upper arm or thigh, 3 mm/day; lower arm or below knee, 1.5 mm/day; and wrist or ankle, 0.5 mm/day. This corresponds roughly to 1.0 mm/day or 1.0 in/month, although regeneration in mild axonotmetic injuries may be faster (McQuarrie 1985; Miller and Kline 1982).

Conservative vs. Surgical Management

Severity of loss distal to the lesion and its electrodiagnostic confirmation and judgment about presence of gross continuity provide the best prognosis concerning spontaneous regeneration. Conservative treatment has been the mainstay for some wartime missile injuries involving nerve. Tinel (1917) in World War I reported greater than 50% spontaneous recovery and characterized surgical intervention before 6 months following injury as impatience. Foerster's retrospective analysis of World War II missile injuries placed spontaneous recovery at 68% (Sunderland 1968), while Omer's (1982)

prospective study of Vietnam injuries reported a 69% recovery with conservative management. However, serious GSWs involving brachial plexus, but usually with elements left in continuity, are more likely to require surgery because one or more elements often do not reverse loss by the early months (Kline 1989).

The series by Nagano and colleague's (1984) on conservative treatment of brachial plexus stretch lesions reported functional recovery in the upper limb in more than 40% of cases. Unfortunately, this is not true if loss is complete in the distribution of two or more roots. Alnot (1984) relates 60% spontaneous recovery in infraclavicular brachial plexus lesions; and in Narakas's (1981) large series of traction injuries to brachial plexus, 46% showed some spontaneous functional return. While conservative management in certain cases can lead to recovery, many stretch lesions to the plexus do not recover without direct repair or other substitutive or reconstructive procedures (Gilbert et al. 1988; Kline et al. 1986; Millesi 1984).

Operative Factors

Primary Vs. Secondary Repair for Transected Nerves

The role of timing of repair for divided nerves has been debated throughout the history of peripheral nerve surgery. It is impossible to perform a controlled, double-blind study in patients, and there is even disagreement in the literature about the definition of primary vs. secondary repair. Primary repair can be defined as nerve repair before associated soft-tissue wounds have healed. To be practical, primary repair is usually performed within 72 hours of injury. Such repair is indicated for sharply transected nerves or plexus elements. The major disadvantage of primary repair for more blunt transections is the difficulty in detecting acutely the extent of neural damage. Delaying the procedure for several weeks allows exact delineation of the extent of injury and subsequent accurate proximal and distal resection of the stumps. Delay also permits elective surgery and may take advantage of a conditioning effect. Research supports an enhanced axonal regeneration rate when preceded several weeks by an axonal lesion (Bisby and Pollick 1983; McQuarrie 1985). Thus, delayed surgical repair may capitalize on the soma's natural peak for biochemical readiness for regeneration. Conversely, factors favoring primary repair include normal anatomy with little scar, minimal stump fibrosis and retraction, and overall decreased delay in restitution of myoneural function (Kline and Hackett 1975; McNamara et al. 1987; Seddon 1972).

When patients are well selected for both primary and secondary repairs, results are similar if the secondary repairs are not delayed too long (Braun 1982; Millesi 1980; Seddon 1943; Snyder 1981; Sunderland 1968; Zachary and Holmes 1946). Primary repair is indicated for sharp transections and for injuries to small peripheral nerves, such as the digital nerves. Otherwise, delayed repair is advantageous (Ducker 1985; Kline 1981; McGillicuddy 1985).

Technique of Nerve Repair

Extent of Mobilization

Prognosis is improved if mobilization can decrease suture line tension and allow end-to-end anastomosis or use of relatively short grafts. The rich and longitudinal intraneural vascular supply of nerves prevents detrimental ischemia, even when a nerve is mobilized from axilla to wrist. Most research indicates that mobilization can improve the chances for adequate repair providing intraneural vessels are not destroyed (Kline et al. 1972; Lundborg 1975; Sunderland 1968).

Lesion In-Continuity: Intraoperative Neurography

The introduction of intraoperative recording has enhanced the surgeon's ability to judge incontinuity lesions, which account for 60% or more of civilian peripheral nerve injuries. Recordings allow the surgeon to determine a lesion's physiological completeness, incom-

pleteness, or degree of regeneration and thus its proper treatment, whether that is neurolysis, split repair, or resection. Erroneous judgment concerning a lesion's regenerative potential based only on physical inspection is avoided (Kline 1983; Kline and Hackett 1985; Kline and Hudson 1985). As a result, regenerating lesions are spared resection while those that are not can be resected with certainty. This approach also permits more through resection of lesions not transmitting. The latter improves prognosis with repair because the leading cause of failure of repair is inadequate resection.

Neurolysis

Some authors have advocated neurolysis when a nerve is surrounded by scar. The actual role neurolysis plays in regeneration is debatable. In many cases in which significant improvement occurs after neurolysis, recovery may be spontaneous if the nerve is not operated on (Kline and Hudson 1989; Woodhall and Beebe 1956). External neurolysis performed carefully should not devascularize nerve or decrease the prognosis (Frykman et al. 1981; Kline et al. 1972; Millesi 1981; Woodhall and Beebe 1956). In one large series of missile injuries, Omer (1982) reported improvement after neurolysis in 60%. If intraoperative stimulation and recording studies show an NAP across a lesion in continuity with complete distal loss in the first 9 months after injury, the recovery rate with neurolysis alone is 90% or better (Kline 1989) (Table 28-1).

Extent of Gap and Tension

Gap is the distance remaining between two stumps when a lesion in continuity is resected or when the neuroma and glioma forming after transection of nerve are trimmed back to healthy tissue. Nerve stump retraction due to the inherent elasticity of nerve also is a factor in producing a gap. When resection of contused stumps or of an in-continuity lesion is necessary, reanastomosis may produce tension at the suture line when the gap is closed. Millesi (1979; 1986) has proposed that tension is the most deleterious factor in nerve repair and

Table 28-1. LSUMC* Operative Results† Based on 378 Serious Upper
Extremity Lesions‡

Nerve Involved/Operation	Lesion Level		
	Upper Arm	Elbow	Forearm or Wrist
Radial nerve/neurolysis			
Grade 3 or better	100	90	94
Grade 4 or better	90	83	89
Radial nerve/repair			
Grade 3 or better	72	81	79
Grade 4 or better	60	62	60
Median nerve/neurolysis			
Grade 3 or better	90	93	93
Grade 4 or better	78	87	90
Median nerve/repair			
Grade 3 or better	68	75	81
Grade 4 or better	45	64	68
Ulnar nerve/neurolysis			
Grade 3 or better	90	92	100
Grade 4 or better	79	83	92
Ulnar nerve/repair			
Grade 3 or better	41	69	56
Grade 4 or better	15	31	40

* LSUMC, Louisiana State University Medical Center.
† Percent with functional return using the grading system in Table 28-5.
‡ Does not include brachial plexus lesions, tumors, or nerve entrapment cases.

has recommended grafts for all defects longer than 2 cm. The critical gap length that precludes end-to-end anastomosis varies according to the nerve involved, level of the lesion, ability to make up length by mobilization and transposition, and the acuteness of the injury (Kline 1983; Millesi 1982). Although some studies have been unable to demonstrate the harmful effect of tension, others have attempted to correlate various levels of tension with different degrees of regeneration (Bratton et al. 1979; Millesi 1986; Orgel 1982; Osterman 1986). Relation between gap and prognosis is not absolute, but Woodhall and Beebe (1956) report a 6% loss of function for each centimeter of gap that had to be repaired. This observation may relate more to the seriousness of the original injury and thus the need for lengthy resection and lengthy grafts than to the gap length itself.

Epineurial Vs. Fascicular Vs. Graft Repair

The advantages of various repair techniques are debated. Even good clinical studies are almost impossible to analyze owing to the multitude of variables involved. Review studies and experimental models have shown nearly identical results between epineurial and fascicular repair (Braun 1982; Kline et al. 1981; McCarroll et al. 1980; McGillicuddy 1985; Orgel 1984). Millesi is a strong proponent for grafts, even with relatively small gaps. While his and other data support an improved prognosis with grafting, evidence also lends credence to a lack of a real advantage with any technique for small gaps (Bratton et al. 1979; Kline and Hudson 1985; McNamara et al. 1987).

The rationale for grafts is largely to decrease tension, and as previously stated, the critical gap necessitating grafts instead of end-to-end anastomosis is variable among surgeons. Most agree that grafts should be no more than 5 mm millimeters in diameter to assure full revascularization. In addition, grafts should be 10% to 15% longer than the gap. In general, short grafts have a better prognosis than long grafts. The main disadvantages of grafts is the added length that axons must transverse to cross the repair site and the additional suture line at the distal graft insertional site. Gaps of more than 4 in are often of poor prognostic outlook, except in some nerves, such as the musculocutaneous, radial, and tibial nerves (Frykman and Colly 1988; Kline and Hudson 1989; Kline 1983; Lundborg 1982; McQuarrie 1985; Sunderland 1968).

Prognostically, epineurial repair is best for both acute repair of sharp, transecting injuries and delayed suture of contused, ragged nerve injuries. Fascicular repair is recommended for sharp, acute distal injuries and partially divided lesions with some fascicles preserved. Interfascicular grafts are best for delayed repair of untidy and lengthy lesions. In summary, prognosis is similar with all three approaches to repair unless large gaps necessitate lengthy grafts.

Neurotization

When there is no distal stump to hook into, burying the proximal stump with or without extension by grafts into muscle is a last resort. Occasionally, recovery has been achieved with neurotization of large muscles, such as biceps or brachialis and gastrocnemius or soleus, but the prognostic value of such a procedure is unknown (Kline and Hudson 1989; McNamara et al. 1987). Proximity of plexus root injury to spinal cord prohibits the use of root(s) as origin for repair in some severe stretch injuries. Nerves other than roots injured have been used to provide lead-out. This type of neurotization uses grafts from the substitutive nerve to bridge the injury site to more distal plexus or peripheral nerves. Some of the nerves or elements used include intercostal nerves, cervical plexus, and accessory, phrenic, and dorsal scapular nerves. Reports indicate that some results are obtained in 30% to 50% of such neurotizations (Hudson and Trammer 1985; McNamara et al. 1987). Unfortunately, these procedures, when they work, can provide only *one* function, such as arm flexion or shoulder abduc-

Table 28-2. Upper and Lower Extremity Repair Results*

Median Nerve Secondary Suture Repair	
Upper arm and elbow level:	36% M3, S3+ or better recovery
Forearm level:	25% M3, S3+ or better recovery
Wrist level:	43.6% M3, S3+ or better recovery

Ulnar Nerve Secondary Suture Repair	
Upper arm and elbow level:	20% M3, S3 or better recovery
Forearm level:	25% M3, S3 or better recovery
Wrist level:	44.5% M3, S3 or better recovery

Radial Nerve Secondary Suture Repair	
All levels:	54% M4 or better recovery

Common Peroneal Nerve Primary and Secondary Suture Repair	
All levels:	34.7% dorsiflexion recovery

Tibial Nerve Primary and Secondary Suture Repair	
All levels:	78.7% plantar flexion recovery
	61.7% sole sensation recovery

* Seddon, H. J. Surgical disorders of the peripheral names, Baltimore: Williams and Wilkins; 1972.

† M3 = Return of function in both proximal and distal muscles of such degree that all important muscles are sufficiently powerful to act against resistance.

M4 = Return of function as M3 but in addition all synergic and independent functions are possible.

S3 = Return of superficial cutaneous pain and tactile sensibility throughout autonomous area with disappearance of any previous response.

S3+ = Return to S3, but also some recovery of two point discrimination within the autonomous area

Table 28-3. Median, Ulnar, and Digital Nerve Repairs*

Median Nerve	
High level suture repair:	30% M4 or better recovery
	17% S3+ or better recovery
Low level suture repair:	45% M4 or better recovery
	33.5% S3+ or better recovery
Combined High and Low Graft Repair:	33% M4 or better recovery
	26% S3+ or better recovery
Ulnar Nerve	
High level suture repair:	17% M4 or better recovery
	20% S3+ or better recovery
Low level suture repair:	32% M4 or better recovery
	34.7% S3+ or better recovery
Combined High and Low Graft repair:	43% M4 or better recovery
	21% S3+ or better recovery
Digital Nerve	
All level suture repair:	59% S3 or better recovery
All level graft repair:	29% S3 or better recovery

* Dellon, A. Results of nerve repair in the hand. In: Evaluation of sensibility and reeducation of sensation in the hand. Baltimore: Williams and Williams Co.; 1981.
† M3 = Return of function in both proximal and distal muscles of such degree that all important muscles are sufficiently powerful to act against resistance.
  M4 = Return of function as M3 but in addition all synergic and independent functions are possible.
  S3 = Return of superficial cutaneous pain and tactile sensibility throughout autonomous area with disappearance of any previous response.
  S3+ = Return to S3, but also some recovery of two point discrimination within the autonomous area

Table 28-4. Louisiana State University Medical Centers Grading Systems

Individual Muscle Grades		
Grade	Evaluation	Description
0	Absent	No contraction
1	Poor	Trace contraction
2	Fair	Movement against gravity only
3	Moderate	Movement against gravity and some (mild) resistance
4	Good	Movement against moderate resistance
5	Excellent	Movement against maximal resistance

Sensory Grades		
Grade	Evaluation	Description
0	Absent	No response to touch, pin, or pressure
1	Bad	Testing gives hyperesthesia or paraesthesia; deep-pain recovery in autonomous zones
2	Poor	Sensory response sufficient for grip and slow protection; sensory stimuli mislocalized with over-response
3	Moderate	Response to touch and pin in autonomous zones; sensation localized and not normal; with some over-response
4	Good	Response to touch and pin in autonomous zones; response localized but not normal; no over-response
5	Excellent	Normal response to touch and pin in entire field, including autonomous zones

Table 28-5. Criteria for Grading Whole-Nerve Injury*

0 (absent)	No muscle contraction; absent sensation
1 (poor)	Proximal muscles contract but not against gravity; sensory grade 1 or 0
2 (fair)	Proximal muscles contract against gravity; distal muscles do not contract; sensory grade if applicable was usually 2 or lower.
3 (moderate)	Proximal muscles contract against gravity and some resistance; some distal muscles contract against at least gravity; sensory grade usually 3
4 (good)	All proximal and some distal muscles contract against gravity and some resistance; sensory grade 3 or better
5 (excellent)	All muscles contract against moderate resistance; sensory grade 4 or better

* Louisiana State University Medical Center's system.

tion. It is difficult to provide multiple functions in an extremity without direct repair. Future developments probably include movement of muscles from one extremity to another and their revascularization and reinnervation at the transfer site.

Postoperative Factors

Wound healing with or without infection, physical therapy, patient attitude, and sensory reeducation can alter the course of functional recovery and prognosis. Infection may damage suture lines, increase scar, and prolong regenerative times. Physical therapy and a lack of fear in using the affected limb can maintain and improve functional muscle and promote effective substitutive maneuvers. As with any disease, psychological factors can enhance or limit recovery, while sensory re-education can increase the patient's functional efficiency. In summary, all postoperative factors should be managed by the surgeon to help achieve a maximal rehabilitative prognosis and as full a reintegration into society as possible.

Follow-Up

It is difficult to standardize assessments of recovery. What one physician calls useful recovery may be labeled as moderate, good, or superior by another (Dellon 1981; Woodhall and Beebe 1956). The functional importance of sensory regeneration varies among nerves. Although relatively unimportant in radial and peroneal nerves, sensation has special significance in median and tibial nerve injuries. Assessment of full recovery of motor and sensory function may take 3 to 5 years. Thus, investigator differences, individual nerve variations regarding

relative sensory and motor importance, and the necessity of meticulous long-term follow-up make comparison of recovery complex and at times frustrating. The modification of Highet's scheme by Zachary and Holmes (1946) has been used to provide a statistical prognosis of recovery in selected nerve injuries (Tables 28-2 and 28-3). Data from the LSUMC panel of patients also have been correlated (Tables 28-4 through 28-7).

Conclusion

This chapter has reviewed the multiple prognostic factors that affect regeneration in peripheral nerve injuries. Although an understanding and interest in each variable is helpful to the physician, their predictive value is limited by their immense interdependence and simultaneous independence. All factors must be weighed to allow helpful prognostic decisions in the medical and surgical management of each individual case.

Table 28-6. LSUMC* Operative Results Based on 378 Serious Upper Extremity Lesions Irrespective of Level and Method of Injury†

Type of Injury	Grade 3 or Better Return
Transections	
Primary suture	78%
Secondary suture	70%
Grafting	63%
In continuity	
Neurolysis	92%
Suture	75%
Grafting	66%

* LSUMC, Louisiana State University Medical Center.

† Does not include brachial plexus lesions, tumors, or nerve entrapment cases.

‡ See Table 28-5 for grading system.

Table 28-7. Comparison of Results* of Brachial Plexus Surgery by Procedure and Lesion Etiology

Operation	Lesion Etiology					
	GSW Complete Loss	GSW Incomplete Loss	Contusion or Stretch Complete Loss	Contusion or Stretch Incomplete Loss	Laceration Complete Loss In-Continuity	Laceration Without Continuity
Neurolysis	96	100	84	94	100	—
Suture	65	100	50	—	73	—
Graft	47	—	49	—	80	—
Split repair	100	—	100	—	100	—
Repair impossible	20	—	0	—	12.5	—
Primary suture						78
Secondary suture						86
Secondary graft						50

* Percent Grade 3 or better recovery (see Table 28-5).

Results are better for upper and middle trunk and lateral and posterior cord elements, and worse for lower trunk and medial cord elements.

Adapted from Kline, D. G.; Judice, D. Operative management of selected brachial plexus lesions. J. Neurosurg. 58:631–649; 1983.

Furthermore, refinement of treatment modalities and additional investigation into the anatomical, physiological, and biochemical features of neural regeneration are paramount to enhance success. Only then can the substantial morbidity of peripheral nerve injury be surmounted.

References

Alnot, J. Y. Intraclavicular lesions. Clinics in Plastic Surgery 11:127–131; 1984.

Bateman, J. E. Results and assessment of disability in iatrogenic nerve injuries. In: Trauma to nerves in limbs. W.B. Sanders; Philadelphia: 1962: p. 285–305.

Bisby, M. A.; Pollick, B. Increased regeneration rate in peripheral nerve axons following double lesions: enhancement of the conditioning lesion phenomenon. Neurobiol. 14:467–472; 1983.

Bratton, B. R.; Kline, D. G.; Coleman, W.; Hudson, A. R. Experimental interfascicular nerve grafting. J. Neurosurg. 51:323–332; 1979.

Braun, R. M. Epineurial nerve structure. Clin. Orthop. Rel. Res. 163:50–56; 1982.

Cabaud, H. E.; Rodkey, W. G.; Nemeth, T. J. Progressive ultrastructural changes after peripheral nerve transection and repair. J. Hand Surg. 7:353–365; 1982.

Clark, W. K. Surgery for injection injuries of peripheral nerves. Surg. Clin. North. Am. 52:1325–1328; 1972.

Cragg, B. G.; Thomas, P. K. Changes in conduction velocity and fiber size proximal to peripheral nerve lesions. J. Physiol. 157:315–327; 1961.

Dellon, A. L. Results of nerve repair in the hand. In: Evaluation of sensibility and re-education of sensation in the hand. Baltimore: Williams and Wilkins; 1981: p. 193–201.

Doyle, J. R. Factors affecting clinical results of nerve suture. In: Jewett, D. L.; McCarroll, H. R. Jr., eds. Nerve repair and regeneration. St. Louis: C.V. Mosby Co; 1980: p. 263–266.

Ducker, T. B. Pathophysiology of peripheral nerve trauma. In: Wilkins, R. H.; Rengachary, S. S., eds. Neurosurgery. New York: McGraw-Hill; 1985: p. 1812–1816.

Frykman, G. K.; Adams, J.; Bowen, W. W. Neurolysis. Orthop. Clin. North Am. 12:325–342; 1981.

Frykman, G. K.; Colly, D. Interfascicular nerve grafting. Orthop. Clin. North Am. 19:71–80; 1988.

Gentili, F.; Hudson, A. R. Peripheral nerve injuries: types, causes, grading. In: Wilkins, R. H.; Rengachary, S. S., eds. Neurosurgery. New York: McGraw-Hill; 1985: p. 1802–1812.

Gilbert, A.; Razaboni, R.; Amar-Khodja, S. Indications and results of brachial plexus surgery in obstetrical palsy. Orthop. Clin. North Am. 19:91–105; 1988.

Groff, R. A.; Houtz, S. J. Recovery and regeneration. In: Manual of diagnosis and management of peripheral nerve injuries. J.B. Lippincott Co.; Philadelphia: 1945: p. 33–35.

Hardy, R. W.; Jr.; Wilbourn, A. J. Thoracic outlet syndromes. In: Wilkins, R. H.; Rengachary, S. S., eds. Neurosurgery. New York: McGraw-HIll; 1985: p. 1767–1770.

Highet, W. B. Grading of motor and sensory recovery in nerve injuries. In: Peripheral nerve injuries, MRC special report series no 282. London: Her Majesty's Stationery Office; 1956.

Horowitz, N.; Rizzoli, H. Postoperative complications in neurosurgical practice, recognition, prevention, management. Baltimore: Williams & Wilkins; 1967.

Hudson, A. R.; Trammer, B. Brachial plexus injuries. In: Wilkins, R. H.; Rengachary, S. S., eds. Neurosurgery. New York: McGraw-Hill; 1985: p. 1817–1832.

Hudson, A. R.; Kline, D. G.; MacKinnon, S. E. Entrapment neuropathies. In: Postoperative complications of extracranial neurological surgery. Baltimore: Williams & Wilkins; 1987: p. 260–282.

Kline, D. G.; Hackett, E.; Davis, G. Effect of mobilization on the blood supply and regeneration of injured nerves. J. Surg. Res. 12:254–266; 1972.

Kline, D. G.; Hudson, A. R.; Bratton, B. R. Experimental study of fascicular nerve repair with and without epineural closure. J. Neurosurg. 54:513–520; 1981.

Kline, D. G. Timing for exploration of nerve lesions and evaluation of the neuroma in-continuity. Clin. Orthop. 163:42–49; 1981.

Kline, D. G.; Hudson, A. R. Acute injuries of peripheral nerves. In: Youmans, J.; ed. Neurologic surgery. W.B. Saunders; Philadelphia: 1989: p. 2423–2510.

Kline, D. G. Microsurgery of peripheral nerves: selection for and timing of the operation. In: Furnas, D., ed. Clinical frontiers in plastic surgery. St. Louis: C. V. Mosby; 1984: p. 341–349.

Kline, D. G.; Judice, D. J. Operative management of selected brachial plexus lesions. J. Neurosurg 58:631–649; 1983.

Kline, D. G.; Hudson, A. R. Complications of nerve injury and nerve repair. In: Greenfield, L. J., ed. Complications in surgery and trauma. Philadelphia: J.B. Lippincott; 1984: p. 695–708.

Kline, D. G.; Hurst, J. Prediction of recovery from peripheral nerve injury. Neurology and Neurosurgery Updated Series 5:2–8; 1984.

Kline, D. G.; Hackett, E. R. Management of the neuroma in continuity. In: Wilkins, R. H.; Rengachary, S. S., eds. Neurosurgery. New York: McGraw-Hill; 1985: p. 1833–1845.

Kline, D. G.; Hackett, E. R. Reappraisal of timing for exploration of civilian peripheral nerve injuries. Surgery 78:54–65; 1975.

Kline, D. G. Diagnostic determinants for management of peripheral nerve lesions. In: Rand, R., ed. Microneurosurgery. 3rd ed., St. Louis: C.V. Mosby; 1985: p. 707–726.

Kline, D. G.; Hudson, A. R. Selected recent advances in peripheral nerve injury research. Surg. Neurol. 24:371–376; 1985.

Kline, D. G.; Hackett, E. R.; Happel, L. H. Surgery for lesions of the brachial plexus. Arch. Neurol. 43:170–181; 1986.

Kline, D. G.; Hudson, A. R. Nerve injuries. In: Horowitz, N.; Rizzoli, H., eds. Postoperative complications of extracranial neurological surgery. Baltimore: Williams & Wilkins; 1987: p. 245–259.

Kline, D. G. Civilian gunshot wounds to brachial plexus. J. Neurosurg. 70:166–174; 1989.

Kutz, J. E.; Shealy, G.; Lubbers, L. Interfascicular nerve repair. Orthop. Clin. North Am. 12:277–286; 1981.

Lindsay, W. K.; Walker, F. G.; Farmer, A. W. Traumatic peripheral nerve injuries in children. Plast. Reconstr. Surg. 30:462–468; 1962.

Lundborg, G. Structure and function of the intraneural microvessel as related to trauma, edema formation, and nerve function. J. Bone Joint Surg. 57A:938–1950; 1975.

Lundborg, G. Regeneration of peripheral nerves—a biological and surgical problem. J. Plast. Reconst. Surg. 19:38–44; 1982.

Lundborg, G. Nerve regeneration and repair: A review. Acta. Orthop. Scand. 58:145–169; 1987.

McCarroll, H. R.; Jr.; Rodkey, W. G.; Cabaud, H. E. Results of suture of cat ulnar nerves: a comparison of surgical techniques. In: Jewett, D. L.; McCarroll, H. R., Jr., eds. Nerve repair and regeneration. St. Louis: C.V. Mosby Co; 1980: p. 228–234.

McGillicuddy, J. E. Techniques of nerve repair. In: Wilkins, R. H.; Rengachary, S. S., eds. Neurosurgery. New York: McGraw-Hill; 1985: p. 1871–1881.

McNamara, M. J.; Garrett, W. E.; Seaber, A. V.; Goldner, J. L. Neurorrhaphy, nerve grafting and neurotization: a functional comparison of nerve reconstruction techniques. J. Hand Surg. 12A:354–360; 1987.

McQuarrie, I. G. Clinical signs of peripheral nerve regeneration. In: Wilkins, R. H.; Rengachary, S. S. eds. Neurosurgery. New York: McGraw-Hill; 1985: p. 1881–1884.

Miller, C. E.; Kline, D. G. Acute exploration and repair of peripheral nerve injury. Contemporary Neurosurgery 4:19:1–6; 1982.

Miller, R. G. AAEE Mimeograph #28: injury to peripheral motor nerves, Muscle Nerve 10:698–710; 1987.

Millesi, H. Microsurgery of peripheral nerves. World. J. Surg. 3:67–79; 1979.

Millesi, H. Interfascicular nerve repair and secondary repair with nerve grafts. In: Jewett, D. L.; McCarrol, H. R., Jr., eds. Nerve repair and regeneration. C.V. Mosby; St. Louis: 1980: p. 299–319.

Millesi, H. Interfascicular nerve grafting. Orthop. Clin. North Am. 12:287–301; 1981.

Millesi, H. Reappraisal of Nerve Repair. Surg. Clin. North Am. 61:321–340; 1981.

Millesi, H. Peripheral nerve injuries. Scand. J. Plast. Reconstr. Surg. 19:25–37; 1982.

Millesi, H. Nerve grafting. Clin. Plast. Surg. 11:105–113; 1984.

Millesi, H. Brachial plexus injuries—management and results clinics. Plast. Surg. 11:115–120; 1984.

Millesi, H. The nerve gap. Hand Clinics 2:651–663; 1986.

Mijaoto, Y. Experimental study of results of nerve suture under tension vs nerve grafting. Plast. Reconstr. Surg. 64:540–549; 1979.

Muller, H.; Grubel, G. Factors influencing peripheral nerve suture results. Arch. Ortho. Traum. Surg. 102:51–55; 1983.

Nagano, A.; Tsuyama, N.; Hara, T.; Singioka, H. Brachial plexus injuries. Arch. Ortho. Trauma Surg. 102:172–178; 1984.

Narakas, A. Brachial plexus surgery. Ortho. Clin. North. Am. 12:303–323; 1981.

Omer, G. E. Past experience with epineurial repair: primary, secondary, and grafts. In: Jewett, D. L.; McCarroll, H. R., Jr., eds. Nerve repair and regeneration. St. Louis: C.V. Mosby Co; 1980: p. 267–276.

Omer, G. E. Results of untreated peripheral nerve injuries. Clin. Orthop. Rel. Res. 163:15–19; 1982.

Orgel, M. G. Experimental studies with clinical application to peripheral nerve injury. Clin. Orthop. Rel. Res. 163:98–106; 1982.

Orgel, M. G. Epineural vs perineural repair of peripheral nerves. Clin. Plast. Surg. 11:101–104; 1984.

Osterman, A. L.; Bedner, J. M.; Bora, F. W. Jr.; Pleasure, D. E.; Urbaniak, J. R. Peripheral nerve grafts: the relationship of axonal growth cone and biomechanical properties. J. Hand Surg 11A:189–195; 1986.

Pollock, L. J.; Davis, L. The results of peripheral nerve surgery. In: Peripheral nerve injuries. New York: Hoeber; 1933: p. 545–561.

Rengarchary, S. S. Entrapment neuropathies. In: Wilkins, R. H.; Rengachary, S. S., eds. Neurosurgery. New York: McGraw-Hill; 1985: p. 1771–1795.

Rosen, J. M. Concepts of peripheral nerve repair. Ann. Plast. Surg. 7:165–171; 1981.

Salisbury, R. E.; Dingeldein, G. P. Peripheral nerve complications following burn injury. Clin. Orthop. Rel. Res. 163:92–97; 1982.

Seddon, H. J. Three types of nerve injury. Brain 66:237–288; 1943.

Seddon, H. J. Results of repair of nerves. In: Surgical disorders of the peripheral nerves. Baltimore: Williams & Wilkins; 1972: p. 299–315.

Sedel, L. The management of supraclavicular lesions. Clin. Plast. Surg. 11:121–126; 1984.

Snyder, C. C. Epineurial repair. Orthop. Clin. North Am. 12:267–276; 1981.

Spinner, M. Factors affecting return of function following nerve injury. In: Injuries to the major branches of peripheral nerves of the forearm. 2nd ed. Philadelphia: W. B. Saunders; 1978: p. 42–51.

Stellini, L. Interfascicular autologous grafts in the repair of peripheral nerve: eight years experience. Br. forum Plast. Surg 35:478–482; 1982.

Sunderland, S. Nerves and nerve injuries. Baltimore: Williams & Wilkins; 1968.

Sunderland, S. The anatomic foundation of peripheral nerve repair techniques. Orthop. Clin. North. Am. 12:245–266; 1981.

Tinel, J. Prognosis and treatment of peripheral nerve lesions. In: Joll, C. A., ed. Nerve wounds. London: Ballance Tindall & Cox; 1917: p. 297–299.

Tupper, J. W.; Oriek, J. C.; Mattick, L. R. Fascicular nerve repairs. Orthop. Clin. North Am. 19:57–69, 1988.

Visser, P. A.; Hermreck, A. S.; Pierce, G. E.; Thomas, J. H. Creighton, A. H. Prognosis of nerve injuries during acute trauma to peripheral arteries. Am. J. Surg. 140:596–599; 1980.

Williams, H. B.; Jabaley, M. E. The importance of internal anatomy of the peripheral nerves to nerve repair in the forearm and hand. Hand Clinics 2:689–707; 1986.

Woodhall, B.; Beebe, G. W. eds. Peripheral nerve regeneration: a follow-up study of 3,656 World War II injuries. Washington, DC: U.S. Government Printing Office; 1956: p. 115–201.

Zachary, R. B.; Holmes, W. Primary suture of nerves. Surg. Gynecol. Obstet. 82:632–651; 1946.

29

Amyotrophic Lateral Sclerosis

STANLEY H. APPEL AND LOUISE V. APPEL

The term motor neuron disease often is used interchangeably with amyotrophic lateral sclerosis (ALS), but ALS is only one type of motor neuron disease. The loss of lower motoneuron function manifests in signs and symptoms that are clinically categorized as progressive muscular atrophy (PMA) or spinal muscular atrophy (SMA). The presence of upper motoneuron deficits alone is termed primary lateral sclerosis (PLS). When a patient presents with symptoms and signs of bulbar difficulty manifested by impairment of speech and swallowing, then he or she is considered to have progressive bulbar palsy (PBP). The combination of upper and lower motoneuron signs and symptoms with bulbar involvement is characteristic of ALS.

Of these disorders, ALS is the most common, PBP is the next most common, PMA or SMA is less common, and PLS is the least common. In one series (Caroscio et al. 1987), approximately 80% of the patients with motor neuron disease had ALS, 10% had PBP, 7% had PMA, and 2% had PLS. These percentages are not an accurate reflection of the prevalence of the different motor neuron diseases because of referral bias, namely, the potential understatement of cases of SMA because patients are being referred to an ALS clinic. The evaluation of referrals of patients with ALS at Baylor College of Medicine in the last 8 years has shown that at least 90% of more than 600 patients with motor neuron disease had ALS. In the authors' clinic, almost all cases with PBP had clinical and pathological evidence of diffuse involvement of upper and lower motoneurons and thus could be considered to have ALS. Furthermore, with the data provided by Caroscio and colleagues (1987), the prognoses of patients diagnosed as ALS were not statistically different from the prognoses of patients with PBP; however, patients with PMA and PLS had a much more benign course.

Because this chapter deals primarily with ALS, PMA, SMA, and PLA are excluded. Nevertheless, it is important to note that many cases of SMA are inherited and are relatively predictable. This statement applies most specifically to the type I SMA, which appears in infants and has been called acute Werdnig-Hoffmann disease. The manifestations may appear *in utero* when kicking movements of the fetus may be markedly decreased or absent. Other abnormalities noticed at birth or within the first few days are decreased muscle tone, decreased respiratory function, and a relatively weak cry. Approximately 25% of all cases of SMA represent this acute infantile form. Typically, patients die within the first few years.

When the disease manifests between the acute infantile ages and adolescence, it is referred to as intermediate or type II SMA. In these cases, the prognosis is more variable, and the course is difficult to predict. Even patients with relatively compromised respiratory function may survive until adulthood. Later onset, juvenile SMA is called type III or Kugelberg-Welander disease. This disorder typically presents between ages 5 and 15, although it may begin earlier or later. Weakness usually progresses slowly with some patients able to ambulate 30 years after the onset of illness; others require a wheelchair in less than 10 years.

It is critical that other conditions associated with slowly progressive proximal weakness, such as limb girdle dystrophy, be ruled out by appropriate electrical and morphological analyses. A more difficult task is to assess prognosis in patients who present later in life with predominantly lower motoneuron involvement. The overlap of patients with motor neuropathy and with lower motoneuron syndromes confounds the issue. An additional source of confusion is patients who present early with lower motoneuron signs and later develop evidence of upper motoneuron and bulbar compromise characteristic of ALS. Their prognosis is more difficult to estimate and will depend on the rate of upper motoneuron and bulbar dysfunction. Nevertheless, most patients with motor neuropathy and lower motoneuron compromise who subsequently develop bulbar symptoms follow a less devastating course than patients with more classical ALS.

ALS is a relentless, incapacitating neuromuscular disease with an unknown cause and with no known therapy to slow its course. The incidence of disease is one to two per 100,000, and the prevalence five to seven per 100,000 (Bobowick and Brody 1973; Jokelainen 1976; Kahana et al. 1976). The disease is approximately twice as common in men as in women and has a mean age of onset of 57 years in most series (Appel et al. 1986). Although many specific etiologies, including viral, toxic, endocrine, and genetic, have been associated with upper and lower motoneuron dysfunction, the majority of ALS cases are sporadic and of unknown etiology. The disease usually begins with focal weakness and progresses with worsening disability of limbs or bulbar musculature. In a study of 58 patients (Appel et al. 1986), compromised speech was the initial manifestation of ALS in 12 patients (21%), and weakness of the extremities was the initial manifestation in 38 patients (66%). Of the 38 patients presenting with limb weakness, 44% presented with weakness or fatigability in the right extremities, 36% in the left extremities, and 20% in both extremities. Twelve percent of the patients presented with sensory complaints, primarily pain or parasthesias (Appel et al. 1986). The pattern of progression of neurological deterioration followed a characteristic course. When the difficulty started in the right lower extremity, the next area of involvement was usually the left lower extremity. When the onset was one upper extremity, the next area of involvement in most patients was the contralateral upper extremity. For patients whose onset was bulbar, the next area of involvement was usually an upper extremity. Furthermore, even in patients diagnosed with PBP with no symptomatic involvement of upper or lower extremities, most patients showed electrical or morphological evidence of limb involvement.

To assess the rate of progression of neurological deterioration, the authors devised a rating scale that provides a quantitative estimate of the clinical status of the patient and of disease progression (Appel et al. 1987). The general clinical experience had been to anticipate variability in the musculature involved and in the progression of the disease. For example, a patient may experience rapid deterioration in speech and swallowing and have minimal change in extremity strength, rendering strength tests meaningless in evaluating disease progression. In contrast, pulmonary function alone may be less accurate in patients for whom extremity strength is compromised and bulbar function is spared. The rating scale includes assessment of swallowing and speech function, respiratory function, muscle strength in upper and lower extremities, and function of upper and lower extremities. The total ALS score consists of five group scores: bulbar, respiratory, muscle strength, lower-extremity function, and upper-extremity function. The bulbar group is composed of the swallowing and speech subgroups, each of which is ranked according to 5 degrees of severity; for example, normal swallowing is considered the ability to eat a general diet. Very severe dysfunction means the patient requires a feeding tube or gastrostomy. In the speech subgroup, "normal" indicates clear speech, while maximal dysfunction is characterized by aphonia. With respiration, scoring is graded into five steps based on changes of forced vital capacity expressed as a percent of predicted response. Muscle strength is assessed using the Medical Research Council system of grading, and tests of grip strength and lateral pinch strength. Lower-extremity function is assessed with timed functions and observed functions. Upper-extremity function is evaluated in four tests of timed functions (e.g., assembly of peg units, turning over blocks) and observed functions (e.g., ability to abduct the arms and independence in dressing and feeding).

With this method of monitoring patients for several years, it was apparent that the total ALS score changed in a linear fashion with time. Linear regression analysis of the change in ALS scores of 74 patients with time yielded a median correlation coefficient of 0.956, supporting the linearity of progression. Figure 29–1 illustrates the total ALS score for three patients following a rapid, an intermediate, or a slow progression of the disease for 10 to 38 months. Using a quantitative testing battery, Andress and coworkers (1986) and Munsat and associates (1987) followed 51 patients with ALS for a minimum of 12 months and found that their

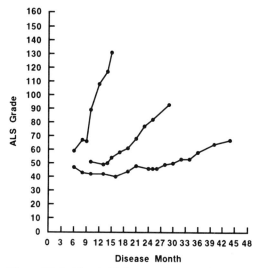

Figure 29-1. Total ALS score for 3 patients progressing at different rates (Appel 1987).

rates of deterioration were linear, with no plateaus or improvement. Using the total ALS score, progression differed markedly in different patients. When the slope of the regression line was calculated (i.e., the point change per 28 days), there was greater than a 20-fold difference in the rate of progression from the slowest to the most rapid course (Figure 29–2). During 1 year, 19% of the patients progressed at a slow rate, changing less than 13 points; 77% of patients progressed at an intermediate rate of 13 to 48 points; and 34% of patients progressed at a rapid rate, changing more than 40 points. The study by Jablecki and colleagues (1989) also documented differing rates of progression in different patients with ALS. In their study, they noted a 60-fold difference between the fastest and slowest rates of progression of 194 patients followed for 8 years. In the patients in the authors' study, the difference was not due to lower motoneuron disease in the slow category and lower and upper motoneuron disease with severe bulbar involvement in the fast category. This is because all patients in the analysis had both lower and upper motoneuron involvement with minimal bulbar compromise at entry into the study.

Bonduelle (1975) reports that ALS patients presenting with bulbar symptoms and

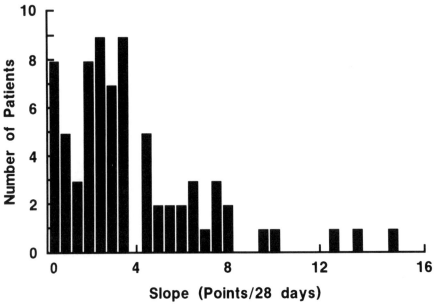

Figure 29-2. The point change per month for patients with ALS documenting the markedly different rates of progression in different patients (Appel 1987).

signs deteriorated more quickly than patients presenting initially with limb weakness. Subsequent studies have confirmed the finding but have noted that bulbar presentation is more common in older patients (Kristensen and Melgaard 1977; Rosen 1978). In fact, if survival in patients presenting with bulbar signs was corrected for age of onset, there was no longer a statistical difference in prognosis of these two groups of patients (Daube 1985; Mulder 1984). The studies by Jablecki and associates (1989) suggest that older patients have a shorter survival, regardless of whether the initial presentation is of bulbar or spinal onset. However, in 318 ALS patients in Israel, patients with onset of disease with bulbar signs had a shorter life expectancy (2.2 years) even when corrected for age and sex (Gubbay et al. 1985). Longer survival in Guamanian ALS patients did depend on an early age at onset and male sex (Reed et al. 1975), but neither bulbar nor extremity onset showed a meaningful pattern of association with the duration of illness. In most studies, regardless of whether bulbar or extremity signs occur first, the rapid progression of bulbar signs is associated with a poor prog-

nosis. The rapid onset of bulbar signs in patients presenting initially with extremity weakness resulted in death within 1.5 to 2 years in 67% of all spinal cases in the series by Boman and Meurman (1967).

A follow-up of the authors' initial group of patients has permitted an estimate of the reliability of the own predictions. Sixty-five patients (47 men and 18 women) were evaluated and graded on seven occasions in a 12-month period between 1984 and 1986. The average number of points per month during that period was used to predict the number of months until the patient reached a terminal stage, which is 135 points in this scoring scheme. The cases were then analyzed to determine the number of months until the patient died, became ventilator dependent, became terminally ill, or was totally dependent at 135 points. Thirteen of 65 patients were still functioning more than 5 years after the evaluation. However, because the patients had been experiencing symptoms before the evaluation, the data had to be analyzed, taking into consideration the number of months that the patients had been experiencing symptoms before they were seen in the clinic. Nineteen of 65 patients (29%)

lived at least 5 years from the onset of first symptoms. Six of the 65 died or became ventilator dependent during the sixth year, and 20% were still functioning during the same period. Seven of the 13 patients who are still functioning are women and only six are men. Because of the overall 2.6:1 ratio of the patients in the study, the over-representation of women surviving 5 years suggests that at least some women may have a more benign course. Six of 47 (12.8%) of the men had a course longer than 5 years, and seven of 18 (39%) women followed such a course.

Fourteen of 65 patients (21.5%) were totally dependent or died 12 months after they were first evaluated in the clinic. Ten men and four women (a ratio of 2.5:1) comprised this group. When the length of time before death or ventilator dependency was dated from the onset of first symptoms, 11 of 65 (17%) followed a course of less than 2 years. Seventeen percent of patients lived 25 to 36 months after first symptoms, 25% lived 37 to 48 months, and 12% lived 49 to 60 months.

The follow-up analysis confirms the belief that at least 20% of ALS patients may follow a benign course lasting more than 5 years from the onset of symptoms. As noted previously, at the time of initial examination, these patients had upper and lower motoneuron involvement and minimal bulbar difficulty. Thus, a prognostic statement excludes patients who might have followed a more benign course because their disease was restricted to lower motoneuron or upper motoneuron deficits alone. Twenty percent to 30% of the ALS patients followed a more rapid course, resulting in death or total dependency in less than 2 years from the onset of symptoms. The data also suggest that the ALS scoring system can predict the progression of disease and prognosis after following the paients for 6 months to 1 year. A detailed analysis of the ability to predict prognosis and survival after scoring patients for several months is being conducted.

In almost all cases, the events leading to death are associated with deteriorating respiratory function, which causes increased susceptibility to infection. Difficulty in swallowing can also lead to aspiration and can impair respiration. Early gastrostomy may avoid many of the complications of swallowing. Tracheostomy and ventilatory assistance may be important life supports, but issues of quality of life and financial means should be evaluated carefully; decisions must be made by the patient and family, based on the unique circumstances of each situation. In the absence of ventilatory support, patients succumb to impaired breathing.

To date, no specific therapy has changed the course of disease progression in ALS. Some provocative data involving immunosuppression are being evaluated in double-blind studies (Appel et al. 1988), and preventing the effects of excitotoxic amino acids also is under examination. The mainstay of current treatment remains symptomatic therapy aimed at improving the quality of daily living and delaying the complications of respiratory and swallowing dysfunction.

Acknowledgment

This study was made possible by the Muscular Dystrophy Association support of our MDA ALS Clinical and Research Center. We are grateful to our ALS team and to the many ALS patients and their families who gave of themselves to further our understanding this devastating condition.

References

Andres, P. L.; Hedlund, W.; Finison, L.; Conlon, T.; Felmus, M.; Munsat, T. L. Quantitative motor assessment in amyotrophic lateral sclerosis. Neurology 36:937–941; 1986.

Appel, S. H.; Stockton-Appel, V.; Stewart, S. S.; Kerman, R. H. Amyotrophic lateral sclerosis-associated clinical disorders and immunological evaluations. Arch. Neurol. 43:234–238; 1986.

Appel, S. H.; Stewart, S. S.; Appel, V.; Harati, Y.; Mietlowski, W.; Weiss, W.; Belendiuk, G. W. A double-blind study of the effectiveness of cyclosporine in amyotrophic lateral sclerosis. Arch. Neurol. 45:381–386; 1988.

Appel, V.; Stewart, S. S.; Smith, G.; Appel, S. H. A rating scale for amyotrophic lateral sclerosis: description and preliminary experience. Ann. Neurol. 22:328–333; 1987.

Bobowick, A. R.; Brody, J. A. Epidemiology of motor neuron disease. N. Engl. J. Med. 288:1047–1055; 1973.

Boman, K.; Meurman, T. Prognosis of amyotrophic lateral sclerosis. Acta. Neurol. Scandinav. 43:489–498; 1967.

Bonduelle, M. Amyotrophic lateral sclerosis. In: Vinken, R. J.; Bruyn, G. W., eds. Handbook of clinical neurology Volume 22. New York: American Elsevier Company; 1975: p 281–338.

Caroscio, J. T.; Mulvihill, M. N.; Sterling, R.; Abrams, B. Amyotrophic lateral sclerosis–its natural history. Neurol. Clin. 5:1–8; 1987.

Daube, J. R. Electrophysiologic studies in the diagnosis and prognosis of motor neuron diseases. In: Aminoff, ed. Symposium of electrodiagnosis. Volume 3. Philadelphia: W. B. Saunders; 1985: p 477–493.

Gubbay, S. S.; Kahana, E.; Zilber, N.; Cooper, G.; Pintov, S.; Leibowitz, Y. Amyotrophic lateral sclerosis. A study of its presentation and prognosis. J. Neurol. 232:259–300; 1985.

Jablecki, C. K.; Berry, C.; Leach, J. Survival prediction in amyotrophic lateral sclerosis. Muscle Nerve 12:833–841; 1989.

Jokelainen, M. The epidemiology of amyotrophic lateral sclerosis in Finland: a study based on the death certificates of 421 patients. J. Neurol. Sci. 20:55–63; 1976.

Kahana, E.; Alter, M.; Feldman, S. Amyotrophic lateral sclerosis: a population study. Neurology 212:205–213; 1976.

Kristensen, O.; Melgaard, B. Motor neuron disease. Acta. Neurol. Scandinav. 56:299–308; 1977.

Mulder, D. W. Motor neuron disease. In: Dyck, P. J.; Thomas, P. K.; Lambert, S. H.; Bunge, R., eds. Peripheral neuropathy. Philadelphia: W. B. Saunders; 1984: p 1525–1536.

Munsat, T.; Andres, P. L.; Bumside, S.; et al. The natural history of amyotrophic lateral sclerosis. Ann. Neurol. 18:157; 1987.

Reed, D. M.; Brody, J. A.; Holden, E. M. Predicting the duration of Guam amyotrophic lateral sclerosis. Neurology 25:277–288; 1975.

Rosen, A. D. Amyotrophic lateral sclerosis. Arch. Neurol. 35:638–642; 1978.

30

Developmental Anomalies

RICHARD E. GEORGE AND HAROLD J. HOFFMAN

Two percent of children are born with congenital malformations and 60% of these involve the central nervous system (CNS) (Murphy 1937). Malformations of the CNS are an important source of morbidity and mortality, frequently depriving the person of intellect and physical agility. With appropriate management, some malformations can be ameliorated and patients may lead rewarding and meaningful lives. Parents of children born with developmental anomalies suffer extreme emotional and psychological turmoil after the birth of their child. The decisions they make as to the treatment of their children are heavily weighted on information provided by treating physicians and their associates (Freeman 1984). The prognosis of many of these conditions has improved with recent advances in medical and surgical therapy. It is imperative that physicians discussing developmental anomalies with parents provide them with accurate and current information as to the prognosis of these conditions.

Spinal Malformations

Myelomeningocele

Myelomeningocele, a term used synonymously with spina bifida cystica, spina bifida aperta, myelocele, and open neural tube defect, consists of a developmental anomaly in which there is a defect of fusion of the posterior neural arch of one or more vertebrae, accompanied by malformation of the underlying neural tube (neural placode). The placode is surrounded by and attached to rudimentary dura and leptomeninges (Humphreys 1985). Myelomeningoceles account for 46% of CNS malformations (Laurence and Tew 1971).

The natural history of patients born with myelomeningoceles is poor; most patients die in infancy (Table 30-1). Those who survive without treatment, however, tend to be the least affected. The majority of survivors have normal IQs and are ambulatory (Table 30-2 and 30-3). Delayed treatment of hydro-

411

Table 30-1. Myelomeningocele Mortality

Reference	Period	Number of Patients	Alive at Follow-Up	Percent Survival	Length of Follow-Up
		Natural History			
Laurence and Tew 1971	1956–1962	274	37	14	6–12 years
Shurtleff et al. 1974	Prior 1965	52	?	~30	Less than 20 years
		Unselected Series			
Lorber 1971	1959–1963	323	134	41	7–12 years
Lorber 1971	1967–1968	201	125	62	2–4 years
Lorber 1972	1962–1964	200	103	52	7–9 years
Ames and Schut 1972	1963–1968	171	138	81	3–8 years
Laurence 1974	1964–1966	104	48	46	7–9 years
Hunt 1983	1963–1971	117	68	58	12–19 years
McLone and Naidich 1989	1975–1978	100	85	85	8–12 years
		Selected Series			
Stark and Drummond 1973	1965–1971	30 (56)	20 (21)	67 (38)	5 years
Lorber and Salfield 1981	1971–1976	42 (113)	36 (36)	86 (32)	3–9 years
Charney et al. 1985	1978–1982	52 (110)	48 (90)	92 (82)	10 months

Figures in parentheses indicate the results of the entire series, including those patients selected for no treatment.

cephalus, however, has been shown to lower the intellectual potential of survivors (Badell-Riberra et al. 1966). Infection, hydrocephalus, and later renal complications are the major causes of death in these patients (Laurence 1974).

Sharrard and colleagues (1963) demonstrated in Sheffield a decreased mortality with early repair of the myelomeningocele and treatment of hydrocephalus. Subsequently, many centers began treating all newborns with myelomeningoceles (see Table 30-1). Although mortality was reduced, many survivors were left severely debilitated (see Table 30-2). Lorber (1971) reviewed a series of 524 aggressively treated infants. He documented improved survival but also demonstrated a high rate of mental retardation and severe physical handicap. Only 6.5% of patients in his series would be considered competitive in society. He advocated four adverse criteria that predicted a poor outcome: (1) severe paraplegia, (2) gross cranial enlargement, (3) kyphosis, and (4) other severe congenital defects or birth injuries. In 1972, Lorber reviewed an additional 270 patients and added a fifth adverse criteria, that of a thoracolumbar lesion.

Following Lorber's publication, several clinical series were published in which patients were selected for aggressive management or supportive care on the basis of Lor-

Table 30-2. Myelomeningocele IQ

Reference	Period	Number of Patients	Number With Normal IQ	Percent Normal	Length of Follow-Up
		Natural History			
Laurence and Tew 1971	1956–1962	28	20	71	6–12 years
		Unselected Series			
Lorber 1971	1959–1963	134	89	66	7–12 years
Lorber 1972	1962–1964	103	58	56	7–9 years
McLone et al. 1985	1975–1978	86	63	73	5–9 years
		Selected Series			
Lorber and Salfield 1981	1971–1976	36	32	89	3–9 years
O'Brien and McLanahan 1981	1971–1979	46	37	80	2–10 years

Table 30-3. Myelomeningocele Ambulatory Status

Reference	Number of Survivors	Percentage of Functional Ambulators
Natural History		
Laurence 1974	36	56
Unselected Series		
Lorber 1971	134	51
Ames and Schut 1972	115	79
Lorber 1972	103	42
Laurence 1974	46	35
Reigel 1988	358	84
Selected Series		
Stark and Drummond 1973	60	75
Lorber and Salfield 1981	36	78
O'Brien and McLanahan 1981	46	46

ber's or other similar adverse criteria. The survivors of the "selected" series tended to have a better outcome than those in "unselected" series (see Tables 30-2 and 30-3); however, their overall mortality was quite high (see Table 30-1). Advancements in shunt technology, the institution of clean intermittent catheterization, and aggressive treatment of secondary complications have improved the results of recent series of unselectively treated patients (see Tables 30-1, 30-2, and 30-3). The adverse criteria of Lorber may indicate a higher probability of neurological or physical disability; however, most authors feel that with current technology, these problems can be ameliorated and no exclusionary criteria should be applied to the treatment of myelomeningocele patients (Ames and Schut 1972; McLaughlin et al. 1982; McLone 1986; Soare and Raimondi 1977). Most centers in North America treat patients with myelomeningoceles in an unselective fashion (Marlin 1990).

Current mortality rates for patients born with myelomeningoceles are 11% to 20% (see Table 30-1). The most common cause of early mortality in recent series is the Chiari malformation, and shunt malfunction is an important cause of late mortality (McLone and Naidich 1989a). Patients with higher le-

sions tend to have a higher mortality than those with lower lesions (Ames and Schut 1972; Humphreys 1985; Lorber 1961; Shurtleff et al. 1975). With the advent of clean intermittent catheterization, renal failure is an infrequent cause of late mortality (McLone et al. 1985).

The incidence of hydrocephalus in patients with myelomeningocele is 80% to 90% (Humphreys 1985; Lorber 1972; McLone et al. 1985; Stark and Drummond 1973). The majority of these patients require cerebrospinal fluid (CSF) shunt insertions. The frequency of hydrocephalus increases with higher lesions (Ames and Schut 1972; Humphreys 1985; Hunt and Holmes 1975). Aqueductal stenosis is the cause of the hydrocephalus in 75% of patients (Gilbert et al. 1986; Stein and Schut 1979).

The prognosis for normal intellectual function is good (see Table 30-2). Severe retardation is present in 10% to 15% of survivors (McLone 1986). Patients with hydrocephalus requiring shunting tend to have lower IQs than those without (Halliwell et al. 1980; Hunt and Holmes 1975; Laurence and Tew 1971; Lonton 1982; Lorber 1972; Mapstone et al. 1984; Soare and Raimondi 1977). Severe hydrocephalus documented at birth, however, does not necessarily indicate a low intellectual potential (Cull and Wyke 1984; McLone 1986; Raimondi and Soare 1974; Soare and Raimondi 1977). The etiology of the decreased intellect in shunted patients is not clear, but it may be related to shunt infection, other shunt complications, or associated cerebral malformations (Gilbert et al. 1986; Hunt and Holmes 1975; Lorber 1972; Mapstone, Rekate et al. 1984; McLone et al. 1982; McLone et al. 1985). Intellect does not appear to be related to the number of shunt revisions (Dennis et al. 1981; Raimondi and Soare 1974). Patients with thoracolumbar lesions typically have lower IQs. This may relate to the higher incidence of hydrocephalus and its related complications (Hunt and Holmes 1975; Lonton 1982; Shurtleff et al. 1975; Soare and Raimondi 1977). Myelodysplastic patients frequently have difficulty with perceptual–motor skills (Soare and Raimondi 1977) and memory function (Cull and Wyke 1984; Reigel 1989). Their verbal skills often are highly developed, leading to the "cocktail party personality," as described under the hydrocephalus section (Humphreys 1985; Ingram and Naughton 1962; Laurence and Tew 1971).

Between 40% and 85% of survivors of myelomeningocele are functional ambulators (see Table 30-3). The prognosis for ambulation depends on the level of the lesion (Welch and Winston 1987). Those with thoracolumbar lesions are functional ambulators in 0% to 33% of cases, while those with sacral lesions are able to walk in most cases (Hoffer et al. 1973; Humphreys 1985; Welch and Winston 1987). The combination of a hip flexion deformity, pelvic obliquity, and scoliosis is likely to preclude walking, as is the presence of any one of these factors to a severe degree (Stillwell and Menelaus 1983). As patients become older and their body weight increases, the percentage of ambulatory patients declines (Humphreys 1985; McLone et al. 1985; Welch and Winston 1987). Many children find that they are more mobile in wheelchairs. Devotion of time to school rather than to physiotherapy also may contribute to the lower frequency of ambulators in older age groups (Welch and Winston 1987).

Patients with myelodysplasia may have impaired motor function of the upper extremities. Wallace (1973) documented neurologically abnormal upper extremities in 69% of 225 children with myelomeningoceles. Of the group with impaired arm function, 58% had significant cerebellar and pyramidal dysfunction that impaired their mobility. Children with myelodysplasia are slower than other children in performing motor tasks. This impairment may be secondary to ataxia, pyramidal tract dysfunction, or perceptuomotor problems (Anderson and Plewis 1977).

Only 3% to 10% of patients with myelodysplasia have normal urinary continence (Ames and Schut 1972; Lorber 1972; Miyazaki 1972; Smith 1972). With clean intermittent catheterization every 2 to 3 hours

and pharmacotherapy, approximately 85% of patients will be "socially continent" and will not emit offensive urinary odors (McLone and Naidich 1989a). Hydronephrosis and pyelonephritis, formerly a frequent cause of late mortality and morbidity, are uncommon with clean intermittent catheterization (McLone et al. 1985). Fecal continence can be obtained in most patients with a regular bowel program (McLone et al. 1985).

Patients with myelomeningoceles have an associated hindbrain malformation known as a Chiari II malformation (Emery and MacKenzie 1973). This developmental anomaly consists of displacement of part of the lower vermis, pons, and medulla oblongata and elongation of the fourth ventricle into the cervical canal (Emery and MacKenzie 1973). The Chiari malformation may become symptomatic during infancy or later in life. Two thirds of children who develop symptoms from their Chiari malformations do so before the age of 3 months (Bamberger-Bozo 1987).

Patients who become symptomatic from the Chiari malformation in infancy typically exhibit swallowing difficulties with prolonged and poor feeding, apneic episodes, stridor, bronchial aspiration, and arm weakness (Bell et al. 1987; Carmel 1983; Hoffman et al. 1975; Park et al. 1983). On examination, they are found to have stridor, facial weakness, depressed or absent gag reflexes, sternocleidomastoid weakness (head lag), nystagmus, weakness and fasciculations of the tongue, fixed retrocollis, weak or absent cries, spastic upper extremity weakness with persistent fisting, increased deep tendon reflexes, and increased tone (Oakes 1985; Park et al. 1983; Salam and Adams 1978; Venes et al. 1986). Most myelodysplastic children have mild symptoms related to their Chiari malformation during infancy. Approximately one third become severely symptomatic and 11% of all myelomeningocele patients die from this (McLone and Naidich 1989a; Park et al. 1983). Patients with vocal cord paralysis, arm weakness, or cardiopulmonary dysfunction have the highest mortality. A shunt malfunction may exacerbate Chiari symptoms, which must be excluded before considering a decompressive procedure for the Chiari malformation (McLone and Naidich 1989a; Park et al. 1983). The value of surgical decompression in the management of symptomatic patients is unclear. Some authors believe that the symptoms related to the Chiari malformation are secondary to malformed cranial nerve nuclei and that decompressive procedures have little effect on survival (McLone and Naidich 1989a). Others believe that the decompensation of previously asymptomatic children shows a compressive etiology (Park et al. 1983). Constrictive transverse dural bands with ischemic changes in the underlying brain are found beneath the arch of C1 in 41% of patients at the time of surgery, also suggesting a compressive etiology for the deterioration of these patients (Park et al. 1983). Holliday and coworkers (1985) demonstrated that brain stem auditory evoked potentials may improve significantly following decompression of the malformation, a finding that cannot be accounted for by immaturity of the CNS. The decompressive procedure recommended for Chiari II malformations is a cervical laminectomy that extends below the level of the herniated vermian tissue. The dura is opened. In patients with hydromyelia, the fourth ventricle is communicated with the subarachnoid space. The central canal may be plugged with muscle or a silastic stent may be placed between the fourth ventricle and the subarachnoid space. A patulous duraplasty is then constructed. The mortality following surgical decompression is 30% to 50% (Bell et al. 1987; Charney et al. 1987; Park et al. 1983). Charney and colleagues (1987) found that infants who had stridor alone at the time of decompression fared better than infants with apnea, cyanotic spells, or dysphagia. Park and associates (1983) also found a lower mortality rate in infants who underwent decompression at an earlier stage of their disease. Eight percent of Park's patients required long-term tracheostomies. Although there is no agreement concerning

the treatment of infants symptomatic from the Chiari malformation, the authors favor early cervical decompression in patients with severe or persistent symptoms.

Symptomatic older children and adolescents with Chiari type II malformations present with symptoms of spasticity, scoliosis, progressive weakness of the arms, atrophy of the intrinsic hand muscles, appendicular or truncal ataxia, nystagmus, pain, and headaches (Oakes 1985; Park et al. 1985). After the onset of symptoms, all patients in the study by Park and coworkers (1985) deteriorated relentlessly. The prognosis of patients undergoing decompression for Chiari II malformations that become symptomatic in childhood or adolescence is good. Two thirds of these patients improve their strength and spasticity and one third stabilize their disease (Bell et al. 1987; Park et al. 1985; Venes et al. 1986). None of the 12 patients who underwent decompression in the study by Park and colleagues (1985) had reversal of their scoliotic curvature.

Hydrosyringomyelia, also known as hydromyelia, is the dilatation of the central canal of the spinal cord with CSF that is or has been continuous with the ventricle. The incidence of this condition in myelodysplasia is 29% to 77% (Cameron 1957; Emery and Lendon 1972; McLone and Naidich 1989a). Hydrosyringomyelia is discussed later.

Scoliosis and kyphosis frequently are found in myelodysplastic patients. In a review of 250 cases, Piggott (1980) reported that spinal curves severe enough to warrant corrective surgery were expected in 50% of patients and only 10% maintained completely undeformed spines. The higher the level of the neurological or bony defect, the greater the probability of spinal deformity.

A late complication of repair of myelomeningoceles is retethering of the spinal cord. Usually, this is heralded by symptoms of gait changes, pain, progressive orthopedic foot deformities, ascending motor loss, and scoliosis (Epstein 1983; McLone and Naidich 1989a; Reigel 1983). The pain may occur at the site of the previous repair or radiate into the legs, especially with exercise. Recurrent hip dislocations commonly are seen in these patients (Reigel 1983). The condition probably occurs in many patients but becomes clinically manifest in only 3% (Epstein 1983; McLone and Naidich 1989a). In a recent large series of patients who underwent untethering for this problem, 87% to 100% experienced relief or improvement of pain following release of the tethered cord, 70% showed improvement in lower extremity orthopedic deformities, 64% had improvement or decrease in spasticity, 57% to 76% displayed motor improvement, 53% showed improvement in scoliosis, and 25% saw improvement in bladder or bowel function (McLone and Naidich 1989a; Reigel 1983). Improvement in bladder function was greatest in those who had recently lost function. Patients must be monitored closely for loss of function from retethering of their spinal cords, and release of the tethering should be undertaken as soon as the diagnosis is established.

Patients with myelodysplasia have an increased frequency of seizures. In a study of 190 patients with myelomeningoceles, Chadduck and Adametz (1988) found that the incidence of seizures was related to ventriculoperitoneal shunting. Patients without shunts had a 2% incidence of seizures while those with shunts who had never required a shunt modification had a 9% incidence. The incidence of seizures rose to 22% if patients had at least one shunt revision and 47% if they experienced a shunt infection. Bartoshesky and coworkers (1985) report a 23% incidence of seizures in 111 patients.

Ocular symptomatology is common in patients with myelodysplasia. Strabismus was present in 44% to 82% of patients with myelomeningoceles with hydrocephalus (Clements and Kaushal 1970; Rothstein et al. 1974; Stanworth 1970). The strabismus appears to be related to the presence of hydrocephalus because squint was much less common in patients without hydrocephalus (Clements and Kaushal 1970; Rothstein et al. 1974).

There is a higher incidence of systemic malformations in patients with myelomeningocele. Bamforth and Baird (1989) found that 6% have associated malformations, including renal anomalies, cleft lip or palate, tracheoesophageal fistula, and diaphragmatic hernia.

Because of recent advances in the care of patients with myelodysplasia, the authors favor an aggressive approach to their management with closure of all lesions within 24 hours and shunting at the onset of progressive hydrocephalus. These patients are best followed in a comprehensive clinic with specialists in pediatrics, neurosurgery, neurology, urology, orthopedics, physical medicine, and social services available to care for the children. The patients require close monitoring for complications, such as the Chiari malformation, hydromyelia, retethering of the spina cord, urological deterioration, and progressive spinal deformities. With close follow-up, the majority of these patients will live useful lives, able to contribute to society.

Spinal Meningocele

A spinal meningocele is a cystic protrusion of the meninges through a defect in the vertebra. The contents of the cyst are limited to CSF, meninges, and skin. The neural elements typically are not involved in the malformation, although occasionally there are associated intraspinal lesions at the site of the meningocele or at other locations (Doral and Guthkelch 1961; Lorber 1972; Tryfonas 1973).

The natural history of patients born with spinal meningoceles is difficult to assess because most patients have their meningoceles repaired at an early age. Laurence (1964) studied patients treated prior to the advent of CSF shunt technology. In this series, all 39 patients with meningoceles survived, 24 without neurological abnormality and four with only minor disability. Sixteen patients had hydrocephalus; all had thoracic or cervical lesions. He noted that 11 of the 17 patients with thin-walled lesions developed hydrocephalus or were physically impaired. This was noted in only five of the 22 patients with normal skin covering their lesions. Doran and Guthkelch (1961) report a similarly favorable outcome in their series of 61 patients treated in the era before shunt availability.

With the development of modern shunt technology, the outlook for patients with spinal meningoceles continues to improve. In a recent series, the survival rate was more than 80% and normal IQs were found in almost all survivors (Laurence 1974; Laurence and Tew 1971; Lorber 1972). Significant physical disabilities were found in only 10% of patients and were more common in patients with associated intraspinal malformations (Laurence and Tew 1971; Lorber 1972).

The authors favor an aggressive approach to the management of infants with spinal meningoceles. The lesions should be closed early, especially if they have a thin skin coverage. The patients should be screened for other intraspinal anomalies and hydrocephalus. They require close follow-up for late urological complications.

Lipomyelomeningocele

Lipomyelomeningocele, a term used synonymously with lumbosacral lipoma, spinal lipoma, and lipoma of the cauda equina, describes a subcutaneous lipomatous mass typically present in the lumbosacral region that extends through a midline defect in the lumbodorsal fascia, vertebral neural arch, and dura, and attaches to an elongated and tethered spinal cord (Hoffman et al. 1985). Most patients with lipomyelomeningoceles have subcutaneous masses, and half have associated cutaneous anomalies, such as hypertrichosis, capillary hemangiomas, skin dimples, dermal sinuses, and skin tags (Bruce and Schut 1979; Hoffman et al. 1985; Pang 1986; Pierre-Kahn et al. 1986). Approximately 20% of patients have congenital neurological deficits (Hirsch and Pierre-Kahn 1988; Pierre-Kahn et al. 1986).

The natural history of lipomyelomeningoceles is one of progressive neurological deterioration with lower extremity sensorimotor deficits, bladder and bowel dysfunction, foot deformities, spinal deformities, trophic changes, and back pain (Bruce and Schut 1979; Chapman and Davis 1983; Dubowitz et al. 1965; Hoffman et al. 1985; Lassman and James 1967; McLone et al. 1983; Pang 1986; Pierre-Kahn et al. 1986). The incidence of worsening in children who were initially normal on examination ranges between 36% and 88% (Bruce and Schut 1979; Chapman and Davis 1983; Hirsch and Pierre-Kahn 1988; Hoffman et al. 1985; Lassman and James 1967; Pang 1986; Pierre-Kahn et al. 1986). By late adolescence, the majority of patients have neurological deficits (Bruce and Schut 1979; Hirsch and Pierre-Kahn 1988; Hoffman et al. 1985; McLone et al. 1983; Pierre-Kahn et al. 1986). Deterioration typically occurs during periods of rapid growth or weight gain (Hirsch and Pierre-Kahn 1988; Hoffman et al. 1976; Pang 1986). Deterioration usually is insidious; however, 16% to 25% of patients deteriorate acutely (Dubowitz et al. 1965; Hirsch and Pierre-Kahn 1988; Pierre-Kahn et al. 1986). Minor trauma occasionally can precipitate rapid neurological deterioration from which complete recovery may not occur (Hirsch and Pierre-Kahn 1988; Pang 1986). Patients have deteriorated as early as 3 to 6 weeks of age (Dubowitz et al. 1965; McLone et al. 1983) or in late adulthood (Lassman and James 1967; Pierre-Kahn et al. 1986). When function is lost, its return following surgery seldom is complete, especially for urological function (Bruce and Schut 1979; Chapman and Davis 1983; Dubowitz et al. 1965; McLone and Naidich 1986; Pang 1986; Pierre-Kahn et al. 1986). The cause of neurological deterioration is believed to be secondary to spinal cord tethering (Kang et al. 1987; Pang 1986; Yamada et al. 1981), although other etiologies, such as compression from the lipoma and trauma from the dural bands and lamina, also are postulated (Dubowitz et al. 1965; Hirsch and Pierre-Kahn 1988; McLone et al. 1983; Pierre-Kahn et al. 1986).

Surgical treatment with modern operative techniques, including evoked potential monitoring, magnification, surgical ultrasonic aspirators, and the CO_2 laser, have provided excellent results. Results of several recent series are shown in Table 30-4. There has been no operative mortality in 252 patients. The overall chance of a patient worsening

Table 30-4. Lipomyelomeningocele Outcome

Author	Status of Patients	Number of Patients	Mortality	Deteriorated	No Change	Improved
Hirsch and Pierre-Kahn 1988	Normal	27	0%	7%	93%	—
Hirsch and Pierre-Kahn 1988	Progressive	23	0%	0%	43%	57%
Hirsch and Pierre-Kahn 1988	Congenital	14	0%	0%	93%	7%
Bruce and Schut 1979	All	40	0%	5%	75%	20%
Hoffman et al. 1985	Normal	23	0%	23%	77%	—
Hoffman et al. 1985	Abnormal	34	0%	14%	60%	26%
McLone and Naidich 1986	Normal	23	0%	0%	100%	—
McLone and Naidich 1986	Abnormal	27	0%	0%	60%	40%
Lassman and James 1967	Normal	8	0%	0%	100%	—
Lassman and James 1967	Abnormal	18	0%	6%	44%	50%
Chapman and Davis 1983	All	17	0%	0%	100%	0%
Overall	All	252	0%	5.9%	75%	19.1%
	Normal	85	0%	8.2%	91.8%	0%
	Abnormal	110	0%	5.5%	58.2%	36.3%

Normal indicates a normal neurological examination prior to surgery; progressive indicates a deteriorating neurological course prior to surgery; congenital indicates a congenital malformation; abnormal indicates abnormal neurological examination prior to surgery.

from surgery is 5.9%; remaining unchanged, 75%; and improving, 19.1%. Patients with neurological abnormalities have a 5.5% chance of worsening, a 58.2% chance of remaining unchanged, and a 36.3% chance of improvement. The incidence of late neurological deterioration following appropriate surgery is 4% to 14% (Hirsch and Pierre-Kahn 1988; Lassman and James 1967; McLone et al. 1983; Pierre-Kahn et al. 1986). Improvements in urological function were less than improvements in motor function (Bruce and Schut 1979; Dubowitz et al. 1965; McLone and Naidich 1986; Pang 1986). Foot deformities frequently progressed despite appropriate surgery (Bruce and Schut 1979; Chapman and Davis 1983; Lassman and James 1967; Pang 1986; Pierre-Kahn et al. 1986).

The authors advocate surgery for this lesion when the patient is 2 months of age or at the time of diagnosis if the lesion is detected later in life. Patients who undergo the repair of a lipomyelomeningocele require careful follow-up of their urological function and monitoring of their neurological examination, because they may experience re-tethering of their spinal cord as described for myelomeningocele repairs.

Diastematomyelia

Diastematomyelia is derived from two Greek words, *"diastema"* meaning cleft and *"myelos"* meaning medulla. It is an embryological defect that is characterized by the division of the spinal cord or cauda equina into two separate portions. The splitting may or may not include the investing dura (Humphreys et al. 1983; Sheptak 1978). Frequently, there is an associated septum of fibrous tissue, cartilage, or bone enclosed in a dural sheath that arises from the vertebral body and transfixes the spinal cord. This malformation almost always is associated with anomalies such as hypoplastic or unsegmented vertebral bodies or hemivertebrae at one or more levels (Hendrick 1971; Winter et al. 1974). The most common site of occurrence is in the lower thoracic or lumbar region (Hendrick 1971) and the most common level is upper lumbar (French 1982;

Hood et al. 1980; Humphreys et al. 1983). There is an increased rate of occurrence of this malformation in women (Hendrick 1971; Humphreys et al. 1983).

Cutaneous manifestations frequently are seen in patients with diastematomyelia (Guthkelch 1974; Humphreys et al. 1983; Sheptak 1978). Hypertrichosis is found in the majority of patients, while capillary hemangiomas, lipoma, or dimpling are less common (Guthkelch 1974; Humphreys et al. 1983; Sheptak 1978). There also is an association of this malformation with myelodysplasia in about one fourth of patients (Guthkelch 1974; Humphreys et al. 1983).

Most patients who become symptomatic from diastematomyelia do so during early childhood (Hendrick 1971; Humphreys et al. 1983; Sheptak 1978). Patients may present with kyphoscoliosis, back stiffness, foot or toe deformities, trophic skin changes, motor weakness of the ankle, gait disturbance, wasting of the calf or foot musculature, absent ankle reflexes, sensory loss, sphincter impairment, or pain (either back or radicular) (Hendrick 1971; Humphreys et al. 1983).

The natural history of this malformation is not well documented. Guthkelch (1974) studied 37 children with diastematomyelia. Thirteen of the 17 patients who were managed expectantly required surgery. Hilal and coworkers (1974) note that patients had an increased tendency for scoliosis when the condition was untreated.

Treatment of diastematomyelia consists of an excision of the septum, using care to avoid injury to the surrounding neural structures. When the lesion is associated with a tight filum terminale, the filum is divided at surgery. Failures following surgery typically are secondary to inadequate removal of the spur or incomplete removal of the dural sheath. If the patient requires spinal instrumentation or fusion, it is important to correct the diastematomyelia before the spinal instrumentation because patients may acutely deteriorate following correction of their scoliosis.

The operative correction of diastematomyelia is prophylactic. Of patients undergoing surgery for this condition, two thirds are unchanged and one third show minor improvements (Guthkelch 1974; Hendrick 1971; Hood et al. 1980; Sheptak 1978; Winter et al. 1974). Improvements commonly seen following surgery are relief of pain, improved gait, and improved sphincter control (Humphreys et al. 1983). Scoliosis frequently progresses despite surgery. Seven out of 11 patients presenting with spinal deformities required spinal instrumentation (Humphreys et al. 1983). Late neurological deterioration is unusual following prophylactic surgery (Guthkelch 1974; Humphreys et al. 1983; Sheptak 1978).

The authors advocate prophylactic surgery for this condition at 2 months of age or as soon as the diagnosis is established if diagnosed later in life. The filum should be divided at the time of the initial surgery if it tethers the spinal cord.

Tight Filum Terminale

The tight filum terminale syndrome results when the spinal cord is tethered from a shortened and thickened filum terminale (Hendrick et al. 1983; McLone and Naidich 1989b). Progressive neurological deterioration results from traction on the conus medullaris by the filum terminale with resultant ischemic injury to the spinal cord. The filum typically measures greater than 2 mm in these patients (McLone and Naidich 1989b). There is a high incidence of this syndrome in infants with anal stenosis or an imperforate anus (McLone and Naidich 1989b).

Patients with this syndrome typically present during periods of rapid growth with weakness and sensory deficits in the lower extremities, bladder dysfunction (usually incontinence), scoliosis, and back or leg pain. Half have cutaneous stigmata, such as a skin dimple, capillary hemangioma, or abnormal hair growth. Orthopedic deformities, such as kyphoscoliosis, pes cavus foot, or a tight Achilles tendon, are common. The onset of symptoms usually is subtle and progressive (Hendrick et al. 1983).

Virtually all patients with this anomaly have spina bifida, usually between L4 and S1. A normal spine x-ray virtually excludes this diagnosis (Hendrick et al. 1983). Magnetic resonance imaging (MRI) or myelography documents the tip of the conus medullaris below L2 in 86% of patients.

The natural history of this condition has not been documented by long-term follow-up. However, because of the progressive pretreatment history, a progressive course may be anticipated (Piatt and Hoffman 1987).

Surgery for this condition consists of a laminectomy with division of the filum terminale. If a filum terminale is absent, the spinal cord is divided from the dura below the take-off of the lowest nerve root.

The outlook for patients undergoing surgery for the tethered cord syndrome is good. Pain responds well to untethering with 92% to 100% of patients free of pain on follow-up (Hendrick et al. 1983; Pang and Wilberger 1982; Piatt and Hoffman 1987). Motor weakness improved or resolved in 72% of 65 patients with this symptom and in no case did weakness progress (Hendrick et al. 1983). Hendrick and colleagues (1983) report that 44% of patients with sensory deficits returned to normal and the remainder improved in their sensation. They also note that 60% of 30 patients with bladder dysfunction returned to normal following untethering and 27% improved. Others have found less encouraging results (Pang and Wilberger 1982; Yoneyama et al. 1985). Patients with the most severe lower urinary dysfunction are the least likely to improve (Hendrick et al. 1983; Pang and Wilberger 1982; Yoneyama et al. 1985). Early diagnosis and adequate release are the keys to successful management. The outcome of patients with kyphoscoliosis is less encouraging. Although McLone and Naidich (1989b) found that one third of patients with this problem improved following surgery, Hendrick and associates (1983) note that all of their patients with this problem required spinal instrumentation and fusion.

Because treatment of this problem is simple, effective, and generally uncomplicated, an aggressive attitude toward the selection of patients for operation is reasonable. A normal infant who presents with a characteristic cutaneous lesion warrants evaluation and treatment because the development and progression of bladder dysfunction is difficult to detect (McLone and Naidich 1989b; Piatt and Hoffman 1987).

Spinal Dermal Sinus Tract

A dermal sinus tract consists of a stalk of fibrous connective tissue, surrounding a lumen, lined by dermis. The stalk begins on the skin with a dermal dimple or cutaneous opening. The cutaneous opening may be no larger than a pinhole. There may be associated cutaneous hemangiomas, subcutaneous dermoid cysts, abnormal hair, or superficial infection. The tract may extend to the dura or penetrate the dura and attach to the conus medullaris. An intradural dermoid cyst may be associated with the lesion (Cheek and Laurent 1985; Wright 1971).

Spinal dermal sinus tracts may be found at any location along the spine above the midsacrum, but they are most common in the lumbosacral region (Cheek and Laurent 1985; Wright 1971). They do not appear to occur in the coccygeal or sacrococcygeal region (Cheek and Laurent 1985). These lesions should not be confused with pilonidal cysts, which occur in the sacrococcygeal region and appear later in life, nor with coccygeal dimples, which are cutaneously lined and have no connection with the CNS.

The natural history of these lesions is not well described. Patients typically are asymptomatic or have symptoms secondary to infections (meningitis or intraspinal abscesses), aseptic meningitis due to rupture of an associated dermoid cyst, or mass effect from a dermoid cyst (Cheek and Laurent 1985; Mount 1949; Wright 1971). Patients usually undergo surgical excision of the tract at the time of diagnosis to prevent complications from these lesions.

The outlook for patients afflicted with these lesions depends on whether supervening infections have occurred or whether an associated dermoid cyst has ruptured. If these complications have not ensued, the excision of the entire sinus tract and any associated dermoid cyst generally is accomplished easily. If the patient has experienced complicating infections or rupture of an associated dermoid cyst, the excision of the entire lesion may not be possible, and many of these patients have permanent sequelae (Cheek and Laurent 1985; Gerlach 1978; Mount 1949; Wright 1971). Because of the potential complications associated with spinal dermal sinus tracts, the authors favor excision of these tracts as soon as the diagnosis is established.

Hydrosyringomyelia

Hydromyelia is a pathological condition characterized by the accumulation of spinal fluid within the central canal of the spinal cord. Syringomyelia, on the other hand, consists of the accumulation of fluid within abnormal cavities in the spinal cord. Although pure forms of either of these conditions may exist, most patients have a combination of the two. The term hydrosyringomyelia is better suited to describe the pathological findings of asymmetric cavitation of the central canal lined by ependymal and glial tissue (Eggers and Hamer 1979).

There are two major classes of hydrosyringomyelia: communicating and noncommunicating (Barnett et al. 1973; Logue and Edwards 1981; Peerless and Durward 1983). In the communicating hydrosyringomyelia, there is a direct communication of the hydrosyringomyelia with the ventricular system. CSF is directed through a patent obex into the central canal. This type of hydrosyringomyelia frequently is seen with developmental anomalies of the posterior fossa, such as Chiari or Dandy-Walker malformations. In noncommunicating hydrosyringomyelia, the posterior fossa is normal and the connection of the hydrosyringomyelia with the ventricular system is not evident. Noncommunicating hydrosyringomyelia typically is associated with neoplasms, previous trauma, spinal arachnoiditis, or is idiopathic (Barnett et al. 1973; Peerless and Durward 1983).

Hydrosyringomyelia typically presents

with a suspended and dissociated sensory loss, motor weakness, amyopathy, spasticity, central pain, scoliosis, neurogenic arthropathies (Charcot's joints), and eventually trophic lesions (Oakes 1985; Park et al. 1985; Peerless and Durward 1983; Salam and Adams 1978; Schlesinger et al. 1981). The sensory disturbances usually involve the hands and arms and are secondary to disruption of the crossing spinothalamic tracts. The loss of pain and temperature may be unnoticed until the patient develops painless burns and chronic ulcerations of the skin. The motor weakness may manifest as a loss of fine hand movements or paraparesis. The pain experienced with hydrosyringomyelia may present as dysesthesias, deep pain between the scapulae, headache, or posterior cervical pain, and it may be exacerbated by Valsalva's maneuver (Oakes 1985).

The natural history of hydrosyringomyelia is described by Boman and Iivanainen (1967). They note a relatively slow progression of symptoms in all patients; however, 50% of patients remained stable in their disease for more than 10 years. Other authors also have noted periods of stability interposed with intervals of deterioration (Levy et al. 1983; Rhoton 1976).

The treatment of hydrosyringomyelia depends on the underlying etiology. Patients with shunt dependent hydrocephalus must have their shunt function assessed before treatment of the hydrosyringomyelia because a shunt malfunction may present as symptomatic hydrosyringomyelia (Hoffman et al. 1987; Oakes 1985). Most authorities recommend that patients with Chiari malformations undergo a decompressive procedure with the fourth ventricle opened and either the obex plugged or a silastic catheter placed into the fourth ventricle (Hoffman et al. 1987; Oakes 1985). Patients who fail to achieve lasting relief from this procedure or patients with noncommunicating hydrosyringomyelia should undergo a shunting procedure. Syringopleural or syringoperitoneal shunts provide better relief of symptoms than syringosubarachnoid shunts (Bar-

baro et al. 1984; Matsumoto and Symon 1989; Peerless and Durward 1983).

The outlook of patients with hydrosyringomyelia treated surgically is reasonably good. Postoperatively, 56% to 81% of patients treated with a decompression and plugging of the obex or a syringostomy improved or were stable (Hoffman et al. 1987; Matsumoto and Symon 1989; Peerless and Durward 1983). In patients who did not achieve relief with a decompression, shunting improved their symptoms in 50% to 86% (Hoffman et al. 1987; Matsumoto and Symon 1989).

The authors recommend that patients with symptomatic communicating hydrosyringomyelia secondary to the Chiari malformation undergo a decompressive procedure with either plugging of the obex or placement of a fourth ventricular subarachnoid shunt. Patients who do not achieve relief with this procedure or those with noncommunicating hydrosyringomyelia should have a syringopleural or syringoperitoneal shunt inserted.

Cranial Malformations

Anencephaly

Anencephaly, also known as exencephaly and extracranial dysencephaly, is a common congenital malformation thought to be caused by failure of fusion of the anterior neuropore or rupture of a previously closed encephalic tube (Giroud 1977). Children born with this malformation have neither a cranial vault nor scalp covering the exposed, partially destroyed brain.

Anencephalic children are born prematurely 53% of the time (Frezal et al. 1964), and in 68% of cases, they are stillborn (Giroud 1977). Other malformations frequently are associated with anencephaly; the most common is spina bifida (17%) (Frezal et al. 1964). Anencephalic children typically die within days of being born. The longest reported survival was 16 days (Ballantyne 1904).

There is no treatment for anencephaly;

the condition is uniformly lethal. Genetic counseling is indicated for the family of an affected child because there is a 3% chance that future pregnancies will be similarly affected (Giroud 1977).

Hydranencephaly

Hydranencephaly is caused by a variety of disorders that have a common result of marked reduction in the cerebrum, which is largely replaced by CSF in a normal-sized cranium (Halsey et al. 1977). The term hydranencephaly encompasses a spectrum of diseases that vary from destruction of major portions of the cerebrum, diencephalon, and upper brain stem to milder forms in which the brain stem and diencephalon are intact. Infants presenting with loss of the diencephalon and possibly the upper brain stem may be liveborn but have decerebrate posturing, myoclonus, generalized flaccidity, and respiratory failure (Halsey et al. 1977). These infants typically die shortly after birth. Infants who are less affected may have a relatively normal appearance at birth. The persistence of the Moro, grasp, and stepping reflexes after 6 months is typical in these children. They may have sleep–wake cycles, spontaneous vowel production, and respond to parental voice by smiling. They never obtain head control, roll over, or sit. Less affected infants may survive for more than 2 years (Halsey et al. 1971), and the authors have had one child with hydranencephaly survive to the age of 6.

Management of children afflicted with this disorder is limited to symptomatic treatment. Many develop hydrocephalus with progressive cranial enlargement. Insertion of a ventriculoperitoneal shunt may facilitate nursing care in this situation. Seizures are common in hydranencephalic children and are frequently difficult to control (Halsey et al. 1977).

Holoprosencephaly

Holoprosencephaly, also referred to as arhinencephaly, designates a spectrum of disorders involving malformations of the face and incomplete organogenesis of the prosencephalon into cerebral hemispheres (DeMyer 1977). DeMyer (1975; 1977) has extensively classified the cerebral and facial manifestations of this disorder. Alobar holoprosencephaly, the most severe form of this disorder, is characterized by a cerebrum without cleavage into lobes or an interhemispheric fissure. Semilobar holoprosencephaly is characterized by a brain that is slightly less severely disturbed having an interhemispheric fissure posteriorly but with broad continuity of the frontoparietal cortex across the midline anteriorly. Lobar holoprosencephaly, the least severe form of the malformation, is classified into two types, type A and B. The type A form of lobar holoprosencephaly has a well-developed interhemispheric fissure but maintains continuity of the cingulate gyrus across the midline. The type B form has only arhinencephaly or hypoplastic olfactory bulbs in a completely hemispherized cerebrum.

The facial malformations associated with holoprosencephaly are characterized by aplasia or hypoplasia of the midline facial bones of frontonasal prominence origin. These facial anomalies have been characterized by DeMyer (1975; 1977) and range from cyclopia, the most severe malformation, to intermaxillary facies, the least severe. Cyclopia is characterized by a single median bony orbit that usually contains an eye. The face may not have a nose or, more commonly, may have a fleshy, penis-like tube known as a proboscis. The proboscis may arise in the midline just above the median orbit, or it may be more lateral. The mouth may or may not be present. Ethmocephaly is characterized by a face with hypotelorism, two separate eyes, and a proboscis attached between the two eyes. Cebocephaly has a face with hypotelorism, two separate eyes, and a proboscis-like nose with a single aperture attached medially below the eyes. Median cleft lip facies exhibit orbital hypotelorism, a flat nasal bridge with nasal alae but no septum, and a median cleft of the upper lip. Finally, intermaxillary rudiment facies have orbital hypotelorism, bilateral cleft lips, and a hypoplastic intermaxillary segment. The nasal septum is incompletely formed.

The severity of the facial deformity in holoprosencephaly predicts the severity of the cerebral malformation and provides a useful guide to prognosis (DeMyer 1975; Demyer 1977). Patients with cyclopia have extreme alobar holoprosencephaly, typically lacking a pituitary gland, and virtually all die in the neonatal period. Patients with ethmocephaly, cebocephaly, and median cleft facies usually have alobar holoprosencephaly; however, the cerebrum is less distorted. They may have a pituitary gland and may live for weeks to months. Patients with intermaxillary rudimentary facies frequently have lobar holoprosencephaly with fully developed pituitary glands, and they may live for years.

The severity of the neurological disturbance parallels the severity of the holoprosencephaly. Patients with alobar holoprosencephaly have no conscious awareness of their environment, and they do not speak, achieve head control, turn over, or acquire any developmental milestones (DeMyer 1975; Demyer 1977; Yokota et al. 1984). They frequently have failure to thrive, poor resistance to infection, and convulsions (Demyer 1977; Yokota et al. 1984). Patients with semilobar and lobar holoprosencephaly are less severely affected but frequently exhibit moderate to severe retardation, spasticity, and failure to thrive (Demyer 1977; Khan et al. 1970; Kobori et al. 1987; Yokota et al. 1984).

The management of patients with holoprosencephaly depends on the severity of their cerebral malformation. Patients with alobar holoprosencephaly should have palliative care and are probably best managed in the hospital or in a custodial care facility. Those with semilobar or lobar holoprosencephaly may survive for extended periods and should receive symptomatic treatment of their epilepsy and hydrocephalus. Families of children born with holoprosencephaly should receive genetic counseling because there is a 6% risk of recurrence in subsequent offspring (Roach et al. 1975).

Cephaloceles

A cephalocele is a congenital herniation of intracranial contents through a cranial defect (McLaurin 1977). Herniations containing brain are known as encephaloceles or, more precisely, encephalomeningoceles. Cephaloceles that contain meninges and no brain tissue are known as cranial meningoceles. Cephaloceles may be classified as occipital, parietal, frontal, sincipital (nasofrontal), and basal. The frequency of cephaloceles varies in different regions of the world. Occipital cephaloceles are much more common in the Western hemisphere, while nasofrontal cephaloceles predominate in Southeast Asia (Matson 1969; Suwanwela and Hongsaprabhas 1966). Matson (1969), reporting on 265 cephaloceles seen in Boston, found that 74% were located in the occipital region, 13% were parietal, 6% were frontal, 5% were nasal (sincipital), and 2% were nasopharyngeal (basal).

Laurence and Tew (1971) describe the natural history of 23 infants born with cephaloceles managed in South Wales between 1956 and 1962. Eight of these infants survived (35%) and only four were considered normal (17%). The mean IQ of the survivors was 56. Hydrocephalus was noted in 50%.

Treatment of cephaloceles depends on the presence or absence of hydrocephalus and the location of the lesion. Patients presenting with hydrocephalus require shunt insertion prior to or at the time of the cephalocele repair. Lesions in the occipital and suboccipital regions should have imaging studies (preferably MRI) prior to closure because the brain stem may herniate into the encephalocele sac (Chapman and Caviness 1988; Smith and Huntington 1977). Occipital and parietal lesions are approached through a transverse incision. The stalk of the lesion is defined, the cephalocele is opened, and the contents are inspected. Viable cerebral tissue is preserved at the time of repair, even if this requires a duraplasty. Frontal, sincipital, and basal lesions are approached through a bifrontal craniotomy. These lesions frequently are accompanied by cranio-facial anomalies, and an approach by a plastic surgery and neurosurgery team is advantageous. The cerebral tissue is inspected and viable tissue is preserved. The cranial defect is reconstructed and associated facial deformities are repaired.

The survival following repair of cephaloceles is significantly improved over the natural history of the disease. Simpson and associates (1984), reporting on a series of 74 patients treated between 1955 and 1983, show a survival rate of 84%. Mealey and colleagues (1970) found a 63% survival in 40 occipital cephaloceles treated between 1948 and 1967, and Guthkelch (1970) achieved a survival rate of 64% in 74 patients treated between 1948 and 1965. Guthkelch (1970) found a higher survival in the 36 patients with meningoceles (88%) than in 38 patients with encephalomeningoceles (39%). Patients with associated hydrocephalus fared worse than nonhydrocephalics.

Overall, 25% to 73% of patients can be expected to be intellectually normal (Ingraham and Swan 1943; Lorber 1967; Simpson et al. 1984). The most important predictor of a good outcome following surgery is the amount of cerebral tissue in the lesion. The frequency of cranial meningoceles and encephaloceles is approximately even (Barrow and Simpson 1966; Guthkelch 1970). Simpson and coworkers (1984) found that nine out of 20 patients with encephalomeningocele were totally dependent on others for care, and one child had a significant mental defect. Patients with occipital meningoceles, on the other hand, were normal or only slightly disabled. Guthkelch (1970) found that 84% of children with meningoceles were intellectually normal, but only 53% of patients with encephalomeningoceles were normal. Other authorities have found similar good outcomes for patients with cranial meningoceles (Lorber 1967).

Hydrocephalus is an important determinant of outcome. Hydrocephalus is present in 16% of Guthkelch's 32 surviving patients with meningoceles (Guthkelch 1970). The incidence of mental retardation was 7% in

patients without hydrocephalus and 60% in those with hydrocephalus. In this same series, 53% of 15 patients surviving with encephalomeningoceles had hydrocephalus. The incidence of mental retardation increased from 43% to 50% in those with hydrocephalus. Others also have found hydrocephalus to be detrimental to the patient's outcome (Matson 1969; Simpson et al. 1984).

The prognosis for sincipital encephalomeningoceles is slightly better than for occipital encephalomeningoceles. Simpson and colleagues (1984) found that 12 out of 20 patients with these lesions were normal or had slight disability. Suwanwela and Hongsaprabhas (1966) report normal IQs in 57% of their patients with sincipital encephalomeningoceles who could undergo neuropsychological evaluations.

The authors recommend that patients born with cephaloceles undergo imaging studies to assess for hydrocephalus and to define the extent of cerebral herniation into the lesion. Patients with hydrocephalus should have a ventriculoperitoneal shunt inserted prior to repair of the lesion or contaminant with the repair. Closure of the cephalocele is indicated in most patients except rare patients born in a preterminal state or with other major associated malformations. Surgery is recommended even for children with large lesions to aid in their nursing care.

Iniencephaly

Iniencephaly is a craniospinal malformation characterized by imperfect formation of the base of the skull, especially around the foramen magnum, rachischisis, and retroflexion of the entire spine, forcing the face of the fetus to look upwards (Ballantyne 1904). Systemic malformations are noted in 84% of affected infants (David and Nixon 1976).

Iniencephalic infants typically are stillborn or die shortly after birth (Nishimura and Okamoto 1977). However, three iniencephalic children reported by Sherk and associates (1974) have survived for up to 17 years.

Table 30-5. Dandy-Walker Syndrome—Associated Malformations

Reference	Number of Patients	Hydro-cephalus	Percent	Agenesis of Corpus Callosum	Percent	Encephalocele	Percent
Maria et al. 1987	20	—	—	2	10	2	10.0
Raimondi et al. 1984	37	33	89.2	15	40.5	2	5.4
Hirsch et al. 1984	40	36	90	3	7.5	7	17.5
Fischer 1973	27	—	—	—	—	0	0.0
Udvarhelyi and Epstein 1975	12	12	100	—	—	0	0.0
Carmel et al. 1977	18	16	88.9	5	27.8	0	0.0
James et al. 1979	10	10	100	—	—	0	0.0
Sawaya and McLaurin 1981	23	22	95.7	4	17.4	1	4.3
Tal et al. 1980	21	—	—	4	19	2	9.5
Total	208	129	92.1	33	20.8	14	6.7

Dandy-Walker Syndrome

The Dandy-Walker malformation is a complex developmental anomaly that has as its essential features aplasia or hypoplasia of the cerebellar vermis and a cystic dilatation of the fourth ventricle (Dandy and Blackfan 1914; Hirsch et al. 1984; Taggart and Walker 1942; Tal et al. 1980). Hydrocephalus is a frequent accompaniment, but it is not an essential component of the malformation (Carmel et al. 1977; Hirsch et al. 1984; Tal et al. 1980). Hydrocephalus is seen in more than 90% of patients exhibiting the syndrome (Table 30-5). The hydrocephalus usually develops after birth and its cause may be due to aqueductal, posterior fossa foraminal, or an incisural blockage of CSF flow (Hirsch et al. 1984). Patients with Dandy-Walker syndrome frequently have an enlarged posterior fossa with elevation of the torcula and lack patency of the outlet foramina of the fourth ventricle (Brown 1977; Dandy and Blackfan 1914; Hart et al. 1972; Hirsch et al. 1984; Sawaya and McLaurin 1981; Taggart and Walker 1942).

The Dandy-Walker malformation frequently is associated with other CNS and systemic malformations. Agenesis of the corpus callosum and occipital encephaloceles occur in 2% and 7% of patients, respectively (see Table 30-5). Other CNS malformations observed include cerebral and cerebellar gyrational anomalies (Golden et al. 1987; Hart et al. 1972), aqueductal stenosis (Golden et al. 1987; Hart et al. 1972), heterotopias (Hart et al. 1972), subdural hygromas (Carmel et al. 1977), spina bifida (Golden et al. 1987; Hart et al. 1972; Tal et al. 1980), syringomyelia (Hart et al. 1972), microcephaly (Golden et al. 1987; Hart et al. 1972), dermoid cysts (Raimondi et al. 1984), porencephaly (Raimondi et al. 1984; Sawaya and McLaurin 1981), and the Klippel-Feil deformity (Hart et al. 1972; Sawaya and McLaurin 1981). Systemic anomalies frequently are noted. Cardiac anomalies include the patent ductus arteriosus (Hirsch et al. 1984; Sawaya and McLaurin 1981), atrial septal defect (Golden et al. 1987), ventricular septal defect (Sawaya and McLaurin 1981; Tal et al. 1980), atrioventricularis communis (Golden et al. 1987), dextrocardia (Hirsch et al. 1984), and tetralogy of Fallot (Tal et al. 1980). Other reported anomalies are pulmonary hypoplasia (Golden et al. 1987), cleft palate (Golden et al. 1987; Hart et al. 1972; Sawaya and McLaurin 1981; Tal et al. 1980), polydactyly or syndactyly (Carmel et al. 1977; Hart et al. 1972; Hirsch et al. 1984), facial angiomas (Carmel et al. 1977; Hirsch et al. 1984; Sawaya and McLaurin 1981), polycystic kidneys (Raimondi et al. 1984; Tal et al. 1980), hydronephrosis (Golden et al. 1987), and omphalocele (Golden et al. 1987).

The mortality rate of the Dandy-Walker malformation has improved (Table 30-6). In early reports, the mortality ranged from 25% to 50%. With improved surgical techniques, this has been lowered to 0% to 26%. Approximately one half of patients with the Dandy-Walker syndrome have an IQ of 80 or higher and have normal neurological examinations (Table 30-7). The etiology of the mental retardation in this condition is unclear but is believed to be related to the as-

Table 30-6. Dandy-Walker Syndrome—Mortality

Author	Period	Total	Deaths	Percent
Maria et al. 1987	1966–1983	20	2	10
Raimondi et al. 1984	1963–1977	37	8	21.6
Hirsch et al. 1984	1969–1982	40	5	12.5
Fischer 1973	1947–1972	27	11	40.7
Udvarhelyi and Epstein 1975	1964–1971	12	0	0
Carmel et al. 1977	1962–1977	18	5	27.8
James et al. 1979	1958–1974	10	4	40
Sawaya and McLaurin 1981	1956–1979	23	6	26.1
Tal et al. 1980	1950–1978	21	10	47.6
Total		208	51	24.5

Table 30-7. Dandy-Walker Syndrome—Morbidity

Reference	Period	Total Number	Number With IQ ≥ 80	Percent	Total Number	Neurologically Normal	Percent
Maria et al. 1987	1966–1983	19	14	73.7	19	16	84.2
Raimondi et al. 1984	1963–1977	33	7	21.2	29	11	37.9
Hirsch et al. 1984	1969–1982	27	16	59.3	—	—	—
Fischer 1973	1947–1972	16	8	50.0	—	—	—
Udvarhelyi and Epstein 1975	1964–1971	12	6	50.0	12	6	50.0
Carmel et al. 1977	1962–1977	13	6	46.2	—	—	—
James et al. 1979	1958–1974	6	2	33.3	—	—	—
Sawaya and McLaurin 1981	1956–1979	14	7	50.0	—	—	—
Tal et al. 1980	1950–1978	11	3	27.3	11	3	27.3
Carmel et al. 1977	1961–1980	12	9	75.0	12	7	58.3
Total		163	78	47.9	83	43	51.8

sociated hydrocephalus and to the associated CNS malformations (Carmel et al. 1977; Fischer 1973; Golden et al. 1987; Hirsch et al. 1984; James et al. 1979; Raimondi et al. 1984; Sawaya and McLaurin 1981). Children with mental retardation frequently exhibit abnormalities on neurological testing (Fischer 1973; Golden et al. 1987). Ataxia, spasticity, and loss of fine motor control often are seen in treated children (Carmel et al. 1977; Golden et al. 1987; Tal et al. 1980). Seizures are noted in 15% of patients with the Dandy-Walker syndrome (Hirsch et al. 1984).

The recommended treatment for the Dandy-Walker malformation is a cyst-peritoneal shunt (Hirsch et al. 1984; Maria et al. 1987; Raimondi et al. 1984; Sawaya and McLaurin 1981). Rarely, patients exhibit aqueductal stenosis with associated hydrocephalus and require a lateral ventricular catheter to be incorporated into the system (Hirsch et al. 1984; Maria et al. 1987; Raimondi et al. 1969). Lateral ventriculoperitoneal shunts are contraindicated as the sole means of treatment owing to the associated risk of upward herniation with foraminal impaction and death (James et al. 1979; Raimondi et al. 1969). Posterior fossa exploration with membrane stripping rarely is successful and carries a significant mortality (Carmel et al. 1977; Fischer 1973; Hirsch et al. 1984; James et al. 1979; Raimondi et al. 1984; Sawaya and McLaurin 1981).

Chiari Malformations

Chiari (1986) classified three hindbrain herniations associated with hydrocephalus. He also described a fourth group of malformations, cerebellar hypoplasia, which is of questionable relationship to the others. Chiari's type I malformation consists of variable displacement of tonsils below the plane of the foramen magnum. Type II malformations, also called Arnold-Chiari malformations, involve caudal displacement of the inferior portion of the cerebellar vermis, medulla, and fourth ventricle into the cervical canal. They are almost always associated with myelodysplasia and are described in the section entitled *Myelomeningocele.*

Type III malformations result from herniation of portions of the cerebellum and in some instances, the brain stem, into a high cervical or suboccipital encephalomeningocele. These malformations are rare and usually are incompatible with life (Carmel 1982).

Hydrosyringomyelia is seen in 20% to 32% of patients with Chiari I malformations (Appleby et al. 1969; Banerji and Millar 1974; Paul and Lye 1983). The hydrosyringomyelia is secondary to obstruction of the outflow pathways of CSF from the fourth ventricle. CSF is diverted into the central canal of the spinal cord. Hydrosyringomyelia is the only anomaly commonly seen with Chiari I malformations (Barkovich 1990).

Chiari I malformations usually present in older children or adults with spastic motor weakness, scoliosis, truncal ataxia, suspended sensory losses typical of hydrosyringomyelia, headaches, facial weakness, downbeating nystagmus, abducens and oculomotor palsies, and deafness (Dure et al. 1989; Dyste and Menezes 1988; Levy et al. 1983; Longridge and Mallinson 1985). The headaches usually are felt in the suboccipital region and can be exacerbated by straining. Unusual presentations include atrophy of the hand muscles, apnea, muscular spasms, deep burning neck or shoulder pain, or torticollis (Dure et al. 1989; Dyste and Menezes 1988).

The natural history of Chiari I malformations is not well described. Conway (1967) described six patients with progressive symptoms over a 2 to 15 year span. Five of the six patients had worsening of their symptoms during the 1.5 to 9 years of follow-up. The patient who was unchanged had been followed only 3 months. Levy and coworkers (1983) studied 127 patients undergoing surgery for Chiari malformations. In the preoperative period, they noted long stretches of clinical stability interspersed by intervals of progression. No patient spontaneously improved.

The surgical management of patients with symptomatic Chiari I malformations consists of a cervical laminectomy to the caudal extent of the tonsillar herniation. If the patient has associated hydrosyringomyelia, the fourth ventricle may be opened and the obex plugged with a piece of muscle (Park et al. 1986). Alternatively, the fourth ventricle may be shunted into the subarachnoid space with a silastic catheter (Oakes 1985; Paul and Lye 1983). The posterior fossa is closed with a water-tight duraplasty.

The outcome of children and adolescents undergoing decompressive procedures for Chiari I malformations is good. Dyste and colleagues (1988) report improvement in 87.5% of patients undergoing surgery and clinical stabilization in the remaining 12.5%. Thirty-seven percent of their patients regained a normal neurological status. Scoliosis typically is stabilized or reversed in these patients. Similar results were reported by others (Dauser et al. 1988; Dure et al. 1989). The outlook was better in patients who underwent a decompression after a shorter symptom duration (Dyste and Menezes 1988).

The prognosis in adults undergoing decompression for Chiari I malformations is worse than for children (Dyste et al. 1989). Levy and associates (1983), in a review of 127 patients, note that 46% achieved long-term improvement of their condition, 26% were unchanged, and 28% were worsened. Similar results have been seen by others (Paul and Lye 1983; Saez et al. 1976; Schlesinger et al. 1981). Patients with a central cord picture had the worst outlook (Levy et al. 1983; Paul and Lye 1983; Saez et al. 1976).

The authors recommend that patients with Chiari I malformations exhibiting progressive symptoms undergo a decompressive procedure with a patulous duraplasty. Those with hydrosyringomyelia should have the fourth ventricle opened and the central canal plugged with muscle. Alternatively, a silastic stent may be placed between the fourth ventricle and the subarachnoid space.

Cranial Dermal Sinus Tract

A cranial dermal sinus tract, like a spinal dermal sinus tract, consists of a stalk of fibrous connective tissue surrounding a lumen lined by dermis. The stalk begins on the skin of the occipital or nasal region with a dermal dimple or cutaneous opening. There may be cutaneous hemangiomas, subcutaneous dermoid cysts, superficial infection, or abnormal hair growth associated with the tract. In the occipital region, the tract extends in a caudal direction (Cheek and Laurent 1985; Matson 1969). If the tract enters the skull, it does so inferior to the torcula (Cheek and Laurent 1985). The tract may extend to the dura or penetrate it and form an intradural dermoid cyst (Cheek and Laurent 1985; Logue and Till 1952; Matson 1969). In the nasal region, the sinus tract invariably occurs at the juncture of the middle and distal thirds of the nasal dorsum in the midline (Cheek and Laurent 1985). The sinus tract extends superiorly and may end extracranially or extend intracranially through a bifid crista or enlarged foramen cecum (Cheek and Laurent 1985).

Symptoms described with dermal sinus tracts include the periodic discharge of infected material onto the skin, a subcutaneous mass, mass effect with cerebellar or frontal lobe findings, hydrocephalus, rupture of the dermoid cyst with aseptic meningitis, infection with bacterial meningitis, or intraparenchymal abscess (Cheek and Laurent 1985; Logue and Till 1952; Matson 1969; Mount 1949). All infants and children with *Staphylococcus aureus* meningitis warrant a careful inspection for a dermal sinus tract (Matson 1969).

The natural history of these lesions is not well described. Once the diagnosis is established, patients typically undergo surgical excision of the tract to prevent complications.

The surgical treatment of these lesions consists of complete excision of the entire tract and associated cysts (Matson 1969). In the occipital region, a midline incision is placed and the sinus tract is followed caudally until it terminates. A careful search for intracranial extension is made. Careful preoperative imaging is mandatory to define the full extent of the tract and to look for associated intracranial lesions. Frequently, the sinus tract connects to an intracranial dermoid cyst (Logue and Till 1952). If the patient has an associated dermoid cyst, it should be excised. Nasal dermal sinus tracts require a combined rhinotomy–craniotomy. The nasal portion of the sinus tract is excised first. If the lesion extends intracranially, the patient then undergoes a craniotomy with excision of the intracranial extension.

The results of surgery for cranial dermal sinus tracts depend on the presence of infection. If infection supervenes, the outcome is unsatisfactory. Matson (1969) reported poor results were reported in four patients who were only diagnosed after severe infection had occurred. Three of these patients died, and the fourth required shunting for residual hydrocephalus. In the other 13 patients, the outcome was excellent. Others have reported similar results (Logue and Till 1952; Mount 1949).

The authors recommend that patients diagnosed with cranial dermal sinus tracts have careful neuroimaging to define the full scope of the lesion. When this is completed, the lesion should be excised entirely (Cheek and Laurent 1985; Logue and Till 1952; Matson 1969; Mount 1949).

Hydrocephalus

Hydrocephalus is the dilation of the ventricular system caused by an imbalance in the production and absorption of CSF. Most cases of hydrocephalus are caused by impaired absorption of CSF, although overproduction of CSF may contribute to the hydrocephalus in the case of a choroid plexus papilloma. Hydrocephalus arises from three basic causes: (1) congenital anomalies, (2) neoplasms, and (3) inflammation. Congenital causes most commonly responsible for hydrocephalus include aqueductal stenosis, myelodysplasia with the Chiari malformation, Dandy-Walker malformations, arachnoid cysts, encephaloceles, and incompetent arachnoid villi. Neoplasms associated with hydrocephalus include posterior fossa tumors, craniopharyngiomas, optic nerve or hypothalamic astrocytomas, pineal region neoplasms, and choroid plexus papillomas. Inflammatory causes of hydrocephalus include infectious etiologies, such as bacterial meningitis, toxoplasmosis, cytomegalic inclusion disease, congenital syphilis, cysticercosis, and posthemorrhagic hydrocephalus secondary to trauma or intraventricular hemorrhage in premature infants (De Lange 1977a; McCullough 1989).

The hydrocephalic patient's prognosis depends on the etiology of the hydrocephalus (Bangash et al. 1986; Billard et al. 1985; Guthkelch and Riley 1969; Hadenius et al. 1962; Laurence and Coates 1962; McCullough 1989). Patients with myelodysplasia, encephaloceles, and Dandy-Walker malformations have complex developmental anomalies in which the hydrocephalus is only one component. Their prognosis depends on the severity of the associated malformations (Di Rocco 1987; Laurence and Coates 1962) and was discussed previously in the appropriate sections. Neoplastic processes may cause hydrocephalus secondary to the obstruction of CSF outflow pathways, but the outlook for these patients primarily is determined by their disease and the therapy that they receive. Likewise, patients with hydrocephalus secondary to infectious processes or trauma may incur severe neurological injury on the basis of their primary insult and their prognosis depends on the severity of this insult (Billard et al. 1985; Laurence and Coates 1962).

In large series studying the history of untreated hydrocephalus, half of the patients died during the follow-up period. (Hadenius et al. 1962; Laurence and Coates 1962; Yashon et al. 1965). Laurence and Coates (1962) determined that the chance of a hydrocephalic patient seen soon after birth reaching adulthood was between 20% and 23%. The major causes of death were cardiorespiratory arrest secondary to increased intracranial pressure, meningitis in patients with open neural tube defects, and bronchopneumonia.

Hagberg and Sjorgen (1966) found that patients with untreated hydrocephalus developed a "chronic brain syndrome," characterized by poor motor performance, poor intellectual development, visual difficulties, hypothalamic dysfunction, and characteristic behavioral disturbances. Normal motor function was found in only 37% of their patients, slight to moderate disability in 44%, and severe disability in 19%. The lower extremities tended to be more severely affected than the upper extremities, and 75% of their patients could not walk by 18 months of age. Ataxia, the predominant neurological sign, was found in 65% of patients and was combined with spastic diplegia in 17%. Laurence and Coates (1962) found similar percentages of motor deficits in their patients. Seizures occurred in 21% of untreated patients (Hagberg and Sjorgen 1966).

Hagberg and Sjorgen (1966) found that 37% of untreated patients had an IQ greater than 90, 38% had IQs between 50 and 90, and 25% were severely retarded with IQs less than 50. Laurence and Coates (1962) found remarkably similar deficits; 38% of their patients had IQs greater than 85, 35% had IQs between 50 and 85, and 27% had IQs less than 50. Both groups commented that patients with the greatest disability had lower IQs. Yashon and associates (1965) noted that verbal performance was higher in untreated hydrocephalic children than nonverbal performance.

Visual difficulties were noted in most untreated hydrocephalic patients. Squint was noted in 60% (Hagberg and Sjorgen 1966) and poor vision or blindness in 13.5% to 21% (Hagberg and Sjorgen 1966; Laurence and Coates 1962). A dysplastic body build was seen in 37% of patients. This was characterized by excessive accumulation of fat on the hips, buttocks, breasts, and lower abdomen and was interpreted as due to hypothalamic dysfunction (Haberg and Sjorgen 1966).

Behavioral problems were documented in 49% to 64% of patients with untreated hydrocephalus (Hadenius et al. 1962; Hagberg and Sjorgen 1966). Many patients exhibited timid and dependent personalities, although occasionally, patients had aggressive behavior (Ingram and Naughton 1962). Ingram and Naughton (1962) note an increased verbal output in many untreated hydrocephalic patients. This hyperverbal syndrome has been termed the "chatter-box" or "cocktail-party" syndrome and was found in 25% of untreated patients (Hagberg and Sjorgen 1966). Children with this syndrome love to chatter and to imitate and pronounce words. Although they appear to be bright, these children tend to think illogically and do not comprehend the meaning of the stream of words that they are using. Their IQ scores typically are much lower than expected.

Laurence and Coates (1967) conclude that all patients with progressive hydrocephalus should be shunted. They recommend that cases that appear to be on the verge of arresting but have not arrested also should be shunted because they could not reliably determine which patients experience a spontaneous arrest of their hydrocephalus. When complete arrest has taken place, however, they caution that patients do not require intervention because most stabilize or improve with time. They note that these patients require close follow-up because the hydrocephalus can occasionally become active again (Laurence and Coates 1967).

Most observers agree with these recommendations (Di Rocco 1987; Gilmore 1990; Hagberg and Naglo 1972). Patients with transient hydrocephalus (as in many cases of intraventricular hemorrhage in preterm infants) may be treated medically with carbonic anhydrase inhibitors, osmotic diuretics, and serial lumbar punctures for communicating hydrocephalus because this may obviate the need for shunting (Gilmore 1990). However, once it is apparent that the hydrocephalus is progressive, a shunt should be inserted. Improved surgical techniques and the deleterious effects of late operation have lead to increased numbers of infants treated with shunt insertion and fewer managed with a "wait and see" policy (Di Rocco 1987; Hagberg and Naglo 1972).

Early series comparing the outcome of treated and untreated hydrocephalic patients documented increased survival when patients were shunted (Foltz and Shurtleff 1963; Shurtleff et al. 1975), although the mortality rates in these series were still high (25% to 50%) (Foltz and Shurtleff 1963; Overton and Snodgrass 1965; Shurtleff et al.

1975; Stark et al. 1974). Improvements in shunt equipment and operative technique have lowered mortality rates. Most recent series of patients treated with ventricular shunting procedures document 5- to 10-year survival rates between 80% and 95% (Amacher and Wellington 1984; Di Rocco 1987; Fernell et al. 1986; Hoffman et al. 1982; Keucher and Mealy 1979; Lorber and Zachary 1968; Mazza et al. 1980; Milhorat 1987). Of the patients who die, many die of their primary disease (frequently neoplasms) or have other conditions which worsen their prognosis (Amacher and Wellington 1984; Hoffman et al. 1982). The number of surgical procedures has little influence on mortality (Di Rocco 1987). Patients with primary cerebral malformations and tumors have high mortality rates and those with infection or posthemorrhagic hydrocephalus have a better outlook (Amacher and Wellington 1984; Bangash et al. 1986).

Studies that compared treated and untreated hydrocephalic patients documented that shunting preserved the intellectual potential of the children and lowered their incidence of physical impairment (Fernell et al. 1988; Foltz and Shurtleff 1963). Foltz and Shurtleff (1963) note that shunting not only increased the survival rates of patients from 22.2% to 61.4%, but also increased the percentage of competitive survivors from 25% to 55%. Of their untreated patients, 54% required institutionalization, and only 16% of treated patients required this care. Fernell and colleagues documented significant decreases in ataxia (68% to 8%), squint (63% to 32%), and "cocktail-party" behavior (27% to 3%) in patients who had been shunted compared to their unshunted patients (Fernell et al. 1988).

The outcome of surgical treatment of hydrocephalus depends on a number of factors, including the general condition of the patient and brain at the time of surgery, type and etiology of hydrocephalus, timing of surgery, complications, supervision of shunt function, and socioeconomic environment (Di Rocco 1987). Recent series of shunted patients document normal IQs in more than two thirds of survivors (Amacher and Wel-

lington 1984; Billard et al. 1985; Hoffman et al. 1982; Lorber and Zachary 1968; Milhorat 1987; Overton and Snodgrass 1965), and more than 50% are judged to be competitive in society (Foltz and Shurtleff 1963; Guidetti et al. 1969).

Despite the fact that most well-shunted hydrocephalic patients have normal intelligence, they may still have neuropsychological impairment (Billard et al. 1985; Dennis et al. 1981; Miller and Sethi 1971; Prigatano et al. 1983). The verbal IQ scores of these children tend to be higher than their performance scores (Badell-Riberra et al. 1966; Bangash et al. 1986; Tew and Lawrence 1975) and detailed neuropsychological evaluations demonstrated that these patients have impaired fine motor skills, visuospatial problem solving, spatial orientation, difficulties in tactile shape recognition, and difficulties in matching stimuli presented to either hand, implying poor transfer of information between hemispheres (split brain phenomenon) (Billard et al. 1985; Miller and Sethi 1971; Prigatano et al. 1983; Tew and Lawrence 1975). They have difficulties distinguishing key features of a stimulus from irrelevant background (Miller and Sethi 1971), and they have deficiencies on measures of verbal reasoning, syntactical comprehension of language, and speed of information processing (Miller and Sethi 1971; Prigatano et al. 1983). Difficulties with verbal and nonverbal memory also have been demonstrated (Cull and Wyke 1984; Prigatano et al. 1983). It is important that parents and teachers of hydrocephalic patients are provided with this information because the poor scholastic performance these children exhibit may not be due to a lack of motivation, but rather to poor retention of verbal material. Shunting may prevent mental retardation from hydrocephalus; however, it does not restore normal neuropsychological function.

Neurological deficits have been found in many patients with well-shunted hydrocephalus. Fernell and colleagues found that 53% of their patients had abnormalities on detailed neurological testing, and 69% were delayed in acquiring gross motor skills. Speech was impaired in 30% of their patients, and a delay in acquiring speech was seen in 31% (Fernell et al. 1988). The frequency of epilepsy (25%) is similar in treated and untreated hydrocephalic patients (Fernell et al. 1988).

Patients with hydrocephalus are at risk for visual impairment. Increased intracranial pressure may cause transient amaurosis, visual field deficits, pregeniculate, or postgeniculate blindness and optic atrophy. Temporal lobe herniation may compress the posterior cerebral arteries, leading to occipital lobe infarction. Pressure on the midbrain may cause oculomotor palsies or gaze disturbances (Bangash et al. 1986; Di Rocco 1987). Squint occurred in 50% of patients with congenital hydrocephalus and increased from 38% to 74% if shunt complications ensued (Stanworth 1970). Patients with treated hydrocephalus also have been found to have a high incidence of hyperopia (20% of eyes examined), myopia (13%), astigmatism (25.5%), and nystagmus 18% (Mankinen and Mustonen 1987).

Certain factors have been identified that correlate highly with poor intellectual development. Patients with prolonged delays between the diagnosis of hydrocephalus and placement of shunts tend to have worse intellectual outcomes (Dennis et al. 1981; Etches et al. 1987; Foltz and Shurtleff 1963; Lorber 1968; Raimondi and Soare 1974). Shunt infections and ventriculitis also have been found to lower IQ (Mapstone et al. 1984; McLone et al. 1982). Patients with hydrocephalus secondary to myelodysplasia or perinatal hemorrhage tend to have good intellectual outcomes (Amacher and Wellington 1984; Raimondi and Soare 1974); however, patients with meningitis, toxoplasmosis, X-linked hydrocephalus, porencephaly, and primary cerebral malformations were generally severely affected (Amacher and Wellington 1984; Billard et al. 1985; Fernell et al. 1988; Halliday et al. 1986; Jansen 1988; Renier et al. 1988). Ocular abnormalities, motor impairment, or seizures have been correlated with lower performance IQs (Dennis et al. 1981). Frequent hospitalizations of the hydrocephalic patient, separation from their families, stress from treatment, and loss of school time deprive these children of the necessary skills on which psychometric tests depend (McLone et al. 1982).

The prognosis for eventual intelligence is not dependent on head size or cortical mantle size at the time of diagnosis (Amacher and Wellington 1984; Hadenius et al. 1962; Hagberg and Sjorgen 1966; Laurence and Coates 1962; Lonton 1984; Lorber 1968; Yashon et al. 1965). Patients with cortical mantles of less than 1 cm may not only have normal IQs, but also may obtain university educations (Hadenius et al. 1962; Tew and Lawrence 1975; Yashon et al. 1965). The cause of disability from hydrocephalus is more than a reduction in the pallium caused by ventricular distension; it is most likely related to the underlying etiology of the hydrocephalus (Laurence and Coates 1962).

Preterm infants who sustain intraventricular hemorrhages and develop posthemorrhagic hydrocephalus represent a subgroup of children whose prognosis depends not only on their hydrocephalus, but also on the severity of the initial hemorrhage. Papile and colleagues (1978) classified intraventricular hemorrhages based on their severity on cranial ultrasound. Grade I bleeds were subependymal without intraventricular extension; grade II hemorrhages extended into the ventricles, but did not demonstrate ventriculomegaly; grade III bleeds demonstrated intraventricular extension with ventriculomegaly; and grade IV hemorrhages exhibited intraventricular and intraparenchymal extension. Volpe (1987) reviewed approximately 800 cases of infants with intraventricular hemorrhage and found that with mild hemorrhage (grade I), the mortality rate was 15% and the rate of progressive ventricular dilatation was 5%; for moderate bleeds (grade II), the rates were 20% and 25%, respectively; for severe hemorrhages (grade III), 40% and 55%; and for severe hemorrhages with intracerebral involvement (grade IV), 60% and 80%. Volpe (1987) reviewed 400 cases of intraventricular hemorrhage and found that 15% of patients with mild hemorrhages, 30% of those with moderate bleeds, 40% of those with severe bleeds, and 90% of those with intracerebral extension, developed major neurological sequelae (spastic diplegia or quadraplegia, often with intellectual deterioration). Similar results have been reported by others (Boynton et al. 1986; Catto-Smith et al. 1985; Costello et al. 1988; Krishnamoorthy et al. 1984; Papile et al. 1987; Williamson et al. 1982).

Factors indicative of a poor outcome were a birth weight less than 1500 g, gestational age of less than 30 weeks, and a head circumference less than 27 cm (Cheek and Desmond 1983; Liechty et al. 1983; Williamson et al. 1982). Seizures also were found to predict a poor outcome (Boynton et al. 1986). Although some authors found that infants with posthemorrhagic hydrocephalus requiring shunt insertion had the same incidence of major handicaps as those without hydrocephalus (Papile et al. 1987; Shankaran et al. 1989), other authors have found a higher incidence of handicaps among patients requiring shunting (Costello et al. 1988; Sasidharan et al. 1986). Among the children with shunts, Shankaran and associates (1989) found a higher incidence of sequelae when lack of ventricular decompression was noted immediately after shunt insertion and when shunt infections occurred (Shankaran et al. 1989). Many of the adverse sequelae that were noted were more often attributable to cerebral ischemia and infarction than to the hydrocephalus (Krishnamoorthy et al. 1984; Shankaran et al. 1989).

Mothers who have hydrocephalic infants diagnosed prenatally present a perplexing prognostic dilemma. The predicted outcome for these infants is particularly important in advising parents about the termination of pregnancy in infants diagnosed early. The management of the delivery also is dependent on the predicted outcome of the infant diagnosed later in gestation (cesarean section vs. vaginal delivery with decompressive obstetrical maneuvers). In studies in which the diagnosis of hydrocephalus was made before birth, the mortality was between 24% and 85% (Chevernak et al. 1984; Pretorius et al. 1985; Serlo et al. 1986). These mortality rates, however, largely reflect termination of the pregnancies and methods used to deliver the babies. Studies of infants born with overt hydrocephalus indicate that the mortality is less than predicted above. Renier and coworkers (1988) reported on 108 patients with overt hydrocephalus at birth. The calculated survival rates at 5 and 10 years were 71% and 62%, respectively. Of the 75 surviving children, 28% have IQs more than 80 and 50% have IQs less than 60. Only 29% of school-aged children have normal academic performances. No neurological disabilities were found in 41% of their patients and 46% were without significant behavioral problems. The outcome was best when there were no associated malformations, no shunt infections, and when hydrocephalus was due to aqueductal stenosis (excluding X-linked hydrocephalus and toxoplasmosis). The outlook for these children was worse than for a comparable group of 492 patients with postnatal hydrocephalus. These patients had a 10-year survival rate of 88% and 58% had IQs of 80 or more (Renier et al. 1988). Mealy and colleagues (1973) report similar results in a group of 79 patients with overt hydrocephalus at birth. McCullough and Balzer-Martin (1982) had better results in a group of 26 treated patients (86% survival at 6 years and normal IQs in 53%); however, 11 of the 37 patients with overt hydrocephalus were not treated.

In advising prospective mothers of hydrocephalic children, the authors recommend that infants with uncomplicated hydrocephalus be carried as close to term as possible and then be delivered in the safest manner possible (usually by cesarean section). The infants should have shunts placed soon after birth. Mothers whose infants have severe CNS malformations (hydranencephaly or holoprosencephaly) or major systemic malformations should be advised of the grim prognosis for their infants and management decisions made accordingly.

The authors recommend that patients with progressive hydrocephalus be treated with ventriculoperitoneal shunts as soon as possible. Transient hydrocephalus, such as that in preterm infants after intraventricular hemorrhages, should have a trial of medical treatment with carbonic anhydrase inhibitors, osmotic diuretics, and serial lumbar punctures to determine if the hydrocephalus will abate prior to placement of a shunt. Children who demonstrate the need for a shunt should have this procedure without any lengthy delays. Patients diagnosed as having "arrested" hydrocephalus require close follow-up with frequent clinical examinations, CT scans, and neuropsychological evaluations. Some of these patients may develop recrudescence of the hydrocephalus late in their course or may suffer a sudden disastrous attack of intracranial hypertension (Di Rocco 1987; Di Rocco et al. 1977; James et al. 1979; Laurence and Coates 1967).

Complications from shunt insertion include shunt infection, shunt obstruction, seizures, subdural hematomas, craniosynostosis, intraventricular hemorrhage, slit ventricle syndrome, and erosion of skin over shunt reservoir (De Lange 1977b; McComb 1990). Complications unique to ventriculoperitoneal shunts include ascites, pseudocyst formation, perforation of the visceral or abdominal wall, intestinal obstruction, spread of infection or neoplasm from the ventricle to the abdomen, and increased incidence of hernias (De Lange 1977b; McComb 1990). A ventriculopleural shunt complication includes fluid accumulations in the pleural cavity (Hoffman et al. 1983). Vascular shunt complications may include subclavian and vena cava obstruction, mural thrombosis, bacterial endocarditis, cardiac arrhythmias, perforation of the cardiac wall with cardiac tamponade, shunt nephritis, embolization of the catheter into the pulmonary vasculature, and chronic pulmonary thromboembolization that could eventually lead to pulmonary hypertension and cor pulmonale (Becker and Nulsen 1968; Keucher and Mealy 1979; McComb 1990).

Ventriculoatrial shunts initially were used by many authors in the early 1960s. As time passed, it was learned that ventriculoatrial shunts required more frequent revisions than the ventriculoperitoneal shunts and had a higher incidence of complications, includ-

ing fatal pulmonary hypertension (Di Rocco 1987; Hoffman et al. 1982; Keucher and Mealy 1979). Ventriculoperitoneal shunts are now used by most neurosurgeons caring for children with hydrocephalus.

References

Amacher, A.; Wellington, J. Infantile hydrocephalus: long-term results of surgical therapy. Child's Brain 11:217–229; 1984.

Ames, M. D.; Schut, L. Results of treatment of 171 consecutive myelomeningoceles—1963 to 1968. Pediatrics 50:466–470; 1972.

Anderson, E.; Plewis, I. Impairment of motor skills in children with spina bifida cystica and hydrocephalus. Br. J. Psychol. 68:61–70; 1977.

Appleby, A.; Foster, J. B.; Hankinson, J.; Hudgson, P. The diagnosis and management of the Chiari malformation. Brain 91:131–140; 1969.

Badell-Riberra, A.; Shulman, K.; Paddock, N. The relationship of nonprogressive hydrocephalus to intellectual functioning in children with spina bifida cystica. Pediatrics 37:787–793; 1966.

Ballantyne, J. W. The embryo. Manual of antenatal pathology and hygiene. Edinburgh: W. Green and Sons; 1904.

Bamberger-Bozo, C. The Chiari II malformation. Malformations. In: Myrianthopoulos, N. C., ed. Handbook of clinical neurology. Amsterdam: Elsevier Science Publishers; 1987: p. 403–412.

Bamforth, S. J.; Baird, P. A. Spina bifida and hydrocephalus: a population study over a 35-year period. Am. J. Hum. Genet. 44:225–232; 1989.

Banerji, N. R.; Millar, J. H. D. Chiari malformations presenting in adult life. Brain. 97:157–168; 1974.

Bangash, I.; D'Souza, B.; Barbosa, E. Hydrocephalus—Effects on cerebral function. In: Hoffman, H.; Epstein, F., ed., Disorders of the developing nervous system: diagnosis and treatment. Boston: Blackwell Scientific Publications; 1986: p. 581–588.

Barbaro, N. M.; Wilson, C. B.; Gutin, P. H.; Edwards, M. S. B. Surgical treatment of syringomyelia. Favorable results with syringoperitoneal shunting. J. Neurosurg. 61:531–538; 1984.

Barkovich, A. J. Congenital malformations of the brain. Pediatric Neuroimaging 1:77–121; 1990.

Barnett, H. J. M.; Foster, J. B.; Hudgson, P. Syringomyelia. In: Walton, N. J., ed. Major problems in neurology. Philadelphia: W. B. Saunders; 1973: p. 1–318.

Barrow, N.; Simpson, D. A. Cranium bifidum:

investigation, prognosis and management. Aust. Paediat. 2:20–26; 1966.

Bartoshesky, L. E.; Haller, J.; Scott, R. M.; Wojick, C. Seizures in children with meningomyelocele. Am. J. Dis. Child. 139:400–402; 1985.

Becker, D.; Nulsen, F. Control of hydrocephalus by valve-regulated venous shunt: Avoidance of complications in prolonged shunt maintenance. J. Neurosurg. 28:215–226; 1968.

Bell, W. O.; Charney, E. B.; Bruce, D. A.; Sutton, L. N.; Schut, L. Symptomatic Arnold-Chiari malformation: review of experience with 22 cases. J. Neurosurg. 66:812–816.

Billard, C. J.; Santini, J.; Gillet, P.; Nargeot, M. C.; Adrien, J. L. Long-term intellectual prognosis of hydrocephalus with reference to 77 children. Pediatr. Neurosci. 12:219–225; 1986.

Boman, K.; Iivanainen, M. Prognosis of syringomyelia. Acta. neur. scand. 43:61–68; 1967.

Boynton, B.; Boynton, C.; Merritt, T.; Vaucher, Y.; James, H.; Bejar, R. Ventriculoperitoneal shunts in low birth weight infants with intracranial hemorrhage: neurodevelopmental outcome. Neurosurgery 18:141–145; 1986.

Brown, J. R. The Dandy-Walker syndrome. In: Handbook of Clinical Neurology. Vol. 30 Congenital Malformations of the Brain and Skull. Vinken, P. J. and Bruyn, G. W. eds. Amsterdam: North Holland; 1977: p. 623–664.

Bruce, D. A.; Schut, L. Spinal lipomas in infancy and childhood. Child's Brain 5:192–203; 1979.

Cameron, A. H. The Arnold-Chiari and other neuroanatomical malformations associated with spina bifida. J. Pathol. Bacteriol. 73:195–211; 1957.

Carmel, P. W. The Arnold-Chiari Malformation. In: Section of Pediatric Neurosurgery of the American Association of Neurological Surgeons, ed. Pediatric neurosurgery. Surgery of the developing nervous system. New York: Grune & Stratton; 1982: p. 61–77.

Carmel, P. W. Management of the Chiari malformations in childhood. Clin. Neurosurg. 30:385–406; 1983.

Carmel, P. W.; Antunes, J. L.; Hilal, S. K.; Gold, A. P. Dandy-Walker syndrome: clinicopathological features and reevaluation of modes of therapy. Surg. Neurol. 8:132–138; 1977.

Catto-Smith, A.; Yu, V.; Bajuk, B.; Astbury, J. Effect of neonatal periventricular haemorrhage on neurodevelopmental outcome. Arch. Dis. Child. 60:8–11; 1985.

Chadduck, W.; Adametz, J. Incidence of seizures in patients with myelomeningocele: a multifactorial analysis. Surg. Neurol. 30:281–285; 1988.

Chapman, P.; Caviness, V. Subtorcular occipital encephaloceles. In: Marlin, A., ed. Concepts in pediatric neurosurgery. Basel: S. Karger; 1988: p. 86–96.

Chapman, P. H.; Davis, K. R. Surgical treatment of spinal lipomas in childhood. In: Raimondi, A. J., ed. Concepts in pediatric neurosurgery. Basel: S. Karger; 1983: p. 178–190.

Charney, E. B.; Rorke, L. B.; Sutton, L. N.; Schut, L. Management of Chiari II complications in infants with myelomeningoceles. J. Pediatr. 111:364–371; 1987.

Charney, E. B.; Weller, S. C.; Sutton, L. N.; Bruce, D. A.; Schut, L. B. Management of the newborn with myelomeningocele: time for a decision-making process. Pediatrics 75:58–64; 1985.

Cheek, W.; Desmond, M. Intraventricular hemorrhage and hydrocephalus in the preterm infant. In: Raimondi, A., ed. Concepts in pediatric neurosurgery. Basel: Karger; 1983: p. 125–132.

Cheek, W. R.; Laurent, J. P. Dermal sinus tracts. In Humphreys, R. F., ed. Concepts in pediatric neurosurgery. Basel: S. Krager; 1985: p. 63–75.

Chevernak, F.; Duncan, C.; Ment, L.; Hobbins, J.; McLure, M.; Scott, D.; Berkowitz, R. Outcome of fetal ventriculomegaly. Lancet 2:179–181; 1984.

Chiari, H. Ueber Veranderungen des Kleinhirns, des Pons und der Medulla oblongata in Folge von genitaler Hydrocephalie des Grosshirns. Denkschr. Akad. Wiss. Wein. 63:71–116; 1896.

Clements, D.; Kaushal, K. A study of the ocular complications of hydrocephalus and meningomyelocele. Trans, Ophthal. Soc. U.K. 90:383–390; 1970.

Conway, L. Hydrodynamic studies in syringomyelia. J. Neurosurg. 27:501–514; 1967.

Costello, A. M.; Hamilton, P. A.; Baudin, J.; Townsend, J.; Bradford, B. C.; Stewart, A. L.; Reynolds, E. O. Prediction of neurodevelopmental impairment at four years from brain ultrasound appearance of very preterm infants. Dev. Med. Child. Neur. 30:711–722; 1988.

Cull, C.; Wyke, M. A. Memory function of children with spina bifida and shunted hydrocephalus. Dev. Med. Child. Neur. 26:177–183; 1984.

Dandy, W. E.; Blackfan, K. D. Internal hydrocephalus. An experimental, clinical and pathological study. Am. J. Dis. Child. 8:406–482; 1914.

Dauser, R. C.; DiPietro, M. A.; Venes, J. L. Symptomtic Chiari I malformation in childhood: a report of 7 cases. Pediatr. Neurosci. 14:184–190; 1988.

David, J. T.; Nixon, A. Congenital malformations associated with anencephaly and inien-cephaly. J. Med. Genet. 13:263–265; 1976.

De Lange, S. A. Progressive hydrocephalus. In: Handbook of Clinical Neurology. Vol. 30 Congenital Malformations of the Brain and Skull. Myrianthopoulos, N. C., ed. Amsterdam: North-Holland Publishing; 1977: p. 525–563.

De Lange, S. A. Treatment of hydrocephalus. In: Handbook of Clinical Neurology. Vol. 30 Congenital Malformations of the Brain and Skull. Myrianthopoulos, N. C., ed. Amsterdam: North-Holland Publishing; 1977: p. 565–607.

DeMyer, W. Median facial malformations and their implications for brain malformations. Birth Defects, Orig. Article Series 11(7):155–181; 1975.

Demyer, W. Holoprosencephaly. (Cyclopia-arhinencephaly) In: Handbook of Clinical Neurology. Vol. 30 Malformations. Vinken, P. J.; Bruyn, G. W., eds. Amsterdam: Elsevier-North Holland Publishing; 1977: p. 225–244.

Dennis, M.; Fitz, C. R.; Netley, C. T.; Sugar, J.; Harwood-Nash, D. C. F.; Hendrick, E. B.; Hoffman, H. J.; Humphreys, R. P. The intelligence of hydrocephalic children. Arch. Neurol. 38:607–615; 1981.

Di Rocco, C. The treatment of infantile hydrocephalus. Boca Raton, Fl: CRC Press, Inc.; 1987.

Di Rocco, C.; Caldarelli, M.; Maira, G.; Rossi, G. The study of cerebrospinal fluid dynamics in apparently "arrested" hydrocephalus in children. Child's Brain. 3:359–374; 1977.

Doran, P. A.; Guthkelch, A. N. Studies in spina bifida cystica. I. General survey and reassessment of the problem. J. Neurol. Neurosurg. Psychiat. 24:331–345; 1961.

Dubowitz, V.; Lorber, J.; Zachary, R. B. Lipoma of the cauda equina. Archs. Dis. Childh. 40:207–213; 1965.

Dure, L. S.; Percy, A. K.; Cheek, W. R.; Laurent, J. P. Chiari type I malformation in children. J. Pediatr. 115:573–576; 1989.

Dyste, G. N.; Menezes, A. H. Presentation and management of pediatric Chiari malformations without myelodysplasia. Neurosurgery 23:589–597; 1988.

Dyste, G. N.; Menezes, A. H.; VanGilder, J. C. Symptomatic Chiari malformations. An analysis of presentation, management, and long-term outcome. J. Neurosurg. 71:159–168; 1989.

Eggers, C.; Hamer, J. Hydrosyringomyelia in childhood—clinical aspects, pathogenesis, and therapy. Neuropaediatrie. 10:87–99; 1979.

Emery, J.; MacKenzie. Medullo-cervical dislocation deformity (Chiari II deformity) related to neurospinal dysraphism (meningomyelocele). Brain 96:155–162; 1973.

Emery, J. L.; Lendon, R. G. Clinical impli-cations of cord lesions in neurospinal dysraphism. Dev. Med. Child. Neur. 14(suppl. 27):45–51; 1972.

Epstein, F. Meningomyelocele: "pitfalls" in early and late management. Clin. Neurosurg. 30:366–384; 1983.

Etches, P.; Chir, B.; Ward, T.; Bhui, P.; Peters, K.; Robertson, C. Outcome of shunted post-hemorrhagic hydrocephalus in premature infants. Pediatr. Neurol. 3:136–140; 1987.

Fernell, E.; Hagberg, B.; Hagberg, G. Epidemiology of infantile hydrocephalus in Sweden: a clinical follow-up study in children born at term. Neuropediatrics 19:135–142; 1988.

Fernell, E.; Hagberg, B.; Hagberg, G.; Wendt, L. V. Epidemiology of infantile hydrocephalus in Sweden: I. Birth prevalence and general data. Acta. Pediatr. Scand. 75:975–981; 1986.

Fischer, E. G. Dandy-Walker syndrome: an evaluation of surgical treatment. J. Neurosurg. 39:615–621; 1973.

Foltz, E.; Shurtleff, D. Five-year comparative study of hydrocephalus in children with and without operation (113 cases). J. Neurosurg. 20:1064–1068; 1963.

Freeman, J. Early management and decision-making for the treatment of myelomeningocele. Pediatrics 73:564–566; 1984.

French, B. N. Midline fusion defects and defects of formation. In: Youmans, J. R., ed. Neurological surgery. Philadelphia: W. B. Saunders; 1982: p. 1236–1380.

Frezal, J.; Kelley, J.; Guillemot, M. L.; Lamy, M. Anencephaly in France. Am. J. Hum. Genet. 16:336–350; 1964.

Gerlach, J. Dermal sinuses and dermoids. In: Handbook of Clinical Neurology. Vol. 32 Congenital Malformations of the Spine and Spinal Cord. Vinken, P. J.; Bruyn, G. W., eds. Amsterdam: North-Holland Publishing; 1978: p. 449–463.

Gilbert, J. N.; Jones, K. L.; Rorke, L. B.; Chernoff, G. F.; James, H. E. Central nervous system anomalies associated with meningomyelocele, hydrocephalus, and the Arnold-Chiari malformation: Reappraisal of theories regarding the pathogenesis of posterior neural tube closure defects. Neurosurgery 18:559–564; 1986.

Gilmore, H. Medical treatment of hydrocephalus. In: Scott, R., ed. Hydrocephalus. Baltimore: Williams and Wilkins; 1990: p. 37–46.

Giroud, A. Anencephaly. In: Handbook of Clinical Neurology. Vol. 30 Congenital Malformations of the Brain and Skull. Vinken, P. J.; Bruyn, G. W., eds. Amsterdam: Elsevier-North Holland Publishing; 1977: p. 173–208.

Golden, J. A.; Rorke, L. B.; Bruce, D. A. Dandy-Walker syndrome and associated anomalies. Pediatr. Neurosci. 13:38–44; 1987.

Guidetti, B.; Occhipinti, E.; Riccio, A. Ventriculo-atrial shunt in 200 cases of nontumoral hydrocephalus: Remarks on diagnostic criteria, postoperative complications and long-term results. Acta. Neurochir. 21:295–308; 1969.

Guthkelch, A. Occipital cranium bifidium. Arch. Dis. Child. 45:104–109; 1970.

Guthkelch, A.; Riley, N. Influence of etiology on prognosis in surgically treated infantile hydrocephalus. Arch. Dis. Child. 44:29–35; 1969.

Guthkelch, A. N. Diastematomyelia with median septum. Brain 97:729–742; 1974.

Hadenius, A.; Hagberg, B.; Hyttnäs-Bensch, K.; Sjörgren, I. The natural prognosis of infantile hydrocephalus. Acta. paediat. scand. 51:117–118; 1962.

Hagberg, B.; Sjorgen, I. The chronic brain syndrome of infantile hydrocephalus. A follow-up study of 63 spontaneously arrested cases. Am. J. Dis. Child. 112:189–196; 1966.

Hagberg, G.; Naglo, A. The conservative management of infantile hydrocephalus. Acta. paediat. scand. 61:165–177; 1972.

Halliday, J.; Chow, C.; Wallace, D.; Danks, D. X linked hydrocephalus: a survey of a 20 year period in Victoria, Australia. J. med. Genet. 23:23–31; 1986.

Halliwell, M. D.; Carr, J. G.; Pearson, A. M. The intellectual and educational functioning of children with neural tube defects. Z. Kinderchir. 31:375–381; 1980.

Halsey, J. H.; Allen, N.; Chamberlin, H. R. The morphogenesis of hydranencephaly. J. neurol. Sci. 12:187–217; 1971.

Halsey, J. H.; Allen, N.; Chamberlin, H. R. Hydranencephaly. In: Handbook of Clinical Neurology Vol. 30. Congenital Malformations of the Brain and Skull. Vinken, P. J.; Bruyn, G. W., eds. Amsterdam: Elsevier-North Holland Publishing; 1977: p. 661–680.

Hart, M. N.; Malamud, N.; Ellis, W. G. The Dandy-Walker syndrome. Neurology 22:771–780; 1972.

Hendrick, E. B. On diastematomyelia. Prog. neurol. Surg. 4:277–288; 1971.

Hendrick, E. B.; Hoffman, H. J.; Humphreys, R. P. The tethered spinal cord. Clin. Neurosurg. 30:457–463; 1983.

Hilal, S. K.; Marton, D.; Pollack, E. Diastematomyelia in children. Radiographic study of 34 cases. Radiology 112:609–621; 1974.

Hirsch, J. F.; Pierre-Kahn, A.; Renier, D.; Sainte-Rose, C.; Hoppe-Hirsch, E. The Dandy-Walker malformation: a review of 40 cases. J. Neurosurg. 61:515–522; 1984.

Hirsch, J. F.; Pierre-Kahn, A. Lumbosacral lipomas with spina bifida. Child. Nerv. Syst. 4:354–360; 1988.

Hoffer, M. M.; Feiwell, E.; Perry, R.; Perry, J.; Bonnett, C. Functional ambulation in patients with myelomeningocele. J. Bone Joint Surg. 55A:137–148; 1973.

Hoffman, H.; Hendrick, E.; Humphreys, R. Management of hydrocephalus. Monogr. neural Sci. 8:21–25, 1982.

Hoffman, H.; Hendrick, E.; Humphreys, R. Experience with ventriculopleural shunts. Child's Brain 10:404–413; 1983.

Hoffman, H. J.; Hendrick, E. B.; Humphreys, R. P. Manifestations and management of Arnold-Chiari malformation in patients with myelomeningocele. Child's Brain 1:255–259; 1975.

Hoffman, H. J.; Hendrick, E. B.; Humphreys, R. P. The tethered spinal cord: its protean manifestations, diagnosis and surgical correction. Child's Brain 2:145–155; 1976.

Hoffman, H. J.; Neill, J.; Crone, K. R.; Hendrick, E. B.; Humphreys, R. P. Hydrosyringomyelia and its management in childhood. Neurosurgery 21:347–51; 1987.

Hoffman, H. J.; Taecholarn, C.; Hendrick, E. B.; Humphreys, R. P. Management of lipomyelomeningoceles: Experience at the Hospital for Sick Children, Toronto. J. Neurosurg. 62:1–8; 1985.

Holliday, P. O., III; Pillsbury, D.; Kelly, D. L. J.; Dillard, R. Brain stem auditory evoked potentials in Arnold-Chiari malformation: possible prognostic value and changes with surgical decompression. Neurosurgery 16:48–53; 1985.

Hood, R. W.; Riseborough, E. J.; Nehme, A.; Micheli, L. J.; Strand, R. D.; Neuhauser, E. B. D. Diastematomyelia and structural spinal deformities. J. Bone Joint Surg. 62A:520–528; 1980.

Humphreys, R. P. Spinal dysraphism. In: Wilkins, R. H.; Rengachary, S. S., eds. Neurosurgery. New York: McGraw-Hill Book Co.; 1985: p. 2041–2052.

Humphreys, R. P.; Hendrick, E. B.; Hoffman, H. J. Diastematomyelia. Clin. Neurosurg. 30:436–456; 1983.

Hunt, G. M. A study of deaths and handicap in a consecutive series of spina bifida treated unselectively from birth. Z. Kinderchir. 38(suppl. II):100–102; 1983.

Hunt, G. M.; Holmes, A. E. Some factors relating to intelligence in treated children with spina bifida cystica. Dev. Med. Child. Neurol. 17(suppl. 35):65–70; 1975.

Ingraham, F. D.; Swan, H. Spina bifida and cranium bifidum: a survey of five hundred forty six cases. N. Engl. J. Med. 228:559–563; 1943.

Ingram, T.; Naughton, J. Paediatric and psychological aspects of cerebral palsy associated with hydrocephalus. Dev. Med. Child. Neurol. 4:287–292; 1962.

James, H. E.; Kaiser, G.; Schut, L.; Bruce, D. A. Problems of diagnosis and treatment in the Dandy-Walker Syndrome. Child's Brain 5:24–30; 1979.

Jansen, J. Etiology and prognosis in hydrocephalus. Child's Nerv. Syst. 4(5):263–267; 1988.

Kang, J. K.; Kim, M. C.; Kim, D. S.; Song, J. U. Effects of tethering on regional spinal cord blood flow and sensory-evoked potentials in growing cats. Child's Nerv. Syst. 3:35–39; 1987.

Keucher, T.; Mealy, J. Jr. Long-term results after ventriculoatrial and ventriculoperitoneal shunting for infantile hydrocephalus. J. Neurosurg. 50:179–186; 1979.

Khan, M.; Rozdilsky, B.; Gerrard, J. W. Familial holoprosencephaly. Dev. Med. Child. Neur. 12:71–76; 1970.

Kobori, J. A.; Herrick, M. K.; Urich, H. Arhinencephaly. The spectrum of associated malformations. Brain 110:237–260; 1987.

Krishnamoorthy, K.; Kuehnle, K.; Tordes, I.; DeLong, G. Neurodevelopmental outcome of survivors with posthemorrhagic hydrocephalus following Grade II neonatal intraventricular hemorrhage. Ann. Neurol. 15:201–204; 1984.

Lassman, L. P.; James, C. C. M. Lumbosacral lipomas: critical survey of 26 cases submitted to laminectomy. J. Neurol. Neurosurg. Psychiat. 30:174–181; 1967.

Laurence, K.; Coates, S. The natural history of hydrocephalus. Detailed analysis of 182 unoperated cases. Arch. Dis. Child. 37:345–362; 1962.

Laurence, K.; Coates, S. Spontaneously arrested hydrocephalus. Results of the re-examination of 82 survivors from a series of 182 unoperated cases. Dev. Med. Child. Neurol. 9(suppl. 13):4–13; 1967.

Laurence, K. M. The natural history of spina bifida cystica. Arch. Dis. Child. 39:41–57; 1964.

Laurence, K. M. Effect of early surgery for spina bifida cystica on survival and quality of life. Lancet. 1:301–304; 1974.

Laurence, K. M.; Tew, B. J. Natural history of spina bifida cystica and cranium bifidum cysticum: major central nervous system malformations in South Wales, Part IV. Arch. Dis. Child. 46:127–138; 1971.

Levy, W. J.; Mason, L.; Hahn, J. F. Chiari malformation presenting in adults: A surgical experience in 127 cases. Neurosurgery 12:377–390; 1983.

Liechty, E.; Bull, M.; Bryson, C.; Kalsbeck, J.; Jansen, R.; Lemons, J.; Schreiner, R. Developmental outcome of very low birth weight infants requiring a ventriculo-peritoneal shunt. Child's Brain. 10:340–349; 1983.

Logue, V.; Edwards, M. R. Syringomyelia and its surgical treatment—an analysis of 75 patients. J. Neurol. Neurosurg. Pyschiat. 44:273–284; 1981.

Logue, V.; Till, K. Posterior fossa dermoid cysts with special reference to intracranial infection. J. Neurol. Neurosurg. Psychiat. 15:1–12; 1952.

Longridge, N. S.; Mallinson, A. I. Arnold-Chiari malformtion and the otolaryngologist: place of magnetic resonance imaging and electronystagmography. Laryngoscope 95:335–339; 1985.

Lonton, A. In infantile hydrocephalus how much brain mantle is needed for normal development? Dilemmas in the management of the neurological patient. New York: Churchill Livingstone; 1984.

Lonton, A. P. Prediction of intelligence in spina bifida neonates. Z. Kinderchir. 37:172–174; 1982.

Lorber, J. Systemic ventriculographic studies in infants born with meningomyelocele and encephalocele, the incidence and development of hydrocephalus. Arch. Dis. Child. 36:381–389; 1961.

Lorber, J. The prognosis of occipital encephalocele. Dev. Med. Child. Neurol. 13:75–86; 1967.

Lorber, J. The results of early treatment of extreme hydrocephalus. Dev. Med. Child. Neurol. 16(suppl.):21–29; 1968.

Lorber, J. Results of treatment of myelomeningocele. An analysis of 524 unselected cases, with special reference to possible selection for treatment. Dev. Med. Child. Neurol. 13:279–303; 1971.

Lorber, J. Spina bifida cystica. Results of treatment of 270 consecutive cases with criteria for selection for the future. Arch. Dis. Child. 47:854–873; 1972.

Lorber, J.; Salfield, S. A. W. Results of selective treatment of spina bifida cystica. Arch. Dis. Child. 56:822–830; 1981.

Lorber, J.; Zachary, R. Primary congenital hydrocephalus. Long-term results of a controlled therapeutic trial. Arch. Dis. Child. 43:516–527; 1968.

Mankinen, H. A.; Mustonen, E. Ophthalmic changes in hydrocephalus. A follow-up examination of 50 patients treated with shunts. Acta. ophthal. (Copenh) 65(1):81–86; 1987.

Mapstone, T.; Rekate, H.; Nulsen, F.; Dixon, M.; Glaser, N.; Jaffe, M. Relationship of cerebral spinal fluid shunting and IQ in children with meningomyelocele. Child's Brain. 11:112–118; 1984.

Maria, B. L.; Zinreich, S. J.; Carson, B. C.; Rosenbaum, A. E.; Freeman, J. M. Dandy-Walker syndrome revisited. Pediatr. Neurosci. 13:45–51; 1987.

Marlin, A. E. The initial treatment of the child with myelomeningocele. In: Marlin, A. E., ed. Concepts in pediatric neurosurgery. Basel: Karger; 1990: p. 7–14.

Matson, D. D. Neurosurgery in infancy and childhood. Springfield: Charles C Thomas; 1969.

Matsumoto, T.; Symon, L. Surgical treatment of syringomyelia—Current results. Surg. Neurol. 32:258–265; 1989.

Mazza, C.; Pasqualin, A.; Da Pian. R. Results of treatment with ventriculoatrial and ventriculoperitoneal shunt in infantile nontumoral hydrocephalus. Child's Brain. 7:1–14; 1980.

McComb, J. Techniques for CSF diversion. In: Scott, R., ed. Hydrocephalus. Baltimore: Williams and Wilkins; 1990: p. 47–65.

McCullough, D. Hydrocephalus: etiology, pathologic effects, diagnosis, and natural history. In: Section of Pediatric Neurosurgery of the American Association of Neurological Surgeons., ed. Pediatric neurosurgery; surgery of the developing nervous system. Philadelphia: W. B. Saunders; 1989: p. 180–199.

McCullough, D.; Balzer-Martin, L. Current prognosis in overt neonatal hydrocephalus. J. Neurosurg. 57:378–383; 1982.

McLaughlin, J. F.; Shurtleff, D. B.; Lemire, R. J.; Hayden, P. W.; Stuntz, J. T. A process for care selection in myelodysplasia. Z. Kinderchir. 37:161–164; 1982.

McLaurin, R. L. Cranium bifidum and cranial cephaloceles. In: Handbook of Clinical Neurology. Vol. 30 Congenital Malformations of the Brain and Skull, Part 1. Vinken, P. J.; Bruyn, G. W., eds. Amsterdam: North-Holland 1977: p. 209–218.

McLone, D. G. Treatment of myelomeningocele: arguments against selection. Clin. Neurosurg. 33:359–370; 1986.

McLone, D. G.; Czyzewski, D.; Raimondi, A. J.; Sommers, R. C.; Central nervous system infections as a limiting factor in the intelligence of children with myelomeningocele. Pediatrics 70:338–342; 1982.

McLone, D. G.; Dias, L.; Kaplan, W. E.; Sommers, M. W. Concepts in the management of spina bifida. In: Humphreys, R. P., ed. Concepts in pediatric neurosurgery. Basel: S. Karger; 1985: p. 97–106.

McLone, D. G.; Mutluer, S.; Naidich, T. P. Lipomeningoceles of the conus medullaris. In: Raimondi, A. J., ed. Concepts in pediatric neurosurgery. Basel: S. Karger; 1983: p. 170–177.

McLone, D. G.; Naidich, T. P. Laser resection of fifty spinal lipomas. Neurosurgery 18:611–615; 1986.

McLone, D. G.; Naidich, T. P. Myelomeningocele: outcome and late complications. In: Section of Pediatric Neurosurgery of the American Association of Neurological Surgeons, ed. Pediatric neurosurgery: surgery of the developing nervous system. Philadelphia: W. B. Saunders; 1989: p. 53–70.

McLone, D. G.; Naidich, T. P. The tethered spinal cord. In: Section of Pediatric Neurosurgery of the American Association of Neurological Surgeons., ed. Pediatric neurosurgery: surgery of the developing nervous system. Philadelphia: W. B. Saunders; 1989: p. 71–96.

Mealey, J., Jr.; Dzenitis, A.; Hockey, A. The prognosis of encephaloceles. J. Neurosurg. 32:209–218; 1970.

Mealey, J., Jr.; Gilmor, R.; Bubb, M. The prognosis of hydrocephalus overt at birth. J. Neurosurg. 39:348–355; 1973.

Milhorat, T. H. Hydrocephaly. In: Handbook of Clinical Neurology. Vol. 6 Malformations. Myrianthopoulos, N. C., ed. New York: Elsevier Science 1987: p. 285–300.

Miller, E.; Sethi, L. The effect of hydrocephalus on perception. Dev. Med. Child. Neurol. 13(suppl. 25):77–81; 1971.

Miyazaki, K. Urological problems in children with spina bifida cystica and sacral defects. Paraplegia. 10:37–43; 1972.

Mount, L. A. Congenital dermal sinuses. JAMA 139:1263–1268; 1949.

Murphy, D. P. The etiology of congenital malformations in light of biologic statistic. Am. J. Obstet. Gynecol. 34:890–897; 1937.

Nishimura, H.; Okamoto, N. Iniencephaly. In: Handbook of Clinical Neurology. Vol. 30 Congenital Malformations of the Brain and Skull. Vinken, P. J.; Bruyn, G. W., eds. Handbook of clinical neurology. Amsterdam: Elsevier-North Holland Publishing; 1977: p. 257–268.

O'Brien, M. S.; McLanahan, C. S. Review of the neurosurgical management of myelomeningocele at a regional pediatric medical center. In: American Society for Pediatric Neurosurgery, ed. Concept in pediatric neurosurgery. Basel: S. Karger; 1981: p. 202–215.

Oakes, J. W. Chiari malformations, Hydromyelia, Syringomyelia. In: Wilkins, R. H.; Rengachary, S. S., ed. Neurosurgery. New York: McGraw-Hill Book Co; 1985; p. 2102–2124.

Overton, M.; Snodgrass, S. Ventriculo-venous shunts for infantile hydrocephalus. A review of five year's experience with this method. J. Neurosurg. 23:517–521; 1965.

Pang, D. Tethered cord syndrome. In: Hoffman, H. J., ed. Neurosurgery: state of the art reviews. Philadelphia: Hanley & Belfus, Inc; 1986: p. 45–79.

Pang, D.; Wilberger, J. E., Jr. Tethered spinal cord in adults. J. Neurosurg. 57:32–47; 1982.

Papile, L.; Burstein, J.; Burstein, R.; Koffler, H. Incidence and evolution of subependymal and intraventricular hemorrhage: A study of infants with birth weights less than 1500 grams. J. Pediat. 92:529–534; 1978.

Papile, L.; Munsick-Bruno, G.; Schaeffer, A. Relationship of cerebral intraventricular hemorrhage and early childhood neurologic handicaps. J. Pediat. 103:273–277; 1987.

Park, T.; Cail, W.; Maggio, W.; Mitchell, D. Progressive spasticity and scoliosis in children with myelomeningocele: Radiological investigation and surgical treatment. J. Neurosurg. 62:367–375; 1985.

Park, T. S.; Hoffman, H. J.; Cail, W. S. Arnold-Chiari malformations. Manifestations and management. In: Hoffman, H. J., ed. Advances in pediatric neurosurgery. Philadelphia: Hanley & Belfus, Inc; 1986: p. 81–99.

Park, T. S.; Hoffman, H. J.; Hendrick, E. B.; Humphreys, R. P. Experience with surgical decompression of the Arnold-Chiari malformation in young infants with myelomeningocele. Neurosurgery 13:147–152; 1983.

Paul, K.; Lye, H. Arnold-Chiari malformation: review of 71 cases. J. Neurosurg. 58:183–187; 1983.

Peerless, S. J.; Durward, Q. J. Management of syringomyelia: a pathophysiological approach. Clin. Neurosurg. 30:531–576; 1983.

Piatt, J. H.; Hoffman, H. J. The tethered spinal cord with focus on the tight filum terminale. A review. II. Clinical presentations, diagnostic investigations, radiological features, urological investigations, electrophysiological evaluation, results. Neuro-orthopedics 4:1–11; 1987.

Pierre-Kahn, A.; Lacombe, J.; Pichon, J.; Giudicelli, T.; Reinier, D.; Sainte-Rose, C.; Perrigot, M. Hirsch, J. Intraspinal lipomas with spina bifida. Prognosis and treatment in 73 cases. J. Neurosurg. 65:756–761; 1986.

Piggott, H. The natural history of scoliosis in myelodysplasia. J. Bone Joint Surg. 62B:54–58; 1980.

Pretorius, D.; Davis, K.; Manco-Johnson, M. L.; Manchester, D.; Meier, P.; Clewell, W. Clinical course of fetal hydrocephalus: 40 cases. A.J.N.R. 6:23–27; 1985.

Prigatano, J.; Zeiner, H.; Pollay, M.; Kaplan, R. Neuropsychological functioning in children with shunted uncomplicated hydrocephalus. Child's Brain. 10:112–120; 1983.

Raimondi, A. J.; Samuelson, G.; Yarzagaray, L.; Norton, T. Atresia of the foramina of Luschka and Magendie. The Dandy-Walker syndrome. J. Neurosurg. 31:202–216; 1969.

Raimondi, A. J.; Sato, K.; Shimoji, T. The Dandy-Walker malformation. Basel: S. Karger; 1984.

Raimondi, A. J.; Soare, P. L. Intellectual development in shunted hydrocephalic children. Am. J. Dis. Child. 127:664–671; 1974.

Reigel, D. H. Tethered spinal cord. In: Humphreys, R. P., Concepts in pediatric neurosurgery. Basel: S. Karger; 1983: p. 142–164.

Reigel, D. H. Myelomeningocele: operative treatment and results—1987. In: Marlin,

A. E., ed. Concepts in pediatric neurosurgery. Basel: S. Karger; 1988: p. 41–50.

Reigel, D. H. Spina bifida. In: Section of Pediatric Neurosurgery of the American Association of Neurological Surgeons, ed. Pediatric neurosurgery: surgery of the developing nervous system. Philadelphia: W. B. Saunders; 1989: p. 35–52.

Renier, D.; Sainte, R. C.; Pierre, K. A.; Hirsch, J. F. Prenatal hydrocephalus: outcome and prognosis. Child. Nerv. Syst. 4(4):213–222; 1988.

Rhoton, A. L. Microsurgery of Arnold-Chiari malformation in adults with or without hydromyelia. J. Neurosurg. 45:473–483; 1976.

Roach, E.; DeMyer, W.; Palmer, K.; Conneally, M.; Merritt, A. Holoprosencephaly: birth data, genetic and demographic analysis of 30 families. Birth Defects, Orig. Article Ser. 11(2):294–313; 1975.

Rothstein, T.; Romano, P.; Shoch, D. Meningomyelocele. Am. J. Ophthalmol. 77:690–693; 1974.

Saez, R.; Onofrio, B.; Yanagihara, T. Experience with Arnold-Chiari malformation, 1960 to 1970. J. Neurosurg. 45:416–422; 1976.

Salam, M. Z.; Adams, R. D. The Arnold-Chiari malformation. In: Handbook of Clinical Neurology. Vol. 32 Congenital Malformations of the Spine and Spinal Cord. Vinken, P. J.; Bruyn, G. W., eds. Handbook of clinical neurology. Amsterdam: North-Holland 1978: p. 99–110.

Sasidharan, P.; Marquez, E.; Dizon, E.; Sridhar, C. V. Developmental outcome of infants with severe intracranial-intraventricular hemorrhage and hydrocephalus with and without ventriculoperitoneal shunt. Child. Nerv. Syst. 2:149–152; 1986.

Sawaya, R.; McLaurin, R. L. Dandy-Walker syndrome. Clinical analysis of 23 cases. J. Neurosurg. 55:89–98; 1981.

Schlesinger, E.; Antunes, J.; Michelsen, J.; Louis, K. Hydromyelia: Clinical presentation and comparison of modalities of treatment. Neurosurgery 9:356–365; 1981.

Serlo, W.; Kirkinen, P.; Jouppila, P.; Herva, R. Prognostic signs in fetal hydrocephalus. Child. Nerv. Syst. 2:93–7; 1986.

Shankaran, S.; Koepke, T.; Woldt, E.; Bedard, M. P.; Dajani, R.; Eisenbrey, A. B.; Canady, A. Outcome after posthemorrhagic ventriculomegaly in comparison with mild hemorrhage without ventriculomegaly. J Pediat. 114:109–114; 1989.

Sharrard, W. J. W.; Zachary, R. B.; Lorber, J.; Bruce, A. M.; A controlled trial of immediate and delayed closure of spina bifida cystica. Arch. Dis. Child. 38:18–22; 1963.

Sheptak, P. E. Diastematomyelia-diplomyelia.

In: Congenital Malformations of the Spine and Spinal Cord. Vinken, P. J.; Bruyn, G. W., eds. Handbook of clinical neurology. Amsterdam: North-Holland Publishing; 1978: p. 239–254.

Sherk, H. H.; Shut, L.; Chung, S.; Iniencephalic deformity of the cervical spine with Klippel-Feil anomalies and congenital elevation of scapula. J. Bone Joint Surg. 56A:1254–1259; 1974.

Shurtleff, D.; Kronmal, R.; Foltz, E. Follow-up comparison of hydrocephalus with and without myelomeningocele. J. Neurosurg. 42:61–68; 1975.

Shurtleff, D. B.; Hayden, P. W.; Loeser, J. D.; Kronmal, R. A. Myelodysplasia: Decision for death or disability. N. Engl. J. Med. 291:1005–1011; 1974.

Simpson, D.; David, D.; White, J. Cephaloceles: treatment, outcome, and antenatal diagnosis. Neurosurgery 15:14–21; 1984.

Smith, E. D. Urinary prognosis in spina bifida. J. Urol. 108:815–817; 1972.

Smith, M.; Huntington, H. Inverse cerebellum and occipital encephalocele. Neurology 27:246–251; 1977.

Soare, P. L.; Raimondi, A. J. Intellectual and perceptual-motor characteristics of treated myelomeningocele children. Am. J. Dis. Child. 131:199–204; 1977.

Stanworth, A. Squint in hydrocephalus. An analysis of cases. Strabismus '69 St. Louis: C. V. Mosby; 1970.

Stark, G.; Drummond, M.; Poneprasert, S.; Robarts, F. Primary ventriculoperitoneal shunts in treatment of hydrocephalus associated with myelomeningocele. Arch. Dis. Child. 49:112–117; 1974.

Stark, G. D.; Drummond, M. Results of selective early operation in myelomeningocele. Arch. Dis. Child. 48:676–683; 1973.

Stein, S. C.; Schut, L. Hydrocephalus in myelomeningocele. Child's Brain 5:413–419; 1979.

Stillwell, A.; Menelaus, M. Walking ability in mature patients with spina bifida. J. Pediatr. Orthop. 3:184–190; 1983.

Suwanwela, C.; Hongsaprabhas, C. Fronto-ethmoidal encephalomeningocele. J. Neurosurg. 25:172–182; 1966.

Taggart, J. K., Jr.; Walker, A. E. Congenital atresia of the foramens of Luschka and Magendie. Arch. Neurol. Psychiat. 48:583–612; 1942.

Tal, Y.; Freigang, B.; Dunn, H. G.; Durity, F. A.; Moyes, P. D. Dandy-Walker syndrome: analysis of 21 cases. Dev. Med. Child. Neurol. 22:189–210; 1980.

Tew, B.; Lawrence, K. The effects of hydrocephalus on intelligence, visual perception, and school attainment. Dev. Med. Child. Neurol. 17(suppl. 35):129–134; 1975.

Tryfonas, G. Three spina bifida defects in one child. J. Pediatr. Surg. 8:75; 1973.

Udvarhelyi, G. B.; Epstein, M. H. The so-called Dandy-Walker syndrome: analysis of 12 operated cases. Child's Brain. 1:158–182; 1975.

Venes, J. L.; Black, K. L.; Latack, J. T. Preoperative evaluation and surgical management of the Arnold-Chiari II malformation. J. Neurosurg. 64:363–370; 1986.

Volpe, J. Intracranial hemorrhage, periventricular-intraventricular hemorrhage of the premature infant. Neurology of the newborn. Philadelphia: W. B. Saunders; 1987.

Wallace, S. The effect of upper limb function on mobility of children with myelomeningocele. Dev. Med. Child. Neurol. 15(suppl. 29):84–91; 1973.

Welch, K.; Winston, K. R. Spina bifida. In: Handbook of Clinical Neurology. Vol. 6 Malformations. Myrianthopoulos, N. C., ed. Handbook of clinical neurology. Amsterdam: Elsevier Science; 1978: p. 477–508.

Williamson, W.; Desmond, M.; Wilson, G.; Andrew, L.; Garcia-Prats, J. A. Early neurodevelopmental outcome of low birth weight infants surviving neonatal intraventricular hemorrhage. J. Perinatal Med. 10:34–41; 1982.

Winter, R. B.; Haven, J. J.; Moe, J. H.; Lagaard, S. M. Diastematomyelia and congenital spinal deformities. J. Bone Joint Surg. 56A:27–39; 1974.

Wright, R. L. Congenital dermal sinuses. Prog. Neurol. Surg. 4:175–191; 1971.

Yamada, S.; Zinke, D. E.; Sanders, D. Pathophysiology of tethered cord syndrome. J. Neurosurg. 54:494–503; 1981.

Yashon, D.; Jane, J.; Sugar, O. The course of severe untreated infantile hydrocephalus. Prognostic significance of the cerebral mantle. J. Neurosurg. 23:509–516; 1965.

Yokota, A.; Oota, T.; Matsukado, Y. Dorsal cyst malformations. Part I. Clinical study and critical review on the definition of holoprosencephaly. Child's Brain. 11:320–341; 1984.

Yoneyama, T.; Fukui, J.; Ohtsuka, K.; Komatsu, H.; Ogawa, A. Urinary tract dysfunctions in tethered spinal cord syndrome: improvement after surgical untethering. J. Urol. 133:999–1001; 1985.

31

Neurofibromatosis

VINCENT M. RICCARDI

A discussion of the prognosis of neurofibromatosis (NF) is a challenge. First, there is more than one type of neurofibromatosis; therefore, the prognoses of more than one disorder must be considered, that is, the neurofibromatoses. Second, each type of neurofibromatosis is extremely variable in its manifestations, so a discussion about prognosis concerns the individual types of manifestations and their cumulative importance for smaller subgroups. This chapter deals with three primary categories of NF: NF-1, NF-2, and atypical NF (Riccardi 1982; Riccardi 1989a; Riccardi and Eichner 1986). For each of these categories, the prognosis for the major complications of the respective disorders will be considered.

NF-1

NF-1 is also known as von Recklinghausen's NF (Riccardi 1981; Riccardi and Eichner 1986). The trait is autosomal dominant, and the genetic locus is the proximal long arm of chromosome 17, specifically, chromosome band 17q11.2 (Collins et al. 1989; Fountain et al. 1989; O'Connell et al. 1989). The defining features of NF-1 include multiple café-au-lait spots, multiple

neurofibromas, and multiple iris Lisch nodules (Lewis and Riccardi 1981). Additional features of the disorder that can adversely influence its prognosis are itemized in Table 31-1. Some of these features are considered in purely anatomical terms (e.g., optic pathway glioma), while some are considered in functional terms (e.g., seizures).

Overview

NF-1 is potentially serious because patients with NF-1 can be compromised seriously in many ways, and this compromise may be delayed for years or even decades (Reynolds and Pineda 1988).

NF-1 as a Cause of Chronic Major Morbidity

NF-1 can cause serious long-term handicaps in many ways. Neurological compromise is the most frequent cause. Paraspinal neurofibromas, particularly in the cervical region where frequently they are associated with quadriparesis or quadriplegia, are among the most important causes of chronic morbidity. At least a portion of intracranial tumors, particularly astrocytomas of the optic pathway or the posterior fossa, can lead to seri-

Table 31-1. Features of NF-1 and Their Severities

Severity Grade	1	2	3	4
Optic pathway gliomas			+	+
Other astrocytomas				
Intracranial			+	+
Spinal			+	+
Malignancy				
Neurofibrosarcoma				+
Other malignancies				+
Skeletal				
Bony bowing or pseudarthrosis		+	+	+
Sphenoid wing dysplasia		+	+	+
Kyphoscoliosis		+	+	+
Short stature		+	+	
Other		+	+	
Congenital glaucoma			+	+
Vascular				
Cerebrovascular			+	+
Gastrointestinal		+	+	+
Renovascular hypertension			+	+
Intellectual compromise		+	+	+
Speech deficits		+	+	
Neurofibromas				
Cosmetic compromise	+	+	+	+
Regional hypertrophy			+	+
Neurological compromise			+	+
Seizures			+	+

ous disability. Cerebrovascular disturbances or hydrocephalus also cause long-term functional compromise. Peripheral neuropathies, resulting from direct involvement by neurofibromas or from surgical complications, are common. A variety of pain syndromes usually associated with large neurofibromas or spinal arachnoid cysts may be a source of major clinical distress. Skeletal dysplasias or hypertrophic overgrowth of congenital plexiform neurofibromas are the second most frequent cause of major disabilities. Among the skeletal dysplasias, kyphoscoliosis, sphenoid wing dysplasia, or tibial pseudarthrosis account for a significant portion of long-term physical handicaps. Congenital plexiform neurofibromas that lead to chronic disability include those involving the upper midline of the body from the retropharyngeal region to the lower reaches of the mediastinum and those involving the limb girdles or face.

NF-1 as a Cause of Death

Neurofibrosarcomas, intracranial tumors, cervical paraspinal neurofibromas, kypho-scoliosis, or pheochromocytomas can lead to a premature death in patients with NF-1.

NF-1 Associated With a Normal or Near-Normal Life

Although NF-1 is severe, many patients with NF-1 lead a wholesome and relatively normal life. Moreover, the serious complications considered here are distributed over the entire NF-1 population; one patient will not develop all of the problems.

The following discussion considers anatomical and functional compromises resulting from expression of the NF-1 mutation in terms of the likelihood that they are present and the respective levels of seriousness.

In general, low likelihood indicates a frequency of 0.5% to 5%; moderate likelihood indicates a frequency of 5% to 15%; and high likelihood indicates a frequency of 20% or more.

Mild compromise is a small degree of disturbance that is considered a nuisance or can be readily obviated or masked; it corresponds to severity grade II (Riccardi and Eichner 1986). Moderate compromise is a disturbance that is substantial but amenable to control or remediation; it corresponds to severity grade III (Riccardi and Eichner 1986). Serious compromise is a disturbance that is responsive only to extreme treatment measures and then generally incompletely; it corresponds to severity grade IV (Riccardi and Eichner 1986).

The frequencies of abnormalities considered here are derived primarily from the published data of Huson and coworkers (1988; 1989a; 1989b); Samuelsson, Samuelsson (1989), and Riccardi (1989a, 1989b); Riccardi and Eichner (1986); Riccardi (1981); and the unpublished experience of the author, based on his involvement with 860 patients with NF-1 enrolled in the Baylor NF Program. Occasionally, more focused studies have been used (e.g., Listernick et al. [1989]). The hypothesis of Miller and Hall (1978) that the severity of NF-1 is influenced by inheritance of the mutation from an affected mother has been disproven (Riccardi and Wald 1987), simplifying the data that follow.

High Likelihood and Mild Compromise

The defining features of NF-1, café-au-lait spots, neurofibromas, and Lisch nodules, are likely to be present in all patients with the disorder if they live long enough. However, a patient rarely is so free of neurofibromas that at least some element of moderate compromise from these is completely absent. On the other hand, the true frequency of mildly affected patients (i.e., minimal or severity grade I) is not known with precision. Among the reasons for this is ascertainment bias and concern for genetic heterogeneity (e.g., if the disorder is limited to café-au-lait spots, the disorder might be NF-6, not NF-1) (Riccardi and Eichner 1986). Other considerations, such as the "predecessor syndrome" (Riccardi and Lewis 1988) or somatic mosaicism (Edwards 1989), are not considered here.

Various types of headache affect 50% to 75% of all patients with NF-1. There are three general types: (1) Approximately half of the headaches associated with NF-1 are coincidental, for example, the "tension headache" and those associated with chronic nasal sinus congestion. (2) A more migraine type of headache, suggesting a vascular pathogenesis, is characteristic of NF-1 in all age groups. It accounts for most of the remaining headaches associated with NF-1 and can be severe; therefore, it raises concern about anatomical lesions within the cranial vault. (3) Infrequently, headaches actually may be a symptom of a brain tumor, hydrocephalus, or cerebrovascular compromise. The prognosis for the first two types of headaches is relatively good; they are readily managed and are associated with minimal to mild morbidity. The third type of headache, which is not always easily distinguished from the other two, may indicate pathological processes that significantly alter the patient's long-term prognosis. Knowledge of a normal cranial neuroimaging study allows the clinician to suggest a cause of majority of headaches occurring in patients with NF-1.

High Likelihood and Moderate Compromise

Learning disabilities occur in at least 40% of patients with NF-1, and the frequency probably is closer to 60% (Denckla 1987; Eliason 1986; 1988; Riccardi 1984; Riccardi and Eichner 1986; Varnhagen et al. 1988). Although the burden of this clinical problem can be relatively modest, it is the most common focus of attention in the NF clinical program. It tends to be extremely disruptive of the affected child's household and social life. The primary problems involve poor attention span, impulsivity, poor visual-motor skills, and a diminished information-processing capability. In addition, there is often generalized incoordination (Dunn and Roos 1988; Riccardi and Eichner 1986). Delayed gross motor development and infantile muscular hypotonia appear to be reasonably accurate harbingers of this type of NF-1 school performance problem. IQ scores generally are within the normal range; however, the mean IQ scores of children with NF-1 are shifted to the left, generally 90 ± 15. The

IQ scores of adults with NF-1 have a mean of 100 ± 15 (Riccardi 1984). In general, the performance of adult patients with NF-1 tends to be better than might have been predicted by their childhood school performance problems. Nonetheless, some continued compromise persists into adulthood, according to Samuelsson and Riccardi (1989a; 1989b).

Many types of speech impediments, without any feature accounting for all patients, are seen in at least 30% of patients with NF-1. Velopharyngeal incompetence appears to be a common specifiable problem (Pollack and Shprintzen 1981), and involvement of the oral, oropharyngeal, or nasopharyngeal structures by the direct or indirect effects of strategically located neurofibromas frequently contribute to this problem. The speech defect often is characterized as hypernasal, breathy, and monotonous (Riccardi and Eichner 1986). Compromise of expressive and receptive prosody skills are a major element of this clinical problem. Language skills also may be affected.

Cosmetic defects of varying degrees probably develop at some time in the life of at least half of all patients with NF-1. They may result from three mechanisms. (1) Localized or segmental hypertrophic growth of plexiform neurofibromas may be a cause; cranial or facial tumors are especially important (Reed et al. 1986), but particularly large tumors of the trunk, buttocks, or limbs also may have a major cosmetic impact. (2) Cutaneous neurofibromas, because of the large numbers or strategic (e.g., facial) location, account for significant cosmetic compromise. (3) Skeletal deformities, including those associated with kyphoscoliosis, sphenoid wing dysplasia, or tibial pseudarthrosis, also can lead to significant cosmetic disfigurement. Particularly for the neurofibroma-associated distortions, surgery is one approach to treatment; however, it is generally unsatisfactory if the perceived goal is restoration of a totally normal appearance.

A heavy psychosocial burden (i.e., an excessive emotional and social discomfiture resulting from the stigmatization of a chronic and progressive disease, the dread of an unknown future, and the cosmetic and other handicaps) is present in a large proportion of patients with NF-1. Teenagers and young adults are particularly likely to experience this complication of NF-1, even if transiently. An excess of unmarried or childless men with moderately severe NF-1 (compared to similarly affected women) probably reflects the magnitude of this aspect of the disorder (Crowe et al. 1956; Riccardi and Eichner 1986). Discussions with knowledgeable clinicians, input from astute social workers or family counselors, and psychotherapy can help alleviate this problem, however (Roback et al. 1981).

Moderate Likelihood and Serious Compromise

Optic pathway gliomas occur in 15% of patients with NF-1 routinely screened using computerized neuroimaging techniques (Lewis et al. 1984; Listernick et al. 1989). Of those patients with optic gliomas, approximately one third are symptomatic from these tumors; the clinical problems range from unilateral visual acuity decreases (i.e., associated with a unilateral optic nerve glioma) (Coyle et al. 1988; Duffner and Cohen 1988) to total blindness, hydrocephalus, precocious puberty, and diencephalic syndrome (Adornato and Berg 1977), and other endocrine disturbances (i.e., associated with chiasmal tumors). Although some controversy remains concerning the efficacy of treatment (e.g., with x-rays) of patients detected before irreversible damage is done [Gould et al. 1987; Hoyt and Baghdassarian 1969; Imes and Hoyt 1986; Weiss et al. 1987), an increased consensus suggests that early detection and properly timed treatment can be important in enhancing the prognosis for patients with this complication of NF-1 (Chung and McCrary 1988; Easley et al. 1988; Packer et al. 1988). Without question, optic pathway gliomas are among the most important of the early childhood complications of NF-1; therefore, no opportunity for minimizing their collective morbidity should be discounted (Cohen and Duffner 1983; Cohen et al. 1986). Although there has been some suggestion that the prognosis is better for patients with an optic glioma as part of NF-1 compared to similar isolated tumors (Stern et al. 1979; 1980), the evidence is not entirely consistent (Alvord and Lofton 1988; Borit and Richardson 1982). The author feels that the apparent "benefit" of NF-1 is the consequent detection of early or asymptomatic gliomas.

Kyphoscoliosis affects approximately 5% of patients with NF-1. In virtually all instances, the onset of the spinal distortion occurs in the first decade of life. Most often the process becomes apparent between ages 6 and 10, but occasionally an earlier onset is noted. Especially for the latter group, a localized or diffuse vertebral dysplasia is the underlying basis for the more serious types of disfigurement (Cimino et al. 1986; Yaghmai 1986). A useful clinical sign of the propensity to develop clinically important, progressive kyphoscoliosis is a distortion of the hair pattern along the adjacent to the midline of the back, a finding some call the Riccardi sign (Flannery and Howell 1987). When the kyphoscoliotic process has been detected, the prognosis can be improved greatly by prompt and aggressive treatment, primarily using spinal fusion, often with internal fixation, such as Harrington rods or similar devices (Chaglassian et al. 1976). External bracing generally only temporizes, but this might be useful as part of a strategy to allow maximum vertical growth before surgery is performed. Only some times are neurofibromas associated with the kyphoscoliotic process, but when they are present, the surgical approach and follow-up must take them into account.

Neurofibrosarcomas occur in approximately 6% of patients with NF-1 (Riccardi and Eichner 1986), although some authors portray the risk as higher (Brasfield and Das Gupta 1972) or lower (Hope and Mulvihill 1981). Ordinarily, neurofibrosarcomas do not occur until the end of the first decade or thereafter, and they tend to cluster in the second, third, and fourth decades (D'Agostino et al. 1963a; 1963b; Riccardi et al. 1988). An analysis of 450 patients with NF-1 presenting to the Baylor NF Program for other reasons demonstrated that nine developed a neurofibrosarcoma, and five were aged 10 to 17 years. The most consistent clinical indicator of a neurofibrosarcoma is pain. Therefore, any new development of pain that cannot be explained must be considered to indicate a neurofibrosarcoma until proven otherwise. Two additional clinical indicators are the sudden appearance of a new mass or the rapid growth of a previously quiescent neurofibroma, and the unexplained development of a focal neurological deficit. The prognosis for patients with NF-1 and a neurofibrosarcoma is bleak unless amputation can completely remove the tumor from the patient's body. In general, local surgery, irradiation, and chemotherapy only temporize at best (Bolton et al. 1989; Goldman et al. 1977). Neurofibrosarcomas represent one of the most important causes of untimely death among young patients with NF-1. Multiple primary lesions are not uncommon (Brasfield and Das Gupta 1972; Leslie and Cheung 1987), and the use of research laboratory techniques (e.g., cytogenetic or DNA analyses) to distinguish metastases from second primaries may help to make decisions about management (Riccardi and Elder 1986).

Low Likelihood and Serious Compromise

Congenital glaucoma probably occurs in 0.5% to 1% of patients with NF-1 (Bost et al. 1985; Grant and Walton 1968). All at-risk newborns must be examined with this feature of the disorder in mind. A protruding ocular globe, frank buphthalmos, or even persistent crying may be helpful clinical signs. Loss of vision in the involved eye is likely, and enucleation may be necessary.

The frequency of pheochromocytomas is generally overestimated among patients with NF-1 (Healy and Mekalatos 1958; Nakamura et al. 1986]. The actual frequency is approximately 0.5% to 1%. Nonetheless, pheochromocytomas are critically important because unappreciated tumors can lead to an untimely, and probably unnecessary, death (e.g., in association with otherwise minor or trivial surgery) (Riccardi and Eichner 1986). Routine screening of patients with

NF-1 using measurements of catechol-amines in 24-hour urine specimens has not been helpful (Riccardi and Eichner 1986). However, even the slightest clinical suspi-cion (e.g., excessive sweating, palpitations, episodic headaches) should lead to a prompt and vigorous effort to discount or confirm the presence of a pheochromocytoma.

Seizures of all types occur in approxi-mately 5% of patients with NF-1. Although most often they are readily managed with anticonvulsant medications, they are listed as "serious complications" to emphasize that for some patients, especially with a neo-natal or infantile onset, seizures may be as-sociated with a serious prognosis. Myo-clonic jerks (infantile spasms with hypsarrhythmia) fall into this category (Goldberg et al. 1985; Terada et al. 1986). In addition, the later onset of seizures may in-dicate the development of hydrocephalus or the progression of a previously quiescent as-trocytoma.

Tibial pseudarthrosis occurs in 0.5% to 1% of patients with NF-1. This lesion usu-ally is unilateral, but bilateral cases are well documented. Conservative treatment with casting and bracing may be sufficient treat-ment, but often surgical treatment (DeBoer et al. 1988) or amputation is required. The recent availability of the Ilizarov bone-lengthening technology (Dal Monte and Donzelli 1987; Green 1988) may significantly improve the prognosis for patients with the more serious forms of this complication of NF-1. Less serious forms of this problem, involving bowing of the tibia, are associated with a much better prognosis. In addition, similar lesions can occur in many other sites, including any of the long bones and clavicles (Ali and Hooper 1982; Bayne 1985). Recently, the author observed three young patients with inherited NF-1 and tib-ial pseudarthrosis who had a paucity of café-au-lait spots. The diagnosis would not have been made based on the café-au-lait spots alone.

Sphenoid wing dysplasia occurs in ap-proximately 2% to 3% of patients with NF-1. Severity ranges from trivial findings on radiographic studies to serious problems, including facial disfigurement and pulsating enophthalmos associated with herniation of a portion of the brain into the orbit (DeVil-liers 1982; Grenier et al. 1984). It is impor-tant to recognize that this is a dysplastic le-sion of the bone (i.e., it is not the result of a local plexiform neurofibroma, which also may be present) and that it will progres-sively worsen. Surgical reconstruction of the posterior orbital wall ultimately may be indicated.

Plexiform neurofibromas are congenital lesions, but their detection in infancy may require a high level of clinical suspicion or radiographic studies (of the chest and abdo-men) (Bourgouin et al. 1988). The presence of overlying hyperpigmentation may facili-tate detection, and if the hyperpigmentation reaches the midline of the body, involve-ment of the neuraxis should be presumed. Conservatively, 2% to 3% of patients with NF-1 ultimately will have serious problems resulting from such tumors. Those involving the second and third branches of the trigemi-nal nerve may have a particularly adverse effect in terms of cosmetic disfigurement and compromise of ocular and oral struc-tures and functions. Plexiform neurofibro-mas involving the retropharyngeal regions and upper mediastinum may seriously com-promise the upper respiratory tree and de-glutition, and they may be associated with kyphoscoliosis, thoracic wall tumor growth, or both. Similar tumors involving the limb girdles and proximal limbs may lead to joint destruction and limited use of the limb, and they may require amputation, often in the teen or young adult years, for effective treatment (Match and Leffert 1987). Deeper involvement of the intrathoracic space or the retroperitoneal space must be consid-ered for such tumors. Plexiform neurofibro-mas with massive overgrowth involving the distal extremities also are a significant cause of major compromise of ambulation or man-ual dexterity and may require disfiguring surgery or amputation.

Paraspinal neurofibromas (Burk et al. 1987; Castelein and MacEwen 1984; Levy et al. 1986; Lewis and Kingsley 1987), particularly in the cervical region, may have a grave prognosis (Adekeye et al. 1984). Massive cervical paraspinal neurofibromas are among the most common causes of serious long-term morbidity in NF-1. Destruction of adjacent vertebrae (Ferner et al. 1988), centripetal growth into and around the spinal cord, and centrifugal growth with consequent compromise of nerve roots and proximal nerve trunks can lead to quadriparesis (or quadriplegia) combined with peripheral neuropathies. The course is slowly but inexorably progressive and frequently leads to premature death. Involvement of the thoracic and lumbosacral regions with neurofibromas can lead to paraparesis (or paraplegia) in a similar manner. Chronic, relatively intense pain is more likely to accompany the lumbosacral lesions than those higher up.

Cerebral and posterior fossa astrocytomas (distinct from astrocytomas of the optic pathways) are a definite part of the NF-1 tumor spectrum, but the frequency is unclear (perhaps as high as 0.5%) (Miller 1975). Moreover, the consequences of such tumors are variable. However, it is because little is known about the natural history of these tumors that they are a problem. In some instances, for unknown reasons, they may become aggressive and lead to serious brain compromise and death (Hochstrasser et al. 1988; Sorensen et al. 1986). It is critically important to distinguish tumors (i.e., lesions showing a breakdown of the blood–brain barrier and a "mass effect") from the hyperintense T_2-weighted signal foci seen frequently on magnetic resonance imaging (MRI) scans of the brains of patients with NF-1. These foci have no apparent prognostic significance (Bognanno et al. 1988; Brown et al. 1987; Duffner et al. 1989; Dunn and Roos 1988).

Hydrocephalus may be seen in approximately 0.5% of patients with NF-1 (Riviello et al. 1988). Although occasionally hydrocephalus may be the result of a demonstrable brain tumor (usually a chiasmal glioma) or Chiari malformation (Afifi et al. 1988), usually the pathogenesis is unclear. Often there is stenosis of the aqueduct of Sylvius, but even then the reason for the stenosis is not apparent. Prompt recognition and treatment can minimize the potential for a seriously worsened prognosis.

Moderate Likelihood and Moderate Compromise

Plexiform neurofibromas in the sites considered previously and elsewhere can alternatively be associated with much less severe outcomes. Although at first, the difference seems to be a matter of size of the neurofibroma, actually the determinants of growth and size of such tumors are unknown, except for the potential contribution of trauma (Riccardi 1987; 1989b). Surgery for cosmetic purposes or to reduce burdensome mass is frequently required and often more than once.

Short stature, other than that associated with, for example, chiasm optic gliomas, occurs in 15% or more of patients with NF-1 (Riccardi and Haeberlin 1989). Alone, this aspect of NF-1 is not particularly serious, but when added to school performance problems, various types of disfigurement, and other clinical tribulations, short stature can be an important aspect of the overall clinical picture. On the other hand, short stature does not necessarily implicate a chiasm glioma or other neuroanatomical basis; more likely; the short stature is another manifestation of the mutation's direct effect on skeletal tissue.

Other skeletal abnormalities, including cranial vault defects (Mann et al. 1983), nondystrophic scoliosis, pectus excavatum, angulation deformities at the knees (genu varum, genu valgum), ankle valgus, and pes planus are relatively common in patients with NF-1. At least 20% have one or more of the abnormalities itemized. Although most are relatively trivial in terms of their clinical impact, some patients may require surgical treatment or prolonged bracing.

Renovascular hypertension occurs in a significant portion of patients with NF-1 (Craddock et al. 1988; Elias et al. 1985; Finley and Dabbs 1988; Lassmann 1988; Pollard et al. 1989). Although the exact frequency of this feature is not known, it is included in this section because the frequency may be higher than 5% and because this feature of NF-1 should be suspected in all age groups, particularly the early and middle teenage years. In addition, pregnancy may be a time when renovascular hypertension presents as an element of NF-1 (Edwards et al. 1983). Also the prognosis of the disorder may be greatly improved by timely surgical intervention (Baxi et al. 1981; Gardiner et al. 1988; Mallmann and Roth 1986).

Pruritus, excessive itching of moderate to major severity, is manifest in more than 5% of patients with NF-1 (Burton and Hanfield-Jones, 1986; Riccardi 1987).

Low Likelihood and
Moderate Compromise

Seizures were considered in the section "Low Likelihood and Serious Compromise." Seizures also are considered here, however, because those with less severe significance tend to present much later, either in late childhood or early adulthood.

Precocious puberty occurs in NF-1 with a frequency of no more than 0.5%, and its presence virtually always indicates the presence of a progressing optic chiasm glioma (Laue et al. 1985; Tertsch et al. 1979). Delay in the onset of progression of puberty is less often a complication of NF-1.

Cerebrovascular disturbances as part of the clinical picture of NF-1 are probably more common than appreciated (DeKersaint et al. 1980). However, an estimate of 0.5% is probably consistent with its clinical detectability. In general, when recognized on the basis of clinical problems, the overall prognosis for the disorder is considerably worse than when no such problems are seen.

Spinal arachnoid cysts in patients with NF-1 (Erkulvrawtr et al. 1979; Kaiser et al. 1986; O'Neill et al. 1983) are more likely to occur in association with the various types of scoliosis, and particularly with lordoscoliosis (Dickson 1985). Accurate frequency figures are not available. However, in a portion of patients with this feature of NF-1,

there may be substantial amounts of localized back pain that has no alternative explanation.

Rare or Occasional Problems With
Variable Consequences

Carcinoid tumors are increasingly recognized as a feature of NF-1, and they may add to a negative prognosis (Dawson et al. 1984; Dayal et al. 1986; Fernandez et al. 1988).

Visceral involvement may, in general, be of two types: that due to the presence of neurofibromas or other tumors within the parenchyma of the various viscera or that due to a more diffuse, less specific pathological process. Genitourinary involvement may take a variety of forms (Blum et al. 1985; Brooks and Scally 1985). Gastrointestinal involvement generally consists of neurofibromas on the serosal surface or within the gut wall (Bhattacharyya and Jamieson 1988; Bridgen et al. 1982; Cosgrove and Fischer 1988); serious, life-threatening hemorrhage may be a consequence (Cox et al. 1988). Alternatively, vascular compromise may lead to postprandial pain and other complications (Brunner et al. 1974; Cameron et al. 1982; DuToit 1987). It is important to distinguish between gastrointestinal involvement of NF-1 from the distinct form of NF known as "gastrointestinal NF" (Heimann et al. 1988; Verhest et al. 1988). Hepatic involvement has been posited, but the relationship of the NF-associated liver involvement is not always clear (Curry and Gray 1972; Ettinger and Freeman 1979; Lederman et al. 1987; Meyer et al. 1982). Diffuse interstitial pulmonary fibrosis is part of NF-1 (Burkhalter et al. 1986; Porterfield et al. 1986; Stark et al. 1988), but this complication has not been seen in the 860 patients with NF-1 enrolled in the Baylor NF Program. Cardiac involvement in terms of congenital lesions (Neiman et al. 1974) or diffuse myocardial hypertrophy (Fitzpatrick and Emanuel 1988) has never been substantiated (Lin and Garver 1988). However, myocardial neurofibromas have been documented (McAllister and Fenoglio 1978).

Spinal astrocytomas are rarely seen in NF-1; they are a much more consistent feature of NF-2. However, when such a tumor occurs in NF-1, it can be extensive and difficult or impossible to treat effectively, creating a serious prognosis for the affected patient.

Hyperparathyroidism (Freimanis et al. 1984; Hoppe et al. 1986) and parathyroid adenomas (Chakrabarti et al. 1979) also have been said to occur frequently in NF-1, but the data are not convincing.

Glomus tumors, particularly involving the tips of one or more fingers, is a definite, although rare and not well-known, feature of NF-1. Surgery is required for adequate pain relief.

Other malignancies also may adversely influence the prognosis of some patients with NF-1 (Brasfield and Das Gupta 1972; Hope and Mulvihill 1981). The most intriguing of these is juvenile chronic myelogenous leukemia (Bader and Miller 1978; Clark and Hutter 1982; Fiorillo et al. 1984); because it does fit neatly into the neurocrisotpathy pathogenesis model (Bolande 1981), it deserves special attention from a research standpoint.

NF-2

NF-2, previously known as bilateral acoustic NF, is different from NF-1. Its clinical burden is much more determined by central nervous system (CNS) tumors (which, in turn, are different from those seen in NF-1) and by paraspinal neurofibromas (Martuza and Eldridge 1988); its autosomal dominant locus is on a different chromosome, the long arm of chromosome 22 (Rouleau et al. 1987; Seizinger et al. 1986; Wertelecki et al. 1988). The prognosis of patients with NF-2 also is different from that of NF-1 (Fickel 1989; Martuza and Eldridge 1988). Patients with NF-2 are spared many of the risks and concerns outlined for NF-1. Thus, those who escape the serious effects of the CNS tumors or the paraspinal neurofibromas generally have a mild course and a good prognosis. However, patients with NF-2 who become symptomatic with

these tumors are likely to be seriously compromised and have a relatively poor prognosis.

The majority of patients with NF-2 have few or no problems during the first decade of life. However, by the end of the third decade, one or more tumors become manifest. The posterior subcapsular cataracts may be detected with increasing frequency as the patient with NF-2 ages, but in any case, they are unlikely to disturb vision (Kaiser-Kupfer et al. 1989; Pearson-Webb et al. 1986). Patients with NF-2 have few café-au-lait spots or cutaneous neurofibromas, which tend to contribute relatively little to the disorder's overall morbidity.

Schwannomas involve the CNS of patients with NF-2 in three ways. First, the definitive lesion is the vestibular schwannoma, more commonly known as an acoustic neuroma. In NF-2, vestibular schwannomas virtually always are bilateral but not necessarily simultaneous in terms of presentation or as a cause of clinical problems (Curati et al. 1986a; 1986b; House 1986; Martuza and Ojemann 1982; Mazzoni 1987; Stack et al. 1988). Most often, progressive growth leads to compression of the auditory portion of the eighth cranial nerve, resulting in varying degrees of deafness—the larger the tumor, the more profound the deafness. In addition, surgical manipulation completes or adds to the deafness and frequently results in an ipsilateral facial palsy. Second, a schwannoma also may develop on the fifth cranial nerve (McCormack et al. 1988). Third, schwannomas may develop within the substance of the spinal cord and on dorsal nerve roots, leading to focal spinal cord deficits.

Meningiomas may develop virtually anywhere within the cranial vault or along the length of the spinal cord, and a tumor at either site may lead to the patient's presentation. Frequently, more than one meningioma complicates the clinical picture for a patient with NF-2. These tumors may be among the most troublesome for patients with NF-2.

Spinal astrocytomas are a hallmark of NF-2, in contradistinction to NF-1. In the latter disorder, astrocytomas of the optic pathway are the rule, while such lesions are virtually unheard of in NF-2.

Ependymomas represent a distinctive lesion that is part of the NF-2 complex. They may occur within the cranial vault or within the spinal cord, but in either case, they lead to serious compromise and death.

Paraspinal neurofibromas often are an unappreciated feature of NF-2 (Levy et al. 1986). Nonetheless, they are a major aspect of this disorder and may lead to serious motor and sensory deficits, and at times, they may be associated with local kyphoscoliosis.

Atypical NF

Riccardi (1982; 1989a) and Riccardi and Eichner (1986) have suggested a variety of alternative types of NF (Calzavara et al. 1988). One point of such a categorization has been to clarify that the clinical course and prognosis of alternative types of NF are not typical of NF-1 or NF-2. Consider for example, a form of NF characterized by café-au-lait spots only. The outlook for that form of NF does not fit the molds of NF-1 and NF-2. This also is true for disorders that overlap with NF incompletely, such as multiple schwannomatosis (Purcell and Dixon 1989; Shishiba et al. 1984).

Treatment Strategies

Surgery is the mainstay of treatment for NF-1 and NF-2. Decreasing each tumor's bulk and the effects of its impingement on local structures are the primary goals of surgery (Kelly et al. 1988; Pickard and Rose 1988). This approach may have dramatic results in the short term, but the long-term results are less than satisfactory. Laser technology for surgical approaches (Katalinic 1987; Roenigk and Ratz 1987) probably offers no unique advantages.

Medicinal approaches to NF are less than satisfactory. Only one approach has been suggested to treat benign tumors, and this involves the use of the mast cell blocker,

ketotifen, to treat neurofibromas (Claman 1987; Riccardi 1987). The medical treatment of other benign tumors, such as optic gliomas (Packer et al. 1988) or acoustic neuromas and other schwannomas, has been wanting (Jahrsdoerfer and Benjamin 1988). Similarly, the medical treatment of malignant tumors (primarily neurofibrosarcomas) that complicate NF-1 and NF-2 is less than satisfactory (Goldman et al. 1977).

Radiation treatment of gliomas probably is effective in some cases (Chung and McCrary 1988; Easley et al. 1988), but otherwise, the use of this modality is probably without complete scientific merit for patients with NF. Possible exceptions might be some types of radiosensitive forms of meningiomas, acoustic neuromas, and neurofibrosarcomas.

Remediation to overcome some of the physical and social handicaps of NF may be especially useful. Programs to assist in school performance, such as prostheses and hearing aids, and communication devices, such as computers, may be especially useful (Brody 1989).

Dialogue for treatment include approaches such as genetic counseling; each person with NF-1 or NF-2 is told of the 50% risk for transmitting the mutant gene to each offspring and that prenatal diagnosis is available for selected patients with NF-1 (i.e., those with multiple generations affected) (Crandall et al. 1988). Educational counseling, clarifying the various types of NF-related problems that may arise, is especially important (Powell 1988a; 1988b). Support groups, such as the Texas NF Foundation; Neurofibromatosis, Inc.; and the Acoustic Neuroma Association may be especially helpful to patients and families dealing with the tribulations and trials of NF. Social counseling may be of particular benefit (Messner et al. 1985; Messner and Smith 1986). Psychotherapy also may be useful. Enhancing self-esteem and sense of control may be specific goals in this regard. Participation in research projects of centers such as the Baylor NF Program (Riccardi 1988) may help some patients and families deal with their frustrations with their disorders.

Acknowledgments

This work was supported by the Texas NF Foundation, the Lyle S. Thompson Memorial Fund, the Masie Creelman Fund, the Walker Family Foundation Fund, and the Salam wa Sahha Fund.

References

Adekeye, E. O.; Abiose, A.; Ord, R. A. Neurofibromatosis of the head and neck: clinical presentation and treatment. J. Maxillofac. Surg. 12:78–85; 1984.

Adornato, B.; Berg, B. Diencephalic syndrome and von Recklinghausen's disease. Ann. Neurol. 2:159–160; 1977.

Afifi, A. K.; Dolan, K. D.; Van Gilder, J. C.; Fincham, R. W. Ventriculomegaly in Neurofibromatosis 1: association with Chiari-I malformation. Neurofibromatosis 1:299–305; 1988.

Ales, J. E.; Alvarez de Mon, M; Yebra, M. Intestinal neurofibromatosis in a 10 year-old patient. Gastroenterol. Hepatol. 7:208–211; 1984.

Ali, M. S.; Hooper, G. Congenital pseudarthrosis of the ulna due to neurofibromatosis. J. Bone. Joint. Surg. 64:600–602; 1982.

Alvord, E. C., Jr.; Lofton, S. Gliomas of the optic nerve or chiasm: outcome by patient's age, tumor site and treatment. J. Neurosurg. 68:85–98; 1988.

Bader, J. L.; Miller, R. W. Neurofibromatosis and childhood leukemia. J. Pediatr. 92:925–929; 1978.

Baxi, R.; Epstein, H. Y.; Abitbol, C. Percutaneous transluminal renal artery angioplasty in hypertension associated with neurofibromatosis. Radiology 139:583–584; 1981.

Bayne, L. G. Congenital pseudarthrosis of the forearm. Hand Clin. 1:457–465; 1985.

Bhattacharyya, R.; Jamieson, G. G. Recurrent perforation complicating intestinal neurofibromatosis. Aust. N.Z. J. Surg. 58:749–751; 1988.

Blum, M. D.; Bahnson, R. R.; Carter, M. F. Urologic manifestations of von Recklinghausen neurofibromatosis. Urology 26:209–217; 1985.

Bognanno, J. R.; Edwards, M. K.; Lee, T. A.; Dunn, P. W.; Roos, K. L.; Klatte, E. C. Cranial MR imaging in neurofibromatosis. A. J. R. 151:381–388; 1988.

Bolande, R. P. The quintessential neurocristopathy: pathogenetic concepts and relationships. Adv. Neurol. 29:67–75; 1981.

Bolton, J. S.; Vauthey, J. N.; Farr, G. H. Jr.; Sauter, E. I.; Bowen, J. C. III; Kline, D. G. Is limb-sparing surgery applicable to neurogenic sarcomas of the extremities? Arch. Surg. 124:118–121; 1989.

Borit, A.; Richardson, E. P. Jr. The biological and clinical behavior of pilocytic astrocytomas of the optic pathways. Brain 105:161–187; 1982.

Bost, M.; Mouillon, M.; Romanet, J. P.; Deiber, M.; Navoni, F. Congenital glaucoma and von Recklinghausen's disease. Pediatrie 40:207–212; 1985.

Bourgouin, P. M.; Shepard, J. A. O.; Moore, E. H.; McCloud, T. C. Plexiform neurofibromatosis of the mediastinum: CT appearance. Am. J. Roentgen. Rad. Ther. Nuc. Med. 151:461–463; 1988.

Brasfield, R. D.; Das Gupta, T. K. Von Recklinghausen's disease: a clinicopathological study. Ann. Surg. 175:86–104; 1972.

Brigden, M. L.; Hosie, R. T.; McCauley, W. D. Sigmoid mesenteric tumour as a manifestation of von Recklinghausen's disease. Can. J. Surg. 25:451–452; 1982.

Brody, H. The great equalizer: PCs empower the disabled. P. C. Computing 2(7):82–93; 1989.

Brooks, P. T.; Scally, J. K. Bladder neurofibromas causing ureteric obstruction in von Recklinghausen's disease. Clin. Radiol. 36:537–538; 1985.

Brown, E. W.; Riccardi, V. M.; Mawad, M.; Handel, S.; Goldman, A.; Bryan, R. N. Magnetic resonance imaging of optic pathways in patients with neurofibromatosis. A. J. N. R. 8:1031–1036; 1987.

Brunner, H.; Stacher, G.; Bankl, H.; Grabner, G. Chronic mesenteric arterial insufficiency caused by vascula neurofibromatosis. Am. J. Gastroenterol. 62:442–447; 1974.

Burk, D. L. Jr.; Brunberg, J. A.; Kanal, E.; Latchaw, R. E. Spinal and paraspinal neurofibromatosis: surface coil MR imaging at 1.5T. Radiology 162:797–801; 1987.

Burkhalter, J. L.; Morano, J. U.; McCay, M. B. Diffuse interstitial lung disease in neurofibromatosis. South. Med. J. 79:944–946; 1986.

Burton, J. L.; Handfield-Jones, S. Neurofibromatosis and generalized pruritus. Clin. Exper. Dermatol. 11:427; 1986.

Calzavara, P. G.; Carlino, A.; Manganoni, A. M.; Anzola, G. P.; Paolini, M. P. Segmental neurofibromatosis: case report and review of the literature. Neurofibromatosis 1:318–322; 1988.

Cameron, A. J.; Pairolero, P. C.; Stanson, A. W.; Carpenter, H. A. Abdominal angina and neurofibromatosis. Mayo Clin. Proc. 57:125–128; 1982.

Castelein, R. M.; MacEwen G. D. A dumbbell

(hourglass) neurofibroma of the spine in a patient with von Recklinghausen's disease. A case report with twelve year follow up. Arch. Orthop. Trauma. Surg. 102:216–220; 1984.

Chaglassian, J. H.; Riseborough, E. J.; Hall, J. E. Neurofibromatous scoliosis: natural history and results of treatment in thirty-seven cases. J. Bone Joint Surg. 58A:695–702; 1976.

Chakrabarti, S.; Murugesan, A.; Arida, E. J. The association of neurofibromatosis and hyperparathyroidism. Am. J. Surg. 137:417–420; 1979.

Chung, S. M.; McCrary, J. A. III. Management of pregeniculate anterior visual pathway gliomas. Neurofibromatosis 1:240–247; 1988.

Cimino, P. M.; Roberts, J. M.; King, A. G.; Burke, S. W.; Larocca, S. H. Dystrophic scoliosis and neurofibromatosis. Is myelogram indicated? Orthop. Trans. 10:580; 1986.

Claman, H. N. New hope for neurofibromatosis—the mast cell connection. J. A. M. A. 258:823; 1987.

Clark, R. D.; Hutter, J. J. Jr. Familial neurofibromatosis and juvenile chronic myelogenous leukemia. Hum. Genet. 60:230–232; 1982.

Cohen, M. E.; Duffner, P. K. Visual-evoked responses in children with optic gliomas, with and without neurofibromatosis. Child's Brain 10:99–111; 1983.

Cohen, M. E.; Duffner, P. K.; Kuhn, J. P.; Seidel, F. G. Neuroimaging in neurofibromatosis. Ann. Neurol. 20:444; 1986.

Collins, F. S.; Ponder, B. A. J.; Seizinger, B. R.; Epstein, C. J. The von Recklinghausen neurofibromatosis region on chromosome 17—Genetic and physical maps come into focus. Am. J. Hum. Genet. 44:1–5; 1989.

Cosgrove, J. M.; Fischer, M. G. Gastrointestinal neurofibroma in a patient with von Recklinghausen's disease. Surgery 103:701–703; 1988.

Coyle, J. T.; Seiff, S. R.; Hoyt, W. F. Orbital optic glioma in neurofibromatosis. Arch. Ophthalmol. 106:718–723; 1988.

Cox, J. G. C.; Royston, C. M. S.; Sutton, D. R. Multiple smooth muscle tumors in neurofibromatosis presenting with chronic gastrointestinal bleeding. Postgrad. Med. J. 64:149; 1988.

Craddock, G. R. Jr.; Challo, V. R.; Dean, R. W. Neurofibromatosis and renal artery stenosis: a case of familial incidence. J. Vasc. Surg. 8:489–494; 1988.

Crandall, K. A.; Edwards, J. G.; Riccardi, V. M. Attitudes of individuals affected with neurofibromatosis toward prenatal diagnosis. Am. J. Hum. Genet. 43:A165; 1988.

Crowe, F. W.; Schull, W. J.; Neel, J. V. A clinical, pathological, and genetic study of multiple neurofibromatosis. Springfield, IL: Charles C. Thomas; 1956.

Curati, W. L.; Graif, M.; Kingsley, D. P. E.; King, T.; Scholtz, C. L.; Steiner, R. E. MRI in acoustic neuroma: a review of 35 patients. Neuroradiology 28:208–214; 1986a.

Curati, W. L.; Graif, M.; Kingsley, D. P. E.; Niendorf, H. P.; Young, I. R. Acoustic neuromas: Gd-DTPA enhancement in MR imaging. Radiology 158:447–451; 1986b.

Curry, B.; Gray, N. Visceral neurofibromatosis. an unusual case of obstructive jaundice. Br. J. Surg. 59:739–741; 1972.

D'Agostino, A. N.; Soule, E. H.; Miller, R. H. Primary malignant neoplasms of nerves (malignant neurilemmomas) in patients with manifestations of multiple neurofibromatosis (von Recklinghausen's disease). Cancer 16:1003–1014; 1963a.

D'Agostino, A. N.; Soule, E. H.; Miller, R. H. Sarcomas of peripheral somatic tissue associated with multiple neurofibromatosis (von Recklinghausen's disease). Cancer 16:1015–1027; 1963b.

Dal Monte, A.; Donzelli, O. Tibial lengthening according to Ilizarov in congenital hypoplasia of the leg. J. Pediatr. Orthop. 7:135–138; 1987.

Dawson, B. V.; Kazama, R.; Paplanus, S. H. Association of carcinoid with neurofibromatosis. South Med. J. 77:511–513; 1984.

Dayal, Y.; Tallberg, K. A.; Nunnemacher, G.; DeLellis, R. A.; Wolfe, H. J. Duodenal carcinoids in patients with and without neurofibromatosis. A comparative study. Am. J. Surg. Pathol. 10:348–357; 1986.

DeBoer, H. H.; Verbout, A. J.; Nielsen, H. K.; van der Eijken, J. W. Free vascularized fibular graft for tibial pseudarthrosis in neurofibromatosis. Acta. Orthop. Scand. 59:425–429; 1988.

DeKersaint Gilly, A.; Zenthe, L.; Dabouis, G.; Mussini, J. M.; Lajat, Y.; Robert, R.; Picard, L. Abnormalities of the intracerebral vasculature in a case of neurofibromatosis. J. Neuroradiol. 7:193–198; 1980.

Denckla, M. B. Cognitive impairments in neurofibromatosis. Dysmorphol. Clin. Genet. 1:49–57; 1987.

DeVilliers, J. C. Neurofibromatous orbitocranial dysplasia in childhood. S. Afr. J. Surg. 20:137–144; 1982.

Dickson, R. A. Thoracic lordoscoliosis in neurofibromatosis: treatment by a Harrington rod with sublaminar wiring. Report of two cases. J. Bone Joint Surg. 67:822–823; 1985.

Duffner, P. K.; Cohen, M. E. Isolated optic nerve gliomas in children with and without neurofibromatosis. Neurofibromatosis 1:201–211; 1988.

Duffner, P. K.; Cohen, M. E.; Seidel, F. G.; Shucard, D. W. The significance of MRI abnormalities in children with neurofibromatosis. Neurology 39:373–378; 1989.

Dunn, D. W.; Roos, K. L. RI evaluation of learning disability and incoordination in neurofibromatosis. Neurofibromatosis 2:1–5, 1988.

DuToit, D. F. Gastric hemorrhage in a patient with neurofibromatosis. A case report. S. Afr. Med. J. 71:730–731; 1987.

Easley, J. D.; Scharf, L.; Chou, J. L.; Riccardi, V. M. Controversy in the management of optic pathway gliomas. 29 patients treated at the Baylor College of Medicine from 1967 through 1987. Neurofibromatosis 1:248–251; 1988.

Edwards, J. H. Familiarity, recessivity and germline mosaicism. Ann. Hum. Genet. 53:33–47; 1989.

Edwards, J. N.; Fooks, M.; Davey, D. A. Neurofibromatosis and severe hypertension in pregnancy. Br. J. Obstet. Gynecol. 90:528–531; 1983.

Elias, D. L.; Ricketts, R. R.; Smith, R. B. Renovascular hypertension complicating neurofibromatosis. Am. J. Surg. 51:97–106; 1985.

Eliason, M. J. Neurofibromatosis: implications for learning and behavior. J. Dev. Behav. Pediatr. 7:175–179; 1986.

Eliason, M. J. Neuropsychological patterns: neurofibromatosis compared to developmental learning disorders. Neurofibromatosis 1:17–25; 1988.

Erkulvrawatr, S.; Gammal, T. E.; Hawkins, J.; Green, J. B.; Srinivasan, G. Intrathoracic meningoceles and neurofibromatosis. Arch. Neurol. 36:557–559; 1979.

Ettinger, L. J.; Freeman, A. I. Hepatoma in a child with neurofibromatosis. Am. J. Dis. Child. 133:528–531; 1979.

Fernandez, M. T.; Puig, L.; Capella, G.; Bordes, R.; de Moragas, J. M. Von Recklinghausen neurofibromatosis with carcinoid tumor and submucous leiomyomas of the duodenum. Neurofibromatosis 1:294–298; 1988.

Ferner, R. E.; Honovar, M.; Gullan, R. W. A spinal neurofibroma presenting as atlanto-axial subluxation in von Recklinghausen neurofibromatosis (NF-1). Neurofibromatosis 2:43–46; 1988.

Fickel, G. Acoustic Neuroma Association annual meeting report: NF-2 update. Neurofibromatosis 2:57–66; 1989.

Finley, J. L.; Dabbs, D. J. Renal vascular smooth muscle proliferation in neurofibromatosis. Hum. Pathol. 19:107–110; 1988.

Fiorillo, A.; Pettinato, G.; Migliorati, R.; Fiore, M.; Menna, G. Neurofibromatosis and chronic myeloid leukemia. Clin. Pediatr. 23:678, 1984.

Fitzpatrick, A. P.; Emanuel, R. W. Familial neurofibromatosis and hypertrophic cardiomyopathy. Br. Heart J. 60:247–251; 1988.

Flannery, D. B.; Howell, C. G. Confirmation of the Riccardi sign. Proc. Greenwood Genet. Ctr. 6:161; 1987.

Fountain, J. W.; Wallace, M. R.; Bruce, M. A.; Seizinger, B. R.; Menon, A. G.; Gusella, J. F.;

Michels, V. V.; Schmidt, M. A.; Dewald, G. W.; Collins, F. S. Physical mapping of a translocation breakpoint in neurofibromatosis. Science 244:1085–1087; 1989.

Freimanis, M. G.; Rodgers, R. W.; Samaan, N. A. Neurofibromatosis and primary hyperparathyroidism. South Med. J. 77:794–795; 1984.

Gardiner, G. A. Jr.; Freedman, A. M.; Shlansky-Goldberg, R. Percutaneous luminal angioplasty: delayed response in neurofibromatosis. Radiolgoy 169:79–80; 1988.

Goldberg, A.; Kohelet, D.; Mundel, G. Congenital neurofibromatosis presenting as neonatal cerebral damage with hypsarrhythmia. Harefuah 108:332–333; 1985.

Goldman, R. L.; Jones, S. L.; Heusinkveld, R. S. Combination chemotherapy of metastatic malignant schwannoma with vincristine, adriamycin, cyclophosphamide, and imidazole carboxamide. Cancer 39:1955–1958; 1977.

Gould, R. J.; Hilal, S. K.; Chutorian, A. M. Efficacy of radiotherapy in optic gliomas. Pediatr. Neurol. 3:29–32; 1987.

Grant, W. M.; Walton, D. S. Distinctive gonioscopic findings in glaucoma due to neurofibromatosis. Arch. Ophthalmol. 79:127–134; 1967.

Green, S. A. Ilizarov esternal fixation. Technical and anatomic considerations. Bull. Hosp. Jt. Dis. Orthop. Inst. 28:28–35; 1988.

Grenier, N.; Guibert Tranier, F.; Nicholau, A.; Caille, J. M. Contribution of computerized tomography to the study of spheno-orbital dysplasia in neurofibromatosis. J. Neuroradiol. 11:201–211; 1984.

Healy, F. H. Jr.; Mekalatos, C. J. Pheochromocytoma and neurofibromatosis. N. Engl. J. Med. 258:540–546; 1958.

Heimann, R.; Verhest, A.; Verschraegen, J.; Grosjean, W.; Draps, J. P.; Hecht, F. Hereditary intestinal neurofibromatosis. I. A distinctive genetic disease. Neurofibromatosis 1:26–32; 1988.

Hochstrasser, H.; Boltshauser, E.; Valavanis, A. Brain tumors in children with neurofibromatosis. Neurofibromatosis 1:233–239; 1988.

Hope, D. G.; Mulvihill, J. J. Malignancy in neurofibromatosis. Adv. Neurol. 29:33–55; 1981.

Hoppe, L. B.; Collicott, P. E.; Stivrins, T. J. Von Recklinghausen's neurofibromatosis and primary hyperparathyroidism: a case report and literature review. Nebr. Med. J. 71:435–437; 1986.

House, J. W.; Waluch, V.; Jackler, R. K. Magnetic resonance imaging in acoustic neuroma diagnosis. Ann. Otol. Rhinol. Laryngol. 95:16–20; 1986.

Hoyt, W. F.; Baghdassarian, S. A. Optic glioma of childhood: Natural history and rationale for conservative management. Br. J. Ophthalmol. 53:793–798; 1969.

Huson, S. M.; Compston, D. A. S.; Clark, P.; Harper, P. S. A genetic study of von Recklinghausen neurofibromatosis in South East Wales. I. Prevalence, fitness, mutation rate and effect of parental transmission on severity. J. Med. Genet. 26:704–711; 1989a.

Huson, S. M.; Compston, D. A. S.; Harper, P. S. A genetic study of von Recklinghausen neurofibromatosis in South East Wales. II. Guidelines for genetic counseling. J. Med. Genet. 27:712–721; 1989b.

Huson, S. M.; Harper, P. S.; Compston, D. A. S. Von Recklinghausen neurofibromatosis: A clinical and population study in South East Wales. Brain 111:1355–1381; 1988.

Imes, R. K.; Hoyt, W. F. Childhood chiasmal gliomas: update on the fate of patients in the 1969 San Francisco Study. Br. J. Ophthalmol. 70:179–182; 1986.

Jahrsdoerfer, R. A.; Benjamin, R. S. Chemotherapy of bilateral acoustic neuromas. Otolaryngol. Head Neck Surg. 98:273–282; 1988.

Kaiser, M. C.; De Slegte, R. G.; Crezee, F. C.; Valk, J. Anterior cervical meningoceles in neurofibromatosis. A. J. N. R. 7:1105–1110, 1986.

Kaiser-Kupfer, M. I.; Freidlin, V.; Datiles, M. B.; Edwards, P. A.; Sherman, J. L.; Parry, D.; McCain, L. M.; Eldridge, R. The association of posterior capsular lens opacity with bilateral acoustic neuromas in patients with neurofibromatosis type 2. Arch. Ophthalmol. 107:541–544; 1989.

Katalinic, D. Therapy of neurofibromatosis with the argon laser. Lasers Surg. Med. 7:128–135; 1987.

Kelly, D. L. Jr.; Britton, B. H.; Branch, C. L. Jr. Cooperative neuro-otologic management of acoustic neuromas and other cerebellopontine angle tumors. South. Med. J. 81:557–561; 1988.

Lassmann, G. Vascular dysplasia of arteries in neurocristopathies: a lesson for neurofibromatosis. Neurofibromatosis 1:281–293; 1988.

Laue, L.; Comite, F.; Hench, K.; Loriaux, D. L.; Cutler, G. B. Jr.; Pescovitz, O. H. Precocious puberty associated with neurofibromatosis and optic gliomas. Treatment with luteinizing hormone releasing hormone analogue. Am. J. Dis. Child 139:1097–1100, 1985.

Lederman, S. M.; Martin, E. C.; Laffey, K. T.; Lefkowitch, J. H. Hepatic neurofibromatosis, malignant schwannoma, and angiosarcoma in von Recklinghausen's disease. Gastroenterology 92:234–239; 1987.

Leslie, M. D.; Cheung, K. Y. P. Malignant transformation of neurofibromas at multiple sites in a case of neurofibromatosis. Postgrad. Med. J. 63:131–133; 1987.

Levy, W. J.; Latchaw, J.; Hahn, J. F.; Sawhny, B.; Bay, J.; Dohn, D. F. Spinal neurofibromas:

a report of 66 cases and a comparison with meningiomas. Neurosurgery 18:331–334; 1986.

Lewis, R. A.; Riccardi, V. M. Von Recklinghausen neurofibromatosis: prevalence of iris hamartomata. Ophthalmology 88:348–354; 1981.

Lewis, R. A.; Riccardi, V. M.; Gerson, L. P.; Whitford, R.; Axelson, K. A. Von Recklinghausen neurofibromatosis: II. Incidence of optic nerve gliomata. Ophthalmology 91:929–935; 1984.

Lewis, T. T.; Kingsley, D. P. Magnetic resonance imaging of multiple spinal neurofibromata—neurofibromatosis. Neuroradiology 29:562–564; 1987.

Lin, A. E.; Garver, K. L. Cardiac abnormalities in neurofibromatosis. Neurofibromatosis 1:146–152; 1988.

Listernick, R.; Charrow, J.; Greenwald, M. J.; Esterly, N. B. Optic glioma in children with neurofibromatosis type 1. J. Pediatr. 114:788–792; 1989.

Mallmann, R.; Roth, F. J. Treatment of neurofibromatosis associated renal artery stenosis with hypertension by percutaneous transluminal angioplasty. Clin. Exper. Theory Pract. 8:893–899; 1986.

Mann, H.; Kozic, Z.; Medinilla, O. R. Computed tomography of lambdoid calvarial defect in neurofibromatosis. Neuroradiology 25:175–176; 1983.

Martuza, R. L.; Eldridge, R. Neurofibromatosis 2 (Bilateral Acoustic Neurofibromatosis). N. Engl. J. Med. 318:684–688; 1988.

Martuza, R. L.; Ojemann, R. G. Bilateral acoustic neuromas: clinical aspects, pathogenesis, and treatment. Neurosurgery 10:1–12; 1982.

Match, R. M.; Leffert, R. D. Massive neurofibromatosis of the upper extremity with paralysis. J. Hand Surg. (Am) 12:718–722; 1987.

Mazzoni, A. Pitfalls in the diagnosis of acoustic neuroma. The ABR-CT protocol. Adv Otorhinolaryngol 37:91–92; 1987.

McAllister, H. A. Jr.; Fenoglio, J. J. Jr. Tumors of the cardiovascular system. In: Atlas of tumor pathology, Second Series, Fascicle 15. Washington, D.C.: Armed Forces Institute of Pathology, 1978: p. 70–71.

McCormick, P. C.; Bello, J. A.; Post, K. D. Trigeminal schwannoma. Surgical series of 14 cases with review of the literature. J. Neurosurg. 69:850–860; 1988.

Messner, R. L.; Messner, M. R.; Lewis, S. J. Neurofibromatosis: a familial and family disorder. J. Neurosurg. Nurs. 17:221–229; 1985.

Messner, R. L.; Smith, M. N. Neurofibromatosis: relinquishing the masks; a quest for quality of life. J. Adv. Nurs. 11:459–464; 1986.

Meyer, G. W.; Griffiths, W. J.; Welsh, J.; Cohen, L.; Johnson, L.; Weaver, M. J. Hepato-biliary involvement in von Recklinghausen's disease. Ann. Intern. Med. 97:722–723; 1982.

Miller, M.; Hall, J. G. Possible maternal effect on severity of neurofibromatosis. Lancet 2:1071–1074; 1978.

Miller, N. R. Optic nerve glioma and cerebellar astrocytoma in a patient with von Reck-linghausen's neurofibromatosis. Am. J. Oph-thalmol. 79:582–588; 1975.

Nakamura, H.; Koga, M.; Sato, B.; Noma, K.; Morimoto, Y.; Kishimoto, S. Von Reck-linghausen's disease with pheochromocytoma and nonmedullary thyroid cancer. Ann. Intern. Med. 105:796–797; 1986.

Neiman, H. L.; Mena, E.; Holt, J. F.; Stern, A. M.; Perry, B. L. Neurofibromatosis and congenital heart disease. Am. J. Roentgenol. 122:146–149; 1974.

O'Connell, P.; Leach, R.; Cawthon, R. M.; Culver, M.; Stevens, J.; Viskochil, D.; Four-nier, R. E. K.; Rich, D. C.; Ledbetter, D. H.; White, R. Two NF-1 translocations map within a 600-kilobase segment of 17q11.2. Science 244:1087–1088; 1989.

O'Neill, P.; Whatmore, W. J.; Booth, A. E. Spi-nal meningoceles in association with neurofi-bromatosis. Neurosurgery 13:82–84; 1983.

Packer, R. J.; Bilaniuk, L. T.; Cohen, B. H.; Braffman, B. H.; Obringer, A. C.; Zimmer-man, R. A.; Siegel, K. R.; Sutton, L. N.; Savino, P. J.; Zackai, E. H.; Meadows, A. T. Intracranial visual pathway gliomas in children with neurofibromatosis. Neurofibromatosis 1:212–222; 1988.

Packer, R. J.; Sutton, L. N.; Bilaniuk, L. T.; Radcliffe, J.; Rosenstock, J. G.; Siegel, K. R.; Bunin, G. R.; Savino, P. J.; Bruce, D. A.; Schut, L. Treatment of chiasmatic/hypotha-lamic gliomas of childhood with chemother-apy: an update. Ann. Neurol. 23:79–85; 1988.

Pearson-Webb, M. A.; Kaiser-Kupfer, M. I.; Eldridge, R. Eye findings in bilateral acoustic (central) neurofibromatosis: Association with presenile lens opacities and cataracts, but ab-sence of Lisch nodules. N. Engl. J. Med. 315:1553–1554; 1986.

Pickard, L. R.; Rose, J. E. Avoidable complica-tions of resection of major nerve trunk neurofi-bromas and schwannomas. Neurofibromatosis 1:43–49; 1988.

Pollack, M. A.; Shprintzen, R. J. Velopharyn-geal insufficiency in neurofibromatosis. Int. J. Pediatr. Otorhinolaryngol. 3:257–262; 1981.

Pollard, S. G.; Hornick, P.; Macfarlane, R.; Caine, R. Renovascular hypertension in neuro-fibromatosis. Postgrad. Med. J. 65:31–33; 1989.

Porterfield, J. K.; Pyeritz, R. E.; Traill, T. A. Pulmonary hypertension and interstitial fibro-sis in von Recklinghausen neurofibromatosis. Am. J. Med. Genet. 25:531–535; 1986.

Powell, P. P. An overview of childhood Von Recklinghausen neurofibromatosis for parents. Neurofibromatosis 1:50–53; 1988a.

Powell, P. P. Schematic representation of von Recklinghausen neurofibromatosis (NF-1): an aid for patient and family education. Neurofi-bromatosis 1:164–165; 1988b.

Purcell, S. M.; Dixon, S. L. Schwannomatosis: an unusual variant of neurofibromatosis or a distinct clinical entity? Arch. Dermatol. 125:390–393; 1989.

Raney, B.; Schnaufer, L.; Ziegler, M.; Chatten, J.; Littman, P.; Jarrett, P. Treatment of chil-dren with neurogenic sarcoma. Experience at the Children's Hospital of Philadelphia, 1958–1984. Cancer 59:1–5; 1987.

Reed, D.; Robertson, W. D.; Rootman, J.; Douglas, G. Plexiform neurofibromatosis of the orbit: CT evaluation. AJNR 7:259–263; 1986.

Reynolds, R. L.; Pineda, C. A. Neurofibromato-sis: review and report. J. Am. Dent. Ass. 117:735–737; 1988.

Riccardi, S. L.; Powell, P. P.; Riccardi, V. M. Documentation of excessive frequency of neurofibrosarcoma (NFS) in NF-1. Am. J. Hum. Genet. 43:A31; 1988.

Riccardi, V. M. Von Recklinghausen neurofibro-matosis. N. Engl. J. Med. 305:1617–1627; 1981.

Riccardi, V. M. Neurofibromatosis: clinical het-erogeneity. Curr. Probl. Cancer 7(2):1–34; 1982.

Riccardi, V. M. Neurofibromatosis as a model for investigating hereditary vs. environmental factors in learning disabilities. The developing brain and its disorders. Tokyo: University of Tokyo Press; 1984.

Riccardi, V. M. Mast cell stabilization to de-crease neurofibroma growth: preliminary ex-perience with ketotifen. Arch. Dermatol. 123:1011–1016; 1987.

Riccardi, V. M. Guidelines for organizing a com-prehensive neurofibromatosis program. Neu-rofibromatosis 1:105–119; 1988.

Riccardi, V. M. Neurofibromatosis: a spectrum of disorders. In: Wetterberg, L., ed. Genetics of neuropsychiatric disease. London; MacMil-lan Press; 1989a: p. 235–248.

Riccardi, V. M. Trauma and wound-healing in the pathogenesis of birth defects. Proc. Green-wood Genet. Ctr. 8:152–153; 1989b.

Riccardi, V. M.; Eichner, J. E. Neurofibromato-sis: phenotype, natural history, and pathogen-esis. Baltimore: Johns Hopkins University Press, 1986.

Riccardi, V. M.; Elder, D. W. Multiple cytoge-netic aberrations in neurofibrosarcomas complicating neurofibromatosis. Cancer Genet. Cytogenet. 23:199–209; 1986.

Riccardi, V. M.; Haeberlin, V. Disjoining of height and head circumference in patients with NF-1: implications for CNS pathogenesis. Am. J. Hum. Genet. 44:A60; 1989.

Riccardi, V. M.; Lewis, R. A. Penetrance of von Recklinghausen neurofibromatosis: A distinction between predecessors and descendants. Am. J. Hum. Genet. 42:284–289; 1988.

Riccardi, V. M.; Wald, J. S. Discounting an adverse maternal effect on neurofibromatosis severity. Pediatrics 79:386–393; 1987.

Riviello, J. J.; Marks, H. G.; Lee, M. S.; Mandell, G. A. Aqueductal stenosis in neurofibromatosis: a report of two cases. Neurofibromatosis 1:312–317, 1988.

Roback, H. B.; Kirshner, H.; Roback, E. Physical self-concept changes in a mildly facially disfigured neurofibromatosis patient following communication skill training. Int. J. Psychiatr. Med. 11:137–143; 1981.

Roenigk, R. K.; Ratz, J. L. Carbon dioxide laser treatment of cutaneous neurofibromas. J. Dermatol. Surg. Oncol. 13:187–190; 1987.

Rouleau, G. A.; Wertelecki, W.; Haines, J. L.; Hobbs, W. J.; Trofatter, J. A.; Seizinger, B. R.; Martuza, R. L.; Superneau, D. W.; Conneally, P. M.; Gusella, J. F. Genetic linkage of bilateral acoustic neurofibromatosis to a DNA marker on chromosome 22. Nature 329:246–248; 1987.

Samuelsson, B.; Samuelsson, S. Neurofibromatosis in Gothenburg, Sweden. I. Background, study design, and epidemiology. Neurofibromatosis 2:6–22; 1989.

Samuelsson, B.; Riccardi, V. M. Neurofibromatosis in Gothenburg, Sweden. II. Intellectual compromise. Neurofibromatosis. 2:78–83; 1989a.

Samuelsson, B.; Riccardi, V. M. Neurofibromatosis in Gothenburg, Sweden. III. Psychiatric and social aspects. Neurofibromatosis. 2:84–106; 1989b.

Seizinger, B. R.; Martuza, R. L.; Gusella, J. F. Loss of genes on chromosome 22 in tumorigenesis of human acoustic neuroma. Nature 322:644–647; 1986.

Shishiba, T.; Niimura, M.; Ohtsuka, F.; Tsuru, N. Multiple cutaneous neurilemmomas as a skin manifestation of neurilemmomatosis. J. Am. Acad. Dermatol. 10:744–754; 1984.

Sorensen, S. A.; Mulvihill, J. J.; Nielsen, A. Nation-wide follow-up of Recklinghausen neurofibromatosis: Survival and malignant neoplasms. N. Engl. J. Med. 314:1010–1015; 1986.

Stack, J. P.; Ramsden, R. T.; Antoun, N. M.; Lyle, R. H.; Isherwood, I; Jenkins, J. P. Magnetic resonance imaging of acoustic neuromas: The role of gadolinium-DTPA. Br. J. Radiol. 61:800–805; 1988.

Stark, P.; Cheng, G. J.; Hildebrandt-Stark, H. E. Pulmonary parenchymal and pleural fibrosis as an expression of Recklinghausen's neurofibromatosis. Radiologe 28:231–232; 1988.

Stern, J. D.; DiGiacinto, G. V.; Housepian, E. M. Neurofibromatosis and optic glioma: clinical and morphological correlations. Neurosurgery 4:524–528; 1979.

Stern, J.; Jakobiec, F. A.; Housepian, E. M. The architecture of optic nerve gliomas with and without neurofibromatosis. Arch. Ophthalmol. 98:505–511; 1980.

Terada, H.; Mimaki, T.; Takiyama, N.; Tagawa, T.; Tanaka, J.; Itoh, N.; Yabuuchi, H. A case of infantile spasms with multiple neurofibromatosis. Brain Dev. 8:145–147, 1986.

Tertsch, D.; Schon, R.; Ulrich, F. E.; Alexander, H.; Herter, U. Pubertas precox in neurofibromatosis of the optic chiasma. Acta. Neurochir. 28:413–415; 1979.

Varnhagen, C. K.; Lewin, S.; Das, J. P.; Bowen, P.; Ma, K.; Klimek, M. Neurofibromatosis and psychological processes. Develop. Behav. Pediatr. 9:257–265; 1988.

Verhest, A.; Heimann, R.; Verschraegen, J.; Vamos, E.; Hecht, F. Hereditary intestinal neurofibromatosis. II. Translocation between chromosomes 12 and 14. Neurofibromatosis 1:33–36; 1988.

Weiss, L.; Sagerman, R. H.; King, G. A.; Chung, C. T.; Dubowy, R. L. Controversy in the management of optic nerve glioma. Cancer 59:1000–1004; 1987.

Wertelecki, W.; Rouleau, G. A.; Superneau, D. W.; Forehand, L. W.; Williams, J. P.; Haines, J. L.; Gusella, J. F. Neurofibromatosis 2: Clinical and DNA linkage studies of a large kindred. N. Engl. J. Med. 319:278–283; 1988.

Yaghmai, I. Spine changes in neurofibromatosis. Radiographics 6:261–285; 1986.

32

Progressive Genetic–Metabolic Diseases

ISABELLE RAPIN

Many neurologists view genetic–metabolic and degenerative diseases of the brain, most of which ultimately have a fatal outcome, as a discouraging part of their field. The number of different diseases and variants of diseases identified is large and growing. The number of diseases in which the molecular biology is understood is exploding, but this number remains much smaller than the diseases in which it is not. Furthermore, in most cases, the lag between elucidation of the biology of a disease and devising a rational and effective treatment is frustratingly long.

This chapter does not attempt to discuss all of these conditions, many of which are diseases of infants and young children that are well covered in recent textbooks of neurology and child neurology (Adams and Lyon 1982; Baraitser 1989; Menkes 1989; Rowland 1989; Swaiman 1989). Nyhan and Sakati's (1987) Diagnostic Recognition of Genetic Disease is an excellent source, especially concerning disorders of amino acid and organic acid metabolism.

Recent advances in genetics have greatly complicated nosology. For example, inac-

tivity of an enzyme may result from any of several allelic point mutations, all of which result in the synthesis of an inactive or less active enzyme or, less often, in the failure of its synthesis. Enzyme activity also may result from the lack of a stabilizing or activator protein. Readers are referred to Mc-Kusick (1990) for up-to-date information concerning the genetics of particular diseases, including brief clinical sketches and pertinent literature references, and to Scriver and coworkers (1989) for detailed biochemical information.

The tables in this chapter are lists of diseases and their prognoses, with no attempt to cover their clinical or biochemical basis. Information about prognosis provided in this chapter should be read with caution; what is believed to be true today is likely to require modification tomorrow.

The author intends to focus on diseases for which there is a specific and effective treatment and to mention some symptomatic and palliative treatments that affect prognosis and longevity of many other diseases because patients can now be maintained almost indefinitely in a vegetative state. These palliative treatments raise many ethical questions about the appropriateness of therapeutic interventions.

Factors That Influence Prognosis

The pace of the illness, the extent of disease in the nervous system, and the severity of its systemic manifestations are important factors influencing prognosis. In general, diseases that are severe enough to cause symptoms at birth or in infancy and for which no treatment is available tend to have the most dismal prognosis and reduce the child to a vegetative state soonest. The endpoint of disease is probably best thought of as regression to a vegetative state because with the use of antibiotics to treat infections, feeding gastrostomies with fundal plication, and meticulous pulmonary toilet it is possible to maintain many patients in a vegetative state for a decade or more.

It is relatively easy to prognosticate in the case of a second affected child in a family because usually recessive disease runs a fairly predictable course among siblings. Because genetic heterogenity is so prevalent, the utmost caution must be exercised when providing prognostic information to patients and their families if the patient does not present with the most classical clinical picture. Heterogeneity highlights the critical importance of obtaining fibroblasts from patients for later study and securing autopsies in undiagnosed patients because this often leads to the discovery of new unsuspected diseases.

Symptomatic and Palliative Treatments

Health Maintenance

Adequate nutrition and general care regularly determine the longevity of seriously affected persons. Some of them cannot be maintained at home and are cared for in institutional settings. The quality of care in these facilities has clearly improved as a result of widely publicized scandals. Fewer patients develop decubitus ulcers, they receive more habilitative services, and their hygiene and nutrition are superior to what

they typically were in the past (Rubin and Crocker 1989). There is a strong trend toward deinstitutionalization in an attempt to relieve crowding and save resources and as a result of greater recognition of the rights of the handicapped for an optimal quality of life. Even severely affected persons, especially when they are children, are maintained at home, often with the help of visiting nurses and the provision of needed appliances. Many of the patients are bused to day programs in the community where attempts are made to occupy them meaningfully, teach them self-help skills, provide them with recreation and socialization, and relieve their caretakers. The nutrition of home-based patients tends to be better than that of the institutionalized, not only because devoted family members spend adequate time feeding them, but because feeding by nasogastric tube or gastrostomy, with or without fundal plication to prevent aspiration, is regularly offered when oral feeding becomes too difficult. Physiotherapy used to minimize joint contractures and increased weight bearing, even strapped to a board, have decreased the prevalence of spontaneous fractures, a significant hazard for immobilized patients. Regular health maintenance visits to a physician is another measure that has had favorable effects on longevity. Whether longevity is a blessing for patients who have reached a vegetative state and for their families is a debatable issue.

Treatment of Intercurrent Illnesses

Most patients with genetic–metabolic diseases of the brain do not die of the disease but of an intercurrent illness. This used to be pneumonia or some other unrecognized infectious illness. Availability and liberal prescription of broad-spectrum antibiotics for nonspecific febrile episodes has minimized infectious causes of death; however, aspiration pneumonia remains a significant and often lethal hazard.

Treatment of Seizures

Many diseases, especially those affecting the gray matter and certain subcortical nuclei, carry a substantial risk of seizures and myoclonus. Vigorous treatment of these seizures with appropriate drugs, monitoring of anticonvulsant blood levels, and avoidance of oversedation and therapy with multiple drugs have substantially altered prognosis, although death during unobserved status epilepticus remains fairly common.

Treatment of Hydrocephalus

Some diseases, in particular Hurler's disease (mucopolysaccharidosis [MPS] type I), some of the other MPSs, and Alexander's disease, cause true hydrocephalus, as opposed to hydrocephalus *ex vacuo,* by producing aqueductal stenosis or involvement of the leptomeninges. Treatment of the hydrocephalus improves the quality of the patient's life by increasing alertness and regaining or preventing loss of some motor and cognitive skills compromised by hydrocephalus.

Treatment of Spasticity and Musculoskeletal Deformities

Many of the diffuse genetic metabolic diseases of the brain produce increasingly severe spasticity. Immobility results in joint contractures. Drugs, such as baclofen and diazepam, have a modestly beneficial effect on spasticity, more so than dantrolene, which tends to increase weakness. Diazepam often has a sedative effect as well, which may or may not be desirable. Physiotherapy and passive range of motion of joints retards the development of painful and fixed joint contractures but cannot prevent them altogether. Surgery to relieve some of these contractures, to improve mobility or facilitate perineal care, and to decrease lung compression by scoliosis may be justified in patients with slowly progressive disease. Awareness and in some cases surgical correction of potentially lethal skeletal complications, such as atlanto-occipital instability and cord compression in some of the MPSs may greatly affect prognosis.

Treatment of Pain

Most of the metabolic diseases of the nervous system do not produce pain, which should be pointed out to patients' families. The lancinating pain of the neuropathy of Fabry's disease responds to phenytoin and carbamazepine. Some severely spastic vegetative patients who lie in opisthotonos may be fretful and seem to gain some relief from muscle relaxants. Passive range of movement retards the development of painful joint contractures, and frequent turning helps prevent skin decubiti.

Treatment of Psychic Pain

Families of progressively neurologically impaired people of all ages, and nondemented affected older children and adults, experience anguish and often are chronically depressed. It is critical to provide them with emotional support, including psychiatric counseling when indicated; practical help in caring for the affected family member; and free time, for example with respite home attendants or nursing home placement. Introducing families with similar problems to one another and to support groups often is particularly helpful. One must also remember to help families cope with the financial demands of the illness.

With illness for which there is no treatment, admitting ignorance and bringing families up-to-date with new scientific developments often helps to alleviate their pain and frustration. Parent groups, such as those of patients with leukodystrophies, mucopolysaccharidoses, Rett syndrome, and others, have not only raised money for research, but have disseminated information about their rare disease and have goaded investigators to work on it. They have facilitated research by agreeing to undergo skin biopsies and blood tests and by helping to recruit new patients into patient registries. Parents often are the most militant proponents of experimental treatments, such as bone marrow transplantation or dietary trials, and of postmortem examination to provide tissue for further study.

Specific Treatments

Dietary Manipulation

Rarely, avoidance of a nutrient that cannot be metabolized, together with its increased excretion, provision of pharmacological doses of a cofactor (such as a vitamin) to

Table 32-1. Diseases for Which There is an Effective Dietary Treatment

Disease	Treatment	Effectiveness*
Phenylketonuria (PKU)	Phenylalanine restriction	Satisfactory
Atypical PKU (several types)	Phenylalanine restriction + B_{12}, levodopa, 5-hydroxytryptophan	Fair to satisfactory
Maple syrup urine disease (classic and variants)	Branched chain amino acid restriction, thiamine	Fair
Homocystinuria (several variants)	Methionine restriction, pyridoxine, cobalamin	Fair to satisfactory
Hartnup	Nicotinic acid	Satisfactory
Lowe's	Vitamin D + alkali	Fair
Organic acidurias	Protein restriction	Poor to satisfactory
Methylmalonic acidemia, B_{12} responsive	Cobalamin	Satisfactory
Multiple carboxylase deficiency	Biotin	Satisfactory
Biotinidase deficiency	Biotin	Satisfactory
Urea cycle defects	Protein restriction + Na benzoate and phenylacetate, arginine	Fair
Galactosemia	Galactose restriction	Satisfactory
Glycogenosis type II	Continuous feeding, cornstarch	Fairly good
Pyruvate dehydrogenase deficiency	Thiamine, lipoic acid	Fair
Pyridoxine dependency	Pyridoxine	Satisfactory
Abetalipoproteinemia	Vitamin E	Satisfactory
Refsum	Phytanic acid restriction + plasmapheresis	Satisfactory
Adrenoleukodystrophy	Glycerol trioleate + restriction of VLCFA†	Still experimental
Cerebrotendinous xanthomatosis	Chenodeoxycholic acid	Promising
Wilson's	Copper restriction and chelation	Satisfactory

* Results of treatment depend on the genetic variant, the age at which therapy was started, and compliance.
† VLCFA = Very long chain fatty acids.

accelerate or bypass a defective reaction, or a combination of such approaches has totally altered prognosis (Table 32-1). In some cases, for example, restriction of phenylalanine in phenylketonuria (PKU), and restriction of dietary copper and copper chelation in Wilson's disease, treatment is so successful that it prevents the appearance of the disease altogether, provided a diagnosis is made before irreversible damage has taken place (Scheinberg and Sternlieb 1984). Widespread neonatal testing for treatable disorders of amino acid metabolism, galactosemia, biotinidase deficiency, and congenital hypothyroidism has dramatically decreased the prevalence of these causes of severe neurological morbidity. Physician education has heightened awareness of acute lethal disorders causing hyperammonemia and metabolic acidosis in neonates. Prognosis has improved for some of the urea cycle defects by combining a low-protein diet with administration of sodium benzoate and phenylacetate or citrulline to promote the excretion of nitrogeneous waste using an alternate pathway to the urea cycle. Modest success has been achieved in some affected patients, although many of the survivors display subnormal intelligence. Much greater success has been achieved in symptomatic female carriers of ornithine transcarbamylase deficiency (Rowe et al. 1986). Prenatal treatment, by providing large doses of cobalamine to mothers known to carry a fetus with vitamin B_{12}-responsive methylmalonic aciduria, can prevent the birth of an already severely affected child (Ampola et al. 1975).

Enzyme Replacement Therapy

This approach, which is still experimental, depends on the delivery of sufficient amounts of a missing enzyme to the affected tissue. Genetic engineering has brightened the prospect for synthesis of some enzymes in therapeutically realistic amounts (Friedman 1989). Problems such as rapid turnover, the development of antibodies to the enzyme, and the need for particular carbohydrate markers to target the enzyme to specific tissues must be addressed. Enzyme replacement therapy for diseases of the nervous system is further complicated by the blood–brain and blood–nerve barriers. Experimental attempts to overcome this problem have included infusion of hyperosmolar agents to open the barrier temporarily, encasement of quanta of the enzyme in lipid-soluble coats (liposomes), and infusion of enzyme directly into the cerebrospinal fluid (CSF) using an indwelling catheter with a subcutaneous refillable reservoir. These techniques have not yet achieved clinical application.

Transplantation

Another approach to enzyme replacement is the engrafting of enzyme-producing tissue. This costly and demanding procedure, which carries a significant morbidity because of the requirement for long-term immunosuppression, may be justified for diseases that have not yet produced severe cognitive impairment. It also may be justified on a research basis in more advanced cases, if patient and family give their informed consent. It is unlikely to become a viable option unless significant progress is achieved (Krivit and Paul 1986).

Liver transplantation has been life-saving in a few patients with Wilson's disease presenting with fulminant liver failure. These patients are among the best candidates for transplantation because often they have not yet developed central nervous system (CNS) damage and because there is an effective treatment for Wilson's disease. They still require a copper-restricted diet and copper-chelating agents, so grafting represents organ replacement and not enzyme replacement in this disease (Rakela et al. 1986). This may not be true is in Fabry's disease although results have been variable. Renal failure is alleviated, but the new kidney does not always provide adequate enzyme replacement and eventually may become diseased as well. However, prolonged survival with improved quality of life has been achieved in some patients (Farragiana et al. 1981; Friedlaender et al. 1987).

Bone marrow transplantation from a histocompatible donor, often a family member, and peritoneal transplantation of fetal membranes have been tried in Gaucher's disease, the metachromatic and Krabbe leukodystrophies, MPS, and other metabolic diseases (Krivit and Paul 1986). This demanding approach is attractive for systemically symptomatic diseases and has had modest success even in some diseases affecting the brain. Bone marrow grafting does not overcome the problem of the blood–brain barrier unless enzyme-producing macrophages or other enzymatically competent cells penetrate the leptomeninges and brain and produce sufficient enzyme to restore function and prevent irreversible neurological damage. Transplantation is unlikely to be effective in conditions such as infantile Tay-Sachs disease in which damage to the brain is already substantial before 20 weeks' gestation, and in which enzyme replacement would have to be provided prenatally to improve prognosis.

Before recommending grafting for metabolic diseases, it is important to weigh the high financial cost, burden of suffering for the child, and questionable efficacy of transplants in mitigating CNS disease when faced with the "sobering morbidity and mortality rates" of bone marrow transplantation (Clark 1990; Rappeport and Ginns 1984). Transplantation must be considered experimental; it is being evaluated in a few specialized centers. It has been publicized widely, and unfortunately, this has raised false hopes for parents to severely affected patients. Table 32-2 summarizes available information on some diseases for which transplantation has been tried, and it shows the results achieved so far.

Molecular Approaches

Recent research on genetic–metabolic diseases has progressed rapidly from ultrastructural and histochemical identification of the affected substrate to its chemical isolation, elucidation of its metabolic pathway, and discovery of the missing enzyme responsible for the disease. Molecular biology now makes it possible to identify the specific DNA defect responsible for inactivity of an enzyme. What was thought to be a single disease, for example classical infantile Tay-

Table 32-2. Transplantation Results

Disease	Type of Transplantation	Results
Fabry	Kidney	Successful in some but not all cases
Niemann-Pick type A	Liver	Ineffective
Niemann-Pick type B	Bone marrow, amniotic membranes	Decreased visceral storage
Swedish neuronopathic Gaucher	Bone marrow	Decreased visceral storage, no effect on CNS?
Hurler, Hunter, Sanfilippo	Bone marrow	Visceral and joint improvement, CNS unimproved or stabilized?
Morquio's, Maroteaux-Lamy	Bone marrow	Modest visceral and joint improvement
Adrenoleukodystrophy	Bone marrow	Ineffective?
Wilson's disease	Liver	Effective in acute liver failure

Sachs disease, turns out to comprise at least four genetically distinct conditions. In addition, there are a dozen other phenotypic variants of GM_2 gangliosidosis caused by other molecular defects affecting hexosaminidases A and B (see below). Correcting the defect at the DNA level by using reverse transcriptase to introduce the correct DNA sequence into the genome has been accomplished in tissue culture and tried in animals. Whether it can be used to prevent or correct human disease remains to be determined.

The Lysosomal Storage Diseases

Lysosomes are cellular organelles with a very acid pH, in which catabolism of many cellular constituents takes place. Lack of one or more lysosomal enzymes results in the storage within lysosomes of undegradable products of metabolism. This storage may result in swelling and distortion of the geometry and function of cells or cell death. In some cases, for example infantile Gaucher disease and Krabbe disease, there also is an accumulation of products toxic to neurons when the block of a pathway activates an alternate degradative pathway. One of the prototypical groups of these disorders, the glycogen storage diseases, is not discussed here because they are covered in the chapter on muscle diseases.

There is no truly effective treatment for the neurological manifestations of these diseases, but prenatal diagnosis and carrier detection are available for most of the lysosomal diseases. Symptomatic treatment must not be neglected (e.g., seizures in many diseases, hydrocephalus in Hurler's and Alexander's diseases, spinal instability in Morquio's disease, and painful crises in Fabry's disease using phenytoin and carbamazepine).

The Sphingolipidoses

Table 32-3 belies the complexity of these disorders (Johnson 1987). For example, there are at least three enzymatic variants of infantile GM_2 gangliosidosis: classic infantile Tay-Sachs disease (lack of hexosaminidase A because of the lack of the hexosaminidase alpha chain coded on chromosome 15), Sandhoff disease (lack of hexosaminidases A and B because of lack of the hexosaminidase beta chain coded on chromosome 5), and the AB variant (hexosaminidase is present, but there is deficiency of an activator protein coded on chromosome 5). There are even more genetic than enzymatic variants because different point mutations may inactivate the same protein, for example, in classic Tay-Sachs disease of Ashkenazi Jewish and French Canadian infants (Myerowitz and Hogikyan 1986). Thus, enzymatic and genetic heterogeneity underlies the varied phenotypes associated with the storage of a particular glycolipid, in this case GM_2 ganglioside.

In general, the earlier the onset of the clinical disease, the shorter and the more severe its course. This correlation is far from perfect, however. For example, siblings from an Ashkenazi Jewish family with GM_2

Table 32-3. Sphingolipidoses

Disease	Variants	Prognosis
GM$_2$ gangliosidosis	Infantile (Tay-Sachs disease)	Death in the preschool years
	Juvenile	Death in adolescence
	Adolescent and adult	Survival for variable number of years
	Chronic	Survival for decades
GM$_1$ gangliosidosis	Infantile	Death in late infancy
	Childhood	Death in adolescence
	Adult	Variable (several mutants probable)
	Dystonic	Survival into young adulthood
Fabry disease		Death in adulthood
Niemann-Pick disease	Group A (infantile)	Death in the preschool years
	Group C (heterogeneous)	Varies from early death to prolonged survival
	Group D (Nova Scotia)	Death in the second decade
	Sea-blue histiocyte (?)	Variable (heterogeneous)
Gaucher disease	Infantile	Death in infancy
	Juvenile neuronopathic	Death in childhood or adolescence
Farber disease		Death in late infancy or early childhood
Krabbe disease	Infantile	Death in infancy
	Late-onset forms	Death after several years
Metachromatic leukodystrophy	Late infantile	Death in the school years
	Juvenile	Death in adolescence
	Adult	Death after protracted course
	Multiple sulfatase deficiency	Death in early childhood

gangliosidosis became symptomatic in the preschool years, but two of three affected patients are in their 40s, have normal intelligence without seizures or retinal involvement, despite severe signs of motor, sensory, basal ganglia, and cerebellar involvement. There are even a few asymptomatic (possibly presymptomatic) adult carriers of hexosaminidase deficiency. Prognosis is difficult in adult variants of the sphingolipidoses because current knowledge consists mostly of single case reports of unusual phenotypes. Not all the sphingolipidoses produce a dementing illness. Some variants of the gangliosidoses resemble motor neuron disease; others resemble dystonia; and adult metachromatic leukodystrophy masquerades as schizophrenia. This variability highlights the need to consider unsuspected metabolic disorders in patients with even very slow but unexplained neurological regression and dictates caution in offering prognostic information.

Early attempts at enzyme replacement, even injected directly into the CSF, have been ineffective (Desnick et al. 1973). Grafting, despite its theoretical appeal and high visibility, has not yet lived up to its reputa-

tion, with the exception of Fabry's disease in which improved renal function has followed kidney transplantation in some cases (Bannwart 1982; Clement et al. 1982; Farragiana et al. 1981). Success with the Swedish neuronopathic variant of Gaucher's disease (Ringden et al. 1988) and slowing of the expected progression of disease in a patient with late infantile metachromatic leukodystrophy (Krivit et al. 1990) are encouraging.

No effective treatment has been devised for any of these disorders, but prevention by prenatal diagnosis and carrier detection in high-risk Ashkenazi groups has greatly decreased the number of infants with Tay-Sachs disease. It is available to families of diagnosed probands with other diseases.

The Mucopolysaccharidoses

The MPSs (Table 32-4) are due to the lack of cleavage in lysosomes of sulfated sugars from acid mucopolysaccharides or glycosaminoglycans. Patients excrete a variety of MPS degradation products in their urine, and their enzymatic deficiencies are well understood. As a group, they are characterized by prominent involvement of connective tissue, including bones, cartilage, heart valves,

Table 32-4. Mucopolysaccharidoses

Type	Eponym (Syndrome)	Age of Onset	Prognosis
IH	Hurler	Infancy	Death in first decade
I H/S	Hurler-Scheie	Childhood	Survival to adulthood
I S	Sheie	Adulthood	Survival for decades
II A	Severe Hunter	Early childhood	Death in adolescence
II B	Mild Hunter	Preschool	Survival to adulthood
III A–D	Sanfilippo	Early childhood	Death in teens or early adulthood
IV A and B	Morquio	Early childhood	Survival to teens or adulthood
VI A	Maroteaux-Lamy	Early childhood	Death in teens or early adulthood
VI B and C	Maroteaux-Lamy	Later childhood	Survival for decades
VII	Sly	Early childhood	Survival to adulthood

ligaments, and in some variants, the cornea and leptomeninges. There is prominent hepatosplenomegaly in Hurler disease, while visceral involvement is less striking in other variants. Accumulation of gangliosides in neurons, including retinal and cochlear neurons, contributes to the dementia, visual and hearing loss of Hurler, severe Hunter's, and Sanfilippo syndromes. Compression of the spinal cord in Maroteaux-Lamy and Morquio diseases and entrapment of peripheral nerves because of storage of MPS in ligaments are prominent in many variants. Hydrocephalus, especially common in Hurler and Maroteaux-Lamy diseases, is due to meningeal storage.

The cause of death often is heart failure, especially in Hurler's disease. It may be respiratory failure due in part to thickening of the walls of the airway in Hunter's disease. Sudden death due to compression of the cervical cord by an unstable odontoid peg occurs in Morquio disease and may take place during anesthesia in any of the MPSs if special precautions are not taken.

The drastic differences in phenotype and prognosis resulting from allelic mutations responsible for Hurler and Scheie diseases are due to different effects of these mutations on the properties of the product protein, the enzyme L-iduronidase. In contrast, there are at least four enzymatically and genetically distinct but clinically similar variants of Sanfilippo syndrome, all of which result in accumulation of the same substrate; they cause less severe connective tissue involvement than Hurler disease, the most severe variant of all the MPSs.

Aggressive treatment of symptoms, such as hydrocephalus, nerve entrapments, and spinal problems, and the wearing of hearing aids enhances the quality of life of patients with MPS. Grafting of fibroblasts has not been proven useful (Gibbs et al. 1983). Bone marrow transplantation improves systemic signs, but so far has had only modest effects on the signs of CNS compromise. Recommendation for early transplantation in nondemented children who suffer from a severe variant may be reasonable. The morbidity and mortality and the cost of transplantation are so high that this is unreasonable in milder variants. Much more time is required before one can reasonably discuss the cost–benefit ratio of this treatment modality.

The Mucolipidoses

This group of diseases is characterized by the storage of glycoproteins in various tissues, while, as is the case in the MPS, glycolipids accumulate in neurons (Table 32-5).

The sialidoses are a heterogeneous group of diseases that range from rapidly lethal congenital forms with ascites and renal failure to the chronic cherry-red spot-myoclonus syndrome, which allows survival to middle age without dementia but has a devastating and virtually untreatable myoclonus. One of the variants, in which there is lack of both sialidase and beta galactosidase, produces a dysostosis reminiscent of MPS. This form, which is especially preva-

Table 32-5. Mucolipidoses

Type	Variants	Age of Onset	Prognosis
Sialidosis	Congenital	Birth	Death in early infancy
	Severe infantile	Birth	Death in late infancy
	Nephrosialidosis	Infancy	Death in childhood
	Mucolipidosis I	Infancy	Death in childhood
	Cherry-red spot-myoclonus	School-age	Survival to adulthood
Galactosialidosis	Infantile	Infancy	Death in infancy
	Japanese–Goldberg	Childhood	Survival to adulthood
Mucolipidosis II	I-cell	Birth	Death in infancy or early childhood
Mucolipidosis III	Pseudopolydystrophy	Early childhood	Survival to adulthood
Mucolipidosis IV		Infancy	Survival to adulthood
Sala disease		Infancy	Survival to adulthood
Fucosidosis	Severe (I and II)	Infancy	Death in childhood
	Mild (III)	Childhood	Survival to early adulthood
Alpha-mannosidosis	Several types	Early childhood	Variable, death in childhood to survival to adulthood
Beta-mannosidosis		Childood	Survival to adulthood
Aspartylglucosaminuria		Early childhood	Survival to adultood

lent in Japan, is crippling but allows survival into adulthood. Similar to Fabry's disease and one of the variants of fucosidosis, the skin is involved, with a characteristic angiokeratomatous rash in the inguinal region.

Mucolipidoses (ML) II and III were confused at first with the MPS, which they resemble because of their clinical features and by the presence of inclusions in fibroblasts, but they differ because of the lack of MPS in the urine. These two diseases are due to deficiency of enzymes that attach phosphomannosyl residues which are required for a number of cytosolic enzymes to enter lysosomes. Several distinct mutations have been found to produce each of these diseases, which are therefore genetically heterogeneous. Prognosis is dismal in ML II, but ML III is much less serious.

ML IV is a disease of Ashkenazi Jews in which clouding of the cornea presents early but for unclear reasons, fluctuates in intensity. There is severe mental and motor defi-

ciency but no clinical storage in the viscera or the skeleton. It has a protracted course and may be missed because it is mistaken for undifferentiated mental deficiency.

There is no specific treatment for any of these diseases. Again, bone marrow transplantation has been tried but in too few patients to enable an appraisal of its efficacy. Prenatal diagnosis and carrier detection are possible for most.

The Ceroid Lipofuscinoses

Table 32-6 summarizes three well-defined childhood variants of Batten disease. There is probably more than one adult variant (Berkovic et al. 1988), and there are many single cases or small series that do not fit into the classical childhood types. Prominent macular degeneration is regularly the presenting sign of the juvenile variant, in which signs of dementia, motor deficit, and seizures may not appear for 4 to 6 years after loss of central vision. Blindness also

Table 32-6. Ceroid Lipofuscinoses

Variants	Age of Onset	Prognosis for Survival	Severity of Illness
Infantile (Finnish)	Infancy	Death in first decade	Devastated in late infancy
Late infantile	Preschool	Death in first decade	Devastated in 2–3 years
Adolescent	Early school-age	Death in teens or 20s	Devastated in late teens
Adult	Adulthood	Death within a decade	Severe

characterizes the infantile and late infantile variants, which usually are fatal before the end of the first decade. Blindness is not a feature of adult cases. The late infantile type is characterized by disabling myoclonus, while the infantile type, which is particularly frequent in Finland, produces such severe neuronal devastation that the cortex becomes virtually aneuronal and the EEG flat within 3 to 4 years. Patients with the juvenile variant may remain ambulatory and able to attend school until their late teens, although perhaps one fourth die in their teens after a more rapidly dementing course with prominent seizures.

Diagnosis often can be suspected clinically, but a rational classification awaits identification of the enzymatic deficit, which is not known in any of these disorders. They are tentatively classified among the lysosomal diseases because of the membrane-bound fluorescent inclusions found in neurons and many peripheral tissues, including muscle and skin. Diagnosis is helped by the ultrastructural appearance of these inclusions, which, although not pathognomonic, tend to be granular in the infantile variant, curvilinear in the late infantile variant, and have a fingerprint pattern in the juvenile variant.

Prenatal diagnosis based on the finding of inclusions in chorionic villi and amniocytes has been accomplished in the infantile and late infantile variants. Unfortunately, often a family is complete before the oldest af-fected sibling with the juvenile variant becomes overtly symptomatic, thus jeopardizing prenatal diagnosis.

There is no effective treatment except attempting to control seizures and providing children with the juvenile variant with an education suitable for the blind. Clonazepam and valproate are the drugs of choice in the late infantile variant. Phenytoin should be avoided as long as possible in the juvenile variant because it seems to increase the patient's ataxia.

The Peroxisomal Diseases

Peroxisomes are subcellular organelles that contain a variety of enzymes, including oxidases whose end product is H_2O_2, which is quite toxic and is broken down by catalase. There are two main varieties of peroxisomal diseases (Table 32-7). In the first, there is a deficiency in the biogenesis of the organelle, which is either absent or small but reduced in number, with resultant deficiency of several peroxisomal enzymes. In the second, the organelle is present, but there is a single peroxisomal enzyme deficiency (Moser et al. 1989).

The most common peroxisomal disease is juvenile X-linked adrenoleukodystrophy (ALD). It is characterized by the accumulation of very long chain fatty acids (VLCFA) that can be seen ultrastructurally as long, flat leaflets in the adrenal gland, the testis, Schwann cells, and oligodendroglia. The

Table 32-7. Peroxisomal Diseases

Diseases	Age of Onset	Prognosis for Survival	Severity
	Diseases With Peroxisomal Structural Defects		
Zellweger	Birth	Death in infancy	Devastated
Neonatal adrenoleukodystrophy	Birth	Death in infancy or preschool	Severe
Infantile Refsum	Birth	Survival into childhood	Severe
Pipecolatacedemia	Infancy	Death in infancy or preschool	Severe
	Diseases With Single Peroxisomal Enzyme Defects		
Adrenoleukodystrophy	Childhood (ALD)*	Death in teens	Severe
	Adulthood (AMN)[+]	Prolonged, variable in AMN	Moderate to severe
Pseudo-Zellweger	Birth	Death in infancy	Devastated
Refsum	Adolescence	Survival to adulthood	Severe

* ALD = Adrenoleukodystrophy
[+] AMN = Adrenomyeloneuropathy

classic neurological picture is one of dementia insidiously starting before 10 years in a boy who subsequently develops cortical blindness and deafness, long tract signs, and in some cases, seizures (Moser et al. 1984). Computed tomography (CT) and magnetic resonance imaging (MRI) show large areas of demyelination that spread from the occipital region forward and less often from the frontal region backwards. Virtually all of these boys die within 1 decade or less of the diagnosis, often after several years in a vegetative state.

Some affected family members of children with classic ALD do not become symptomatic until adulthood when they present with signs of a chronic myelopathy and neuropathy called adrenomyeloneuropathy (AMN). Occasionally, individuals remain asymptomatic or develop only adrenal insufficiency. Reasons for this extremely variable prognosis are unknown, but members of the same family, presumably with the same mutation, may have entirely different courses. Obligate carrier women often develop a mild to moderate spastic paraparesis in middle life and rarely a progressive dementia.

As a result of research by the father of a boy with this disease, an experimental treatment is undergoing clinical trial (Moser et al. 1987). It consists of curtailing the intake of food containing long chain fatty acids and administering glycerol trioleate and C_{22} oils; this results in a dramatic decrease in the blood levels of VLCFA (Moser et al. 1989). It improved nerve conduction velocity in an adult with the AMN variant. Whether it will retard the development of the disease in presymptomatic people may be hard to determine. Not surprisingly, it has not improved boys with advanced disease.

Infantile ALD is an autosomal recessive disease that is named because of the storage of VLCFA. It is close in phenotype with other infantile peroxisomal diseases that are only beginning to be sorted out clinically and biochemically. Some are being tried on the experimental VLCFA-lowering diet.

Zellweger disease is the most severe of the peroxisomal diseases and the most common after ALD (Kelley 1983; Moser and Goldfisher 1985). The children are severely hypotonic and have a characteristic facies and visceromegaly. No treatment has helped so far, and the children die in infancy, as do children with a Zellweger-like phenotype and a deficit of thiolase, another peroxisomal enzyme (Goldfisher et al. 1986; Schram et al. 1987).

Refsum disease is due to the lack of the enzyme phytanic acid oxidase, presumed to be a peroxisomal enzyme. The phenotype varies depending on how much phytanic acid, which comes from chlorophyll, was included in the diet. It usually appears in adolescence as a chronic, slowly progressive ataxic polyneuritic gait defect and is associated with pigmentary degeneration of the retina and deafness. It is reversible over a period of months using a phytanic-restricted diet. Plasmapheresis can be used to hasten recovery and allows patients to end their monotonous diet (Dickson et al. 1989).

The Mitochondrial Cytopathies

These disorders are not discussed here because they are covered in Chapter 26, even though many patients present with signs of an encephalopathy. Their clinical and biochemical basis is complex, and their course is extraordinarily varied. Some may become specifically treatable as their nosology is unravelled.

The Aminoacidurias and Other Organic Acidurias

The most effective dietary therapies have been achieved in this group of disorders; however, as is shown in Tables 32-1, 32-8, and 32-9, much remains to be done. The book by Nyhan and Sakati (1987) is a particularly useful reference on the aminoacidurias and organic acidurias and their management. Patients with these rare disorders and their many variants must be referred to centers with extensive experience in their treatment for death or severe damage to the brain to be avoided. These centers have the resources to study patients at the metabolic and molecular level. This is essential be-

Table 32-8. Aminoacidurias and Other Disorders of Nitrogen Metabolism*

Disease	Age of Onset	Prognosis for Survival (untreated)	Severity (untreated)	Effectiveness of Treatment
Phenylketonuria	Birth	Survive to adulthood	Severe	Satisfactory
PKU variants	Variable	Survival	Variable, some severe	Depends on the variant
Maple syrup urine	Birth	Death in infancy	Devastated	Reasonably good
MSUD variants	Variable	Variable	Mild to severe	Satisfactory
Homocystinuria	Childhood	Survive to adulthood	Moderate	Moderately good
Homocystinuria B$_6$ responsive	Infancy	Most survive to adulthood	Moderate to severe	Moderately good
Homocystinuria variants	Infancy	Survive to adulthood	Moderate to severe	Depends on variant, some good
Nonketotic hyperglycinemia	Birth	Death in infancy	Devastated	Poor
Hartnup	Childhood	Survive to adulthood	Moderate, intermittent	Satisfactory
Urea cycle defects	Early infancy	Death in early infancy	Severe	Poor to fair
Lowe	Birth	Survive to teens or adulthood	Severe	Fair
Lesch-Nyhan	Infancy	Survive to teens or early adulthood	Severe	Poor
Spongy degeneration	Infancy	Death in childhood	Severe	None
Alexander	Childhood	Death in childhood	Severe	None
Alexander variants	Variable	Variable	Variable, severe	None

* See Table 1 for effectiveness of treatment.

Table 32-9. Organic Acidemias

Disease	Age of Onset	Prognosis for Survival (untreated)	Severity	Effectiveness of Treatment
Glutaric acidemia	Early infancy	Death in infancy	Severe (most variants)	Poor
Isovaleric acidemia	Early infancy	Death in infancy	Severe (most variants)	Poor in severe neonatal form, better in intermittent form
Propionic I & II acidemia (ketotic hyperglycinemia)	Early infancy	Death in infancy	Severe	Poor to fair
Methylmalonic acidemia	Early infancy	Death in infancy	Severe	Poor
Methylmalonic B12 responsive	Early infancy	Death in infancy	Severe	Satisfactory
Multiple carboxylase deficiency	Early infancy	Death in infancy	Severe	Satisfactory
Biotinidase deficiency	Infancy	Death in early childhood	Severe	Satisfactory
Dicarboxylic acidemia	Variable	Variable, intermittent	Variable	Variable

cause there are many variants of the classic disorders, and their prognosis and management may differ entirely. These centers also can provide prenatal diagnosis for many disorders.

Except for Lesch-Nyhan disease, Lowe disease, and ornithine transcarbamylase deficiency, which are X-linked recessive, all are inherited as autosomal recessive traits. The most severe diseases, such as urea cycle defects, some organic acidurias, and maple syrup urine disease (MSUD), rapidly produce a severe encephalopathy with coma and death. Affected children usually are normal at birth and become symptomatic soon after feeding has started. In critically ill neonates, it may not be enough to stop feedings and administer glucose with pharmacological doses of vitamins that act as coenzymes (e.g., thiamine, riboflavine, cobalamine, and biotin). It may be necessary to perform an exchange transfusion or dialysis while awaiting the results of analyses of blood gases, serum ammonia, gas liquid chromatography, and mass spectrometry of the blood and urine. Long-term results in such extreme cases often are unsatisfactory, with the child surviving with a profoundly damaged brain. Early diagnosis requires a high degree of suspicion for a disorder other than neonatal asphyxia in a lethargic or comatose infant with or without seizures.

In the case of PKU and MSUD, preventive treatment consists of a synthetic diet deficient in phenylalanine in the case of PKU and of branched amino acids in the case of MSUD. How long to continue the diet is controversial. For PKU, the original recommendation was up to age 4 when major tracts are myelinated, but several reports (Nyhan and Sakati 1989) indicate that IQ may decline modestly and some behavior problems may emerge when the diet is discontinued. In a national collaborative study of 125 10-year-olds with PKU, there was a significant relationship between blood phenylalanine levels between ages 3½ and 10 years and cognitive outcome variables, suggesting that the diet should be continued until 10 years of age (Michals et al. 1988). Avoidance of profound mental deficiency

and achievement of normal or near-normal intelligence using diet in promptly diagnosed PKU has led to mandatory neonatal screening programs for this and other metabolic disorders, such as MSUD, galactosemia, biotinidase deficiency, and hypothyroidism. In the severe form of MSUD, dietary restriction may have to be continued indefinitely. Patients with milder forms of MSUD may not need treatment but are at risk for an acute encephalopathy at the time of intercurrent febrile illnesses.

Women homozygous for PKU need to return to a phenylalanine-restricted diet before conception to avoid bearing retarded children, because dietary restriction started in the first or second trimester of pregnancy does not prevent mother's high phenylalanenemia from damaging the fetus' brain (Mabry et al. 1966). It is uncertain whether this is true of other disorders treated successfully enough for women to procreate.

The management of infants with hyperammonemia has been successful using protein restriction and sodium benzoate, phenylacetate, citrulline, and arginine to activate an alternate pathway for the excretion of waste products of nitrogen metabolism (Batshaw 1984). Treated children are at high risk for developing hyperammonemic coma during the course of intercurrent infections. Outcome in the disorders of the urea cycle is variable, with significant mortality and prevalence of mental deficiency. Female carriers for ornithine transcarbamylase deficiency who present with migraine headaches, ill-defined episodes of lethargy, or frank hyperammonemic coma, also require treatment (Batshaw et al. 1980).

Lesch-Nyhan disease, a disorder of purine metabolism, can result from many different mutations inactivating hypoxanthine guanine phosphoribosyltransferase (Gibbs et al. 1989). Prognosis in the classic infantile variant is poor; all affected boys manifest a movement disorder in infancy that precludes ambulation, most have significant mental deficiency, and all mutilate themselves, often to a grotesque degree (Nyhan and Sakati 1987). With treatment with allopurinol, death from uricemic nephropathy may be postponed. Gouty arthropathy is alleviated with colchicine. These treatments do not affect the basic defect or its neurological manifestations. Carbidopa and tetrabenazine seem to provide a modest amelioration of the movement disorder (Jankovic et al. 1988). Variants of the disease have hyperuricemia and deafness without the movement disorder or self-mutilation; they are due to other purine enzyme deficiencies.

In Lowe syndrome, vigorous treatment of the renal defect with alkali and of rickets with vitamin D enables many boys to survive to adulthood. This treatment does not prevent the cataracts, which are present at birth, or mental deficiency.

There is no treatment for invariably fatal disorders such as nonketotic hyperglycinemia and spongy degeneration of the brain (Canavan disease). Sodium benzoate lowers glycine levels in the blood and CSF and may provide some control for the intractable seizures and myoclonus of nonketotic hyperglycinemia, but it does not seem to influence the children's profound mental deficiency (Wolff et al. 1986). Survival in a vegetative state may extend to the second decade, especially in Canavan disease.

Prognosis is variable in Alexander disease, a condition of unknown cause and for which no treatment exists. It may present as a chronic illness in preschool children, in adolescents, or in adults in whom bulbar signs predominate. In general, the earlier the signs of the illness, the more severe its course, which may extend for 5 years in children, but lasts a decade or more in adults.

Other Disorders of Lipid Metabolism

This is a heterogeneous group of disorders, of which a few are now treatable (Table 32-10). This is not the case with Wolman disease, a lysosomal storage disease of early infancy in which death occurs before 1 year of age, or with Pelizeaeus-Merzbacher (P-M) disease, an X-linked trait producing striking nystagmus in infancy and a leuko-

dystrophy with long tract and cerebellar signs and moderate mental deficiency. The course is very slowly progressive in P-M but motor disability is severe, and most patients do not survive beyond the teens or young adulthood.

Marinesco-Sjögren syndrome is another rare, untreatable, chronic disease of early childhood, producing ataxia, mental deficiency, cataracts, and failure to grow. It may be a lysosomal disease. Sjögren-Larsson syndrome, which is most prevalent in northern Sweden and presents with spastic quadriplegia precluding ambulation, mental deficiency, retinal changes, and ichthyosis, is said to improve with dietary restriction of fat and supplementation with medium chain triglycerides.

Cerebrotendinous xanthomatosis does not become symptomatic until the teens at the earliest and often not until middle age. It is often mistaken for multiple sclerosis or some other ataxic syndrome, but tendinous xanthomas and cataracts indicate otherwise. Untreated, most patients die of myocardial infarction or bulbar palsy in middle life, following decades of a slowly progressive deterioration affecting peripheral nerves, the spinal cord, and the brain. Treatment with chenodeoxycholic acid produces substantial improvement in all the manifestations of the disease, if it is started before they become severe (Berginer et al. 1984).

Early treatment with large doses of vitamin E prevents most of the neurological symptoms of abetalipoproteinemia (Bassen-Kornzweig [B-K] disease) and reverses some of them. There is no treatment for Tangier disease, in which a neuropathy does not become manifest until middle adulthood. In contrast, B-K produces ataxia, a neuropathy, and retinal degeneration in adolescence, which gradually progresses over a decade.

Other Disorders

Table 32-11 lists a miscellaneous series of diseases, most of which are not yet understood and therefore cannot be diagnosed in utero. With the exception of kinky hair dis-

Table 32-10. Other Disorders of Lipid Metabolism

Disease	Age of Onset	Prognosis for Survival (untreated)	Severity (untreated)	Treatment	Effectiveness
Wolman disease	Early infancy	Death before 1 year	Devastating	—	—
Pelizeus-Merzbacher	Infancy	Death in teens, young adult	Moderate to severe	—	—
Orthochromatic leukodystrophy	Childhood	Death within a decade	Severe	—	—
Marinesco-Sjogren	Infancy	Death in childhood (?)	Severe	—	—
Sjogren-Larsson	Early childhood	Survival to adulthood	Severe	Low fat, MCT oil	Fair
Cerebrotendinous xanthomatosis	Teens, adulthood	Survival for decades	Moderate to severe	Chenodeoxycholic acid	Promising
Tangier disease	Teens, adulthood	Survival for decades, early atherosclerosis	Mild to moderate	—	—
Abetalipoproteinemia	Early childhood	Survival to adulthood	Mild to moderate	Vitamin E	Satisfactory

Table 32-11. Other Disorders

Disease	Age of Onset	Prognosis for Survival (untreated)	Severity (untreated)	Specific Treatment	Effectiveness
Wilson's	Childhood to early adulthood	Variable	Severe	Low copper diet, chelation	Satisfactory
Kinky hair	Early infancy	Death in infancy or early childhood	Devastated	—	—
Lafora	Mid childhood	Death within a decade	Severe	—	—
Baltic myoclonus	Mid childhood	Survival to adulthood	Moderate to severe	—	—
Neuroaxonal dystrophy	Infancy	Death in childhood	Severe	—	—
Hallervorden-Spatz	Childhood or later	Most survive to adulthood	Severe	—	—
Ataxia-telangiectasia	Early childhood	Survival to teens, adulthood	Moderate to severe	—	—
Cockayne syndrome	Infancy	Death in childood or teens	Severe	—	—
Xeroderma pigmentosum	Infancy	Most survive to adulthood	Severe	—	—
Familial dysautonomia	Infancy	Survival to early adulthood	Moderate to severe	Bethanecol	Fair

ease, they are inherited as autosomal recessive traits.

Prognosis in Wilson disease has been revolutionized by the introduction of a low-copper diet and by the administration of copper-chelating agents. Treated before irreversible cavitation of the basal ganglia and before death from fulminant hepatic failure, Wilson's disease is compatible with indefinite survival, regression of all neurological signs, and even successful pregnancy (Scheinberg and Sternlieb 1984). Liver transplantation occasionally has been successful in children with fulminant hepatic failure unresponsive to more conservative treatments.

Copper malabsorption syndrome (kinky hair disease), an X-linked trait, is at the other extreme and has resisted all efforts at providing copper by systemic infusion. The infants are symptomatic from birth and die in infancy or are in a vegetative state in the preschool years.

Of the two forms of familial myoclonus, Lafora disease is more severe, resulting in death within 6 to 10 years, while Baltic myoclonus or Unverricht-Lundborg syndrome is compatible with survival into adulthood. No curative treatment is available, and the myoclonus does not necessarily respond to klonopin or valproate. Dementia is more severe and occurs earlier in Lafora disease.

The biochemical basis of infantile neuroaxonal dystrophy (NAD) and Hallervorden-Spatz disease (H-SD) is unknown. They share the occurrence of axonal spheroids, which in NAD are found in peripheral nerve terminals and throughout the CNS, while in H-SD, they predominate in the basal ganglia. No treatment is known for either disease. Children with NAD succumb in a vegetative state during the school years, while those with H-SD, who may not become symptomatic until their teens, may survive for a decade or more and may not be severely demented despite their devastating motor deficit. The combination of spasticity and dystonia is characteristic of both age groups. Variants with an earlier or later onset and a more or less severe course exist.

Ataxia telangiectasia presents in the preschool years with ataxia. Neurological deterioration is slowly progressive and is compatible with survival into young adulthood if immunological incompetence does not result in respiratory failure following severe sinopulmonary infection or in a lymphoma or other malignancy. Infusion of fresh frozen plasma to raise the level of immunoglobulin A (IgA) may be beneficial in the 90% of immunologically incompetent patients, but it has no effect on the neurological symptomatology. These patients often have multiple endocrine deficiencies, including diabetes mellitus, and are exquisitely sensitive to radiation and radiomimetic drugs, which precludes effective therapy of the malignancies that kill 20% of these patients, often in their teens (Morrell et al. 1986). Blacks are three times as likely as whites to die of a malignancy. Variability in the manifestations of the disease is accounted for in part by genetic heterogeneity, with some patients surviving into their 50s with only a moderate neurological handicap. Heterozygote carriers are at substantially increased risk of harboring a malignancy, especially breast cancer.

Cockayne syndrome is a severe untreatable disease producing extreme dwarfing with microcephaly, mental deficiency, ataxia, spasticity, neuropathy, deafness, and visual loss. Death tends to occur in the second decade, although occasionally patients survive to the end of the third decade. Genetic heterogeneity may account for a variable prognosis. A very severe form of Cockayne's disease is associated with xeroderma pigmentosum and results in early death. Both diseases and ataxia telangiectasia are due to defects in DNA repair. Xeroderma pigmentosum is a genetically heterogeneous dermatologic condition characterized by extreme sensitivity to ultraviolet light resulting in multiple carcinomas and melanomas of the skin and, in 18% of cases with ataxia, neuropathy, mental deficiency, hearing loss, and endocrine abnormalities (Kraemer et al. 1987). Probability of survival is 90% at age 13, 80% at age 28, 70% at age 40. Death occurs an average of 30 years earlier than in the general population. Cancer is responsible for one third of the deaths.

Prognosis in familial dysautonomia, a recessive disease of Ashkenazi Jews, has been greatly improved by vigorous treatment of the gastrointestinal and hypertensive crises and hyperthermia that threatened survival, even of infants. Among 227 patients described by Axelrod and Abularrage (1982), one third were 20 years or older. Long-term survival into the late 30s and early 40s and even pregnancy occur in optimally managed patients in whom dysautonomic crises are vigorously treated with chlorpromazine, bethanecol, and diazepam. While survivors may be normally intelligent, they tend to be somewhat less so than their siblings and tend to have difficult personalities.

References

Adams, R. D.; Lyon, G. Neurology of hereditary metabolic diseases of children. New York: McGraw-Hill; 1982.

Ampola, M. G.; Mahoney, M. J.; Nakamura, E.; Tanaka, K. Prenatal therapy of a patient with vitamin-B_{12} responsive methylmalonic acidemia. N. Engl. J. Med. 293:313–317; 1975.

Axelrod, F. B.; Abularrage, J. J. Familial dysautonomia: a prospective study of survival. J. Pediatr. 101:234–236; 1982.

Baraitser, M. The genetics of neurological disorders. 2nd ed. Oxford: Oxford University Press; 1989.

Bannwart F. Morbus Fabry: Licht- und elektronmikroskopischer Befund 12 Yahre nach erfolgreicher Nierentransplantation. Schweiz. Med. Wschr. 112:1742–1747; 1982.

Batshaw, M. L. Hyperammonemia. Curr. Prob. Pediat. 14(11):1–69; 1984.

Batshaw, M. L.; Roan, Y.; Jung, A. L.; Rosenburg, L. A.; Brusilow, S. W. Cerebral dysfunction in asymptomatic carriers of ornithine transcarbamylase deficiency. N. Engl. J. Med. 302:482–485; 1980.

Berginer, V. M.; Salen, G.; Shefer, S. Long-term treatment of cerebrotendinous xanthomatosis with chenodeoxycholic acid. N. Engl. J. Med. 311:1649–1652; 1984.

Berkovic, S. F.; Carpenter, S.; Andermann, F.; Andermann, E.; Wolfe, L. S. Kuf's disease: a critical reappraisal. Brain 111:27–62; 1988.

Clark, J. G. The challenge of bone marrow transplantation. Mayo Clin. Proc. 65:111–114; 1990.

Clement, M.; McGonigle, R. J. S.; Monkhouse, P. M.; Renal transplantation in Anderson-Fabry disease. J. Roy. Soc. Med. 75:557–560; 1982.

Desnick, R. J.; Bernlohr, R. W.; Krivit, W., eds. Enzyme replacement therapy in genetic disease. Baltimore: Williams and Wilkins; 1979.

Dickson, N.; Mortimer, J. G.; Faed, J. M.; Pollard, A. C.; Styles, M.; Peart, D. A. A child with Refsum disease: successful treatment with diet and plasma exchange. Dev. Med. Child Neurol. 31:92–97; 1989.

Farragiana, T.; Churg, J.; Grishman, E.; Strauss, L.; Prado, A.; Bishop, D. F.; Schuchman, E.; Desnick, R. J. Light and electron-microscopic histochemistry in Fabry's disease. Am. J. Pathol. 103:926–931; 1981.

Friedlaender, M. M.; Kopolovic, J.; Rubinger, D.; Silver, J.; Drukker, A.; Ben-Gershon, Z.; Durst, A. L.; Poportzer, M. M. Renal biopsy in Fabry's disease eight years after successful renal transplantation. Clin. Nephrol. 27:206–211; 1987.

Friedman, T. Progress toward human gene therapy. Science 24:1275–1281; 1989.

Gibbs, D. A.; Spellacy, E.; Tompkins, R.; Watts, R. W.; Thowbray, J. F. A clinical trial of fibroblast transplantation for the treatment of mucopolysaccharidoses. J. Inher. Metab. Dis. 6:62–81; 1983.

Gibbs, R. A.; Ngyuen, P. N.; McBride, L. J.; Koepf, S. M.; Caskey, C. T. Identification of mutations leading to the Lesch-Nyhan syndrome by automated direct DNA sequencing of in vitro amplified cDNA. Proc. Natl. Acad. Sci. USA 86:1919–1923; 1989.

Goldfisher, S. L.; Collins, J.; Rapin, I.; Neumann, P.; Neglia, W.; Spiro, A. J.; Ishii, T.; Roels, F.; Vameca, J.; Van Hoof, F. Pseudo-Zellweger syndrome: deficiencies in several peroxisomal oxidative activities. J. Pediatr. 108:25–32; 1986.

Jankovic, J.; Caskey, T. C.; Stout, J. T.; Butler, I. J. Lesch-Nyhan syndrome: a study of motor behavior and cerebrospinal fluid neurotransmitters. Ann. Neurol. 23:466–469; 1988.

Johnson, W. G. Neurological disorders with hexosaminidase deficiency. In: Moss, A. J., ed. Pediatrics update. New York: Elsevier Science; 1987: p. 91–104.

Kelley, R. I. The cerebrohepatorenal syndrome of Zellweger, morphologic and metabolic aspects. Am. J. Med. Genet. 16:503–517; 1983.

Kolodny, E. H. Enzyme and gene therapy for storage diseases. In: Swann, J. W., ed. Disorders of the developing nervous system: changing views on their origins, diagnoses, and treatments. New York: Alan R. Liss; 1988: p. 229–246.

Kraemer, K. H.; Lee, M. M.; Scotto, J. Xeroderma pigmentosum: cutaneous, ocular, and neurologic abnormalities in 830 published cases. Arch. Derm. 123:241–250; 1987.

Krivit, W.; Paul, N. W., eds. Bone marrow transplantation for treatment of lysosomal storage disease. New York: Alan R. Liss; 1986.

Krivit, W.; Shapiro, E.; Kennedy, W.; Lipton, M.; Lockman, L.; Smith, S.; Summers, C. G.; Wenger, D. A.; Tsai, M. Y.; Ramsay, N. K. C.; Kersey, J. H.; Yao, J. K.; Kaye, E. Treatment of late infantile metachromatic leukodystrophy by bone marrow transplantation. N. Engl. J. Med. 322:28–32; 1990.

Mabry, C. C.; Denniston, J. C.; Coldwell, J. G. Mental retardation in children of phenylketonuric mothers. N. Engl. J. Med. 275:1331–1336; 1966.

McKusick, W. A. Mendelian inheritance in man. Catalog of autosomal dominant, autosomal recessive, and X-linked phenotypes. 9th ed. Baltimore: Johns Hopkins University Press; 1990.

Menkes, J. H. Textbook of child neurology. 4th ed. Philadelphia: Lea and Febiger; 1989.

Michals, K.; Azen, C.; Acosta, P.; et al. Blood phenylalanine levels and intelligence of 10-year-old children with PKU in the National Collaborative Study. J. Am. Diet. Assoc. 88:1226–1229; 1988.

Morrell, D.; Chromartie, E.; Swift, M. Mortality and cancer incidence in 263 patients with ataxia-telangiectasia. J. N. C. I. 77:89–92; 1986.

Moser, A. B.; Borel, J.; Odone, A.; Naidu, S.; Cornblath, D.; Sanders, D. B.; Moser, H. W. A new dietary therapy for adrenoleukodystrophy: biochemical and preliminary clinical results in 36 patients. Ann. Neurol. 21:240–249; 1987.

Moser, H. W.; Goldfisher, S. L. The peroxisomal disorders. Hosp. Pract. 20(9):61–70; 1985.

Moser, H. W.; Mihalik, S.; Watkins, P. A. Adrenoleukodystrophy and other peroxisomal disorders that affect the nervous system, including new observations on L-pipecolic acid oxidase in primates. Brain Dev. 11:89–90; 1989.

Moser, H. W.; Moser, A. B.; Singh, I.; O'Neill B. P. Adrenoleukodystrophy: survey of 303 cases: biochemistry, diagnosis, and therapy. Ann. Neurol. 16:628–641; 1984.

Myerowitz, R.; Hogikyan, N. D. Different mutations in Ashkenazi Jewish and non-Jewish French Canadians with Tay-Sachs disease. Science 232:1646–1648; 1986.

Nyhan, W. L.; Sakati, N. O. Diagnostic recognition of genetic disease. Philadelphia: Lea and Febiger; 1987.

Rakela, J.; Kurtz, S. B.; McCarthy, J. T.; Ludwig, J.; Ascher, N. L.; Bloomer, J. R.; Claus, P. L. Fulminant Wilson's disease treated with post dilution hemofiltration and orthotopic liver transplantation. Gastroenterology 90:2004–2007; 1986.

Rappeport, J. M.; Ginns, E. I. Bone-marrow transplantation in severe Gaucher's disease. N. Engl. J. Med. 311:84–88; 1984.

Ringdén, O.; Groth, C.-G.; Erikson, A.; Bäckman, L.; Grangvist, S.; Månsson, J. E.; Svennerholm, L. Long-term follow-up of the first successful bone marrow transplantation in Gaucher disease. Transplantation 46:66–70; 1988.

Rowe, P. C.; Newman, S. L.; Brusilow, S. W. Natural history of symptomatic partial ornithine transcarbamylase deficiency. N. Engl. J. Med. 314:541–547; 1986.

Rowland, L. P., ed. Merritt's textbook of neurology, 8th ed. Philadelphia: Lea and Febiger; 1989.

Rubin, I. L.; Crocker, A. C. Developmental disabilities: delivery of medical care for children and adults. Philadelphia: Lea & Febiger; 1989.

Scaggiante, B.; Pineschi, A.; Sustersich, M.; Andolina, M.; Agosti, E.; Romeo, D. Successful therapy of Niemann-Pick disease by implantation of human amniotic membrane. Transplantation 44:59–61; 1987.

Scheinberg, I. H.; Sternlieb, I. Wilson's disease. Philadelphia: W. B. Saunders; 1984.

Schram, A. W.; Goldfisher, S. L.; van Roemund, C. W. T.; Human peroxisomal 3-oxacyl-coenzyme A thiolase deficiency. Proc. Natl. Acad. Sci. 84:2494–2496; 1987.

Scriver, C. R.; Beaudet, A. L.; Sly, W. S.; Valle, D., eds. The metabolic basis of inherited disease. 6th ed. New York: McGraw-Hill; 1989.

Swaiman, K. F. Pediatric neurology: principles and practice. St Louis: C. V. Mosby; 1989.

Wolff, J. A.; Kulovich, S.; Yu, A. L.; Qiao, C. N.; Nyhan, W. L. The effectiveness of benzoate in the management of seizures in nonketotic hyperglycinemia. Am. J. Dis. Child. 140:596–602; 1986.

33

Spinocerebellar Degenerations

DAVID A. STUMPF

Many ataxia patients are given erroneous prognostic information. Several factors contribute to these errors. First, the diagnosis may be incorrect. Predictions about the course of illness depend on the correct diagnosis; with spinocerebellar degenerations, this can be a major problem. Unfortunately, early in the course when predictions may have their greatest value, the diagnosis is most problematic. This complex subject is beyond the scope of this chapter, and the reader is referred to other sources for the author's perspective on classification (Stumpf 1978; 1985; 1990). Second, the prognosis is not known early in the course of most ataxia cases. Even when the diagnosis is clear, often there is a variation in the clinical course from patient to patient. Third, the neurologist may pay insufficient attention to systemic disease. Spinocerebellar degenerations are identified as progressive disorders of the nervous system. Other organ systems also may be affected, and their involvement often produces the major morbidity and fatal complications.

Pseudodegeneration

Not all that deteriorates is degenerative or progressive. Several patients have presented with a fixed neurological deficit that functionally deteriorates during the adolescent growth spurt when the center of gravity rapidly shifts upward. These patients have had anomalies of the cerebellum (hypoplasia) or hypomyelination of peripheral nerves. Thus, they have ataxia or areflexia clinically and small cerebellum on scan or reduced nerve conduction velocity. Using magnetic resonance imaging (MRI), the smooth cerebellar surface indicates an anomaly, rather than atrophy, of a previously normal-sized cerebellum. These patients had previously been given grim prognoses with diagnoses such as Friedreich's ataxia. In each case, after the growth spurt, they recompensated and remained functional, albeit clumsy, individuals.

These patients are important for two reasons. First, one needs to be aware of this problem to counsel properly. Second, they indicate that even in a true degenerative disorder, linear skeletal growth may make the deterioration appear more severe than it will be later in life.

Friedreich's Disease

Friedreich's disease (FD), classically defined (Geoffroy et al. 1976), is a rapidly progressive disorder beginning before puberty and rendering the individual wheelchair-bound by the early 20s or before. Life span is shortened in FD; median survival is 35 years. Survival rates at 10, 20, and 30 years are 96%, 80%, and 61%, respectively. Age at onset does not predict survival. Men fare worse than women (Leone et al. 1988). Earlier deaths generally are due to cardiomyopathy, which produces life-threatening arrhythmias or congestive failure.

Cardiomyopathy is present in all FD patients. Echocardiography is the most precise test and generally detects the hypertrophic myocardium at the time of neurological diagnosis. As the cardiomyopathy progresses, the ventricle dilates and the ventricular wall thins. The heart can be very volatile in FD; neurologists should advocate that their patients seek cardiology consultations. Cardiologists, accustomed to dealing with more common ischemic disorders, frequently have clinical instincts or styles that are too cavalier for the FD patient. The author has seen several patients in life-threatening crises because of rapid changes in medications that seemed reasonable but were not in FD patients.

Diabetes mellitus develops clinically in 20% of FD patients; an additional 20% have diabetic glucose tolerance curves. Insulin dependency generally develops late in the course. The diabetes in FD is not merely the glucose intolerance seen in other neuromuscular disorders with decreased muscle mass. It behaves like idiopathic, type I diabetes mellitus (Schoenle et al. 1989) and is attributable to a degeneration of pancreatic beta cells.

Quality of life in FD is frequently compromised by scoliosis, which nearly all patients develop. Many patients develop a 60-degree curve, greatly impairing comfort in a chair and respiratory function. Scoliosis in FD is generally regarded as neuromuscular, implying ''don't operate.'' However, FD scoliosis behaves clinically much like idiopathic scoliosis (Labelle et al. 1986). Thus, severe curves develop when the scoliosis develops before the adolescent growth spurt. Later

developing curves tend to be less. Bracing is less successful in FD (Cady and Bobechko 1984), and early operation can greatly enhance the quality of later life. Prognostic information is critical in evaluating the virtues of surgery in a patient. The surgery is a major undertaking. Patients with severe cardiac disease or other poor prognostic features are less acceptable candidates for surgery. Diabetic crises can be precipitated by operation, can be avoided if they are anticipated; a glucose tolerance curve before surgery can be helpful.

FD is one of the most tragic of the neurological disorders, striking the young in their bodies but not in their minds. Young Friedreich's patients show determination, particularly in adolescence. Plans and dreams, often sought through valiant efforts, unfortunately are not realized. Patients are unemployable, except in the most sheltered environments. The psychological toll of FD is devastating. Nearly all patients are suicidal, and occasionally they are successful. Accurate prognostic counseling may produce more realistic planning and less frustration, hopefully obviating the need for later psychotherapy.

In clinics that see many patients, it is clear that some families—indistinguishable on the basis of neurological features—have a better prognosis. Unfortunately, they are not common, approximately 5% to 10% of American families. An allelic or different genetic defect, rather than an effect of a modifier gene, is suggested by the similarly slower course in siblings, which can be helpful in identifying this more insidious form. These patients progress much slower, remaining ambulatory into the late 20s or longer and survive into their 60s. Genetic studies also support the allelic hypothesis (Keats et al. 1989).

Dominant Hereditary Ataxia

Dominant ataxia is probably a heterogeneous condition, based on genetic linkage data. However, clinical variation, even within a family, is substantial. Anticipation, an earlier age of onset with subsequent generations, is observed in some families. As a result, approximately 5% to 10% of patients have onset at less than 10 years of age, and they interrupt the transmission of the disorder in a family or a branch of the family. Generally, the earlier the onset, the worse the prognosis. At one extreme is the patient with onset at 1 year of age and death after an additional year. At the opposite extreme is onset in the seventh decade and little debility before death from another cause after many years. On average, patients survive approximately 15 years after the initial diagnosis.

Early diagnosis provides valuable genetic counseling insights. Examinations of presymptomatic, at-risk individuals may reveal ophthalmoplegia, ataxia, or other signs highly suggestive of the disorder. CT or MRI scans are abnormal in some at-risk individuals; these changes may correlate with subsequent development of the disease (Aita 1988), but further studies are necessary to document this.

The clinical picture many evolve with time. Parkinsonian features (bradykinesia and rigidity) may develop after several years. This may respond favorably to L-dopa. Lower motor neuron findings and dementia may occur late in the course.

Other Hereditary Ataxias

There are many recessively inherited disorders producing ataxia (Stumpf 1990). Studies of prognosis often are limited to survival data. Systemic organ difficulties often are limiting factors for survival. It is particularly important to detect treatable disorders through metabolic testing (Stumpf 1987). In these patients, the course can be arrested and at times, reversed by treatment.

Sporadic Ataxia

Multisystem atrophy (MSA) is the usual condition producing a sporadic spinocerebellar degeneration. Degeneration occurs—in variable degrees in each patient—in the

cerebellum, pyramidal tract, basal ganglia, and intermediolateral cell column in the cord. The clinical picture can include ataxia, spastic paraparesis, bradykinesia, and genitourinary dysfunction or orthostatic hypotension. The prognosis is related to these symptoms. Some patients have ataxia or spastic paraparesis with no other symptoms and tend to have the best prognosis; long-term survival is possible.

When the autonomic symptoms predominate early in the course, it can be difficult to distinguish MSA from a pure autonomic dysfunction (PAF). This distinction is important because PAF has a much better prognosis. Approximately 75% of MSA patients present with ataxia and then develop orthostasis within an average of 4 years. The other 25% begin with orthostasis and develop neurological symptoms within an average of 2 years. Hypotension is associated with a more debilitating degenerative course. The overall survival is 8 to 9 years in MSA patients (Polinsky and Nee 1989).

Bradykinesia may respond to L-dopa, improving the quality of life; this therapy may, however, precipitate hypotensive episodes.

References

Aita, J. A. Cranial CT and olivopontocerebellar atrophy. J. Neurol. 235:(suppl. 1):S40; 1988.

Cady, R. B.; Bobechko, W. P. Incidence, natural history, and treatment of scoliosis in Friedreich's ataxia. J. Pediatr. Orthopedics 4:673–676; 1984.

Geoffroy, G.; Barbeau, A.; Breton, G.; Lemieux, B.; Aube, M.; Leger, C.; Bouchard, J. P. Clinical description and roentgenologic evaluation of patients with Friedreich's ataxia. Can. J. Neurol. Sci. 3:279–286; 1976.

Keats, B. J.; Ward, L. J.; Shaw, J.; Wickremasinghe, A.; Chamberlain, S. "Acadian" and "classical" forms of Friedreich's ataxia are most probably caused by mutations at the same locus. Am. J. Med. Genet. 33:266–268; 1989.

Labelle, H.; Tohmé, S.; Duhaime, M.; Allard, P. Natural history of scoliosis in Friedreich's ataxia. J. Bone Joint Surg. 68A:564–572; 1986.

Leone, M.; Rocca, W. A.; Rosso, M. G.; Mantel, N.; Schoenberg, B. S.; Schiffer, D. Friedreich's disease: survival analysis in an Italian population. Neurology 38:1433–1438; 1988.

Polinsky, R. J.; Nee, L. E. Autonomic failure: nosology, clinical evolution and prognosis. Ann. Neurol. 26:121A; 1989.

Schoenle, E. J.; Boltshauser, E. J.; Baekkeskov, S.; Landin Olsson, M.; Torresani, T.; von Felten, A. Preclinical and manifest diabetes mellitus in young patients with Friedreich's ataxia: no evidence of immune process behind the islet cell destruction. Diabetologia 32:378–381; 1989.

Stumpf, D. A. Friedreich's Ataxia and other hereditary ataxias. In: Tyler, H. R.; Dawson, D. M., eds. Current neurology, Vol. 1. Boston: Houghton-Mifflin; 1978: p 86–111.

Stumpf, D. A. Treatment of neurologic degenerative disorders. Int. Pediatrics. 2(2):109–114; 1987.

Stumpf, D. A. The inherited ataxias. Neurol. Clin. 2:47–57; 1985.

Stumpf, D. A. Cerebellar diseases. Neurology Raven Press; New York: 1990.

Stumpf, D. A. Cerebellar Diseases. In: Comprehensive Neurology. Rosenberg, R. N.; ed. New York: Raven Press; 1991; p. 833–843.

34

Epilepsy

L. JAMES WILLMORE

Initial assessment and formulation of treatment plans for a patient with epilepsy require establishment of goals, review of expected outcome, and identification of criteria for success. A patient's seizure disorder is classified after accumulation of intake data, including details of ictal behavior, interictal neurological function, and electroencephalographic (EEG) patterns. Once classification is complete, treatment can be selected based on this categorization. When treatment is started, the patient will have overt and subtle anxieties about the potential for seizure recurrence. Because of this anxiety, the treating physician must be ready to provide reassurance that the events associated with epilepsy and the accompanying social and medical dilemmas are controllable. However, for the patient's confidence to be maintained, the physician must establish realistic expectations. Knowledge of prognosis will aid in communication about the potential outcome of therapy and will establish realistic goals.

One method of monitoring seizure management is to compare expected prognosis with observed efficacy of treatment. Medical management based on monitoring blood-level, reviewing seizure frequency, and assessing duration of seizure-free interval requires comparison of a patient against standards for a specific seizure classification. If treatment fails to achieve expected levels of control or improvement in seizures, then changes may be needed. Thus, change in medication regimen, improvement in the data base used in classification, or reassessment of risk factors independent of seizure type may be appropriate. The risk factors for identification of patients with treatment-resistant epilepsy include psychological impairment, presence of lesional epilepsy, and occurrence of certain epilepsy syndromes.

Knowledge of prognosis may govern decisions about discontinuation of treatment with antiepileptic medications. Although initial assessment should attempt to match the standards for control for a specific seizure type, drug-induced remission invariably leads to questions about duration of therapy and the potential for total remission of the clinical disorder. This review attempts to establish guidelines for communication, identify nonepileptic risk factors associated with failure to match standards of control, and list situations in which an accurate classification and formulation of prognosis may depend on serial assessment and require patience in establishing an opinion about ultimate outcome.

Definitions

Prognosis is a knowledge of the course of a disease and the outlook afforded by this knowledge. Understanding the prognosis of epilepsy requires an understanding that this rubric collects clinically similar but nosologically separate disorders linked by the occurrence of seizures of cerebral origin. When applied to epilepsy, remission implies abatement of overt manifestations without cure; the original disorder is present, but symptoms are controlled by continued use of mediation. A true remission occurs when patients do not experience seizures and are not receiving medication.

Classification of Seizures and Epilepsy

Seizures may be characterized by clinically observed behaviors with accompanying EEG patterns. The altered behavior observed during a seizure is characteristically sudden, circumscribed, and involuntary with alterations in consciousness, posture, motor function, or sensation symptomatic of altered brain function. Abnormal electrical discharges accompany the altered behavior. Epilepsy exists if a patient is free of metabolic or acutely active brain diseases while experiencing recurrent spontaneous seizures. The International Classification of Epileptic Seizures (1981) established criteria useful for identifying and characterizing specific seizures experienced by patients. The use of classification has an impact on drug selection, formulation of prognosis, and management planning.

Seizures initially are classified as generalized or partial in pattern. Seizures generalized at onset include those with physiological involvement of both hemispheres, impairment of consciousness, and bilateral motor manifestations. Generalized seizures are further divided into convulsive and nonconvulsive types. Nonconvulsive seizures include both typical and atypical absence seizures. Patterns of convulsive seizures include clonic, tonic, and tonic–clonic behaviors. Partial seizures arise from a localized region of neurons. Symptoms reflect components of normal function with sensory experiences, motor manifestations, or psychic perceptions. The seizure is identified as complex if consciousness is impaired. Automatisms, or parodies of situational behaviors, form a component of complex partial seizures.

An epilepsy syndrome is characterized by observing a group of signs and symptoms that occur together and with an interrelationship that distinguishes an epileptic disorder as unique (Commission 1989). Important components of a syndrome include occurrence of a specific type or pattern of seizure, family history of epilepsy, age at onset, and presence or absence of abnormal neurological or psychological findings. The severity, chronicity, or diurnal pattern of seizures, and occasionally impairments of development, mental function, or life span, also infuence the identity of an epilepsy syndrome. The presence of EEG abnormalities and specific ictal EEG components are adjunctive information when characterizing syndromes.

Prognosis of Seizures and Epilepsy Syndromes

Generalized and Partial Seizure Disorders

Prognosis for medical control of generalized and partial seizures has developed during transition in methodology for assessment, change in epilepsy classification, introduction of new drugs, and more rational use of older agents (Mattson et al. 1985). The ability to assure compliance by measurement of serum levels of antiepileptic drugs, establishment of an accurate classification using assessment of seizures with simultaneous EEG video monitoring, and addition of carbamazepine and valproate to the drug armamentarium render early observations about prognosis less pessimistic.

Although classification of seizures depends on observations of behavior with adjunctive use of ictal and interictal EEG patterns, tonic–clonic convulsive behaviors may be a manifestation of a primary seizure disorder or a secondarily generalized phenomena as a component of partial seizures. Prior to development of monitoring technilogies that allow accurate differentiation of these two seizure types, investigators considered patients with convulsive seizures as belonging to a homogenous category. However, assessment of the distribution of seizure types in patients older than 15 years shows that the majority of those with convulsive seizures have a secondarily generalized pattern (Keranen et al. 1988).

Referral patterns to specialty centers bias observations about treatment efficacy and prognosis. Institutionalized patients or patients being treated in a referral center have a 2-year terminal remission of approximately 30% (Rodin 1969). Retrospective assessment of patients in the Mayo Clinic record linkage provides a more optimistic view, with 5-year remissions found in at least 70% of patients followed for 20 years. Of these patients, 50% were in true remission because they had undergone medication discontinuation (Annegers et al. 1979).

Patients with new-onset seizures treated with a single drug with an endpoint of relapse of seizures had a good outcome in 65% (Beghi and Tognoni 1988). Unsuccessful control occurred in patients with longer duration of epilepsy, greater numbers of seizures before institution of treatment, a known etiology, epileptiform patterns on EEG assessment, and the presence of a partial seizure disorder. Prospective assessment of patients identified with at least two tonic–clonic or partial seizures and treatment with a single major antiepileptic drug showed 73% with 1-year remission by 2 years of follow-up, 88% by 4 years, and 92% by 8 years (Elwes et al. 1984; Shorvon and Reynolds 1982). Although an overall remission of seizures may approach 60% (Okuma

and Kumashiro 1981) poor seizure control was associated with impaired social adjustment, multiple types of seizures with abnormalities on EEG, delay on onset of treatment beyond 1 year, and frequent seizures (Holowach et al. 1972).

Complex partial seizures (CPS) appear to have a less favorable prognosis for control when compared to numbers of patients with generalized seizures achieving complete control (Loiseau et al. 1983). Good outcome of treatment of CPS is influenced by normal mentation, short duration of illness, and low seizure frequency (Chao et al. 1962; Lindsay et al. 1979). Although early estimates for control were in the range of 20%, especially when compared to patients with generalized tonic–clonic seizures (GTCS) (Juul-Jensen 1963; Lennox and Lennox 1960; Rodin 1969), some observers report 40% of patients seizure free for more than 4 years (Currie et al. 1971). Some studies show that 63% of patients with CPS alone can achieve control with medical management (Schmidt et al. 1983). However, if CPS is associated with concomitant monthly GTCS, only 44% achieve control. Schmidt (1984) found that patients at high risk for development of intractable CPS included those with clusters of seizures (more than one seizure per day), an aura at the onset of seizures, and psychiatric disease. Schmidt (1984) included 82 patients referred for chronic CPS to a seizure clinic. Drug trials were performed using high-dose carbamazepine, phenytoin, and barbiturates to the point of clinical toxicity. In this especially difficult patient group, high-dose therapy resulted in complete control in 22%, while 38% had increased or no change in seizure frequency. Approximately 30% remain intractable (Kotagal et al. 1987; Reynolds et al. 1983).

Epilepsy Syndromes

Since the repertoire of seizure manifestations is limited or stereotyped, similar seizure types may be components of several syndromes with widely divergent prognoses. Clear discernment of the specific nosological category and formulation of prognosis in the childhood syndromes often require repeated assessment, careful evaluation of development, and review of treatment responses. Hasty formulation of prognosis before time to observe may undermine accuracy if either extreme of outcome eventuates. Although some syndromes are identified as benign or progressive, these terms usually are applied to ultimate outcome of intellectual function and survival. Although some syndromic seizures are benign in their impact on behavior, lifelong treatment with antiepileptic drugs may be required.

Syndromes Without Associated Behavioral Impairment

Benign Childhood Epilepsy With Centrotemporal Spikes

This common syndrome is characterized by hemifacial brief and partial motor seizures that are usually nocturnal and occasionally generalized. The pattern is associated with somatosensory symptoms. Patients develop normally and do not have any focal changes on contrast studies. A genetic predisposition and male predominance also is found. The EEG shows sleep activation of high-voltage centrotemporal spikes with a blunt configuration and a slow wave. Recovery with complete remission is usual in more than 90% of patients by the middle teenage years (Beaussart 1972; Dreifuss 1989; Loiseau et al. 1988; Lombroso 1967).

Absence Epilepsy (Pyknolepsy)

The prevalence of childhood absence epilepsy is estimated from 2.3% (Livingston et al. 1965) to 8% (Cavazzuti 1980). Prognosis of absence is formulated by observing previously normal children with seizure onset before puberty, with absence episodes as the initiating seizure type. The associated EEG abnormality of 3-Hz spike and wave occurs in the milieu of a normal background pattern (Penry et al. 1975).

Although remission rates for absence epilepsy occasionally are optimistic, long-term follow-up shows that development of tonic–clonic seizures reduces the numbers of patients reaching medication-free remission.

Remissions of 79.3% are reported. However only 44% of patients sustain remission among all patients evaluated who are older than 15 years and are at least 15-years-old after onset (Currier et al. 1963). Cessation of seizures seems to be favored by short duration of illness, implying the need for early identification of patients and rapid institution of treatment.

Categorization of patients with absence as complete control without antiepileptic medications, complete control on medication, and developing GTCS reveals complete remission in 29%, while almost 70% of patients need continued mediation administration (Sato et al. 1983). Assessment of factors associated with a good prognosis using univariate analysis showed normal IQ and no history of GTCS associated with favorable prognosis. Multivariate analysis showed an IQ greater than 90 as the most favorable indicator of a good prognosis.

Tonic–clonic seizures complicate the course of absence epilepsy in approximately 50% of patients (Currier et al. 1963; Livingston et al. 1965; Loiseau 1985). Risk factors for development of tonic–clonic seizures include later onset of absence (Lennox and Lennox 1960), difficulty in controlling absence seizures, and abnormal background activity on EEG (Penry et al. 1975; Sato et al. 1976). Prior to development of broad-spectrum drugs, there was some support for identification of at-risk patients and administration of an additional drug in a prophylactic fashion to prevent tonic–clonic seizures (Livingston et al. 1965). This dual drug approach has been supplanted by the use of valproate, a broad-spectrum medication with combined efficacy against absence and GTCS. Discussion of expectations with parents and patients should include information about the possibility of developing tonic–clonic seizures.

Childhood Epilepsy With Occipital Paroxysms

There are two variants of this rare syndrome. Patients may have partial seizures characterized by eye deviation and vomiting, usually at night, with evolution to hemiconvulsions or generalized convulsive seizures (Gastaut 1982). A late-onset variant (Panayiotopoulos 1989) has a diurnal pattern with visual symptoms that include illusions, hallucinations, phosphene, and amaurosis often followed by hemiclonic seizures or automatisms. Migraine headache or a family history of migraine often is observed. A focal pattern of repetitive occipital spikes or sharp and slow waves with high-amplitude spike waves in the occipital and posterior temporal areas is found on the EEG. Fixation-off sensitivity and attenuation of spikes with opening of the eyes are characteristic. Prognosis generally is excellent, and seizures occur in isolation; complete remission in 1 to 2 years and younger than age 12 is usual (Panayiotopoulos 1989).

Benign Neonatal Familial Convulsions

These neonates have clonic or apneic seizures with a dominantly inherited pattern. No specific changes are found on EEG (Bjerre and Corelius 1968; Dreifuss 1989). Although rare, some patients develop seizures during later development.

Benign Neonatal Convulsions

Frequent and repeated clonic or apneic seizures occur in the first week of life. The interictal EEG shows alternating sharp theta waves. No later consequence is known (Brown 1973).

Benign Myoclonic Epilepsy of Infancy

These children have brief episodes of myoclonus each day. There is a family history of convulsions or epilepsy. The EEG shows normal background for age but with generalized spike waves occurring in brief bursts, especially during the early stages of sleep (Dravet et al. 1985).

Juvenile Absence Epilepsy

This syndrome is characterized by absence seizures with onset near puberty. There is equal sex distribution, and seizure frequency on any day is low. Tonic–clonic seizures are especially frequent on awakening. The morphology of the spike-wave pattern is of faster frequency than childhood absence. Response to therapy is excellent (Dreifuss 1989; Wolf 1985).

Juvenile Myoclonic Epilepsy (Impulsive Petit Mal)

These patients are identified by the occurrence of characteristic myoclonic jerks, GTCS, a family history of seizures, and photosensitivity. Precipitants include sleep deprivation, stress, and alcohol use. The myoclonic jerks are prominent in the morning. The correct diagnosis is important because the seizures are usually well controlled with valproate. The prognosis for normal intellectual function is usually good; however, the outlook for a drug-free remission is poor (Asconape and Penry 1984; Delgado-Escueta and Enrile-Bascal 1984).

Syndromes With Less Favorable Behavioral Outcomes

Infantile Spasms (West Syndrome)

This syndrome is characterized by arrest of psychomotor development in an infant younger than 1 year with infantile spasms and the pattern of hypsarrhythmia on EEG. Prognosis generally is poor, especially if the etiology can be identified (Jeavons et al. 1973; Seki et al. 1976). Lack of response to a treatment trial with ACTH or steroid administration also heralds a poor outcome (Glaze et al. 1988; Kellaway et al. 1979; Lacy and Penry 1979; Lombroso 1983; Matsumoto et al. 1981; Sneed et al. 1989).

The EEG pattern of hypsarrhythmia is a characteristic of this syndrome. Patients are divided into groups depending on the ability to identify a specific etiology causing this syndrome. Most patients have neurological injury with superimposition of seizures rather than seizures causing lack of development of intellectual function.

Diagnosis by clinical pattern and characteristic EEG changes requires a therapeutic trial with adrenocorticotropic hormone (ACTH). Steroid responsiveness tends to differentiate subgroups with better overall prognosis. The outcome for becoming seizure-free ranges from 40% to 60% of patients. However, the prognosis for attaining normal intellectual function is poor. Less than 25% of patients with infantile spasms attain normal IQ (Jeavons et al. 1973; Seki et al. 1976). Some reports find better intellectual function in patients with idiopathic disease (Matsumoto et al. 1981; Riikonen et al. 1982).

Debate over the exact dose of ACTH to be administered and the timing of steroid treatment tends to confound interpretation of reports. Most investigators suggest institution of immediate treatment with ACTH, with better intellectual development in patients treated within 1 month of diagnosis (Pollack et al. 1979; Sneed et al. 1989).

Lennox-Gastaut Syndrome

This syndrome is characterized by onset between 1 and 8 years of age. The behavioral seizures include myoclonic, partial, generalized tonic–clonic, axial, atonic, and absence seizures. EEG shows abnormal background, with slow spike waves and multifocal changes. Mental retardation is observed in more than 80% of patients, and seizures often are intractable with only sporadic patients experiencing remission of epilepsy (Beaumanoir 1985). The evolution of this syndrome results in a chronic disorder with impairment of intellectual and psychological function. Seizures tend to remit and exacerbate. Remission of seizures and functional psychological performance is rare (Beaumanoir 1985; Dreifuss 1989).

Childhood epilepsy syndromes with poor prognosis for seizure control and intellectual outcome occasionally present with variants in which the patient has a poor response to medication but has a benign course. Patients with features of benign focal epilepsy of childhood, but with intractable myoclonic–astatic epilepsy, have been identified within the clinical pattern of Lennox-Gastaut (Deonne et al. 1986).

Epilepsy With Myoclonic–Astatic Seizures

Seizure patterns are similar to those observed in the Lennox-Gastaut syndrome, with astatic or myoclonic–astatic seizures of high frequency; status epilepticus also is common. Onset is in childhood, with a familial tendency and normal development prior to the onset of seizures. Tonic seizures that occur late in the course of the illness are associated with unfavorable outcome (Doose et al, 1970; Dreifuss 1989).

Epilepsy With Myoclonic Absences

These patients have tonic contractions associated with absences and severe bilateral rhythmical clonic jerks. The patient maintains awareness of the jerks. Seizures often are intractable, with mental deterioration and some tendency to evolve to other less favorable syndromes (Tassinari 1985).

Early Myoclonic Encephalopathy

This syndrome is associated with poor outcome and early death or evolution to infantile spasms. No specific etiologic factor has been identified. Onset is usually before 3 months of age. The seizures are characterized by the occurrence of fragmentary myoclonus, erratic partial seizures, massive myoclonia, or tonic spasms. Suppression–burst patterns on EEG evolve into hypsarrhythmia (Aicardi 1985; Dreifuss 1989).

Neonatal Seizures

Patterns of clinical manifestations are subtle; they involve tonic, horizontal deviation of the eyes, with presence or absence of jerking; blinking of the eyelids or eyelid flutter; sucking, smacking, or other patterns of buccal–lingual movements; swimming or bicycle pedaling movements; or apneic spells (Dreifuss 1989). Patients may have posturing in tonic extension or multifocal clonic seizures. Clonus may migrate, and they may have strictly focal clonic seizures. EEG shows a suppression–burst pattern. The presence of tonic seizures portends a poor prognosis because of the underlying pathogenesis, such as intraventricular hemorrhage. Myoclonic patterns in the neonatal period have a poor prognosis because of evolution into the myoclonic encephalopathy syndrome (Dreifuss 1989; Mizrahi and Kellaway 1987). Of patients with neonatal seizures who survive, 20% to 30% have congenital malformations or develop seizures later in life (Brown 1973; Cavazzuti et al. 1984; Holden et al 1982; Nelson and Ellenberg 1986; Schulte 1966).

Severe Myoclonic Epilepsy in Infancy

These children develop normally until seizures begin during the first year of life. The seizure pattern is generalized or has a unilateral clonic pattern precipitated by fever. Secondary myoclonic seizures and partial seizures develop. EEG shows focal and generalized spike waves and polyspikes with photosensitivity. There is a characteristic family history of febrile and nonfebrile seizures. Patients have retarded psychomotor development with ataxia, pyramidal dysfunction, and eventually interictal myoclonus. The seizures are intractable to all forms of treatment (Dravet et al. 1985).

Epilepsy With Continuous Spike Waves During Slow-Wave Sleep

These patients have atypical absence when awake and partial or generalized seizures during sleep. EEG shows diffuse and continuous spike waves during slow-wave sleep. Absence of tonic seizures portend a good prognosis with eventual abatement of seizures (Tassinari et al. 1985).

Acquired Epileptic Aphasia (Landau-Kleffner syndrome)

These patients have generalized convulsive or partial seizures with rapid reduction of spontaneous speech and verbal auditory agnosia. The EEG shows multifocal spikes and spike-and-wave discharges. Seizures tend to remit younger than age 15, as do the EEG charges. Remission of the aphasia depends on the age of onset and on rehabilitative efforts (Landau and Kleffner 1957).

Continuous Partial Seizures

Patients have partial seizures with motor manifestations and later develop myoclonic jerks arising from the same anatomical region. The EEG is normal except for focal spikes and slow waves. Prognosis is related to the underlying pathogenic lesion and usually is not associated with progression in clinical pattern. The childhood form of this syndrome, with focal seizures involving motor systems, progressive motor deficits, and mental deterioration, may be of viral etiology (Dreifuss 1989; Rasmussen et al. 1958).

Progressive Myoclonic Epilepsy

Although generalized or focal myoclonic seizures are characteristic of several childhood syndromes, progressive neurological dysfunction with impairments of pyramidal or extrapyramidal motor systems or visual systems identifies patients with this form of myoclonic epilepsy (Berkovic et al. 1986; Roger 1985). Causes of progressive myoclonic epilepsy include Unverricht-Lundborg disease, Lafora-Body Disease, neuronal ceroid lipofuscinosis, sialidosis, and mitochondrial encephalomyopathy (Berkovic et al. 1986). Because benign childhood syndromes occasionally are difficult to differentiate from progressive myoclonic seizures, repetitive examination with assessment of neurological deficits identifies patients with diseases of poor prognosis. The outcome depends on the underlying disease; the progressive myoclonic epilepsies are associated with a uniform fatal outcome.

Sudden Death

Questions about effects of epilepsy on life span and the impact of chronic medication (Reynolds 1975) use may have practical implications for patient management. A review of coroner's cases (Terrence et al. 1975) reveals common factors in sudden death among epileptics. Most deaths occur while patients are in bed, and 6% to 30% occur during sleep. At autopsy, few patients had therapeutic blood levels of prescribed antiepileptic drugs; 50% had no detectable levels (Jay and Leestma 1981). Seizures were not always a factor in the events immediately associated with death (Hirsch and Martin 1971); cardiac arrhythmia has been implicated (Leestma et al. 1984; 1989). The prevalence of sudden death has been estimated to be between 1 of 2000 and 1 of 900 patients with epilepsy.

Remission and Discontinuation of Antiepileptic Drugs

Although treatment-induced remission with continued use of antiepileptic medications is reached in 50% to 70% of all patients (Reynolds 1987), the ability to achieve drug-free remission appears to depend on age (Pestre et al. 1987). Children have a relatively good prognosis for a drug-free remission, with an average relapse rate during a trial off of medication of 35% (Blom et al. 1978; Chadwick 1985; Chadwick and Reynolds 1985; Holowach et al. 1972; Janz 1987; Thurston et al. 1982). Patients who are seizure-free for 4 to 5 years may be offered an opportunity to see if a remission can be sustained without medication. At least 50% of eligible adults in one series declined such a trial because of social and situational difficulties, including the need to drive an automobile and fear of seizure recurrence (Overweg et al. 1987). An exacerbation of seizures during a trial off of medication may be associated with an inability to achieve the pretrial levels of complete control (Todt 1984). Approximately 50% to 65% of adult patients with 2- to 5-year treatment remission may sustain control off drugs (Callaghan et al. 1988; Overweg et al. 1987).

The rate of discontinuation of medication in a remitted patient seems to be independent of pharmacologic principles. Children with medication dosage reductions conducted for 6 to 12 months remain in remission in a greater percentage than those with drug tapering over shorter periods of time (Arts et al. 1988; Todt 1984). However, shorter taper times may prove to be effective as well (Tennison et al. 1989). Identification of patients at risk for recurrence should allow some refinement of selection of controlled patients who might be eligible for a trial off of medication. Patients who need indefinite treatment include those with juvenile myoclonic epilepsy.

Decisions to withdraw antiepileptic medications should be governed by knowledge of factors that favor success. If the pattern of seizure was generalized tonic–clonic in nature, then successful withdrawal is more likely. A drug-free remission is favored by short duration of seizures prior to induction of remission, a long period of control (Oller-Daurella 1999; Thurston 1982), a normal EEG or normalization of the EEG over the treatment time (Shinnar et al. 1985; Todt 1984), and absence of focal or diffuse cerebral lesions. Slow drug withdrawal also favors sustaining remission.

The Single Seizure

A patient with an initial isolated seizure presents a dilemma regarding knowledge of prognosis. The decision is not made in isolation; full and informed participation by the patient is critical. Social variables, including loss of driving privileges and the situational embarrassment of recurrent seizures, influence the decision to treat a patient with an isolated seizure. Naturally, the critical variable is identification of the event as isolated or initial. Careful assessment of historical information may show that a patient has had events in the past that, on knowledgeable scrutiny, are revealed to be seizures (Annegers et al. 1979).

If a seizure is the first unprovoked event, then risk factors for recurrence must be assessed, drug risks reviewed (Reynolds 1975), and the patient's cooperation enlisted in the decision to treat or not treat with antiepileptic drugs. Recurrence is most common within the first 6 months after the signal seizure (Elwes et al. 1985). More than 50% of patients who have a recurrence do so within 6 months. If a patient is a willing and informed participant, then it is prudent to withhold antiepileptic drugs until a second seizure occurs. This assumes that risk factors are not present (Hirtz et al. 1984; Johnson et al 1972). The recurrence rate after a first unproved seizure varies from 36% to 77%. If care is taken to be sure that the signal seizure is the first episode, then 36% of patients have a second seizure.

Risk factors for recurrence following an initial seizure include patients with known prior neurological injury or lesions, a positive family history of epilepsy in a sibling, a persistent but transient neurological deficit (Todd's paralysis), or an EEG with generalized epileptiform discharges (Annegers et al. 1986; Hauser et al. 1982). The risk of recurrence after two seizures approaches 80% to 90%; such a patient must be treated with antiepileptic medications (Camfield et al. 1985; Cleland et al. 1981; Reynolds 1987).

Prognosis of Patients Treated With Surgery

Patients selected for surgical intervention usually meet criteria that include intractable seizures arising from brain regions amenable to respective procedures (Engel 1987). Intractability is defined as a condition of enduring seizures in spite of treatment with all major antiepileptic drugs administered to the point of clinical toxicity. Thus patients treated using the upper limits of published blood levels for antiepileptic drugs as a guide have not had an adequate trial of medical therapy on a given drug. Other criteria include the ability to identify a focus of epileptiform discharge that is consistently associated with the patient's clinical seizure, localization of dysfunction by neuropsychological testing, and delineation of focal structural or metabolic abnormalities by MRI or positron emission tomography (Theodore et al. 1986).

Surgical procedures for epilepsy include *en bloc* resection of the temporal lobe, removal of extratemporal foci by regional re-section, partial hemispherectomy, and corpus callosum sectioning (CCS). The criteria for use of the latter procedure differs from the goals for temporal lobectomy or focal resection (Walczak et al. 1990; Wieser 1988). The usual intent of focal resection is complete control of seizures to the point of cure without continued need for antiepileptic drugs. Hemispherectomy is effective for complete seizure control while CCS is a palliative procedure used with the intent to eliminate seizures causing injuries or to reduce the incidence of a specific type of seizure (Spencer et al. 1988).

The outcome for these surgical procedures is shown in Table 34-1. The data are taken from large centers applying protocols for performance of surgical treatment. The efficacy of hemispherectomy is high, although the numbers of patients qualifying for this procedure is low (Rasmussen and Villemure 1989). Temporal lobectomy results in complete control of seizures in 55% of patients. The reason that resection of foci not within the temporal lobe is less effective than temporal resection is unknown, although methodology to localize such lesions and possible substrate differences regarding the tendency to develop epilepsy in these patients may be operant. Although callosal section appears much less efficacious than the other procedures, the intent usually is not complete seizure control but palliation (Spencer et al. 1988). CCS often is used when hemispherectomy cannot be performed (Spencer 1988).

Application of surgical treatment for epilepsy requires careful selection of patients,

Table 34-1. Survey of Outcome

	Hemispherectomy	Temporal Lobectomy	Ext T	CCS
Patients	88	2,336	825	197
Seizure free	77.3%	55.5	43.2	5.0%
Improved	18.2%	27.7	27.8	71.0
Not improved	4.5%	16.8	29.1	23.9

Modified from Engel, J., Jr. Outcome with respect to epileptic seizures. In: Surgical treatment of the epilepsies. New York: Raven Press; 1987: p 553–571 with permission.

(Ext T, extra temporal resection; CCS, corpus callosum section)

using interictal and ictal recordings of EEG combined with careful assessment to localize foci by several modalities. Prognosis is good for cure in carefully selected patients (Lindsay et al. 1984).

Conclusion

Nonictal influences on prognosis include the presence of structural neurological disease, psychiatric disorders, and situational handicaps that interfere with drug compliance. Long duration of epilepsy before institution of treatment, as observed by Gowers in 1895, the presence of a symptomatic seizure disorder; and a family history of epilepsy also infuence prognosis. Although speculative, early treatment and achievement of complete control of seizures may have an influence on the interval between seizures because the time between seizures appears to shorten with increase in the number of tonic–clonic seizures.

Planning should involve an outline of a sequence of drugs to be tried and information for the patient regarding symptoms that may signal toxicity, because dosing will be governed by clinical responses and by measurement of antiepileptic blood levels. High-dose monotherapy, with avoidance of duplication of drug types and sequencing, should improve chances that an efficacious drug can be identified. Because addition of a second drug is not always helpful, sequencing of the major antiepileptic drugs should be completed before assuming monotherapy is not efficacious. Although efforts to control seizures should not be rationed, planning and communication will be aided by identification of patients at high risk for treatment failure.

References

Aicardi, J. Early myoclonic encephalopathy. In: Roger, J.; Dravet, C.; Bureau, M.; Dreifuss, F. E.; Wolf, P. eds. Epileptic syndromes in infancy, childhood and adolescence. London: John Libbey; 1985: p. 89–99.

Annegers, J. F.; Hauser, W. A.; Shirts, S. B.; Kurland, L. T. Factors prognostic of unprovoked seizures after febrile convulsions. N. Engl. J. Med. 316:493–498; 1987.

Annegers, J. F.; Hauser, W. A.; Elveback, L. R. Remission of seizures and relapse in patients with epilepsy. Epilepsia 20:729–737; 1979.

Annegers, J. F.; Shirts, S. B.; Hauser, W. A.; Kurland, L. T. Risk of recurrence after initial unprovoked seizure. Epilepsia 27:43–50; 1986.

Arts, W. F. M.; Visser, L. H.; Loonen, M. C. B.; Tjiam, A. T.; Stroink, H.; Stuurman, P. M.; Poortvliet, D. C. J. Follow-up of 146 children with epilepsy after withdrawal of antiepileptic therapy. Epilepsia 29:244–250; 1988.

Asconape, J.; Penry, J. K. Some clinical and EEG aspects of benign juvenile myoclonic epilepsy. Epilepsia 25:108–114; 1984.

Beaumanoir, K. A. The Lennox-Gastaut syndrome. In: Roger, J.; Dravet, C.; Bureau, M.; Dreifuss, F. E.; Wolf, P. eds. Epileptic syndromes in infancy, childhood and adolescence. London: John Libbey; 1985: p. 89–99.

Beaussart, M. Benign epilepsy of children with rolandic (centro-temporal) paroxysmal foci. A clinical entity. Study of 221 cases. Epilepsia 13:795–811; 1972.

Beghi, E.; Tognoni, G. Prognosis of epilepsy in newly referred patients: a multicenter prospective study. Epilepsia 29:236–243; 1988.

Berkovic, S. F.; Andermann, F.; Carpenter, S.; Wolfe, L. S. Progressive myoclonus epilepsies: specific causes and diagnosis. N. Engl. J. Med. 315:297–306; 1986.

Bjerre, I.; Corelius, E. Benign familial neonatal convulsions. Acta. Paediatr. Scand. 57:557–561; 1968.

Blom, S.; Heijbel, J.; Bergfors, P. G. Incidence of epilepsy in children: a follow-up study three years after the first seizure. Epilepsia 19:343–350; 1978.

Brown, J. K. Convulsions in the newborn period. Dev. Med. Child. Neurol. 15:823–846; 1973.

Callaghan, N.; Garrett, A.; Goggin, T. Withdrawal of anticonvulsant drugs in patients free of seizures for 2 years. A prospective study. N. Engl. J. Med. 318:942–946; 1988.

Camfield, P. R.; Camfield, C. S.; Dooley, J. M.; Tibbles, J. A. R.; Fung, T.; Garner, B. Epilepsy after a first unprovoked seizure in childhood. Neurology 35:1657–1660; 1985.

Cavazzuti, G. B. Epidemiology of different types of epilepsy in school-age children of Modena, Italy. Epilepsia 21:57–62; 1980.

Cavazzuti, G. B.; Ferrari, P.; Lalla, M. Follow-up study of 482 cases with convulsive disorders in the first year of life. Dev. Med. Child. Neurol. 26:425–437; 1984.

Chadwick, D. The discontinuation of antiepileptic therapy. In: Meldrum, B. M.; Pedley, T. A., eds. Recent advances in epilepsy, vol.

2. Edinburgh: Churchill Livingston; 1985: p. 111–124.

Chadwick, D.; Reynolds, E. H. When do epileptic patients need treatment? Starting and stopping medication. Br. Med. J. 290:1885–1888; 1985.

Chao, D.; Sexton, J. A.; Pardo, S. S. Temporal lobe epilepsy in children. J. Pediatr. 60:686–693; 1962.

Cleland, P. G.; Mosquera, I.; Steward, W. P.; Foster, J. B. Prognosis of isolated seizures in adult life. Br. Med. J. 283:1364; 1981.

Commission on Classification and Terminology of the International League Against Epilepsy. Proposal for revised classification of epileptics and epileptic syndromes. Epilepsia 30:389–399; 1989.

Commission on Classification and Terminology of the International League Against Epilepsy. Proposal for revised clinical and electroencephalographic classification of epileptic seizures. Epilepsia 22:489–501; 1981.

Currier, R. D.; Kooi, K. A.; Saidman, L. J. Prognosis of pure petit mal. A follow-up study. Neurology 13:959–967; 1963.

Currie, S.; Heathfield, K. W. G.; Henson, R. A.; Scott, D. F. Clinical course and prognosis of temporal lobe epilepsy: a survey of 666 patients. Brain 94:173–190; 1971.

Delgado-Escueta, A. V.; Enrile-Bascal, F. Juvenile myoclonic epilepsy of Janz. Neurology 34:285–294; 1984.

Deonne, T.; Ziegler, A. L.; Despland, P. A. Combined myoclonic-astatic and "benign" focal epilepsy of childhood ("Atypical benign partial epilepsy of childhood"). A separate syndrome? Neuropediatrics 17:144–151; 1986.

Doose, H.; Gerken, H.; Leonhardt, R.; Volzke, E.; Volz, C. Centrencephalic myoclonic-astatic petit mal. Clinical and genetic investigations Neuropediatrics 2:59–78; 1970.

Dravet, C.; Bureau, M.; Roger, J. Severe myoclonic epilepsy in infants. In: Roger, J.; Dravet, C.; Bureau, M.; Dreifuss, F. E.; Wolf, P., eds. Epileptic syndromes in infancy, childhood and adolescence. London: John Libbey; 1985: p. 58–67.

Dravet, C.; Bureau, M.; Roger, J. Benign myoclonic epilepsy in infants. In: Roger, J.; Dravet, C.; Bureau, M.; Dreifuss, F. E.; Wolf, P., eds. Epileptic syndromes in infancy, childhood and adolescence. London: John Libbey; 1985: p. 121–129.

Dreifuss, F. E. Pediatric epilepsy syndromes: an overview. Clev. Clin. J. Med. 56:s166–s171; 1989.

Elwes, R. D. C.; Johnson, A. L.; Shorvon, S. D.; Reynolds, E. H. The prognosis for seizure control in newly diagnosed epilepsy. N. Engl. J. Med. 311:944–947; 1984.

Elwes, R. D. C.; Chesterman, P.; Reynolds, E. H. Prognosis after a first untreated tonic-clonic seizure. Lancet 2:752–753; 1985.

Engel, J., Jr. Outcome with respect to epileptic seizures. In: Surgical treatment of the epilepsies. New York: Raven Press; 1987: p. 553–571.

Gastaut, H. A new type of epilepsy: benign partial epilepsy of childhood with occipital spike-waves. In: Akimoto, H.; Kazamatsuri, H.; Seine, M., eds. Advances in epileptology: XIIIth Epilepsy International symposium. New York: Raven Press; 1982: p. 18–25.

Glaze, D. G.; Hrachovy, R. A.; Frost, J. D., Jr.; Kellaway, P.; Zion, T. E. Prospective study of outcome of infants with infantile spasms treated during controlled studies of ACTH and prednisone. J. Pediatr. 112:389–396; 1988.

Gowers, W. R. Epilepsy and other chronic convulsive diseases: their causes, symptoms and treatment. New York: Dover Publications; Reprinted 1964. First Published 1885 by William Wood and Company, London.

Hauser, W. A.; Anderson, V. E.; Loewenson, R. B.; McRoberts, S. M. Seizure recurrence after a first unprovoked seizure. N. Engl. J. Med. 307:522–528; 1982.

Hauser, W. A. Epidemiology of epilepsy in children. Cleveland Clin. J. Med. 56:s185–s194; 1989.

Hirsch, C. S.; Martin, D. L. Unexpected death in young epileptics. Neurology 21:682–690; 1971.

Hirtz, D. G.; Ellenberg, J. H.; Nelson, K. B. The risk of recurrence of nonfebrile seizures in children. Neurology 34:637–641; 1984.

Holden, K. R.; Mellits, E. D.; Freeman, J. M. Neonatal seizures. I. Correlations of prenatal and perinatal events with outcomes. Pediatrics 70:165–176; 1982.

Holowach, J.; Thurston, D. L.; O'Leary, J. Prognosis of childhood epilepsy. N. Engl. J. Med. 286:169–174; 1972.

Janz, D. When should antiepileptic drug treatment be terminated? In: Advances in epileptology, vol. 16. New York: Raven Press; 1987: p. 365–372.

Jay, G. W.; Leestma, J. E. Sudden death in epilepsy: a comprehensive review of the literature and proposed mechanism. Acta. Neurol. Scand. 63(suppl. 82):1–66; 1981.

Jeavons, P. M.; Bowser, B. D.; Dimitrakoudi, M. Long-term prognosis of 150 cases of "West Syndrome." Epilepsia 14:153–164; 1973.

Johnson, L. C.; DeBolt, W. L.; Long, M. T.; Ross, J. J.; Sassin, J. F.; Arthur, R. J.; Walter, R. D. Diagnostic factors in adult males following initial seizures. Arch. Neurol. 27:193–197; 1972.

Juul-Jensen, P. Epilepsy. A clinical and social

analysis of 1020 adult patients with epileptic seizures. Copenhagen: Munksgaard; 1963.

Kellaway, P.; Hrachovy, R. A.; Frost, J. D.; Zion, T. Precise characterization and quantification of infantile spasm. Ann. Neurol. 6:214–218; 1979.

Keranen, T.; Sillanpaa, M.; Piekkinen, P. J. Distribution of seizure types in an epileptic population. Epilepsia 29:1–7; 1988.

Kotagal, P.; Rothner, A. D.; Erenberg, G.; Curse, R. P.; Wyllie, E. Complex partial seizures of childhood onset: a five-year follow-up study. Arch. Neurol. 44:1177–1180; 1987.

Lacy, J. R.; Penry, J. K. Infantile Spasms. New York: Raven Press; 1976.

Landau, W. M.; Kleffner, F. R. Syndrome of acquired aphasia with convulsive disorder in children. Neurology 7:523–530; 1957.

Leestma, J. E.; Kalelkar, M. B.; Teas, S. S.; Jay, G. W.; Hughes, J. R. Sudden unexpected death associated with seizures: analysis of 66 cases. Epilepsia 25:84–88; 1984.

Leestma, J. E.; Walczak, T.; Hughes, J. R.; Kalelkar, M. B.; Teas, S. S. A prospective study on sudden unexpected death in epilepsy. Ann. Neurol. 26:195–203; 1989.

Lennox, W. G.; Lennox, M. A. Epilepsy and related disorders, vol. I. Boston: Little, Brown and Co; 1960.

Lindsay, J.; Ounsted, C.; Richards, P. Long-term outcome in children with temporal lobe seizures: I. Social outcome and childhood factors. Dev. Med. Child Neurol. 21:285–298; 1979.

Lindsay, J.; Ounsted, C.; Richards, P. Long-term outcome in children with temporal lobe seizures. V. Indications and contra-indications for neurosurgery. Dev. Med. Child Neurol. 26:25–32; 1984.

Livingston, S.; Torres, I.; Pauli, L. L.; Rider, R. V. Petit mal epilepsy. Results of a prolonged follow-up study of 117 patients. J.A.M.A. 194:227–232; 1965.

Loiseau, P. Intractable epilepsy: prognostic evaluation. In: Schmidt, D.; Morselli, P. L., eds. Intractable Epilepsy. Experimental and clinical aspects. New York: Raven Press; 1986: p. 227–236.

Loiseau, P. Childhood absence epilepsy. In: Roger, J.; Dravet, C.; Bureau, M.; Dreifuss, F. E.; Wolf, P. eds. Epileptic syndromes in infancy, childhood and adolescence. London: John Libbey; 1985: p. 106–120.

Loiseau, P.; Dartigues, J. F.; Pestre, M. Prognosis of partial epileptic seizures in the adolescent. Epilepsia 24:472–481; 1983.

Loiseau, P.; Duche, B.; Cordova, S.; Dartigues, J. F.; Cohadon, S. Prognosis of benign childhood epilepsy with centrotemporal spikes: a follow-up study of 168 patients. Epilepsia 29:229–235; 1988.

Lombroso, C. T. Sylvian seizures and midtemporal spike foci in children. Arch. Neurol. 17:52–59; 1967.

Lombroso, C. T. A prospective study of infantile spasms: clinical and therapeutic correlations. Epilepsia 24:135–158; 1983.

Matsumoto, A.; Watanabe, K.; Negoro, T.; Sugiura, M.; Iwase, K.; Hara, K.; Miyazaki, S. Long-term prognosis after infantile spasms: a statistical study of prognostic factors in 200 cases. Dev. Med. Child Neurol. 23:51–65; 1981.

Mattson, R. H.; Cramer, J. A.; Collins, J. F.; Smith, D. B.; Delgado-Escueta, A. V.; Browne, T. R.; Williamson, P. D.; Treiman, D. M.; McNamara, J. O.; McCutchen, C. B.; Homan, R. W.; Crill, W. E.; Lubozynski, M. F.; Rosenthal, N. P.; Mayersdorf, A. Comparison of carbamazepine, phenobarbital, phenytoin, and primidone in partial and secondarily generalized tonic–clonic seizures. N. Engl. J. Med. 313:145–151; 1985.

Mizrahi, E.; Kellaway, P. Characterization and classification of neonatal seizures. Neurology 37:1837–1844; 1987.

Nelson, K. B.; Ellenberg, J. H. Antecedents of seizure disorders in early childhood. Am. J. Dis. Child. 140:1053–1062; 1986.

Okuma, T.; Kumashiro, H. Natural history and prognosis of epilepsy: report of a multi-institutional study in Japan. Epilepsia 22:35–53; 1981.

Oller-Daurella, L.; Pamies, R.; Oller, F. V. Reduction or discontinuance of antiepileptic drugs in patients seizure free for more than 5 years. In: Janz, D., ed. Epileptology. Stuttgart: Thieme-Verlag; 1976: p. 218–227.

Overweg, J.; Binnie, C. D.; Oosting, J.; Rowan, A. J. Clinical and EEG prediction of seizure recurrence following antiepileptic drug withdrawal. Epilepsy Res. 1:272–283; 1987.

Panayiotopoulos, C. P. Benign childhood epilepsy with occipital paroxysms: a 15-year prospective study. Ann. Neurol. 26:51–56; 1989.

Penry, J. K.; Porter, R. J.; Dreifuss, F. E. Simultaneous recording of absence seizures with videotape and electroencephalography: a study of 374 seizures in 48 patients. Brain 98:427–440; 1975.

Pestre, M.; Loiseau, P.; Larrieu, E.; Dartigues, J. F.; Cohadon, S. Withdrawal of antiepileptic drug therapy in adolescent epileptic patients. In: Wolf, P.; Dam, M.; Janz, D.; Dreifuss, F. E., eds. Advances in epileptology, vol. 16. New York: Raven Press; 1987: p. 395–400.

Pollack, M. A.; Zion, T. E.; Kellaway, P. Long term prognosis of patients with infantile spasms following ACTH therapy. Epilepsia 20:255–260; 1979.

Rasmussen, T.; Villemure, J. G. Cerebral hemispherectomy for seizures with hemiplegia.

Cleve. Clin. J. Med. 56(suppl. 1):S62–S68; 1989.

Rasmussen, T. E.; Olszewsik, J.; Lloyd-Smith, D. Focal cortical seizures due to chronic localized encephalitides. Neurology 8:435–445; 1958.

Reynolds, E. H. Early treatment and prognosis of epilepsy. Epilepsia 28:97–106; 1987.

Reynolds, E. H. Chronic antiepileptic toxicity: a review. Epilepsia 16:319–352; 1975.

Reynolds, E. H.; Elwes, R. D. C.; Shorvon, S. D. Why does epilepsy become intractable? Lancet 2:952–954; 1983.

Reynolds, E. H. Early treatment and prognosis of epilepsy. Epilepsia 28(2):97–106; 1987.

Reynolds, E. H. The initiation of anticonvulsant drug therapy. In: Pedley, T. A.; Meldrum, B. S., eds. Recent advances in epilepsy, vol. 2. Edinburg: Churchill Livingston; 1984: p. 101–110.

Riikonen, R. A long-term follow-up of 214 children with the syndrome of infantile spasms. Neuropediatrics 13:14–23; 1982.

Rodin, E. A. The prognosis of patients with epilepsy. Springfield, Ill: Charles C Thomas; 1969.

Roger, J. Progressive myoclonic epilepsy in childhood and adolescence. In: Roger, J.; Dravet, C.; Bureau, M.; Dreifuss, F. E.; Wolf, P., eds. Epileptic syndromes in infancy, childhood and adolescence. London: John Libbey; 1985: p. 302–310.

Rose, A. L.; Lombroso, C. T. Neonatal seizure states. Pediatrics 45:404–425; 1970.

Sato, S.; Dreifuss, F. E.; Penry, J. K.; Kirby, D. D.; Palesch, Y. Long-term follow-up of absence seizures. Neurology 33:1590–1595; 1983.

Saunders, M.; Marshall, C. Isolated seizures: an EEG and clinical assessment. Epilepsia 16:731–733; 1975.

Schmidt, D. Single drug therapy for intractable epilepsy. J. Neurol. 229:221–226; 1983.

Schmidt, D.; Tsai, J. J.; Janz, D. Generalized tonic–clinic seizures in patients with complex partial seizures: Natural history and prognostic relevance. Epilepsia 24:43–48; 1983.

Schmidt, D. Two antiepileptic drugs for intractable epilepsy. J. Neurol. Neurosurg. Psychiat. 45:1119–1124; 1982.

Schmidt, D. Prognosis of chronic epilepsy with complex partial seizures. J. Neurol. Neurosurg. Psychiat. 47:1274–1278; 1984.

Schulte, F. J. Neonatal convulsions and their relation to epilepsy in early childhood. Dev. Med. Child Neurol. 8:381–392; 1966.

Seki, T.; Kawahara, Y.; Hirose, M. The long-term prognosis of infantile spasms—The present condition of cases in infantile spasms followed in school age. Folia. Psychiatr. Neurol. Japan 30:297–306; 1976.

Shinnar, S.; Vining, E. P. G.; Mellits, E. D.; D'Souza, B. J.; Holden, K.; Baumgardner, R. A.; Freeman, J. M. Discontinuing antiepileptic medication in children with epilepsy after two years without seizures. N. Engl. J. Med. 313:976–980; 1985.

Shorvon, S. D.; Reynolds, E. H. Early prognosis of epilepsy. Br. Med. J. 285:1699–1701; 1982.

Shorvon, S. D. The temporal aspects of prognosis in epilepsy. J. Neurol. Neurosurg. Psychiat. 47:1157–1165; 1984.

Sneed, O. C.; Benton, J. W.; Hosey, L. C.; Swann, J. W.; Spink, D.; Martin, D.; Rej, R. Treatment of infantile spasms with high-dose ACTH: efficacy and plasma levels of ACTH and cortisol. Neurology 39:1027–1031; 1989.

Spencer, S. S. Corpus callosum section and other disconnection procedures for medically intractable epilepsy. Epilepsia 29(suppl. 2):S85–S99; 1988.

Spencer, S. S.; Spencer, D. D.; Williamson, P. D.; Sass, K.; Novelly, R. A.; Mattson, R. H. Corpus callosotomy for epilepsy. I. Seizure effects. Neurology 38:19–24; 1988.

Tassinari, C. A.; Bureau, M. Epilepsy with myoclonic absences In: Roger, J.; Dravet, C.; Bureau, M.; Dreifuss, F. E.; Wolf, P., eds. Epileptic syndromes in infancy, childhood and adolescence. London: John Libbey; 1985: p. 121–129.

Tassinari, C. A.; Bureau, M.; Dravet, C.; Dalla Bernardina, B.; Roger, J. Epilepsy with continuous spikes and waves during slow sleep. In: Roger, J.; Dravet, C.; Bureau, M.; Dreifuss, F. E.; Wolf, P., eds. Epileptic syndromes in infancy, childhood and adolescence. London: John Libbey; 1985: p. 194–204.

Tennison, M. B.; Greenwood, R. S.; Lewis, D. V.; Benoit, S. E. Rate of taper of antiepileptic drugs and the risk of seizure recurrence (abstract). Ann. Neurol. 26:439; 1989.

Terrence, C. F.; Wisotzkey, H. M.; Perper, J. A. Unexpected, unexplained death in epileptic patients. Neurology 25:594–598; 1975.

Theodore, W. H.; Dorwart, R.; Holmes, M.; Porter, R. J.; DiChiro, D. Neuroimaging in refractory partial seizures: comparison of PET, CT, and MRI. Neurology 36:750–759; 1986.

Thomas, M. H. The single seizure: its study and management. J.A.M.A. 169:457–459; 1959.

Thurston, J. H.; Thurston, D. L.; Hixon, B. B.; Keller, A. J. Prognosis in childhood epilepsy. Additional follow-up of 148 children 15 to 23 years after withdrawal of anticonvulsant therapy. N. Engl. J. Med. 306:831–836; 1982.

Todt, H. The late prognosis of epilepsy in childhood; results of a prospective follow-up study. Epilepsia 25:137–144; 1984.

Tsai, J. J.; Schmidt, D. Seizure relapse in epilepsy with complex partial seizures. In: Wolf, P.; Dam, M.; Janz, D.; Dreifuss, F. E., eds.

Advances in epileptology, vol. 16. New York: Raven Press; 1987: p. 405–407.

Walczak, T. S.; Radtke, R. A.; McNamara, J. O.; et al. Anterior temporal lobectomy for complex patrial seizures: Evaluation, results, and long-term follow-up in 100 cases. Neurology 40:413–418; 1990.

Wieser, H. G. Selective amygdalo-hippocampec-tomy for temporal lobe epilepsy. Epilepsia 29(suppl. 2):S100–S13; 1988.

Wolf, P. Juvenile absence epilepsy. In: Roger, J.; Dravet, C.; Bureau, M.; Dreifuss, F. E.; Wolf, P., eds. Epileptic syndromes in infancy, childhood and adolescence. London: John Libbey; 1985: p. 242–246.

35

Vasculitis and Collagen Vascular Disease

STEPHEN E. NADEAU

The value of studies of prognosis in vasculitis and collagen vascular disease is limited considerably by ascertainment bias and the relative lack of controlled studies of therapeutic agents.

The potential effects of ascertainment bias are most significant in disorders such as rheumatoid arthritis (RA) and systemic lupus erythematosus (SLE), in which a great range in disease severity exists and many patients are only intermittently or mildly symptomatic. The potential effects of ascertainment bias are less important in disorders such as polyarteritis nodosa (PAN) or Wegener's granulomatosis (WG), which tend to exist only in a severe, often life-threatening form that almost invariably leads to hospitalization.

The absence of controlled studies most adversely affects estimates of the value of therapeutic interventions in chronic, fluctuating disorders with low to moderate acute morbidity and mortality, such as RA and SLE. On the other hand, controlled studies have not been necessary to demonstrate the efficacy of certain treatments for fulminant disorders such as WG; untreated WG is almost uniformly fatal, whereas administration of cyclophosphamide induces lasting remissions in 90% of patients.

Despite these limitations, considerable data are available that are of some use to the clinician. The disorders that are considered have been reviewed comprehensively (Nadeau and Watson 1990); only studies of prognosis are discussed here.

Polyarteritis Nodosa

The 5-year survival rate in untreated patients with PAN is 13%, and 40% to 50% of deaths occur in the first 3 months (Frohnert and Sheps 1967; Sack et al. 1975). Mortality may be higher in older patients and in patients with renal or intestinal involvement (Cohen et al. 1980; Frohnert and Sheps 1967; Leib et al. 1979; Sack et al. 1975). Five-year survival rates of steroid-treated patients in large series have been reported to be 48% to 57% (Cohen et al. 1980; Frohnert and Sheps 1967; Medical Research Council 1960; Sack et al. 1975). Leib and coworkers (1979) report a 5-year survival rate of 12% in untreated patients, 53% in steroid-treated patients, and 80% in patients treated with steroids combined with an immunosuppressive agent, usually azathioprine, in a dose of 100 to 325 mg/day. In a study by Fauci and colleagues (1979), 14 of 17 patients achieved complete remissions using steroids combined with cyclophosphamide in a dose of 1 to 2 mg/kg/day. In two of three patients treated with azathioprine in this group, disease progression was halted only when cyclophosphamide was substituted. The relative value of these two drugs remains a source of controversy.

Plasmapheresis has been reported to be successful in the treatment of vasculitis (Goldman et al. 1979), but data are not adequate to assess its potential impact on mortality.

Limited data are available on the prognosis of certain neurological features. Moore and Fauci (1981) found that the constellation of seizures and encephalopathy was invariably associated with acute disease and resolved quickly and reliably with definitive treatment with cyclophosphamide and steroids. In this same study, eight patients had mononeuritis multiplex, of whom two ultimately returned to normal, one recovered full function but retained some signs of neuropathy, and four improved markedly but remained disabled. Of three patients with complete quadraplegia, two became independent in activities of daily living and one became functionally normal. Chang and associates (1984) found that 85% of patients with vasculitic mononeuropathy multiplex showed meaningful improvement, and 20% returned to normal.

Limited Forms

Polyarteritis nodosa may be limited largely to the skin, in which case nerves and muscles of the lower extremities also may be involved. It also may predominantly affect peripheral nerves, producing mononeuritis multiplex or a symmetric polyneuropathy, in which case there is often muscular involvement that is largely silent clinically. In both of these syndromes, the clinical course usually is far more indolent than in systemic PAN, and the prognosis is relatively benign. Diaz-Perez and Winkelman (1974) reported a single death among 23 patients with cuta-

neous PAN, 48% with neuromuscular involvement during a follow-up of 7 months to 27 years (median, 7 years). Dyck and co-workers (1987) report only three deaths, all from unrelated disease, among 20 patients with isolated vasculitic neuropathy treated with a variety of agents for up to 35 years (median follow-up, 11.5 years). In some instances, isolated vasculitic neuropathy may be associated with malignancy (Harati and Niakan 1986; Vincent et al. 1986), in which case outcome is adversely affected.

Allergic Granulomatosis

Allergic granulomatosis is a variant of PAN characterized by microvascular and small artery involvement, granulomas, eosinophilia, and pulmonary involvement. Chumbley and associates (1977), in the single longitudinal series available, report a 5-year survival rate of 62% with steroid treatment.

Cerebral Vasculopathy With Drug Abuse

Many reports have appeared in the literature of acute stroke, intracerebral hemorrhage, and the anterior spinal artery syndrome in young adults using or abusing drugs, such as amphetamine, ephedrine, phenylpropanolamine, cocaine, and heroin. A cerebral vasculitis frequently has been imputed from the finding of diffuse segmental constrictions on cerebral angiograms, coupled with one report of several cases of systemic PAN presumably related to methamphetamine abuse (Citron et al. 1970). Virtually all available data suggest that this concept of a drug-related, isolated central nervous system (CNS) vasculitis is invalid. The pathology is not characteristic of vasculitis and more closely resembles the vasculopathy associated with aneurysmal subarachnoid hemorrhage (Nadeau and Watson 1990). All patients display features of a single, sudden, severe CNS vascular injury; systemic manifestations characteristically are absent, the sedimentation rate is usually normal, and there is no evidence of an ongoing pathological process. Survival figures are consistent with a single event related to drug-induced vasospasm or hypertension and not with far more malignant, ongoing, inflammatory processes, such as PAN or cerebral granulomatous angiitis. Among 88 cases in 35 reports, 40 patients had intracranial hemorrhage, 27 (68%) of whom survived, seven having received a craniotomy and one, cytotoxic drugs; 15 patients had cerebral infarcts, 14 (93%) of whom survived; 13 patients had myelopathy, 11 (85%) of whom survived; and 20 patients had uncharacterized intracranial events, all of whom survived (Nadeau and Watson 1989). As expected, many patients were left with significant neurological deficits.

Wegener's Granulomatosis

Mean survival in untreated WG is 5 months (Walton 1958). A life table analysis of data from this series indicates 6-month, 1-year, and 2-year survival rates of 36%, 16%, and 6%, respectively. Average survival among patients treated with steroids in 12.5 months (Hollander and Manning 1967). Life table analysis indicates 6-month, 1-year, and 2-year survival rates of 59%, 53%, and 11%, respectively. Cyclophosphamide in conjunction with steroids is now universally regarded as the treatment of choice. Fauci and colleagues (1983) at the National Institutes of Health (NIH) have had the greatest experience with this treatment. In a group of 85 patients, complete remission was achieved in 93%; patients who did not go into remission died of active disease. Mean duration of remission was 48 months (range 7 months to 13.2 years), and 23 had been off therapy for a mean of 35 months. Twenty five had relapsed and required retreatment. Of 11 patients begun on azathioprine, only one went into remission; remission was achieved in the remainder when they were switched to cyclophosphamide. No one else has accumulated such a large series of patients, but several investigators have reported results comparable to those of Fauci and coworkers (Pinching et al. 1983; Reza et al. 1975). No data are available on the efficacy of plasmapheresis, although it is frequently used.

The major causes of death in WG are acute fulminant disease and chronic renal failure.

Data on the response of neurological disease to treatment are scant. Two investigators have remarked on the dramatic reversal of major CNS deficits in five patients in response to definitive treatment (Novack and Pearson 1971; Sahn and Sahn 1976).

Temporal Arteritis

Community-based studies indicate that temporal arteritis (TA), as it is currently managed, has no statistically significant effect on mortality (Bengtsson and Malmvall 1981; Boesen and Sorensen 1987; Gouet et al. 1985; Huston et al. 1978). Comparable data are not available from the presteroid era. Mayo Clinic studies indicate a 5-year survival rate of 74% and a 10-year rate of 53% (Huston et al. 1978). On the other hand, the subpopulation of TA patients referred to neurology and ophthalmology clinics may have a slightly worse prognosis: Graham and associates (1981) found that women in their clinics had a statistically significant reduction in survival. Not surprisingly, causes of death in TA patients are virtually identical to those in age-matched controls. Huston and colleagues (1978) found that in Olmsted County, Minnesota, only one of 21 deaths in TA patients could be attributed to the disease. Deaths attributable to TA are due to myocardial infarction, stroke, or a ruptured abdominal aortic aneurysm (Save-Soderbergh et al. 1986).

The most severe neurological consequence of TA, stroke, occurs so infrequently that incidence figures have not been calculated; almost certainly, less than 1% of patients are affected. However, substantially higher stroke incidence rates may be noted among TA patients referred to neurology or ophthalmology clinics. For example, Graham and coworkers (1981) recorded eight deaths within 6 weeks of presentation in their series of 90 patients, four due to brain stem infarction, three with postmortem documentation of giant-cell arteritis of the vertebral arteries.

The major neuro-ophthalmological consequence of TA, visual loss, occurs frequently; however, it is not possible to calculate the precise risk of this complication because life table methods have never been used, and many reports reflect the substantial ascertainment bias associated with referral populations, particularly when referrals are to ophthalmology or neurology clinics. Unilateral or bilateral loss of vision occurred in up to 50% of patients in older series (Goodman 1979). For example, among Mayo Clinic referrals between 1931 and 1959, 42% became blind in one or both eyes (Hollenhorst et al. 1960). Hamilton and associates (1971) report that 50% of inadequately treated patients with unilateral loss of vision became totally blind. In more recent series, the incidence of visual loss has decreased to the 10% range. Again, the figures for the Mayo Clinic referral population are representative: among 166 biopsy-proven patients seen between 1980 and 1983, 14 (8%) suffered permanent visual loss (Caselli et al. 1988). This marked improvement in visual prognosis reflects the efficacy of steroids and is the principle evidence of that efficacy. It also reflects a trend toward earlier diagnosis and initiation of effective treatment.

Community-based studies avoid most of the ascertainment bias characteristic of clinical series and permit much more valid estimates of the incidence of visual loss. In a Mayo Clinic community study, seven (17%) patients experienced permanent visual impairment, two of whom became blind (Huston et al. 1978). In a community study based in Goteburg, Sweden, nine (7%) patients experienced permanent visual impairment, and two became blind (Bengtsson and Malmvall 1981). Visual loss rarely is an initial feature of TA, but it is usually an early feature (Fig. 35-1). In the Swedish study, acute visual loss was the initial manifestation in only one patient, and it followed other disease features by up to 6 months in the other eight patients. In the Mayo Clinic community study (Huston et al. 1978), all instances of visual loss occurred before the diagnosis of TA was made. Thus, because of the efficacy of steroids in preventing visual loss, the nature and timing of medical intervention are crucial in determining risk of visual impairment.

Steroids constitute definitive treatment of TA. A prospective, randomized study of daily vs. alternate-day steroids has demonstrated that alternate-day steroids are significantly less effective in suppressing signs and symptoms of arteritis (Hunder et al. 1975). Experienced clinicians have not found it necessary to use other immunosuppressive agents in the treatment of TA (Healey and Wilske 1983), but a recent controlled study

Figure 35-1. Estimated duration of symptoms of temporal arteritis before loss of vision occurred: 73 cases from 1931 to 1959 at the Mayo Clinic. Exact timing of steroids was not specified. Steroids were first introduced in October, 1949, but did not become widely available until the mid-1950s. (From Hollenhorst et al. 1960, with permission.)

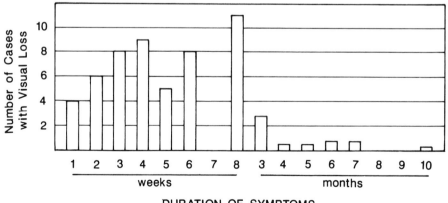

DURATION OF SYMPTOMS

Table 35-1. Incidence of Complications of Steroid Therapy in Temporal Arteritis*

	Allsop and Gallagher (1981)	Spiera and Davison (1978)	Huston et al. (1978)	Beevers et al. (1973)	Gouet et al. (1985)	von Knorring (1979)
Number of patients	63	22	33	44	87	53
Percent with side effects	13	45	61	59	49	19
Number of cases with side effects						
Osteoporosis	3			4		
Vertebral collapse	2	5	11		18	2
Diabetes	2	3	1	2	6	4
Cushingoid features			17	13	17	2
Peptic ulcer disease	2	2		1	4	
Myopathy		3	7	4	14	
Psychosis		3				
Subcapsular cataracts			2		3	1
Fluid retention requiring diuretics				13		
Hypertension or aggravated hypertension				10	15	
Infection					11	1

* Steroid regimens were not specified in detail.

demonstrated that azathioprine is effective in TA and permits a significant reduction in steroid dosage (DeSilva and Hazleman 1986).

Steroids generally are ineffective in reversing visual impairment. In a literature review, Schneider and coworkers (1971) found that vision improved in only 15% of patients after steroid treatment.

Precise data do not exist on the risks of steroid side effects associated with various drug regimens in patients with different risk profiles, but several reports have provided clinically useful information in this regard (Table 35-1).

Takayasu's Arteritis

Considerable data are available on the prognosis of Takayasu's arteritis but there are no valid studies of its natural history. Fraga and associates (1972) report a 2-year mortality rate of 75% based on their review of 135 cases in the early literature, but this figure is seriously confounded by ascertainment bias. In contemporary series, some patients have not received any treatment, but most have been given steroids, a few have had cytotoxic agents, and many have undergone surgery. Ishikawa and colleagues (1983) have provided the most extensive data. Ishikawa (1981) reports both 5- and 10-year survival rates of 90%. Hall and associates (1985), in a Mayo Clinic referral series, re-

port a 5-year survival rate of 94%. Stroke incidence rates are not available, but 7% to 11% of patients in clinical series have experienced strokes (Lupi-Herrera et al. 1977; Paloheimo 1967; Ueda et al. 1969).

Ishikawa (1981) further examined prognosis as a function of clinical profile. Patients were classified according to the presence or absence of four complications generally reflecting the severity of arterial disease: (1) Takayasu's retinopathy, (2) secondary hypertension, (3) aortic regurgitation, and (4) an aortic or arterial aneurysm. Patients with no complications, or a single complication in mild or moderate degree (the mild group), were compared with patients with two or more complications, or a single complication in severe degree (the severe group). Ten-year survival in the mild group was 100%. Five- and 10-year survivals in the severe group were both 74%. Ishikawa (1981) also determined the numbers of patients who survived and had not experienced one or more events that were either life-threatening or had a major impact on quality of life, including stroke, acute pulmonary edema, dissecting aortic aneurysm, unilateral or bilateral blindness, and massive hemoptysis or hematemesis. Overall, event-free 5- and 10-year survival rates were 87% and 80%, respectively. In the mild group, both 5- and 10-year event-free survival rates were 97%. In the severe group, they were

72% and 59%, respectively. In this series, deaths occurred exclusively among patients who presented with marked restriction of physical activity due to easy fatiguability, dyspnea, recurrent syncope, blurred vision, or headache. In a later study, Ishikawa and colleagues (1983) classified patients according to status of cervical vessels. Patients with 70% or greater stenosis of three or less major cervical systems (left and right carotid, left and right vertebral or subclavian—the mild group) were compared with patients with 70% or greater stenosis of three systems and greater than 50% stenosis of the fourth (the severe group). A retinal artery systolic pressure of less than 40 mm and evidence of ischemic retinopathy such as microaneurysms were found almost exclusively in the severe group, and all patients in this group were so affected. Mortality rates were similar in the two groups but rates of stroke or blindness were 25% and 33%, respectively, in the severe group, and 6% and 0%, respectively, in the mild group.

Major causes of death in Takayasu's arteritis include congestive heart failure, renal failure, stroke, and aneurysmal rupture (Lupi-Herrera et al. 1977; Nakao et al. 1967; Ishikawa 1981; Fraga et al. 1972).

Hall and associates (1985) report that 74% of their Mayo Clinic patients were working full time 5 years after diagnosis.

Japanese patients with histocompatibility antigen allotype Bw52–Dw12 appear to experience more fulminant disease with more rapid progression of arterial signs (Numano et al. 1982).

There is a universal consensus as to the efficacy of steroids even without controlled studies or suitable historical controls. Fraga and colleagues (1972) report that steroids induced a prompt fall of the sedimentation rate and resolution of systemic signs and symptoms as well as headache and weakness, and they note a halt in the progression of arterial stenoses and even the reappearance of once absent pulses. Shelhamer and coworkers (1985) found that steroids halted progression in eight of their 16 patients with active disease at NIH. Hall and associates (1985) report that eight of 16 patients with

absent pulses experienced eventual return of pulses.

Only one study is available on the efficacy of cytotoxic agents. Shelhamer and colleagues (1985) found that all seven steroid-resistant patients experienced remission with treatment with cyclophosphamide 2 mg/kg/day, although two of these patients subsequently developed late progression of arterial disease.

Surgery is generally felt to be warranted in patients with refractory renovascular hypertension, arterial aneurysms, or features of cerebral hypoperfusion (recurrent syncope or presyncope, ischemic retinopathy). Mortality in older series was prohibitive (Kimoto 1979), mainly due to intracerebral or subarachnoid hemorrhage—apparently due to inadequate perioperative control of blood pressure—or to arterial dehiscence when patients were operated on during active phases of their disease. Lagneau and colleagues (1987) report one operative death among 39 patients; Robbs and coworkers (1986) report one death among 25 operated; Shelhamer and associates (1985) show no deaths among seven patients undergoing 13 procedures; and Hall and colleagues (1985) show no deaths among 12 operated patients.

Systemic Lupus Erythematosus

Mortality rates among patients with SLE followed at major medical centers have fallen steadily in the last 50 years, and recent publications report 10-year survival rates in excess of 80% (Table 35-2). Several factors can be identified to account for this remarkable improvement in prognosis: changes in ascertainment; advances in general medical care; and the use of specific treatments, particularly steroids.

Ascertainment of increasing numbers of patients with mild disease can be credited to the development of better diagnostic tests, particularly the lupus erythematous (LE) prep and antinuclear antibody (ANA) tests, and to improved diagnostic skills among practicing physicians (Albert et al. 1979, Wofsy 1987). Patients in early studies averaged nearly twice the number of American

Table 35-2. Five- and 10-Year Survival From Time of Diagnosis of Systemic Lupus Erythematosus

Period	Study	Center	Percent Surviving	
			5 Years	10 Years
1938–1953	Jessar et al. (1953)	Columbia–Presbyterian	30.7	—
1949–1959	Kellum and Haserick (1964)	Cleveland Clinic	69.4	53.7
1955–1961	Leonhart (1966)	Sweden	70.5	51.0
1957–1968	Urman and Rothfield (1977)	New York University	72.4	63.4
1958–1969	Fernandez-Herlihy (1972)	Lahey Clinic	89.2	78.0
1963–1969	Estes and Christian (1971)	Columbia–Presbyterian	76.9	59.1
1968–1976	Urman and Rothfield (1977)	University of Connecticut	91.4	84.7
1970–1975	Feinglass et al. (1976)	Johns Hopkins	94.4	81.6
1965–1978	Ginzler et al. (1982)	Multicenter Study	86.0	76.0

Rheumatism Association (ARA) criteria for the diagnosis of lupus, suggesting that they were more severely ill (Albert et al. 1979). The potential impact of this trend on survival data is illustrated in a recent Mayo Clinic study: 5- and 10-year survival among patients with definite SLE (four or more ARA criteria) were 75% and 63%, respectively; comparable figures among patients with suspected SLE (two or three criteria) were 95% and 89%, respectively (Michet et al. 1985). The latter two figures are not significantly different than expected for a normal population.

Important advances in medical care have included antibiotics, diuretics, better drugs for the management of hypertension and congestive heart failure, and recently, dialysis. The impact of these measures on lupus survival is uncertain, but careful review of available data suggests that multiple factors, including steroid treatment, have contributed to improved survival. This is discussed further in the section entitled, Steroid Treatment.

Causes of Death

Dubois and associates (1974) have provided the most extensive data on causes of death. In the presteroid era, uremia and CNS disease accounted for 26% of deaths, and infections accounted for another 16%. In the 1963 to 1973 period, 14% of patients died of uremia, 8% of CNS disease, and 18.5% of infections. In the multicenter study of outcome in SLE (Rosner et al. 1982), 31% of deaths

were due to active SLE (18% nephritis), and 33% were due to infection. Ten percent of deaths were directly due to CNS disease, and CNS disease contributed to death in another 4.5%. Six of 222 patients had a terminal thrombotic thrombocytopenic purpura-like syndrome. There were no deaths from malignancy.

Factors Influencing Survival

Demographic Characteristics

Lupus is overwhelmingly a disease of women, and it is far more prevalent among blacks. However, most studies have suggested that neither race nor sex influence survival (Dubois et al. 1974; Estes and Christian 1971; Ginzler et al. 1982; Rosner et al. 1982; Siegel et al. 1969; Wallace et al. 1981). Studenski and colleagues (1987), using a Cox proportional hazards model* to evaluate risk factors for death in their Duke University population, found that nonwhite race and low socioeconomic status contributed independently to mortality. However, there is reason to suspect that referral patterns in this study were biased against blacks with mild disease. Ginzler and associates (1982), in the multicenter cooperative study of outcome, also found reduced survival among blacks, but there was no significant difference in survival between publicly

---

* The Cox procedure couples time-dependent analysis of survival with capabilities for simultaneous and stepwise multivariate logistic regression.

Table 35-3. Survival Among Patients With Systemic Lupus Erythematosus as a Function of Age at Diagnosis

Study	Group	Percent Surviving 5 Years	10 Years
Fish et al. (1977)	Age less than 20	—	86.0
Walravens and Chase (1976)	Age less than 20		
	With antimetabolite treatment	78.9	73.6
	Without antimetabolite treatment	60.9	60.9
Meislin and Rothfield (1968)	Age less than 20		
	Renal disease	44.3	30.4
	No renal disease	59.2	59.2
Wallace et al. (1981)	Age less than 16		
	With nephritis	73	58
	Without nephritis	100	100
	Age 16–50		
	With nephritis	81	68
	Without nephritis	94	88
	Age >50		
	With nephritis	83	48
	Without nephritis	86	81

funded blacks and whites, or between privately funded blacks and whites. Privately funded patients in general had a better prognosis. In a recent report of Dubois' enormous private practice population in Los Angeles (609 patients), survival rates were similar in blacks and whites (Wallace et al. 1981).

Children with SLE have a higher incidence of fever, weight loss, cardiac disease, and renal disease (Nadeau and Watson 1990), but most series suggest that survival is similar to that in older patients, with the possible exception of children with nephritis (Table 35-3).

Disease Manifestations

The single factor that has the most profound influence on survival in lupus is the presence of renal disease (Fish et al. 1977; Lee et al. 1977; Wallace et al. 1981). In Dubois' private practice group, 5- and 10-year survival rates were 80% and 65%, respectively, in patients with nephritis, and 93% and 88%, respectively, in those without (Wallace et al. 1981).

Sthoeger and coworkers (1987), in their study of 49 male patients with SLE in Tel Aviv, found that survival among those with vasculitis was significantly worse. Five- and 10-year figures in patients with vasculitis were 66% and 54%, respectively, and in those without vasculitis, 97% and 97%, respectively.

A number of studies suggest that the presence of neuropsychiatric manifestations is associated with a worse prognosis (Table 35-4). However, no data suggest that neuropsychiatric manifestations *per se* affect survival; more likely, their presence is a reflection of more severe disease generally. This conclusion is supported by the results of the multicenter study of outcome; survival rates among 1103 patients were reduced among those with seizures ($p < .05$) and among those with psychosis ($p < .05$). However, in a stepwise linear regression, qualitative urine protein, highest serum creatinine, and lowest hematocrit were significantly associated with mortality in declining order of importance, and neither seizures nor psychosis were independent predictors of survival (Ginzler et al.)

Patients with SLE exhibit individual propensities for the production of particular antibodies that help to determine the expression of clinical disease. Among these, antibodies to native, double-stranded DNA,

Table 35-4. The Influence of Neuropsychiatric Manifestations on Survival in Systemic Lupus Erythematosus

Study	Neuropsychiatric Manifestations	Percent Surviving 5 Years	Percent Surviving 10 Years
Feinglass et al. (1976)	Present	92	81
	Absent	95	82
Gibson and Meyers (1976)	Present	79	50*
	Absent	90	83*
Lee et al. (1977)	Present	81.7	—
	Present or absent	91	—
Sthoeger et al. (1987)	Psychosis or convulsions present	61.9	40.9
	Psychosis or convulsions absent	91.8	91.8†

* Not a life table analysis.

† Difference between groups significant ($p < .05$.)

present in about 40% of SLE patients, have been particularly associated with glomerulonephritis and generally more severe disease. Antibodies to Sm, an RNAse-resistant extractable nuclear protein, which are present in 20% to 25% of SLE patients, also are associated with more severe disease, renal and CNS involvement, and vasculitis. Beaufils and coworkers (1983) report four deaths among 12 patients with antibody to Sm antigen in their small series, and no deaths among 22 patients without, Sm antigen ($p < .02$, simple chi-square analysis).

Treatment

Steroids

There has never been a controlled trial of steroids in SLE. The immediate and dramatic response of disease manifestations to steroid treatment seen in virtually all patients provides irrefutable evidence of their efficacy in reducing morbidity. Controversy has centered on the influence of steroids on disease outcome, and more specifically, on the possible contribution of steroids to the decline in mortality observed during the past 50 years. By far, the best analysis of this issue is by Albert and associates (1979). They performed a meta-analysis of 52 studies selected from the literature between 1932 and 1977 that included unselected patient populations and contained adequate data.

When average disease duration until death, survival by series, and 1- and 5-year survival were plotted by year of study (Fig. 35-2), a steady, linear improvement was noted. In particular, there was no change in slope after 1956, the year when steroids became widely available; this was confirmed using the technique of piece-wise linear regression. This constancy of slope strongly suggests that the improvement in survival is multifactorial in origin and cannot be attributed to any single intervention. In an analysis of 142 patients followed at Massachusetts General Hospital between 1922 and 1966, disease-specific mortality was related to the presence of nephritis and measures of disease severity and steroid use had no independent predictive value. However, in a stratified survival curve analysis, steroid use was associated with a significant improvement in survival, but only among the patients with the most severe disease (Fig. 35-3). Because it is likely that the most severely affected patients were treated with steroids, the steroid-treated group might have been expected to do worse unless steroids were truly effective; thus, Figure 35-3C provides reasonably strong evidence of a significant favorable effect of steroid use on outcome limited to patients with severe disease.

Pulse-steroid therapy (e.g. methylprednisolone 1 g/day for 3 days) has recently

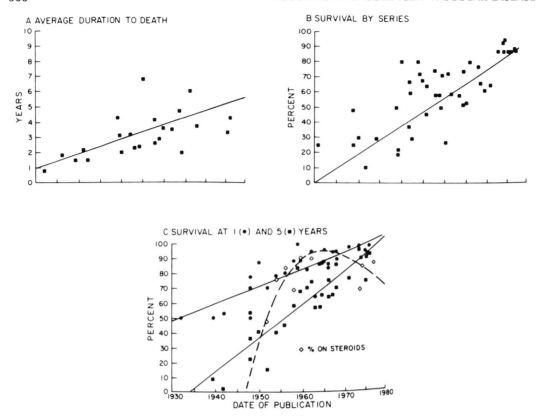

Figure 35-2. Analysis of survival statistics culled from the literature of systemic lupus erythematosus. **A.** The average duration of disease until death in each series is plotted against the publication date of that series. The regression line is drawn by the method of least squares and has a correlation coefficient, r, of 0.64. **B.** Plot of the percent survival in each paper against the publication date (r = 0.76). **C.** The survival at one (r = 0.83) and five (r = 0.86) years of disease is plotted against the date of publication of the series. C also depicts the percent of patients receiving high dose steroids in each of the respective series against the publication date. (From Albert et al. 1979, with permission.)

become popular as adjunctive therapy for severe manifestations of SLE. This therapy appears to be reasonably safe, although a 3% incidence of infections has been reported (Garrett and Paulus 1980) and several instances of sudden death have been reported (Adelman et al. 1986). Edward and colleagues (1987) conducted a prospective randomized trial of pulse therapy vs. prednisone 100 mg/day in 21 patients. Outcome, rated on a four-point scale, was actually slightly worse in the pulse-treated group.

Other Immunosuppressive Agents

There has long been agreement on the steroid-sparing effects of immunosuppressive drugs, particularly azathioprine and cyclophosphamide. On the other hand, the ef-

fect of such drugs on disease outcome has been a source of major controversy and innumerable studies. In a meta-analysis of 250 patients with lupus nephritis in eight prospective randomized studies, Felson and Anderson (1984) found that patients treated with prednisone and azathioprine or cyclophosphamide did significantly better than patients treated with prednisone alone in terms of degree of renal deterioration, incidence of end-stage renal disease, nephritis-related mortality, and overall mortality. In a prospective study of 107 patients with lupus nephritis at NIH, five treatment regimens were tested: prednisone; azathioprine; oral cyclophosphamide; oral cyclophosphamide and azathioprine; and intravenous cyclosphosphamide. The endpoint was renal fail-

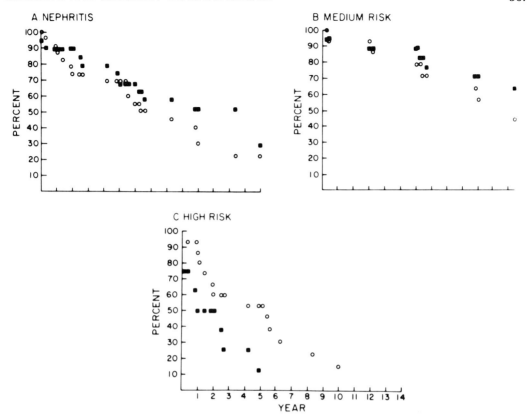

Figure 35-3. Stratified survival curve analysis of the Massachusetts General Hospital lupus patients who entered after 1950. **A.** The subgroup of patients with nephritis was examined. Steroid treated patients (○) were compared to untreated patients (■) and found not to be significantly different in survival ($p = 0.247$). **B.** The difference in survival between steroid treated and untreated patients within the medium risk stratum was examined and also found not to be significant ($p = 0.766$). **C.** The difference in survival between steroid treated and untreated patients within the high risk category was examined and found to be significant ($p = 0.015$) with improved survival in the steroid treated group. P values are strata specific Mantel-Hansell chi-square tests of association. Relative risk of death in untreated versus treated patients in the high risk group is 3.76 at one year of disease, 1.75 at two years, 1.87 at three years, and 1.77 overall. (From Albert et al. 1979, with permission).

ure (Austin et al. 1986). All regimens containing immunosuppressive agents were superior to prednisone alone, but results were statistically significant only for intravenous cyclophosphamide. However, there were no statistically significant differences in survival (Klippel 1987), and immunosuppressive agents, particularly cyclophosphamide, were associated with a high incidence of side effects. There was a 25% to 33% incidence of herpes zoster, 14% to 17% incidence of hemorrhagic cystitis, 11% to 17% incidence of neoplasia, and 18% to 71% incidence of premature ovarian failure. These data provide strong support for the efficacy of adjuvant immunosuppressive agents in lupus nephritis, but they indicate considerable caution in their use.

There are no data bearing directly on the use of these agents for other disease manifestations. Their potential value in lupus vasculitis may be inferred from experience with PAN (see above), but this application has not been studied specifically.

Prognosis of Neurological Manifestations

There has been no systematic study of the prognosis of neurological manifestations of lupus and the data presented here are largely anecdotal.

The acute dementia of lupus typically is characterized by attentional impairment with impaired alertness, distractibility, delirium, or coma. In many patients, features of psychosis are prominent. These disorders, thought to be caused by the cytotoxic effects of antibodies directed against some component of neurons, generally respond well to high-dose steroids (Nadeau and Watson 1990). However, Denburg and colleagues (1987; Carbotte et al. 1986) have raised the possibility that patients recovering from acute CNS disease may be left with significant residua that are largely subclinical. They found, using an extensive neuropsychological test battery, that 61% of patients tested within 4 weeks of acute illness and 72% of patients tested more than 4 weeks after the acute illness had significant impairment. The profile of impairment was consistent with a disorder primarily of frontal lobe and memory function. These investigators also found that 35% of lupus patients without a history of neuropsychiatric disease were impaired on neuropsychological testing. The interpretation of these results is clouded by the fact that the investigators included lupus patients with all types of CNS disorders, not just encephalopathy. Goodwin and Goodwin (1979) used a simple mental status test (the Cognitive Capacity Screening Examination) in the long-term follow-up of 22 lupus patients, 10 with a history of "cerebritis." They found that test scores were inversely correlated with duration of disease (r = −0.59, p < .01). The limited quality of these studies precludes firm conclusions, but these investigations do raise the possibility that lupus patients with encephalopathy are left with significant residua following acute recovery, that many patients without a history of encephalopathy have impaired mental function, and that many patients may experience insidious chronic deterioration of mental function that is largely subclinical.

Cranial nerve palsies in SLE have been correlated with the presence of microinfarcts in the brain stem due to lupus vasculopathy (Ellis and Verity 1979; Johnson and Richardson 1968). That they are frequently transient in nature presumably reflects the capability for recovery from small, ischemic, mainly demyelinative lesions.

The early course of ischemic strokes in SLE does not appear to differ from that of patients with stroke due to atheromatous disease. However, 60% to 100% of SLE patients with stroke have circulating antiphospholipid antibodies, either lupus anticoagulant or anticardiolipin antibodies, or both. There is convincing evidence that these antibodies are the cause of most strokes and deep venous thrombosis and pulmonary embolism in lupus (Nadeau and Watson 1990). Recurrent strokes have been reported, but there has been no quantitative assessment of risk of recurrence (Levine and Welch 1987). There have not been any controlled studies of anticoagulants for any thrombotic complications of antiphospholipid antibodies, but Asherson and associates (1985) did report that all of six patients with high titer anticardiolipin antibody completing routine courses of warfarin for deep venous thrombosis experienced recurrent venous thrombosis (five) or myocardial infarction (one) within 12 weeks of being taken off warfarin. Largely anecdotal data suggest that steroids and immunosuppressive agents have little impact on the titer of antiphospholipid antibodies (Nadeau and Watson 1990).

Libman-Sacks endocarditis probably is not associated with increased risk of stroke (Galve et al. 1988).

Lupus myelopathy is fatal in 33% of reported cases. In the review by Andrianakos and coworkers (1975), the mortality rate was 50%, but since then, it has dropped to 11% (DeBrun-Fernandez et al. 1987). Twenty percent of cases have shown significant improvement (Al-Husaini and Jamal 1985). Improvement occurred in six of 12 patients treated with prednisone in a dose of at least 40 mg daily beginning within 24 hours of onset of illness; improvement occurred in only one of 23 patients who received no steroids (seven) or received steroids more than 24 hours after illness onset. These results, significant at the p < .01 level (chi-square statistic), suggest that steroids may be beneficial if initiated early.

Lupus chorea, which is thought to be etiologically related to circulating antiphospholipid antibodies, is a self-limited syndrome. Bruyn and Padberg (1984) reviewed the literature and found that among 50 cases, chorea lasted less than 2 weeks in 20, 2 to 4 weeks in nine, 1 to 3 months in 13, 4 to 6 months in five, and up to 3 years in three. There were up to four recurrences in 10 patients. No data provide any support for the use of steroids to treat lupus chorea.

Side Effects of Medications

An exhaustive review of side effects of medications used in SLE is not provided here, but some consideration of potential long-term consequences of steroids and immunosuppressive agents used in the context of rheumatologic disease is appropriate (see also Perry 1982; Pirofsky and Bardana 1977). Aseptic necrosis of the femoral heads and risks of infection and neoplasia are discussed.

Aseptic necrosis occurs exclusively in patients treated with steroids, but there is no correlation between incidence and duration of steroid use, total dose, average dose, or peak dose. Also, no clinical or demographic features have identified patients at increased risk for this complication (Dimant et al. 1978). Dimant and coworkers (1978) found a 9% incidence of aseptic necrosis in their series, but the projected incidence for the course of lupus was 30%.

Ginzler and associates (1978) have performed the most extensive study of infectious risks of steroids and azathioprine in lupus. They found that the rate of bacterial infection increased from 10.3 to 86.8 per 100 patient years as steroid dosage increased from 0 to the 41 to 100 mg/day range. The rate of opportunistic infections (tuberculosis, fungi, cytomegalovirus, pneumocystis) increased from 0.8 per 100 patient years on no steroids to 7.8 on 21 to 40 mg of prednisone per day and 42.1 on 41 to 100 mg of prednisone per day. The incidence of herpes zoster did not correlate with steroid dosage. Most cases were associated with azathioprine use. On the other hand, azathioprine did not contribute to an increased risk of nonzoster infection except when excessive dosage led to myelosuppression. In the multicenter study of outcome in SLE (Rosner et al. 1982), infectious mortality correlated with peak steroid dose but not with use of either azathioprine or cyclophosphamide. SLE appears to be associated with increased risk of infection (Staples et al. 1974), and this risk is potentiated by the presence of nephritis and uremia (Ginzler et al. 1978).

Fauci and associates (1983) note nine episodes of herpes zoster in seven of 65 patients with WG treated with cyclophosphamide, suggesting that this drug is similar to azathioprine in its potentiation of zoster.

Table 35-5. Survival in Scleroderma*

| | | Observed Percent Surviving | | | |
| | | From Diagnosis | | From First Symptom | |
Study	Center	5 Years	10 Years	5 Years	10 Years
Wynn et al. (1985)	Indiana and Hershey	—	—	68.9	51.2
Tuffanelli and Winkelmann (1961)	Mayo Clinic	70.3†	58.9†	—	—
Bennett et al. (1971)	Hammersmith Hospital, Great Britain	73	50	—	—
Medsger et al. (1971)	Pittsburgh	47.7 (95.9)	—	—	—
Medsger and Masi (1973)	Veterans Administration	41.6 (92.7)	—	—	—
Barnett et al. (1988)	Melbourne, Australia	74	85.9	77.8	—
Zarafonetis et al. (1988)	University of Michigan	81.4	69.4	—	—

* Figures in parentheses indicate expected survival.

† Not a life table analysis.

Kinlen (1985) investigated long-term risk of cancer relative to a control population among 1634 patients with a variety of immune diseases (RA, lupus nephritis, glomerulonephritis, Crohn's disease) treated with cyclophosphamide, azathioprine, or chlorambucil for more than 3 months. The relative risk of cancer of any type was 1.6 among treated patients. The relative risk for non-Hodgkin's lymphoma was 10.9, squamous cell carcinoma of the skin 5.0, bladder carcinoma 3.7, and hepatoma 9.1. Excess bladder cancer was exclusively associated with cyclophosphamide.

Scleroderma (Progressive Systemic Sclerosis)

Reported 5-year survival rates among scleroderma patients from time of diagnosis range from 42% to 81% (Table 35-5). The reasons for this great variability are not precisely known, but ascertainment bias is undoubtedly important because of the chronicity of scleroderma in general and of scleroderma confined to acral areas in par-

ticular. Mortality is greatest during the first year after diagnosis, but this may be an artifact related to the acute illness that precipitated the hospitalization during which the diagnosis was made.

Measurement of survival from the time of appearance of first symptom (see Table 35-5) is useful for comparing certain clinical subtypes of scleroderma (e.g. limited vs. diffuse forms) when one of those subtypes (limited) frequently has a prolonged early course marked by minimal symptomatology (Table 35-6). However, this method of measurement introduces a serious "immortality" artifact because no deaths are counted from time of first symptoms to time of presentation; any patients who died in this period would never have presented for diagnosis.

Factors Influencing Survival

Demographic Characteristics

Multiple studies have shown that older age at diagnosis has a major adverse effect on survival, even taking into account the re-

Table 35-6. Survival in Scleroderma by Organ Involvement

Study	Group	From Diagnosis 5 Years	From Diagnosis 10 Years	From First Symptom 5 Years	From First Symptom 10 Years
Barnett (1978)	Acrosclerotic	73.4	59.2	—	—
	Diffuse	40.4 (4 yr.)	—	—	—
Tuffanelli and Winkelman (1962)	Acrosclerotic	74*	62	—	—
	Diffuse	17*	15	—	—
Medsger et al. (1971)	No lung, heart, kidney disease	63.4†	—	—	—
	Lung, no heart or kidney disease	35.3	—	—	—
	Heart, no kidney disease	25.9	—	—	—
Medsger and Masi (1973)	No lung, heart, kidney disease	60.4†	—	—	—
	Lung, no heart or kidney disease	34.2	—	—	—
	Heart, no kidney disease	22.1	—	—	—
Barnett et al. (1988)	Type I‡	78.7	—	92.9	85.2
	Type II	77.0	—	91.0	73.7§
	Type III	47.4	—	55.0	30.9§

* Not a life table analysis.

† Difference between groups significant, $p < .01$

‡ Type I: Skin changes isolated to fingers.
  Type II: Skin changes beyond fingers but confined to extremities.
  Type III: Diffuse

§ Type I–III and Type II–III differences significant, $p < .001$.

Table 35-7. Survival in Scleroderma by Age*

| | | Observed Percent Surviving | | | |
| | | From Diagnosis | | From First Symptom | |
Study	Group	5 Years	10 Years	5 Years	10 Years
Bennett et al. (1971)	Age <40	95	70	—	—
	Age >40	50	30	—	—
Medsger et al. (1971)	Age <45	57.7 (98.5)	—	—	—
	Age >45	41.4 (93.6)†	—	—	—
Medsger and Masi (1973)	Age <50	54.1 (97.1)	—	—	—
	Age >50	29.3 (86.4)‡	—	—	—
Barnett et al. (1988)	Age <40	—	—	91.8 (99.2)	89.5 (98.3)
	Age >40	—	—	79.6 (94.5)§	63.4 (88.8)
Zarafonetis et al. (1988)	Age <40	87.8	77.8	—	—
	Age >40	77.2	64.0†	—	—

* Figures in parentheses indicate expected survival.
† Difference between age groups significant, $p < .001$.
‡ Difference between age groups significant, $p < .01$.
§ Difference between age groups not significant when difference in expected survival taken into account.

Table 35-8. Survival in Scleroderma by Sex*

| | | Observed Percent Surviving | | | |
| | | From Diagnosis | | From First Symptom | |
Study	Group	5 Years	10 Years	5 Years	10 Years
Medsger et al. (1971)	Men	35.8 (93.5)	—	—	—
	Women	53.8 (97.0)†	—	—	—
Barnett (1978)	Men	—	—	69.5	55.6
	Women	—	—	87.5	77.4†
Zarafonetis et al. (1988)	Men	69.2	54.6	—	—
	Women	84.2	72.7‡	—	—

* Figures in parentheses indicate expected survival.
† Difference between groups significant, $p < .001$.
‡ $p = .057$

Table 35-9. Survival in Scleroderma by Race*

| | | Observed Percent Surviving From Diagnosis | |
| | | 5 Years | 10 Years |
Study	Group		
Medsger et al. (1971)	White	49.6 (94.8)	—
	Black	37.2 (93.4)†	—
Medsger and Masi (1973)	White	44.7 (94.4)	—
	Black	40.2 (91.3)‡	—
Zarafonetis et al. (1988)	White	82.4	70.7
	Black	74.4	59.0

* Figures in parentheses indicate expected survival.
† Difference between groups significant, $p < .05$, first year only.
‡ Difference between groups not significant.

duced expected survival of older patients without scleroderma (Table 35-7). Men have a significantly lower survival rate than women (Table 35-8). Medsger and co-workers (1971) found that blacks had a significantly worse prognosis during the first year following diagnosis, but survival differences between blacks and whites beyond the first year were not significant. Other investigators have not found significant differences between survival among blacks and whites (Table 35-9).

Organ Involvement

Several investigators have shown that scleroderma carries a much better prognosis when the pathology is confined to acral areas (see Table 35-6). Barnett and colleagues (1988) have shown that the extent of skin involvement provides an easily measurable index of disease severity that has high prognostic value (see Table 35-6). Skin involvement extending beyond the metacarpal–phalangeal joints correlates strongly with visceral involvement, and most studies have shown that skin involvement *per se* has no effect on survival (Medsger et al. 1971; Medsger and Masi 1973). However, Zarafonetis and associates (1988) found, using a Cox proportional hazards model, that extent of skin involvement remained a significant independent predictor of reduced survival, even when pulmonary and cardiac involvement and the presence of hypertension were controlled for.

Of all visceral involvement, renal disease carries the worst prognosis. Renal disease typically presents in the form of scleroderma renal crisis (SRC), which produces malignant hypertension and rapidly progressive renal failure. In a review of their experience with 68 patients, Traub and coworkers (1983) found that all patients whose hypertension was inadequately treated died within 4 months with a mean survival of 23 days. Seventy-one percent of those who received intensive antihypertensive therapy, dialysis, or both, died within 1 year. There is some evidence that angiotensin-converting enzyme inhibitors (e.g., captopril) may be particularly effective in controlling the hypertension of SRC (Traub et al. 1983).

After renal disease, cardiac and pulmonary disease (fibrosis), in declining order of importance, have the greatest impact on survival. Medsger and colleagues (1971), in their study of Pittsburgh patients, and Medsger and Masi (1973), in their study of Veterans Administration hospital patients, found that subpopulations with pulmonary and cardiac disease had significantly reduced survival. Several investigators (Bennett et al. 1971; Kostis et al., 1988; Wynn et al. 1985), also using univariate survival curve analysis, found that various clinical features of cardiac or pulmonary involvement, such as an S3 gallop, electrocardiographic abnormalities, reduced maximum midexpiratory flow rate, and cardiomegally on chest radiograph, were associated with significantly reduced survival. Wynn and associates (1985), in a study of 62 patients using a Cox proportional hazards model, found that only age and the presence of an S3 gallop were significantly related to survival. Zarafonetis and coworkers (1988), using a Cox regression on data from 390 patients, found that hypertension, cardiac involvement, pulmonary disease, and extent of skin involvement, in declining order of predictive power, were significantly associated with reduced survival; renal involvement just failed to reach statistical significance, presumably because of the small number of patients affected.

The only common neurological manifestation of scleroderma is myopathy, which is present in 27% to 96% of patients (Nadeau and Watson 1990). Cranial neuropathies, mainly trigeminal, are uncommon, and polyneuropathy is rare. None of these disease manifestations has been shown to have any effect on mortality. However, West and associates (1981) have noted a high incidence of cardiomyopathy with associated fascicular blocks, ventricular arrhythmias, and congestive heart failure, in the small subgroup of myopathy patients with inflammatory myositis.

Autoantibodies

Anticentromere antibody (ACA) is found in 43% of patients with limited scleroderma (Steen et al. 1988). Frequently, these patients have the syndrome of calcinosis, Raynaud's phenomenon, esophageal dysmotility, sclerodactyly, and telangiectasias (CREST), and visceral involvement is minimal. ACA rarely is found in diffuse scleroderma (disseminated skin involvement often accompanied by visceral disease). One third of patients with diffuse scleroderma and 18% of patients with limited scleroderma have an antibody known as anti-Scl-70. The presence or absence of these antibodies does not appear to predict survival when the clinically defined distribution of disease is taken into account.

Morbidity

A single study of morbidity is available. Barnett and colleagues (1988) report the functional status of 75 of the 89 living patients from their series of 177 patients from 1953 to 1983. Sixty-seven percent were functionally normal or performed normal activities with minor inconvenience; 23% were capable of only modified work and recreation; and 10% were severely disabled. The subgroup of patients with sclerodactyly only did better: 76% were normal or minimally disabled, and 4% were severely disabled. Deformity of the small joints of the hands was a major contributor to functional impairment.

Treatment

Two drugs have been used extensively in the primary treatment of scleroderma: d-penicillamine and para-aminobenzoic acid (PABA). Neither has been tested in a prospective, randomized fashion, but available data nonetheless provide considerable support for their use.

The best and most extensive study of d-penicillamine has been by Medsger's group in Pittsburgh (Steen et al. 1982). These investigators reviewed their experience with 118 patients with scleroderma of less than 3 years' duration, survival for more than 1 year, and no renal involvement within 6 months of presentation. Seventy-three had received d-penicillamine for at least 6 months and were compared with the remaining 45. The total skin score, a measure of dermal involvement, declined by 25% or

more in 34 of 56 (61%) of d-penicillamine patients vs. nine of 28 (32%) of untreated patients ($p < .05$). Three d-penicillamine patients (4%) and nine (20%) untreated patients developed renal disease ($p = .01$). Five-year survival in the d-penicillamine group was 88%; in the untreated group, it was 66% ($p < .05$). Neither colchicine nor immunosuppressive agents appeared to alter outcome, although few patients were treated with these agents. The results of this study become more persuasive when it is recognized that d-penicillamine tends to be used in patients with more severe and rapidly progressive disease, because it is in these patients that the morbidity associated with the drug is most readily justified. Side effects of d-penicillamine included pruritus (1.5%), rash (14.3%), fever (2.7%), altered taste (2.7%), mouth ulcers (2.7%), leukopenia less than 4000 (2.3%), proteinuria more than 400 mg/24 hours (9.3%), and myasthenia gravis (1.5%); 47.5% of patients experienced side effects (Steen et al. 1986). Side effects were mainly related to the rate of increase of d-penicillamine. There is a strong association between d-penicillamine-induced myasthenia gravis and HLA antigens DR1 and DR5 (Garlepp et al. 1983; Steen et al. 1986).

The only extensive trial of PABA has been by Zarafonetis and associates (1988) at the University of Michigan. The chief impetus for using PABA has been to ameliorate the dermal manifestations of scleroderma. In an open study, these investigators found that 5- and 10-year survival rates from time of diagnosis were 70% and 57%, respectively, in untreated patients; 74% and 62%, respectively, in inadequately treated patients; and 88% and 77%, respectively, in adequately treated patients ($p < .01$). The fact that PABA is cleared through the kidneys and therefore contraindicated in patients with impaired renal function might lead to the suspicion that the apparent benefit of the drug was an artifact of patient selection. This suspicion has some support from a Cox proportional hazards analysis that included pulmonary, cardiac, renal, and dermal involvement; hypertension; and PABA treatment. The negative association of PABA with mortality dropped to the $p = .08$ significance level in this analysis (Zarafonetis et al. 1988). On the other hand, in a separate analysis that excluded patients with hypertension or renal involvement, PABA was significantly associated with improved survival ($p = .048$).

No data support the routine use of steroids in scleroderma, but several investigators have reported a consistently favorable response of scleroderma myositis to steroids (Clements et al. 1978; Mimori 1987; West et al. 1981).

In the single reported series of scleroderma patients with vasculitis (Oddis et al. 1987), both patients with dermal leukocytoclastic vasculitis responded well to steroids, and all five patients with PAN responded to steroids in combination with immunosuppressive agents (cyclophosphamide, nitrogen mustard, or azathioprine).

Topical glyceryl trinitrate (Franks 1982) and nifedipine (Rodehoffer et al. 1983; Smith and McKendry 1982) have been shown to be effective in the treatment of Raynaud's phenomenon in double-blind, crossover studies.

No treatment has benefitted cranial neuropathy or polyneuropathy due to scleroderma (Nadeau and Watson 1990).

Rheumatoid Arthritis

Mortality

Studies of survival in RA patients have shown consistently that they have increased mortality (Tables 35-10, 35-11, 35-12) (Symmons 1988). Only one study, a community-based investigation from the Mayo Clinic (Linos et al. 1980), has failed to find significantly reduced survival, and this study has been criticized for the inclusion of patients with very mild disease whose diagnosis was not confirmed by a rheumatologist and who may not have had RA. The literature is succinctly summarized by Rasker and Cosh (1981): "The generally agreed reduction of life span in patients with RA is mainly attributable to a minority of patients, about one third in number, who have relatively severe disease and die 10 or more years prematurely as a result." This conclusion is supported by the finding that patients with classical RA (seven or more of 11 diagnostic criteria) (Arnett et al. 1988; Ropes et al. 1958) have a higher mortality rate than those with definite RA (five or six diagnostic criteria) (see Table 35-10) (Rasker and Cosh 1981; Vollertsen et al. 1986). Also, inpatients generally have a higher mortality rate than patients in the general population (see Table 35-11) (Allebeck et al. 1981; Allebeck 1982). Nevertheless, every study is susceptible to at least some ascertainment bias because every RA patient must experience symptoms of sufficient severity and duration to seek medical consultation, and greater disease severity may be required to make the diagnosis of RA or to precipitate a referral to a major diagnostic center or to a specialist. Symmons and coworkers (1986) demonstrated the potential impact of such ascertainment bias by comparing survival

Table 35-10. Survival in Rheumatoid Arthritis by Diagnostic Classification*

Study	Study Population	Group	Observed Percent Surviving			
			5 Years	10 Years	15 Years	20 Years
Rasker and Cosh (1981)	Outpatient presenting less than 1 year after onset	Definite	96	81	73	—
		Classical	92	77	58	48†
Mitchell et al. (1986)	Outpatient–inpatient	ARA I	96	89	81	—
		Class II–IV	88	76	62	—
Vollertsen et al. (1986)	Mayo Clinic Referrals	Definite	87 (87)	77 (77)	62 (65)	—
		Classical	75 (92)	62 (79)	45 (67)	—
		Vasculitis	61 (90)	—	—	—

* Figures in parentheses indicate expected survival.

† Between groups difference significant, $p = .01$

Table 35-11. Survival in Rheumatoid Arthritis (RA) by Sex and Referral Base*

Study	Study Population	Group	Observed Percent Surviving			
			5 Years	10 Years	15 Years	20 Years
Allebeck et al. (1981)	Community Study	Men	80 (94)	60 (90)†	—	—
		Women	89 (97)	79 (90)†	—	—
Vandenbroucke et al. (1984)	Outpatient‡	Men	96 (99)	75 (98)	55 (81)	—
		Women	94 (99)	83 (94)	66 (81)	—
Mutru et al. (1985)	Outpatient–inpatient	Men	79 (85)	55 (68)§	—	—
	<5 yrs. after onset	Women	85 (94)	69 (84)§	—	—
Mitchell et al. (1986)	Outpatient–inpatient	All RA:				
		Men	88	73	60	—
		Women	92	84	73	—
		Definite RA:				
		Men	89 (93)	73 (82)	61 (68)	—
		Women	93 (96)	85 (91)	73 (83)	—
Uddin et al. (1970)	Outpatient–inpatient	Men	74 (84)	56 (67)‖	—	—
		Women	82 (95)	58 (79)‖	—	—
Symmons et al. (1986)	Outpatient–inpatient	<5 yrs. after onset				
		Men	88	82	71	—
		Women	93	91	80	—
		>5 yrs. after onset				
		Men	84	64	55	—
		Women	81	78	56	—
Monson and Hall (1976)	Inpatient	Men	95	88	80	65¶
		Women	94	88	79	68¶
Allebeck (1982)	Inpatient	Men	67 (86)†			
		Women	70 (89)†			

* Figures in parentheses indicate expected survival.
† Relative risk of death or standardized mortality ratio, significant, $p < .001$.
‡ 50% lupus erythematosus-prep positive.
§ Case control study: figures for controls in parentheses; difference from controls significant, $p < .0001$.
‖ Difference from expected significant, $p < .01$.
¶ Survival rates relative to U.S. population.

among patients presenting within 5 years of disease onset with survival of patients presenting more than 5 years after onset. The latter group, which was more likely to be referred because of disease complications, had greater mortality (see Table 35-11).

Mitchell and colleagues (1986), using a stepwise Cox regression, found that age, ARA class, male sex, the presence of proteinuria, number of active joints, and the log of the latex fixation titer were significantly associated with mortality (see Tables 35-10, 35-11, 35-12). Life expectancy was reduced an average of 4 years in men and 10 years in women. Survival is most reduced among patients with disease onset early in life (Allebeck 1982; Cobb et al. 1953; Monson and Hall 1976) (see Table 35-12). Gordon and colleagues (1973) found that the mortality on long-term follow-up was twice as high in a group of hospitalized patients with one or more extra-articular features (e.g., nodules, vasculitis, pulmonary fibrosis, noncompressive neuropathy) as in those without such features. Vollertsen and associates (1985) found that RA patients with vasculitis have reduced survival (see Table 35-10), but when the data are corrected for the selection artifact introduced by referral distance, there is no longer a significant difference in survival between patients with RA and vasculitis and patients with classical RA. Clinical features of rheumatoid vasculitis associated with a poor prognosis include neuropathy, weight loss, histological evidence of vasculitis on rectal biopsy (Scott et al. 1982), high immunoglobulin G (IgG), IgM and rheumatoid factor titers, gangrene of digits and extremities, ischemic bowel lesions, and cardiac involvement (Geirsson et al. 1987).

Table 35-12. Survival in Rheumatoid Arthritis by Age at Diagnosis*

Study	Study Population	Group	Observed Percent Surviving			
			5 Years	10 Years	15 Years	20 Years
Mitchell et al. (1986)	Outpatient–inpatient	Age <50	99	95	88	—
		Age >50	85	71	54	—
Cobb et al. (1953)	Inpatient	Onset <25	96 (99)	87 (97)	83 (95)	80 (94)
		Onset 25–49	94 (99)	84 (95)	75 (89)	69 (86)
		Onset 50+	89 (95)	75 (80)	56 (61)	43 (51)

* Figures in parentheses indicate expected survival.

Causes of Death

The major causes of death in RA are cardiovascular disease (42.1%), cancer (14.1%), infection (9.4%), renal disease (7.8%), respiratory disease (7.2%), RA itself (5.3%), and gastrointestinal disease (4.2%) (mainly peptic ulcer disease due to treatment). The figure for respiratory disease is inflated by the inclusion of respiratory infection under respiratory disease in the International Classification of Disease (Pincus and Callahan 1986). An excess of RA deaths are due to infection and to renal and gastrointestinal disease (Allebeck 1982; Cobb et al. 1953; Fries et al. 1985; Mitchell et al. 1986; Monson and Hall 1976; Mutru et al. 1985; Uddin et al. 1970; Vandenbrouke et al. 1984). Mutru and coworkers (1985), in their 1000-patient case-control study, found an excess of deaths due to vascular disease among male RA patients, and Allebeck and associates (1982) found an excess of cardiovascular deaths among RA patients of either sex discharged from Stockholm hospitals. Mitchell and colleagues (1986) report that 10% of deaths were directly attributable to RA, its complications, or its treatment, and Monson and Hall (1976) report a markedly increased standardized mortality ratio for diseases of bones and joints. RA is more likely to be the direct cause of death among younger patients (Rasker and Cosh 1981). Most studies, including that of Mutru and colleagues (1986), have not found an excess of deaths due to malignancy (Allebeck et al. 1981; Allebeck 1982; Fries et al. 1985; Mitchell et al. 1986; Uddin et al. 1970) but Monson and Hall (1976) report a standardized mortality

ratio of 1.36 for cancer, and Prior and co-workers (1984) report a ratio of 1.6. In the latter study, the excess cancer mortality was attributable to a significantly increased incidence of non-Hodgkin's lymphoma and leukemia. Neurological disease does not contribute significantly to excess mortality (Mitchell et al. 1986) but is associated with more severe disease.

Disability

Sherrer and associates (1986), in their longitudinal study of more than 1000 RA patients in Saskatchewan, found that disease progression, as measured by a disability index, ratings of radiological changes, and functional class, tended to be maximal in the first 10 to 15 years. After this time, clinical stabilization tended to occur. In a stepwise linear regression analysis, the entry characteristics that had significant predictive power relative to the disability index were (in declining order of importance) age, radiological grade, initial functional class, female sex, presence of rheumatoid nodules, and disease duration.

Treatment

The 1% prevalence of RA has provided great impetus and ample opportunity for the study of therapeutic agents. A large number of uncontrolled and controlled studies have been performed, and space does not permit an adequate discussion here of even the controlled trials. Therefore, this literature is only summarized, and, with few exceptions, only controlled studies are considered.

The overriding goal of clinical investigators has been to find therapeutic agents that alter the long-term course of the disease. Unfortunately, developing reliable measures of long-term outcome has proven very difficult (Gabriel and Luthra 1988), and the only measure that has found substantial acceptance is degree of radiographic progression. Even this measure has problems, and detecting differences in radiographic outcome requires long-term follow-up. Consequently, few adequate studies are available. Most studies have used outcome measures that are more easily quantified and susceptible to rapid change, such as pain (often scored on a visual analog scale), clinical score rated by patient or physician, grip strength, duration of morning stiffness, number of tender or swollen joints, circumference of proximal interphalangeal joints, number of new rheumatoid nodules, 50-foot walk time, steroid requirement, sedimentation rate, and C-reactive protein level. There is little question that these are valid measures of acute disease activity and morbidity, but their relationship to long-term outcome is unclear. Several investigators have noted that radiographic progression may continue even with good clinical response to a drug (Gabriel and Luthra 1988). Furthermore, as steroid trials in the 1950s demonstrated, a drug may be acutely beneficial and reduce radiological progression at the same time that its side effects lead to increased disability, hospitalization, and mortality (Fries 1983).

Multiple prospective, randomized studies using measures of disease activity have shown that salicylates, nonsteroidal anti-inflammatory drugs (NSAIDs) (indomethacin, sulindac, ibuprofen, fenprofen, naproxen, tolmetin, mefenamic acid, tolfenamic acid) (Simon and Mills 1980), the antimalarials chloroquine and hydroxychloroquine, gold salts, d-penicillamine, sulfasalazine, methotrexate, azathioprine, and cyclophosphamide are effective in the treatment of RA in comparison with placebo or salicylates and NSAIDs (Baker and Rabinowitz 1986; Bunch and O'Duffy 1980; Gabriel and Luthra 1988; Hochberg 1986; Maksymowych and Russell 1987; Pinals et al. 1986;

St. Clair and Polisson 1986; Tugwell et al. 1987; Wilke and Mackenzie 1986). Some comparative studies have shown antimalarials to be less effective than gold, while others have shown them to be equal in effect to gold salts, azathioprine, and d-penicillamine (Maksymowych and Russell 1987). Several studies have shown sulfasalazine to have less than (Bax and Amos 1985; Situnayake et al. 1987) or equal efficacy to gold salts or d-penicillamine (Neumann et al. 1983; Pullar et al. 1983). Whatever sulfasalazine lacks in efficacy tends to be compensated for by a lower incidence of serious side effects compared with gold salts or d-penicillamine. Multiple studies have shown the oral gold-containing drug, auranofin, to be similar to parenteral gold salts (gold sodium thiomalate, aurothioglucose) in efficacy but superior in terms of incidence of side effects (Blodgett and Pietrusko 1986; Hochberg 1986). Several studies have shown gold salts and d-penicillamine to be roughly equivalent in efficacy and incidence of serious side effects (Bunch and O'Duffy 1980; Hochberg 1986). Controlled studies generally have shown methotrexate to be at least equal, and probably superior, to the purine analogs, azathioprine and 6-mercaptopurine in efficacy (Hamdy et al. 1987; Sambrook et al. 1986; Tugwell et al. 1987). Limited data suggest that azathioprine and cyclophosphamide are more effective than gold and that cyclophosphamide is more effective than azathioprine (Bunch and O'Duffy 1980).

Scott and Bacon (1984) used bolus cyclophosphamide and pulse methylprednisolone in an open trial in 21 of 45 patients with rheumatoid vasculitis. Although the patients in the cyclophosphamide group had more severe initial disease, a greater incidence of neuropathy, and more frequent evidence of arteritis on biopsy, they experienced more frequent healing of ulcers and neuropathy, a lower incidence of relapse, fewer serious complications, and a lower mortality rate relative to those who received conventional therapy (mainly second line agents, prednisone, azathioprine, or chlorambucil).

The extensive evidence of the efficacy of

pharmacotherapy in suppressing acute manifestations of RA notwithstanding, only gold salts and cyclophosphamide have been unequivocally demonstrated to retard radiographic progression (Iannuzzi et al. 1983). No conclusions can be drawn from studies of other agents.

There is strong evidence from a large number of well-executed, randomized, controlled trials of the efficacy of all of the drugs used routinely for relieving acute manifestations of RA, and there is limited evidence of reduction of radiographic progression by gold and cyclophosphamide. However, the practical use of antirheumatic drugs is dictated in considerable part by their frequent and potentially serious side effects. Among the most noteworthy are gastrointestinal upset and bleeding and reduced glomerular filtration with NSAIDs; nausea, vomiting, dizziness, and headache with sulfasalazine; retinopathy with antimalarials; taste loss, nausea, rash, stomatitis, proteinuria, and thrombocytopenia with d-penicillamine; rash, stomatitis, thrombocytopenia, neutropenia, marrow aplasia, and nephropathy with gold; elevated hepatic transaminases, hepatic cirrhosis, pancytopenia, pneumonitis, stomatitis, and diarrhea with methotrexate; myelosuppression, hepatotoxicity, nausea, vomiting, and increased risk of cancer with azathioprine; and myelosuppression, alopecia, cystitis, sterility, herpes zoster, and increased risk of neoplasia with cyclophosphamide (see the section entitled Side Effects of Medications under Systemic Lupus Erythematosus). Neurological complications have been reported with most of these agents, but they are rare (Nadeau and Watson 1990). Rheumatologists consider salicylates and NSAIDs to be the mainstay of therapy for RA. Gold and d-penicillamine have been the traditional second-line agents. Antimalarials have long been out of favor because of their retinal toxicity, but recent data suggest that the incidence of this complication is negligible with currently accepted doses. Sulfasalazine is a drug engineered for use in RA in the 1940s when RA was thought to be an infectious disease. Because of an early adverse report, it fell out of

favor, but recent data provide strong support for this drug as a second-line agent, particularly because of its fairly benign side-effect profile. Methotrexate traditionally has been considered a third-line agent because of its grouping with antineoplastic drugs, but it has proven to be highly effective and remarkably well tolerated in the accepted weekly oral dosage. This experience, and the lack of any demonstrable association with neoplasia, have moved this drug well into the group of second-line agents. Cyclophosphamide and azathioprine remain agents of last resort because of their serious side effects.

The effectiveness of steroids in suppressing acute manifestations of RA was demonstrated in several trials in the 1950s (Gabriel and Luthra 1988), but unacceptable long-term complications have led to the current pattern of usage, namely 1- to 2-week courses of moderate-dose prednisone to suppress acute flares of disease. In this context, pulse steroids (i.e., 1 g methylprednisolone intravenously for 1–3 days) have enjoyed considerable popularity. Controlled studies have demonstrated the effectiveness of this regimen acutely and chronically (given on a monthly basis); they also have shown that oral prednisone, 1 g, is as effective and well tolerated, and that lower doses, (e.g., 320 mg) are just as effective (Liebling et al. 1981; Radia and Furst 1988; Smith et al. 1988). Low-dose steroid therapy (5–7.5 mg prednisone per day) has been investigated and limited data suggest that it may retard radiographic progression (Masi 1983; St. Clair and Polisson 1986). Halla and coworkers (1984), in an open study of 31 patients, report that rheumatoid myositis was responsive to low-dose prednisone (less than 12.5 mg/day).

Controlled trials have supported other treatment modalities, including total lymphoid irradiation (Zvaifler 1987) and drugs including chlorambucil (Cannon et al. 1985), nonathymulin (Amor et al. 1987), fish oil (Kremer et al. 1987), cyclosporin A (Dougados et al. 1988; Førre et al. 1987), and gamma-interferon (Veys et al. 1988). All of these are considered investigational. Apheresis has not been effective in controlled trials (St. Clair and Polisson 1986).

Disease of the Cervical Spine

The only neurological manifestations of RA for which prognostic data are available are those due to cervical spine involvement. Disease of the cervical spine in RA patients parallels that in other joints. Four patterns of major cervical pathology have been delineated:

1. Atlantoaxial subluxation, defined by separation of the anterior arch of C1 from the dens by more than 3 mm and caused by erosion of the transverse ligament (the most common clinically significant abnormality)
2. Vertical subluxation of the odontoid through the ring of the atlas and ultimately into the foramen magnum (atlantoaxial impaction), caused by lysis of the C1–C2 facet joints, the occipital condyles, or fracture of C1
3. Backward subluxation of the atlas on the axis, usually caused by erosion of the odontoid in conjunction with laxity of the C1–C2 facet joints (the least common abnormality)
4. Subaxial subluxation, often referred to as stairstep subluxation because of its propensity to occur at multiple levels (the second most common abnormality (Nadeau and Watson 1990)

Radiographic Abnormalities

Wolfe and associates (1987), in their population study, found atlantoaxial subluxation in 7.8% of patients in their first decade of RA and in 38.4% of patients in their fourth decade (Table 35-13). These figures correlate roughly with clinical studies (Table 35-14) that generally include only patients with more advanced disease. Atlantoaxial subluxation can be expected to progress in 17% to 41% of patients during the course of 5 years (see Table 35-14).

Atlantoaxial impaction occurs in 4% to 33% of RA patients, depending to some extent on the means of measurement and to a great extent on the duration of RA (see Table 35-14). It may progress indefinitely until the odontoid projects well into the foramen magnum. In most cases of atlantoaxial subluxation in which there is apparent improvement, it is illusory because impaction has occurred and the improvement reflects the fact that the body of C2 has replaced the odontoid within the ring of C1 (Mathews 1974; Weissman et al. 1982; Winfield et al. 1981).

Backward subluxation of the atlas on the axis accounts for 6.7% of atlantoaxial subluxations (Weissman et al. 1982). On the other hand, erosion of the odontoid in RA is common and closely related to the duration of the disease (see Table 35-13). No adequate studies exist on the natural history of backward subluxation.

The prevalence of subaxial subluxations in patients with late RA is not much greater than in patients in their first decade of disease (see Table 35-13). Subaxial subluxations progress in 5 to 10 years in 16% to 31% of patients (see Table 35-14). Atlantoaxial subluxation coexists in up to 73% of

Table 35-13. Percent of Rheumatoid Arthritis Patients With Selected Radiographic Features of Cervical Spine Disease as a Function of Duration of Disease in a Population Study of Rheumatoid Arthritis

	Decade of Disease			
	1	2	3	4
Atlantoaxial subluxation	7.8	13.4	48.9	38.4
Subaxial subluxation	35.7	33.8	48.9	40.4
Odontoid erosion	32.9	35.7	76.8	70.1

From Wolfe, B. K.; O'Keeffe, D.; Mitchell, D. M.; Tchang, S. P. K. Rheumatoid arthritis of the cervical spine: early and progressive radiographic features. Radiology 165:145–148; 1987, with permission.

Table 35-14. Progression of Selected Radiographic Features of Cervical Spine Disease in Clinical Studies of Rheumatoid Arthritis Patients*

Study	n	Duration Follow-Up (Years)	Feature	Percent Initially	Percent at Follow-Up	Percent With Progression
Isdale and Conlon (1971)	171	6	AAS	22	46	32
			SAS	8	15	—
Smith et al. (1972)	55	5–15	AAS	100	—	35
			SAS	100	—	16
			AI	7	15	—
Mathews (1974)	54	5	AAS	24	—	17
			AI	8	33	—
Pellicci et al. (1981)	106	5	AAS, SAS or AI	43	70	—
			AAS	26	39	—
			SAS	5	9	—
			AI	1	2	—
			AAS + SAS	8	12	—
			AAS + AI	3	5	—
			AAS + SAS + AI	0	3	—
Weissman et al. (1982)	144	5	AAS	97	—	41
Winfield et al. (1981)	100†	7	AAS	4	12	—
			SAS	12	24	—
			AI	0	3	—
Redlund-Johnell and Pettersson (1985)	49	2+	SAS	100	—	31

* Differences in initial prevalence of radiographic abnormalities reflect differences in patient populations studied.

† All patients had RA of less than one year duration at start of study.

AAS = atlantoaxial subluxations; SAS = subaxial subluxation; AI = atlantoaxial impaction.

these patients and impaction in up to 50% (see Table 35-14). (Redlund-Johnell and Pettersson 1985).

RA patients with cervical disease tend to have more severe RA and therefore have a diminished life expectancy, but cervical involvement *per se* does not diminish longevity (Pellicci et al. 1981; Smith et al. 1972).

Neurological Abnormalities

In general, cervical disease is remarkably well tolerated by RA patients, notwithstanding their often alarming radiological appearance. Pain is common if not universal, but serious neurological disease is unusual. In the study by Winfield and coworkers (1981) of 100 patients with RA of less than 1 year's duration, subluxations eventually developed in 33 in the course of 7 years of follow-up, but no patient developed neurological signs (Table 35-15). Pellicci and colleagues (1981), found that only 15% of their patients with radiographic abnormalities had neurological symptoms, and one had signs at the start of their study (see Table 35-14). A mean of 5 years later, 16% of

asymptomatic and 36% of symptomatic patients had progressed to neurological stages II and III, degrees of impairment that would usually prompt consideration of surgery. Neurological stage II is defined as long tract signs associated with subjective weakness and stage III as long tract signs associated with objective weakness. Weissman and associates (1982) found that cord compression occurred significantly more often in men.

The nature of the cervical disease strongly affects the probability of neurological manifestations. With atlantoaxial subluxation, long tract signs are fairly common (see Table 35-15), but significant myelopathy (i.e. stage II or stage III) is unusual. There is only a rough correlation between the degree of subluxation and presence of myelopathy (Fig. 35-4), reflecting variations in intrinsic spinal canal size and the presence of other mechanisms and loci of cord compression. Myelopathy reliably occurs when both the anterior and posterior subarachnoid spaces are obliterated in flexion (Stevens et al. 1986). Rheumatoid pannus posterior to the odontoid contributes significantly to spinal

Table 35-15. Neurological Progression in Rheumatoid Arthritis (RA) Patients With Cervical Spine Disease*

Study	n	Duration of Follow-Up (Years)	Radiographic Feature	Neurological Feature	Percent With Neurological Feature		Percent With Progression
					Initial	Final	
Mathews (1974)	54	5	Any subluxation	Long tract signs	6.6	16.6	—
			AAS	Long tract signs	16.0	33.0	—
			AI	Long tract signs	17.0	50.0	—
Pellicci et al. (1981)	74	5	Any subluxation	Stage 0 to Stage I	85†	—	21
				Stage 0 to Stage II, III	85	—	16
				Stage I to Stage II, III	15	—	36
Winfield et al. (1981)	33‡	7	Any subluxation	Neurological sign	0	0	—
Weissman et al. (1982)	144	5	AAS	Myelopathic weakness	3	10	—
Smith et al. (1972)	150	5–15	Any subluxation	Myelopathy	13	15	—

* Differences in initial prevalence of neurological abnormalities reflect differences in patient population studied.

† Percent with initial stage at beginning of study.

‡ All patients had RA of less than 1 year's duration at start of study.

AAS = atlantoaxial subluxation; AI = atlantoaxial impaction.

Stage 0, no neurological symptoms or signs; Stage I, neurological symptoms without signs; Stage II, subjective weakness with hyperreflexia and dysesthesia; Stage III, objective weakness and long tract signs.

From Ranawat, C. S.; O'Leary, P.; Pellicci, P.; Tsairis, P.; Marchisello, P.; Dorr, L. Cervical spine fusion in rheumatoid arthritis. J. Bone Joint Surg. 61(A):1003–1010; 1979, with permission.

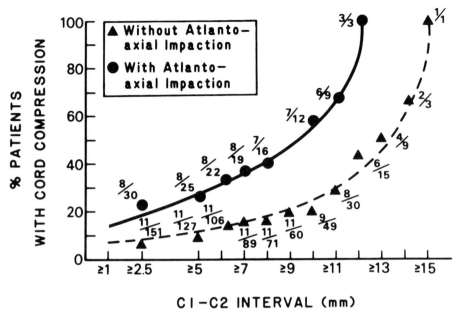

Figure 35-4. Effect of atlantoaxial impaction on the development of spinal cord compression. The presence of atlantoaxial impaction increases the risk of developing neurologic sequelae given a particular amount of C1-C2 subluxation. (From Weissman et al. 1982, with permission.)

canal stenosis in up to 40% of patients (Breedveld et al. 1987; Menezes et al, 1985; Stevens et al. 1986). Myelopathy also occurs in the absence of obliteration of the subarachnoid space in flexion and correlates with degree of atlantoaxial mobility, suggesting that traction of the cord across the bulge of protruding odontoid and pannus significantly contributes to cord injury (Stevens et al. 1986). In spite of all these sources of complexity, Weissman and associates (1982) found in their large study that 9 mm of separation between the anterior arch of C1 and the dens seemed to represent a threshold below which myelopathy rarely occurred (see Fig. 35-4). The prevalence of myelopathic weakness in this study of RA patients with atlantoaxial subluxation increased over the course of 5 years from 3% to 10%.

Atlantoaxial impaction significantly reduces the degree of atlantoaxial subluxation necessary to produce myelopathy (see Fig. 35-4), and its development is one of the major causes of rapidly progressive myelopathy (see Table 35-14). Pellicci and co-workers (1981) report that in three of the

seven operated patients in their series, impaction precipitated the rapid deterioration that led to surgery. Weissman and colleagues (1982) found that a patient with impaction had a 21% chance of developing cord compression, compared to an 8% chance for a patient without impaction ($p <$ .01); 45% of myelopathic patients had impaction vs. 20% without myelopathy. A number of cases of sudden death have been reported with impaction; death typically occurs when, as a consequence of collapse of a lateral mass of C1 or an occipital condyle, the dens is thrust upward into the medulla (Nadeau and Watson 1990). Such observations have led most authorities to recommend surgery as soon as neurological signs appear in the patient with impaction.

Neurological symptoms have been reported in 29% of patients with backward atlantoaxial subluxation, but myelopathy is rare (Redlund-Johnell 1984; Weissman et al.) Redlund-Johnell (1984) found that only patients who also had impaction were symptomatic. The prognosis is not well documented.

Neurological symptoms have been reported in 22% of patients with subaxial subluxation and are most common in patients with a sagittal canal diameter of less than 13 mm and in patients with reduced height of the cervical spine due to vertebral body compression at one or more levels (Redlund-Johnell and Pettersson 1985). Notwithstanding the absence of adequate documentation of the neurological prognosis of subaxial subluxation, most authorities recommend surgery as soon as neurological signs appear (i.e., stage II neurological impairment) because rapid progression frequently occurs at this point (Nadeau and Watson 1990).

Treatment of Cervical Disease

Cervical collars may help to assure patients and relieve pain, but there is no evidence that they halt the progression of neurological manifestations or protect against neurological disaster (Pellicci et al. 1981; Smith et al. 1972).

Surgery generally is indicated to relieve intractable pain, to treat patients with stage III neurological impairment, and to treat patients who have stage II impairment in the presence of rapid neurological progression, impaction, or subaxial subluxation. The approach that is most widely used and has stood the test of time is the Gallie procedure (Table 35-16), which consists of posterior wiring and fusion with iliac bone chips. The occiput is included in the fusion mass when there is irreducible atlantoaxial subluxation. Some surgeons use internal fixation with methylmethacrylate ("bone cement") in all cases to reduce the duration of postoperative immobilization, while others reserve methacrylate for patients with eroded or osteoporotic lamina. Crockard and coworkers (1986) have advocated transoral resection of the odontoid and adjacent pannus in conjunction with posterior fusion in all patients with atlantoaxial subluxation, while others reserve this procedure for irreducible impactions. Anterior fusion has been used very little and has not been successful. Laminectomy is used selectively in patients with myelographic block.

Operative mortality ranges as high as 20%

in reported series, but results in recent studies have been better (see Table 35-16). Neurological improvement has been reported in 72%. Among eight series of stage II and III patients in which results of long-term follow-up have been documented, 50% of patients have lived longer than 7 years (Fig. 35-5) and 78% have remained free of neurological relapse during this time (Fig. 35-6). No studies exist that permit a valid comparison between the results of surgery and conservative management. Many clinicians have worried that high cervical fusion might accelerate the progression of subaxial subluxation. Redlund-Johnell and Pettersson (1985) concluded from their experience with 26 patients, seven operated, that the rate of progression was not affected.

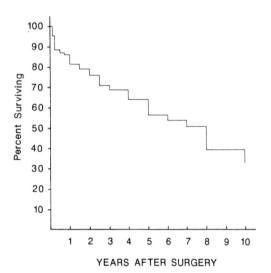

Figure 35-5. Survival following cervical surgery in 138 patients with RA. One hundred twenty were neurologic stage III and 26 were stage II at the time of surgery (Classification scheme of Ranawat et al. 1979). Ninety patients had atlantoaxial subluxation, 26 subaxial, and 20 combined atlantoaxial and subaxial. Eight patients underwent anterior fusion, and 130 posterior fusion, in many including the occiput. In addition, laminectomy was performed in a small number of patients and internal fixation with methacrylate was used in a small number of patients. Fifteen had simultaneous posterior fusion and tarnsoral odontoidectomy. Data from Meijers et al. (1984), Conaty and Mongan (1981), Crellin et al. (1970), Crockard et al. (1986), Bryan et al. (1982), Brattstrom and Granholm (1976), Ranawat et al. (1979), and Ferlic et al. (1975).

Table 35-16. Immediate Results of Cervical Operations in Rheumatoid Arthritis Patients With Stage II (n = 41) or Stage III (n = 198) Neurological Manifestations*

Study	n	Type of Subluxation	Operative Deaths (Percent)	Percent With Improvement (Worsening)	Nature of Surgery
Cregan (1966)	9	SA three, AA six	1 (11)	67	Posterior fusion
Crellin et al. (1970)	5	AA	2 (40)	40	Posterior fusion
	3	SA	0	100	Anterior fusion
	1	AA + SA	0	0	Posterior fusion
	2	Myelographic block	0	0	Laminectomy
Ferlic et al. (1975)	10	SA one, AA seven SA + AA two	2 (20)	50	Posterior fusion ± laminectomy
Brattström and Granholm (1976)	17	AA, four with AI	0	29	Posterior fusion with methacrylate + laminectomy
Ranawat et al. (1979)	5	SA	0	20 (40)	Anterior fusion
	15	AA	2 (13)	40 (7)	Posterior fusion
	5	SA, two with AI, one with AA	0	60 (20)	Posterior fusion
Conaty and Mongan (1981)	6	AA	0	67	Posterior fusion
	6	Irreducible AA, four with AI	1 (17)	33	Posterior fusion
	6	SA	1 (17)	50	Posterior fusion
	4	AA + SA	1 (25)	50	Posterior fusion
Bryan et al. (1982)	8	AA + AI two, AA + SA two, AA + AI + SA four	0	75	Posterior fusion + methacrylate
Meijers et al. (1984)	34	AA 20, SA five, AA + SA nine	3 (9)	97	Posterior fusion ± laminectomy
Menezes et al. (1985)	43	AI	0	100	Posterior fusion methacrylate in 28, transoral decompression in seven; ± laminectomy
Lipson (1985)	4	Posterior AA	0	100	Posterior fusion ± methacrylate
Crockard et al. (1986)	14	AA	0	93	Posterior fusion ± transoral decompression
Zoma et al. (1987)	28	AA eight, AI three, SA nine, AA + SA eight	2 (7)	60	One anterior, 27 posterior fusions ± laminectomy ± methacrylate
Fehring and Brooks (1987)	14	AA	1 (7)	71	Posterior fusion ± methacrylate
Total	239		16 (7)	72	

* See Table 35–14 for staging scheme.

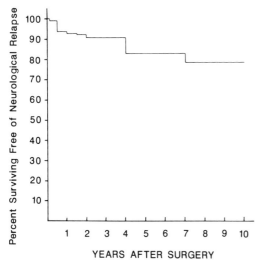

Figure 35-6. Survival free of neurologic relapse following cervical surgery in 138 patients with RA. Only relapses attributable to cervical disease were counted. Many relapses reflected pathology at a level different than that treated in the initial operation. See Figure 35-5 for details of the patient population. Data from Meijers et al. (1980), Conaty and Mongan (1981), Crellin et al. (1970), Crockard et al. (1986), Bryan et al. (1982), Brattstrom and Granholm (1976), Ranawat et al. (1979), and Ferlic et al. (1975).

The frequently high mortality rate associated with cervical operations, and the failure of one third or more of RA patients to improve following surgery should prompt considerable caution in approaching operative intervention. Many patients are in poor condition and do not tolerate surgery. Furthermore, the surgery presents special problems and should be performed only by persons with experience with RA patients. Most surgical failures are attributable to failure of bony fusion because of difficulty in maintaining adequate postoperative fixation, graft resorption, and failure of graft union (pseudarthrosis formation).

RA patients tolerate halo immobilization poorly, developing extensive skin and soft-tissue breakdown and infections. Consequently, many surgeons have resorted to prolonged preoperative and postoperative immobilization in a circular bed using head tongs; unfortunately, this results in hospitalizations of up to 3 or 4 months (Conaty and Mongan 1981; Crellin et al. 1970; Meijers et

al. 1984). One approach to this problem, pioneered by Brattström and Granholm (1976), has been to use internal fixation with methyl methacrylate. This permits mobilization of patients within days using only a form-fitted hard cervical collar (Bryan et al. 1982; Menezes et al. 1985). Another approach has been the combined transoral decompression of the odontoid and retro-odontoid pannus and because the posterior spinous elements and posterior cervical–occipital fusion developed by Crockard and colleagues (1986; Stevens et al. 1986). These investigators have reduced the length of hospital stay to an average of 2 weeks by completely obviating preoperative traction and markedly reducing the duration of postoperative immobilization. Despite a high rate of graft failure (50%), most patients do well because the extensive operative decompression prevents recurrent cord compression despite the redevelopment of some degree of subluxation.

Achieving adequate graft union has been a problem because of difficulty in maintaining adequate immobilization for long enough are frequently osteoporotic, eroded, and fragile in RA patients. This is particularly true of the arch of C1, which is small and provides minimal surface to which bone grafts or methacrylate can adhere. Fusion failure rates of 25% to 50% have been reported after operation for atlantoaxial subluxation (Conaty and Mongan 1981; Fehring and Brooks 1987; Ranawat et al. 1979; Stevens et al. 1986). One approach to this problem has been to include the occiput in the cervical fusion mass, but Zoma and associates (1987) report a reduction of fusion failure only from 44% to 38% by doing this, and use of this approach is inconsistent. Other alternatives include the combined anterior decompression–posterior cervico-occipital fusion technique of Crockard and colleague (1986). This has achieved a high clinical success rate despite a high radiographic failure rate (Stevens et al. 1986). The use of methacrylate, which has been associated with a high rate of both clinical and radiographic success when the occiput is included in the fusion mass, is another alternative; however, there is a cost of a modest delay in healing and increase in infection rate (Brattström and Granholm 1976; Bryan et al. 1982; Menezes 1985).

Sjögren's Syndrome

There are no satisfactory data regarding the prognosis of Sjögren's syndrome. The prognosis of secondary Sjögren's syndrome (i.e., in association with RA, SLE, or scleroderma) is probably similar to that of the underlying disease. However, this inference has not been formally tested.

Several drugs have been tested in prospective randomized studies for their efficacy in relieving symptoms of exocrine dysfunction in Sjögren's syndrome: xerophthalmia and xerostomia. Short trials of mucosolvan (but not bromhexine, of which it is thought to be a metabolite) (Manthorpe et al. 1984; Prause et al. 1984), nandrolone decanoate (Drosos et al. 1988), cyclosporin A (Drosos et al., 1986), and n-acetylcystein (Walters et al. 1986) have produced significant clinical improvement without altering objective measures of salivary or lacrimal function.

Mixed Connective Tissue Disease

There are no satisfactory data regarding the prognosis of mixed connective tissue disease (MCTD). The prevailing clinical impression is that disease outcome is similar to that of the disorder it most closely resembles. For example, when features of SLE predominate, the expected prognosis is that of lupus; however, renal disease, a major source of morbidity and mortality in lupus, is uncommon in MCTD (Nadeau and Watson 1990). When features of scleroderma predominate, the expected prognosis is that of scleroderma.

Acknowledgments

I am most grateful to Linda Pritchett who assisted in preparing the manuscript.

References

Adelman, D. C.; Saltiel, E.; Klinenberg, J. R. The neuropsychiatric manifestations of systemic lupus erythematosus: an overview. Semin. Arthritis Rheum. 15:185–199; 1986.

Albert, D. A.; Hadler, N. M.; Ropes, M. W. Does corticosteroid therapy affect the survival of patients with systemic lupus erythematosus? Arthritis Rheum. 22:945–953; 1979.

Al-Husaini, A.; Jamal, G. A. Myelopathy as the main presenting feature of systemic lupus erythematosus. Eur. Neurol. 24:94–106; 1985.

Allebeck, P. Increased mortality of rheumatoid arthritis. Scand. J. Rheumatol. 11:81–86; 1982.

Allebeck, P.; Ahlbom, A.; Allander, E. Increased mortality among persons with rheumatoid arthritis, but where RA does not appear on death certificate. Scand. J. Rheumatol. 10:301–306; 1981.

Allsop, C. J.; Gallagher, P. J. Temporal artery biopsy in giant-cell arteritis. Am. J. Surg. Pathol. 5:317–323; 1981.

Amor, B.; Dougados, M.; Mery, C.; Dardenne, M.; Bach, J. F. Nonathymulin in rheumatoid arthritis: two double blind, placebo controlled trials. Ann. Rheum. Dis. 46:549–554; 1987.

Andrianakos, A. A.; Duffy, J.; Suzuki, M.; Sharp, J. T. Transverse myelopathy in systemic lupus erythematosus. Report of three cases and review of the literature. Ann. Intern. Med. 83:616–624; 1975.

Arnett, F. C.; Edworthy, S. M.; Bloch, D. A.; McShane, D. J.; Fries, J. F.; Cooper, N. S.; Healey, L. A.; Kaplan, S. R.; Liang, M. H.; Luthra, H. S.; Medsger Jr., T. A.; Mitchell, D. M.; Neustadt, D. H.; Pinals, R. S.: Schaller, J. B.; Sharp, J. T.; Wilder, R. L.; Hunder, G. G. The American Rheumatism Association 1987 revised criteria for the classification of rheumatoid arthritis. Arthritis Rheum. 31:315–323; 1988.

Asherson, R. A.; Chan, J. K. H.; Harris, E. N.; Gharavi, A. E.; Hughes, G. R. V. Anticardiolipin antibody, recurrent thrombosis, and warfarin withdrawal. Ann. Rheum. Dis. 44:823–825; 1985.

Austin, H. A.: Klippel, J. H.; Balow, J. E.; De Riche, N. G. H.; Steinberg, A. D.; Plotz, P. H.; Decker, J. L. Therapy of lupus nephritis. Controlled trial of prednisone and cytotoxic drugs. N. Engl. J. Med. 314:614–619; 1986.

Baker, D. G.; Rabinowitz, J. L. Current concepts in the treatment of rheumatoid arthritis. J. Clin. Pharmacol. 26:2–21; 1986.

Barnett, A. J. Scleroderma (progressive systemic sclerosis). Progress and course on a personal series of 118 cases. Med. J. Aust. 2:129–134; 1978.

Barnett, A. J.; Miller, M. H.; Littlejohn, G. O. A

survival study of patients with scleroderma diagnosed over 30 years (1953–1983): the value of a simple cutaneous classification in the early stages of the disease. J. Rheumatol. 15:276–283; 1988.

Bax, D. E.; Amos, R. S. Sulphasalazine: a safe, effective agent for prolonged control of rheumatoid arthritis. A comparison with sodium aurothiomalate. Ann. Rheum. Dis. 44:194–198; 1985.

Beaufils, M.; Kouki, F.; Mignon, F.; Camus, J-P.; Morel-Maroger, L.; Richet, G. Clinical significance of anti-Sm antibodies in systemic lupus erythematosus. Am. J. Med. 74:201–205; 1983.

Beevers, D. G.; Harpur, J. E.; Turk, K. A. D. Giant cell arteritis: the need for prolonged treatment. J. Chron. Dis. 26:571–584; 1973.

Bengtsson, B. A.; Malmvall, B. E. The epidemiology of giant cell arteritis including temporal arteritis and polymyalgia rheumatica. Arthritis Rheum. 24:899–904; 1981.

Bennett, R.; Bluestone, R.; Holt, P. J. L.; Bywaters, E. G. L. Survival in scleroderma. Ann. Rheum. Dis. 30:581–588; 1971.

Blodgett, R. C.; Pietrusko, R. G. Long-term efficacy and safety of auranofin: a review of clinical experience. Scand. J. Rheumatol. 63:67–78; 1986.

Boesen, P.; Sorensen, S. F. Giant cell arteritis, temporal arteritis, and polymyalgia rheumatica in a Danish community. A prospective investigation 1982–1985. Arthritis Rheum. 30:294–299; 1987.

Brattström, H.; Granholm, L. Atlanto-axial fusion in rheumatoid arthritis. Acta Orthop. Scand. 47:619–628; 1976.

Breedveld, F. C.; Algra, P. R.; Vielvoye, C. J.; Cats, A. Magnetic resonance imaging in the evaluation of patients with rheumatoid arthritis and subluxations of the cervical spine. Arthritis Rheum. 30:624–629; 1987.

Bruyn, G. W.; Padberg, G. Chorea and systemic lupus erythematosus. A critical review. Eur. Neurol. 23:278–290; 1984.

Bryan, W. J.; Idglis, A. E.; Sculco, T. P.; Ranawat, C. S. Methylmethacrylate stabilization for enhancement of posterior cervical arthrodesis in rheumatoid arthritis. J. Bone Joint Surg. 64(A):1045–1050; 1982.

Bunch, T. W.; O'Duffy, J. D. Disease-modifying drugs in progressive rheumatoid arthritis. Mayo Clin. Proc. 55:161–179; 1980.

Cannon, G. W.; Jackson, C. G.; Samuelson, C. O.; Ward, J. R.; Williams, H. J.; Clegg, D. O. Chlorambucil therapy in rheumatoid arthritis: clinical experience in 28 patients and literature review. Semin. Arthritis Rheum. 15:106–118; 1985.

Carbotte, R. M.; Denburg, S. D.; Denburg, J. A. Prevalence of cognitive impairment in systemic lupus erythematosus. J. Nerv. Ment. Dis. 174:357–364; 1986.

Caselli, R. J.; Hunder, G. G.; Whisnant, J. P. Neurologic disease in biopsy-proven giant cell (temporal) arteritis. Neurology 38:352–359; 1988.

Chang, R. W.; Bell, C. L.; Hallett, M. Clinical characteristics and prognosis of vasculitic mononeuropathy multiplex. Arch. Neurol. 41:618–621; 1984.

Chumbley, L. C.; Harrison, E. G.; DeRemee, R. A. Allergic granulomatosis and angiitis (Churg-Strauss syndrome). Mayo Clin. Proc. 52:477–484; 1977.

Citron, B. P.; Halpern, M.; McCarron, M.; Lundberg, G. A.; McCormick, R.; Pincus, I. J.; Tatter, D.; Haverback, B. J. Necrotizing angiitis associated with drug abuse. N. Engl. J. Med. 283:1003–1011; 1970.

Clements, P. J.; Furst, D. E.; Campion, D. S.; Bohan, A.; Harris, R.; Levy, J.; Paulus, H. E. Muscle disease in progressive systemic sclerosis. Arthritis Rheum. 21:62–71; 1978.

Cobb, S.; Anderson, F.; Bauer, W. Length of life and cause of death in rheumatoid arthritis. N. Engl. J. Med. 249:553–556; 1953.

Cohen, R. D.; Conn, D. L.; Ilstrup, D. M. Clinical features, prognosis, and response to treatment in polyarteritis. Mayo Clin. Proc. 55:146–155; 1980.

Conaty, J. P.; Mongan, E. S. Cervical fusion in rheumatoid arthritis. J. Bone Joint Surg. 63A:1218–1227; 1981.

Cregan, J. C. F. Internal fixation of the unstable rheumatoid cervical spine. Ann. Rheum. Dis. 25:242–252; 1966.

Crellin, R. Q.; MacCabe, J. J.; Hamilton, E. B. D. Severe subluxation of the cervical spine in rheumatoid arthritis. J. Bone Joint Surg. 52(B):244–251; 1970.

Crockard, H. A.; Pozo, J. L.; Ransford, A. O.; Stevens, J. M.; Kendall, B. E.; Essigman, W. K. Transoral decompression and posterior fusion for rheumatoid atlanto-axial subluxation. J. Bone Joint Surg. 68(B):350–356; 1986.

DeBrun-Fernandes, A. J.; Lucena-Fernandes, M.; Neto, M. L.; Cossermelli, W. Myelitis in systemic lupus erythematosus. Arthritis Rheum. 30:238–239; 1987.

Denburg, J. A.; Carbotte, R. M.; Denburg, S. D. Neuronal antibodies and cognitive function in systemic lupus erythematosus. Neurology 37:464–467; 1987.

DeSilva, M.; Hazleman, B. L. Azathioprine in giant cell arteritis/polymyalgia rheumatica: a double blind study. Ann. Rheum. Dis. 45:136–138; 1986.

Diaz-Perez, J. L.; Winkelmann, R. K. Cutaneous periarteritis nodosa. Arch. Dermatol. 110:407–414; 1974.

Dimant, J.; Ginzler, E. M.; Diamond, H. S.;

Schlesinger, M.; Marino, C. T.; Weiner, M.; Kaplan, D. Computer analysis of factors influencing the appearance of aseptic necrosis in patients with SLE. J. Rheumatol. 5:136–141; 1978.

Dougados, M.; Awada, H.; Amor, B. Cyclosporin in rheumatoid arthritis: a double blind, placebo controlled study in 52 patients. Ann. Rheum. Dis. 47:127–133; 1988.

Drosos, A. A.; Skopouli, F. N.; Costopoulos, J. S.; Papadimitriou, C. S.; Moutsopoulos, H. M. Cyclosporin A (CyA) in primary Sjögren's syndrome: a double blind study. Ann. Rheum. Dis. 45:732–735; 1986.

Drosos, A. A.; Van Vliet-Dascalopoulou, E.; Andonopoulos, A. P.; Galanopoulou, V.; Skopouli, F. N.; Moutsopoulos, H. M. Nandrolone decanoate (deca-durabolin) in primary Sjögren's syndrome: a double blind pilot study. Clin. Exp. Rheumatol. 6:53–57; 1988.

Dubois, E. L.; Wierzchowiecki, M.; Cox, M. B.; Weiner, J. M. Duration and death in systemic lupus erythematosus. An analysis of 249 cases. J. A. M. A. 227:1399–1402; 1974.

Dyck, P. J.; Benstead, T. J.; Conn, D. L.; Stevens, J. C.; Windebank, A. J.; Low, P. A. Nonsystemic vasculitic neuropathy. Brain 110:843–854; 1987.

Edwards, J. C. W.; Snaith, M. L.; Isenberg, D. A. A double blind controlled trial of methylprednisolone infusions in systemic lupus erythematosus using individualized outcome assessment. Ann. Rheum. Dis. 46:773–776; 1987.

Ellis, S. G.; Verity, M. A. Central nervous system involvement in systemic lupus erythematosus: a review of neuropathologic findings in 57 cases, 1955–1977. Semin. Arthritis Rheum. 8:212–221; 1979.

Estes, D.; Christian, C. L. The natural history of systemic lupus erythematosus by prospective analysis. Medicine 50:85–95; 1971.

Fauci, A. S.; Haynes, B. F.; Katz, P.; Wolff, S. M. Wegener's granulomatosis: prospective clinical and therapeutic experience with 85 patients for 21 years. Ann. Intern. Med. 98:76–85; 1983.

Fauci, A. S.; Katz, P.; Haynes, B. F.; Wolff, S. M. Cyclophosphamide therapy of severe systemic necrotizing vasculitis. N. Engl. J. Med. 301:235–238; 1979.

Fehring, T. K.; Brooks. A. L. Upper cervical instability in rheumatoid arthritis. Clin. Orthop. 221:137–148; 1987.

Feinglass, E. J.; Arnett, F. C.; Dorsch, C. A.; Zizik, M.; Stevens, M. B. Neuropsychiatric manifestations of systemic lupus erythematosus: diagnosis, clinical spectrum, and relationship to other features of the disease. Medicine 55:323–339; 1976.

Felson, D. T.; Anderson, J. Evidence for the superiority of immuno-suppressive drugs and prednisone over prednisone alone in lupus nephritis. N. Engl. J. Med. 311:1528–1533; 1984.

Ferlic, D. C.; Clayton, M. L.; Leidholt, J. D.; Gamble, W. E. Surgical treatment of the symptomatic unstable cervical spine in rheumatoid arthritis. J. Bone Joint Surg. 57(A):349–354; 1975.

Fernandez-Herlihy, L. Systemic lupus erythematosus: a clinical and prognostic analysis of 120 cases. Lahey Clin. Found. Bull. 21:49–62; 1972.

Fish, A. J.; Blau, E. B.; Westberg, N. G.; Burke, B. A.; Vernier, R. L.; Michael, A. F. Systemic lupus erythematosus within the first two decades of life. Am. J. Med. 62:99–117; 1977.

Førre, Ø.; Bjerkhoel, F.; Salvesen, C.; Berg K. J.; Rugstad, H. E.; Saclid, G.; Mellbye, O. J.; Käss, E. An open, controlled, randomized comparison of cyclosporine and azathioprine in the treatment of rheumatoid arthritis: a preliminary report. Arthritis Rheum. 30:88–92; 1987.

Fraga, A.; Mintz, G.; Valle, L.; Flores-Izquierdo, G. Takayasu's arteritis: frequency of systemic manifestations (study of 22 patients) and favorable response to maintenance steroid therapy with adrenocorticosteroids (12 patients). Arthritis Rheum. 15:617–624; 1972.

Franks, A. G. Topical glyceryl trinitrate as adjunctive treatment in Raynaud's disease. Lancet 1:76–77; 1982.

Fries, J. F. Toward an understanding of patient outcome measurement. Arthritis Rheum. 26:697–704; 1983.

Fries, J. F.; Bloch, D.; Spitz, P.; Mitchell, D. M. Cancer in rheumatoid arthritis: a prospective long-term study of mortality. Am. J. Med. 78(suppl. 1A):56–59; 1985.

Frohnert, P. P.; Sheps, S. G. Long-term follow-up study of periarteritis nodosa. Am. J. Med. 43:8–14; 1967.

Gabriel, S. E.; Luthra, H. S. Rheumatoid arthritis: can the long-term outcome be altered? Mayo Clin. Proc. 63:58–68; 1988.

Galve, E.; Candell-Riera, J. C.; Pigrau, C.; Permanyer-Miralda, G.; Garcia-DelCastillo, H.; Soler-Soler, J. Prevalence, morphologic types, and evolution of cardiac valvular disease in systemic lupus erythematosus. N. Engl. J. Med. 319:817–823; 1988.

Garlepp, M. J.; Dawkins, R. L.; Christiansen, F. T. HLA antigens and acetylcholine receptor antibodies in penicillamine induced myasthenia gravis. Br. Med. J. 286:338–340; 1983.

Garrett, R.; Paulus, H. Complications of intravenous methylprednisolone pulse therapy. Arthritis Rheum. 23:677; 1980.

Geirsson, A. J.; Sturfelt, G.; Truedsson, L. Clinical and serological features of severe vasculitis in rheumatoid arthritis: prognostic implications. Ann. Rheum. Dis. 46:727–733; 1987.

Gibson, T.; Myers, A. R. Nervous system involvement in systemic lupus erythematosus. Ann. Rheum. Dis. 35:398–406; 1976.

Ginzler, E.; Diamond, H.; Kaplan, D.; Weiner, M.; Schlesinger, M.; Seleznick, M. Computer analysis of factors influencing frequency of infection in systemic lupus erythematosus. Arthritis Rheum. 21:37–44; 1978.

Ginzler, E. M.; Diamond, H. S.; Weiner, M.; Schlesinger, M.; Frees, J. F.; Wasner, C.; Medsger, Jr., T. A.; Ziegler, G.; Klippel, J. H.; Hadler, N. M.; Albert, D. A.; Hess, E. V.; Spencer-Green, G.; Grayzel, A.; Worth, D.; Hahn, B. H.; Barnett, E. V. A multicenter study of outcome in systemic lupus erythematosus. I. Entry variables as predictors of prognosis. Arthritis Rheum. 25:601–611; 1982.

Goldman, J. A.; Casey, H. L.; McIlwain, H.; Kirby, J.; Wilson, C. H.; Muller, S. B. Limited plasmapheresis in rheumatoid arthritis with vasculitis. Arthritis Rheum. 22:1146–1150; 1979.

Goodman, B. W. Temporal arteritis. Am. J. Med. 67:839–852; 1979.

Goodwin, J. S.; Goodwin, J. M. Cerebritis in lupus erythematosus. Ann. Intern. Med. 90:437–438; 1979.

Gordon, D. A.; Stein, J. L.; Broder, I. The extra-articular features of rheumatoid arthritis. A systematic analysis of 127 cases. Am. J. Med. 54:445–452; 1973.

Gouet, D.; Marechaud, R.; Alcalay, M.; Berq-Girandon, B.; Bontoux, D.; LeFevre, J.-P.; Gil, R.; Risse, J.-F.; Sudre, Y. Survival in giant cell arteritis: a 14-year survey of 87 patients. J. Rheumatol. 12:1209–1210; 1985.

Graham, E.; Holland, A.; Avery, A.; Ross Russell, R. W. Prognosis in giant-cell arteritis. Br. Med. J. 282:269–271; 1981.

Hall, S.; Barr, W.; Lie, J. T.; Stanson, A. W.; Kazmier, F. J.; Hunder, G. G. Takayasu arteritis. A study of 32 North American patients. Medicine 64:89–99; 1985.

Halla, J. T.; Koopman, W. J.; Fallahi, S.; Oh, S. J.; Gay, R. E.; Schrohenloher, R. E. Rheumatoid myositis. Clinical and histologic features and possible pathogenesis. Arthritis Rheum. 27:737–743; 1984.

Hamdy, H.; McKendry, R. J. R.; Mierens, E.; Liver, J. A. Low-dose methotrexate compared with azathioprine in the treatment of rheumatoid arthritis. A twenty-four-week controlled clinical trial. Arthritis Rheum. 30:361–368; 1987.

Hamilton, C. R.; Shelley, W. M.; Tumulty, P. A. Giant cell arteritis: including temporal arteritis and polymyalgia rheumatica. Medicine 50:1–27; 1971.

Harati, Y.; Niakan, E. The clinical spectrum of inflammatory angiopathic neuropathy. J. Neurol. Neurosurg. Psychiat. 49:1313–1316; 1986.

Healey, L. A.; Wilske, K. R. Treatment of resistant giant cell arteritis. Arthritis Rheum. 26:930; 1983.

Hochberg, M. C. Auranofin or d-penicillamine in the treatment of rheumatoid arthritis. Ann. Intern. Med. 105:528–535; 1986.

Hollander, D.; Manning, R. T. The use of alkylating agents in the treatment of Wegener's granulomatosis. Ann. Intern. Med. 67:393–398; 1967.

Hollenhorst, W. W.; Brown, J. R.; Wagener, H. P.; Shick, R. M. Neurologic aspects of temporal arteritis. Neurology 10:490–498; 1960.

Hunder, G. G.; Sheps, S. G.; Allen, G. L.; Joyce, J. W. Daily and alternate day corticosteroid regimens in treatment of giant cell arteritis. Comparison in the prospective study. Ann. Intern. Med. 82:613–618; 1975.

Huston, K. A.; Hunder, G. G.; Lie, J. T.; Kennedy, R. H.; Elveback, L. R. Temporal arteritis. A 25-year epidemiologic, clinical and pathologic study. Ann. Intern. Med. 88:162–167; 1978.

Iannuzzi, L.; Dawson, N.; Zein, N.; Kushner, I. Does drug therapy slow radiographic deterioration in rheumatoid arthritis? N. Engl. J. Med. 309:1023–1028; 1983.

Isdale, I. C.; Conlon, P. W. Atlanto-axial subluxation. A six-year follow-up report. Ann. Rheum. Dis. 30:387–389; 1971.

Ishikawa, K. Survival and morbidity after diagnosis of occlusive thromboaortopathy (Takayasu's disease). Am. J. Cardiol. 47:1026–1032; 1981.

Ishikawa, K.; Uyama, M.; Asayama, K. Occlusive thromboaortopathy (Takayasu's disease): cervical arterial stenoses, retinal arterial pressure, retinal microaneurysms and prognosis. Stroke 14:730–735; 1983.

Jessar, R. A.; Lamont-Havers, R. W.; Ragan, C. Natural history of lupus erythematosus disseminatus. Ann. Intern. Med. 38:717–731; 1953.

Johnson, R. T.; Richardson, E. P. The neurological manifestations of systemic lupus erythematosus. A clinical-pathological study of 24 cases and review of the literature. Medicine 47:337–369; 1968.

Kellum, R. E.; Haserick, J. R. Systemic lupus erythematosus: a statistical evaluation of mortality based on a consecutive series of 299 patients. Arch. Intern. Med. 113:92–99; 1964.

Kimoto, S. The history and present status of aortic surgery in Japan, particularly for aortitis syndrome. J. Cardiovasc. Surg. 20:107–126; 1979.

Kinlen, L. J. Incidence of cancer in rheumatoid arthritis and other disorders after immunosuppressive treatment. Am. J. Med. 78(suppl. 1A):44–49; 1985.

Klippel, J. H. Morbidity and mortality. In: Balow, J. E.; moderator. Lupus nephritis. Ann. Intern. Med. 106:79–94; 1987.

Kostis, J. B.; Siebold, J. R.; Turkevich, D.; Massi, A. T.; Grau, R. G.; Medsger, T. A.; Steen, V. D.; Clements, P. J.; Szydlo, L.; D'Angelo, W. A. Prognostic importance of cardiac arrhythmias in systemic sclerosis. Am. J. Med. 84:1007–1015; 1988.

Kremer, J. M.; Jubiz, W.; Michalek, A.; Rynes, R. I.; Bartholomew, L. E.; Bigaquette, J.; Timchalk, M.; Beeler, D.; Lininger, L. Fish-oil fatty acid supplementation in active rheumatoid arthritis. A double-blinded, controlled, crossover study. Ann. Intern. Med. 106:497–503; 1987.

Lagneau, P.; Michel, J. B.; Vuong, P. N. Surgical treatment of Takayasu's disease. Ann. Surg. 205:157–166; 1987.

Lee, P.; Urowitz, M. B.; Bookman, A. A. M.; Koehler, B. E.; Smythe, H. A.; Gordon, D. A.; Ogryzlo, M. A. Systemic lupus erythematosus. A review of 110 cases with reference to nephritis, the nervous system, infections, aseptic necrosis and prognosis. Q. J. Med. 46:1–32; 1977.

Leib, E. S.; Restivo, C.; Paulus, H. E. Immunosuppressive and cortico-steroid therapy of polyarteritis nodosa. Am. J. Med. 67:941–947; 1979.

Leonhart, T. Long-term prognosis of systemic lupus erythematosus. Acta Med. Scand. 445(suppl.):440–442; 1966.

Levine, S. R.; Welch, K. M. A. The spectrum of neurologic disease associated with antiphospholipid antibodies. Lupus anticoagulants and anticardiolipin antibodies. Arch. Neurol. 44:876–883; 1987.

Leibling, M. R.; Leib, E.; McLaughlin, K.; Blocka, K.; Furst, D. E.; Nyman, K.; Paulus, H. E. Pulse methylprednisolone in rheumatoid arthritis. A double-blind crossover trial. Ann. Intern. Med. 94:21–26; 1981.

Linos, A.; Worthington, J. W.; O'Fallon, W. M.; Kurland, L. T. The epidemiology of rheumatoid arthritis in Rochester, Minnesota: a study of incidence, prevalence, and mortality. Am. J. Epidemiol. 111:87–98; 1980.

Lipson, S. J. Cervical myelopathy and posterior atlanto-axial subluxation in patients with rheumatoid arthritis. J. Bone Joint Surg. 67(A):593–597; 1985.

Lupi-Herrera, E.; Sánchez-Torres, G.; Marcushamer, J.; Mispireta, J.; Horwitz, S.; Vela, J. E. Takayasu's arteritis. Clinical study of 107 cases. Am. Heart J. 93:94–103; 1977.

Maksymowych, W.; Russell, A. S. Antimalarials in rheumatology: efficacy and safety. Semin. Arthritis Rheum. 16:206–221; 1987.

Manthorpe, R.; Petersen, S. H.; Prause, J. U. Mucosolvan in the treatment of patients with primary Sjögren's syndrome. Results from a double-blind cross-over investigation. Acta. Ophthalmol. 62:537–541; 1984.

Masi, A. T. Low dose glucocorticoid therapy in rheumatoid arthritis (RA): transitional or selected add-on therapy? J. Rheumatol. 10:675–678; 1983.

Mathews, J. A. Atlanto-axial subluxation in rheumatoid arthritis. A 5-year follow-up study. Ann. Rheum. Dis. 33:526–531; 1974.

Medical Research Council. Treatment of polyarteritis nodosa with cortisone: results after three years. Br. Med. J. 1:1399–1400; 1960.

Medsger, T. A.; Masi, A. T. Survival with scleroderma—II: A life-table analysis of clinical and demographic factors in 358 males U.S. veteran patients. J. Chron. Dis. 26:647–660; 1973.

Medsger, T. A.; Masi, A. T.; Rodnan, G. P.; Benedek, T. G.; Robinson, H. Survival with systemic sclerosis (scleroderma). A life-table analysis of clinical and demographic factors in 309 patients. Ann. Intern. Med. 75:369–376; 1971.

Meijers, K. A. E.; Kremer, H. P. H.; Luyendijk, W.; Onvlee, G. J.; Thomeer, R. T. W. M. Cervical myelopathy in rheumatoid arthritis. Clin. Exp. Rheumatol. 2:239–245; 1984.

Meislin, A. G.; Rothfield, N. Systemic lupus erythematosus in childhood. Analysis of 42 cases, with comparative data on 20 adult cases followed in the community. Pediatrics 42:37–49; 1968.

Menezes, A. H.; VanGilder, J. C.; Clark, C. R.; El-Khoury, G. Odontoid upward migration in rheumatoid arthritis. An analysis of 45 patients with "cranial settling." J. Neurosurg. 63:500–509; 1985.

Michet, C. J.; McKenna, C. H.; Elveback, L. R.; Kaslow, R. A.; Kurland, L. T. Epidemiology of systemic lupus erythematosus and other connective tissue disease in Rochester, Minnesota, 1950 through 1979. Mayo Clin. Proc. 60:105–113; 1985.

Mimori, T. Scleroderma-polymyositis overlap syndrome. Clinical and serologic aspects. Int. J. Dermatol. 26:419–425; 1987.

Mitchell, D. M.; Spitz, P. W.; Young, D. Y.; Bloch, D. A.; McShane, D. J.; Fries, J. F. Survial, prognosis, and causes of death in rheumatoid arthritis. Arthritis Rheum. 29:706–714; 1986.

Monson, R. R.; Hall, A. P. Mortality among arthritics. J. Chron. Dis. 29:459–467; 1976.

Moore, P. M.; Fauci, A. S. Neurologic manifestations of systemic vasculitis. A retrospective and prospective study of the clinicopathologic features and responses to therapy in 25 patients. Am. J. Med. 71:517–524; 1981.

Mutru, O.; Laakso, M.; Isomäki, H.; Koota, K. Ten year mortality and causes of death in patients with rheumatoid arthritis. Br. Med. J. 290:1797–1799; 1985.

Nadeau, S. E.; Watson, R. T. Neurologic manifestations of vasculitis and collagen vascular syndromes. In: Joynt, R. J., Clinical neurology. Philadelphia: J. B. Lippincott; 1990:p. 1–166.

Nakao, K.; Ikeda, M.; Kimata, S-I.; Nitani, H.; Miyahara, M.; Ishimi, Z.; Hashiba, K.; Takeda, Y.; Ozawa, T.; Matsushita, S.; Kuramochi, M. Takayasu's arteritis. Clinical report of eighty-four cases and immunological studies of seven cases. Circulation 35:1141–1155; 1967.

Neumann, V. C.; Grindulis, K. A.; Hubball, S.; McConkey, B.; Wright, V. Comparison between penicillamine and sulphasalazine in rheumatoid arthritis: Leeds-Birmingham trial. Br. Med. J. 287:1099–1102; 1983.

Novack, S. N.; Pearson, C. M. Cyclophosphamide therapy in Wegener's granulomatosis. N. Engl. J. Med. 284:938–942; 1971.

Numano, F.; Ohta, N.; Susazuki, T. HLA and clinical manifestations in Takayasu disease. Jpn. Circ. J. 46:184–189; 1982.

Oddis, C. V.; Eisenbeis, C. H.; Reidbord, H. E.; Steen, V. D.; Medsger, T. A. Vasculitis in systemic sclerosis: association with Sjögren's syndrome and the CREST syndrome variant. J. Rheumatol. 14:942–948; 1987.

Paloheimo, J. A. Obstructive arteritis of Takayasu's type. Acta Med. Scand. 468(suppl.):1–31; 1967.

Pellicci, P. M.; Ranawat, C. S.; Tsairis, P.; Bryan, W. J. A prospective study of the progression of rheumatoid arthritis of the cervical spine. J. Bone Joint Surg. 63(A):342–350; 1981.

Perry, M. C. Chemotherapy, toxicity and the clinician. Semin. Oncol. 11:1–4; 1982.

Pinals, R. S.; Kaplan, S. B.; Lawson, J. G.; Hepburn, B. Sulfasalazine in rheumatoid arthritis. A double-blind, placebo-controlled trial. Arthritis Rheum. 29:1427–1434; 1986.

Pinching, A. J.; Lockwood, C. M.; Pussell, B. A.; Rees, A. J.; Sweny, P.; Evans, D. J.; Bowley, N.; Peters, D. K. Wegener's granulomatosis: observations on 18 patients with severe renal disease. Q. J. Med. 52:435–460; 1983.

Pincus, T.; Callahan, L. F. Taking mortality in rheumatoid arthritis seriously—predictive markers, socioeconomic status and comorbidity. J. Rheumatol. 13:841–845; 1986.

Pirofsky, B.; Bardana, E. J. Immunosuppressive therapy in rheumatic disease. Med. Clin. North Am. 61:419–437; 1977.

Prause, J. U.; Frost-Larsen, K.; Høj, L.; Isager, H.; Manthorpe, R. Lacrimal and salivary se-

cretion in Sjögren's syndrome: the effect of systemic treatment with bromhexine. Acta Ophthalmol. 62:489–497; 1984.

Prior, P.; Symmons, D. P. M.; Scott, D. L.; Brown, R.; Hawkins, C. F. Causes of death in rheumatoid arthritis. Br. J. Rheumatol. 23:92–99; 1984.

Pullar, T.; Hunter, J. A.; Capell, H. A. Sulphasalazine in rheumatoid arthritis: a double blind comparison of sulphasalazine with placebo and sodium aurothiomalate. Br. Med. J. 287:1102–1104; 1983.

Radia, M.; Furst, D. E. Comparison of three pulse methylprednisolone regimens in the treatment of rheumatoid arthritis. J. Rheumatol. 15:242–246; 1988.

Ranawat, C. S.; O'Leary, P.; Pellicci, P.; Tsairis, P.; Marchisello, P.; Dorr, L. Cervical spine fusion in rheumatoid arthritis. J. Bone Joint Surg. 61(A):1003–1010; 1979.

Rasker, J. J.; Cosh, J. A. Cause and age at death in a prospective study of 100 patients with rheumatoid arthritis. Ann. Rheum. Dis. 40:115–120; 1981.

Redlund-Johnell, I. Posterior atlanto-axial dislocation in rheumatoid arthritis. Scand. J. Rheumatol. 13:337–341; 1984.

Redlund-Johnell, I.; Pettersson, H. Subaxial antero-posterior dislocation of the cervical spine in rheumatoid arthritis. Scand. J. Rheumatol. 14:355–363; 1985.

Reza, M. J.; Dornfeld, L.; Goldberg, L. S.; Bluestone, R.; Pearson, C. M. Wegener's granulomatosis. Long-term followup of patients treated with cyclophosphamide. Arthritis Rheum. 18:501–506; 1975.

Robbs, J. V.; Human, R. R.; Rayarutnam, P. Operative treatment of nonspecific aortoarteritis (Takayasu's arteritis). J. Vasc. Surg. 3:605–616; 1986.

Rodehoffer, R. J.; Rommer, J. A.; Wigley, F.; Smith, C. R. Controlled double-blind trial of nifedipine in the treatment of Raynaud's phenomenon. N. Engl. J. Med. 308:880–883; 1983.

Ropes, M. W.; Bennett, G. A.; Cobb;, S.; Jacox, R.; Jessar, R. A. 1958 revision of diagnostic criteria for rheumatoid arthritis. Bull. Rheum. Dis. 9:175–176; 1958.

Rosner, S.; Ginzler, E. M.; Diamond, H. S.; Weiner, W.; Schlesinger, M.; Fries, J. F.; Wasner, C.; Medsger, Jr., T. A.; Ziegler, G.; Klippel, J. H.; Hadler, N. M.; Albert, D. A.; Hess, E. V.; Spencer-Green, G.; Grayzel, A.; Worth, D.; Hahn, B. H.; Barnett, E. V. A multicenter study of outcome in systemic lupus erythematosus. II. causes of death. Arthritis Rheum. 25:612–617; 1982.

Sack, M.; Cassidy, J. T.; Bole, G. G. Prognostic factors in polyarteritis. J. Rheumatol. 2:411–420; 1975.

Sahn, E. E.; Sahn, S. A. Wegener granulomatosis with aphasia. Arch. Intern. Med. 136:87–89; 1976.

Sambrook, P. N.; Champion, G. D.; Brown, C. D.; Cohen, M. L.; Compton, P.; Day, R. O.; deJager, J. Comparison of methotrexate with azathioprine or 6-mercaptopurine in refractory rheumatoid arthritis: a life table analysis. Br. J. Rheumatol. 25:372–375; 1986.

Save-Soderbergh, J.; Malmvall, B-E.; Andersson, R.; Bengtsson, B-A. Giant cell arteritis as a cause of death. Report of nine cases. J.A.M.A. 255:493–496; 1986.

Schneider, H. A.; Weber, A. A.; Ballen, P. H. The visual prognosis in temporal arteritis. Ann. Ophthalmol. 3:1215–1230; 1971.

Scott, D. G. I.; Bacon, P. A. Intravenous cyclophosphamide plus methylprednisolone in treatment of systemic rheumatoid vasculitis. Am. J. Med. 76:377–384; 1984.

Scott, D. G. I.; Bacon, P. A.; Elliott, P. J.; Tribe, C. R.; Wallington, T. B. Systemic vasculitis in a district general hospital 1972–1980: clinical and laboratory features, classification and prognosis in 80 cases. Q.J. Med. 51:292–311; 1982.

Shelhamer, J. H.; Volkman, D. J.; Parrillo, J. E.; Lawley, T. J.; Johnston, M. R.; Fauci, A. S. Takayasu's arteritis and its therapy. Ann. Intern. Med. 103:121–126; 1985.

Sherrer, Y. S.; Bloch, D. A.; Mitchell, D. M.; Young, D. Y.; Fries, J. F. The development of disability in rheumatoid arthritis. Arthritis Rheum. 29:494–500; 1986.

Siegel, M.; Gwon, N.; Lee, S. L.; Rivero, I.; Wong, W. Survivorship in systemic lupus erythematosus: relationship to race and pregnancy. Arthritis Rheum. 12:117–125; 1969.

Simon, L. S.; Mills, J. A. Nonsteroidal antiinflammatory drugs. N. Engl. J. Med. 302:1179–1185, 1237–1243; 1980.

Situnayake, R. D.; Grindulis, K. A.; McConkey, B. Long term treatment of rheumatoid arthritis with sulphasalazine, gold, or penicillamine: a comparison using life-table methods. Ann. Rheum. Dis. 46:177–183; 1987.

Smith, C. D.; McKendry, R. J. R. Controlled trial of nifedipine in the treatment of Raynaud's phenomenon. Lancet 2:1299–1301; 1982.

Smith, M. D.; Ahern, M. J.; Roberts-Thomson, P. J. Pulse steroid therapy in rheumatoid arthritis: can equivalent doses of oral prednisone give similar clinical results to intravenous methylprednisolone? Ann. Rheum. Dis. 47:28–33; 1988.

Smith, P. H.; Benn, R. T.; Sharp, J. Natural history of rheumatoid cervical luxations. Ann. Rheum. Dis. 31:431–439; 1972.

Spiera, H.; Davison, S. Long-term follow-up of polymyalgia rheumatica. Mt. Sinai J. Med. 45:225–229; 1978.

Staples, P. J.; Gerding, D. N.; Decker, J. L.; Gordon, R. S. Incidence of infection in systemic lupus erythematosus. Arthritis Rheum. 17:1–10; 1974.

St. Clair, E. W.; Polisson R. P. Therapeutic approaches to the treatment of rheumatoid disease. Med. Clinic North Am. 70:285–303; 1986.

Steen, V. D.; Blair, S.; Medsger, T. A. The toxicity of d-penicillamine in systemic sclerosis. Ann. Intern. Med. 104:699–705; 1986.

Steen, V. D.; Medsger, T. A.; Rodnan, G. P. D-penicillamine therapy in progressive systemic sclerosis (scleroderma). Ann. Intern. Med. 97:652–659; 1982.

Steen, V. D.; Powell, D. L.; Medsger, T. A. Clinical correlations and prognosis based on serum autoantibodies in patients with systemic sclerosis. Arthritis Rheum. 31:196–203; 1988.

Stevens, J. M.; Kendall, B. E.; Crockard, H. A. The spinal cord in rheumatoid arthritis with clinical myelopathy: a computed myelographic study. J. Neurol. Neurosurg. Psychiat. 49:140–151; 1986.

Sthoeger, Z. M.; Geltner, D.; Rider, A.; Bentwich, Z. Systemic lupus erythematosus in 49 Israeli males: a retrospective study. Clin. Exp. Rheumatol. 5:233–240; 1987.

Studenski, S.; Allen, N. B.; Caldwell, D. S.; Rice, J. R.; Polisson, R. P. Survival in systemic lupus erythematosus. A multivariate analysis of demographic factors. Arthritis Rheum. 30:1326–1332; 1987.

Symmons, D. P. M. Mortality in rheumatoid arthritis. Br. J. Rheumatol. 27(suppl. I):44–54; 1988.

Symmons, D. P. M.; Prior, P.; Scott, D. L.; Brown, R.; Hawkins, C. F. Factors influencing mortality in rheumatoid arthritis. J. Chron. Dis. 39:137–145; 1986.

Traub, Y. M.; Shapiro, A. P.; Rodnan, G. P.; Medsger, T. A.; McDonald, R. H.; Steen, V. D.; Osial, T. A.; Tolchin, S. F. Hypertension and renal failure (scleroderma renal crisis) in progressive systemic sclerosis. Medicine 62:335–352; 1983.

Tuffanelli, D. L.; Winkelmann, R. K. Systemic scleroderma. A clinical study of 727 cases. Arch. Dermatol. 84:359–371; 1961.

Tuffanelli, D. L.; Winkelmann, R. K. Diffuse systemic scleroderma. A comparison with acrosclerosis. Ann. Intern. Med. 57:198–203; 1962.

Tugwell, P.; Bennett, K.; Gent, M. Methotrexate in rheumatoid arthritis. Indications, contraindications, efficacy, and safety. Ann. Intern. Med. 107:358–366; 1987.

Uddin, J.; Kraus, A. S.; Kelly, H. G. Survivorship and death in rheumatoid arthritis. Arthritis Rheum. 13:125–130; 1970.

Ueda, H.; Morooka, S.; Ito, I.; Yamaguchi, H.;

Takeda, T.; Saito, Y. Clinical observation of 52 cases of aortitis syndrome. Jpn. Heart J. 10:277–288; 1969.

Urman, J. D.; Rothfield, N. F. Corticosteroid treatment in systemic lupus erythematosus: survival studies. J.A.M.A. 238:2272–2276; 1977.

Vandenbroucke, J. P.; Hazevoet, H. M.; Cats, A. Survival and cause of death in rheumatoid arthritis: a 25-year prospective followup. J. Rheumatol. 11:158–161; 1984.

Veys, E. M.; Mielants, H.; Verbruggen, G.; Grosclaude, J.-P.; Meyer, W.; Galazka, A. Schindler, J. Interferon gamma in rheumatoid arthritis-double blind study comparing human recombinant interferon gamma with placebo. J. Rheumatol. 15:570–574; 1988.

Vincent, D.; Dubas, F.; Haux, J. J.; Godeau, P.; Lhermitte, F.; Buge, A.; Castaigne, P. Nerve and muscle microvasculitis in peripheral neuropathy: a remote effect of cancer? J. Neurol. Neurosurg. Psychiatry 49:1007–1010; 1986.

Vollertsen, R. S.; Conn, D. L.; Ballard, D. J.; Ilstrup, D. M.; Kazmar, R. E.; Silverfield, J. C. Rheumatoid vasculitis: survival and associated risk factors. Medicine 65:365–375; 1986.

von Knorring, J. Treatment and prognosis in polymyalgia rheumatica and temporal arteritis. A ten-year survey of 53 patients. Acta Med. Scand. 205:429–435; 1979.

Wallace, D. J.; Podell, T.; Weiner, J.; Klinenberg, J. R.; Forouzesh, S.; Dubois, E. L. Systemic lupus erythematosus—survival patterns. Experience with 609 patients. J.A.M.A. 245:934–938; 1981.

Walravens, P. A.; Chase, H. P. The prognosis of childhood systemic lupus erythematosus. Am. J. Dis. Child. 130:929–933; 1976.

Walters, M. T.; Rubin, C. E.; Keightley, S. J.; Ward, C. D.; Cawley, M. I. D. A double blind, cross-over, study of oral N-acetylcysteine in Sjögren's syndrome. Scand. J. Rheumatol. 61(suppl.):253–258; 1986.

Walton, E. W. Giant-cell granuloma of the respiratory tract (Wegener's granulomatosis). Br. Med. J. 2:265–270; 1958.

Weissman, B. N. W.; Aliabadi, P.; Weinfeld, M. S.; Thomas, W. H.; Sosman, J. L. Prognostic features of atlantoaxial subluxation in rheumatoid arthritis patients. Radiology 144:745–751; 1982.

West, S. G.; Killian, P. J.; Lawless, O. J. Association of myositis and myocarditis in progressive systemic sclerosis. Arthritis Rheum. 24:662–667; 1981.

Wilke, W. S.; Mackenzie, A. H. Methotrexate in rheumatoid arthritis. Current status. Drugs 32:103–113; 1986.

Winfield, J.; Cooke. D.; Brooke, A. S.; Corbett, M. A prospective study of the radiological changes in the cervical spine in early rheumatoid disease. Ann. Rheum. Dis. 40:109–114; 1981.

Wofsy, D. New approaches to treating systemic lupus erythematosus. West. J. Med. 147:181–186; 1987.

Wolfe, B. K.; O'Keeffe, D.; Mitchell, D. M.; Tchang, S. P. K. Rheumatoid arthritis of the cervical spine: early and progressive radiographic features. Radiology 165:145–148; 1987.

Wynn, J.; Fineberg, N.; Matzer, L.; Cortada, X.; Armstrong, W.; Dillon, J. C.; Kinney, E. L. Prediction of survival in progressive systemic sclerosis by multivariate analysis of clinical features. Am. Heart J. 110:123–127; 1985.

Zarafonetis, C. J. D.; Dabich, L.; Negri, D.; Skovronski, J. J.; DeVol, E. B.; Wolfe, R. Retrospective studies in scleroderma: effect of potassium para-aminobenzoate on survival. J. Clin. Epidemiol. 41:193–205; 1988.

Zoma, A.; Sturrock, R. D.; Fisher, W. D.; Freeman, P. A.; Hamblen, D. L. Surgical stabilization of the rheumatoid cervical spine. A review of indications and results. J. Bone Joint Surg. 69(B):8–12; 1987.

Zvaifler, N. J. Fractionated total lymphoid irradiation: a promising new treatment of rheumatoid arthritis? Yes, no, maybe. Arthritis Rheum. 30:109–114; 1987.

36

Neurosarcoidosis

BARNEY J. STERN

Neurological disease occurs in 5% of patients with sarcoidosis (Stern et al. 1985). Central (CNS) or peripheral nervous system dysfunction is the presenting feature of sarcoidosis in one half or more of the patients with neurosarcoidosis (Oksanen 1986; Pentland et al. 1985; Stern et al. 1985). Furthermore, approximately three fourths of sarcoidosis patients who are destined to develop neurological disease do so within the first 2 years of their systemic illness (Stern et al. 1985). Unfortunately, it is not known why only a small proportion of patients with sarcoidosis have neurological problems or how to predict which patients will encounter neurological difficulties.

Therefore, neurosarcoidosis is a diagnostic consideration in patients with sarcoidosis who develop neurological signs or symptoms and in patients without known sarcoidosis who develop an illness consistent with neurosarcoidosis. Sarcoidosis presents with a variety of neurological problems as outlined in Table 36-1.

Two thirds of patients with neurosarcoidosis have a monophasic illness with minimal residual neurological deficits (Luke et al. 1987; Oksanen 1986; Pentland et al. 1985). These patients typically have a cranial mononeuropathy. Approximately one third of patients have a remitting–relapsing or chronically progressive illness that is associated with substantial morbidity and a mortality approximating 10% (James and Sharma 1967; Luke et al. 1987). These patients often have CNS mass lesions, hydrocephalus, or a relatively diffuse encephalopathy/vasculopathy (Luke et al. 1987; Oksanen 1986; Pentland et al. 1985). Patients with recurrent aseptic meningitis, cranial polyneuropathies, myopathy, or neuropathy can have a protracted course but usually survive their illness albeit with considerable morbidity at times (Ando 1985; Luke et al. 1987). It is not possible to confidently identify which patients will have a relapsing course (Chen and McLeod 1989), but those who are destined to do so tend to have recurrent disease presenting with similar neurological manifestations (Luke et al. 1987). Patients with either an acute or more chronic neurological presentation do not differ in their prognosis (Chapelon et al. 1990).

The true natural history of neurosarcoido-

Table 36-1. Manifestations
of Neurosarcoidosis

Cranial Neuropathy
Meningeal Disease
   Aseptic meningitis
   Dural mass
Myopathy
Neuropathy
Hydrocephalus
Parenchymatous Disease
   Endocrinopathy
   Mass lesion(s)
   Encephalopathy/vasculopathy
   Seizures

sis is not well known. Therefore, it is difficult to report specific numbers from large clinical series regarding the prognosis of the various neurological manifestations of sarcoidosis. Also, reports from referral centers probably reflect a skewed patient population. Most patients with neurological complications of sarcoidosis are treated with corticosteroids or rarely with other therapies, such as radiation or immunosuppressive agents. Published observations of treated patients may, however, approximate the natural history of the disease; Zaki and associates (1987) have noted no significant differences in the outcome of 183 patients with pulmonary sarcoidosis randomized to either corticosteroid therapy or placebo. Corticosteroids are effective treatment for the *symptoms* of sarcoidosis (Israel 1987), but whether corticosteroids affect the ultimate prognosis of neurosarcoidosis remains to be determined (Chapelon et al. 1990). Other therapeutic modalities are needed to alter the outcome of patients with refractory disease.

Patients with CNS sarcoidosis who die tend to do so at a young age. In the pathological series reviewed by Manz (1983), the mean age at death of patients with CNS sarcoidosis ranged from 24 to 38.2 years. The mean duration of neurological disease ranged from 2.3 to 4.7 years before death. Even allowing for the bias inherent in a pathological analysis, these data reinforce the potentially fulminant course of CNS sarcoidosis.

Cranial Neuropathy

Peripheral facial palsy, the most common neurological complication of sarcoidosis, has a good prognosis (Colover 1948; James and Sharma 1967; Silverstein et al. 1965; Wiederholt and Siekert 1965), with improvement in more than 80% of patients (Stern et al. 1985; Oksanen 1986). Patients typically recover over several weeks, although residual weakness may remain (Matthews 1979).

Eighth nerve dysfunction, whether auditory or vestibular, can recover or have a less fortunate outcome, characterized by sudden deafness, fluctuating or progressive symptomatology with persistent hearing loss, tinnitus, or vestibular dysfunction (Souliere et al. 1991).

Unilateral or bilateral optic neuropathy, often accompanied by chiasmatic infiltration, can lead to progressive visual loss (Gelwan et al. 1988). Optic atrophy can develop insidiously and serial neuro-ophthalmological assessments are necessary to monitor the clinical course.

Any of the other cranial nerves can be involved in CNS sarcoidosis. The clinical outcome typically is benign (Chapelon et al. 1990), especially if only one or two nerves are involved and disease is limited to the cranial nerves. If CNS parenchymal disease or hydrocephalus is present, the prognosis seems to parallel the course of these problems rather than the cranial neuropathy.

Patients with a single cranial neuropathy have been treated with corticosteroids, although it is unclear whether this alters the generally benign clinical course. Patients with optic or eighth nerve disease often require protracted corticosteroid therapy to maintain function.

Meningeal Disease

Symptomatic acute aseptic meningitis usually is a self-limited problem (Chapelon et al. 1990), but it may be recurrent. Typically, patients are offered corticosteroid therapy. A chronic cerebrospinal fluid (CSF) pleocytosis is not usually symptomatic, but hypoglycorrachia, elevated immunoglobulin G (IgG) index, oligoclonal bands, elevated CSF angiotensin converting enzyme (ACE) level, and elevated CSF CD4 : CD8 ratio would suggest continuing CNS inflammation. The relevance of these findings to the clinical outcome is unclear. If a patient is clinically stable or asymptomatic, the author has not tried to normalize the CSF with corticosteroid therapy but has titrated therapy to the development of new signs or symptoms. Rigorous data to support this approach are lacking.

A dural granulomatous mass requires a more cautious prognosis. These lesions may respond to therapy and not recur following withdrawal of corticosteroid therapy, or they may develop into a chronic problem and exhibit a chronic remitting–relapsing or progressive course. In the latter case, the patient's outcome is similar to that of patients with CNS parenchymatous disease.

Myopathy

Patients with myopathy have a guarded prognosis (Chapelon et al. 1990). An acute myositis can resolve after several months of corticosteroid therapy. More commonly, there is a chronic, indolent myopathy that can be remitting–relapsing or chronically progressive and debilitating (Matthews 1979; Wolfe et al. 1987). Long-term corticosteroid treatment is routinely used and can be helpful.

Neuropathy

A mononeuropathy implies a good prognosis, whereas the outcome of patients with a mononeuritis multiplex or generalized sensory, motor, or sensorimotor neuropathy is guarded (Zuniga et al. 1991). A Guillain-Barré syndrome can occur with good recovery (Zuniga et al. 1991; Miller et al. 1991). A trial of corticosteroid therapy is indicated.

Hydrocephalus

The prognosis of hydrocephalus is related to the underlying cause. Hydrocephalus can be due to active granulomatous inflammation involving the choroid plexus or ependymal lining of the ventricles or causing a periventricular parenchymal mass or meningitis. Hydrocephalus can result from an end-stage fibrotic process obstructing CSF flow. Patients can develop acute life-threatening increased intracranial pressure, or they may have a subacute or even asymptomatic presentation.

Patients with an active inflammatory component to their hydrocephalus may improve, at least temporarily, with corticosteroid therapy (Lundh and Wikkelso 1987). However, concurrent parenchymal disease and the hydrocephalus often require continuing therapy (Foley et al. 1989). Often a ventriculoperitoneal shunt is required because the patient becomes intolerant of the high-dose corticosteroid needed to maintain CSF flow or the hydrocephalus is progressive in spite of high corticosteroid doses. A patient with symptomatic hydrocephalus due to end-stage fibrosis would not be expected to respond to medical management with corticosteroids and therefore requires a ventriculoperitoneal shunt.

Once a shunt is placed, the patient is at risk not only from the underlying inflammatory disease and the concurrent medication needed to keep this under control, but also from shunt complications, such as infection and obstruction. Blockage of the shunt can be related to a high CSF protein or spread of granulomatous inflammation onto and within the shunt tubing (Stern, unpublished data, 1985).

With all the complications associated with hydrocephalus due to sarcoidosis, it is clear why these patients constitute a particularly high-risk group (Chapelon et al. 1990).

CNS Parenchymatous Disease

The most common CNS parenchymal lesions involve the hypothalamus and cause neuroendocrinological dysfunction and vegetative symptoms. Once altered, it is unusual for hormonal function to normalize (Scott et al. 1987). However, hyperprolactinemia can resolve (Nakao 1978), and rarely normal gonadotropin function can return (Hidaka et al. 1987).

Patients with one or more CNS granulomatous mass lesions have a guarded prognosis. Untreated, these patients usually progressively deteriorate. However, although resolution of the disease can be achieved in some patients (Chapelon et al. 1990), many patients have continuing disease characterized by remissions and exacerbations that are linked to their corticosteroid dose. Other patients exhibit a progressive deterioration in spite of "optimal" therapy. Surgical intervention is best reserved for diagnostic purposes or for patients who deteriorate in spite of aggressive medical management (Peeples et al. 1991). Radiation therapy and immunosuppressive agents other than corticosteroids occasionally are used in refractory patients and may improve neurological function.

A sarcoidosis-related diffuse encephalopathy/vasculopathy can present as a monophasic illness and resolve, but more commonly, it exhibits a remitting–relapsing or chronically progressive course in spite of intense medical management.

Convulsions suggest the presence of parenchymal brain disease if there is no obvious metabolic perturbation accounting for the seizures. Typically the seizures are not difficult to treat, but the underlying parenchymal disease often is the limiting factor regarding patient outcome (Krumholz et al. 1988). Therefore, seizures are a potential marker for cerebral sarcoidosis and are poor prognostic indicators (Delaney 1980).

As previously mentioned, patients with CNS parenchymatous disease involving the brain or spinal cord account for much of the morbidity and mortality attributed to neurosarcoidosis, especially if there is associated hydrocephalus. Patients are at risk not only for progressive neurological deterioration, but also therapy can pose significant hazards. Complications of prolonged corticosteroid administration, such as infection and osteoporosis (Rizzato et al. 1987), are particular problems.

Conclusion

Neurosarcoidosis can exhibit a broad spectrum of manifestations. In an analogous manner, the prognosis is varied and linked not only to the extent of neurological disease, but also to the complications that arise from therapeutic interventions. Fortunately, most patients with sarcoidosis do well, but 1% to 2% of all patients with sarcoidosis encounter considerable difficulties that challenge their fortitude and the skills of their physicians.

References

Ando, D. G.; Lynch, J. P.; Fantone, J. C. Sarcoid myopathy with elevated creatine phosphokinase. Am. Rev. Respir. Dis. 131:298–300; 1985.

Chapelon, C.; Ziza, J. M.; Piette, J. C.; Levy, Y.; Raguin, G.; Wechsler, B.; Bitker, M. O.; Blétry, O.; Laplane, D.; Bousser, M. G.; Godeau, P. Neurosarcoidosis: Signs, course and treatment in 35 confirmed cases. Medicine 69:261–276; 1990.

Chen, R. C. Y.; McLeod, J. G. Neurological complications of sarcoidosis. Clin. Exp. Neurol. 26:99–112; 1989.

Colover, J. Sarcoidosis with involvement of the nervous system. Brain. 71:451–475; 1948.

Delaney, P. Seizures in sarcoidosis: A poor prognosis. Ann. Neurol. 7:494; 1980.

Foley, K. T.; Howell, J. D.; Junck, L. Progression of hydrocephalus during corticosteroid therapy for neurosarcoidosis. Postgrad. Med. J. 65:481–484; 1989.

Gelwan, M. J.; Kellen, R. I.; Burde, R. M.; Kupersmith, M. J. Sarcoidosis of the anterior visual pathway: successes and failures. J. Neurol. Neurosurg. Psychiatry 51:1473–1480; 1988.

Hidaka, N.; Takizawa, H.; Miyachi, S.; Hisatomi, T.; Kosuda, T.; Sato, T. Case report: A case of hypothalamic sarcoidosis with hypopituitarism and prolonged remission of hypogonadism. Am. J. Med. Sci. 294:357–363; 1987.

Israel, H. L. Corticosteroid treatment of sarcoidosis—Who needs it? N.Y. State J. Med. 87:490; 1987.

James, D. G.; Sharma, O. P. Neurological complications of sarcoidosis. Proc. R. Soc. Med. 60:1169–1170; 1967.

Krumholz, A.; Stern, B. J.; Stern, E. G. Clinical implications of seizures in neurosarcoidosis. Arch. Neurol. 48:842–844; 1991.

Luke, R. A.; Stern, B. J.; Krumholz, A.; Johns, C. J. Neurosarcoidosis: The long-term clinical course. Neurology 37:461–463; 1987.

Lundh, T.; Wikkelso, C. Sarcoidosis with hydrocephalus: report of a case successfully treated with a ventriculo-peritoneal shunt and methylprednisolone pulse therapy. Acta. Neurol. Scand. 76:365–368; 1987.

Manz, H. J. Pathobiology of neurosarcoidosis and clinicopathologic correlation. Can. J. Neurol. Sci. 10:50–55; 1983.

Matthews, W. B. Neurosarcoidosis. Handbook of clinical neurology. New York: Elsevier–North Holland, Inc., 1979.

Miller, R.; Sheron, N.; Semple, S. Sarcoidosis presenting with an acute Guillain-Barré syndrome. Postgrad. Med. J. 65:765–767; 1989.

Nakao, K.; Noma, K.; Sato, B.; Yano, S.; Yamamura, Y.; Hibano, T. Serum prolactin levels in eighty patients with sarcoidosis. Eur. J. Clin. Inves. 8:37–40; 1978.

Oksanen, V. Neurosarcoidosis: clinical presentations and course in 50 patients. Acta. Neurol. Scand. 73:283–290; 1986.

Peeples, D. M.; Stern, B. J.; Violet, J.; Sahni, K. S. Germ cell tumors masquerading as central nervous system sarcoidosis. Arch. Neurol. 48:554–556; 1991.

Pentland, B.; Mitchell, J. D.; Cull, R. E.; Ford, M. J. Central nervous system sarcoidosis. Q. J. Med. 56:457–465; 1985.

Rizzato, G.; Fraioli, P. Natural and cortico-steroid-induced osteoporosis in sarcoidosis: Prevention, treatment, follow up, and reversibility. Sarcoidosis. 7:89–92; 1990.

Scott, I. A.; Stocks, A. E.; Saines, N. Hypothalamic/pituitary sarcoidosis. Aust. N. Z. J. Med. 17:243–245; 1987.

Silverstein, A.; Feuer, M. M.; Siltzbach, L. E. Neurologic sarcoidosis. Arch. Neurol. 12:1–11; 1965.

Souliere, C. R.; Kava, C. R.; Barrs, D. M.; Bell, A. F. Sudden hearing loss as the sole manifestation of neurosarcoidosis. Otolaryngol. Head Neck Surg. 105:376–381; 1991.

Stern, B. J.; Krumholz, A.; Johns, C.; Scott, P.; Nissim, J. Sarcoidosis and its neurological manifestations. Arch. Neurol. 42:909–917; 1985.

Wiederholt, W. C.; Siekert, R. G. Neurologic manifestations of sarcoidosis. Neurology 15:1147–1154; 1965.

Wolfe, S. M.; Pinals, R. S.; Aelion, J. A.; Goodman, R. E. Myopathy in sarcoidosis: Clinical and pathologic study of four cases and review of the literature. Semin. Arthritis Rheumatism 16:300–306; 1987.

Zaki, M. H.; Lyons, H. A.; Leilop, L.; Huang, C. T. Corticosteroid therapy in sarcoidosis. A five-year, controlled follow-up study. N. Y. State J. Med. 87:496–499; 1987.

Zuniga, G.; Ropper, A. H.; Frank, J. Sarcoid peripheral neuropathy. Neurology 41:1558–1561; 1991.

37

Paraneoplastic Diseases

MICHAEL J. GLANTZ AND S. CLIFFORD SCHOLD, JR.

Paraneoplastic diseases of the nervous system are identifiable, nonmetastatic neurological syndromes that occur in the setting of systemic malignancy (Henson and Urich 1982). These conditions are rare, and they should be considered only after the more common causes of neurological dysfunction in patients with cancer (Table 37-1) have been excluded (Patchell and Posner 1985). These "remote effects" of cancer on the nervous system can be organized according to the anatomical site of predominant involvement (Table 37-2). Some disorders are incompletely described or are recognized in only a few case reports. Enough is known in other cases to permit a discussion of prognosis of the paraneoplastic syndrome (PNS) and of the coexisting cancer.

Paraneoplastic syndromes of the nervous system affect prognosis in several ways. First, the remote effect may lead to death or severe disability. Second, recognition of the PNS may precede identification of the underlying malignancy, permitting early diagnosis and perhaps more successful treatment of the tumor. At times, the activity of the PNS may parallel that of the tumor, providing a clinical "marker" and allowing early recognition of disease reactivation. Finally, the presence of a PNS may be associated with an altered history of the underlying malignancy. This last consideration reflects a complex cellular and biochemical interaction between tumor and host that affects the length and quality of the host's survival (Abeloff 1987; Dickson and Lippman 1987; Sporn and Roberts 1985).

Paraneoplastic Syndromes of the Brain

Paraneoplastic encephalomyelitis is an inflammatory disease of the central gray matter, characterized by perivascular lymphocytic infiltrates, astrocytosis, and neuronal loss (Dropcho and Whitaker 1989; Henson et al. 1965). The most severely affected structures dictate the specific nomenclature and the signs and symptoms of this syndrome (Hundforth et al. 1983). Thus, in limbic encephalitis, the cingulate cortex, hippocampus, and amygdaloid nuclei are involved, and seizures, marked impairment in recent memory, depression, anxiety, and confusion progressing to dementia constitute the clinical syndrome (Brennan and Craddock 1983; Carr 1982; Corsellis et al. 1968; Drop-

Table 37-1. Causes of Neurological Dysfunction in Patients With Cancer

Metastatic nervous system disease
Metabolic derangements
Nutritional deficiencies
Opportunistic infections
Complications of therapy
Paraneoplastic syndromes

cho and Whitaker 1989; Garcia et al. 1987; Halperin et al. 1981; Henson et al. 1965; Henson and Urich 1982; Kaplan and Stabashi 1974; Markham and Abeloff 1982; Zangemeister et al. 1978). Involvement of the medullary and (to a lesser extent) pontine nuclei produces a brain stem encephalitis with corresponding cranial nerve dysfunction. The cerebrospinal fluid (CSF) generally shows an elevated protein concentration and a lymphocytosis, and the EEG shows bitemporal (or generalized) slowing with epileptiform activity in limbic encephalitis. Age at onset and sex reflect the epidemiology of the underlying tumor—usually oat cell carcinoma of the lung, less commonly non–small-cell lung cancer, breast, testicular, ovarian, gastric, or uterine cancer or Hodgkin's disease. In one third of the cases, presentation of the neurological disease precedes recognition of the tumor (Brennan and Craddock 1983; Carr 1982;

Table 37-2. Anatomy of Paraneoplastic Nervous System Disorders

Brain
   Limbic encephalitis
   Brain stem encephalitis
   Subacute cerebellar degeneration
   Opsoclonus–myoclonus syndrome
   Cerebrovascular disease
   Retinopathy
Spinal Cord
   Motor neuronopathy
   Subacute necrotizing myelopathy
Peripheral Nerves and Nerve Roots
   Subacute sensory neuronopathy (dorsal root
     ganglionitis)
   POEMS syndrome
Neuromuscular Junction
   Lambert-Eaton myasthenic syndrome
Muscle
   Dermatomyositis
   Carcinoid myopathy

Corsellis et al. 1968; Dropcho and Whitaker 1989; Halperin et al. 1981; Hensen et al. 1965; Kaplan and Stabashi 1974; Markham and Abeloff 1982; Zangemeister et al. 1978). The onset of symptoms (except seizures) is subacute or gradual, and a course of 2 to 24 months (average 10.5 months) is typical. Death usually is secondary to the cancer, but incapacity due to dementia or bulbar dysfunction (in brain stem encephalitis) also contributes. There are sporadic claims of recovery or arrest of disease either spontaneously or in response to treatment of the underlying cancer (Brennan and Craddock 1983; Carr 1982). With one reported exception, corticosteroids are of no benefit. When oat cell lung cancer is the associated malignancy, the tumor frequently is small (sometimes diagnosed only at autopsy), and survival (in at least one series) appears to be prolonged (Dropcho and Whitaker 1989; Shapiro 1976).

In nearly 60% (55 of 97 evaluable cases in the literature) of patients, paraneoplastic cerebellar degeneration also presents before recognition of an associated cancer (Anderson et al. 1988; Brain and Wilkinson 1965; Croft and Wilkinson 1969; Dropcho and Whitaker 1989; Greenberg 1984; Greenlee and Brashear 1983; Greenlee and Lipton, 1986; Trotter et al. 1976). Intervals of 1 month to 8 years between the onset of the neurological illness and the diagnosis of cancer have been reported, with a mean interval of 20 months. Oat cell, breast, and ovarian cancers and Hodgkin's disease are most commonly associated. Dramatic, selective loss of cerebellar Purkinje's cells and frequently perivascular mononuclear infiltrates are identified at autopsy. Clinically, diffuse cerebellar dysfunction, with prominent dysarthria, gait ataxia, and eye movement abnormalities (nystagmus, ocular dysmetria, occasionally opsoclonus) develop over days to weeks. Anti-Purkinje cell antibodies (APCA) to several partially characterized protein antigens have been identified in serum or spinal fluid in approximately half of patients with paraneoplastic cerebellar degeneration (PCD); usually they are associated with breast and ovarian tumors and occasionally with other malignancies (Anderson et al. 1988; Anderson et al. 1988; Dropcho and Whitaker 1989; Greenlee et al. 1986; Greenlee and Lipton 1986). In both antibody-positive and -negative patients, the CSF often is abnormal (approximately 75% of cases). A mild lymphocytic pleocytosis, elevated protein concentration, oligoclonal bands, and an increased immunoglobulin G (IgG) index are the usual findings. Cerebellar atrophy on computed tomography (CT) or magnetic resonance imaging (MRI) appears late in the course of the disease (Greenberg 1984). A few reports of spontaneous arrest (rarely reversal) of this condition and rare instances of apparent response to vitamin supplementation, plasmapheresis, and treatment of the underlying tumor are known (Anderson et al. 1988c; Auth and Chodoff 1957; Eekhoff 1985; Greenberg 1984; Greenlee and Lipton 1986; Martin and Griffith 1971; Paone and Jeyasingham 1980; Roberts 1967; Trotter et al. 1976). In two patients, remission of the PNS followed successful antineoplastic therapy, and recurrent cerebellar symptoms heralded the return of the tumor (Anderson et al. 1988d). Typically (at least 70% of case reports), the syndrome progresses to significant disability, independent of age, sex, tumor type, CSF findings, or course of the underlying cancer. The presence or absence of an APCA does not predict response to plasmapheresis (Anderson et al. 1988d; Greenlee et al. 1986; Jaeckle et al. 1985); however, the disease in antibody-negative patients is less fulminant, less severe, and less likely to present before the identification of an underlying tumor (Anderson et al. 1988d; Jaeckle et al. 1985). Antibody-positive patients are more often women, and there are no examples of spontaneous improvement in this group.

Between 2% and 3% of children with neuroblastoma are afflicted with a striking paraneoplastic syndrome called myoclonic encephalopathy (Altman and Baehner 1976; Rosen et al. 1984). Gait and limb ataxia, myoclonus, and opsoclonus constitute a clinical triad. There are no consistent pathological abnormalities in the few autopsied cases, and elevated urinary or serum catecholamines do not influence the presentation or course of this disorder. Affected children are almost always 2 years old or younger at the onset of the PNS (83% of reported cases), and the symptoms nearly always precede recognition of the associated tumor (96% of cases) by 1 week to 4 years (average 26 months) (Altman and Baehner 1976; Bray et al. 1969; Moe and Nellhaus 1970; Rosen et al. 1984; Senelick et al. 1975). Spontaneous remissions and relapses occur, and treatment with ACTH mitigates the neurological disorder in at least two thirds of patients.

Successful treatment of the tumor is associated with resolution of the PNS in up to 75% of cases. Nevertheless, residual ataxia and mental retardation are common (82% of patients). The tumor, when accompanied by myoclonic encephalopathy, generally is more benign, histologically (ganglioneuroma) and clinically (Evans stages I, II or IV-S) and, even for tumors of the same stage, is associated with a notably better prognosis (Altman and Baehner 1976).

A related syndrome, consisting of opsoclonus, myoclonus (predominantly truncal), ataxia, and encephalopathy has been described as a complication of oat cell cancer of the lung (more than half of the cases), non–small-cell lung cancer, and tumors of the breast, female genital tract, bladder, thyroid, and bone (Anderson et al. 1988d). Onset is acute to subacute (2 weeks to 2 months) and precedes the diagnosis of cancer (by 2 to 12 weeks, average 3.5 weeks) in approximately 50% of patients. The CSF formula is similar to that in PCD, but autopsy findings and antibody studies generally have been unrevealing. Complete or partial remissions or a remitting and relapsing course occur in nearly two thirds of patients, and some responses may have been related to treatment with clonazepam, thiamine, or effective cancer chemotherapy. Steroids have not been clearly beneficial.

Cerebrovascular disease is common with systemic cancer, and three disorders—nonbacterial thrombotic endocarditis (NBTE), disseminated intravascular coagulation

(DIC), and cerebral venous thrombosis— can be considered paraneoplastic diseases (Graus et al. 1985b; Patchell et al. 1985a). As a group, these disorders constitute the third most common neurological complication of cancer (after metastases and metabolic encephalopathy), and they differ from all the other PNS because they occur almost exclusively after the diagnosis of cancer is well established.

Evidence of NBTE is seen at autopsy in 1% to 2% of patients with tumors and as many as 7.7% of bone marrow transplant recipients (Graus et al. 1985b; Jerman and Fick 1986; Koorker et al. 1976; Lune and Bodey 1978; MacDonald and Robbine 1957; Patchell et al. 1985a; Patchell et al. 1985b; Regan and Okazaki 1974; Wiznitzer et al. 1984). Adenocarcinomas of the gastrointestinal (GI) tract and non–small-cell lung cancers are the most frequent underlying tumors. DIC occurs in approximately 1% of patients with a malignancy, primarily lymphomas and leukemias (Collins et al. 1975; Graus et al. 1985; Schwartzman and Hill 1982). Both NBTE and DIC typically occur with widespread systemic disease; DIC often occurs with concomitant sepsis. Despite treatment (aspirin and occasionally valve replacement for NBTE; heparin anticoagulation in DIC), significant neurological deficits are the rule, and death occurs in days to several weeks in nearly all patients. In contrast, the prognosis for patients with cerebral venous thrombosis often is good (Graus et al. 1985b; Hickey et al. 1982; Priest et al. 1980; Sigsbee et al. 1979). Lymphoma, leukemia, and occasionally breast or ovarian cancer are commonly associated tumors, and the superior saggital sinus usually is involved. Chemotherapy regimens incorporating asparaginase also have been implicated (Priest et al. 1980), and widespread, often refractory disease is the usual setting. After the abrupt onset of headache, signs of increased intracranial pressure, seizures, and focal neurological deficits, gradual recovery often rewards attentive supportive care (15 of 26 [58%] of well-documented cases). A gradual onset with few or no focal signs is less common and may betoken a poor outcome. The role of anticoagulation remains unsettled.

Paraneoplastic Syndromes of the Spinal Cord

Numerous case reports describe a paraneoplastic motor neuronopathy affecting the lower motor neurons and occurring in patients with Hodgkin's and non-Hodgkin's lymphomas (Schold et al. 1979; Walton et al. 1968). The clinical course is marked by generalized or asymmetric muscle weakness, atrophy, fasciculations, and reduced deep tendon reflexes. The lower extremities often are more severely affected than the upper extremities. Bulbar and sensory involvement is exceptional. Pathologically, anterior horn cell loss without associated inflammation is seen. Some loss of cells in Clarke's column also may occur. The onset of neurological symptoms followed the diagnosis of hematologic malignancy by 2 to 7 months in 10 of 13 reported cases. Neurological improvement, sometimes a complete resolution of symptoms, occurred from 3 months to 3 years in eight of these cases, independent of the course of the cancer. Treatment with steroids has been ineffective; however, treatment of the underlying malignancy with vincristine may produce a temporary exacerbation of symptoms.

A rare, fulminant spinal cord disorder has been reported in the setting of lymphoma, leukemia, and carcinomas of the lung, breast, prostate, kidney, stomach, thyroid, and ovary. Subacute necrotizing myelopathy is characterized pathologically by indiscriminant spinal cord necrosis, most prominently in the midthoracic region (Gieron et al. 1987; Handforth et al. 1983; Henson and Urich 1982; Ojeda 1984; Richter and Moore 1968). The brain and brain stem occasionally are involved. Vasculopathy and inflammatory changes are absent. In one third of cases, the onset of pain, ascending lower extremity weakness, paresthesias, and sphincter dysfunction precede the diagnosis of cancer. Red blood cells, white blood cells, and an elevated protein concentration usually are seen in the CSF. Myelography and CT of the spine are reliably normal. One case report (with autopsy confirmation of the diagnosis) describes increased T_2 signal on MRI without swelling or mass effect, corresponding to pathologically involved areas of the cervical spinal cord (Gieron et al. 1987). The course is relentlessly progressive, and death, usually directly related to the neurological syndrome, occurs in an average of 5 weeks. The pathophysiology is unknown, and no known treatment alters the natural history.

Paraneoplastic Syndromes of Peripheral Nerves and Nerve Roots

Subacute sensory neuropathy (SSN), a pure sensory neuropathy that is analogous to the paraneoplastic inflammatory disorders of more rostral CNS structures (limbic system, brain stem), is now well described (Croft and Wilkinson 1969; Henson and Urich 1982; Horwich et al. 1977). This condition is a dorsal root ganglionitis, resulting in loss of posterior root ganglia and posterior columns and wallerian degeneration of the sensory component of peripheral nerves. In half of affected patients, there is clinical or pathological evidence of a similar inflammatory process concurrently involving motor neurons, brain stem, cerebellum, or limbic structures. Pain, paresthesias, and a sensory ataxia dominate the clinical picture. The face and trunk typically are spared, and sphincter dysfunction is rare. CSF protein concentration is elevated in three fourths of patients. Small-cell lung cancer is the most commonly associated tumor, although other lung cancers and cancers of the breast, ovary, esophagus, stomach, colon, and bladder also are represented. An antinuclear antibody (anti-Hu) that reacts with all CNS neuronal nuclei has been described in some patients with oat cell cancer and subacute sensory neuropathy (Anderson et al. 1988c; Budde-Steffen et al. 1988; Graus et al. 1985a; Graus et al. 1986) (and in several patients with oat cell cancer and other PNSs with or without SSN [Anderson et al. 1988; Graus et al. 1987]). Antibody from five patients reacted with tumor obtained from one of the patients (Budde-Steffen et al. 1988). The diagnostic, therapeutic, and prognostic implications of these finding have not been fully established, but the anti-Hu antibody appears to be a specific marker for many of the paraneoplastic syndromes associated with small-cell lung cancer. Clinically, the signs and symptoms of SSN usually precede the diagnosis of cancer (two thirds of the cases) by 1 to 46 months (average 12.5 months). The course is subacute, with peak disability attained over several weeks to months. Spontaneous arrest of the disease process after months to 1 year leaves most patients moderately to severely impaired. The course of SSN (except in two case reports [Sagar and Read 1982]) is unrelated to the associated cancer and is unaltered by steroids, vitamins, immunosuppression, or antineoplastic therapy. Survival depends on the virulence of the underlying malignancy.

A prominent sensorimotor polyneuropathy, distinguished by concurrent dysfunction in multiple organ systems, and associated with osteosclerotic plasmacytomas, also has been reported (Bardwick et al. 1980; Kelly et al. 1983; Nakanishi et al. 1984; Resnick et al. 1981). In these patients, polyneuropathy, organomegaly, multiple endocrinopathies, M-proteins, skin changes (hyperpigmentation, anasarca), and papilledema (POEMS) may be seen. Features of this PNS precede (by 2 to 24 months on average) the diagnosis of a plasmacytoma in three fourths of patients. The neurological component of the syndrome is the sole (51%) or prominent (14%) presenting symptom in most patients. A distal, symmetric, sensorimotor polyneuropathy with gradual spread proximally is seen. Except for papilledema (present in more than 60% of cases), cranial nerves are spared. CSF protein concentration characteristically is elevated, and nerve conduction velocities are markedly slowed. The neuropathy (and the other manifestations of POEMS syndrome) frequently (in nearly 40% of published cases) improves or resolves following successful treatment of the tumor with local irradiation, surgery, steroids, or antineoplastic chemotherapy. Cross-reactivity between the tumor-derived abnormal protein and a component of peripheral myelin represents the likely pathogenetic mechanism of this PNS (Mendell et al. 1985).

Paraneoplastic Syndromes of the Neuromuscular Junction

The Lambert-Eaton myasthenic syndrome (LEMS) is an autoantibody-mediated neuromuscular disorder that occurs as a remote effect of cancer in one half to two thirds of affected patients and without a known cancer in the remainder (Engel 1984; Lambert et al. 1956; Lennon et al. 1982; O'Neill et al. 1988; Autoantibody activity in Lambert-Eaton myasthenic syndrome 1988). When cancer is present, small-cell cancer of the lung is most common, and the tumor becomes clinically evident within 2 years of onset of the PNS is nearly all cases (O'Neill et al. 1988). Cancers of the breast, prostate, kidney, stomach, and rectum also have been reported but may reflect chance associations. Younger women are less likely and older men are more likely to harbor an associated malignancy. In both groups, an increased incidence of other autoantibodies, systemic autoimmune disorders, and the HLA-B8 haplotype are seen (Newsom-Davis 1985; O'Neill et al. 1988). The onset of symptoms usually presages the diagnosis of tumor by weeks or months. Proximal muscle weakness, myalgias, paresthesias, loss of deep tendon reflexes, dry mouth, blurred vision, sphincter dysfunction, and impotence are typical clinical manifestations (Brown and Johns 1974; Engel 1984; Heath et al. 1988; Henson and Urich 1982; Lambert et al. 1956; Lennon et al. 1982; Rubenstein et al. 1979; Autoantibody activity in Lambert-Eaton myasthenic syndrome, 1988). Facilitation of strength with maximal voluntary muscle contraction often is difficult to demonstrate clinically, but the electrophysiological correlate, an incremental response to high-frequency repetitive stimulation, is a cornerstone of diagnosis. Autoantibodies directed against antigenic

determinants on the presynaptic voltage-dependent calcium channel probably play a decisive pathophysiological role (Kim and Neher 1988; Lang et al. 1981; Autoantibody activity in Lambert-Eaton myasthenic syndrome 1988). Although a progressive course usually is described, treatment with guanidine hydrochloride (Lambert 1956) or 3,4-diaminopyridine (Lundh et al. 1984; Murray and Newsom-Davis 1981), frequently mitigates the symptoms. Corticosteroids or a combination of corticosteroids and immunosuppressive agents often provide relief (Dau and Denys 1982; Lang et al. 1981; Newsom-Davis and Murray 1984; Streeb and Rothner 1981; Vroom and Engle 1969), perhaps more consistently in patients also afflicted with cancer. Occasionally, reports describe resolution or remission following chemotherapy, irradiation, or resection of the associated tumor (Anderson et al. 1953; Clamon et al. 1984; Jenkyn et al. 1980; Katter et al., 1985). Plasmapheresis also is effective, particularly in patients with an associated malignancy (Dau and Denys 1982; Lang et al. 1981; Newsom-Davis and Murray 1984; Arnason and Areen et al. 1986). Prognosis, at least in patients with oat cell cancer, is improved if LEMS also is present. A more benign clinical course, fewer and less frequent CNS metastases, and significantly longer survivals are reported (de la Monte et al. 1984).

Paraneoplastic Syndromes of Muscle

Controversy persists regarding the status of polymyositis and dermatomyositis as paraneoplastic conditions. Most authors agree that polymyositis alone is associated with malignancy only by chance, while cancer underlies dermatomyositis in 7% to 40% of cases (Callen 1984; Callen 1987; Henson and Urich 1982; Lakhan et al. 1986; Patchell et al. 1985a). This association is further strengthened by excluding patients younger than 50. Clinically, cutaneous manifestations include a heliotrope rash and periungual erythema or telangiectasia. Symmetrical, predominently proximal muscle weakness and atrophy are the neurological hallmarks. Muscle tenderness is less common. Some series report an increased incidence of dysphagia (Pearson 1963) and respiratory embarassment (Talbott 1980) in patients with cancer-associated dermatomyositis. Laboratory abnormalities include elevated muscle enzymes (approximately 95% of cases) and erythrocyte sedimentation rate (50% of cases). Electromyography (EMG) demonstrates brief, low-amplitude polyphasic waves; increased insertional activity; and fibrillation potentials. None of these findings influence prognosis. Gastrointestinal, breast, and ovarian cancers are the most common associated malignancies, but reports have implicated other neoplasms, including lymphomas and leukemias. In up to one half of patients, the onset of dermatomyositis precedes the diagnosis of the cancer by 1 to 2 years. Spontaneous arrests and remissions occur occasionally, and improvement following successful surgery or chemotherapy directed at the underlying malignancy is common. Treatment with corticosteroids produces improvement in most patients (Pearson 1966; Rowland et al. 1977). The addition of other immunosuppressive agents may enhance the response (Bunch et al. 1980; Jacobs 1977; Metzger et al. 1974). At least three fourths of patients achieve a significant clinical and laboratory response with these measures. Refractory patients have been treated successfully with plasmapheresis and total lymphoid irradiation (Bohan et al. 1977; Callen et al. 1987; Carpenter et al. 1977). A poor response to immunosuppression suggests an underlying malignancy (Budde-Steffen et al. 1988). Dysphagia and profound weakness at the time of presentation also are poor prognostic signs (Benbassat et al. 1985; Bohan et al. 1977; Carpenter et al. 1977). A rare type of paraneoplastic myopathy is specifically associated with carcinoid tumors (Berry et al. 1974; Henson and Urich 1982; Lederman et al. 1985; Swash et al. 1975). The clinical picture is similar to polymyositis. However, the disorder develops long after the diagnosis of cancer has been established, and it

may respond to treatment with serotonin antagonists.

Summary

Many paraneoplastic neurological syndromes, mediated indirectly by a variety of extraneural malignancies, have been identified. The prognosis of the paraneoplastic syndrome and its associated neoplasm may be related in one of several ways (Table 37-3). The neurological disorder may follow a relentlessly progressive course, oblivious to the course of the underlying tumor. In this group are most cases of subacute cerebellar degeneration, subacute necrotizing myelopathy, encephalomyelitides (limbic, brain stem, dorsal root), and the neurological consequences of DIC and NBTE. In other cases, successful antineoplastic therapy results in improvement of the neurological syndrome. The neuropathy of POEMS is the best example. Sometimes spontaneous resolution occurs or specific therapy is available for the remote effect. This occurs in myoclonic encephalopathy, paraneoplastic motor neuronopathy, cerebral venous thrombosis, LEMS, dermatomyositis, and carcinoid myopathy. Often, recognition of the PNS suggests the existence (and sometimes the type) of an occult tumor and provides the opportunity for earlier, more effective cancer treatment. In the final scenario, the presence of a paraneoplastic disorder is associated with a more benign course of the underlying malignancy. Myoclonic encephalopathy and some of the syndromes seen with oat cell cancer (such as LEMS) fall into this category.

The cases marked by inexorable progression underscore the incomplete knowledge of the pathogenesis of the remote effect. When direct therapeutic intervention is possible or when treatment of the underlying cancer improves the paraneoplastic syndrome, a known or suspected immune-mediated process generally is at work. In these reversible syndromes, few or no histological changes are found at biopsy or autopsy.

Table 37-3. Clinical Course of Neurological Paraneoplastic Syndromes

Paraneoplastic Syndrome	Onset Relative To Cancer Diagnosis	Usual Outcome	Improvement With Cancer Treatment
Encephalomyelitis	Precedes (33%) average 11 months	Progressive neurological dysfunction	Rare
Paraneoplastic cerebellar degeneration	Precedes (60%) average 20 months	Progressive neurological dysfunction	Occasional
Myoclonic encephalopathy	Precedes (96%) average 26 months	Remission (spontaneous and after treatment with ACTH) with residual deficits	Common
Opsoclonus–myoclonus	Precedes (50%) average 4 weeks	Complete or partial remissions, relapsing–remitting course in ⅔	Occasional
Nonbacterial thrombotic endocarditis, disseminated intravascular coagulation	Follows in nearly all cases	Dysfunction progressing to death	Rare
Cerebral venous thrombosis	Follows in nearly all cases	Neurological recovery in more than half	Occasional
Spinal motor neuronopathy	Follows (75% of cases)	Stabilization and gradual improvement	No
Subacute necrotizing myelopathy	Precedes (33%)	Progressive dysfunction leading to death (average 5 weeks)	No
Subacute sensory neuropathy	Precedes (66%) average 12.5 months	Progresses to stable, usually marked dysfunction	No
Myeloma-associated neuropathy (POEMS)	Precedes (75%) (2–24 months)	Progressive until underlying neoplasm treated	Common
LEMS	Typically precedes; 50% to 66% ultimately develop cancer	Progressive weakness without intervention	Occasional
Dermatomyositis	Precedes (50%) (1–2 years)	Progressive weakness without intervention	Common
Carcinoid myopathy	Follows	Progressive weakness without intervention	No

Cases in which the presence of a paraneoplastic neurological disorder is associated with improved tumor prognosis remain perplexing. Shared antigens between the tumor (usually a neural crest derivative) and the host's nervous system tissue may create a situation in which a vigorous immune response effectively controls the cancer with a disabling remote effect (Bell and Seetharam 1977; Bell and Seetharam 1979; Bell et al. 1976; Souhami et al. 1979). A better understanding of this relationship may in the future offer greater control over the outcomes of the tumor and the paraneoplastic syndrome.

Acknowledgments

This work supported in part by grant T32–NS07304.

References

Abeloff, M. D. Paraneoplastic disorders: a window on the biology of cancer. N. Engl. J. Med. 317:1598–1600; 1987.

Altman, A. J.; Baehner, R. L. Favorable prognosis for survival in children with coincident opso-myoclonus and neuroblastoma. Cancer 37:846–852; 1976.

Anderson, N. E.; Budde-Steffen, C.; Wiley, R. G.; Thurman, L.; Rosenblum, M. K.; Nadeau, S. E.; Posner, J. B. A variant of the anti-Purkinje cell antibody in a patient with paraneoplastic cerebellar degeneration. Neurology 38:1018–1026; 1988a.

Anderson, N. E.; Budde-Steffen, C.; Rosenblum, M. K.; Graus, F.; Ford, D.; Synek, B. J. L.; Posner, J. B. Opsoclonus, myoclonus, ataxia, and encephalopathy in adults with cancer: a distinct paraneoplastic syndrome. Medicine 67:100–109; 1988b.

Anderson, H. J.; Churchill-Davidson, H. C.; Richardson, A. T. Bronchial neoplasm with myasthenia: prolonged apnoea after administration of succinylcholine. Lancet 1291–1293; 1953.

Anderson, N. E.; Rosenblum, M. K.; Graus, F.; Wiley, R. G.; Posner, J. B. Autoantibodies in paraneoplastic syndromes associated with small cell lung cancer. Neurology 38:1391–1398; 1988c.

Anderson, N. E.; Rosenblum, M. K.; Posner, J. B. Paraneoplastic cerebellar degeneration: clinical-immunological correlations. Ann. Neurol. 24:559–567; 1988d.

Arnason, B. G.; Areen, J.; et al., Consensus Conference. The utility of therapeutic plasmapheresis for neurological disorders. JAMA 256:1333–1337; 1986.

Auth, T. L.; Chodoff, P. Transient cerebellar syndrome from extracerebral carcinoma. Neurology 7:370–372; 1957.

Autoantibody activity in Lambert-Eaton myasthenic syndrome, Lancet 1:920; 1988.

Bardwick, P. A.; Zvaifler, N. J.; Gill, G. N.; Newman, D.; Greenway, G. D.; Resnick, D. L. Plasma cell dyscrasia with polyneuropathy, organomegaly, endocrinopathy, M protein, and skin changes: the POEMS syndrome. Report on two cases and a review of the literature. Medicine 59:311–322; 1980.

Bell, C. E.; Seetharam, S. Identification of the Schwann cell as a peripheral nervous system cell possessing a differentiation antigen expressed by a human lung tumor. J. Immunology 118:826–831; 1977.

Bell, C. E.; Seetharam, S. Expression of endodermally derived and neural chest-derived differentiation antigens by human lung and colon tumors. Cancer 44:13–18; 1979.

Bell, C. E.; Seetharam, S.; McDaniel, R. C. Endodermally-derived and neural crest-derived differentiation antigens expressed by a human lung tumor. J. Immunol. 116:1236–1243; 1976.

Benbassat, J.; Gefel, D.; Larholt, K.; Sukenik, S.; Morgenstern, V.; Zlotnick, A. Prognostic factors in polymyositis/dermatomyositis: a computer-assisted analysis of ninety-two cases. Arthritis Rheum. 28:249–255; 1985.

Berry, E. M.; Maunder, C.; Wilson, M. Carcinoid myopathy and treatment with cyproheptadine (periactin). Gut 15:34–38; 1974.

Bohan, A.; Peter, J. B.; Bowman, R. L.; Pearson, C. M. A computer-assisted analysis of 153 patients with polymyositis and dermatomyositis. Medicine 56:255–286; 1977.

Brain, R.; Wilkinson, M. Subacute cerebellar degeneration associated with neoplasms. Brain 88:465–478; 1965.

Bray, P. F.; Ziter, F. A.; Lahey, M. E.; Myers, G. G. The coincidence of neuroblastoma and acute cerebellar encephalopathy. J. Pediatr. 75:983–990; 1969.

Brennan, L. V.; Craddock, P. R. Limbic encephalopathy as a non-metastatic complication of oat cell lung cancer. Am. J. Med. 75:518–520; 1983.

Brown, J. C.; Johns, R. J. Diagnostic difficulties encountered in the myasthenia syndrome sometimes associated with carcinoma. J. Neurol. Neurosurg. Psychiatry 37:1214–1224; 1974.

Budde-Steffen, C.; Anderson, N. E.; Rosenblum, M. K.; Posner, J. B. Expression of an antigen in small cell lung carcinoma lines detected by antibodies from patients with paraneoplastic dorsal root ganglionitis. Cancer Res. 48:430–434; 1988.

Bunch, T. W.; Worthington, J. W.; Combs, J. J.; Ilstrup, M. S.; Engle, A. G. Azathioprine with prednisone for polymyositis. Ann. Intern. Med. 92:365–369; 1980.

Burton, G. V.; Bullard, D. E.; Walther, P. J.; Burger, P. C. Paraneoplastic limbic encephalopathy with testicular carcinoma: a reversible neurologic syndrome. Cancer 62:2248–2251; 1988.

Callen, J. P. Myositis and malignancy. Clin. Rheum. Dis. 10:117–131; 1984.

Callen, J. P. Dermatomyositis. In: Callen, J. P.; Meckler, R. J., eds. Neurologic clinics: neurocutaneous disorders. Philadelphia: W. B. Saunders; 1987:p. 379–403.

Carpenter, J. R.; Bunch, T. W.; Engel, A. G.; O'Brien, P. Survival in polymyositis: corticosteroids and risk factors. J. Rheumatol. 4:207–214; 1977.

Carr, I. The ophelia syndrome: memory loss in Hodgkin's disease. Lancet 1:844–845; 1982.

Clamon, G. H.; Evans, W. K.; Shepherd, F. A.; Humphrey, J. G. Myasthenic syndrome and small cell cancer of the lung: variable response to antineoplastic therapy. Arch. Intern. Med. 144:99–1000; 1984.

Collins, R. C.; Al-Mondhiry, H.; Chernik, N. L.; Posner, J. B. Neurologic manifestations of intravascular coagulation in patients with cancer: a clinicopathologic analysis of 12 cases. Neurology 25:795–806; 1975.

Corsellis, J. A. N.; Goldberg, G. J.; Norton, A. R. "Limbic encephalitis" and its association with carcinoma. Brain 91:481–497; 1968.

Croft, P. B.; Wilkinson, M. The course and prognosis in some types of carcinomatous neuromyopathy. Brain 92:1–8; 1969.

Dau, P. C.; Denys, E. H. Plasmapheresis and immunosuppressive drug therapy in the Eaton-Lambert syndrome. Ann. Neurol. 11:570–575; 1982.

de la Monte, S.; Hutchins, G. M.; Moore, G. W. Paraneoplastic syndromes and constitutional symptoms in prediction of metastatic behavior of small cell carcinoma of the lung. Am. J. Med. 77:851–856; 1984.

Dickson, R. B.; Lippman, M. E. Estrogenic regulation of growth and polypeptide growth factor secretion in human breast carcinoma. Endocr. Rev. 8:29–43; 1987.

Dorfman, L. J.; Forno, L. S. Paraneoplastic encephalomyelitis. Acta. Neurol. Acand. 48:556–574; 1972.

Dropcho, E. J.; Whitaker, J. N. Cerebellar ataxia as a paraneoplastic syndrome. Hosp. Prac. 69–84; 1989.

Eekhof, J. L. A. Remission of a paraneoplastic cerebellar syndrome. Clin. Neurol. Neurosurg. 87:133–134; 1985.

Engel, A. G. Myasthenia gravis and myasthenic syndromes. Ann. Neurol. 16:519–534; 1984.

Fawcett, D. P.; McBrien, M. P. Transitional cell carcinoma of the renal pelvis presenting with peripheral neuropathy. Br. J. Urol. 49:202; 1977.

Garcia, J. H.; Heinsohn, M. E.; Guptka, K. L.; Faught, R. R.; Bonnin, J. M. Encephalopathy in a 68 year old man. S. Med. J. 80:1277–1284; 1987.

Gieron, M. A.; Margraf, L. R.; Korthals, J. K.; Gonzalvo, A. A.; Murtagh, R. F.; Hvizdala, E. V. Progressive necrotizing myelopathy associated with leukemia: clinical, pathologic, and MRI correlation. J. Child. Neurol. 2:44–49; 1987.

Graus, F.; Cordon-Cardo, C.; Posner, J. B. Neuronal antinuclear antibody in sensory neuronopathy from lung cancer. Neurology 35:538–543; 1985a.

Graus, F.; Elkon, K. B.; Cordon-Cardo, C.; Posner, J. B. Sensory neuronopathy and small cell lung cancer: antineuronal antibody that also reacts with the tumor. Am. J. Med. 80:45–52; 1986.

Graus, F.; Elkon, K. B.; Lloberes, P.; Ribalta, T.; Torres, A.; Ussetti, P.; Valls, J.; Obach, J.; Agusti-Vidal, A. Neuronal antinuclear antibody (anti-Hu) in paraneoplastic encephalomyelitis simulating acute polyneuritis. Acta. Neurol. Scand. 75:249–252; 1987.

Graus, F.; Rogers, L. R.; Posner, J. B. Cerebrovascular complications in patients with cancer. Medicine 64:16–35; 1985b.

Greenberg, H. S. Paraneoplastic cerebellar degeneration. J. Neurooncol. 2:377–382; 1984.

Greenlee, J. E.; Brashear, H. R. Antibodies to cerebellar Purkinje cells in patients with paraneoplastic cerebellar degeneration and ovarian carcinoma. Ann. Neurol. 14:609–613; 1983.

Greenlee, J. E.; Brashear, H. R.; Jaeckle, K. A.; Stroop, W. G. Anticerebellar antibodies in sera of patients with paraneoplastic cerebellar degeneration: studies of antibody specificity

and response to plasmapheresis. Ann. Neurol. 20:139; 1986.

Greenlee, J. E.; Lipton, H. L. Anticerebellar antibodies in serum and cerebrospinal fluid of a patient with oat cell carcinoma of the lung and paraneoplastic cerebellar degeneration. Ann. Neurol. 19:82–85; 1986.

Halperin, J. J.; Richardson, E. P.; Ellis, J.; Ross, J. S.; Wray, S. H. Paraneoplastic encephalomyelitis and neuropathy: report of a case. Arch. Neurol. 38:773–775; 1981.

Handforth, A.; Nag, S.; Dharp, D.; Robertson, D. M. Paraneoplastic subacute necrotic myelopathy. Can. J. Neurol. Sci. 10:204–207; 1983.

Heath, J. P.; Ewing, J. P.; Cull, R. E. Abnormalities of autonomic function in the Lambert Eaton myasthenic syndrome. J. Neurol. Neurosurg. Psychiatry 51:436; 1988.

Henson, R. A.; Hoffman, H. L.; Urich, H. Encephalomyelitis with carcinoma. Brain 88:449–464; 1965.

Henson, R. A.; Urich, H. Cancer and the nervous system: the neurological manifestations of systemic malignant disease. Oxford: Blackwell Scientific Publications; 1982:p. 422–426.

Hickey, W. F.; Garnick, M. B.; Henderson, I. C.; Dawson, D. M. Primary cerebral venous thrombosis in patients with cancer—a rarely dignosed paraneoplastic syndrome: report of three cases and review of the literature. Am. J. Med. 73:740–750; 1982.

Horwich, M. S.; Cho, L.; Porro, R. S.; Posner, J. B. Subacute sensory neuropathy: a remote effect of carcinoma. Ann. Neurol. 2:7–19; 1977.

Jacobs, J. C. Methotrexate and azathioprine for treatment of childhood dermatomyositis. Pediatrics 59:212–218; 1977.

Jaeckle, K. A.; Graus, F.; Houghton, A.; Cardon-Cardo, C.; Nielsen, S. L.; Posner, J. B. Autoimmune response of patients with paraneoplastic cerebellar degeneration to a Purkinje cell cytoplasmic protein antigen. Ann. Neurol. 18:592–600; 1985.

Jenkyn, L. R.; Brooks, P. L.; Forcier, R. J.; Maurer, L. H.; Ochoa, J. Remission of Lambert-Eaton syndrome and small cell anaplastic carcinoma of the lung induced by chemotherapy and radiation therapy. Cancer 46:1123–1127; 1980.

Jerman, M. R.; Fick, R. B. Nonbacterial thrombotic endocarditis associated with bone marrow transplant. Chest 90:919–922; 1986.

Kaplan, A. M.; Stabashi, H. H. Encephalitis associated with carcinoma: central hypoventilation syndrome + cytoplasmic inclusion bodies. J. Neurol. Neurosurg. Psychiatry 37:1166–1176; 1974.

Katter, S.; Dhingra, H. M.; Farha, P. Primary therapy for small cell lung cancer reversing the

Eaton Lambert syndrome. S. Med. J. 78:197–198; 1985.

Kelly, J. J.; Kyle, R. A.; Miles, J. M.; Dyck, P. J. Osteosclerotic myeloma and peripheral neuropathy. Neurology 33:202–210; 1983.

Kim, Y. I.; Neher, E. IgG from patients with Lambert-Eaton syndrome blocks voltage-dependent calcium channels. Science 239:450–458; 1988.

Klingele, T. G.; Burde, R. M.; Rappazzo, J. A.; Isserman, M. J.; Burgess, D.; Kantor, O. Paraneoplastic retinopathy. J. Clin. Neuro. Ophthalmol. 4:239–245; 1984.

Koorker, J. C.; MacLean, J. M.; Sumi, S. M. Cerebral embolism, marantic endocarditis and cancer. Arch. Neurol. 33:260–264; 1976.

Lakhanpal, S.; Bunch, T. W.; Ilstrup, D. M.; Melton, L. J., III. Polymyositis-dermatomyositis and malignant lesions: does an association exist. Mayo. Clin. Proc. 61:645–653; 1986.

Lambert, E. H. Defects of neuromuscular transmission in syndromes other than myasthenia gravis. Ann. N. Y. Acad. Sci. 135:367–384; 1966.

Lambert, E. H.; Eaton, L. M.; Rooke, E. D. Defect of neuromuscular condition associated with malignant neoplasms. Am. J. Physiol. 187:612–613; 1956.

Lang, B.; Newsom-Davis, J.; Wray, D.; Vincent, A. Autoimmune etiology for myasthenic (Eaton-Lambert) syndrome. Lancet 2:224–226; 1981.

Lederman, R. J.; Bukowski, R. M.; Nickerson, P. Carcinoid myopathy. Neurology 35:165; 1985.

Lennon, V. A.; Lambert, E. H.; Whittingham, S.; et al. Autoimmunity in the Lambert-Eaton myasthenic syndrome. Muscle Nerve 5:521–525; 1982.

Lundh, H.; Nilsson, O.; Rosen, I. Treatment of Lambert-Eaton syndrome: 3,4-diaminopyridine and pyridostigmine. Neurology 34:1324–1330; 1984.

Lune, M.; Bodey, G. P.; Nonbacterial thrombotic endocarditis in cancer patients: comparison of characteristics of patients with and without disseminated intravascular coagulation. Med. Pediatr. Oncol. 4:149–157; 1978.

MacDonald, R. A.; Robbine, S. L. The significance of nonbacterial thrombotic endocarditis: an autopsy and clinical study of 78 cases. Ann. Intern. Med. 46:255–273; 1957.

Markham, M.; Abeloff, M. D. Small cell lung cancer and limbic encephalitis. Ann. Intern. Med. 96:785; 1982.

Martin, E. S.; Griffith, J. F. Myoclonic encephalopathy and neuroblastoma: report of a case with apparent recovery. Am. J. Dis. Child. 122:257; 1971.

Mendell, J. R.; Sahenk, Z.; Whitaker, J. N.;

Trapp, B. D.; Yates, A. J.; Griggs, R. C.; Quarles, R. H. Polyneuropathy and IgM monoclonal gammopathy: studies on the pathogenic role of anti-myelin-associated glycoprotein antibody. Ann. Neurol. 17:243–254; 1985.

Metzger, A. L.; Bohan, A.; Goldberg, L. S.; Bluestone, R. Polymyositis and dermatomyositis: combined methotrexate and corticosteroid therapy. Ann. Intern. Med. 81:182–189; 1974.

Moe, P. G.; Nellhaus, G. Infantile polymyoclonia-opsoclonus syndrome and neural crest tumors. Neurology 20:756–764; 1970.

Murray, N. H.; Newsom-Davis, J. Treatment with oral 4-aminopyridine in disorders of neuromuscular transmission. Neurology 31:265–271; 1981.

Nakanishi, T.; Sobue, I.; Toyokura, Y.; Nishitani, H.; Kuroiwa, Y.; Satoyoshi, E.; Tsubaki, T.; Igata, A.; Ozaki, Y. The Crow-Fukase syndrome: a study of 102 cases in Japan. Neurology 34:712–720; 1984.

Newsom-Davis, J. Lambert-Eaton myasthenic syndrome. Springer Seminars in Immunopathology 8:129–140; 1985.

Newsom-Davis, J.; Murray, N. M. P. Plasma exchange and immunosuppression drug therapy in the Lambert-Eaton myasthenic syndrome. Neurology 34:480–485; 1984.

Ojeda, V. J. Necrotizing myelopathy associated with malignancy: a clinicopathologic study of two cases and literature review. Cancer 53:1115–1123; 1984.

O'Neill, J. H.; Murray, N. M. F.; Newsom-Davis, J. The Lambert-Eaton myasthenic syndrome: a review of 50 cases. Brain 111:577–596; 1988.

Paone, J. F.; Jeyasingham, K. Remission of cerebellar dysfunction after pneumonectomy for bronchogenic carcinoma. N. Engl. J. Med. 302:156–157; 1980.

Patchell, R. A.; Posner, J. B. Neurologic complications of systemic cancer. In: Vick, N. A.; Bigner, D. D., eds. Neurologic clinics. Symposium on neuro-oncology. Philadelphia: W. B. Saunders; 1985:p. 729–750.

Patchell, R. A.; White, C. L.; Clark, A. W.; Beschorner, W. E.; Santos, G. W. Neurologic complications of bone marrow transplantation. Neurology 35:300–306; 1985a.

Patchell, R. A.; White, C. L.; Clark, A. W.; Beschorner, W. E.; Santos, G. W. Nonbacterial thrombotic endocarditis in bone marrow transplant patients. Cancer 55:631–635; 1985b.

Pearson, C. M. Patterns of polymyositis and their responses to treatment. Ann. Intern. Med. 59:827–838; 1963.

Pearson, C. M. Polymyositis. Annu. Rev. Med. 17:63–82; 1966.

Priest, J. R.; Ramsay, N. K. C.; Latchaw, R. E.; Lockman, L. A.; Hasegawa, D. K.; Coates, T. D.; Coccia, P. F.; Edson, J. R.; Nesbit, M. E.; Krivit, W. Thrombotic and hemorrhagic strokes complicating early therapy for childhood acute lymphoblastic leukemia. Cancer 46:1548–1554; 1980.

Reagan, T. J.; Okazaki, H. The thrombotic syndrome associated with carcinoma: a clinical and neuropathologic study. Arch. Neurol. 31:390–395; 1974.

Resnick, D.; Greenway, G. D.; Bardwick, P. A.; Zvaifler, N. J.; Gill, G. N.; Newman, D. R. Plasma-cell dyscrasia with polyneuropathy, organomegaly, endocrinopathy, M-protein, and skin changes: the POEMS syndrome. Distinctive radiographic abnormalities. Radiology 140:17–22; 1981.

Richter, R. B.; Moore, R. Y. Non-invasive central nervous system disease associated with lymphoid tumors. Johns Hopkins Med. J. 122:271–283; 1968.

Roberts, H. J. V. A case of spinocerebellar degeneration associated with carcinoma of the prostate. Med. J. Aust. 1:391–393; 1967.

Rosen, E. M.; Cassady, J. R.; Frantz, C. N.; Kretschmar, C.; Levey, R.; Sallan, S. E. Neuroblastoma: The joint center for radiation therapy/Dana-Farber Center Institute/Children's Hospital Experience. J. Clin. Oncol. 2:719–732; 1984.

Rowland, L. P.; Clark, C.; Olarte, M. Therapy for dermatomyositis and polymyositis. Adv. Neurol. 17:63–97; 1977.

Rubenstein, A. E.; Horowitz, S. H.; Bender, A. N. Cholinergic dysautonomia and Eaton-Lambert syndrome. Neurology 29:720–723; 1979.

Sagar, H. J.; Read, D. J. Subacute sensory neuropathy with remission: an association with lymphoma. J. Neurol. Neurosurg. Psychiatry 45:83–85; 1982.

Schold, S. C., Jr.; Cho, L.; Somasundarum, M.; Posner, J. B. Subacute motor neuronopathy: a remote effect of lymphoma. Ann. Neurol. 5:271–287; 1979.

Schwartzman, R. J.; Hill, J. B. Neurologic complications of disseminated intravascular coagulation. Neurology 32:791–797; 1982.

Senelick, R. C.; Bray, P. F.; Lahey, M. E.; Van Dyk, J.; Dale, G. Neuroblastoma and myoclonic encephalopathy: two cases and a review of the literature. J. Pediatr. Surg. 75:983–990; 1975.

Shapiro, W. R. Remote remote effects of neoplasms on the central nervous system: encephalopathy. Adv. Neurol. 15:101–117; 1976.

Sigsbee, B.; Deck, M. D. F.; Posner, J. B. Nonmetastatic superior sagittal sinus thrombosis complicating systemic cancer. Neurology 29:139–146; 1979.

Souhami, R. L.; Beverley, P. C. L.; Bobrow, L. G. Antigens of small-cell lung cancer. Lancet 2:325–326; 1979.

Sporn, M. B.; Roberts, A. B. Autocrine growth factors and cancer. Nature 313:745–747; 1985.

Streeb, E. W.; Rothner, A. D. Eaton-Lambert myasthenic syndrome: long-term treatment of three patients with prednisolone. Ann. Neurol. 10:488–453; 1981.

Swash, M.; Fox, K. P.; Davidson, A. R. Carcinoid myopathy: serotonin-induced muscle weakness in man? Arch. Neurol. 32:572–574; 1975.

Talbott, J. H. Acute dermatomyositis-polymyositis and malignancy. Semin. Arthritis Rheum. 6:305–360; 1980.

Thirkill, C. E.; Roth, A. M.; Keltner, J. L. Cancer-associated retinopathy. Arch. Ophthalmol. 105:372–375; 1987.

Trotter, J. L.; Hendin, B. A.; Osterland, C. K. Cerebellar degeneration with Hodgkin's disease: an immunological study. Arch. Neurol. 33:660–661; 1976.

Vroom, F. O.; Engle, W. K. Non-neoplastic steroid-responsive Lambert-Eaton myasthenic syndrome. Neurology 19:281; 1969.

Walton, J. N.; Tomlinson, B. E.; Pearce, G. W. Subacute "poliomyelitis" and Hodgkin's disease. J. Neurol. Sci. 6:435–445; 1968.

Wiznitzer, M.; Packer, R. J.; August, C. S.; Burkey, E. D. Neurological complications of bone marrow transplantation in childhood. Ann. Neurol. 16:569–576; 1984.

Zangemeister, W. H.; Schwendemann, G.; Colmant, H. J. Carcinomatous encephalopathy in conjunction with encephalomyeloradiculitis. J. Neurol. 218:63–71; 1978.

38

Exposure to Industrial Toxins

CHRISTOPHER G. GOETZ

In an industrial and predominantly urban society, the consequences of factory-produced toxins and fumes have been increasingly recognized. Whereas environmentalists must be concerned with the toxic effects on plant and ecological balance, neurologists must focus on the specific effects of industrial chemicals and gases on the human nervous system. In spite of increasing concern, however, the neurological sequelae of acute and chronic exposure to industrial products are incompletely studied. This chapter concentrates on the neurological prognosis of victims exposed to various industrial chemicals by inhalation, skin exposure, and ingestion. Most cases of toxic exposure are inadvertent, but some specific agents are prominent substances of volitional abuse. The topics cover the major organic solvents, gases, and pesticides.

Methyl Alcohol

Methanol is used as a solvent, a combustible, a component of antifreeze, and an adulterant of alcoholic beverages. Although the compound is only mildly toxic, its oxidation products, formaldehyde and formic acid, in-

duce a severe metabolic acidosis and account for most of the signs of methanol intoxication. To a large extent, the prognosis of the intoxicated patient depends on the degree and duration of acidosis, although both formaldehyde and formate probably have neurotoxic effects independent of acidosis (Fink 1943; Roe 1969). The amount of methanol needed to cause serious effects varies with the individual and may largely depend on concomitant ethanol ingestion. Methanol and ethanol share the same degrading enzyme, alcohol dehydrogenase, so ethanol intoxication tends to diminish the rapid production of toxic formaldehyde and formic acid from methanol.

Because the oxidation and excretion of methanol is slow, toxic signs may not develop for 12 to 48 hours, especially in a patient with cointoxication with ethanol. Acute toxicity may last for several days, and persistent neurological sequelae are well described (Kaplan and Levreault 1944). The visual apparatus, central nervous, gastrointestinal, and respiratory systems are affected by methanol.

Early toxic symptoms are nausea, vomiting, generalized weakness, severe abdominal pain, vertigo, and headache. Symptoms similar to ethanol intoxication are restlessness, incoordination, delirium, and hallucinations. Visual disturbance and ocular abnormalities are frequent; visual loss may start within the first hours after ingestion or may be delayed for several days. Amblyopia, scotomas, or total blindness occur, and the pupils become dilated and fail to react to light. Ocular pain is common. Ophthalmoscopic examination may reveal injection of the discs, with blurring of the margins and pericapillary and macular edema (Krolman and Pidde 1968). Increasing severity of the poisoning may lead to stupor, coma, tonic muscle contractions, hyperactive reflexes, opisthotonos, and generalized convulsions. In severely acidotic patients, death occurs from respiratory failure (Krolman and Pidde 1968).

Survival prognosis depends mainly on the degree of acidosis and the effectiveness of treatment. Only in unusual cases does the narcotic effect of the methanol or toxicity directly relate to formaldehyde or formic acid mortality (Ritchie 1975). In the largest clinical group of 323 patients reported with methanol exposure to bootleg whiskey, 6.2% died. When acidosis accompanied the methanol intoxication, the mortality rate was higher; when plasma CO_2 values were less than 20 mEq, 19% died, and when CO_2 values were less than 10 mEq, 50% died (Bennett et al. 1953). Coma and seizures do not always indicate a hopeless prognosis (Schneck 1979). Typical shallow slow respiration, tonic limb contractions, and opisthotonic posturing are considered a likely fatal triad (Roe 1955). In regard to vision, formate appears to be the primary toxic agent (Martin-Amat et al. 1978), and levels of serum formate correlate well with acute signs of neurotoxicity (Osterloh et al. 1986). For prognosis, the absence of pupillary light reflexes is generally associated with a low likelihood of survival (Benton and Calhoun 1952). Patients with severe retinal edema but retained pupillary responses usually are left with some degree of permanent visual loss. Most patients, however, have partial or complete recovery of visual acuity, and the recovery may occur within the first hours after correction of acidosis. If vision has not returned to normal within 6 days, however, full recovery is deemed unlikely (Roe 1953).

In survivors of methanol intoxication, long-term visual impairment and extrapyramidal signs may be prominent. Bradykinesia, low voice volume, masked facies, and tremor occur. Dementia and additional motor signs with increased or decreased reflexes may accompany this clinical picture. In two patients with parkinsonism, bilateral symmetric infarctions of the frontal and central white matter and putamen were found. In the more severely affected patient, levodopa therapy was attempted but without success. One brain was examined at autopsy and cystic degeneration of the putamen and subcortical white matter accompanied widespread neuronal damage to the brain stem and spinal cord. The prominent putaminal necrosis seen in these patients may have related to decreased venous outflow through the Rosenthal's veins or to a possible accumulation of formic acid and formaldehyde in the putamen (McLean et al. 1980). An alternative presentation is dystonia and severe hypokinesis with putaminal necrosis but without other parkinsonian signs (LeWitt and Martin 1988). In the latter example, dopaminergic drugs were moderately helpful in ameliorating symptoms.

In an attempt to improve prognosis, a three-part approach to treatment was developed, involving ethanol, bicarbonate, and in severe cases dialysis. Ethanol administration saturates the alcohol dehydrogenase enzyme and retards the conversion of methanol into its toxic byproducts (Lawrence and Haggerty 1971). Massive and rapid alkalinization with bicarbonate may be necessary to correct acidosis, and it must be carried out for several days to avoid relapse (Rumack 1976). Dialysis is advocated when the methanol blood concentration is more than 50 mg/dl, and more recent work suggests that serum formate concentrations may be a better criterion for hemodialysis (Osterloh et al.

1986). The rapidity of therapeutic intervention is important, at least for the protection of vision. The longer the initial visual loss is present prior to the onset of therapy, the less likely full vision will be regained (Krolman and Pidde 1968).

Because methanol presumably remains in the body for a long time, transient correction of acidosis may lead to clinical relapse. In the review by Bennett and coworkers (1953) of more than 300 patients, many were treated in the emergency room with only transiently assured correction of acidosis. On discharge from the emergency room, patients were instructed to consume bicarbonate as outpatients. Six cases returned in relapse: three had mild and easily correctable symptoms; two arrived in a moribund state 12 and 24 hours after discharge; and one had mild symptoms but collapsed suddenly and died. For these reasons, persistent observation for several days is indicated. Bennett and colleagues (1953) also observed that acidotic patients must be treated aggressively regardless of how clinically well they appear. Six acidotic patients who were rational and conversing on admission suddenly became comatose and died within minutes. Therefore, intensive observation of all acidotic patients would presumably improve prognosis (Bennett et al. 1953).

Ethylene Glycol

Ethylene glycol is commonly used as an antifreeze for automobiles, as a solvent, and as a constituent in the manufacturing of explosives. The lethal dose approximates 100 ml (Hunt 1932). Ingestion of this solvent results in 40 to 60 deaths per year (Vale et al. 1976).

Symptoms of toxicity may appear early or may be delayed for several hours. Initially, the patient becomes restless and agitated. This is followed by somnolence, stupor, coma, and convulsions (Peterson et al. 1981). Death results from respiratory or cardiac failure. Abducens or soft-palate paralysis, aphasia, nystagmus, and fecal or urinary incontinence have been observed (Berger and Ayyar 1981). Hypocalcemia can occur and may be responsible for episodes of tetany.

Factors that influence prognosis are undetermined. Treatment consists chiefly of symptomatic measures, respiratory support, correction of severe metabolic acidosis, and control of hypocalcemia. As with methyl alcohol toxicity, ethylene glycol intoxication can be treated with ethyl alcohol infusions reaching 5 to 10 g/hour. Hemodialysis has been associated with improvement in gait disorders and focal neurological signs, specifically visual and speech (Berger and Ayyar 1981).

Isopropyl Alcohol

Isopropyl alcohol is a solvent used in the manufacturing of perfumes, cosmetics, and various lacquers and varnishes. It also is commonly used as rubbing alcohol and in tepid sponges to reduce elevated body temperatures. The toxicity and narcosis effect of isopropyl alcohol is approximately twice that of ethanol. Dizziness and headache, followed by ataxia, depression, and narcosis usually occur in acute isopropyl alcohol intoxication. At higher levels, patients may present in coma. When hypotension occurs, a poor prognosis is suggested (Adelson 1962). Because isopropyl alcohol has a longer duration of action than ethanol and its major metabolite, acetone, also is a central nervous system (CNS) depressant, hemodialysis is considered a life-saving procedure in severely intoxicated patients (Rosansky 1982). Approximately 27 g of isopropyl alcohol can be removed per hour with hemodialysis, which is 50 times the rate of removal achieved by urinary excretion.

Benzene

The hydrocarbon benzene is one of the most important of the industrial poisons, ranking second only to tetrachlorethane in the production of serious intoxications. Benzene is widely used in industry, although its use has been curtailed because of toxic reactions. It is used in the processing of rubber, motor fuel, dyes, leather, and paints. Workers in chemical laboratories and in coal tar distilleries are subject to exposure to this toxin.

Acute benzene intoxication can follow oral ingestion in suicide attempts, but commonly it occurs in workers who have been cleaning or repairing leaks in benzene tanks or who have been exposed to toxic concentrations of this substance in the air (Dinberg 1945).

Excessive ventilation of benzene vapor produces euphoria and an alcohol-like inebriation, associated with giddiness, tinnitus, headache, drowsiness, and vomiting. The gait becomes ataxic and muscular twitching appears with severe intoxication. Finally, convulsive seizures and unconsciousness occur. In very severe intoxications, there may be almost immediate loss of consciousness and respiratory depression.

In patients who survive acute effects, recovery usually is complete, although headache, chest pain, shortness of breath, anorexia, nausea, and vomiting may be present for days or weeks. No specific risk factors have been identified to affect prognosis in these acute intoxications.

Chronic benzene intoxication has been observed during benzene treatment for leukemia. CNS symptoms occur as a result of anemia and from direct toxic effects of the compound. Pyramidal tract involvement, hyperesthesia, ataxia, paresthesias, paraplegia, retrobulbar neuritis, peripheral neuropathy, and seizures have been observed (Klein 1950). Recovery in nonfatal cases takes several months.

The best studied of the neurological complications related to chronic benzene intoxication is peripheral neuropathy. However, because toluene often is mixed with benzene, the specificity of benzene neuropathy is not fully established. Six exposed Turkish shoemaker–leather workers were studied neurologically and electrophysiologically and showed slowed conduction velocities and clinical signs of atrophy (Baslo and Aksoy 1982). The severity of residual neurological dysfunction and the duration of nonexposure, that is the period from last ben-

zene exposure to the neurological evaluation, were inversely correlated. In fact, the two patients with the longest freedom from exposure (as high as 96 months) had minimal or no neurological abnormalities. The degree of pancytopenia associated with benzene exposure did not correlate with neurological prognosis (Kahn and Muzyka 1973).

Toluene (Methylbenzene)

Toluene is one of the most widely used solvents and is used as a paint and lacquer thinner, a cleaning and dying agent, and a constituent of fuels. It is commonly used in histology laboratories and has been estimated to be the most widely abused solvent (King 1983). Neurological signs occur after exposure to as little as 100 ppm, and with concentrations of 800 ppm, signs may persist for several days. After acute exposure, there is an early exhilaration, followed by fatigue, mild confusion, ataxia, and dizziness, intentional exposure by sniffing glue with the head in a plastic bag may result in blood levels of 6.5 mg/dl. Psychotic behavior, unconsciousness, and death have been reported following such exposure (Winek et al. 1968). In most cases, however, single acute intoxication episodes are associated with no neurological residua.

After chronic toluene abuse, the prognosis is more serious. Tremulousness, unsteadiness, emotional lability, and insomnia are prominent features, and jaw jerk, snout, and other primitive reflexes have been reported in chronically exposed patients months after cessation of exposure (Browning 1937). Cerebellar ataxia with and without optic nerve dysfunction also had been reported with cognitive impairment (Barnes 1967; King 1983). In a 20-patient study of solvent (mainly toluene) abusers, after 4 weeks of solvent abstinence, cognitive impairment persisted in 60%, pyramidal signs in 50%, cerebellar signs in 45%, and brain stem or cranial nerve impairment in 25% (Hormes et al. 1986; Rosenberg et al. 1988). Brain stem auditory evoked responses (BAERs) and magnetic resonance imaging

(MRI) scans also showed persistent abnormalities in chronic abusers (Rosenberg et al. 1988). Irreversible optic neuropathy can occur, although even severe visual compromise (4/200) can reverse to 20/30 after weeks of complete abstinence from chronic toluene exposure (Keane 1978).

Additionally, a patient may present with a polyneuropathy, symmetric weakness, and atrophy after chronic toluene exposure. The long-term outcome of these patients, however, has not been well described.

n-Hexane and Methyl Butyl Ketone

n-Hexane is a component of numerous glues and, like toluene, is a source of major inhalation abuse. The substance is associated with a euphoric effect in patients with acute exposure and a pronounced peripheral neuropathy after chronic intoxication. Although many glues contain a combination of n-hexane and toluene, the independent role of n-hexane has been indicated in cases in which the neuropathy did not develop until the glue sniffer changed to a glue containing n-hexane (Towfighi et al. 1976). The neurotoxic product is probably not hexane, but 2,5 hexadione (2,5, HD), which also is a metabolite of methyl butyl ketone (MBK) (Spencer et al. 1980). Workers exposed to n-hexane first report headache, nausea, and anorexia. Two to 6 months after initial exposure, victims may demonstrate a neuropathy, including symmetric, predominantly distal, motor deficits and frequent sensory symptoms that may progress for up to 3 months after exposure ceases. Although clinically detectable neurological abnormalities and sensory symptoms have been reported to persist for 2 to 3 years after exposure ceases, partial or complete recovery is the rule (Bravaccio et al. 1981; Spencer and Schaumburg 1985b).

MBK is a component of paint thinners, cleaning agents, and solvents for printing. Numerous outbreaks of polyneuropathy have been reported (Teitelbaum 1977). The sensory polyneuropathy is insidious and painless, usually beginning several months after continued chronic exposure. Autonomical and cranial nerve dysfunction also occurs.

As with n-hexane neuropathy, there is no recognized form of treatment for MBK neurotoxicity. The administration of prednisolone and large doses of vitamin B have been tried without effect (Yamamura 1969). Removing the patient from the exposure source generally is associated with a good prognosis, although for the first few weeks or months, the disease may intensify. Only rarely have patients exposed to either hexacarbon shown signs of long-term CNS damage (Spencer and Schaumburg 1985b).

Gasoline

The composition of gasoline is extremely varied. Originally, it was composed largely of butanes, hexanes, and pentanes, but various other hydrocarbons have been added. Because of its antiknock qualities, tetraethyl lead is present in most gasoline, and triorthocresyl phosphate occasionally is found as well. With the inhalation of toxic quantities of gasoline, inebriation, blurred vision, incoordination, confusion, excitement, and occasionally delirium with visual hallucinations may occur. Voluntary gasoline abusers report that 15 to 20 breaths of gasoline vapors cause a 5 to 6 hour euphoric intoxication (Hartman 1988). These symptoms are followed by depression, headache, lethargy, trembling, staggering, and nausea. In severe cases, symptoms progress to dyspnea, cyanosis, and convulsions. Death has occurred in cases of extreme exposure, but generally recovery is followed by an alcoholic-like "hangover." In a 50-patient study of chronic gasoline sniffers, 92% had abnormal neurological examinations at the time of gasoline sniffing, but all except one case resolved after 8 weeks. Lead chelation was instituted in 39 of these patients. Twenty-seven of the 46 patients had an abnormal EEG with very low voltage and an excess of diffuse slow activity; 10 of the 15 EEGs performed 8 to 12 weeks later were normal. Common neurological signs that resolved were brisk deep tendon reflexes, intention tremor, and ataxia. However, cases of persistent seizure disorders, ataxia, and dementia have been reported (Browning 1937; Valpey et al. 1978).

The precise toxic component in gasoline remains undetermined, but some toxicity may relate to solvents and others to tetraethyl lead. Chelation has been used successfully and has been reported to improve long-term clinical outcome, although this has not been tested in controlled trials (Valpey et al. 1978). A chronic encephalopathy can be progressive in patients who persist in gasoline sniffing (Valpey et al. 1978).

Turpentine and Other Solvent Mixtures

Turpentine is widely used in the paint and varnish industry and has been used medically as an expectorant. Intoxication can occur after inhalation of fumes in industry or after ingestion (accidentally or in attempted suicide). With acute intoxication, patients develop headache and dizziness, followed by confusion, excitement, and delirium. Patients may become ataxic and exhibit visual disturbances. These resolve without sequelae in most cases. In more severely intoxicated cases, renal failure and gastrointestinal irritation occur, along with encephalopathy that may progress to coma and generalized convulsions. In cases in which pneumonia and poor oxygenation may develop, a longstanding posthypoxic encephalopathy also may develop.

Two major toxic syndromes occur in association with chronic mixed solvent exposure: encephalopathy and neuropathy. Cerebellar and myopathic changes have been described but are less common (Pedersen et al. 1980). Chronic encephalopathic symptoms include fatigue, concentration difficulties, impaired memory, and sleep disruption (Eskelinen et al. 1986). Specific neuropsychological impairment of verbal memory and slowed emotional reactivity have been reported. In 80 patients exposed to mixed solvents and followed for a mean of 5.8 years after initial chronic exposure, objec-

tive clinical signs of CNS dysfunction and evidence of progression (Juntunen 1982) persisted in most cases. In many cases, no neurological abnormality was detected at initial diagnosis, but at follow-up, cerebellar or pyramidal involvement was documented. Of the 10 patients who demonstrated signs of peripheral neuropathy at diagnosis, the long-term prognosis will split in approximately equal numbers between improvement ($N = 4$), no change ($N = 4$), and deterioration ($N = 3$). In a second study specifically aimed at follow-up prognosis, 26 patients chronically exposed to solvents who had shown signs of encephalopathy were removed from the solvents for 2 years and reexamined neurologically, neuropsychologically, and by CT scans. In general, the encephalopathy persisted in spite of removal of the solvent exposure. In three patients, neurological function deteriorated (Bruhn et al. 1981).

Other population studies have compared painters or floor layers with workers not exposed to mixed solvents, such as carpenters and construction workers. These studies show more depression, slower reaction times, and more impaired visual constructive tasks in the solvent-exposed groups (Ekberg et al. 1986; Lindstrom 1982). Patients who voluntarily abuse solvents by soaking a cloth and inhaling the fumes also have been studied for chronic encephalopathy lasting beyond the period of toxic exposure; results of persistent disability were similar (Berry et al. 1977).

In another follow-up study in which investigators reexamined 26 of their original 50 painters chronically exposed to solvent mixtures for an average of 28 years, subjective complaints of headaches and dizziness declined for most patients, but no one improved in neuropsychological, neurological, or neuroradiological status after 2 years without solvent exposure. Specifically, if cortical atrophy or other objective signs of neurological dysfunction developed, recovery was slow or absent. Exposed workers without objective signs of structural abnormalities or neurological dysfunction appeared to be much more likely to recover (Hartman 1988).

Several studies suggest that chronic exposure, usually longer than 3 years, of mixed solvents (specifically constituent not always determined) is associated with an encephalopathy that progresses during exposure and tends to plateau but may progress after removal from the source of intoxication (Spencer and Shaumberg 1985).

A second neurological syndrome seen in solvent mixtures is referred to as Huffer's neuropathy (Procktop et al. 1974). This disorder was originally described in patients who had inhaled lacquer thinner over several years but became acutely ill when the solvent mixture was changed to another composition. A severe and predominantly motor neuropathy developed and continued to progress for up to 2 months after cessation of exposure. The weakness involved proximal and distal muscles and the muscles of respiration and facial expression. Some patients required respiratory support. Areflexia was the rule, although sensory loss was mild. Most patients showed improvement in motor function after exposure ceased, but many showed persistent deficits. Electromyographical studies demonstrated severely slowed conduction velocities and signs of acute denervation.

Tetrachlorethane

This solvent is the most dangerous of the chlorinated hydrocarbons and its use often is restricted because of toxicity. It has been marketed commercially under names such as Alanol, cellon, Novania, and Westron. Tetrachlorethane exerts a prolonged narcotic effect, resulting in unconsciousness, loss of corneal reflexes, cyanosis, and death, often within 12 hours. More commonly, however, tetrachlorethane poisoning results from chronic exposure to excess concentrations of the vapor in the air. In such cases, headache, giddiness, anorexia, nausea, and other nonspecific complaints predominate (Browning 1937). Good prognostic studies have not been conducted with

this agent. The treatment of toxicity is aimed chiefly at protection of the liver from parenchymal damage (Fairhall 1949).

Trichlorethylene and Perchloroethylene

Trichlorethylene (TCE) is an important organic solvent used occasionally in medicine as an anesthetic but more widely used in the dry cleaning industry. Deliberate inhalation of TCE induces a rapid state of euphoria, and many workers have become addicted to the fumes (Granjean et al. 1955). Sudden death associated with bronchial constriction, pulmonary edema, and myocardial irritability have been reported infrequently (Seage and Bruns 1971). Neurologically, cranial and peripheral neuropathies are the most striking feature of TCE toxicity. There is a propensity for trigeminal involvement, both sensory and motor, with facial nerve involvement also a characteristic (Feldman 1979). Neuro-ophthalmological complications, including retrobulbar neuropathy, optic atrophy, and ocular motor disturbance may occur, as well as cerebellar and extrapyramidal dysfunction in the form of tremor and ataxia.

In the first 24 hours after exposure, the entire face usually becomes anesthetic, but by 4 to 8 weeks, the peripheral face generally has returned to normal. Twenty to 30 weeks after exposure, the anesthetic area further retracts to involve only the eyes, nose, and mouth. By 80 weeks, only spotty hypalgesia is noted (Feldman 1979).

In addition to the well-documented cranial neuropathies of TCE exposure, less specific neurasthenic symptoms, including fatigue, insomnia, anxiety, and emotional giddiness, have been reported that may last for several weeks (Spencer and Schaumburg 1985).

Perchloroethylene (PCE) has been widely substituted for TCE in the dry cleaning industry and produces similar clinical problems. Acute intoxication signs, usually dizziness, headache, nausea, and vomiting, clear after removal of the patient from the PCE fumes. After months or years of exposure, an encephalopathy has been described consisting of irritability, labile affect, memory loss, and mild confusion. Even 1 year after cessation of chronic exposure, long-lasting changes in personality and memory loss have been reported (Gold 1969).

Carbon Tetrachloride

Carbon tetrachloride is widely used for manufacturing refrigerants and insecticides (Hardin 1954). Following oral ingestion, the intoxicated patient develops nausea and vomiting, often associated with abdominal pain and diarrhea. The acute symptoms usually are more severe following inhalation. In the 77-patient study review by Hardin (1954), neurological symptoms, such as headache, dizziness, and convulsions, were the sixth, 10th, and 12th most frequent symptoms after nausea, vomiting, abdominal pain, edema, and jaundice. Staggering gait, confusion, and unconsciousness also may occur during intoxication but were less common (Kazantzis and Bomford 1960). Stewart and Witts (1944) examined 78 industrial workers in an English factory who were exposed intermittently to carbontetrachloride vapors. More than 50% of the workers were discharged from work because of symptoms attributable to gaseous exposure, including insomnia, irritability, nausea and vomiting, abdominal pains, and diarrhea. Objective evidence of hepatic or renal failure was rarely found, and no long-term follow-up information after exposure cessation was reported. Late sequelae of acute exposure can include optic atrophy with amblyopia (listed 14th on Hardin's list), generalized convulsions, and delirium terminating in death. In such cases, the patient usually is the victim of severe hepatic and renal failure (Morgan et al. 1949). After chronic carbon tetrachloride exposure, persisting states of mental confusion, polyneuropathy, and parkinsonism have been described, although specific population studies have not been performed (Melamed and Levy 1977).

Factors that may alter the toxicity of carbon tetrachloride have been studied. First, ingestion of alcohol concomitant to or preceding exposure to the toxin increases the victim's susceptibility to poisoning (Morgan et al. 1949). Furthermore, obesity, diabetes, and prior liver or renal disease also appear to lower the threshold to the toxic effects of carbon tetrachloride (Council of Industrial Health 1946). Evidence from animal studies suggests that phenobarbital may negatively affect prognosis because the drug increases microsomal enzyme activity leading to greater quantities of toxic metabolites. Oxidizing substances, such as vitamin E and selenium, protect against carbon tetrachloride intoxication in animals, although these two therapies have not been studied extensively in humans. Hemodialysis has been recommended in cases of exposure (Fogel 1983).

Any exposure concentration of more than 100 ppm can be associated with severe liver or kidney damage. When hepatic necrosis exists, long-standing cirrhotic changes develop in surviving patients. Hepatoma also has developed in some cases and occasionally has been ascribed to the carbon tetrachloride exposure (Louria and Bogden 1980).

Carbon Monoxide

Carbon monoxide, an odorless and nonirritating gas, is the most abundant air pollutant in the lower atmosphere. Acute poisoning accounts for more than 3000 accidental or suicidal deaths each year in the United States. The effects of chronic exposure to carbon monoxide, due to inhalation of polluted air, are of increasing concern to scientists and clinicians. The threshold limit of carbon monoxide is 50 ppm, causing a carboxyhemoglobin saturation level of 8% to 10% after an 8-hour exposure. Levels of carboxyhemoglobin that exceed 50% saturation are considered life-threatening, and levels of 70% to 75% usually are fatal. Such saturation levels may occur by exposure to very high acute concentrations or by exposure to relatively lower concentrations for a chronic period. The threshold limit assumes that the carbon monoxide comes from a single source and does not take into account endogenous factors or smoking, which also can increase the victim's baseline saturation.

Mild neurological effects include headache, dizziness, and visual disturbances; more severe toxic reactions include convulsive disorder, pyramidal and extrapyramidal signs, cerebral blindness, and deafness. Such symptoms may be transient to persistent, appearing immediately, days, or weeks after intoxication. Myositis and calcification in the muscles with mild necrosis also may occur; these are related to the direct effects of the toxin on myoglobin and the crushing effect of the patient's own body on muscles.

Behavioral and psychiatric alterations, including irritability and violent behavior, personality disturbances, inappropriate euphoria, confusion, and impaired judgment may occur. In a follow-up study, 3 years after carbon monoxide poisoning, 13% of patients showed gross neuropsychiatric damage, 33% showed a deterioration of personality, and 43% demonstrated impaired memory.

More recent concern over the chronic effects of carbon monoxide focus on patients with continued industrial or urban exposure to the gas. Cigarette smokers also fit into this potential category. Six patients with leaking exhaust systems in their automobiles who were chronically self-exposed to carbon monoxide (McClain and Becker 1975). Over weeks they developed an insidious syndrome of somnolence and mental decline with gait difficulty, incoordination, and slurred speech. When the exhaust system was corrected, their carboxyhemoglobin levels declined and their symptoms resolved. A more focal neurological syndrome reported after acute and chronic exposure to carbon monoxide is parkinsonism (Westerman et al. 1965). Some patients show diffuse neurological involvement associated with parkinsonian features, while others demonstrate purer parkinsonism. This clinical syndrome usually is associated with globus pallidal lucency on the CT scan and pallidal atrophy histologically. Patients may respond to levodopa or to anticholinergic drugs, but the response is unpredictable.

If toxic levels of carboxyhemoglobin are determined, the diagnosis of carbon monoxide poisoning is safely made. However, the level of carbon monoxide does not indicate the severity of the poisoning because a patient removed from the intoxicated environment and allowed to breathe fresh air will show a rapid decline in the carboxyhemoglobin levels.

The prognosis for patients with significant carbon monoxide poisoning is difficult to determine immediately after the patient is discovered. Patients with carboxyhemoglobin levels as low as 29% have died after a carbon monoxide poisoning, while others with much higher levels have survived.

Personality change and memory impairment are a particular concern; in a 3-year follow-up study of 63 patients who survived acute carbon monoxide intoxication, eight (12.7%) showed gradual behavioral improvement, 21 (33.3%) showed deterioration, and 27 (43%) reported memory impairment. Personality deterioration and memory impairment correlated, and the degree of depressed consciousness at hospital admission correlated significantly with the development of neuropsychiatric sequelae (Smith and Brandon 1973).

Treatment is well established and improves survival. Upon discovery of the patient with carbon monoxide poisoning, the physician should immediately remove the patient from the contaminated environment. Fresh air and inactivity are recommended to maintain the tissue demands for oxygen at an absolute minimum. Resuscitation in the contaminated environment may be dangerous to the rescuer and to the patient. If the patient cannot breathe, artificial respiration is indicated. One hundred percent oxygen should be started as soon as possible through an oral or nasal mask. The half-life of carboxyhemoglobin is 40 minutes after the administration of 100% oxygen.

Hyperbaric oxygen reduces the half-life of carboxyhemoglobin to less than 25 minutes. The following are recommendations for hyperbaric oxygen: 46 minutes of 100% oxygen at 3 atm absolute pressure, followed by 2 atm absolute pressure for 2 hours or until the proper carboxyhemoglobin level is achieved (Cohen 1975; Sauerhoff and Michaelson 1973). Hyperbaric oxygen is considered the treatment of choice for carbon monoxide poisoning, and if possible, it should be administered to every patient who presents with signs and symptoms of severe intoxication. In premature infants, the toxicity of oxygen must be considered and weighed against the likelihood of the diagnosis in questionable cases.

To decrease the tissue demand for oxygen, hypothermia also has been advocated, although no controlled studies have determined its effects on prognosis. Patients covered in ice are maintained for 8 to 12 hours at a temperature of 30° to 32°C. Shivering, which increases the metabolic rate, should be prevented and often can be controlled by administering 25 mg of chlorpromazine. If seizures develop, they may be treated with steroids, such as dexamethasone, 16 to 20 mg/day.

Nitrous Oxide

Volitional inhalation of nitrous oxide has gained increasing popularity, especially among dentists and other health-care workers (Layzer et al. 1978). The gas also is available in various compressed dispenser cartridges and may account for "whipped cream dispenser" polyneuropathy (Sahenk et al. 1978). Nitrous oxide neuropathy may be a predominantly distal polyneuropathy or may follow a radicular pattern. Also, cases of mixed or primarily sensory neuropathy have been described. Cessation of exposure has been associated with gradual clinical improvement, although as late as 6 months after removal from nitrous oxide, mild neuropathic symptoms may persist (Layzer et al. 1978; Sahenk et al. 1978).

In addition to neuropathy, myelopathic signs of upper motor neuron disease also may exist, provoking a picture resembling subacute combined degeneration due to vitamin B_{12} deficiency. In fact, nitrous oxide has been shown to inactivate certain B_{12}-dependent enzymes (Chanarin 1982). The gait difficulty, sensory ataxia, paresthesias,

and impotency regress at least partially after removal from exposure (O'Donoghue 1985).

Methyl Chloride

Methyl chloride is primarily a foaming agent for plastics. Acute exposures are associated with a toxic confusional state, involving slurred speech, ataxia, and occasionally convulsions (Kegel et al. 1929). Headache, dizziness, and personality changes also may occur. These changes, however, usually revert to normal, although they may take several weeks or months to clear (MacDonald 1964). During the first months after acute exposure, symptoms have recurred after apparent recovery and without further exposure (Scharnweber et al. 1974).

The most complete follow-up of patients exposed to acute intoxication of methyl chloride is a report by Gudmundsson (1977), who reevaluated patients 13 years after a refrigeration leak. Of the 15 original patients, four died, one within 24 hours of the episode, and two others committed suicide 11 and 18 months later. The fourth patient was assessed as 75% disabled because of severe neurological and psychiatric disturbances and died 10 years after the accident at age 34. Of the 11 survivors in 1976, 10 were examined; six had abnormal neurological examinations, including peripheral neuropathy ($N = 2$), tremor ($N = 3$), paralysis of visual accommodation ($N = 2$), and single cases of muscle atrophy, primative reflexes, and habit spasms. The degree of disability was less than the evaluation 20 months after the accident, suggesting that with time, very slow improvement occurred. Only one patient showed complete regression of symptoms between the 20-month follow-up and the 7-year follow-up (Gudmundsson 1977).

After chronic methyl chloride exposure, patients may develop an insidious form of dementia and ataxia. Peripheral neuropathy, more common with many other industrial toxins, is less conspicuous with methyl chloride (Scharnweber et al. 1974). Removal of the patients is associated with significant improvement, although some patients' long-term personality changes persisted, and 3 months after cessation of chronic exposure, some patients still showed tremor and nervousness (Scharnweber et al. 1974).

Carbon Disulfide

Carbon disulfide is used to manufacture viscose, rayon, cellophane, and adhesives. Acute fulminant intoxications can result in toxic psychosis, agitated delirium, and permanent sequelae of mental impairment (Gordy and Trumper 1940). The more important syndrome follows chronic exposure and involves the development of progressive mental impairment with cranial and peripheral neuropathies. Mental changes include irritability, personality changes, and emotional lability. A characteristic cranial nerve dysfunction involving the selective loss of corneal reflexes without facial sensory complaints also has been reported. The peripheral neuropathy is relatively mild, with mixed sensory and motor involvement and predominantly distal distribution. An unusual parkinsonian presentation has been reported after carbon disulfide intoxication and was marked by characteristic resting tremor, rigidity, and bradykinesia. It was associated with peripheral neuropathy and marked mental impairment (Allen 1979). In spite of extensive studies of patients exposed to chronic carbon disulfides, specific prospective examinations of neurological and neuropsychological function after cessation of exposure have not been reported in detail.

Ethylene Oxide

Ethylene oxide is a gas used as a precursor to various other industrial chemicals, including ethylene glycol. Generally considered innocuous, this gas recently has been implicated in human toxic peripheral neuropathies, cerebellar dysfunction, and encephalopathy. Signs of acute toxicity involving the nervous system are agitation and confusion, headaches, and drowsiness. With excessive doses, convulsions and death may occur (O'Donoughue 1985). After chronic exposure, a variety of syndromes can develop (Gross et al. 1979). Of four men exposed for 2 months to ethylene oxide at levels of more than 700 ppm, one developed repeated motor seizures, and the other three had a sensory motor peripheral neuropathy. Clinically, the latter patients showed generalized fatigue, weakness of the distal muscles, ataxia, and reduction or loss of the deep tendon reflexes. One man had additional cranial neuropathy, difficulty with memory, and slurred speech. All showed marked improvement within 2 weeks of cessation of exposure (Gross et al. 1979).

Organophosphates

These chemicals are powerful inhibitors of acetylcholinesterase and pseudocholinesterase (Arlien-Søberg 1985). In humans, the former enzyme is found in nervous tissue, specifically in the brain, spinal cord, and myoneural junctions at preganglionic and postganglionic parasympathetic synapses and preganglionic and some postganglionic sympathetic nerve endings. Excess acetylcholine causes overstimulation and then depolarization blockade of cholinergic transmission (Goetz 1985).

Most reports of major organophosphate poisoning have been due to parathion or methylparathion. Clinically, intoxication may range from latent, asymptomatic poisoning to a life-threatening illness. Toxicity usually is monitored with serum cholinesterase activity. Decreases of 10% to 50% may not be clinically detectable. When levels are moderately depressed (20% of normal), sweating, cramps, tingling of the extremities, and mild bulbar weakness with fasciculation may occur. At 10% of normal, consciousness becomes depressed, and myosis with no pupillary response to light occurs. The patient may become cyanotic from respiratory weakness and pooled secretions may obstruct the airway. Although symptoms generally abate after removal of the causative agent, the resolution may slow; headaches and weakness may persist for 1 month. Eye discomfort has been reported 5 months after exposure (Midtling et al. 1985).

Pralidoxime usually is used intravenously to reverse the acute cholinergic alterations. The major action of this drug is to reactivate organophosphate-inhibited acetylcholinesterase activity. The compound removes the phosphate group bound to the esteratic site (Lotti 1982). Because pralidoxime has a quaternary structure, it does not penetrate the blood–brain barrier well and therefore does not reach high CNS concentrations. However, it has controlled seizures, possibly through its peripheral effect on respiratory and skeletal muscles (Namba et al. 1971).

After acute organophosphate intoxication with mortality, death almost always occurs within the first 24 hours in untreated cases and within 10 days in treated cases (Namba et al. 1971). If anoxia does not develop, complete recovery should be expected within 10 days, although electroencephalogram (EEG) abnormalities can be detected weeks after exposure (Wislicki 1960). Occasionally, mild persistent residua are reported; for example, 25 of 600 acutely exposed victims complained of persistent neurobehavioral symptoms, including confusion, concentration difficulties, muscle aches, and mild diffuse weakness lasting for weeks. After chronic exposure, residua may be seen more predictably; in a 16-patient study of chronic (months) exposure to organophosphates, eight patients had impaired memory, seven had depression, and five had "schizophrenic reactions" lasting up to 1 year after cessation of exposure (Gershon and Shaw 1961). Extensive follow-up data are not available; however, in Australia, no increase in mental hospital admissions was noted in a region where organophosphates were widely used compared to areas where they were rarely used (Stoller et al. 1965).

Neuropathy is another prominent toxic sign of organophosphate exposure and has been reported as a progressive disorder after both acute and chronic poisoning. In such cases, the acute cholinergic toxicity is followed after several days by a delayed, progressive weakness, including foot drop, absent ankle jerk, and muscle weakness.

Persisting neuropathy can develop as late as 3 months after exposure ceases (Petry 1952). Pyramidal signs can accompany this syndrome (Senanayake 1982). Examples of responsible compounds include tri-aryl phosphate, nipafox, trichlorfon, and methamidophos. There is no correlation between the potency of acute cholinergic effects and the likelihood of chronic neuropathy (Cherniak 1082). However, for alkyl or mixed alkyl–aryl esters with the structure of R1R2P(O or S)X, factors that increase the relative risk for delayed neurotoxicity compared with acute toxicity include phosphonate or phosphoramidates, rather than phosphates and long chain length (four or five carbon atoms) or hydrophobicity of R1 and R2.

The pathophysiology of the neuropathy does not appear to relate to effects on the cholinergic system. Instead, it relates to a two-step phosphorylation and "aging" of the protein called neuropathy target esterase (NTE) (Lotti 1986).

Dichlorodiphenyltrichloroethane and Organochloride Insecticides

The chlorinated hydrocarbon insecticides, of which, dichlorodiphenyltrichloroethane (DDT) is the prototype, share numerous toxicological properties, although certain individual variations exist. In general, these compounds are toxic to the nervous system and induce tremor and convulsions. DDT is associated with acute and chronic neurotoxicity. Acutely, patients notice a metallic taste in the mouth, and within 1 hour, they experience dryness of the mouth and extreme thirst. The patient may experience drowsiness or insomnia, eye burning, and a gritty feeling of the eyelids. Muscle spasms, tremor, and stiffness of the jaw may develop later. After chronic low-dose exposure, weakness of the arms may occur, or weakness may be relatively isolated to distal muscles and cause bilateral wrist drop (Garrett 1947). Mononeuropathy, optic neuropathy, and polyneuropathy have been described with chronic intoxication (Committee on Pesticides 1951). When these symptoms result from exposure of the skin to DDT, they may be limited to the limb that was in direct contact with the toxin (Mackerras and West 1946). With chronic high-dose exposure, generalized convulsions, coma, and death may ensue (Globus 1948).

After acute exposure to DDT, motor signs generally abate; however, even months later, objective disability can persist in the form of weakness and poor coordination (Taylor and Calabrese 1979). In cases of retrobulbar neuritis, recovery has been reported after a few months (Campell 1952).

Chlordecone (Kepone) has not been reported to induce seizures regularly but causes a dramatic irregular postural tremor that is maximal when the limbs are extended but not specifically enhanced by actions. In more severe cases, the tremor is evident at rest. Additionally, opsoclonia is associated with chlordecone exposure and occasionally pseudotumor cerebri (Sanborr et al. 1979; Taylor et al. 1978). In the chlordecone outbreak in Virginia, more than half of the active employees of the affected plant were moderately ill. A few patients required hospitalization for months, but severe toxic signs generally abated when patients were removed from the toxic environment. However, 4 years later, several workers continued to show incapacitating tremor (Taylor et al. 1978).

Paraquat

Prognosis related to acute paraquat poisoning has been specifically addressed by Bismuth and colleagues (1982) in a 28-patient study. Eleven patients survived, representing a mortality rate of approximately 60%. To determine which factors likely related to survival, they analyzed sex, route of exposure, exposure doses, and preexisting medical conditions in the survivors and in those who died. They determined that paraquat inhalation is benign because the aerosolized particles do not attain significant alveolar penetration and because patients survived without sequelae when exposed by this route. Percutaneous penetration was more dangerous, but only occasionally were severe intoxications caused by this route. Oral ingestion was the most dangerous, with a mortality rate of 70%. The minimal lethal dose is 35 mg/kg, so one mouthful of a 20% solution may kill a victim. A 50% mortality occurred with one mouthful and 92% mortality with two mouthfuls. In the group with the lower exposure dose, death occurred from pulmonary fibrosis 5 to 31 days after intoxication, whereas the group ingesting the higher dose died within 48 hours from circulatory collapse. In 13 cases, gastric content prior to intoxication could be determined. Only one of five patients poisoned immediately after eating a meal died, whereas seven of eight who were intoxicated at a time unrelated to eating died. Gastric endoscopy was useful in predicting survival because nine of 14 patients with ulcerations died, and no patient without ulcerations died. Finally, and possibly most important, the plasma level of paraquat was helpful in determining outcome; at 4, 6, 10, 16, and 24 hours after ingestion, patients with levels at or below 2.0, 0.6, 0.3, 0.2, and 0.1 mg/L were likely to survive. Renal failure and pulmonary function results had no prognostic value.

Therapeutic interventions that influence prognosis include rapid removal of paraquat from the gastrointestinal tract, although the efficacy of this therapy has not been fully established. Forced renal diuresis does not enhance removal of the herbicide from the blood (Paillard, private communication), and because the pulmonary accumulation of paraquat is so rapid, hemodialysis and hemoperfusion have not regularly improved survival. Long-term neurological prognosis in the seven survivors of this study was not specified.

Agent Orange—Dioxin

Tetrachlordibenzo-P-dioxin (TCDD), has been the source of intense controversy in industrial medicine. A large-scale dioxin intoxication occurred in Seveso, Italy, in 1976, although prior episodes had been reported. Dioxin exposure has been associated with skin lesions, including chloracne, multiple irregularities in blood chemistries, and neurological and psychiatric alterations (Togoni and Bonaccorsi 1982).

Dioxin is a solid substance, insoluble in water and only slightly soluble in fats and chlorinated solvents. It is a contaminant of such herbicides as Agent Orange. It is highly heat stable and exerts its biological effects at extremely low concentrations. Its half-life in soil is estimated to be approximately 1 year, and microbial degradation rarely occurs. Although there are significant differences in species susceptibility to the toxin, the toxic effect occurs slowly and progresses several days or weeks after a dose. Since the Seveso accident, long-term follow-up has been conducted on intoxicated patients from this group and from a group of 80 patients from Czechoslovakia (Pazderova-Vejlupkova 1981). The earliest symptoms following intoxication are the gradual development of chloracne, a global feeling of sickness, fatigue, and weakness. Usually this occurs days or weeks after the initial exposure, but in some patients, these consistent findings did not occur until several months after work with dioxin was completed. In the Czechoslovakian series, 23% of patients exposed to chronic dioxin showed initial evidence by clinical examination or electromyographic study of polyneuropathy. In most cases, the neuropathy progressed and then stabilized after 4 years. In addition, four patients showed signs of facial weakness as evidence of a peripheral VII cranial nerve lesion. Encephalopathy occurred initially in 7% of the patients and progressed in chronic follow-up to include 9% of those studied. Early psychiatric signs included neurotic symptoms, neurasthenia, and depression occurring in 83% of patients. The latter syndrome tended to decrease in severity and was only present in 58% in long-term follow-up. Other studies have questioned any direct behavioral consequence of dioxin exposure (Hartman 1988).

Importantly, these neuropathic symptoms did not occur initially. Polyneuropathy in some patients occurred only in the third or fourth year after exposure had ceased. The origin of this abnormality is unclear because many of the patients had abnormal glucose tolerance tests and other metabolic abnormalities, including hypercholesterolemia, hyperlipemia, and high pre-beta fractions. In contrast, the psychiatric problems were seen early and may have related to exogenous influences, including fear of death, disfigurement from the cutaneous manifestations, and concerns about job permanency. In a series of patients who were examined by the author (unpublished) and who were exposed to dioxin and other chemicals in an American work setting, 20 to 46 (43%) had an intention tremor, most of the postural type, and more than half had evidence of peripheral neuropathy. Further studies have documented action dystonia that began months or years after chronic exposure started and stabilized thereafter (Klawans 1987).

References

Adelson, L. Fatal intoxication with isopropyl alcohol (rubbing alcohol). Am. J. Clin. Pathol. 38:144–151; 1962.

Allen, N. Solvents and other industrial organic compounds. In: Vinken, P. J.; Bruyn, G., eds. Handbook of clinical neurology. Amsterdam: North Holland; 1979:p. 361–375.

Arlien-Søberg, P. Chronic effects of organic solvents on the central nervous system and diagnostic criteria. In: Chronic effects of organic solvents on the central nervous system and diagnostic criteria. Copenhagen: World Health Organization; 1985:p. 218–226.

Barnes, R. Poisoning by the insecticide chlordane. Med. J. Aust. 1:972–974; 1967.

Baslo, A.; Aksoy, M. Neurologic abnormalities in chronic benzene poisoning. Environ. Res. 27:457–465; 1982.

Bennett, I. L.; Cary, F. H.; Mitchell, G. L. Acute methyl alcohol poisoning: a review based on experiences in an outbreak of 323 cases. Medicine 32:431–463; 1953.

Benton, C. D., Jr.; Calhoun, F. P., Jr. The ocular effects of methyl alcohol poisoning: Report of a catastrophe involving three hundred and twenty persons. Trans. Am. Acad. Ophthalmol. Otolaryngol. 56:875–885; 1952.

Berger, J. R.; Ayyar, O. R. Neurological complication of ethylene glycol intoxication. Arch. Neurol. 38:724–726; 1981.

Berry, G. J.; Heaton, R. K.; Kirby, M. W. Neuropsychological deficits of chronic inhalant abusers. In: Rumack, B. H.; Temple, A. R., eds. Management of the poisoned patient. Princeton: Science Press; 1977:p. 9–31.

Bismuth, C.; Garnier, R.; Dally, S.; Fournier, P. E. Prognosis and treatment of paraquat poisoning: a review of 28 cases. J. Toxicol.—Clin. Toxicol. 19(5):461–474; 1982.

Bravaccio, F.; Ammendola, A.; Barruffo, L.; Carlomagno, S. H-reflex behavior in glue (n-hexane) neuropathy. Clin. Toxicol. 18:1369–1375; 1981.

Browning, E. Toxicity of industrial organic solvents. London: His Majesty's Stationery Office; 1937.

Bruhn, P.; Arlien-Søborg, A.; Gyldensted, C.; Christensen, E. L. Prognosis in chronic toxic encephalopathy: a two-year follow-up study in 26 house painters with occupational encephalopathy. Acta. Neurol. Scand. 64:259–272; 1981.

Campbell, A. M. G. Neurological complications associated with insecticides and fungicides. Br. Med. J. 2:415–417; 1952.

Chanarin, I. The effects of nitrous oxide on cobalamins, folates, and other related events. CRC. Crit. Rev. Toxicol. 10:179–213; 1982.

Cherniak, M. G. Organophosphorus esters and polyneuropathy. Ann. Intern. Med. 104:264–266; 1985.

Cohen, M. M. Neural Toxins. In: Cohen, M. D., ed. Biochemistry of neural disease. Hagerstown, MD: Harper & Row; 1975:p. 220–244.

Committee on Pesticides. Pharmacology and toxicologic aspects of DDT (chlorophenothane, USP). J.A.M.A. 145:728–729; 1951.

Council of Industrial Health. The recognition and treatment of carbon tetrachloride poisoning. J.A.M.A. 132:786–792; 1946.

Dinberg, M. C. Benzene poisoning. Can. Med. Assoc. J. 52:176–179; 1945.

Ekberg, K.; Barregard, L.; Hagberg, S.; Sallsten, S. Chronic and acute effects of solvents on central nervous system functions in floor-layers. British Journal of Industrial Medicine 43:101–106; 1986.

Eskelinen, L.; Luisto, M.; Tenkanen, K.; Mattel, O. Neuropsychological methods in the differentiation of organic solvent intoxication from certain neurological conditions. J. Clin. Exp. Neuropsychol. 8:239–256; 1986.

Fairhall, L. T. Industrial toxicology. Baltimore: Williams & Wilkins; 1949.

Feldman, R. G. Trichloroethylene. In: Vinken, P. J.; Bruyn, G. W., eds. Handbook of clinical neurology. Amsterdam: North Holland; 1979:p. 457–477.

Fink, W. H. The ocular pathology of methylalcohol poisoning. Am. J. Ophthalmol. 26:694–698; 1943.

Fogel, R. P. Carbon tetrachloride poisoning treated with hemodialysis and total parenteral nutrition. Can. Med. Assoc. J. 114:560–561; 1983.

Garrett, R. M. Toxicity of DDT for man. J. Med. Assoc. State Ala. 17:75–78; 1947.

Gershon, S.; Shaw, F. H. Psychiatric sequelae of chronic exposure to organophosphorus insecticides. Lancet 1:1371–1376; 1961.

Globus, J. H. Histopathologic observations on central nervous system in so-treated monkeys, dogs, cats, and rats. J. Neuropathol. Exp. Neurol. 7:418–422; 1948.

Goetz, C. G. Neurotoxins in clinical practice. New York: SP Medical and Scientific Books; 1985.

Gold, J. H. Chronic perchloroethylene poisoning. Can. Psychiat. Assoc. J. 14:627–630; 1969.

Gordy, S. T.; Trumper, M. Carbon disulfide poisoning. Ind. Med. 9:231–182–184; 1940.

Grandjean, E.; Haas, H. K.; Knoepel, V.; Munchinger, H.; Rosenmund, H.; Turrian, P. A. Investigations into the effects of exposure to trichloroethylene in mechanical engineering. Br. J. Intern. Med. 12:131–138; 1955.

Gross, J. A.; Haas, M. L.; Swift, T. R. Ethylene oxide neurotoxicity: report of four cases and review of the literature. Neurology 29:978–984; 1979.

Gudmundsson, G. Methyl chloride poisoning 13 years later. Arch. Environ. Health 32:236–240; 1977.

Hardin, B. L. Jr. Carbon tetrachloride poisoning: a review. Ind. Med. 23:93–96; 1954.

Hartman, D. E. Neuropsychological toxicology: identification and assessment of human neurotoxic syndromes. New York: Pergamon Press; 1988.

Hormes, J. T.; Filley, C. M.; Rosenberg, N. L. Neurologic sequelae of chronic solvent vapor abuse. Neurology 36:698–702; 1986.

Hunt, R. Toxicity of ethylene and prophylene glycol. Ind. Eng. Chem. 24:361–363; 1932.

Juntunen, J. Neurological examination and assessment of the syndromes caused by exposure to neurotoxic agents. In: Gilioli, R.; Cassitto, M. G.; Foa, V., eds. Neurobehavioral methods in occupational health. New York: Pergamon Press; 1982:p. 3–10.

Kahn, H.; Muzyka, V. Chronic effect of benzene on porphyrin metabolism. Work Environ. Health 10:140–143; 1973.

Kaplan, A.; Levreault, G. V. Methyl alcohol poisoning: report of 42 cases. U.S. Navy Med. Bull. 44:1107–1109; 1944.

Kazantzis, G.; Bomford, R. R. Dyspepsia due to inhalation of carbon tetrachloride vapor. Lancet 1:360–364; 1960.

Keane, J. R. Toluene optic neuropathy. Ann. Neurol. 4:390–394; 1978.

Kegel, A. H.; McNally, W. E.; Pope, A. S. Methyl chloride poisoning from domestic refrigerators. J.A.M.A. 93:353–358; 1929.

King, M. Long-term neuropsychological effects of solvent abuse. In: Cherry, N.; Waldron, H. A., eds. The neuropsychological effects of solvent exposure. Havant, Hampshire: Colt Foundation; 1983:p. 75–84.

Klawans, H. L. Dystonia and tremor following exposure to 2,3,7,8-Tetrachlorodibenzo-p-dioxin. Movement Disorders 2:255–261; 1987.

Klein, H. Zur pathologischen Histologie nach akuter Benzinvergiftung. Dtsch Z Ges Gerichtl Med. 40:76–79; 1950.

Krolman, G. M.; Pidde, W. J. Acute methyl alcohol poisoning. Can. J. Ophthalmol. 3:270–278; 1968.

Lawrence, R. A.; Haggerty, R. J. Household agents and their potential toxicity. Modern Treatment 8:511–514; 1971.

Layzer, R. B.; Fishman, R. A.; Schaefer, J. A. Neuropathy following abuse of nitrous oxide. Neurology 28:504–508; 1978.

LeWitt, P. A.; Martin, S. D. Dystonia and hypokinesia with putaminal necrosis after methanol intoxication. Clin. Neuropharmacol. 2:161–167; 1988.

Lindstrom, K.; Marbelin, T. Personality and long-term exposure to organic solvents. Neurobehav. Toxicol. 2:89–100; 1982.

Lotti, M. Pralidoxine in parathion poisoning. J. Toxicol. Clin. Toxicol. 19:121–124; 1982.

Lotti, M. Biological monitoring for organophosphate-induced delayed polyneuropathy. Tox. Lett. 33:167–172; 1986.

Louria, D. B.; Bogden, J. D. The dangers from limited exposure to carbon tetrachloride. CRC Crit. Rev. Toxicol. 72:177–188; 1980.

MacDonald, J. D. C. Methyl chloride intoxication: report of 8 cases. J. Occup. Med. 6:81–89; 1964.

Mackerras, I. M.; West, R. F. K. DDT poisoning in man. Med. J. Aust. 1:400–414; 1946.

Martin-Amat, G.; McMartin, K. E.: Hayreh, M. S.; Tephly, R. R. Methanol poisoning: ocular toxicity produced by formate. Toxicol. Appl. Pharmacol. 45:201–205; 1978.

McClain, R. M.; Becker, B. A. Teratogenicity, fetal toxicity and placental transfer of lead nitrate in rats. Toxicol. Appl. Pharmacol. 31:72–78; 1975.

McLean, D. R.; Jacobs, H.; Mielke, B. W. Methanol poisoning. Ann. Neurol. 8:161–167; 1980.

Melamed, E.; Levy, S. Parkinsonism associated with chronic inhalation of carbon tetrachloride. Lancet 1:1015–1117; 1977.

Midtling, J. E.; Barnett, P. G.; Coye, M. J.; Clinical management of field worker organophosphate poisoning. West J. Med. 142:514–518; 1985.

Morgan, E. C.; Wyatt, J. P.; Sutherland, R. B. An episode of carbon tetrachloride poisoning with renal complications. Can. Med. Assoc. J. 60:145–148; 1949.

Namba, T.; Nolte, C.; Jackrel, J.; Grob, D. Poisoning due to organophosphate insecticides. Am. J. Med. 50:475–492; 1971.

O'Donoghue, J. L. Neurotoxicity of industrial and commercial chemicals (2 volumes). Boca Raton: CRC Press; 1985.

Osterloh, J. D.; Pond, S. M.; Grady, S.; Becker, C. E. Serum formate concentrations in methanol intoxication as a criterio for hemodialysis. Ann. Intern. Med. 104:200–203; 1986.

Pazderova-Vejlupkova, J. Development and prognosis of chronic intoxication by tetrachlordibenzo-p-dioxin in men. Arch. Environ. Health 36:5–11; 1981.

Pederson, W.; Nygaard, E.; Nielson, O. Solvent-induced occupational myopathy. Occup. Med. 22:603–607; 1980.

Peterson, C. D.; Collins, A. J.; Himes, J. M.; Bullock, M. L.; et al. Ethylene glycol poisoning. N. Engl. J. Med. 304:21–23; 1981.

Petry, H. Polyneuritis durch E 605. Zbl Arbeitsmed 1:86. Cited in Arch. Indust. Hyg. 6:461–470; 1952.

Procktop, L. D.; Alt, M.; Tison, J. Huffer's neuropathy. J. A. M. A. 229:1083–1084; 1974.

Ritchie, J. M. The aliphatic alcohols. In: Gilman, A. G.; Goodman, L. S., eds. The pharmacological basis of therapeutics. New York: MacMillan Publishing Co.; 1975:p. 372–386.

Roe, O. Past, present and future fight against methanol blindness and death. Trans. Ophthalmol. Soc. U.K. 89:235; 1969.

Roe, O. The metabolism and toxicity of methanol. Pharmacol. Rev. 7:399–412; 1955.

Roe, O. Clinical investigations of methyl alcohol poisoning with special reference to the pathogenesis and treatment of amblyopia. Acta. Med. Scand. 113:558–608; 1953.

Rosansky, S. J. Isopropyl alcohol poisoning treated with hemodialysis: Kinetics of isoprophyl alcohol and acetone removal. J. Toxicol. Clin. Toxicol. 19(3):265–270.

Rosenberg, N. L.; Spitz, M. C.; Filley, C. M.; Schaumburg, H. H. Central nervous system effects of chronic toluene inhalation: Clinical, brainstem evokes response, and MRI studies. Neurology 38(suppl.):113–120; 1988.

Rumack, B. H. Methanol. In: Rumack, B. H., ed. Poisindex. Denver: Micromedex; 1976.

Sahenk, Z.; Mendell, J. R.; Couri, D.; Nachman, J. Generalized polyneuropathy: Inhalation of N_{20}. Neurology 28:485–492; 1978.

Sanborr, G. E.; Selhorst, J. B.; Calabrese, V. P.; Taylor, J. R. Pseudotumor cerebri and insecticide intoxication. Neurology 29:1222–1228; 1979.

Sauerhoff, M. W.; Michaelson, I. A. Hyperactivity and brain catecholamines in lead-exposed developing rats. Science 182:1022–1024; 1973.

Senanayake, N. Acute polyneuropathy after poisoning by a new organophosphate insecticide. N. Engl. J. Med. 306:155–170; 1982.

Scharnweber, H. C.; Spears, G. N.; Cowles, S. R. Chronic methyl chloride intoxication in six industrial workers. J. Occup. Med. 16:112–118; 1974.

Schneck, S. A. Methyl alcohol. In: Vinken, P. J.; Bruyn, G. W., eds. Handbook of clinical neurology: intoxications of the nervous system. Amsterdam: North-Holland Publishing Company; 1979:p. 351–360.

Seage, A. J.; Bruns, M. W. Pulmonary edema following exposure to trichlorethylene. Med. J. Aust. 2:484–487; 1971.

Smith, J. S.; Brandon, S. Morbidity from acute carbon monoxide poisoning at three-year follow-up. Br. Med. J. 1:318–321; 1973.

Spencer, P. S.; Couri, D.; Schaumburg, H. H. N-hexane and methyl n-butyl ketone. In: Spencer, P. S.; Schaumburg, H. H., eds. Experimental and clinical neurotoxicology. Baltimore: Williams and Wilkins; 1980: p. 456–475.

Spencer, P. S.; Schaumburg, H. H. Organic solvent neurotoxicity. Facts and research needs. Scand. J. Work Environ. Health (Suppl.):53–60; 1985a.

Spencer, P. S.; Schaumburg, eds. Experimental and clinical neurotoxicology. Baltimore: Williams and Wilkins; 1985b.

Stewart, A.; Witts, L. J. Chronic carbon tetrachloride intoxication. Br. J. Indust. Med. 1:11–40; 1944.

Stoller, A.; Krupinski, J.; Christophers, A. J.; Blanks, G. K. Organophosphorus insecticides and major mental illness. An epidemiological investigation. Lancet 1:1387–1389; 1965.

Taylor, J. R.; Calabrese, V. P. Organochlorine and other insecticides. In: Vinken, P. J.; Bruyn, G. W., eds. Handbook of clinical neurology: intoxications of the nervous system. Amsterdam: North-Holland; 1979:p. 391–455.

Taylor, J. R.; Selhorst, J. B.; Houff, S. A.; Martinez, A. J. Chlordecone intoxication in man. Neurology 28:626–629; 1978.

Togoni, C.; Bonaccorsi, A. Epidemiological problems with TCDD (A Critical Review). Drug Metab. Rev. 13:447–469; 1982.

Towfighi, J.; Gonatas, N. K.; Pleasure, D.; Cooper, H. S.; McCree, L. Glue sniffer's neuropathy. Neurology 26:238–242; 1976.

Vale, J. A.; Widdop, B.; Bluett, N. H. Ethylene glycol poisoning. Postgrad Med. J. 52:598–603; 1976.

Valpey, R.; Sumi, S. M.; Compass, M. K.; Goble, G. J. Acute and chronic progressive encephalopathy due to gasoline sniffing. Neurology 281:507–511; 1978.

Westerman, M. P.; Pfitzer, E.; Ellis, L. D.; Jensen, W. N. Concentration of lead in bones in plumbism. N. Engl. J. Med. 273:1246–1248; 1965.

Winek, C. L.; Wecht, C. H.; Collom, W. D. Toluene fatality from glue sniffing. Pa. Med. 71:81–84; 1968.

Wislicki, L. Differences in the effect of oximes on striated muscle and respiratory centre. Arch. Int. Pharmacodyn. Ther. 129:443–448; 1960.

Yamamura, Y. N-hexane polyneuropathy. Folia Psychiatrica et Neurologica Japonica 23:45–48; 1969.

39

Heavy Metals and Neurological Disease

ANTHONY J. WINDEBANK

The metals discussed in this section are not known to have biological functions. They are all toxic to some extent, a predominant clinical effect of many is nervous system toxicity. The type of toxicity depends on the type and the chemical state of the metal. Route of ingestion or absorption also may be important in determining toxicity. All of the toxic effects of metals are reversible to some extent, and therefore it is important to determine whether a specific neurological manifestation can be linked to a metal exposure or an elevated body fluid level of a particular metal. The ubiquitous "urine heavy metal screen" obtained for every patient with peripheral neuropathy must detect hundreds of false-positive results for every true metal-induced neuropathy. Detailed reviews of the diagnostic strategies for metal toxin-induced neurological disease are available elsewhere (Windebank et al. 1984; Windebank 1987). For this review, it is assumed that a well-established diagnosis of metal-induced neurological disease has been made.

Arsenic

Acute arsenic poisoning is almost always due to ingestion of inorganic arsenic salts. Many cases are intentional poisoning. After a single significant dose of arsenic, the patient usually develops nausea, vomiting, and diarrhea within hours. This may proceed rapidly to death due to circulatory collapse. If the dose is not lethal, the gastrointestinal symptoms may subside within a few days. Massive but not fatal poisoning may result in a transient generalized encephalopathy with somnolence, stupor, or delirium. This delirious state may be accompanied by paranoid ideation (Danan et al. 1984; Windebank et al. 1984). These central effects clear rapidly if repeated exposure does not occur.

Skin changes, including palmar erythema and scaling, appear within days of exposure and often accompany the onset of peripheral nerve symptoms. The appearance of neuropathic symptoms may be delayed for several weeks after a single acute exposure (LeQuesne and McLeod 1977). The peripheral neuropathy may progress for days to weeks and then stabilize. It is characterized by the distal to proximal progression of pain, sensory loss, and weakness that begins in the lower limbs and spreads to the upper limbs. Maximum deficit usually is established after approximately 6 weeks.

Because of the temporal profile, the initial clinical diagnosis often is acute inflammatory demyelinating polyradiculoneuropathy

571

(Guillain-Barré syndrome). Arsenic neuropathy is distinguished by the degree of distal pin, the proximal to distal progression, and the skin and nail changes. The diagnosis should be further suspected if the cerebrospinal fluid (CSF) protein level is close to normal (<100 mg/dl) and the electrophysiological picture suggests an axonal neuropathy.

The diagnosis is established by the observation of elevated arsenic excretion in the urine. Blood arsenic levels increase acutely but decrease rapidly, with a half-life of 6 hours. Urine levels may be extremely high (>10,000 μg/24 hours) for the first few days after ingestion but may fall to normal within several weeks (half-time for excretion is 6–8 days). If the patient is seen weeks after the putative exposure, then estimation of hair or nail arsenic content is useful. This estimation is particularly useful in the patient with a borderline elevation of urine arsenic levels (25–200 μg/24 hours). Such levels may be induced by ingestion of seafood during or immediately before urine collection. The arsenic from seafood is pentavalent, relatively nontoxic, and not deposited in the keratin of hair and nails.

If arsenic poisoning is diagnosed within a few days of exposure, treatment with chelating agents, such as British antilewisite (BAL) or EDTA, may be considered. However, if the acute systemic illness has resolved and the neuropathy is not progressing, there is probably no advantage to trying to increase the already rapid excretion rate.

The prognosis for recovery of the neuropathy depends on the severity of nerve damage. Electrophysiological and biopsy studies confirm that the major pathological process in nerve is axonal degeneration (Dyck et al. 1968; LeQuesne and McLeod 1977). The process of recovery occurs over months to years. Maximal recovery may not be achieved for 3 to 4 years, and in severe cases, it is incomplete.

Lead

There are three main types of lead-induced neurological disease: (1) lead encephalopa-thy in infants due to ingestion of inorganic lead compounds; (2) lead encephalopathy in adults due to ingestion of organic lead compounds; and (3) lead neuropathy due to ingestion of inorganic lead compounds.

Lead Encephalopathy in Infants

This encephalopathy occurs in infants who are just becoming mobile and is almost invariably associated with pica. The lead is ingested from paint flakes and dust in older houses. For unknown reasons, the encephalopathy usually presents during the early summer months. Prodromal symptoms include gastrointestinal upset and headache, which may be followed rapidly by behavioral alterations, lethargy, confusion, and coma. Physical signs during the acute illness include confusion, ataxia, and convulsions with increased intracranial pressure. The individual rate of progression through the symptoms appears to depend on the quantity of lead ingested. The mortality for children who develop frank encephalopathy with raised intracranial pressure has been high (10% to 50%) (Sachs et al. 1970). It is probable that with modern intensive care techniques and intracranial pressure monitoring, this mortality rate is lower; however, there are no recent studies to confirm this.

Acute treatment involves general support, aggressive management of raised intracranial pressure, and measures directed toward promoting lead excretion. In children with encephalopathy, BAL should be given intramuscularly (75 mg/m^2 every 4 hours) for 4 to 5 days, accompanied by intravenous EDTA (1500 mg/m^2 per day) for the same period (Piomelli et al. 1984). Convulsions should be treated with phenobarbital. Many of the children are anemic and iron deficient. This enhances the absorption of lead and should be investigated and treated appropriately. Subclinical renal impairment usually recovers without long-term sequellae.

Recovery from the acute illness is good, but 20% of children who survive may have a chronic seizure disorder or some impairment of intellectual function. Long-term follow-up is extremely important to identify and manage these potential deficits. Also, it is important to monitor blood lead levels serially. The lead exposure is environmental and tends to be more prevalent in children with fewer socioeconomic advantages. These children are at risk for significant reexposure.

Lead Encephalopathy in Adults

Inorganic lead salts in old paint and in environmental dust due to lead supplemented gasoline do not produce encephalopathy in adults because ionic lead does not readily cross the blood–brain barrier. Organic lead compounds, such as tetraethyl lead, rapidly enter the brain after ingestion and produce a severe encephalopathy that has many features resembling the childhood disease. It is managed in the same way. Most cases are due to intentional ingestion, and if large quantities are absorbed, the prognosis is poor.

Lead Neuropathy

Classic lead neuropathy due to chronic exposure to inorganic salts of lead is extremely rare. This clinical entity was common when silver was mined without protective industrial hygiene. Lead compounds commonly were found with silver deposits. The reduced use of lead in paints and plumbing materials also has contributed to the disappearance of this type of neuropathy. Occasionally, cases are seen, and the criteria for diagnosis have been described in detail elsewhere (Windebank 1987).

The neuropathy may be unusual because the upper limbs may be affected first with prominent asymmetric weakness of the wrist and finger extensor muscles. Peroneal weakness may be the first sign in children who develop lead neuropathy (Sachs et al. 1970; Seto and Freeman 1964). Identification and removal of the source of exposure are the most important aspects of treatment. Improvement of the neuropathy is the rule, but it may not be complete in severe cases.

Thallium

Thallium was used therapeutically as a depilatory in the treatment of ringworm in chil-

dren. Munch (1934) pointed out that the incidence of neurological side effects was extremely high. These salts are no longer used therapeutically, but they are still available in pest control products. As with arsenic, most cases of thallium intoxication are due to poisoning or intentional self-administration. The clinical effects of thallium administration are similar to those of arsenic poisoning. Severe systemic illness occurs within hours of ingestion. This may involve vomiting, diarrhea, renal impairment, circulatory collapse, and abnormalities of liver function. If the patient survives this acute illness, neuropathic symptoms begin distally after several days and spread proximally over days to weeks. The most distinctive feature of thallium poisoning is hair loss, which begins 2 to 3 weeks after ingestion. This hair loss occurs over the whole body.

The best treatment of thallitoxicosis is unclear. The acute systemic illness often overshadows the neurological manifestations. Thallium appears to be less toxic than arsenic to sensory neurons *in vitro* (Windebank 1986). This may explain why death from systemic collapse occurs rather than survival with peripheral neuropathy (Cavanagh et al. 1974; Davis et al. 1981). Use of chelating agents has not been helpful. There is some evidence that Tl^+ interferes with K^+ metabolism, and some authors have suggested the use of intravenous potassium to accelerate thallium excretion. There is no good evidence that this helps, and it may be impractical or dangerous with renal failure. A more rational and safer therapy involves oral or nasogastric administration of potassium ferric ferrocyanide II (Prussian blue). This compound is not absorbed by the gut, but the K^+ is replaced by Tl^+, which is strongly bound and thus unabsorbable. This prevents further absorption of thallium and may lead to increased excretion through the gastrointestinal tract.

Very few cases of thallium poisoning are recognized before death, and it is probable that cases are not diagnosed even at autopsy (Cavanagh et al. 1974). If the patient survives the acute systemic illness and neurological signs develop, these may not be reversible. Good evidence suggests that thallium produces both peripheral and central (posterior column) degeneration in humans (Cavanagh et al. 1974). This central injury is unlikely to recover.

Gold

Gold salts are used in the treatment of rheumatoid arthritis. Evidence from animal (Pollock et al.) and human studies (Meyer et al. 1978; Katrack et al. 1980) show that these salts may produce a distal symmetric neuropathy in some people. It may be difficult to distinguish from neuropathy caused by the underlying rheumatoid disease. The neuropathy may be accompanied by electrophysiological evidence of myokymia. Observation of this phenomenon is distinctive for gold intoxication. If the diagnostic question cannot be resolved, then the therapy should be discontinued. Gold-induced neuropathy appears to be readily reversible.

Mercury

The type of clinical disorder produced by mercury intoxication depends on the molecular form of the element. Chronic exposure to very low doses of short-chain alkyl mercury compounds (ethyl mercury and methyl mercury) results in prominent irreversible central nervous system damage. This includes constriction of visual fields, ataxia, dysarthria, decreased hearing, tremor, and mental impairment. Several major outbreaks of organic mercury poisoning have occurred as a result of eating grain treated with mercury-containing fungicides. The first major epidemic was recognized when inorganic mercury salts in industrial effluent were discarded in Minamata Bay, Japan. These compounds were metabolized to organic mercurials by marine microorganisms, and then they entered the food chain, resulting in hundreds of deaths and thousands of cases of severe permanent neurological disability (Kurland et al. 1960).

Elemental mercury exposure usually occurs through inhalation of mercury vapor. Metallic mercury is quite volatile at room temperature and, in an enclosed space, will produce toxic levels of vapor in the air. Chronic exposure may produce depression or other mood changes, tremor, and occasionally a progressive motor neuropathy (Swaiman and Flagler 1971; Windebank et al. 1984). All of these manifestations are reversible if the source of exposure is eliminated.

Other Metals

Aluminum in the water supply has been implicated as a toxin in some individuals on chronic hemodialysis. These patients present with a rapidly progressive encephalopathy. Early signs include episodes of speech arrest, myoclonus, dysarthria, and ataxia. Once the role of aluminum is recognized, reduction in body aluminum load results in some improvement in the symptoms of "dialysis dementia." The incidence of this disorder has decreased since dialysis units started to monitor aluminum levels. Occasionally, patients with relatively normal aluminum balance present with the same clinical picture. Either the disease is multifactorial or other metabolic derangements can produce the same disorder.

Bismuth has been reported to produce an encephalopathy. Bismuth salts have been used for years in the treatment of diarrhea. They were most popular in France and Australia, from which come the most extensive series of case reports (Buge and Rancourel 1975; Burns et al. 1974; Supino-Viterbo et al. 1977). The patients usually were given insoluble bismuth salts, such as bismuth subgallate, by mouth but would develop high blood bismuth levels. There were various reasons suggested for the solubilization of the salts in the gastrointestinal tract. The encephalopathy that presented with myoclonic jerks, dysarthria, and ataxia has many features in common with aluminum encephalopathy. Most patients appeared to recover when the bismuth was excreted. In an experimental study, Basinger and colleagues (1983) demonstrated that D-penicillamine is the chelating agent that most rapidly promotes bismuth excretion.

Cadmium produces hemorrhagic lesions

in the brain and spinal ganglia of experimental animals. It is not known to produce any nervous system disease in humans. Itai-itai is a disease produced by chronic cadmium poisoning. It involves severe bone pain that may produce gait disorders but is not due to neurological involvement (Tischner 1980).

Manganese toxicity occurs almost exclusively in people who work in manganese mining, ore extraction, or smelting. There have been a number of detailed reviews concerning the metabolism and clinical manifestations produced by exposure to the dust of this metal (Mena 1979; Piscator 1979). The major features are pulmonary irritation, behavioral changes, mood changes, and an extrapyramidal syndrome. The latter is characterized by loss of postural reflexes, rigidity, masked facies, dysarthria, tremor, and occasionally dystonic posturing. The early manifestations may be reversed by removing the patient from the exposure. Once the extrapyramidal syndrome is established, it usually is not reversible. It is responsive to treatment with L-dopa (Mena 1979).

Tin is not toxic in metallic or inorganic forms. Organic tin compounds, however, produce central and peripheral nervous system disorders in humans and experimental animals. The most usual toxic forms are the short-chain alkyl derivatives (e.g., dimethyl and trimethyl tin and tetraethyl tin). These compounds have widespread industrial use in plastic manufacturing and as catalysts. In animals, organotin compounds produce central nervous system demyelination in low doses and peripheral nervous system demyelination in high doses (Blake et al. 1981). A large epidemic of organotin poisoning occurred in France in 1953. Approximately 210 people took a tin-containing remedy (probably triethyl tin) for treatment of skin infections (Alajouanine et al. 1958). One hundred people died, and autopsy demonstrated widespread intramyelinic edema (Cossa and Radermecker 1958; Foncin and Gruner 1979). Trimetyl tin intoxication (Besser et al. 1987) appears to cause neuronal damage in the central nervous system, producing a limbic–cerebellar syndrome.

Severe intoxication produces death. Patients with significant neurological deficit from trimethyl or triethyl tin do not recover, while those with only mild behavioral changes improve to normal. Besser and colleagues (1987) treated three of their six patients with plasma exchange, but it is not clear if this alters the course of the disease.

Acknowledgments

The expert secretarial assistance of Linda A. Goldbeck is most gratefully acknowledged.

References

Alajouanine, T.; Derobert, L.; Thieffry, S. Etude clinique d'ensemble de 120 cas d'intoxication par les sels organiques d'etain. Rev. Neurol. (Paris) 98:85–96; 1958.

Basinger, M.; Jones, M.; McCroskey, S. Antidotes for acute bismuth intoxication. J. Toxicol. Clin. Toxicol. 20(2):159–165; 1983.

Besser, R.; Krämer, G.; Thümler, R.; Bohl, J.; Gutmann, L.; Hopf, C. Acute trimethyltin limbic-cerebellar syndrome. Neurology 37:945–950; 1987.

Blake, W. D.; Krigman, M. R.; Thomas, D. J.; Mushak, P.; Morell, P. Effect of triethyl tin on myelination in the developing rat. J. Neurochem. 36:44–52; 1981.

Burns, R.; Thomas, W.; Barron, V. J. Reversible encephalopathy possibly associated with bismuth subgallate ingestion. Br. Med. J. 1:220–223; 1974.

Buge, A.; Rancure, G. Les encéphalopathies aigues myocloniques par les sels oraux de bismuth. Rev. Méd. (Paris) 24:1668–1674; 1975.

Cavanagh, J. B.; Fuller, N. H.; Johnson, H. R. M.; Rudge, P. The effects of thallium salts, with particular reference to the nervous system changes. Q. J. Med. 43:293–319; 1974.

Cossa, P.; Radermecker, J. Encéphalopathies toxiques au Stalinon (aspects anatomocliniques et electroencephalographiques). Rev. Neurol. (Paris) 98:97–109; 1958.

Danan, M.; Dally, S.; Conso, F. Arsenic-induced encephalopathy. Neurology 34:1524; 1984.

Davis, L. E.; Standefer, J. C.; Kornfeld, M.; Abercrombie, D. M.; Butler, C. Acute thallium poisoning: toxicological and morphological studies of the nervous system. Ann. Neurol. 10:38–44; 1981.

Dyck, P. J.; Gutrecht, J. A.; Bastron, J. A.; Karnes, W. E.; Dale, A. J. D. Histologic and teased fiber measurements of sural nerve in disorders of lower motor and primary sensory neurons. Mayo Clin. Proc. 43:81–123; 1968.

Foncin, J. F.; Gruner, J. E. Tin neurotoxicology. In: Vinken, P. J.; Bruyn, G. W., eds. Handbook of clinical neurology. Amsterdam: Elsevier Science Publishers; 1979: p. 279–290.

Katrak, S. M.; Pollock, M.; O'Brien, C. P.; Nukada, H.; Allpress, S.; Calder, C.; Palmer, D. G.; Grennan, D. M.; McCormack, P. L.; Laurent, M. R. Clinical and morphological features of gold neuropathy. Brain 103:671; 1980.

LeQuesne, P. M.; McLeod, J. G. Peripheral neuropathy following a single exposure to arsenic. J. Neurol. Sci. 32:437–451; 1977.

Mena, I. Manganese poisoning. In: Vinken, P. J.; Bruyn, G. W., eds. Handbook of clinical neurology; Amsterdam: Elsevier Science Publishers; 1979: p. 217–237.

Meyer, M.; Haecki, M.; Ziegler, W.; Forster, W.; Schiller, H. H. Autonomic dysfunction and myokymia in gold neuropathy. In: Canal, N.; Pozza, G., eds. Peripheral neuropathies. Proceedings of the international symposium on peripheral neuropathy, Milan, 1978. Amsterdam: Elsevier Science Publishers; 1978: p. 475–480.

Munch, J. C. Human thallotoxicosis. J.A.M.A. 102:1929–1934; 1934.

Piscator, M. Manganese. In: Friberg, L.; Nordberg, G. F.; Vouk, V. B. eds. Handbook on the toxicology of metals; Amsterdam: Elsevier Science Publishers; 1979: p. 485–501.

Sachs, H. K.; Blanksma, L. A.; Murray, E. F.; O'Connell, M. J. Ambulatory treating of lead poisoning: report of 1,155 cases. Pediatrics 46:389–396; 1970.

Seto, D. S. Y.; Freeman, J. M. Lead neuropathy in childhood. Am. J. Dis. Child. 107:337–342; 1964.

Spino-Viterbo, V.; Sicard, C.; Risvegliato, M.; Rancure, G.; Buge, A. Toxic encephalopathy due to ingestion of bismuth salts: clinical and EEG studies of 45 patients. J. Neurol. Neurosurg. Psychiatry 40:748–752; 1977.

Swaiman, K. F.; Flagler, D. G. Penicillamine therapy of the Guillain-Barré syndrome caused by mercury poisoning. Neurology 21:456–457; 1971.

Tischner, K. Cadmium. In: Spencer, P.; Schaumburg, H. eds. Experimental and clinical neurotoxicology; Baltimore: Williams and Wilkins; 1980: p. 348–355.

Windebank, A. J. Specific inhibition of myelination by lead in vitro; comparison with arsenic, thallium, and mercury. Exp. Neurol. 94(1):203–212; 1986.

Windebank, A. J. Peripheral neuropathy due to chemical and industrial exposure. In: Matthews, W. B., ed. Handbook of clinical neurology; Amsterdam: Elsevier Science Publishers; 1987: p. 263–292.

Windebank, A. J.; McCall, J. T.; Dyck, P. J. Metal neuropathy. In: Dyck, P. J.; Thomas, P. K.; Lambert, E. H.; Bunge, R. P., eds. Peripheral neuropathy; Philadelphia: W. B. Saunders; 1984: p. 2133–2161.

40

Neurological Effects of Chronic Alcoholism

ELLIOTT L. MANCALL

The abuse of alcohol, particularly ethyl (grain) alcohol, leads directly or indirectly to a remarkable range of neurological disorders with widely varying severities and outlooks (Adams and Victor 1989; Charness et al. 1989; Victor and Adams 1953). Many of these disorders unfortunately are treatable to only a limited extent. Although rarely fatal, they are responsible for substantial morbidity and at times permanent physical or mental incapacity.

Disorders of the central (CNS) and peripheral (PNS) nervous systems and of muscle as observed in the chronic alcoholic may be divided into those due to the direct toxic effect of ethyl alcohol, those reflecting withdrawal from alcohol, the "abstinence syndromes," and those due to concomitant and usually protracted nutritional depletion. Manifestations of the direct toxicity of alcohol may appear in occasional or accustomed drinkers if a sufficient amount of alcohol has been consumed. On the other hand, abstinence syndromes tend to appear particularly in the periodic (binge or spree) drinker, while nutritional disorders are most prominent in the steady and chronic drinker. These distinctions are useful clinically, despite their somewhat artificial character, and they are used in this chapter.

Many other neurological disorders that are not classifiable in this way occur with special frequency in an alcoholic population. Among these are the effects of craniocerebral trauma, most importantly subdural hematoma, and the many complications of liver failure, including inter alia, hepatic encephalopathy, and myelopathy; these are covered elsewhere in this book. The fetal alcohol syndrome represents a special case of direct ethanol toxicity and will be reviewed in that category. Only disorders complicating the preferential use of ethyl alcohol are considered. The consequences of ingestion of other alcohols, by accident or design, are beyond the scope of this presentation. The management of alcoholism, in a generic sense, is far removed from the parameters of this discussion. The interested reader is referred to other reviews, such as that by Peachey and Annis (1984), for a contemporary critique of the issues involved.

Toxic Effects of Ethyl Alcohol

Inebriation

The clinical manifestation of acute alcoholic intoxication or inebriation requires little elaboration. Alterations of mood, judgement, and alertness; blurring of vision; vertigo; nystagmus; dysarthria; incoordination of the limbs; and truncal ataxia are among the more common expressions of acute intoxication. Factors influencing the onset and severity of this syndrome include not only the amount of alcohol consumed, but also the rate of consumption, the degree of habituation or tolerance, the concomitant use of other drugs and food, genetic predisposition, and psychological attitude and expectations at the time of drinking. The extent of habituation, defining the difference between the occasional and the accustomed drinker, is of particular importance but is poorly understood. Modifications in transport of alcohol across the intestinal mucosa, induction of hepatic enzyme systems, changes in neurotransmitter interrelationships, and morphological alterations of the neuronal lipoprotein membrane complex with resultant modifications in ion flux are among the many mechanisms that have been implicated as responses to acute or chronic ingestion of alcohol. However, the precise pathophysiological mechanism(s) basic to the development of tolerance remains unclear.

Several specific issues require particular attention concerning the inebriated person. Most important is the development of stupor and coma. Miles (1922) has pointed out that in the unaccustomed drinker, stupor appears at a blood-alcohol level of 300 mg/dl, deep coma supervenes at a level of 400 mg/dl, and death from respiratory depression occurs when the level approaches 500 mg/dl. The accustomed drinker may tolerate these high levels more effectively. In any case, the development of coma demands urgent therapeutic intervention (Mancall and Silver 1986). Hemodialysis may be life-saving. The use of the opiate antagonist naloxone has been reported to reverse coma in such patients (Lyon and Antony 1982; Mackenzie 1979); unfortunately, this is not universal, and in some studies the use of naloxone has failed to reverse the signs of intoxication (Lancet editorial 1983). Gastric lavage may be helpful in acute cases. Nonspecific stimulants, such as caffeine, are of dubious benefit. Hypertonic glucose, always administered with thiamine to preclude the development of Wernicke's encephalopathy, may be beneficial but only in the few cases in which hypoglycemia is documented.

The "alcoholic blackout" occasionally is encountered following acute ingestion of alcohol. This may be defined as a self-limited and brief period of failure of short-term memory recording, similar clinically to transient global amnesia. The explanation for this defect is not clear. Such episodes, which often are recurrent, generally are evidence of a serious state of alcohol abuse. They require no specific therapy.

Although alcohol ordinarily is considered a sedative or depressing drug, its ingestion, even in small amounts, is sometimes followed by the development of a paradoxical state of excitement with aggressive and often violent behavior, referred to as pathological intoxication. The cause of this idiosyncratic response to ethanol is not known. There is no specific therapy, but a quiet environment and, when needed, restraints and parenteral haloperidol have a place in management. The period of agitation tends to be brief, usually ending within a few hours after a period of deep sleep. Suicide attempts are not infrequent and the patient must be monitored with this in mind.

Myopathy

Acute alcoholic myopathy (rhabdomyolysis) most frequently appears shortly after a period of unusually excessive drinking. Although total agreement is lacking (Laureno 1979), it is widely held (Rubin 1979; Urbano-Marquez et al. 1989) that this acute disorder reflects the direct toxicity of alcohol on skeletal and cardiac muscle. The syndrome is characterized clinically by muscle weakness and pain, involving particularly the proximal muscles of the limbs, symmetrically or asymmetrically, and generally associated with severe muscle tenderness and sometimes myoedema. Myoglobinuria and elevated serum creatine kinase levels typically are encountered. Muscle biopsy demonstrates severe necrosis of muscle fibers, involving particularly type 1 fibers, with subsequent active muscle regeneration. There is no specific therapy. The disorder ordinarily is self-limited if the patient remains abstinent. Unfortunately, return of muscle function is not always complete, especially in people with very severe and widespread muscle destruction. Variable degrees of weakness and atrophy may persist indefinitely.

A more chronic form of myopathic disorder also is encountered in the alcoholic (Mancall and McEntee 1966), again involving primarily the proximal musculature but associated with less dramatic necrotic changes within muscle, a predominant type 2 fiber atrophy, absence of myoglobinuria, and only a modest rise in serum creatine kinase levels. It is uncertain whether this more indolent myopathy also reflects the direct toxicity of ethanol or, alternatively, chronic malnutrition. In this more chronic form, continued abstinence and improved nutrition ordinarily lead to restoration of function. The importance of an associated neuropathy in the pathogenesis of the acute and chronic forms of myopathy should not be underestimated.

Fetal Alcohol Syndrome

The offspring of alcoholic women are known to exhibit a constellation of neurological abnormalities, including growth and intellectual retardation, hyperactivity, microcephaly, craniofacial anomalies, cleft palate, dislocations of the hip and a variety of other arthropathies, and many other congenital anomalies affecting the skin, heart, and urogenital system. Although the precise pathogenetic mechanisms underlying fetal alcohol syndrome (FAS) remain incompletely established, it appears to be a dose-dependent phenomenon, reflecting the quantity of alcohol consumed during pregnancy by the mother. It occurs in more than one third of the offspring of serious alcoholics; the children of women who drink less may evidence portions of this syndrome, called fetal alcohol effects. The possible etiologic significance of maternal undernutrition, smoking, and concomitant drug usage remains to be clarified, but it is likely that they play a less

important role than that of ethanol itself (Charness et al. 1989).

The 10-year follow-up study by Streissguth and coworkers (1985) indicates that growth deficiency persists in FAS despite some catch-up phenomena, although the typically emaciated appearance of the younger child may disappear by adolescence. Some modifications of the craniofacial anomalies have been observed, and cardiac defects, such as atrial or ventricular septal defects and patent ductus arteriosus, assume less significance or resolve spontaneously. Osseous or dental anomalies may worsen and require surgical intervention. Defective cognitive development and attention defects remain serious problems, often attended by major social and emotional dislocations. The degree of intellectual impairment in any child appears to correlate with the severity of the malformations and the degree of growth retardation; the best ultimate predictive factor appears to be the severity of the maternal alcoholism.

Effects of Withdrawal

Following withdrawal, whether relative or complete, a sequence of neurological disorders appears, beginning with tremulousness and hallucinations, proceeding to seizures, and in approximately 15% of cases, to overt delirium tremens (Victor and Adams 1953). The first of these appears relatively early after reduction or elimination of ethanol, whereas delirium tends to develop later in withdrawal. Although common in combination, each of these disorders may be encountered independently; because the therapeutic and prognostic implications of each disorder vary, it is appropriate to treat them separately.

The use of psychoactive substances has been widely advocated in the management of alcohol withdrawal in general (Holloway et al. 1984; Mancall and Silver 1986) and is said to have a significant impact on outcome. Benzodiazepines, particularly chlordiazepoxide (Librium), are most widely used and are claimed to effect prognosis significantly by reducing the morbidity and perhaps the mortality associated with these disorders. Unfortunately, chlordiazepoxide has a relatively long half-life (approximately 10 hours), and its metabolites, which have variable half-lives, are biologically active and are characterized by unpredictable but slow elimination; hepatic disease prolongs elimination even further. The resulting potential for benzodiazepine toxicity is significant. Oxazepam (Serax) has a shorter half-life and does not have active metabolites; therefore, it is often preferred in these circumstances. Unfortunately, the majority of benzodiazepines are available only for oral or intravenous use. Neither of these routes is ideal in an agitated negative patient in withdrawal. Because lorazepam (Ativan) is suitable for intramuscular administration, it has been particularly favored by some investigators (Rosenbloom 1986).

Other drugs advocated for use in withdrawal include diazepam (Valium), chlorpromazine (Thorazine), meprobamate, and paraldehyde. Because phenothiazines such as chlorpromazine alter seizure thresholds, they must be used with great caution. The prominence of adrenergic signs in withdrawal, reportedly related to elevated spinal fluid and cerebral levels of norepinephrine (Hawley et al. 1981), has prompted the use of propranolol (Inderal) as well.

Alcoholic Tremulousness

Developing as soon as 6 to 8 hours after withdrawal, alcoholic tremulousness is the earliest and most benign of the abstinence syndromes. The patient exhibits prominent postural tremors of the limbs, with concomitant tremors of the face, tongue, and head. Autonomic manifestations, such as tachycardia and facial flushing, occur along with insomnia, excitement, anorexia, retching, and heightened sensitivity to external stimuli. This disorder requires no specific therapy, but its appearance should alert the observer to the potential for more serious abstinence syndromes. If the patient remains abstinent, the somatic tremors subside without incident within several days, as do the epiphenomena of tachycardia, flush-

ing, and agitation. Many patients experience a curious inner tremor persisting for weeks; an excessive startle response also may linger.

It remains unclear whether this acute but transient movement disorder bears any relationship to the chronic postural tremor exhibited by some alcoholics or to essential tremor (Neiman et al. 1990).

Alcoholic Hallucinosis

Within 12 to 18 hours of withdrawal, as many as 25% of patients exhibit vivid visual, auditory, or mixed hallucinations. Hallucinations are vivid and real to the patient, who often responds in an appropriate manner to these sensory experiences. They usually subside uneventfully without therapy within several days if the patient remains abstinent. In a small proportion of cases, however, the patient lapses into a stage of chronic auditory hallucinosis and may ultimately exhibit features of chronic paranoid schizophrenia. Whether this state of chronic hallucinosis is an independent process or represents a pre-existing but previously inapparent psychosis that has been unmasked by alcohol withdrawal remains unclear.

Withdrawal Seizures

In 12% of hospitalized alcoholics (Victor and Adams 1953), generalized tonic–clonic seizures appear within 18 to 36 hours of withdrawal. Also known as alcoholic epilepsy or rum fits, these tend to occur in isolation or in a brief flurry of several seizures. Status epilepticus may occur but is very uncommon. When focal features are observed, either clinically or electroencephalographically, the possibility of other causes of late-onset seizures must be considered. The risk of developing such seizures apparently increases with repetitive bouts of detoxification (Lechtenberg and Worner 1990). Seizures also may develop during a period of active drinking, and it has been proposed that the relationship of seizures to alcohol use is dose-dependent. It has been suggested that seizures in these circumstances are due to the direct toxicity of alcohol rather than to withdrawal (Ng et al. 1988; Simon 1988); there is little general support for this theory.

The biological mechanism(s) underlying the appearance of withdrawal seizures remains unclear. There is clearly heightened epileptogenic potential during the critical period of withdrawal, as documented, for example, by unusual sensitivity to metrazol and to photic driving. Hypomagnesemia and respiratory alkalosis developing early in withdrawal and, at least in the latter instance, reflecting rebound hyperactivity of the respiratory centers of the brain stem have been implicated. Increased concentration of the excitatory neurotransmitter glutamate, increased number of postsynaptic glutamate-receptor binding sites, depressed inhibitory neurotransmitter gamma-aminobutyric acid (GABA), and altered G proteins and calcium channels also have been proposed to have significant pathogenetic roles (Chernoff 1989; Simon 1988).

Seizures occurring during abstinence are self-limited and generally require no specific therapeutic intervention. If status epilepticus develops, appropriate emergency management is warranted. The routine, chronic use of anticonvulsants to prevent withdrawal seizures in alcoholic patients is unnecessary because these occur only during periods of withdrawal. They commonly follow a period of binge drinking, during which time, the patient is not likely to be therapeutically compliant; they are not likely to occur at other times. The use of benzodiazepines during withdrawal has been considered beneficial in reducing the frequency of seizures. The administration of an intravenous loading dose of phenytoin, followed by regular oral dosages of this agent through withdrawal, also has been advocated.

Delirium Tremens

This syndrome of disordered sensory perceptions, hallucinations, agitation, confusion, insomnia, fever, and autonomic overactivity occurs in 5% to 15% of patients in withdrawal. It generally appears later than the other abstinence syndromes, with a peak at 72 hours after withdrawal. Its onset may be heralded by tremors, hallucinations, and seizures. Delirium tends to begin abruptly and is relatively short-lived; it ends suddenly and in less than 72 hours in more than 80% of the cases (Victor and Adams 1953).

A significant mortality rate is associated with delirium tremens, ranging up to 15% in most reported series. No single cause of death can be identified; intercurrent infections; electrolyte and other metabolic abnormalities; renal, adrenal, or hepatic failure; trauma; and acute peripheral vascular collapse have been implicated in small numbers of cases. Along with appropriate supportive management, early and vigorous use of benzodiazepines substantially reduces the morbidity and the mortality associated with this disease.

Nutritional Disorders

In an alcoholic who drinks steadily for months or years, malnutrition becomes a major clinical threat. In the urban United States, nutritional disorders of the nervous system in adults appear most commonly in the nutritionally depleted alcoholic population. The alcoholic uses the "empty" calories of the ethanol to displace required nutrients from the diet. Although an adequate dietary history is difficult to obtain in the best circumstances and laboratory documentation of such a deficiency even more difficult, one can infer the presence of undernutrition by observing weight loss with atrophy of muscle and subcutaneous tissue, glossitis, cheilosis, angular stomatitis, and nonspecific laboratory abnormalities, such as hypochromic anemia and hypoproteinemia. However, such evidence is indirect at best, and firm documentation, especially of a specific vitamin insufficiency, is extremely problematic. Many patients manage remarkably well for extended periods of time despite a meager diet; they may develop neurological symptoms only when burdened with a superimposed metabolic demand, a sequence of events most convincingly documented in the case of Wernicke's encephalopathy.

Among these diseases, several (polyneuropathy, Wernicke-Korsakoff syndrome, amblyopia, pellagra) are of undoubted nutritional origin, with the specific deficiency reasonably well defined. The essential dietary ingredient lacking is a member of the B-vitamin group. Two others (cerebellar degeneration and Marchiafava-Bignami disease) are clearly due to undernutrition, but the nutritional lack has not been established. The role of malnutrition in the pathogenesis of central pontine myelinolysis remains a matter of dispute.

Nutritional Polyneuropathy

Commonly referred to as alcoholic neuropathy or neuritic beriberi, nutritional polyneuropathy is the most common nutritional disorder encountered in the alcoholic population. In the series of Victor and Adams (1953), 70% of their patients were found to have features of peripheral neuropathy, although many were asymptomatic. Some investigators (Behse and Buchthal 1977) have stressed the importance of alcohol *per se* in pathogenesis, with or without attendant malnutrition, but there is little doubt about the fundamental importance of malnutrition in this regard. The use of the term neuritic beriberi connotes a pure thiamine deficiency, but such specificity may not be warranted in these circumstances.

Weakness, numbness, paresthesiae, and pain are the most common complaints in symptomatic individuals. The onset is insidious, and the evolution is subacute. The lower extremities invariably are involved initially; the upper extremities are involved as the disease progresses. Variable motor, sensory, and reflex changes are found on examination, generally in a distal and symmetrical distribution. In milder cases, sensory changes predominate; however, as the disease becomes more severe, motor alterations appear with progressive weakness and atrophy (Hawley et al. 1982). Cranial nerve palsies may appear and autonomic changes are frequent. Chronic skin ulcers and Charcot-like arthropathies develop, and signs of other nutritional diseases of the nervous system are common. Pain, painful paresthesiae, and dysesthesiae are among the most disabling components of the illness.

Abstinence, improved nutrition, and the supplemental use of parenteral vitamins, particularly thiamine, are fundamental to the therapy of this disorders. Analgesics may be required to control pain. Carbamazepine or amytriptyline often are of major assistance, but sympathetic blocks sometimes are required in very severe burning. The judicious use of physiatric techniques is an essential therapeutic ingredient.

Despite adequate therapy, recovery from nutritional polyneuropathy is slow and incomplete; however, some patients experience complete recovery with abstinence, even without vitamin supplementation (Hillbom and Wennberg 1984). Many patients note at least some improvement in sensory complaints relatively soon; at least mild degrees of sensory loss and loss of the tendon reflexes may persist indefinitely (Hawley et al. 1982). In more severely involved patients, recovery is much slower and often unsatisfactory, with significant and sometimes incapacitating sensory and motor residuals observed for months or years.

Wernicke-Korsakoff Syndrome

Commonly considered two different disorders, Wernicke's encephalopathy and Korsakoff's psychosis in fact represent the acute and chronic phases of a single illness due to thiamine deficiency (Victor et al. 1989). The assumption that there is a definite link between the two disorders is based on the observations that the typical psychological changes of Korsakoff's psychosis may be present from the onset of Wernicke's disease or may emerge during treatment; that residual features of a prior bout of Wernicke's encephalopathy often are noted on careful examination of patients with Korsakoff's psychosis; and that the distribution of the pathological changes in the two conditions is identical, the differences noted being explicable on the basis of differences in chronology.

Wernicke's encephalopathy presents in an acute or subacute manner. Characteristic clinical features include the following: An abnormal mental status, variable in expression, occurs. In some patients, apathy, listlessness, a short attention span, and confusion appear; an hallucinatory–confusional state occurs in others; and in others, a serious memory disorder exists. Ophthalmoplegia, especially bilateral sixth nerve palsies, develop. Virtually any combination of altered ocular movements, including gaze palsies, may be encountered. Nystagmus generally is present in horizontal and vertical planes. Finally, ataxia of the trunk and gait occurs, but with little affection of the limbs. Autonomic changes may appear, and although beriberi heart disease is unusual in these patients, acute peripheral vascular collapse may account for sudden death. In the series of patients with Wernicke's encephalopathy by Adams and Victor (1953), a mortality of 17% was observed, generally attributable to sepsis or hepatic failure.

The lesions of Wernicke's encephalopathy follow a regular and stereotypical pattern; particularly affected are the brain stem, cerebellum, and hypothalamus. The lesions, characteristically comprising subtotal tissue necrosis without inflammatory response, are typically found symmetrically disposed in the mammillary bodies, beneath the walls of the third ventricle, in the medial dorsal nucleus of the thalamus, in the periaqueductal grey matter of the mesencephalon, beneath the floor of the fourth ventricle in the pons and medulla, and in the superior vermis of the cerebellum. In terms of clinicopathological correlation, it is logical to consider the lesions in the periaqueductal grey matter and pontine tegmentum responsible for the ophthalmoplegias, affection of the vestibular complex for nystagmus, changes in the vermis for ataxia, and the lesions in the mammillary bodies and thalamus for the alterations in mentation.

Following the parenteral administration of thiamine, the ophthalmoplegias improve within a few hours, and complete resolution occurs in less than 1 week in most patients (Victor et al. 1989). In a smaller number of patients, particularly those exhibiting gaze palsies, 4 weeks of therapy may be required for full recovery. More time is required for resolution of ataxia and nystagmus. Vertical nystagmus disappears in most patients within 4 weeks; horizontal nystagmus, however, remains as a permanent sequella of the disease in more than half of the patients. Truncal and gait ataxia similarly improve with continued treatment with thiamine. Only 38% recover completely, even with 2 months or more of therapy; incomplete recovery is observed in 37% and no improvement in ataxia in fully 27% of the cases. It is thus clear that even prompt institution of parenteral therapy with thiamine does not necessarily result in complete recovery, particularly concerning nystagmus and ataxia.

The alterations in the mental status of Wernicke's encephalopathy respond unpredictably to thiamine. Patients who are simply confused appear to recover completely within several weeks. However, as the sensorium clears, many of these patients exhibit the amnestic dementia that is present in some from the onset and that is the typical and fundamental change of Korsakoff's psychosis.

Korsakoff's psychosis generally is characterized as an amnestic dementia with a severe disorder of memory comprising anterograde and retrograde amnesia. Cognitive functions are relatively better preserved but often are impaired to some extent. Alterations in behavioral patterns also are typical; apathy is the most prominent change. Neurological examination often demonstrates residuals of a previous perhaps unsuspected bout of Wernicke's encephalopathy, particularly horizontal nystagmus and gait ataxia.

The distribution of the pathological changes in Korsakoff's psychosis is identical to that encountered in Wernicke's encephalopathy. As noted, the lesions in the mammillary bodies and perhaps especially the thalamus appear most consistently related to the memory disorder.

In 186 patients who survived an acute bout of Wernicke's encephalopathy studied by Victor and colleagues (1989), 84% exhibited features of Korsakoff's psychosis. In 104 of these patients who were assessed for recovery of mental function for 2 months or more, four groups of roughly equal size could be discerned: 26% demonstrated no recovery; 28%, slight; 25%, significant; and only 21%, complete recovery from the amnestic state. Some patients required continued treatment for many months before gradual, and at times remarkably complete, recovery ensues. McEntee and associates (1980; 1984) have observed that the adrenergic agonist clonidine improves memory performance in Korsakoff's psychosis, and they speculate that damage to adrenergic neural systems in paraventricular brain stem and hypothalamic structures is fundamental to the memory deficit in this disorder.

Apart from the well-defined amnestic syndrome of Korsakoff, a number of chronic alcoholics exhibit a more global dementia, generally referred to as alcoholic dementia. This may represent the clinical expression of the cerebral convolutional atrophy observed pathologically in alcoholics, an atrophic process that appears at least partially reversible with abstinence as determined by magnetic resonance imaging (MRI) and computed tomography (CT) studies (Schroth et al. 1988). It may, however, reflect neurological disorders ranging from pellagra to subdural hematoma to multi-infarct dementia to Alzheimer's disease. Lishman's contention (1981; 1986) that alcoholic dementia is a neglected and perhaps subclinical form of Wernicke's encephalopathy has few adherents.

Nutritional Amblyopia

Also referred to as nutritional retrobulbar neuritis or tobacco–alcohol amblyopia, nutritional amblyopia is well established as a result of deficiency of one or several B vitamins, perhaps particularly thiamine, riboflavin, or cobalamine (Victor et al. 1960). This is an acutely or subacutely evolving disorder characterized by progressive and bilateral impairment of visual acuity and of color vision. Examination confirms bilateral impairment of visual acuity and of color vision, always with symmetrical central, cecocentral, or paracentral scotomata. In the early

stages of the disease, the fundi appear normal, although optic atrophy may develop later. Pathologically one observes a symmetrical degeneration of the visual conducting system in the optic nerves, chiasm, and tracts, with involvement primarily of the papillomacular fiber bundles. It is clearly distinct from the retinal degeneration of methyl alcohol intoxication.

Treatment with B vitamins, when introduced promptly, results in complete return of visual acuity and fields. If the disease is not recognized and treatment is delayed, permanent blindness and optic atrophy result. At times, irreversible visual loss develops with remarkable speed.

Pellagra

Pellagra is due to a deficiency of nicotinic acid or its immediate precursor, tryptophan. The fully developed syndrome of diarrhea, dermatitis, and dementia as recognized in the early part of the century is uncommonly encountered today, particularly in the West. The alcoholic population represents the most consistent reservoir of this disorder (Serdaru et al. 1988). Neurological alterations include irritability, confusion, insomnia, dementia, and amnesia. Extrapyramidal or cerebellar signs occur, and occasionally the optic nerves are affected. A polyneuropathy is common, and features of a myelopathy may appear. Widespread chromolytic changes may be found pathologically throughout the neuraxis, most prominently in Betz's cells of the motor cortex, and symmetrical degenerative changes may be found in the posterior and lateral columns of the spinal cord. Treatment with niacin often is unsatisfactory, and it has been suggested (Victor and Adams 1956) that pyridoxine deficiency and possibly deficiencies of other B vitamins may play a significant pathogenetic role.

Nicotinic acid deficiency encephalopathy is a term applied to a poorly understood syndrome involving elderly patients and characterized by confusion, stupor, and rigidity. Infantile reflexes, such as forced grasping and sucking, also are common. The disorder responds promptly to the administration of nicotinic acid. Neglected in medical writings for many years, it has recently resurfaced (Lishman, 1981) as another, generally overlooked, form of abnormal mentation in alcoholics.

Cortical Cerebellar Degeneration

"Alcoholic" cerebellar degeneration is a frequent complication of chronic alcoholism (Victor et al. 1959) and represents the most common of the acquired cerebellar degenerations of adulthood. It develops after long-standing drinking and undernutrition. Ordinarily, patients present with ataxia of the trunk and the gait and incoordination of the lower extremities as demonstrated, for example, with the heel–knee–shin test. The symptoms are insidious in onset and slowly progressive; evolution of the disease often takes weeks to months.

On examination, patients demonstrate a serious ataxia of the trunk and of gait and incoordination of the legs. The upper extremities are spared, and little if any dysarthria or other "cerebellar signs" are seen. The pattern of clinical abnormalities remains remarkably fixed in extent, without regard to duration of the illness.

Just as the clinical symptoms and signs of alcoholic cerebellar degeneration are restricted in distribution, the pathological changes also are anatomically restricted in most cases. Alterations predominate in the superior vermis and the anterior and superior portions of the cerebellar hemispheres; occasionally, changes are found elsewhere in the cerebellar cortex, and secondary changes may be observed in the vicinity of the deep cerebellar nuclei and in the olivary complex of the medulla. All neurocellular elements of the cerebellar cortex may be involved, but Purkinje cells appear particularly vulnerable. Pathological features of other nutritional diseases, especially Wernicke's encephalopathy, may be encountered.

This form of cortical cerebellar degeneration is of nutritional origin. Its appearance has been well documented with serious nu-tritional depletion without the use of alcohol (Mancall and McEntee 1966). The coexistence of pathological features of Wernicke's encephalopathy in many cases and the prominence of cerebellar ataxia in otherwise uncomplicated instances of Wernicke's encephalopathy suggest that the two diseases are intimately linked. It is possible that thiamine deficiency is the basis for both disorders (Adams and Victor 1989).

Abstinence, improved nutrition, and supplemental vitamins contribute to modest improvement in the cerebellar deficit in these patients; usually an extended period of time is required, and such improvement possibly reflects improvement in other disorders that may contribute to ataxia, such as polyneuropathy. Unfortunately, such improvement is extremely limited, and all patients are left with significant cerebellar ataxia indefinitely.

Marchiafava-Bignami Disease

This rare disorder, once thought to be a pathological curiosity, is increasingly identified using the CT and MRI (Kawamura et al. 1985). Originally, it was thought to occur in middle aged or elderly Italian men who were addicted to drinking crude red wine; with time, it has become abundantly clear that the most important pathogenetic factor is severe and chronic undernutrition (Koeppen and Barron 1978). The specific dietary lack has not been identified.

Unlike other nutritional disorders of the nervous system, the features of Marchiafava-Bignami disease are not stereotypical. The clinical sympatomatology includes dementia, psychosis, asphasia, seizures, tremor, paralysis, hypertonia, and coma. The course of the disease often is phasic. Some patients have exhibited complete remissions, but overall, the disease is progressive, usually leading to death in less than 2 years. Pathological changes comprise symmetrical degeneration of myelin with relative preservation of axons in the corpus callosum and other commissural bundles. The cerebral white matter may be involved, and similar lesions are found in the middle cerebral peduncles, the optic chiasm, and the posterior columns of the spinal cord.

No specific treatment is available for this disease, which appears to lead to death in all recognized cases, with rare exceptions (Leventhal et al. 1965).

Central Pontine Myelinolysis

Central pontine myelinolysis (CPM) (Adams et al. 1959) is a dramatic disease characterized clinically by the rapid development of a flaccid quadriplegia with bulbar palsy, with preserved consciousness and sensation. Movements of the eyelids and globes are intact in most instances. Nystagmus has been noted. If the patient survives for more than several days, spasticity and hyperreflexia may ensue. It is likely that mentation is intact, although difficulties in communication make cognitive function extremely difficult to assess. Neuroradiological and pathological studies have documented the existence of asymptomatic cases; the size of the pontine lesion is apparently the principle determinant of symptomatic expression.

The classic pathological change is symmetrical demyelination in the basis pontis, beginning in the midline and spreading symmetrically to either side. Axons traversing the lesion are relatively well preserved, as are the intrinsic nerve cells of the basis pontis. Extrapontine lesions have been described in areas as diverse as the subcortical white matter, basal ganglia and thalamus, cerebellar white matter, and spinal cord; the term extrapontine myelinolysis is sometimes applied to these cases.

Although originally considered a reflection of undernutrition in chronic and debilitated alcoholics in whom the disease was first described, it has become apparent that neither alcohol *per se* nor malnutrition play an exclusive or essential role in pathogenesis. In many cases of CPM, significant hyponatremia is documented, and in many instances, the disease appears *de novo* following rapid correction of such hyponatremia (Ayus et al. 1985; Laureno and Karp 1988; Norenberg et al. 1982; Sterns et al. 1986; Tomlinson et al. 1976). It has been suggested (Ayus et al. 1987) that an increase in serum sodium to normal or hypernatremic levels in the first 48 hours and a change in the serum sodium concentration of more than 25 ml/L in the first 48 hours are important precipitating characteristics. Rapid correction of severe vasopressin-induced hyponatremia in experimental animals has been followed by the development of similar lesions (Laureno 1983). However, the majority of patients with CPM are seriously ill and often are extremely debilitated and undernourished before they develop neurological disease. Some have undergone significant hypoxic episodes; therefore, it is possible that, although not universally applicable, such might represent ancillary factors in pathogenesis. The suggestion that the demyelination reflects an osmotic shift rather than the serum sodium level per se recently has been supported by the observation by McKee and associates (1988) of serum hyperosmolality in all patients who developed CPM in association with severe burns.

The prognosis for survival in patients with CPM who become symptomatic is extremely poor. In asymptomatic cases, the disease does not recognizably contribute to morbidity or mortality.

References

Adams, R. D.; Victor, M.; Mancall E. L. Central pontine myelinolysis. A hitherto undescribed disease occurring in alcoholic and malnourished patients. Arch. Neurol. 81:154; 1959.

Adams, R. D.; Victor, M. Principles of neurology. 4th ed. New York: McGraw Hill; 1989.

Ayus, J. C.; Krothapalli, R. K.; Arieff, A. I. Changing concepts in treatment of severe symptomatic hyponatremia. Am. J. Med. 78:897–902; 1985.

Ayus, J. D.; Krothapalli, R. K.; Arieff, A. I. Treatment of symptomatic hyponatremia and its relations to brain damage. N. Engl. J. Med. 317:1190–1195; 1987.

Behse, F.; Buchthal, F. Alcoholic neuropathy: clinical, electrophysiological, and biopsy findings. Ann. Neurol. 2:95; 1977.

Charness, M. E.; Simon, R. P.; Greenberg, D. A. Ethanol and the nervous system. N. Engl. J. Med. 321:442; 1989.

Hawley, R. J.; Kurtzke, J. F.; Armbrustmacher;

Saini, N.; Manz, H. The course of alcoholic–nutritional peripheral neuropathy. Acta. Neurol. Scand. 66:582; 1982.

Hawley, R. J.; Major, L. F.; Schulman, E. A.; Lake, R. CSF levels of norepinephrine during alcohol withdrawal. Arch. Neurol. 38:289; 1981.

Hillbom, M.; Wennberg, A. Prognosis of alcoholic peripheral neuropathy. J. Neurol. Neurosurg. Psychiatry 47:699; 1984.

Holloway, H. C.; Hales, R. E.; Watanabe, H. K. Recognition and treatment of acute alcohol withdrawal syndromes. Psychiatr. Clin. North Am. 7:729; 1984.

Kawamura, M.; Shiota, J.; Yagishita, T.; Hirayama, K. Marchiafava-Bignami disease: computed tomographic scan and magnetic resonance imaging. Ann. Neurol. 18:103; 1985.

Koeppen, A. H.; Barron, K. D. Marchiafava-Bignami disease. Neurology 28:290; 1978.

Lancet editorial. Naloxone for ethanol intoxication. Lancet 2:145; 1983.

Laureno, R. Letter to the Editor. N. Engl. J. Med. 22:1239; 1979.

Laureno, R. Central pontine myelinolysis following rapid correction of hyponatremia. Ann. Neurol. 13:232–242; 1983.

Laureno, R.; Karp, B. I. Pontine and extrapontine myelinolysis following rapid correction of hyponatremia. Lancet 1:1439; 1988.

Lechtenberg, R.; Worner, T. M. Seizure risk with recurrent alcohol detoxification. Arch. Neurol. 47:535; 1990.

Leventhal, C. M.; Baringer, J. R.; Arnason, B. G. A case of Marchiafava-Bignami disease with clinical recovery. Trans. Am. Neurol. Assoc. 90:87; 1965.

Lishman, W. A. Cerebral disorder in alcoholism: syndromes of impairment. Brain 104:1; 1981.

Lishman, W. A. Alcoholic dementia: a hypothesis. Lancet 1:1184; 1986.

Lyon, L. J.; Antony, J. Reversal of alcoholic coma by naloxone. Ann. Intern. Med 96:464; 1982.

Mackenzie, A. I. Naloxone in alcoholic intoxication. Lancet 1:733; 1979.

Mancall, E. L.; McEntee, W. J. Alterations of the cerebellar cortex in nutritional encephalopathy. Neurology 15:303; 1966.

Mancall, E. L.; McEntee, W. J.; Hirschhorn, A. M.; Gonyea, E. F. Proximal muscular weakness and atrophy in the chronic alcoholic. Neurology 16:301; 1966.

Mancall, E. L.; Silver, P. Alcohol intoxication and withdrawal. In: Johnson, R., ed. Current therapy in neurological disease. Philadelphia: B. C. Decker.

McEntee, W. J.; Mair, R. G. Memory enhancement in Korsakoff's psychosis by clonidine. Ann. Neurol. 7:466; 1980.

McEntee, W. J.; Mair, R. G.; Langlais, P. J. Neurochemical patterns in Korsakoff's psychosis: implications for other cognitive disorders. Neurology 34:648; 1984.

McKee, A. C.; Winkelman, M. D.; Banker, B. Q. Central pontine myelinolysis in severely burned patients: Relationship to serum hyperosmolality. Neurology 38:1211–1217; 1988.

Miles, W. R. The comparative concentrations of alcohol in human blood and urine at intervals after ingestion. J. Pharmacol. Exp. Ther. 20:265; 1922.

Neiman, J.; Lang, A. E.; Fornazzari, L.; Carlen, P. L. Movement disorders in alcoholism: a review. Neurology 40:741; 1990.

Ng, S. K. C.; Hauser, W. A.; Brust, J. C. M.; Susser, M. Alcohol consumption and withdrawal in new-onset seizures. N. Engl. J. Med. 319:666; 1988.

Norenberg, M. D.; Leslie, K. O.; Robertson, A. S. Association between rise in serum sodium and central pontine myelinolysis. Ann. Neurol. 11:128–135; 1982.

Peachey, J. E.; Annis, H. Pharmacologic treatment of chronic alcoholism. Psychiatr. Clin. North Am. 7:745; 1984.

Rosenbloom, A. J. Optimizing drug treatment of alcohol withdrawal. Am. J. Med. 81:901; 1986.

Rubin, E. Alcoholic myopathy in heart and skeletal muscle. N. Engl. J. Med. 301:28; 1979.

Schroth, G.; Naegele, T.; Klose, U.; Mann, K.; Petersen, D. Reversible brain shrinkage in abstinent alcoholics, measured by MRI. Neuroradiology 30:385; 1988.

Serdaru, M.; Hausser-Hauw, C.; Laplane, D.; Buge, A.; Castaigne, P.; Goulon, M.; Lhermitte, F.; Hauw, J. J. The clinical spectrum of alcoholic pellagra encephalopathy. A retrospective analysis of 22 cases studied pathologically. Brain 111:829–842; 1988.

Simon, R. P. Alcohol and seizures. N. Engl. J. Med. 319:715; 1988.

Sterns, R. H.; Riggs, J. E.; Schochet, S. S. Osmotic demyelination syndrome following correction of hyponatremia. N. Engl. J. Med. 314:1535–1542; 1986.

Streissguth, A. P.; Clarren, S. K.; Jones, K. Natural history of the fetal alcohol syndrome: a 10 year follow-up of eleven patients. Lancet 2:85; 1985.

Tomlinson, B. E.; Pierides, A. M.; Bradley, W. G. Central pontine myelinolysis: Two cases with associated electrolyte disturbance. Q. J. Med. 45:373; 1976.

Urbano-Marquez, A.; Estruch, R.; Navarro-Lopez, F.; Grau, J. M.; Mont, L.; Rubin, E. The effects of alcoholism on skeletal and cardiac muscle. N. Engl. J. Med. 320:409; 1989.

Victor, M.; Adams, R. The effect of alcohol on the nervous system. Proc. Assoc. Res. Nerv. Ment. Dis. 32:526; 1953.

Victor, M.; Adams, R. D. Neuropathology of experimental vitamin B-6 deficiency in monkeys. Am. J. Clin. Nutr. 4:346; 1956.

Victor, M.; Adams, R. D.; Collins, G. H. The Wernicke-Korsakoff syndrome and related neurologic disorders due to alcoholism and malnutrition. Philadelphia: F. A. Davis; 1989.

Victor, M.; Adams, R. D.; Mancall, E. L. A re-stricted form of cerebellar degeneration occur-ring in alcoholic patients. Arch. Neurol. 1:579–688; 1959.

Victor, M.; Mancall, E. L.; Dreyfuss, D. M. Defi-ciency amblyopia in the alcoholic patient: a clinicopathological study. Arch. Ophthalmol. 64:1; 1960.

41

Primary Headache Disorders

JOEL R. SAPER

During the past 10 years, clinical and pathogenetic perspectives on migraine and chronic headaches have changed considerably. With this change has come a major reconsideration of fundamental attitudes toward the head pain problem: who suffers from it, why they suffer, and when and how head pain should be treated (Lance 1982; Raskin 1988; Saper 1986; 1983; 1987).

In 1962, the Ad Hoc Committee on the Classification of Headache (1962) presented what was to become the standard classification for head pain disorders for the next 20 years. Based on the premise of a clear distinction between migraine and tension headaches (TH), the report focused on symptom-specific etiologies: vasculature in migraine and musculature in tension headache. However, many authorities have expressed the view that it is impossible to reconcile the myriad phenomena of migraine, TH, and cluster headache (CH) with the basic foundations of this classification (Raskin 1988). The emphasis on peripheral phenomena (blood vessels and muscles) and the separation of migraine and tension headache have been challenged.

Moreover, several authors (Cohen 1978; Featherstone 1984; Lance 1982; Raskin 1988; Saper 1986; 1983) cite data that they believe support the view that migraine and tension headaches are physiologically related entities with varied symptomatic expressions, reflecting a central or central–peripheral disturbance of neuroreceptor and neurovascular function. The central hypothesis is supported by several factors, including the physiological events that are purported to occur during headache: the symptom overlap between migraine and TH; the general acknowledgement that the clinical phenomena, including pain, cannot be satisfactorily or entirely explained by the disturbances of the vascular or muscular structures alone; and conflicting vascular flow studies that appear to challenge traditional views on the order of events in migraine (Cohen 1978; Dalessio et al. 1961; Dalessio 1980; Featherstone 1984; Lance 1982; Olesen 1986; 1987; Raskin 1988; Saper 1983; 1986). Changing views on the presumed mechanism of action of well-known therapeutic agents has contributed to the altered point of view (Raskin 1988).

These and other factors highlight the fact that the exact pathogenesis of chronic headache is not established, creating varying and divergent views among credible authorities as to the primary pathogenesis of headaches. Raskin (1988), Lance (1982), the author (Saper 1986), and others believe that the disturbances that cause migraine and the other chronic headache disorders reside primarily in the central brain stem–hypothalamic region, involving alterations in ascending and descending serotonergic systems. Through undetermined phenomena, these disturbances result in what is now considered chronic headache. Moreover, it is believed that a genetic predisposition to these disturbances exists (Lance 1982; Raskin 1988; Saper 1986). Hypothetically, this predisposition might be activated by a variety of assaults (perturbations) (Lance 1988), including head trauma, infection, emotional despair, and the overuse of analgesics (discussed later).

In addition to general disagreement regarding the pathogenesis of the primary headache disorders, the absence of generally applied terminology, accepted criteria, and classification brings even greater confusion and denies a reasonable perspective with which to interpret the studies of headache. No major international effort to reclassify headache is underway. Although progress is evident in addressing the failures of previous classifications, early drafts demonstrate many of the same difficulties that plagued earlier attempts.

For these and other reasons, determining prognosis and natural history of chronic headache is difficult. One can make reasonable, though tentative, generalizations, but for the most part, precise outcome data and natural history observations are lacking. Well-considered speculation and personal experience, however, can be of practical value and must suffice until more definitive information is available.

Migraine

Migraine is the most well known of the chronic headache disorders. References to it may be found in writings dating back thousands of years. Historically, the term "classic migraine" has been used to describe the syndrome of headache accompanied by premonitory neurological symptoms, including sensory, motor, or visual disturbances. "Common migraine" has been used to describe those migrainous attacks in which no focal neurological disturbances are noted. It is now clear that common migraine, the more frequently occurring of the two, can be accompanied by numerous neurological symptoms prior to and during the attack, involving hypothalamic, autonomic, and behavioral symptomatology (Lance 1982; Raskin 1988; Saper 1983).

Attacks of migraine vary in presentation, location, duration, and quality. Agreement as to what constitutes a typical migraine attack is lacking. Pain may be unilateral or bilateral and is usually, but not necessarily, associated with gastrointestinal (GI) complaints, light sensitivity, scalp tenderness, and general malaise. The pain may vary from mild to incapacitating. Some migraine attacks occur in the absence of pain altogether. The frequency of attacks also is variable, with some patients reporting one attack in a lifetime and others experiencing a migraine transformation, evolving from intermittent attacks to daily pain.

Prevalence

According to the most recent studies, the annual prevalence of migraine in the Western world is up to 30% of women and 17% of men, aged 24 to 34 years. A decline of 10% and 5%, respectively, occurs in those aged 75 years and older (Raskin 1988; Waters 1986). These figures may, however, underestimate the true incidence and prevalence of migraine. It is now believed that "ordinary headaches," not generally considered migraine, may be of migraine origin (Raskin 1988). Thus, data on prevalence, onset, and natural history must be interpreted with caution.

Genetics

Most patients with migraine, currently estimated to be 50% to 90%, describe a family history of migraine (Dalsgaard-Nielson 1965; Lance 1966; Lance 1982; Raskin 1988). This is compared to family history of headache in 10% to 20% of people without headache. In one study (Dalsgaard-Nielsen 1965), 90% of 100 women with migraine had a first-degree relative with migraine, and 73% had at least one parent with migraine, usually the mother.

Though the mode of inheritance is not clearly established, the transmission of migraine from patient to child is most likely through an autosomal dominant pattern with incomplete penetrance (Ad Hoc Committee 1962; O'Hare et al. 1981; Parrish and Stevens 1977; Raskin 1988).

Currently, no biological markers have identified the presence of the genetic predisposition. The tendency may lower the threshold to environmental (external) or internal activation (perturbation), which may help to explain the process by which headache onset is determined.

Age of Onset

Most patients with migraine experience their first attack younger than age 40 (Selby and Lance 1962). Migraine has been reported as young as 2 to 3 years of age (Barlow 1984). Of children with migraine, 20% to 35% had their first attack younger than age 5 (Congdon and Forythe 1979; Prensky 1976).

Boys and girls experience migraine approximately equally (perhaps slightly more in boys than girls) (Prensky 1976). In adults, 70% to 80% of the migraine population consists of women (Raskin 1988; Saper 1983).

Although onset can occur at any age, less than 10% of patients have the onset after 40 to 50 years of age (Selby and Lance 1962).

Natural History and Prognosis

Migraine in General

Contrary to the widely held view that migraine tends to cease, it is likely that migraine continues throughout the lives of many or most of its sufferers. Moreover, a large but uncertain amount of people with intermittent migraine develop daily chronic

headache and neck pain patterns later in life (Raskin 1988; Saper 1986; 1983). Several authorities now believe that migraine transforms or evolves to a more progressive disorder. This perspective is based on the premise that "ordinary" headaches and neck pain (tension headache) are of migraine origin and can occur daily. If true, the natural history of migraine as an intermittent illness must be reconsidered. In this case, migraine may have variations that will substantially impact the determination of the natural history of this condition.

Nonetheless, using traditional criteria by which migraine is considered an intermittent illness with diagnostic criteria based on that premise, a study (Whitty and Hockaday 1968) demonstrated that 63 of 92 people experiencing migraine for 16 to 69 years were still experiencing attacks after 15 to 20 years of evaluation. Forty-four patients suggested that their attacks were less severe. Of 18 patients 64 years or older, half continued to have headaches with only a slight tendency for improvement (Raskin 1988). In 40 patients who had experienced menopause before follow-up, more than half (22) noted major alterations in their headaches. Six experienced worsening, two noted improvement, and two noted cessation of attacks. In 12 patients, there were not enough data to allow for a determination. Thus, of the 28 postmenopausal female patients whose data were sufficient to draw conclusions, 24 were the same or worsened, and four improved or were no longer experiencing headaches.

Independent of treatment variables, migraine and the symptoms that accompany migraine generally follow an extremely variable course, but in a large percentage of patients, symptoms continue throughout a lifetime (Raskin 1988; Saper 1983). Attacks may cease in early life and recur several decades later or may begin in infancy and continue throughout the entire life span (Graham 1968).

Hormonal Influences

It is clear that various hormonal milestones in a woman's life can affect, usually adversely, the migraine pattern. Current theories suggest that alterations in brain serotonin function or serotonergic and noradrenergic receptors in the central nervous system (CNS) may represent the mechanism by which gonadal hormonal cycles in women affect migraine (Biegon et al. 1983; Quay and Meyer 1978; Raskin 1988; Saper 1983).

Menstruation. Up to 60% of women with migraine report an attack just before, during, or after menstruation, and approximately 14% experience headache only at this time (Epstein et al. 1975; Raskin 1988; Saper 1983).

Somerville (1972a) notes that menstrual headache can be delayed by sustaining the level of estradiol. Oral administration of estrogen only during the premenstrual period, however, does not prevent menstrual migraine (Somerville 1975). According to Raskin (1988), once the mechanism of migraine has been prompted by falling estrogen levels, an increase in tissue levels of estrogen cannot effectively prevent the development of an attack.

Pregnancy. Pregnancy has a variable effect on migraine (Ekbom 1981). Migraine may begin for the first time during pregnancy, and in patients with preexisting migraine, it may worsen during pregnancy, usually during the first trimester (Callaghan 1968). However, it is likely that during the second and third trimesters, patients with preexisting migraine experience improvement (Somerville 1972b).

Following delivery (usually between the third and sixth day), headache is reported in up to 40% of women, most of whom had migraine prior to their pregnancies (Stein et al. 1984). Severe preeclampsia predisposes to the development of postpartum migraine (Rubin and McCabe 1984). Rapidly falling estrogen and progesterone levels or alterations in serotonin metabolism are considered critical factors.

Oral Contraceptives. Before the development of low estrogen content oral contra-

ceptives, their use adversely influenced the frequency and severity of migraine attacks (Ramcharan et al. 1980; Saper 1983). It is now believed that women using low-dose preparations are less likely to develop or suffer exacerbations of migraine (Raskin 1988). However, because there is a considerable delay from the first administration of oral contraceptives to the exacerbation or onset of headache, it may be difficult to determine the accuracy of this correlation. Most women who experience adverse influences from oral contraceptives do so within the first months of usage (Ryan 1978). Data based on higher-dose estrogen preparations show that 30% to 70% of women improve after the oral contraceptives are discontinued (Dennerstein et al. 1978; Kudrow 1975). Some women, however, experience improvement in headache following the administration of oral contraceptives (Raskin 1988).

Generally, authorities believe that an increased risk of cerebrovascular disease occurs in migraine-prone women who take oral contraceptives (Saper 1983).

Menopause. Current opinion is that no predictable and generalized change in the headache pattern of migraine occurs after menopause (Epstein 1975; Whitty and Hockaday 1968), and in observations in one study (Nattero 1982), less than half experienced worsening after menopause. Exogenous hormonal administration during or after menopause may variably improve migraine (Dennerstein et al. 1978; Greenblatt and Brunetau 1974; Shoemaker et al. 1977) or worsen headaches.

The use of pure, synthetic estrogen instead of conjugated, animal-derived preparations may be advantageous (Raskin 1988).

Hysterectomy and Oophorectomy. Generally, hysterectomy with or without oophorectomy offers no benefit for migraine sufferers (Alvarez 1940; Raskin 1988; Utian 1974). Anecdotal exceptions occasionally are noted. Following oophorectomy, estrogen replacement therapy generally ensues and complicates the ability to obtain adequate data on the effect of the oophorectomy.

Childhood Migraine: Natural History

The natural history of childhood migraine varies. Of 67 patients between the ages of 13 and 21 years studied 6 years after initial evaluation, attacks of headache had ceased in more than half. Improvement was noted in 34%, while in 15%, headaches clinically remained the same or worsened (Bille 1962).

In another study of 58 patients (mean age 21 years) studied 9 to 14 years after initial diagnosis, absence of headache was noted in 33% and improvement in 47% (Hinricks and Keith 1965). Significantly more men than women improved.

In general, childhood migraine may have a more favorable natural history and course than adult-onset migraine, and this may be particularly true for men compared to women (Sillanpää 1983). That women dominate the adult population of migraine sufferers may be due to the adverse influence of hormones in women or a more favorable prognosis in boys (Raskin 1988).

Migraine in childhood may be different clinically than in adulthood. A shorter duration of attacks is common; vomiting and abdominal pain are more likely; and a higher incidence of convulsions in childhood between episodes of headache is noted (Prensky 1976). Vertigo, motion sickness, and lightheadedness are more likely between attacks in children than in adults (Fenichel 1967). Sleep disturbances and bed-wetting, night terrors, and sleep walking also are more likely (Raskin 1988).

Children with cyclic vomiting seem more likely to develop migraine later in life (Hammond 1974; Raskin 1988; Prensky 1976). It is unclear whether this is the same process or related to episodic abdominal pain in children, which often occurs in conjunction with vomiting and fever (Raskin 1988). Most authorities believe migraine is more common than epilepsy as a cause of recurring abdominal pain in children. Central neuronal excitability is likely to be a factor common to the pathogenesis of migraine and epilepsy (Raskin 1988).

Headaches in children may be more readily controlled than in adults. A generally better response to anticonvulsant treatment is noted (Raskin 1988).

Transformation of Migraine

Most patients with recurring migraine experience attacks one to four times per month. Up to approximately 50% of patients with acute migraine attacks also experience mild to moderate headaches or neck pain intermittently or continuously. The features of these mild headaches are not by traditional diagnostic criteria categorized as migraine, but often they respond to antimigraine medications (Barrie et al. 1968; Raskin 1988; Saper 1983) and are thought by several authorities to be migraine.

Raskin and Appenzeller (1980) describe a clinical continuum that they consider primary to the basic migraine–tension headache predisposition, emphasizing that the general clinical features thought to be specific and characteristic of one entity frequently are encountered in another. The authors note the absence of compelling clinical and biological distinctions between these two conditions, suggesting the likelihood that these disorders occur at the ends of a continuum in which quantitative rather than etiologically qualitative differences are basic.

Mathew and coworkers (1982) describe the evolution of intermittent migraine to daily migraine. They consider chronic daily headache (tension headache, ordinary headache, neck pain) to be migraine and use the term "migraine transformation." Saper and colleagues (1983; 1986) in 1982 and 1983 expanded this concept. In analyzing the data of more than 500 patients with daily chronic headache prior to first evaluation, they demonstrated the following: a female to male ratio identical to that of migraine (3:1); the onset of headache as intermittent migraine years earlier; a frequent association of depression and sleep disturbance; and a family history of headache in more than 90% of women and 84% of men. A higher than expected incidence of substance abuse also was noted. More than 70% of patients used daily analgesics.

No identifiable event could explain consistently the clinical transformation from intermittent migraine in these patients to daily chronic headache, and the authors expressed the opinion that this phenomenon represented a spontaneous biological transition (transformation) or progression of the illness. Moreover, the pathogenesis was speculated to be a disturbance of the neurotransmitter–receptor systems in the brain, and this transformation or progression could result from internal (genetic predispositional factors) or external adverse factors.

The European literature had identified the same clinical phenomena using different terms, defining common migraine as an "evolutive disease," progressing to the status of what is called "continuous migraine with interparoxysmal headache" (Genazzani et al. 1984; Nappi 1981; 1985; Saper 1982; 1986).

Thus, from a prognosis and natural history perspective, it appears that a substantial percentage of patients with intermittent migraine evolve to a daily chronic headache disturbance (prospective studies are required to determine the amount). Antecedents capable of determining this transformation have not been identified with certainty. A family history of headache, depression, or other general pain disturbance; substance abuse; a history (in the headache patient) of sleep disturbance; excessive analgesic or ergotamine tartrate usage; and depression are noted so frequently that they may eventually be identified as important markers for this potential transformation. Pathogenetically, some authors consider endorphin disturbances critical to the transformation process, while others consider most important the monoamine system and the interplay between monoamine and endorphin physiology in the modulation of pain (Lance 1982; Raskin 1988; Saper 1983; 1986; 1987a).

It is conceivable that patients predisposed to migraine have genetically altered neuro-physiological mechanisms in the pain-modulating zone of the brain stem compared to non–headache-suffering people. Perturbation (adverse influences) by traumatic, chemical (analgesic overuse) (Saper 1986; 1987b), psychiatric, ischemic, or other influences prompts the transformation (Raskin 1988).

Migraine as a Cause of Stroke

Ischemia traditionally has been considered the cause of the neurological events in classic migraine, but recent studies suggest that central neuronal disturbances are more likely to be the primary factor (Bousser et al. 1985; Lance 1982; Raskin 1988).

From a prognostic point of view, uncertainty exists regarding the risks of permanent sequelae in patients with ocular, hemiplegic, aphasic, hemianopic, and retinal dysfunctions in migraine (Rascol et al. 1982; Raskin 1988). When present, arterial occlusion resulting in cerebral infarction has been identified most commonly within the posterior cerebral circulation.

It appears that permanent stroke-like sequelae from migraine are rare. When they occur, often they duplicate the symptoms that had been occurring transiently over the years or represent the first neurological event experienced in association with migraine (Raskin 1988).

Migraine and Brain Hemorrhage

Cole and Aubé (1987) reported on four middle-aged women suffering severe intracerebral hemorrhage during migraine. Shuaib and colleagues (1989) have published a fifth case. This author has evaluated and treated two additional cases (report in preparation).

Permanent and at times potentially fatal cerebral consequences can occur as a result of migraine, but these appear to occur relatively rarely. It is not possible to identify prognostic markers or risk factors with certainty. When possible, avoiding and controlling known stroke risks, such as smoking, oral contraceptives, hyperlipidemia, and prolonged use of vasoconstrictive therapies (chronic ergotamine administration), may be useful, but their absolute relevance cannot be determined.

Transient Migraine Accompaniments: Later-Life Migraine

Fisher (1980) describes 120 patients who developed episodic neurological impairment without identifiable cerebrovascular disease on angiography. Most of the patients experienced visual disturbances, commonly lasting 5 to 30 minutes. All patients were older than 40 years, and approximately half had a history of recurring migraine prior to developing the syndrome. No permanent neurological impairment developed. After a round of attacks, the episodic symptoms did not recur. Approximately 50% of patients experienced headache with the neurological events. In those who eventually died of other causes, autopsy showed no significant cerebral atherosclerosis.

Fisher (1980; 1986) calls these events "transient migraine accompaniments" and distinguishes these from transient ischemic events. These late-life migraine accompaniments, as he has called this phenomenon, appear to carry a good prognosis.

Migraine and Treatment: Prognostic Implications

It is estimated that 75% to 90% of patients with recurring migraine can be helped significantly (Edmeads 1988; Lance 1982; Raskin 1988; Saper 1983; 1986). Treatment requires the elimination or modification of provoking (activating) factors and the implementation of adequate pharmacotherapy. Treatment of the acute attack includes a variety of therapeutic agents ranging from analgesics to ergotamine tartrate. Drug absorption from the GI tract can be enhanced by pretreatment with 10 to 20 mg of oral metoclopramide (Saper 1983).

Preventive therapy uses a wide range of medications, including beta-blocking agents, ergot derivatives (methysergide), tricyclic antidepressants, nonsteroidal anti-inflammatory agents, monoamine oxidase inhibitors, and many others. A common mechanism of these agents may be an influence on central neurophysiology, particularly that involving serotonergic mechanisms (Raskin 1988).

The treatment of patients with refractory and chronic headache must be individually determined. Many patients with frequent migraine headaches experience the adverse influence that comes from chronic overuse of analgesics. The "rebound" phenomenon or "toxic headache" concept is based on the premise that too frequent use of ergotamine tartrate (more than 2 days per week) or analgesics (more than 2 days per week) will, in an undetermined way, result in a medication–headache cycle. The next headache (usually within 24–72 hours) is followed by an abstinence response (headache and accompaniments) from the medication on which dependency has been established (Rapoport 1986; Rapoport et al. 1986; Saper 1983; 1986; 1987; Saper and Jones 1986).

Hospital Treatment and Outcome

Patients with resistant migraine disorders require aggressive treatments that may best be administered during hospitalization. In 1979, the first inpatient specialty unit for headache was developed in Ann Arbor, Michigan (VanMeter et al. 1983). Currently, several other units, modeled after this program, have been established. These programs, using the team approach to aggressive and comprehensive therapy, provide an opportunity for many of the factors related to refractoriness in the headache population to be addressed, including substance abuse, behavioral disturbances, life-style factors, and other general health considerations. More importantly, the hospital treatment programs provide an important opportunity for the aggressive implementation of pharmacotherapy. Intravenous protocols (Raskin 1986), treatment of dehydration, treatment of withdrawal, dietary manipulation, and evaluation of sleep disturbances, sleep apnea, and chronobiological disturbances can be facilitated.

Except for a short-term study by Lake and Saper (1988), prospective studies on the use of inpatient programs are not available.

One retrospective study suggests a beneficial effect (Diamond et al. 1986). Many patients considered untreatable in outpatient settings can be effectively treated in a hospital-controlled setting, and it appears (tentatively) that many patients have long-term beneficial effects. However, recurrence and exacerbations of acute, severe headaches are not uncommon. Hospital specialty unit treatment may be able to establish a baseline of headache control that, with effective continuity of care and regular treatment maintenance, can offer acceptable management to a population of patients considered refractory and often dependent on analgesics. Moreover, by interrupting the pain cycle, the crisis status of these patients can be altered, and prevention of iatrogenic illness may be possible. Nonetheless, an uncertain percentage of patients remain refractory to even the most aggressive treatment efforts.

Cluster Headache

CH occurs primarily in men. A prevalence of less than 70 cases per 100,000 has been established (D'Alessandro et al. 1986; Raskin 1988). The condition usually begins after the age of 20 but may have onset after age 70, and variants of the condition may occur in childhood. Each attack of typical cluster headache lasts from one-half hour to approximately 2 hours, with most attacks lasting 45 to 60 minutes (Kudrow 1980; Saper 1983). Headaches may develop at the same time each day, or they may be random. The cyclical pattern of the attacks (bouts of headaches for 1 to several months followed by remission for years at a time) has been characterized as the cluster (Kudrow 1980).

Episodic CH is the most common and well-known form and is characterized by recurring bouts or clusters of headaches, which then enter remission. Most patients with episodic CH will experience approximately one to two bouts per year. The duration of cycles or bouts ranges from 1 month to 2 years in 80% of cases and from 6 months to 2 years in 60% (Kudrow 1980). In some cases, headaches may remain in remission for 25 years. The median duration of remission is similar for men and women, with a median of 7 to 12 months (Kudrow 1980). Sixty-seven percent of patients have remissions of 1 year or less and in 81%, 2 years or less.

Approximately 10% of patients with CH will experience a chronic headache form. Chronic CH is divided into a primary and a secondary form. Primary chronic CH is characterized by recurring headache events, which, from their onset, do not enter a period of remission. The secondary chronic form is characterized by an evolution from episodic to chronic form. Absence of remission (interim) for 1 year or more justifies the term. The chronic phase may last up to 4 to 5 years (Ekbom 1986). The later the first onset of the episodic form, the greater the chance of evolution into chronic form (Kudrow 1980).

Long-term natural history data are incomplete. Most cases eventually cease spontaneously (Hornabrook 1964; Kudrow 1980). Hereditary factors are less important in CH than they are in migraine (Kudrow 1980). The mechanism of CH is not understood, but hypothalamic and neurovascular peripheral factors seem to be important (Raskin 1988). The influence of chronobiological (hypothalamic pacemaker) factors may be directly relevant to the periodicity and mechanism of CH.

Treatment consists of the administration of a wide range of medications and other therapies that are detailed in standard references (Lance 1982; Raskin 1988; Saper 1983; 1986).

Treatment for CH generally is successful in controlling the attacks but requires persistence, diligence, innovation, and compassion. There is no evidence that effective control of a particular cycle of headaches has an impact on the development of future cycles or on the earlier termination of that particular cycle.

Patients should discontinue smoking, all alcoholic products and avoid daytime napping. Discontinuance of smoking and alcohol is particularly difficult. It is suggested that a predisposition to these habituating tendencies is present (Kudrow 1980; Saper 1983; 1984; 1986).

Resistant cases are best managed in the hospital where aggressive control of provoking factors, implementation of intravenous (Raskin 1986) and other pharmacotherapeutic agents, and general support can be provided while preventive programs and protective measures are implemented. Suicidal ideation and depression are encountered in many patients with CH. A relationship to cyclical depression on a central basis may exist.

For refractory cases, various surgical procedures are available (Ekbom et al. 1987; Onofrio and Campbell 1986; Mathew and Hurt 1988; Solomon and Apfelbaum 1986). Surgical treatment is not recommended except in the most severe and intractable instances and when all traditional treatments, including hospital care, have been tried. Surgical outcome by qualified and experienced teams demonstrate that surgical treatment is effective in more than 50% of patients (Mathew and Hurt 1988; Onofrio and Campbell 1986).

Chronic Paroxysmal Hemicrania

Several variants of CH are described, the most well known of which is chronic paroxysmal hemicrania (CPH) (Sjaastad and Dale 1974; Sjaastad et al. 1986; 1980). Initially thought to be primarily an illness of young women, it also occurs in men. Attacks bear a relationship to CH, but attack frequency of more than 10 per day is common, which is not generally encountered in CH. Pregnancy tends to bring relief (Sjaastad et al. 1980). Treatment of CPH is achieved effectively with indomethacin, but attacks often return on discontinuance. Prolonged remissions have been reported (Jensen and Joensen 1982; Raskin 1988; Sjaastad and Antonaci 1987).

Conclusion

A reliable and scholarly scientific approach to data collection, evaluation, and treatment of headache is emerging, and it is more ac-

ceptable and valuable than in the past. Unfortunately, general acceptance of a classification scheme, diagnostic criteria, and standard terminology is years away. Until this occurs, dependable data on the epidemiology and natural history of primary headaches remain elusive, and conclusions will be tentative.

References

Ad Hoc Committee on Classification of Headache. Arch. Neurol. 6:173–176; 1962.

Alvarez, W. C. Can one cure migraine in women by inducing menopause? Mayo Clin. Proc. 15:380–382; 1940.

Baldi, E.; Salmon, S.; Anselmi, B.; et al. Intermittent hypoendorphaemia in migraine attacks. Cephalgia 2:77–81; 1982.

Barlow, D. F. Headaches and migraine in childhood. London: Spastics International; 1984.

Barrie, M. A.; Fox, W. R.; Weatherall, M.; Wilkinson, M. I. P. Analysis of symptoms of patients with headaches and their response to treatment with ergot derivatives. Q. J. Med. 37:319–336; 1968.

Biegon, A.; Reches, A.; Snyder, L.; McEwen, B. S. Serotonergic and noradrenergic receptors in the rat brain: modulation by chronic exposure to ovarian hormones. Life Sci. 32:2015–2021; 1983.

Bille, B. Migraine in school children. Acta. Paediatr. 51(suppl. 136):141–151; 1962.

Bousser, M. G.; Baron, J. C.; Chiras, J. Ischemic strokes in migraine. Neuroradiology 27:583–587; 1985.

Callaghan, N. The migraine syndrome in pregnancy. Neurology 18:197–199; 1968.

Cohen, M. J. Psychophysiological studies of headache: is there a similarity between migraine and muscle contraction headache? Headache 18:189–196; 1978.

Cole, A. J.; Aubé, M. Late onset migraine with intracerebral hemorrhage: a recognizable syndrome. Neurology 37(suppl. 1):238; 1987.

Congdon, P. J.; Forythe, W. I. Migraine in childhood: A study of 300 children. Dev. Med. Child Neurol. 21:209–216; 1979.

Dalessio, D. J.; Camp, W. A.; Goodell, I. I.; Wolff, H. G. Studies on headache. The mode of action of UML-491 and its relevance to the nature of vascular headache of the migraine type. Arch. Neurol. 4:235; 1961.

Dalessio, D. J. Wolff's headache and other head pain. New York: Oxford University Press; 1980.

Dalsgaard-Nielsen, T. Migraine and heredity. Acta. Neurol. Scand. 41:287–300; 1965.

D'Alessandro, R.; Gamberini, G.; Benassi, G.;

et al. Cluster headache in the Republic of San Marino. Cephalalgia 6:159–162; 1986.

Dennerstein, L.; Laby, B.; Burrows, G. D.; et al. Headache and sex hormone therapy. Headache 18:146–153; 1978.

Diamond, S.; Freitag, F. G.; Solomon, G. D. Impact of inpatient headache treatment on subsequent medical care. Headache 28:317; 1986.

Edmeads, J. Emergency management of headache. Headache 28:675–679; 1988.

Ekbom, K. Chronic migrainous neuralgia. In: Rose, F. C., ed. Handbook of clinical neurology. Vol. 48. Amsterdam: Elsevier Science Publishing; 1986: p. 247–255.

Ekbom, K.; Lindgren, L.; Nilsson, B. Y.; et al. Retrogassarian glycerol injection in the treatment of chronic cluster headache. Cephalalgia 7:21–27; 1987.

Ekbom, K.; Waldenlind, E. Cluster headache in women: evidence of hypofertility headaches in relationship to menstruation and pregnancy. Cephalalgia 31:27–33; 1981.

Epstein, M. T.; Hockaday, J. M.; Hockaday, T. D. R. Migraine and reproductive hormones throughout the menstrual cycle. Lancet 1:543–548; 1975.

Featherstone, H. J. Migraine and muscle contraction headaches: a continuum. Headache 25:194–198; 1984.

Fenichel, G. M. Migraine as a cause of benign paroxysmal vertigo of childhood. J. Pediatr. 71:114–115; 1967.

Fisher, C. M. Late life migraine accompaniments as a cause of unexplained transient ischemic attacks. Can. J. Neurol. Sci. 7:9–17; 1980.

Fisher, C. M. Late life migraine accompaniments—further experience. Stroke 17:1033–1042; 1986.

Genazzani, A. R.; Nappi, G.; Facchinetta, F.; et al. Progressive impairment of CSF B-EP levels in migraine sufferers. Pain 18:127–133; 1984.

Graham, J. R. Migraine: clinical aspects. In: Vincen, P. J.; Bruyn, G. W., eds. Handbook of clinical neurology. Vol. 5. New York: John Wiley and Sons: 1968: p. 45–58.

Greenblatt, R. B.; Brunetau, D. W. Menopausal headache—psychogenic or metabolic. J. Am. Geriatr. Soc. 22:186–190; 1974.

Hammond, J. The late sequelae of recurrent vomiting of childhood. Dev. Med. Child. Neurol. 16:15–22; 1974.

Hinricks, W. L.; Keith, H. M. Migraine in childhood: a follow-up report. Mayo Clin. Proc. 40:593–596; 1965.

Hornabrook, R. W. Migrainous neuralgia. N. Z. Med. J. 63:774–779; 1964.

Jensen, N. B.; Joensen, P. Chronic paroxysmal hemicrania: continued remission of symptoms after discontinuation of indomethacin. Cephalalgia 2:163–164; 1982.

Kudrow, L. Cluster headache: mechanisms and management. New York: Oxford University Press; 1980.

Kudrow, L. The relationship of headache frequency to hormone use in migraine. Headache 15:36–40; 1975.

Lake, A. E.; Saper, J. R. Prospective outcome evaluation of an accredited inpatient headache program (abstract). Headache 28:315–316; 1988.

Lance, J. W. Mechanism and management of headache. 4th ed. Boston: Buttersworth; 1982.

Lance, J. W.; Anthony, M. Some clinical aspects of migraine: the prospective survey of 500 patients. Arch. Neurol. 15:356–361; 1966.

Mathew, N. T.; Hurt, W. Percutaneous radiofrequency trigeminal gangliorhizolysis in intractable cluster headache. Headache 28:328–331; 1988.

Mathew, N. T.; Reuveni, U.; Perez, F. Transformed or evolutive migraine. Headache 27:102–106; 1987.

Mathew, N. T.; Staubits, E.; Nigam, M. B. Transformation of episodic migraine into daily headache: analysis of factors. Headache 22:66–68; 1982.

Nappi, G.; Faccinetti, F.; Martignoni, E.; et al. Plasma and CSF endorphin levels in primary and sympatomatic headache. Headache 25:141–144; 1985.

Nappi, G.; Micieli, G.; Sandrini, G.; et al. Headache temporal patterns: towards a chronobiological model. In: Nappi, G.; Sjaastad, O., eds. Chronobiological correlates of headache. Oslo: Universitets Forlaget; 1981: p. 21–30.

Nattero, G. Menstrual headache. Adv. Neurol. 33:215–326; 1982.

O'Hare, J. A.; Feely, M. J.; Callaghan, N. The clinical aspects of familial hemiplegic migraine in two families. Ir. Med. J. 74:291–295; 1981.

Olesen, J. The ischemic hypothesis of migraine. Arch. Neurol. 44:321–322; 1987.

Olsesen, J. The pathophysiology of migraine. In: Rose, F. C., ed. Handbook of clinical neurology. Vol. 48. Amsterdam: Elsevier Science Publishers; 1986: p. 59–83.

Onofrio, B. M.; Campbell, J. K. Surgical treatment of cluster headache. Mayo Clin. Proc. 61:537–544; 1986.

Parrish, R. M.; Stevens, H. Familial hemiplegic migraine. Minn. Med. 70:709–715; 1977.

Prensky, A. L. Migraine and migrainous variants in pediatric patients. Pediatr. Clin. North Am. 23:461–471; 1976.

Quay, W. B.; Meyer, D. C. Rhythmicity in periodic functions of the central nervous system. In: Essman, W. B., ed. Serotonin in health and disease. Vol. 2. New York: Spectrum; 1978: p. 159–204.

Ramcharan, S.; Pellegrin, E. A.; Roy, R. M.;

et al. The Walnut Creek contraceptive drug study—A prospective study of the side effects of oral contraceptives. J. Reprod. Med. 25(suppl.):346–372; 1980.

Rapoport, A. Analgesic rebound. Top. Pain Management 1:8, 29–32; 1986.

Rapoport, A.; Weeks, R. E.; Sheftell, F. D.; et al. The analgesic washout period: a critical variable of evaluation of headache treatment efficacy (abstract). Neurology 36(suppl):100–101; 1986.

Rascol, A.; Clanet, M.; Rascol, O. Cerebrovascular accidents complicating migraine attacks. In: Rose, E. C.; Avery, W. K., eds. Cerebral hypoxy in the pathogenesis of migraine. London: Pittman Publishers; 1982: p. 110–115.

Raskin. N. H. Headache. 2nd ed. New York: Churchill-Livingstone; 1988.

Raskin, N. H. Repetitive intravenous dihydroergotamine as therapy for intractable migraine. Neurology 36:995–997; 1986.

Raskin, N. H.; Appenzeller, O. Headache. Philadelphia: W. B. Saunders; 1980.

Rubin, P. C.; McCabe, R. Post-partum migraine in severe eclampsia. Lancet 2:285–286; 1984.

Ryan, R. E. Sr. A controlled study of the effect of oral contraceptives on migraine. Headache 17:250–252; 1978.

Saper, J. R. Changing perspectives on chronic headache. Clin. Pain 2:19–28; 1986.

Saper, J. R. Drug treatment of headache: changing concepts and treatment strategies. Semin. Neurol. 7:178–191; 1987a.

Saper, J. R. Ergotamine dependency: a review. Headache 27:435–438; 1987.

Saper, J. R. Headache disorders: current concepts and treatment strategies. Littleton, MA: Wright, PSG; 1983.

Saper, J. R. Non-headache disorders and characteristics of cluster headache patients. In: Mathew, N. T., ed. Cluster headache. New York: Spectrum Publications; 1984: p. 31–44.

Saper, J. R.; Jones, J. M. Ergotamine tartrate dependency: features and possible mechanisms. Clin. Neuropharmacol. 9:244–256; 1986.

Selby, G.; Lance, J. W. Observations on 500 cases of migraine in allied vascular headaches.

J. Neurol. Neurosurg. Psychiatry 23:23–32; 1962.

Shuaib, A.; Metz, L.; Hing, T. Migraine in intracerebral hemorrhage. Cephalgia 9:59–61; 1989.

Shoemaker, E. S.; Forney, J. P.; McDonald, P. C. Estrogen treatment of post-menopausal women. JAMA 238:1524–1530; 1977.

Sillanpää, M. Changes in the prevalence of migraine and other headaches during the first seven years of school. Headache 23:15–19; 1983.

Sjaastad, O.; Dale, I. Evidence for a new ? treatable entity. Headache 14:105–108; 1974.

Sjaastad, S.; Aasly, J.; Fredriksen, T.; et al. Chronic paroxysmal hemicrania. On the autonomic involvement. Cephalalgia 6:113–124; 1986.

Sjaastad, S.; Antonaci, F. Chronic paroxysmal hemicrania: longlasting remission in the chronic stage. Cephalalgia 7:203–205; 1987.

Sjaastad, S.; Apfelbaum, R.; Caskey, W.; et al. Chronic paroxysmal Hemicrania: clinical manifestations. A review. Ups. J. Med. Sci. 31(suppl):27–33; 1980.

Solomon, S.; Apfelbaum, R. I. Surgical decompression of the facial nerve in the treatment of chronic cluster headache. Arch. Neurol. 43:479–482; 1986.

Somerville, B. W. The role of estradiol withdrawal in the etiology of menstrual migraine. Neurology 22:355–365; 1972a.

Somerville, B. W. A study of migraine in pregnancy. Neurology 22:824–828; 1972b.

Somerville, B. W. Estrogen withdrawal in migraine, I & II. Neurology 25:239–244, 245–250; 1975.

Stein, G.; Morton, J.; Marsh, A.; et al. Headaches after childbirth. Acta. Neurol. Scand. 69:74–79; 1984.

Utian, W. H. Estrogen headache and oral contraceptives. S. Afr. Med. J. 48:2105–2108; 1974.

VanMeter, M.; Saper, J. R.; Ross, M. Inpatient treatment of intractable headaches: an outcome study (abstract). Headache 23:144; 1983.

Waters, W. E. Headache (series in clinical epidemiology). London: Croom Helm; 1986.

Whitty, C. W. M.; Hockaday, J. M. Migraine: a follow-up study of 92 patients. Br. Med. J. 1:735–736; 1968.

42

Reflex Sympathetic Dystrophy

STUART M. WEIL

The term "reflex sympathetic dystrophy" encompasses a combination of symptoms, including pain, hyperesthesia, vasomotor, and dystrophic changes. A variety of terms have been used to describe these symptoms, including causalgia "major and minor," Sudeck's atrophy, mimocausalgia, shoulder-hand syndrome, sympathalgia algoneurodystrophy, and others. Reflex sympathetic dystrophy was first recognized as a clinical syndrome in 1864 by Weir Mitchell as a result of gunshot injuries to the peripheral nerves. It resulted in burning pain, progressive and trophic changes in the affected limb. (Mitchell 1864). It is now recognized that a variety of injuries can cause reflex sympathetic dystrophy, including blunt trauma, penetrating trauma, myocardial and cerebral infarction, degenerative joint disease, frostbite, and burns. The common mechanism by which these etiologies result in the same or similar symptoms is largely unknown.

Symptoms may be manifest a few hours after the injury, but they usually occur gradually over a period of days to weeks. The syndrome of reflex sympathetic dystrophy is characterized by progression, which has been divided into several stages for purposes of understanding the process. The length of each stage may vary considerably, ranging from weeks to years. Regardless of the way in which reflex sympathetic dystrophy is defined, it has generally been recognized that it may spread beyond the local area of injury, particularly if treatment is not initiated early or effectively (Rowlingson 1983).

Three stages are recognized as comprising the syndrome of reflex sympathetic dystrophy. The acute stage begins at the time of injury and usually lasts several weeks. It is characterized by spontaneous aching or burning pain that is restricted to a particular vascular or neural territory. During this stage, hyperpathia, pain characterized by over-reaction after stimulation to pain, may occur. There also may be the sensation of hyper or hypoesthesia or dysesthesia. Early in the course, the skin may appear warm, dry, and red. Later, the skin can become cold and cyanotic with local edema and increased hair and nail growth (Merskehy et al. 1979). Range of motion also may be decreased in the affected extremity. Approximately 3 to 6 months after injury, a dystrophic change may begin. Edematous tissue can become indurated, and the skin can become cool with cyanosis. Hair loss may occur, and nails may become ridged, cracked, and brittle. Constant pain radiating from the site of injury may occur, and the hyperpathia is more pronounced than in the acute phase. Radiographic evauation may demonstrate osteoporosis in the affected limb. A third stage of reflex sympathetic dystrophy is called atrophic. This phase usually occurs more than 6 months after the time of injury when pain becomes less prominent. Irreversible trophic changes may occur, however, including thinning of the skin and subcutaneous tissue. There may be tapering of digits with fascial thickening and flexion contractures. Sweating may be increased or decreased, and marked bony demineralization and ankylosis can be demonstrated.

Reflex sympathetic dystrophy is defined by signs and symptoms rather than by cause. Because of the wide variability of the syndrome, prognosis has been difficult to define. It is known, however, that pain persists for more than 6 months in 85% of patients and for more than 1 year in 25% of patients (Bonica 1980). It is widely accepted that prognosis is unrelated to the degree of nerve injury, the age of the patient, or associated soft-tissue injury (Sunderland 1978). The most important factor in determining prognosis of reflex sympathetic dystrophy appears to be the use of early aggressive treatment, often regardless of the type of treatment initiated (Sunderland 1978).

As previously mentioned, the diagnosis of reflex sympathetic dystrophy is based almost entirely on clinical observation. Patients usually report a history of burning pain that occurs hours or days after the precipitating injury. Pain can begin within a few hours to 1 week after the injury in more than 80% of cases (Bonica 1980); however, some patients report the onset of pain up to 1 month or longer after the injury (Baker et al. 1969). Pain is severe and often has a profound effect on an individual's lifestyle. When the injury is limited to an extremity,

the pain almost always begins in the distribution of the injured nerve. The pain eventually may spread proximally and can involve an entire extremity.

Several findings may be useful in diagnosing reflex sympathetic dystrophy. The skin may appear smooth and glossy; there may be cutaneous vasoconstriction and subcutaneous atrophy. The combination of these findings often is considered diagnostic of reflex sympathetic dystrophy (Horowitz 1984). Several laboratory tests have been used in the diagnosis of reflex sympathetic dystrophy, primarily aimed at assessing skin temperature (Lofstrum et al. 1980) or osteoporosis (Kozin et al. 1981). The most useful diagnostic test actually may be therapeutic. The response to sympathetic blocks, although not 100%, may help to make the diagnosis (White J. C., Sweet W. H., 1969).

Developing an effective and efficient treatment plan for reflex sympathetic dystrophy, like determining prognosis in the management of the disease, can be difficult. A variety of treatments have been used, and many appear to be useless. A few of the approaches available for management of reflex sympathetic dystrophy include sympathetic blockade, anticonvulsants, narcotic analgesics, tricyclic antidepressants, anti-inflammatory agents, and calcium channel blockers. Regardless of the treatment chosen, early and aggressive initiation of treatment is essential.

Treatment of reflex sympathetic dystrophy relies on interruption of sympathetic fibers, which may be accomplished in a variety of ways. Sympathetic blocks performed at the stellate ganglion or the lumbar sympathetic ganglia can be useful for diagnostic and short-term therapeutic purposes. It has been reported that the duration of pain relief may actually outlast the duration of anesthetic agent used. This has been reported by many authors, especially if the block is performed within a few days of injury (Miller et al. 1980). Success rates as high as 50% to 60% have been reported (Wang et al. 1985). Response is a function of the timing of the block in relation to the onset of pain, the skill of the individual performing the block, and the completeness of the blockade attained (Lofstrom et al. 1980). If patients receive temporary pain relief from local anesthetic blocks, a more permanent sympathectomy may be indicated. These permanent or surgical sympathectomies may be performed in the upper thoracic or lumbar region, depending on the distribution of pain. Before performing surgical sympathectomy, a series of sympathetic blocks may be used to alleviate pain. It has been reported that up to 70% of patients demonstrate pain relief in a 3-year period if the blocks are initiated within 6 months of the onset of symptoms (Wang et al. 1985). However, only 50% of patients treated 6 to 12 months after the onset of symptoms reported improvement at a 3-year follow-up.

The response rate for surgical sympathectomy is variable. Results have ranged from 12% to 97% (Payne 1986). In follow-up periods ranging from 3 to 8 years, pain may recur in as many as 33% of patients. Even if patients have recurrent pain, many of them may still report improvement over preoperative condition (Schumacker 1985).

Rather than performing a sympathectomy surgically, a pharmacological sympathectomy also may be performed. Guanethidine and reserpine have been used intravenously to produce regional sympathetic blockade by replacing norepinephrine from sympathetic efferent fibers. Pharmacological sympathectomy can be used in patients who have had persistent pain after an unsuccessful sympathectomy or in patients who are poor surgical candidates. Up to 80% of patients report pain relief at a 15-week follow-up (Hannington-Kiff 1974).

Pharmacological sympathectomy also can be performed using oral sympatholytic agents. Beta-adrenergic blockers (propranolol) or alpha-adrenergic blockers (guanethidine, prazosin, and phenoxybenzamine) have been used for pain management. Propranolol has not been shown to be useful in the management of reflex sympathetic dystrophy, while the alpha-adrenergic blockers have been useful (Scadding et al. 1982). Pra-

zosin has been useful in one case report, while phenoxybenzamine has been the most useful. In one study, 40 patients with causalgia after military injuries had complete resolution of pain for follow-up periods ranging from months to 6 years. There was no recurrence of pain after treatment (Ghostine et al. 1984). This work has been independently confirmed (Haddad 1985).

Other medications have been used in the management of reflex sympathetic dystrophy that have a wide mechanism of action. Anti-inflammatory agents, including steroidals and nonsteroidals, may be helpful in the management of these patients (Kozin et al. 1976). Tricyclic antidepressants, including amitriptyline and others, have been used in the management of reflex sympathetic dystrophy. There are no controlled clinical trials to determine their effectiveness, but anecdotal experience suggests that the tricyclic may be useful (Botney and Fields 1983).

The use of narcotic analgesics in the management of reflex sympathetic dystrophy is controversial. Addiction has been reported, and adequate pain relief is not usual (Magora et al. 1980).

Anticonvulsants have been used in a number of pain syndromes, including reflex sympathetic dystrophy. Twenty-three of 28 patients with post-traumatic neuralgia experienced pain relief with a variety of anticonvulsant agents (Swerdlow 1984). Calcium channel blockers also have been used. Although no double-blind clinical trials have been reported, in an uncontrolled study, relief from pain was obtained in up to 60% of patients (Prough 1985). Other adjuncts also are available for management of reflex sympathetic dystrophy. Probably the most useful method is physical therapy, which can be used to treat and minimize dystrophic changes in muscles and joints. Patients may be instructed on home physical therapy, which allows the patient to be actively involved in therapy. Electrical stimulation, including transcutaneous electrical nerve stimulation, has been useful in isolated cases (Meyers 1972). More recently use of dorsal column stimulation has been found to also be affective (Robaina et al. 1989). Patients were followed for a mean of 27 months and 90% of them had good or excellent pain relief. Moreover, these patients were then allowed to undergo physical therapy, which they had not been able to do prior to placement of the dorsal column stimulator. This study specifically evaluated the efficacy of dorsal column stimulators for the upper extremity of reflex sympathetic dystrophy, but also suggests that it may be affective for symptoms in the lower extremities. Trigger-point injections also have been performed, but data are not available regarding its usefulness specifically for reflex sympathetic dystrophy.

Other surgical procedures also have been proposed for the management of reflux sympathetic dystrophy. Neuroablative procedures, including neurolysis, neurectomy, rhizotomy, cordotomy, thalamotomy, cingulotomy, and mesencephalic tractotomy, have been proposed, and all have been unsuccessful on any reasonable follow-up periods (Sunderland 1978).

Management is the most important factor in the prognosis of reflex sympathetic dystrophy. When initially evaluating a patient, it may be necessary to admit him or her to a pain unit for diagnosis and therapy. Patients with reflex sympathetic dystrophy should be treated as early as possible because early treatment is the most important factor in determining good outcome. Once the diagnosis of reflex sympathetic dystrophy is made, sympathetic blockade should be initiated promptly. The type of blockade may be individualized to the chronicity and intensity of pain. For recent-onset causalgia, lumbar or cervical sympathetic block with local anesthetic agents may be the treatment of choice. In patients with a contraindication to this type of treatment, intravenous guanethidine or reserpine or oral sympatholytic agents may be initiated instead. Adjunctive modalities, including electrical stimulation and physical therapy, should be instituted at the same time as sympathetic blockade. Moreover, dorsal column stimulation has been shown to be affective in not only reducing pain, but also improving func-

tional capacity. Some patients may respond permanently to these adjunctive modalities and not require further sympathetic blocks or surgical intervention.

References

Baker, A. G.; Wingarner, F. G. Causalgia: a review of twenty-eight treated cases. Am. J. Surg. 117:690–694; 1969.

Bonica, J. J. Sympathetic nerve blocks for pain diagnosis and therapy. Vol. I. New York: Beron Laboratories; 1980.

Botney, M.; Fields, H. L. Amitriptyline potentiates morphine analgesia by direct action on the central nervous system. Ann. Neurol. 13:160–164; 1983.

Ghostine, S. Y.; Comair, Y. G.; Turner, D. M.; Kassell, N. F.; Azar, C. G. Phenoxybenzamine in the treatment of causalgia: report of forty cases. J. Neurosurg. 60:1263–1268; 1984.

Haddad, A. Phenoxybenzamine for treatment of causalgia. J. Neurosurg. 62:157–159; 1985.

Hannington-Kiff, J. G. Intravenous regional sympathetic block with guanethidine. Lancet 1:1019–1020; 1974.

Horowitz, S. H. Iatrogenic causalgia: classification, clinical findings, and legal ramifications. Arch. Neurol. 41:821–824; 1984.

Kozin, F.; McCarthy, D. J.; Sims, J.; Genant, H. The reflex sympathetic dystrophy syndrome. Am. J. Med. 60:321–331; 1976.

Kozin, F.; Ryan, L. M.; Carerra, G. F.; Soin, J. S.; Wortmann, R. L. The reflex sympathetic dystrophy syndrome. Scintigraphic studies, further evidence for the therapeutic efficacy of systemic corticosteroids and proposed diagnostic criteria. Am. J. Med. 70:23–30; 1981.

Lofstrom, J. B.; Lloyd, J. W.; Cousins, M. J. Sympathetic neural blockade of the upper and lower extremity. In: Cousins, M. J.; Bridenbaugh, P. O., eds. Neural blockade in clinical anesthesia and management of pain. Philadelphia: J. B. Lippincott; 1980: p. 620–624.

Magora, F.; Oshwang, D.; Shoor, J.; Katzenelson, R.; Cotev, S.; Davidson, J. T. Observations on extra-dural morphine analgesia in various pain conditions. Br. J. Anaesth. 52:247–252; 1980.

Merskehy, H.; Albe-Fessard, D. G.; Bonica, J. J. Pain terms: a list with definitions and notes on usage. Pain 6:249–252; 1979.

Meyers, G. A.; Fields, H. L. Causalgia treated by elective large fiber stimulation of peripheral nerve. Brain 95:163–168; 1972.

Miller, R. D.; Munger, W. L.; Powell, P. E. Chronic pain and local anesthetic neural blockade. In: Cousins, M. J.; Bridenbaugh, P. O., eds. Neural blockade in clinical anesthesia and management of pain. Philadelphia: J. B. Lippincott; 1980: p. 616–636.

Mitchell, S. W.; Morehouse, G. R.; Keen, W. W. Gunshot wounds and other injuries to the nerves. Philadelphia: J. B. Lippincott; 1864.

Payne, R. Neuropathic pain syndromes, with special reference to causalgia and reflex sympathetic dystrophy. Clin. J. Pain 2:59–73; 1986.

Prough, D. S.; McLeskey, C. H.; Boehlin, C. G. Efficacy of oral nifedipine in the treatment of reflex sympathetic dystrophy. Anesthesiology 62:796–799; 1985.

Robaina, F. J. Dominguez, M.; Diaz, M.; Rodriguez, J. L.; DeVera, J. A. Spinal cord stimulation for the relief of chronic pain, in vasospastic disorders of the upper limbs. Neurosurgery 24:63–67; 1989.

Rowlingson, J. C. The sympathetic dystrophies. Int. Anesthesiol. Clin. 21:117–129; 1983.

Scadding, J. W.; Wall, P. D.; Parry, C. W.; Brook, D. M. Clinical trial of propranolol in post-traumatic neuralgia. Pain 45:190–194; 1982.

Schumacker, H. B. A personal overview of causalgia and other reflex dystrophies. Ann. Surg. 201:278–289; 1985.

Sunderland, S. Nerves and nerve injuries. 2nd ed. New York: Churchill Livingston; 1978.

Swerdlow, M. Anticonvulsant drugs and chronic pain. Clin. Neuropharmacol. 7:51–82; 1984.

Wang, J. K.; Johnson, K. A.; Ilstrup, D. M. Sympathetic blocks for reflex sympathetic dystrophy. Pain 23:13–17; 1985.

White, J. C.; Sweet, W. H. Pain and the neurosurgeon. Springfield: Charles C Thomas; 1969.

43

Postherpetic Neuralgia

C. PETER N. WATSON

The efficient physician is the man that successfully amuses his patients while nature effects a cure. *Philosophical Dictionary,* Voltaire (1694–1778)

Many therapies have been recommended for the treatment of postherpetic neuralgia (PHN). Most of these therapies have not been generally used and may simply serve, as Voltaire said, "to amuse . . . patients while nature effects a cure." Many studies of these therapies have been uncontrolled trials in populations of unknown age with postherpetic pain of unspecified duration. The chief difficulty is that the pain associated with herpes zoster (HZ) and PHN improves with time, and this gives a false impression of the efficacy of a remedy. This amelioration is most dramatic in the early weeks after the onset of the skin lesions but also is seen to a lesser extent in the ensuing months and years. This information about the pain's natural course is important for three reasons. First, it helps to reassure patients that improvement can occur even with pain of long duration. Second, it becomes possible to interpret more knowledgeably

the results of clinical trials. Finally, the researcher can more effectively design studies to assess the efficacy of a particular treatment. The clinical features, pathology, and pathogenesis of PHN are not discussed in this chapter nor are prophylaxis and treatment discussed in detail. For these, the reader is referred to review articles (Loeser 1986; Portenoy et al. 1986, Watson and Evans 1986).

Definition

It is important to discriminate between the acute pain of HZ and true PHN. A reasonable definition of PHN is neuropathic pain that persists in the area affected by HZ after the usual time that the skin lesions heal (usually 4 weeks). At this time, most patients with HZ will be pain-free. One can then arbitrarily define PHN as pain occurring more than 1 month after HZ onset. As this chapter shows, resolution of pain continues to occur spontaneously in a considerable number of patients between 1 and 3 months; therefore, to ensure a chronic, fairly stable pain state for study design purposes it may be preferable to choose patients with pain lasting more than 3 or 6 months.

The Natural History of Postherpetic Pain in All Age Groups

When PHN is defined as pain persisting more than 1 month after HZ, the incidence in all age groups has been 9% (Ragozzino et al. 1982), 9.7% (Burgoon et al. 1957), and 14.3% (Hope-Simpson 1975) in different studies. Higher assessments have been made by De Moragas and Kierland (1957) and Rogers and Tindall (1971). Perhaps differences are related to whether the inquiry was directed at any sort of discomfort or rather significant persistent pain.

The number of patients suffering PHN for more than 1 year has been estimated to be 22% (Ragozzino et al. 1982) and 49% (De Moragas and Kierland 1957) of the number affected at 1 month. Thus, of 100 patients of all ages with HZ, approximately 10 will have pain at 1 month and two to five at 1 year or more.

The author's data support continued improvement with extended periods of time. A population of 132 patients with PHN was followed at intervals of 3 months, 6 months, and yearly thereafter for periods of up to 10 years after the initial pain clinic visit. At each interval, 35% to 50% of patients were doing well and having no pain or mild pain. Between 40% and 60% of patients doing well at each following assessment had the disorder for 1 year or more when first seen, so even long-lasting pain can improve with time. A disturbing group of 26 patients (20%) initially responded well to treatment, but with time, they lost this effect and were repeatedly documented as having poor results despite retreatment with the initially successful methods and others.

Effect of Age, Sex, Affected Dermatome, Other Diseases, and Antibody Titers on the Risk and Duration of PHN

Although the overall tendency is for PHN to improve, there is a strong relationship between age and the incidence and severity of the disorder. De Moragas and Kierland (1957) found an increasing incidence of PHN lasting 1 month or more when patients were studied by age groups. The incidence was 33% between ages 40 and 49, increasing with each decade to 74% at age 70 and older. Also severity was increased (as measured by duration) with age, so the prevalence of pain more than 1 year after HZ was 7% in the group aged 40 to 49, climbing to 47.5% in those older than 70 years. Other authors also have found this age-related increase in incidence (Burgoon et al. 1957; Glynn 1987; Rogers and Tindall 1971). Rogers and Tindall (1971) estimate the incidence of PHN lasting more than 4 weeks as 46.9% in a group older than 60 years and only 15.9% in those younger than 60 years.

A number of studies have indicated that women suffer PHN more commonly than men (Glynn 1987; Hope-Simpson 1975; Lewith et al. 1983; Russell et al. 1957; Watson et al. 1988b). Although Hope-Simpson (1975) thought this was a true increase, Watson and colleagues (1988b) and Glynn (1987) studied a larger number of patients and found that this female predominance simply reflects general population statistics.

There is an increased tendency for PHN to affect the first division of the trigeminal nerve (VI) and thoracic, especially midthoracic, dermatomes (Glynn 1987; Watson et al. 1988b). There is a corresponding lesser incidence in cervical and lumbosacral areas. This dermatomal distribution is similar to observations in HZ (Burgoon et al. 1957, Hope-Simpson 1975; Ragozzino et al. 1982) and also may reflect the centripetal distribution of varicella. PHN in VI has been reported to occur with increased frequency (Harding et al. 1987) and to last longer (De Moragas and Kierland 1957; Hope-Simpson 1975; Watson et al. 1988b). An increasing incidence of this long-duration pain with trigeminal involvement has been found with increasing age (De Moragas and Kierland 1957).

Although an increased risk of HZ has been associated with leukemia, lymphoma, and chemotherapy, no increased incidence of PHN has been reported with these conditions. Diabetes mellitus may predispose to HZ and increase the likelihood of PHN (Brown 1976; McCulloch et al. 1982). Higa and associates (1988) have reported that the severity of HZ could be defined by the maximum antibody titer to varicella-zoster virus and that independent of age, this had the greatest influence in the duration of treatment of acute herpetic pain by sympathetic blocks.

The Prevention of Postherpetic Neuralgia

Many therapeutic approaches to the treatment of HZ have been claimed to prevent progression to PHN. Most have been used with small numbers of patients and for short periods of follow-up.

Controlled trials appear to support the use of oral corticosteroids. Elliot (1964) treated 16 patients older than 50 years with HZ, using prednisone or cortisone and compared them with 10 controls. No treated patient had pain for more than 4 weeks. Two control patients had pain at 8 and 13 months. Eaglstein and coworkers (1970) randomized 35 patients with HZ in a double-blind fashion to oral triamcinolone or placebo in patients older than 60 years. Eleven (73%) of 15 of the control group developed PHN more than 2 months from the pain onset; three (30%) of 10 in the steroid-treated group experienced this. No difference in PHN between the two groups was present at 1 year follow-up (one case in each group). Keczkes and Basheer (1980) randomized 40 patients to oral prednisolone or carbamazepine (CBZ) in a nonblinded trial. They found that 13 (65%) of 20 patients in the CBZ-treated group developed PHN lasting 2 or more months vs. three (30%) of 10 patients in the steroid-treated group. Four patients in the CBZ group had pain for 1 year or more, whereas none in the steroid group had pain for more than 6 months.

A recent randomized, double-blind, placebo-controlled trial (Esmann et al. 1987) compared prednisolone and oral acyclovir with placebo and acyclovir in HZ and found no difference in PHN at 6 months follow-up.

Acyclovir was given to prevent dissemination of HZ. Despite this, it seems reasonable, based on the other trials, to use moderate doses of an agent such as prednisolone (60 mg daily) at the onset of the skin lesions with gradual dose reduction for 2 weeks in the nonimmunosuppressed patient. Dissemination of the virus does not appear to be a risk with this approach (Eaglstein et al. 1970, Keczkes and Basheer 1980). There is a need for further study of more patients treated with steroids in this condition to settle the disagreement regarding efficacy and to assess their effect on the long-term natural history of PHN.

There have been conflicting reports of the effectiveness of sympathetic blocks used at the stage of HZ to prevent the development of PHN (Colding 1969; 1973; Dan et al. 1983; 1985; Riopelle et al. 1984; Yanagida et al. 1987). Some of the reasons for this are the lack of controls, the failure to state patients' ages, and insufficient follow-up of adequate numbers of patients.

Galbraith (1983) reports a randomized, double-blind, placebo-controlled trial of amantadine hydrochloride for the acute illness in a population with a mean age of 70 years. The time taken to achieve total pain relief with this agent was reduced by half. A double-blind, placebo-controlled study of levodopa and benserazide (Kernbaume and Hauchecorne 1981) claimed a reduction in PHN in patients aged 65 years and older. Although a difference was detected at 21 days, no such effect was present 60 days after the onset of the rash. Intramuscular interferon alpha reduced the severity and duration of PHN in patients with malignant neoplasms in two randomized, placebo-controlled, double-blind studies (Merigan et al. 1978; 1981). Intravenous vidarabine reduced the duration but not the incidence of PHN in a double-blind, placebo-controlled study of HZ in immunosuppressed patients (Whitley et al. 1982). The authors conclude that because of its potential toxicity, the drug should be reserved for immunocompromised patients. Well-controlled studies (Balfour 1986; Bean et al. 1982; 1983; Esmann et al. 1982; Peterslund et al. 1981) have shown that acyclovir did not reduce the incidence or severity of PHN in immunocompetent (Bean et al. 1982; 1983) or immunocompromised (Balfour et al. 1983) adults. However, used intravenously, it shortened the period of acute pain and accelerated cutaneous healing (Bean et al. 1982; 1983; Esmann 1987; Peterslund et al. 1981). A randomized study that compared acyclovir with vidarabine in immunocompromised patients found no difference in the incidence of PHN 28 days or more after the rash (Shepp et al. 1986).

A randomized, placebo-controlled, double-blind trial of intramuscular adenosine monophosphate (AMP) has markedly reduced PHN (Sklar et al. 1985). The number of patients in this trial was small and there was concern about the toxicity of AMP in humans and that the severity of disease was not comparable among the groups (Sherlock and Corey 1985).

Treatment of Postherpetic Neuralgia

This chapter does not review exhaustively the treatment of PHN. For this, the reader is referred elsewhere (Loeser 1986; Portenoy et al. 1986; Watson and Evans 1986). Because of the natural history of improvement with time, particularly soon after HZ onset, definition of PHN and selection of a control population are important for evaluating any treatment approach; only controlled trials with adequately defined study populations are considered here.

Two randomized, double-blind, placebo-controlled trials support the analgesic effect of amitriptyline (AT) in PHN (Max et al. 1988; Watson et al. 1982). This effect was seen with mean doses of 65 mg (Max et al. 1988) and 75 mg (Watson et al. 1988b) in approximately two thirds of patients (Watson et al. 1982). The drug was thought to have an analgesic action independent of its antidepressant effect. After 12 months, 12 (55%) of 22 patients were maintaining a good response on AT (Watson et al. 1982). After 5 years, 10 (52%) of 19 were doing well, but only four were on AT. Long-term data de-

rived from 132 patients indicate that of patients doing well at follow-up, at least 50% continued to use antidepressants (AT, nortriptyline, maprotiline) at 3, 6, and 12 months and 2 and 3 years from the first pain clinic visit.

Capsaicin (8-methyl-N-vanillyl-6-nonenamide) is the pungent principle of hot paprika or chili peppers and other plants of the nightshade family. The chemical structure is similar to the oil of cloves, an old remedy for toothache, which produces a long-lasting trigeminal anesthesia. Capsaicin selectively stimulates and then blocks unmyelinated sensory afferents from skin and mucous membranes (Jansco et al. 1967; 1977; Otsuka 1976; Yaksh et al. 1979). Many of these contain substance P, an excitatory peptide neurotransmitter and other neurotransmitters, such as calcitonin gene-related peptide and somatostatin. It is thought that capsaicin relieves pain by releasing and then depleting substance P. Two open-label studies support an analgesic action of capsaicin ointment in PHN. One study involved 12 patients and found that nine (75%) experienced "substantial relief of pain" (Bernstein et al. 1987). It was not clear how many patients actually had a good or satisfactory response. The author has conducted an open-label trial of 33 patients and found a good response in 56% of the 23 patients completing the trial, with some significant change in 79% (Watson et al. 1988). Burning after application was a significant problem, resulting in premature termination of the study in one third of the initial study population. Follow-up at 3 months after completion of the trial indicated that pain returned in most patients after treatment was stopped. The following guidelines should be observed for the use of this agent:

1. It must be used four to five times daily. (If used less, it may not effectively deplete substance P.)
2. It should be used for 4 weeks. (Patients may not get relief for 2 or 3 weeks.)
3. Patients should be seen regularly to encourage persistence and deal with capsaicin-induced burning.

4. If burning occurs, it may be tolerated by the previous application of 5% lidocaine ointment, initially covering a small area of skin, and by regular analgesics.

5. Hands should be washed after use to avoid inadvertent eye exposure, and the ointment should be kept above the eyebrow and away from the eye if used for ophthalmic nerve involvement.

Although a number of surgical approaches to PHN have been suggested, results generally have been unsatisfactory. Dorsal root entry zone (DREZ) lesions have been reported to be of some use. Three articles (Friedman et al. 1984; Friedman and Nashold 1984; Nashold et al. 1983) report the beneficial effect of DREZ lesions in PHN in 50% to 66% of patients with followup of 6 to 25 months. Many of these patients had pain for at least 5 months and were refractory to many medical approaches (Friedman et al. 1984; Friedman and Nashold 1984). Significant complications can result from this operation, including leg weakness; altered, disagreeable sensation below the lesion; and back discomfort at the level of the surgery. Because of this and the overall efficacy, this approach should be considered a last resort for refractory patients.

Conclusions

In most patients, postherpetic pain improves with time. This occurs most dramatically soon after HZ onset but continues to occur during ensuing months and years. A small number of patients initially may respond to therapy and then become intractable with time. Some remain so for years from the onset of the disorder. A direct relationship is present between advancing age and increasing risk of developing persistent pain.

Studies of corticosteroids suggest that they are safe and may exert a preventive effect, at least for the short term in the non-immunosuppressed patient. Long-term effects on the natural history are more difficult to assess and further studies are necessary

to conclusively determine if steroids are effective. Although sympathetic blockade is widely believed to be effective for acute pain with HZ, controlled trials are lacking, and the effect in preventing PHN has not been demonstrated clearly. Amantadine hydrochloride has been shown by one controlled trial to be effective, and this regimen (100 mg twice daily for 1 month) is a reasonable, safe alternative if steroids are contraindicated. Antiviral agents in the immunocompetent patient have not been shown to prevent PHN (acyclovir), to have potential toxicity (vidarabine, AMP), or are impractical because of unavailability (interferon).

With established PHN, the only controlled trials support the use of low-dose antidepressant therapy in altering at least the short-term natural course of the disorder from a moderate or severe to a mild intensity in approximately two thirds of patients. Other therapies must be regarded as unproven or of questionable value. Surgical treatment for the most part is unhelpful; however, DREZ lesions may be considered in medically intractable patients who are good surgical risks.

References

Balfour, H. Acyclovir therapy for herpes zoster: advantages and adverse effects. J.A.M.A. 255:387–388; 1986.

Balfour, H.; Bean, B.; Laskin, O. L.; et al. Acyclovir halts progression of herpes zoster in immunocompromised patients. N. Engl. J. Med. 380:1453; 1989.

Bean, B.; Braun, C.; Balfour, H. H. Acyclovir therapy for acute herpes zoster. Lancet 2:118–121; 1982.

Bean, B.; Aeppli, D.; Balfour, H. H. Acyclovir in shingles. J. Antimicrob. Chemother. 12(suppl. B):123–127; 1983.

Bernstein, J. E.; Bickers, D. R.; Dahl, M. V.; Roshal, J. Y. Treatment of chronic postherpetic neuralgia with topical capsaicin. J. Am. Acad. Dermatol. 17:93–96; 1987.

Brown, G. R. Herpes zoster. Correlation of age, sex, distribution, neuralgia, and associated disorders. South Med. J. 69:576–578; 1976.

Burgoon, C. F.; Burgoon, J. S.; Baldridge, G. D. The natural history of herpes zoster. J.A.M.A. 164:256–269; 1957.

Golding, A. The effect of sympathetic blocks on

herpes zoster. Acta. Anaesthesiol. Scand. 13:113–141; 1969.

Colding, A. Treatment of pain: organization of a pain clinic, treatment of herpes zoster. Proc. R. Soc. Med. 66:541–543; 1973.

Dan, K.; Higa, K.; Noda, B. Nerve block for herpetic pain. In: Fields, H. L.; Dubner, R.; Cervero, F., eds. Advances in pain research and therapy. Vol. 9. New York: Raven Press; 1985: p. 831–838.

Dan, K.; Higa, K.; Tanaka, K.; Mori, R. Herpetic pain and cellular immunity. In: Yokota, T.; Dubner, R., eds. Current topics in pain research and therapy. Proceedings of the International symposium on pain. Amsterdam: Excerpta Medica; 1983: p. 293–305.

De Moragas, J. M.; Kierland, R. R. The outcome of patients with herpes zoster. Arch. Dermatol. 75:193–196; 1957.

Eaglstein, W. H.; Katz, R.; Brown, J. A. The effects of corticosteroid therapy on the skin eruption and pain of herpes zoster. J.A.M.A. 211:1681–1683; 1970.

Elliott, F. A. Treatment of herpes zoster with high doses of prednisone. Lancet 2:610–611; 1964.

Esmann, V.; Ipsen, J.; Peterslund, N. A.; Sayer-Hansen, K.; Schonheyder, H.; Juhl, H. Therapy of acute herpes zoster with acyclovir in the nonimmunocompromised host. Am. J. Med. 73:320–325; 1982.

Esmann, V.; Kroon, S.; Peterslund, N. A.; et al. Prednisolone does not prevent postherpetic neuralgia. Lancet 1:126–128; 1987.

Friedman, A. H.; Nahold, B. S.; Overlmann-Levitt, J. DREZ lesions for postherpetic neuralgia. J. Neurol. 60:1258–1262; 1984.

Friedman, A. H.; Nashold, B. S. DREZ lesions for postherpetic neuralgia. Neurosurgery 15:969–970; 1984.

Galbraith, A. W. Treatment of acute herpes zoster with amantadine hydrochloride (Symmetrel). Br. Med. J. 4:693–695; 1983.

Glynn, C. A study of postherpetic neuralgia and its treatment. Pain Clin. 1:(4):237–246; 1987.

Harding, S. P.; Lipton, J. R.; Wells, J. C. D. Natural history of herpes zoster ophthalmicus predictors of postherpetic neuralgia and ocular involvement. Br. J. Opthalmol. 71:353–358; 1987.

Higa, K.; Dan, K.; Manabe, H.; Noba, D. Factors infuencing the treatment of acute herpetic pain with sympathetic nerve block: importance of severity of herpes zoster assessed by the maximum antibody titres to varicella zoster virus in otherwise healthy patients. Pain 32:147–157; 1988.

Hope-Simpson, R. E. Postherpetic neuralgia. J. R. Coll. Gen. Pract. 25:571–575; 1975.

Jansco, N.; Jansco-Gabor, A.; Szolcsany, J. Direct evidence for neurogenic inflammation and its prevention by dernervation and by pretreatment with capsaicin. Br. J. Pharmacol. 31:138–151; 1967.

Jansco, G.; Kiraly, E.; Jansco-Gabor, A. Pharmacologically induced selective degeneration of chemosensitive primary sensory neurons. Nature 270:741–743; 1977.

Jessell, T. M.; Iversen, L. L.; Cuello, A. C. Capsaicin-induced depletion of substance P from primary sensory neurones. Brain Res. 152:132–188; 1978.

Keczkes, K.; Basheer, A. M. Do corticosteroids prevent postherpetic neuralgia? Br. J. Dermatol. 102:551–555; 1980.

Kernbaum, S.; Hauchecorne, J. Administration of levodopa for relief of herpes zoster pain. J.A.M.A. 246:132–134; 1981.

Lewith, G. T.; Field, F.; Machin, D. Acupuncture versus placebo in postherpetic pain. Pain 17:361–368; 1983.

Loeser, J. D. Herpes zoster and postherpetic neuralgia. Pain 25:149–164; 1986.

Max, M. B.; Schafer, R. N. C.; Culnane, M.; Amitriptyline but not lorazepam relieves postherpetic neuralgia. Neurology 38:1427–1432; 1988.

McCulloch, D. K.; Fraser, D. M.; Duncan, L. P. J. Shingles in diabetes mellitus. Practitioner 226:531–532; 1982.

Merigan, T. C.; Gallagher, J. G.; Pollard, R. B.; Short course human leukocyte interferon in treatment of herpes zoster in patients with cancer. Antimicrob. Agents Chemother. 19:193–195; 1981.

Merigan, T. C.; Rand, K. H.; Pollard, R. B.; et al. Human leukocyte interferon for the treatment of herpes zoster in patients with cancer. N. Engl. J. Med. 298:981–987; 1978.

Nashold, B. S., Jr.; Ostdahl, R. H.; Bullitt, E.; et al. Dorsal root entry zone lesions: a new neurosurgical therapy for deafferentation pain. In: Bonica, J. J.; Albefessard, D., eds. Advances in pain research and therapy. New York: Raven Press; 5:738–750; 1983.

Otsuka, M.; Konski, S. Release of substance P-like immunoreactivity from isolated spinal cord of newborn rat. Nature 164:83; 1976.

Peterslund, N. A.; Seyer-Gansen, K.; Ipsen, J.; Esmann, V.; Schonheyder, H.; Juhl, H. Acyclovir in herpes zoster. Lancet 2:827–831; 1981.

Portenoy, R. K.; Duma, C.; Foley, K. M. Acute herpetic and postherpetic neuralgia: clinical review and current management. Ann. Neurol. 20:651–664; 1986.

Ragozzino, M. W.; Melton, L. J.; Kirland, L. T.; et al. Population based study of herpes zoster and its sequelae. Medicine 21:310–316; 1982.

Riopelle, J. M.; Naraghi, M.; Grush, K. P.

Chronic neuralgia incidence following local anesthetic therapy for herpes zoster. Arch. Dermatol. 120:747–750; 1984.

Rogers, R. S.; Tindall, J. P. Geriatric herpes zoster. J. Am. Geriatr. Soc. 19:495–503; 1971.

Russell, W. R.; Espir, M. L. E.; Morganstern, F. S. Treatment of postherpetic neuralgia. Lancet 1:242–245; 1957.

Shepp, D. H.; Dandliker, P. S.: Meyers, J. D. Treatment of varicella zoster virus infection in severely immunocompromised patients: a randomized comparison of acyclovir and vidarabine. N. Engl. J. Med. 314:208–212; 1986.

Sherlock, C. H.; Corey, L. Adenosine monophosphate for the treatment of varicella zoster infections, a large dose of caution. J.A.M.A. 253:1444–1445; 1985.

Sklar, S. H.; Blue, W. T.; Alexander, E. J.; et al. Herpes zoster, the treatment and prevention of neuralgia by adenosine monophosphate. J.A.M.A. 253:1427–1430; 1985.

Watson, C. P.; Evans, R. J.; Reed, K.; et al. Amitriptyline versus placebo in postherpetic neuralgia. Neurology 32:670–673; 1982.

Watson, C. P. N.; Evans, R. J. Postherpetic neuralgia: a review. Arch. Neurol. 43:836–840; 1986.

Watson, C. P. N.; Evans, R. J.; Watt, V. R. Postherpetic neuralgia and topical capsaicin. Pain 33:333–340; 1988a.

Watson, C. P. N.; Evans, R. J.; Watt, V. R.; Birkett, N. Postherpetic neuralgia: 208 cases. Pain 35:289–297; 1988b.

Whitley, R. J.; Soong, S. J.; Dolin, R. Early vidarabine therapy to control the complications of herpes zoster in immunosuppressed patients. N. Engl. J. Med. 307:971–975; 1982.

Yaksh, T. L.; Farb, D. H.; Leeman, S. E.; Jessell, T. M. Intrathecal capsaicin depletes substance P in the rat spinal cord and produces prolonged thermal analgesia. Science 206:481–483; 1979.

Yanagida, H.; Suwa, K.; Corssen, G. No prophylactic effect of early sympathetic blockade on postherpetic neuralgia. Anesthosiol. 66:73–76; 1987.

44

Trigeminal Neuralgia

GERHARD H. FROMM

Demographic Factors

Trigeminal neuralgia can occur at any age but predominantly afflicts people older than 50 years, with the peak age of onset in the sixth decade (Harris 1940; Rushton and MacDonald 1957). Women are affected more commonly than men with a ratio of 1.6 : 1.0 (Penman 1968). The maxillary and mandibular divisions of the trigeminal nerve are much more frequently involved than the ophthalmic division (Penman 1968; Selby 1984). There also is a tendency for the right side of the face to be involved more often than the left (Harris 1940; Rushton and Mac-Donald 1957; Yoshimasu et al. 1972), and there have been occasional reports of a familial occurrence (Harris 1940; Yoshimasu et al. 1972).

The combination of trigeminal neuralgia and multiple sclerosis occurs more frequently than by chance. It appears that in these patients, the trigeminal neuralgia is caused by plaques of demyelination in the pons at the trigeminal root entry zone (Jensen et al. 1982; Kerr 1970; Lazar and Kirkpatrick 1979; Olafson et al. 1966). These patients are younger and are more

likely to have bilateral trigeminal neuralgia than patients with only trigeminal neuralgia (Brisman 1987a; 1987b; Rushton and Olafson 1965).

Symptoms

The typical attack of trigeminal neuralgia consists of a brief paroxysm of jabbing, burning, or electric shock-like pain in the distribution of one or more branches of the trigeminal nerve. These paroxysms of pain usually are triggered by touching an area of orofacial skin or mucous membrane (Lance 1982; Selby 1984; White and Sweet 1969). However, trigeminal neuralgia does not always start with typical attacks. Some patients report a dull and longer-lasting pain in the upper or lower jaw for months to years before developing typical trigeminal neuralgia in the same branch of the trigeminal nerve (Fromm et al. 1990; Mitchell 1980; Symonds 1949).

Another characteristic of trigeminal neuralgia is a tendency for exacerbations and remissions. In the series reported by Rushton and MacDonald (1957), 78 of 155 patients had one or more symptom-free intervals lasting 6 months or longer, and 38 had remissions of 12 months or longer. However, the natural history of this disorder is further characterized by a progressive increase in frequency, severity, and duration of the exacerbations with time. Some patients eventually have innumerable attacks every day with no let-up.

Trigeminal neuralgia also is unique because it does not respond to conventional analgesic drugs. Even opiates are only marginally effective at very high doses, causing marked sedation. On the other hand, as discussed later, trigeminal neuralgia does respond to some antiepileptic and antispastic drugs that do not otherwise have analgesic properties.

Signs

The neurological examination is normal in trigeminal neuralgia (Lance 1982; Selby 1984; White and Sweet 1969). The diagnostic evaluation of these patients also should include a magnetic resonance imaging (MRI) study (Yuh et al. 1988) because some patients appear to have typical trigeminal neuralgia without neurological deficit, but actually have a cerebellopontine angle tumor (Bullitt et al. 1986; Friedman et al. 1982; Nguyen et al. 1986; Richards et al. 1983; Tanaka et al. 1987; Zorman and Wilson 1984).

Treatment

Although trigeminal neuralgia is among the most excruciatingly painful disorders, it is nonfatal. It should therefore be managed by means that carry the least risk of morbidity or mortality. Medical treatment should be used first, and surgical intervention reserved for patients who fail to respond to medical therapy or develop unacceptable side effects.

No medical or surgical treatment modality has been developed that is capable of permanently stopping the painful paroxysms in all patients. This is probably due in part to the progressive increase in severity of tri-

geminal neuralgia described previously. Only 50% of patients continue to respond to carbamazepine after 5 to 16 years of treatment (Taylor et al. 1981), while at most, 30% continue to respond to phenytoin after 2 years of treatment (Albrecht and Krump 1954; Bergouignan 1957; 1958; White and Sweet 1969). Similarly, 30% to 50% of patients experience a recurrence of trigeminal neuralgia within 18 months to 5 years after glycerol rhizotomy (Beck et al. 1986; Burchiel et al. 1988; Dieckman et al. 1987), and the recurrence rate after radiofrequency rhizotomy ranges from 21% to 65% (Burchiel et al. 1981; Ferguson et al. 1981; Sweet 1986). There are recurrences even after microvascular decompression or suboccipital rhizotomy (Burchiel et al. 1981; Ferguson et al. 1981; Sweet 1986), but the recurrence rate is much lower. Seventy percent of patients experience acceptable pain relief 8.5 years after microvascular decompression (Burchiel et al. 1988).

The greater safety of baclofen makes it the drug of choice for the initial treatment of trigeminal neuralgia, even though it does not appear as effective as carbamazepine (Fromm et al. 1984). It should be started at a dose of 5 to 10 mg three times per day and then increased by 10 mg/day every other day until the patient is pain free or side effects occur (Fromm 1991; Fromm and Terrence 1987). Patients with severe trigeminal neuralgia may need to take baclofen at 3- to 4-hour intervals because of the short biological half-life of this drug.

Because trigeminal neuralgia often is characterized by exacerbations and remission, the dose of baclofen should be tapered gradually after the patient has been pain free for several weeks. If the painful attacks do not recur, the patient can then remain off medication until the next exacerbation occurs. Baclofen should never be discontinued abruptly after long-term administration because hallucinations or seizures may occur (Hyser and Drake 1984; Lees et al. 1977; Stien 1977; Terrence and Fromm 1981). If withdrawal symptoms occur, they are treated by reinstating the previous dose of baclofen and then gradually reducing it by 5 to 10 mg/day each week.

Carbamazepine should be tried if baclofen is ineffective or causes unacceptable side effects. The starting dose should be 100 to 200 mg twice per day. This dose can then be increased by 200 mg/day every other day until the patient is pain free or side effects occur. Most patients will have to take carbamazepine three times per day because of its relatively short biological half-life with chronic use. Fortunately, aplastic anemia is a rare complication of carbamazepine therapy. However, leukopenia, neutropenia, or thrombocytopenia occur in approximately 2% of patients (Hart and Easton 1982). A complete blood count should therefore be performed before commencing treatment, and then at 2-week intervals for the first 2 months (Fromm 1990; Fromm and Terrence 1987). Thereafter, the blood count should be checked at 3-month intervals.

As with baclofen, an attempt to gradually taper off the carbamazepine should be made after the patient has been pain free for several weeks to see if the trigeminal neuralgia is in remission and medication is no longer required.

Due to the progressive increase in severity of trigeminal neuralgia with time, many patients eventually become refractory to baclofen and carbamazepine monotherapy. At that point, the combined administration of baclofen and carbamazepine often controls the attacks (Fromm et al. 1984). Baclofen can be combined with phenytoin in patients who cannot tolerate carbamazepine. The usual dose of phenytoin is 300 to 400 mg/day. Occasionally, patients require a combination of all three drugs: baclofen, carbamazepine, and phenytoin.

Patients may suffer a severe exacerbation of trigeminal neuralgia with extremely frequent excruciating attacks that preclude taking anything by mouth. This can be helped by the intravenous injection of a loading dose of 1000 mg phenytoin. The phenytoin should be administered by slow infusion at a rate no greater than 25 mg/min. Continuous electrocardiogram (EKG) monitoring and frequent blood pressure determinations are required during the infusion, and the injection must be discontinued if arrhythmia or hypotension occur. Intravenous administration of phenytoin is contraindicated in the presence of sinus bradycardia, sinoatrial block, second or third degree atrioventricular block, or severe myocardial insufficiency.

Patients whose trigeminal neuralgia progresses to the extent that none of the medications help become candidates for surgical treatment. The choice between a major procedure (microvascular decompression) or a minor procedure (radiofrequency or glycerol gangliolysis and rhizotomy) depends on the patient's preference, age, and general medical condition. As previously stated, no procedure is effective in all cases. Furthermore, it is not possible to predict accurately who will benefit from each operation. Microvascular decompression offers the possibility of a much longer duration of pain relief (Burchiel et al. 1988) and lack of sensory deficits, but it entails major surgery and longer hospitalization (Burchiel et al. 1981; Ferguson et al. 1981; Fritz et al. 1988; Hanakita and Kondo 1988; Sweet 1986). Radiofrequency or glycerol gangliolysis and rhizotomy involve significantly fewer risks and a much shorter hospital stay, but the pain recurs much sooner and patients frequently experience facial dysaesthesias postoperatively (Burchiel et al. 1988; Burchiel et al. 1981; Ferguson et al. 1981; Sweet 1986). It also is important to remember that there is a wide range in the success rate and incidence of complication reported by various centers for each of these procedures. This variation probably is related to differing levels of experience and expertise. The choice of a neurosurgeon who is skilled in a procedure is at least as important as the choice of the procedure itself.

Finally, patients who experience a recurrence of their trigeminal neuralgia after surgery may respond to another trial of medical therapy. Baclofen, carbamazepine, and phenytoin should therefore be tried once more at that point.

References

Albrecht, K.; Krump, J. Diagnose, Differential-diagnose und Behandlungsmoglichkeiten der Trigeminusneuralgie. Munch. Med. Wschr. 96:1037–1039; 1954.

Beck, D. W.; Olson, J. J.; Urig, E. J. Percutaneous retrogasserian glycerol rhizotomy for treatment of trigeminal neuralgia. J. Neurosurg. 65:28–31; 1986.

Bergouignan, M. Quelques developpements recents de la therapeutique medicale dans la nevralgie essentielle du trijumeau: interet de certaines medications anti-epileptiques. Gaz. Med. France 64:1571–1577; 1957.

Bergouignan, M. Quinze ans d'essais therapeutiques dans la nevralgie essentielle du trijumeau. Rev. Neurol. 98:414–416; 1958.

Brisman, R. Trigeminal neuralgia and multiple sclerosis. Arch. Neurol. 44:379–381; 1987a.

Brisman, R. Bilateral trigeminal neuralgia. J. Neurosurg. 67:44–48; 1987b.

Bullitt, E.; Tew, J. M.; Boyd, J. Intracranial tumors in patients with facial pain. J. Neurosurg. 64:865–871; 1986.

Burchiel, K. J.; Steege, T. D.; Howe, J. F.; Loeser, J. D. Comparison of percutaneous radiofrequency gangliolysis and microvascular decompression for the surgical management of tic douloureux. Neurosurgery 9:111–119; 1981.

Burchiel, K. J.; Clarke, H.; Haglund, M.; Loeser, J. D. Long-term efficacy of microvascular decompression in trigeminal neuralgia. J. Neurosurg. 69:35–38; 1988.

Dieckmann, G.; Bockermann, V.; Heyer, C.; Henning, J.; Roesen, M. Five-and-a-half years' experience with percutaneous retrogasserian glycerol rhizotomy in treatment of trigeminal neuralgia. Appl. Neurophysiol. 50:401–413; 1987.

Ferguson, G. G.; Brett, D. C.; Peerless, S. J.; Barr, H. W. K.; Girvin, J. P. Trigeminal neuralgia: a comparison of the results of percutaneous rhizotomy and microvascular decompression. Can. J. Neurol. Sci. 8:207–214; 1981.

Friedman, A. H.; Wilkins, R. H.; Kenan, P. D.; Olanow, C. W.; Dubois, P. J. Pituitary adenoma presenting as facial pain: report of two cases and review of the literature. Neurosurgery 10:742–745; 1982.

Fritz, W.; Schafer, J.; Klein, H. J. Hearing loss after microvascular decompression for trigeminal neuralgia. J. Neurosurg. 69:367–370; 1988.

Fromm, G. H. Medical treatment of trigeminal neuralgia. In: Fromm, G. H.; Sessle, B. J., eds. Trigeminal neuralgia: current concepts regarding pathogenesis and treatment. Boston: Butterworths; 1991:p. 131–144.

Fromm, G. H.; Terrence, C. F. Medical treatment of trigeminal neuralgia. In: Fromm, G. H., ed. Medical and surgical management of trigeminal neuralgia. Mount Kisko, NY: Futura Publishing Co.; 1987:p. 61–70.

Fromm, G. H.; Terrence, C. F.; Chattha, A. S. Baclofen in the treatment of trigeminal neuralgia: double-blind study and long-term follow-up. Ann. Neurol. 15:240–244; 1984.

Fromm, G. H.; Graff-Radford, S. B.; Terrence, C. F. Pre-trigeminal neuralgia. Neurology 40:1493–1495; 1990.

Hanakita, J.; Kondo, A. Serious complications of microvascular decompression operations for trigeminal neuralgia and hemifacial spasm. Neurosurgery 22:348–352; 1988.

Harris, W. An analysis of 1,433 cases of paroxysmal trigeminal neuralgia (trigeminal tic) and the end-results of gasserian alcohol injection. Brain 63:209–224; 1940.

Hart, R. G.; Easton, J. D. Carbamazepine and hematological monitoring. Ann. Neurol. 11:309–312; 1982.

Hyser, C. L.; Drake, M. E., Jr. Status epilepticus after baclofen withdrawal. J. Natl. Med. Assoc. 76:533–538; 1984.

Jensen, T. S.; Rasmussen, P.; Reske-Nielsen, E. Association of trigeminal neuralgia with multiple sclerosis: clinical and pathological features. Acta. Neurol. Scand. 65:182–189; 1982.

Kerr, F. W. L. Peripheral versus central factors in trigeminal neuralgia. In: Hassler, R.; Walker, A. E., eds. Trigeminal neuralgia: pathogenesis and pathophysiology. Stuttgart: Georg Thieme; 1970:p. 180–190.

Lance, J. W. Mechanisms and management of headache. 4th ed. London: Butterworths; 1982.

Lazar, M. L.; Kirkpatrick, J. B. Trigeminal neuralgia and multiple sclerosis: demonstration of the plaque in an operative case. Neurosurgery 5:711–717; 1979.

Lees, A. J.; Clarke, C. R. A.; Harrison, M. J. Hallucinations after withdrawal of baclofen. Lancet 1:858; 1977.

Mitchell, R. G. Pre-trigeminal neuralgia. Br. Dent. J. 149:167–170; 1980.

Nguyen, M.; Maciewicz, R.; Bouckoms, A.; Poletti, C.; Ojemann, R. Facial pain symptoms in patients with cerebellopontine angle tumors: a report of 44 cases of cerebellopontine angle meningioma and a review of the literature. Clin. J. Pain 2:3–9; 1986.

Olafson, R. A.; Rushton, J. G.; Sayre, G. P. Trigeminal neuralgia in a patient with multiple sclerosis: an autopsy report. J. Neurosurg. 24:755–759; 1966.

Penman, J. Trigeminal neuralgia. In: Vinken, P. J.; Bruyn, G. W., eds. Handbook of clinical neurology. Vol. 5. Amsterdam: North Holland; 1968:p. 296–322.

Richards, P.; Shawdon, H.; Illingworth, R. Operative findings on microsurgical exploration of the cerebello-pontine angle in trigeminal neuralgia. J. Neurol. Neurosurg. Psychiatry 46:1098–1101; 1983.

Rushton, J. G.; MacDonald, H. N. A. Trigeminal neuralgia: special consideration of nonsurgical treatment. J. A. M. A. 165:437–440; 1957.

Rushton, J. G.; Olafson, R. A. Trigeminal neuralgia associated with multiple sclerosis: report of 35 cases. Arch. Neurol. 13:383–386; 1965.

Selby, G. Diseases of the fifth cranial nerve. In: Dyck, P. J.; Thomas, P. K.; Lambert, E. H.; Bunge, R., eds. Peripheral neuropathy. 2nd ed. Philadelphia: W. B. Saunders; 1984:p. 1224–1265.

Stien, R. Hallucinations after sudden withdrawal of baclofen. Lancet 2:44–45; 1977.

Sweet, W. H. The treatment of trigeminal neuralgia (tic douloureux). N. Engl. J. Med. 315:174–177; 1986.

Symonds, Sir C. Facial pain. Ann. R. Coll. Surg. Engl. 4:206–212; 1949.

Tanaka, A.; Takaki, T.; Maruta, Y. Neurinoma of the trigeminal root presenting as atypical trigeminal neuralgia: diagnostic values of or-bicularis oculi reflex and magnetic resonance imaging. Neurosurgery 21:733–736; 1987.

Taylor, J. C.; Brauer, S.; Espir, M. L. E. Long-term treatment of trigeminal neuralgia with carbamazepine. Postgrad. Med. J. 57:16–18; 1981.

Terrence, C. F.; Fromm, G. H. Complications of baclofen withdrawal. Arch. Neurol. 38:588–589; 1981.

White, J. C.; Sweet, W. H. Pain and the neurosurgeon. Springfield, IL: C. C. Thomas; 1969:p. 123–178.

Yoshimasu, F.; Kurland, L. T.; Elveback, L. R. Tic douloureux in Rochester, Minnesota, 1945–1969. Neurology 22:952–956; 1972.

Yuh, W. T. C.; Wright, D. C.; Barloon, T. J.; Schultz, D. H.; Sato, Y.; Cervantes, C. A. MR imaging of primary tumors of trigeminal nerve and Meckel's cave. Am. J. Neurorad. 9:665–670; 1988.

Zorman, G.; Wilson, C. B. Outcome following microsurgical vascular decompression or partial sensory rhizotomy in 125 cases of trigeminal neuralgia. Neurology 34:1362–1365; 1984.

45

Whiplash Syndrome

RANDOLPH W. EVANS

The term "whiplash injury" is used synonymously with acute or chronic cervical sprain or cervical myofascial pain syndrome (Evans 1992). This term was introduced by Dr. Harold Crowe in 1928. While many say that the term is unscientific, whiplash has become well entrenched in the medical literature and the lay-vocabulary. Whiplash describes the typical mechanism of the hyperextension-flexion injury which occurs when an occupant of a motor vehicle is hit from behind by another vehicle. The National Safety Council estimated that there were 3,280,000 rear-end collisions in the United States in 1989. The worldwide incidence is equally substantial. The precise number of whiplash injuries per year has not been determined, but a rough estimate for the United States is more than 1 million.

Rear-end collisions are responsible for about 85% of all whiplash injuries (Hohl 1974a; Deans et al. 1986). Neck pain develops in 56% of patients involved in a front- or side-impact accident (Deans et al. 1986). Seventy-three percent of patients wearing a seatbelt develop neck pain as compared to 53% of those not wearing seatbelts (Deans et al. 1987). However, proper use of headrests can reduce the incidence of neck pain in rear-end collisions by 24% (Nygren 1984; Morris 1989). Whiplash syndrome has a female gender predominance, especially in the 20 to 40 year age group with an overall male:female ratio of about 30% to 70% (Hohl 1974a; Balla 1980; Pearce 1989).

According to one estimate, 94% of patients with late whiplash syndrome see more than one specialist (Balla 1980). Therefore, it is not surprising that the whiplash type of injury is one of the most common problems that neurologists and neurosurgeons evaluate and treat. In addition, because a significant percentage of these injuries results in lawsuits, insurance companies, attorneys, and courts frequently request prognostic information. Many clinicians estimate prognosis based on the perceived severity of the accident, symptoms, objective findings on examination and testing, duration and extent of treatment, credibility of the patient and concerns about compensation neurosis or malingering, their own clinical experience, and for some, doubts about whether chronic whiplash syndrome actually exists. For many, the prognostic literature has not been read or is considered unimportant in comparison to the other factors used in making prognostic judgments. In addition, specialists have a widespread cynicism about treating anxious and depressed patients with chronic or late whiplash syndrome who never seem to get well despite multiple physician opinions, extensive testing, and endless treatments.

Evidence of Organicity

The acute whiplash syndrome, for which symptoms include neck stiffness, aching, and headaches, is well accepted by the public and the medical profession. However, as symptoms persist for months or years, as with chronic or late whiplash syndrome, doubt increases and the patient is suspected of having an emotional problem (The Lancet editorial 1991). This situation is very similar to doubts about the post-concussion syndrome both now and in the late nineteenth century when the controversy over functional vs. organic etiology was about both the post-concussion syndrome and the whiplash type injury ("railway spine") (Trimble 1981; also see "The Post-concussion Syndrome" chapter in this book). Although the case for organicity is not quite as strong as the one for the post-concussion syndrome, the arguments are still quite compelling.

Pathology

There is evidence of structural damage from whiplash-type injuries. Experimentally caused acceleration/extension injuries in primates have demonstrated multiple injuries including muscle damage, rupture of the anterior longitudinal and other ligaments, avulsion of disc from vertebral body, retropharyngeal hematoma, intralaryngeal and esophageal hemorrhage, cervical sympathetic nerve damage, and even various brain injuries including hemorrhages and contusions of brain and brain stem (MacNab 1964; Wickstrom et al. 1967; Ommaya et al. 1968). A recent magnetic resonance imaging (MRI) study performed on selected patients within 4 months of whiplash injuries has confirmed some of these observations (Davis et al. 1991). Findings included ruptures of the an-

terior longitudinal ligament, horizontal avulsion of the vertebral end plates, separation of the disc from the vertebral end plate, occult fractures of the anterior vertebral end plates, acute posterolateral cervical disc herniations, focal muscular injury of the longus colli muscle, posterior interspinous ligament injury, and prevertebral fluid collections. Keith (1985) has described injury of the second cervical ganglion and nerve as the cause of unilateral neck and suboccipital area pain with decreased sensation in the C2 dermatome as a cause of symptoms in a small minority of patients following whiplash injury.

Myofascial Pain

A majority of patients with whiplash injuries may have injury only to muscle, ligaments, and connective tissue—myofascial injury. Although myofascial pain syndromes associated with trigger points and areas of referred pain have been well described with increasing documentation (Simons 1988), acceptance by many in the medical community is not forthcoming. Although myofascial pain is the most common cause of neck and back pain, the surgical community has been fixated with disc disease since the description of "Rupture of the intervertebral disc with involvement of the spinal canal" by Mixter and Barr in 1934. Unfortunately, the equally important observations of Kellgren's in his 1938 paper, "A preliminary account of referred pain arising from muscle," have been overlooked. The improvement in patients with referred pain from trigger points after local anesthetic injection is as impressive today as in 1938.

Vestibular Dysfunction

Many patients complain of dizziness and vertigo following whiplash injuries. Toglia (1976) reported abnormal vestibular tests in 309 patients with whiplash injuries and dizziness: latent nystagmus in 29% of the patients, abnormal calorics in 57%, and abnormal rotatory tests in 51%. Hinoki (1985) reported abnormal equilibrium tests indicating that overexcitation of cervical and lumbar proprioceptors due to hypertonicity of the soft supporting tissues may cause some cases of vertigo. He also reports abnormal optokinetic nystagmus in other patients, suggesting brain stem and cerebellar dysfunction. Chester (1991) has described dysfunction of the semicircular canals, otolith structures, and occasionally perilymph fistulas by a variety of tests including moving platform posturography and electronystagmography (ENG) studies.

Psychological Factors

Prognostic studies, discussed in the next section, demonstrate complaints equally in claimants and non-claimants that persist long after legal settlement. Psychological factors such as neurosis are commonly cited as the cause of persisting symptoms. However, Radanov and coworkers (1991) performed a recent prospective study of 78 consecutive patients demonstrating that psychosocial factors, negative affectivity, and personality traits were not significant in predicting duration of symptoms. Older age of patients, initial neck pain intensity, and injury-related cognitive impairment were significant factors predicting illness behavior. Additionally, other recent studies demonstrate cognitive impairment in a small minority of patients sustaining whiplash injuries (Yarnell and Rossie 1988; Olsnes 1989; Kischka et al. 1991). Deficits are noted in tests of attention, concentration, cognitive flexibility, and memory.

Prognostic Studies

Studies on the prognosis of whiplash injuries are fraught with methodological differences including selection criteria of patients, patient attrition rates, prospective and retrospective designs, duration of follow-up, and treatment used. Some studies do not discuss the treatment modalities used, if any. Although the majority of patients studied probably have only soft tissue injuries, testing such as cervical myelography, computed axial tomography (CAT) scan, or MRI studies have not been uniformly performed to exclude disc disease, cervical stenosis, and ra-

diculopathy due to spondylosis. In general, patients are placed into study groups based on the relatively mild nature of the complaints and physical findings, as compared to patients with cervical fractures, myelopathy, and so forth.

Onset, Frequency, and Duration of Symptoms

Sixty-two percent of patients presenting to the hospital following a motor vehicle accident complain of neck pain (Deans et al. 1987). The onset of neck pain occurs in 65% of patients within 6 hours, within 24 hours in an additional 28%, and within 72 hours in the remaining 7% (Greenfield and Ilfeld, 1977; Deans et al. 1987).

The following six studies evaluated the persistence of neck pain and headaches (Table 45-1). Greenfield and Ilfeld (1977) reported a short-term prospective study of 179 consecutive patients seen in a private practice. All of the patients were involved in litigation. Patients were treated with isometric neck and shoulder exercises. Poor responders (24% of the total patients) also received cervical traction. After an average length of treatment of 7.4 weeks, 37% of the patients were asymptomatic, 47% were improved, and 16% showed no significant recovery.

Deans and coworkers (1987) retrospectively evaluated 137 patients 1 to 2 years after presentation to the hospital following a motor vehicle accident. The following percentages of patients had neck pain for the following durations: 18%, 1 week; 18%, 1 week to 1 month; 13%, 1 to 3 months; 8%, 4 to 6 months; and 1%, 7 to 12 months. The pain lasted more than 1 year in 26.3% with 22.6% experiencing occasional pain, and 3.7% severe continuous pain. By comparison, 7.2% of a control population had neck pain.

Maimaris, Barnes, and Allen (1988) performed a retrospective follow-up for approximately 2 years of 102 patients with whiplash injuries. At the end of 2 years, one-third of patients were still symptomatic with neck pain reported in 89%, neck stiffness in 40%, shoulder pain in 37%, headache in 26%, interscapular pain in 29%, and re-

Table 45-1. Percentage Of Patients With Persistence Of Neck Pain And Headaches After A Whiplash Injury

	1 week	1 month	2 months	3 months	6 months	1 year	2 years	10 years
Neck Pain	92%[†]	64%[†]	63%[*]	51%[†]	43%[†]	26%[†]	29%[‡] 44%[*] 81%[∥] 90%[¶]	74%[#]
Headaches		82%[**]		73%[**]			9%[‡] 37%[§] 37%[∥] 70%[¶]	33%[#]

[*] Greenfield and Ilfeld (1977).
[†] Deans et al. (1987).
[‡] Maimaris, Barnes, and Allen (1988).
[§] Norris and Watt (1983). Subgroup of patients with subjective symptoms only after a mean follow up of 19.7 months.
[∥] Norris and Watt (1983). Subgroup with subjective symptoms and a reduced cervical spine range of movement after a mean follow up of 23.9 months.
[¶] Norris and Watt (1983). Subgroup with subjective symptoms, reduced cervical range of movement, and objective neurological loss after a mean follow up of 24.7 months.
[#] Gargan and Bannister (1990).
[**] Balla and Karnaghan (1987).

ferred symptoms in 40%. Of the other two-thirds who were symptom-free by 2 years, 88% were symptom-free within 2 months.

Norris and Watt (1983) performed a long-term prospective study of 61 patients presenting to the emergency room who were involved in rear-end collisions. Patients were divided into three groups depending on initial symptoms and signs. Of patients with subjective symptoms only and no abnormalities on physical examination and after a mean follow-up of 19.7 months, 44% had neck pain, 37% had headaches, and 17% had paresthesias of upper extremities. Patients with symptoms and a reduced range of movement of the cervical spine but no abnormal neurological signs were followed for a mean of 23.9 months: 81% had neck pain, 37% had headaches, and 29% had paresthesias. Patients with symptoms, a reduced range of cervical movement, and evidence of objective neurological loss were followed for a mean of 24.7 months: 90% had persisting neck pain, 70% reported headaches, and 60% had paresthesias.

Gargan and Bannister (1990) followed up 43 of the 62 patients reported by Norris and Watt (1983) for a mean of 10.8 years. Twelve percent recovered completely, 28% had in-

strusive symptoms, and 12% had severe residual symptoms.

Headaches

Headaches following whiplash injuries are predominantly of the muscle contraction type often associated with greater occipital neuralgia. Occasionally whiplash can precipitate chronic migraine headaches in patients without prior migraines (Winston 1987; Weiss 1991).

Balla and Karnaghan (1987) studied whiplash headache in prospective and retrospective studies. In the prospective study of 180 patients with whiplash injury seen within 4 weeks of the accident, 82% complained of headaches, which were occipitally located in 46%, generalized in 34%, and other locations in 20%. Fifty percent of the patients had pain more than half the time. At 12 weeks, the headache persisted in 73% with one-third having pain more than half of the time.

In the retrospective study of 100 patients, 90% were seen within 6 months and 3 years of the accident. Eighty percent were still having headaches which occurred once a week or more in 40% and were constant in more than half the cases. Half of the patients awoke with headaches and in two thirds, the headaches were more prominent in the morning.

Return to Work and Chronic Disability

In a retrospective analysis of over 5000 cases of whiplash injury, 26% were not able to return to normal activities at 6 months (Balla 1988). In a retrospective study of 102 consecutive patients seen in the emergency department, patients in the good prognostic group (66% of total) had an average time off work of 2 weeks with a maximum of 16 weeks (Maimaris et al. 1988). One third of these patients had no time off at all. In the poorer prognostic group patients, 20% had no time off work, 9% did not return to work by 2 years, and the average time off work was 6 weeks.

In a series of consecutive medico-legal cases, 79% of patients returned to work by 1 month, 86% by 3 months, 91% by 6 months, and 94% by 1 year (Pearce 1989). Nygren (1984) reported that permanent medical disability occurred in 9.6% of patients involved in rear-end collisions and 3.8% involved in front- or side-impact accidents.

Prognostic Variables

Older age of patients was not related to the degree of recovery in a 7-week follow-up (Greenfield 1977). Older age was prognostically significant in follow-ups performed for 6 months (Radanov et al. 1991), 2 years (Maimaris et al. 1988), and 10 years (Gargan and Bannister 1990). The majority of patients who develop the chronic or late whiplash syndrome are between the ages of 21 and 50 (Balla 1980).

The presence of interscapular or upper back pain (Greenfield and Ilfeld 1977; Maimaris et al. 1988), occipital headache (Maimaris et al. 1988), multiple symptoms or paresthesias at presentation (Hohl 1974a; Maimaris et al. 1988; Gargan and Bannister 1990), reduced range of movement of the cervical spine (Norris and Watt 1983), and the presence of an objective neurological deficit (Norris and Watt 1983; Maimaris et al. 1988) predict a less favorable recovery.

Upper-middle compared to lower and higher occupational categories have an increased incidence of symptoms persisting for greater than 6 months. (Balla 1980). Symptoms present 2 years after injury are still present 10 years after injury (Gargan and Bannister 1990).

The duration of symptoms was similar in patients involved in rear-end collisions as compared to other types of collisions (Pennie and Agambar 1991). There is only a minimal association of a poor prognosis with the speed or severity of the collision and the extent of vehicle damage (Kenna and Murtagh 1987). In high-speed collisions, hyperextension injuries can be lessened since the back of the seat is often broken.

Radiographic Findings

Asymptomatic Cervical Spondylosis and Disc Disease

Cervical spondylosis occurs with increasing frequency with age and is often asymptomatic. Irvine and coworkers (1965) took cervical spine x-rays of 10% of the patients on a general practice list in a mining area. Spondylosis occurred in men and women, respectively, at the following ages: 20–29 years, 13%, 5%; 30–39 years, 36%, 11%; 40–49 years, 66%, 46%; 50–59 years, 87%, 73%; 60–69 years, 98%, 91%; and 70 and older, 100%, 96%.

Degenerative disc disease also occurs with increasing frequency with age and is often asymptomatic. Friedenberg and Miller (1963) evaluated cervical spine x-rays of asymptomatic persons, 80 men and 80 women. The following percentages of patients had one or more degenerative discs at the following ages: 30–40 years, 6%; 40–50 years, 25%; 50–60 years, 64%, and 60–70 years, 75%. The incidence was slightly higher and the changes slightly more severe in men than in women. Degeneration occurred most frequently at the C5–6 and C6–7 levels. There was a close correlation between degeneration of the intervertebral disc and proliferative changes at the posterolateral margins of the bodies and the intervertebral foramina.

Teresi and coworkers (1987) studied 100 patients referred for MRI examinations of the larynx without symptoms of cervical spine disease. Disc protrusions were present in 20% of patients 45 to 54 years of age and in 57% of patients older than 64 years. Posterolateral protrusions were seen in 9%. Spinal cord impingement was observed in 16% of patients 45 to 64 years of age and 26% of patients over the age of 64. Spinal cord compression occurred in 7% of cases solely due to disc protrusion with a reduction of spinal cord area up to 16%. With this significant incidence of asymptomatic spondylosis and disc disease, it can be difficult in the individual case to attribute radiographic abnormalities to a whiplash injury unless recent prior films are available for comparison.

Trauma May Accelerate Spondylosis and Disc Disease

There is some evidence suggesting that trauma and whiplash injuries can accelerate the development of cervical spondylosis and degenerative disc disease. Irvine and colleagues (1965) found an increased incidence of spondylosis in miners younger than 40 years who were doing the heaviest work and in people with a history of a serious head or neck injury.

Hohl (1974) performed a 7-year follow-up cervical spine series study of patients with whiplash injuries and a mean age of 30. Thirty-nine percent had developed degenerative disc disease at one or more levels as compared to the expected incidence in this population of 6% found in the Friedenberg and Miller study (1963). Hohl and Hopp (1978) performed a similar 7-year follow-up study of patients who had preexisting degenerative changes at the time of a whiplash injury. New degenerative change occurred at another level in 55%, although there was no correlation between the development of degenerative changes and continued symptoms. In the 10-year follow-up study previously described (Gargan and Bannister 1990), degenerative spondylosis was more common at all ages than in age- and sex-matched controls. This was particularly evident in the 30–40 age group, where spondylosis occurred in 33% compared to 10% of controls (Watkinson 1990).

Cervical Spine X-ray Findings and Prognosis

Cervical spine x-rays at or shortly after the time of the injury provide some prognostic information. Preexisting degenerative osteoarthritic changes portend a poorer prognosis (Norris and Watt 1983; Miles et al. 1988). Abnormal cervical spine curves have been variably reported as prognostic (Norris and Watt 1983) and not prognostic (Maimaris et al. 1988) of a poor outcome. An angular deformity of the cervical spine when not associated with spinal cord injury was found to carry a good prognosis whereas prevertebral soft tissue swelling was found to have no prognostic significance (Miles et al. 1988). Cervical stenosis is a risk factor for the development of myelopathy after whiplash injuries both with cervical spine fractures and/ or dislocations (Eismont et al. 1984) and without them (Epstein et al. 1988).

Litigation And Symptoms

Many lay people and professionals consider prolonged neck pain after a whiplash injury to be a form of compensation neurosis or malingering. A few studies have looked at the effect of litigation on symptoms (Table 45-2).

In Memphis, Tennessee, Gotten (1956) interviewed 100 patients 1 to 26 months after settlement of claims for injury. Fifty-four percent of the patients had no significant symptoms, 34% reported minor discomfort on damp, cloudy days or with exercises or lifting, and 12% continued to have severe symptoms. He concluded, "Once the psychoneurotic symptoms had developed, they persisted for many months and were refractory to treatment, being finally resolved to a great extent by settlement of the litigation."

In Toronto, Ontario, MacNab (1964) surveyed patients 2 or more years after settlement of litigation. Of the 145 patients of the original cohort of 266 available for follow-

Table 45-2. Persistence of Neck Symptoms After Settlement of Litigation

Study	Number of Patients	Selection of Patients	Time from Injury to Settlement	Length of Follow-up after Settlement	Neck Complaints
Gotten (1956) Memphis, TN	100	Neurosurgery office practice	Not reported	1–26 months	54% no "appreciable" symptoms 34% minor symptoms 12% severe symptoms
MacNab (1964) Toronto, Ontario	145	Orthopedic office practice	Not reported	2 or more years	83% most with minor symptoms
Schutt and Dohan (1968) Newark, NJ	7 all women	Employees of RCA Plant	Not reported	Not reported Followup 6–26 months after injury	71%
Hohl (1974) Los Angeles, CA	102 total not stated	Orthopedic office practice patients without cervical degenerative changes	within 6 months	4½+ years	17%
	not stated		after 18 months	3½+ years	62%
Norris and Watt (1983) Sheffield, England		Prospective study of consecutive ER presentations			
Symptoms, no signs	14		17.25 ± 11.9 months	35.8 ± 8.4 months	50% improved 50% no change
Symptoms, reduced range of motion	14		15.9 ± 11.2 months	43.4 ± 9.4 months	64% no change 36% improved
Symptoms, reduced range of cervical movement, and objective neurological loss	8		27.6 ± 6.5 months	43 ± 10 months	50% no change 25% improved 25% worse
Maimaris et al (1988) Leicester, England	10	Retrospective study of consecutive ER presentations	Average 9 months	15–20 months	100%

up, 121 were still having symptoms which for most were minor rather than a cause of significant disability. He also noted that in patients with concomitant injuries, the broken wrist or sprained ankle would heal as expected and yet the neck pain would persist. "It is difficult to understand why the patients' traumatic neurosis should be confined solely to their necks and not be reflected in continuing disability in relation to other injuries sustained at the same time. Moreover, if the symptoms resulting from an extension-acceleration injury of the neck are purely the result of a litigation neurosis, it is difficult to explain why 45% of the pa-

tients should still have symptoms 2 years or more after settlement of their court action."

Schutt and Dohan (1968) studied 74 women with whiplash injuries in New Jersey. Symptoms and litigation status were assessed 6 to 26 months after the accident. Of the 5% with litigation pending, 75% had persisting symptoms. Of the 9.5% with litigation settled, 71% had persisting symptoms. Of the 23% with no litigation at all, 82% had persisting symptoms. The remaining 9.5% had an unknown litigation status.

Norris and Watt (1983) reported a series of 61 patients with whiplash injuries from rear-end collisions in Sheffield, England.

The average time from injury to settlement was 17.25 months for patients with symptoms but no signs. At a mean follow-up of 35.8 months, 50% had persisting symptoms and 50% had improved. Of the patients with symptoms and a reduced range of movement of the cervical spine but no neurological signs, 64% had not improved and the other 36% had improved after a mean follow-up of 43.4 months and an average time from injury to settlement of 15.9 months. Of the patients with symptoms, a reduced range of cervical movement, and evidence of objective neurological loss, 25% had improved, 50% had not changed, and 25% had worsened at a mean follow-up of 43 months and an average time from injury to settlement of 27.6 months. Symptoms present 2 years after injury were still present 10 years after injury (Gargan and Bannister 1990).

Maimaris, Barnes, and Allen (1988) performed a retrospective study of 102 consecutive patients presenting to the emergency room with whiplash injuries with a followup of about 2 years. "The average time for settlement was 9 months, yet all these patients continued to have symptoms for between 2 and 2.5 years after injury. These results suggest that litigation does not influence the natural progression of symptoms." Pennie and Agambar (1991) found no significant difference in recovery between claimants and nonclaimants.

In summary, most of these studies indicate that the end of litigation does not signal the end of symptoms for many symptomatic patients; they are not cured by a verdict (Mendelson 1982). The patients who exaggerate or malinger are a distinct minority.

Treatment

Patients with whiplash injuries undergo an astounding array of treatments that have been poorly studied in randomized prospective studies. Physicians and healers are unwitting partners in health care (Murray and Rubel 1992). Medications prescribed include nonsteroidal anti-inflammatory medications, muscle relaxants, analgesics, and tricyclic antidepressants. Soft collars and exercises are recommended. Surgery is sometimes advised for questionably causative bulging discs and spondylosis. Physical therapy modalities include heat, cold, traction, massage, ultrasound, stretch and spray, electrical stimulation, acupressure, and transcutaneous electrical nerve stimulation (TENS). Trigger point injections and greater occipital nerve blocks are performed with local anesthetic agents and at times steroids. Acupuncture, osteopathic manipulation, and chiropractic adjustments (occasionally under general anesthesia) are performed. Psychotherapy and biofeedback are also utilized.

These treatments are provided without data demonstrating efficacy, duration or intensity of treatment, or guidelines for patient selection. The costs are substantial, particularly in the United States where these services have been relatively overvalued. Often, these treatments are provided by well meaning practitioners. In a small minority of cases, practitioners are more concerned with their financial interests in the treatments rather than with improvement of the patient's symptoms and function. Some plaintiffs and their attorneys encourage additional treatment attempting to magnify the nature of the injury. Adequately controlled prospective studies are greatly needed to help provide a scientific consensus for management (Newman 1990).

Randomized Studies

Three recent studies compare active vs. passive treatments. Mealy and colleagues (1986) performed a randomized study of 62 patients, comparing active to passive treatment. The active treatment group received neck mobilization using the Maitland technique of passive mobilization, daily active exercises of the cervical spine, and local heat. The standard treatment group was treated with a soft cervical collar and 2 weeks of rest before gradual mobilization. Simple analgesics were used for both groups. After 8 weeks, the active treatment group showed greater improvement as measured by a linear analog pain scale and active cervical movement.

McKinney and associates (1989) performed a singleblind prospective randomized trial of 247 consecutive patients presenting to the emergency room with an acute whiplash injury. All patients were give a soft cervical collar and placed on an analgesic, codydramol. The first treatment group was given general advice about mobilization after an initial rest period of 10 to 14 days. The second group received physical therapy 3 times per week for 6 weeks consisting of hot and cold applications, shortwave diathermy, hydrotherapy, traction, and active and passive repetitive movements of the neck using Maitland exercises. They also received posture instructions. The third group was instructed to use heat sources at home and was instructed in mobilizing exercises that were demonstrated. Results were obtained by use of a visual analog pain scale and cervical movement. After 2 months, the patients given specific instructions in mobilization did just as well as the patients receiving 6 weeks of physical therapy. Both of these groups did better than the group given just the soft cervical collar, rest, and analgesia.

Pennie and Agambar (1990) performed a prospective trial of management of 135 adults. All patients were provided soft collars initially, then randomly assigned to a standard or active treatment group. Standard treatment was 2 weeks of rest in either a soft or molded foam collar with slight flexion followed by a program of active exercises. At 6 to 8 weeks, the patients who had deteriorated or did not improve were referred for physical therapy. The active treatment consisted of intermittent halter traction twice a week and instruction in neck and shoulder exercises. There was no significant difference between the collar group and the active traction group in both the 6–8-week or 5-month surveys.

Conclusion

Gay and Abbott (1953) studied 50 patients who suffered a whiplash injury and concluded, "Characteristically, these patients were more disabled and remained handicapped for longer periods than was anticipated, considering the mild character of the accident." Additionally, "The symptoms from this condition were so tenacious and the rate of recovery so slow that it was felt that every effort should be made by public authorities to prevent this kind of suffering." Gotten (1956) reviewed the chronic cases and stated, ". . . in some instances there were indications that the injury was being used by the patient as a convenient lever for personal gain."

After 40 years of prognostic studies, similar polarization of opinion still exists. Porter (1989) states, "Pain, suffering, and disability after acute neck sprains may be reduced by doctors recognizing that these injuries, especially those that occur after rearend impacts, may cause long term disability." Conversely, Pearce states, "Few topics provoke so much controversy or heated opinion, based on so little fact as whiplash injuries. In emergency departments, orthopaedic, neurological and rheumatological clinics, and not least in the courts, this common syndrome is shrouded in mystery and creates clinical insecurity in those who attempt to explain its mechanism, its prognosis and treatment. These problems are compounded in medico-legal practice where the potential rewards of successful litigation may colour the clinical picture. Most victims of whiplash injury have, however, sustained no more than a minor sprain to the soft tissues and unusually severe or protracted complaints may demand explanations which lie outside the fields of organic and psychiatric illness."

Certainly, a small percentage of patients are malingering or are instead motivated by a not-so-hidden agenda. Premorbid psychopathology can be accentuated by pain to further obscure the picture. However, if the patient is considered credible, the physician should not be deterred from making a diagnosis of chronic whiplash syndrome just because the diagnosis is often one of exclusion based on subjective complaints without demonstrable pathophysiology. Using similar criteria, clinicians diagnose most migraine and tension type headaches on a daily basis without any reservations. Rather than debating whether chronic whiplash syndrome exists, the challenge for future prognostic studies is to determine the relative efficacy of the various treatments available. Perhaps then, unnecessary treatments and expenditures could be reduced and eliminated and more efficacious treatments developed.

References

Balla, J. I. The late whiplash syndrome. Aust. N. Z. J. Surg. 50:610–614; 1980.

Balla, J.; Karnaghan, J. Whiplash headache. Clin. Exp. Neurol. 23:179–182; 1987.

Chester, J. B. Whiplash, postural control, and the inner ear. Spine 16:716–720; 1991.

Davis, S. J.; Teresi, L. M.; Bradley, W. G.; Ziemba, M. A.; Bloze, A. E. Cervical spine hyperextension injuries: MR findings. Radiology 180:245–251; 1991.

Deans, G. T.; McGalliard, J. N.; Rutherford, W. H. Incidence and duration of neck pain among patients injured in car accidents. Br. Med. J. 292:94–95; 1986.

Deans, G. T.; McGalliard, J. N.; Kerr, M.; Rutherford, W. H. Neck sprain—a major cause of disability following car accidents. Injury 18:10–12; 1987.

Eismont, F. J.; Clifford, S.; Goldberg, M.; Green, B. Cervical sagittal spinal canal size and spine injury. Spine 9:663–666; 1984.

Epstein, N.; Epstein, J. A.; Benjamin, V.; Ransohoff, J. Traumatic myelopathy in patients with cervical spinal stenosis without fractures or dislocation—methods of diagnosis, management, and prognosis. Spine 5:489–496; 1980.

Evans, R. W. Some observations on whiplash injuries. Neurologic Clinics 10(4); 1992 (in press).

Friedenberg, Z. B.; Miller, W. T. Degenerative disc disease of the cervical spine. A comparative study of asymptomatic and symptomatic patients. J. Bone Joint Surg. 45A:1171–1178; 1963.

Gargan, M. F.; Bannister, G. C. Long-term prognosis of soft tissue injuries of the neck. J. Bone Joint Surg. 72B:901–903; 1990.

Gay, J. R.; Abbott, K. H. Common whiplash injuries of the neck. J.A.M.A. 152:1698–1704; 1953.

Gotten, N. Survey of 100 cases of whiplash injury after settlement of litigation. J.A.M.A. 162:865–867; 1956.

Greenfield, J.; Ilfeld, F. W. Acute cervical strain: evaluation and short term prognostic factors. Clin. Orthop. 122:196–200; 1977.

Hinoki, M. Vertigo due to whiplash injury: a neurotological approach. Acta. Otolaryngol. (Stockh.) Suppl. 419:9–29; 1985.

Hohl, M. Soft tissue injuries of the neck in automobile accidents: factors influencing prognosis. Journal of Bone and Joint Surgery 56A:1675–82; 1974.

Hohl, M.; Hopp, E. Soft tissue injuries of the neck. II. Factors influencing prognosis (abstr). Orthop. Transact. 2:29; 1978.

Irvine, D. H.; Fisher, J. B.; Newell, D. J.; Klukvin, B. N. Prevalence of cervical spondylosis in a general practice. Lancet 1:1089–1092; 1965.

Keith, W. S. "Whiplash"—Injury of the 2nd cervical ganglion and nerve. Can. J. Neurol. Sci. 13:133–137; 1986.

Kellgren, J. H. A preliminary account of referred pain arising from muscle. B.M.J. 1:325–327; 1938.

Kenna, C.; Murtagh, J. Whiplash. Austr. Family Physician 16:727, 729, 733, 736; 1987.

Kischka, U.; Ettlin, Th.; Heim, S.; Schmid, G. Cerebral symptoms following whiplash injury. Eur Neurol 31:136–140; 1991.

The Lancet editorial. Neck injury and the mind. Lancet 338:728–729; 1991.

McKinney, L. A.; Dornan, J. O.; Ryan, M. The role of physiotherapy in the management of acute neck sprains following road-traffic accidents. Arch. Emerg. Med. 6:27–33; 1989.

MacNab, I. Acceleration injuries of the cervical spine. J. Bone Joint Surg. 46A:1797–1799; 1964.

MacNab, I. The whiplash syndrome. Orthop. Clin. North Am. 2:389–403; 1971.

Maimaris, C.; Barnes, M. R.; Allen, M. J. Whiplash injuries of the neck: a retrospective study. Injury 19:393–6; 1988.

Mealy, K.; Brennan, H.; Fenelon, G. C. C. Early mobilization of acute whiplash injuries. Br. Med. J. 292:656–657; 1986.

Mendelson, G. Not "cured by a verdict." Effect of legal settlement on compensation claimants. Med. J. Aust. 2:132–134; 1982.

Miles, K. A.; Maimaris, C.; Finlay, D.; Barnes, M. R. The incidence and prognostic significance of radiological abnormalities in soft tissue injuries to the cervical spine. Skeletal Radiol 17:493–496; 1988.

Mills, H.; Horne, G. Whiplash—man-made disease? N. Z. Med. J. 99:373–374; 1986.

Mixter, W. J.; Barr, J. S. Rupture of the intervertebral disc with involvement of the spinal canal. N.E.J.M. 211:210–216; 1934.

Morris, F. Do head-restraints protect the neck from whiplash injuries? Arch. Emergency Med. 6:17–21; 1989.

Murray, R. H.; Rubel, A. J. Physicians and healers—unwitting partners in health care. N.E.J.M. 326:61–64; 1992.

Norris, S. H.; Watt, I. The prognosis of neck injuries resulting from rear-end vehicle collision. J. Bone Joint Surg. 65B:608–611; 1983.

Nygren, A. Injuries to car occupants: some aspects of the interior safeties of cars. Acta. Otolaryngol. Suppl. (Stockh.) 395:1–164; 1984.

Olsnes, B. T. Neurobehavioral findings in whiplash patients with long-lasting symptoms. Acta. Neurol. Scand. 80:584–588; 1989.

Ommaya, A. K.; Faas, F.; Yarnell, P. Whiplash injury and brain damage—an experimental study. J.A.M.A. 204:285–289; 1968.

Pearce, J. M. S. Whiplash injury: a re-appraisal. J. Neurol. Neurosurg. Psychiatry 52:1329–1331; 1989.

Pennie, B. H.; Agambar, L. J. Whiplash injuries: a trial of early management. J. Bone Joint. Surg. 72B:277–279; 1990.

Porter, K. M. Neck sprains after car accidents: a common cause of long term disability. Br. Med. J. 298:973–974; 1989.

Radanov, B. P.; Dvorak, J.; Valach, L. Psychological changes following whiplash injury of the cervical vertebrae. Schweiz Med. Wochensch 119:536–543; 1989.

Radanov, B. P.; Stefano, G. D.; Schnidrig, A.; Ballinari, P. Role of psychosocial stress in recovery from common whiplash. Lancet 338:712–715; 1991.

Schutt, C. H.; Dohan, F. C. Neck injury to women in auto accidents: a metropolitan plague. J.A.M.A. 206:2689–2692; 1968.

Simons, D. G. Myofascial pain syndromes: where are we? Where are we going? Arch. Phys. Med. Rehabil. 69:207–212; 1988.

Teresi, L. M.; Lufkin, R. B.; Reicher, M. A.; Moffit, B. J.; Vinuela, F. V.; Wilson, G. M.; Bentson, J. R.; Hanafee, W. N. Asymptomatic degenerative disk disease and spondylosis of the cervical spine: MR imaging. Radiology 164:83–88; 1987.

Toglia, J. U. Acute flexion-extension injury of the neck. Electronystagmographic study of 309 patients. Neurology 26:808–814, 1976.

Trimble, M. R. Post-traumatic neurosis. Chichester: John Wiley and Sons; 1981.

Watkinson, A. F. Correspondence: whiplash injury. Br. Med. J. 301:983; 1990.

Weiss, H. D.; Stern, B. J.; Goldberg, J. Post-traumatic migraine: chronic migraine precipitated by minor head or neck trauma. Headache 31:451–456; 1991.

Wickstrom, J.; Martinez, J.; Rodriguez, R. Cervical sprain syndrome and experimental acceleration injuries of the head and neck. In Selzer, M. L.; Gikas, P. W.; Huelke, D. F. (eds.): Proc. Prevention of Highway Accidents Symposium. University of Michigan, 1967, p. 182–187.

Winston, K. Whiplash and its relationship to migraine. Headache 27:452–457, 1987.

Yarnell, P. R.; Rossie, G. V. Minor whiplash head injury with major debilitation. Brain Injury 2:255–258; 1988.

46

Neoplasms

VICTOR A. LEVIN AND RICHARD P. MOSER

Cerebral Gliomas

Gliomas represent the largest group of primary central nervous system (CNS) neoplasms. Histologically, they vary in malignancy from the juvenile pilocytic astrocytomas of childhood to the most aggressive, glioblastoma multiforme. While most gliomas are astrocytomas, oligodendrogliomas, ependymomas, and various combinations of two or more cell types occur in this broad grouping. Among the gliomas, each subtype and level of malignancy is associated with age-specific prevalence patterns.

Treatment

Treatment is multimodal, incorporating surgery, radiation therapy, and chemotherapy is selected cases. Surgery is necessary for histological diagnosis, grading of malignancy, and debulking. While radiation therapy is beneficial for high-grade gliomas, the benefit is controversial for low-grade gliomas.

The use of chemotherapy is more controversial. This is primarily because adjuvant chemotherapy following radiation is given for 1 year or longer and is associated with continual morbidity during treatment. In addition, except for a limited number of earlier trials by the U.S. Brain Tumor Study Group (BTSG) and the Radiation Treatment Oncology Group (RTOG), adjuvant chemotherapy usually is not currently randomized against radiation therapy alone.

Controlled clinical trials have demonstrated the efficacy of many drugs when combined with irradiation as adjuvant therapy. Efficacy has been shown for carmustine (BCNU), lomustine (CCNU), PCNU, procarbazine, streptozotocin, and the combination of lomustine, procarbazine, and vincristine. Lesser activity has been shown for some of the newer agents, such as aziridinylbenzoquinone and spiromustine. The consensus has been that adjuvant chemotherapy following surgery and radiation therapy for glioblastoma and anaplastic astrocytomas modestly increases both time to tumor progression and survival (FazeKas 1977; Kramer 1983; Leibel et al. 1975; Levin et al. 1989; Marsa et al. 1975; Sheline et al. 1964; Walker et al. 1978; Walker et al. 1979).

Prognosis

Survival is generally well correlated to histological diagnosis, although patient age at the time of tumor onset can be an extremely powerful deterent to long-term survival regardless of age. Between these two views, however, there is overlap because younger patients usually have lower grade gliomas.

While gliomas occur more frequently in men than in women, survival is unaffected by sex. Similarly, while gliomas are less common among blacks than other races, survival appears to be unrelated to race. Aside from occupational patterns, survival is unaffected by the geographical location of patients. Typically, patients with a long history of symptoms, particularly seizures, live longer after histological diagnosis than those with shorter histories.

Tumor volume closely reflects tumor cell burden, and postsurgical tumor volume correlates inversely with survival for gliomas (Andreou et al. 1983; Levin et al. 1980). For patients receiving postsurgery radiation and chemotherapy, there is an approximate two-fold difference in survival for patients whose tumors are in the lower 10% vs. those with tumors in the upper 90% in volume (Levin et al. 1980). For this reason, gross total surgical resection was a dominant factor favoring longer survival in a large series of patients with low-grade astrocytomas treated at the Mayo Clinic (Laws et al. 1984). Most large cooperative group trials of radiation–chemotherapy regimens show a correlation between the extent of surgical resection and subsequent survival in patients with more malignant astrocytomas (Levin et al. 1990; Walker et al. 1978; Walker et al. 1979). In addition, a review by Salcman and associates (1982) of older literature reporting on the results of treatment of more than 600 patients with such malignant gliomas who received only surgical treatment confirms this correlation.

The benefit of postoperative radiation therapy for low-grade astrocytomas is shown by 5-year survival rates of 13% to 19% for incomplete surgical resection vs. 41% to 46% with the addition of postoperative radiotherapy (Fazekas 1977; Leibel et al. 1975). Laws and colleagues (1984) report a 5-year survival rate of 49% for patients who received at least 40 Gy versus 34% ($p = 0.05$) for those with lesser doses or no irradiation. Ten-year survival rates of 6% to 35% have been reported, depending on extent of resection and whether radiotherapy was used following resection (Bloom 1982; Laws et al. 1984; Leibel et al. 1975).

For more anaplastic gliomas, radiation therapy clearly is beneficial. The most telling randomized clinical trial in which patients with malignant gliomas were randomized to receive or not receive postoperative radiotherapy was carried out by the BTSG (Walker et al. 1978). Of 222 patients, 90% had glioblastoma multiforme and 9% anaplastic astrocytoma. The median survival was 14 weeks without radiotherapy and 36 weeks with it ($p = 0.001$). The 12-month survival rates were 24% with radiotherapy and 3% without. Combining data from several BTSG clinical trials demonstrated that 60 Gy in 6 to 7 weeks yielded a better survival than 50 Gy in 5 to 6 weeks (Walker et al. 1979). No additional improvement in survival was noted when the total dose was escalated from 60 to 70 Gy.

For anaplastic gliomas, an RTOG–Eastern Cancer Oncology Group (ECOG) prospective randomized trial of 626 patients found a median survival time of 28 months for anaplastic astrocytoma 8 months and 15% survival at 18 months for glioblastoma multiforme. Comparable differences also have been reported for anaplastic astrocytoma (or malignant astrocytoma) and glioblastoma multiforme using data from older retrospective studies (Kramer 1983; Marsa et al. 1975; Sheline 1975).

Chemotherapy programs are not reviewed here, but it is apparent that some regimens are better than others. For example, postradiation BCNU was compared to the PCV combination (Levin et al. 1990).

The last analysis showed for glioblastoma multiforme patients, that 50% of patients survived 50 weeks with PCV vs. 59 weeks for BCNU; 25% survived 94 and 71 weeks, respectively. This was more significant ($p = .009$), however, for anaplastic tumors. Fifty percent of PCV patients survived 157 weeks vs. 82 weeks for BCNU patients; 25% of patients were alive at more than 320 weeks with PCV, but only 214 weeks with BCNU.

Brain Stem Gliomas

Gliomas of the brain stem are usually intrinsic or intrinsic with exophytic components in the fourth ventricle, peripontine cisterns, or in both locations. The intrinsic lesions can be diffuse and infiltrative or focal; the latter have a better prognosis (Edwards et al. 1989). In addition, patients with exophytic components in the fourth ventricle do better than those with infiltrative central lesions. As with cerebral gliomas, lower grade anaplastic tumors do better than higher grade anaplastic tumors.

Treatment

Surgery for brain stem tumors is of limited value. By virtue of location, these tumors rarely are totally excised. Surgery for exophytic tumors appears to be somewhat beneficial (Epstein and Wisoff 1988). Diffuse pontine tumors frequently can be biopsied with the risk of creating transient or permanent neurological deterioration. Ventriculoperitoneal shunting for obstructive hydrocephalus can provide symptomatic improvement in patients in whom the tumor blocks the fourth ventricle.

Radiation remains the standard therapy for all brain stem tumors that are more malignant than juvenile pilocytic astrocytoma. The use of chemotherapy to treat patients if radiation therapy fails is palliative.

Prognosis

The poor results obtained with conventional radiation dose schedules, with or without chemotherapy (Fulton et al. 1981; Jenkin 1983; Levin et al. 1984), has led to newer hyperfractionation radiation treatments that deliver higher total doses. In one study, 100 cGy were given twice daily for a total dose of 72 Gy (Edwards et al. 1989). Results from this hyperfractionation program found a 50% increase in survival among children when compared to other regimens (Table 46-1).

Table 46-1. Brain Stem Glioma Therapy

Treatment	Median Survival	Reference
Radiation therapy alone 50 to 60 Gy	35	32
Radiation therapy followed by CCNU+VCR+PRED	44	32
5-FU+CCNU prior to RT+HU+MISO	44	38
Hyperfractionated radiation therapy at 72 Gy	64/92*	26

* ≤18 yrs/>18 yrs at treatment

Adjuvant chemotherapy has not been shown to be advantageous, although chemotherapy for recurrent or progressive brain stem gliomas has demonstrated some palliative benefit (Fulton et al. 1981; Rodriguez et al. 1988).

Cerebellar Astrocytomas

In general, these tumors have a better prognosis than cerebral or brain stem gliomas. They occur at a younger age and are more likely to be low grade (Allen et al. 1986). Sometimes these tumors are associated with neurofibromatosis.

Treatment

The treatment of choice is gross total surgical resection regardless of histology; if this fails, extensive subtotal resection is advocated. For tumors more malignant than juvenile pilocytic astrocytoma, radiation therapy is advocated because it substantially improves the survival for patients with incompletely resected highly anaplastic gliomas (Allen et al. 1986; Chamberlain et al. 1990).

Chemotherapy has been used infrequently, so it is hard to evaluate its efficacy. For highly anaplastic gliomas that recur or progress following radiation therapy, chemotherapy can provide palliation.

Prognosis

Cure is achievable but not guaranteed for patients with fully resected low-grade cerebellar astrocytomas. Five-year survival rates of 91% in children with cerebellar astrocytomas have been reported (Duffner et al. 1986). Chemotherapy at recurrence can provide palliation with relapse-free survivals in 50% of patients at 18 months and 25% of patients at 32 months or more (Chamberlain et al. 1990).

Oligodendrogliomas

While most oligodendrogliomas occur in the cerebral hemispheres, approximately 10% disseminate through the cerebrospinal fluid (CSF) pathways and have a worse prognosis (Levin et al. 1989). Even though oligodendrogliomas can be grossly resected, they frequently recur in the operative site and require reoperation.

Treatment

No prospective controlled studies of radiation or chemotherapy for oligodendrogliomas have been published. Five- and 10-year survival rates of 85% and 55%, respectively, can be achieved for patients who receive radiotherapy and 31% and 25% for those who do not (Bullard et al. 1987; Chin et al. 1980). A more recent review (Wallner et al. 1988) yields 5- and 10-year actuarial survival rates of 78% and 56% for patients with pure oligodendroglioma who had received at least 45 Gy and 54% and 18% for those who were not irradiated. The survival curve for patients with irradiated mixed tumors (i.e., oligodendroglioma–astrocytoma) was virtually identical to that for the irradiated pure tumors.

Chemotherapy for recurrent or progressive disease appears to be of some benefit (Cairncross and Macdonald 1988; Levin et al. 1989). A nitrosurea-based chemotherapy program, such as the combination of CCNU, procarbazine, and vincristine, can produce a median relapse-free survival of approximately 1.4 years (Levin et al. 1989).

Prognosis

Like astrocytomas, oligodendrogliomas vary in malignancy and survival is closely

coupled with histology (Chin et al 1980; Ludwig et al. 1986). Those designated grades A and B do better than those considered grades C and D (Ludwig et al. 1986); some believe that a three-tiered histological grading system is sufficient (Bullard et al. 1987; Chin et al. 1980).

Ependymoma

Ependymomas occur intracranially above the tentorium cerebelli, infratentorially, and along the spinal axis. The ratio of spinal axis to intracranial ependymomas is approximately 1:2 (Helseth and Mrk 1989). In a recent series of 62 patients, 35% had supratentorial, 33% had infratentorial, and 30% had intramedullary spinal cord tumors. These groups had mean ages of 17, 7, and 41 years, respectively, at the time of first symptoms (Rawlings et al. 1988). Extension into the subarachnoid space occurs in up to 50% and carries a worse prognosis. The anaplastic ependymomas are more likely to disseminate through the CSF pathways. Differentiated or grade 2 (World Health Organization classification) ependymomas are twice as frequent as anaplastic ependymomas (grade 3) (Ernestus et al. 1989).

Treatment

Surgery is the major treatment modality for both cranial and spinal axis tumors. Following complete surgical resection of differentiated ependymomas, radiation normally is not given. Following recurrence and for incompletely resected tumors, radiation therapy is advocated. For the treatment of anaplastic ependymomas following surgery, radiation therapy is required; chemotherapy as postradiation therapy awaits validation.

The use of chemotherapy is experimental. Palliation with BCNU (Levin et al. 1989) and dibromodulcitol (Levin et al. 1984) has been observed. Activity in an adjuvant setting with radiation therapy has been noted in children (Kun et al. 1988).

Prognosis

The 5- and 10-year survival rates for those who received more than 45 Gy is 61% to

75% and 58%, respectively (Shaw et al. 1987; Wallner 1986). For anaplastic ependymomas, 5-year survival rates range from 10% to 50%. The 5-year survival rate for patients with spinal axis ependymoma was 89% in contrast to 24% for patients with intracranial ependymoma (Helseth and Mrk 1989).

The risk of CNS dissemination along CSF pathways from intracranial ependymomas depends on the site of origin and the grade of malignancy, with 50% incidence occurring in cases with high-grade lesions situated in the posterior fossa (Bloom et al. 1990). Survivals at 5, 10, and 15 years in 51 children were 51%, 40%, and 31%, respectively.

Although some authors argue that histology does not influence survival (Ross and Rubinstein 1989), most believe that marked survival differences exist (Nazar et al. 1990). The 5-year survival rate without recurrence was 57.4% in grade 2 ependymomas and 24.1% in grade 3 ependymomas (DiMarco et al. 1988).

Primary CNS Lymphoma

Most primary CNS lymphomas are considered to be B-cell and of the histiocytic (large cell or large cell immunoblastic) type (Hochberg and Miller 1988). These tumors occur in immunocompromised hosts and are high in renal transplant patients (Schneck and Penn 1971) and in acquired immune deficiency syndrome (AIDS) patients (Payan et al. 1984; Rosenblum et al. 1988). These tumors are increasing at an alarming rate and may shortly become the most common neurological neoplasm because of the increase in sporadic occurrence and in the AIDS population (Hochberg and Miller 1988). Three percent of AIDS patients will develop this tumor either prior to AIDS diagnosis or during their course of treatment. In addition to acquired immunosuppression, patients with inherited disorders of the immune system are predisposed to the development of CNS lymphoma.

Typically, these tumors involve multiple areas of the neuraxis, the eye, and multiple intracranial sites without obvious evidence of systemic lymphoma (Hochberg and Miller 1988).

Treatment

Because a tissue diagnosis is essential and surgery is not curative, patients may receive only a computed tomography (CT)–stereotaxic biopsy. The mainstay of therapy is radiation. For grossly confined intracranial tumors, whole-brain radiation therapy is indicated; for more widespread disease associated with leptomeningeal spread, craniospinal axis irradiation is indicated. Chemotherapy at initial diagnosis and for recurrent or progressive disease is widely used, although a clear advantage of one regimen over another is lacking (Hochberg and Miller 1988; Murray et al. 1986; Neuwelt et al. 1986).

Prognosis

Radiation therapy is of proven value, although the improvement may last only 12 to 24 months (Hochberg and Miller 1988; Murray et al. 1986; Neuwelt et al. 1986). The 1- and 5-year survival rates are 66% and 7%, respectively. AIDS-related CNS lymphomas appear to respond to irradiation similar to other CNS lymphomas; however, AIDS patients are much more likely to die of other causes.

Positive CSF cytology or intracranial and spinal lesions carry a poorer prognosis.

Primitive Neuroepithelial Tumors

Primarily a disease of early childhood, primitive neuroepithelial tumors (PNET) represent a controversial nosology of primitive tumors. Some authors divide them into the following classifications: medulloepithelioma, neuroblastoma, spongioblastoma, ependymoblastoma, pineoblastoma, and medulloblastoma. With the exception of the medulloblastoma, primitive neuroepithelial tumors are rare. This discussion of PNETs is restricted to tumors that are located in the cerebral hemisphere and are composed of predominantly undifferentiated neuroepithelial tumor with or without glial or neuronal differentiation.

Treatment

The initial therapy for primitive neuroecto-dermal tumors is surgical bulk reduction whenever feasible. It is well documented that PNETs metastasize and that these patients should be staged and their tumors treated like medulloblastoma and given craniospinal axis irradiation. Most specialists treat these patients as "poor risk" medulloblastomas. Because this is an uncommon tumor, no controlled chemotherapy trials exist, although chemotherapy before and after radiation frequently is given.

Prognosis

The prognosis for these patients is poor, regardless of treatment or stage. Prediction of survival is lacking, but few patients survive even 4 years (Tomita et al. 1988). Berger and colleagues (1983) reviewed the results of treatment for 11 patients with cerebral neuroblastoma and found that of six patients with cystic tumors, none had recurrence, while four of the five solid-tumor patients had recurrences.

Medulloblastoma

Initially localized to the cerebellum, these tumors can spread down the spinal axis and outside the neuraxis. Some studies indicate that up to 30% of cases have positive cytology or myelographic evidence of spinal metastasis at diagnosis (Allen et al. 1986; Bloom 1982). Medulloblastoma occurs in children and adults with a ratio of approximately 2 : 1 in some series.

Treatment

Surgery should be aggressive and seek to remove as much tumor as possible. Conventional radiation therapy is given to the craniospinal axis with a "boost" to the posterior fossa (Bloom 1982; Bloom et al. 1982).

Chemotherapy before and after radiation therapy is being evaluated in an effort to improve survival (Levin et al. 1988; Loeffler et al. 1988; Packer et al. 1988). Medulloblasto-mas are responsive to wide variety of antineoplastic agents, including vincristine, nitrosourea, procarbazine, dibromodulcitol, cyclophosphamide, methotrexate, cisplatin, and various drug combinations (Allen et al. 1986; Friedman and Schold 1985; Levin et al. 1983; vanEys et al. 1988). Many patients treated with craniospinal irradiation have reduced bone marrow reserves, which complicates chemotherapy.

Prognosis

Prognosis is defined, to an extent, by staging criteria. Poor risk is defined as (1) less than a 75% resection; (2) metastasis to spinal cord, cerebrum, leptomeninges, or seeding of the cerebellum; (3) positive CSF cytology 2 weeks after surgery; (4) invasion of the brain stem; and (5) age less than 4 years (Allen et al. 1986; Bloom et al. 1982; Levin et al. 1988). Poor risk associated with an age of less than 4 years is due to the fact that radiation therapists do not treat with full doses of craniospinal irradiation at this age.

Based on similar factors, the 5-year survival of poor-risk patients with craniospinal irradiation with or without chemotherapy is approximately 25% to 40% (Allen et al., 1986; Bloom et al., 1982; Evans et al., 1990; Park 1983). Good-risk patients, on the other hand, have 5-year survivals of 66% to 70% (Allen et al. 1986; Bloom et al. 1982; Evans et al. 1990; Levin et al. 1988). Results of multi-institutional randomized trials conducted by the Children's Cancer Study Group (CCSG) demonstrated 5-year event-free survival rates of 59% with and 50% without chemotherapy.

Long-term childhood survivors can develop neuropsychological dysfunction, decreased cognition, impaired growth of the spine, and pituitary–hypothalamic dysfunction. Growth hormone deficiency is the most frequent endocrine dysfunction after irradiation.

Metastatic Cancer

Widespread dissemination is the life-limiting factor for most patients afflicted with cancer. Control of the hematogenous spread or local extension into other vital organ systems determines the prognosis. While the brain metastases frequently are one more manifestation of end-stage cancer, their presence significantly may affect the quality and length of survival.

For each type of primary cancer, a pattern of central nervous system involvement can be defined. In gynecology, ovarian and cervical cancers rarely metastasize to the brain (<1%) and when detected, durable remissions still can be achieved. Brain metastases in stage IV melanoma are seen in up to 72% of patients. In breast cancer, 26% of patients will develop metastatic disease within 10 years of diagnosis. Brain metastasis is documented in 16% of patients with metastatic cancer, but this represents only 4.2% of the entire population of patients with a breast primary. While the synchronous discovery of the lung primary and brain metastasis is common, breast and colorectal cancers rarely present with central nervous system signs and symptoms as the initial manifestation of disease.

Treatment

The treatment plan is dictated by the need to provide immediate palliation of neurological signs and symptoms, the desire to achieve durable remissions, and the status of the patient's systemic disease and overall performance. Surgery is indicated for diagnosis when in doubt. Resection of a symptomatic tumor provides immediate palliation and may enhance the durability of subsequent therapy. Whole-brain irradiation has been the standard treatment of brain metastases and when combined with steroids, may provide adequate symptomatic control. Patients who are bed-ridden at the time of brain metastases are not likely benefit from any form of therapy.

Prognosis

Patient selection will markedly enhance the length of survival in patients who undergo specific treatments. For example, Patchell and associates (1986) report a 19-month me-

dian survival in lung cancer patients who were operated on for a single brain metastasis. All of these patients had no active systemic disease, but they represented only 9% of the patients seen with metastatic lung cancer. Thus, their impact on the expected survival of the entire group was modest. At M. D. Anderson Cancer Center, the median survival for patients who underwent craniotomy for removal of one or more metastases was slightly less than 12 months. However, durable remissions lasting more than 3 years were observed in up to 20% of patients (Moser and Johnson 1989). In a randomized trial designed to study the role of surgery in the treatment of single metastases, surgical resection followed by radiation resulted in longer survivals, fewer recurrences, and better quality of life than similar patients treated with radiotherapy alone (Patchell et al. 1990). Thus, in patients with a good performance status and potentially treatable systemic disease, surgery, radiation, and in some situations chemotherapy can result in both immediate palliation and durable symptom-free remissions.

Germ Cell Tumors—Germinomas

Germinomas are the most common intracranial germ cell tumor and are histologically indistinguishable from testicular seminoma. They represent 0.2% of all primary intracranial tumors documented in the United States during the years 1980 and 1985 (Mahaley et al. 1989). This low number contrasts to a much higher reported incidence in Japan and certain regions of Germany where 4.5% to 10% of tumors are called germinomas. Men (73.1%) were more likely to develop a germinoma, and the mean age at diagnosis was 19.8 years (Dehner 1983). Most germinomas arise in the midline axis from the suprasellar cistern to the pineal region. Local leptomeningeal dissemination and spinal metastases are seen.

Treatment

The role of surgery other than for biopsy and tissue diagnosis remains controversial

(Moser and Backlund, 1984). Though the concept of surgical cytoreduction remains attractive, resection exposes the patient to the operative risks and danger of regional tumor dissemination. Germinomas are exquisitely sensitive to both ionizing radiation and chemotherapy. The presence of a relatively homogeneous, well-circumscribed, extra-axial, enhancing pineal region mass in a young man is so characteristic of a germinoma that diagnostic radiotherapy can be justified. In such a situation, the patient receives 20 Gy over 2 weeks. A reduction of more than 50% in the tumor diameter is consistent with the diagnosis of this highly radiosensitive tumor, and a complete course of radiation (craniospinal up to 22 Gy and tumor region to 50 Gy) is administered. Patients should have a ventricular drain placed if acutely ill from obstructive hydrocephalus, and the CSF should be analyzed for human chorionic gonadotropin (HCG) and alpha fetoprotein (AFP). If a definite reduction is not seen, surgical resection should be considered the next step. In this era of microsurgical technique, a limited needle biopsy can be reserved for situations in which open exploration is not feasible. Accurate staging of the disease requires CSF cytology and spinal imaging, using either magnetic resonance imaging (MRI) or myelography with CT. Chemotherapy is not commonly used as the initial treatment for pure germinomas, although a regimen of cytoxin and *cis*-platinum is highly effective in the treatment of systemic germinomas and has been shown to eradicate brain metastases (Allen et al. 1987).

Prognosis

The reported 5-year survival from the recent Commission on Cancer survey was 67.7%. Survival following radiation alone or combined with initial surgery have ranged from 60% to 79% (Amendola et al. 1984; Jenkin et al., 1978; Packer et al. 1984; Sano and Matsutani 1981; Sung et al. 1978). The outcome is adversely affected by the extent of tumor dissemination within the neuroaxis.

Germ Cell Tumors—Teratoma

Of primary intracranial tumors, 0.5% are teratomas, but in children, the incidence is higher (Zulch 1986). The majority of tumors are found along the midline. Teratomas are heterogeneous on imaging studies, reflecting the cysts, cartilage, hair, teeth, and other products of the germ cell layers present. Sarcomatous or malignant germ cell tissue is present in the teratoid malignant tumors (Rubinstein 1971).

Treatment

Complete surgical removal should be attempted if possible. The benign (mature) teratoma may have an indolent clinical course. If symptomatic mass effect is present, radiation therapy will not reduce the tumor burden (Obrador et al. 1976; Stein 1979). If malignant germ cell tumor (choriocarcinoma, embryonal carcinoma, or endodermal sinus) is present within a teratoma, it should be treated aggressively with both radiation and chemotherapy.

Prognosis

In contrast to pure germinomas, the 5-year survival for the teratoma patient is only 35% (Jennings et al. 1977). Although complete surgical removal and long-term survival are well known, the persistence of tumor mass effect in some patients and the presence of malignant tissue in others probably accounts for the mortality encountered.

Germ Cell Tumors—Embryonal Carcinoma, Choriocarcinoma, Yolk Sac tumor

These tumors are rare and represent the malignant correlate of the embryonal pluripotent stem cells and extraembryonic derivatives. They are found most often in the pineal region, and specific serologic and immunohistochemical markers (HCG or AFP) can be demonstrated (Edwards et al. 1988).

Treatment

These tumors require aggressive multimodality therapy. Complete surgical excision should be attempted. In most situations, the patient should receive local radiation and systemic chemotherapy, using agents most active against the same corresponding systemic cancer. Children younger than 4 years should be considered for chemotherapy alone and radiation reserved for relapse.

Prognosis

A dismal prognosis exists for these patients, which is particularly true for choriocarcinoma. The median survival is generally less than 1 year in most anecdotal series.

Pineal Cell Tumors—Pineocytoma

Pineocytomas are rare tumors that represent less than 1% of intracranial tumors. They are seen in all age groups and are equally distributed in men and women. Neoplasms arising from the pineal parenchyma represent 20% to 30% of pineal region tumors (Donat et al. 1978). Neoplastic pineal parenchymal cells are capable of differentiating into tumors with astrocytic and ganglionic components (Rubinstein 1972).

Treatment

The tumors tend to be well encapsulated, and complete macroscopic resection frequently can be achieved. Durable remissions have been reported following surgery (Lapras 1984). Radiation therapy has been used to treat pineocytomas. (Disclafani et al. 1989). No published reports document responses to chemotherapy.

Prognosis

In patients with documented pineocytoma, long-term survival was seen only in patients whose tumors were completely resected. In other series, most patients did not survive more than 2 years after diagnosis, although the presence of astrocytic or neuronal differentiation may be associated with a better prognosis. However, Disclafani and colleagues (1989) report on six cases treated with 45 to 54 Gy radiation. Survival ranged up to 29 years with a median of 9 years.

Pineal Cell Tumors—Pineoblastoma

Of pineal parenchymal neoplasms, this tumor is the least differentiated and is included with other primitive neuroectodermal tumors. It occurs most frequently in the first 2 decades of life, and there is no sexual predilection.

Treatment

As with other primitive neuroectodermal tumors, gross total removal may enhance the durability of remissions. Surgery alone is never definitive. Craniospinal radiation is indicted for children older than 3 years, and additional chemotherapy should be considered. In younger children, chemotherapy should be attempted prior to radiation.

Prognosis

Few long-term survivors have been documented. The outcome appears significantly worse than with other primitive neuroectodermal tumors such as medulloblastoma. This may reflect the greater difficulty in achieving complete tumor resections.

Pituitary Adenoma—Nonsecreting Pituitary Tumors

Nonsecreting adenomas comprise 25% to 30% of pituitary tumors. Patients most commonly present with macroadenomas, producing symptoms related to mass effect on chiasm, hypothalamus, and adjacent pituitary tissue.

Treatment

The primary therapy for these tumors is surgical resection and postoperative radiotherapy for residual tumor. Transsphenoidal decompression is the treatment of choice. Conventional radiotherapy is appropriate when residual disease is present (Pistenma et al. 1975).

Prognosis

Improvement in neurological functions compromised by the tumor mass will occur in the majority of patients following surgical decompression. Radiation therapy may influence the development of hypopituitarism if not already present. Progressive tumor growth has been documented in approximately 20% of patients who were irradiated because of residual tumor postoperatively. With the exception of the few patients with very aggressive and destructive tumors of the skull base, durable, disease-free survival is the rule (Ebersold et al. 1986).

Pituitary Adenoma—Secreting Pituitary Tumors

These tumors secrete prolactin, thyrotropin, gonadotropin, growth hormone, and ACTH. Hormone secretion from the pituitary gland depends on the intimate association between the hypothalamus and pituitary (Challa et al. 1985). The pathogenesis of pituitary tumors remains unknown. There is evidence for and against the hypothesis that secretory tumors arise as the result of hypothalamic dysregulation (Kovacs and Horvath 1987). For each of the hormones, 80% to 90% appear to arise de novo within the pituitary. The prolactinomas are most common, while the thyrotropin-secreting tumors are rare (Kleinberg et al. 1977). Gonadotropin cell adenomas occur frequently and are generally seen in middle-aged men.

Treatment

For prolactinomas, medical therapy with dopamine agonist generally is safer and more effective than surgery or pituitary irradiation (Parl et al. 1986). Even very large tumors can shrink dramatically in response to bromocriptine (Pelkonen et al. 1981). The other secreting tumors are best treated by transsphenoidal excision and postoperative irradiation if invasive residual tumor is present (Jennings et al. 1977; Tyrrell et al. 1978). Dopamine agonist and somatostatin analogues may have a role in growth hormone-secreting tumors (Lamberts et al. 1985). Heavy-particle irradiation and stereotactic radiosurgery have been used in the treatment of ACTH-producing tumors (Kjellberg and Kliman 1980). Metyrapone

and aminoglutethimide also have been used in the medical therapy of Cushing's disease (Kreiger 1979).

When a patient has evidence of a pituitary tumor, a serum T4 and TSH level must be obtained (Hamilton et al. 1970). In primary hypothyroidism, secondary pituitary enlargement will resolve spontaneously when thyroid replacement is administered.

Prognosis

Long-term survival is the rule for patients with functioning pituitary tumors. Severe systemic sequelae from acromegaly, Cushing's disease, hyperthyroidism, and treatment-induced panhypopituitarism can be seen. These conditions can have a profound effect on the quality of life, but with good medical management, a satisfactory level of function can be maintained.

Tumors of Meningeal and Related Tissues—Meningioma

These include meningotheliomatous, fibrous, transitional, psammous, angiomatous, hemangioblastic, hemangiopericytic, papillary, and anaplastic tumors. In the American College of Surgeons' brain tumor survey, meningiomas and anaplastic meningiomas represented 21.9% and 1.2% of tumors reported, respectively (Mahaley et al. 1989). The mean age was 59, and 68% were women. The tumor occurs in children and adolescents in approximately 2% of cases. Meningiomas are found most commonly along the falx cerebri, spender ridges, parasellar region, and cerebral convexities. Tumors less frequently arise in the optic nerve sheath, foramen magnum, cerebellopontine angles, pineal region, and ventricular system. Meningiomas in bone and paranasal sinuses rarely are seen (Batsakis 1979). Meningiomas generally are firm and well demarcated. When they invade the pia mater, edema is seen in the adjacent brain. To some degree, meningiomas can be graded. Meningotheliomatous, fibrous, transitional, psammomatous, angiomatous, hemangioblastic meningiomas are considered grade I or benign (Kepes 1982). Hemangiopericytic and papillary meningiomas are more prone to malignancy, though any histologic subtype may transform into a malignant phenotype (Ludwin et al. 1975).

Treatment

Surgical removal is the treatment of choice for symptomatic tumors. It provides a diagnosis, immediate palliation, and potential for cure. Recurrences are seen in approximately 10% of tumors in which a gross total removal was performed. This may represent residual microscopic disease in most cases and possible multicentric origin in a few. The *en plaque* meningiomas of the skull base rarely are resected completely. Angiomatous and anaplastic meningiomas have a higher recurrence rate. Radiation therapy should be used in incomplete tumor resections in which the likelihood of symptomatic recurrence is expected (Carella et al. 1982). In anaplastic and angiomatous meningiomas, radiation should be considered as part of the primary therapy. Chemotherapy is recommended for recurrent tumors following surgery and radiation therapy. The doses and agents used are those most active against other soft-tissue sarcomas.

Prognosis

The 5-year survival for meningiomas and anaplastic meningiomas is 91% and 61% respectively (Mahaley et al. 1989). In the incompletely resected tumors, radiation therapy suppresses regrowth in approximately 50% of patients. There is no prognostic difference between the benign subtypes (Skullerud and Loken 1974).

Tumors of Meningeal and Related Tissues—Meningeal Sarcomas

These include fibrosarcoma, polymorphic cell sarcoma, meningeal sarcomatosis, rhabdomyosarcoma, mesenchymal chondrosarcoma, and malignant fibrous histiocytoma. Frank sarcomatous neoplasms of meningeal origin comprise approximately 1.2% of intracranial tumors (Zulch 1986). In addition to the fibrosarcomas, the polymorphic cell sarcomas, and the primary meningeal sarcomatosis, others are associated with the blood vessels within the brain parenchyma (Kernohan and Uhlheim 1962; Lukes et al. 1983). Meningeal sarcomas, particularly fibrosarcomas, may arise as a result of previous radiation therapy (Gonzales-Vitale et al. 1976).

Treatment

In all cases, complete surgical removal is the treatment of choice. Radiation therapy to the involved region may reduce the rate of recurrence. Chemotherapy may play a role in the management of these tumors. Sarcomas can respond to these systemic agents and to the standard front-line drugs most active against the systemic version of the tumor.

Prognosis

Most of these tumors are so rare that no valid information is available regarding patient outcomes. Long-term survivors have been reported for most of the sarcomas mentioned (Simpson et al. 1986). This most likely reflects the fact that some of these tumors are well circumscribed and amenable to curative resection. Based on the experience with other organ system sarcomas, multimodality therapy, including aggressive surgery, radiation, and chemotherapy, results in improved survival.

Tumors of Nerve Sheath Cells—Neurilemoma and Neurofibroma

The acoustic nerve is the most common site for development of a neurilemoma. Rarely, the trigeminal and glossopharyngeal nerves are involved as a solitary finding. Other cranial nerve and bilateral acoustic nerve involvement are generally associated with neurofibromatosis type 2. Tumors can arise in the dorsal spinal nerve roots and grow large in the paraspinal soft tissue. Recent studies have demonstrated a consistent chromosomal abnormality. The Commission on Cancer reports that in 1980 and 1985, 3.5% of primary intracranial tumors were neurilemomas, and the mean age at diagnosis was 50 (Mahaley 1989). The tumor occurs more frequently in women (57.2%) and has been seen in young children. Neurofibromas are generally multiple and associated with Recklinghausen's disease. They are peripheral nerve tumors that can grow intracranially or intraspinally.

Treatment

Surgical resection is the treatment of choice in most situations. Because the tumors are histologically benign and may behave in a clinically indolent manner, no treatment may be indicated for the asymptomatic elderly or frail patient. Small tumors, generally less than 2 cm, have been treated using single-dose, highly focused ionizing radiation (stereotactic radiosurgery).

Prognosis

The expected outcome for most patients is quite favorable. The 5-year survival for patients with neurilemoma operated in 1980 was 96.3%. The few deaths seen are usually the result of large recurrent tumors in which the surgical complication and mortality rates are much higher. Hearing loss is the most common early neurological finding. Modern diagnostic imaging allows for the detection of smaller lesions, resulting in minimal surgical morbidity and better preservation of cranial nerve functions.

Malformative Tumors—Craniopharyngioma

Craniopharyngiomas are most frequently associated with childhood and adolescence, but they may remain asymptomatic until late adulthood (Colmant and Noltenius 1988). These tumors are thought to arise from the vestigial remnants of Rathke's pouch (other names for this tumor have included tumors of Rathke's cleft or the hypophyseal duct). They arise exclusively in the sellar region and may be contained solely within the sella or more frequently in the suprasellar space. The tumors may expand as predominately solid or cystic lesions. The cyst may be multiloculated and may be responsible for the majority of mass effect. Though they may expand into the sphenoid sinus, they

usually grow into the suprasellar region, displacing the optic apparatus and effacing the ventricular system.

Craniopharyngiomas generally grow slowly, although cyst enlargement can be quite dramatic even over a few months. The solid cellular component consists of stratified squamous epithelium arranged in various bands, cords, or papillary structures surrounding the vascularized connective tissue stroma (Russell and Rubinstein 1977). Histologically apparent malignant transformation is not associated with this type of tumor, but aggressive biological activity is seen.

Treatment

The optimal management of these tumors remains controversial. Complete tumor removal of small intrasellar and suprasellar tumors can be resected with minimal morbidity and a durable remission can be expected. Larger tumors that distort the floor of the third ventricle or expand into its lumen can elicit a reactive gliosis and become enmeshed in the ventricular wall. In this situation, removal frequently is associated with significant hypothalamic dysfunction. Cyst drainage and removal of all solid tumor without stripping all of the cyst wall from adjacent brain tissue and vessels will provide immediate decompression and restoration of neurological function and greatly reduce the surgical morbidity. The addition of ionizing radiation in older children and adults may maximize the opportunity for a durable symptom free remission (Amacher 1980).

Prognosis

The biological behavior of these tumors generally is indolent. They may be found as an incidental mass and followed for years with no symptoms or radiographic progression. As surgical morbidity declines, improved postoperative endocrine support, in cases of panhypopituitarism, can permit a near-normal life. Despite their benign histology, death or significant disability from progressive tumor and the cumulative toxicity of treatment are frequent. Radical tumor resections are associated with recurrences in at least 23% of cases (Shapiro et al. 1979). Local control rates after complete resection, subtotal resection, or incomplete resection and postoperative irradiation are 70%, 26%, and 75%, respectively (Wen et al. 1989). The 10-year survival rates are better for patients treated with surgery and radiation (approximately 70%) than with surgery alone (50%) (Sung et al. 1981).

References

Allen, J. C.; Bloom, J.; Ertel, I.; Evans, A.; Hammond, D.; Jones, H.; Levin, V.; Jenkin, D.; Sposto, R.; Wara, W. Brain tumors in children: current cooperative and institutional chemotherapy trials in newly diagnosed and recurrent disease. Semin. Oncol. 13:110–122; 1986.

Allen, J. C.; Kim, J. H.; Packer, R. J. Neoadjuvant chemotherapy for newly diagnosed germ cell tumors of the central nervous system. J. Neurosurg. 67:65–70; 1987.

Amacher, A. L. Craniopharyngioma: the controversy regarding radiotherapy. Child's Brain 6:57–64; 1980.

Amendola, B. E.; McClatchey, K.; Amendola, M. A. Pineal region tumors: analysis of results. Int. J. Radiat. Oncol. Biol. Phys. 10:991–997; 1984.

Andreou, J.; George, A. E.; Wise, A.; de Leon, M.; Krichoff, I. I.; Ransoff, J.; Foo, S. H. CT prognostic criteria of survival after malignant glioma surgery. Am. J. Neuroradiol. 4:488–490; 1983.

Batsakis, J. G. Tumors of the head and neck, clinical and pathological considerations, 2nd ed. Baltimore, Maryland: Williams and Wilkins; p. 347; 1979.

Berger, M. S.; Edwards, M. D.; Wara, W. M.; Levin, V. A. Primary cerebral neuroblastoma: long-term follow-up review and therapeutic guidelines. J. Neurosurg. 59:418–423; 1983.

Bloom, H. J.; Glees, J.; Bell, J. The treatment and long-term prognosis of children with intracranial tumors: a study of 610 cases, 1950–1981. Int. J. Radiat. Oncol. Biol. Phys. 18:723–745; 1990.

Bloom, H. J. G. Intracranial tumors: response and resistance to therapeutic endeavors 1970–1980. Int. J. Radiat. Oncol. Biol. Phys. 8:1083–1113; 1982.

Bloom, H. J. G. Medulloblastoma in children: increasing survival rates and further prospects. Int. J. Radiat. Oncol. Biol. Phys. 8:2023–2027; 1982.

Bloom, H. J. G.; Thornton, H.; Schweisguth, O. SIOP medulloblastoma and high grade ependymoma therapeutic clinical trials: Preliminary results (1975–1981). In: Raybaud, C.; Clement, R.; Lebreuli, G.; Bernard, J. L. eds. Pediatric oncology. Amsterdam: Excerpta Medica; p. 309; 1982.

Bullard, D. E.; Rawlings, C. E., III; Phillips, B.; Cox, E. B.; Schold, S. C., Jr.; Burger, P.; Halperin, E. C. Oligodendroglioma. An analysis of the value of radiation therapy. Cancer 60:2179–2188; 1987.

Cairncross, J. G.; Macdonald, D. R. Successful chemotherapy for recurrent malignant oligodendroglioma. Ann. Neurol. 23:360–364; 1988.

Carella, R. J.; Ransohoff, J.; Newall, J. Role of radiation therapy in the management of meningioma. Neurosurgery 10:332–339; 1982.

Challa, V. R.; Marschall, R. B.; Hopkins, M. B., III; Kelly, D. L. Jr.; Civantos, F. Pathobiologic study of pituitary tumors: report of 62 cases with a review of the recent literature. Hum. Pathol. 16:873–884; 1985.

Chamberlain, M. C.; Silver, P.; Levin, V. A. Poorly differentiated gliomas of the cerebellum. A study of 18 patients. Cancer 65:337–340; 1990.

Chin, H. W.; Hazel, J. J.; Kim, T. H.; Webster, J. H. Oligodendrogliomas. I. A clinical study of cerebral oligodendrogliomas. Cancer 45:1458–1466; 1980.

Colmant, H. J.; Noltenius, H. Tumors of the central nervous system. In: Noltenius, H. ed. Human oncology, pathology and clinical characteristics. Baltimore: Urban and Schwarzenberg; p. 749–849; 1988.

Dehner, P. L. Gonadal and extragonadal germ cell neoplasia of childhood. Human Pathol. 14:493–511; 1983.

DiMarco, A.; Campostrini, F.; Pradella, R.; Reggio, M.; Palazzi, M.; Grandinetti, A.; Garusi, G. F. Postoperative irradiation of brain ependymomas. Analysis of 33 cases. Acta. Oncol. 27:261–267; 1988.

Disclafani, A.; Hudgins, R. J.; Edwards, M. S.; Wara, W.; Wilson, C. B.; Levin, V. A. Pineocytomas. Cancer. 63:302–304; 1989.

Donat, J. F.; Okazaki, H.; Gomez, M. R.; Reagen, T. J.; Baker, H. L., Jr.; Laws, E. R., Jr. Pineal tumors: A 53-year experience. Arch. Neurol. 35:736–740; 1978.

Duffner, P. K.; Cohen, M. E.; Myers, M. H.; Heise, H. W. Survival of children with brain tumors: SEER Program, 1973–1980. Neurology 36:597–601; 1986.

Ebersold, M. J.; Quast, L. M.; Laws, E. R.; Scheithauert, B.; Randall, R. V. Long-term results in transsphenoidal removal of nonfunctioning pituitary adenomas. J. Neurosurg. 64:713–719; 1986.

Edwards, M. S.; Hudgins, R. J.; Wilson, C. B.; Levin, V. A.; Wara, W. M. Pineal region tumors in children. J. Neurosurg. 68:689–697; 1988.

Edwards, M. S.; Wara, W. M.; Urtasun, R. C.; Prados, M.; Levin, V. A.; Fulton, D.; Wilson, C. B.; Hannigan, J.; Silver, P. Hyperfractionated radiation therapy for brain-stem glioma: a phase I–II trial. J. Neurosurg. 70:691–700; 1989.

Epstein, F.; Wisoff, J. H. Intrinsic brainstem tumors in childhood: surgical indications. J. Neurooncol. 6:309–317; 1988.

Ernestus, R. I.; Wilcke, O.: Schroder, R. Intracranial ependymomas: prognostic aspects. Neurosurg. Rev. 12:157–163; 1989.

Evans, A. E.; Jenkin, R. D. T.; Sposto, R.; Ortega, J. A.; Wilson, C, B.; Wara, W.; Ertel, I. J.; Kramer, S.; Chang, C. H.; Leikin, S. L.; Hammond, G. D. The treatment of medulloblastoma. J. Neurosurg. 72:572–582; 1990.

Fazekas, J. T. Treatment of grade I and II brain astrocytomas. The role of radiotherapy. Int. J. Radiat. Oncol. Biol. Phys. 2:661–666; 1977.

Friedman, H. S.; Schold, S. C. Jr. Rational approaches to the chemotherapy of medulloblastoma. Neurol. Clin. 3:843–853; 1985.

Fulton, D. S.; Levin, V. A.; Wara, W. M.; Edwards, M. S. Chemotherapy of pediatric brain stem tumors. J. Neurosurg. 54:721–725; 1981.

Gonzales-Vitale, J. C.; Slavin, R. E.; McQueen, J. D. Radiation induced intracranial malignant fibrous histiocytoma. Cancer 37:2960–2963; 1976.

Hamilton, C. R., Jr.; Adams, L. C.; Maloof, F. Hyperthyroidism due to the thyrotropin-producing pituitary chromophobe adenoma. N. Engl. J. Med. 280:1077–1080; 1970.

Helseth, A.; Mrk, S. J. Primary intraspinal neoplasms in Norway, 1955 to 1986. A population-based survey of 467 patients. J. Neurosurg. 71:842–845; 1989.

Hochberg, F. H.; Miller, D. C. Primary central nervous system lymphoma. J. Neurosurg. 68:835–853; 1988.

Jenkin, D. Posterior fossa tumors in childhood: radiation treament. Clin. Neurosurg. 30:203; 1983.

Jenkin, R. D.; Simpson, W. J.; Keen, C. W. Pineal and suprasellar germinomas. Results of radiation treatment. J. Neurosurg. 48:99–107; 1978.

Jennings, A. S.; Liddle, G. W.; Orth, D. N. Results of treating childhood Cushing's disease with pituitary irradiation. N. Engl. J. Med. 297:957–962; 1977.

Jennings, M. T.; Gelman, R.; Hochberg, F. Intracranial germ-cell tumors: natural history and pathogenesis. J. Neurosurg. 63:155–167; 1985.

Kepes, J. J. Meningiomas: biology, pathology, and differential diagnosis. New York: Masson; 1982.

Kernohan, J. W.; Uhlheim, A. Sarcomas of the brain. Springfield, IL: Charles C. Thomas; 1962.

Kjellberg, R. N.; Kliman, B. Radiosurgery therapy for pituitary adenoma. In: Post, K. D., Jackson, I. M. D; Reichlin, S. eds. The pituitary adenoma. New York: Plenum; p. 459–478; 1980.

Kleinberg, D. L.; Noel, G. L.; Frantz, A. G. Galactorrhea: a study of 235 cases, including 48 with pituitary tumors. N. Engl. J. Med. 296:588–600; 1977.

Kovacs, K.; Horvath, E. Pathology of Pituitary Tumors. In: Endocrinology and metabolism clinics of North America, 16:3, Philadelphia: W. B. Saunders; 1987: p. 529–551.

Kreiger, D. T. Pharmacological therapy of Cushing's disease and Nelson's syndrome. In: Linfoot, J. A. ed. Recent advances in the diagnosis and treatment of pituitary tumors. New York: Raven Press; 1979: p. 337–340.

Kun, L. E.; Kovnar, E. H.; Sanford, R. A. Ependymomas in children. Pediatr. Neurosci. 14:57–63; 1988.

Lamberts, S. W.; Uitterlinden, J. P.; Verschoor, L.; van Dongen, K.J.; del Pozo, E. Long-term treatment of acromegaly with the somatostatin analogue SMS 201–995. N. Engl. J. Med. 313:1576–1580; 1985.

Lapras, C. Surgical therapy of pineal region tumors. In: Neuwelt, E. ed. Diagnosis and treatment of pineal region tumors. Baltimore: Williams & Wilkins; 1984: p. 289–299.

Laws, E. R., Jr.; Taylor, W. F.; Bergstrahl, E. J.; Okazaki, H.; Clifton, M. B.; The neurosurgical management of low-grade astrocytoma. Clin. Neurosurg. 33:575–588; 1986.

Laws, E. R.; Taylor, W. F.; Clifton, M. B.; Okazaki, H. Neurosurgical management of low-grade astrocytoma of the cerebral hemispheres. J. Neurosurg. 61:665–673; 1984.

Leibel, S. A.; Wara, W. M.; Sheline, G. E.; Townsend, J. J; Baldrey, E. B. The treatment of meningiomas in childhood. Cancer. 37:2709–2712; 1976.

Leibel, S. A.; Sheline, G. E.; Wara, W. M.; Boldrey, E. B.; Nielsen, S. L. The role of radiation therapy in the treatment of astrocytomas. Cancer 35:1551–1557; 1975.

Levin, V. A.; Edwards, M. S. B.; Gutin, P. H.; Vestnys, P.; Fulton, D.; Seager, M.; Wilson, C. B. Phase II evaluation of dibromodulcitol in the treatment of recurrent medulloblastoma, ependymoma, and malignant astrocytoma. J. Neurosurg. 61:1063–1068; 1984.

Levin, V. A.; Edwards, M. S.; Wara, W. M.; Allen, J.; Ortega, J.; Vestnys, P. S. 5-fluorouracil and CCNU followed by hydroxyurea, misonidazole and irradiation for brain stem gliomas: a pilot study of the Brain Tumor Research Center and the Childrens Cancer Group. Neurosurgery. 14:679–681; 1984.

Levin, V. A.; Hoffman, W.; Heilbron, D. C.; Norman, D.; Wilson, C. B. Prognostic significance of the pretreatment CT scan on time to progression for patients with malignant gliomas. J. Neurosurg. 52:642–647; 1980.

Levin, V. A.; Rodriguez, L. A.; Edwards, M. S.; Wara, W.; Liu, H. C.; Fulton, D.; Davis, R. L.; Wilson, C. B.; Silver, P. Treatment of medulloblastoma with procarbazine, hydroxyurea, and reduced radiation doses to whole brain and spine. J. Neurosurg. 68:383–387; 1988.

Levin, V. A.; Shelin, G. E.; Gutin, P. H. Neoplasms of the central nervous system. In: DeVita, V. T., Jr.; Hellman, S.; Rosenberg, S. A. eds. Cancer. Principles and practice of oncology, 3rd ed. Philadelphia: J. B. Lippincott, Co.; 1989: p. 1557–1579.

Levin, V. A.; Silver, P.; Hannigan, J.; Wara, W. M.; Gutin, P. H.; Davis, R. L.; Wilson, C. B. NCOG 6G61 Final Report: superiority of post-radiotherapy adjuvant chemotherapy with CCNU, procarbazine, and vincristine (PCV) over BCNU for anaplastic gliomas. Int. J. Radiat. Oncol. Phys. Biol. 18:321–324; 1990.

Levin, V. A.; Vestnys, P. S.; Edwards, M. S.; Wara, W. M.; Fulton, D.; Barger, G.; Seager, M.; Wilson, C. B.; Improvement in survival produced by sequential therapies in the treatment of recurrent medulloblastoma. Cancer 51:1364–1370; 1983.

Loeffler, J. S.; Kretschmar, C. S.; Sallan, S. E.; LaVally, B. L.; Winston, K. R.; Fischer, E. G.; Tarbell, N. J.; Pre-radiation chemotherapy for infants and poor prognosis children with medulloblastoma. Int. J. Radiat. Oncol. Biol. Phys. 15:177–181; 1988.

Logothetis, C.; Samuels, M.; Trindade, A. The management of brain metastases in germ cell tumors. Cancer 49; 12–18; 1982.

Ludwig, C. L.; Smith, M. T; Godfrey, A. D.; Armbrustmacher, V. W. A clinicopathologic study of 323 patients with oligodendrogliomas. Ann. Neurol. 19:15–21; 1986.

Ludwin, S. K.; Rubinstein, L. J.; Russell, D. S. Papillary meningioma: a malignant variant of meningioma. Cancer 36:1363–1373; 1975.

Lukes, A.; Wollmann, R.; Stefannson, K. Meningeal sarcomatosis and multiple astrocytomas. Arch. Neurol. 40:179–182; 1983.

Mahaley, M. S.; Mettlin, C.; Natarajan, N.; Laws, E. R.; Peace, B. National Survey of patterns of care of brain tumor patients. J. Neurosurg. 71:826–836; 1989.

Marsa, G. W.; Goffinet, D. R.; Rubinstein, L. J.; Bagshaw, M. A. Megavoltage irradiation in the treatment of gliomas of the brain and spinal cord. Cancer 36:1681–1989; 1975.

Moser, R. P.; Backlund, E. O. Stereotactic techniques in the diagnosis and treatment of pineal region tumors. In: Neuwelt, E. ed. Diagnosis and treatment of pineal region tumors. Baltimore: Williams & Wilkins; 1984: p. 236–253.

Moser, R. P.; Johnson, M. L. Surgical management of brain metastases: how aggressive should we be? Oncology 3:123–134; 1989.

Murray, K.; Kun, L.; Cox, J. Primary malignant lymphoma of the central nervous system. J. Neurosurg. 65:600–607; 1986.

Nazar, G. B.; Hoffman, H. J.; Becker, L. E.; Jenkin, D.; Humphreys, R. P.; Hendrick, E. B. Infratentorial ependymomas in childhood: prognostic factors and treatment. J. Neurosurg. 72:408–417; 1990.

Neuwelt, E. A.; Frenkel, E. P.; Gumerlock, M. K.; Braziel, R.; Dana, B.; Hill, S. A. Developments in the diagnosis and treatment of primary CNS lymphoma. Cancer 58:1609–1620; 1986.

Obrador, S.; Soto, M.; Gutierrez-Diaz, J. A. Surgical management of tumors of the pineal region. Acta. Neurochir. 34:159–171; 1976.

Packer, R. J.; Sutton, L. N.; Rorke, L. B.; Rosenstock, J. G.; Zimmerman, R. A.; Littman, P.; Bilaniuk, L. T.; Bruce, D. A.; Schut, L. Intracranial embryonal cell carcinoma. Cancer 54:520–524; 1984.

Packer, R. J.; Siegel, K. R.; Sutton, L. N.; Evans, A. E.; DAngio, G.; Rorke, L. B.; Bunin, G. R.; Schut, L. Efficacy of adjuvant chemotherapy for patients with poor-risk medulloblastoma: a preliminary report. Ann. Neurol. 24:503–508; 1988.

Park, T. S.; Hoffman, H. J.; Hendrick, E. B.; Humphreys, R. P. Experience with surgical decompression of the Arnold-Chiari malformation in young infants with myelomeningocele. Neurosurgery 13:147–152; 1983.

Park, T. S.; Hoffman, H. J.; Hendrick, E. B.; Humphreys, R. P. Medulloblastoma: clinical presentation and management. Experience at the Hospital For Sick Children, Toronto, 1950–1980. J. Neurosurg. 58:543–552; 1983.

Parl, F. F.; Cruz, V. E.; Cobb, C. A.; Bradley, C. A.; Aleshire, S. L. Late recurrence of surgically removed prolactinomas. Cancer 57:2422–2426; 1986.

Patchell, R. A.; Cirrincione, C.; Thaler, H. T.; Galicich, J. H.; Kim, J.-H.; Posner, J. B. Single brain metastases: surgery plus radiation or radiation alone. Neurology 36:447–453; 1986.

Patchell, R. A.; Tibbs, P. A.; Walsh, J. W.; Dempsey, R. J.; Maruyama, Y.; Kryscio, R. J.; Markesbery, W. R.; MacDonald, J. S.; Young, B. A randomized trial of surgery in the treatment of single metastases to the brain. N. Engl. J. Med. 322:494–500; 1990.

Payan, M. J.; Gambarelli, D.; Routy, J. P.; Choux, R.; Blanc, A. P.; Alliez, B.; Toga, I. Primary lymphoma of the brain associated with AIDS. Acta. Neuropathol. 64:78–80; 1984.

Pelkonen, R.; Grahne, B.; Hirvonen, E.; Karonen, S.; Salmi, J.; Tikkanen, M.; Valtonen, S. Pituitary function in prolactinoma: effect of surgery and postoperative bromocriptine therapy. Clin. Endocrinol. 14:335–348; 1981.

Pistenma, D. A.; Goffinet, D. R.; Bagshaw, M. A.; Hanbery, J. W.; Eltringham, J. R. Treatment of chromophobe adenomas with megavoltage irradiation. Cancer 35:1574–1582; 1975.

Rawlings, C. E. III; Giangaspero, F.; Burger, P. C.; Bullard, D. E. Ependymomas: a clinicopathologic study. Surg. Neurol. 29:271–281; 1988.

Rodriguez, L.; Prados, M.; Fulton, D.; Edwards, M. S. B.; Silver, P.; Levin, V. A. Treatment of recurrent brain stem glioma and other CNS tumors with 5-fluorouracil, CCNU, hydroxyurea, and 6-mercaptopurine. Neurosurg. 22:691–693; 1988.

Rosenblum, M. L.; Levy, R. M.; Bredesen, D. E.; So, Y. T.; Wara, W.; Ziegler, J. L. Primary central nervous system lymphomas in patients with AIDS. Ann. Neurol. 23(suppl):13–16; 1988.

Ross, G. W.; Rubinstein, L. J. Lack of histopathological correlation of malignant ependymomas with postoperative survival. Neurosurg. 70:31–36; 1989.

Rubinstein, L. J. Sarcomas of the nervous system. In: Minckler, J. ed. Pathology of the nervous system, Vol. 2. New York: McGraw-Hill Book Company; 1971: p. 2144–2164.

Rubinstein, L. Tumors of the central nervous system. Atlas of tumor pathology, 2nd Series, Fascicle 6. Washington DC: AFIP.

Russell, D. C.; Rubinstein, L. J. Pathology of tumors of the nervous system, 4th ed. Baltimore: Williams and Wilkins; 1977.

Salcman, M.; Kaplan, R. S.; Ducher, T. B.; Adbo, H.; Montgomery, E. Effect of age and reoperation on survival on the combined modality treatment of malignant astrocytoma. Neurosurg. 10:454–463; 1982.

Sano, K.; Matsutani, M. Pinealoma (germinoma) treated by direct surgery and postoperative irradiation. Child's Brain 2:81–97; 1981.

Schneck, S. A.; Penn, I. De novo brain tumors in renal transplant recipients. Lancet 1:983–986; 1971.

Shapiro, K.; Till, K.; Grant, D. N. Craniopharyngiomas in childhood: a rational approach to treatment. J. Neurosurg. 50:617–623; 1979.

Shaw, E. G.; Evans, R. G.; Scheithauer, B. W.; Ilstrup, D. M.; Earle, J. D. Postoperative ra-

diotherapy of intracranial ependymoma in pediatric and adult patients. Int. J. Radiat. Oncol. Biol. Phys. 13:1457–1462; 1987.

Sheline, G. E. Radiation therapy of primary tumors. Semin. Oncol. 2:29; 1975.

Sheline, G. E.; Boldrey, E.; Karlsberg, P.; Therapeutic considerations in tumors affecting the central nervous system: oligodendrogliomas. Radiology 82:84; 1964.

Silverberg, K.; Boring, C. C.; Squires, T. S. Cancer Statistics, 1990. Ca-A J. Clin. 40:9–28; 1990.

Simpson, R. H.; Phillips, J. I.; Miller, P.; Hagen, D.; Anderson, J. E. M. Intracerebral malignant fibrous histiocytoma. Clin. Neuropath. 5:185–189; 1986.

Skullerud, K.; Loken, A. C. The prognosis in meningioma. Acta. Neuropath. (Berl) 29:337–344; 1974.

Smith, M. T.; Ludwig, C. L.; Godfrey, A. D.; Armbrustmacher, V. W. Grading of oligodendrogliomas. Cancer 52:2107–2114; 1983.

Stein, B. M. Supracerebellar–infratentorial approach to pineal tumors. Surg. Neurol. 11:331–337; 1979.

Sung, D.; Harisiadis, L.; Chang, C. H. Midline pineal tumors and suprasellar germinomas: highly curable by radiation. Radiology 128:745–751; 1978.

Sung, D. I.; Chang, C. H.; Harisiades, L.; Carmel, P. W. Treatment results of craniopharyngiomas. Cancer 47:847–852; 1981.

Tomita, T.; McLone, D. G.; Yasue, M. Cerebral primitive neuroectodermal tumors in childhood. J. Neurooncol. 6:233–243; 1988.

Tyrrell, J. B.; Brooks, R. M.; Fitzgerald, P. A.; Cofoid, P. B.; Forshaw, P. H.; Wilson, C. B. Selective trans-sphenoidal resection of pituitary adenomas. N. Engl. J. Med. 298:753–758; 1978.

van Eys, J.; Baram, T. Z.; Cangir, A.; Bruner, J. M.; Martinez-Prieto, J. Salvage chemotherapy for recurrent primary brain tumors in children. J. Pediatr. 113:601–606; 1988.

Walker, M. D.; Alexander, E., Jr.; Hunt, W. E.; MacCarty, C. S.; Mahaley, M. S., Jr.; Mealey, J., Jr.; Norrell, H. A.; Owens, G.; Bansohoff, I.; Wilson, C. B.; Gehan, E. A.; Strike, T. A. Evaluation of BCNU and/or radiotherapy in the treatment of anaplastic gliomas: A cooperative clinical trial. J. Neurosurg. 49:333–343; 1978.

Walker, M. D.; Strike, T. A.; Sheline, G. E. An analysis of dose-effect relationship in the radiotherapy of malignant gliomas. Int. J. Radiat. Oncol. Biol. Phys. 5:1725–1731; 1979.

Wallner, K. E.; Gonzales, M.; Sheline, G. E. Treatment of oligodendrogliomas with or without postoperative irradiation. J. Neurosurg. 68:684–688; 1988.

Wallner, K. E.; Wara, W. M.; Sheline, G. E.; Davis, R. L. Intracranial ependymomas: results of treatment with partial or whole brain irradiation without spinal irradiation. Int. J. Radiat. Oncol. Biol. Phys. 21:1937–1941; 1986.

Wen, B. C.; Hussey, D. H.; Staples, J.; et al. A comparison of the roles of surgery and radiation therapy in the management of craniopharyngiomas. Int. J. Radiat. Oncol. Biol. Phys. 16:17–24; 1989.

Zulch, K. J. Brain tumors. Their biology and physiology. 3rd ed. New York: Springer; 1986.

47

Cancer Pain

EUGENIE A. M. T. OBBENS

Chronic cancer pain may shorten a patient's life expectancy. It often is accompanied by depression, anxiety, guilt, and fear, and it even may lead to suicide. Even though pain is certainly not an inevitable part of cancer, the fear of pain often causes unnecessary distress. With a better understanding of such pain and of existing and new treatment modalities, pain relief is now possible for most patients.

Incidence of Pain

Approximately one third of cancer patients actively on treatment have significant pain. This number is as high as 80% in patients with terminal cancer (Bonica 1985). Pain is especially common in malignancies such as bone, lung, and pancreatic cancer, sarcoma, and metastatic cancer (Bonica 1985).

In cancer patients, pain is most often caused by the tumor. Of 86 consecutive patients evaluated at the Pain Clinic of Memorial Sloan-Kettering Cancer Center, 53 (62%) had pain caused by tumor involvement, while 24 patients (28%) had pain directly related to the cancer treatment (Foley 1982). In the remaining nine patients, the

pain was due to other causes, such as musculoskeletal pain and postherpetic neuralgia.

In children, malignancies often present with pain. Of 92 children who presented with a variety of newly diagnosed malignancies, 72 reported pain, and in 57 patients (62%), pain was one of the initial symptoms (Miser et al. 1987a). In most of these patients, the pain resolved rapidly once appropriate treatment was instituted. Of 139 children and young adults undergoing treatment at the Pediatric Branch of the National Cancer Institute, pain was present in 48% of in-patients and in 27% of all out-patients (Miser et al. 1987b). The pain was predominantly treatment-related in both groups, either alone (46%) or with pain due to tumor (14%). Oral mucositis due to chemotherapy and total body irradiation were major causes of morbidity in children.

Pain Due to Tumor

Bone Tumors

One of the most common causes of pain in cancer patients is involvement of bone by primary or metastatic tumors. In a literature survey by Bonica (1985), 70% to 80% of patients with bone tumors suffered from pain. Bone metastases are especially common in patients with prostate cancer, but they also are found often in breast cancer, lung cancer, and other malignancies. In children, two thirds of all cases of tumor pain were due to bone tumors (Miser et al. 1987b).

Bone metastases mainly develop in the proximal bones, with a high prevalence to the spine. In autopsy studies on bone metastases, the vertebral column was involved in 80% of patients with multiple bone metastases (Torma 1957; Walther 1948). Deformity of the vertebra may cause nerve root compression, which will then cause radicular pain in addition to the earlier occurring local pain.

Tumor involvement of the epidural space of the spinal canal is suggested when the pain worsens when laying flat, coughing or sneezing, and increased intra-abdominal pressure. Pain is a presenting symptom in more than 90% of patients with epidural spinal cord compression due to tumor (Gilbert et al. 1978; Greenberg et al. 1980). In spinal cord compression, irradiation in combination with high doses of dexamethasone has resulted in pain relief in 82% of patients at completion of treatment, while intravenous dexamethasone at diagnosis may cause immediate pain relief as well (Greenberg et al. 1980). This entity carries a poor prognosis because it often develops at the end of the disease. In a prospective study, less than 50% of patients were alive at 2 months follow-up after treatment (Obbens et al. 1984).

Other pain syndromes caused by bone metastases include metastases to the base of the skull. They often present with pain, either locally or referred to the vertex, temples, or neck and shoulders (Greenberg et al. 1981). Rib or other long bone tumors also can cause referred pain, such as pain in the knee due to a tumor in the proximal femur.

Bone pain usually responds well to local irradiation. In a review of 16 nonrandomized studies of external beam irradiation for bone pain, a mean overall response rate in terms of pain relief was 86% (range, 73% to 100%) (Hoskin 1988). In the four studies that re-

ported long-term follow-up, persisting pain relief ranged from 91% of responding patients at 1-year follow-up in one study, to four out of 10 surviving patients at 4 years in another study (Garmatis and Chu 1978; Penn 1976). The likelihood of response did not appear to be influenced by the histology of the primary tumor.

Other modalities of anticancer treatment, such as chemotherapy and hormonal therapy, may result in pain relief. The pain response rate usually correlates with the objective tumor response rate for the specific therapy and malignancy, although no good data are available. In advanced breast cancer, hormonal manipulation through bilateral oophorectomy, adrenalectomy, hypophysectomy, chemical ablation, or hormone replacement rarely achieves more than 40% tumor response rate; however, with androgen or progestin therapy, pain relief has been achieved in up to 90% of patients with bone metastases from breast cancer (Hoy 1989; Ramirez and Levin 1984). For bone metastases from prostate cancer, estrogen therapy or orchiectomy provide pain relief in 35% to 70% of patients (Hoy 1989). In general, bone pain responds well to analgesic medication. In patients with mild pain, nonopioid analgesics, such as acetaminophen or nonsteroidal anti-inflammatory drugs, often are adequate. If these do not control the pain, codeine or another weak opioid analgesic may be added. For severe pain, an opioid agonist, such as morphine, is the drug of choice.

A methodical approach for the use of these drugs in a stepwise fashion has been outlined in the World Health Organization Guidelines for Cancer Pain Relief (World Health Organization 1986). With this approach, tested in Japan and Italy in patients with a variety of tumors and pain etiologies, up to 93% of patients achieved adequate pain control (Takeda 1986; Ventafridda et al. 1987).

For patients in whom pain cannot be controlled sufficiently with oral narcotic analgesics, the epidural, intrathecal, or intraventricular route may be a good alternative method of administration. Of 20 cancer patients treated with intraventricular morphine sulfate, 10 were able to continue this treatment with good relief of pain until death 1 to 17 months later, while in eight patients, treatment was discontinued because of side effects or inadequate pain control (Obbens et al. 1987). Side effects that limit the effectiveness of spinal opioids include infection, urinary retention, nausea and vomiting, and itching (Moulin and Coyle 1986).

Alternative treatments consisting of calcitonin or diphosphonates have shown promise in the management of malignant bone pain, but more studies are needed to assess response rates (Hanks 1988).

As with any pain due to tumor, sudden worsening of bone pain suggests new or rapidly expanding disease and requires another evaluation and treatment plan.

Nerve Involvement

In addition to pain due to compression of the spinal cord or nerve roots, pain can be caused by involvement of other nervous structures, such as peripheral nerves, nerve plexuses, or leptomeninges. Pain in the involved arm and shoulder may be the initial symptom of tumor infiltrating or compressing the brachial plexus. This was demonstrated in a retrospective review of 78 patients, 70% of whom had breast or lung cancer (Kori et al. 1981). Pain was the presenting symptom in 81% of patients. The pain typically increased with movement in the shoulder. In 46% of patients treated, radiation therapy resulted in significant pain relief, but no improvement in weakness or sensory dysfunction occurred. Other studies report significant pain relief with radiation therapy in 49% to 73% of patients and suggest that symptom palliation is dose-dependent. As with other nerve pain, this pain often does not respond well to analgesics. The often severe chronic pain also is inadequately relieved with most surgical procedures, such as dorsal rhizotomy, hypophysectomy, or limb amputation, but good pain relief was reported by Kori and coworkers (1981) in the three patients in their study who underwent a percutaneous cordotomy.

The lower brachial plexus or the C8 and T1 nerve roots may be compressed by tumors in the apex of the lung, known as Pancoast tumor or superior pulmonary sulcus syndrome. Pain in the shoulder and medial aspect of the arm often is accompanied by weakness of the hand muscles and Horner's syndrome. In one study (Sundaresan et al. 1987), pain was a prominent feature at presentation in 93% of patients and was present for a median duration of 5 months prior to diagnosis. Pain was sufficiently severe that all patients required narcotic analgesics for pain relief. Of their 30 patients, 20 had a brachial plexopathy. Eight of these 20 also had evidence of spinal invasion with epidural spinal cord compression. The latter group, as well as those with extensive local invasion of the brachial plexus, carried the poorest prognosis, probably because such invasion tends to occur late in the disease. With surgical resection and preoperative irradiation, they found a high correlation between pain relief and tumor response after treatment. Initial therapy for Pancoast's tumor often results in pain relief. With radiation therapy, pain relief has been reported in up to 100% of patients after doses of 60 Gy (Ampil 1985). After initial surgery, pain relief has varied from 58% to 90% of patients (Kanner et al. 1982). In most of these patients, however, the pain returns as the tumor progresses.

Compression or invasion of the lumbar plexus may be seen in patients with locally extensive pelvic tumors, such as prostate, rectal, bladder, testicular, and cervical cancer (Jaeckle et al. 1985; Thomas et al. 1985). Most patients have pain as their first symptom of plexus compression, followed weeks to months later by weakness and sensory dysfunction. The pain may be localized to the pelvic or lumbosacral region or radiate down one leg. Specific cancer treatment, such as radiation therapy, surgery, or chemotherapy, has not been effective for pain control. In one study, such treatment resulted in a subjective improvement of pain in only 15% of patients (Jaeckle et al. 1985). When tumor invades the leptomeninges, 70% of patients have radicular or local spinal symptoms, including paresthesias in one or more extremities and pain in the back or neck (Wasserstrom et al. 1982). This pain is fairly resistant to narcotic analgesics, but it will respond to spinal axis irradiation or intrathecal chemotherapy.

Pain due to nerve damage is the most difficult to treat, and in general, it does not respond well to drug treatment. Often, it is not controlled by narcotic analgesics alone but may respond better to corticosteroids (Twycross 1988). A combination of drug and nondrug measures is often required to obtain partial relief.

Certain spinal procedures, such as rhizotomy or chordotomy, may be indicated in chronic unilateral pain. In 20 patients with lumbosacral or pelvic tumors, a chordotomy resulted in dramatic pain relief that lasted an average of 12 weeks (Macaluso et al. 1988). In a group of 45 patients with Pancoast's syndrome, 20 (45%) were pain-free until death as a result of a chordotomy (Ischia et al. 1985).

Tumor Invading Gastrointestinal Organs

Pain due to tumor invasion of the gastrointestinal tract may be especially severe with involvement of the stomach or colon. Usually a cramping abdominal pain, it may radiate to the groin, back, or scapula. Rectal cancer often causes pain in the lumbosacral spine and legs or in the perineal region. In a retrospective review of 150 patients with adenocarcinoma of the rectum in the years 1967 to 1982 who were suffering from severe chronic pain, the pain was located in the lumbosacral spine or leg in 47% of patients (Watson and Evans 1986). Twenty-two percent had pain in the perineum, 20% had rectal pain, and 11% had abdominal pain. The effect of their treatment is difficult to evaluate because only 53 patients were treated with narcotic analgesics, dosages of which were not mentioned. Of 53 patients who underwent a chordotomy, 77% had long-term pain relief, with a median time of 5 months, and one third of patients were pain-free until death. A phenol block, performed in 43 patients, achieved pain relief in 53% for a median time of 8 weeks. Radiation therapy was very effective in relieving pain in the six patients who were able to receive additional irradiation.

Cancer of the pancreas often causes severe deep midabdominal pain that may radiate to the back. Bonica (1985) found pain to be a prominent symptom in 79% of patients with advanced pancreatic cancer. This pain is notoriously unrelieved by specific antitumor treatment and also may not respond well to narcotic analgesics. Good pain relief can be achieved by a neurolytic celiac plexus block (Lebovits and Lefkowitz 1989). Of 136 patients with pancreatic cancer treated by Brown and colleagues (1987), long-term pain relief was obtained in 116 (85%). When blocks were repeated for recurrence of pain after an initial successful block, pain relief was achieved in 81% of patients.

Pain Due to Treatment

Not only the cancer, but also the treatment for cancer may result in chronic pain. This can be due to any of the three major treatment modalities: surgery, radiation therapy, and chemotherapy.

Postsurgical Pain

The most classic example of chronic pain due to surgery are postthoracotomy and postmastectomy pain. In these syndromes, the pain usually develops within a few months after surgery. This deafferentation pain has a dysesthetic quality and is associated with sensory loss in the area of the scar. Patients who have undergone a mastectomy also develop pain and numbness in the posterior arm and axilla. Because the pain often worsens with movement, a frozen shoulder may develop. The incidence of these pain syndromes appears to be less than 1% (Kanner, unpublished data), but no good epidemiological data are available. As in other deafferentation syndromes, pain management often is disappointing. The pain does not respond well to morphine, and only a minority of patients have obtained pain relief through ablative neurosurgical procedures, such as cordotomy or stereotactic intracerebral lesions (Tasker and Dostrovsky 1989).

Extensive resection in the head and neck region also may result in persistent pain, often of a lancinating quality. This is thought to be due to damage of postauricular and cervical nerve branches and may respond to carbamazepine.

Phantom pain is a persistent pain in the amputated body part where pain before amputation often was a major symptom. This usually involves a limb, but it also may involve other organs, such as rectum or breast. The limb still may be perceived as whole, often in an awkward position. This pain appears to be of central origin, probably caused by a decrease in tonic inhibition due to loss of sensory input. Although no treatment modality is known to give consistent pain relief, patients may respond to a variety of treatments, such as tricyclic antidepressants, narcotic analgesics, beta-blocking drugs, acupuncture, or dorsal root entry zone lesions (Marsland et al. 1982; Saris et al. 1985; Urban et al. 1986).

Postradiation Pain

Radiation therapy sometimes results in chronic pain through the nerve damage it can cause. Although peripheral nerves are relatively resistant to radiation therapy, doses of more than 60 Gy gradually may damage underlying nerves or plexuses through obliteration of small vessels and fibrosis. In general, symptoms do not appear until months or years after the treatment. Pain often is preceded by a slowly progressive weakness and sensory loss in the distribution of the involved nerves or plexus.

The most common example is radiation-induced brachial plexopathy, often seen in patients with breast cancer whose radiation ports include the supraclavicular fossa. Severe pain is less of a problem than dysesthesias, heaviness, lymphedema, and weakness. In the series by Kori and associates (1981) of 22 patients, only four (18%) presented with pain, although pain eventually was reported in 65% and became a major problem at some point in 35% of patients.

This pain can be severe; Lederman and Wilbourn (1984) describe a patient who underwent an amputation for incapacitating pain and weakness in the arm due to radiation-induced nerve damage. As with metastatic plexopathy, pain relief has not been very effective in this group of patients.

Radiation therapy to the lower abdomen and pelvis may damage the lumbosacral plexus. This most often presents with indolent lower extremity weakness, in contrast to lumbosacral plexopathy due to tumor, in which pain is the most prevalent early symptom. The weakness may be bilateral and usually progresses slowly or stops. Pain initially occurs only in approximately 10% of patients, but eventually it is present in half (Thomas et al. 1985). Pain treatment has been unsatisfactory for this syndrome.

Painful paresthesias and burning or shooting pains in the neck developing years after irradiation and encompassing the spinal cord may herald the onset of a progressive radiation myelopathy. Later in the disease, aching, burning, or shooting pains at the level of the irradiated spinal segment often are experienced (Palmer 1972; Reagan et al. 1968). Although there is no effective treatment for radiation myelopathy, pain and weakness may temporarily improve with corticosteroids.

Pain Due to Chemotherapy

Pain as a complication of chemotherapy most often is connected with peripheral neuropathy. This mainly occurs with repeated administration of vinka alkaloids and less frequently with platinum. In addition to a progressive distal weakness and sensory loss in a stocking-glove distribution, the patient often suffers from paresthesias and dysesthesias. With vincristine, the severity of the neuropathy appears to be dose-related (Kaplan and Wiernik 1982; Young and Posner 1980). The time till return of normal function and disappearance of paresthesias depends on the severity of symptoms at discontinuation of the drug. In cases of severe vincristine neuropathy, resolution of symptoms may take many months and may remain incomplete. This is true for *cis*-plati-num-induced neuropathy. The latter also may be exacerbated if treatment with *cis*-platinum is resumed (Becher et al. 1980). There is no effective treatment to hasten the recovery of these drug-induced neuro-pathies (Jackson et al. 1986; Kaplan and Wiernik 1982). However, tricyclic antide-pressants such as amitriptyline may help to relieve the pain.

References

Ampil F. L. Radiotherapy for carcinomatous brachial plexopathy. Cancer 56:2185–2188; 1985.

Becher, R.; Schutt, P.; Osieka, R.; Schmidt, C. G. Peripheral neuropathy and ophthalmo-logic toxicity after treatment with cisdichloro-diaminoplatinum II. J. Cancer Res. Clin. On-col. 96:219–221; 1980.

Bonica, J. J. Treatment of cancer pain: current status and future needs. In: Fields, H. L.; Dubner, R.; Cervero, F.; eds. Advances in pain research and therapy. Vol. 9. New York: Raven Press; 1985: p. 589–616.

Brown, D. L.; Bulley, K.; Quiel, E. L. Neuro-lytic celiac plexus block for pancreatic cancer pain. Anesth. Analg. 66:869–873; 1987.

Foley, K. M. Clinical assessment of cancer pain. Acta. Anaesthesiol. Scand. 74(suppl.):91–96; 1982.

Garmatis, C. J.; Chu, F. The effectiveness of ra-diation therapy in the treatment of bone metas-tases from breast cancer. Radiology 16:235–237; 1978.

Gilbert, R. W.; Kim, J. H.; Posner, J. B. Epidu-ral spinal cord compression from metastatic tu-mor: diagnosis and treatment. Ann. Neurol. 3:40–51; 1978.

Greenberg, H. S.; Deck, M. D. F.; Vikram, B.; Chu, F. C. H.; Posner, J. B. Metastasis to the base of the skull: clinical findings in 43 pa-tients. Neurology 31:530–537; 1981.

Greenberg, H. S.; Kim, J.-H.; Posner, J. B. Epi-dural spinal cord compression from metastatic tumor: results with a new treatment protocol. Ann. Neurol. 8:361–366; 1980.

Hanks, G. W. The pharmacological treatment of bone pain. Cancer Surv. 7:87–101; 1988.

Hoskin, P. J. Scientific and clinical aspects of radiotherapy in the relief of bone pain. Cancer Surv. 7:69–86; 1988.

Hoy, A. M. Radiotherapy, chemotherapy and hormone therapy; treatment for pain (1989). In: Wall, P. D.; Melzack R., eds. Textbook of pain. Edinburgh: Churchill Livingstone: 1989: p. 966–978.

Ischia, S.; Ischia, A.; Luzzani A.; Toscano D.; Steele A. Results up to death in the treatment of persistent cervico-thoracic (Pancoast) and thoracic malignant pain by unilateral percuta-neous cervical cordotomy. Pain 21:339–355; 1985.

Jackson, D. V.; Pope, E. K.; McMahan, R. A.; Cooper, M. R.; Atkins, J. N.; Callahan, R. D.; Paschold, E. H.; Grimm, R. A.; Hopkins, J. O.; Muss, H. B.; Richards, F.; Stuart, J. J.; White, D. R.; Zekan, P. J.; Cruz, J. M.; Spurr, C. L.; Capizzi, R. L. Clinical trial of pyridox-ine to reduce vincristine neurotoxicity. J. Neu-rooncol. 4:37–41; 1986.

Jaeckle, K. A.; Young, D. F.; Foley, K. M. The natural history of lumbosacral plexopathy in cancer. Neurology 35:8–15; 1985.

Jensen, T. S.; Krebs, B.; Nielsen, J.; Rasmus-sen, P. Immediate and long-term phantom limb pain in amputees: Incidence, clinical charac-teristics and relationship to pre-amputation limb pain. Pain 21:267–278; 1985.

Kanner, R. M.; Martini, N.; Foley, K. M. Inci-dence of pain and other clinical manifestations of superior pulmonary sulcus (Pancoast) tu-mors. In: Bonica, J. J.; Ventafridda, V.; Pagni, C. A., eds. Advances in pain research and therapy. Vol. 4. New York: Raven Press; 1982: p. 27–39.

Kaplan, R. S.; Wiernik, P. H. Neurotoxicity of antineoplastic drugs. Semin. Oncol. 9:103–130; 1982.

Kori, S. H.; Foley, K. M.; Posner, J. B. Brachial plexus lesions in patients with cancer. Neurol-ogy 31:45–50; 1981.

Lebovits, A. H.; Lefkowitz, M. Pain manage-ment of pancreatic carcinoma: a review. Pain 36:1–11; 1989.

Lederman, R. J.; Wilbourn, A. J. Brachial plex-opathy: recurrent cancer or radiation? Neurol-ogy 34:1331–1335; 1984.

Macaluso, C.; Foley, K. M.; Arbit, E. Cordot-omy for lumbosacral, pelvic, and lower ex-tremity pain of malignant origin: safety and effi-cacy. Neurology 38(suppl. 1):110, 1988.

Marsland, A. R.; Weekes, J. W. N.; Atkinson, R. L.; Leong, M. G. Phantom limb pain: a case for beta blockers? Pain 12:295–297; 1982.

Miser, A. W.; Dothage, J. A.; Wesley, R. A.; Miser, J. S. The prevalence of pain in a pediat-ric and young adult cancer population. Pain 29:73–83; 1987b.

Miser, A. W.; McCalla, J.; Dothage, J. A.; Wes-ley, M.; Miser, J. S. Pain as a presenting symp-tom in children and young adults with newly diagnosed malignancy. Pain 29:85–90; 1987a.

Moulin, D. E.; Coyle, N. Spinal opioid analge-sics and local anesthetics in the management of chronic cancer pain. J. Pain Sympt. Manag. 1:79–86; 1986.

Obbens, E. A. M. T.; Kim, J. H.; Thaler, H.; Deck, M. D. F.; Posner, J. B. Metronidazole as a radiation enhancer in the treatment of metastatic epidural spinal cord compression. J. Neurooncol. 2:99–104; 1984.

Obbens, E. A. M. T.; Hill, C. S.; Leavens, M. E.; Ruthenbeck, S. S.; Otis, F. Intraventricular morphine administration for control of chronic cancer pain. Pain 28:61–68; 1987.

Posner, J. B. Metronidazole as a radiation enhancer in the treatment of metastatic epidural spinal cord compression. J. Neurooncol. 2:99–104; 1984.

Palmer, J. J. Radiation myelopathy. Brain 95:109–122; 1972.

Penn, C. R. H. Single dose and fractionated palliative irradiation of osseous metastases. Clin. Radiol. 27:405–408; 1976.

Ramirez, L. F.; Levin, A. B. Pain relief after hypophysectomy. Neurosurgery 14:499–504; 1984.

Reagan, T. J.; Thomas, J. E.; Colby, M. Y. Chronic progressive radiation myelopathy. J.A.M.A. 203:106–110; 1968.

Saris, S. C.; Iacono, R. P.; Nashold, B. S. Dorsal root entry zone lesions for post-amputation pain. J. Neurosurg. 62:72–76; 1985.

Sundaresan, N.; Hilaris, B. S.; Martini, N. The combined neurosurgical-thoracic management of superior sulcus tumors. J. Clin. Oncol. 5:1739–1745; 1987.

Takeda, F. Results of field-testing in Japan of the WHO draft interim guidelines on relief of cancer pain. Pain Clinics I:83–89; 1986.

Tasker, R. R.; Dostrovsky, J. O. Deafferentiation and central pain. In: Wall, P. D.; Melzack, R. eds. Textbook of pain. Edinburgh: Churchill Livingstone; 1989: p. 154–180.

Thomas, J. E.; Cascino, T. L.; Earle, J. D. Differential diagnosis between radiation and tumor plexopathy of the pelvis. Neurology 35:1–7; 1985.

Torma, T. Malignant tumours of the spine and the spinal extradural space. Acta. Chir. Scand. 225(suppl.):15–176; 1957.

Twycross, R. G. Opioid analgesics in cancer pain: current practice and controversies. Cancer Surv. 7:29–53; 1988.

Urban, B. J.; France, R. D.; Steinberger, E. K.; Scott, D. L.; Maltbie, A. A. Long-term use of narcotic/antidepressant medication in the management of phantom limb pain. Pain 24:191–196; 1986.

Ventafridda, V.; Tamburini, M.; Caraceni, A.; De Conno, F.; Naldi, F. A validation study of the WHO method for cancer pain relief. Cancer 59:851–856; 1987.

Wasserstrom, W. R.; Glass, J. P.; Posner, J. B. Diagnosis and treatment of leptomeningeal metastases from solid tumors: experience with 90 patients. Cancer 49:759–772; 1982.

Walther, H. E. Krebsmetastasen. Basel: Bruno Schwabe; 1948.

Watson, C. P. N.; Evans, R. J. Intractable pain with cancer of the rectum. Pain Clinics 1:29–34; 1986.

World Health Organization. Cancer pain relief. Geneva: World Health Organization; 1986.

Young, D. F.; Posner, J. B. Nervous system toxicity of the chemotherapeutic agents. In: Vinken, P. J.; Bruyn, G. W., eds. Handbook of clinical neurology. Vol. 39. Amsterdam: North Holland Publishing Company; 1980: p. 91–129.

48

Sleep Disorders

SIMON J. FARROW

The etiology and natural history of most sleep disorders are uncertain, so that diagnostic criteria, prognosis and the benefits of available treatment are difficult to assess.

Narcolepsy

For instance, there is disagreement concerning criteria for the diagnosis of narcolepsy (Burton et al. 1989). The clinical syndrome of idiopathic narcolepsy may be easily recognizable and clearly associated with the electroencephalographically confirmable abnormality of sleep-onset rapid eye movement sleep (Mitler et al. 1979; Rechtschaffen et al. 1963; Rosa et al. 1979); but many subjects do not report all the major symptoms and there seems to be a spectrum in which the signs and symptoms of narcolepsy shade uneasily into those of other causes of excessive sleepiness. It is not clear to what extent a combination of the clinical presentation, the close association between narcolepsy and the HLA-DR2 and DQW1 histocompatibility antigens (Billiard et al. 1986; Guilleminault and Grumet 1986; Honda et al. 1986; Honda et al. 1989; Langdon et al. 1986; Poirier et al. 1986), and a rigorous analysis of polysomnographic data (Carnellas et al. 1989; Nahmias and Karetzky 1989; Rhoads et al. 1989) will make possible a clear distinction between idiopathic narcolepsy and other forms of excessive daytime sleepiness. Estimates of prevalence and prognosis are likely to be significantly affected by variation in diagnostic criteria.

Prevalence has been variously estimated at between 0.02% and 0.07% (Coleman et al. 1982; Dement et al. 1972; Dement et al. 1973; Nevsimalova-Brŭhova and Roth 1972). Both the narcolepsy syndrome (Baraitser and Parkes 1978; Dement et al. 1972; Sours 1963; Yoss and Daly 1960) and other types of excessive daytime sleepiness (Honda et al. 1986; Kessler et al. 1974) are much more common in the families of narcoleptics, but the mode of inheritance is unclear. Parkes (1985) has estimated that between a quarter and a half of all narcoleptics have an affected close relative.

Onset of the syndrome is usually said to be in the teens or early 20s but has been reported as earlier in upto 6% (Roth 1980) or even 20% (Navelet et al. 1976) of cases. It has been stated that onset over the age of fifty years accounts for less than 5% of all cases (Parkes 1985).

Development of the syndrome can be abrupt (Roth 1980), but there may be a prodromal period of between one and twenty years in upto 10% of narcoleptics (Parkes et al. 1974). The course of the illness is often progressive at first, with gradual appearance of more symptoms. In one study of 70 patients, mean age of onset of sleepiness, cataplexy, sleep paralysis and hypnagogic hallucinations was reported as 22, 28, 29 and 27 years respectively (Billiard et al. 1983). Subjects frequently seem to reach a plateau of severity, but its level is not predictable. Fluctuations in the severity of the symptoms is common and appears to be related to a wide variety of circumstances (Roth 1980). Prolonged remissions may occur but they are rare and the etiology is uncertain. Cataplexy and sleep paralysis may disappear in some patients (Parkes 1985). It does not appear possible to predict the course of the illness in any individual case.

The effects of narcolepsy on the individual vary from inconvenient to incapacitating and narcoleptics have been reported as being involved in more than twice as many sleep related automobile accidents as controls (Aldrich 1989). But there appear to be no other developments of the syndrome as such and there is no significant evidence of a relationship between idiopathic narcolepsy and the development of any other disease.

There is no evidence that any available treatment alters the course of the illness.

The symptoms may be relieved by behavioral modification.

Pharmaceutical treatment of the sleepiness has traditionally been with stimulant drugs. Many narcoleptics report good long term relief at a constant dose. Side effects may be difficult to tolerate. The use of stimulants for the treatment of narcolepsy has been reviewed by Parkes (1985) and Guilleminault (1989). Effective treatment of motor symptoms with tricyclics and other antidepressants has been reported (Schmidt et al. 1977; Shapiro 1975). Mitler (1989), in a well organized trial, demonstrated improvement towards control values on some measures of performance and sleepiness with pemoline, methylphenidate amphetamine, and viloxazine (a propranolol derivative)

and relief of cataplexy with protriptylene and viloxazine. Relief of cataplexy with viloxazine was also reported from a multi center study (Godbout et al. 1989).

About two thirds of the patients in one study reported continuing symptomatic relief with codeine (Fry et al. 1986).

Prolonged relief of symptoms has been reported in upto 75% of patients treated with the sedative gamma-hydroxybutyrate (Mamelak et al. 1986; Scrima et al. 1989), but laboratory findings are less certain (Mamelak et al. 1986).

There appears to be little information upon which to base predictions of the effect of specific drugs in any particular individual.

Sleep Apnea

On the other hand the conflict between the reduction of motor activity associated with sleep and the mammalian system of tidal ventilation is such that disturbances of the multiple-loop systems which control ventilation during sleep are common. The hard-to-define (Association of Sleep Disorders Centers 1979; Bornstein 1982; Henderson-Smart and Cohen 1986; Parkes 1985) but often easily recognizable sleep apnea–hypopnea syndrome may be associated with immaturity, anatomical variation, developmental anomaly, or many aquired conditions which affect breathing. Investigators have begun to define the effect of different factors on occurrence and prognosis.

The incidence of apnea of prematurity is particularly related to gestational age at birth and degree of development (Booth et al. 1983; Heldt 1988; Henderson-Smart et al. 1983; Henderson-Smart and Cohen 1986), although other "triggering" and predisposing factors are important (Butcher-Puech et al. 1985; Thach 1985). Untreated, even mild apnea may be associated with subsequent deterioration. In one study, mortality was 70% for the sickest infants and 23% for those whose treatment was initially postponed because the apnea was considered mild (Jones and Lukeman 1982).

Treatment with stimulants or positive airway pressure (Miller et al. 1985), singly or in combination, is generally considered effective. In one study of caffeine alone (Anwar et al.), episodes of prolonged apnea were reported to disappear in all infants treated. In another, using theophylline alone (Muttitt et al. 1988), a response was noted in up to 77%. Ment and colleagues (1985) did not find evidence that methylxanthine therapy for apnea of prematurity had an adverse effect on cognitive development at 18 months. No particular advantage of any specific methylxanthine has been demonstrated (Brouard et al. 1985). It has been suggested that there is some advantage to treatment with a stimulant as opposed to positive airway pressure (Ryan et al. 1989) and that there is no significant difference between continuous and intermittent airway pressure (Sims et al. 1985).

The tone of some follow-up studies is less optimistic (Sims et al. 1985). Cardiorespiratory abnormalities lasting for at least several months after discharge from the hospital were recorded in 16% of infants in one study (Rosen et al. 1986). Reports of developmental sequelae vary from 20% with major handicaps at 9 to 24 months (Jones and Lukeman 1982) to only two with an abnormal developmental quotient when examined after 16 months of age, out of 25 babies found to be apneic at less than 31 weeks postmenstrual age (Levitt et al. 1988). It is not clear that this major difference can be attributed to the effects of improved methods of treatment. Results of surveys vary with the methodology (Thach 1985; Jones and Lukeman 1982). Moreover, it is difficult to define the effect of sleep apnea in the high frequency of residual nervous system dysfunction associated with prematurity.

Apneic episodes and periodic breathing occur during sleep in normal full-term infants (Ellingson et al. 1982; Hoppenbrouwers et al. 1977). A definition of abnormal breathing during sleep in infants is difficult and has been discussed by Guilleminault (1987). Acceptable criteria of significance may be bradycardia or hypoxemia associated with ventilatory abnormality.

Sleep apnea in full term infants that is not clearly a part of another illness is usually associated with a "narrow upper airway" or an abnormality of nervous system development. As in the Pierre Robin syndrome, these may be related. Serious sequelae of untreated infantile sleep apnea are common and include cor pulmonale, pulmonary hypertension, failure to thrive and developmental delay. Brouillette (1982) reported cor pulmonale in 55% of patients, permanent neurological damage in 9%, failure to thrive in 27% and other disturbances in 23%. Deykin and coworkers (1984) reported significant impairment in gross motor development and mild cognitive deficiencies in formerly apneic children when compared with age-similar siblings. Mauer and colleagues (1983) reported that none of 12 children with disordered breathing during sleep and demonstrated polysomnographic abnormalities had cor pulmonale, pulmonary hypertension, or alveolar hyperventilation.

Although the sleep apnea narrow upper airway syndrome of infancy and childhood is associated with many primary diagnoses as reviewed by Thach (1985), treatment consists primarily of surgical relief of the obstruction (Frank et al. 1983) or the use of devices to maintain airway patency during sleep. Surgery produced relief in 20 out of 22 children in one series (Brouillette et al. 1982) and 12 out of 12 in another (Mauer et al. 1983). Airway patency may also be obtained with a nasopharyngeal airway or positive airway pressure. Long term follow-up data appear to be limited.

Infantile sleep apnea not associated with a narrow upper airway or intercurrent illness is usually related to abnormality of the central nervous system. Specific treatment for identifiable conditions, such as syringomyelia, hydrocephalus and Arnold-Chiari malformation, has been reported as producing improvement of the associated hypoventilation or apnea (Balk et al. 1985; Weese-Mayer et al. 1988). However, successful treatment of sleep apnea associated with Arnold-Chiari malformation has also been reported for the use of caffeine (Davis et al. 1989). Treatment of idiopathic apnea of infancy with stimulants has been used as a general approach (Anwar et al. 1986; Guilleminault 1987; Finer et al. 1984), but it may be ineffective or have unacceptable side effects. Tracheostomy and mechanical ventilation during sleep may produce relief in infants with central apnea in whom stimulants are not effective (Guilleminault 1987). Reports of long-term follow-up are not extensive and seem disappointing. In one study, children with congenital central hypoventilation syndrome all had severely abnormal ventilation during sleep when followed for two to fourteen years (Oren et al. 1987).

In spite of the clear association between apnea of infancy and near-miss sudden infant death syndrome (Guilleminault 1987; Thach 1985), useful predictors of concurrence have not been demonstrated (Oren et al. 1987; Oren et al. 1989; Ward et al. 1986).

Irregular breathing during sleep occurs in normal adults (Krieger et al. 1989). Definitions of abnormal breathing during sleep have been arbitrary in the absence of data relating the development of daytime behavioral abnormality or of other pathology to the severity of the breathing disturbance.

However, disturbance of breathing during sleep sufficiently severe to be described as sleep apnea is common (Block et al. 1979; Coleman et al. 1982). Estimates of prevalence vary from 0.01% to 1% in the general population in the United Kingdom (Parkes 1985), through 1.3% among Swedish men aged 30 to 69 (Gislason et al. 1988), to 6% in the industrial population in Israel to (Lavie 1983) 10% of middle-aged men in a German sample (Peter et al. 1989). Sleep-related breathing disorders have been detected in upto 50% of individuals in some samples of elderly Americans (Ancoli-Israel et al. 1985; Carskadon and Dement 1981; Krieger et al. 1983; McGinty et al. 1982). Reports on the rate of progression have conflicted (Bliwise et al. 1989; Mason et al. 1989).

Sleep apnea in adults is associated with insomnia and hypersomnolence (Association of Sleep Disorders Centers 1979) which may be dangerously severe (Aldrich et al. 1987; Aldrich 1989; Richter and Plattner 1989). Mood disorders and paranoid behavior in apneic patients seem common in clinical practice, but objective correlation has been difficult to establish (Millman et al. 1989). Cognitive deficits have been associated with sleep apnea in adults (Bedard et al. 1989; Greenberg et al. 1987) and in the elderly (Bliwise et al. 1989).

Sleep apnea is prevalent in patients with hypertension (Fletcher et al. 1985; Lavie et al. 1984), and hypertension in patients with sleep apnea (Guilleminault et al. 1980; Kales et al. 1985; Rauscher et al. 1989), but both are associated with the common denominators of obesity and male sex and it is not clear that the sleep apnea is causative or contributory (Fletcher et al. 1985; Rauscher et al. 1989). A suggested causative relationship between sleep apnea and diurnal pulmonary hypertension has not been clearly demonstrated (Weitzenblum et al. 1988). However, hemodynamic changes associated with the apneic episodes are well documented (Marrone et al. 1989; McGinty et al. 1988; Tilkian et al. 1976), as are cardiac arrhythmias (Bolm-Audorf et al. 1984; Otsuka et al. 1987; Shepard et al. 1985), and there is some evidence that there is a relationship between the severity of the sleep apnea syndrome and the cardiovascular pathology (Cirignotta et al. 1989; McGinty et al. 1988; Otsuka et al. 1987; Shepard et al. 1985; Weitzenblum et al. 1988).

Available data suggest that survival is reduced for those with untreated sleep apnea as opposed to those whose sleep apnea is treated aggressively and to those sample members found not to have sleep apnea. These data also appear to imply that survival is inversely related to severity of the sleep apnea syndrome (Bliwise et al. 1988; He et al. 1988; Partinen et al. 1988; Thorpy et al. 1987; Thorpy et al. 1989). Actuarial assessment of reduced survival in people with sleep apnea is beginning to become available (He et al. 1988; Partinen et al. 1988; Thorpy et al. 1987).

Adult sleep apnea has been described in association with a wide variety of different pathologies, so that the effects of treatment on prognosis vary with the associated condition. Conventionally, it has been classified according to the degree of apparent airway obstruction demonstrated polysomnographically.

Treatment designed to relieve airway obstruction is often demonstrably effective in the short term. Long term compliance with mechanical treatment is reported to be from about 50% to over 80% (Browman et al. 1989; Dickins et al. 1989; Kribbs et al. 1989; Krieger and Kurtz 1988; Nino-Murcia et al. 1989; Samelson 1989). So far, follow up suggests that positive airway pressure devices are persistently effective with continued use (Nino-Murcia et al. 1989) and that the tongue retaining device (the oral prosthesis for which follow-up studies seem most extensive) may also remain effective (Cartwright et al. 1988; Samuelson 1989). Tracheostomy may be effective for long term relief of apnea in selected patients (Thorpy et al. 1987; He et al. 1988; Partinen et al. 1988). Pharyngeal surgery alone has a relatively poor success rate. In one study (He et al. 1988), the cumulative survival of the group treated with pharyngeal surgery alone was not different from the survival of untreated patients with moderate sleep apnea. Other studies report a success rate of from 50% to much less (Aubert-Tulkens et al. 1989; Katsantonis et al. 1988). A 67% success rate at 6 months has been reported for mandibular surgery in patients for whom these more conservative procedures were not successful (Riley et al. 1989).

Successful treatment of "central" sleep apnea with obstruction relieving devices has been reported for patients in whom a small upper airway appeared to be a significant etiological factor (Guilleminault et al. 1987). Purely central sleep apnea may be treated with mechanical ventilation or diaphragmatic stimulation. Reports of successful pharmaceutical treatment of "central sleep apnea," have been generally anecdotal and have been reviewed by White and coworkers (1989). A recent report has suggested statistically significant treatment of selected patients with prednisone (Kimura et al. 1989).

Although weight loss is clearly an important aspect of the treatment of sleep apnea in obese patients (Riley et al. 1989), obtaining reliable data on the long-term success of this form of treatment appears as difficult as the treatment itself.

Restless Legs and PLMS

Restless legs and periodic leg movements of sleep, or nocturanl myoclonus, are frequently mentioned as possibly associated disorders which prolong the onset or reduce the quality of sleep (Adams and Victor 1985; Coleman et al. 1982; Montplaisir and Godbout 1989; Parks 1985; Weitzman and Rowland 1984). However, the criterion for diagnosis of restless legs appears to be the clinician's decision that, in the absence of any more objectively definable pathology, the patient's symptoms resemble those described by Ekbom (1960) and others (Brodeur et al. 1988; Hening et al. 1986; Telstad et al. 1984). Screening tests, alternative diagnoses and general appreciation of the syndrome vary widely with the discipline of the researcher (Blattler and Muhlemann, 1982; Montplaisir et al. 1985; Rudiger 1980). Estimates of prevalence of 5% or greater (Bornstein 1961; Ekbom 1960) depend on reports of patients with some aspects of the syndrome only, which emphasizes the diagnostic confusion. Anecdotal reports suggest that idiopathic restless legs syndrome may occur at any age, may vary in intensity through the life of the patient, may remit completely and is usually not associated with the development of other pathology. Treatment was reviewed by Gibbs and Lees (1986). More recent reports have not altered the impression that no single treatment is credibly reliable, although various treatments seem effective sometimes.

Often, there is an association between complaints of disagreeable sensation in the lower legs which can only be relieved by walking and the polygraphically demonstrable periodic leg movements of sleep.

There is a spectrum of increasingly complex patterns of leg muscle activity which may occur during sleep (Parkes 1986). Whether defined quite specifically to include only prominent repetitive contractions of the anterior tibial compartment muscles

(Coleman and Guilleminault 1982) or to include several types of "muscle activity in the legs" associated with disturbed sleep, (Martman and Scrima 1986) periodic leg movements of sleep are common among patients studied polygraphically in sleep laboratories (Coleman et al. 1980; Coleman et al. 1982) and have been associated with complaints of insomnia and excessive daytime sleepiness (Coleman et al. 1980; Rosenthal et al. 1984; Saskin et al. 1985). However, they are present in the normal population (Lugaresi et al. 1986; Montagna et al. 1988) and were found in 34% of a sample of elderly people originally identified by telephone survey and up to 60% of individuals over sixty recruited from a senior citizens' center and studied for 3 nights (Mosko et al. 1988). Also, the intensity of periodic leg movements in any individual may vary considerably from night to night (Bliwise et al. 1988; Mosko et al. 1988). Leg movements are reported in association with many other sleep disorders (Coleman 1980), particularly narcolepsy and sleep apnea (Broughton et al. 1985; Fry et al. 1989; Hartman and Scrima 1986; Mosko et al. 1988; Pearce et al. 1989) and, indeed, as becoming more common in patients whose sleep apnea is treated with nasal controlled positive airway pressure (Fry et al. 1989; Pearce et al. 1989). So that although intermittent leg movements in sleep become much more common with age (Ancoli-Israel et al. 1987; Coleman et al. 1981), and although, in severe form, they appear to be associated with disturbed sleep, neither the etiology, the effects, nor the associated morbidity, if any, are sufficiently clear to provide useful prognostic data, still less to indicate general treatment with sedative drugs.

Sleepwalking and Night Terrors

Sleep-walking and night terrors are examples of more prolonged and organized motor sequences occurring in sleep. Occasional sleep-walking is common in children. Reported incidence in large samples varies from 10% to 40% (Cirignotta et al. 1983; Fisher and Wilson 1987; Klackenberg 1987). More than two or three incidents a year is much less common (Fisher and Wilson 1987; Gass and Strauch 1984; Klackenberg 1987). The sleep-walking usually begins between the ages of 4 and 7, is most prevalent between the ages of 10 and 12, and becomes much less frequent thereafter. In adults, the reported frequency of at least occasional sleep-walking varies from 1.5% to 5% (Bixler et al. 1979; Cirignotta et al. 1983; Davies et al. 1942), but probably less than 1% sleep-walk more than once a year (Cirignotta et al. 1983). Strong family histories (Kales et al. 1980) and a higher concordance in monozygotic than dizygotic twins (Bakwin 1970) have been noted. There is no clear evidence that sleep-walking is the harbinger of any other abnormal behavior or pathology.

Night terrors seem to occur more frequently in families in which sleep-walking is prevalent, and also possibly to have a hereditary component (Hallstrom, 1972; Kales et al., 1980). They are reported as occurring in up to 6% of children (Beltramini and Hertzig 1983; Gass and Strauch 1984; Klackenburg 1987) and are more likely than sleep-walking to begin in adolescence or adult life.

Sleep-walking and night terrors in adults come to medical attention when they cause distress or danger. Severe personal injury to or death of the subject or a third party have been described on a case-by-case basis (Broughton and Warnes 1989; Hartman 1983; Oswald and Evans 1985; Scott 1988).

The distinction between sleep-walking and night terrors becomes blurred in adults as does the distinction between apparent arousals from non rapid eye movement sleep and similar behaviors with other possible etiologies (Broughton and Warnes 1989; Maselli et al. 1988). Also, there is an association with other behavioral abnormalities and the drugs used to treat them (Allen 1983; Glassman et al. 1986; Kales et al. 1980; Regestein and Reich 1985; Scott 1988), so that useful statistical data are not available.

Persistent reduction in the unwanted behavior has been claimed for a variety of psychotherapeutic, behavioral, and pharmacological treatments (Cooper 1987; Guilleminault 1987; Koe 1989; Lask 1988; Maselli et al. 1988). Controlled trials have not been performed.

Hypersomnolence

Some patients whose illness does not meet an acceptable definition of narcolepsy and whose sleep is not disturbed by chronic alterations of circadian rhythm, sleep apnea, excessive motor activity, pain or affective disorder are persistently or intermittently excessively sleepy.

Some of these patients are eventually diagnosed as having idiopathic narcolepsy (Parkes 1985).

Excessive sleepiness, distinguishable from stupor, has been reported in association with otherwise definable abnormalities of the nervous system, including gliosis (Erlich and Itabashi 1986), inflammation (Parkes 1985; Poirier et al. 1987), prescribed non "hypnotic" drugs (Thachil et al. 1987), mitochondrial disease (Kotagal 1985), and tumor (Parkes 1985) and may improve following appropriate treatment (Korfali 1988; Rubenstein and Gray 1988). It has also been measured in association with conditions in which the pathophysiology is less obvious, for instance, infectious mononucleosis (Guilleminault et al. 1986). The prognosis remains unclear.

Some people are persistently, excessively sleepy and do not fit any of these categories. For some, the difficulty seems to contain components of sleep related breathing disorder, periodic movements of sleep, or the combination of fibromyalgic symptoms and abnormal sleep EEG that Moldofsky (1986) has described as a rheumatic modulation disorder. These patients seem to be quite common. Their symptoms often are stable or only slowly progressive and resistant to attempts to treat the individual components. They may remain functional members of the community for many years. Other patients better fit the description of idiopathic CNS hypersomnia as discussed by Guilleminault (1989). This group seems more often to suffer from severe and disabling sleepiness, and often to develop tolerance to available stimulants.

Episodic extreme sleepiness has often been described under the eponym of Kleine-Levin syndrome. Reports have been reviewed by Critchley (1962) and others (Fresco et al. 1971; Parkes 1985). The frequency of attacks varies. Apparently complete remission is reported in some patients (Billiard 1981; Guilleminault et al. 1986). Published accounts include descriptions of clinically different syndromes, and recent investigational techniques suggest a variety of etiologies which may affect the hypothalamus or associated areas (Gadoth et al. 1987; Korfall et al. 1988; Testa et al. 1987). It is likely that prognosis depends on the specific pathology. Some therapeutic success has been reported for lithium (Hart 1985; Will et al. 1988).

Insomnia

Complaints of disturbed sleep are common (Bixler et al. 1979; Hammond 1964). Thirty-five percent of one sample reported "trouble sleeping in the past year" (National Institute of Mental Health, consensus development conference, 1984). In a recent study (Ford and Kamerow 1989), 10% of 7954 respondents in a community sample reported insomnia. But this word has no generally agreed meaning and is used in recent classifications of sleep disorders to denote symptoms, objective findings and nosological entities (Association of Sleep Disorders Centers 1979; International classification of sleep disorders 1990). Consequently, there are few meaningful data concerning prognosis in untreated insomnia.

The prognosis for cure or relief of insomnia associated with psychiatric illness (Ford and Kamerow 1989), other neurodegenerative processes (Von Economo 1930), diseases which prevent the bodily relaxation associated with sleep and the sleep disorders themselves (including those reviewed in the preceding sections of this chapter) is related to that for treatment of the underlying condition.

Advances in the treatment of supposedly idiopathic insomnia have come from better appreciation of underlying pathology. For instance, elimination of cow's milk from the diet of infants with chronic insomnia who were found to have immunologic reaction to cow's milk was associated with improved sleep (Khan et al., 1989). Bright light therapy for circadian rhythm disorders was associated with improved sleep in night-workers (Czeisler 1990) and with both improved sleep and higher serum melatonin levels in workers suffering from midwinter insomnia in the sub-arctic region (Hansen et al. 1987).

There is no strong evidence that any currently available pharmacological preparation is effective in the long term treatment of insomnia perceived as a nosological entity.

Some success has been reported for behavioral treatment of insomnia if subjects and codependents are sufficiently motivated (Akerstedt and Torsvall 1978; Ferber 1989). For instance, behavioral treatment of the phase delay syndrome, was reported as successful for all patients in one brief series (Czeisler et al. 1981). Another study demonstrated statistically significant improvement at follow-up in a non homogeneous group of subjects whose persistent insomnia was treated by restriction of time in bed (Spielman et al. 1987). Successful treatment of insomnia by psychoanalysis (Williams 1987) and by hypnosis (Cochrane 1989) has been reported on an individual basis. Long term follow-up data for behavioral treatment of insomnia are not available.

References

Adams, R. D.; Victor, M. Principles of Neurology. 3rd ed. New York: McGraw-Hill; 1985.

Akerstedt, T.; Torsvall, L. Experimental changes in shift schedules—their effects on well-being. Ergonomics 21(10):849–856; 1978.

Aldrich, M. S.; Aldrich, C. K.; Kehn, T. Automobile accidents: relation to excessive daytime sleepiness. Sleep res. 16:296; 1987.

Aldrich, M. S. Automobile accidents in patients with sleep disorders. Sleep 12(6):487–494; 1989.

Allen, R. M. Attenuation of drug-induced anxiety dreams and pavor nocturnus by benzodiazepins. J. Clin. Psychiatry 44(3):106–108; 1983.

Ancoli-Israel, S.; Kripke, D. F.; Mason, W.; Kaplan, O. J. Sleep apnea and periodic movements in an aging sample. J. Gerontol. 40(4):419–425; 1985.

Ancoli-Israel, S.; Mason, W.; Kaplan, O.; Kripke, D. F. Stability of apneas and leg jerks in the elderly. Sleep Res. 16:297; 1987.

Anwar, M.; Mondestin, H.; Mojica, N.; Nova, R.; Graff, M.; Hiatt, M.; Hegyi, T. Effect of caffeine on pneumogram and apnoea of infancy. Arch. Dis. Child. 61(9):891–895; 1986.

Association of Sleep Disorders Centers, Dignostic Classification of Sleep and Arousal Disorders. 1st ed. Sleep Disorders Classification Committee, Roffwarg, H. P., chairman. Sleep 2:39–41, 65–69; 1979.

Aubert-Tulkens, G.; Hamoir, M.; Van den Eeckhaut, J.; Rodenstein, D. O. Failure of tonsil and nose surgery in adults with long-standing sleep apnea syndrome. Arch. Intern. Med. 149(9):2118–2121; 1989.

Bakwin, H. Sleep-walking in twins. Lancet. 2(670):446–447; 1970.

Balk, R. A.; Hiller, F. C.; Lucas, E. A.; Scrima, L.; Wilson, F. J.; Wooten, V. Sleep apnea and the Arnold-Chiari malformation. Am. Rev. Respir. Dis. 132(4):929–930; 1985.

Baraitser, M.; Parkes, J. D. Genetic study of narcoleptic syndrome. J. Med. Genet. 15(4):254–259; 1978.

Bedard, M. A. Montplaisir, J.; Rouleau, I.; Malo, J. Neuropsychological deficits in obstructive sleep apnea syndrome (OSAS): Effect of sleep disruption, daytime somnolence and oxygen desaturation. Sleep Res. 18:199; 1989.

Beltramini, A. U.; Hertzig, M. E. Sleep and bedtime behavior in preschool-aged children. Pediatrics 71(2):153–158; 1983.

Billiard, M. The Kleine-Levin syndrome. In: Koella, W. P., ed. Sleep 1980. Basel: S Karger; 1981: p. 124–127.

Billiard, N.; Besset, A.; Cadilhac, J. The clinical and polygraphic development of narcolepsy. In: Guilleminault, C.; Lugaresi, E., eds. Sleep/wake disorders: natural history, epidemiology and long term evolution. New York: Raven Press; 1983: p. 171–185.

Billiard, M.; Seignalet, J.; Besset, A.; Cadilhac, J. HLA-DR2 and narcolepsy. Sleep 9:149–152; 1986.

Bixler, E.; Kales, A.; Soldatos, C. R.; Kales, J. D.; Healey, S. Prevalence of sleep disorders in the Los Angeles metroplitan area. Am. J. Psychiatry 136(10):1257–1262; 1979.

Blattler, W.; Muhlemann, M. Restless legs and nocturnal leg spasms—forgotten facts in diagnosis—new facts for therapy. Schweiz Med Wochenschr 112(4):115–117; 1982.

Bliwise, D. L.; Bliwise, N. G.; Partinen, M.; Pursley, A. M.; Dement, W. C. Sleep apnea and mortality in an aged cohort. Am. J. Public Health 78(5):544–547; 1988.

Bliwise, D. L.; Carskadon, M. A.; Dement, W. C. Nightly variation of periodic leg movements in sleep in middle aged and elderly individuals. Arch. Gerontol. Geriatr. 7(4):273–279; 1988.

Bliwise, D. L.; Ingham, R. H.; Nino-Murcia, G.; Pursley, A. M.; Dement, W. C. Five year follow-up of sleep related respiratory disturbances and neuropsychological variables in elderly subjects. Sleep res. 18:202; 1989.

Block, A. J.; Boysen, P. G.; Wynne, J. W.; Hunt, L. A. Sleep apnea, hypopnea and oxygen desaturation in normal subjects. A strong male predominance. N. Engl. J. Med. 300(10):513–517; 1979.

Bolm-Audorff, U.; Kohler, U.; Becker, E.; Fuchs, E.; Mainzer, K.; Peter, J. H.; von Wichert, P. Nocturnal cardiac arrhythmias in the sleep apnea syndrome. Dtsch Med Wochenschr 109(22):853–856; 1984.

Booth, C. L.; Morin, V. N.; Waite, S. P.; Thomas, E. B. Periodic and non-periodic sleep apnea in premature and fullterm infants. Dev. Med. Child. Neurol. 25(3):283–296; 1983.

Bornstein, B. Restless legs. Psychiatr. Neurol. 141:165–201; 1961.

Bornstein, S. K. Guilleminault, C., ed. Sleeping and waking disorders: indications and techniques. Boston: Butterworths; 1982.

Brodeur, C.; Montplaisir, J.; Godbout, R.; Marinier, R. Treatment of restless legs syndrome and periodic movements during sleep with L-dopa: a double-blind, controlled study. Neurology 38(12):1845–1848; 1988.

Brouard, C.; Moriette, G.; Murat, I.; Flouvat, B.; Pajot, N.; Walti, H.; deGamarra, E.; Relier, J. P. Comparative efficacy of theophylline and caffeine in the treatment of idiopathic apnea in premature infants. Am. J. Dis. Child. 139(7):698–700; 1985.

Broughton, R.; Tolentino, M. A.; Krelina, M. Excessive fragmentary myoclonus in NREM sleep: a report of 38 cases. Electroencephalogr. Clin. Neurophysiol. 61(2):123–133; 1985.

Broughton, R.; Warnes, H. Violence and Sleep: eleven cases. Sleep Res. 18:205; 1989.

Brouillette, R. T.; Fernbach, S. K.; Hunt, C. E. Obstructive sleep apnea in infants and children. J. Pediatr. 100(1):31–40; 1982.

Browman, C. P.; Newman, J. C.; Winslow, D. H. Nasal CPAP therapy for obstructive sleep apnea: determinants of long term compliance. Sleep res. 18:206; 1989.

Burton, S. A.; Dement, W. C.; Ristanovic, R. K.; eds. Recent developments in the diagnosis and treatment of narcolepsy. Chicago: Associations Plus, Inc.; 1989: p. 3–14, 27–33.

Butcher-Puench, M. C.; Henderson-Smart, D. J.; Holley, D.; Lacey, J. L.; Edwards, D. A. Relation between apnoea duration and type and neurological status of preterm infants. Arch. Dis. Child. 60(10):953–958; 1985.

Canellas, F.; Nougier, J.; Tafti, M.; Billiard, M. Nighttime and daytime sleep in idiopathic central nervous system hypersomnolent subjects in comparison with narcoleptic subjects. Sleep Res. 18:211; 1989.

Carskadon, M. A.; Dement, W. C. Respiration during sleep in the aged human. J. Gerontol. 36(4):420–423; 1981.

Cartwright, R.; Stefoski, D.; Caldarelli, D.; Kravitz, H.; Knight, S.; Lloyd, S.; Samelson, C. Toward a treatment logic for sleep apnea: the place of the tongue retaining device. Behav. Res. Ther. 26(2):121–126; 1988.

Cirignotta, F.; Zucconi, M.; Mondini, S.; Lenzi, P. L.; Lugaresi, E. In: Guilleminault, C.; Lugaresi, E., eds. Sleep/wake disorders. Natural history, epidemiology and long term evolution. New York: Raven Press; 1983: p. 237–241.

Cirignotta, F.; Zucconi, M.; Mondini, S.; Gerardi, R.; Lugaresi, E. Cerebral anoxic attacks in sleep apnea syndrome. Sleep 12(5):400–404; 1989.

Cochrane, G. The use of indirect hypnotic suggestions for insomnia arising from generalized anxiety: a case report. Am. J. Clin. Hypn. 31(30):199–203; 1989.

Coleman, R. M.; Pollak, C. P.; Weitzman, E. D. Periodic movements in sleep (nocturnal myoclonus): relation to sleep disorders. Ann. Neurol. 8(4):416–421; 1980.

Coleman, R. M.; Miles, L. E.; Guilleminault, C. C.; Zarcone, V. P. Jr.; van den Hoed, J.; Dement, W. C. Sleep-wake disorders in the elderly: polysomnographic analysis. J. Am. Geriatr. Soc. 29(7):289–296; 1981.

Coleman, R. M.; Guilleminault, C., ed. Sleeping and waking disorders: indications and techniques. Boston: Butterworths; 1982.

Coleman, R. M.; Roffwarg, H. P.; Kennedy, S. J.; Guilleminault, C.; Cinque, J.; Cohn, M. A.; Karacan, I.; Kupfer, D. J.; Lemmi, H.; Miles, L. E.; Orr, W. C.; Phillips, E. R.; Roth, T.; Sassin, J. F.; Schmidt, H. S.; Weitzman, E. D.; Dement, W. C. Sleep-wake disorders based on a polysomnographic diagnosis. A national cooperative study. J.A.M.A. 247(7):997–1003; 1982.

Cooper, A. J. Treatment of coexistent night-terrors and somnambulism in adults with imipramine and diazepam. J. Clin. Psychiatry 48(5):209–210; 1987.

Critchley, M. Periodic hypersomnia and megaphagia in adolescent males. Brain 85:628–656; 1962.

Czeisler, C. A.; Richardson, G. S.; Coleman, R. M.; Zimmerman, J. C.; Moore-Ede, M. C.; Dement, W. C.; Weitzman, E. D. Chronotherapy: resetting the circadian clocks of patients with delayed sleep phase insomnia. Sleep 4(1):1–21; 1981.

Czeisler, C. A.; Johnson, M. P.; Duffy, J. F.; Brown, E. N.; Ronda, J. M.; Kronauer, R. E. Exposure to bright light and darkness to treat physiologic maladaptation to night work. New

England Journal of Medicine 18:1253–1259; 1990.

Davies, E.; Hayes, M.; Kirman, B. H. Somnambulism. Lancet 1:186; 1942.

Davis, J. M.; Zirman, R.; Aranda, J. V. Use of caffeine in the treatment of apnea associated with the Arnold-Chiari malformation. Dev. Pharmacol. Ther. 12(2):70–73; 1989.

Dement, W. C.; Zarcone, V.; Varner, V.; et al. The prevalence of narcolepsy. Sleep Res. 1:148; 1972.

Dement, W. C.; Carskadon, M. A.; Ley, R. The prevalence of narcolepsy. Sleep Res. 2:147; 1973.

Deykin, E.; Bauman, M. L.; Kelly, D. H.; Hsieh, C. C.; Shannon, D. Apnea of infancy and subsequent neurologic, cognitive, and behavioral status. Pediatrics 73(5):638–645; 1984.

Dickins, Q. S.; Jenkins, N. A.; Chambers, G. W.; Schweitzer, P. K.; Walsh, J. K. Long-term nasal CPAP use. Sleep Res. 18:223; 1989.

Ekbom, K. A. Restless legs syndrome. Neurology 10:868–873; 1960.

Ellingson, R. J.; Peters, J. F. Nelson, B. Respiratory pauses and apnea during daytime sleep in normal infants during the first year of life: longitudinal observations. Electroencephalogr. Clin. Neurophysiol. 53(1):48–59; 1982.

Erlich, S. S.; Itabashi, H. H. Narcolepsy: a neuropathologic study. Sleep 9(1):126–132; 1986.

Ferber, R. A. Behavioral "insomnia" in the child. Psychiatr. Clin. North Am. 10:641–653; 1987.

Finer, N. N.; Peters, K. L.; Duffley, L. M.; Coward, J. H. An evaluation of theophylline for idiopathic apnea of infancy. Dev. Pharmacol. Ther. 7(2):73–81; 1984.

Fisher, B. E.; Wilson, A. E. Selected sleep disturbances in school children reported by parents: prevalence, interrelationships, behavioral correlates and parental attributions. Percept. Mot. Skills 64 (3 pt 2):1147–1157; 1987.

Fletcher, E. C.; DeBehnke, R. D.; Lovoi, M. S.; Gorin, A. B. Undiagnosed sleep apnea in patients with essential hypertension. Ann. Intern. Med. 103(2):190–195; 1985.

Ford, D. E.; Kamerow, D. B. Epidemiologic study of sleep disturbances and psychiatric disorders. An opportunity for prevention? J.A.M.A. 262(11):1479–1484; 1989.

Frank, Y.; Kravath, R. E.; Pollak, C. P.; Weitzman, E. D. Obstructive sleep apnea and its therapy: clinical and polysomnographic manifestations. Pediatrics 71(5):737–742; 1983.

Fresco, R.; Giudicelli, S.; Poinso, Y.; Tatossian, A.; Mouren, P. Kleine-Levin syndrome, recurrent hypersomnia of male adolescents. Ann. Med. Psychol. 1(5):625–668; 1971.

Fry, J. M.; Pressman, M. R.; DiPhillipo, M. A.;

Forst-Paulus, M. Treatment of narcolepsy with codeine. Sleep 9(1):269–274; 1986.

Fry, J. M.; DiPhillipo, M. A.; Pressman, M. R. Periodic leg movements in sleep following treatment of obstructive sleep apnea with nasal continuous positive airway pressure. Chest 96(1):89–91; 1989.

Gadoth, N.; Dickerman, Z.; Bechar, M.; Laron, Z.; Lavie, P. Episodic hormone secretion during sleep in Kleine-Levin syndrome: evidence for hypothalamic dysfunction. Brain Dev. 9(3):309–315; 1987.

Gass, E.; Strauch, I. The development of sleep behavior between three and eleven years in Proceedings of the 7th European sleep congress. Munich. 1984.

Gibb, W. R.; Lees, A. J. The restless legs syndrome. Postgrad. Med. J. 62(727):329–333; 1986.

Gislason, T.; Almqvist, M.; Eriksson, G.; Taube, A.; Boman, G. Prevalence of sleep apnea syndrome among Swedish men—an epidemiological study. J. Clin. Epidemiol. 41(6):571–576; 1988.

Glassman, J. N.; Darko, D.; Gillin, J. C. Medication-induced somnambulism in a patient with schizoaffective disorder. J. Clin. Psychiatry 47(10):523–524; 1986.

Godbout, R.; Poirier, G.; Montplaisir, J. New treatments for narcolepsy. In: Burton, S. A.; Dement, W. C.; Ristanovic, R. K., eds. Recent developments in the diagnosis and treatment of narcolepsy. Chicago: Associations Plus, Inc.; 1989: p. 79–81.

Greenberg, G. D.; Watson, R. K.; Deptula, D. Neuropsychological dysfunction in sleep apnea. Sleep 10(3):254–262; 1987.

Guilleminault, C.; Cummiskey, J.; Dement, W. C. Sleep apnea syndrome: recent advances. Adv. Intern. Med. 26:347–372; 1980.

Guilleminault, C.; Grumet, C. HLA-DR2 and narcolepsy: not all narcoleptic-cataplectic patients are DR2. Hum. Immunol. 17:1–2; 1986.

Guilleminault, C.; Mondini, S. Mononucleosis and chronic daytime sleepiness. A long-term follow-up study. Arch. Intern. Med. 146(7):1333–1335; 1986.

Guilleminault, C. Disorders of arousal in children: somnambulism and night terrors. In: Guilleminault, C., ed. Sleep and its disorders in children. New York: Raven Press; 1987: p. 243–252.

Guilleminault, C. Sleep apnea in the full-term infant. In: Guilleminault C., ed. Sleep and its disorders in children. New York: Raven Press; 1987: p. 195–211.

Guilleminault, C.; Quera-Salva, M. A.; Nino-Murcia, G.; Partinen, M. Central sleep apnea and partial obstruction of the upper airway. Ann. Neurol. 21(5):465–469; 1987.

Guilleminault, C.; Kryger, M. H.; Roth, T.; Dement, W. C., eds. Principles and practice of sleep medicine. Philadelphia: W. B. Saunders; 1989.

Guilleminault, C. Idiopathic central nervous system hypersomnia. In: Kryger, M. H.; Roth, T.; Dement, W. C., eds. Principles and practice of sleep medicine. Philadelphia: W. B. Saunders; 1989: p. 347–350.

Hallstrom, T. Night terrors in adults through three generations. Acta. Psychiatr. Scand. 48(4):350–352; 1972.

Hammond, E. C. Some preliminary findings on physical complaints from a prospective study of 1,064,004 men and women. Am. J. Public Health 54:11–23; 1964.

Hansen, T.; Bratlid, T.; Lingjarde, O.; Brenn, T. Midwinter insomnia in the subarctic region: evening levels of serum melatonin and cortisol before and after treatment with bright artificial light. Acta. Psychiatr. Scand. 75(4):428–434; 1987.

Hart, E. J. Kleine-Levin syndrome: normal CSF monoamines and response to lithium therapy. [letter] Neurology 35(9):1395–1396; 1985.

Hartman, E. Two case reports: night terrors with sleepwalking—a potentially lethal disorder. J. Nerv. Ment. Dis. 171(8):503–505; 1983.

Hartman, P. G.; Scrima, L. Muscle activity in the legs (MAL) associated with frequent arousals in narcoleptics, nocturnal myoclonus and sleep apnea (OSA) patients. Clin. Electroencephalogr. 17(4):181–186; 1986.

He, J.; Kryger, M. H.; Zorick, F. J.; Conway, W.; Roth, T. Mortality and apnea index in obstructive sleep apnea. Experience in 385 male patients. Chest 94(1):9–14; 19

Heaf, D. P.; Helms, P. J. Dinwiddie, R.; Matthew, D. J. Nasopharyngeal airways in Pierre Robin syndrome. J. Pediatr. 100(5):698–703; 1982.

Heldt, G. P. Development of stability of the respiratory system in preterm infants. J. Appl. Physiol. 65(1):441–444; 1988.

Henderson-Smart, D. J.; Pettigrew, A. G.; Campbell, D. J. Clinical apnea and brain-stem neural function in preterm infants. N. Engl. J. Med. 309(7):353–357; 1983.

Henderson-Smart, D. J.; Cohen, G. Apnoea in the newborn infant. Aust. Paediatr. J. 22(suppl. 1):63–66; 1986.

Hening, W. A.; Walters, A.; Kavey, N.; Gidro-Frank, S.; Cote, L.; Fahn, S. Dyskinesias while awake and periodic movements in sleep in restless legs syndrome: treatment with opioids. Neurology 36(10):1363–1366; 1986.

Honda, Y.; Juji, T.; Matsuki, K.; Naohara, T.; Satake, M.; Inoko, H.; Someya, T.; Harada, S.; Doi, T. HLA-DR2 and Dw2 in narcolepsy and in other disorders of excessive somnolence without cataplexy. Sleep 9:133–142; 1986.

Honda, Y.; Matsuki, K.; Juji, T.; Inoko, H. Recent progress in HLA studies and a genetic model for the development of narcolepsy. In: Burton, S. A.; Dement, W. C.; Ristanovic, R. K., eds. Recent developments in the diagnosis and treatment of narcolepsy. Chicago: Associations Plus Inc.; 1989: p. 27–33.

Hoppenbrouwers, T.; Hodgman, J. E.; Harper, R. M.; Hofmann, E.; Sterman, M. B.; McGinty, D. J. Polygraphic studies of normal infants during the first six months of life: III. Incidence of apnea and periodic breathing. Pediatrics 60(4):418–425; 1975.

International classification of sleep disorders: Diagnostic and coding manual. Diagnostic Classification Steering Committee. Thorpy, M. J., Chairman. Rochester, Minnesota: American Sleep Disorders Association; 1990.

Jones, R. A.; Lukeman, D. Apnoea of immaturity. 2. Mortality and handicap. Arch. Dis. Child. 57(10):766–768; 1982.

Kahn, A.; Mozin, M. J.; Rebuffat, E.; Sottiaux, M.; Muller, M. F. Milk intolerance in children with persistent sleeplessness: a prospective double-blind crossover evaluation. Pediatrics 84(4):595–603; 1989.

Kales, J. D.; Kales, A.; Soldatos, C. R.; Caldwell, A. B.; Charney, D. S.; Martin, E. D. Night terrors. Clinical characteristics and personality patterns. Arch. Gen. Psychiatry 37(12):1413–1417; 1980.

Kales, A.; Cadieux, R. J.; Bixler, E. D.; Soldatos, C. R.; Vela-Bueno, A.; Misoul, C. A.; Locke, T. W. Severe obstructive sleep apnea—I: Onset, clinical course, and characteristics. J. Chronic. Dis. 38(5):419–425; 1985.

Kales, A.; Soldatos, C. R.; Bixler, E. O.; Ladda, R. L.; Charney, D. S.; Weber, G.; Schweitzer, R. K. Hereditary factors in sleepwalking and night terrors. Br. J. Psychiatry 137:111–118; 1980.

Katsantonis, G. P.; Schweitzer, P. K.; Branham, G. H.; Walsh, J. K. Management of obstructive sleep apnea: comparison of various treatment modalities. Laryngoscope 98(3):304–309; 1988.

Kessler, S.; Guilleminault, C.; Dement, W. C. A family study of fifty REM narcoleptics. Arch. Neurol. Scand. 50:503–512; 1974.

Kimura, H.; Tatsumi, K.; Kunitomo, F.; Okita, S.; Tojima, H.; Kouchiyama, S.; Masuyama, S.; Shinozaki, T.; Honda, Y.; Kuriyama, T. Progesterone therapy for sleep apnea syndrome evaluated by occlusion pressure responses to exogenous loading. Am. Rev. Respir. Dis. 139(5):1198–1206; 1989.

Klackenberg, G. Incidence of parasomnias in children in a general population. In: Guilleminault, C., ed. Sleep and its disorders in children. New York: Raven Press: 1987: p. 99–113.

Koe, G. G. Hypnotic treatment of sleep terror disorder: a case report. Am. J. Clin. Hypn. 32(1):36–40; 1989.

Korfali, E.; Askoy, K.; Safi, I. Slit ventricle syndrome presenting with paroxysmal hypersomnia in an adult: case report. Neurosurgery 22(3):594–595; 1988.

Kotagal, S.; Archer, C. R.; Walsh, J. K.; Gomez, C. Hypersomnia, bithalamic lesions and altered sleep architecture in Kearns-Sayre syndrome. Neurology 35(4):574–577; 1985.

Kribbs, N. B.; Dinges, D. F.; Schuett, J. S.; Pack, A. I.; Kline, L. R. Prevalence and effects of intermittent CPAP use in obstructive sleep apnea. Sleep Res. 18:251; 1989.

Krieger, J.; Turlot, J. C.; Mangin, P.; Kurtz, D. Breathing during sleep in normal young and elderly subjects: hypopneas, apneas and correlated factors. Sleep 6(2):108–120; 1983.

Krieger, J.; Kurtz, D. Objective measurement of compliance with nasal CPAP treatment for obstructive sleep apnea syndrome. Eur. Respir. J. 1(5):436–438; 1989.

Langdon, N.; Lock, C.; Welsh, K.; Vergani, D.; Dorow, R.; Wachtel, H.; Palenschat, D.; Parkes, J. D. Immune factors in narcolepsy. Sleep 9:143–148; 1986.

Lask, B. Novel and non-toxic treatment for night terrors. Br. Med. J. 297(6648):592; 1988.

Lavie, P. Sleep apnea in industrial workers. In: Guilleminault, C.; Lugaresi, E., eds. Sleep/wake disorders: natural history, epidemiology and long term evolution. New York: Raven Press; 1983: p. 127–135.

Lavie, P.; Ben-Yosef, R.; Rubin, A. E. Prevalence of sleep apnea syndrome among patients with essential hypertension. Am. Heart. J. 108(2):373–376; 1984.

Levitt, G. A.; Mushin, A.; Bellman, S.; Harvey, D. R. Outcome of preterm infants who suffered neonatal apnoeic attacks. Early Hum. Dev. 16(2–3):235–243; 1988.

Lugaresi, E.; Cirignotta, F.; Coccagna, G.; Montagna, P. Nocturnal myoclonus and restless legs syndrome. Adv. Neurol. 43:295–307; 1986.

Mamelak, M.; Scharf, M. B.; Woods, M. Treatment of narcolepsy with gamma-hydroxybutyrate. A review of clinical and sleep laboratory findings. Sleep 9:285–289; 1986.

Marrone, O.; Bellia, V.; Ferrara, G.; Milone, F.; Romano, L.; Salvaggio, A.; Stallone, A.; Bonsignore, G. Transmural pressure measurements. Importance in the assessment of pulmonary hypertension in obstructive sleep apneas. Chest 95(2):338–342; 1989.

Maselli, R. A.; Rosenberg, R. S.; Spire, J. P. Episodic nocturnal wanderings in non-epileptic patients. Sleep 11(2):156–161; 1988.

Mason, W. J.; Ancoli-Israel, S.; Kripke, D. F. Apnea revisited: a longitudinal follow-up. Sleep 12(5):423–429; 1989.

McGinty, D.; Littner, M.; Beahm, E.; Ruiz-Primo, E.; Young, E.; Sowers, J. Sleep related breathing disorders in older men: a search for underlying mechanisms. Neurobiol. Aging 3(4):337–350; 1982.

Mauer, K. W.; Staats, B. A.; Olsen, K. D. Upper airway obstruction and disordered nocturnal breathing in children. Mayo Clin. proc. 58(6):349–353; 1983.

McGinty, D.; Beahm, E.; Stern, N.; Littner, M.; Sowers, J.; Reige, W. Nocturnal hypotension in older men with sleep-related breathing disorders. Chest 94(2):305–311; 1988.

Ment, L. R.; Scott, D. T.; Ehrenkranz, R. A.; Duncan, C. C. Early childhood developmental follow-up of infants with GMH/IVH: effect of methylxanthine therapy. Am. J. Perinatol. 2(3):223–227; 1985.

Miller, M. J.; Carlo, W. A.; Martin, R. J. Continuous positive airway pressure selectively reduces obstructive apnea in preterm infants. J. Pediatr. 106(1):91–94; 1985.

Millman, R. P.; Fogel, B. S.; McNamara, M. E. Carlisle, C. C. Depression as a manifestation of obstructive sleep apnea: reversal with nasal continuous positive airway pressure. J. Clin. Psychiatry 50(9):348–351; 1989.

Mitler, M. M.; Van den Hoed, Jr.; Carskadon, M. A.; Richardson, G.; Park, R.; Guilleminault, C.; Dement, W. C. REM sleep episodes during the multiple sleep latency test in narcoleptic patients. Electroencephalogr. Clin. Neurophysiol 46:479–481; 1979.

Mitler, M. M. New developments in the therapy of narcolepsy. In: Burton, S. A.; Dement, W. C.; Ristanovic, R. K., eds. Recent developments in the diagnosis and treatment of narcolepsy. Chicago: Associations Plus, Inc.; 1989: p. 69–78.

Moldofsky, H. Sleep and musculoskeletal pain. Am. J. Med. 81(3A):85–89; 1986.

Montagna, P.; Liguori, R.; Zucconi, M.; Sforza, E.; Lugaresi, A.; Cirignotta, F.; Lugaresi, E. Physiological hypnic myoclonus. Electroencephalogr. Clin. Neurophysiol. 70(2):172–176; 1988.

Montplaisir, J.; Godbout, R.; Boyhen, D.; DeChamplain, J.; Young, S. N.; Lapierre, G. Familial restless legs with periodic movements in sleep: electrophysiologic, biochemical, and pharmacologic study. Neurology 35(1):130–134; 1985.

Montplaisir, J.; Godbout, R. Restless legs syndrome and periodic movements during sleep. In: Kryger, M. H.; Roth, T.; Dement, W. C., eds. Principles and practice of sleep medicine. Philadelphia: W. B. Saunders; 1989: p. 402–409.

Mosko, S. S.; Dickel, M. J.; Ashurst, J. Night-to-night variability in sleep apnea and sleep related periodic leg movements in the elderly. Sleep 11(4):340–348; 1988.

Muttitt, S. C.; Tierney, A. J.; Finer, N. N. The dose response of theophylline in the treatment of apnea of prematurity. J. Pediatr. 112(1):115–121; 1988.

Nahmias, J.; Karetzky, M. Narcolepsy versus idiopathic CNS hypersomnolence: a comparision of patient and poly somnographic characteristics. Sleep Res. 18:275; 1989.

Navelet, Y.; Anders, T.; Guilleminault, C. Narcolepsy in children. In: Guilleminault, C.; Dement, W. C.; Passouant, P., eds. Narcolepsy. New York: Spectrum; 1976: p. 571–584.

Nevsimalova-Brühova, S.; Roth, B. Heredofamilial aspects of narcolepsy and hypersomnia. Schweiz. Arch. Neurol. Neurochir. Psychiatr. 110(1):45–54; 1972.

Nino-Murcia, G.; McCann, C. C.; Bliwise, D. L.; Guilleminault, C.; Dement, W. C. Compliance and side effects in sleep apnea patients treated with nasal continuous positive airway pressure. West. J. Med. 150(2):165–169; 1989.

Oren, J.; Kelly, D. H.; Shannon, D. C. Familial occurrence of sudden infant death syndrome and apnea of infancy. Pediatrics 80(3):355–358; 1987.

Oren, J.; Kelly, D. H.; Shannon, D. C. Long-term follow-up of children with congenital central hypoventilation syndrome. Pediatrics 80(3):375–380; 1987.

Krieger, J.; Kryger, M. H.; Roth, T.; Dement, W. C., eds. Principles and practice of sleep medicine. Philadelphia: W. B. Saunders; 1989.

Oren, J.; Kelly, D. H.; Shannon, D. C. Pneumogram recordings in infants resuscitated for apnea of infancy. Pediatrics 83(3):364–368; 1989.

Oswald, I.; Evans, J. On serious violence during sleep-walking. Br. J. Psychiatry 147:688–691; 1985.

Otsuka, K.; Sadakane, N.; Ozawa, T. Arrhythmogenic properties of disordered breathing during sleep in patients with cardiovascular disorders. Clin. Cardiol. 10(12):771–782; 1987.

Parkes, J. D.; Fenton, G.; Struthers, G.; Curzon, G.; Kantamaneni, B. D.; Buxton, B. H.; Record, C. Narcolepsy and cataplexy. Clinical features, treatment and cerebrospinal fluid findings. Q. J. Med. 43(172):525–536; 1974.

Parkes, J. D. The parasomnias. Lancet 2(8514):1021–1025; 1986.

Parkes, J. D. Sleep and its disorders. London: W. B. Saunders; 1985.

Partinen, M.; Jamieson, A.; Guilleminault, C. Long-term outcome for obstructive sleep apnea syndrome patients. Mortality Chest 94(6):1200–1204; 1988.

Pearce, J. W.; Kapuniai, L. E.; Crowell, D. H. Periodic leg movements in OSA: a comparison before and during treatment with nasal CPAP. Sleep Res. 18:282; 1989.

Peter, J. H.; Amend, G.; Cassel, W.; Fett, I.; Riess, M.; Schneider, H.; von Wichert, A. Morbidity and prevalence of snoring and sleep apnea in patients of an outpatient department and field study. Sleep Res. 18:283; 1980.

Poirier, G.; Montplaisir, J.; Dumont, M.; Duquette, P.; Decary, F.; Pleines, J.; Lamoureux, G. Clinical and sleep laboratory study of narcoleptic symptoms in multiple sclerosis. Neurology 37(4):693–695; 1987.

Poirier, G.; Montplaisir, J.; Decary, F.; Momege, D.; Lebrun, A. HLA antigens in narcolepsy and idiopathic central nervous system hypersomnolence. Sleep 9:153–158; 1986.

Rauscher, H.; Popp, W.; Vollmann, A.; Ritsenka, L.; Zwick, H. Obstructive sleep apnea—a risk factor for arterial hypertension. Wien Klin Wochenschr 101(6):200–203; 1989.

Rechtschaffen, A.; Wolpert, E. A.; Dement, W. C.; Mitchell, S. A.; Fisher, C. Nocturnal sleep of narcoleptics. Electroencephalogr. Clin. Neurophysiol. 15:599–609; 1963.

Regestein, Q. R.; Reich, P. Agitation observed during treatment with newer hypnotic drugs. J. Clin. Psychiatry 46(7):280–283; 1985.

Rhoads, N. P.; Hayes, B.; Guilleminault, C.; Nino-Murcia, G.; Dement, W. C. The narcoleptic tetrad and objective polysomnographic tests. Sleep Res. 18:294; 1989.

Richter, H. J.; Plattner, E. A. Obstructive sleep apnea and falling asleep at the wheel. Sleep Res. 18:296; 1989.

Riley, R. W.; Powell, N. B.; Guilleminault, C. Inferior mandibular osteotomy and hyoid myotomy suspension for obstructive sleep apnea: a review of 55 patients. J. Oral Maxillofac. Surg. 47(2):159–164; 1989.

Rosa, R.; Kramer, M.; Foright, P. Narcolepsy: symptom frequency and associated disorders. Sleep Res. 8:213; 1979.

Rosen, C. L.; Glaze, D. G.; Frost, J. D. Jr. Home monitor follow-up of persistent apnea and bradycardia in preterm infants. Am. J. Dis. Child 140(6):547–550; 1986.

Rosenthal, L.; Roehrs, T.; Sicklesteel, J.; Zornick, F.; Wittig, R.; Roth, T. Periodic movements during sleep, sleep fragmentation and sleep-wake complaints. Sleep 7(4):326–330; 1984.

Roth, B. Narcolepsy and Hypersomnia. Basel: S Karger; 1980.

Rubinstein, I.; Gray, T. A.; Moldofsky, H.; Hofstein, V. Neurosarcoidosis associated with hypersomnolence treated with corticosteroids and brain irradiation. Chest 94(1):205–206; 1988.

Rudiger, E. Casuistic contribution to the Ekbom–syndrome. Wien Med Woochenschr 130(17):563–565; 1980.

Ryan, C. A.; Finer, N. N.; Peters, K. L. Nasal intermittent positive-pressure ventilation offers no advantages over nasal continuous positive pressure airway pressure in apnea of prematurity. Am. J. Dis. Child. 143(10):1196–1198; 1989.

Samelson, C. F. A survey of the effectiveness of the tongue retaining device for the control of snoring and/or obstructive sleep apnea. Sleep Res. 18:299; 1989.

Saskin, P.; Moldofsky, H.; Lue, F. A. Periodic movements in sleep and sleep-wake complaint. Sleep 8(4):319–324; 1985.

Schmidt, H. S.; Clark, R. W.; Hyman, P. R. Protriptyline: an effective agent in the treatment of the narcolepsy-cataplexy syndrome and hypersomnia. Am. J. Psychiatry 134(2):183–185; 1977.

Scott, A. I. Attempted strangulation during phenothiazine-induced sleep-walking and night terrors. Br. J. Psychiatry 153:692–694; 1988.

Scrima, L.; Hartman, P. G.; Johnson, F. H. Jr.; Hiller, F. C. Efficacy of gamma-hydroxybutyrate versus placebo in treating narcolepsy-cataplexy: double-blind subjective measures. Biol. Psychiatry 26(4):331–343; 1989.

Shapiro, W. R. Treatment of cataplexy with clomipramine. Arch. Neurol. 32(10):653–656; 1975.

Shepard, J. W., Jr.; Garrison, M. W.; Grither, D. A.; Dolan, G. F. Relationship of ventricular ectopy to oxyhemoglobin desaturation in patients with obstructive sleep apnea. Chest 88(3):335–340; 1985.

Sims, M. E.; Yau, G.; Rambhatla, S.; Cabal, I.; Wu, P. Y. Limitations of theophylline in the treatment of apnea of prematurity. Am. J. Dis. Child. 139(6):567–570; 1985.

Sours, J. A. Narcolepsy and other disturbances in the sleep-waking rhythm: a study of 115 cases with review of the literature. J. Nerv. Ment. Dis. 137:225–242; 1963.

Spielman, A. J.; Saskin, P.; Thorpy, M. J. Treatment of chronic insomnia by restriction of time in bed. Sleep 10(1):45–56; 1987.

Telstad, W.; Sorensen, O.; Larsen, S.; Lillevold, P. E.; Stensrud, P.; Nyberg-Hansen, R. Treatment of the restless legs syndrome with carbamazepine: a double blind study. Br. Med. J. [Clin Res] 288(6415):444–446; 1984.

Testa, S.; Opportuno, A.; Gallo, P.; Tavolato, B. A case of multiple sclerosis with an onset mimicking the Kleine Levin syndrome. Ital. J. Neurol. Sci. 8(2):151–155; 1987.

Thachil, J.; Zeller, J. R.; Kochar, M. S. Hypersomnolence with beta-adrenergic blockers. Chest 92(5):943–944; 1987.

Thach, B. T. Sleep apnea in infancy and childhood. Med. Clin. North. Am. 69(6):1289–1315; 1985.

Thorpy, M. J.; Ledereich, P. S.; Glovinsky, P. B.; Burack, B.; Rozycki, D. L.; McGregor, P.; Partinen, M.; Sher, A. E. Survival of patients with obstructive sleep apnea. Sleep Res. 16:444; 1987.

Thorpy, M. J.; Ledereich, P. S.; Burack, B.; McGregor, P. A. Increased mortality in women with obstructive sleep apnea. Sleep Res. 18:315; 1989.

Tilkian, A. G.; Guilleminault, C.; Schroeden, J. S.; Lehrman, K. L.; Simmons, F. B.; Dement, W. C. Hemodynamics in sleep-induced apnea. Studies during wakefulness and sleep. Ann. Intern. Med. 85(6):714–719; 1976.

von Economo, C. Sleep as a problem of localization. J. Nerv. Ment. Dis. 71:249–259; 1930.

Ward, S. L.; Keens, T. G.; Chan, L. S.; Chipps, B. E.; Carson, S. H.; Deming, D. D.; Krishna, V.; MacDonald, H. M.; Martin, G. I.; Meredith, K. S.; et al. Sudden infant death syndrome in infants evaluated by apnea programs in California. Pediatrics 77(4):451–458; 1986.

Weese-Mayer, D. E.; Brouillette, R. T.; Naidich, T. P.; McLone, D. G.; Hunt, C. E. Magnetic resonance imaging and computerized tomography in central hypoventilation. Am. Rev. Respir. Dis. 137(2):393–398; 1988.

Weitzenblum, E.; Krieger, J.; Apprill, M.; Vallee, E.; Ehrhart, M.; Ratomaharo, J.; Oswald, M.; Kurtz, D. Daytime pulmonary hypertension in patients with obstructive sleep apnea syndrome. Am. Rev. Respir. Dis. 138(2):345–349; 1988.

Weitzman, E. D.; Rowland, L. P., ed. Merritt's textbook of neurology. 7th ed. Philadelphia: Lea and Febiger; 1984.

White, D. P.; Kryger, M. H.; Roth, T.; Dement, W. C., eds. Principles and practice of sleep medicine. Philadelphia: W. B. Saunders; 1989.

Will, R. G.; Young, J. P.; Thomas, D. J. Kleine-levin syndrome: report of two cases with onset of symptoms precipitated by head trauma. Br. J. Psychiatry 152:410–412; 1988.

Williams, M. Reconstruction of an early seduction and its after effects. J. Am. Psychoanal. Assoc. 35(1):145–163; 1987.

Yoss, R. E.; Daly, D. D. Hereditary aspects of narcolepsy. Trans. Am. Neurol. Assoc. 85:239–240; 1960.

49

Pregnancy

JAMES O. DONALDSON

Pregnancy can alter the natural history and complicate the treatment of many neurological diseases. Alternatively, some neurological diseases change the management of pregnancy and childbirth. Common clinical situations are discussed in this chapter. The reader interested in a specific condition should consult the discussion of that disease elsewhere in this book.

Epilepsy

The prognosis of epilepsy during pregnancy can be estimated from the degree of seizure control before pregnancy (Knight and Rhind 1973). Almost all women who convulse at least once each month can be expected to have a higher frequency of convulsions during pregnancy. If a woman with epilepsy has not convulsed in 9 months before becoming pregnant, the risk of a seizure during pregnancy is only 25%. If she has been seizure-free for 2 years, the risk drops to 10% (Schmidt et al. 1983). A history of catamenial seizures does not predict the effect of pregnancy (Holmes 1988).

The major factor responsible for increased seizure frequency is altered drug metabolism (Leppik and Rask 1988). The distribution and hepatic metabolism are increased for almost all drugs. In the case of phenytoin, gastrointestinal absorption is diminished and near term, the portion not bound to serum proteins increases. The renal excretion of phenobarbital increases. Furthermore, there may be poor compliance with the prescribed regimen. Thus, during pregnancy and the puerperium, blood levels of anticonvulsants fluctuate. Maintaining blood levels with a range that is therapeutic for each patient can prevent seizures. Little can be done to prevent high estrogen concentrations and, near term, sleep deprivation from decreasing seizure thresholds.

The outcome of pregnancy depends more on socioeconomic status, maternal age and parity, and regular prenatal care than on maternal epilepsy. The only obstetrical complication of epilepsy consistently found in studies is third-trimester vaginal bleeding (Hiilesmaa et al. 1985).

The most serious threat from epilepsy is status epilepticus. Although some women deliver normal babies after status epilepticus, approximately one third of mothers and one half of fetuses do not survive (Teramo and Hiilesmaa 1982). A single tonic–clonic maternal seizure is followed by a decreased fetal heart rate for 20 minutes or more (Yerby 1987). A similar response occurs after an eclamptic seizure but not during or after temporal lobe seizures.

Infants of epileptic women have a higher risk of major birth defects. The relative was 1.25, 1.6, 1.87, and 2.2 in four series (Donaldson 1989). In the best controlled study, a Norwegian survey of 3879 infants at risk, the absolute risk increased from 3.5% to 4.4% (Bjerkedal 1982). The risk of orofacial clefts is increased approximately fivefold (Friis et al. 1986). Initially, the incidence of congenital heart disease was thought to be increased; however, more recent studies have not confirmed this (Friis and Hauge 1985). The discrepancy is accounted for by trimethadione, which is now considered a human teratogen.

The only major malformations specifically related to anticonvulsants are defects in closure of the neural tube, which occur in 1% to 2% of fetuses exposed to valproic acid in the first trimester (Lammer et al. 1987). Significant neural tube defects can be detected in the second trimester by ultrasonography and are suggested by an elevated concentration of alphafetoprotein in the amniotic fluid.

Minor dysphoric facial features have been described in various "fetal anticonvulsant syndromes." None of the features are specific to any anticonvulsant. Hypoplasia of the fingernails and the distal phalanges occurs in 15% to 30% of infants exposed *in utero* to phenytoin (Kelly 1984) and, according to one unconfirmed study, in 26% of infants exposed to carbamazepine *in utero* (Jones et al. 1989). Usually, these minor malformations are outgrown, becoming less recognizable as the child matures.

The increased risk of birth defects may be mitigated by monotherapy, if possible, and by the pregestational administration of a multiple vitamin containing folic acid. Two studies have shown that infants of epileptic women with normal concentrations of folate have a lower risk of birth defects (Baile and Lewenthal 1984; Dansky et al. 1987).

Cerebrovascular Disease

Subarachnoid Hemorrhage

The prognosis of spontaneous subarachnoid hemorrhage is serious, regardless of whether the patient is pregnant. The incidence of the initial rupture increases with each trimester of pregnancy for cerebral aneurysms and for aneurysms at other sites in the body (Wiebers 1988). Arteriovenous malformations are more likely to bleed initially during the second trimester and during delivery. For both conditions, intrapartum rebleeding is common and probably is related to Valsalva's maneuver, which almost always accompanies severe labor pains.

If possible, the aneurysm should be clipped or the arteriovenous malformation excised during pregnancy. Normal vaginal delivery can then ensue without special risks. Surgery for unruptured aneurysms can be postponed until the puerperium with the probable exception of giant aneurysms.

If the lesion cannot be cured, vaginal delivery is a risk. Some physicians advise vaginal delivery for multiparous women whose aneurysms bled early in pregnancy. Panting and adequate regional anesthesia can help to prevent bearing down. However, primiparous women with aneurysms and almost all women known to have cerebral arteriovenous malformations should be delivered by cesarean section late in pregnancy.

Cerebral Ischemia

Pregnancy increases the risk of cerebral ischemia approximately four- to ten-fold (Wiebers 1985). The first postpartum week is particularly dangerous; as many ischemic strokes occur at this time as during the second half of pregnancy.

The prognosis and treatment often are dictated by the underlying cause of the stroke or transient ischemic attack (TIA). For this reason, an aggressive, careful work-up, including high-quality angiography is indicated. Approximately one fourth of pregnancy-associated ischemic events are due to premature arteriosclerotic disease. Approximately one half are due to a long list of sometimes rare conditions, including moyamoya disease, sickle hemoglobinopathies, vasculitis, paradoxical emboli, subacute bacterial endocarditis, lupus anticoagulant, mitral valve prolapse, peripartum cardiomyopathy, thrombotic thrombocytopenic purpura, and metastatic choriocarcinoma. No explanation is found for the remaining one fourth.

The management of unexplained single and multiple TIAs during pregnancy is debatable. No definitive literature exists. The author's clinical judgment is to anticoagulate women after the second TIA. Heparin is the preferred anticoagulant during pregnancy because it does not cross the placenta. Similarly unknown is the risk of a cerebral ischemic event during pregnancy if there is a history of ischemic events while taking an oral contraceptive or during a previous pregnancy. The lack of case reports is evidence of its rarity. Thus, prophylactic anticoagulation is not recommended during the subsequent pregnancy.

Cerebral Vein Thrombosis

Puerperal aseptic cerebral phlebothrombosis is a well-recognized entity, although rarely encountered in Europe and North America. It is inexplicably more common in India (Srinivasan 1988). Magnetic resonance imaging (MRI) may be diagnostic. If not, digital subtraction angiography is an excellent method for examining cerebral veins and dural sinuses and for establishing the diagnosis. Computed tomography (CT) is an excellent way to detect intracerebral hemorrhage, which helps to determine management and predict prognosis. If the clot propagates quickly, preventing adequate drainage by the collateral veins, hemorrhagic venous infarction occurs. The presence of intracerebral bleeding is a strong contraindication to anticoagulation, which could prevent further propagation of the clot. Thus, early diagnosis is important.

The prognosis of puerperal cerebral vein thrombosis is very good or very bad. Most survivors recover function, probably because unlike an arterial stroke, oxygenated blood can reach the affected brain. The risk of death has not been assessed because CT scanning has helped to decide whether to anticoagulate. Previously 20% to 30% of patients died. Cerebral venous thrombosis has recurred following a subsequent pregnancy, but the risk is low and probably does not justify prophylactic postpartum anticoagulation with its risk of uterine hemorrhage.

Eclampsia

The cerebral manifestations of the toxemia of pregnancy are not confined to seizures, but this has been the traditional clinical criterion differentiating eclampsia from presumably less severe forms of toxemia. Furthermore, cerebral lesions, clinically expressed as visual hallucinations, cortical blindness, and sometimes coma, can occur before the first seizure. Cerebral edema has been demonstrated on CT scans before seizures. Because eclampsia may be pathophysiologically defined as hypertensive encepalopathy, usually in a previously normotensive woman, it may be better to define the beginning of eclampsia as when the mean arterial blood pressure exceeds the upper limit of autoregulation of cerebral perfusion by blood pressure (Donaldson 1988). Nevertheless, the traditional distinction has merit because the hypoxia and acidosis accompanying a seizure can quickly escalate the seriousness of an already sick patient.

Approximately 40% of the deaths due to the toxemia of pregnancy are caused by cerebral pathology—primarily intracerebral and intraventricular hemorrhages and cerebral edema. A few patients who do not die are left in a chronic vegetative state. Most of the remainder recover without sequelae, notably epilepsy (Sibai et al. 1985).

Immunogenic Neuromuscular Disease

The immune system is affected by pregnancy. One reason is the immunosuppressive effect of fetally produced alphafetoprotein, which dissipates quickly after delivery. Often, autoimmune diseases remit during pregnancy, only to flare after birth. Numerous patients do not follow this pattern, so it is unwise to predict the course for any patient.

The fetal effect of maternal autoimmune diseases can be due to the type of immune response in the disease. In humans, only immunoglobulin G (IgG) crosses the placenta. Thus, IgG-mediated maternal diseases (e.g., Graves' disease, idiopathic thrombocytopenic purpura (ITP), and myasthenia gravis) could provoke fetal and neonatal diseases if enough antibody crosses the placenta. IgM and immune complexes do not cross the placenta, and cannot cause a neonatal disease. An insignificant amount of IgG is transferred by breast milk. However, in rats, IgG does not cross the placenta but is transferred in milk.

Plasmapheresis can be used during pregnancy to treat myasthenia gravis and acute Guillain-Barré syndrome (Parry and Heiman-Patterson 1988). Years before plasmapheresis was used in myasthenia gravis, the safety of this procedure during pregnancy was established for prevention of erythroblastosis fetalis in infants of women with high Rh antibody titers.

Myasthenia Gravis

Before early thymectomy, during pregnancy approximately one third of myasthenic women worsened, one third improved, and one third stayed the same. Approximately 40% relapsed postpartum. Following early thymectomy, significantly fewer patients worsen during pregnancy (Eden and Gall 1983). The author advises women with generalized myasthenia gravis to have a thymectomy and to allow their disease to stabilize during the following year before attempting to become pregnant. Women with restrictive respiration due to severe myasthenia gravis have more problems as the diaphragm becomes elevated in late pregnancy. Thymectomy during pregnancy may be indicated for some women.

At least 12% of infants of mothers with generalized myasthenia gravis develop transient neonatal myasthenia gravis, the onset of which may not occur until the fourth day of life (Morel et al. 1988). The delayed neonatal onset and the absence of fetal myas-

thenia are presumably caused by fetally produced alpha-fetoprotein, which blocks antiacetylcholine receptor antibody from binding to its antigen (Abramsky et al. 1979).

Polymyositis

The prognosis of polymyositis (subacute inflammatory myopathy) during pregnancy is poor for mother and the fetus. The maternal disease worsens. More than one half of pregnancies terminate with spontaneous abortion, stillbirth, and neontal death (Rosenzweig et al. 1989).

Polymyositis in young women is commonly associated with systemic lupus erythematosus, scleroderma, or another collagen-vascular disease. Fetal loss is greater than normal among women with polymyositis, regardless of any associated disease.

Guillain-Barré Syndrome

The Guillain-Barré Syndrome (acute inflammatory polyneuropathy) occurs randomly throughout pregnancy. Its course is not influenced by pregnancy. Infants are unaffected (Parry and Heiman-Patterson 1988).

Chronic Inflammatory Polyneuropathy

Chronic inflammatory polyneuropathy often worsens during the third trimester and after giving birth (McCombe et al. 1987). The infants are healthy.

Prepartum Mononeuropathy

Bell's Palsy

Sir Charles Bell first noted a relationship between pregnancy and idiopathic unilateral facial paralysis. The incidence during the third trimester is approximately 10 times higher than expected (McGregor et al. 1987). Cases appearing the week before delivery usually resolve completely carly in the puerperium. There is no effect on the course of pregnancy.

Carpal Tunnel Syndrome

Nocturnal acroparesthesiae are common among pregnant women, but median neuropathy is relatively infrequent. Usually, it presents in the second half of pregnancy and resolves spontaneously 6 to 8 weeks after childbirth. Unless weakness indicates a need for surgery, most women experience adequate symptomatic relief from nocturnal splinting of the wrist (Ekman-Ordeberg et al. 1987).

Meralgia Paresthetica

Painful stinging paresthesiae in the area innervated by the lateral femoral cutaneous nerve present after the 30th week of gestation and resolve within 3 months after delivery (Massey 1988).

Maternal Obstetric Palsy

An intrapelvic peripheral nerve or nerve trunk can be compressed during labor by the fetal head (Massey 1988). Cephalopelvic disproportion, dystocia, and primiparity are common features. The most common maternal obstetrical palsy is foot drop caused by compression of the lumbosacral trunk (L4, L5) by the fetal brow as it crosses the brim of pelvis. This paralysis is almost always unilateral. It may be closely mimicked by compression of the lateral peroneal nerve between leg holders and the fibular head. Femoral neuropathy may be unilateral or bilateral, and it may coexist with an obturator neuropathy. This neuropathy is rare because obstetricians are less reluctant to deliver by cesarean section any woman whose labor has arrested in a transverse lie.

The prognosis depends on whether myelin sheaths are distorted or axons are crushed. Mixed lesions occur. Electromyography performed approximately 3 weeks after childbirth can detect denervation potentials. If only neuropractic injury occurs, complete recovery is expected within 6 weeks after the delivery.

Prevention of an intrapelvic entrapment neuropathy during a subsequent pregnancy requires an assessment of the underlying cephalopelvic disproportion, the size of the baby, and the severity of the first neuropathy. A trial of labor may be reasonable, but if dystocia develops, a cesarean section usu-

ally is done, especially for women who had axonal degeneration.

Pseudotumor Cerebri

Pseudotumor cerebri can develop in the third, fourth, or fifth month of pregnancy; it may last for a few months and spontaneously remit, or it can persist until after childbirth and then dissipate (Digre et al. 1984). Pseudotumor cerebri can recur with subsequent pregnancies. Women who have pseudotumor cerebri when they become pregnant typically worsen.

The outcome of pregnancy is good; Labor and delivery are normal, and the babies are healthy. Epidural anesthesia can be used. Treatment consists of curbing excessive weight gain and monitoring visual function while awaiting the expected spontaneous remission. Serial cerebrospinal fluid drainage and corticosteroid therapy may be needed, depending on the clinical situation.

Migraine

Most patients with classic migraine are headache-free during pregnancy (Reik 1988); however, 15% to 25% worsen. Migraine with symptoms attributable to angiospasm in the distribution of a middle cerebral artery is not infrequent. Fortunately beta-adrenergic blockers can be used for prophylaxis if migraine is frequent (Rubin 1981). Ergot is not recommended during pregnancy and lactation.

Muscle contraction headaches, including bruxism, usually indicate unresolved psychological and situational problems. Psychiatric consultation may be appropriate for pregnant women with severe, frequent tension headaches because they may be harbingers of postpartum depression.

Multiple Sclerosis

The natural history of multiple sclerosis is punctuated irregularly by exacerbations. The relapse rate decreases with each successive trimester of pregnancy; however, for the entire "pregnancy year" (including 3 postpartum months), the exacerbation rate is at least as high as for a nonpregnancy year (Birk et al. 1988). Approximately 40% of women have an exacerbation during the 6 postpartum months. This rate is approximately the same for mothers who breast-feed as for those who choose to bottle-feed their babies (Nelson et al. 1988).

For most women, multiple sclerosis does not affect the management and outcome of pregnancy (Birk et al. 1988). The frequency of spontaneous abortion is normal, but more women elect to induce abortion. The bulk of a gravid uterus may complicate the management of a neurogenic bladder and increase the risk of cystitis. Regional and epidural anesthesia can be used during childbirth, but traditionally spinal anesthesia has been avoided.

Tumors

Brain tumors appear to grow faster during pregnancy (Roelvink et al. 1987). Estrogen receptors have been demonstrated in meningiomas, neurofibromas, and, to a lesser extent, gliomas. Some tumors shrink after childbirth. There is a high mortality rate for malignant and intraventricular supratentorial tumors and almost all infratentorial tumors except acoustic neuromas.

Intrasellar pituitary microadenomas usually do not become symptomatic during pregnancy, whereas approximately one third of macroadenomas (diameter less than 10 mm) and extrasellar adenomas cause headache and restriction of visual fields (Molitch 1985). Thus, macroadenomas usually are treated with surgery before pregnancy.

References

Abramsky, O.; Brenner, T.; Lisak, R. P.; Zeidman, A.; Beyth, Y. Significance in neonatal myasthenia gravis of inhibitory effect of amniotic fluid on binding of antibodies to acetylcholine receptor. Lancet 2:1333–1335; 1979.

Baile, Y.; Lewenthal, H. Effect of folic acid supplementation on congenital malformations due to anticonvulsant drugs. Europ. J. Obstet. Gynecol. Reprod. Biol. 18:211–216; 1984.

Birk, K.; Smeltzer, S. C.; Rudick, R. Pregnancy and multiple sclerosis. Semin. Neurol. 8:205–213; 1988.

Bjerkedal, T. Outcome of pregnancy in women with epilepsy, Norway, 1967 to 1978: congenital malformations. In: Janz, D.; Dam, M.; Richens, A.; Bossi, L.; Helge, H.; Schmidt, D. Epilepsy, pregnancy, and the child. New York: Raven Press; 1982: p. 289–295.

Blandfort, M.; Tsuboi, T.; Vogel, F. Genetic counseling in the epilepsies. I. Genetic risks. Hum. Genet. 76:303–331; 1987.

Dansky, L. V.; Andermann, E.; Rosenblatt, D.; Sherwin, A. L.; Andermann, F. Anticonvulsants, folate levels, and pregnancy outcome: a prospective study. Ann. Neurol. 21:176–182; 1987.

Dirge, E. B.; Varner, M. W.; Corbett, J. J. Pseudotumor cerebri and pregnancy. Neurology 31:877–880; 1984.

Donaldson, J. O. Eclamptic hypertensive encephalopathy. Semin. Neurol. 8:230–233; 1988.

Donaldson, J. O. Neurology of pregnancy. 2nd ed. London: W. B. Saunders; 1989.

Eden, R. D.; Gall, S. A. Myasthenia gravis and pregnancy: a reappraisal of thymectomy. Obstet. Gynecol. 62:328–333; 1983.

Ekman-Ordeberg, G.; Salgeback, S.; Ordeberg, G. Carpal tunnel syndrome in pregnancy, a prospective study. Acta. Obstet. Gynecol. Scand. 66:133–235; 1987.

Friis, M. L.; Hauge, M. Congenital heart defects in live-born children of epileptic parents. Arch. Neurol. 42:374–376; 1985.

Friis, M. L.; Holm, N. V.; Sindrup, E. H.; Fogh-Anderson, P.; Hauge, M. Facial clefts in sibs and children of epileptic patients. Neurology 36:346–350; 1986.

Hiilesmaa, V. K.; Bardy, A.; Teramo, K. Obstetric outcome in women with epilepsy. Am. J. Obstet. Gynecol. 152:499–504; 1985.

Holmes, G. L. Effects of menstruation and pregnancy on epilepsy. Semin. Neurol. 8:234–239; 1988.

Kelly, T. E. Teratogenicity of anticonvulsant drugs III: Radiographic hand analysis of children exposed in utero to diphenylhydantoin. Am. J. Med. Genet. 19:445–450; 1984.

Knight, A. H.; Rhind, E. G. Epilepsy and Pregnancy: a study of 153 pregnancies in 59 patients. Epilepsia 16:99–110; 1975.

Jones, K. L.; Lacro, R. V.; Johnson, K. A.; Adams, J. Pattern of malformations in the children of women treated with carbamazepine during pregnancy. N. Engl. J. Med. 320:1661–1669; 1989.

Leppik, I. E.; Rask, C. A. Pharmacokinetics of antiepileptic drugs during pregnancy. Semin. Neurol. 8:240–246.

Lammer, E. J.; Sever, L. E.; Oakley, G. P. Teratogen update: valproic acid. Teratology 35:465–473; 1987.

Massey, E. W. Mononeuropathies in pregnancy. Semin. Neurol. 8:193–196; 1988.

McCombe, P. A.; McManis, P. G.; Frith, J. A.; Pollard, J. D.; McLeod, J. G. Chronic inflammatory demyelinating polyradiculoneuropathy associated with pregnancy. Ann. Neurol. 21:102–104; 1987.

McGregor, J. A.; Guberman, A.; Amer, J.; Goodlin, R. Idiopathic facial nerve paralysis (Bell's Palsy) in late pregnancy and the puerperium. Obstet. Gynecol. 69:435–438; 1987.

Molitch, M. D. Pregnancy and the hyperprolactinemic women. N. Engl. J. Med. 312:1364–1370; 1985.

Morel, E.; Eymard, B.; Vernet-der Garabedian, B.; Pannier, C.; Dulac, O.; Bach, J. F. Neonatal myasthenia gravis: A new clinical and immunologic appraisal of 30 cases. Neurology 38:138–142; 1988.

Nelson, L. M.; Franklin, G. M.; Jones, M. C. Risk of multiple sclerosis exacerbation during pregnancy and breast-feeding. J.A.M.A. 259:2441–3443; 1988.

Parry, G. J.; Heiman-Patterson, T. D. Pregnancy and autoimmune neuromuscular disease. Semin. Neurol. 8:197–204; 1988.

Reik, L. Headaches in pregnancy. Semin. Neurol. 8:187–192; 1988.

Roelvink, N. C. A.; Kamphorst, W.; van Alphen, H. S. M.; Rao, B. R. Pregnancy-related primary brain and spinal cord tumors. Arch. Neurol. 44:209–215; 1987.

Rosenzweig, B. A.; Rotmensch, S.; Binette, S. P.; Phillipe, M. Primary idiopathic polymyositis and dermatomyositis complicating pregnancy: diagnosis and management. Obstet. Gynecol. Surv. 44:162–170; 1989.

Rubin, P. C. Beta-blockers in pregnancy. N. Engl. J. Med. 305:1323–1326; 1981.

Schmidt, D.; Canger, P.; Ayanzini G.; Battino, D.; Cusi, C.; Beck-Mannagetta, G.; Koch, S.; Rating, D.; Janz, D. Change in seizure frequency in pregnant epileptic women. J. Neurol. Neurosurg. Psychiatry 46:751–755; 1983.

Sibai, B. M.; Spinato, J. A.; Watson, D. L.; Lewis, J. A.; Anderson, G. D. Eclampsia IV. Neurological fundings and future outcome. Am. J. Obstet. Gynecol. 152:184–192; 1985.

Srinivasan, K. Puerperal cerebral venous and arterial thrombosis. Semin. Neurol. 8:222–225; 1988.

Teramo, K.; Hiilesmaa, V. K. Pregnancy and fetal complications on epileptic pregnancies: review of the literature. In: Janz, D.; Dam, M.; Richens, A.; Bossi, L.; Helge, H.; Schmidt, D. Epilepsy, pregnancy, and the child. New York: Raven Press; 1982: p. 53–59.

Wiebers, D. O. Ischemic cerebrovascular complications of pregnancy. Arch. Neurol. 42:1106–1113; 1985.

Wiebers, D. Subarachnoid hemorrhage in pregnancy. Semin. Neurol. 8:226–229; 1988.

Yerby, M. S. Problems and Management of the pregnant woman with epilepsy. Epilepsia 28(suppl. 3):S29–S36; 1987.

50

Behavioral and Cognitive Disorders

ANDREW KERTESZ

The prognosis of behavioral disorders in neurology directly depends on their etiology. Considerable experience has been accumulated with various syndromes after strokes and trauma. Much of the information concerning the aphasic syndromes is related to the natural history of recovery from stroke. Even though each etiology is treated separately, there is a certain degree of commonality related to severity, lesion size, location, and other biological factors.

Aphasia

Aphasic syndromes are common in neurology, and knowledge of their prognosis is essential for practicing neurologists and researchers. In addition to the practical aspects of recovery from aphasia, the theoretical aspects are important for understanding the organization of language in the brain and recovery mechanisms in general. Language is a uniquely human attribute that is processed mainly in the left hemisphere, although right hemisphere language capacity maybe important for recovery. The factors in recovery from aphasia have been investigated extensively. Some are valid for other

behavioral disorders as well, and only the differences are pointed out in this chapter. The factors listed are inter-related, and at times, their interaction is as important as their distinctiveness. They are discussed in their approximate order of importance for prognosis.

Severity

A major prognostic factor in recovery is the initial severity of aphasia (Godfrey and Douglass 1959; Kertesz and McCabe 1977; Schuell et al. 1964). Even though some findings indicate that the most severely affected patients or global aphasics show little gain regardless of whether they are treated (Sarno et al. 1970), several other studies demonstrate that some of these patients recover surprisingly well and that other factors, such as lesion size, location, and etiology, may make a great deal of difference (Kertesz 1979; Kertesz et al. 1989a). Severely affected patients have a lot of room for improvement. Although mildly affected patients often recover completely, the amount of gain is small. This "ceiling" effect should be considered when actual recovery rates are used in studies for prognosis (Kertesz and McCabe 1977). Recovery rates and outcome measures are affected by initial severity in different ways.

Etiology

Another major factor in recovery is etiology. Post-traumatic aphasics recover much better than patients with vascular disease (Butfield and Zangwill 1946). In one study (Kertesz and McCabe 1977), complete recovery was demonstrated in more than half of the post-traumatic patients, and recently, the Vietnam veterans study of aphasia outcome reached similar conclusions (Ludlow et al. 1986). Some of the patients from this head injury study remained nonfluent after 30 years, particularly those who had extensive cortical and subcortical lesions. Penetrating wounds with a large amount of brain tissue destruction are not comparable to most instances of closed-head injury, which have good prognosis. Even patients with global aphasia can recover to a mild, anomic state after a closed head injury (Kertesz and McCabe 1977). Persisting, severe aphasia is unusual after a closed-head injury, even though dysarthria can be incapacitating. Memory loss and nonverbal cognitive deficits are the major problems in survivors of severe motor vehicle accidents (Levin et al. 1982). The scatter of the extent of recovery is greater in trauma than in stroke because of the combination of contusion and concussion and because of the variability of penetrating head injuries in contrast to the stereotypical occurrence of vascular occlusions in the same arterial territory.

Aphasia from subarachnoid hemorrhage also is variable in the rate of recovery, which is related to whether the patient sustained hemorrhage or an infarction and to the eventual tissue destruction. The prognosis is predictable to some extent by the initial severity of the aphasia (Kertesz and McCabe 1977). Severe, persisting jargon aphasia and global aphasia can be seen following ruptured middle cerebral artery aneurysms in which the cerebral spasm produces extensive lesions. If only a hematoma occurs, absorption and recovery usually are good.

Type of Aphasia

Type of aphasia interacts with severity. However, aside from the severity factor, there is a difference between various aphasia types, which is because various language components differ in their rate of recovery. Head (1926) also observed that various types of aphasia recover at different rates. Subsequent investigations indicated that motor or expressive aphasia improves the most (Butfield and Zangwill 1946; Kertesz and McCabe 1977; Messerli et al. 1976; Weisbenburg and McBride 1935). Although Vignolo (1964) showed that expressive disorders have a poor prognosis, Basso and coworkers (1982) did not find a difference between fluent and nonfluent aphasic patients in their recovery. This variation in conclusion reflects problems in classification. Vignolo's study (1964) was heavily weighted toward global aphasics with poor recovery. Other studies have included aphasics with milder expressive difficulty, which produces a better prognosis. When the severity is taken into consideration and aphasia types with equal severity are compared, relatively small differences are seen between fluent and nonfluent patients. This may be the reason that Basso and colleagues (1982) did not demonstrate differences between aphasia types. However, when studying a large aphasic population, Broca's aphasics recovered somewhat better than Wernicke's aphasics of equivalent severity (Kertesz 1979; 1988).

When considering recovery from various types of aphasia, the methods of classification, methods of measurement, and relationship to severity must be considered carefully before comparing different studies. Most clinicians distinguish between Broca's and Wernicke's aphasics or use the expressive–receptive dichotomy, but for the purpose of prognosis, global aphasics (those who have poor or nonexisting comprehension and very poor output) should be distinguished from Broca's aphasics who may be nonfluent but have reasonable comprehension. In fact, this preservation of comprehension that makes the group that often is called Broca's aphasia, or by some speech pathologists "verbal apraxia," has a good overall prognosis.

Apart from the exceptional instances of recovery in global aphasia discussed later, most global patients remain severely impaired, but approximately half recover enough comprehension to be reclassified as a Broca's aphasic. Broca's aphasics have an intermediate outlook that is evenly divided between fair and good recovery, and in these cases, lesion size and location can help to predict the final outcome. Residual Broca's aphasia often is classified as anomic aphasia, except for the remaining phonological errors (Crary and Kertesz 1988). Severely affected Wernicke's aphasics often retain fluent jargon for many months, and those persisting after 1 year usually remain the same. However, some lose the phonemic jargon, and the language deficit becomes semantic jargon with verbal substitutions, eventually, anomia develops. Wernicke's aphasia also may develop toward conduction aphasia with difficulty with repetition but with improved comprehension. Anomic, conduction, and transcortical aphasics have good prognosis, and 62.5% of the conduction aphasics, 50% of transcortical aphasics, and 48% of anomic patients have full recovery by 1 year. A common end-stage of recovery is anomic aphasia, and mostly word-finding difficulty in spontaneous speech is the most significant residual symptom.

The Rate and Stages of Recovery

Most patients show a great deal of recovery in the first 2 weeks. This first stage of recovery is related to the absorption of hemorrhage, cellular debris, and edema; to recovery from electrolyte disturbance and cellular reaction, ionic imbalance, and membrane failure; and possibly to the re-establishment of the circulation in the ischemic penumbra (Astrup et al. 1981). Second stage recovery, which takes place months or years after injury, remains largely unexplained. Although axonal regrowth and collateral sprouting are important mechanisms in the peripheral and sometimes in the central nervous system

(CNS), large destructive lesions in humans probably are compensated by the reorganization of intact structures. This second stage process shows the greatest improvement in the first 3 months after injury.

There is considerable agreement in the literature that the steepest recovery in aphasia occurs in the first 2 or 3 months (Basso et al. 1975; Culton 1969; Kertesz and McCabe 1977; Sarno and Levita 1971; Vignolo, 1964). After 6 months, a plateau occurs. Typical recovery curves are asymptotic, and their smoothness depends on the number of examinations performed in the study. Reproducible recovery curves can be obtained by examining the patients in the acute stage and at 3 months, 6 months, and 1 year follow-ups.

Many studies use outcome measures instead of recovery rates because of the difficulty of maintaining patients in a follow-up study. If there is a great deal of attrition, the recovery curve is irregular with dips at later stages. This often is related to the milder patients being discharged and the more severe patients remaining in the rehabilitation setting where they are available for assessment.

There is relatively little, if any, spontaneous recovery after 1 year (Butfield and Zangwill 1946; Kertesz and McCabe 1977; Sands et al. 1969; Vignolo 1964). There are, however, reports of improvement after therapy for many years after the injury (Broida 1977; Marks et al. 1957; Schuell et al. 1964; Smith et al. 1972), but most of these are uncontrolled studies. Some aphasics continue to improve for several years, but others decline to some extent after therapy is discontinued, according to a retrospective study by Hanson and associates (1989).

The author estimated the extent of final recovery at 1 year in an aphasic population and found that 25% had gained enough to be considered fully recovered. We took a standard deviation below the mean, using the control population as the cut-off point between normals and aphasics (Kertesz 1979). One fourth of the overall population of aphasics remains severely effected. One fourth are fair, and one fourth have good recovery.

Complete recovery often is relative, but fluent, slightly paraphasic speech and a slight amount of word finding difficulty may be acceptable for most individuals but not for those who have higher education and depend on the full capacity of language and verbal cognition.

Language Components

Various language functions show different rates of recovery, but most depend on the actual test or type of patients tested. Altogether, a correlation exists between the recovery rates of the subtests, but at times, the dissociations are important. There is a general agreement that patients with a comprehension deficit recover the most (Lomas and Kertesz 1978; Prins et al. 1978; Vignolo 1964). The recovery of single-word comprehension tends to be good even in patients with severe, initial deficits (Sarno and Levita 1971). Selnes and associates (1983) found that only very extensive left-hemisphere lesions had incomplete recovery of single-word comprehension, and damage to Wernicke's area did not preclude this return. They also indicated that sentence comprehension was less likely to recover when Wernicke's area was affected. They suggested that single-word comprehension recovers even in larger lesions because this is a function the right hemisphere is capable of assuming. There is convergent evidence from split-brain patients and hemispherectomies that the right hemisphere has the capacity to comprehend nouns in the auditory and visual modality, especially when they are concrete and imageable (Zaidel 1976). Aspects of comprehension recovery were further explored by Gainotti and Monteleone (1988). They found that the degree of recovery was greater in naming and in semantic–lexical comprehension than in the other comprehension tasks concerning syntax and phonology.

Naming recovers considerably, although a residual deficit of word finding and object naming tends to persist in most residual aphasias even when comprehension and speech output are recovered (Kertesz 1979).

Among recovered patients, anterior lesions tend to produce residual phonologic paraphasias, and posterior lesions, produce semantic paraphasias (Knopman et al. 1984). A residual repetition deficit was related to damage to Wernicke's area. Lesions outside of that area have good prognosis for the resolution of repetition deficit (Selnes et al. 1985). Repetition recovered to a greater extent than naming in aphasics (Lomas and Kertesz 1978). The evolution of error types in expressive performance was examined by Crary and Kertesz (1988). Broca's aphasics tended to produce phonological and word omission errors when they recovered. They usually are reclassified as anomic aphasia at this stage, but the type of error distinguishes them from Wernicke's aphasics who have residual semantic and omission errors.

Writing usually is more severely affected than speech, and recovery of writing is often less than recovery of oral language function (Kertesz 1979; 1988).

Recovery from reading disorders has been studied by Newcombe and coworkers (1975) who tried to quantitate errors made by eight patients. They found an asymptotic curve that plateaus around 1 year. The recovery of reading was parallel with the recovery of aphasia and was maximum in the first 3 months in the author's laboratory (Kertesz 1979). Recovery of calculation was similar to the recovery of reading in these studies.

Lesion Size

Autopsy correlations provide frequent evidence that lesion size is negatively related to the extent of recovery. This is well known to clinicians and to experimenters who have measured lesion size and the effect of recovery or the ability of animals to relearn tasks. These give rise to the well-accepted principle of "mass effect" (Lashley 1938). However, lesion size is only one of the many complex factors; therefore, it can be misleading to look at an image of a lesion and try to establish the prognosis from the size alone. Modern neuroimaging has provided an opportunity to study lesion characteristics at the time of the stroke. Outcome measures seem to relate best to lesion size (Kertesz et al. 1979; Knopman et al., 1983; Selnes et al., 1983). Some studies indicate that the language area of approximately 60 sq cm is a critical mass, and larger lesions result in relatively less recovery (Knopman et al. 1983). Lesions smaller than 60 cc, even though they may produce a fairly typical initial deficit, will likely recover well. Some residual deficit, however, often remains with small lesions.

Recovery rates also show a negative correlation with lesion size, except for the recovery rate for comprehension, which often shows no, or positive, correlation with lesion size. This can be understood best by looking at studies of recovery from modalities; comprehension is often the most improved component of speech (Knopman et al. 1983; Lomas and Kertesz 1978). Patients with large lesions who had poor comprehension (severe Broca's aphasia or global aphasia) often show substantial improvement in comprehension. Patients with smaller lesions, such as anomic aphasics, already have good comprehension; therefore, there is less room for recovery (a ceiling effect). The large lesions with more recovery and the small lesions with little change create a paradoxically positive correlation. This can be eliminated to some extent by covarying the recovery rates with the initial severity (Kertesz 1988).

Lesion Location

Lesion location interacts closely with lesion size in recovery because brain structures are not equipotential to compensate for language and cognitive deficit. Certain crucial language areas are important for recovery, and when these are damaged, little or no recovery takes place. The study of lesion location and recovery has progressed recently because *in vivo* imaging techniques allow for reasonable follow-up studies in a larger population.

Mohr and associates (1978) emphasize that large lesions with posterior extension recover poorly from Broca's aphasia, while lesions only in Broca's area have good prognosis. Involvement of the central structures, such as the precentral and postcentral gyri of the face, tongue, and mouth area, in addition to Broca's area contribute to the persistence of deficit (Kertesz et al. 1979; Mohr et al. 1978; Selnes et al. 1983). Subcortical involvement should be also considered in the prognosis of Broca's aphasia. Limited damage to subcortical structures alone rarely cause persisting aphasia. Extensive subcortical white-matter lesions on the other hand may. Bonhoeffer (1914) postulated that larger white-matter lesions undercut Broca's area, preventing access. Large periventricular white-matter lesions with involvement of the subcallosal fasciculus were thought to be important factors in persisting nonfluent aphasia (Naeser et al. 1989). We also found a few entirely subcortical white-matter lesions were found that caused persistent global aphasia (Kertesz et al. 1989a). Large lesions that involve a contiguous cortical and subcortical portion of the speech areas have poor prognosis. Those with sparing of the parietotemporal area will likely recover comprehension and will have functional communication even though residual difficulties with articulation and fluency may remain. Global aphasics with an anterior and posterior lesion but a spared central area tend to have a good prognosis (Ferro 1983; Tranel et al. 1987; Van Horn and Hawes 1982).

The author evaluated the predictive value of lesion location in Broca's aphasics by determining the structures that were significantly involved in persisting cases (Kertesz et al. 1989a). The structures that were involved in both the recovered and persisting cases were the inferior frontal gyrus, especially the pars opercularis; the precentral gyrus and the anterior insula. Involvement of the putamen and caudate was twice as frequent in the persistent cases. It appears that there are four major structures contributing to the network of elaborating articulate speech: Broca's area, the anterior insula, the striatum, and the inferior central cortex. If all four are involved, the prognosis is poor for recovery of fluency. Fairly large lesions with sparing any of these four may still show good recovery.

The structural correlates of prognosis in Wernicke's aphasia show that lesion size is important, but poor prognosis also was associated with involvement of the supramarginal and the angular gyrus (Kertesz et al. 1989b). Subcortical involvement did not seem to be as important as in Broca's aphasia for the prognosis. Naeser and colleagues (1987) also studied Wernicke's aphasics and found that lesions in more than half of Wernicke's cortical area had poor recovery of auditory comprehension 1 to 2 years after stroke onset. They defined Wernicke's areas as the posterior two thirds of the superior temporal gyrus.

Cerebral blood flow (CBF) and positron emission tomography (PET) studies of cerebral metabolism have not added much practical predictive information. One CBF study showed no significant change, while clinical recovery occurred in severe aphasics (Demeurisse et al. 1983). Patients who improve more show more activation in the left hemisphere. This is a consequence of the size of the lesion, which correlates with the CBF changes. Another CBF study showed more than 60% hemispheric flow in patients with good recovery (Nagata et al. 1986).

The variable ability to transfer language to the right hemisphere has been suggested since the time of Wernicke. Clinicians have thought that the right hemisphere must be responsible for the recovery of very large lesions on the left side (Nielsen 1946). Left hemispherectomies and callosal sectioned patients also indicated some degree of right hemisphere language function, which may play an important role in recovery. Dichotic listening tests showed increasing left ear advantage with recovery (Johnson et al. 1977; Pettit and Noll 1979). This increasing left ear advantage was observed in a nonfluent group, whereas a paradoxical decrease was observed in the fluent group (Castro-Caldas et al. 1980).

Age

Often, it is believed that younger patients recover better (Eisenson 1949; Vignolo 1964; Wepman 1951). This impression may be created by including post-traumatic aphasics in a population that compares patients with a much younger mean age than that of the stroke patients. A more homogenous population of stroke patients fails to show a correlation with age (Culton 1971; Kertesz and McCabe 1977; Sarno and Levita 1971; Smith et al., 1972). Some elderly patients show remarkable recovery, and some younger patients are affected most. However, in childhood aphasia, the issue of the plasticity of a younger organism enters the determination of prognosis. The plasticity in immature animals is called the Kennard principle and is based on the dramatic recovery shown in ablation studies (Kennard 1936). It is often stated on the basis of hemispherectomies and studies of childhood aphasia (Basser 1962) that a child will recover from an aphasic deficit by substitution until they reach a critical age. Just when this occurs, is somewhat controversial. However, most people believe that puberty represents an important factor (Lenneberg et al. 1967). Not only is it more difficult to learn a new language after puberty, but it appears that brain plasticity also diminishes and a lesion will result in more persisting aphasia after puberty. Recent studies indicate that some lesions produce fairly persisting aphasia in children as well (Aram et al. 1985; Woods and Carey 1979; Woods and Teuber 1978). Previously, the uniformly good prognosis in childhood aphasia was related to the etiology to some extent as infectious cases and post-traumatic cases have recovered better. Children with Keffler-Landau syndrome tend to be severely affected with epilepsy, and the aphasic disturbance also tends to persist. This syndrome has several etiologies, and at times, a progressive encephalitis produces progressive symptoms.

Sex

Suggestions that women have more bilateral language representation (McGlone 1977) implicate the possibility that women may recover from aphasia better than men. Studies looking at recovery, however, did not show any sex difference (Basso et al. 1982; Kertesz 1988; Kertesz and McCabe 1977).

Handedness

Subirana (1969) and Gloning and coworkers (1969) suggested that left-handers recover better from aphasias than right-handers. However, the data were not based on objective language examination, and subsequent studies could not confirm this. Gloning and colleagues (1969) also suggested that left-handers are likely to become aphasic regardless of which hemisphere is damaged. The author's experience is the opposite; there were less left handers in the aphasic population than expected from the general population (Kertesz 1979). If left handers had more bilateral language distribution, one would expect less aphasia among left handers because either hemisphere could take over.

Intelligence and Education

Premorbid intelligence and education often influence prognosis positively. However, the study of Keenan and Brassel (1974) indicate that health, employment, and age had little, if any, prognostic value. Sarno and associates (1970) also demonstrate that recovery was not influenced by age, education, occupational status, and preillness language proficiency. Intelligence *per se* is difficult to estimate premorbidly. Higher educational achievement, which may or may not be related to higher premorbid intelligence, is considered by some to be favorable for recovery (Darley 1972; Smith 1971; Wepman 1951).

Nonverbal Cognition

The recovery of right hemisphere syndromes of visuospatial deficit, neglect, anosognosia, motor impersistence, and prosopagnosia were examined by Hier and associates (1983). Age and sex did not influence recovery significantly in this older stroke group. Hemorrhages and smaller lesions tended to recover faster. Sparing of the frontal lobe was important for the recovery of visuospatial deficit and neglect.

Neglect

The prognosis of neglect is not as well determined as verbal deficit. Lawson (1962) described some improvement in two cases of left visuospatial neglect for up to 2 years. Gainotti (1968) reported evidence of improvement from the initial severe involvement to complete recovery by 3 years. Persisting cases of neglect also have been described (Zarit and Kahn 1974). Campbell and Oxbury (1976) examined stroke patients in the acute stage, 3 to 4 weeks and 6 months after stroke. They gave the patients tests of copying and drawing, block design, picture completion, and cube counting. Although neglect seemed to recover, by 6 months, those who had neglect still had a visuospatial deficit. The most significant recovery from visuospatial deficit occurs in the first 3 months, and plateau occurs subsequently (Kertesz 1979). Left spatial neglect contributed significantly to the prognosis of hemiplegia (Denes et al. 1982).

Apraxia

Recovery from ideomotor apraxia was better with anterior lesions (Basso et al. 1987). Although less than one third of their patients were followed beyond 1 year, they found that improvement occurred beyond 6 months. Age, sex, lesion size, and type of aphasia did not seem to be related significantly to outcome. Posterior temporoparietal brain regions were more often involved in persistent apraxia. The recovery from apraxia was documented in 50 aphasics, and it showed that global aphasics have better recovery in praxis than in the language scores (Kertesz 1979). The initial recovery of praxis may not be as rapid as that of language, but dissociations occur in both directions, resulting in some patients with aphasia who have no apraxia and vice versa.

Cognitive Recovery from Head Injury

In severe head injury, cognitive recovery depends on the extent of injury, above all the duration of coma, the presence or absence of penetrating injury, and age (Bond 1975). Injuries are regarded as severe when post-traumatic amnesia (PTA) exceeds 24 hours and very severe if PTA is longer than 1 week. Recovery of orientation coincides well with the re-establishment of continuous memory. Mandleberg and Brooks (1975) showed less initial impairment and better recovery from verbal subtests than performance on the Wechsler Adult Intelligence Scale (WAIS) in severe closed head injury in adults. Recovery of nonverbal subtests continued after about 3 years, although a plateau also was reached at 13 months. The outcome and prognosis of head injuries is detailed in other chapters.

Cortical Blindness and Visual Agnosia

The prognosis of cortical blindness usually is good, and recovery often occurs within days. Cortical blindness, which may be associated with denial of blindness (Anton's syndrome), may improve towards visual agnosia (impaired recognition of objects), which then further recovers towards optic aphasia, and with further recovery, a residual hemianopia, or hemianopia with pure alexia, may be seen. A large number of cases of cortical blindness was analyzed by Aldrich and associates (1987), who concluded that the prognosis in cortical blindness is poor when caused by a stroke, and bioccipital abnormalities on the CT scan also are associated with poor prognosis.

Therapy and Prognosis

The literature on the treatment of language disorders is probably the most extensive, although the rehabilitation of other cognitive problems, such as neglect or memory, recently has become more common. Few well-controlled, scientifically sound studies exist. There are many reasons for this, most of them social. It is difficult to obtain untreated controls in localities where treatment is available (usually where such studies are feasible). Accurate measurements of cognitive deficits are also difficult to standardize.

One recent, large-scale, multicenter study of speech therapy in aphasia indicated improvement in the treated group that was sig-

nificantly greater than in the untreated group in a crossover design using the same population at different stages of their recovery (Wertz et al. 1986). Another study found no effect (Lincoln et al. 1984). Some studies found no difference between aphasics treated by speech pathologists or by volunteers (David et al. 1982; Meikle et al. 1979; Shewan and Kertesz 1984). Retrospective studies that showed the beneficial effect of therapy had significant problems in patient selection, such as including more severely affected patients in the untreated groups and failing to control for time-from-onset (Basso et al. 1979). Other studies using special techniques of therapy, such as melodic intonation (Sparks et al. 1974), visual communication therapy (Gardner et al. 1976; Glass, 1973), visual action therapy, and sign language (Helm-Estabrooks et al. 1982; Skelly et al. 1974) described only a few selected patients.

Therapy for apraxia of speech is considered one of the most successful models of speech therapy (Rosenbek et al. 1973). Recently, the psycholinguistic approach to aphasia therapy has attempted to develop interventions on the basis of processing models (Lesser 1987). The use of microcomputers may influence the prognosis of aphasia, but so far, only uncontrolled descriptions of such use have been available (Colby et al. 1981; Dean 1987; Katz 1986; Kinsey 1986). The role of microcomputers probably is most efficient in the supplementary treatment of reading (Katz 1987). Pharmacotherapy of aphasia with bromocriptine has been attempted, but only case descriptions are available (Albert et al. 1988). Poststroke depression occurs in up to 20% of patients with left frontal injury and in about 15% after right hemispheric stroke (Robinson et al. 1984). A trial of antidepressants may improve prognosis and the success of rehabilitation.

The treatment for right hemisphere deficits, such as neglect, has been described (Diller and Gordon 1981). Perceptual remediation of visual scanning, somatosensory awareness, and size estimation improved prognosis (Gordon et al. 1985). Computer retraining of cognitive deficits also has been attempted (Bracy 1983; Ben-Yishay et al. 1985). Cognitive rehabilitation has attracted attention, but it still lacks a solid empirical basis. Memory rehabilitation has become a separate entity in the last decade (Sohlberg and Mateer 1989; Wilson, 1987).

References

Albert, M. L.; Bachman, D. L.; Morgan, A.; Helm-Estabrooks, N. Pharmacotherapy for aphasia. Neurology 38:877–879; 1988.

Aldrich, M. S.; Alessi, A. G.; Beck, R. W.; Gilman, S. Cortical blindness: Etiology, diagnosis, and prognosis. Ann. Neurol. 21:149–158; 1987.

Aram, D. M.; Ekelman, B. L.; Rose, D. F.; Whitaker, H. A. Verbal and cognitive sequelae following unilateral lesions acquired in early childhood. J.C.E.N. 7:55–78; 1985.

Astrup, J.; Siesjo, B. K.; Symon, L. Thresholds in cerebral ischemia—the ischemic penumbra. Stroke 12:723–725; 1981.

Basser, L. S. Hemiplegia of early onset and the faculty of speech, with special reference to the effects of hemispherectomy. Brain 85:427–460; 1962.

Basso, A.; Capitani, E.; Vignolo, L. A. Influence of rehabilitation on language skills in aphasic patients: a controlled study. Arch. Neurol. 36:190–196; 1979.

Basso, A.; Capitani, E.; Moraschini, S. Sex differences in recovery from aphasia. Cortex 18:469–475; 1982.

Basso, A.; Capitani, E.; Della Sala, S.; Laiacona, M.; Spinnler, H. Recovery from ideomotor apraxia—a study on acute stroke patients. Brain 110:747–760; 1987.

Ben-Yishay, Y.; Piasetsky, E. G.; Rattok, J. A systematic method for ameliorating disorders in basic attention. In: Meirer, M. J.; Diller, L.; Benton, A. L., eds. Neuropsychological rehabilitation. London: Churchill Livingstone; 1985.

Bond, M. R. Assessment of psychosocial outcome after severe head injury. In: Ciba Foundation Symposium 34 (new series). Outcome of severe damage to the central nervous system. Elsevier: Amsterdam; 1975.

Bonhoeffer, K. Klinischer und anatomischer Befund zur Lehre von der Apraxie und der 'motorischen Sprachbahn'. Monatsschr. Psychiatr. Neurol. 35:113; 1914.

Bracy, O. L. Computer based cognitive rehabilitation. Cognitive Rehab. 1:7; 1983.

Broida, H. Language therapy effects in long term aphasia. Arch. Phys. Med. Rehabil. 58:248; 1977.

Butfield, E.; Zangwill, O. L. Re-education in aphasia: a review of 70 cases. J. Neurol. Neurosurg. Psychiatry 9:75; 1946.

Campbell, D. C.; Oxbury, J. M. Recovery from unilateral visuo-spatial neglect. Cortex 12:303–312; 1976.

Castro-Caldas, A.; Silveira Botelho, M. Dichotic listening in the recovery of aphasia after stroke. Brain Lang. 10:145–151; 1980.

Colby, K. M.; Christinaz, D.; Parkinson, R. C.; Graham, S.; Karpf, C. A word finding computer program with a dynamic lexical-semantic memory for patients with anomia using an intelligent speech prosthesis. Brain Lang. 14:272–281; 1981.

Crary, M. A.; Kertesz, A. Evolving error profiles during aphasia syndrome remission. Aphasiology 2:67–78; 1988.

Culton, G. L. Spontaneous recovery from aphasia. J. Speech Hearing Res. 12:825; 1969.

Culton, G. L. Reaction to age as a factor in chronic aphasia in stroke. J. Speech Hearing Disord. 30:3; 1971.

Darley, F. L. The efficacy of language rehabilitation in aphasia. J. Speech Hearing Disord. 37:3; 1972.

David, R.; Enderby, P.; Bainton, D. Treatment of acquired aphasia: Speech therapists and volunteers compared. J. Neurol. Neurosurg. Psychiatry 45:957; 1982.

Dean, E. C. Short report. Microcomputers and aphasia. Aphasiology 1:267–270; 1987.

Demeurisse, G.; Verhas, M.; Capon, A.; Paternot, J. Lack of evolution of the cerebral blood flow during clinical recovery of stroke. Stroke 14:77–81; 1983.

Denes, G.; Semenza, C.; Stoppa, E.; Lis, A. Unilateral spatial neglect and recovery from hemiplegia: a follow-up study. Brain 15:21; 1982.

Diller, L.; Gordon, W. A. Rehabilitation and clinical neuropsychology. In: Boll, T. J., Filskov, S. B., eds. Handbook of clinical neuropsychology. Toronto: John Wiley & Sons; 1981.

Eisenson, J. Prognostic factors related to language rehabilitation in aphasic patients. J. Speech Hearing Disord. 14:262; 1949.

Ferro, J. Global aphasia without hemiparesis. Neurology 33:1106; 1983.

Gainotti, G. Les manifestations de negligence et d'inattention pour l'hemispace. Cortex. 4:64–91; 1968.

Gainotti, G.; Monteleone, D. Spontaneous recovery of various aspects of language comprehension and of naming in aphasia. In: Hoofien, D.; Vakil, E.; Grosswasser Z., eds. Rehabilitation of the brain injured. London: Freund Publishing House, Ltd.; 1988.

Gardner, H.; Zurif, E. G.; Berry, T.; Baker, E. Visual communication in aphasia. Neuropsychologia 14:275; 1976.

Glass, A. V.; Gazzaniga, M. S.; Premack, D. Artificial language training in global aphasia. Neuropsychologia 11:95; 1973.

Gloning, I.; Gloning, K.; Haub, G.; Quatember, R. Comparison of verbal behaviour in right-handed and non-right-handed patients with anatomically verified lesion to one hemisphere. Cortex 5:53–62; 1969.

Gloning, I.; Trappl, R.; Heiss, W. D.; Quatember, R. Prognosis and speech therapy in aphasia. In: Hook, O.; Sarno, M. T., eds. Neurolinguistics. 4. Recovery in aphasics. Amsterdam: Swets & Zeitlinger, B. V.; 1976.

Godfrey, C. M.; Douglass, E. The recovery process in aphasia. Can. Med. Assoc. J. 80:618–624; 1959.

Gordon, W. A.; Hibbard, M. R.; Egelko, S. Perceptual remediation in patients with right brain damage. Arch. Phys. Med. Rehabil. 66:353; 1985.

Hanson, W. R.; Metter, E. J.; Riege, W. H. The course of chronic aphasia. Aphasiology 3:19–29; 1989.

Head, H. Aphasia and kindred disorders of speech. Cambridge: Cambridge University Press; 1926.

Helm-Estabrooks, N.; Fitzpatrick, P. M.; Barresi, B. Visual action therapy for global aphasia. J. Speech Hearing Disord. 47:385; 1982.

Hier, D. B.; Mondlock, J.; Caplan, L. R. Recovery of behavioral abnormalities after right hemisphere stroke. Neurology 33:345–350; 1983.

Johnson, J.; Sommers, R.; Weidner, W. Dichotic ear preference in aphasia. J. Speech Hearing Res. 20:116; 1977.

Katz, R. C. Aphasia treatment and microcomputers. San Diego: College Hill Press; 1986.

Katz, R. C. Reply: common ground. Aphasiology 1:171; 1987.

Keenan, S. S.; Brassel, E. G. A study of factors related to prognosis for individual aphasic patients. J. Speech Hearing Disord. 39:257; 1974.

Kennard, M. A. Age and other factors in motor recovery from precentral lesions in monkeys. Am. J. Physiol. 115:138–140; 1936.

Kertesz, A. Aphasia and associated disorders. New York: Grune and Stratton; 1979.

Kertesz, A. What do we learn from recovery? In: Waxman, S. G., ed. Advances in Neurology, 47, Functional recovery in neurological disease. New York: Raven Press; 1988: p. 279–292.

Kertesz, A.; McCabe, P. Recovery patterns and prognosis in aphasia. Brain 100:1; 1977.

Kertesz, A.; Harlock, W.; Coates, R. Computer tomographic localization, lesion size and prognosis in aphasia. Brain Lang. 8:34; 1979.

Kertesz, A.; Dennis, S.; Polk, M. The role of lesion size and location in recovery from global and Broca's aphasia. Abstract presented at the International Neuropsychological Society meeting. Vancouver: British Columbia; 1989a.

Kertesz, A.; Polk, M.; Dennis, S.; McCabe, P. The structural determinants of recovery in Wernicke's aphasia and comprehension deficit. Abstract presented at the American Academy of Neurology. Chicago: 1989b.

Kinsey, C. Microcomputers speech therapy for dysphasic adults. A comparison with two conventionally administered tasks. Br. J. Disord. Commun. 21:125; 1986.

Knopman, D. S.; Selnes, O. A.; Niccum, N.; Rubens, A. B. A longitudinal study of speech fluency in aphasia: CT scan correlates of recovery and persistent nonfluency. Neurology 33:1170–1178; 1983.

Knopman, D. S.; Selnes, O. A.; Niccum, N.; Rubens, A. B. Recovery of naming in aphasia: relationship to fluency, comprehension, and CT findings. Neurology 34:1461–1470; 1984.

Lawson, I. R. Visual-spatial neglect in lesions of the right cerebral hemisphere: a study in recovery. Neurology 12:23; 1962.

Lashley, L. S. Factors limiting recovery after central nervous lesions. J. Nerv. Ment. Dis. 88:733–755; 1938.

Lenneberg, E. Biological foundations of language. New York: Wiley; 1967.

Lesser, R. Cognitive neuropsychological influences on phasia therapy. Aphasiology 1:189–200; 1987.

Levin, H. S.; Benton, A. L.; Grossman, R. G. Neurobehavioral consequences of closed head injury. New York: Oxford University Press; 1982.

Lincoln, N. B.; McGuirk, E.; Mulley, G. P.; Lendrem, W.; Jones, A. C.; Mitchell, J. R. A. The effectiveness of speech therapy for aphasic stroke patients: a randomized controlled trial. Lancet 1:1197; 1984.

Lomas, J.; Kertesz, A. Patterns of spontaneous recovery in aphasic groups: A study of adult stroke patients. Brain Lang. 5:388–401; 1978.

Ludlow, C.; Rosenberg, J.; Fair, C.; Buck, D.; Schesselman, S.; Salazar, A. Brain lesions associated with nonfluent aphasia fifteen years following penetrating head injury. Brain 109:55–80; 1986.

Mandleberg, I. A.; Brooks. D. N. Cognitive recovery after severe head injury. 1. Serial testing on the Wechsler Adult Intelligence Scale. J. Neurol. Neurosurg. Psychiatry 38:1121; 1975.

Marks, M. M.; Taylor, M. L.; Rusk, L. A. Rehabilitation of the aphasic patient: a survey of three years' experience in a rehabilitation setting. Neurology 7:837; 1957.

McGlone, J. Sex differences in the cerebral organization of verbal functions in patients with unilateral lesions. Brain 100:775–793; 1977.

Meikle, M.; Wechsler, E.; Tupper, A.; Benninson, M.; Butler, J.; Mullhall, D.; Stern, G. Comparative trial of volunteer and professional treatments of dysphasia after stroke. Br. Med. J. 2:87; 1979.

Messerli, P.; Tissot, A.; Rodriguez, J. Recovery from aphasia: Some factors of prognosis. In: Neurolinguistics. 4. recovery in aphasics. Amsterdam: Swets & Zeitlinger, 1976.

Mohr, J. P.; Pessin, M. S.; Finkelstein, S.; Funkenstein, H. H.; Duncan, G. W.; Davis, K. R. Broca aphasia: Pathologic and clinical. Neurology 28:311–324; 1978.

Naeser, M.; Helm-Estabrooks, N.; Haas, G.; Auerbach, S.; Srinivasan, M. Relationship between lesion extent in 'Wernicke's area' on computed tomographic scan and predicting recovery of comprehension in Wernicke's aphasia. Arch. Neurol. 44:73–82; 1987.

Naeser, M.; Palumbo, C. L.; Helm-Estabrooks, N.; Stiassny-Eder, D.; Albert, M. L. Severe nonfluency in aphasia. Role of the medial subcallosal fasciculus and other white matter pathways in recovery of spontaneous speech. Brain 112:1–38; 1989.

Nagata, K.; Yunoki, K.; Kabe, S.; Suzuki, A.; Araki, G. Regional cerebral blood flow correlates of aphasia outcome in cerebral hemorrhage and cerebral infarction. Stroke 17:417–423; 1986.

Newcombe, F.; Hiorns, R. W.; Marshall, J. C.; Adams, C. B. T. Acquired dyslexia: patterns of deficit and recovery. In: Ciba Foundation Symposium 34 (new series). Outcome of severe damage to the central nervous system. Amsterdam: Elsevier/Excerpta Medica/North-Holland; 1975.

Nielsen, J. M. Agnosia, apraxia, aphasia. New York: Hoeber; 1946.

Pettit, J.; Noll, J. Cerebral dominance in aphasia recovery. Brain Lang. 7:191–200; 1979.

Prins, R. S.; Snow, C. E.; Wagenaar, E. Recovery from aphasia: spontaneous speech versus language comprehension. Brain Lang. 6:192–211; 1978.

Robinson, R. G.; Kubos, K. L.; Starr, L. B.; Rao, K.; Price, T. R. Mood disorders in stroke patients—importance of location of lesion. Brain 104:81–93; 1984.

Rosenbek, J. C.; Lemme, M. L.; Ahern, M. B.; Harris, E. H.; Wertz, R. T. A treatment for apraxia of speech in adults. J. Speech Hear. Disord. 38:462; 1973.

Sands, E.; Sarno, M. T.; Shankweiler, D. Long-term assessment of language function in aphasia due to stroke. Arch. Phys. Med. Rehab. 50:202; 1969.

Sarno, M. T.; Levita, E. Natural course of recovery in severe aphasia. Arch. Phys. Med. Rehab. 52:175; 1971.

Sarro, M. T.; Silverman, M.; Levita, E. Psycho-social factors and recovery in geriatric patients with severe aphasia. J. Am. Geriatr. Soc. 18:405; 1970.

Schuell, A.; Jenkins, J. J.; Pabon, J. Aphasia in adults. New York: Harper & Row; 1964.

Selnes, O. A.; Knopman, D. S.; Niccum, N.; Rubens, A. B. CT scan correlates of auditory comprehension deficits in aphasia: a prospective recovery study. Ann. Neurol. 13:553–566; 1983.

Selnes, O. A.; Knopman, D. S.; Niccum, N.; Rubens, A. B. The critical role of Wernicke's area in sentence repetition. Ann. Neurol. 17:549–557; 1985.

Shewan, C. M.; Kertesz, A. Effects of speech and language treatment on recovery from aphasia. Brain Lang. 23:272–299; 1984.

Skelly, M.; Schinksi, L.; Smith, R.; Furst, R. S. American Indian Sign (AMERIND) as a facilitator of verbalization for the oral verbal apraxia. J. Speech Hearing Disord. 39:445; 1974.

Smith, A. Objective indices of severity of chronic aphasia in stroke patients. J. Speech Hearing Disord. 36:167; 1971.

Smith A.; Chamoux, R.; Leri, J.; London, R.; Muraski, A. Diagnosis, intelligence and rehabilitation of chronic aphasics. Ann Arbor: University of Michigan Department of Physical Medicine and Rehabilitation; 1972.

Sohlberg, M. M.; Mateer, C. A. Introduction to cognitive rehabilitation. New York: Guildford Press; 1989.

Sparks, R.; Helm, N. A.; Albert, M. L. Aphasia rehabilitation resulting from melodic intonation therapy. Cortex 10:303–316; 1974.

Subirana, A. Handedness and cerebral dominance. In: Vinken, P. J.; Bruyn, G. W., eds. Handbook of clinical neurology. Amsterdam: North-Holland; 1969.

Tranel, D.; Biller, J.; Damasio, H.; Adams, H. P.; Cornell, S. H. Global aphasia without hemiparesis. Arch. Neurol. 44:304–308; 1987.

Van Horn, G.; Hawes, A. Global aphasia without hemiparesis: A sign of embolic encephalopathy. Neurology 32:403–406; 1982.

Vignolo, L. A. Evolution of aphasia and language rehabilitation: A retrospective exploratory study. Cortex 1:344; 1964.

Weisbenburg, T.; McBride, K. E. Aphasia: A clinical and psychological study. New York: Commonwealth Fund; 1935.

Wepman, J. M. Recovery from aphasia. New York: Ronald Press; 1951.

Wertz, R. T.; Weiss, D. G.; Aten, J. L.; Brookshire, R. H.; Garcia-Bunuel, L.; Holland, A. L.; Kurtzke, J. F.; LaPointe, L. L.; Milianti, F. J.; Brannegan, R.; Greenbaum, H.; Marshall, R. C.; Vogel, D.; Carter, J.; Barnes, N. S.; Goodman, R. Comparison of clinic, home, and deferred language treatment of aphasia. A

Veterans Administration cooperative study. Arch. Neurol. 43:653–658; 1986.

Wilson, B. A. Rehabilitation of memory. New York: Guildford Press, 1987.

Woods, B. T.; Carey, S. Language deficits after apparent clinical recovery from childhood aphasia. Ann. Neurol. 6:405–409; 1979.

Woods, B. T.; Teuber, H. L. Changing patterns of childhood aphasia. Ann. Neurol. 3:273–280; 1978.

Zaidel, E. Auditory vocabulary of the right hemisphere following brain bisection or hemidecortication. Cortex 12:187–211; 1976.

Zarit, S. H.; Kahn, R. L. Impairment and adaptation in chronic disabilities: spatial inattention. J. Nerv. Dis. 159:63; 1974.

Index

Figures are denoted by f; tables by t.